FROM ANIMALS TO ANIMATS 8

FROM ANIMALS TO ANIMATS 8

Proceedings of the Eighth International Conference on Simulation of Adaptive Behavior

edited by
Stefan Schaal, Auke Ijspeert, Aude Billard, Sethu Vijayakumar,
John Hallam, and Jean-Arcady Meyer

A Bradford Book

The MIT Press
Cambridge, Massachusetts
London, England

ISSN 1089-4365
ISBN 978-0-262-69341-7 (pb.:alk.paper)

The MIT Press is pleased to keep this title available in print by manufacturing single copies, on demand, via digital printing technology.

CONTENTS

Preface

THE ANIMAT APPROACH TO ADAPTIVE BEHAVIOR

State of the Artificial Rat Psikharpax 3
David Filliat, Benoit Girard, Agnès Guillot, Mehdi Khamassi, Loïc Lachèz, and Jean-Arcady Meyer

Interactions of Environment, Behavior, and Synaptic Patterns in a Neuro-Robotic Model 13
William H. Alexander and Olaf Sporns

Comparative Cognitive Robotics: Using Autonomous Agents as Empirical Models of Animal Learning 23
Roul Sebastian John and Christian W. Werner

Morphology and Computation 33
Chandana Paul

Constructing an Animat Using 505 Sub-minds from 234 Different Authors 39
Ciarán O'Leary, Mark Humphrys, and Ray Walshe

May We Have Your Attention: Analysis of a Selective Attention Task 49
Eldan Goldenberg, Jacob Garcowski, and Randall D. Beer

Formalization of Recognition, Affordances and Learning in Isolated or Interacting Animats 57
Philippe Gausiier, Jean Christophe Baccon, Ken Prepin, Jacqueline Nadel, and Laurence Hafemeister

Gallery Structure Formation in *Conelyborus scarabaeoides* 67
Luc Steels

PERCEPTION

Towards a Theory of How Bats Navigate through Foliage 77
Jonas Reijniers and Herbert Peremans

From Bat Pinnae to Sonar Antennae: Augmented Obliquely Truncated Horns as a Novel Parametric Shape Model 87
Rolf Müller and John C. T. Hallam

Evolving Adaptive, High-Dimensional, Camera-Based Speed Sensors 96
Ralf Salomon

Adaptive Object Tracking with an Anthropomorphic Robot Head 104
Hyundo Kim, Boris Lau, and Jochen Triesch

Morphology and Learning—A Case Study on Whiskers 114
Miriam Fend, Roland Abt, Marco Diefenbacher, Simon Bovet, and Martin Krafft

Simulating Whisker Sensors—On the Role of Material Properties for Morphology, Behavior, and Evolution 122
Simon Bovet, Miriam Fend, and Rolf Pfeifer

Spatiotemporal Processing of Whisker Input Supports Texture Discrimination in a Brain-based Device 130
Anil K Seth, Jeffrey L. McKinstry, Gerald M. Edelman, Jeffrey L. Krichmar

A Biomimetic Whisker for Texture Discrimination and Distance Estimation 140
DaeEun Kim and Ralf Möller

MOTOR CONTROL

A Simple, Adaptive Locomotion Toy-System 153
Jonas Buchli and Auke Jan Ijspeert

GasNets and Other Evolvable Neural Networks Applied to Bipedal Locomotion 163
Gary McHale and Phil Husbands

Adaptive Locomotion in a Complex Environment: Simulation of Stick Insect Gap Crossing Behavior 173
Bettina Blaesing

A Simulation Model of the Locomotion Controllers for the Nematode *Caenorhabditis elegans* 183
John Bryden and Netta Cohen

Planning the Sequencing of Movement Primitives 193
Marcelo Kallmann, Robert Bargmann, and Maja Matarić

Biologically Inspired Adaptive Dynamic Walking of a Quadruped Robot 201
Hiroshi Kimura, Yasuhiro Fukuoka, and Avis H. Cohen

ACTION SELECTION AND BEHAVIORAL SEQUENCES

Action Selection and Mental Transformation Based on a Chain of Forward Models 213
Heiko Hoffmann and Ralf Möller

Self-Growth of Basic Behaviors in an Action Selection Based Agent 223
Olivier Buffet, Alain Dutech, and François Charpillet

Optimal and Non-Optimal Compromise Strategies in Action Selection 233
Frederick L. Crabbe

Using Hormonal Feedback to Modulate Action Selection in a Competitive Scenario 243
Orlando Avila-García and Lola Cañamero

NAVIGATION

A Spiking Neuron Model of Head-Direction Cells for Robot Orientation 255
Thomas Degris, Loïc Lachèze, Christian Boucheny, and Angelo Arleo

Navigation and Cognitive Map Formation Using Aperiodic Neurodynamics 264
Derek Harter and Robert Kozma

Adaptive Motor Primitive and Sequence Formation in a Hierarchical Recurrent Neural Network 274
Rainer W. Paine and Jun Tani

LEARNING AND ADAPTATION

Competitive-Cooperative-Concurrent Reinforcement Learning with Importance Sampling 287
Eiji Uchibe and Kenji Doya

Estimating Future Reward in Reinforcement Learning Animats using Associative Learning 297
George D. Konidaris and Gillian M. Hayes

Hebbian Reinforcement Learning in a Modular Dynamic Network 305
Emmanuel Daucé

Learning Human-like Movement Behavior for Computer Games 315
Christian Thurau, Christian Bauckage, and Gerhard Sagerer

On Ashby's Homeostat: A Formal Model of Adaptive Regulation 324
J. Michael Herrmann, Michael Holicki, and Ralf Der

Toward a Dynamical Systems Analysis of Neuromodulation 334
Christopher Buckley, Seth Bullock, and Netta Cohen

Homeostatic Plasticity in Recurrent Neural Networks 344
Hywel Williams

EVOLUTION

Evolving Walking: The Anatomy of an Evolutionary Search 357
Chad W. Seys and Randall D. Beer

The Force Model: Concept, Behavior, Interpretation 364
Ralf Salomon

Evolving a Neurocontroller through a Process of Embryogeny 373
Diego Federici

COLLECTIVE AND SOCIAL BEHAVIORS

Coordination and Behaviour Integration in Cooperating Simulated Robots 385
Gianluca Baldassarre, Domenico Parisi, and Stefano Nolfi

Effects of the Topology of Social Networks on Information Transmission 395
Jason Nobel, Simon Davy, and Daniel W. Franks

Evolving Functional Self-Assembling in a Swarm of Autonomous Robots 405
Vito Trianni, Elio Tuci, and Marco Dorigo

How Can a Life-like Agent Evoke Its Emotions For Users?—A Method for Inducing Users' Natural Behaviors to Establish "Mutual Adaptation" 415
Takanorio Komatsu

Hamelin: A Model for Collective Adaptation Based on Internal Stimuli 425
Vincent Thomas, Christine Bourjot, Vincent Chevrier, and Didier Desor

BASC, a Bottom-up Approach to Automated Design of Spatial Coordination 435
Fabien Flacher and Olivier Sigaud

The More Radical, the Better: Investigating the Utility of Aggression in the Competition among Different Agent Kinds 445
Matthias Scheutz and Paul Schermerhorn

Imitation in Embodied Agents Results in Self-organization of Behavior 455
Bart Jansen, Tom ten Thij, Tony Belpaeme, Bart De Vylder, and Bart de Boer

ADAPTIVE BEHAVIOR IN LANGUAGE AND COMMUNICATION

A Connectionist Approach to Learn Association between Sentences and Behavioral Patterns of a Robot 467
Yuuya Sugita and Jun Tani

Imitation is not Enough for Lexicon Learning 477
Jason Fleischer and Jonathan Shapiro

The Sensorimotor Bases of Linguistic Structure: Experiments with Grounded Adaptive Agents 487
Angelo Cangelosi

APPLIED ADAPTIVE BEHAVIOR

Evolving Opponents for Interesting Interactive Computer Games 499
Georgios N. Yannakakis and John Hallam

Evolvable Recovery Membranes in Self-monitoring Aerospace Vehicles 509
Peter Wang and Mikhail Prokopenko

Learning to Behave: Adaptive Behavior for Planetary Surface Rovers 519
Terry Huntsberger and Hrand Aghazarian

Author Index 529

PREFACE

This book regroups the articles that were presented at the Eighth International Conference on Adaptive Behavior (SAB2004), held in Los Angeles, USA, on July 13-17 2004. The objective of the biennial SAB conference is to bring together researchers from a wide range of backgrounds including ethology, theoretical biology, neuroscience, psychology, artificial life, machine learning, and robotics, to further our understanding of the behaviors and underlying mechanisms that allow natural and artificial animals ("animats") to adapt and survive in uncertain environments.

Adaptive behavior research is distinguished by its focus on the modeling and creation of complete animal-like systems, which – however simple at the moment – may be one of the best routes to understanding intelligence in natural and artificial systems. This conference is part of a long series that started with the first SAB conference, which was held in Paris in September 1990, and was followed by conferences in Honolulu 1992; Brighton 1994, Cape Cod 1996, Zurich 1998, Paris 2000, and Edinburgh 2002. Like most of its predecessors, SAB2004 is a single-track conference with several poster sessions in order to favor a friendly atmosphere with multiple interactions and discussions among participants. As a new feature, last-minute-results poster sessions were introduced at SAB2004. The short abstracts associated with those posters are not published in this volume.

The 53 papers published here were selected from 104 submissions after a two-pass review process to ensure high and consistent overall quality. Articles cover all main areas in animats research, including motor control, perception, action selection, navigation, collective behavior, learning, and evolution. The authors focus on well-defined models, computer simulations, and robotic demonstrations that help characterize and compare various organizational principles or architectures capable of inducing adaptive behavior in real animals or synthetic agents. We hope that these articles will provide stimulating reading material, with a good overview of the latest developments in this exciting field.

The conference and its proceedings would not exist without the substantial help of a wide range of people. Foremost, we would like to thank the members of the Program Committee who thoughtfully reviewed all the submissions and provided detailed suggestions on how to improve the articles. The committee members were:

Ronald Arkin, *Georgia Institute of Technology, Atlanta, USA*
Christian Balkenius, *Lund University, Lund, Sweden*
Randy Beer, *Case Western University, Cleveland, USA*
Luc Berthouze, *Neuroscience Research Institute (AIST), Tsukuba, Japan*
Aude Billard, *Swiss Federal Institute of Technology (EPFL), Lausanne, Switzerland*
Andrea Bonarini, *Politecnico di Milano, Milano, Italy*
Lashon Booker, *The MITRE Corporation, USA*
Joanna Bryson, *University of Bath, Bath, UK*
Larry Bull, *University of the West of England, Bristol, UK*
Seth Bullock, *University of Leeds, Leeds, UK*
Jose M. Carmena, *Duke University, Durham, USA*
Philippe Codognet, *University of Paris VI, Paris, France*
Thomas Collett, *University of Sussex, Brighton, UK*

Holk Cruse, *University of Bielefeld, Bielefeld, Germany*
Kerstin Dautenhahn, *University of Hertfordshire, Hatfield, UK*
Yiannis Demiris, *Imperial College London, London, UK*
Ezequiel DiPaolo, *University of Sussex, Brighton, UK*
Marco Dorigo, *IRIDIA - Université Libre de Bruxelles, Brussels, Belgium*
Michael Dyer, *University of California (UCLA), Los Angeles, USA*
Örjan Ekeberg, *Royal Institute of Technology, Stockholm, Sweden*
Dario Floreano, *Swiss Federal Institute of Technology (EPFL), Lausanne, Switzerland*
Philippe Gaussier, *Cergy Pontoise University, Cergy Pontoise, France*
Agnes Guillot, *University of Paris VI, Paris, France*
Bridget Hallam, *University of Southern Denmark, Odense, Denmark*
John Hallam, *University of Southern Denmark, Odense, Denmark*
Inman Harvey, *University of Sussex, Brighton, UK*
Gillian Hayes, *University of Edinburgh, Edinburgh, UK*
Phil Husbands, *University of Sussex, Brighton, UK*
Auke Jan Ijspeert, *Swiss Federal Institute of Technology (EPFL), Lausanne, Switzerland*
Yasuo Kuniyoshi, *University of Tokyo, Tokyo, Japan*
Anders Lansner, *Royal Institute of Technology, Stockholm, Sweden*
Pier Luca Lanzi, *Politecnico di Milano, Milano, Italy*
Wolfgang Maass, *Graz University of Technology, Graz, Austria*
Hanspeter Mallot, *Max-Planck Institute for Biological Cybernetics, Tuebingen, Germany*
Chris Melhuish, *University of the West of England, Bristol, UK*
Jean-Arcady Meyer, *University of Paris VI, Paris, France*
Francesco Mondada, *Swiss Federal Institute of Technology (EPFL), Lausanne, Switzerland*
Ryohei Nakano, *Nagoya Institute of Technology, Nagoya, Japan*
Chrystopher L. Nehaniv, *University of Hertfordshire, Hatfield, UK*
Ulrich Nehmzow, *University of Essex, Colchester, UK*
Jason Noble, *University of Leeds, Leeds, UK*
Stefano Nolfi, *Institute of Cognitive Sciences and Technologies-CNR, Rome, Italy*
Domenico Parisi, *Institute of Psychology, Rome, Italy*
Rolf Pfeifer, *University of Zurich, Zurich, Switzerland*
Tony Prescott, *University of Sheffield, Sheffield, UK*
Herbert L. Roitblat, *University of Hawaii, USA*
Stefan Schaal, *University of Southern California, Los Angeles, USA*
Olivier Sigaud, *University of Paris VI, Paris, France*
Olaf Sporns, *Indiana University, Bloomington, USA*
Luc Steels, *Sony Computer Science Laboratory, Paris, France and University of Brussels (VUB), Brussels, Belgium*
Wolfgang Stolzmann, *DaimlerChrysler AG, Germany*
Jun Tani, *Brain Science Institute RIKEN, Saitama, Japan*
Tim Taylor, *University of Edinburgh, Edinburgh, UK*
Peter Todd, *Max Planck Institute for Human Development, Berlin, Germany*
Richard Vaughan, *Simon Fraser University, Burnaby, Canada*
Sethu Vijayakumar, *University of Edinburgh, Edinburgh, UK*

Barbara Webb, *University of Edinburgh, Edinburgh, UK*
Myra S. Wilson, *University of Wales, Aberystwyth, UK*
Brian Yamauchi, *iRobot Corporation, Burlington, USA*
Tom Ziemke, *University of Skovde, Skovde, Sweden*

We are also indebted to our sponsors:

- The International Society for Adaptive Behavior (ISAB)
- Hewlett Packard
- K-Team
- The University of Southern California (USC)
- The Swiss Federal Institute of Technology, Lausanne (EPFL),
- The University of Edinburgh
- The University of Southern Denmark
- Université de Pierre et Marie Curie (Paris 6)

The enthusiasm and hard work of numerous individuals was essential to the conference's success. Above all we would like to thank members of our laboratories, and in particular those of the Computational Learning and Motor Control laboratory at USC, for their help with the local arrangements.

Finally, once again, we would like to warmly thank Jean Solé for the artistic conception of the SAB2004 poster and the proceedings cover.

We invite readers to enjoy and profit from the articles in this book, and look forward to the next conference, SAB2006.

Stefan Schaal, Auke Ijspeert, Aude Billard, Sethu Vijayakumar, John Hallam, and Jean-Arcady Meyer

THE ANIMAT APPROACH TO ADAPTIVE BEHAVIOR

State of the artificial rat Psikharpax

David Filliat*,* , Benoit Girard*,**, Agnès Guillot*, Mehdi Khamassi*,**, Loïc Lachèze*, Jean-Arcady Meyer***

jean-arcady.meyer@lip6.fr

*AnimatLab/LIP6, 8 rue du capitaine Scott, 75015 Paris, France

** LPPA, Collège de France, 11 place Marcelin Berthelot, 75005 Paris, France

*** DGA/Centre Technique d'Arcueil, 16 bis Av. Prieur de la Cote d'Or, 94114 Arcueil, France

Abstract

This paper describes the current state of advancement of the Psikharpax project, which aims at producing an artificial rat equipped with control architectures and mechanisms that reproduce as nearly as possible those that have been widely studied in the natural rat.

The article first describes the navigation system of Psikharpax, which is inspired from the anatomy and physiology of dedicated structures in the rat's brain, like the hippocampus and the postsubiculum. Then, it defines the animat's action-selection system, which aims at replicating other structures, the basal ganglia. It also explains how navigation and action-selection capacities have been combined thanks to the interconnection of two different loops in the basal ganglia: a ventral loop that selects the direction of motion, and a dorsal loop that selects other behaviors, like feeding or drinking. Finally, preliminary results on the implementation of learning mechanisms in these structures are also presented.

1. Introduction

As everybody knows, Psikharpax was the King of the Rats -- i.e., an intelligent and adaptive character -- in the Batrachomyomachy, a parody of Iliad written in Greek verses and (falsely) attributed to Homer. The name means “crumb robber”. It has been given to a modern project that brings together several academic partners -- the Laboratoire de Physiologie de la Perception et de l'Action (LPPA), in Paris; the Laboratoire d'Informatique et de Microélectronique (LIRMM), in Montpellier; the Center for Neuromimetic Systems of the Swiss Federal Institute of Technology, in Lausanne; the Adaptive Behaviour Research Group, in Sheffield; and the AnimatLab, in Paris -- as well as two private companies -- BEV.S.A. and Wany.S.A. It aims at equipping an artificial rat with a bio-inspired behavioral control architecture that should hopefully afford it some of the capacities of autonomy and adaptation that characterize the natural rat. In particular, Psikharpax will be endowed with internal needs - such as hunger, rest, or curiosity - which it will try to satisfy in order to survive within the challenging environment of a laboratory populated with humans and, possibly, other robots. To this end, it will sense and act on its environment in pursuit of its own goals and in the service of its needs, without help or interpretation from outside the system.

Both fundamental and applied objectives are targeted by the project. In the first place, in the perspective of assessing the coherency and completeness of current knowledge about the rat's nervous system and about the mechanisms that contribute to its adaptive capacities, this project aims at integrating these mechanisms in a robot that may be confronted with the same situations as those a real rat may encounter in a laboratory or in nature. Secondly, this project should also help in assessing the operational value of these adaptive capacities in the situations where an artificial agent has to "survive" or fulfil its mission, without human assistance, and in a more or less unpredictable environment.

This paper is centred on the AnimatLab's contribution to this project and will describe the software developments that will contribute to endowing Psikharpax with integrated capacities for navigation, action selection and learning. It will capitalize on both simulations and robotic implementations. In the latter case, only commercial robots or simple home-made machines are concerned. However, the paper will end with a short presentation of parallel developments that will produce the final robotic platform on which the control software will ultimately be implemented.

2. General situation of the project

It is clear from recent reviews (Bar-Cohen and Breazeal, 2003; Holland and McFarland, 2001; Webb and Consi, 2001) that, although many research efforts have been devoted to the design of biomimetic sensors or effectors for robots, relatively little work has been done on control system architectures, and what has been done has focused primarily on invertebrate models. Only a few groups are currently building biomimetic robot control architectures modelled on mammalian nervous systems and, moreover, their efforts are often centred on isolated behaviors, like locomotion in cats (Patla et al., 1985) or feeding in mice (Guillot and Meyer, 1987), which are not dealt with in an integrated perspective.
To the best of our knowledge, the scope of the Psikharpax project is unique. First, it draws inspiration from a vertebrate instead of an invertebrate. Second, it aims at designing both biomimetic sensors and control architectures. Third, because it capitalizes on a dedicated robotic platform, it will integrate a variety of sensors, actuators and control systems making it possible to assess its adaptive capacities in much more challenging circumstances than those that characterize seemingly comparable biomimetic robotic approaches (Capi et al., 2002; Gaussier et al., 2000; Hafner et al., 2003; Montes-Gonzalez, 2000; Touretzky and Saksida, 1997) – a point to which we will return at the end of this paper.

3. Experimental results

To demonstrate the range of adaptive behaviors that a rodent-like control architecture affords a robot, we want to endow our artificial rat with the following capacities:

- to explore its environment and build a cognitive map of it;
- to use this map to localize both itself and places where rewards and punishments have been experienced;
- to learn which behaviors and objects in the environment generate which emotions and fulfil which motivations;
- to use its motivational system to select a current goal, like finding something to eat when it is hungry, or finding something to drink when it is thirsty;
- to switch to another behavior when specific emotions, like fear, are generated by specific events;
- to adjust its energy balance, notably by initiating rest periods.

Basically, this plan calls on two major control systems, respectively responsible for Psikharpax's navigation and action selection capacities. It also presupposes that these systems be integrated with various motivations and emotions in a coherent whole, and that it also exhibit appropriate learning mechanisms. The following paragraphs describe significant steps already made in these directions.

3.1. Navigation

Numerous simulation models – see Trullier et al. (1997) for a review – call upon so-called *place cells* and *head direction cells* to implement navigation systems that are inspired from the anatomy and physiology of dedicated structures in the rat's brain, like the *hippocampus* and the *postsubiculum*.

Figure 1. Schematics of allothetic data used for navigation. The broken line joins the points detected by the robot's sonar sensors in eight absolute directions. The rectangles arranged in a circle indicate the mean grey-level perceived in the corresponding direction by the camera.

The model currently used in Psikharpax's developments implements a *multiple-hypothesis tracking* navigation strategy, maintaining a set of hypotheses about the robot's position that are all updated in parallel (Filliat and Meyer, 2003; Meyer and Filliat, 2003).

It serves to build a dense topological map (Filliat and Meyer, 2002), in which nodes store the allothetic data that the robot can perceive at the corresponding places in the environment. These data correspond to mean grey-levels perceived by a directional camera in each of 36 surrounding directions, and to sonar data providing distances to obstacles in eight directions (Figure 1).

A link between two nodes memorizes how far and in which direction the corresponding places are positioned relatively to each other, as measured by the robot's idiothetic sensors, i.e., by its odometry (Figure 2, left).

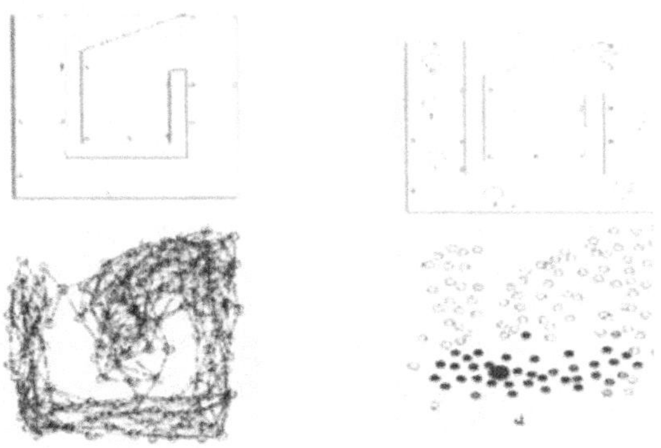

Figure 2. Left: The topological map (bottom) created by the robot when it explores an unknown environment (top). Right: This map may be used by the robot to localize itself because the activity distribution of the nodes in the map changes as the robot moves through successive places a, b,…g in the environment (top). Thus, when the robot is at place d, a blob of activity in the map surrounds the node that corresponds to this place (bottom). The grey level of each small node in the map indicates its activity, ranging from 0 for white nodes to 1 for black nodes. Larger black dots indicate the nodes which were successfully recognized.

The robot's position is represented by an activity distribution over the nodes, the activity level of a given node representing the probability that the robot is currently located at the corresponding position (Figure 2, right). This navigation model has been implemented on a Pioneer 2 mobile robot and proved to be efficient at exploring an unknown laboratory (Figure 3). From scratch, the robot succeeds in exploring its environment and in localizing itself accurately within it. Moreover, the system makes it possible for an external user to designate a goal-place in the map, from which a spreading activation algorithm computes the shortest path joining the goal to the robot's current position. When the robot follows this path, if it encounters an unexpected obstacle, the spreading activation algorithm may be triggered again to check if an alternative way to the goal exists. If so, the robot will follow the corresponding detour to fulfil its mission.

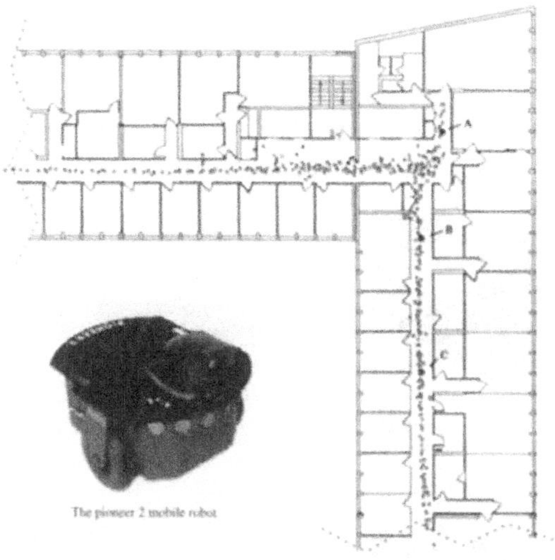

Figure 3. An example of a map created while a Pioneer 2 robot explored an unknown laboratory. The map is superimposed on an architectural sketch of the environment. When the robot is located in place A, if it is manually moved to place B, it will temporarily hesitate between two possible positions, B and C, that look very similar. However, after a few moves beyond place B , it will correctly identify its position.

3.2. *Action selection*

To survive, a rat must be able to solve the so-called *action selection problem* – i.e., it must be able to decide at every moment what to do next in service of its needs.

Some of the circuits that are involved in this task are known to be located in *basal ganglia-thalamus-cortex* loops and have inspired the *GPR model* (according to the authors Gurney, Prescott and Redgrave, 2001a & b) that is implemented in Psikharpax's control architecture (Girard et al., 2002; Girard, 2003). Basically, this model (Figure 4) assumes that the numerous segregated channels observed in the basal ganglia each correspond to a discrete motor action (the granularity of which is still not deciphered) that is inhibited by default and thus prevented from being executed. Inputs to these channels are so-called *salience, a* kind of *common currencies* for actions that take into account both internal and external perceptions to assess the relevance of each action with respect to the robot's needs. A positive feedback loop involving the thalamus serves to introduce some persistence in such assessments. Two parallel *selection* and *control* circuits within the basal ganglia act to modulate interactions between channels.

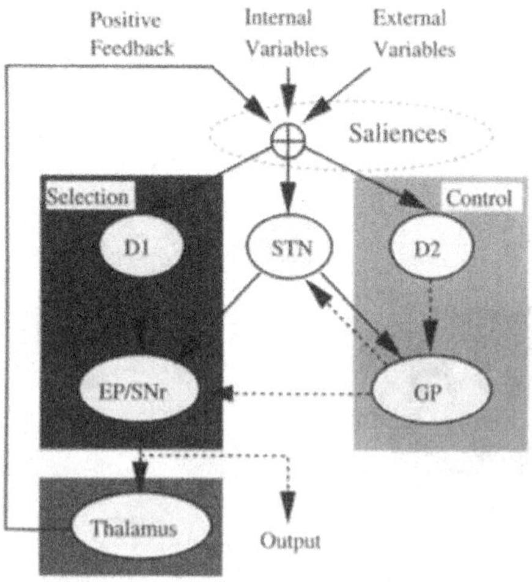

Figure 4. A single channel within the basal ganglia: the GPR model. D1 and D2 are striatal neurons with different dopamine receptors. STN is sub-thalamic nucleus. EP/SNr designates both the entopeduncular nucleus and the substantia nigra reticulata. GP is the globus pallidus. Solid arrows represent excitatory connections, dotted arrows represent inhibitory connections.

Finally, at the output of these circuits, the action that is less inhibited by others is selected and allowed to be executed by the motor system (Figure 5).

Experimental results demonstrate the model's ability to promote survival in the sense that it permanently keeps two *essential variables* (Ashby, 1952) above minimal levels: *Potential Energy* (obtained via "feeding") and *Energy* (converted from Potential Energy via "resting"). Moreover, the model avoids dithering or interferences between actions - thanks to saliences' discrepancies enhanced by the control circuit - and demonstrates some advantages with respect to a simpler version in which the selected action would be the one with the greatest salience - thanks to the conjugated action of the control circuit and the thalamic loop.

This model has been implemented in a Lego robot whose task was to efficiently select between four actions – wandering, avoiding obstacles, "feeding" and "resting" – in order to "survive" in an environment where it could find "food" and "rest" places (Girard et al., 2003) (Figure 6).

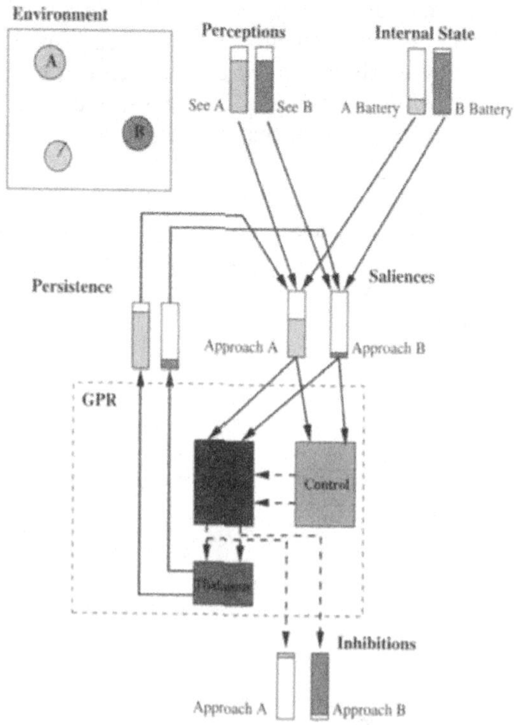

Figure 5. An illustration of the action selection circuit involving two possible actions: move-toward-A or move-toward-B. Solid arrows represent excitatory connections, dotted arrows represent inhibitory connections. The saliences associated with the two corresponding actions and channels depend upon both external and internal perceptions. In this case, although A and B objects are both perfectly perceived, because the A-battery level is lower than that of the B-battery, the action move-toward-A gets less inhibited than the other, and the robot accordingly decides to move towards the A object.

Figure 6. Left: The environment showing "food" (A) and "rest" (B) places. Right: A Lego robot equipped with light sensors (A) and bumpers (B).

However, the robot's survival depended on its chances of getting to the right place at the right moment, i.e., to a food place when its *Potential Energy* level was low, or to a rest place when it lacked *Energy*. Obviously, additional adaptive capacities would depend on the robot's capacity to record the position of such places on its map and to use this map to reach such places when needed. This has been made possible thanks to a model combining navigation and action selection capacities.

3.3. *Combining navigation and action selection*

The connection of the previously-described navigation and action selection models and their implementation on a simulated robot were inspired by recent hypotheses concerning the role of dedicated structures within the basal ganglia – like the *nucleus accumbens* in particular – and the interplay of *basal ganglia-thalamus-cortex* loops in the rat's brain (Girard, 2003; Girard et al., in press). The corresponding model (Figure 7) basically involves two such loops: a *ventral loop* that selects locomotor actions, like moving to the north or the east, and a *dorsal loop* that selects non-locomotor actions, like feeding or resting. Each of these loops has been modeled as a GPR system like the one previously described. The STN of the dorsal loop provides the interconnection between them because it sends excitatory projections to the output of the ventral loop. As a consequence, when the dorsal loop is active and triggers some non-locomotor action, the excitatory signal that is sent towards the ventral loop raises the inhibition level of every locomotor action and prevents it from being selected. Hence the robot cannot move and eat at the same time.

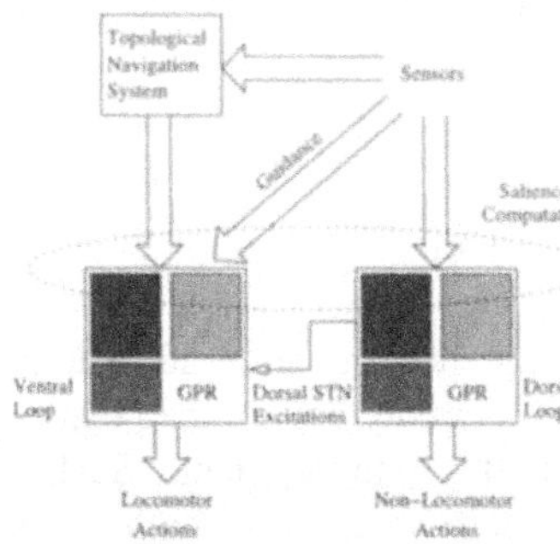

Figure 7. Interconnection of the ventral and dorsal loops in the basal ganglia. The ventral loop selects locomotor actions, the dorsal loop selects non locomotor actions. The latter subsumes the former.

Saliences in the ventral loop depend upon direction profiles that are generated by two different navigation strategies, i.e., a simple *guidance* strategy and a more elaborate *topological navigation* strategy (Trullier et al., 1997). This makes it possible for the robot to be attracted either by an object that it directly perceives or to move towards a region where such an object is located in its map. The latter possibility puts some constraints on action selection because the robot is committed to regularly returning to previously mapped areas in its environment in order to check the accuracy of the current map.

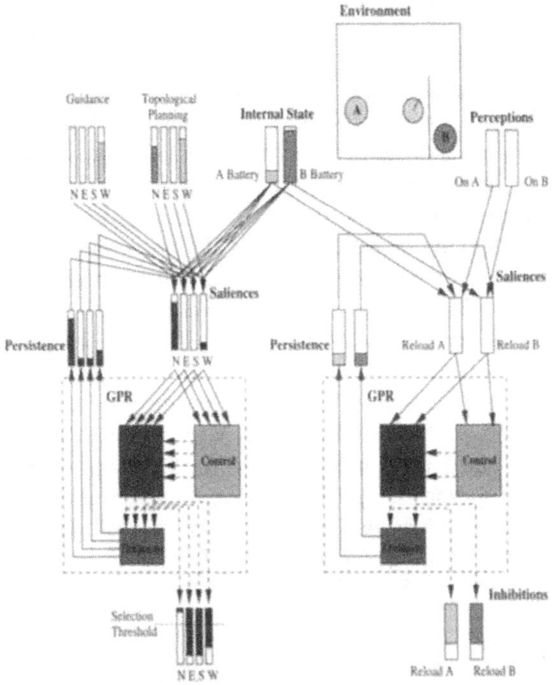

Figure 8. The model integrating navigation and action selection calls upon two basal ganglia-thalamus-cortex loops. Each loop is managed by a GPR model, and the coordination between loops is provided by the subthalamic nucleus of the dorsal loop, which is connected to the ventral loop (connection not shown here). The dorsal loop selects one of the two possible reloading actions, the ventral loop selects one of the 36 directions of movement (simplified here to four cardinal directions). Inhibitory connections are represented by dotted arrows, excitatory connections by solid arrows. In the current situation, only locomotor actions are selected because external perceptions are not strong enough to suppress the inhibition of reloading actions. When the robot gets close to the B object, this fact will be detected by its sensors, a reloading action will be triggered and an excitatory signal will be sent to the ventral loop in order to inhibit further locomotor actions.

This need is expressed by a *Disorientation* variable managed by the model, which increases when the robot enters unexplored areas, decreases when it returns to known areas, and affects the computation of saliences. Saliences in the dorsal loop depend upon both internal and external perceptions (Figure 8).

This model has been implemented in a version that manages 36 locomotor actions – i.e., moving in each of the 36 possible directions – and two non-locomotor actions – i.e., reloading actions that change the robot's *Energy* and *Potential Energy* levels.

The robot simulated in the environment on the left of Figure 9 survives successfully because it uses its map to navigate between places *E* and *Ep* where it can reload its *Energy* and *Potential Energy* levels. Likewise, in the environment on the right of Figure 9, assuming that place *Ep1* is the only one that the robot has previously encountered and recorded on its map, if it decides to move towards that place to reload its *Potential Energy* and if it detects on its way the close presence of another food place like *Ep2*, it will give up navigating towards *Ep1* and will opportunistically divert to *Ep2*. Then, having consumed the corresponding "food", it will record the position of *Ep2* on its map. Thus, next time it needs to reload its *Potential Energy*, it will have the choice of navigating towards *Ep1* or *Ep2*.

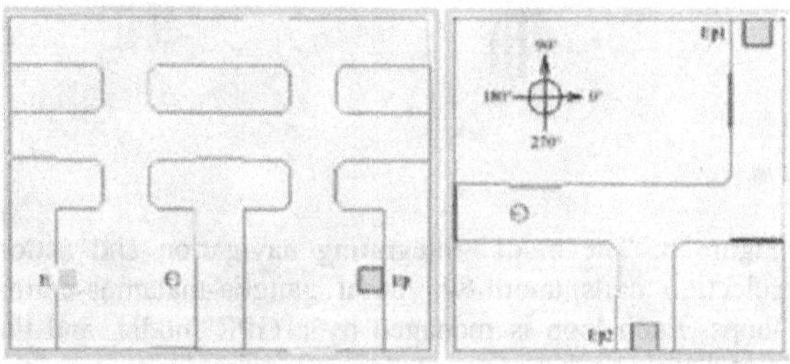

Figure 9. Two environments used to test the connection of navigation and action selection models. E: "rest" place, Ep: "food" place.

In the environment of Figure 10, the robot has the choice between two trajectories leading to a "food" place. The first one is shorter but entails passing through a "dangerous" place. The second one is longer, but safer. The robot is able to decide to navigate through the longer path when its *Potential Energy* level is not low enough to compromise its survival by a long journey, but it chooses the shorter path in the opposite case, at the risk of facing the potential danger recorded on its map.

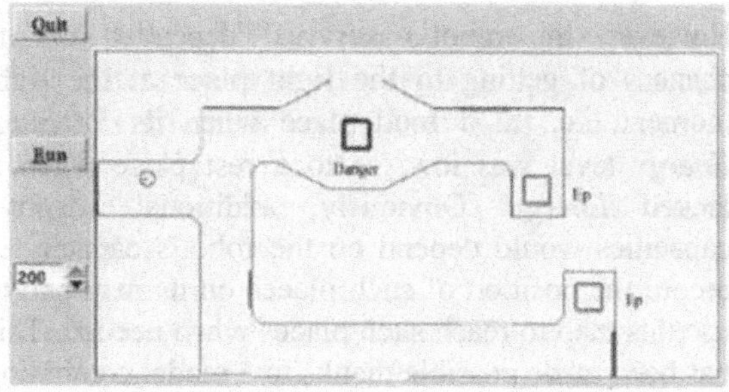

Figure 10. Two trajectories leading to "food" places (Ep). One is shorter than the other but entails passing through a dangerous place (Danger).

In the complex and challenging environment of Figure 11, the simulated robot autonomously survives, thanks to the numerous adaptive mechanisms and behaviors it has been endowed with (Girard, 2003).

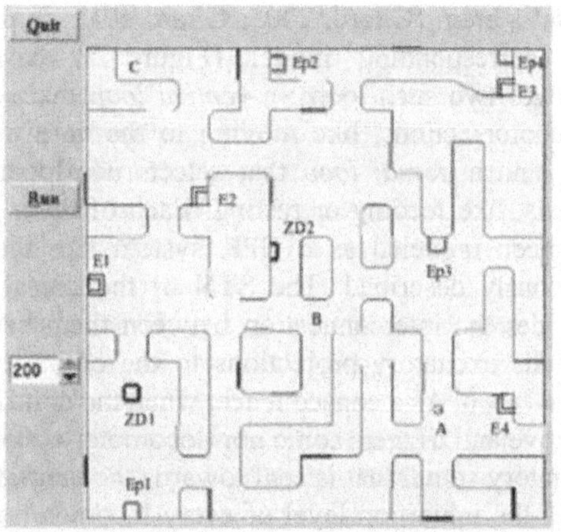

Figure 11. A complex environment with four "rest" places (E), four "food" places (Ep) and two dangerous places (ZD).

3.4. *Learning*

In an unknown environment, a rat is able to explore it and to incrementally build a map that describes the topology of this environment. Such *associative learning,* which combines both allothetic and idiothetic data, has been implemented in the navigation model described above.

However, a rat is also able to improve its behavior over time through *reinforcement learning*, i.e., thanks to adaptive mechanisms that raise its chances of exhibiting behaviors leading to rewards and that lower those of behaviors leading to punishments. Concerning action selection, a recently debated hypothesis (Barto, 1995; Houk et al., 1995; Schultz et al., 1997) postulates that such mechanisms could be mediated by dopamine signals within so-called *actor-critic* architectures (Figure 12).

Within such architectures, an action-selection module plays the role of an actor, while a critic module calls upon both the episodic reinforcement signal r_t occasionally generated by the robot's actions and a dopamine signal that is assumed to evaluate the difference gP_t-P_{t-1} between currently expected and future rewards. This estimate is used in the actor module to adapt the way saliences are computed and used to select the most appropriate action, i.e., the action the most likely to maximize the reward that it will generate.

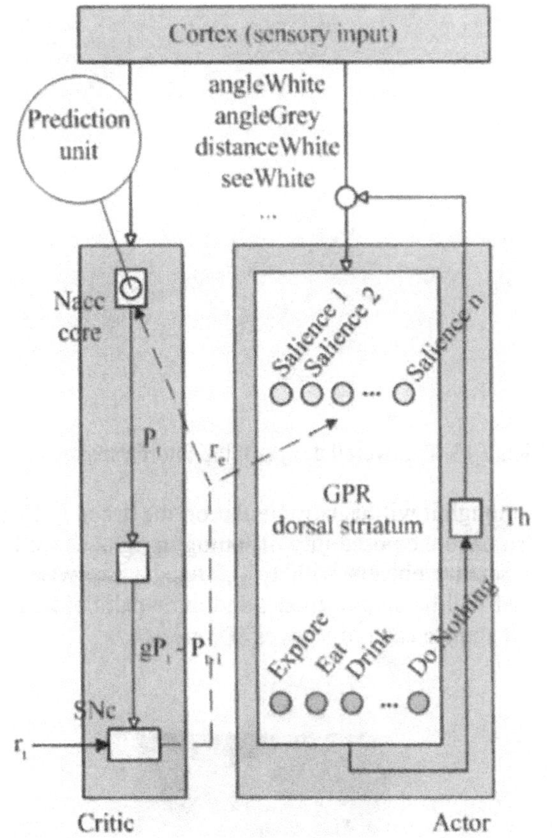

Figure 12. The actor-critic model of reinforcement learning. The actor module is a GPR model that is segregated in different channels, with saliences as inputs and actions as outputs. The critic module (involving the nucleus accumbens core (Nacc core) and the substantia nigra compacta (SNc)) propagates towards the actor module an estimate r_e of the instantaneous reinforcement triggered by the selected action.

This type of model has been implemented in a simulated robot that must learn in a plus-maze, and through successive trials, which movement to perform in order to get to the end of an arm where a door may provide access to a reward - i.e., some water to drink (Figure 13). This kind of learning corresponds to a simple Stimulus-Response association. At every trial, one lamp out of four is lighted indicating the door behind which the reward is accessible. When the robot succeeds in getting such reward, the corresponding lamp is turned off and the robot must learn to return to the center of the maze, where the lighting of another lamp will designate another reward place. This setting reproduces an experiment on real rats (Albertin et al., 2000) and helps to interpret the corresponding results.

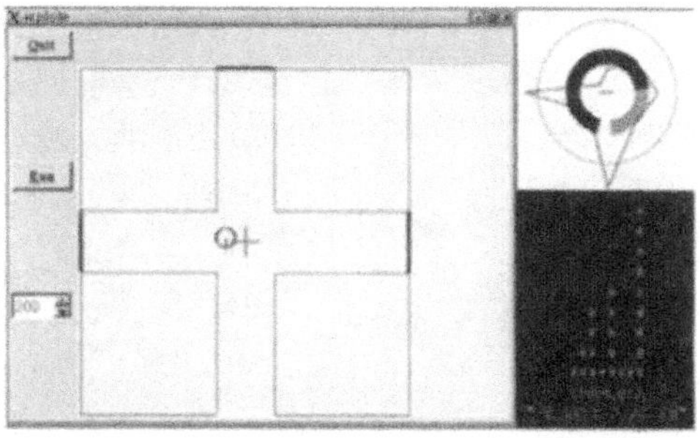

Figure 13. Left: the robot in the plus maze environment. A lighted lamp indicates which door (in white) leads to reward. The other doors do not lead to reward and are shown in black. Upper right: the robot's visual perceptions. Lower right: activation level of each channel in the actor module.

Two different critic modules, respectively adapted from Houk, Adams and Barto (1995) and from Baldassare (2002), have been implemented and connected to the same actor module, i.e., the action selection model described in Section III.2 above (Khamassi et al., in press). The main difference between these modules is that the first module (Critic 1) calls upon only one unit to predict an instantaneous reward, while the second (Critic 2) calls upon two units, on the one hand, and takes into account the time course of the rewarding stimulus, on the other hand.

The corresponding results were compared on the basis of two criteria: the zone in the environment where the robot has already learned something - i.e., where it is able to select the appropriate actions to get to the reward - and the way the prediction errors decrease at reward location along successive trials - information that is essential to propagate learning to other regions of the environment. It thus turns out that Critic 2 gives better results than Critic 1 (Figure 14) but that the corresponding module still has to be improved, for instance through the use of a greater number of prediction units, or the use of one controller per critic unit (Doya et al., 2000), to let learning extend to the whole experimental environment and to speed up the corresponding process.

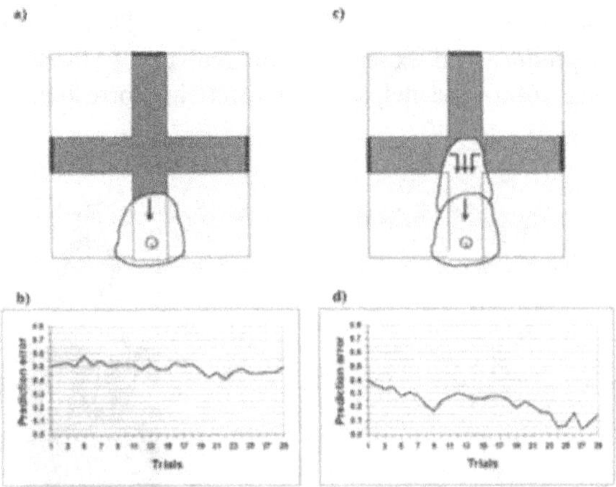

Figure 14. Reinforcement learning results obtained with an actor-critic architecture in which two different critic modules have been tested (left: Critic 1; right: Critic 2). Depending on which module is used, the correct actions to perform are learned in specific regions of the maze (a & c: white areas in the plus maze) and the improvement patterns of the prediction error at reward location over successive trials are different (b & d).

Referring to the beginning of Section 3 above, it thus appears that almost all the capacities that we wanted to endow our artificial rat with have started to be implemented, even if this animat currently exhibits very few motivations and emotions. In 2004, these capacities will be integrated on a Pekee platform that will make a first assessment of the Psikharpax endeavour possible.

4. The future robot

A more advanced version of Psikharpax is under construction. The morphology and sensori-motor equipment of this 50cm-long robot will be as closely inspired from the real rat as possible (Figure 15). In particular, it will be endowed with three sets of allothetic sensors: a two-eyed visual system, an auditory system calling upon two electronic cochleas, and a haptic system made of 50 whiskers on each side of its head.

Psikharpax will also be equipped with three sets of idiothetic sensors: a vestibular system reacting to linear and angular accelerations of its head, an odometry system monitoring the length of its displacements, and capacities to assess its current energy level. Sensor fusion will be accomplished through the use of GVPP, a biomimetic chip dedicated to low-level real-time signal processing that already serves robot vision (Gourichon et al., 2002). The robot's capacity to move and act in the environment will be afforded by several motors and actuators. In particular - despite the fact that such a device is not really biomimetic - two wheels will allow Psikharpax to move at a maximum speed of a few meters per second.

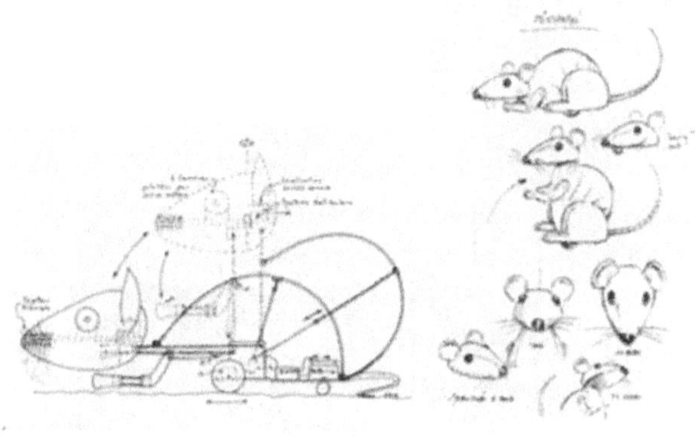

Figure 15. The overall design of the future Psikharpax.

Although it will usually lie flat on the ground, it will also have the possibility of setting upright, as well as of seizing objects with two forelegs. Likewise, its head will be able to rotate, and three pairs of motors will actuate each of its eyes (Figure 16).

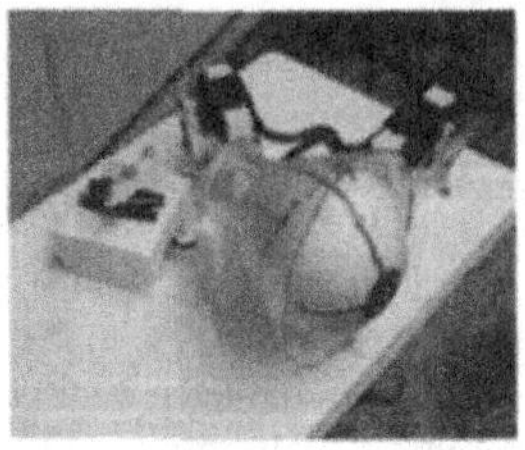

Figure 16. An eye equipped with a camera and a log-polar sensor, which is actuated by three motors. The whole device obeys the Listing's law (von Helmholtz, 1955).

Finally, several low-level reflexes will connect Psikharpax's sensors to its actuators, thus making it possible, for instance, to keep looking at an object even when its head is moving, and to avoid an obstacle detected by its whiskers or by its visual or auditory systems.

5. Directions of future work

Being able to integrate the past (through its recorded map), the present (through its sensors) and the future (through its planning capacities), Psikharpax is a nice example of a *motivationally autonomous animat* (McFarland and Bösser, 1993) whose survival may depend on a huge number of perceptions and actions that must be carefully integrated. Future work will be devoted to increasing this number as much as possible, in order to delineate the limits of the controllers described above, an attitude that is at the opposite of that of traditional robotics, when the chances of threatening the survival of a costly device are deliberately held as low as possible. Clearly, the control of such a complex sensori-motor system is a great challenge that goes beyond any other attempt at making an animat do something more than merely moving around. To successfully tackle this challenge will entail extending the capacities of the controllers described herein by letting them manage more perceptions -- notably a sense of smell -- more actions, more motivations and more emotions. It will also probably involve adding other biomimetic control structures, among which the cerebellum and the prefrontal cortex will probably prove to be of utmost importance, as well as other learning processes than the S-R one currently implemented, notably those that would afford capacities for goal-directed behaviors and behavioral sequence chunking (Dayan, 2001; Graybiel, 1998).

6. Conclusions

Considerable research effort has been already devoted to the Psikharpax project and the corresponding results have been summarized in this paper. In particular, several adaptive mechanisms have been designed that make navigation, action selection and learning possible. When a dedicated version of the Psikharpax platform becomes available, the adaptive mechanisms that have been studied so far will be implemented and integrated on it. This will probably generate new and interesting control challenges, from which useful contributions to biology, cognitive sciences and robotics are to be expected.

Acknowledgments

This research has been granted by the LIP6 and the Project *Robotique et Entités Artificielles* (ROBEA) of the Centre National de la Recherche Scientifique, France.

References

Albertin, S. V., Mulder, A. B., Tabuchi, E., Zugaro, M. B. & Wiener, S. I. (2000). Lesions of the Medial Shell of the Nucleus Accumbens Impair Rats in Finding Larger Rewards, but Spare Reward-Seeking Behavior. *Behavioral Brain Research.* 117(1-2):173-83.

Ashby, W. (1952). *Design for a Brain.* Chapman and Hall.

Baldassarre, G. (2002). A Modular Neural-Network Model of the Basal Ganglia's Role in Learning and Selecting Motor Behaviours. *Journal of Cognitive Systems Research*, 3(1):5-13.

Bar-Cohen, Y. and Breazeal, C. (2003) (Eds). *Biologically Inspired Intelligent Robots.* SPIE Press.

Barto, A.G. (1995). Adaptive Critics and the Basal Ganglia. In Houk et al., (Eds). *Models of Information Processing in the Basal Ganglia,* pp 215-232. The MIT Press.

Capi G., Uchibe E., Doya K. (2002). Selection of Neural Architecture and the Environment Complexity. *The 3rd International Symposium on Human and Artificial Intelligence Systems: Dynamic Systems Approach for Embodiment and Sociality*, pp 231-237. Fukui, Japan.

Dayan. P. (2001). Motivated Reinforcement Learning. *NIPS,* 14: 11-18. The MIT Press.

Doya, K.; Samejima, K.; Katagiri, K.; Kawato, M. (2002) Multiple Model-based Reinforcement Learning. *Neural Computation.* 14, 6: 1347-69

Filliat D. and Meyer, J.-A. (2003). Map-based Navigation in Mobile Robots - I. A review of localization strategies. *Journal of Cognitive Systems Research.* 4,4, 243-282.

Filliat, D. and Meyer, J.-A. (2002). Global Localization and Topological Map Learning for Robot Navigation. In Hallam et al. (Eds). *From Animals to Animats 7. Proceedings of the Seventh International Conference on Simulation of Adaptive Behavior.* pp 131-140. The MIT Press.

Gaussier, P., Leprêtre, S., Quoy, M., Revel, A., Joulain, C., Banquet, J.P. (2000). Experiments and Models about Cognitive Map Learning for Motivated Navigation. In Demiris and Birk, (Eds). *Interdisciplinary approaches to robot learning. Robotics and Intelligent Systems.* 24 : 53-94. World Scientific.

Girard, B. (2003) *Intégration de la navigation et de la sélection de l'action dans une architecture de contrôle inspirée des ganglions de la base.* PhD. Thesis, University of Paris 6, France.

Girard, B., Cuzin, V., Guillot, A., Gurney, K., Prescott, T. (2002). Comparing a Bio-inspired Robot Action Selection Mechanism with Winner-takes-all. In Hallam et al. (Eds). *From Animals to Animats 7. Proceedings of the Seventh International Conference on Simulation of Adaptive Behavior* pp. 75-84. The MIT Press.

Girard, B., Cuzin, V., Guillot, A., Gurney, K. and Prescott, T. (2003). A Basal Ganglia inspired Model of Action Selection Evaluated in a Robotic Survival Task. *Journal of Integrative Neuroscience,* 2(22), 179-200.

Girard, B., Filliat, D., Meyer, J-A., Berthoz A., and Guillot, A. (In Press). An Integration of two Control Architectures of Action Selection and Navigation inspired by Neural Circuits in the Vertebrates: the Basal Ganglia. *Proceedings of the 8th Neural Computation and Psychology Workshop. Progress in Neural Processing.* World Scientific.

Gourichon, S., Meyer, J.-A. and Pirim, P. (2002). Using colored snapshots for short-range guidance in mobile robots. *International Journal of Robotics and Automation: Special Issue on Biologically Inspired Robotics.* 17, 4:154-162.

Graybiel, A. M. (1998). The Basal Ganglia and Chunking of Action Repertoires. *Neurobiology of Learning and Memory,* 70:119-136.

Gurney, K.N., Prescott, T.J. and Redgrave, P. (2001a). A Computational Model of Action Selection in the Basal Ganglia. I. A new functional anatomy. *Biological Cybernetics.* 84, 401-410.

Gurney, K.N., Prescott, T.J. and Redgrave, P. (2001b). A Computational Model of Action Selection in the Basal Ganglia. II. Analysis and simulation of behaviour. *Biological Cybernetics.* 84, 411-423.

Hafner, V. V., Fend, M., Lungarella, M., Pfeifer, R., König, P., Körding, K. P. (2003), Optimal coding for naturally occurring whisker deflections. *Proceedings of the Joint International Conference on Artificial Neural Networks and Neural Information Processing (ICANN/ICONIP).* pp. 805-812. Springer Lecture Notes in Computer Science,.

von Helmholtz, H. (1955). *Treatise on Physiological Optics.* The Thoemmes Press.

Holland, O. and McFarland, D. (2001). *Artificial Ethology.* Oxford University Press.

Houk, J.C., Adams, J.L. and Barto, A.G. (1995). A Model of how the Basal Ganglia generate and Use Neural Signals That Predict Reinforcement. In Houk et al., (Eds). *Models of Information Processing in the Basal Ganglia.* The MIT Press.

Khamassi, M., Girard, B., Berthoz, A. and Guillot, A. (In Press). Comparing three Critic Models of Reinforcement Learning in the Basal Ganglia Connected to a Detailed Actor in a S-R Task. *Proceedings of the 8th Intelligent Autonomous Systems Conference.* IOS Press.

McFarland, D.J. & Bösser, T. (1993) *Intelligent Behavior in Animals and Robots.* Cambridge: The MIT Press.

Meyer, J.-A. and Filliat D. (2003). Map-based navigation in mobile robots - II. A review of map-learning and path-planning strategies. *Journal of Cognitive Systems Research.* 4, 4: 283-317.

Montes Gonzales, F., Prescott, T.J., Gurney, K., Humphries, M. and Redgrave, P. (2000). An Embodied Model of Action Selection Mechanisms in the Vertebrate Brain. In Meyer et al. (Eds). *From Animals to Animats 6. Proceedings of the Sixth International Conference on Simulation of Adaptive Behavior* pp. 157-166. The MIT Press.

Schultz, W., Dayan, P. & Montague, P. R. (1997). A Neural Substrate of Prediction and Reward. *Science,* 275:1593-1599.

Touretzky, D. and Saksida. L. (1997). Operant Conditioning in Skinnerbots. *Adaptive Behavior,* 5, 3/4 :219-247.

Trullier, O., Wiener, S., Berthoz, A. and Meyer, J.A. (1997). Biologically-based artificial navigation systems: review and prospects. *Progress in Neurobiology.* 51, 483-544.

Webb, B. and Consi, T.R. (Eds) (2001). *Biorobotics: Methods and Applications.* The MIT Press.

Interactions of Environment, Behavior, and Synaptic Patterns in a Neuro-Robotic Model

William H. Alexander
Department of Psychology
Indiana University
Bloomington, IN 47405
wialexan@indiana.edu

Olaf Sporns
Department of Psychology
Indiana University
Bloomington, IN 47405
osporns@indiana.edu

Abstract

We investigate the changing patterns of behavior of an autonomous robot embedded in a variety of environments, and the consequences of alterations in the environment and of the robot's behavior on the development of a neural model of neuromodulation. Our experimental environments consist of varying ratios of appetitive and aversive objects. In environments with a high ratio of uniformly distributed rewarding objects, the behavior of the robot is observed to affect the distribution of objects throughout the environment. Additionally, the alteration of the environment in this manner influences the behavior of the robot. As the interaction of the robot with objects in the environment produces clusters of appetitive objects, the robot tends to spend a proportionally longer amount of time near such clusters. The mutual influence of environment and behavior is also observed to alter the development of connections within the neural model used for signaling the prediction of a reward. The development of correct predictions about the occurrence of reward depends on both the behavior of the robot and the properties of the environment in which the robot is placed.

1. Introduction

Recent research in robotics has focused on the vital role of the environment in shaping behavior and neural development (Lungarella & Pfeifer, 2001; Weng et al., 2001; Almassy et al., 1998; Chiel & Beer, 1997). In such work, specific environmental properties (e.g., the fluid dynamics of water, or the statistical properties of a visual image) are necessary to support a given behavior or the development of specific neural properties. While the view of cognition as being fundamentally *embodied* is increasingly acknowledged (Verschure et al., 2003), study of embodied cognition may place emphasis on the morphological and physical properties of a robot (Pfeifer & Scheier, 1999), or on collections of simple rules that reproduce certain behaviors when placed in an environmental context (Mataric, 1997).

A particular approach to the study of embodied cognition emphasizes detailed models of specific neural components. A popular subject for research in this area has been the role of neuromodulation in the nervous system. Functional effects of neuromodulatory systems have been implemented in computational models as influences on neuronal response functions, learning rates, and synaptic plasticity (Servan-Schreiber et al., 1990; Hasselmo, 1995; Pennartz, 1996; Fellous and Linster, 1998; Doya, 2000; Hasselmo et al., 2002). Neuromodulatory systems are similar to value systems (Friston et al., 1994; Sporns et al., 2000, Rucci et al., 1997) in that they both serve as mediators of environmental saliency, and support long-lasting behavioral changes as a result of interaction with salient stimuli. In the context of embodied cognition, value systems are responsive to the occurrence of specific sensory stimuli, often as a result of an agent's behavior.

In previous work (Sporns & Alexander, 2002; Alexander & Sporns, 2003), we implemented a neural model of the primate midbrain dopamine system based on observed neurophysiology and anatomy. The model was tested both in computer simulations and embedded in an autonomous robot. Dopamine neurons in the ventral tegmental area of the midbrain have a characteristic pattern of activity related to processing reward information (Schultz, 1998; Montague et al., 1997). One interpretation of the dopamine signal is that it signals the unexpected occurrence or absence of a reward. In the course of learning trials during which a stimulus precedes the delivery of a reward, the dopamine signal is "transferred" to the onset of the stimulus. Dopamine neurons show a phasic increase in activity at the appearance of the stimulus, while no such activation occurs when the reward is delivered as usual. When the reward is delivered earlier than usual, dopamine neurons again respond. Absence of the reward at the expected time results in a transient decrease in baseline activity, with an increase if the reward is delivered later than usual. Our model was able to reproduce all aspects of the dopamine signal, both in computer simulation as well as under controlled-timing experiments with the robot.

In addition to replicating these characteristics, however, an additional goal was to investigate the activity of the model in a freely behaving agent under various environmental conditions (Alexander & Sporns, 2003). Little experimental data is available on the activity of the dopamine system in animals under unconstrained, autonomous behavior. A particular advantage of using neural models embedded in a behaving robot is the ability to continuously monitor the states of all units within the model, and to connect internal states with overt behaviors over the course of complete experimental runs. Previous research examined the development of synaptic weights involved in the generation of predictions during unconstrained behavior. In environments with a high number of rewarding objects, the ability of the model to generate and maintain predictions about the temporal relationship between a stimulus and the occurrence of a reward was severely disrupted. For environments with few appetitive objects, the model was able to more reliably predict the timing between a stimulus and a reward.

Additional observations suggested various structural properties of the environment (e.g., composition and distribution of objects) affected both the behavior of the robot as well as the development of the neural model. These effects are not inherent to the environment, or pre-specified in the programming of the robot or neural model. Rather, they are best viewed as the product of the interaction of the robot with the environment over a period of time. In this paper, we investigate the progressive, simultaneous alteration of the behavior of an autonomous robot and environments containing varying ratios of appetitive to aversive objects. Changes to structural properties of the environment depend on repeated interaction of the robot with objects in the environment. These changes, in turn, are observed to affect further behavior of the robot. The continuous, mutual influence of environment and behavior is discussed in terms of its impact on the development of the neural model.

2. Methods

Previous publications (Sporns & Alexander, 2002; Alexander & Sporns, 2003) detail the specifics of the robot, environment and neural model. For the present paper, we limit our discussion to salient features of the robot, model, and environment.

2.1 General

Data presented in this paper were collected using a robotic platform (Khepera) in a circular enclosure approximately 1 meter in diameter. The neural model was implemented in MatLab 6.5 (Mathworks, Natick, MA) running on a Linux 7.4 workstation (ASL, Newark, CA). The robot was interfaced to the workstation via a flexible wire bundle mounted at the top of the enclosure. A camera also mounted at the top of the enclosure was used to obtain overhead RGB images of the robot and environment during experimental runs at the rate of approximately 1 image every 2 seconds.

2.2 Environment

The environment consisted of a white floor and white walls lined with a strip of IR-reflective paper. Objects in the environment were 1 inch black cubes with a single colored face at the top. The colored face was either red or blue. Except for their color, objects were indistinguishable from one another. All blue objects were coated along their sides with conductive carbon paint, while all red objects were electrically non-conductive. Objects were light enough to be easily manipulated by the robot's gripper unit.

2.3 Robot

The robot, Monad, is a wheeled, circular robotic platform (approximately 5 cm in diameter) capable of rotation and translation through the environment at a constant speed. The robot is equipped with an arm/gripper unit with one degree of freedom that may be used to interact with objects encountered in the environment. Resistivity sensors located in the gripper measure electrical conductivity in objects with which the robot interacts. For the present paper, high conductivity is interpreted to be an aversive stimulus, while low conductivity is appetitive. A CCD camera turret mounted on the robot provided RGB images to the neural model, while IR sensors mounted along the base of the platform assisted in the detection and avoidance of obstacles.

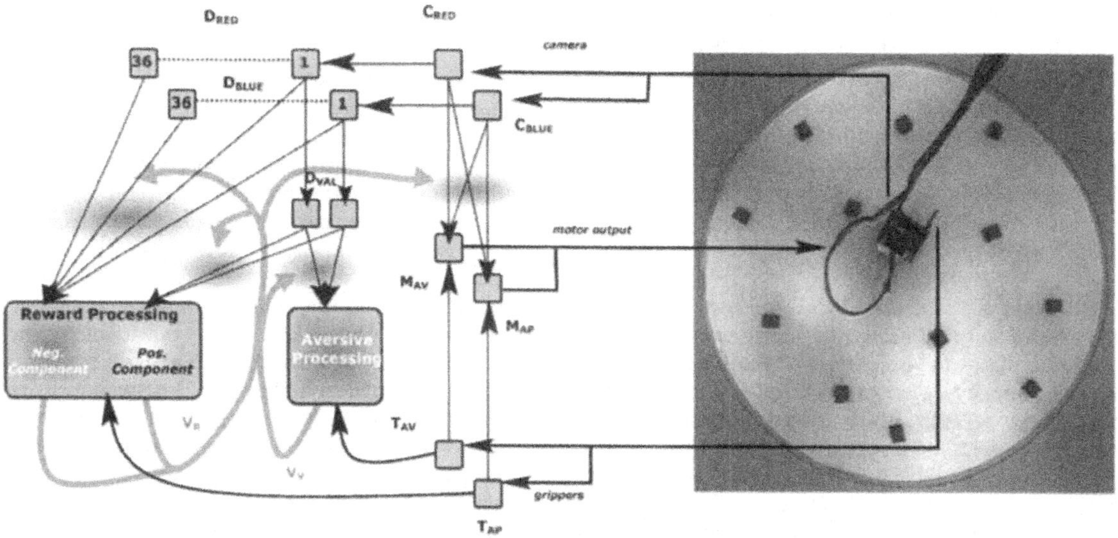

Figure 1: (Left) Schematic Diagram of Monad's neural model; for more details see Alexander and Sporns (2003) and Sporns and Alexander (2002). Gray boxes represent neural areas, black arrows represent interconnections, and gray arrows represent neuromodulatory projections (value signal). The model receives sensory input via a CCD camera and conductivity sensors. (Right) Monad in its environment. The environment contains a mixture of appetitive and aversive objects distributed throughout the environment. Sensory information is delivered to the neural model, and motor commands are received via a flexible wire bundle.

2.4 Neural model

The neural model is designed to incorporate anatomical and functional properties of the mammalian midbrain dopamine system. Figure 1 shows the connectivity of units within the model. RGB images supplied by the onboard camera activate color-sensitive units (C_{RED} and C_{BLUE}) in the model sensitive to the presence of red or blue stimuli. During interaction with an object, readings from the robot's resistivity sensor activate units (T_{AP} and T_{AV}) in the model corresponding with the occurrence of an appetitive or aversive event. The model is also used to control the movement of the robot. Activity in motor output units M_{AP} and M_{AV} controls object approach and avoidance.

At the onset of a trial, the model's color-selective units become active, signaling the presence of a salient (red or blue) external stimulus. The signal from these units is transformed into a representation of the stimulus in two ways. Activity in C_{RED} or C_{BLUE} is propagated to one of two units (depending on color) labeled D_{VAL}. D_{VAL} remains active during the course of a trial, representing the appearance of a stimulus for a set amount of time (~36 iterations of the model). Activity in C_{RED} or C_{BLUE} also drives activity in the groups of units labeled D_{RED} and D_{BLUE}. These groups, composed of 36 units each, essentially constitute a delay chain; the activity of C_{RED} drives activity in the first unit of D_{RED}, which then activates the second unit, and so on. Through their activity, units in the delay chains maintain a continuous temporal representation of the external stimulus. Units in the delay chain become active a certain number of iterations of the model after the stimulus onset (from 1 to 36 iterations), and their activity has a fairly broad temporal profile, such that there is significant overlap in the activity of neighboring cells within the delay chains.

Units associated with the temporal stimulus representation project to one of two components of the neural model that we relate to the primate VTA. The activity of a unit in the first component, S_{S1}, signals the occurrence of an event that is primarily rewarding, or the occurrence of an event that is predictive of a primary reward. Essentially, this component of the model represents positive errors in prediction, and its activity depends on projections from T_{AP} and D_{VAL}. Activity in T_{AP} signals the occurrence of an event that is primarily rewarding. D_{VAL} projects to the positive component via a single modifiable connection. Initially, activity in D_{VAL} has no effect on activity in the first component. As this connection is strengthened in the course of learning, however, D_{VAL} is sufficient to activate units in this component, indicating the prediction of an appetitive event.

Temporal representation units in D_{RED} and D_{BLUE} project to the second component, the activity of which indicates a negative error in prediction. When a primary reward reliably follows the onset of a stimulus at a consistent time, connections from the temporal units to the negative component are selectively strengthened, constituting a prediction about the timing of a rewarding event. When a reward that is predicted to occur at a specific time is withheld, the activity of a neural unit in the negative component of the model, S_{S2}, signals a negative prediction error.

A key feature of this model is the use of value-dependent learning rules used to update all modifiable connections within the model. Value-dependent learning constitutes a form of traditional Hebbian learning rules, with the addition of a term, V_R, which is a signed scalar variable. The value signal is derived from the activity of the positive and negative components of the neural model such that $V_R = S_{S1}-S_{S2}$. Learning occurs, then, only when a positive or negative prediction error is made.

2.5 Behavior

Monad's behavior is defined by several behavioral modes, and at each point in time, one of these modes is active. In the absence of salient visual stimuli, Monad's behavior is to *navigate* through the environment at a speed of approximately 4 cm/s while sensing IR reflectance and avoiding obstacles. Upon detection of a high contrast object, Monad *approaches* the object under the guidance of the visual system. When the object is proximal, Monad attempts to *interact* with the object. Interaction with an object consists of lowering the arm, rotating slightly to align the object with the gripper's sensors and closing the gripper, followed by a single iteration during which the electrical conductivity of the object is registered. The stereotypic behavior of the robot interacting with the object causes the object to shift its position slightly (~7 cm). After the interaction, Monad releases the object and *withdraws*, rotating away from the object and raising its arm. In the course of learning, the stereotypic behavior described above may be altered. After learning that certain objects are aversive, Monad's behavior will be to *avoid* the object, rather than to approach; the robot will reverse direction, turn 90 degrees and resume navigation. When an object is learned to be rewarding, Monad's response is to immediately approach and interact with the object.

2.6 Environmental Measurements

Images from the overhead camera were used to collect data on the position of the robot and of objects within the environment. During experimental runs, a single 640x480 RGB image (e.g., Fig. 1) was acquired approximately every two seconds. In addition, a time stamp was applied to the image, and used to synchronize environmental data with states of the neural model. A single pixel corresponds with an actual distance of approximately 2.6mm.

Overhead data was used to derive two measures of the structural properties of the environment. The first measure describes the distribution of rewarding objects throughout the environment. The average pair-wise distance is calculated for *N* appetitive objects as:

$$\frac{1}{N(N-1)}\sum_{i=1}^{N}\sum_{j=1,j\neq i}^{N} dist(i,j)$$

The second measure recorded the number of appetitive objects within 10 cm of the robot.

3. Results

Previous studies of the present model have investigated the functional characteristics of the neuromodulatory system (Sporns & Alexander, 2002; Alexander & Sporns, 2003). The focus of the experiments reported in this paper is on the influence of environmental properties and behavior on the development of the model. To this end, four manipulations of the environment were imposed, varying the ratio of appetitive objects to aversive objects. The first environment consisted of 12 objects, all of which had a red face and were appetitive. The objects were distributed uniformly throughout the environment by the experimenter prior to the start of an experimental run. The second environment consisted of 8 appetitive objects and 4 aversive (blue) objects. Again, the objects were distributed uniformly through the environment, with aversive and appetitive objects intermixed. The third and fourth environmental manipulations consisted of ratios of 6 appetitive objects to 6 aversive objects, and 4 appetitive objects to 8 aversive objects , respectively. The environment containing 12 appetitive objects is referred to as a "high-ratio" environment, while the environment with 4 appetitive objects is "low-ratio," indicating the ratio of appetitive to aversive objects in each environment. Other types of environments used are intermediate. Three experimental runs, lasting 20,000 iterations each (approximately 80 minutes in real time), were conducted for each of the four conditions. For all runs, Monad was allowed to behave in an autonomous manner with minimal intervention by the experimenter.

3.1 Development of the Environment

One of the goals of the present research is to describe the alteration of environmental properties caused by the activity of the robot. Previous observations have noted the tendency of objects within the environment to cluster together over the course of an experiment (Sporns & Alexander, 2003). While the interaction of the robot with objects in the environment results in progressive clustering of appetitive objects, this clustering is not a programmed behavior of the robot, but emerges from the continuous interaction of the robot and environment. Fig. 2a shows this alteration in the environment for a condition with a high number of appetitive objects (12) and a low number (0) of aversive objects (a "high-ratio" environment). The frame on the left shows the roughly uniform distribution of objects at the beginning of an experimental run, while the right frame shows the final distribution for the same run. Several appetitive objects have been shifted into a cluster in the center of the enclosure, with a few isolated objects. In contrast, for conditions with a low number of appetitive objects (4) and a high number (8) of aversive objects ("low-ratio"), appetitive and aversive objects remain well separated from one another (Fig. 2b).

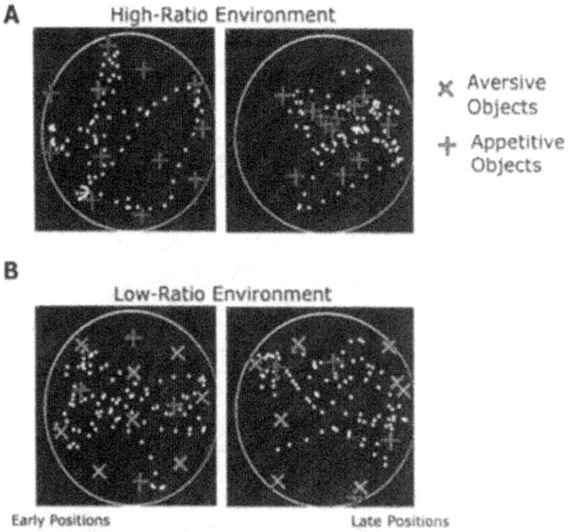

Figure 2: (A, left) The beginning of an experimental run in an environment with a high number of appetitive objects. Initially, the distribution of objects is uniform, and the robot traverses large areas of the environment. (Right) Near the end of the run, the objects are clustered near the center of the environment, and the robot's area of movement is limited. (B, left and right) The beginning and end of an experimental run in an environment with few appetitive objects. Objects remain widely distributed throughout the environment and the robot's motion is not restricted by clusters of rewarding objects.

Fig. 3a,b shows the change in this average measure over three runs per condition for all four conditions. Initially, the average pair-wise distance is fairly high for all conditions. For high-ratio conditions, however, the distance tends to decrease over the course of an experimental run. For low-ratio conditions, the distance tends to fluctuate widely between a low and high value. This kind of fluctuation is also observed to a lesser extent in the intermediate and high-ratio conditions. Due to the relative sparseness of appetitive objects in low-ratio conditions, the alteration of the position of one object naturally has a greater influence on our distance measure than in conditions with a higher number of objects. Additionally, clusters of appetitive objects may be formed and destroyed several times in the course of an experimental run, resulting in periodic increases in the average pair-wise distance between objects as Monad interacts with individual objects within a cluster. Such interactions may cause a cluster to decompose entirely, shift position in the environment, or re-form in two or more separate clusters.

3.2 Behavioral Changes

In addition to describing the development of the environment, of special interest are the consequences of alterations in the structural properties of the environment for the behavior of the robot. Fig. 2 shows the positions of the robot in the environment for approximately four minutes at the beginning (left) and end (right) for high-ratio (top) and low-ratio (bottom) experimental runs. Initially, in both cases, the robot tends to traverse a wide area of the environment. As the objects in the high-ratio environment cluster, the robot tends to spend more time near such clusters. In contrast, the presence of several aversive objects in the low-ratio environment results in restricting the areas of the environment the robot enters due to the aversive nature of the conditioned response. Fig. 4 shows the increasing in the frequency with which the robot is proximal (defined here as within 10 cm of the robot) to one or more appetitive objects for a representative experimental run in a high-ratio environment, as well as the lack of such tendencies for a run in a low-ratio environment. For high-ratio conditions, appetitive objects are typically close to the robot due in part to the overall density of appetitive objects within the environment. However, during autonomous behavior, the number of instances that the robot is proximal to appetitive objects, as well as the number of objects that are proximal, tends to increase. In low-ratio environments, no consistent increase in the proximity of appetitive objects is observed. These effects coincide with the progressive decrease in the average distance between objects; as objects tend to cluster together, the robot is more often proximal to single and multiple objects simultaneously.

As a consequence of this increase in the robot's proximity to appetitive objects, the pattern of interactions between Monad and objects is altered. Figure 5a shows this alteration for individual runs in a high and low-ratio environment. In high-ratio environments, there are on average a higher number of interactions than in low-ratio environments (approximately 180 to 65, respectively). This is due in part to the number of appetitive objects; clearly, more opportunities for interaction are afforded in environments with a high number of such objects. Successive interactions in high-ratio environments tend to occur in rapid bursts of activity, and are more temporally separated in low-ratio environments. Additionally, the timing of successive interactions changes as a consequence of alterations in the properties of the environment. In early stages of experimental runs (figure 5b, left frames), prior to significant development of the environment, interactions are temporally spread out for both high and low ratio environments. In later stages of the run, this timing has changed significantly in the high-ratio environment, as the robot interacts with clustered objects in rapid succession. The change is less obvious in the low-ratio environment, as appetitive objects tend to remain well separated from one another.

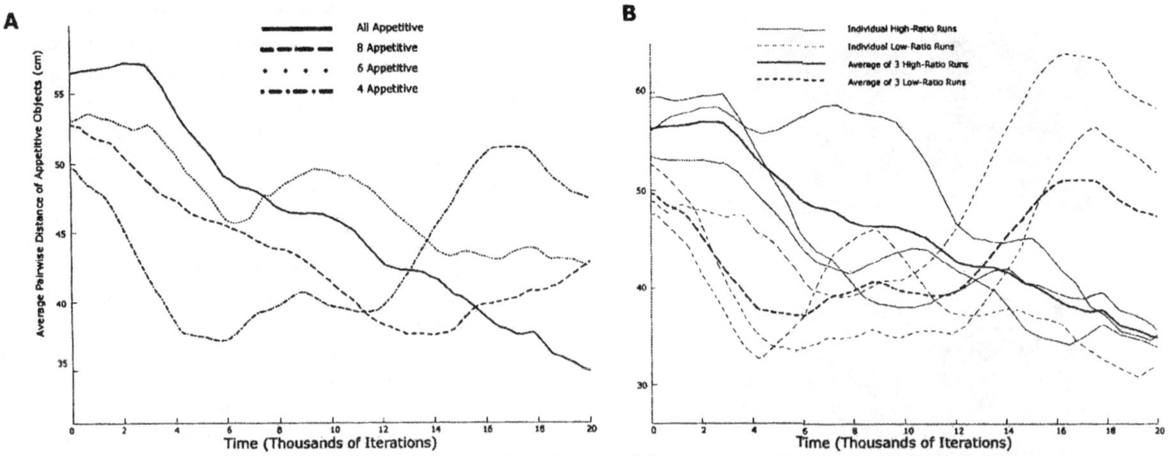

Figure 3: (A) the average pair-wise distance of appetitive objects of 3 experimental runs for high-ratio, low-ratio, and intermediate conditions. A consequence of the robot's behavior in environments with a high-ratio of appetitive to aversive objects is the consistent decrease in the average pair-wise distance of rewarding objects. (B) Average pair-wise distance for individual runs in high-ratio and low-ratio environments, showing a similar trend.

3.3 Neural Development

An additional goal of these experiments is to connect the co-development of the environment and robot behavior with patterns of development of synaptic connections within the neural model. Learning in the model is modulated by a signed value signal that acts on widespread targets within the model. Synaptic connections from D_{RED} to the negative component of the value system code for temporal relationships between the onset of a stimulus and the occurrence of a reward. The development of a reliable prediction requires consistent timing between the onset of a stimulus and an appetitive event. Changes in environmental contingencies and patterns of behavior affect the temporal relationship of a stimulus and reward, and this in turn influences the development of connections coding for this relationship.

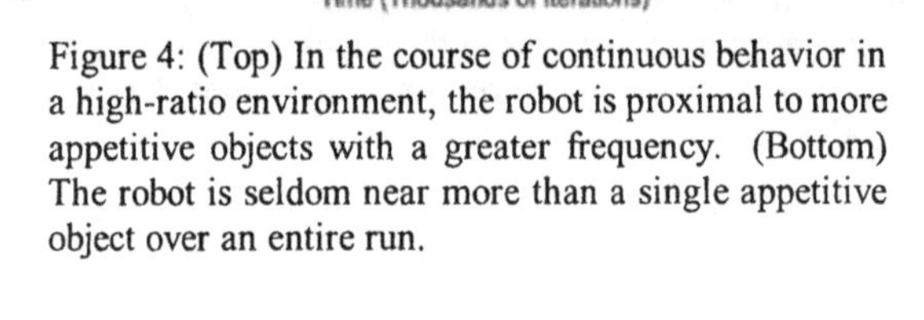

Figure 4: (Top) In the course of continuous behavior in a high-ratio environment, the robot is proximal to more appetitive objects with a greater frequency. (Bottom) The robot is seldom near more than a single appetitive object over an entire run.

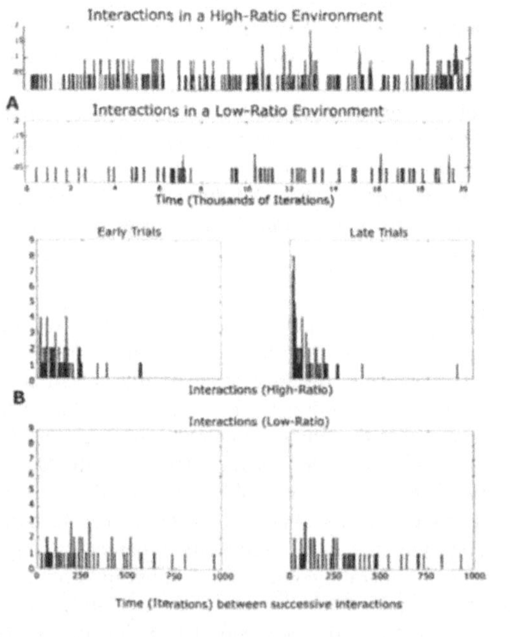

Figure 5: (A) Moving average of appetitive events for selected runs. Interactions with rewarding objects occur more often in high-ratio than in low ratio environments. Additionally, interactions in highly rewarding environments tend to occur in rapid succession. (B) The influence of the robot on structural properties of the environment further alters these patterns of interaction. The robot plays a key role in altering the statistical qualities of its sensory input.

To examine the pattern of synaptic development in the model in relation to global environmental variables, we relate the cumulative instances of synaptic change (i.e., the value signal is negative or positive) to their timing in relation to the onset of a reward-predicting stimulus. Fig. 6 shows this accumulation for low-ratio (left) and high-ratio conditions (right). Values in the graph greater than zero signify the number of learning instances for which synaptic connections increased in strength, while values less than zero indicate the number of learning instances for which modifiable connections were weakened. Each bin represents the amount of time (number of iterations) after the onset of a stimulus.

For the low-ratio environment, instances of learning tend to occur around a period in time after the onset of a stimulus consistent with a stereotyped behavioral routine of object detection, approach, and interaction. A few trials result in the occurrence of a value signal at later times. Generally, these late interactions are caused by the appearance of additional (appetitive or aversive) objects in Monad's field of view, resulting in a delayed or interrupted approach. Occasionally, Monad will interact with the same object multiple times during the same trial due to obstacle avoidance behaviors that cause the object to disappear and then reappear as Monad attempts to maneuver. Instances of learning in which the value signal is negative generally occur around the same point in time as positive value signals. These negative prediction errors are caused by an inherent "jitter" in the timing of approach behavior, as well as by the interference of multiple objects (aversive or appetitive) with approach behavior.

For high-ratio environments, the learning associated with a positive value signal in response to a rewarding event again generally occurs around the time needed for characteristic approach and interaction. However, proportionally more instances of learning occur at longer time periods following the onset of a stimulus due to multiple interactions within the same trial. A comparison between instances of learning early in the trial and late in the trial can be made by taking the total number of instances of positive value signals for times early in a trial (bins 1 to 18) and dividing by the sum of positive value signals later (19 to 36), constituting a ratio of early instances of learning to later instances. For environments with few appetitive objects, this ratio is 11.15, while it is 2.83 for high-ratio environments. The tendency for the learning to occur at later times in individual trials can be explained by the development of clusters of appetitive objects in high-ratio environments. Interactions with objects in such clusters often occur in rapid succession, frequently during the same trial. The occurrence of multiple rewards within the same trial drives learning events for time periods associated with a relatively long delay from the onset of a stimulus.

In addition to comparisons of learning instances based on global variables such as the ratio of appetitive to aversive objects, it is of particular interest to compare learning that may be more directly related to local environmental conditions. While the environment as a whole may contain a large number of rewarding objects, the area immediately surrounding the robot during a trial may be locally sparse with appetitive objects. Similarly, a globally sparse environment may occasionally have a locally dense concentration of rewarding objects. Fig. 7 shows the occurrence and within-trial timing of appetitive events for trials occurring in locally sparse (left) and dense (right) areas of the environment, cumulative over all conditions. Since the model's generation of value signals is dependent prior developmental history, we use the actual occurrence of a reward, rather than instances of learning, in this context.

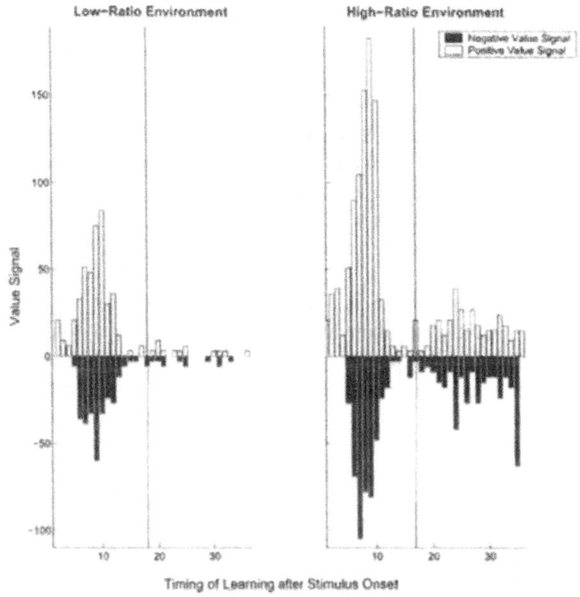

Figure 6: (Left) Instances of positive and negative value signals typically occur around a fixed time after the onset of a stimulus in environments with few appetitive objects. The consistent delivery of a reward after the onset of a stimulus is necessary to reproduce properties of the midbrain dopamine system. (Right) Multiple interactions during a single trial result in proportionally more instances of learning at later delay periods in environments with a large number of appetitive objects.

For areas of the environment that are locally sparse across all conditions, occurrences of rewarding events have the same general profile as discussed for all instances of learning in a low-ratio environment. Rewards in locally sparse environments typically occur around a point of time consistent with normal approach and interaction. For rewarding events that occur in a locally dense area of the environment, the occurrence of a reward after the onset of a reward-predicting stimulus again has a profile similar to that of learning instances in high-ratio environments. For the most part, rewards in such areas of the environment occur on a time scale appropriate for typical approach. However, due to the density of objects in the immediate area, multiple interactions may occur in the course of a single trial, resulting in rewarding events at later times in a single trial.

Again, we compare the number of appetitive events in the first 18 iterations of all trials to the number of appetitive events in the last 18 iterations. This ratio for locally sparse areas is 7.68, and is 2.65 for locally dense areas.

4. Discussion

The experiments presented in this paper investigate how environment, behavior and synaptic patterns interact and co-develop over time. As an autonomous robot is placed in an environment, e.g. containing a number of appetitive objects, its own actions can result in progressive alterations of both the spatial distribution of appetitive objects (e.g., the average distance from one object to other objects) which in turn can affect the behavior of the robot (e.g., in terms of its proximity to appetitive objects). In addition, such alterations affect the ability of the neural model to generate precise predictions regarding the temporal relationship between a reward-predicting stimulus and the occurrence of a reward.

For environments with a high density of appetitive objects, the behavior of the robot alters the environment in specific ways, producing clusters of objects. In turn, these alterations influence the behavior of the robot; clusters of appetitive objects exert an attractive force on the robot, resulting in restricted motion of the robot throughout the environment and changes in the patterns of interaction with objects. Low-ratio environments also produce alterations in behavior; however, these changes depend more on the initial distribution of aversive objects than on the continuous restructuring of the environment due to the activity of the robot. In such environments, the movement of the robot is restricted by the presence of aversive objects rather than by clusters of rewarding object. Additionally, the patterns of interaction remain relatively unchanged; appetitive objects are more often encountered one at a time, rather than in groups.

The characteristic activity of the primate midbrain dopamine system (Schultz, 1998) is highly dependent on the precise timing between a stimulus and reward. Previous studies of the present model (Alexander & Sporns, 2003; Sporns & Alexander, 2002) have demonstrated the model's ability to replicate salient aspects of the dopamine response under conditions in which the timing of reward delivery is carefully controlled. This consistent temporal relationship between the stimulus and reward is observed in the current experiments for conditions with a low number of appetitive objects, and for object interactions in locally sparse areas of the environment. In contrast, for conditions with a high number of appetitive objects, and for locally dense areas, this relationship is more variable, resulting in disruption of the model's ability accurately to predict the precise timing of the occurrence of a reward.

Analysis of the within-trial timing of rewarding events and instances of learning in which synaptic weights increase provides additional insight into the effects of environmental properties on neural development. While global properties such as the number of appetitive objects appear to have a large influence on the pattern of synaptic weights (e.g., fig. 6), this influence is more readily interpretable in the context of interactions between the robot and the local environment. An ongoing concern of research in embodied cognition is the identification of properties of the environment that directly influence brain and behavior (Clark, 1996).

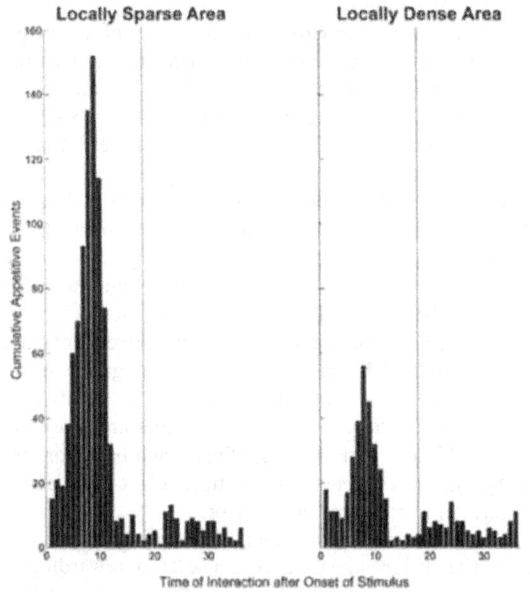

Figure 7: The cumulative occurrences of a primary reward are plotted in relation to the time after the onset of a stimulus at which they occurred and according to the local properties of the environment. In areas of the environment with very few appetitive objects proximal to the robot, rewards are typically delivered around a time associated with normal approach and interaction for the robot. When the area around the robot is cluttered with appetitive objects, however, this consistency is lost.

A potential implication of these observations is that the "normal" function of the midbrain dopamine system depends on properties of the environment that can be regarded as predictive of reward. Rewarding events occur more frequently and at less predictable times when the robot is proximal to clusters of multiple appetitive objects regardless of the global properties of the environment, and are more predictable when the robot is in a locally sparse area of the environment. Rapid sequences of rewarding interactions may constitute a disruption of normal reward processing. Alternatively, the inability of the model to maintain appropriate predictions for such sequences may indicate additional environmental cues important for reward processing not encompassed by the model. Cues such as the number of proximal objects and their positions relative to an agent, in addition to the temporal representation of a stimulus, might act to modify the prediction of a reward.

The focus of this paper is on the development of environment and behavior primarily in terms of rewarding objects within the environment. The presence of several aversive objects in low-ratio environments represents a significant complication in the experiments presented in this paper. The expression of an avoidance behavior may potentially affect the environmental properties discussed above. Further research along these lines should examine the potential interactions between competing behaviors, especially as it pertains to neural development.

In this paper, we discuss the simultaneous co-development of a neural model of a neuromodulatory system, the behavior of an autonomous robot, and the characteristics of an environment in which it is embedded. We find that the development of synaptic weights depends not only on the initial properties of the environment, but also on progressive changes within the environment that are due to the behavioral activity of the robot. These observations point to the need to incorporate environmental and behavioral influences into an account of neural development. Our work suggests future lines of empirical research aimed at understanding the action of the midbrain dopamine system during autonomous behavior. Our work also supports the importance of regarding neural function and cognition as fundamentally embodied.

Acknowledgements

This work was supported by NIH/NIDA grant 1R21DA15647-01.

References

Alexander, W. and Sporns, O. (2003). An embodied model of reward conditioning. Adaptive Behavior, 11, 143-159.

Almassy, N., Edelman, G.M., and Sporns, O. (1998) Behavioral constraints in the development of neuronal properties: A cortical model embedded in a real world device. *Cerebral Cortex*, 8, 346-361.

Chiel, H. J., & Beer, R. D. (1997). The brain has a body: Adaptive behavior emerges from interactions of nervous system, body and environment. Trends in Neurosciences, 20, 553–557.

Clark, A. (1996). *Being There: Putting Brain, Body, and World Together Again.* Cambridge, MA, The MIT Press

Doya, K., (2000) Metalearning, neuromodulation, and emotion. In: *Affective Minds*, Hatano, G., Okada, N., and Tanabe, H. (eds.), pp. 101-104, Elsevier.

Friston, K.J., Tononi, G., Reeke, Jr., G.N., Sporns, O., and Edelman G.M. (1994) Value-dependent selection in the brain: Simulation in a synthetic neural model. *Neuroscience*, 59, 229-243.

Fellous, J.-M., and Linster, C. (1998) Computational models of neuromodulation. *Neural Computation*, 10, 771-805.

Hasselmo, M.E., Wyble, B.P., and Fransen, E. (2002) Neuromodulation in mammalian nervous systems. In: Arbib, M. (ed.). *Handbook of Brain Theory and Neural Networks, 2nd Edition*. Cambridge, MA: MIT Press.

Hasselmo, M.E. (1995) Neuromodulation and cortical function: Modeling the physiological basis of behavior. *Behavioral Brain Resarch*, 67, 1-27.

Lungarella, M., Pfeifer, R. (2001) Robots as cognitive tools: information theoretic analysis of sensory-motor data. In *Proceedings of the 2001 IEEE-RAS International Conference on Humanoid Robots.*

Mataric, M. (1997). Studying the role of embodiment in cognition. Cybernetics and Systems, 28, 457-70.

Montague, P.R., Dayan, P., and Sejnowski, T.J. (1996) A framework for mesencephalic dopamine systems based on predictive hebbian learning. Journal of *Neuroscience*, 16, 1936-1947.

Pennartz, C.M.A. (1996) The ascending neuromodulatory systems in learning by reinforcement: comparing computational conjectures with experimental findings. *Brain Research Reviews*, 21, 219-245.

Pfeifer, R., and Scheier, C. (1999) *Understanding Intelligence*. MIT Press, Cambridge, MA.

Reeke, G.N., Jr., O. Sporns, and G.M. Edelman (1990) Synthetic neural modeling: The "Darwin" series of recognition automata. *Proc. IEEE* 78, 1498-1530.

Rucci, M., Tononi, G., and Edelman, G.M. (1997) Registration of neural maps through value-dependent learning: Modeling the alignment of auditory and visual maps in the barn owl's optic tectum. *Journal of Neuroscience*, 17, 334-352.

Schultz, W. (1998) Predictive reward signal of dopamine neurons. *Journal of Neurophysiology*, 80, 1-27.

Servan-Schreiber, D., Printz, H., and Cohen, J.D. (1990) A network model of catecholamine effects: gain, signal-to-noise ratio, and behavior. *Science*, 249, 892-895.

Sporns, O., Almassy, N., and Edelman, G.M. (2000) Plasticity in value systems and its role in adaptive behavior. *Adaptive Behavior*, 8, 129-148.

Sporns, O. (2002) Embodied cognition. In: *Handbook of Brain Theory and Neural Networks, 2nd Edition.*, Arbib, M. (ed.), Cambridge, MA: MIT Press.

Sporns, O., and Alexander, William H. (2002) Neuromodulation and plasticity in an autonomous robot. *Neural Networks* 15, 761-774.

Verschure, P.F.M.J., Voegtlin, T. & Douglas, R.J. (2003) Environmentally mediated synergy between perception and behaviour in mobile robots. Nature, 425: 620-624.

Comparative cognitive robotics: Using autonomous robots as empirical models of animal learning

Roul Sebastian John
University of Osnabrück
Institute of Cognitive Science
Katharinenstr. 24
D-49069 Osnabrück
Germany
rjohn@uos.de

Christian W. Werner
Heinrich-Heine-University of Düsseldorf
C. and O. Vogt Institute of Brain Research
Universitätsstr. 1
D-40225 Düsseldorf
Germany
wernerc@uni-duesseldorf.de

Abstract

In this article, we explore a methodology to use autonomous mobile robots as empirical models in the comparative psychology of learning and adaptation. Starting from the successful example of biorobotics which uses robots as models in biology, we propose an extended methodology called "comparative cognitive robotics". Instead of biology and ethology, we cooperate with comparative psychology in the process of constructing empirical robot models. Like in biorobotics, the robot models and the animals to be modeled are tested in the same experiments within the same experimental environments, applying the same means of behavioral analysis. We use the experimentally controlled environments of comparative psychology to test both animal subjects and robot models. This way, we do not need to resort to the abstracted description of a human observer to provide input to our model or to interpret its output, which might otherwise bias our results. As an application of our new methodology, we describe the construction of an empirical autonomous robot model of visual discrimination learning in chickens and report the results of the first comparisons between the robot and the animal performance in the same behavioral experiments. Due to emergent properties of its interaction with the environment, our robot model does not have to use abstracted representations of features or stimulus categories, or associative strengths altered by learning rules. It can thus be considered to be more parsimonious than previous, non-robotic approaches to model discrimination learning.

1. Autonomous robots as empirical models

1.1 Biorobotics

In the last ten to twenty years, autonomous robots have successfully been applied as empirical models in biology. Robot models proved to be helpful to study simple behavioral patterns in animals and to better understand the role of morphology in adaptive behavior. The research program of so-called "biorobotics" is characterized by a close collaboration between empirical biological research and the construction of robot models. The robot model and the animals are tested under similar conditions and their performance is analyzed using similar performance measures. The behavioral data generated by the model animal and by the robot model is compared and is taken as an indication of the quality of the model. Empirical experiments with the animal and the construction of the robot model inspire each other.

For two very instructive examples of biorobotics, see the model of stick insect walking presented in Dean, Kindermann, Schmitz et al., 1999 and the model of ant navigation described in Lambrinos, Möller, Labhart et al., 2000; for a comprehensive overview of biorobotics and further examples, see Webb, 2001.

Biorobotics itself has been an extension of the methodology of computational neuroethology which works with simulated models. For an overview of the methodology of computational neuroethology, see e.g. Beer 1990, and the special issue of *Adaptive Behavior* on computational neuroethology, Vol. 3 No. 4.

1.2 *From biorobotics to comparative cognitive robotics*

Though very successful, the methodology of biorobotics as it is currently applied also faces some important restrictions. Like the simulated models of computational neuroethology, the autonomous robot models of biorobotics are restricted in the sense that …

- Only one species or a small group of species is modeled.
- Rather fixed behavioral mechanisms are modeled, which show only very simple forms of adaptation or learning.
- The models often focus on a neuronal level of explanation. But as explanations on this level are not (yet?) available for more complex phenomena, this very much restricts the range of phenomena that can be studied.

Not all of these restrictions necessarily hold at the same time for all models in biorobotics. For example, a model might only partly be implemented as simulated neural networks, while other parts take the form of modules that deliver information in an appropriate form without going into detail how this information might be computed in the brain. At the same time, these three restrictions are of course also closely connected. For example, the focus on a neuronal level of explanation leads to a restriction of the complexity of behavioral patterns that can be studied. And very specialized behavioral patterns might only be found in a limited range of species.

The general restriction of biorobotics is its focus on biology, or (neuro-)ethology. This science is an important source of knowledge on the way animals realize their outstanding autonomy. But there are further sources of empirical data on animal behavior outside of biology that should to be taken into account to achieve the long-term goal to build autonomous robots.

1.3 *Comparative cognitive robotics*

While cognitive scientists still mostly use computer-based and mathematical models, we would like to illustrate how the modeling approaches of cognitive science can successfully be extended by the additional use of autonomous robot models. We are aiming at developing empirical autonomous robot models. With the term “empirical models”, we denote models that are created to reproduce data gathered from a natural system by implementing a theory on processes that might generate the behavior measured by this data. Biorobotics is a very successful approach towards the implementation of empirical autonomous robot models in biology. We take biorobotics as our example to derive an extended methodology that can also be applied in cognitive science. Instead of biology, we choose the comparative psychology of learning and adaptation as our collaboration partner in the construction of autonomous robot models. Comparative psychology aims at finding common mechanisms of learning and adaptation in a wide variety of species which can be as different as monkeys, pigeons, dolphins, rats, and even bees or humans. This research gives insight into possible evolutionary steps in the development of complex cognitive abilities as well as into mechanisms which seem to be common to a large variety of species (for an overview, see e.g. Domjan, 1998; MacPhail, 1987; Pearce & Bouton, 2001; Roitblat, 1995; Roitblat & von Fersen, 1992; Shanks, 1995; Staddon, 2000, 2001). By choosing comparative psychology rather than biology as our collaboration partner, we overcome the restriction of biorobotics to rather fixed behavioral patterns that are specific for single species and/or should best be understood on a neuronal level of explanation. As will be illustrated below, we instead focus on psychological, yet “subsymbolic” models of behavioral control, which are implemented in autonomous robots. These autonomous robots are again evaluated as empirical models by closely collaborating with empirical research and by testing the robot models and the model animals under the same or very similar conditions. But this time, the environments used to test the robot models and the animals stem from the empirical research of comparative psychology, not from biology or (neuro-)ethology (for an extensive discussion of this methodology see John, 1998).

To summarize, the main properties of comparative cognitive robotics (which have partly been inherited from biorobotics) are:

(1) As a collaboration partner in constructing autonomous robots, we choose the comparative psychology of learning and adaptation.

(2) We are mainly interested in those phenomena of learning and adaptation that can be found in a wide variety of species.

(3) Empirical research with the model animal and the construction of the robot model work hand in hand. Work on the model inspires new experiments and new theories, and the empirical findings are used to update the model.

(4) The model animal and the robot model are tested in the same or similar environments commonly used in comparative psychology, with the same or similar means of analysis. The match between those measurements is taken as an indication of the quality of the model.

The term “comparative cognitive robotics” is meant to indicate that we are interested in mechanisms which could also form the basis of phenomena of complex cognition and that we are using methods and experimentally controlled environments of comparative psychology.[1]

[1] The term “cognitive robotics” has been introduced earlier for application-oriented autonomous robot research that is using complex (“cognitive”) algorithms and for robots illustrating principles that might later be used to develop models of cognitive abilities. This research in cognitive robotics has to be distinguished from our approach towards building autonomous robots as empirical models of phenomena studied with methods of comparative psychology. Research groups for cognitive robotics can e.g. be found at the University of Toronto, the MIT or the Fraunhofer Institute for Autonomous Intelligent Systems (AIS) in Sankt Augustin, Germany.

2. Experiments on visual discrimination

As one application of our new methodology, we have implemented an autonomous robot model of visual discrimination learning in chickens (*Gallus gallus* fd). In this article, we will mainly focus on the implementation and on the level of description used for our model and only give a very brief overview of our animal experiments. For a comprehensive discussion of the psychological theory and the detailed results of the animal experiments, please refer to Werner et al., 1999, 2001, 2003, to appear.

2.1 Theoretical background

Being able to treat different things differently is a key to any sensible ("intelligent") behavior, in humans, animals, and robots. Learning to differentiate between just a few simple objects (or patterns) according to their visual appearance is called visual discrimination. Learning to discriminate a multitude of complex visual patterns and transferring this ability to new patterns of the same kind is called visual categorization. Visual discrimination and categorization is an important ability of every intelligent system, and also forms the basis of "higher" forms of cognition (e.g. denoting visual categories by symbols).

An important question concerning visual discrimination is which information on visual stimuli enters the learning process. In principle, there are two extreme possibilities: Visual stimuli might first be analyzed and decomposed into high-level "features". During learning, associations are then formed between representations of these stimuli parts and—e.g.—information on reinforcement (see e.g. Cook, Riley & Brown, 1992; Riley, 1984). Such a type of processing is commonly called "analytic processing". An advantage of processing visual stimuli analytically could be that the different features of a stimulus could selectively be taken into account during learning, e.g. "relevant" features could be "attended to" while "irrelevant" features would be "ignored". However, analyzing a stimulus into its features is a costly process. And at least in theory, the number of features that could be constructed from a raw image and could later be found relevant in learning is endless. Thus, another possibility is to assume that the stimuli are not represented as separated features during learning. Rather, a representation of the whole pattern of stimulation (an "exemplar") is assumed to enter the process of learning (see e.g. D'Amato & Salmon, 1984; Grant & MacDonald, 1984; Pearce, 1994, 2002). Such a type of processing is commonly called "holistic processing". Of course it is also possible that both types of processing occur, e.g. dependent on the type of stimulus that is perceived. Some stimuli can easily be decomposed. Such stimuli are called "separable compound stimuli" as their "compounds" can easily be separated in an analytic visual process. For other stimuli, the visual dimensions are so closely integrated that it seems hard or even impossible to separate them. Such stimuli are called "integral compound stimuli". It is assumed that separable compound stimuli are processed analytically, while integral compound stimuli are rather processed holistically (see e.g. Kemler Nelson, 1993; Monahan & Lockhead, 1977; Riley, 1984).

But is it really necessary to assume two different types of processing? Or is it possible to account for the same empirical findings by assuming just the simplest type of processing, i.e. holistic processing? This is the central question we address with our empirical animal research and our autonomous robot model.

Experiments carried out earlier already did indicate that just a holistic type of processing could suffice to account for visual discrimination in chickens. Stimuli with a multitude of feature dimensions were constructed. Only one of these dimensions was relevant to predict reinforcement. However, it could be shown that the animals' behavior was controlled by all feature dimensions. The animals seemingly were not able to focus on single dimensions while ignoring the others. The principal components analysis of the behavioral data further revealed that the animals' performance could not be explained by a prototype model. The model which best accounted for the data was an exemplar model (Werner & Rehkämper, 1999, 2001).

2.2 The animal experiment

This finding was backed up in a systematic study combining both separable and integral compound stimuli in several training and transfer phases that also involved a reversal of reinforcement for single features of the stimuli (Werner et al., 2003, to appear). The stimuli used in our new study were compositions of the color features green and red and of two possible line orientation features: vertical and horizontal. Stimuli did either show only one of these features (a black vertical or horizontal line on transparent background or a purely red or green stimulus), or a compound of both features. Compounds could either be separable (black vertical or horizontal bar on red or green background) or integral (red or green bar on transparent background, the bar being either vertical or horizontal). A black-and-white illustration of the stimuli used and of the schedule of the experiment is shown in fig. 1.[2]

Uncolored bars on colored background are commonly considered to be separable compound stimuli, as they can easily be separated into parts. They are assumed to be rather processed analytically. On the other hand, colored bars on transparent background inseparably integrate the dimensions of color and line orientation, and are thus commonly seen as integral compound stimuli, which are rather processed holistically (see Leith & Maki, 1977).

[2] Please note that the different hatchings in this black-and-white reproduction correspond to the two colors red and green that were used, respectively. Further note that we show the different stimuli in circles because they were projected on round pecking discs. Being round is thus not to be considered as a feature that could help to distinguish between reinforced and non-reinforced stimuli.

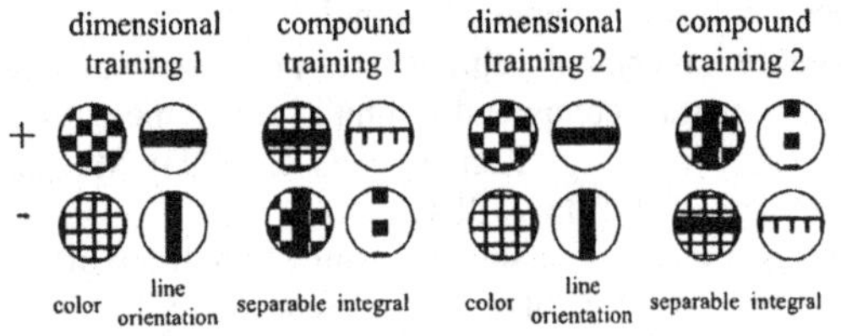

Figure 1: Stimulus material and training sequence of a visual discrimination experiment. Values of the color dimension were red and green, lines could be oriented either horizontally or vertically. "+" and "-" indicate reinforcement and non-reinforcement.

Nine adult chicken hens (about two years of age) of the breed Japanese Bantam (Chabo) served as subjects. All chickens were trained in a sequence of four training phases: two with four different dimensional stimuli (dimensional training 1 and 2) and two with four different compound stimuli (compound training 1 and 2). The reinforced dimensions were balanced across groups of chickens. In one daily session, each hen was trained for twenty trials. The four different stimuli of the current training phase were presented five times in a session. Two stimuli were presented at a time, each one on a separate pecking disk. One stimulus always was a reinforced one, while pecking on the other would not lead to reinforcement. The first trial started with a 60-sec delay followed by the onset of stimulus presentation, which was response-terminated after a delay of 2 sec. A single key-peck on one of the two discs was sufficient to start either food-access for 3 sec, if the choice was correct, or a pause of the same duration, if it was incorrect. The next trial followed after another pause of 5 sec. Each hen was trained until 90% of the pecks were correct during a single session, for both groups of stimuli that were presented. That is, one session e.g. in dimensional training had to reveal 90% correct choices for color *and* for line orientation stimuli; 90% correct choices just for one of these groups did not fulfill the learning criterion. If the learning criterion was fulfilled, the respective hen entered the next training phase. As illustrated in fig.1, a half reversal of reinforcement of one dimension occurred from one training phase to the next. For example for those chickens for which red and horizontal were the reinforced dimensions in dimensional training 1, red was not reinforced in compound training 1, while horizontal still indicated reinforcement. This sequence of half-reversals resulted in a full-reversal of both dimensions of the whole compound stimuli between compound training 1 and 2.

Further details on the experiment, its theoretical background and a detailed analysis of the empirical data can be found in a separate article (Werner et al., to appear). The main results can be summarized as follows:

- In the dimensional training phases, color stimuli were learned better than line orientation stimuli, i.e. the percentage of correct pecks was higher in each session, and the level of 90% correct was reached faster for color stimuli than for line orientation stimuli.
- In compound training, the separable compound stimuli were learned better than the integral compound stimuli.
- The half-reversal of reinforcement of the color dimension from dimensional training to compound training resulted in a decreased discrimination performance for separable compound stimuli, whereas the half-reversal of reinforcement of the line orientation dimension resulted in a decreased performance for integral compound stimuli.
- The resulting full reversal of reinforcement for both stimulus dimensions between compound training 1 and compound training 2 led to an additional decrease of performance for both types of compound stimuli.

Especially the interaction between the type of half-reversal and a decrease of performance for a specific type of compound stimulus would not be expected in the common feature-based theory (the most advanced such theory in animal psychology which should actually be able to deal with our kind of experiment can be found in Pearce, 1994, 2002). At least the dimensional and the separable compound stimuli would be assumed to be analyzed into and represented by their features, and it should be possible to selectively "attend" only to those features which are relevant for predicting reinforcement. As the reinforcement of one feature dimension is reversed from dimensional training to compound training, the animal should focus on the other dimension, which would still allow to perfectly predict reinforcement. If at all, only integral compound stimuli should show decreased performance, as they are thought to be processed rather holistically, and it could be more difficult to attend to only one of their stimulus dimensions. The empirical results, however, do not match with these predictions at all. They can best be accounted for if we again assume the stimuli to be represented as whole pictorial exemplars rather than by separated features. If such a holistic stimulus representation is assumed, it is clear that dimensional color stimuli (e.g., a completely red stimulus) are more similar to separable compound stimuli (e.g., a black horizontal bar on a red background) than to integral ones (e.g., a red horizontal bar on a transparent background), while dimensional line orientation stimuli (e.g., a black horizontal bar on transparent background) are more similar to integral compound stimuli (e.g., a red horizontal bar on a transparent background) than to separable compound stimuli (e.g., a black horizontal bar on a red background). (From this verbal description alone, this might still not seem obvious. See fig.1 or best sketch a colored version of the stimuli to see why.) This way, the interaction between type of reversal and affected compound stimulus type which seems mysterious from the perspective of a feature-based approach can easily be explained by assuming a holistic stimulus representation. For a detailed analysis of this explanation, and a comparison with other approaches, see Werner et al. 1999, 2001, 2003, to appear. Here, we will focus on the question how an empirical autonomous robot model for this type of experiment was developed in the framework of comparative cognitive robotics.

3. An application of comparative cognitive robotics: Modeling visual discrimination

3.1 Robot and animal environment

First, a robot had to be constructed that could be used in exactly the same experimental environment and in the same experiments as the animals. An illustration of the environment used is shown in fig. 2. Both the chickens and the robot were tested in a conditioning chamber. The stimuli just described were projected on the left and right pecking disks with a carrousel slide projector controlled by a computer. The chickens were moving between both pecking disks and had to select one of them by pecking. Then the next slide was shown. Due to their distance, the chickens could not see both pecking disks at a time, but had to inspect them separately. The robot we constructed thus also had to be able to move between the two pecking disks and to detect the disks with a camera. In addition, it was equipped with a simple "pecking" mechanism. Basic behaviors were programmed to move the robot in the box and to situate it at the appropriate position in front of a pecking disk. The only change made to the environment concerned reinforcement. In the case of the chickens, pecking on the correct stimulus was reinforced by opening a food hopper for a certain amount of time. In addition, a light was switched on to signal the availability of food. In the case of the robot, the food hopper was removed. Instead, only the light signal was used to indicate reinforcement to our robot model.

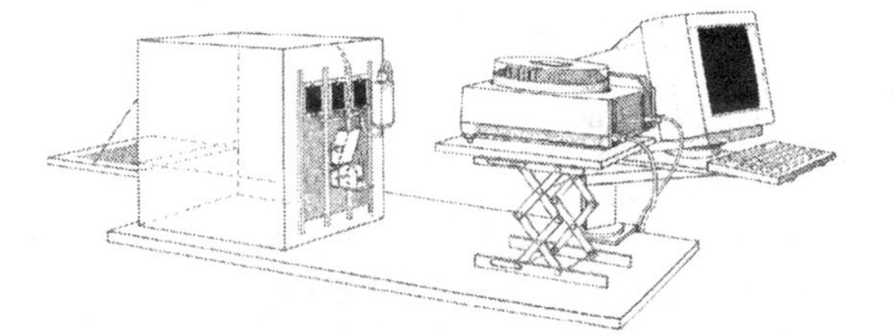

Figure 2: Drawing of the conditioning chamber, carrousel slide projector, and PC-components used to control the experiment.

3.2 Robot construction

A view of the complete robot is shown in fig. 3. The robot was constructed using the LEGO Mindstorms™ kit. A standard web cam was attached to the robot that allowed to switch off the automatic color correction which would otherwise have biased the processing of colored stimuli. The robot was controlled by a separate computer. Moving instructions were sent and data from the bumpers at each side of the robot was received by an IR sender/receiver which replaced the food hopper in the experimental box. The camera data was transferred through a cable also shown in the picture. A more detailed view of the "pecking" mechanism is shown in fig. 4.

Figure 3: Full view of the autonomous mobile robot model.

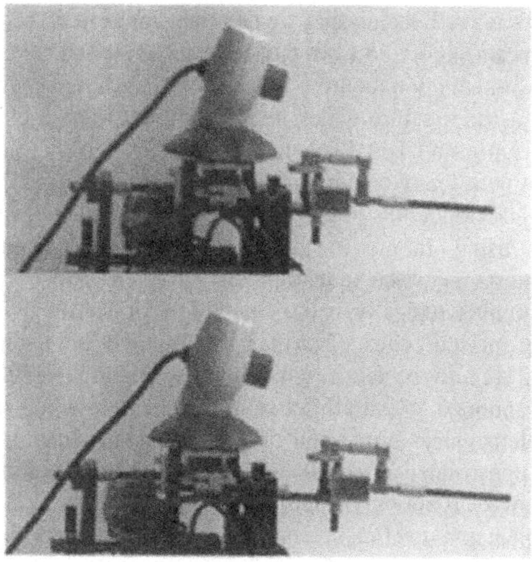

Figure 4: A detailed view of the mechanism used for "pecking", fully retracted (upper half) and fully extended (lower half).

3.3 Holistic stimulus processing: Arguments against using features

Special care had to be taken in the design of the robot control and learning algorithm. The main goal of our model is to test whether holistic processing of the presented stimuli alone can account for the performance of the chickens. If a robot model that is equipped only with a holistic type of processing is showing the same performance as the chickens, in the same experiments, within the same environment, this would be a strong indication that indeed no analytical type of processing has to be assumed to explain the experimental animal data. Thus, no prior analysis and decomposition of the stimuli could be assumed.

A separate argument against the use of features stems from the use of autonomous robot models in general. In simulated models, the input and output is typically coded in features. These features can be expressed by symbols or as the activation of units. If we e.g. wanted to model a learning process that associates a black horizontal bar on red background with reinforcement, the features "horizontal bar" and "red background" would typically be coded separately. If we wanted to use the same approach within our autonomous robot, we would first have to program an additional process that maps the camera input to these separate feature codes. But this seems impractical for a number of reasons: First, algorithms for extracting such visual features are notoriously complex and error-prone, especially if applied in a mobile robot. While the named features might still be extractable, extracting features becomes more and more difficult as the complexity of the proposed features increases. Especially for real-world contexts, the visual processing should thus be kept as simple as possible. Second, if we assume an initial step of visual processing that codes any given input into features, and if our learning algorithm actually operates on these codes, we might as well decide that we are only interested in learning and not so much in how this code is generated. We could then assume the coding process as given and just feed ready-made feature codes into our model (like, e.g., in the model presented in Pearce, 1994, 2002). In this case, no robot with real-world input would be needed. Actually, this is what frequently happens in research labs that start out with using robots to study cognitive phenomena like learning. In the construction of their robot models, researchers typically resort to models of learning that use feature-based codes. But then they realize that such codes could directly be fed into the models, and the use of robots is abandoned.[3] If at all, robots might then only be used as a supplementary illustration of the learning mechanism that was constructed in simulation. In the case of comparative cognitive robotics, robots are not to be used as later supplements to finished simulated models, but they are themselves to be taken as models. Therefore, a different approach towards the development of an algorithm for learning (but also for any other cognitive phenomenon that is modeled) has to be taken.

Finally, a third argument against the use of a feature-code as the input to our learning mechanism is that it is unclear which features should be used. Typically, the features to be used stem from a verbal description of stimulus features that a human observer found relevant. In the case of our experiment, the stimuli that were used could be described as a combination of features ("red vertical bar on transparent background", "black horizontal bar on red background", etc.). This suggests that these named features should be used in a coded input. But actually, there is no guarantee that precisely these features are extracted by every observer, especially when it is non-human, like in our case. And if our model rested on the assumption that a list of relevant features for every situation that might occur is pre-specified, this would contradict our long-term goal to construct a robot that is able to act as autonomously as a natural agent.

These arguments do not speak in general against the use of features within the processes that control our autonomous robot model. But if any such high-level features were needed, they would have to be derived by the robot itself, and not to be pre-specified by a human engineer. In general, an autonomous robot model cannot assume "stimuli" as given. Stimuli are abstractions from a real-world input that the robot has to find by itself. This process of stimulus construction thus has to be part of any autonomous robot model of learning, and it has largely been ignored in existing models that assume a feature-coded representation of given stimulus inputs.

3.4 An exemplar-based learning mechanism

With our autonomous robot model, we wish to test our hypothesis that the learning performance of the animals can be accounted for by assuming only an exemplar-based, holistic type of processing. We are thus using only this kind of processing in our robot. The key assumption is that no abstracted, "higher-level" descriptions of the perceived patterns need to be generated. We therefore decided to start with a most parsimonious mechanism that does not presuppose any further processing of the pixel arrays generated by the camera. A camera picture of a stimulus is stored as an exemplar in a database, together with the information whether pecking on this stimulus picture led to reinforcement or not. Every such exemplar is stored, no exemplar is discarded, deleted or overwritten. When a new stimulus picture is detected in the middle of the camera image, the learning system first searches for similar pictures in its database. If no similar picture was encountered before, the robot (like the chickens) will peck. In the exemplar that is generated it will then be stored in the database whether pecking on this picture led to reinforcement or not. If there are exemplars with similar

[3] One interesting example are the ingenious studies on the origin of language by Luc Steels and his collaborators at the Free University of Brussels. In their so-called "Talking Heads Experiments" (see Steels, 1999 for an extensive overview), simple geometrical figures are identified on camera pictures. While the segregation algorithms and the stimulus dimensions are given, the features in these dimensions are dynamically generated such that a distinctive representation of the geometrical figures is possible. Although this experiment is dependent on a separate source of input, it does not depend on the use of artifacts coupled to a real-world environment. This shows in the fact that the cameras and real boards with geometrical figures are now being replaced by a simulated environment (Vogt, 2003). The use of robots for the study of language is not given up, however. But in their new robotic models, Steels and his collaborators have given up feature-based representations of stimuli and instead use an exemplar-based approach that is very similar to our own (see Steels & Kaplan, 2001), though developed totally independently and for a rather different field of research.

pictures in the database, some of them will indicate that pecking on a similar picture led to reinforcement, while other exemplars represent earlier experiences in which no reinforcement was detected. The probability to peck on the picture that is currently perceived by the robot is then computed following a simple algorithm. First, the sum s1 of the weighed similarity to the current camera image of all stored pictures that are associated with reinforcement is computed. Second, the sum s2 is computed in the same way for all pictures that are associated with non-reinforcement. And third, the probability for pecking is computed as p1 = s1/(s1+s2), while the probability of non-pecking is given by p2=s2/(s1+s2). Thus, the decision whether to peck or not is taken by following a simple implementation of the Luce decision rule that is well established in psychological theory (Luce, 1959; Page, 2000).

As can be seen, similarity plays a central role in our algorithm. As a first approach, we use the simplest similarity measure we could think of:

(1) compare pixel (x,y) of picture 1 with corresponding pixel (x,y) of picture 2
(2) difference := Euclidean distance of the colors of the two pixels in L*a*b* color space
(3) compute mean distance of all corresponding pixels
(4) the most similar picture is the one with the lowest mean distance

The L*a*b* color space was used because it maps the perceived similarity of colors tested in psychophysical experiments onto Euclidean distances (for a description of the L*a*b* color space and a discussion of its limitations see Lew, 2001). To weigh the similarity, a decreasing exponential function was used, following the suggestion of a general law of generalization by Roger Shepard (1987).

You will surely notice that this similarity measure is very sensitive to noise and even to simple shifts of the images. This would commonly seem to be a disadvantage, because even pictures which show "the same stimulus" will not necessarily be regarded as very similar. But note that this view again entails the use of a feature-based description of the stimuli: Two pictures showing a "black vertical bar on red background" would only seem to be "the same stimulus" if we were using a feature-based representation of these stimuli that alludes to their verbal description. On the level of the camera input, however, these two stimuli might very well show a lot of differences. The fact that we did not try to "tame" the real-world nature of the visual input before feeding it into the learning mechanism actually turned out to be an advantage of our model.

3.5 *Results of first experiments*

Our robot model is equipped with just a few basic behaviors to move within the experimental environment and a simple exemplar-based learning mechanism which operates on unprocessed sensory data. Already with these simple assumptions, we were able to obtain a good fit of the behavioral data gathered from the animals and from the robot model in the first experiments. The robot and the animals were tested in the same behavioral experiments within the same experimental environment. The same methods of measurement and of analyzing the data were used. The performance of the chicken hens and the robot model was measured by the percentage of "correct pecks". As can be seen in fig. 5, in the two experimental sessions we have carried out so far, the performance of the model and the hens matched within a range of only around 1% difference.

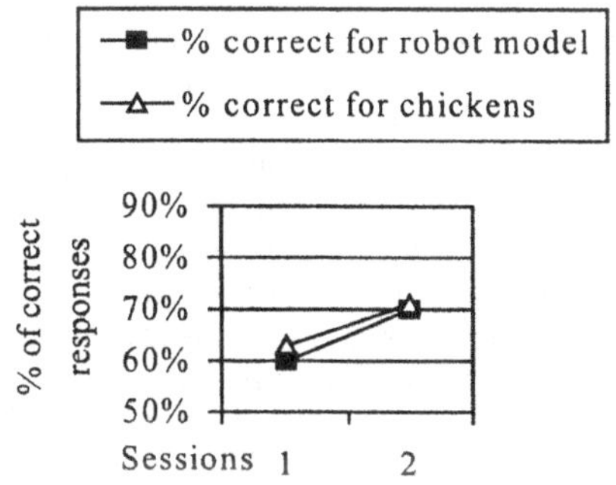

Figure 5: Results of a first comparison of the performance of the chickens and the robot model in the same experiments within the same experimental environment. Performance is measured by the percentage of pecks on the "correct" (i.e. the currently reinforced) stimuli. Data is averaged over nine runs with different experimental conditions (e.g. red and horizontal reinforced vs. green and vertical reinforced). Each session under one experimental condition consisted of 20 trials. All stimuli were shown an equal number of times. For further details on the experiments, see paragraph 2.2.

Furthermore, the "fine structure" of the animal performance was also reflected in our robot data. As reported above, the animals were able to acquire the color discrimination much faster than the discrimination between the two line orientations. As can be seen in fig. 6, the same was found for our robot model.

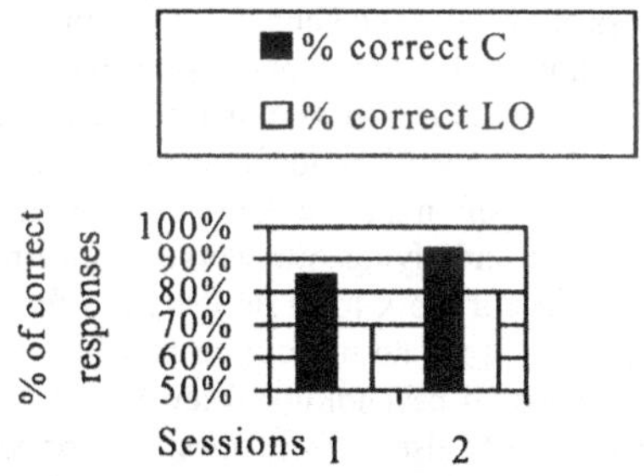

Figure 6: Comparison of the discrimination of color and line orientation stimuli. Like for the chickens, learning the color discrimination was easier for the robot than learning the line orientation discrimination. The data refers to the same experimental sessions as fig.5.

4. Discussion and conclusion

The robot model that we developed is still rather restricted. First, we did not model the process of auto-shaping that normally takes place when animals newly enter an experimental chamber (see Brown & Jenkins, 1968). In this process, the animals are trained to associate the disks with food and will thus peck on them in the following experiments. Our robot can only be compared to experienced chickens which already know that attending to the pictures shown on the pecking disks will be important, like the chickens in our discrimination learning experiments. Second, the similarity measure will have to be further adapted to physiological data on chicken vision. In this first prototype model, we used a color space derived from psychophysical experiments on human vision, which is comparable to chicken vision only within a certain range. Careful experimentation is needed to empirically derive an independent similarity measure for our model animals. This will be a main task in our further research. Third, the full range of experiments carried out with the chickens will have to be replicated with our robot model, to prove that indeed all of the empirical findings can be generated without using a stimulus representation that is based on abstracted features. We do not claim that a more abstract representation of stimuli will never be necessary to account for behavioral phenomena exhibited by animals. We are just interested to see when exactly additional stimulus processing has to be assumed, how complex it actually needs to be, and which phenomena can be explained better with a more parsimonious, holistic exemplar model. Both to obey the principle of parsimony, and to guarantee that the additional processes which might then have to be assumed work in a real-world context.

One common worry about our kind of model is that it does not use networks with units that some people like to call "neurons". But we wanted to relate our model to the very successful tradition of using exemplar-based models in both animal and human learning psychology (see Domjan, 1998, Estes, 1994). Our earlier research had explicitly compared different approaches to model empirical data on visual categorization in chickens, and exemplar-based models turned out to be the most promising ones (see Werner et al., 1999, 2001). Exemplar-based models can easily be mapped on network models (see Page, 2000), but this different kind of implementation would blur the theoretical concepts underlying our work. Concerning the question of relating our work to brain data, the first simple answer is to stress that we do not aim to build models on a neuronal level, but on a psychological level of description. This way, we are able to also address questions for which no data on a neuronal level is currently available. Still, some critics might feel that exemplar-based models should be ruled out right away, because they "do not work like the brain". This is a common thinking error that we should shortly comment on. Storing an exemplar which represents a current input and/or output does not mean that there is a single, localized representation in the brain which corresponds to this exemplar. And comparing exemplars in a psychological model does not mean that there has to be a center with that function in the brain. This would be just one way to map the psychological level of description onto the neuronal level which does not seem to be too plausible. Instead, an "exemplar" on the psychological level could be seen to amount to all processes that happen within a brain when it reacts to some sensory input with a certain behavioral output. Depending on the kinds of inputs and outputs we decide to measure, the processes corresponding to such an exemplar can also be very different. We just need to assume that these processes leave some kind of "trace" within the neuronal hardware, i.e. the hardware is affected by these processes and will change a little bit, however lightly. This change in the neuronal hardware when reacting to a sensory input with a behavioral output is the neuronal correspondence of "storing an exemplar". Furthermore, we could assume that similar reactions to similar sensory inputs lead to similar processes in the brain and that this influences the tendency to show such a behavioral reaction to such an input in the future. Such an influence on the neuronal level is expressed by comparing exemplars in models that operate on a psychological level of description. This is again just one way of relating psychological terms to hypothetical brain mechanisms. But it is obviously not as implausible as the first one. This example should help to see why it would be premature to rule out exemplar-based models as "not brain-like". One other important reason not to use networks for our model is the parsimony principle that we are committed to. "Parsimony" in this case means reducing the number of hypotheses inherent in our model as much as possible.[4] In the case of neural networks, a lot of parameters have to be fixed, like the number of units, the topology of their connections, the way these connections are altered over time (which can be different for different connections, although this option is not often used), etc.. Every such parameter is a hypothesis. Given the large number of units and connections that can usually be found in such networks, which leads to a plethora of parameters that influence the dynamics, using such a model would be a contrast to our ideal of sticking to the parsimony principle.

Although our model will still have to be extended in a number of ways, we think it demonstrates how autonomous robots can successfully be applied as empirical models of cognitive processes in the framework of comparative

[4] NB: This is *not* the same as parsimony in the sense of reducing the use of computer resources! Judging the quality of a model of a natural system based on the amount of computer resources it uses is a bad idea. One problem with such an approach is again that there are many ways to match one level of description onto another level. (In this case: The use of resources by a computer model and the use of resources by the modeled natural system.) In the sense that we use the word, a model that can account for the empirical data by assuming just one simple process is more parsimonious than a model that has to assume several more complex processes, even if the latter uses less computer resources than the first.

cognitive robotics. The model has a number of interesting characteristics. The robot learns to react differently to stimuli that are defined by different features (red vs. green, horizontal vs. vertical). But no features are abstracted and no abstract representation of the stimuli is generated. In a more conventional approach, at least two conceptual steps would be assumed to relate the camera input to an appropriate response. In a first step, all pictures which relate to "the same stimulus" would be mapped to an abstracted stimulus representation, e.g. all pictures showing a green stimulus would be mapped to one or several units that "represent the green stimulus". This mapping process might have been implemented by an engineer, or it might have been acquired by the system itself in several training sessions. In a second conceptual step, this abstracted stimulus representation would be associated with an appropriate behavior, e.g. pecking. The first step might even be omitted, if the stimuli are readily provided by a simulated environment. In our model, however, no such mapping process takes place. Stimulus pictures are directly associated with a behavior and the reinforcement. Most pictures of the purely green stimulus and most pictures of the black vertical bar on transparent background could e.g. be found to be associated with a pecking behavior that was reinforced, but there is no representation within the agent that could be said to represent the concept of "green stimulus" or "black vertical bar stimulus". These concepts only exist on the level of the observer, but they do not exist within the robot model itself. We call this ability to differentially react to different categories of pictures without forming any abstracted representation of these categories within the agent "*categorization without categories*".

A second, related characteristic of our model is the lack of learning rules. In a conventional model that operates with abstracted stimulus representations, the learning problem is reduced to associating such a stimulus representation with the representation of a behavior. This problem is so simple that it could actually be learned in just one or two trials. If the system experiences that pecking on the red stimulus is not reinforced, but pecking on the green stimulus is reinforced, this simple association could be learned instantaneously. But animals normally need a larger number of trials to learn such relations. We would have to introduce a learning rule that slowly increases the "strength" of the association between the stimulus representation and the response so that it matches with the learning curves that show in the empirical data. Even worse, we would have to assume two different learning rules, as learning the discrimination between color stimuli turned out to proceed much faster than learning the discrimination between line orientation stimuli. In the case of our robot model, however, there is no such thing as an abstracted stimulus representation, and thus we also could not use this kind of learning rule. The association between a stimulus picture, a response and a reinforcement always remains the same, there is no "associative strength". But it turned out that the empirical data could fully be explained without learning rules. The real-world nature of the input entails that a number of exemplars of each kind of stimulus will have to be collected before new camera pictures are mapped with a high probability to the "correct response" by our simple learning and robot control mechanism. And it is again because of the interaction of this mechanism with a real-world environment that color stimuli are learned more easily than line orientation stimuli. The former are simply less affected by noise and shifts of the camera image than the latter. The difference in learning of color and line orientation stimuli turned out to be an emergent property of our simple holistic, exemplar-based learning mechanism. But this emergent property only became visible when the mechanism was implemented and tested in a real-world artifact, i.e. an autonomous mobile robot model that was tested in the same environment as the animals.

As our autonomous agent model does not need to assume abstracted features, representations of categories, and associative strengths changed by learning rules, it can be considered to be more parsimonious than earlier models of discrimination learning.

Acknowledgments

The animal research part of this study was supported by the DFG (*Deutsche Forschungsgemeinschaft*). The work on the robot part of this project was supported by the Volkswagen Foundation, the Gottlieb Daimler and Karl Benz Foundation and by a graduate grant of the University of Osnabrück.

At the University of Osnabrück, research described here was carried out as part of a one year Master student project in the first German program for cognitive science. The Master students working in this project called EROSAL ("Empirical RObot Study on Animal Learning") have been: Christa Deiwiks, Aikaterini "Katerina" Gergou, Leonhard Läer, Rüdiger Land, Sascha Lange, Jan Plate, Ulaş Türkmen.

We further thank Inga Tiemann and Julia Cnotka for their support in conducting the visual discrimination study and Christine Opfermann-Rüngeler for the preparation of graphic illustrations and of the stimulus slides.

References

Beer, Randall D. (1990). *Intelligence as Adaptive Behavior: An Experiment in Computational Neuroethology.* San Diego, CA: Academic Press. (Perspectives in Artificial Intelligence, Volume 6)

Brown, P. L., & Jenkins, H. M. (1968). Auto-shaping of the pigeon's key-peck. *Journal of the Experimental Analysis of Behavior* 11, 1-8.

Cook, R. G., Riley, D. A., & Brown, M. F. (1992). Spatial and configural factors in compound stimulus processing by pigeons. *Animal Learning & Behavior* 20, 41-55.

D'Amato, M. R., & Salmon, D. P. (1984). Cognitive processes in cebus monkeys. In: Roitblat, H. L., Bever, T. G., & Terrace, H. S. (Eds.): *Animal cognition.* Hillsdale, NJ: Erlbaum, pp. 149-168.

Dean, J., Kindermann, T., Schmitz, J., Schumm, M., & Cruse, H. (1999). Control of walking in the stick insect: from behavior and physiology to modeling. *Autonomous Robots* 7, 271-288.

Domjan, Michael (1998). *The Principles of Learning and Behavior, 4th Edition.* Pacific Grove, CA: Brooks / Cole.

Estes, W. K. (1994). *Classification and Cognition.* New York, NY; Oxford, UK: Clarendon Press.

Grant, D. S., & MacDonald, S. E. (1984). Matching to element and compound samples in pigeons: the role of sample coding. *Journal of Experimental Psychology: Animal Behavior Processes* 12, 160-171.

John, R. S. (1998). *Methodologische Probleme der Verhaltensbasierten Künstlichen Intelligenz aus kognitionswissenschaftlicher Perspektive.* Magisterarbeit, Arbeitsbereich Computerlinguistik und Künstliche Intelligenz, Universität Osnabrück. [*Methodological problems of behavior-based artificial intelligence from a cognitive science perspective.* Master thesis. Department for Computational Linguistics and Artificial Intelligence, University of Osnabrück, Germany]

Kemler Nelson, D. G. (1993). Processing integral dimensions: the whole view. *Journal of Experimental Psychology: Human Perception and Performance* 19, 1105-1113.

Lambrinos, D., Möller, R., Labhart, T., Pfeifer, R., & Wehner, R. (2000). A mobile robot employing insect strategies for navigation. *Robotics and Autonomous Systems* 30, 39-64.

Leith, C. R., & Maki, W. S. Jr. (1977). Effects of compound configuration on stimulus selection in the pigeon. *Journal of Experimental Psychology: Animal Behavior Processes* 3, 229-239.

Lew, M. S. (Ed.) (2001). *Principles of Visual Information Retrieval.* London, Berlin, Heidelberg: Springer.

Luce, R. D. (1959). *Individual Choice Behaviour: A Theoretical Analysis.* New York: Wiley.

MacPhail, E. M. (1987). The comparative psychology of intelligence. *The Behavioral and Brain Sciences* 10, 649-695.

Monahan, J. S., & Lockhead, G. R. (1977). Identification of integral stimuli. *Journal of Experimental Psychology: General* 106, 94-110.

Page, M. P. A. (2000). Connectionist Modelling in Psychology: A Localist Manifesto. *The Behavioral and Brain Sciences* 23 (4), 443-467.

Pearce, John M. (1994). Similarity and discrimination: a selective review and a connectionist model. *Psychological Review* 101, 587-607.

Pearce, John M. (2002). Evaluation and development of a connectionist theory of configural learning. *Animal Learning and Behaviour* 30 (2002), 73-95.

Pearce, John M., & Bouton, M. E. (2001). Theories of associative learning in animals. *Annual Review of Psychology 52,* 111-139.

Riley, D. A. (1984). Do pigeons decompose stimulus compounds?. In: Roitblat, H. L., Bever, T. G., & Terrace, H. S. (Eds.). *Animal cognition.* Hillsdale, NJ: Erlbaum, pp. 333-350.

Roitblat, Herbert L. (Ed.) (1995). *Comparative Approaches to Cognitive Science.* London, UK; Cambridge, MA: MIT Press.

Roitblat, Herbert L., & von Fersen, L. (1992). Comparative cognition, representations and processes in learning and memory. *Annual Review of Psychology* 43, 671-710.

Shanks, David R. (1995). *The psychology of associative learning.* Cambridge: Cambridge University Press.

Shepard, R. N. (1987). Towards a universal law of generalization for psychological science. *Science* 237, 1317-1323.

Staddon, J. E. R. (2000). *The new behaviorism: Mind, mechanism and society.* Philadelphia, PA: Psychology Press.

Staddon, J. E. R. (2001). *Adaptive Dynamics: The Theoretical Analysis of Behavior.* London, UK; Cambridge, MA: MIT Press.

Steels, Luc (1999). *The Talking Heads Experiments, Volume 1: Words and Meanings.* Special pre-edition for LABORATORIUM, Antwerpen.

Steels, Luc, & Kaplan, Frédéric (2001). AIBO's first words. The social learning of language and meaning. *Evolution of Communication* 4 (1), 3-32.

Vogt, Paul (2003). THSim v3.2: The Talking Heads Simulation Tool. In: Banzhaf, W., Christaller, Th., Dittrich, P., Kim, J. T., & Ziegler, J. (Eds.): *Advances in Artificial Life, Proceedings of the 7th European Conference on Artificial Life, ECAL 2003.* Berlin, Heidelberg, New York: Springer, pp. 535-544.

Webb, Barbara (2001). Can robots make good models of biological behaviour?. *The Behavioral and Brain Sciences* 24 (6), 1033-1050.

Werner, C. W., & John, R. S. (2003). Exemplar-Based Visual Discrimination and Categorization in Chickens and Implementation of an Autonomous Agent Model. In: Schmalhofer, F., Young, Richard M., & Katz, G. (Eds.): *Proceedings of the European Cognitive Science Conference, EuroCogSci03.* Mahwah, NJ; London, UK: Lawrence Erlbaum.

Werner, C. W., & Rehkämper, G. (1999). Discrimination of multidimensional geometrical figures by chickens: categorization and pattern-learning. *Animal Cognition* 2, 27-40.

Werner, C. W., & Rehkämper, G. (2001). Categorization of multidimensional geometrical figures by chickens (*Gallus gallus* f. domestica): fit of basic assumptions from exemplar, feature and prototype theory. *Animal Cognition* 4, 37-48.

Werner, C. W., Tiemann, I., Cnotka, J., & Rehkämper, G. (to appear). Processing of visual compound stimuli in chickens (Gallus gallus f. domestica) is not predictable by constituting features, to appear in: *Animal Cognition.*

Morphology and Computation

Chandana Paul
Artificial Intelligence Laboratory
University of Zurich
CH-8051 Zurich, Switzerland
chandana@ifi.unizh.ch

Abstract

The body of a situated agent plays an important role in adaptive behavior as it enables sensory motor interactions with the environment which can give rise to emergent intelligent behavior. Using the physical dynamics of the body, the agent can perform behaviors with much simpler control than would otherwise be required for the same task. The physical structure of the body, or the *morphology*, determines its possible set of sensory motor interactions as well as its dynamics, and as a consequence, determines the complexity of the control required. This intrinsic relationship between the morphology and required control has been identified as the *morphology and control tradeoff* (7). What has not been previously considered, however, is that the controller is a computational entity whereas the morphology a physical entity, and if a tradeoff exists between them, it may be possible that the morphology can subsume a computational role. It is this computational contribution which may then reduce the computational demands on the controller, giving rise to the common phenomena of the morphology and control tradeoff. This paper introduces this idea and proves that it is indeed possible for the morphology to perform computation which can reduce the computational demands on the controller.
Keywords: embodiment, robots, morphology

1. Introduction

It is known in the study of adaptive behavior that the morphology of an agent's body can be used to achieve simplified control. Brooks showed that embodiment would enable the use of simplified controllers requiring little or no representation (2). Pfeifer has similarly discussed the importance of considering the design of the morphology of a robot, to reduce or simplify the control required and has named this relationship the *morphology and control tradeoff* (7). Braitenberg has shown through a series of simple vehicle designs that the control of behaviors such as obstacle avoidance and light following can be acheived using simple sensory-motor control if the morphology, in particular the placement of sensors and actuators is appropriate (1). In a more biologically inspired system, Cruse (4) gives an example of how mechanics of the body are used in biology, to simplify the coordination of insect legs. Although these studies have highlighted the reduced computational complexity of control as a result of the dynamics of the body, they have not considered that this may be the case because in effect the morphology performs some of the "computation" that the controller would otherwise have to perform. This paper shows on the basis of the thought experiments in Sections 2 and 3 that it is indeed possible for the morphology to perform computation which in effect reduces the computational requirements on the controller.

2. The XOR Robot: a thought experiment

Consider a robot which is controlled by perceptron networks. The perceptron is a well-known neural network, which consists of a single output neuron, with multiple inputs, adjustable synaptic weights and a threshold function. The perceptron convergence theorem, proved by Rosenblatt showed that the perceptron could be used to compute any function, as long as it is linearly separable (5). This means that the perceptron could be used to compute functions such as AND and OR, for example, but not functions such as XOR or XNOR.

Consider now that there are two perceptron networks, with inputs A and B. The first one performs an OR function, and the second an AND function. The networks are connected to actuated parts of a simple robot body (Fig. 1). The first network is connected to M_1, which is a wheel at the center of the base of the robot which causes forward motion. The second network is connected to M_2, which lifts the wheel off the ground. Now when A and B are both off, both networks output 0, which means that the wheel is on the ground but does not move, so the robot is stationary. When only A is active, the AND network outputs a 0, so the wheel remains on the ground. The OR network outputs a 1, and makes the wheel turn, thus causing forward motion. When B is 1, the situation is the same, the AND network outputs a 0 and the OR outputs a 1, causing forward motion. However, when A and B are both on, then the OR network causes M_1 to move, but the AND network causes the wheel to lift off so that it no longer touches the ground so the vehicle is once again stationary. The following table summarizes the behavior of the robot in these four conditions.

A	B	Behavior
F	F	stationary
F	T	moving
T	F	moving
T	T	stationary

This table looks like the truth table of the XOR function. How is this possible?

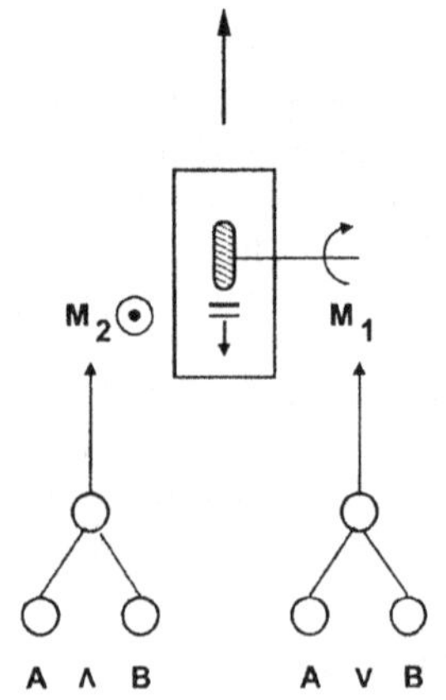

Figure 1: **The XOR Robot** This robot has one wheel, with two actuated degrees of freedom. The motor M_1 is responsible for turning the wheel so that the robot moves forward. The motor M_2 is responsible for lifting the wheel off the ground. Each motor is controlled by a separate perceptron network, which takes as inputs A and B. M_1 is controlled by a network which computes A OR B, and M_2 by a network which computes A AND B. Using only these controllers, the robot is able to display the XOR function in its behavior.

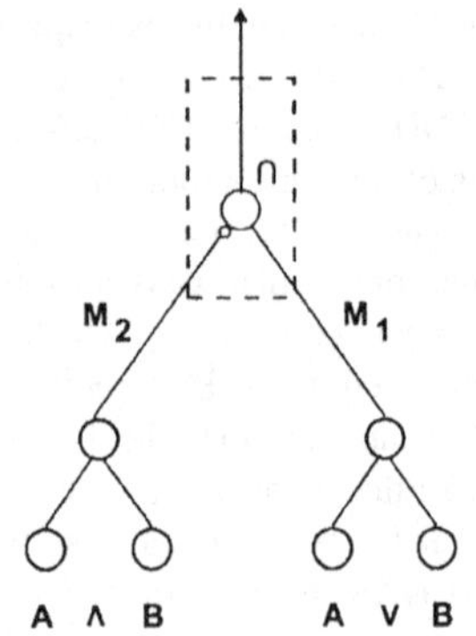

Figure 2: **Computational structure equivalent to the XOR Robot** The body of the XOR robot acts as if it is performing the computational function M_1 AND $\tilde{M}_2$

The explanation is that the robot's behavior is not simply determined by the output of the neural networks. It is also affected by the interaction between the actuated components of the body. The body through its mechanical structure provides

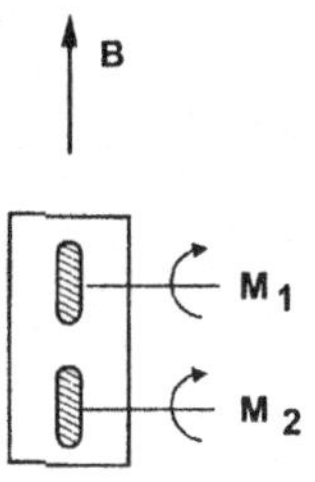

Figure 3: **The OR Robot** This robot has two wheels both of which turn in the same direction when actuated. Each wheel is actuated by one motor which is responsible for turning the wheel. The wheels are also capable of being driven passively.

some additional computational ability, which allows the XOR function to be displayed. In this case, the body provides the computational equivalent of a second layer of neural processing, in which it performs a NOT on its first input, followed by an AND, as shown in Figure 2. The function can be written as $M_1 \wedge \sim M_2 \rightarrow B$. This clearly shows that through its morphology, a robot body can perform a quantifiable computation which reduces the computation required in the controller.

3. The OR Robot

It can be shown that such computational ability is by no means an isolated occurrance. In fact, various simple morphologies can give rise to computation. In order to show this through a second robot example the OR robot is presented. The OR Robot (Fig. 3) is a simple mechanical structure, with a rectangular body and two wheels which are aligned along central longitudinal axis of the body, one behind the other. Each wheel has a motor, which when turned on, rotates the wheel so that the body is propelled in the forward direction. When the motor is off, the wheel can also be passively driven.

In this robot, when both motors are off, the robot is stationary. When only M_1 is active, one wheel turns and passively drives the other wheel so the robot moves forward. Similarly, when only M_2 is active the robot, the robot also moves forward. Finally, if M_1 and M_2 are both active, both wheels turn, and so the robot also moves forward. The behavior of the robot in these four conditions is summarized in the table below.

M_1	M_2	Behavior
F	F	stationary
F	T	moving
T	F	moving
T	T	moving

It is clear that this table is the truth table of the the familiar OR operator from Boolean logic, and that when a controller sends commands to the motors M_1 and M_2, the computational function performed by the controller is augmented by a final OR operation. Utilizing this to acheive the desired over-

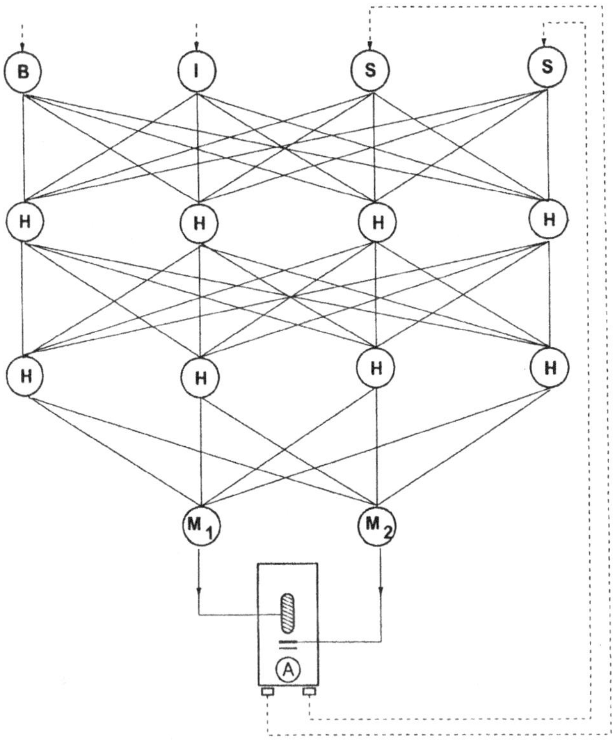

Figure 4: **Conventional neural network architecture for Vaccuum Cleaning Robot** The robot is controlled by a neural network which has an input layer with sensory input nodes, labelled S, input node I and bias node B. There are two hidden layers, each with 4 hidden neurons each, labelled H. The final layer is the output layer which has two output nodes to send motor commands to the robot.

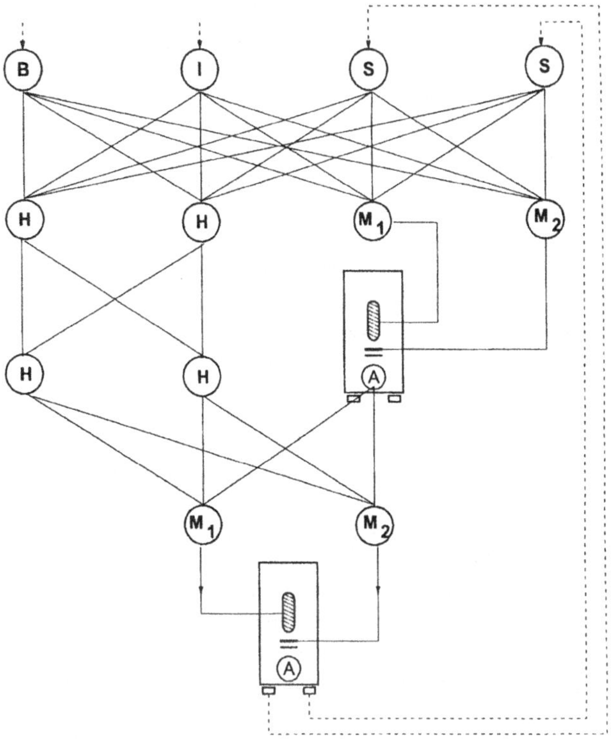

Figure 5: **Morphological computation used in the neural control of the Vaccuum Cleaning Robot** The robot is controlled by a neural network which has an input layer with sensory input nodes, labelled S, input node I and bias node B. There are two hidden layers. The first hidden layer has 2 hidden nodes, represented by H, and two nodes which convey motor commands to the robot, represented by M_1 and M_2. The second hidden layer also has two hidden nodes, labelled H, and a node which conveys the output of the robot accelerometer A. The final layer is the output layer which has two output nodes which send motor commands to the robot.

all computation can reduce the computational requirement of the controller by one logical function.

4. Explicit Computation

Although the computational abilities of the robot morphologies described in Sections 2 and 3 are relatively small they serve as an important proof of concept, that morphology can perform computation which can reduce the computational requirements of the controller.

At first glance, it may be argued that this computation simply exists in the eyes of the observer, and is not a real computation. In this section, adopting the commonsense definition that a computation is *real* if it can be used as a computation, it is shown that the computation performed by the morphology of the XOR robot is in fact real.

Consider this example. The body of the XOR Robot is now to be used as the basis of a real world robotic application such as vaccuum cleaning. It is enhanced with cameras to sense the environment, a microprocessor for computational processing and behavior arbitration, a binary velocity sensor and of course a vaccuum unit. Its task is to survey its initial environment and decide whether to vaccuum or not. If the floor does not look particularly dirty, it can wait. If the floor is dirty, it can decide to start vaccuuming down the long straight corridor.

Such behavior arbitration requires a certain amount of processing. A neural network based controller could be used for the task. This may look like the network shown in Fig. 4. This is a standard neural network architecture with an input layer, two hidden layers and an output layer. The input layer receives inputs from the robots visual sensors S, a higher level signal I, and a bias node B. The inputs are processed through the three layers of the network and result in output motor commands issued to the robot. Each individual node in such a network is a computational sub-unit of processing.

Now since the robot is equipped with a velocity sensor it is possible to measure the behavioral outcome of the robot as a result of motor actuation. Making the simplifying assumptions that the robot cannot overcome inertia and the force of friction when the wheel is not in contact with the ground or not actuated, and that when moving the robot comes to an immediate halt when the motors are switched off due to gear resistance and friction. Thus, the output of the velocity sensor directly depends on whether the wheel is in contact with the ground and actuated and in essence conveys the output of the computation performed by the body.

Using this sensor, the robot body itself could then be used as a computational sub-unit within the controller. For example, we could use it to replace a node in the network, as shown in Fig. 5. As can be seen Fig. 5 is very similar to Fig. 4 except that in the second hidden layer of the network one node has been removed and another has been replaced by the robot body such that it receives inputs from two neurons of the previous hidden layer which actuate the motors, and sends binary outputs measured by the velocity sensor, A. Removal of a neuron obviously does not diminish the computational nature of the network, although it may alter the function. Similarly, replacing a hidden neuron with the robot body, simply changes the function of that computational sub-unit from a standard threshold function on the weighted sum of the inputs, to $M_1 \wedge \sim M_2$. It does not invalidate the controller as a computational entity itself and therefore demonstrates that the transformation performed by the morphology can be used as a traditional computation, and is therefore real.

This example is of course a toy example. Performing the XOR function on a processor would be trivial from the point of view of processing time and energy costs compared to having to move the body. However, in a more complex morphology where a simple motion may perform the equivalent of an extremely complex computation, it may be useful to utilize this ability.

5. Complex Morphology

In the above example, the morphology of the XOR robot performed a simple binary function. served as the basis for the Vaccuum Cleaning Robot. The computation performed by the body of the XOR robot was easy to recognize as a common binary function, and so it could be shown that starting from a morphology which performed a simple computation, a controller architecture could be designed to utilize it. In robots with more complex morphologies such as manipulators or legged robots, which have rotary or prismatic actuated joints and joint angle sensors for sensing, the computation performed is a complex analog function determined by the dynamics of the body. Yet, it is nonetheless a computational function, in the sense of defining a mapping from motor commands as inputs to the physical consequences measured by the sensors as outputs.

Consider for example a 2 DOF robot manipulator with rotary joints and joint angle sensors, which is controlled by a neural network with sensory feedback (Fig. 6). For simplicity we assume that input I is 0 and the weights of the network

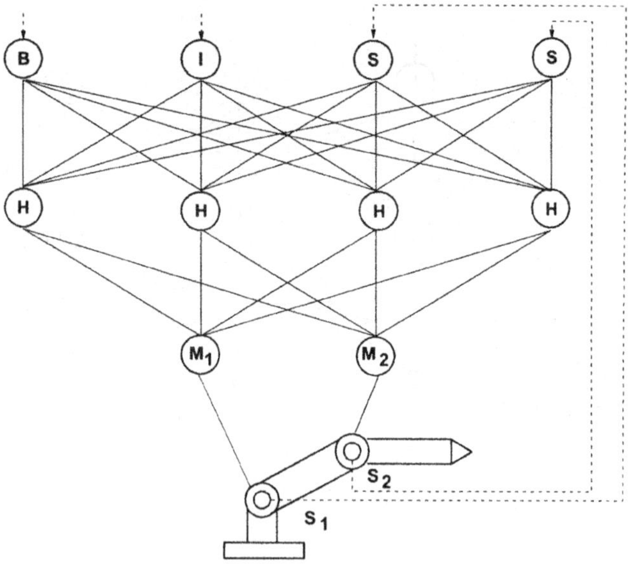

Figure 6: **Control of a robot manipulator** A 2 DOF manipulator is is controlled by a neural network with sensory feedback. The network has an input layer with 2 sensory input nodes, labelled S, an input node I, and a bias node B. The sensory nodes recieve proprioceptive inputs from the manipulator on joint angles. There is one hidden layer with four hidden nodes represented by H. The output layer has two output nodes which send motor commands to the two actuators of the robot, M_1 and M_2

are constant. The controller is designed such that it reads in sensory input from the joint angle sensors, processes the information through the network, sends motor commands, and only then reads in the next value of sensory input for the next iteration of neural processing. It is clear to see that in each such iteration there are two transformations. One is the transformation of the sensory inputs to motor outputs through the neural network. The second is the transformation of motor outputs to sensory inputs through the forward dynamics of the manipulator. Thus, the entire system forms a single computational entity.

Viewing the system from this perspective, it is clear how a morphology and control trade-off can exist in an embodied agent with sensorimotor control. Changing the physical characteristics of the robot, including even simple parameters such as link length or mass distribution, will affect the dynamics of the manipulator. This will, in effect, change the computational relationship between the motor commands and its physical consequences measured by the sensors. In some cases this change may be for the worse, but in others it may be for the better with regards to the task, and reduce the requirements on control.

6. Discussion

6.1 *Morphological Computation*

The thought experiments of Sections 2 and 3 demonstrate that it is theoretically possible for the morphology of an agent to perform a computational function. Such computation which can be latently performed by the morphology can be called *morphological computation*. The knowledge that physical structure can perform such computation is as old as Babbage's design for his Analytical Engine, a massive brass, steam-powered mechanical computer, from 1837. But as computation has grown more siliconized in the 20th century, a conceptual divide has arisen between the computational and physical hardware in robotics and AI. The body is viewed as an effector for the computations performed by the control hardware. This has occluded understanding the basis of the morphology and control trade-off. The main aim of the previous sections has been to remove this occlusion and show that computational properties can arise even in very simple morphologies reducing the requirements on the controller.

6.2 *Explicit morphological computation*

The Vaccuum Cleaning Robot demonstrates that when a robot with latent morphological computation is augmented with a sensor which can sense the behavioral consequences, it makes the computational function defined by the morphology explicit, such that it can be used as a standard computational sub-unit, at any stage of processing.

The fact that this is possible shows that morphological computation is not simply a way of *viewing* the transformation of motor commands to physical behavior as a computational process, but that it is a real computation as is traditionally understood, and can be used accordingly.

6.3 *Using computational and motor functions*

The Vaccuum Cleaning Robot also shows that in a system where appropriate sensing enables the explicit use of morpohological computation, a controller can exploit both the computational and motor functions. This can be acheived by separating the two functions in time, and alternating between them.

6.4 *Duality*

In engineering, it is traditional to think of a single component of a system performing a single function. Thus, in most cases in human engineered technology each component usually has a single function, and if it has more than one function, it is usually used to perform only one of them at a time. This modular approach to design is the basis of human engineering, which leads to a tacit belief in the idea that natural systems may also be designed according to these principles. However, in the case of an intelligent embodied agent, this does not need to be the case. Although, it is difficult for human engineers to conceptualize, it is possible for a single action of an agent to serve both motor and computational functions *at the same time*. In fact, many physical action performed by an embodied agent have the inherent possibility for *duality of function*.

In order to illustrate the concept, and show that it is relevant for natural intelligence, a simple analogy is made to human behavior. Consider the case when there are a large number of coins on a desk which must be counted (more than can be counted through a cursory visual inspection) and they must be put away in a drawer under the desk. Notice that there are two components to the task requirements. One is the computational task of counting the coins, which involves the development of an abstract representation (number) based on the environmental condition. The second is a physical task of relocating the coins from the desktop to the drawer. There are multiple ways of accomplishing this, but one common approach to the task that humans use, is to slide the coins off the desktop into the drawer one by one, counting each one in the process. In this simple behavior, the action of sliding the coin into the drawer has duality of function. One function is computational: it serves to separate the coin which has already been counted from the set of coins which has not been counted. The second function is physical: relocating the coins from the desktop to the drawer. These two functions are seamlessly integrated into one action, without any conscious forethought. The example demonstrates that if the opportunity arises to utilize duality of function, natural intelligence has been designed to be able to exploit it.

The Vaccuum Cleaning Robot, however, was an example of "human design" in that the computational role of the body was separated from the physical role in time. This had the advantage that it highlighted the use of the computational function of the body, by separating it from its physical role. However, theoretically, no such separation is necessary. It should be possible to design systems which use their physical structure for both physical and computational purposes at the same time.

6.5 *Effect of the environment*

The environment plays an important role in morphological computation. In the thought experiments of Sections 2 and 3, the systems were embedded in a single environment. However, when the environment of the system changes, the functions describing the system will also change. This is very clear to see in the examples presented. Consider the case when the OR robot is placed on a downhill incline. Now, the robot moves forward regardless of whether the actuators are active or not. The computational function of the structure changed. Thus, the ecological niche is crucial for the particular characteristics of the morphological computation which arise.

6.6 *Morphology and Control*

It has been understood in the study of adaptive behavior that embodiment can lead to simplified control. However, the reasons for this have not been clear. It has been known that interaction of loosely coupled sensorimotor processes with the dynamics of the body gives rise to *emergent behaviors* which are not explicitly represented in the controller (3)(6)(7). However, the mechanism of emergence itself has not been further investigated. By showing here that the body can perform a computational role, the basis of such emergence can begin to be understood. In some cases at least it is can be understood that the body provides the "computational glue" between loosely coupled sensorimotor processes.

The fact that such computation can be so simply acheived suggests that it is not simply a rare phenomena, but possibly a pervasive characteristic of physical structures. Further scientific investigation is necessary in this area. There also is a long way to go in understanding morphological computation and its relationship to control. Particularly it needs to be understood what characteristics of the morphology affect control requirements, and how. It is likely that such understanding will require the development of new analytical techniques to identify and quantify the computational contributions of a morphology.

If such an understanding can be acheived, however, it will open a new realm of possibilities for designing robots which incorporate "intelligence" in their bodies. By utilizing smart morphologies which are known to optimize dynamics for control, more robust and adaptive agents will be able to be designed.

7. Conclusion

This paper introduced the concept of *morphological computation*, computation performed by the mechanical structure of a robot body. To show the existence of such computation, an example of a robot controlled by perceptron networks was given, which utilizing the structure of its morphology was able to display the XOR function in its behavior. The XOR function being linearly inseparable would have been impossible to acheive with only a perceptron network, and thus proves that the morphology can perform a computational role.

This was followed by the description of a vaccuum cleaning robot which used the computation performed by its body in its own neural control. The example demonstrated that the computation performed by the body is both explicit and real. Furthermore, it explicitly illustrated a situation where the morphology replaced part of the computational control. The illustration sheds new light on how smart morphologies could reduce control.

Future work will focus on developing a theoretical framework for morphological computation which will provide analytical methods for studying the computational role of a robot body. This could be used to guide the design of smart mechanical structures, which will be able to exploit the morphology to reduce control.

References

[1] Braitenberg, V. (1984) Vehicles: Experiments in Synthetic Psychology. *MIT Press* Cambrige, MA

[2] Brooks, R.A (1991) Intelligence without Representation. Artificial Intelligence 47, 139-160

[3] Brooks, R.A (1989) A Robot that Walks; Emergent Behaviors from a Carefully Evolved Network, *Neural Computation*, 1(2):253-262

[4] Cruse, H., Bartling, C., Dean, J., Kindermann, T., Schmitz, J., Scumm, M. and Wagner H. (1996) Coordination in a six-legged walking system: Simple solutions to complex problems by exploitation of physical properties. In *Proc. 4th Int. Conf. on Simulation of Adaptive Behavior*, Cape Cod, USA, 84-93

[5] Haykin, S. (1994) Neural Networks: A comprehensive foundation. *Macmillan College Publishing Company*, New York

[6] Mataric, M. J. (1992) Designing emergent behaviors: From local interactions to collective intelligence", in *From Animals to Animats: International Conference on Simulation of Adaptive Behavior* eds. J-A. Meyer, H. Roitblat, and S. Wilson, MIT Press, 432-441

[7] R. Pfeifer and C. Scheier. (1999) *Understanding Intelligence.* MIT Press, Cambridge, Massachusetts.

Constructing an animat mind using 505 sub-minds from 234 different authors

Ciarán O'Leary* **Mark Humphrys**** **Ray Walshe****

*School of Computing
Dublin Institute of Technology
Kevin St.
Dublin 8
Ireland
www.comp.dit.ie/coleary

**School of Computing
Dublin City University
Glasnevin
Dublin 9
Ireland
computing.dcu.ie/~humphrys
computing.dcu.ie/~ray

Abstract

The World-Wide-Mind (WWM) is a scheme for putting minds (by which we mean the software to drive a real or virtual agent) online for remote re-use by third parties. Although the animat community favours the extension of existing agent minds in an ongoing process, there is currently no easy way to re-use minds built by other authors. If re-use became the norm, it is suggested that larger, more diverse and more complex minds could be built.

This paper describes an ongoing project that uses WWM technology to develop a novel set of minds that incorporate a range of algorithms and techniques. We demonstrate that it is possible to evolve superior minds by the artificial selection and combination of existing online minds.

The project that is described here involved 234 authors who developed 505 different minds for the virtual animal in a re-implementation of Tyrrell's Simulated Environment (Tyrrell, 1993). These minds, which were developed by two classes of undergraduate Computer Science students, were subsequently selected based on their performance in the virtual world, and integrated into larger minds. At the time of writing the most successful mind developed used as components the two sub-minds that most successfully satisfied the animal's two main goals. Since a huge number of combinations of minds is possible, it is important that the work is distributed among a community of researchers. The architecture of the World-Wide-Mind makes this possible.

1. Introduction

The World-Wide-Mind (WWM) (Humphrys and O'Leary, 2002) is a scheme for putting animat minds online for remote re-use by third parties. Although the animat community favours the extension of existing agent minds in an ongoing process (Wilson, 1990), there is currently no easy way to re-use minds built by other authors (Guillot and Meyer, 2000). If re-use became the norm, it is suggested that larger, more diverse and more complex minds could be built.

This paper describes an ongoing project that uses WWM technology to develop a novel set of minds that incorporate a range of algorithms and techniques. We demonstrate that it is possible to evolve superior minds by the artificial selection and combination of existing minds written by diverse authors, in a way that would not previously have been possible.

1.1 Project Description

The ongoing project that is described here has so far involved 234 authors who each developed minds (or *action selection mechanisms*) for the virtual animal in a re-implementation of Tyrrell's Simulated Environment (Tyrrell, 1993). Each author (the majority of whom were undergraduate Computer Science students) is permitted to develop any number of minds, which are then made available online, as web services. Since the minds are online, third parties can select a specific mind and run it in the virtual world, located at w2m.comp.dit.ie. This effectively means plugging the mind into the animal in the virtual world, and letting it select the actions at each point in time. The score that is achieved by each mind in each run is recorded in an online scoreboard that can be viewed by all other authors. The scoreboard had two effects (i) it introduces an element of competition between authors that encourages them to write improved versions of their minds, and (ii) it identifies for authors the best minds to incorporate into their own minds as *sub-minds* or modules.

1.2 $Mind_M$ Services

Authors are actively encouraged to "cheat" or use other author's minds as modules in their own minds. This type of integration is made possible through remote reuse of online components. The manner in which existing minds can be integrated into new minds takes a number of forms. For example an author can select the two minds at the top of the scoreboard and wrap a high level mind around them. The high level mind will make the decision on which sub-mind to call at which point in time, resulting in a division of the problem state-space between the two sub-minds. Obviously this strategy can be employed with any number of sub-minds. Using the terminology introduced in (Humphrys, 2001), this is a $Mind_M$ service.

A number of other strategies involving varying levels of competition between sub-minds are described in (Humphrys and O'Leary, 2002) but have not been tried out as part of this project, for reasons that will be described later.

At the time of writing 505 minds had been written and made available for re-use online. Of these, less than 5% were $Mind_M$ services. In total, the scoreboard contains 4684 runs, meaning that the average number of runs for each mind was 9.28 (however, many minds would have been abandoned by their authors after a very small number of runs). In order to assess each individual mind correctly, a much higher number of runs would be necessary, but very difficult to carry out given the time required (for statistically sound results, Tyrrell (1993) tested each action selection mechanism he implemented 1650 times).

For this reason, we rely on the community of authors, or interested third parties to run the tests. They do so by selecting their mind, or a mind they are interested in and running it in the world. Over time the better minds will rise up the scoreboard at the expense of the poorer minds. These better minds are then more likely to get artificially selected for inclusion in a $Mind_M$ service. In this way, we expect to see the evolution over time of successively improved minds (Darwin, 1859, Ch. 1).

The results outlined here are based on ten $Mind_M$ services developed by the first author. They are accompanied by a description of the strategy used for developing each $Mind_M$ service based on the scores available in the scoreboard. In addition we describe how these $Mind_M$ services, which are also available online, will be subject to the same sort of evolution by artificial selection that the original minds are.

1.3 Roadmap

This paper will proceed as follows. Section 2 provides an overview of the World-Wide-Mind project, including a detailed description of the architecture. Section 3 describes Tyrrell's Simulated Environment (SE), as a re-implementation of this virtual world was used for the research outlined in this paper. Section 4 shows how the implementation of Tyrrell's SE was put online as part of the World-Wide-Mind architecture. This includes a description of the format for the scoreboard that is core to evolving the minds. Section 5 provides a discussion of the ongoing project that started with 234 diverse authors. We explain how their minds were made available online, subsequently run and recorded in the scoreboard. These minds were then selected based on their scores, and integrated into $Mind_M$ services, according to different strategies, as described in section 6. Section 7 provides the results for these $Mind_M$ services. Section 8 provides a discussion of these results. Section 9 describes further work, and section 10 will place our work in context by showing its relationship to other modular forms of artificial minds and agent architectures.

2 World-Wide-Mind

The World-Wide-Mind (WWM) is a scheme for facilitating the development of large minds through publishing virtual worlds and sub-minds online as web services. A web service is effectively a programming language object whose methods are invoked over HTTP, the language of the web rather than being invoked in a local environment. The WWM specification (Humphrys, 2001a) defines three types of entity, a *world service*, a *mind service* and a *client*. Messages sent between entities are composed in the XML-like protocol of the WWM named SOML, the Society of Mind Markup Language (this is an update of the protocol referred to as AIML in (Humphrys and O'Leary, 2002b)).

2.1 World service

A world service is a virtual world (real worlds involving embodied agents are possible, though none are implemented as yet) that supports a set of SOML messages for retrieving the *state* of the world, and for instructing the agent in the world to execute a given *action*. For most problems, the state refers to the current perceptions of the agent in the world, although this could obviously be extended to include any type of information about the world. The action will be some value or symbol that can be interpreted by the agent in the world.

A world service defines its own format for state and action, so that when another entity queries it for its state, the requesting entity must parse the state according to the rules defined for that particular service. Similarly the world service will define its own format for its actions. This removes the requirement for the type of symbolic languages used for agent communication in multi-agent systems and on the Semantic Web, as discussed in (O'Leary and Humphrys, 2002).

2.2 Mind service

A mind service generally represents an action taking mechanism developed for a given type of world service. The mind service will support the necessary SOML messages to receive the state from the world service, parse it according to the format defined for the world service, and then suggest one of the actions possible for that particular world.

Once a mind is created there are three options available to the mind author. Firstly they can put the mind online with no additional information describing the algorithm used by the mind. This effectively means that the mind will be used as a black box and it can only be selected for reuse in a $Mind_M$ service based on its performance.

Secondly, the author of a mind service could publish details on its implementation on an accompanying web site. $Mind_M$ authors who wished to reuse this mind as a component could then design their $Mind_M$ to take advantage of the published attributes of the mind.

Thirdly, minds could propose an action accompanied by some additional information to help in the higher level action selection. Such minds (referred to as $Mind_I$ services in (Humphrys and O'Leary, 2002)) are aware that they may not be the only mind involved in selecting actions for a particular body, and would need to be designed appropriately. For example, Humphrys (1997) describes how sub-minds that use reinforcement learning can calculate values that represent how much they want to be obeyed in particular states.

In the experiment carried out here, minds are not accompanied by web sites describing their algorithm, nor do they propose weights with their chosen actions, and as a result must be reused as black boxes. Future research will address methods for proposing weights in societies of sub-minds that implement heterogeneous algorithms.

2.3 Client

The client is the entity that is responsible for *plugging* a mind into an agent in a world. It does this by selecting a world service at a given URL and selecting a mind service at another URL. It then informs both the world service and the mind service that it intends to start a *run* involving them. Both services will typically create new instances of the algorithm that they represent and return a unique ID for this instance to the client. The world service algorithm implements the problem (e.g. Tyrrell's SE) and the mind service algorithm implements a solution (one author's mind).

The client will use this ID in all communication with both services. Both IDs along with both URLs uniquely identify the subsequent interaction between the entities, which is termed a *run*. A run consists of constant iterations of the following steps:

(i) Client requests the world service to return its state.
(ii) Client takes the world state and sends it to mind service, which then returns the action that it suggests for the given state.
(iii) Client takes the action selected by the mind and sends it to the world service, instructing it to execute this action in the world.

This process continues until the world service sends a message to the client to say that the run has ended. The client will then send a message to the mind to tell it that it is no longer needed for this run.

During a run, the world service does not need to know which mind service is being used to select its actions (although in reality as we'll see, this information is sent so that it can be recorded on the scoreboard). The mind service will normally not need to be sent any information regarding the world service, since it will have been created specifically for that particular problem. The client will only need to know the URL of both services, but will not need to understand anything about the representation of the state or action for the world, since its only function is to pass this data between the services.

2.4 SOML

The Society of Mind Markup Language (SOML) is the XML-like language that defines the general format for the messages that are sent between entities on the WWM. The protocol is defined in full in (O'Leary, 2003). In brief, the following are the core messages:

(i) `newrun`: Can be sent to any type of service to request that a new instance be created.
(ii) `endrun`: Can be sent to any type of service to inform it that a given run has ended. This type of message can also be sent by a world service to a client to inform it that the run has ended in the world.
(iii) `getstate`: Can be sent to a world service to request that its state be returned.
(iv) `getaction`: Can be sent to a mind service to request that an action be suggested for the state that is contained in the request.
(v) `takeaction`: Can be sent to a world service to instruct it to take the action contained in the request.
(vi) `getdescription`: Can be sent to any type of service. The response will typically contain information such as the author of the service, the date it was last modified and so on.
(vii) `getscoreboard`: This request can contain a URL of a given mind and a number (x). When received by a world service, the top x scores achieved by the mind identified by the URL should be returned. If the request does not contain any parameters, all scores should be returned. The format of the score is also defined by the service, as described in the section below.

2.5 Scoreboard

As described above, world services should support a request named `getscoreboard`. This should return the scores achieved by a given mind, or indeed all scores achieved in the world. The world service can define its own format for its score, the only requirement being that the scoreboard should be ordered in some way that places better performing minds closer to the top of the scoreboard than those minds that they outperform.

One possible way of constructing a scoreboard is to record every run in the world, with the best run at the top. Using this approach the size of the scoreboard grows very large very quickly (this is the approach we used, and the scoreboard is already over 2MB in size), and the ordering on the scoreboard may appear misleading. For example, the mind that scored the best run may have been very fortunate in the world that it was given for that particular run, and may perform badly in most other instances of the world.

Another possibility would be to record statistical information on the performance of each mind, so that instead of recording every score, the average and standard deviation of its scores are recorded, and the scores are ordered appropriately, more accurately reflecting the relative performances of the minds.

We used the first format for this experiment. Since all information is recorded in the scoreboard, and is available to others using the `getscoreboard` request, anyone who is interested can perform their own statistical analysis of the scores and select the best mind based on this.

The information that is recorded for each score includes the URL of the mind service, the date and time of the run, the name and contact details (e-mail/URL) of the author of the mind, the name and contact details of the user, or person who started the run and most importantly the score achieved, in a format decided upon by the author of the world service.

3 Problem Domain

At present there are two world services online that conform to the latest version of SOML. The first of these is a simple blocks world implementation, for which twelve mind services have been developed, including two $Mind_M$ services. A description of the problem and the design of the minds is discussed in earlier work (O'Leary and Humphrys, 2003).

A second available world service, which is the problem world at the centre of the work described here, is a re-implementation in Java of Tyrrell's Simulated Environment (SE) (Tyrrell, 1993). The SE will be described in this section. Its implementation as a world service will be dealt with in the next section.

3.1 Tyrrell's Simulated Environment

Tyrrell's PhD thesis examined a number of different action selection mechanisms when applied to a complex, multi-goal problem. In order to do this he wrote a virtual environment that modelled an animal in a heavily populated, dangerous environment. The primary goal of the animal was to mate, as this would ultimately determine how likely it was to pass on its genes to subsequent generations. However, in order to mate the animal needed to survive in the world long enough to be presented with sufficient opportunities to court and mate with other members of its own species.

The world which the animal occupied had fifteen different features, including food and water (both of which could be toxic), a den, cover and shade. It also featured several different types of animal; predators and non-predatory animals that needed to be avoided, prey that could be consumed and animals of the same species that could be mated with. The world experienced changes in weather and also cycled through different times of day.

The animal had its own models of perception and navigation that provided it with information about its surrounding environment and memories of places that it had previously visited. It was also provided with motor control, or the ability to convert a chosen action into the necessary movements of the body. Each of these three systems were error-prone, in that the animal could misperceive features in its immediate environment or *forget* about a feature it had encountered. Also, any action that it chose was subject to a low probability of being incorrectly executed.

3.2 Action Selection

At each time-step the animal was required to choose one of 35 actions to execute. These included 16 different moving actions for travelling at different speeds in different directions, 9 different looking actions that could improve its perceptions for a given direction, or allow it to quickly scan its environment for predators, and various actions for eating, sleeping, resting, cleaning and mating.

The animal should choose the action that is most likely to increase its expected future genetic fitness i.e. the number of times it mates in this case. In order to do this the animal should maintain good health by eating and cleaning. It should also avoid an early death by hiding from predators and looking out for other dangerous places such as cliffs and marshes. In addition it should try to make its way home at nighttime to sleep.

Tyrrell implemented six different algorithms for the animal's action selection in the environment. His own algorithm which he named the *Extended Rosenblatt and Payton* algorithm performed best in the tests he conducted. His results were later improved upon by Bryson with her Edmund algorithm (Bryson, 2000).

3.3 Discussion

It is obvious that the action selection problem for Tyrrell's SE is difficult. All approaches to solving it so far have focussed on building a single algorithm that can account for all the goals that must be satisfied by the animal. What we propose is that rather than designing the best solution from the beginning, we start with a solution and then add to that in subsequent implementations. This would be a difficult task if every author were required to understand every implementation that came before him. We intend to examine if it is possible to integrate solutions that we do not understand completely or at all, that are judged simply on their performance. This may become easier if the algorithms are available for remote re-use, online.

4 Tyrrell's SE online

An implementation of Tyrrell's SE is available online as a WWM world service at w2m.comp.dit.ie. When queried for its state it will return a vector on numbers representing the animal's perceptions and memories and well as some additional information about the environment (it is given all the information that was available to the action selection mechanisms developed by Tyrrell).

It will accept as an action any number from 0 to 34, each representing one of the 35 actions that can be selected by the action selection mechanism. For a detailed explanation of state and action see w2m.comp.dit.ie.

Any mind service that will be created for this implementation of the SE should accept the state, process it and return an action i.e. it will serve as an action selection mechanism, except it will be located somewhere else on the Internet.

4.1 Scoring in the SE

Every run in the world service is recorded on the scoreboard, including the URL of the mind, author and user details as explained. The scores are ranked according to two criteria, firstly the number of times that the animal mated in the environment during that particular run, and secondly according to how long the animal survived in the environment, in time steps.

In addition, various other items are recorded for each run. This information is useful in developing a profile for the mind that was involved in the run, which will be used when we attempt to integrate individual minds into Mind_M services.

This additional information includes (i) the number of times that the animal successfully ate food or drank water, (ii) the number of times that the animal ate toxic food or drank toxic water, (iii) the number of times the animal caught and consumed prey, (iv) the number of times the animal was injured by predators, other animals or by visiting dangerous places and (v) the amount of time that the animal spent in the environment without knowing how to return to its den.

In addition, for each run in the world, the number of times that each of the 35 actions was chosen is recorded. Using all of this information it is possible to infer to some degree the type of strategy that was used by the mind. The more information that is recorded, the easier it is to build up this type of profile, but there are real limits that must be observed. As mentioned earlier, the size of the scoreboard is already over 2MB in size. The more information that is recorded the larger this will grow.

A decision was made to record the data for each run rather than simply maintain statistical information for each mind. This meant that anyone interested in using a particular mind could perform their own analysis of the runs conducted using the mind, and develop their own profile for it. This also meant that the scoreboard was ordered according to each run, rather than the performance of each mind in all its runs collectively.

A HTML version of the scoreboard is available at w2m.comp.dit.ie. The SOML for the scoreboard can be retrieved using the `getscoreboard` request. This will contain all the additional information recorded for each run.

5 Format of experiment

The world service is publicly available for anyone to use (at w2m.comp.dit.ie), and anyone is free to develop a mind service to work with it, once they observe the correct formats for state and action. We now explain how the first set of minds were built.

5.1 Community of Students

In order to get a large number of minds online for the purposes of this research, an assignment was given to two classes of undergraduates studying for an honours degree in Computer Applications at Dublin City University. Altogether, over two hundred students developed mind services for the version of Tyrrell's SE that is online.

In order to focus on the action selection aspect of the assignment rather than the networking (relatively trivial though it is) the students were given a tool kit that they could use to develop the mind and put it online. The majority of the students developed their software in this way, although a minority did implement minds using web technologies such as PHP and Java Servlets.

Students were permitted to create as many different mind services as they wanted over a period of six weeks. At the time of writing (two weeks after the deadline for the assignment) 505 had been put online. Once a mind was put online, it was automatically run in the world to record some account of its performance. Many minds were then run again several times by their authors or others who wanted to observe the performance of the mind, with a view to integrating the mind in their own Mind_M services. Some students developed such minds, although this can only be inferred from the response times of the mind service (it is not unusual for a Mind_M service to take over an hour and a half to perform a run in the world, in particular when it is calling on more than one sub-mind).

5.2 Artificial Selection of Minds

As the experiment continued, the better minds rose to the top of the scoreboard. Over a prolonged period of time, the artificial selections made by authors and users will result in a larger number of runs in the better minds, which will in turn result in their being selected more often for inclusion in

$Mind_M$ services, and consequently the creation of better minds.

In addition, since there is useful information apart from scores recorded on the scoreboard, authors can observe the strategies that are employed by the better minds (in one case, it would appear the *sleeping* was preferred to *eating* in order to live longer) and try to employ these strategies in the minds they are authoring without remotely re-using the original mind.

6 Integration strategies

Using the large set of mind services that had been put online, we hand designed a further ten $Mind_M$ services. The first problem that we are faced with in developing a $Mind_M$ service is in identifying at least two existing minds that could work together. A second problem is in identifying when to select the action proposed by either mind. One simple approach would be to start a run in one mind and let it run for a finite number of steps, then stop it and start another mind which will then run and so on. A problem with this is that everything that the mind may have learned up to that point from its experience of the environment is lost. A second approach might be to start both minds and present both of them with the state of the world at every time-step, letting them all suggest an action, but only select the action proposed by one of them, possibly based on a voting strategy. At least with this strategy all sub-minds get to experience the environment for the entire duration of the run, but now all the minds think that they are being listened to at every point in time, which is clearly not the case.

We describe here a number of different approaches we took to selecting and integrating two minds.

6.1 $Mind_M$ I

The first strategy used for building a $Mind_M$ service was to select the mind that had achieved the best run in the world, i.e. the one at the top of the scoreboard, and examine its entries in the scoreboard. From this we were able to develop a profile of the mind, based on the actions that it had chosen in its runs as well as the additional pieces of information described above.

The profile that was developed for the mind was based on its average performance across all its runs, as well as analysis of individual scores. A summary of its profile is given here

1. Lives for an average of seven days in the world (a day is 500 time steps).
2. Often employs sleeping as a method for living long, and consequently misses out on some mating opportunities.
3. Favours drinking over eating. Does not eat enough.
4. Efficient cleaner.
5. Has not been injured by other animals or dangerous places, possibly because they have not been encountered yet. Has not used its *move fast* actions at all, which implies that it has not had to escape from predators.

Based on this we selected a mind that could compensate for the deficiencies highlighted in the above profile. We identified the failure of the animal to eat enough to be its primary fault, so we analysed the scoreboard to find a mind that regularly selects eating actions, and still occupies a high position on the scoreboard. The mind selected was the one that ate the most on average in a run.

To integrate the two minds, we employed a simple strategy. We started a run in both minds at the beginning of our run. We selected the action of our main mind the majority of the time. We selected the action of our second mind whenever it suggested an eating action.

There are obviously many shortcomings in this approach. If the first mind is pursuing some goal such as *avoid predator*, and the second mind suggests an eating action then the animal will stop to eat, and probably end up eaten himself. However, the second mind we selected is relatively high up on the scoreboard, so it would be fair to assume that it is aware of the presence of the predator as well, and would not select an eating action while a predator was close.

6.2 $Mind_M$ II

A second approach we took to developing a $Mind_M$ service was to select, once again, the mind at the top of the scoreboard, but this time make a different judgement on how to select an *eating* mind for it. In this case we selected the mind that was the most successful eater i.e. the one that ate non-toxic food every time that it selected an eating action. Other minds had selected eating actions when it was incorrect to do so. We used the same approach to integrating the minds i.e. the eater was only listened to when it suggested an eating action.

6.3 $Mind_M$ III

For this mind, we selected as the primary mind the one which had the best average score, rather than the highest individual score. We created a profile of this mind and saw that it was quite similar to the mind described in section 6.1 above. It is not surprising that two successful minds should have a similar strategy, even if this is not the best strategy. One may itself be a $Mind_M$ service that wraps up the other mind, using the actions that it suggested the majority of the time. Alternatively, the author of one of the minds may have observed that the other mind was achieving good scores on the scoreboard. Then using the information in the scoreboard, he may have developed his own profile of the mind and attempted to mimic that strategy in his own implementation. Another reason may of course be that they both arrived at the same design independently.

We then integrated the mind we had selected at this stage with the *successful eater* we had identified for $Mind_M$ II above, using the strategy described earlier.

6.4 $Mind_M$ IV & V

In this case we picked another successful eater that had a better average overall performance in the world than the eater used above. We integrated it with the two main minds we had been using so far to give $Mind_M$ IV and $Mind_M$ V.

6.5 $Mind_M$ VI

In order to score well in the world the animal must survive for a long time and mate as frequently as possible. With this in mind we selected two minds which we felt would satisfy these two goals best. The first mind is the one that survived longest in the world on average, and the second one also lived quite long (more than two days) but mated more frequently. We integrated them in a manner similar to the minds built earlier, obeying the actions of the first mind in all cases except where the second mind proposed a mating related action (*courting* or *mating*).

6.6 $Mind_M$ VII & VIII

Some mind implementations focussed entirely on mating to the cost of all other goals. These minds did not survive long in the world, but mated more frequently in the short period of time that they did survive. This suggested that they were better at identifying good mating opportunities since they were not concerned with other goals. We picked one such mind and integrated it with the main mind used in section 6.5. This was $Mind_M$ VII.

We then selected another frequently mating mind that appeared to have an interest in a limited number of other goals such as eating and cleaning. We integrated this with our long living mind, taking its suggestion whenever it proposed a mating related action, and named it $Mind_M$ VIII.

6.7 $Mind_M$ IX

A $Mind_M$ service can call on, or wrap up, any number of sub-minds that it invokes remotely. The interface to these minds remains the same regardless of how they are implemented, including situations where the sub-minds are themselves $Mind_M$ services. For the construction of $Mind_M$ IX we integrated $Mind_M$ II as described above with a frequent mater. This whole mind then uses three sub-minds to select its actions, with the frequent mater only being listened to when it proposes a mating action.

6.8 $Mind_M$ X

The obvious shortcoming of all the minds described thus far is that a mind that has been identified as being strong in achieving a particular goal is only called upon when it proposes the final consummatory action. This is because it is difficult to identify when a mind is suggesting an appetitive action that will ultimately lead to the achievement of the goal. For example, although we have used sub-minds that we identified as being best for solving the *eating* or *mating* goals, we have not accepted the actions that they proposed when they may have started to solve that particular goal. The reason for this is that we have no way of knowing that they are beginning to solve that goal, since the only information we are provided with from the sub-mind is the action that is proposed.

In an attempt to address this particular issue, we implemented a $Mind_M$ service that performed its arbitration using the *Drives* algorithm that Tyrrell had developed for the Simulated Environment. This algorithm worked by taking all the animal's perceptions and memories and calculating simple *motivational strengths* or *drives* for each of the sub-goals such as finding food, water or mates, or avoiding predators or dangerous places.

For our implementation, we calculated the *drive* for each sub goal and selected a sub-mind based on the winning, or highest drive. Whenever *mating* won the competition we selected a frequently mating mind; if eating won we selected a successful eater, and for all other drives we defaulted to the most successful overall mind.

Clearly, any number of strategies could be used for determining when to switch between sub-minds, as will be discussed in the *Further Work* section below. This $Mind_M$ service was simply a demonstration of one possible strategy.

6.9 Discussion

We have outlined how ten different $Mind_M$ services have been developed from existing components. As can be seen there are two main issues involved in the creation of a $Mind_M$ service using minds in this black box fashion:

(i) *Profiling each mind based on its performance in the world.* The profile should provide sufficient information to be able to identify what goals are handled well by each mind.

(ii) *Deciding when to switch between minds.* In the first nine implementations above, minds were selected and deselected based entirely on an action that could be related to only one goal. When appetitive actions can be associated with more than one goal (e.g. move towards direction of nearest mate) it is difficult to infer that the sub-mind should be given control of the agent.

7 Results

Firstly it should be stated that the implementation of Tyrrell's SE used here is not identical to that used for other published results by Tyrrell (1993) and Bryson (2000). The first author reimplemented the SE using Java, based on the description provided in Tyrrell's thesis and also his freely available C code.

7.1 Tyrrell's Extended Rosenblatt and Payton Algorithm

Tyrrell's algorithm used a hierarchy of nodes that take input from both internal and external stimuli and pass excitation on to nodes at lower levels in the hierarchy. Nodes at the lowest level represented actions, and at every point, the action node with the highest level of activation would be selected for execution. This system allowed all actions to be considered at each timestep, and for separate parts of the creature's mind to contribute excitation to the same action, facilitating the selection of compromise actions, or actions that can contribute to the satisfaction of more than one goal. Using this algorithm, Tyrrell's animal mated an average of **8.09** times in a lifetime.

7.2 Bryson's Edmund Algorithm

Bryson's Edmund algorithm was implemented for several different versions of the SE. This algorithm allowed the creature for focus its attention on the highest priority goal that could be addressed in a particular state. For this reason it was able to dramatically reduce the amount of computation that was required at each timestep, and was also able to improve on the results achieved. Bryson feels that the simplicity of the implementation leads to a higher probability of finding the correct parameters for the system. Using this algorithm, Bryson's animal mated an average of **9.12** times in a lifetime.

7.3 Implementation of Simple Minds

Although the internal design of each of the minds developed for this experiment were not available to the authors of $Mind_M$ services, subsequent assignment submissions by the cohort of students contained descriptions of the algorithms used. Minds tended to be hand designed, priority based algorithms, where the animal checked various goals in order, obeying the first one that could be satisfied. The ordering which was given to the goals, the manner in which an action was selected for a goal, the manner in which it could be determined whether a goal should be addressed in a particular state and the amount of state that was saved between actions determined the relative success of each of the minds. Most minds implemented highly reactive algorithms, although the most successful individual mind maintained its own (more accurate) record of recently visited states, in addition to that provided by the world service.

In total there were 505 minds that were involved in 4684 runs. The minds were scored firstly on the number of times they mated, and secondly on the amount of time-steps they survived in the environment. The overall average number of times that the animal mated across all runs was 1.23 times. The animal lived for an average of 1066 time steps.

The top ten minds mated an average of 3.3 times per run, and lived for an average of 1916 time-steps. For the top one hundred minds these figures were 2.6 and 1686 respectively.

The best performing mind mated an average of 3.65 times and lived for an average of 1378 time-steps. The mind that lived the longest in the environment was able to survive for an average of 4413 time-steps, although its strategy was to find its den and sleep continuously, consequently it never mated. The average lifetime of the longest living mind that was also able to mate was 3348 time-steps, spending 36% of its time asleep.

7.4 $Mind_M$ Results

The results achieved in this implementation of the Simulated Environment do not rival the results published by Tyrrell or Bryson for this particular problem. The scores achieved by the best $Mind_M$ services *did* however beat the scores achieved by all the other minds developed, and did so by selecting and integrating these as sub-minds. This point will be returned to in the *Discussion* section.

The results achieved by these $Mind_M$ services are given in Table 1 below.

	Average Performance			**Best Performance**		
Mind	**Mated**	**Lived**	**Pos.**	**Mated**	**Lived**	**Pos.**
I	2.67	2365	22	8	3196	25
II	3.16	2226	9	12	4768	2
III	2.89	1182	15	10	1936	8
IV	3.67	1276	2	9	1769	12
V	2.2	2421	41	5	2886	116
VI	1.96	2827	61	7	4496	34
VII	3.7	1539	1	20	3916	1
VIII	1.53	226	120	6	248	115
IX	2.24	2776	36	5	4192	127
X	2.46	979	27	7	2013	47

Table 1: The number of mates, time-steps lives and overall positions of each of the ten MindM services.

$Mind_M$ VII performs best on average and also reached the top of the scoreboard with the best individual score.

8 Discussion

From the results that have been shown it is clear that it is possible to create superior minds from existing minds if the correct minds are chosen, and if the integration strategy suits the algorithms that are being executed by each of the minds. In our case, the best performing mind used the mind that had lived the longest in the environment and the most aggressive mater. An aggressive mater will constantly prioritise mating over the pursuit of other goals. The mind that it was integrated with appeared to distribute its attention among multiple goals. When the two worked together they appeared to get the best of both worlds. The mating mind sought out opportunities to mate whenever possible, whereas the remainder of the time the other part of the mind was focussed on ensuring that the creature lived long enough to be presented with sufficient opportunities to mate.

In many cases minds were put together that did not work well, at least with the type of strategy that was used for integration. For example, Mind_M X should have outperformed all other minds because of the care that was taken in switching between the various sub-minds at the correct time. However, in this situation it would seem that all that happened was that the sub-minds were interrupted while pursuing particular goals that the other minds could not pick up on. Even in this situation, the minds that were selected for the mating and eating goals were often not the ones that actually executed those actions. Better minds, such as our winner Mind_M VII use sub-minds that are very different from each-other. It is easier to divide the state-space between these types of minds, as their respective roles are more clearly defined.

Tyrrell (1993) and Bryson (2000) both built action selection mechanisms that outperformed our collection of minds. However, our sub-minds require much more testing, as does the re-implementation of Tyrrell's SE. Also, the action selection needs to be more sophisticated. Minds need to be designed to co-operate with the types of action selection that is taking place in other minds. For example, minds may need to flag appetitive and consummatory actions, perhaps by returning strengths or W-values (Humphrys, 1997) with each action.

9 Further Work

This work is at a relatively early stage. The assignment that was given to the undergraduate students only ended recently, thus limiting the time available to develop more sophisticated Mind_M services. However, it is clear that if minds are available online as services, they effectively exist forever, and can be used by anyone with access to the World-Wide-Web. This should mean that over time, as more runs get executed and more scores are recorded, it will be possible to get a clearer image of the profile of each mind. With more information about the profiles of each of the minds, it should be possible for interested authors to develop better Mind_M services, and consequently get better ideas for how to write minds for the animal. Evolving a mind in this fashion is different to developing it as part of a single project. Evolution requires time and the cumulative effects of much recombination in order to be successful (Dawkins, 1986). In nature this happens by natural selection. In our situation what is required is the conscious selections of a community of interested researchers.

9.1 Integration

Our integration strategies were relatively straightforward, with the exception of the *drives* based mind, which failed to perform well in the tests conducted. We intend to examine related work in behavior/goal selection and agent architectures which can be tried out with our architecture. Some of the work in this area is discussed in the *Related Work* section below.

Earlier publications discussed the possibility of allowing sub-minds exchange numeric weights that express the degree to which one sub-mind *wants* to run at any particular point in time. Such a mind could use Hierarchical Q-Learning (Lin, 1993) or W-Learning (Humphrys, 1997) as outlined in (Humphrys and O'Leary, 2002). While approaches such as these will be tried for certain problems, they may not be appropriate for the research described here, in their current form. Firstly, both learning schemes require specific types of problems. Tyrrell (1993) outlined why his SE is not suitable for these types of algorithms, given the size of the state-space and the requirement for experience of all different types of states. Secondly, each mind service author would need to understand and correctly implement the algorithm for proposing an action with a given weight, which will be possible in certain domains but was not possible here. It may be possible if proxies (Schmidt *et. al.*, 2000) are wrapped around sub-minds that will then implement the required algorithm.

10 Related Work

Bryson (2001) proposes a design methodology named *Behavior Oriented Design (BOD)* that defines an iterative engineering approach to the development of complex, complete agents. BOD stresses the need for an initial decomposition of the agent requirements into a set of sub-systems and then a continuous revision, weighing the level of granularity of each system against the requirement for arbitration between these. Such a cyclic approach could be applied to our own work in the development of a single agent where the requirements are considered in terms of the available sub-minds.

Other modular forms of agent design, including Brooks (1986) and Humphrys (1997) address issues in behavior decomposition and arbitration, while the whole animat approach to AI (Wilson, 1990) is predicated on the idea of starting with simple whole minds to which new components are added over time. This is the approach we took to the development of Mind_M services.

Work in Internet-agent systems is related to our own in that the various components are distributed throughout the network. While we have no initial requirement for the type of symbolic languages (FIPA, 2002) or complex platforms (Poslad et. al., 2000) that are used in this area, we still face similar technical issues in terms of network latency.

Distributed systems programmers develop components and web services that are automatically located and invoked by clients, primarily in the area of e-Commerce. A great deal of work in this area is now focussed on the development of profiles for each of the services that run over the Web and more recently the Semantic Web (DAML-S Coalition, 2002). Such profiles however are written at development time by the creators of the service and are not automatically inferred and updated from the performance of the service.

11. Summary

Complex minds are necessarily composed of a multitude of different algorithms and structures (Minsky, 1986). Such minds are difficult to build because the required expertise is rarely available to a single project or lab. The World-Wide-Mind project aims to make it easier to build these types of minds, by proposing the online reuse of existing mind components.

The work described here has shown how a mind can be evolved over a period of time by utilising the independent expertise and artificial selection of a community of authors.

Acknowledgements

The authors wish to thank the many students from CA3 and CAE3 in Dublin City University whose hard work made this research possible. All author's names can be seen on the scoreboard at w2m.comp.dit.ie; those used here, and the $Mind_M$ services they were used in are Peter Kehoe (version 7 of his mind was used in I, II & V, version 10 in VI, VII, VIII & X), John Dundon (I), Colin Sheridan (II & III), Eugene Gibney (III, IV, VI & X), Lousie O'Nolan (IV & V), Mark Ruddy (version 8 used VII, version 9 in VIII), Andrew Pimlott (IX) and Michael Dowling (X).

The authors also wish to thank two anonymous reviewers for their helpful comments.

References

Brooks, Rodney (1986), A robust layered control system for a mobile robot. *IEEE Journal of Robotics and Automation*

Bryson, Joanna (2000), The Study of Sequential and Hierarchical Organisation of Behaviour via Artificial Mechanisms of Action Selection, *MPhil Dissertation: University of Edinburgh*

Bryson, Joanna (2001), Intelligence by Design: Principles of Modularity and Coordination for Engineering Complex Adaptive Agents, *PhD Thesis, MIT*

DAML Services Coalition (2002), DAML-S: Web Service Description for the Semantic Web, *in proceedings of First International Conference of the Semantic Web (ISWC-02)*

Darwin, Charles (1859), The Origin of Species.

Dawkins, Richard (1986), The Blind Watchmaker: Why the Evidence of Evolution Reveals a Universe Without Design, *W.W. Norton & Company*

FIPA (2002). FIPA ACL message structure specification. *http://www.fipa.org/specs/fipa00061/SC00061G.html*

Guillot, A. and Meyer, J.-A. (2000), From SAB94 to SAB2000: What's New, Animat?, *Proc. 6th Int. Conf. on Simulation of Adaptive Behavior (SAB-00).*

Humphrys, Mark (1997) Action Selection methods using Reinforcement Learning, *PhD Thesis, University of Cambridge*

Humphrys, Mark (2001), The World-Wide-Mind: Draft Proposal, *Dublin City University, School of Computing, Tech Report CA-0301 computing.dcu.ie/~humphrys/WWM/*

Humphrys, Mark and O'Leary, Ciarán (2002), Constructing complex minds through multiple authors, *Proc. 7th Int. Conf. on Simulation of Adaptive Behavior (SAB-02)*

Lin, L-J (1993), Scaling up Reinforcement Learning for robot control, *10th Int. Conf. on Machine Learning*

Minsky, Marvin (1985), The Society of Mind, *Simon and Schuster*

O'Leary, Ciarán and Humphrys, Mark (2002), Lowering the entry level: Lessons from the Web and the Semantic Web for the World-Wide-Mind, *1st Int. Semantic Web Conf. (ISWC-02)*

O'Leary, Ciarán and Humphrys, Mark (2003), Building a hybrid Society of Mind using components from ten different authors, *in Proceedings of Seventh European Conference on Artificial Life.*

O'Leary, Ciarán (2003), SOML Specification, *http://w2m.comp.dit.ie/services/documentation/*

Poslad, S, Buckle, P and Hadingham, R (2000), The FIPA-OS agent platform: Open Source for Open Standards, *in 5th Int. Conf. on the Practical Application of Intelligent Agents and Multi-Agent Technology (PAAM-2000)*

Schmidt, D, Stal, M, Rohnert, H and Buschmann, F (2000), Pattern-Oriented Software Architecture: Patterns for Concurrent and Networked Objects, *Wiley and Sons*

Tyrrell, Toby (1993), Computational Mechanisms for Action Selection, *PhD Thesis, University of Edinburgh, Centre for Cognitive Science*

Wilson, S.W. (1990), The animat path to AI, *Proc. 1st Int. Conf. on Simulation of Adaptive Behavior (SAB-90)*

May We Have Your Attention: Analysis of a Selective Attention Task

Eldan Goldenberg[1], Jacob Garcowski[1] and Randall D. Beer[1,2]
[1]Dept. of Electrical Engineering and Computer Science
[2]Dept. of Biology
Case Western Reserve University
Cleveland, OH 44106
{exg39, jrg13, rxb9}@cwru.edu

Abstract

In this paper we present a deeper analysis than has previously been carried out of a selective attention problem, and the evolution of continuous-time recurrent neural networks to solve it. We show that the task has a rich structure, and agents must solve a variety of subproblems to perform well. We consider the relationship between the complexity of an agent and the ease with which it can evolve behavior that generalizes well across subproblems, and demonstrate a shaping protocol that improves generalization.

1. Introduction

Within the adaptive behavior community, it is widely believed that situated, embodied and dynamical approaches have significant implications for cognitive science (Miller, 1994; Clark, 1997; Harvey, 2000). In order to realize this potential, however, we must create model agents capable of *minimally cognitive behavior*, the simplest behavior that raises issues of genuine cognitive interest. Toward this end, we have previously evolved dynamical "nervous systems" for a variety of minimally cognitive behaviors, including categorical perception, short-term memory and selective attention (Beer, 1996; Slocum et al., 2000).

Our primary motivation for these studies has been to investigate the mechanisms by which the evolved agents operate. For example, a detailed dynamical analysis of an evolved agent capable of visually-guided object discrimination has recently been carried out (Beer, in press). This analysis involved characterizing (1) the dynamics of the entire evolved brain-body-environment system, (2) the dynamics of the agent and environment subsystems and the interactions that give rise to the observed behavior of the coupled system, and (3) the neuronal properties underlying the agent dynamics.

In attempting to apply these techniques to more complex behaviors, we have found it necessary to first understand in detail the structure of the minimally cognitive tasks that we set. While simple tasks may have only one or a small set of obvious best solutions, more complex tasks may involve tradeoffs that interact in complicated ways. These tradeoffs must be understood before a detailed dynamical analysis of agents performing such tasks can be undertaken.

In this paper, we perform a task analysis of the selective attention problem. First, we present a largely unsuccessful attempt to quantify the difficulty of this task. We then demonstrate that this attempt fails because the task has considerably more structure than is initially apparent, containing many different subproblems on which different evolved agents specialize in different ways. Next, we present the beginnings of a taxonomy of these subproblems and their particular difficulties, and show that the overall success of different agents can be understood in these terms. Finally, we demonstrate that, in random trials, these different subproblems occur with frequencies that sometimes differ by orders of magnitude. By manipulating these frequencies, we can put selective pressure on the evolution of agents that perform better on the harder subproblems, or that generalize better across the full range of subproblems.

2. Methods

Agents are trained to handle a selective attention task (Slocum et.al., 2000; Downey 2000). An agent, with limited sensory capability, is required to catch two falling objects. The agent is represented by a circle, and is capable of moving horizontally. The falling objects, also circles, have separate constant velocity, and are constrained such that the agent should be capable of catching both.

An agent is a circle of diameter 30, having 9 proximity sensors of range 205 evenly distributed over a visual angle of $\pi/6$. The world the agent exists in is 400 units wide. The agent can not move outside of this world, however, it also has no way to know that it has been stopped at the edge. The agent's horizontal velocity is proportional to the sum of two opposing motor neurons' output. The constant of proportionality is 5, thus, the maximum velocity of the agent in either direction is 5. The falling objects have diameter 26.

The first object's vertical velocity is in the range [3,4], while the second object's vertical velocity is in the range [1,2]. Both objects horizontal velocities are in the range [–2, 2]. The objects are constrained to start within the agent's field of view, and travel in such a way as it is always possible for the agent to catch both. The faster object will always land first. The second object will land so that it can be reached by the agent traveling at 5α, where α is a scaling factor, set to 0.7, to ensure the agent would be capable of catching both objects. When an object lands, it is removed from the simulation, so that it does not confuse the agent's visual sensors. See Figure 1 below.

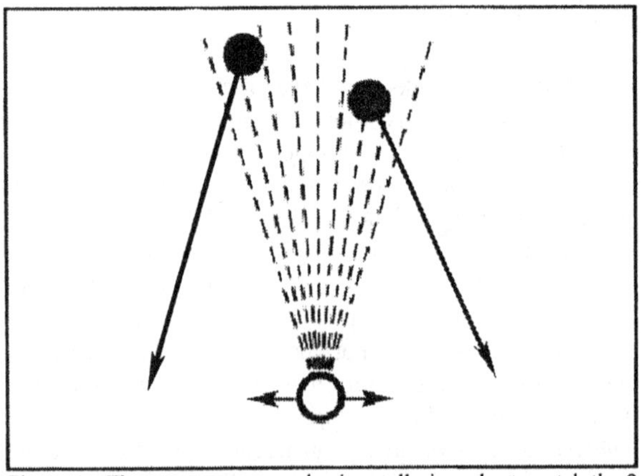

Figure 1: The agent can move horizontally in order to catch the 2 objects which fall from above.

An agent's behavior is controlled by a continuous-time recurrent neural network (CTRNN) with the following state equation:

$$\tau_i \dot{y}_i = -y_i + \sum_{j=1}^{N} w_{ji}\, \sigma(g_j(y_j + \theta_j)) + I_i \quad i = 1, \ldots, N$$

Where y is the state of each neuron, τ is its time constant, w_{ij} is the strength of the connection from the jth to the ith neuron, g is a gain, θ is a bias term, $\sigma(x)=1/(1+e^{-x})$ is the standard logistic activation function, and I represents an external input (zero on all but sensory neurons). A forward Euler method with a step size of 0.1 was used to integrate the CTRNN.

The CTRNN is constrained to be bilaterally symmetric, with 9 sensory neurons, a set even number of interneurons, and 2 motor neurons. Sensory neurons only receive input from the environment, but they connect to all interneurons and motor neurons. The interneurons and motor neurons are fully interconnected (every interneuron or motor neuron is connected to every other interneuron or motor neuron). Agents are constrained to bilateral symmetry both to simplify the problem and to ensure that an agent will act the same in mirror situations. All sensory neurons share the same gain and bias, interneuron and motor neuron biases are constrained to the range of [–5,5], and motor neuron gains are fixed at 5. All gains are positive, and all time constants are greater than 1.

The parameters of agents were evolved using a real-valued genetic algorithm (Mitchell, 1996). An individual agent was represented by a real-value vector of length M. Initially a population of 100 individuals was created by setting their evolvable parameters to random numbers in the range [–1, 1]. The top 2 individuals were copied into the next generation automatically. The remaining individuals were created by mutation. A linear-rank based method was used to select individuals for mutation. A selected parent was mutated by adding a vector whose direction was randomly distributed on an M-dimensional hypersphere and whose magnitude was a Gaussian random variable with mean 0 and variance σ^2, of 1. Agents were trained over 9000 generations.

An agent's performance on any individual trial was (200 – p), where p is the error, measured as the sum of the absolute distances between the center of the agent and the center of the falling objects when they land. Therefore, the maximum performance of an agent on any trial is 200. Agents are ranked based upon their average performance for a set of trials. For purposes of reporting, agent performance will be given as a percentage of the maximum possible.

Agents were trained using a shaping scheme. As evolution continues, the agents were presented with more difficult problems. Each agent was trained on a set of 30 trials. Over time, the trials that the agent performs best on were replaced with others. Whenever the best agent performed better than a threshold T, or 600 generations passed since the last trial change, the trial that it performed best at was replaced with a new trial. The first 5 of these new trials were hand selected. Subsequently, new trials were generated by selecting the first random trial that the agent performs worse than 170 on. T = 198 – n/14 – gen/2500, where n is the number of additional trials added, and gen is the current generation. For work using similar incremental shaping schemes to guide evolution or learning towards flexible behavior, see (Dorigo & Colombetti, 1994; Saksida et.al., 1997).

3. Task Difficulty

How difficult is this selective attention task? A more difficult task should provide for a more interesting minimally cognitive behavior study. A good way to show the difficulty of a task is to consider how complex an agent is needed to solve it. For our task, we trained agents with differing numbers of interneurons. 16 genetic algorithm runs were done for each number of interneurons.

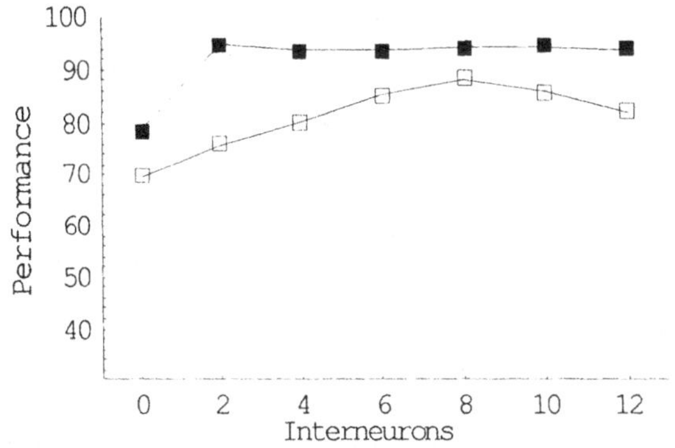

Figure 2: The best and average performance for each set of interneurons. Best performance is solid boxes, average is outlined boxes.

Figure 2 shows an unexpected result: agents with low numbers of interneurons are still capable of performing the task as well as higher numbers of interneurons. The best 2 interneuron agent is able to perform as well as all of the other agents. The best performance only drops when zero interneuron agents (purely reactive agents, with feed forward connections from the sensor neurons to the motor neurons) are considered. An agent with two interneurons should not be able to hold as much internal state as agents with 8, 10, or 12 interneurons, and yet is still capable of performing as well. This seems to go directly against our beliefs that this task requires both short-term memory and selective attention, cognitive abilities that should require more internal state. Previous work on this problem (Downey, 2000) suggested that the ideal number of interneurons would be 10, and yet a 2-interneuron agent can perform as well as a 10 interneuron agent.

The mean values in Figure 2 show a more expected shape. These values suggest that while a 2-interneuron agent is capable of solving the problem, training it to do so is more difficult.

At first, the surprisingly small difference in performance between the best agent of each type would seem to imply that the task is simpler than had previously been assumed (Downey, 2000). We contend that it is misleading to treat this task as uniformly complex. Closer analysis shows that the complexity of the task, and of the agent architecture required to accomplish it, varies considerably from one trial to another. The interaction between the random task parameters makes some trials entirely trivial, but others are more difficult for the agents.

4. A Taxonomy of Subproblems

Detailed analysis of the task parameters and their effects shows that there is a wide variety of subtasks, ranging from the entirely trivial to very difficult cases on which only a small proportion of agents perform well. Previous work (Downey, 2000) considered four subgroups of trials, and found significant variance between the agents' performance across these subgroups. We are incrementally refining this classification, to which end we divided a sample of 100,000 randomly generated trials into 24 mutually exclusive categories according to salient features of the setup.

Comparisons of several agents' performance across these trial categories showed that many could be meaningfully grouped together, and that it was more useful to base these groupings on the demands they make of the agent, or the strategies an agent could use to perform well on them. This produces overlapping groups, which we describe below. After eliminating the easiest trial types, we consider those that require some short-term memory and those that set a selective attention task more difficult than simply ignoring the more distant object from the start of the trial, before looking at some special cases.

Subsequent analysis of the individual agents' performance will focus on their relative strengths and weaknesses, from the perspective of the problems they have to solve in particular trial groups. Rather than presenting a full taxonomy, this section will set out the main groupings that will be used to illustrate the evolved agents' strengths and weaknesses.

4.1 Trivial trial types

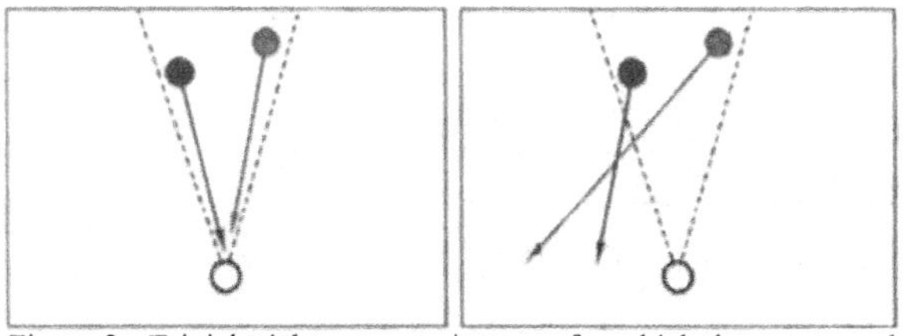

Figure 3: Trivial trial types: an instance for which the agent need not move at all (left) and one for which it need only move reactively (right). *In all figures in this section the agent is shown as an empty circle, its field of vision is between the dashed lines, the objects are shown as filled circles with their trajectories as arrows, and the darker object (which starts to the left) lands first. Here a horizontal line indicates the second object's height when the first lands.*

Some trials make very few demands of an agent, and therefore provide little information when analyzed. An obvious example of this is that it is possible for both objects to land directly on the agent's starting position, in which case the agent will get a perfect score by simply not moving at all. This caricature example is rather rare, accounting for only 5.2% of trials, but there is a more common subgroup of trials which can be dismissed as trivial because they only require purely reactive behavior of the agent. Any trial in which the object that starts closer to the agent is the first to land, and the second object is within the agent's field of vision when the first is caught, is trivial in this sense. Altogether, such trials account for 31% of the total. An example of each of these is shown in Figure 3 above.

4.2 Selective attention problems

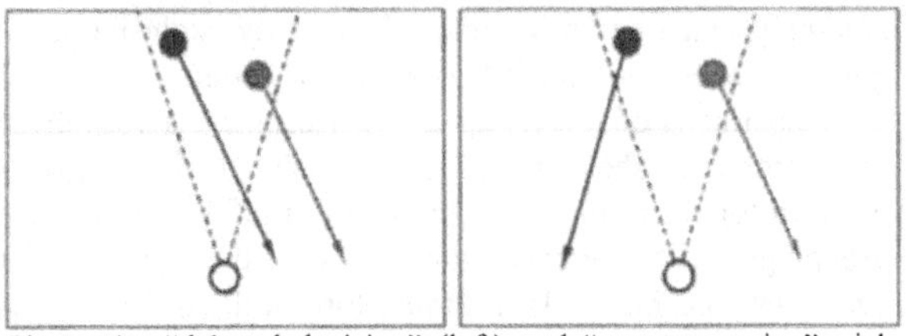

Figure 4: "delayed decision" (left) and "unseen passing" trials (right). *Here the horizontal lines indicate the height at which the darker object overtakes the lighter one.*

Selective attention becomes a more difficult problem when the object that will land first starts further away from the agent than the other. Two examples of this are shown in Figure 4 above. We will refer to trials in which the object that starts further away lands first as "delayed decision" trials, because the agent can not correctly choose which object to catch first until it has observed them moving. These instances make up 48% of all randomly generated trials[1].

A more difficult subset of the "delayed decision" trials occurs when the faster object passes the slower one at a large enough horizontal distance that the agent can not see both. In these cases, the correct choice of which object to follow can no longer be made based on distance information alone; the objects' relative speeds must be considered as well. We call such instances "unseen passing" trials, and they account for only 0.099% of the total.

4.3 Short-term memory problems

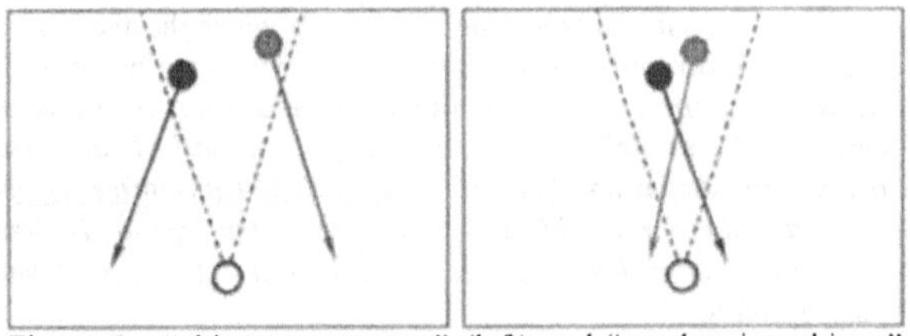

Figure 5: "object permanence" (left) and "overlapping objects" (right) trials. *Here the horizontal line indicates the height of the second object when the first lands.*

Another class of trials requires the agent to store information so that it can find an object again after having to lose sight of it. We consider two such situations, as illustrated by Figure 5 above.

If the second object is not within the agent's field of vision when the first is caught, then the agent must move towards the second object without being able to see it. This requires both that the agent stores information about the unseen object, and that it switches its attention to this internal information in response to catching the first object. These "object permanence" trials account for 39% of the total.

An additional difficulty is presented when the two objects cross horizontally, or start with a horizontal overlap. The poor sensory resolution of the agent means that as long as a group of neighboring rays are all broken, it will not be clear whether they are being broken by one large object or multiple smaller ones, and one object can be entirely occluded by the other. In such cases, which account for 58.6% of the total, it is possible for the agent to follow the correct object at the start but be fooled by the overlap, and switch its attention as a result.

4.4 'Cheat' heuristics and their limitations

For each subset of trials, it is possible for an agent to catch both objects by the using one of various simple 'cheat' heuristics. However, each such heuristic applied uniformly will only solve a small proportion of all trials; reliable performance requires a more flexible set of responses from the agent. For instance, an agent which always follows the object that starts further away will be successful on the delayed decision trials, but its overall performance will be poor because it will fail to catch the first object on the more straightforward trials.

The short-term memory requirement can also be circumvented for a subset of trials by the agent automatically moving in a pre-set direction after catching the first object if the second is not within sight, and switching to reactive behavior once it has moved far enough that it can see the second object again. However, any such behavior will guarantee failure to catch the second object on a proportion of trials (as shown in Figure 8 below). The most successful pre-set heuristic would be to continue in the same direction after the first catch, which will succeed in 78% of "object permanence" trials (making up 30% of all trials), but still guarantees failure in the remaining "object permanence" trials, which account for 8.4% of all random trials.

4.5 Summary

The taxonomy presented above is not a comprehensive classification of all the subtleties of this task, but a step towards this goal. The variety of different subtasks, different combinations of which are involved in each random trial, ensures that agents must have some degree of behavioral flexibility to perform well across all trial types. The following section will illustrate the interaction between individual agents and different trial types.

[1] Such instances account for less than 50% of all trials because randomly generated trials in which the two objects will land too far apart for the agent to catch both are rejected. This rejection criterion affects a slightly higher proportion of parameter sets in which the object that starts higher also moves faster.

5. Individual Variation

The agents we have evolved show a wide variety of behavior patterns, and relative strengths and weaknesses. In this section we present some illustrative examples of the range of agents we have evolved, and how they are differentially affected by the complexities of the task.

We use two different performance metrics to compare the performance of agents on various subsets of trials. The fitness measure described in Section 2 above will be used, but it is supplemented with a binary measure of whether each object is caught by the agent on each given trial, because this can make analysis clearer. A catch is defined as any collision between the agent and the object, without necessarily having to be exactly aligned, and we report the percentage of trials in which the agent catches each or both objects. The agents were tested on the same batch of 100,000 randomly generated trials described in Section 4 above, and their performance was recorded for various subsets of these trials defined by the combinations of parameters.

5.1 A blind spot for the best agent

It is instructive to begin the analysis with the best performing agent that we have generated to date. This agent has 10 interneurons, an overall fitness score of 97%, and catches both objects in 99% of all trials. It performs reliably well across almost all trial categories, catching both objects with 100% reliability in many, and with better than 97% reliability in all but one. However, its performance in the "unseen passing" situation is strikingly poor: earning a fitness score of 27%, and it only catches both objects in 20% of these trials.

In fact, closer analysis shows that this agent almost always (95% of trials) succeeds in catching the second object in these cases; it is the first object that it has trouble with (only catching it in 25% of trials). An example of its typical behavior is illustrated by figure 6 below. It seems that this agent has evolved a strategy of choosing the closer object when it is no longer possible to keep both within sight. In most trials (all but the "unseen passing" type described in Section 4.2 above, which only accounts for 0.099% of the total) this strategy will lead to a correct choice of the first object. It does fail consistently on one subgroup of trials, but because this subgroup is so rare it still allows the agent to score impressively well overall.

This striking weakness in an otherwise highly performing agent shows that the task is not uniformly complex, but rather that special cases provide particular challenges for the agent. Conversely, the trial type on which a particular agent performs worst will not necessarily be the hardest type for all other agents, as the next subsection will show.

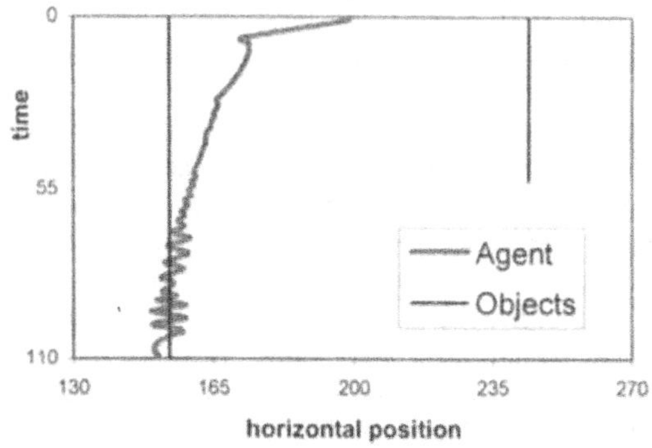

Figure 6: The best agent usually fails to catch the first object in the "unseen passing" situation. *The first object's trace ends when it reaches the agent's level.*

5.2 "Unseen passing" trials are tractable

There are also agents for which the "unseen passing" trials present little or no handicap, but which have weaknesses on other trial types. The best performance on this subgroup of trials was from an agent with only 4 interneurons, which scored almost as well on this subset as on average. The overall fitness for this agent is 91%, and it catches both objects in 92% of all trials; clearly less reliable than the agent described in 5.1 above. However, in the "unseen passing" situation it catches both objects in 89% of trials.

There are other subsets of trials which give this agent more trouble than the "unseen passing" ones. Its worst performance is for a subset that we were not expecting to provide particular difficulties: where the first object starts directly above the agent, and both objects fall diagonally in the same direction, without crossing in the horizontal plane, as illustrated in Figure 7 below. These trials account for 0.74% of the total, and this agent manages to catch both objects in only 76% of such trials.

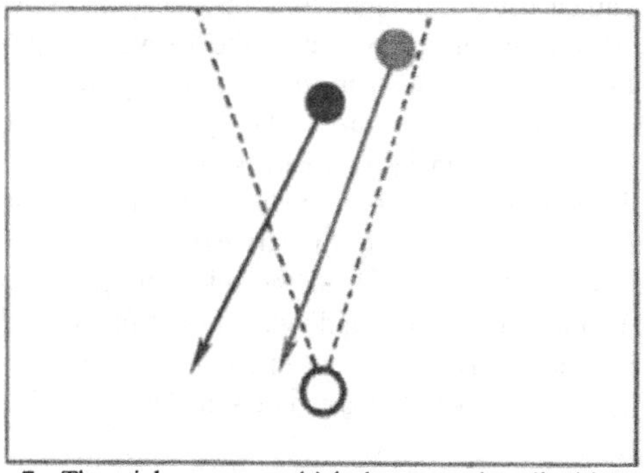

Figure 7: The trial setup on which the agent described in Section 5.2 performs worst.

5.3 Fixed action patterns

Both of the agents described above perform well enough in general that they do seem to be responding to the task in a generalized manner. There are others that show evidence of using a much simpler approach. Another 4-interneuron agent provides a particularly striking example of this.

This agent's overall performance is not impressive: it has a fitness of only 59%, and catches both objects in 67% of all trials. However, there is a subset of nontrivial trials for which its performance is far better than this. For trials which can be solved by rigidly applying the heuristic of "follow the closer object, and then continue to move in the same direction to find the other after catching it" (as shown in Figure 8 (left) below) this agent catches both objects in 97% of trials. By contrast, it only catches both objects in 46% of trials for which it would have to reverse its movement after the first catch (shown in Figure 8 (right)).

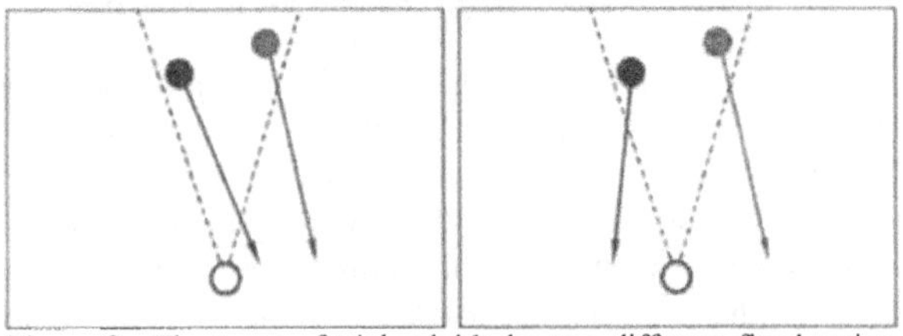

Figure 8: The types of trial soluble by two different fixed action patterns: Continuing in the same direction after catching the first object (left) or reversing direction (right)

Breaking down the data further shows that this agent catches the first object reliably in most trial types (99.9% of those shown in Figure 8 (left), 93% of those shown in Figure 8 (right), and 97% of all trials), so the variation in its performance comes almost entirely from what it does *after* the first catch. It catches the second object with 97% reliability in the "continue in the same direction after the first catch" cases, as compared with only 53% in the "reverse after the first catch" trials, and 70% overall.

A similar pattern, though less pronounced, is shown by 4 other 4-interneuron agents out of the 16 we have evolved, so while it does not account for the majority of results it is clearly not a unique instance. Our speculative explanation is that this is a comparatively easy strategy to evolve within the limitations of the 4-interneuron architecture, and because the trials on which this strategy fails account for a relatively small proportion of the total, the poor performance on these does not drag down overall fitness enough for these agents to always die out in evolution. More analysis needs to be done before this can be stated authoritatively.

5.4 What can be done with 2 interneurons

In general, it does appear harder to evolve agents that perform well with smaller numbers of interneurons. Agents with fewer interneurons also tend to perform more inconsistently across different trial types, indicating that they are more likely to rely on simple 'tricks' to solve the majority of subproblems. However, the best performing 2-interneuron agent performs surprisingly well across all trial types.

This agent has an overall fitness of 94% (as compared with the best agent of all, which was described in 5.1 and scores 97%), and catches both objects in 97% of all trials (as compared to 99% for the best agent). Its weakest performance is on the same "unseen passing" trials that stymie the best agent, but it is actually far less handicapped by this subset, earning a fitness score of 74% and catching both objects for 68% of such instances (as compared to only 20% for the otherwise superior agent described in 5.1).

One possible explanation for this agent's high overall performance is that it has developed a sort of toolkit of simple solutions to specific subproblems, and is able to select the right one in most cases. However, the standard deviation for this agent's reliability across all of the trial subtypes we have analyzed is 6.8 (as compared to 27 for the particularly inconsistent agent described in 5.3 above) indicating overall flexibility of behavior. Furthermore, it is hard to see how an agent with only two interneurons–and therefore only two pathways from visual input to motor output–could have a comprehensive repertoire of 'programmed' behaviors.

All of these data point to this agent having evolved a general strategy that is robust to the various different types of problem presented by different trial parameters, in spite of having such a simple control architecture. However, this is not true of any other 2-interneuron agents in the 16 we have evolved. It seems to be the case that while more interneurons are not necessary to perform well on this task, they make evolving reliable solutions considerably easier. This is borne out by the difference between the mean fitness of agents with 2 interneurons and those with 10, as shown in Figure 2 above; while this good 2-interneuron agent is an outlier, the best 8 or 10-interneuron agents are not far removed from the average performers in their class. This is a finding that needs more analysis and a larger number of evolution runs than time has permitted at this point.

5.5 Summary

The examples discussed above show that the task we are analyzing has a richly varied set of subtasks, which make different demands of our evolved agents. Because they are categorically different subtasks, rather than degrees of difficulty along one dimension, agents can develop idiosyncratic patterns of relative strengths and weaknesses, with a subtask that stymies one agent providing no trouble at all for another. The selective weaknesses of generally high performing agents can be at least partly explained by the rarity 'in the wild' of the trial types on which they fail; they have not evolved an appropriate response to situations that were either never or very rarely encountered during evolution. Section 6 below presents an attempt at testing this hypothesis.

The consistently high performance of the best 2-interneuron agent proves that 2 interneurons are sufficient to develop a generalized response to the task, contrary to our previous expectations. However, the exceptional nature of this agent, contrasted with the high average performance of 10-interneuron agents suggests that 10-interneuron architectures are far more likely to evolve good solutions. It may be that producing such a range of behaviors with such a simple architecture requires a very precise, and therefore brittle, combination of parameters, though this is a tentative explanation that needs more work to support it.

6. Improving Generalization

By modifying the trials an agent sees during training, we can increase performance on any subproblem, hopefully allowing for greater generalization.

An agent is exposed to a surprisingly small number of trials throughout its evolution. An agent will always be evaluated on 30 trials. New trials are added by replacing the trial that the agent does best on. During training, an agent will be introduced to a new trial when it is performing around 99%, or 600 generations after the last new trial was introduced. The performance limit for introducing a new trial is prohibitively high, therefore, it can be assumed that an agent will be introduced to a new trial every 600 generations, so only 14 trials will be added during the 9000 generations of training[2]. The first 35 trials are hand selected to encourage a valid solution to be formed, the remainder are selected from a group of 30 randomly generated trials, where the first one found to cause the current top-performing agent to fail (score below 85%) is added. Informal observation says that it is unusual for more than 30 trials to be generated before the agent fails on one. Thus, a reasonable upper bound for the number of trials an agent will encounter during training is 305 (including the initial 35 trials).

There are a large variety of subproblems, and many are quite rare. An example is the "unseen passing" subproblem, which only occurs 0.099% of the time. With this extreme rarity, the expected number of "unseen passing" trails that an agent will see during training is less than 0.2673.

By changing the trials an agent is exposed to during training, we should be able to increase performance in different subproblems. To test this hypothesis, the shaping schedule was modified to include one "unseen passing" problem. By introducing one "unseen passing" problem early in training we give agents much longer to train on the "unseen passing" subproblem. If an agent was to encounter an "unseen passing" trial during standard shaping, it would most likely be in a later generation, giving the genetic algorithm less time to evolve good solutions. Figure 9 below shows the average performance on the "unseen passing" subproblem for agents trained normally and those trained with the modified shaping. Due to time constraints, only 4 agents could be evolved for each number of interneurons on the modified shaping. 16 agents were evolved for each number of interneuron using standard shaping. The average score for agents with modified shaping is higher than standard agents, showing that an agent's performance on a subtask can be improved by modifying the shaping. While the best performing agents were evolved with the standard shaping, we believe this is an artifact of having 4 times more standard shaping agents.

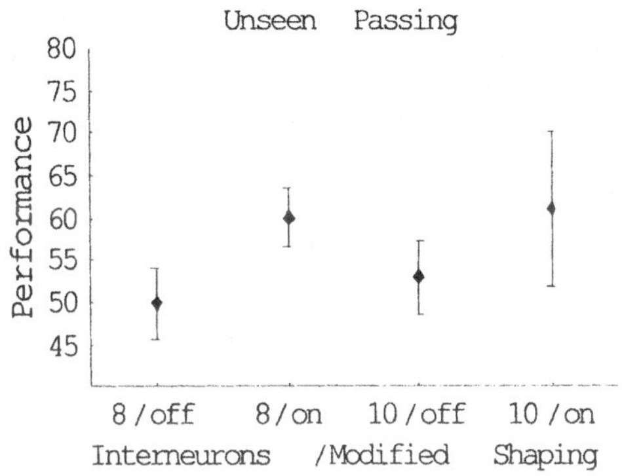

Figure 9: Performance on "unseen passing" subproblem with and without shaping modified to include an "unseen passing" trial. Error bars represent the standard error of the averages.

7. Conclusion

In this paper we have revisited a selective attention task which has previously been considered (Slocum et al., 2000; Downey, 2000) and analyzed the problem space in greater depth. Empirical observation of a selection of agents evolved for this task has shown that their performance does not vary monotonically across parameter sets. Instead, there is a rich substructure within the problem space, consisting of a range of distinct subproblems that agents must solve to perform well.

Our agents have evolved distinct behavioral responses to the variety of subproblems, as shown by the heterogeneous pattern of strengths and weaknesses. Our results to date indicate that agents with more complex control architectures are easier to evolve robust behavior with, but that this is possible even with only 2 interneurons. We have also found that the relative frequency with which agents are exposed to a given subproblem during evolution influences the likelihood that the agent will develop an appropriate response to it, and manipulated this effect to increase the reliability with which we can evolve agents whose behavior generalizes.

[2] Additional trials are actually introduced at generation 601, 1202, 1803... This causes only 14 trials to be added over 9000 generations.

We believe that understanding the dynamics of an evolved agent requires us to also understand the subtleties of the task for which they are evolved. The selective attention task we are using turns out to have a more complex structure than is at first apparent, and analyzing this has helped us to characterize the agents.

There are several directions for future work that we believe will be productive. Our decomposition of the problem space is not yet exhaustive, and extending this will probably allow us to make finer-grained distinctions between behavior patterns. This in turn would support a more detailed analysis of individual agents, linking the internal dynamics of the CTRNNs to their observed behavior. Such an analysis would help us to understand how an agent with as few as 2 interneurons can perform well on subproblems requiring a combination of short-term memory and selective attention.

We have made various tentative claims about the relative difficulty of evolving robust behavior with different numbers of interneurons. Evaluating these claims fully will require a more detailed characterization of the fitness space, which in turn will allow us to further refine the shaping protocol and produce robust behavior more reliably.

Acknowledgments

This work was supported in part by grant EIA-0130773 from the NSF.

References

Beer, R.D. (1996). Toward the evolution of dynamical neural networks for minimally cognitive behavior. In P. Maes, M. Mataric, J. Meyer, J. Pollack and S. Wilson (eds.), *From animals to animats 4: Proceedings of the Fourth International Conference on Simulation of Adative Behavior* (pp. 421-429). MIT Press.

Beer, R.D. (in press). The dynamics of active categorical perception in an evolved model agent. To appear in *Adaptive Behavior*.

Clark, A. (1997). *Being There: Putting Brain, Body, and World Together Again*. MIT Press.

Dorigo, M. and Colombetti, M. (1994). Robot Shaping: Developing Situated Agents through Learning. *Artificial Intelligence* 70(2):321-370.

Downey, D.C. (2000). *An Evolution of Minimally Cognitive Behavior: Short-term memory and selective attention.*. M.S. Thesis. EECS Department, Case Western Reserve University.

Harvey, I. (2000) Robotics: Philosophy of Mind using a Screwdriver. In T. Gomi (ed.) *Evolutionary Robotics: From Intelligent Robots to Artificial Life*, Vol. III (pp. 207-230). AAI Books.

Miller, G.E. (1994) *Artificial life as theoretical biology: how to do real science with computer simulation.* Cognitive Science Research Paper 378. School of Cognitive and Computing Sciences, University of Sussex.

Mitchell, M. (1996). *An Introduction to Genetic Algorithms*. MIT Press.

Saksida, L.M., Raymond, S.M. and Touretzky, D.S. (1997). Shaping robot behavior using principles from instrumental conditioning. *Robotics and Autonomous Systems* 22(3-4):231-249.

Slocum, A.C., Downey, D.C. and Beer, R.D. (2000). Further experiments in the evolution of minimally cognitive behavior: From perceiving affordances to selective attention. In J. Meyer, A. Berthoz, D. Floreano, H. Roitblat and S. Wilson (eds.), *From Animals to Animats 6: Proceedings of the Sixth International Conference on Simulation of Adaptive Behavior* (pp. 430-439). MIT Press.

Formalization of recognition, affordances and learning in isolated or interacting animats

P. Gaussier, J.C Baccon, K. Prepin(*), J. Nadel (*) and L. Hafemeister
Neuro-cybernetic team, Image and Signal processing Lab., ETIS UMR 8051 CNRS
Cergy Pontoise University / ENSEA, 6 av du Ponceau, 95014 Cergy , France
(*) UMR CNRS 7593, Hopital la Pitié Salpétrière, Paris
email: gaussier @ensea.fr

Abstract

In this paper, we summarize the main properties of an algebra useful to describe architectures devoted to the control of autonomous and embodied "intelligent" systems. First, we use this formalism to propose a definition of perception as a potential function built from the integration of the sensori-motor signals. Next, we demonstrate the capability of a very simple architecture to learn to recognize and reproduce facial expressions without the innate capability to recognize the facial expressions of others. The solution relies on the importance of the interactions with another system/agent knowing already a set of emotional expressions. A condition for the learning stability of the proposed architecture is derived. The teacher agent must act as a mirror of the baby agent (and not as a classical teacher). As a result, the proposed architecture is able to learn the task. Finally, we discuss the limitations of the proposed formalism.

1. Introduction

Nowadays hardware and software technologies allow to build more and more complex artifacts. Unfortunately, we are almost unable to compare two control architectures proposed to solve one given problem. Of course, one can try an experimental comparison on a given benchmark but the results focus on the optimality regarding the benchmark (how to deal with unknown or unpredictable events?). We should be able to analyze, compare and predict the behaviors of different control architectures. For instance, we must be able to decide if two architectures belong or not to the same family and can be reduced to a single architecture. On another level, new design principles are proposed to create more "intelligent" systems (Pfeifer and Scheier, 1999) but there is no real formalization of these principles. The only way to correctly understand and use them is to have a long explanation build on examples showing cases of success stories (examples of good robotic architectures). Our situation can be compared to the period before Galileo when people knew objects fall but were unable to relate that to the concept of mass and acceleration in order to predict what will happen in new experiments. We urgently need tools to analyze both natural and artificial intelligent systems. Previous works have focused on mathematical tools to formalize pure behaviorist or reactive systems (Steels, 1994). People have also tried with no real success to measure the complexity (in terms of fractal dimension for instance) of very simple behaviors like an obstacle avoidance (Smithers, 1995). The most interesting tools are dedicated to specific part of our global problem such as learning (see NN literature), dynamical systems (Schöner et al., 1995) or some game theory aspects (Ikegami, 1993). Yet, it remains difficult to overstep the old frame of the cybernetics (Wiener, 1961, Ashby, 1960). Finding the fundamental variables and parameters regarding some particular cognitive capabilities will be a long and difficult work but we believe we have to try to start it. After a short summary of our formalism, we will use it for the learning of attraction basins that characterize place or object recognition. A new formal definition of the perception will be derived. Next, several formal simplification rules based on the optimization of our perception criteria will be proposed. They will be illustrated on a simple theoretical model of the development of the capability to express and recognize more and more complex facial expressions. We will try to discuss, which are the basic mechanisms necessary to allow a naive agent to acquire the capability to understand/read the facial emotions of a teacher agent and to mimic them (so as to become a teacher and to allow turn taking in an emotion expression game).

2. Basic formalism of a CS

We summarize here the basis of our mathematical formalism. Figure 1 shows a typical control architecture for what we will call a cognitive [1] system (CS). The input and output of a CS are represented by vectors in the "bracket" notation[2]. An input or output vector x (column vector of size m) is noted $|x\rangle$ with $|x\rangle \in R^{+m}$ [3] while its transposed vector is noted $\langle x|$. Hence $\langle x|x\rangle$ is a scalar representing the square of $|x\rangle$ norm. The multiplication of a vector $|x\rangle$ by a matrix A is $|y\rangle = A|x\rangle$ with $|y\rangle \in R^n$ for a matrix A of size $n \times m$. A CS is supposed to be

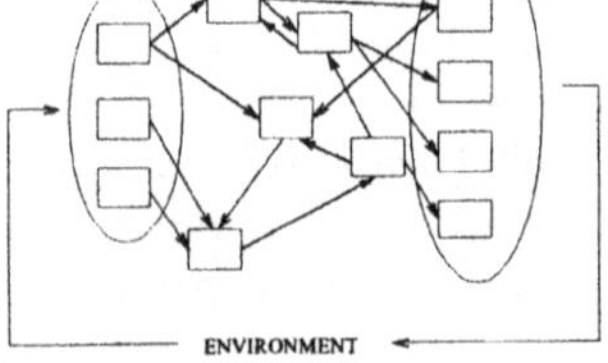

Figure 1: Typical architecture that can be manipulated by our formalism.

made of several elements or nodes or boxes associated with input information, intermediate processes and output (command of actions). We can consider that any element of a CS filters an input vector according to a matrix of weights W and a non linear operator **k**. This operator represents the way to use the W matrix and the pattern of interactions between the elements of the same block. It can be a simple scalar product (or distance measure) or even a more complex operator such as an "If...then...else..." treatment (hard decision making), a pattern of lateral interactions in the case of a competitive structure, a recurrent feedback in the case of a dynamical system, a shifting mechanism, a mechanism to control a focus of the attention... Hence, we can consider these elements as "neurons" even if they can be more complex algorithmic elements in other programming languages. For instance, in the case of a simple WTA[4] box, we can write the WTA output $|y\rangle$ is $wta(A|x\rangle)$ with $|y\rangle = (0, ..., y_j, ...0)$ and $j = ArgMax(q_i)$ and $q_i = \langle A_i|x\rangle$. In the case of a Kohonen map, $|y\rangle = koh(A|x\rangle)$, the main difference is the way the output is computed: $q_i = \sum_j |A_{ij} - x_j|$. To be more precise, we should write $|y\rangle = koh(A, |x\rangle)$. Because, in the general case, an operator can have an arbitrary number of input groups, we will consider the recognition of an input is performed according to the type of its associated weight matrix. For instance, "one to one" input/output connections represented by the general identity weight matrix I is considered as the signature of a reflex pathway (because there is almost no interest to consider "one to one" learnable links). Basically, we distinguish 2 main types of connectivity according to their learning capabilities (learning possible or not): the "one to one" links (see fig. 2a) and the "one to many" connections (see fig. 2b) which are used for pattern matching processes, categorization... or all the other possible filtering. "One to many" connections will be represented in general by a A. In the case of a complex competitive and conditioning structure with 1 unconditional (US) and 2 conditional (CS) inputs, we should write for instance $|y\rangle = c(A_1, |CS_1\rangle, A_2, |CS_2\rangle, I, |US\rangle)$. To avoid too many commas in the operator expression, we simply write $|y\rangle = c(A_1|CS_1\rangle, A_2|CS_2\rangle, I|US\rangle)$ [5]. This allows

[1] The term cognitive must be understood here in the sense of the study of particular cognitive capabilities (cogitare - to think) and not as a positive a priori for any kind of cognitivist approach.

[2] The formalism is inspired from Hilbert space used in quantum mechanics. Nevertheless, in our case it is not an Hilbert space since the operator is not linear...

[3] We consider the components of the different input/output vectors can only be positive/activated or null/inactivated. Negative activities are banned to avoid positive effects when combined with a negative weight matrix.

[4] Winner Takes All.

[5] In previous papers, it was possible to write $|y\rangle = c(A_1|CS_1\rangle + A_2|CS_2\rangle + I|US\rangle)$ but many reviewers complained about the risk of misunderstanding the meaning of the operator +.

to be sure a particular matrix is always associated to the correct input vector but it does not mean the matrix has to be multiplied by the vector (this computation choice is defined by the operator itself). The main

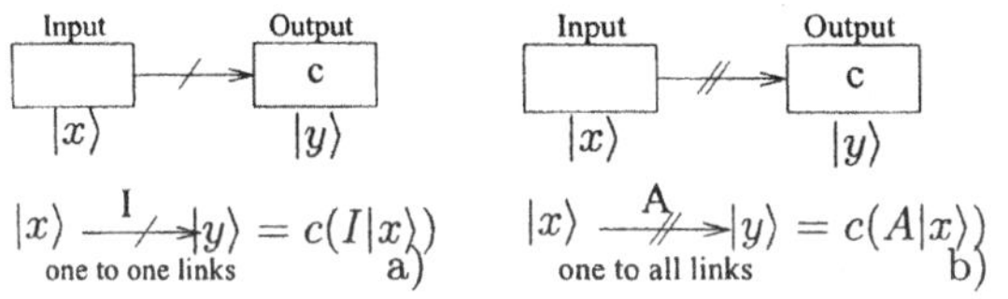

Figure 2: Arrows with one stroke represent "one to one" reflex connections (one input connected to one output in an injective manner). Arrows with labels and 2 parallel strokes represent "one to many" modifiable connections between input and output nodes. a) Unconditional "one to one" connections (used as a reflex link) between two groups. Upper image is the graphical representation and lower image is the formal notation. b) "One to many" connections with a competitive group representing the categorization of the input stimulus at the level of the output group.

difference with classical automata networks is that most of our operators are supposed to adapt or learn online new input/output associations according to their associated learning rule. For instance, in the case of a classical Kohonen rule, we should write $\frac{dA_{ij}}{dt} = koh\_learning(|y\rangle, |x\rangle)$. Hence, 2 equations have to be written for each elementary box: one for the computation of the system output and another one for the weight adaptation (modification of the box memory). In this paper, we will not discuss the interest or defaults of particular learning rules. We will simply suppose the existence of learning rules able to stabilize the weight matrices in the case the system is in a particular "perception state" (this notion will be defined in the next section). So we will not need to write explicitly the learning rule but it will be crucial to remember our operators represent 2 different functions and flow of information moving in opposite directions. The first one will allow to transform sensorial information in an output code while the second one will act on the group memory in order to maintain a certain equilibrium defined by the learning rule (Gaussier, 2001).

3. Stable state of perception

In previous papers, we have shown a homing behavior can be obtained from the competition between few sensori-motor associations learned around a goal location (see PerAc architecture proposed in (Gaussier and Zrehen, 1995, Gaussier et al., 2000)). For one position P in a given environment (or state space), we suppose 2 sensation vectors $|S_r\rangle$ and $|S_g\rangle$ can be defined. First, $|S_r\rangle = f(P)$ represents a coarse information about the motor system "proprioception" (a feedback information from the execution of the motor command) or the direction of the goal (if the goal is in the immediate neighborhood). It can be considered as a reflex or regulatory pathway linking a proprioceptive sensor to the motor command $|Ac\rangle$. Second, $|S_g\rangle = g(P)$ represents a more global information on the environment allowing to build a local but robust distance measure (metric). This measure is learned and performed by a competitive recognition group R ($|R\rangle$ representing its activity). This sensori-motor architecture for homing behavior can be defined by the following equations:

$$\begin{aligned} |R\rangle &= c(A_1|S_g\rangle) \\ |Ac\rangle &= c(A_2|R\rangle, I|S_r\rangle) \end{aligned} \quad (1)$$

which can be represented by the diagram fig. 3. The operator c represents a competitive structure (soft-WTA) able to self-organize itself according to one sensorial data flow and/or to condition one input data flow according to an unconditional flow (both learning and output computation are mixed in the single c notation). To illus-

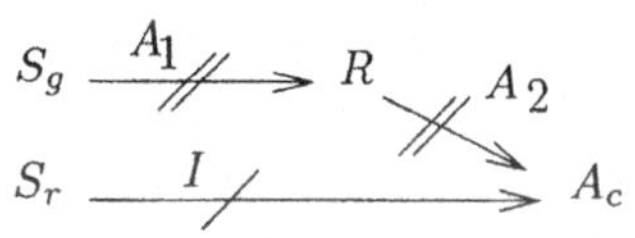

Figure 3: Diagram representing eq. 1: a simple network that can be used to learn how to return to a given place or to focus on a given object in a visual scene.

trate the system functioning and generalize it to object recognition, we simulate a very simple visual system that must learn to recognize a cross (see fig. 5b). We consider our system is recognizing an object if it is able to maintain the object in the center of its fovea or to stabilize its focus on this object. In our example, the system need to learn how to return to the center of the cross when it comes from the right

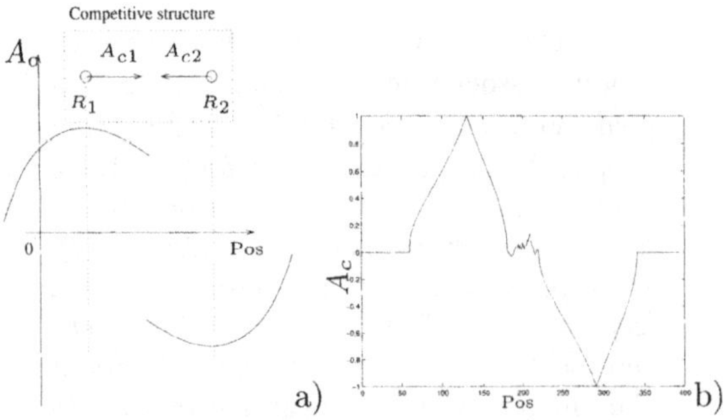

Figure 4: a) Theoretical system actions after learning of 2 sensation/actions associations and their competition according to the system position on the x axis. b) experimental value of the action level in the 1D case of fig. 5. Both max. and min. peaks are associated with the learned locations (perfect correlation between learned and tested views).

or from the left. Hence the system has to learn at least two positions located symmetrically around the goal. In our simulation, the learned locations were centered at the cross extremities (see fig. 5b). The recognition $|R\rangle$ of both learned positions is performed by a simple point to point cross-correlation between the learned local views (centered around the learned positions) and the current view centered on the gaze of the artificial eye. After the competition, the component R_1 or R_2 of $|R\rangle$ is activated according to the level of the local view recognition. The real decision is only taken at the motor level ($|Ac\rangle$ vector [6]) and must be understood according to the global temporal dynamic of the system. Let's notice, the system behavior does not directly depend on the absolute level of recognition of the learned views or places. Only the rank in the competition process matters. Because most of the visual perturbations have the same effect on each elementary recognition, the system will continue to behave correctly until the noise has an effect on the rank in the competition for the view recognition (while classical systems fail when the noise/perturbation oversteps an absolute recognition threshold). Fig. 5 shows few robot trajectories of the gaze direction according to a random initial gaze direction. We can see the system always converges to the center of the object whatever its starting position is. Fig. 5 b) presents a typical trajectory and the vector field in the 2D case (the system has only learned the 4 cross extremities and the associated actions to move in the direction of the center of the cross). Learning more places allows to produce a smoother vectorial field but the global behavior remains the same. Interestingly, in the proposed architecture the goal location is not learned in the N.N. Only the strategy to reach that place is learned. We claim that when our robot is going in the direction of the goal location it is truly "recognizing" that place. It is important to note that in

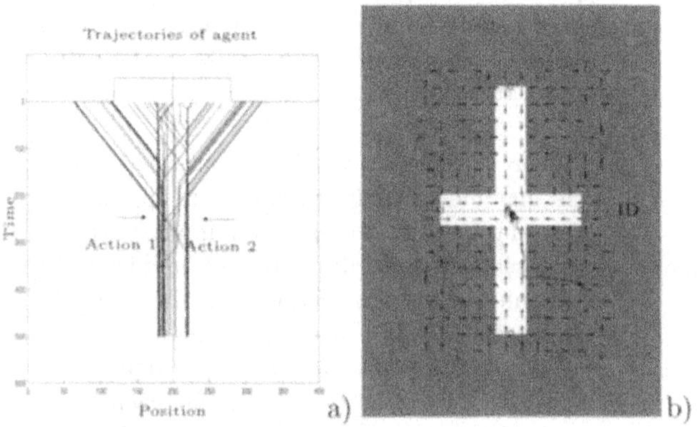

Figure 5: a) Examples of trajectories in the 1D case. The vertical line at position 200 represent the desired final position. b) Image used for the simulations (300x400 pixels, 256 gray-scale). Each small square is the position of the learned local views (101x101 pixels). The arrows indicate the orientation of the movement done at those locations. The larger square is the desired final position (goal). The line shows a sample of trajectory. In this figure only 4 views are used by the agent (one for each extremity of the cross).

the dynamical systems theory, the action is defined as the derivative of a potential field (Kelso, 1995, Schöner et al., 1995). If the field is defined according to a given position $\vec{p}$ in the environment, we will define[7]:

$$|Ac\rangle = -\overrightarrow{grad}\ Per \qquad (2)$$

The perception Per can be seen as a scalar function ψ representing an invariant of the

[6] In the 1 dimension case, we suppose the robot eye can only move on a horizontal line $|Ac\rangle$ vector need 2 components: Ac_1 associated to a movement from the left to the right and Ac_2 for a movement in the opposite direction. The simulated robot actions are represented by the intensity of the speed vector on the 1D axis (the sign representing the direction of the vector - see fig. 4).

[7] We could define : $m|Ac\rangle = -\overrightarrow{grad}\ Per$ with m the "mass" (real or virtual) of the considered system. We suppose the Perception Per takes into account a "mass" linked to the physical body.

system (a kind of energy measure). Hence, the perception can only be defined for an active system and is dependent of the system dynamical capabilities (kind of body, sensors and actuators). Of course, we can also write: $Per(\vec{p}) = -\langle Ac|\vec{p}\rangle = -\int_{\vec{p}+\delta\vec{p}} Ac\ d\vec{r}$. This corresponds to our intuition of the recognition as an attraction basin. The basin shown fig. 6 results from the mathematical integration of the curves proposed fig. 4 to represent the learned sensori-motor associations and their effect according to the system location. The obtained potential field defines an attraction basin with a single minima guiding the agent towards the goal location. The only important constraint is to guaranty that the recognition of the sensori information decreases monotonously over a given neighborhood when moving away from the learned positions. To sum up, the important change

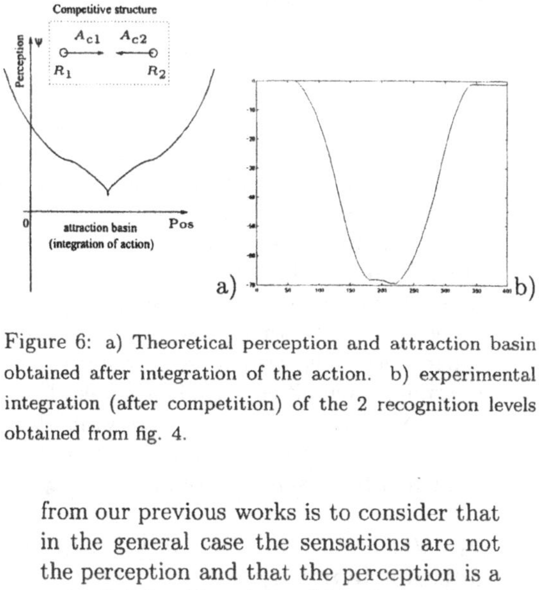

Figure 6: a) Theoretical perception and attraction basin obtained after integration of the action. b) experimental integration (after competition) of the 2 recognition levels obtained from fig. 4.

from our previous works is to consider that in the general case the sensations are not the perception and that the perception is a potential function defined by the conjunction of sensations and actions. Learning to recognize an object or a spatial location can be seen as building an attraction basin or learning some particular affordances (for instance on-line learning of few sensori-motor associations). In this case, we will say a system is in a stable state of perception if it is able to maintain itself in the associated attraction basin. Hence, recognizing an object (from visual, tactile, auditory... informations) can be seen as maintaining the system in a particular dynamical attraction basin (Gibson, 1979).

4. Formal simplification rules

Now, the problem is to be able to simplify a CS architecture in another one (presumably simpler to analyze and to understand). Two architectures will be considered as equivalent if they have the same behavioral attractors. This means we cannot study a control architecture alone. The interactions with the environment must be taken into account. After the learning of a first behavior, the dynamics of the interactions with the environment (the perception state) is supposed to be stabilized. In the present formalism, two types of diagram simplifications will be considered. Simplifications of the first type can be performed at any time and leave the fundamental properties of the system completely unchanged (these are very restrictive simplification rules). Those of the second type only apply after learning stabilization (if learning is possible!). They allow strong simplifications but the resulting system is no more completely equivalent to the departure system. We now present a first example of a simplification rule based on the existence of unconditional and reflex links. If we consider a linear chain of unconditional links between competitive structures such as WTA, the intermediate competitive boxes are useless since they replicate on their output their input information. Hence we can write for instance that if we have: $|b\rangle = c(I|a\rangle)$ and $|d\rangle = c(I|b\rangle)$ then $|d\rangle = c(I|c(I|a\rangle))$ which should be equal to $|d\rangle = c(I|a\rangle)$ because a cascade of competitions leads to an isomorphism between the different output vectors which become equivalent to each other after the self organization of the different groups. So we can deduce the following rule $\mathbf{c}(\mathbf{I}|\mathbf{c}(.)) = \mathbf{c}(.)$. Other static simplification rules can be build in the same way (Gaussier, 2001). Other simplifications can be used to represent the effect of learning. Except for robustness, these simplifications can be introduced to compare different control architectures (or to build more complex controllers). We will suppose that the system is in a stable state of perception or interaction with its environment. That is to say, it exists a time period where the system remains almost unchanged (internal modification must not have an effect on the system behavior). Fig. 7 shows an intuitive representation of the evolution of a system behavior through time. The system behav-

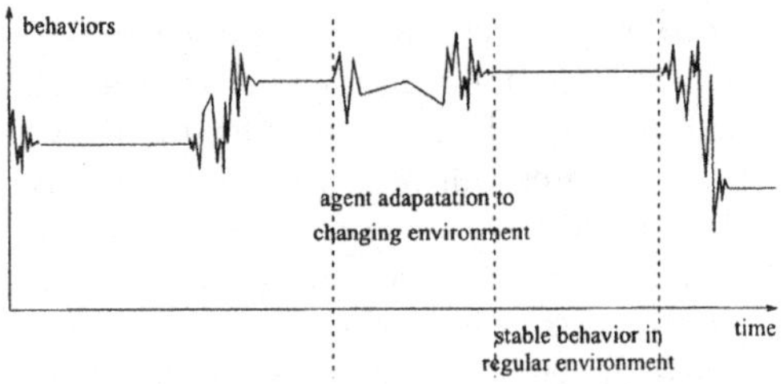

Figure 7: Intuitive representation of what is a stable behavior allowing formal simplifications of the system.

ior can evolve to adapt itself to an environment variation (or to the variation of an internal signal). In this case, it moves from a stable state to an unstable state or transition phase. It is only during the stable phases that the following simplifications can be considered as valid. Hence, we have to highlight a "before learning state" and an "after learning state" since some of the simplifications can be made at any time while some others must necessarily be made in the "after learning state". A very simple exam-

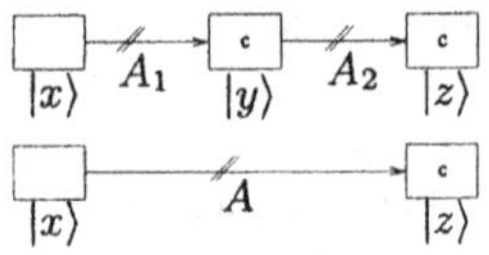

Figure 8: A cascade of competitive or unsupervised classification structures can be simplified in a single competitive or classification box with a possible loss of performance but without a change in the main properties of the architecture.

ple of such a simplification is the case of strict self organized learning group or competitive boxes (c operator) push-pully connected, fig. 8. We have $|y\rangle = c(A_1|x\rangle)$ and $|z\rangle = c(A_2|y\rangle)$ with A_1 and A_2 the matrices to learn the relevant input configurations. So $|z\rangle = \mathbf{c}(\mathbf{A_2}|\mathbf{c}(\mathbf{A_1}|\mathbf{x}\rangle)) = \mathbf{c}(\mathbf{A}|\mathbf{x}\rangle)$ since it is always possible to create a bijection between the activation of a given neuron in a first group and the activation of another neuron in a second group. Both sets of neurons can be considered as equivalents. A more interesting case corresponds to the conditioning learning. The conditioning network (fig. 9 a) should be equivalent "after learning" to the simple network shown fig. 9 b and can be translated by the following equation: $c(I|US\rangle, A|CS\rangle) \approx c(A|CS\rangle)$ where $|US\rangle$ represents the unconditional stimulus and $|CS\rangle$ the conditional stimulus. The simplification "before learning" considers only the reflex pathway: $c(I|US\rangle, A|CS\rangle) \approx c(I|US\rangle)$ (functioning is equivalent in a short time delay but there is no possible adaptation) whereas the other simplification represents the equivalent NN in the "after learning" situation: not equivalent if the environment changes too much and leads the agent to be inadapted. We have shown

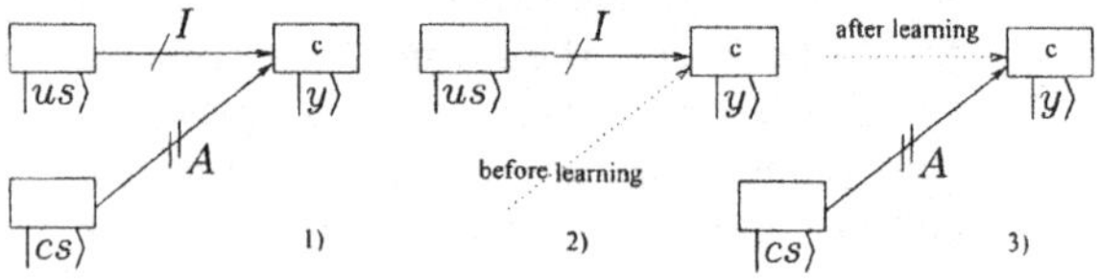

Figure 9: Image 1 is the graphical representation of a conditioning learning $|y\rangle = c(I|US\rangle, A|CS\rangle)$. Image 2 is the graphical representation of the equivalent network before learning and Image 3 after learning $|y\rangle = c(A|CS\rangle)$.

in (Gaussier, 2001) that maximizing the dimensionality (rank) of the perception matrix $\sum_P |Ac\rangle\langle S|$ can be equivalent to the mean square error minimization performed when trying to optimize the conditioning learning between the action proposed by the conditional link and the action proposed by the unconditional link. Hence, learning can be seen as an optimization of the tensor representing the perception. In other words, we can say the proposed simplification rules are relevant if the system is adapted to its environment or if the system perceives its environment correctly according to the capabilities of its own control architecture (learning capabilities). We can notice that $Per = \sum_P \langle Ac|S\rangle = tr(\sum_P |Ac\rangle\langle S|)$ while the "complexity" of the system behavior can be estimated from $rank((\sum_P |Ac\rangle\langle S|)$.

5. Application to social interactions learning

In this section, our goal is to show how our formalism can be applied to analyze a very simple control architecture and justify some psychological models (see (Canamero, 2001) for a discussion on the importance of an emotional system in autonomous agents). At the opposite to the classical pattern recognition approach, we will show that an online dynamical action/perception approach between two interacting systems has very important properties. The system we will consider is composed of two identical agents (same architecture) interacting in a neutral environment (see fig. 10). One agent is supposed

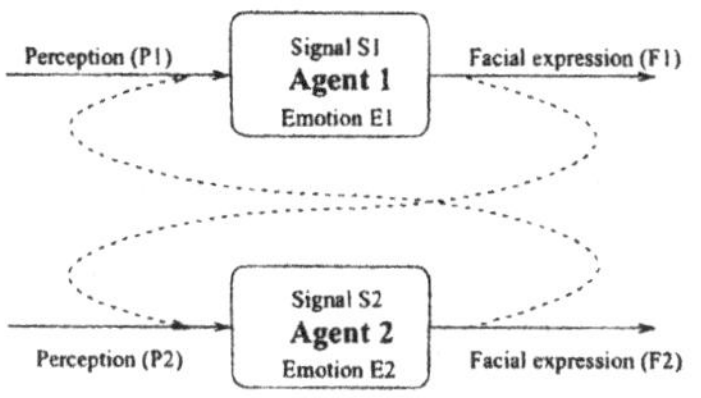

Figure 10: The bidirectional dynamical system we are studying. Both agents face each other. Agent 1 is considered as a newborn and agent 2 as an adult mimicking the newborn facial expressions. Both agents are driven by internal signals which can induce the feeling of particular emotions.

to be an adult with perfect emotion recognition capabilities and also the perfect capability to express an intentional emotion. The second agent will be considered as a newborn without any previous learning on the social role of emotions [8]. First, we will determine the conditions for a stable interaction and show that in this case learning to associate the recognition of a given facial expression with the agent own "emotions" is a behavioral attractor of the global system. We suppose our agents receive some visual signals (P_i perception of agent i). They can learn and recognize them ($|R_i\rangle$ activity). Hence, the perception of a face displaying a particular expression should trigger the activation of a corresponding node in R_i.

$$|R_i\rangle = c\,(A_{i1}|P_i\rangle) \tag{3}$$

c represents a competitive mechanism allowing to select a winner among all the vector components (see section 3). A_{i1} represents the weights of the neurons in the recognition group of the agent i allowing a direct pattern matching. Our agents are also affected by the perception of their internal milieu (hunger, fear etc.). We will call S_i the internal signals linked to physiological inputs such as fear, hunger... "Emotion" recognition E_i depends on the internal milieu. The recognition of a particular internal state will be called an emotional state E_i. We suppose also E_i depends on the visual recognition R_i of the visual signal P_i. At last, the agents can express a motor command F_i corresponding to a facial expression. If one agent can act as an adult, it must have the ability to "feel" the emotion recognized on someone else's face (empathy). At least, one connection between the visual recognition and the group of neuron representing its emotional state must exist. In order to display emotional state, we must also suppose there is a connection from the internal signals to the control of the facial expression. The connection can be direct or through another group devoted to the representation of emotions. For sake of homogeneity, we will suppose that the internal signal activates through an unconditional link the emotion recognition group which activates through an unconditional connection the display of a facial expression (hence it is equivalent to a direct activation of F_i by S_i - see (Gaussier, 2001) for a formal analysis of this kind of properties). Hence, the sum of both flows of informations can be formalized as follow:

$$|E_i\rangle = c\,(I|S_i\rangle, A_{13}|R_i\rangle) \tag{4}$$

At last, we can also suppose the teacher agent can display a facial expression without "feeling" it (just by a mimicking behavior obtain form the recognition of the other facial expression). The motor output of the teacher facial expression then depends on both facial expression recognition and the will to express a particular emotion:

$$|F_i\rangle = c\,(I|E_i\rangle, A_{12}|R_i\rangle) \tag{5}$$

Fig. 11 represents the network associated to the 3 previous equations describing our candidate architecture. In a more realistic architecture, some intermediate links allowing the inhibition of one or another pathway could be added but it is out of the scope of the present paper, which aims at illustrating what can be done with our formalism on a very simple example.

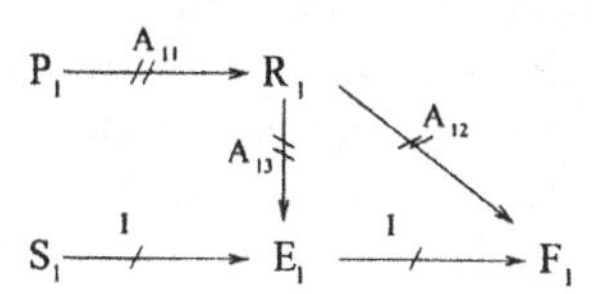

Figure 11: Schematic representation of an agent that can display and recognize "emotions" (notations see fig. 2).

[8]We will have to explain how our agent can recognize and reproduce gestures it cannot see itself perform and by what mechanism it connects the felt but unseen movements of self with the seen but unfelt movements of the other (Meltzoff and Moore, 1997).

5.1 Condition for learning stability

First, we can study the minimal conditions allowing the building of a global behavioral attractor (learning to imitate and to understand facial expression). Fig. 12 represents the complete system with both agents in interaction. It is considered as a virtual net that can be studied in the same way than an isolated architecture thus allowing to deal at the same time with the agent "intelligence" and with the effects of the embodiment and/or the dynamics of the action/perception loops. The following simplifications apply before learning and concern only the unconditional links (see in the previous section the simplification of a conditioning structure before learning). We simply consider the activation of S can induce a reflex activation of a stereotyped facial expression F before (and after) the learning of the correct set of conditioning. The resulting network is shown fig. 13. Next, the linear chains of "one to many" modifiable connections and their associated competitive learning structures can also be simplified since $\mathbf{c}(\mathbf{A}|\mathbf{c}(.) \equiv \mathbf{c}(.)$. We finally obtain the network shown fig. 14 a). It is much simpler on fig. 14 to see the condition of the learning stability. Since, the chosen simplifications allow to obtain a virtual network with learnable bidirectional connections between F_1 and F_2, a condition for the learning stability is that these connection weights remain stable. If S_1 and S_2 are independent, learning cannot be stable since S_1 and S_2 are connected through unconditional links to F_1 and F_2 respectively. The only way to stabilize learning is to suppose S_1 and S_2 are similar enough. Otherwise a lot of "energy" is lost to adapt continuously the connections between F_1 and F_2 (see (Gaussier, 2001) for more details). Because, the agent representing the baby must not be explicitly supervised, a simple solution is to suppose the agent representing the parent is nothing more than a mirror. We obtain the network fig. 14 b) where the architecture allows the system to learn the "internal state"-"facial expression" associations. Hence, we show that from our departure control architecture, learning is only possible if the parent agent (supposed to be the teacher) imitates the baby agent. The roles are switched according to the classical point of view of AI and learning theory. This shows how taking account the dynamics of interactions between two agents can change our way of thinking learning and more generally cognition problems.

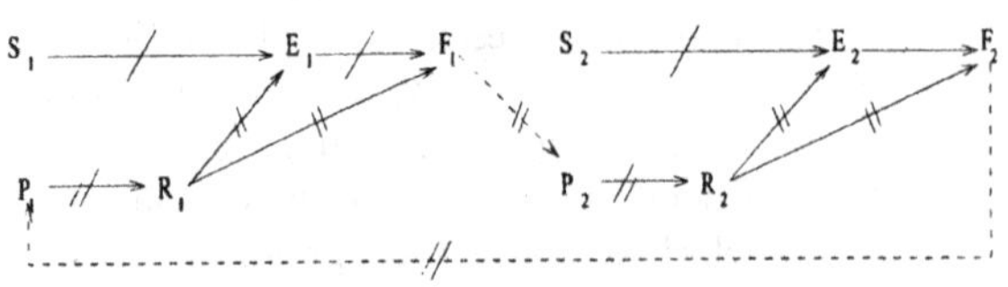

Figure 12: Schematic representation of the global network representing the interaction between 2 identical emotional agents. The dashed links represent the connections from the display of a facial expression to the other agent perception system (effect of the environment).

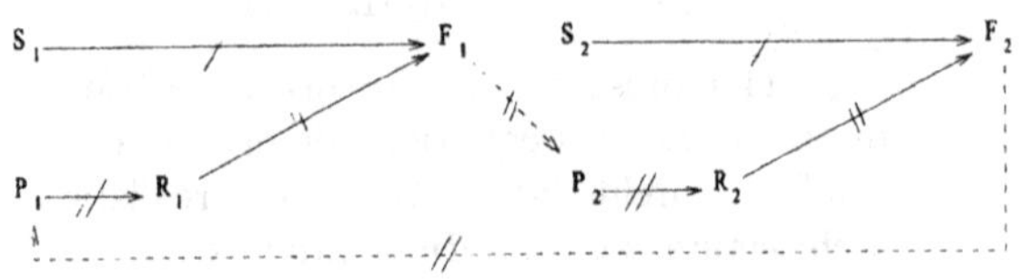

Figure 13: Schematic representation of the simplified network representing the interaction between 2 identical emotional agents (modification of fig. 12)

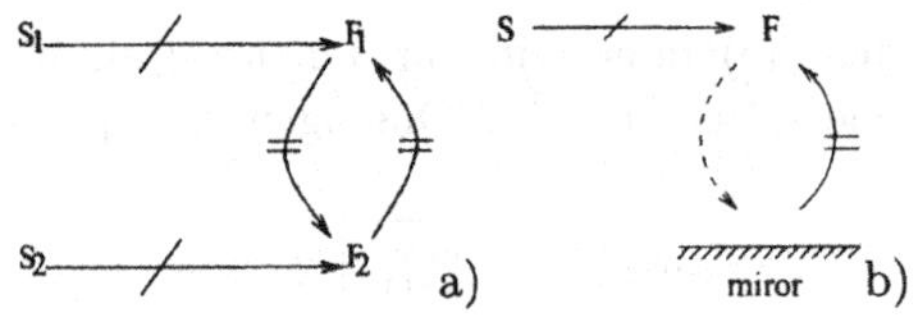

Figure 14: a) Final simplification of the network representing the interaction between 2 identical emotional agents (modification of fig. 13). b) Minimal architecture allowing the agent to learn "internal state"-"facial expression" associations.

5.2 Learning the emotional value of facial expressions

The first simplifications bring us to the conclusion that learning stabilization is possible if the teacher/parent agent acts as an imitator of the baby agent. Now, we will suppose these conditions are respected. From the initial equations of the system, we will derive another set of simplifications in order to prove the beginner (or naive) agent can learn to associate the visual facial expression displayed by the teacher agent to the correct emotional state. We suppose the agent 1 perceptive input P_1 is the result of a linear projection of the facial expression (output) of the agent 2 and vice versa. We will write $|P_1\rangle = B_1|F_2\rangle$ and $P_2 = B_2|F_1\rangle$. Hence, $|R_1\rangle = c(A_{11}|P_1\rangle) = c(A_{11}.B_1|F_2\rangle) = c(A'_{11}|F_2\rangle)$ (with $A'_{11} = A_{11}.B_1$). We can then replace in this new expression of R_1, $|F_2\rangle$ by the result of the computation of the second agent (using eq. 5). We obtain:

$$\begin{aligned}|R_1\rangle &= c(A'_{11}|c(I|E_2\rangle, A_{23}|R_2\rangle))\\ &= c(A'_{11}|c(I|E_2\rangle, A_{23}|c(A_{21}|P_2\rangle)))\end{aligned}$$

On the other side, we have $|P_2\rangle = B_2|F_1\rangle$ so:

$$\begin{aligned}|R_1\rangle &= c(A'_{11}|c(I|E_2\rangle, A_{23}|c(A_{21}\cdot B_2|F_1\rangle)))\\ &= c(A'_{11}|c(I|E_2\rangle, A_{23}|c(A'_{21}|F_1\rangle))) \quad (6)\end{aligned}$$

A'_{21} is defined as the matrix resulting from $A_{21} \cdot B_2$.

All the preceding simplifications could be made at any time (here, it is before learning). The following simplification can be done only after learning (and need the learning stability condition ie the second agent is a mirror of the first one). If the obtention of learning is possible (the error between F_1 and E_2 can be minimized in the means square sense), conditioning learning in eq. 6 should result in: $I|E_2\rangle \approx A_{23}.c(A'_{21}|F_1\rangle)$. if both architectures are identical, since there is no influence of learning on this simplification, we obtain by symmetry: $|E_1\rangle \approx A_{13}.c(A'_{11}|F_2\rangle)$. Then, we can simplify eq. 6.

$$\begin{aligned}|R_1\rangle &\approx c(A'_{11}|c(A_{23}|c(A'_{21}|F_1\rangle)))\\ &\approx c(A'_{123}|F_1\rangle) \quad (7)\end{aligned}$$

(we also have $|R_1\rangle \approx c(A'_{12}|E_2\rangle)$ but we won't use it.) Eq. 7 can be interpreted as the fact the activity of agent 1 visual face recognition is a function of its own facial expression. If we replace the value of F_1 obtained from eq. 5 in eq. 7, we obtain:

$$|R_1\rangle \approx c(A'_{123}|c(I|E_1\rangle, A_{13}|R_1\rangle)) \quad (8)$$

Here again $I|E_1\rangle$ is the reflex link and $A_{13}|R_1\rangle$ the conditional information. The conditional link can learn to provide the same results as the reflex link. If E_1 can be associated to R_1 then we obtain:

$$\begin{aligned}|R_1\rangle &\approx c(A'_{123}|c(I|E_1\rangle))\\ \text{and } |R_1\rangle &\approx c(A'_{123}|E_1\rangle) \quad (9)\end{aligned}$$

This result shows the activity of the face recognition system is a direct function of the agent emotional state (R_1 can be deduce from E_1). In conjunction with the relation linking E_1 to R_1 (eq. 4) we can deduce the agent 1 (baby) has learned to associate the visual recognition of the tested facial expressions to its own internal feeling (E_1). The agent has learned how to connect the felt but unseen movements of self with the seen but unfelt movements of the other. It could be generalized to other movements since we showed in (Gaussier et al., 1998, Andry et al., 2001, Andry et al., 2002) that a simple sensori-motor system is sufficient to trigger low level imitations.

6. Conclusion and discussion

In this paper, we have improved a formalism proposed in (Gaussier, 2001) to define more correctly what perception can be in a situated and embodied agent. The formalism has been used to simplify an "intelligent" system and to analyze some of its properties. We have shown a very simple architecture can learn the bidirectional association between an internal "emotion" and its associated facial expression. To demonstrate this feature, we have first proved that learning is only possible if one of the agents acts as a mirror of another. We have proposed a theoretical model that can be used as a tool not only to understand artificial emotional brains but also natural emotional brains. Let us consider a newborn. She expresses internal states of pleasure, discomfort, disgust, etc, but she is not aware of what she expresses. Within our theoretical framework, we can expect that she will learn main associations between what she expresses and what she experiences through her partners' mirroring of her own expressions. Seeing what she feels will allow the infant to associate her internal state with an external signal (i.e. her facial expression mirrored by someone else). Empirical studies of mother-infant communication support this view. For instance, two-month-old infants facing a non contingent televised mother who mirrors their facial expressions with a delay become wary, show discomfort and stop imitating the mother's facial expressions (Nadel et al., 2004). The primary need of mirroring is also demonstrated by the progressive disappearance of facial expressions in infants born blind. Another prospective benefit of the model is to give a simple developmental explanation of how facial expressions come to inform the growing infant about external events through the facial reading of what those events trigger in others (Feinman, 1992). Finally the model leads to suggest a main distinction between two processes of emotional matching: matching a facial emotion without sharing the emotion expressed: in this case there is a decoupling (Scherer, 1984) between what is felt and what is shown, thus it is pure imitation, and matching a facial emotion with emotional sharing, that is to say feeling what the other expresses through the process of mirroring, a definition of empathy (Decety and Chaminade, 2002).

In our team, we also use this formalism as a programming language to describe different architectures for visual object recognition, visual navigation, planning, visuomotor control of an arm, affordance learning, imitation games... Future works will have to focus on how to manage different learning time constants and how to represent the body/controller co-development. What are the minimal structures that cannot be simplified? What kind of really different operators have to be considered? What kind of invariant has to be added to take into account all the ideas of the animat approach and the embodied cognition? Computational models avoiding the passive aspect of the architectures studied in this paper should be the next step of our work especially to study when internal recurrences or loops in the system architecture can be simplified or not.

Acknowledgements: Different parts of this work have been inspired from fruitful discussions during the workshop on "perceptive suppleances" sponsored by the STIC department of the French CNRS and especially discussions with O. Gapenne, C. Lenay, K. OReagan and J. Stewart. This work is supported by the CNRS (ACI Computational Neurosciences and CNRS team project on "imitation in robotics and development").

References

Andry, P., Gaussier, P., Moga, S., Banquet, J., and Nadel, J. (2001). Learning and communication in imitation: An autonomous robot perspective. *IEEE transactions on Systems, Man and Cybernetics, Part A*, 31(5):431–444.

Andry, P., P.Gaussier, and Nadel, J. (2002). From sensorimotor coordination to low level imitation. In *Second international workshop on epigenetic robotics*, pages 7–15.

Ashby, W. (1960). *Design for a brain.* London: Chapman and Hall.

Canamero, L. (2001). Emotions and adaptation in autonomous agents: A design perspective. *Cybernetics and Systems*, 32(5):507–529.

Decety, J. and Chaminade, T. (2002). Neural correlates of feeling sympathy. *Neuropsychologia*, 42:127–138.

Feinman, S. (1992). *Social referencing and the social construction of the reality in infancy.* Plenum Press, New York.

Gaussier, P. (2001). Toward a cognitive system algebra: A perception/action perspective. In *European Workshop on Learning Robots (EWLR)*, pages 88–100, http://www-etis.ensea.fr/~neurocyber/EWRL2001_gaussier.pdf.

Gaussier, P., Joulain, C., Banquet, J., Lepretre, S., and Revel, A. (2000). The visual homing problem: an example of robotics/biology cross fertilization. *Robotics and autonomous system*, 30:155–180.

Gaussier, P., Moga, S., Quoy, M., and Banquet, J. (1998). From perception-action loops to imitation processes: a bottom-up approach of learning by imitation. *Applied Artificial Intelligence*, 12(7-8):701–727.

Gaussier, P. and Zrehen, S. (1995). Perac: A neural architecture to control artificial animals. *Robotics and Autonomous System*, 16(2-4):291–320.

Gibson, J. (1979). *The Ecological Approach to Visual Perception.* Houghton Mifflin, Boston.

Ikegami, T. (1993). Ecology of evolutionary game strategies. In *ECAL 93*, pages 527–536.

Kelso, J. S. (1995). *Dynamic patterns: the self-organization of brain and behavior.* MIT Press.

Meltzoff, A. and Moore, M. K. (1997). Explaining facial imitation: A theoritical model. *Early Development and Parenting*, 6:179–192.

Nadel, J., Revel, A., Andry, P., and Gaussier, P. (2004). Toward communication: first imitations in infants, low-functioning children with autism and robots. *Interaction Studies*, 5:45–75.

Pfeifer, R. and Scheier, C. (1999). *Understanding intelligence.* MIT press.

Scherer, K. (1984). Emotion as a multicomponent process. *Rev. Person. Soc. Psychol.*, 5:37–63.

Schöner, G., Dose, M., and Engels, C. (1995). Dynamics of behavior: theory and applications for autonomous robot architectures. *Robotics and Autonomous System*, 16(2-4):213–245.

Smithers, T. (1995). On quantitative performance measures of robot behaviour. *Robotics and Autonomous Systems*, 15:107–133.

Steels, L. (1994). A case study in the behavior-oriented design of autonomous agents. In *SAB'94*, pages 445–451.

Wiener, N. (1948, 1961). *CYBERNETICS or Control and Communication in the Animal and the Machine.* MIT Press.

Gallery Structure Formation in Conleyborus scarabaeoides.

Luc Steels

Sony CSL
5, rue Amyot
75005 PARIS
steels@arti.vub.ac.be

University of Brussels
(VUB AI Lab)
Pleinlaan 2
1000 Brussels

Abstract

The paper presents an agent-based model of gallery structure formation by bark beetles. It attempts to show that the remarkably complex patterns observed in nature are an emergent side effect of several basic processes mostly related to the embodied interaction of the beetle with the wood.

1. Introduction

Members of the *Scolytidae* species (bark beetles) construct intriguing structures by excavating slender mines or burrows in the bark of trees or between the bark and sap wood (figure 1). Typically (for example for the *Anisandrus dispar*, the European shot-hole borer), the adult bores a hole in a tree during the spring, constructs a maternal chamber, and then digs a characteristic network of galleries of uniform diameter emanating from the chamber. About 40 eggs are laid and larvae feed on xylophagous fungi (*Ambrosia*) introduced by the female on the walls of the galleries (Grgoire, 1988). The gallery structures are definitely not random and neither completely regular. They are not identical for different subspecies of Scolytidae and show large variation even within the same subspecies or for the same individual.

This paper reports on attempts to understand the nature of these structures. At first sight, they have a calligraphic character and resemble marks in Minoan Linear A and B (Chadwick, 1980). So we may wonder whether any intelligent design or conscious effort to follow specific patterns is implied, similar to the way human cultures have developed various writing systems which are transmitted from one generation to the next. If not, then what are the structural principles underlying these fascinating markings and what are the mechanisms that implement them? The question is of ecological value because bark beetles have a devastating effect on ecosystems in Northern Europe, where they attack fruit trees, and in Canada, where they attack the soft pine population. More generally, research into the mechanisms that give rise to the markings produced by Scolytidae raises deeper questions about the origins of complexity in animal behavior, in particular how far seemingly complex structures necessarily require highly intelligent (and perhaps even conscious) organisms to produce them or whether they can be the result of self-organisation.

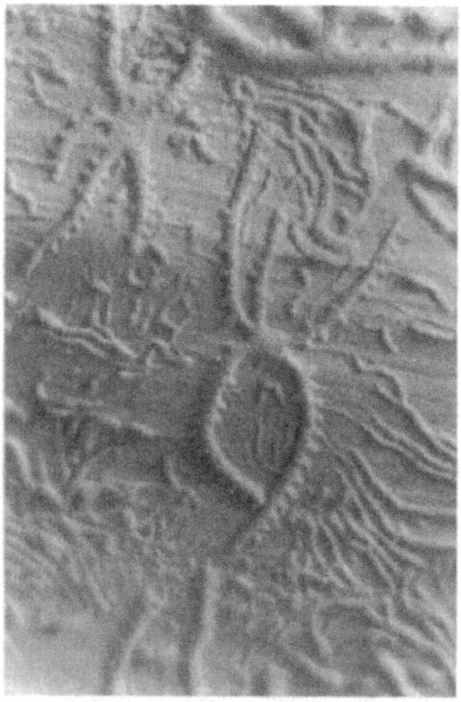

Figure 1: Typical example of markings produced by Scolytidae based on the excavation of galleries in tree wood. (Field data B. Conley, Upstate New York)

2. Method

Science observes natural phenomena and attempts to find first the systematicity and then the underlying principles that could give rise to the phenomena. These principles can be mathematically captured in systems of differential equations, which are then shown, usually through Monte Carlo simulations, to adequately cover the phenomena one is interested in. Researchers in 'artificial life' have pioneered an alternative modeling approach in the form of agent-based models. In this approach, the relevant aspects of the environment are captured in virtual worlds, and the crucial properties of organisms are implemented as behaviors of artificial software constructions known as agents. These agents act like animated robots (sometimes called animats) that operate in a virtual world, similar to the way animated characters like Lara Croft operate in a computer game world. An early example of this approach, inspired by stygmergetic behavior of social insects, can be found in Steels (1990). Contemporary ethologists now routinely use both types of models interchangeably, but in this paper only agent-based models will be employed, mainly because they allow a much better treatment of the very complex phenomena we want to investigate.

The real world is complex at many different scales and not everything can be captured in a model (whether based on differential equations or agents). To be informative, the model should only incorporate what is relevant for an explanation of the phenomena of interest. Moreover the model should not necessarily be completely faithful to reality. As Galileo showed, there is great value in making idealised abstractions, like the ideal pendulum or the frictionless cannonball. These abstractions can be understood thoroughly, and then used to confront richer realities, which are contingent on many additional factors and hence much less amenable to model construction.

I will follow the same approach in this paper and define an 'idealised bark beetle', further called the Conleyborus scarabaeoides (by analogy with bark beetle species that resemble it). The Conleyborus scarabaeoides (commonly named the Conley beetle) is an idealisation in the sense that it has only the behaviors that are relevant for gallery construction. More realistic investigations for naturally occuring species can use this idealised beetle as a first approximation.

This work is in line with earlier work in modeling the adaptive behavior of animals using computational simulations or robots (see e.g. Prescott and Ibbottson (1997)).

3. Hypothesis

The main hypothesis of the investigation is that the markings produced by bark beetles are the emergent side effect of the complex interactions between a number of coupled dynamical systems. In other words, there is no pre-established pattern that the beetles have innately stored nor a culturally established repertoire that they try to copy, similar to the way human beings reproduce signs. The causal factors cannot be reduced to a single source and the markings emerge through self-organisation. This hypothesis is in line with earlier work on how social insects, like termites, honey bees, wasps, or ants, collectively construct complex nests, establish paths from food sources to their nests, or assemble in swarms to migrate or defend themselves against predators (Camazine, et.al., 2001).

In fact, some aspects of *Scolytidae* behavior have already been approached this way. Larvae have been shown experimentally to exhibit aggregation behavior and self-organisation models have been proposed to explain how the aggregation occurs (Deneubourg, et.al., 1990). The larvae use chemical communication which pulls them together due to a positive feedback mechanism. Larvae emit a chemical substance and are themselves attracted to it. So the higher the concentration in a certain area, the more the larvae are attracted to that area, and the concentration increases even further attracting more larvae. The chemical attraction is countered by negative feedback because larvae try to avoid each other. The net effect of both dynamical processes is that larvae stick together and form aggregate paths, particularly under the bark of the tree (figure 2). Notice also that the patterns in figure 2 are very different from those in figure 1. The latter are based on galleries inside the wood. In the remainder of this paper we focus only on the second type of structures, and particularly on patterns similar to those shown in figure 1.

Figure 2: Characteristic aggregation paths of larvae between bark and tree wood. (Author's field data, Barranco de Arure, Gomera, Spain)

In the case of gallery formation in *Scolytidae*, the literature suggests that the intervening coupled dynamical systems include:

1. *The textural properties of the wood.* The beetles make tunnels within the wood and will meet more or less resistance depending on texture. For example, they navigate around branch points where the wood is much harder. It follows that we will see different markings depending on the tree type and even the individual tree.
2. *The reaction of the tree to the presence of larvae.* When tunnels are made, the tree reacts by producing large amounts of resin which can potentially kill the larvae. The larvae react to this by sticking together to avoiding getting overwhelmed.
3. *The embodiment of the beetle (and the larvae).* The size of the galleries is largely determined by the size of the body (figure 3) and the beetles tend to excavate in the same direction because turning the body around is not so easy. So we expect to see gentle curves, rather than 90 degree angles.

Figure 3: Excavation of a tunnel by a larvae. The size of the tunnel is largely determined by body size. (Image from Inra Zoology HYPP)

4. *Repulsion among the beetles.* Beetles keep out of each others' way because they are in competition for the same food sources. Thus when there is an initial hole in a certain location, the next beetle arriving on the same branch will start a new hole at a certain distance to avoid interference with the first. While digging, beetles must avoid getting into other galleries. A tree branch is a limited terrain and can only sustain a limited population of beetles.

5. *Presence of fungi.* In most cases, the beetles do not live off the wood itself, but use fungi as a food source. These fungi are introduced by females and thrive in the dust resulting from tunnel excavation. The fungi population has its own dynamics which is coupled to the dynamics of the larvae population. They co-exist in a symbiotic relation with the beetles.

6. *Beetle life cycle.* Beetles lay eggs, and clearly the number of eggs and where they are deposited in the maternal chamber or in the tunnels determine how the tunnels will further expand. Beetles hibernate in the tunnels and they get further excavated due to beetle traffic along existing paths.

This list is not intended to be exhaustive but illustrates that a complex set of factors interferes in gallery construction. Given all these factors, it is obvious that a model of differential equations is not very realistic. An agent-based model will still be very complex but can be built up step by step and allows the integration of spatio-temporal structures much more easily. It is also obvious that not everything can or needs to be modeled to help us gain insight in the processes that give rise to the markings of interest here. I will here ignore the growth process of the tree, the resin attacks, the chemical communication and aggregation of the larvae, the life-cycle of the beetle, the temperature changes in the environment, and many other factors that all play a role in this ecosystem. The markings produced by the beetles exhibit a bewildering complexity, without being entirely random, and the great challenge is to show that they are nevertheless based on a number of simple underlying principles. These principles must translate into behaviors that can be carried out by a beetle based only on local observation and control and with a minimum of sensing or complex information processing.

When observing a surface as in figure 1, we get only a very partial view of the total ecosystem that plays a role in gallery formation. A tree branch is a three dimensional structure and beetles excavate in all directions. So we will use a three-dimensional cellular automaton as the basic grid to model the world, as opposed to the more common two-dimensional cellular automata used in many 'artificial life' models. Only the surface layer of the automaton will be visualised. To emphasise the three-dimensional character of the galleries, I designed the drawing programs in such a way that structures are drawn with shadows, so that they appear to stick out.

The position of the beetle is defined in terms of <x,y,z> coordinates on the 3D grid. Each beetle also has a characteristic size, modeling the body shape, a tilt plane which is, by default, level with the ground, and an orientation (a 360 degree angle with respect to the tilt plane).

4. Behaviors of the *Conleyborus scarabaeoides*

I now show that gallery formation producing markings with a complexity similar to natural markings can be explained with a few rather straightforward behaviors: tunneling, path retracing, borehole initiation, chamber construction, and spawning of larvae. Each of these behaviors contains random choices which are influenced by the state of the environment and it is through their interaction with the environment that the typical markers emerge.

1. Tunneling

The Conley beetles or their larvae form tunnels by crawling step by step over the 3D grid and removing wood in the process. Given the position of a beetle in x,y,z coordinates, the next adjacent position is computed by translating the angle into a new set of coordinates, and the beetle moves to this position, excavating the area on its path based on its current size (figure 4). Because of the grain-size of the cellular automata, a tunnel cannot be as smooth as for real life beetles.

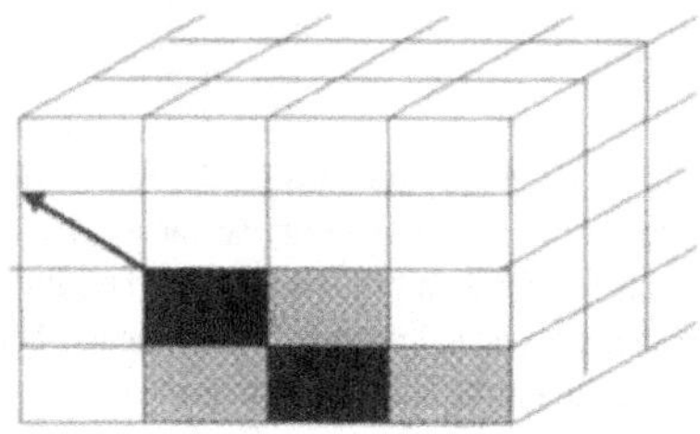

Figure 4: A 3D-Cellular automaton is used to model the wood. The next move of a beetle is calculated based on its orientation and its current position.

A tunnel has an orientation and a size. The size of the tunnel is a side effect of the size of the beetle, which excavates enough to comfortably pass through, i.e. slightly larger than its own size. When a tunnel is bending, slightly more area is escavated to allow the body to make the turn.

The orientation of the tunnel is a side effect of the orientation of the beetle, which is basically constant, with slight, random variation (left and right with respect to the existing orientation). This variation models the resistance encountered by the wood. Variations can accumulate and may then give rise to a bend in the tunnel.

This kind of tunneling behavior already generates the typical single gallery strands which we observe in natural markings (see figure 5, left). Notice the use of shading, producing an apparent depth. The markings produced by the beetles give the impression of paths left by rolling stones, coming down from a mud hill. Note that the tunnel is not a computational object in the simulation but a side effect of beetle behavior.

This specification of tunneling behavior cannot be the whole story however. First of all we also see in natural settings characteristic symbol-like structures consisting of an internal blob (interpreted as the presence of a maternal chamber) with tentacles going up or down similar to a sea anemone (figure 1). This suggests that local orientation in tunneling behavior should occasionally be influenced by a global direction. More research is required to investigate the precise cause or ecological reason for pursuing a global direction. In the Conley beetle simulations discussed here, I have assumed that the global orientation is either guided by gravity or antigravity, depending on the initial orientation: When orientation is above 270 degrees or below 90 degrees, the beetle will progressively seek an orientation of zero degrees, otherwise it tends towards 180 degrees. With this kind of influence on tunneling, the Conley beetles produce tentacle-like tunnels (see figure 5, right). The size of the gallery generally decreases as the beetle approaches the end of the tunnel.

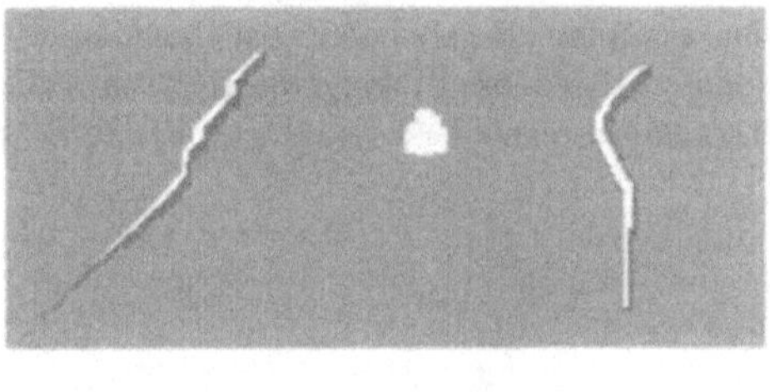

Figure 5: The three main structures produced as a side effect of the tunneling behaviors of Conleyborus scarabaeoides Left, construction of a tunnel in random direction (top to bottom). Middle, maternal chamber excavated by random movements. Right, tentacular structure due to the influence of a global orientation.

A second point is that in natural settings, beetles tend to avoid an already existing gallery, even though this is far from always the case (figure 1). This implies that beetles must sense from a distance whether there is a gallery in the direction in which they are going. Because this is presumably not possible using visible means, we can speculate that it is based on chemical communication, which has been proven to show an important role for other aspects of beetle behavior as well.

Avoiding existing galleries has also been incorporated in the tunneling behavior of Conleyborus scarabaeoides. Around each gallery a 'taboo' zone is dynamically computed as tunnels are excavated. Beetles are able to sense these taboo zones and perform obstacle avoidance behavior around them or halt in front of them.

2. Path Retracing.

Once a tunnel exists, beetles can move through them, retracing the position from where they came. This requires that they can turn around in a gallery, and that they can compute the next move based on a perception of the orientation of the gallery in front of them. Retracing of a path may give rise to additional excavation, slightly enlarging the gallery. Again, the behavior is not necessarily completely successful, and sometimes the Conley beetle may start to excavate from a point somewhere in the middle of a gallery, wrongly believing it has reached the maternal chamber.

3. Borehole Initiation.

A beetle arrives on the bark of a tree and then bores a hole into the branch. The selection of the entry point is partially random, and partially determined by the presence of other holes. In nature, beetles avoid making a new hole too close to an existing one in order not to get into competition right away (figure 6). We therefore must supply the Conley beetle with an entry point selection process, based on the perception of existing holes observed by matching a visual pattern against perceived reality. Once a position is chosen, excavation of a tunnel starts in a random orientation and goes for a randomly chosen distance into the wood.

Figure 6: Holes are randomly spaced but at a certain distance from each other. (Author's Field data, Barranco de Arure, Gomera, Spain)

4. Chamber construction

The Conley beetle constructs a maternal chamber by making random excavating movements around a central position. The number of movements is randomly determined. This results in an amorphous area of undetermined size (see figure 5, middle).

The four behaviors discussed so far are chained: A beetle arrives on a piece of wood and initiates a borehole. After some time, it constructs a maternal chamber, and then it starts making a tunnel in a randomly chosen direction. When the tunnel has a certain depth (again randomly determined) the beetle retraces the path back to the maternal chamber and then chooses a new direction for the next tunnel.

5. Spawning of larvae

The final behavior I had to introduce to to get the characteristic gallery structures of *Scolytidae*, concerns the spawning of the larvae. They mature and then start to construct their own tunnels in the wood. The number of larvae and the times at which they are spawned are randomly determined but they follow a more or less regular pattern. The position where larvae are deposited determines from where a new tunnel starts, and the orientation of the tunnel is basically perpendicular to the orientation of the gallery or chamber structure from where the larvae start to move, but alternating left and right. Beetles and larvae build further on existing gallery structures, thus illustrating the kind of stygmergetic behavior which is very common in social insects.

When the beetle deposits larvae while moving in a tunnel (which may bend slightly), we often see a series of parallel lines, because only the galleries in one perpendicular direction are visible. The galleries in the other perpendicular direction go deeper inside the wood and are not visible on the surface. In most cases we do not see the main gallery in which the adult beetle is moving either, because it is also located below the surface.

When the beetle deposits larvae while moving on the edge of a maternal chamber, we see the characteristic 'symbols' as in natural markings (see figure 1 or figure 7). The beetle is moving around the edge of the maternal chamber and new galleries start at random points on the edges, but then move globally up or down.

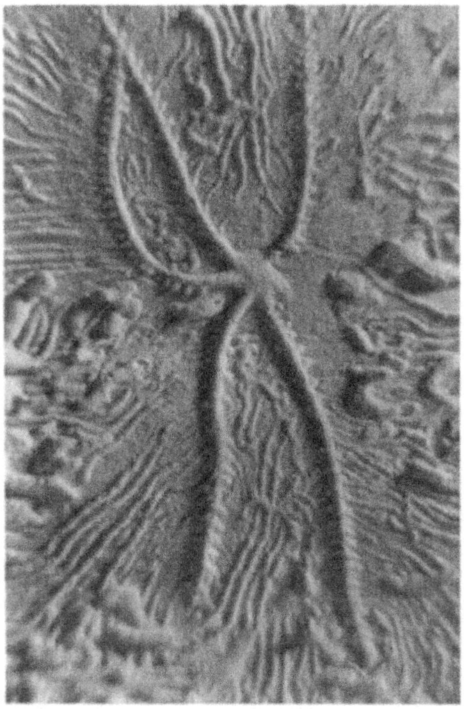

Figure 7: Tentacular structures arise around a maternal chamber producing a characteristic 'symbolic marking.(Field data B. Conley, Upstate New York)

5. Results and Discussion

The results of putting the different behaviors of Conleyborus scarabaeoides in action on a 3D grid are shown in figure 8. This image has been produced by a single invocation of the simulation program where a random number of beetles attack virgin wood, make boreholes and maternal chambers, excavate tunnels, and spawn larvae that then construct their own tunnels. The image has been made slightly less sharper through a 3x3 filter, simulating the effect of taking a photograph of the wood surface. We clearly observe structures that are similar to those observed in the natural settings shown in figures 1 or 7.

More examples of simulation results are shown in figure 9. The comparison with natural markings could potentially be made more precise by a mathematical characterisation of structures in natural and artificial settings. But the analogy is striking to the naked eye.

These simulations therefore give substance to the main claim of the present paper, namely that the markings produced by *Scolytidae* are the emergent side effect of a number of dynamical processes, including the way tunnels are constructed in the wood, the embodiment of the beetle, the intervals with which larvae are spawned, etc. The observed complexity is due to the randomness, influencing decisions at all levels of this ecosystem. The observed order is due to the many constraints inherent in the physical and biological world and a number of default choices, for example, to start a new gallery perpendicular to an existing one and not in a completely random direction. Gallery formation in *Scolytidae* therefore follows the same general principle as other structures formed by social insects. The hypothesis of high intelligence (let alone consciousness), innate knowledge of patterns, or global coordination and planning of behavior is not required.

Returning to the intriguing similarity between the markings of *Scolytidae* and early writing systems such as Minoan Linear A or B, I refer to some of my other work, which has shown that the building blocks of human communication systems, such as the speech sounds of human languages, can also be explained as the emergent side effect of self-organising processes. In the case of speech, there are free choices that language users collectively make, but these are constrained by the physics of the sound medium, the nature and control of the vocal apparatus, the capacities of the human auditory system, and the behavior of neural networks in the brain (Steels and Oudeyer, 2000). Similar research could investigate the origins of writing systems (such as clay tablet markings, Chi-

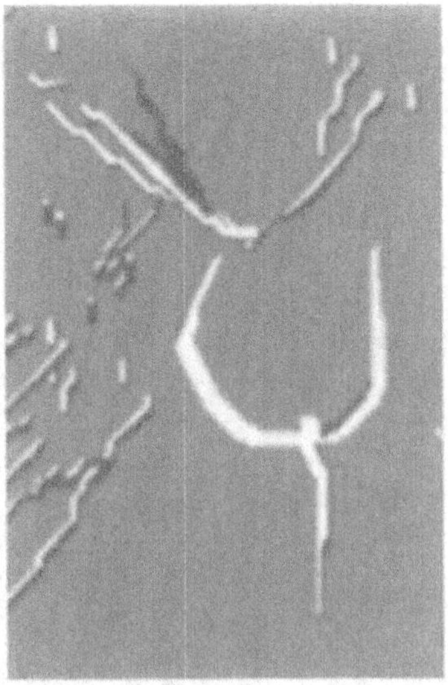

Figure 8: Close-up of markings produced by *Conleyborus scarabaeoides*. We see a typical symbolic structure in the bottom right corner and other galleries around it.

nese characters, or Minoan Linear A or B), by modeling the influence of the physics of the medium, the interaction of available tools with the medium, the nature of human gestures, and the limits and potentials of human visual pattern recognition. Of course, in the case of speech or writing systems, there is a cultural transmission process as well, because the physical signs play a role in collective semiotic processes that are transmitted from one generation to the next (Steels and Kaplan, 2000).

Acknowledgement

I am indebted to Brian Conley whose interest and extensive field work on *Scolytidae* in Upstate New York was a direct inspiration for the research presented here, and to Marleen Wynants for help during my own field work on *Scolytidae* in the Barranco Arrure, Gomera, Spain, as well as early feedback on the simulations. I programmed the 3D cellular automaton, the agent behaviors, and the graphic visualisation directly in Macintosh Common Lisp, although the use of a variant of Logo would have been a viable alternative.

6. Bibliography

Camazine, S., J-L Deneubourg, N.R. Franks, J. Sneyd, G. Theraulaz, and E. Bonabeau. (2001) Self-Organization in Biological Systems. Princeton University Press, Princeton, New Jersey.

Chadwick, J. (1980) The Decipherment of Linear B. Cambridge Univ Press, Cambridge.

Deneubourg, J-L., J.C. Grgoire, and E. Le Fort (1990) Kinetics of the larval gregarious behaviour in the bark beetle Dendroctonus micans. Journal of Insect Behavior 3:169-182.

Gregoire, (1988) The greater European spruce beetle. In: Berryman, A. (ed.) Population Dynamics of Forest Insects. pp. 455-478. Plenum, New York.

INRA HYPP Zoology pages http://www.inra.fr/Internet/Produits/HYPPZ/

Prescott, T.J. and C. Ibbottson (1997) A robot trace maker: modeling the fossil evidence of early invertibrate behavior. Artificial Life, 3, 298-306.

Steels, L. (1990) Cooperation through self-organisation. In: De Mazeau, Y. (ed.) Distributed Artificial Intelligence. North-Holland Pub., Amsterdam.

Steels, L. and P-Y Oudeyer (2000) The cultural evolution of syntactic constraints in phonology. pp. 382-394. In: Bedeau, et.al. (ed.) (2000) Proceedings of Artificial Life VII. The MIT Press, Cambridge Ma.

Steels, L. and Kaplan, F. (2000) Collective learning and semiotic dynamics. pp. 679-688. In: Floreano, D. and Nicoud, J-D. and Mondada, F., (ed.) (2000) Advances in Artificial Life (ECAL 99), Lecture Notes in Artificial Intelligence 1674, Springer-Verlag, Berlin.

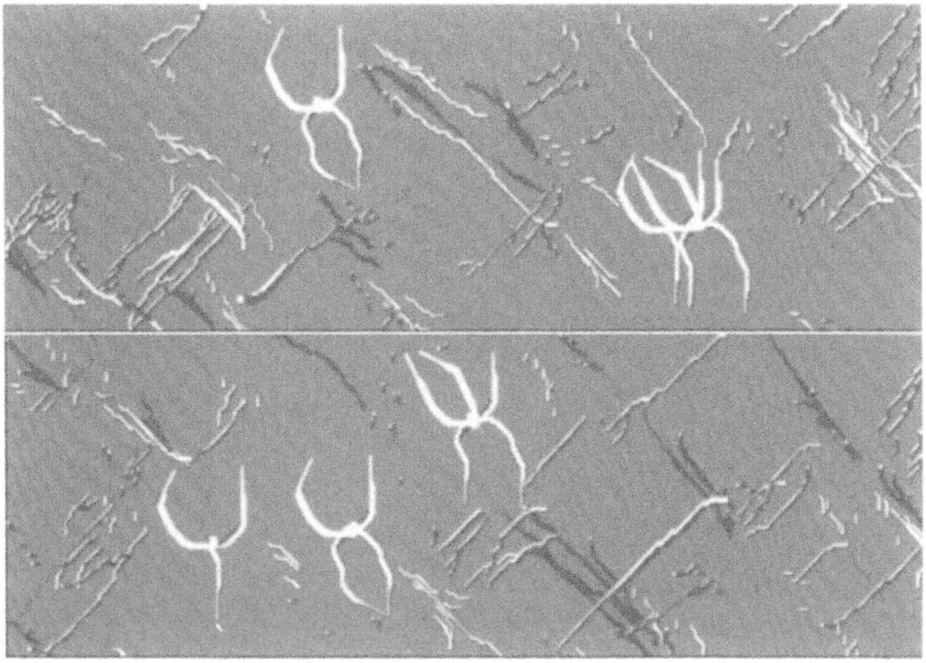

Figure 9: More markings produced by *Conleyborus scarabaeoides*. Each is based on a single run of the simulation with random choices for number of beetles attacking the wood. Roughly parallel lines (as in bottom image, left lower corner) are produced by offshoots from an invisible deeper perpendicular gallery.

PERCEPTION

Towards a theory of how bats navigate through foliage

Jonas Reijniers and Herbert Peremans*
*Departement Milieu, Technologie en Technologiemanagement
Universiteit Antwerpen
Prinsstraat 13
2000 Antwerpen
herbert.peremans@ua.ac.be

Abstract

In this paper we study echolocation by spectral analysis as a possible mechanism used by bats to navigate through their natural environment. We simulate the echo signals received by a bat inspired sonar system confronted with realistic reflectors, i.e. a large number of closely spaced scatterers. We propose a method to localize such natural reflectors on the basis of the time-frequency representation of the returned echo that is calculated by the bat's cochlea. The sonar system is simplified into a piston emitter and two piston receivers, of which the latter can be rotated. The method proposed reconstructs the environment by comparing the returned echoes with predefined templates corresponding with echoes from different angles. We show that the overlap between members of this set of templates can be used as a measure of their accuracy for localization purposes. This method is then applied to both monaural and binaural sonar systems and is shown to allow useful navigation behaviour. The problem of ghost images, i.e. erroneous interpretations of ambiguous position estimates, is shown to be inherent to single measurement based decisions. Finally, we show how active control of the ear configuration allows binaural sonar systems to eliminate ghost images by combining the results from a sequence of measurements, each corresponding with a well chosen ear configuration. Despite the description in this paper being restricted to the two-dimensional situation, this is not essential and we discuss how the mechanism described can be straightforwardly extended to the full 3D case.

1. Introduction

Navigation through a realistic outdoor environment, i.e one containing trees, bushes and other natural reflectors, while solely relying on sonar would seem to be an impossible task if it were not routinely solved by bats. A bat flying in a natural environment, e.g. a forest, is confronted with large numbers of (randomly) distributed scatterers of various sizes, some directed toward the bat, others away from it. Hence the bat receives at both ears a complex echo composed of many interfering echo's, (Müller and Kuc, 2000). Yet, it manages to extract the spatial information it needs in order to manoeuvre with great skill through its environment while at the same time using its sonar to hunt for prey (Popper and Fay, 1995). In spite of the efforts expended in recent years (Thomas et al., 2003), it is still not well understood how bats manage to do this.

Bats can be divided into two broad non-taxonomic groups based on the time-frequency-structure of their echolocation pulses: fm-bats (fm=frequency modulated) echolocation calls are multi-harmonic chirps. In cf-fm-bats, echolocation pulses are often dominated by prolonged cf (=constant frequency) signal portions, although frequency-modulated "heads" or "tails" are always present as well. For bats (Nowak, 1994) that emit broadband calls, i.e. fm-bats, a number of schemes have been proposed that are all to some extent based on how the particular shape of the bat pinna (outer ear) gives rise to systematic variations in the spectrum of the received echo depending on the spatial position of the reflecting target. Kuc (Kuc, 1994) describes a mechanism for bat prey localization and prey capture based on the presence of a fundamental call frequency and at least one harmonic in the bat cry. The SCAT receiver (Saillant et al., 1993, Peremans and Hallam, 1998, Matsuo et al., 2001) has been proposed as a mechanism to convert the output of the bat's cochlea into a time-domain representation, i.e. the impulse response, of the environment. In (Simmons et al., 2002) it is shown how this measured impulse response varies in a systematic manner with the position of the reflecting target. In (Walker et al., 1998, Peremans et al., 1998) a mechanism is described that is based on stereotypical ear movements, as observed for particular bat species and can be applied to both broadband and narrowband target localization. Spatial dependent filtering of the reflected echo by the pinnae seems less useful for narrowband calls, i.e. cf-fm-bats, as most of the call energy is emitted at a constant frequency. However, since the calls of such bats also contain a broadband part the mechanisms described above could be made use of by these bats as well. In addition, Müller et al. (Müller and Schnitzler, 2000) analyze the feasibility of an acoustic flow mechanism based on Doppler-shift and amplitude cues for narrowband localization by cf-fm-bats. Note that such Doppler-shift information is not present for fm-bats as their pulses can be shown to be largely Doppler-shift invariant (Altes and Reese, 1975). Although all of these methods result in realistic localization performance for situations giving rise to small numbers of isolated echoes, they break down when confronted with natural scenes giving rise to large numbers of highly overlapping echoes. In this paper we will focus on the latter situation and investigate the feasibility of direction dependent filtering as a localization strategy in the presence of multiple overlapping echoes.

In Sec. 2. we describe the various components of the model of the bat sonar system as well as the acoustic model describing reflection on natural reflectors that will be used throughout this paper. Next, we consider a simple monaural sonar system consisting of a circular piston which acts both as emitter and receiver. Based on this simple configuration, we develop our approach in Sec. 3. However, due to its simplicity this system displays a number of limitations in realistic localization tasks. Hence, in Sec. 4. we extend the sonar system into a more complex binaural one consisting of two piston receivers that can be rotated. We study how this affects the performance. In Sec.5. we summarize our findings and discuss how the approach that has been restricted, for clarity reasons, to localization in two dimensions (2D) can be extended for a true 3D sonar localization system.

2. Simulation models

In this section we describe and comment on the models that were used in this study.

2.1 Biomimetic sound processing

Mouth and ears In the interest of clarity we limit ourselves to a 2D analysis in this paper. Hence, we approximate the 3D description, see Fig. 1 (a), of the head of a bat, i.e. one transmitter (mouth) placed in front of and somewhat below the two receivers (ears) as shown in Fig. 1 (b) by the 2D description shown in Fig. 1 (c). Furthermore, as the distance to the target L in a navigation task is always large compared to the size of the head ($2d \approx 2-3$cm) we consider all the relevant bearing angles to be equal: $\theta_{l_{ear}} = \theta_{r_{ear}} = \theta$ or equivalently we consider the mouth and ears to coincide. In addition, in the monaural case, Sec. 3. we assume transmitter and receiver to have the same orientation, i.e. $\phi = 0$, whereas

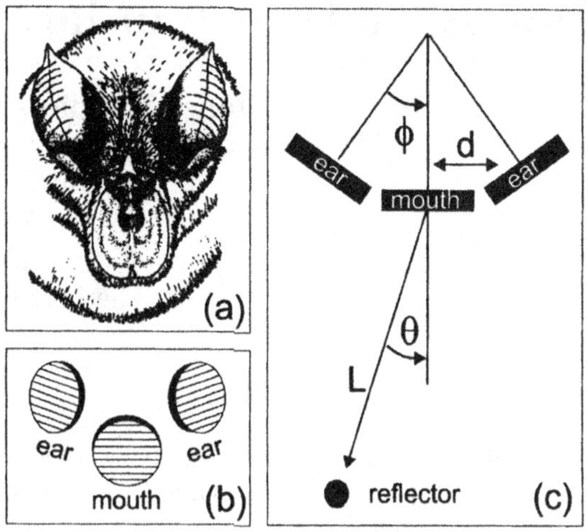

Figure 1: (a) A front view of a bat head (© M. Trappe); (b) Schematic front view of the sonar system model consisting of piston-type transducers for mouth and ears; (c) The 2D version of the sonar system model.

in the binaural case, Sec. 4., the orientation of the two receivers will in general differ from that of the transmitter, as shown in Fig. 1.

We approximate the mouth and ears by circular piston-like transducers, see Sec. 2.2 despite the experimental data (Hartley and Suthers, 1987, Firzlaff and Schuller, 2003) clearly indicating that real bat sonar systems have a considerably more complex directivity. Nevertheless, we introduce this assumption for illustration purposes as it allows us to derive analytic expressions for the directivity. Since none of the conclusions drawn below is dependent on the particular directivity selected, the piston model can be replaced with a more accurate one when and where available.

Echolocation call We consider a broadband call of 2ms duration. The amplitude increases from zero to full strength in 0.5ms, remains constant for 1ms and then decreases again in 0.5ms. The instantaneous frequency of the call sweeps down from 100 to 20kHz. A spectrogram-like representation of this emit signal is shown in Fig. 5 (a). Such call duration and frequency sweep are in accordance with experimental data on bats.

Cochlea The bats auditory system is structured in the same way as that of other mammals (see e.g., (Pickles, 1982)). The transduction stage located in the inner ear (cochlea) performs a joint time-frequency analysis of the incoming signal. A simple model of this analysis is a bandpass filter bank with subsequent demodulation in each channel by a combination of half-wave rectification and lowpass filtering. In fm-bats the layout of the auditory filter bank follows the general mammalian pattern of keeping filter quality constant as center frequency varies.

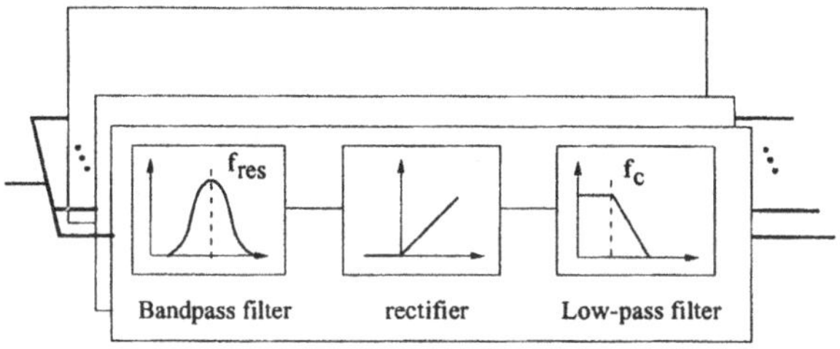

Figure 2: The cochlear model of the bat.

Specifically, our model of the bat cochlea, see Fig. 2, is based upon a gammatone filter bank model (Slaney, 1993). The filter bank consists of 77 frequency channels with central frequencies ranging between [10kHz, 150kHz] and with constant quality factor $Q = 20$. Keeping filter quality constant implies that bandwidth (f_{bw}) is a linear function of center frequency (f_c): $f_{bw} = f_c/Q$. Additionally, overlap between neighbouring filters is kept constant (Slaney, 1993), which results in a wider spacing of filters as bandwidth increases. We have run additional experiments while varying the number of frequency channels, i.e. 9, 16, 33 and 77 frequency channels. The results, e.g. overlap between the templates (see Sec. 3.2.2), were not affected until the number of frequency channels dropped below 16. This seems to suggest that the higher numbers of frequency channels present in a real bat cochlea, $\approx 700 - 3000$, are mostly there to increase redundancy and not so much information content of the signals. The transduction of the movement of the membrane into neural activity, i.e. implemented biologically by the hair cells of the cochlea, is modeled to a first approximation by processing the outputs of the filter bank with a half-wave rectifier and a low-pass filter (Schroeder and Hall, 1974). These last two steps are equivalent to a simple amplitude demodulation scheme that approximately recovers the envelope of the outputs of the bandpass filters in the filter bank. The outputs of these low-pass filters are taken to be the final output of the system. Fig. 5 (a) shows an example of the output of the system when processing an echo consisting of a replica of the call, picked up by the ear, and two echoes from objects at different bearings and distances. From this figure it is clear that the cochlea produces a spectrogram-like output representation.

In the bat's brain the cochlear output is then translated into a spike code and sent down the auditory nerve into the brain for further processing. However, we assume that no essential information is lost in this process so that we can make abstraction of it in the current study.

2.2 *Echo formation*

We now describe the models used to simulate how the sound pulses emitted by the sonar system are transformed into echoes by the environment.

Filter model Only pointlike sound sources radiate sound equally in all directions, sound sources of finite extent instead have a directivity, also called radiation pattern, which for a piston-like transducer is given by (Pierce, 1994)

$$H_p(\theta, f) = \frac{|J_1(ka\sin(\theta))|}{|ka\sin(\theta)|}, \tag{1}$$

with a being the radius of the transducer, $k = 2\pi f/v_{sound}$, and θ the angle between the normal to the transducer surface, i.e. the maximal sensitivity axis, and the direction of interest, see Fig. 1 (c). As this result shows, see Fig. 3 (a), the radiation pattern is both angle and frequency dependent.

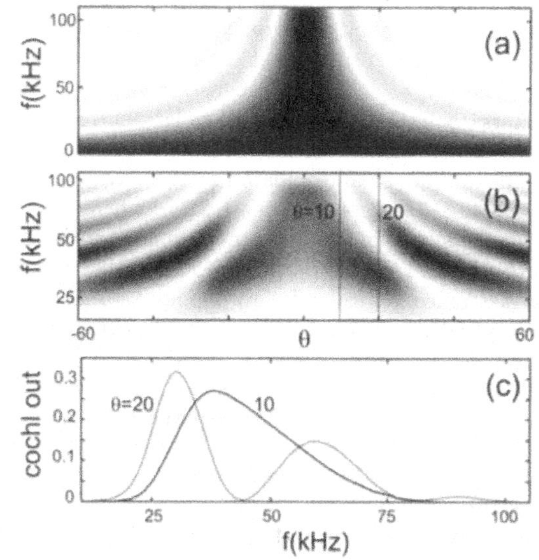

Figure 3: (a) Radiation pattern of a circular piston, $a =$ 1.4cm, as a function of frequency f and bearing angle θ. (b) The set of angular templates $T_\theta(n)$ for the monaural sonar system. (c) The two templates, i.e. normalized cochlear output patterns, corresponding with reflectors at $\theta = 10°$ and $\theta = 20°$ (the two vertical lines in (b)).

Hence, a sonar system does not emit the same call in all directions and objects situated at different bearing angles are sounded by pulses with a different frequency content. Similarly, upon reception the echoes returning from the environment undergo a second filtering caused by the finite spatial extent of the receiver. For the monaural system, as emitter and receiver coincide, this is exactly the same filtering as was done by the emitter (reciprocity).

For the complete echo formation process, assuming a target at position (r, θ) and linear physical phenomena throughout the echo formation and denoting the call spectrum by $S_c(f)$, we can express the echo spectrum $S_e(f)$ (Peremans, 1997) as

$$S_e(f) = H_p(\theta, f) H_{air}(f) H_{refl}(f) H_p(\theta, f) S_c(f), \quad (2)$$

where for the monaural system both the transmitter and the receiver result in identical filter contributions $H_p(\theta, f)$ and the propagation of the sound pulse through the air is modeled by

$$H_{air}(f) = \frac{e^{-\alpha(f)2r} e^{-j2\pi f 2r/v_{sound}}}{r^2}.$$

The latter expression contains spherical spreading, propagation delay and frequency dependent absorption. Note that none of these contributions are angle dependent.

Reflector models The last term $H_{refl}(f)$ to explain in Eq. (2) represents the reflection process occurring at one of the reflecting facets of the natural reflector, e.g. the leaf of a tree. We consider two types of reflectors: pointlike and extended reflectors. The pointlike reflectors model reflecting features of the natural reflector that are small compared to the wavelengths, i.e. $\lambda = 3.4 - 17$mm, of the sound waves whereas the extended reflectors model those that are large with respect to those wavelengths. In the case of pointlike reflectors we take reflection to be due to omnidirectional and frequency independent diffraction, i.e. $H_{refl}(f) = C/r^2$, with C a constant.

The extended reflectors, representing the larger leaves, are modeled by disks with the size of the leaf as diameter. Making use of Huygens' principle (Pierce, 1994) we have investigated the off-axis reflection of an incident planar wave by such a disk, see Fig. 4 (a).
This analysis, as shown in Figs. 4 (b) and (c), clearly indicates that such reflection is approximated accurately by considering the reflecting disk to be a secondary source acting like a piston. However, unlike for a piston source, the maximal sensitivity axis of the leaf's radiation pattern is no longer the normal to the leaf (piston) surface but instead this axis is rotated to coincide with the main reflection direction, i.e. $\theta_i = \theta_r$. If the angle of incidence exceeds $\pi/4$, we assume the backscattering to be diffraction off the edge of the leaf.

Natural environment model We assume the natural environment to consist of trees. A tree is considered to be a set of leaves randomly placed in the xy-plane within a certain radius R of the center of the tree. All leaves of one tree are given the same size and a randomly distributed orientation. For the experiments reported on here, we have used uniform distributions but if additional information is available about the leaf distribution of particular trees one could introduce more specific distributions. As explained above leaves are modeled either as omnidirectional point scatterers (black dots in Fig. 8 (c)) or as extended reflectors (lines in Fig. 9 (c)) with a piston type radiation pattern. Despite the simplistic nature of these scattering models we believe that they allow us to judge to what extent the introduction of additional filtering, different as it may be from our particular model, during reflection affects the performance of the proposed scheme.

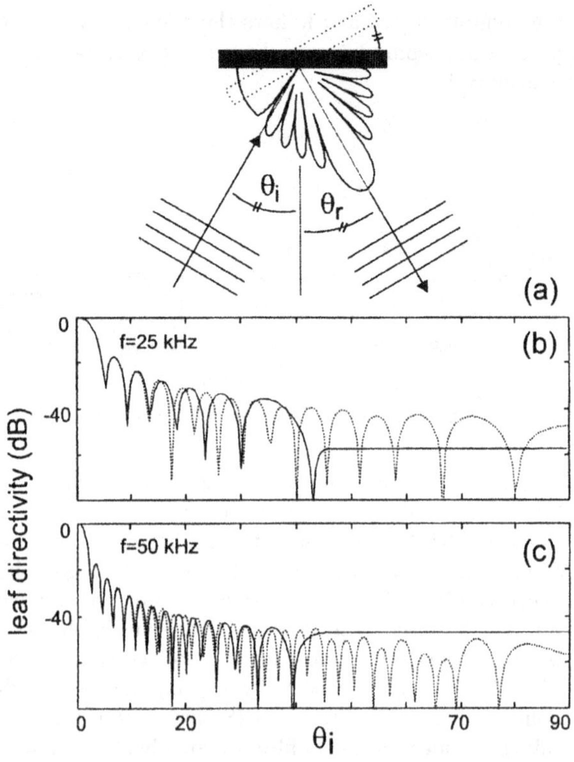

Figure 4: (a) Schematic representation of the reflection of a planar wave incident on a disk (radius=5cm); (b) comparison of results from Huygens' principle (dotted line) and piston with rotated radiation pattern (solid line) at f=25kHz and f=50kHz.

When calculating the composite echo we consider reflection at each leaf independently, and accumulate the returned echoes afterwards. As a result we still have to compensate for the fact that leaves on the outside of the tree will to some extent cover those on the inside. This phenomenon is taken into account by introducing a penetration depth P_d, which we define as the maximal distance inside the tree from which echoes can be returned. This parameter is highly correlated with the density of the foliage: denser foliage results in shorter penetration depths. The effect of taking into account a particular P_d can be ascertained from Figs. 8,9 (c). It is clear that this is only a rough approximation of the multiple scattering phenomena that determine the response of natural reflectors as shown by measurements on real plants (Müller and Kuc, 2000).

3. A simple monaural sonar system

In order to present our approach we first study the simplest situation and take the sonar system to consist of an emitter and a single receiver, i.e. we consider only the left or right half of the bat sonar system Fig. 1 (a). For all realistic situations we can neglect the distance between the emitter and the receiver compared to the distance between the sonar system and the reflector, i.e. $d << L$ in Fig. 1 (c). Hence, we assume emitter and receiver to be identical and coincident. As shown below, this simple monaural system will already allow limited bearing angle determination, although inclusion of a second receiver (ear) and allowing the ears to rotate will improve localization performance significantly.

As a result of the filter operations, see Eq. (1) performed by the sound generation and reception apparatus of the bat the bearing information of a reflector is encoded in the spectrum of the echo. We now show how this property can be used to localize natural reflectors consisting of many closely spaced reflecting facets.

3.1 Calculation of templates

As described above, we make use of Eq. (2) and the cochlear model of Sec. 2.1 to calculate the expected echo $s_e(t)$ and its corresponding cochlear response for a single pointlike reflector placed at different bearings. As echo travel time τ is related in a systematic way with distance, i.e. $\tau = 2r/v_{sound}$, the distance dependencies of the echo signal can be easily compensated for by a time dependent gain compensation. All results shown below assume such compensation for distance dependencies has been applied. Next, we store the cochlear responses in a set of templates. Because of the finite response times of the cochlear frequency channels, these templates are two dimensional objects (frequency vs. time/distance) as is clear from Fig. 5 (a). However, we note that it is sufficient to consider only the maximum filter output, i.e. the values along the white curves in Fig. 5 (a), to estimate the angular information. Indeed, these maxima provide the weights applied to the known responses of the cochlear frequency channels to the call. Note that because of the frequency swept nature of the latter, the different frequency channels of the cochlea attain a maximum at different points in time. However, we assume that the received signal as processed by the cochlea also contains the call so that the systematic variation of frequency over time present in the call can be compensated for in the received echoes, see Fig. 5 (b), and a self-calibrating system results. This calibration has been applied to all the cochlear outputs shown below. For the situation described here, i.e. two echoes from $\theta = 10°$ and $\theta = 20°$, this results in the two cochlear output patterns, i.e. angular templates, shown in Fig. 3 (c).

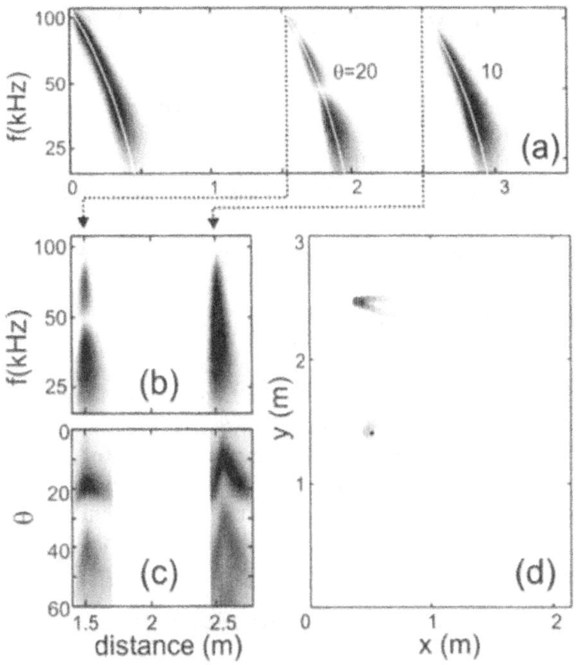

Figure 5: (a) Cochlear time-frequency representation of picked up call + echoes from two spatially well separated reflectors (time axis rescaled into traveled distance axis). (b) Cochlear output after call induced delay variation has been compensated for. (c) The overlap function, overlap(θ, r), calculated for the cochlear output shown in (a). (d) The estimated positions (larger gray dots) and the true positions (black dots) of the reflectors.

In order to build our database of angle dependent templates, we position a point reflector at every angle and calculate the corresponding cochlear output. To remove signal strength dependencies this output is normalized before storing it as a template. If we do this for our simple sonar configuration, we obtain for each angle θ the associated template $T_\theta(n)$ with $n = 1 \cdots N$ and N cochlear frequency channels. The resulting set of templates for the monaural sonar system is shown in Fig. 3 (b). We notice that there is a symmetry around $\theta = 0$ which means that with this monaural system one cannot distinguish left from right. We also notice that with increasing θ increasing numbers of notches are shifted in from the high frequency region.

3.2 Matching with templates

3.2.1 Isolated echoes

Suppose we have a composite echo consisting of two echoes originating from two different point reflectors, positioned at different angles $\theta = 10°, 20°$ and different distances $r = 1.5, 2.5$m. The echo as processed by the cochlea is shown in Fig. 5 (a). To reconstruct the environment we compare the normalized cochlear output pattern $S_{cochlea}(n, r)$ at every distance r with the set of angular templates, and search for the one that resembles the cochlear output the most. The echo corresponding with that particular distance r is then assumed to come from the corresponding direction. Hence, in order to find the template $T_\theta(n)$ which is closest to the cochlear output pattern at distance r we calculate its overlap with each of the angular templates

$$\text{overlap}(\theta, r) = \sum_{n=1}^{N} T_\theta(n) S_{cochlea}(n, r),$$

with N the number of cochlear frequency channels. Because of normalizations overlap(θ, r) lies within the interval $[0, 1]$. This overlap function calculated for the cochlear output shown in Fig. 5 (a) is shown in Fig. 5 (c). In Fig. 5 (d) we plot the estimated angle (circular dots), i.e. the angle corresponding with the maximal overlap, see Fig. 5 (c), for every r

$$\theta^*(r) = \arg(\max_\theta(\text{overlap}(\theta, r))).$$

We use the total, i.e. summed over all frequency channels, power of the cochlear output at that distance r to weigh the importance of the corresponding angle estimate. Hence, we can limit angle estimation to those distances that correspond with sufficiently high cochlear output power. Every dot has a gray-value set by its associated cochlear output power: black corresponds to highest intensity, white to zero intensity. From this example we conclude that this scheme results in accurate position estimates as long as the reflectors are radially separated giving rise to isolated, i.e. non-overlapping, echoes.

However, from Fig. 5 (c) one can see that for each distance r the overlap has multiple local maxima for several angular templates indicating that multiple angular estimates are good candidates to be associated with each distance r. Consequently, because the presence of noise and/or deviations from the theoretic model influences the relative ordering of these maxima, angle estimation by real sonar systems is not guaranteed to return correct angle estimates. Hence, this scheme will suffer from ambiguity and give rise to ghost images that, as we will see in Sec. 4., can only be eliminated by interpreting sequences of echoes. Such ambiguity is to be expected, since our set of angular templates is not an orthogonal one. Indeed, as indicated by Fig. 6 (a), different templates have nonzero overlap. From this analysis it is clear that the suitability of the proposed scheme for angle determination is primarily determined by the set of angular templates: the more their overlap function

$$\text{overlap}(\theta_1, \theta_2) = \sum_{n=1}^{N} T_{\theta_1}(n) T_{\theta_2}(n),$$

resembles the Dirac function $\delta(\theta_1 - \theta_2)$, the better the angle estimation will be.

3.2.2 Overlapping echoes

Above we have analyzed position estimation of a single pointlike scatterer at a certain distance. While this analysis might be appropriate for prey localization on the wing: a single target well separated from background clutter, it is unrealistic in a navigation context. When navigating through a natural environment, e.g. foliage, the received echoes are much more complicated and composed of large numbers of overlapping echoes caused by the different reflecting facets of the natural reflectors, e.g. trees. Applying the approach outlined above to this more complex situation results in less ideal performance, as illustrated by Figs. 8,9 (c). Our experiments indicate that this performance loss can be explained by the following three observations.

(i) As we discussed before, the set of templates is not orthogonal and as a result different templates corresponding with different bearing angles have similar overlap values when compared with the cochlear output. In combination with the next observation this gives rise to the appearance of ghost images. As stated before this problem can be minimized by selecting a set of templates that minimizes overlap between the different templates.

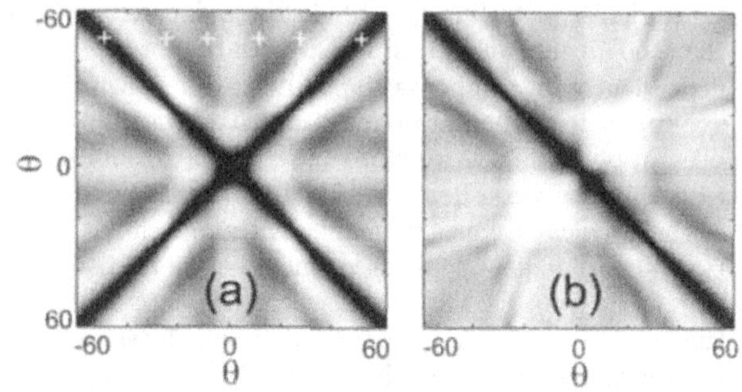

Figure 6: (a) The overlap, overlap(θ_1, θ_2), between the templates associated with the monaural sonar system; (b) The overlap between the templates associated with the binaural sonar system ($\phi = 7°$).

However, as shown by Fig. 6 (a) ambiguities are inherent to the simple monaural sonar system, e.g. the regions (white crosses) indicated on the horizontal line corresponding with $\theta = 55°$. In addition to the left-right ambiguity, giving rise to the symmetric nature of the overlap function, we notice that two other ambiguity regions are present for this bearing. We conjecture that minimizing such ambiguities has been one of the driving forces in the evolution of the, often complexly shaped, bat pinnae that can be observed in nature. Fig. 7 shows a CAD model of such a real pinna that was constructed using a CT-microscanner, (Peremans and Müller, 2003). Also, we show below that using a more complex sonar system can lead to further improvements, see Fig. 6 (b).

Figure 7: The pinna of the Brown long-eared bat (Bat specimen provided by the Dept. of Zoology, University of Erlangen).

(ii) This scheme allows detection of only one angle (reflector) for every distance r. In a real environment this need not be the case, e.g. multiple leaves of the same tree at approximately the same distance but at different angles, as illustrated by the black dots in Fig. 8 (c). The echoes from these reflecting facets will interfere and give rise to a cochlear output pattern that is not so straightforward to classify, see Fig. 8 (a). Indeed, nonlinearities in the cochlear processing make it impossible to compose such cochlear outputs out of linear combinations of the memorized templates. In a particular frequency channel of the cochlea interference occurs whenever the response time of that frequency channel is longer than the delay between two echoes. Hence, we conclude that such interference effects are most pronounced for the low frequency channels as they have the longest response times. As can be seen from Fig. 8 (b) and (c) these interference effects in combination with the ambiguities in the template matching result in a number of incorrectly positioned reflectors. Nevertheless, in our opinion the results shown in Fig. 8 (c) indicate that useful navigation behaviour can still be performed. This can be understood by noting that the strength of an echo is highly correlated with the angle, i.e. smaller angles corresponding with higher strengths, and the template matching, in the presence of interfering echoes, is

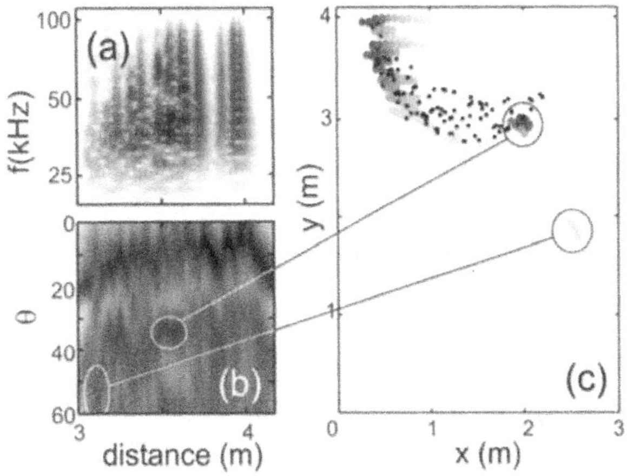

Figure 8: (a) Cochlear output for the scene shown in (c); (b) Overlap between the templates and the cochlear output at each distance r; (c) A tree consisting of multiple pointlike reflecting facets (black dots) and their estimated positions (larger gray dots).

predominantly driven by the strongest echo. Hence, in the presence of multiple overlapping echoes the localization mechanism described above returns preferentially the position estimates of those targets that correspond with the smallest bearing angles. However, from safety considerations, we conclude that such behaviour is advantageous as it guarantees that the parts of the complex reflector that are closest to the flight path, assuming the latter usually coincides with the direction of 'sight', are mapped out most accurately.

(iii) So far, we only considered pointlike scatterers. The situation gets more complicated when we increase the accuracy of our simulation of the natural habitat of a bat by introducing extended reflectors, e.g. a tree consisting of multiple larger leaves (black lines) as shown in Fig. 9 (c).
As such reflectors each have a radiation pattern of their own, see Sec. 2.2, and hence introduce additional filtering, their presence complicates angle estimation even further. This is verified by the increased ambiguity shown in Fig. 9 (b) and the concomitant increase in the number of ghost images as well as the overall decline in localization accuracy seen in Fig. 9 (c). Note that in these circumstances most of the leaves (black lines) generate echoes that are significantly smaller than the ones generated by a few leaves (bold black lines) that are pointing more or less directly toward the sonar system, i.e. the foliage now acts like an acoustic mirror-ball. Since the template matching process will be dominated by these glint producing reflectors we no longer see the outlines of the tree being traced out by the sonar system but instead get a few isolated position estimates corresponding with the favorably oriented leaves of the tree. Note

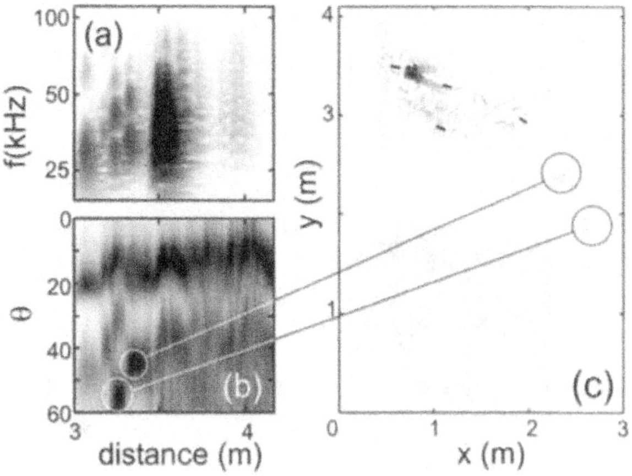

Figure 9: (a) Cochlear output for the scene shown in (c); (b) Overlap between the templates and the cochlear output at each distance r; (c) A tree consisting of multiple extended reflecting facets (black lines) and their estimated positions (larger gray dots). The bold lines indicate the leaves that are facing, error$\leq 5°$, in the direction of the sonar system.

that the latter can be positioned anywhere within the tree and hence should be used with caution when supporting navigation behaviour. On the other hand, we conclude from our simulations that the filtering by the extended reflectors is less pronounced than initially expected because only those reflectors that are more or less directly facing the sonar system, hence introducing only limited filtering (Fig. 3 (a)), generate significant contributions to the composite echo. This explains why the position estimates are still fairly accurate. However, the additional filtering, limited as it might be, does result in increased ambiguity, i.e. more ghost images, as the returned spectra resemble the templates less.

In the next section we will investigate to what extent these ambiguities can be further reduced by introducing a more realistic binaural sonar system.

4. A binaural sonar system

In this section we will study a more realistic, binaural, sonar configuration, which on top of being able to distinguish between left and right will also enable us to develop strategies for eliminating most of the ambiguities present in the output of the monaural system. Our binaural system, see Fig. 1, consists of one (piston) emitter, and two (piston) receivers which can be rotated in the horizontal plane over angles ϕ (left ear) and $-\phi$ (right ear). We still only consider echoes from the far-field, i.e. $d << L$. Hence, apart from the rotation over ϕ, we consider emitter and receivers to coincide like we did for the monaural sensor.

4.1 Ear configuration

To estimate the bearings of reflectors the overall approach is identical to the one described for the monaural case. First, we place a pointlike reflector at all possible bearings and store for each bearing the normalized cochlear outputs of the right and left ear in the corresponding angular template. Next, the bearing estimate is provided by the template that overlaps the most with the normalized cochlear outputs generated by a new measurement. Except that now, since we have two orientable receivers, the set of possible templates is much larger. For every ear-configuration, i.e. orientation of the right and left ear (ϕ_L, ϕ_R), we calculate a new set of angular templates that have to be compared with the cochlear output to determine the reflector bearing. For simplicity, we will restrict the analysis in this paper to the symmetric configurations where both ears are oriented over the same angle $\phi_L = -\phi_R = \phi$. Fig. 10 shows what these binaural templates look like for an ear configuration $\phi = 7°$.

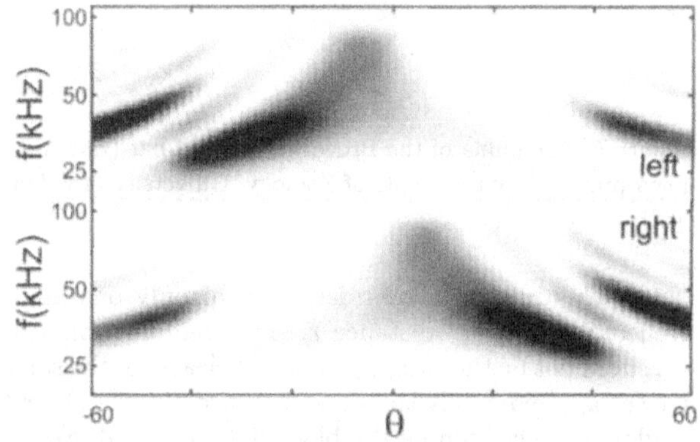

Figure 10: The set of angular templates $T_\theta(n)$ for the binaural sonar system, ear configuration $\phi = 7°$.

Calculating the overlap(θ_1, θ_2) between the binaural templates for this ear configuration, see Fig. 6 (b), clearly shows that this binaural system does not suffer from a left-right ambiguity anymore.

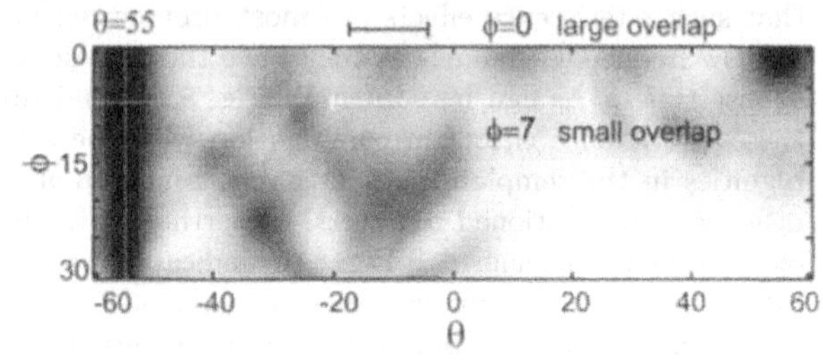

Figure 11: The overlap$(55°, \theta_2)$ for all ear configurations $\phi \in [0°, 30°]$.

To show the effect of the ear configuration on the overlap function, Fig. 11 displays the overlap between the angular template $T_{55^\circ}(n)$ and all other angular templates $T_\theta(n)$, $\forall\theta \in [-60^\circ, 60^\circ]$, for all ear configurations $\phi \in [0^\circ, 30^\circ]$. This result shows that the ambiguity regions of the angular template $T_{55^\circ}(n)$ can be varied significantly by choosing the ear configuration. Similar results can be shown to hold for all angular templates $T_\theta(n)$. Thus we propose that the ear configuration ϕ is an obvious control parameter that can be made use of by an active sensing strategy aimed at eliminating ghost images. Indeed, we conjecture that those bat species that show considerable ear mobility while their sonar system (Walker et al., 1998) is active make use of this extra degree of freedom, i.e. control the ear configuration, so as to reduce the overlap in the region of interest.

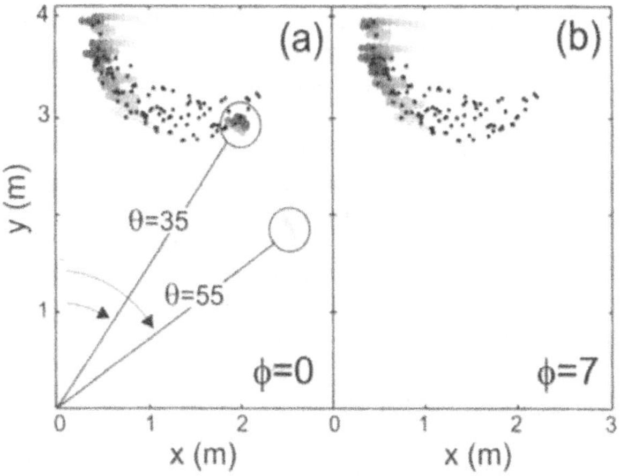

Figure 12: The same tree from Fig. 8 and the estimated positions (larger gray dots) when looking at it with a binaural sonar system, (a) ear configuration $\phi = 0^\circ$ and (b) ear configuration $\phi = 7^\circ$.

4.1.1 Active control of ear configuration

A systematic study of the overlap functions for different ear configurations shows that no single ear configuration results in an orthogonal set of templates. Hence, ghost images will always occur. However, for different ear configurations they will appear at different locations. To investigate how controlling the ear configuration could be used to eliminate ghost images we take another look at the results shown in Fig. 8 as reproduced in Fig. 12(a). The ghost images occur at bearing angles $\theta \approx 55^\circ$ and $\theta \approx 35^\circ$ for ear configuration $\phi = 0^\circ$. To decide whether the objects positioned at $\theta \approx 55^\circ$ are really there we choose a new ear configuration. To do this we calculate for every ϕ the overlap between the $T_{55^\circ}(n)$ template and every other template as shown in Fig. 11. Next, we look for regions of high overlap in the current ear configuration, i.e. $\phi = 0^\circ$ in this example, as these correspond with the rival interpretations for the bearing of the ghost images and choose a new ear configuration that results in a small overlap value for precisely those same ambiguity regions, i.e. $\phi = 7^\circ$ in this example. A new measurement with that new ear configuration results in the position estimates shown in Fig. 12 (b).

Analysis of this result shows that while new ghost images are produced they occur at completely different positions, whereas position estimates corresponding with true reflectors stay the same. This is to be expected as we chose an ear configuration that guarantees that the overlap between the template corresponding with the bearing of the ghost image is maximally different from the templates corresponding with the rival bearings. Hence, a straightforward combination of the results from this sequence of measurements, each corresponding with a different ear configuration, makes it possible to reliably eliminate the ghost images.

5. Conclusions

In this paper we have proposed and analyzed a localization mechanism that allows a binaural sonar system, i.e. a simplified version of the bat's sonar system, to navigate through its natural environment based on the time-frequency representations produced by its left and right cochlea. Although the analysis in this paper has been restricted to the 2D situation, the mechanism described can be straightforwardly extended to the full 3D case. Indeed, we have built a robotic sonar system (Peremans et al., 1998) that uses the mechanism described here to localize (accuracy $\approx 3^\circ$) a single, isolated, pointlike reflector in 3D making use of piston-like transducers. The increase in complexity for going from 2D to 3D localization is very modest: the number of templates required to attain this performance was only $2 \times N \times M \times 30$ with N, M the number of different azimuth and elevation values and 30 the number of frequency channels. If such 3D localization performance is possible with the highly regular radiation patterns of circular piston-like transducers, it will most certainly be possible with the much more irregular radiation patterns associated with bat sonar systems (Hartley and Suthers, 1987, Firzlaff and Schuller, 2003). Indeed, from the analysis given here we conclude that a more irregular radiation pattern will result in angular templates that differ more from each other and hence give rise to much lower overlap and less ambiguity. This could be one explanation why evolution has come up with such very particular shapes, see Fig. 7, for the ears and noses of most bats.

From the analysis given in this paper we further conclude that the single most important difference between a situation in which isolated echoes occur and one in which many highly overlapping echoes occur is the generation of ghost images. However, since the use of simple active control of the ear configuration has been shown to allow the elimination of such ghost images it seems that this spectrum based localization scheme will work equally well when confronted with realistic natural reflectors. Hence, we consider it plausible that bats make use of this simple scheme, possibly in combination with other methods, when navigating through foliage. To test this conjecture we are currently implementing a robotic sonar head, i.e. the CIRCE head, of similar size to a real bat head and consisting of an emission/reception system capable of generating/processing bat vocalizations in real-time and a multi-degree of freedom mechanical system allowing pinnae movement and shape control. This sonar head will allow us to test these theories on a real robot system navigating in real-time through an environment containing realistically complex reflectors.

6. Acknowledgment

This work is supported by the European Union, IST-FET Program, Life-like Perception Systems Initiative (IST-2001-35144) and is part of the CIRCE (Chiroptera Inspired Robotic CEphaloid) project.

References

Altes, R. and Reese, W. (1975). Doppler-tolerant classification of distributed targets– a bionic sonar. *IEEE Transactions on Aerospace and Electronic Systems*, 11(5):708–723.

Firzlaff, U. and Schuller, G. (2003). Spectral directionality of the external ear of the lesser spear-nosed bat, phyllostomus discolor. *Hearing Research*, 181:27–39.

Hartley, D. and Suthers, R. (1987). The sound emission pattern and the acoustical role of the noseleaf in the echolocating bat, carollia perspicillata. *Journal of the Acoustical Society of America*, 82(6):1892–1900.

Kuc, R. (1994). Sensorimotor model of bat echolocation and prey capture. *Journal of the Acoustical Society of America*, 96:1965–1978.

Matsuo, I., Tani, J., and Yano, M. (2001). Model of echolocation of multiple targets. *J Acoust Soc Am*, 110:607–624.

Müller, R. and Kuc, R. (2000). Foliage echoes: a probe into the ecological acoustics of bat echolocation. *J. Acoust. Soc. Am.*, 108(2):836–845.

Müller, R. and Schnitzler, H. (2000). Acoustic flow perception in cf-bats: Extraction of parameters. *J. Acoust. Soc. Am.*, 108(3):1298–1307.

Nowak, R. M. (1994). *Walker's BATS of the world.* The Johns Hopkins University Press, Baltimore.

Peremans, H. (1997). Broad beamwidth ultrasonic transducers for tri-aural perception. *The Journal of the Acoustical Society of America*, 102(3):1567–1572.

Peremans, H. and Hallam, J. (1998). The spectrogram correlation and transformation receiver, revisited. *The Journal of the Acoustical Society of America*, 104(2):1101–1110.

Peremans, H. and Müller, R. (2003). A comprehensive robotic model for neural & acoustic signal processing in bats. In *Proc. of the 1st Int. IEEE EMBS Conf. on Neural Engineering*, pages 458–461, Capri.

Peremans, H., Walker, A., and Hallam, J. (1998). 3d object localisation with a binaural sonarhead, inspirations from biology. In *1998 IEEE Int. Conf. on Robotics and Automation*, pages 2795–2800.

Pickles, J. O. (1982). *An Introduction to the physiology of hearing.* Academic Press, London.

Pierce, A. (1994). *Acoustics.* Acoustical Society of America, New York.

Popper, A. and Fay, R., (Eds.) (1995). *Hearing by Bats.* Springer Verlag.

Saillant, P., Simmons, J., and Dear, S. (1993). A computational model of echo processing and acoustic imaging in frequency-modulated echolocating bats: The spectrogram correlation and transformation receiver. *Journal of the Acoustical Society of America*, 94(5):2691–2712.

Schroeder, M. R. and Hall, J. L. (1974). Model for mechanical to neural transduction in the auditory receptor. *J. Acoust. Soc. Am.*, 55:1055–1060.

Simmons, J., Wotton, J., Ferragamo, M., and Moss, C. (2002). Transformation of external-ear spectral cues into perceived delays by the big brown bat, eptesicus fuscus. *J Acoust Soc Am*, 111:2771–2782.

Slaney, M. (1993). An efficient implementation of the patterson-holdsworth auditory filter bank. Technical Report 35, Apple Computer Inc.

Thomas, J., Moss, C., and Vater, M., (Eds.) (2003). *Echolocation in Bats and Dolphins.* University of Chicago Press.

Walker, V. A., Peremans, H., and Hallam, J. (1998). One tone, two ears, three dimensions: A robotic investigation of pinnae movements used by rhinolophid and hipposiderid bats. *The Journal of the Acoustical Society of America*, 104(1):569–579.

From Bat Pinnae to Sonar Antennae: Augmented Obliquely Truncated Horns as a Novel Parametric Shape Model

Rolf Müller **John C. T. Hallam**
The Maersk Institute, University of Southern Denmark
Campusvej 55, DK-5230 Odense M, Denmark
rolfm@mip.sdu.dk

Abstract

Biosonar systems operating in air (bats) use intricate baffle shapes for shaping the sonar beam pattern. Because of its relative ease of implementation, beamforming with such baffle shapes is also an attractive solution for enhancing the sonar-sensing capabilities of animats. The work presented here introduces a novel family of parametric, biologically inspired, baffle-shapes for sonar sensors: augmented obliquely truncated horns. The structural features of these shapes are: an obliquely truncated horn, a washboard ripple pattern on its inner surface, and a pin-shaped flap ("tragus") partially occluding the aperture through which sound enters the baffle. Shapes generated by this model can be completely controlled with 22 parameters. Numerical methods to generate representations of these shapes, evaluate their acoustic properties, and visualize the results are described. The investigated acoustical system properties are: directivity pattern, shape of iso-gain surfaces in three dimensions and gain contours on planar cuts through the spatial distribution of the gain. The last property (cuts) serves as a representation of the shape of a sonar spotlight shone onto an extended target of relatively low curvature. The results demonstrate that ripple and tragus have a profound impact on the functional properties of the baffle shapes. The most prominent effects include breaking of symmetry and generation of side lobes in different frequency bands. The two features have been shown offer the opportunity to shape the spatial distribution of receiver gain into a large number of profoundly different configurations. Evidence of interactions between the features indicates that a combination of both features may generate system properties inaccessible through any one feature alone. The tool-chain of numerical methods used in the present work establishes a foundation for future investigations of performance measures to be used in optimizing the baffle shapes for a particular task.

1. Introduction

Sonar is a sensory modality quite commonly found in mammals. Approximately a quarter of all species use it, the most notable groups being bats and dolphins. Biosonar is known to be a very capable far sense which meets the informational needs arising from the wide variety of ecological niches occupied by the animals which use it. As part of their multi-layer, feedback-controlled architecture, biosonar systems operating in air typically perform some kind of beamforming operation with intricate baffle shapes surrounding the sites of sound emission (mouth or nose) and reception (ears). Optimization in the course of biological evolution has created a highly variable ensemble of such shapes. Some of this variability is likely attributable to optimization of biosonar function with respect to the specific sets of sensing tasks. Consequently, it appears worthwhile to explore whether and how biomimetic reproductions of the beamforming strategies underlying this shape variability can enhance sonar performance and turn sonar into a far sense which can increase the sensing skills of animats.

The standard technical solution to the beamforming problem is array signal processing (Johnson and Dugeon, 1993). These methods require a grid of individual transducers, the input or output signals of which can be accessed and manipulated individually in a way that mimics different shapes or orientations of the grid. For an animat to perform this kind of operation, it needs to be able to deal with the mechanical load of the array and the parallel signal-preconditioning stages it requires. It must also make available the necessary signal processing capabilities. For sonar, these requirements are not easy to meet on a small animat. Hence, if only one aperture or aperture baffle shape is required at a time, using a physical baffle can be expected to be the solution which offers the greater ease of implementation under these circumstances. Furthermore, while only one shape is available at a given time, this shape need not necessarily remain constant over time. In fact, it could be changed, if the baffle is made from a malleable material and a suitable actuation strategy can be found, to flip it from one functional shape to another.

Sonar systems operated on robots – whether intended to mimic animals or not – typically do not use baffle shapes for beamforming. However, prior exploratory work on this subject has been reported in (Carmena et al., 2001). In this prior work, a small number of reflecting facets was placed around an ultrasonic transducer or simple shapes, like a paraboloid, were used. In the work reported here, we propose a novel parametric model for baffle shapes to be used with ultrasonic receivers. These shapes are inspired by the pinna (outer ear) shapes found in bats, and extend a general, simplified model for the mammalian pinna proposed by (Fletcher and Thwaites, 1988, Fletcher, 1992), the obliquely truncated horn. In the extension of the model, two conspicuous structural features prominent in many bat species are included. The augmentation sacrifices the advantage that the obliquely truncated horn is analytically tractable under simplifying approximations. It retains some of its structural simplicity however, and hence allows a shape to be described by a comparatively small set of parameters. For each set of parameters, the acoustic properties of the resulting shape need to be evaluated numerically. Besides introducing an extended shape model, the reported work employs a different numerical approach, which combines detailed numerical modelling of diffraction by the shape with ready accessibility of far-field system properties and presents the results using three different techniques for visualization and qualitative evaluation. This qualitative evaluation is used to establish that the proposed model can generate interesting functional properties and that its novel structural features also have a strong impact on function hence paving the way to quantitative evaluation and search for parameter values which attain optimization tasks set by sonar tasks.

2. Methods

2.1 Shape Model

The model presented here is an augmented version of the obliquely truncated horn model introduced by (Fletcher and Thwaites, 1988, Fletcher, 1992), with which it shares its basic shape of a single circular cone with specified height and opening angle. In the model described here (s. Fig. 1), two planar cuts are made to generate a conical tube with two open ends, called a "horn". This procedure is repeated with a second iden-

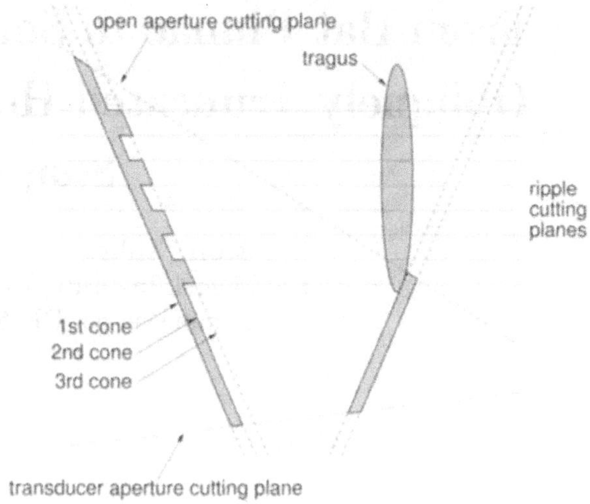

Figure 1: Schematic sketch of the proposed model. Solid black lines delineate the outline of the final pinna shape, gray lines cutting planes and dashed black lines portions cut away.

tical cone, which is slightly shifted along the cone axis prior to cutting. The two cones are taken to be the inner and outer surface of a horn with finite wall thickness. The aperture closer to the cone vertex is chosen to hold the transducer, the other one is the pinna aperture through which sound enters. Hence, the angle at which the cone is cut to create the transducer aperture controls how the pinna axis is oriented with respect to the transducer normal. The horn is than augmented by two structural features, which are highly conspicuous in many bat pinnae: a "tragus" and a "washboard ripple" (s. Fig. 2).

Many bat species have an appendage which partially occludes the pinna aperture. This appendage is called a tragus or an antitragus, depending on which portion of the pinna it originates from. Size, shape and insertion point of these appendages can vary widely between species; they can be either very prominent or almost entirely absent. For the sake of simplicity, the appendage of the model is referred to summarily as a tragus. It is described by a sphere which can be scaled and rotated independently in three dimensional space. It can also be translated, but translations are limited to positions where the tragus touches the inner cone surface.

In order to generate a washboard surface ripple, a third cone generated with its axis collinear to the two previous ones and its apex shifted towards the inside of the pinna. This cone is taken as the inner surface of a new pinna wall layer which is then cut by a stack of parallel planes. The inner wall layer is alternately retained and discarded between the planes of this stack, so that the spacing of the planes equals half of the spa-

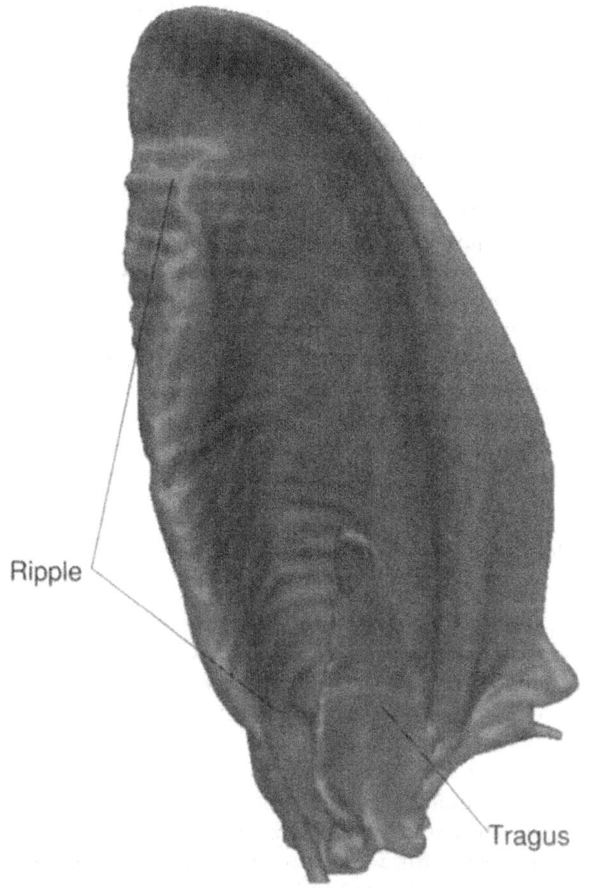

Figure 2: Example of a natural bat pinna shape, taken from the Brown Long-eared Bat (*Plecotus auritus*, Linnaeus, 1758). The pinna shows both structural features (washboard ripple and tragus) discussed here.

tial period of the generated ripple. The amplitude of the ripple is given by the distance between the second and third cone. The ripple generated in this way has a square wave cross-section. It can be smoothed by applying a spatial lowpass filter. The model is described by a total of 22 parameters, which are summarized in Table 1. In this listing the presence or absence of one of the two novel features (tragus and ripple) is not treated as a separate parameter, because a feature can be omitted by choosing appropriate parameters. For example, choosing a ripple amplitude of zeros means "no ripple", choosing a tragus size of zero means "no tragus" and setting the cutoff frequency of the spatial lowpass filter to infinity means "no smoothing of the ripple".

Shape generation was implemented using Boolean combinations of implicit functions' classes provided by the Visualization ToolKit (VTK) described in (Schroeder et al., 2003).

Cone:	
opening angle	1 DF
cone hight	1 DF
wall thickness	1 DF
Aperture Cutting Planes:	
normals	4 DF
intersection points with cone axis	2 DF
Ripple:	
amplitude	1 DF
spatial frequency	1 DF
ripple orientation	2 DF
smoothing	1 DF
Tragus:	
size	3 DF
position	2 DF
orientation	3 DF
Total:	*22 DF*

Table 1: Parameters for augmented obliquely truncated horn model.

2.2 *Numerical Prediction of Acoustical Properties*

A description of the sensitivity (gain) of a given shape for a receiver (a microphone) can be obtained by considering the amplitude of wave field that this receiver would emit if it were used as an emitter (a loudspeaker), since the two will be the same by virtue of the reciprocity principle. This approach is chosen here, because obtaining one outgoing wavefield is computationally more efficient then obtaining the response to many different incoming wavefields impinging from all directions of interest. Therefore, the baffle shapes will be treated as baffles of acoustic sources (loudspeakers) although they are intended for use with microphones. In the model, the entire cross-section of the transducer-opening of the pinna is actuated in phase as a planar piston transducer with an elliptical shape.

With few notable exceptions, bat pinnae are used in the far-field. The far-field of an antenna is defined – regardless of the physical nature of the wave phenomenon considered – as the set of points in space which satisfy

$$r \gg 8\frac{a^2}{\lambda}, \tag{1}$$

where r is the distance of the point to the aperture, a the aperture's "characteristic dimension" (for example, the radius of a circular aperture) and λ the wavelength. If, for example, a pinna's characteristic dimension is $a = 2\,\text{cm}$ and the wavelength is $\lambda = 5\,\text{mm}$, the far-field boundary given in Eq. 1 is 32 cm away from the pinna.

For an harmonic problem, where a continuous excitation by a single monochromatic wave is considered at a time, the complex wave amplitude (sound pressure amplitude) P is function of three spatial dimensions and frequency f; there is no time-dependence. If only points in the far-field are considered and their positions are expressed in spherical coordinates, the expression for P can be separated into a product of two terms. Both terms are dependent on frequency, but as far as their dependence on spatial variables is concerned, one term is a function of the two angles ϕ, θ only and independent of r, whereas the other is dependent on r alone:

$$P(\phi, \theta, r, f) \xrightarrow{r \gg 8\frac{a^2}{\lambda}} D(\phi, \theta, f)\frac{1}{r}e^{-jkr}. \quad (2)$$

The second term ($\frac{1}{r}e^{-jkr}$) describes the propagation of a spherical wave. Since this term is the same for any pinna shape, it can be omitted from the description of a specific pinna, but is almost certainly relevant to evaluating system performance in any given sonar task. The first term $D(\phi, \theta, f)$ is called the directivity and needs to be determined for every pinna shape of interest.

Since there is no analytic approximation which would yield the directivity D as a function of the model parameters listed in Table 1, D has to be obtained numerically. To achieve this goal, diffraction around the pinna has been modelled by a finite-element element approach in this study: finite element meshes consisting of cubic elements were generated from voxel grids which can be thought of as stacked cross-sectional images of the shapes with 1-bit color depth. The voxel grids/cross-sectional images were obtained by sampling the Boolean combinations of implicit functions describing the pinna-shapes. Element vertices were placed at grid voxel locations, if the locations of all vertices corresponding to a possible position of an element were found to have the color value representing air. The finite-element calculations proper (assembly of the master stiffness matrix and solving) were performed using the CFS++ program developed by the Department of Sensor Technology, University of Erlangen-Nürnberg, Germany.

Since the size of the elements into which the air-volume around the pinna is discretized must be considerably smaller than a wavelength, it is not possible to reconcile extending the computational domain into the far-field with maintaining a manageable problem size. This is illustrated by the pinna example given above: if a spatial resolution of 10 elements per wavelength were chosen and wavefield amplitudes at twice the boundary value stated in Eq. 1 were desired for this case, it would result in a problem size of $\sim 17\,000\,000\,000$ points. In order to make accessible far-field properties, a two stage approach has by taken – as described in (Müller and Hallam, 2004) – is taken, where the computational domain of the finite element model is restricted to the vicinity of the pinna surface, and a near-field-to-far-field transformation based on the Kirchhoff Integral is applied to field values obtained on the boundary (Ramahi, 1997). For this approach, the computational domain of the finite element model was wrapped in a layer of two-dimensional absorbing boundary elements, which are capable of mimicking undisturbed outward propagation to a very good approximation (Enquist and Majda, 1977). Complex field amplitudes were obtained for the three outer layers of finite elements on the boundary of the computational domain in order to estimate both the field amplitude Ψ and its derivative $\vec{\mathbf{n}} \cdot \nabla\Psi$ with regard to the surface normal $\vec{\mathbf{n}}$ of the boundary S of the computational domain. Using only these two variables, the field amplitude at any position $\vec{\mathbf{x}}$ outside the computational domain can be extrapolated using the Kirchhoff integral (Jackson, 1999)

$$\Psi(\vec{\mathbf{x}}) = -\frac{1}{4\pi}\int_S \frac{e^{jkr}}{r}\vec{\mathbf{n}} \cdot \left[\nabla\Psi + jk\left(1 + \frac{j}{kr}\right)\frac{\vec{\mathbf{r}}}{r}\Psi\right] ds. \quad (3)$$

To facilitate these computations, the geometry of the computational domain of the finite element model was always chosen to be a cuboid. In this way, computations were kept simple since only six different surface normals had to be considered. Furthermore, discontinuity in the orientation of the surface normal were limited to the 12 edges of the cuboid and were found to be compatible with a sufficiently small overall error in the numerical solution of problems with a known solution (for example, simulation of bipole source yielded root-mean-square errors around or below 5% for more than 10 points per wavelength and a cubic computational domain with edges longer than three wavelengths). The positions $\vec{\mathbf{x}}$ serving as extrapolation targets were chosen on the surface of a sphere with a radius satisfying Eq. 1. The normalized field magnitudes obtained for these points $|\Psi(\vec{\mathbf{x}})|$ were retained as an estimate for the directivity $D(\phi, \theta, f)$.

2.3 Evaluation and Visualization

Directivity D is a function of three independent variables ϕ, θ, f. Since two of these variables are angles describing a position on the surface of a sphere, suitable visualization techniques should be conveying the spatial relationship between the different combinations of ϕ and θ as faithfully as possible. Three approaches were found useful in achieving this goal: cartographic projections, renderings of three-dimensional iso-gain surfaces, and renderings on iso-gain contours on planar cuts. All three techniques generate one representation per frequency value. Therefore, representations for a sample of different frequency values were either tiled next to or superposed on each other to visualize the frequency dimension.

The different cartographic projections available offer a wide array of trade-offs between the various desirable properties of a map (conformality, preservation of distances, directions, equal scale and area-relationships) and the inevitable side effects they have on each other. Here, a Lambert azimuthal equal-area projection (Weisstein, 2002) is shown (s. Fig. 3) because its equal-area property can accurately represent the angular extent of sound beams. At the same time, it allows a large portion of a sphere to be mapped while shape distortions are kept close to the theoretical minimum. This allows presentation of an overview of the directivity pattern over an entire hemisphere. For each frequency under consideration, a separate map is plotted, the colors on the map code for directivity gain.

In conjunction with the attenuation associated with spherical spreading (and other sources of attenuation not considered here), directivity gain also has an effect on the range at which the total gain falls under a given threshold for the first time. The combination of directivity gain and attenuation through propagation losses defines closed volumes inside which the total gain exceeds a given threshold. The boundary surfaces of these volumes (s. Fig. 4) can be calculated from wavefield magnitude data which is re-synthesized everywhere within the space of interest. This re-synthesis is computationally much cheaper than a finite element computation covering the same volume, since it only requires evaluation of the product between directivity magnitude $|D(\phi, \theta, f)|$ and the magnitude of a spherical wave $\frac{1}{r}$. One closed boundary surface is obtained for each frequency; surfaces for different frequencies can be rendered together and marked by different hue or gray values.

The spatial aspect of directivity is particularly interesting in the case of extended sonar targets, where it defines a footprint or graded spotlight which imposes a weighting on the contributions that different elements of the target surface can make to the received echo. It is therefore interesting to know what kind of footprint shapes the different pinna shapes produce when aimed at an extended target in a distance. If the the curvature radius of a target is large compared to the extent of the footprint, the surface of such a target can be approximated by a plane and the footprint obtained for a given gain threshold is a planar cut through the three-dimensional iso-gain surfaces discussed before. Plots (s. Fig. 5) of these contours sacrifice a large portion of the total spatial information present in renderings of the entire iso-gain surfaces, but provide less cluttered visualizations that are directly linked to a common sonar scenario (scanning of an extended target of low curvature).

3. Results

3.1 Shape Generation

The proposed model has augmented the obliquely truncated horn in way that enables the generation of a much greater variety of shapes (s. Fig. 6 for some examples). The implementation of the model using Boolean combinations of implicit functions allows for efficient interfacing both with shape visualization algorithms and mesh generation for a finite element analysis of acoustic properties. Furthermore, meshes can be generated in stereolithography STL-format which is suitable for the manufacture of physical prototypes in a rapid-prototyping process (s. Sec. 3.3 below) Depending on the desired resolution, mesh representations of the shapes for visualization or finite element analysis can be generated within a few seconds up to a few minutes on a standard PC.

3.2 Predicted Acoustical Properties

The acoustical data presented here compares four different shapes generated by the model. All four shapes are based on the same obliquely truncated horn (20° opening angle, height 25 mm, wall thickness 1 mm). The shapes differ in the presence or absence of a tragus and washboard ripple, the two principal features considered here, yielding four different shapes:

- plain obliquely truncated horn (no ripple, no tragus)
- smooth obliquely truncated horn with tragus (no ripple)
- obliquely truncated horn with washboard ripple
- obliquely truncated horn with washboard ripple and tragus

The generated washboard ripple had a period of 6 mm and an amplitude of 2 mm. It was not smoothed, so the cross-section of the ripple was a square wave.

The directivity pattern of shapes which lack a tragus are bilateral symmetric (s. Fig. 3) like the shapes themselves. As reported by (Fletcher, 1992), the plain obliquely truncated horn shows evidence of the direction of highest gain being dependent on center frequency, this is particularly obvious in comparing the 30 and 40 kHz results in Fig. 3, which is again in accordance with Fletcher's findings that this dependence is most pronounced for lower frequencies.

Both added features – tragus and ripple – were found to have a profound impact on the predicted acoustical properties of the shape. An asymmetrically placed and oriented tragus can be seen to break symmetry in the directivity pattern as well. This effect is obvious in all three visualization methods (s. Figs. 3, 4 and 5). Within the studied range of frequencies (25 to 75 kHz in total), the tragus has a greater effect at higher frequencies

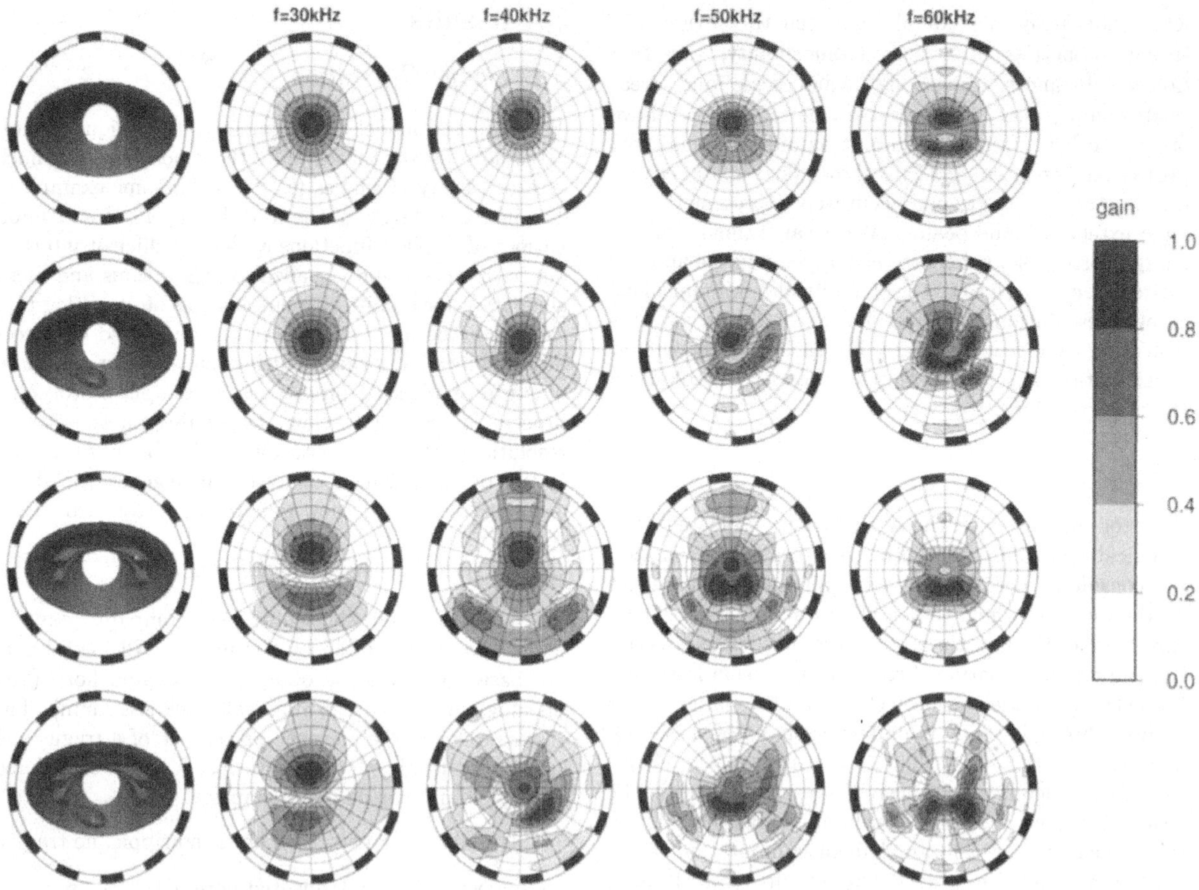

Figure 3: Comparison of the directivities $D(\phi, \theta, f)$ of four different pinna shapes generated from the model. Each row of subgraphs shows the results obtained for one shape. A top view of the shapes is shown on the left of each row. The maps use the Lambert azimuthal equal-area projections, the azimuthal (longitudinal) orientation of the maps is identical to the view of the shapes. The maps cover an entire hemisphere.

where it results in a wider spread of directivity gain over angle (Fig. 3). This spread is partially attributable to the generation of side lobes with a rather small angular extent, which show up as a family of slender beams pointing to the right in Fig. 4. At the lower frequencies (s. 30 kHz in Fig. 3), its impact on the overall distribution of directivity gain over angle is comparatively small.

The introduction of washboard ripple has an even larger impact on the distribution of directivity gain, since it generates a pattern of broad side lobes (s. Figs. 3, 4), which are particularly prominent around the center of the studied frequency band (40 and 50 kHz in Fig. 3).

Tragus and washboard ripple can be seen to interact in a manner, which is not entirely obvious from the effects which they have have in isolation (s. Figs. 3, 4 and 5). When combined with the washboard ripple, the tragus breaks symmetry, directing the higher directivity gain to the opposite side of the pinna and the most dramatic effect on the overall spatial gain distribution is seen at the higher frequencies (s. Figs. 3, 4). The effects of washboard ripple and tragus can be clearly seen to interact on the broad, low-frequency side-lobes generated by the washboard pattern, which are now shifted off the mid-line by the tragus as well (s. Figs. 3, 4).

The changes in the directivity pattern caused by the added model features can be seen to have an impact on the patterns with which large objects would be scanned by the receiver sensitivity (s. Fig. 5). The single orientation of a planar cut shown here cannot be representative for all the patterns which would be found if differently oriented cutting planes are chosen. Hence, an exposition of the effect of the model's added features on the shape of these sonar spotlights has to be postponed until future work on a quantitative characterization of these effects has been carried out. The example shown here, however, suffices to demonstrate that the absence or presence of both features can have a profound impact on the way an extended sonar target will be probed by the spatial sensitivity of the receiver.

3.3 *Physical Renderings*

A manufacturing test has been conducted in order to establish if physical prototypes can be manufactured from these shapes. The material chosen for the test was Nylon 7000, a glass-filled poly-amide powder with high durability, good thermal stability and chemical resistance. The samples (s. Fig. 7) were manufactured in a selective laser-sintering process at the Catholic University of Leuven, Belgium. The prototypes were found to have a satisfactory resolution of surface features and to be able to withstand wear and tear during laboratory use.

4. Discussion and Future Work

Many bat pinna have structural features that are more prominent in these animals than in humans and most other mammals. The two most conspicuous of these features, a washboard ripple pattern and a tragus have been added to augment the obliquely truncated horn, a standard model for the mammalian pinna. With a comparatively small set of parameters (22), this augmented model gives access to a much richer space of structures which is likely to cover many of the functionally relevant features evolved in bats to make them excel in the use of biosonar. Indeed, the data presented here demonstrates that both added features have a profound impact on the functional properties of the pinna, as described by the directivity pattern.

With regard to future research, the work reported here can be seen as step one in a two-step approach. In step one, the model has been defined and it has been established that it can generate interesting behaviors and that its novel features are of prime functional relevance. These findings warrant the cost at which these added features were introduced, namely the loss of approximate analytic tractability and a much larger parameter space, which more was more than doubled in size from 9 to 22 parameters. Future work will build on these findings in step two, where optimization goals will be defined and the parameter space of the model will be searched for locations which achieve them.

The methodological advances made in the development and initial, qualitative, evaluation of the acoustical properties of the model are well suited to enable the work in step two: In order to search the parameter space of the model (for example, by genetic algorithm methods), numerical representations of many different shapes have to be generated and evaluated with regard to the properties of their directivity patterns. The generation of finite element meshes from sets of model parameters has already been implemented in a way which ensures that its computational cost is negligible compared to the cost incurred by the finite element calculations. A parallelized implementation of the finite element computations will be required in order to make the latter step tractable. For the same reason, it is desirable to find the lowest possible resolution for the finite element mesh, which will still allow to predict the major features of the directivity pattern with sufficient accuracy. Making such a judgment not only requires an assessment of the relative differences between the results obtained at different resolutions, but also an estimate of the error between the numerical predictions and physical measurements. Such measurements will be carried out with physical prototypes, for which the manufacturing process has already been selected and tested as demonstrated by the prototype samples shown here. The established visualization methods are based on directivity functions and gain magnitude grids in two and three dimensions. These representations are well suited to form a basis for the development of automated, quantitative evaluation procedures needed for optimization, which will have to assess measures of the distribution of gain over angle, in three-dimensional space, or within the spotlight cast on an extended target. The renderings of these distributions as established here can provide guidance for the choice of these measures.

In nature, bat pinna shapes are integrated into a biosonar system, which offers additional degrees of freedom in positioning and orienting the pinnae, in choosing the time-frequency structure of the sonar pulse as well as in many other aspects of behavior. In order to do justice to the importance of system integration, the pinna shapes discussed here are conceived as part of the development of a complete biomimetic sonar head in the CIRCE project (Peremans and Müller, 2003, Müller and Peremans, pear). Other important lines of work carried out in this project will provide actuation for the pinna, with two rotational degrees of freedom, without significantly exceeding the overall size of a large bat head. Furthermore, future versions of the pinna shapes are in preparation which will be manufactured from a malleable material. These malleable pinna will be actuated to allow for changes in pinna shape as part of an adaptive sensing behavior performed by the animat.

Acknowledgements

Manfred Kaltenbacher and Alexander Streicher of the Department of Sensor Technology, University of Erlangen-Nürnberg, Germany provided the CFS++ source code and crucial advice on the finite element method. Filips Schillebeeckx arranged for the prototype fabrication at the selective laser-sintering facilities of the Catholic University of Leuven, Belgium. Supported by the European Union (IST Program, LPS Initiative, CIRCE Project IST-2001-35144).

References

Carmena, J. M., Kämpchen, N., Kim, D., and Hallam, J. C. T. (2001). Artificial ears for a biomimetic sonarhead: From multiple reflectors to surfaces. *Artificial Life*, 7(2):147–69.

Enquist and Majda (1977). Absorbing boundary conditions for the numerical simulation of waves. *Mathematics of Computation*, 31(139):629–51.

Fletcher, N. H. (1992). *Acoustic Systems in Biology*. Oxford University Press.

Fletcher, N. H. and Thwaites, S. (1988). Obliquely truncated simple horns: idealized models for vertebrate pinnae. *Acustica*, 65:194–204.

Jackson, D. E. (1999). *Classical Electrodynamics*. John Wiley & Sons, New York, 3rd edition.

Johnson, D. H. and Dugeon, D. E. (1993). *Array Signal Processing: Concepts and Techniques*. Prentice Hall, 1st edition.

Müller, R. and Hallam, J. C. T. (2004). Biomimetic smart antenna shapes for ultrasonic sensors in robots. In *Proceedings of the 35th International Symposium on Robotics ISR 2004, 6 pages, to appear.*

Müller, R. and Peremans, H. (to appear). *Neuro-Nanotechnology - Artificial Implants - Neural Prosthesis*, volume 4 of *IEEE Press Series on Biomedical Engineering*, chapter Biomimetic Integration of Neural and Acoustic Signal Processing. Wiley/IEEE Press, 26 pages.

Peremans, H. and Müller, R. (2003). A comprehensive robotic model for neural & acoustic signal processing in bats. In *Proceedings of the 1st International IEEE EMBS Conference on Neural Engineering*, pages 458–461. IEEE EMBS.

Ramahi, O. M. (1997). Near- and far-field calculations in fdtd simulations using kirchhoff surface integral representation. *IEEE T. Antenn. Propag.*, 45(5):753–9.

Schroeder, W., Martin, K., and Lorensen, B. (2003). *The Visualization Toolkit An Object-Oriented Approach To 3D Graphics.* Kitware, Inc., 3rd edition.

Weisstein, E. W. (2002). *CRC Concise Encyclopedia of Mathematics.* CRC Press, 2 edition.

Figure 4: Renderings of three-dimensional iso-gain surfaces for the combined effect of directivity gain magnitude $|D(\phi, \theta, f)|$ and spherical spreading losses $\frac{1}{r}$. Iso-gain contours are rendered for 101 frequencies linearly spaced from 22 to 75 kHz in steps of 0.5 kHz; dark surface colors represent low frequencies, bright colors high frequencies. The shape renderings are provided to show the alignment between the orientations of the shapes and the iso-gain surfaces; they are enlarged 25× with respect to iso-gain surfaces and shifted to side for better visibility.

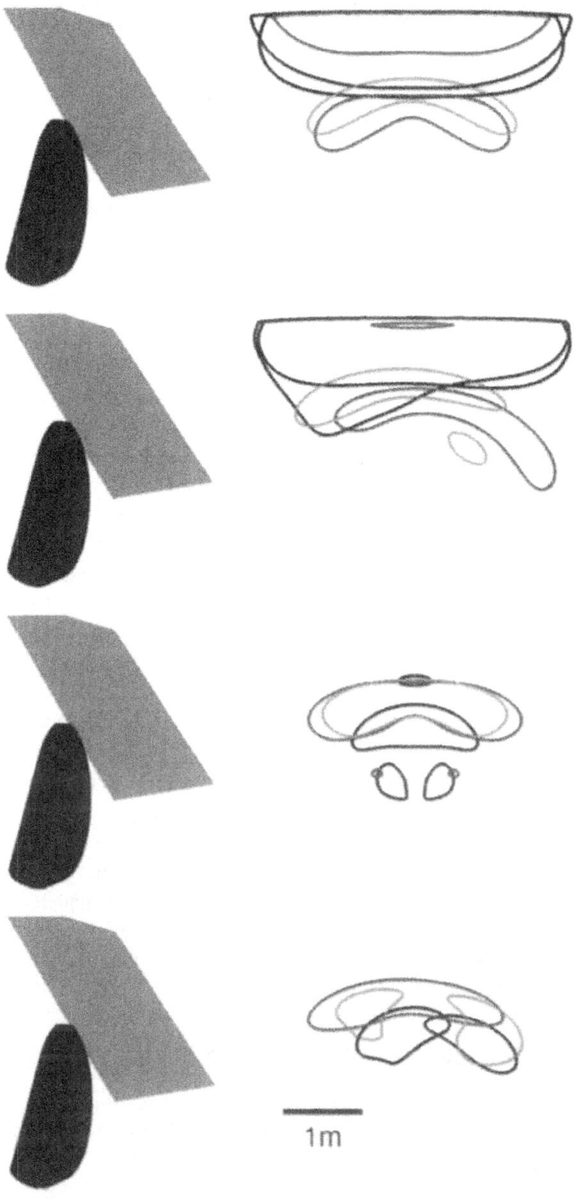

Figure 5: Renderings on iso-gain contours on planar cuts. The shapes are enlarged 100× with respect to depiction of the plane area, and are provided for reference in orientation only. The contours correspond to the same four frequency values which are shown in Fig. 3: 30,40,50 and 60 kHz; color-coded from dark (30 kHz) to bright (60 kHz). Gain values shown are renormalized within the plane, the graph shows -3 dB contours. .

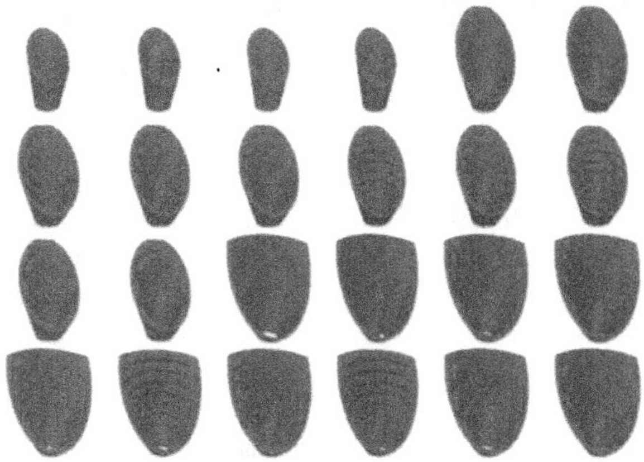

Figure 6: Example shapes generated by the parametric model. The shapes shown here do not cover variations in all model parameters. The parameters varied within the sample are: cone angle, presence or absence of tragus and washboard ripple, spatial frequency and orientation of the ripple as well as cutoff-frequency of the spatial smoothing.

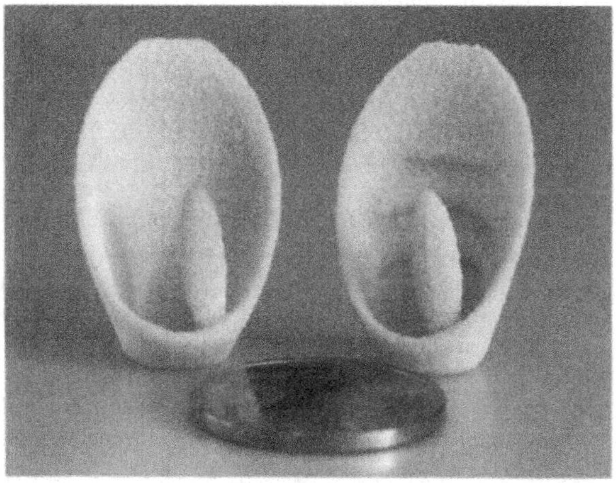

Figure 7: Manufactured samples of shapes generated by the proposed model made from Nylon 7000 in a selective laser-sintering rapid-prototyping process.

Evolving Adaptive, High-Dimensional, Camera-Based Speed Sensors

Ralf Salomon[*]
[*] Department of Electrical Engineering and Information Technology University of Rostock
18051 Rostock, Germany
ralf.salomon@etechnik.uni-rostock.de

Abstract

This paper reviews some attempts that exploit a phenomenon, also known as motion parallax, to estimate the distance of closest approach of a moving object. Despite their success, the existing evolutionary methods lack some desirable properties, such as reasonable scalability and online learning. To overcome these practically-relevant limitations, this paper proposes a new model that is based on Hebbian learning. Due to its scalability and online learning capabilities, this model is especially suited to mobile robots.

1. Introduction

Since the beginning of the 90ies a new research area has emerged, for which Brooks has coined the term new artificial intelligence (new AI for short). New AI aims at understanding (natural) intelligence and its underlying mechanisms by building systems that exhibit "intelligent" behavior (R. A. Brooks, 1991a, 1991b, R. Pfeifer and C. Scheier 1999). These systems are often realized as mobile robots, which are supposed to operate in dynamically changing, partially unknown environments without any human control (that is why they are attributed *autonomous*). New AI prefers a synthetic approach, i.e., understanding by building. In order to reach its research goals, new AI draws a significant amount of inspiration from natural systems. It therefore often investigates (biological) hypotheses and aims at validating them in simulation or on particularly-designed robots.

Even though the ultimate goal is to build physical robots, most attempts resort to simulation for obvious limitations of evolvable hardware available today (see, for example, the conference series *Simulation of Adaptive Behavior*), A very nice exception is the *eyebot* on which Lichtensteiger and Eggenberger have evolved simplified insect eyes (L. Lichtensteiger and P. Eggenberger, 1999). Section 2. briefly explains how Lichtensteiger and Eggenberger used evolutionary algorithms to evolve the eye's morphology that allows the robot to consistently estimate the distance of closest approach. Despite its successes, previous research (L. Lichtensteiger and P. Eggenberger, 1999, R. Salomon and L. Lichtensteiger, 2000) has indicated that this approach is practically limited to about 10-20 sensors, since the evaluation of a particular sensor distributions requires approximately one minute on a physical robot. Evolutionary algorithms normally generate a number λ offspring per generation. Since every offspring has to be evaluated (in real-world experiment), evolutionary methods require a significant amount of experimentation time.

Furthermore, evolutionary algorithms, like backpropagation-type learning procedures, require a *preselected* training set for *offline* learning and thus lack any online adaptation capabilities. By way contrast, living creatures are inherently adaptive, scale very well over their visual system, and apparently do not suffer from overlearning effects (that much). Section 3. proposes a new, simple model that achieves these design goals to a large extent by using a simple Hebbian-based learning rule. Section 3. also discusses some of the model's basic properties.

Sections 4. and 5. discuss the methods and results of some representative experiments. The results show that the proposed learning method is able to train the network model such that it can consistently determine the robot's speed and thus can determine the distance of closest approach. Section 6. concludes with a brief discussion.

2. Background and Previous Research

This section summarizes previous research and includes the description of the robot, its eye, the neural network controller, as well as a phenomenon, called motion parallax.

2.1 The Eyebot

Inspired by biological evidence (T. S. Collett, 1978, N. Franceschini, J. Pichon, and C. Blanes, 1992, G. A. Horridge, 1978), Lichtensteiger and Eggenberger (1999) constructed the *eyebot* (Figure 1) to model the eye of an insect, such as the house fly. It consists of a chassis, an on-board controller, and sixteen independently-controllable facet units, which are all mounted on a common vertical axis. A facet unit (Figure 2) basically consists of the sensor, a thin tube, two cog-wheels, a motor, and a potentiometer. By means of the cog-wheels, the motor can position the facet within a range of about 200 degrees, and the potentiometer provides feedback about its

Figure 1: The "eyebot" consists of a chassis, an on-board controller, and sixteen independently-controllable facet units (see Figure 2), which are all mounted on a common vertical axis.

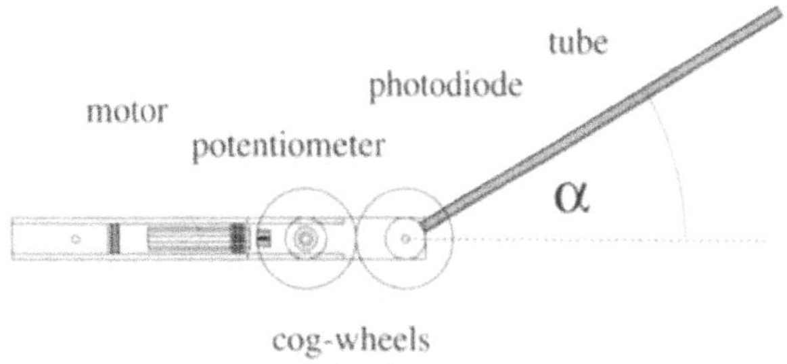

Figure 2: The robot's facet basically consists of the sensor (the photo diode), a thin tube, two cog-wheels, a motor, and a potentiometer. By means of the cog-wheels, the motor can position the facet within a range of about 200 degrees, and the potentiometer provides feedback about its actual position.

actual position, i.e., its angle α_i. The thin opaque tube is used to reduce the sensor's aperture to about two degrees. These tubes are the primitive equivalent to the biological ommatidia (T. S. Collett, 1978, N. Franceschini, J. Pichon, and C. Blanes, 1992, G. A. Horridge, 1978). It should be noted that such a low-cost construction is subject to several imprecisions and tolerances, which might be sensed as noise during operation.

2.2 *Motion Parallax: A Mathematical Description*

Figure 3 sketches how a phenomenon, also known as motion parallax, can be utilized to avoid obstacles. Let us assume that the compound eye, presented as the observer R, is at a fixed position. If the obstacle X is moving at constant speed v, the observer views the obstacle under different angles α_1, α_2, and α_3 at different time steps t_1, t_2, and t_3. Let r denote the vector from observer R to obstacle X and let v_t denote the component of v that is perpendicular to vector r. For the distance d of closest approach, the following relation holds

$$d = r \sin \alpha \,. \tag{1}$$

Since $v_t = v \sin \alpha$, the angular velocity $\omega = \dot{\alpha}$ with which the image of X moves through the visual field of the observer is

$$\omega = \frac{v_t}{r} = \frac{v \sin \alpha}{r} \,. \tag{2}$$

If the agent can estimate the velocity v and can measure both α and ω, it can calculate its distance r to the obstacle at any time. Solving eq. (2) for r and substituting into eq. (1) leads to

$$d = \frac{v}{\omega} \sin^2 \alpha \,. \tag{3}$$

Let us assume that the agent uses some sensors each of which can detect the obstacle if it appears under a particular angle α. The agent can then estimate the angular velocity $\omega = d\alpha/dt \approx \Delta\alpha/\Delta t$ by the change of α per time interval:

$$d = \frac{v}{(d\alpha/dt)} \sin^2 \alpha \approx v \frac{\Delta t}{\Delta \alpha} \sin^2 \alpha \,. \tag{4}$$

Similar to biological systems, an agent with s facets can estimate $\omega \approx \Delta\alpha/\Delta t$ by utilizing a simple neural network, which consists of s input units u^I and s-1 output units u^O. As can be seen in Figure 3, each output unit u_i^O is connected to two input units u_i^I and u_{i+1}^I each of which is in turn connected to one facet with all connections being topology preserving, i.e., neighboring facets signal to neighboring inputs, which in turn connect to the same output unit. Furthermore, each input unit u_i^I has an associated time constant τ_i during which the unit remains active after it has been triggered by an appropriate input.

Each triple u_i^I, u_{i+1}^I, and u_i^O constitutes a motion sensor. An input unit u_i^I is activated by the appearance of a sufficiently-high "dark-to-bright" stimulus. Then, this unit remains active during the decay time τ_i. If also the neighboring input unit u_{i+1}^I becomes active during this time interval, the output unit u_i^O is triggered (due to its "and" operation). If however, the stimulus moves too slowly, the first input neuron u_i^I becomes inactive and the output unit u_i^O is not triggered. Depending on its relative velocity v, the agent may trigger an appropriate avoidance action if the sum of active output units exceeds a critical threshold, which can be, for example, dependent on v (see also eq. (4)). For an agent to be successful, it is essential to avoid obstacles only if necessary, since an agent that constantly avoids obstacles is rather useless. Therefore the parameter τ_i determines a critical speed between two neighboring sensors.

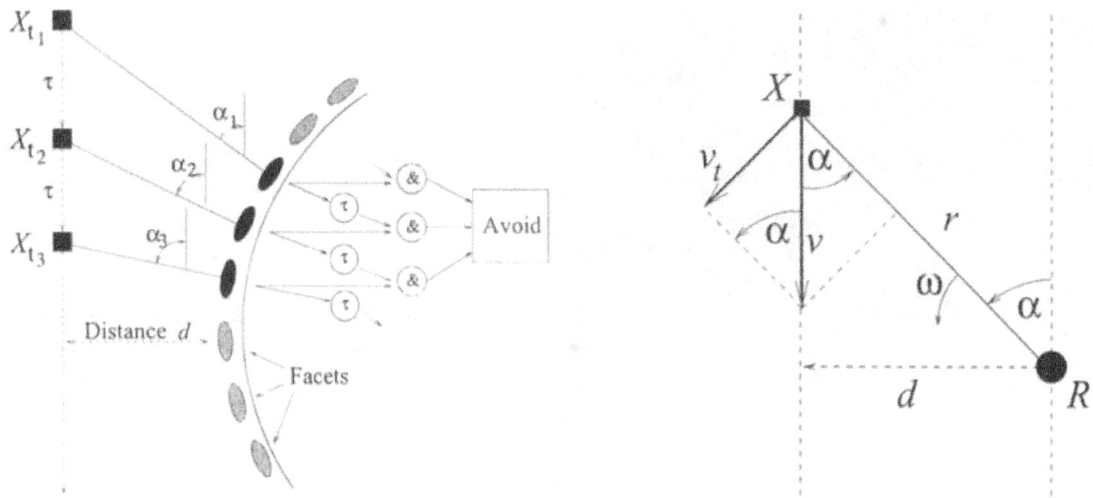

Figure 3: Left: due to their small aperture of about two degrees, the insect eye's facets recognize a moving object at different times $t_1, \ldots t_3$. With some units that maintain their activity during a time period τ after stimulation and output units that perform an and-operation over neighboring input pairs, a neural network is able to detect too small a distance of closest approach to an obstacle. Right: the angular velocity ω under which an obstacle X is seen equals $\omega = v_t/r = v \sin\alpha/r$ and depends on the actual angle α leading to a distance of closest approach $d = (v/\omega)\sin^2\alpha$.

An agent is in principle able to calculate the distance d of closest approach by utilizing only two facets. However, this calculation would be limited to a particular angle of its visual field, which would be too restrictive for real-world mobile robots. By using a compound eye, the agent is able to anticipate potential collisions regardless of the angle under which the object is seen. That is, two compound eyes would provide an agent with an almost 360° view.

2.3 Previous Results

Lichtensteiger and Eggenberger (1999) used a simple neural network with all weights being constant and equal and applied evolutionary algorithms to evolve a suitable sensor distribution, i.e, the different angles α, such that a predefined fitness function $f = \sum_i((t_{i+1} - t_i) - \tau)^2$ minimizes. Even with significantly improved algorithms (R. Salomon and L. Lichtensteiger, 2000), the results clearly indicate that this approach is practically limited to about 10-20 sensors due to the experimentation time and convergence problems.

Alternatively, the sensor distribution could be fixed and an algorithm like backpropagation could be applied to train the individual time delays τ_i. However, this approach as well as using evolutionary algorithms require the preselection of suitable training cases and offline learning, which immediately excludes any suitable online adaptation to changing conditions, such as broken sensors or a tilted camera.

3. The Model

The left-hand-side of Figure 4 shows the general architecture. A high-resolution input device, such as a CCD camera, feeds its activation to a speed sensor module. This module in turn communicates with a motor (sub-) system via reciprocal connections. And the motor system has direct connections to the actual actuators. The remainder of this section is devoted to some of the model's architectural details, including the actual wiring, the learning procedure, as well as some of the model's properties.

3.1 The Architecture

As has already been discussed in the background section, an object passing by the input device activates the input units i for a constant duration σ at different time steps t_i. Now, let I denote units of the input device and let S denote units of the speed sensor module. Every speed unit S_i directly connects to *exactly one* (randomly chosen) input unit I_j with the index j referencing to the speed unit's position (relative to the input device). Furthermore, all speed units also connect the neighborhood of their reference input unit. For example, speed unit S_i connects to the reference input unit I_r and also connects to $I_{r-m} \ldots I_{r+m}$ for some $m > 0$. Except for the reference connection, all signals are delayed by τ and have to pass a gate. The time delays τ may vary across all connections, and the gate has to be activated by the reference unit. That is, a speed sensor S_i receives delayed signals from the input device only during the time window defined by the reference unit I_r.

It should be noted here that all sensor units S_i are treated in the same way. For readability purposes, these neurons are arranged on a grid with columns and rows representing reference position r and speed, respectively. All speed sensors

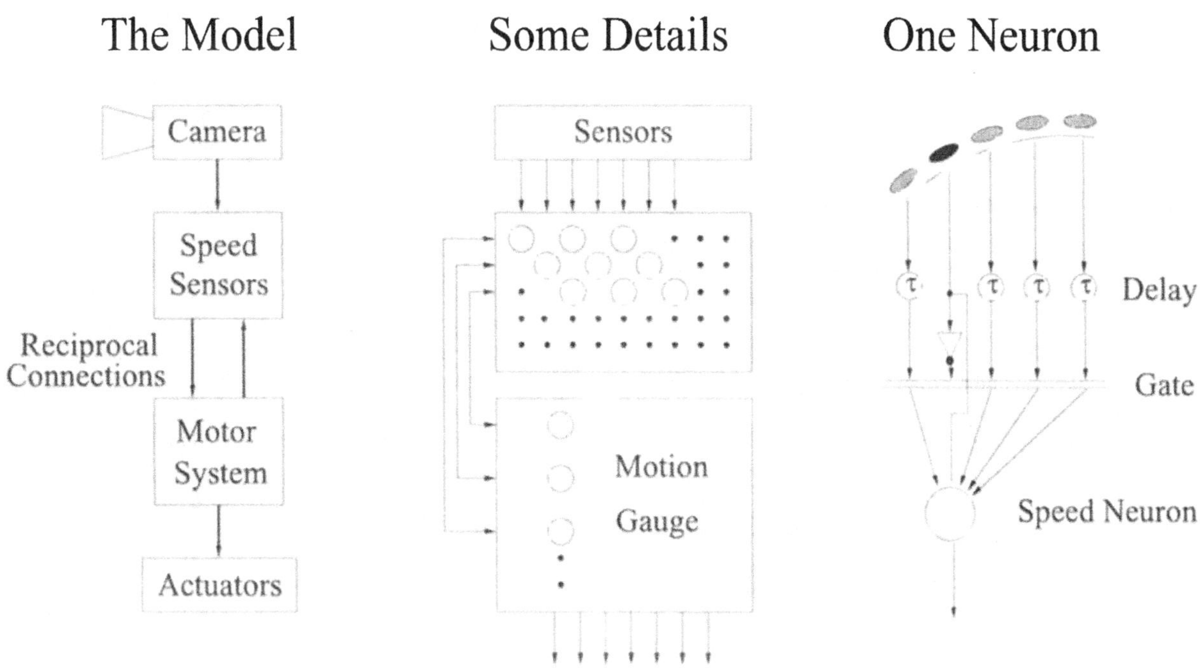

Figure 4: This fi gure sketches the model's architecture with increasingly more details when going from left to right. The left-hand-side shows the overall architecture, in which a high-resolution input device (e.g., a CCD camera) feeds its activation into a speed sensor module. This module communicates with a motor (sub-) system]ia reciprocal connections. The middle part depicts some selected neurons, and in the right-hand-side, the fi gure shows the wiring between a particular speed sensor and its incoming signals it receives from the input device. For further detail, see text.

belonging to the same speed value, communicate with corresponding units in the motor system.

3.2 The Learning Rule

Without loss of generality, the model makes the following four assumptions:

1. The input unit's activation I is either 0 or 1.
2. The gate either passes the incoming activation or sets it to 0.
3. If an input sensor views an object it sets its activation to 1 for the duration σ.
4. The activation of the speed sensor units is bounded by $0 \le S_i \le 1$.

The operation of the speed sensor units is inspired by biological integrate-and-fire neurons. For each connection, S_i first calculates the overlap o_{ij} of its reference signal (I_r) and the incoming delayed signals ($I_{j \neq r}$). More formally, the overlap is given by $o_{ij} = (1/\sigma)\int_t I_r(t)I_j(t)dt$. Due to the division by the signal duration constant σ, the overlap is normalized to $0 \le o_{ij} \le 1$. To sharpen the edges, the overlap is modified by (with ϵ denoting an arbitrarily-chosen, small, positive constant):

$$o'_{ij} \leftarrow \frac{\ln(o_{ij} + \epsilon) - \ln \epsilon}{\ln \epsilon} . \tag{5}$$

If activated by the reference signal I_r, the speed sensor unit S_i then applies the following, rather generic Hebbian-based learning rule (for some properties, see below):

$$w_{ij} \leftarrow w_{ij} + \eta o'_{ij} - (\eta/\alpha) w_{ij} , \tag{6}$$

with w_{ij} denoting the connection (weight) connecting input unit I_j with speed sensor unit S_i, η denoting a learning constant, and α denoting an unlearning constant.

3.3 Model Properties and Behaviors

The learning rule, as defined by eq. (6) has the following properties: due to the unlearning constant α, the connection weight is bounded by $w_{ij} < \alpha o'_{ij}$, and since $o'_{ij} \le 1$, the relation $w_{ij} < \alpha$ holds. Due to this bounding, the repetitive application of the learning rule eq. (6) lets the connections approach this bound, i.e., $\lim_{t \to \infty} w_{ij} = o'_{ij}$. In other words, the network's connections assume values proportional to their incoming overlap values during the time window defined by the reference signal I_r. Without loss of generality, the unlearning constant can be set to α=1.

The learning constant η in eq. (6) determines the slope with which the final values are approached; it does not influence the final values. With smaller η values, the learning procedure consumes more time to saturate but is less sensitive to changing environmental conditions. Large η values have the opposite effect.

Like the high-resolution input device, the speed sensor module features very many units, to which learning applies as described above. The speed sensor module arranges all units on a grid with the x and y axes representing position r and speed, respectively. All units in a specific row are responsible for the same speed and are collectively coupled with corresponding units in the motor system. Within a specific row, all speed sensor units connect to different reference input units.

In the very beginning, all connections are initialized to some, randomly-chosen values. When the robot is moving with a particular speed v, the operating motor system activates the appropriate row of the speed sensors. When these active speed sensor units receive their reference signal (what they do at different time steps), they update their connections w_{ij} by applying learning rule (6). During the initial leaning (exploration) phase, the robot repetitively moves with different speeds in order to train all speed sensor units.

It should be mentioned here that the robot itself selects a particular moving speed by means of its motor system, and that the visual system organizes itself by evaluating visual stimulation it receives as a result of the agent-environment interaction (i.e., exploiting the agent's morphology).

After the initial learning phase, the speed sensor units may also by activated by the following activation rule:

$$S_i = \frac{o'_{ij} w_{ij}}{\sum_j w_{ij}}, \tag{7}$$

with o'_{ij} denoting the modified overlap with the reference signal as described above. As cn be seen from eq. (7), the speed sensor unit integrates all incoming delayed signals during the time window given by the reference signal I_r. This integration is weighted by the w_{ij}'s and then normalized. This normalization has the effect that the parameter α has no effect and that only slightly varies with changing configurations (i.e., signal duration σ and number of connected inputs).

The main idea behind this model is as follows: Since all speed sensor units are treated in the same way and connected only to some local region of the input device, the model allows for simple scaling over very many input units of high-resolution sensor devices.

4. Methods

Unless otherwise stated, all experiments have used the following parameter settings: signal duration σ=2, learning rate η=0.01, unlearning rate α=1, overlap-modification parameter ϵ=0.001 (see, eq. (5)). During the initial training phase, the speed was set to values, such that the object travels about one sensor per time unit.

5. Results

This section presents some results, obtained with a computer simulation. In order to be comprehensible, this section presents figures with only a few units. The presentation of a simple learning process of one speed sensor unit would merely demonstrate the convergence property of equation (6), and is thus omitted here.

Figure 5 demonstrates the model's adaptation capability. After the network has learned an initial input distribution, which has been arbitrarily chosen for illustration purposes, the network can easily adapt to input changes. In the figure, the input values from position 0-15 have been changed, and the adaptation process takes about 140 epochs. It can be seen that the model just adapts to the changed input.

As Section 3. has already discussed, the speed sensor module employs neural populations for different speed values v. Figure 6 shows five different units that are particularly trained for varying speeds $v \in \{0.83, 0.91, 1.00, 1.10, 1.20\}$. In this test case, an object passes by with speed $v = 1.0$, and the figure shows how the units' activations change over time (each tic marks a particular test point). It can be clearly seen that the units' responses decrease with the difference between both the test and its inherent (trained) speed. In this figure, the x and y axes represent time and activation, respectively. It should be noted that the depicted units have been particularly chosen, such that a moving object activates them at different time steps (otherwise it would became impossible to read the figure).

Figure 6 suggests, as has been expected, that the speed sensor units react quite sensitively to a passing object. The next experiment investigates the sensor module's behavior with respect to varying object speeds. To this end, an object is passing the input device several times each time with a different speed $v_1, \ldots, v_m$. For each pass, the system monitors the activation of a set of selected speed sensor units S_i and remembers each unit's maximal activation. These maximal activations are then plotted in a graph. Figure 7 presents such a graph with five selected units. It can be clearly seen that the speeds are v_1=0.77, v_2=0.88, v_3=1.00, v_4=1.14, and v_5=1.30 units per time step. All units exhibit their maximal activation (not necessarily 1 due to eq. (7)) at their particular training speed, and are less active elsewhere. It is obvious that a subsequent layer, e.g., another neural layer or a fuzzy controller, can easily reconstruct the object's/robot's speed.

6. Discussion

This paper has proposed a new model to alleviate a particular deficiency in the area of autonomous agents. The proposed model allows the utilization of a high-resolution input device, such as a CCD camera, by featuring a simple Hebbian-based learning rule. This learning procedure has the following advantages over currently used methods: entire online learning without any need of preselecting any training pattern set, straight forward scaling property by using only local informa-

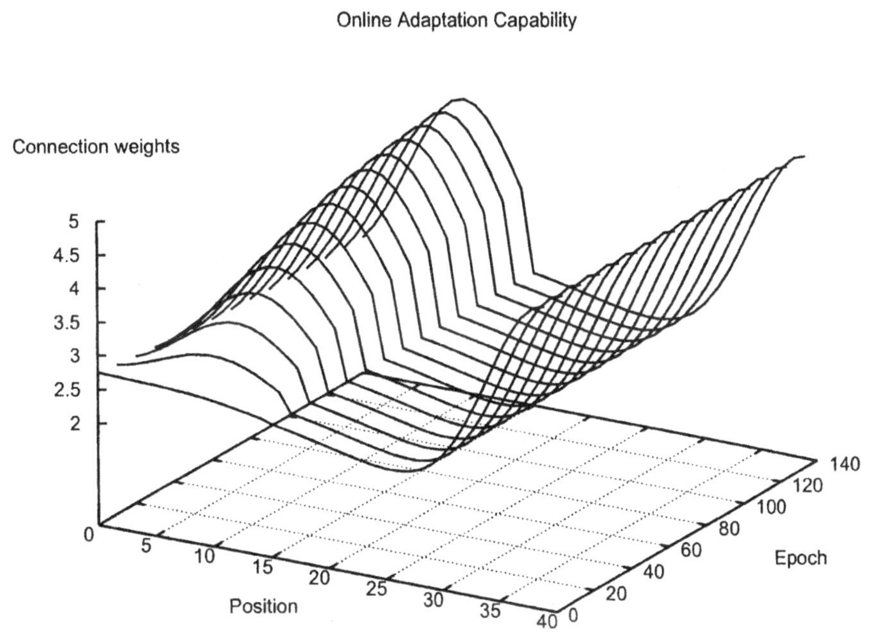

Figure 5: After the network has learned a certain input distribution (arbitrarily chosen for illustration purposes), it can adapt to a new distribution (positions 0-15) within 140 epochs.

tion among units, and no overlearning deficiencies and thus constant applicability with inherent adaptation properties.

The model's design is largely *inspired* by biological observations. However, no claim is made that in turn, any part is biological plausible or that any biological system is equivalently working like the proposed model; it is a just a technical application.

When setting up the system, the speed sensor module may be equipped with as many units as desired, and then each has to connected to a reference input sensor as well as the *reference unit's neighborhood.* In the initial training phase, the speed sensor units may be grouped to populations with each population being responsible for a particular speed.

Future research will be devoted to the following steps: implementation of the model on a physical robot and trying to substitute the gate functionality that is activated by the reference signal, by more natural mechanism; this reference signal is currently required, but substituting it would probably simplify the model.

Acknowledgments

The author gratefully thanks Lukas Lichtensteiger for pointing me to his research and all the experimental details, including the data, the robot, as well as the background literature. The author is also very grateful to the anonymous reviewers for their feedback.

References

R. A. Brooks (1991a). Intelligence Without Reason. *Proceedings of the 12th Intl. Conference on Artificial Intelligence (IJCAI-91),* Morgan Kaufmann, San Mateo, CA, 569-595.

R. A. Brooks (1991b). Intelligence without representation. *Artificial Intelligence,* 47:139-159.

T. S. Collett (1978). Peering – a locust behavior pattern for obtaining motion parallax. *Journal of Experimental Biology,* 76:237-241.

N. Franceschini, J. Pichon, and C. Blanes (1992). From insect vision to robot vision. *Philosophical Transactions of the Royal Society of London B,* 337:283-294.

G. A. Horridge (1978). Insects which turn and look. *Endeavour* 1:7-17.

L. Lichtensteiger and P. Eggenberger (1999). Evolving the Morphology of a Compound Eye on a Robot. *Proceedings of the Third European Workshop on Advanced Mobile Robots (Eurobot '99).* IEEE Piscataway, NJ, pp. 127-134.

R. Pfeifer and C. Scheier (1999). *Understanding Intelligence* MIT Press, Cambridge, MA.

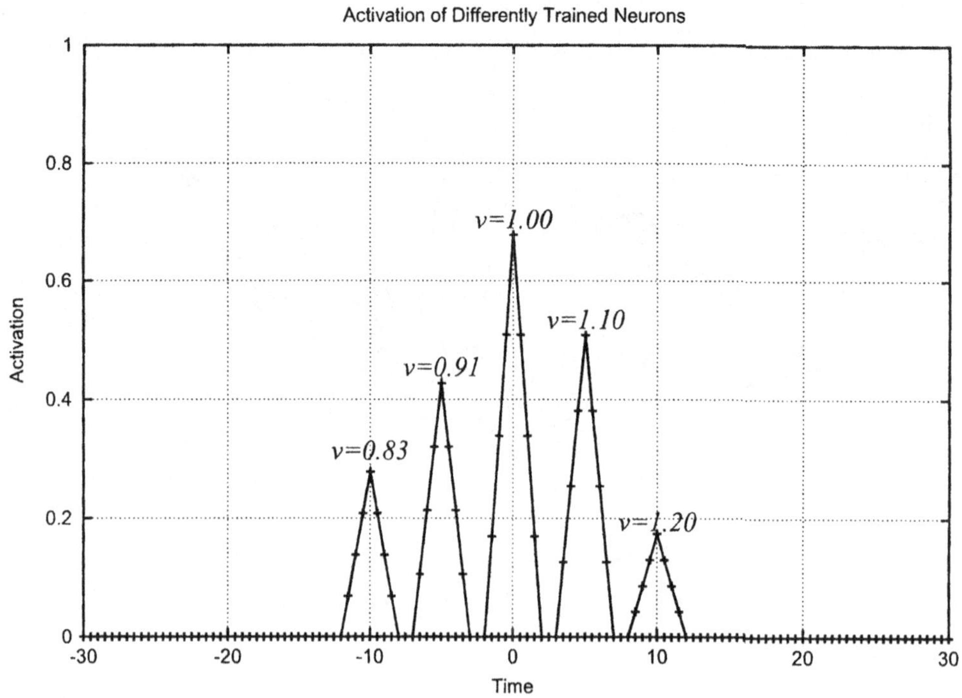

Figure 6: This figure shows the activation of five speed sensor units, each trained with a different speed v, when an object is passing by with v=1 unit per time step.

R. Salomon and L. Lichtensteiger (2000). Exploring different Coding Schemes for the Evolution of an Artificial Insect Eye. *Proceedings of The First IEEE Symposium on Combinations of Evolutionary Computation and Neural Networks.* 10-16.

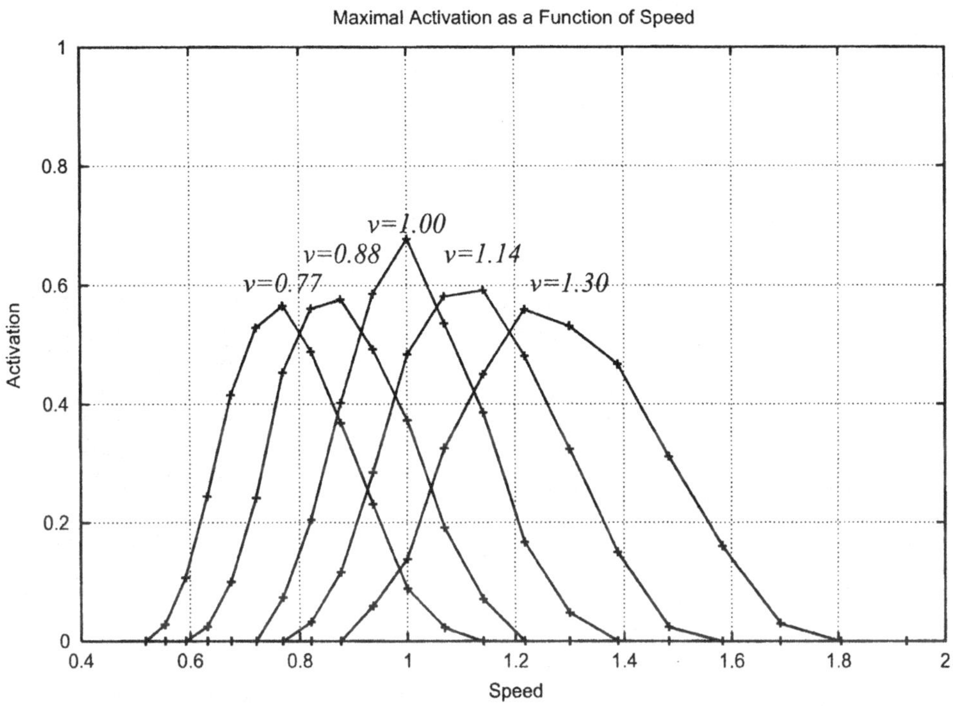

Figure 7: This figure shows how the speed sensor units I react to varying speeds. Each figure point represent the maximal activation of a unit during a complete pass of an object passing the robot.

Adaptive Object Tracking with an Anthropomorphic Robot Head

Hyundo Kim*, Boris Lau, Jochen Triesch***

*Complex Systems & Cognition Lab
University of California, San Diego
hyundo@ucsd.edu, triesch@cogsci.ucsd.edu
http://csclab.ucsd.edu

**Department of Neuroinformatics
Ilmenau Technical University, Germany
boris.lau@stud.tu-ilmenau.de

Abstract

We present a system for model-free, adaptive object tracking with a redundant anthropomorphic robot head. The visual tracking system is based on the idea of "Democratic Integration" to fuse information from several cues in a biologically plausible, self-organized fashion. The motor control system utilizes the idea of "controller partitioning" to exploit complementary capabilities of redundant degrees of freedom in the neck and eyes. A biologically inspired vergence control mechanism ensures coordination of the robot's two eyes. We demonstrate the performance of the overall system in different active tracking scenarios.

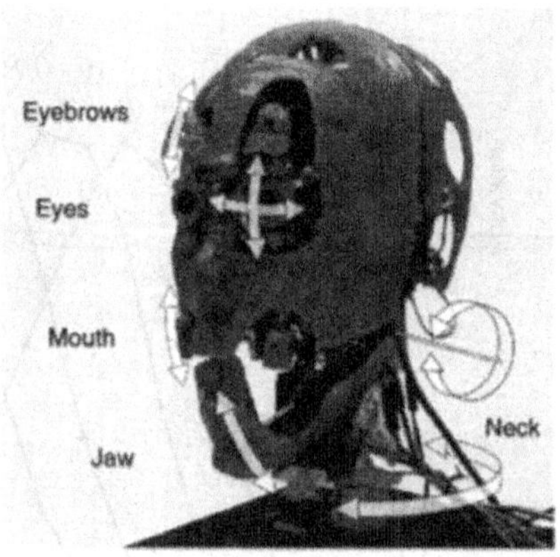

Figure 1: Anthropomorphic robot head with nine DoFs.

1. Introduction

The long term goal of our research is to understand how biological systems can autonomously learn to see. How does the developing primate brain form representations of the objects in its environment? Since the developing visual system actively explores its visual environment, it shapes the statistics of what it is seeing rather than being passively exposed to it. Furthermore, it shapes these statistics in a purposeful manner — even infants tested minutes after birth show preferences for some visual stimuli over others. Also, eye movements are likely to play an important role in the development of invariant representations. Thus we believe that in order to fully understand the post-natal development of the primate visual system it is crucial to take its active and goal-driven nature into account. Our approach to the problem of autonomous visual learning is to build embodied models of these active learning processes. To this end we have developed an anthropomorphic robot head platform, whose kinematic structure closely resembles that of the human neck/eye system (Fig. 1). In this paper, we present a system for active object tracking implemented on the robot head. This system will form the "front end" for a more sophisticated module for learning higher level object representations in future explorations into autonomous learning in actively developing visual systems. The system presented here allows our robot to actively track "interesting" objects in its environment, creating the opportunity to learn models of their appearance and behavior.

Our specific goals for this system were the following. First, the tracking scheme needs to be model-free, i.e. it must not require a detailed prior model of the object to be tracked, because we are interested in the *completely autonomous* learning of representations of unfamiliar objects. Second, despite the lack of a model of either the object or its typical patterns of motion, the system should function robustly and be able to cope with fast object movements over a wide range. Third, the system should operate in real-time, generating motor responses at the rate of the visual input of 30Hz. Finally, the system should be roughly consistent with biological constraints, if it is to be part of a larger model for the learning of object representations in the primate brain.

The remainder of the paper is organized as follows. Section 2 briefly describes our redundant anthropomorphic robot head. Section 3 explains the adaptive visual tracking system. In Sec. 4 we describe how our parti-

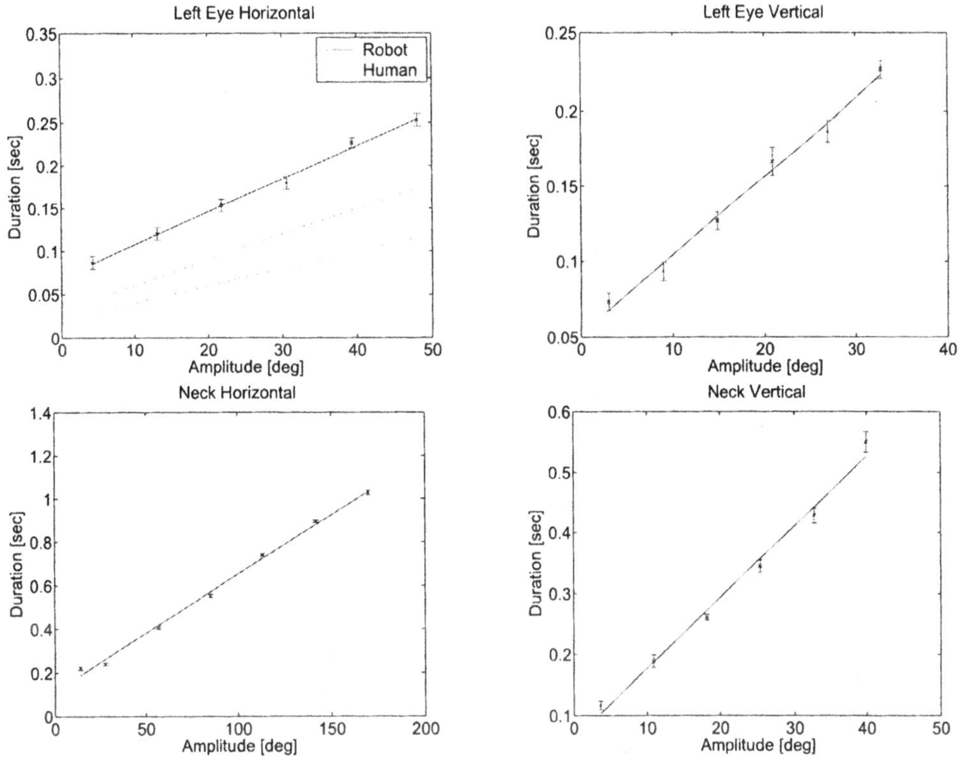

Figure 2: Average time elapsed vs. size of motion for horizontal (top left) and vertical (top right) movements of the left eye and horizontal (bottom left) and vertical (bottom right) movements of the neck. For horizontal eye movements (top left), the dashed lines indicate the range of saccade durations typically observed in humans.

tioned motor control system of the robot exploits the complementary nature of the robot's degrees of freedom (DoFs). Section 5 presents experiments and Sec. 6 discusses our system from a broader perspective.

2. Anthropomorphic Robot Head

We have developed an anthropomorphic robot head that closely mimics the neck-eye system of the human head (Fig. 1). The robot head is intended as a flexible platform for studying autonomous visual learning and human robot interaction. A detailed description will be published elsewhere (Kim et al., 2004). Here we only give a brief overview. The robot head has a total of 9 degrees of freedom (DoFs). Six of them are for the neck and the two eyes: both eyes can independently pan and tilt. The neck can pan and tilt, too. The remaining 3 DoFs are for facial expressions (jaw opening, eyebrow lifting, smiling/frowning) but will not be used in this paper. The robot head is controlled from a host computer (Linux PC, Pentium-4 3.06GHz processor, 2GB RAM).

The robot's eyes are two miniature CCD cameras (Point Grey Research, model FireFly). The cameras are connected to the host computer via two IEEE 1394 interfaces with 400 Mbit/s. Each camera gives 24 bit color images with up to 640x480 resolution at 30 fps. We use a focal length of 4mm, providing a diagonal field of view of 100 degrees.

All DoFs are actuated via servo motors. They provide fast, easy, and fairly accurate control, while being robust and inexpensive. Most DoFs are connected via four bar linkages, introducing a slightly non-linear relation between the rotation of a servo shaft and the rotation of the actuated DoF. Table 1 gives the range of motion and the average angular resolution for the robot's major DoFs. The servos are controlled through two serial control boards (Mini SSC II). The control boards receive commands from the host computer through a 9600bps serial interface allowing us to issue up to 400 motor commands per second. Since we are using 9 DoFs, we can send a command to each servo every 22.5ms — faster than the frame rate of the cameras. Due to the use of very small, light-weight cameras and fast servo motors, our robot can make saccade-like gaze shifts that come close to the speeds of human saccades. Figure 2 compares the durations of the robot's saccades with those of humans (Becker, 1991). The focus of this paper is smooth pursuit-like tracking, however.

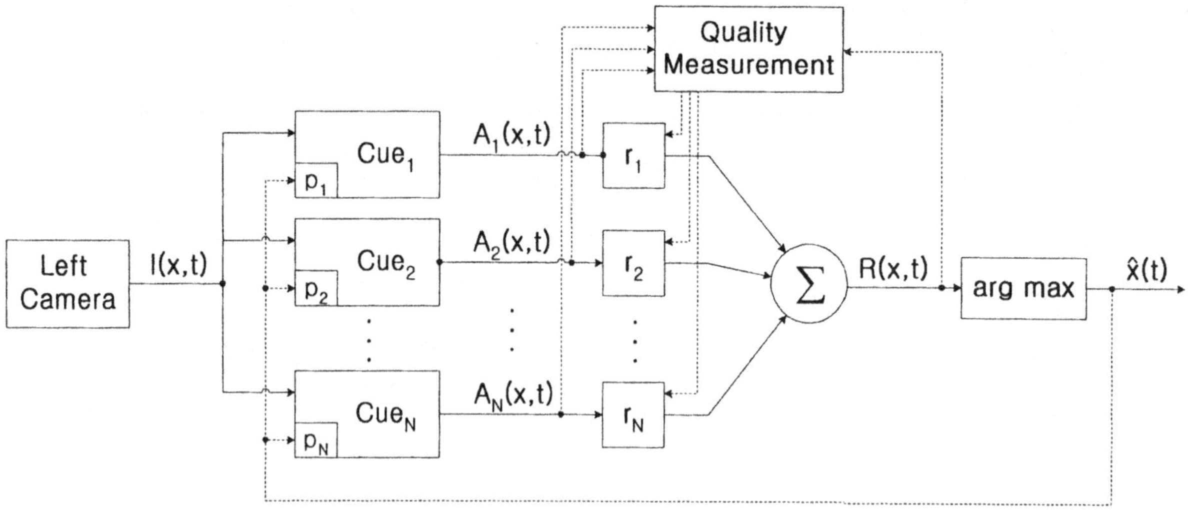

Figure 3: Block diagram of the democratic integration system. Solid lines represent feed-forward information flow, while dashed lines represent adaptation signals.

	Range		Resolution	
	Hor.	Ver.	Hor.	Ver.
Left Eye	42.8°	35.3°	0.37°	0.28°
Right Eye	37.9°	33.4°	0.33°	0.27°
Neck	180°	40.2°	0.71°	0.17°

Table 1: Range of motion and average angular resolution of the neck and eye DoFs.

3. Adaptive Visual Tracking

Given the lack of a detailed model of the target object we need to have a tracking system that evaluates and adapts itself according to the tracking results. We use the Democratic Integration idea (DI), which is a framework for integration of different visual cues in a self-organized manner (Triesch and von der Malsburg, 2001). This system adaptively fuses several cues into one result. There are two ways in which it is adaptive: first, it allows for a re-weighting of cues based on their estimated reliabilities, second, cues adapt object models to better match the fused result. Such re-weighting can also be observed in human object tracking (Triesch et al., 2001).

The two eyes of our robot head play different roles in the tracking system. The dominant left eye uses an adaptive tracking scheme to estimate the object's position. A vergence controller steers the right eye to the location looked at by the left eye, as described in Section 4.

Figure 3 shows a block diagram of the DI system. We are using four cues (shape, color, motion, prediction). A detailed description of the cues is given in the appendix. Our current implementation of the active DI-style tracking system works in real time (30Hz) with an image resolution of 160x120 pixels.

3.1 *Democratic Integration System*

In this system N different cues search for features in the image that match the appearance of the object. Each cue i computes a saliency map $A_i(x,t)$ that measures the similarity of the cue's object model $p_i(t)$ to features extracted from the image $I(x,t)$. High values indicate high similarities. Each cue has a reliability $r_i(t)$ that controls the cue's influence on the result saliency map $R(x,t)$. Concretely, $R(x,t)$ is a weighted sum of the cues' individual saliency maps:

$$R(x,t) = \sum_{i=1}^{N} r_i(t) A_i(x,t). \tag{1}$$

The estimated target position $\hat{x}(t)$ is the point yielding the highest value in the result saliency map, as long as this value exceeds a threshold T (here $T = 0.6$). Otherwise no target is detected:

$$\hat{x}(t) = \arg\max_x \{R(x,t)\}, \text{ if } R(\hat{x},t) > T. \tag{2}$$

3.2 *Adaptation of Reliabilities*

The adaptive parameters in the integration of the cues are the reliabilities $r_i(t)$. They reflect a cue's agreement with the result in the recent past. The agreement is measured with the quality $\tilde{q}_i(t)$. A detailed discussion of different quality metrics can be found in (Triesch and von der Malsburg, 2001). Here we measure the quality by comparing the response in the cues'

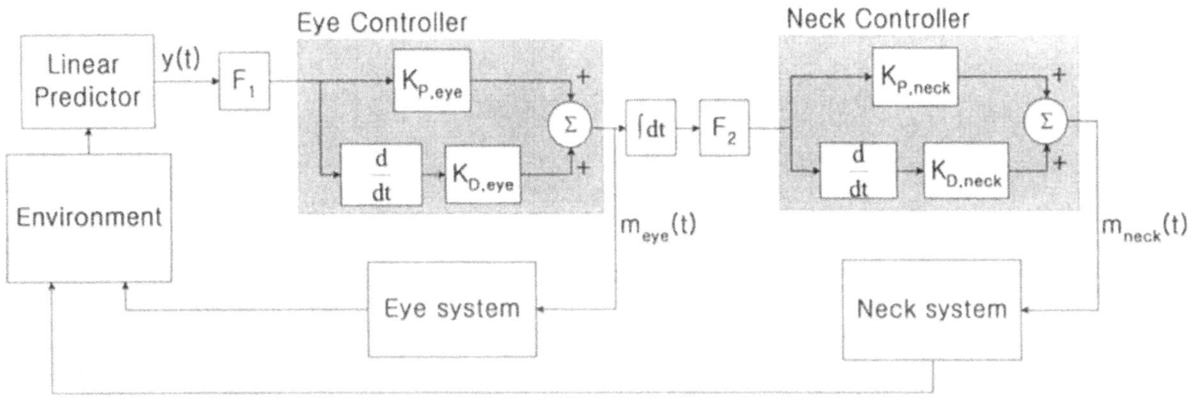

Figure 4: Block diagram of the partitioned controller.

saliency maps at the estimated target position with the average saliency $\bar{A}_i$ of each cue:

$$\tilde{q}_i(t) = \max\left\{0, A_i(\hat{x}, t) - \bar{A}_i(t)\right\}. \quad (3)$$

If a target has been detected, the reliabilities are updated to follow the normalized qualities $q_i(t)$:

$$\begin{aligned} q_i(t) &= \tilde{q}_i(t) / \sum_{k=1}^{N} \tilde{q}_k(t) \\ r_i(t) &= \alpha_{rel} \cdot q_i(t) + (1 - \alpha_{rel}) \cdot q_i(t-1). \quad (4) \end{aligned}$$

If no target has been detected, the reliabilities are adapted towards their initial values. $\alpha_{rel} \in [0, 1]$ is a adaptation constant, controlling the speed of adaptation. A value close to 1 leads to a fast adaption, a small value close to 0 causes slow changes. Due to the normalization of the $q_i(t)$, the $r_i(t)$ are also normalized with their sum converging to one.

The adjustment of the adaptation constant α_{rel} is important: If it is too small the system can't react fast enough to the sudden failure of one or more cues. On the other hand, if α_{rel} is too large, this typically also worsens performance.

3.3 Adaptation of Cues

The adaptive cues try to maximize their agreement with the result saliency map by adapting their object models towards features $f_i(t)$ extracted at the estimated target position. This causes a better agreement with the result in the future, provided that the visual appearance of the tracked object is not changing much from one frame to the next. The shape cue for example uses a small grayscale template (e.g. 7x7 pixel) as object model $p_i(t)$ and searches for image locations that are similar to the template. The process of object model adaption uses similar dynamics as (4). If a target has been detected, the cues' object models are adapted according to:

$$p_i(t) = \alpha_{cue} \cdot f_i(t) + (1 - \alpha_{cue}) \cdot p_i(t-1). \quad (5)$$

If no target has been detected, the cues' object models are adapted towards their initial values. Each cue can have its own adaptation constant α_i, but in our experiments they are all identical: $\alpha_i = \alpha_{cue} \forall i$. The adjustment of the adaptation constant α_{cue} is also important: a constant object model is unlikely to match the object for more than a few frames, but a object model changing too quickly would not carry any information about the stable properties of the object.

3.4 Initialization

As we have mentioned in the Introduction, infants tested shortly after birth already have preferences for some visual stimuli over others. Here, in the initialization of the cues, we incorporate this idea of having preferences for the visual system by setting initial models for the cues. One always needs a priori knowledge about the target to guide initial target selection. For example, a system configured to do face tracking could have its color object model set to skin color initially, or can use an additional face detection cue. Other cues like the motion cue can work without such a prior model and may have high reliabilities right from the start. If we don't have a priori knowledge for a cue we set its reliability to zero and its object model to an arbitrary value. As soon as the other cues agree on a target this cue will be bootstrapped and automatically integrated into the tracking process. It is important to find a set of cues that is sufficient for bootstrapping the system.

4. Motor Control System

The robot head has pan and tilt DoF for both eyes and the neck, such that multiple combinations of pan and tilt angles of the eyes and the neck can provide the same orientation of the camera in space. This redundancy needs to be resolved by the controller. Our solution is based on the idea of controller partitioning (Oh and Allen, 2001), which exploits complementary properties of redundant DoFs.

4.1 Eye-neck control mechanism

Since the robot's cameras are very small, the eyes move faster than the neck, where the servo has to move the entire robot head. On the other hand, the neck has a much wider range of movement. The basic idea behind our partitioned controller is that once the tracked object is out of the center of the image, we use the DoF with the faster response (eye) first to reduce the visual error, and then the slower DoF (neck) to bring the eye back to its center position. In other words, we use the neck DoF to compensate for the small range of the eye DoF.

Inside the partitioned controller, we have 2 PD controllers in a cascade. The first controller receives the deviation $y(t)$ of the target object from the center in pixel coordinates. The second controller takes as input the estimated eye position derived from the output of the first controller. Here we explicitly assume that the horizontal and vertical mapping from image coordinates to motor commands are independent from each other and can be controlled separately. This may not be completely accurate due to a number of reasons but is a good approximation. The first PD controller computes the motor command $m_{eye}(t)$ given the visual error:

$$\begin{aligned} m_{eye}(t) &= K_{P,eye} \cdot y(t) + K_{D,eye} \cdot \dot{y}(t) \quad (6) \\ \dot{y}(t) &= y(t) - y(t-1)\,. \end{aligned}$$

$K_{P,eye}$ is the proportional gain for the eye, and $K_{D,eye}$ the derivative gain.

This motor command is used to estimate the future eye angular deviation from its center position $\hat{x}_{eye}(t)$:

$$\hat{x}_{eye}(t) = \hat{x}_{eye}(t-1) + m_{eye}(t)\,. \quad (7)$$

Here, we assume linearity and no delays to approximate the eye position $\hat{x}_{eye}(t)$ calculated from the previous motor commands. In other words, we make an assumption that the estimated eye angular deviation from its center position has a one-to-one relationship with the given motor command $m_{eye}(t)$. This approximation works reasonably well as demonstrated in Section 5. Then this estimated deviation is fed to the neck controller, where another PD controller is used to make a neck movement $m_{neck}(t)$, that brings the eye back to its center position:

$$m_{neck}(t) = K_{P,neck} \cdot \hat{x}_{eye}(t) + K_{D,neck} \cdot m_{eye}(t)\,. \quad (8)$$

Both controllers ignore inputs that are below a certain threshold. The vision system can detect errors that are smaller than the smallest possible corrective movement. The thresholds will suppress small oscillations around the target position. The first threshold (± 2 pixels for the full image resolution) corresponds to the visual error for the eye controller. The second one (± 1 ticks) corresponds to the angular error for the neck controller.

A block diagram for the complete controller is given in Figure 4. In this figure, F_1 corresponds to the threshold opearation in the eye controller and F_2 to the threshold operation in the neck controller. We also use a simple linear predictor in front of the partitioned controller which is basically the same as the prediction cue described in the appendix. One could use a more sophisticated predictor that estimates the target dynamics to improve the tracking performance, as demonstrated in (Shibata and Schaal, 2001b).

Note that due to the feedback delays in the system, the active tracking will always lag behind the target, showing object following behaviors.

Figure 5: Left and right image pair before (above) and after (below) vergence control.

4.2 Vergence control mechanism

Since tracking is done with only one "dominant" eye, we need a method for ensuring the coordination of both eyes, i.e. a way of controlling the non-dominant eye to look at the same object. We have adopted and extended previous work by Theimer and Mallot for doing vergence control (Theimer and Mallot, 1994). They extract Gabor jets from the center of left and right image and then compute a phase-based distance estimate in pixel coordinates. This estimate can be used to bring the center of the left and right image to the same target without specific knowledge about the object. This method mimics an array of disparity tuned simple cells as could be found in the primary visual cortex.

Our version extends this idea and extracts a grid of Gabor jets from the right image and compares it to a Gabor jet from the left image. Due to space limitations, we will provide a detailed description in a separate paper. Figure 5 shows the left-right image pairs before and after

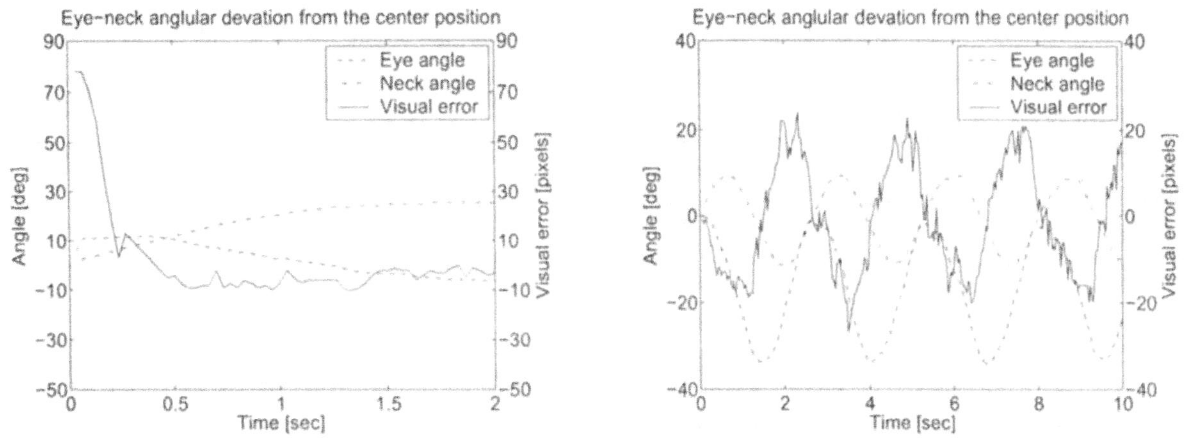

Figure 6: Eye and neck angle deviation from center position due to a step input(left) and a swinging pendulum(right).

the vergence control. A red cross is drawn at the center for comparison.

Alternatively, one could solve this vergence problem in the tracking application by having two separate tracking modules for the left and right eye and then have each module control one eye independently of the other. In this case the neck controller would be driven by the average deviation of the left and right eye from their central positions. As long as the two tracking modules share models, both eyes will track the same object. This tracking approach, although more time consuming, is more biologically plausible than the current scheme. Future work will try to combine two DI systems with the vergence control mechanism presented here.

5. Experiments

5.1 Experimental setups

We have experimented with two different setups to evaluate the performance of our active tracking system. In the first setup, we let the robot head track a tennis ball in order to evaluate the controller's behavior in response to step inputs and oscillatory inputs. For initialization, we supply the system with color models for the color cue only. It is able to track robustly by bootstrapping the other cues. In the second setup, we track the face of a moving person.

For the DI tracking system, the initial reliabilities of the cues are 0.55 for the motion cue, 0.45 for the color cue and 0 for the shape and prediction cues. We have used a template size of 7x7 for both experiments. Blurring in the motion cue is done using a gaussian filter of size 5x5 for the tennis ball tracking experiment and 7x7 for the face tracking (see appendix). Blurring in the color cue is done using a gaussian filter of size 3x3 for tennis ball tracking and 5x5 for face tracking. The adaptation

$K_{P,eye,horizontal}$	0.189	$K_{P,neck,horizontal}$	0.0189
$K_{D,eye,horizontal}$	0.1	$K_{D,neck,horizontal}$	0.1
$K_{P,eye,vertical}$	0.284	$K_{P,neck,vertical}$	0.0474
$K_{D,eye,vertical}$	0.15	$K_{D,neck,vertical}$	0.1
$K_{P,vergence}$	0.075		
$K_{D,vergence}$	0.1		

Table 2: Controller gains used in the experiments.

constant for the dynamics of cue reliability, α_{rel}, is set to 0.1 and the adaptation constant for the dynamics of the cues' object models, α_{cue}, is set to 0.05.

The gains for the controller were set manually after some experimentation (see Table 2) and are the same for all experiments. For tuning the gains, first, we set the gains for the neck controller equal to zero and moved the eyes only. We increased the gains of the eye controller until we have sufficiently fast response with the overshoot below a certain threshold, say ± 5 pixels. Then we fixed the gains of the eye controller and increased the gains for the neck controller till we have sufficiently small overshoot with fast response time.

5.2 Step input

For the first experiment, we let the system track the tennis ball for a while without moving the robot head. Then we switched on the control system to measure the step response, and plotted the eye and neck angular deviation from its center position along with the visual error. We only show data for the horizontal movements of the left (dominant) eye and the neck. The object was placed near the horizontal boundary of the image in order to generate a visual error, close to the maximum of ± 80 pixels (± 35 degrees).

In the left side of Figure 6, the step input is given at

0 s. In other words, we turned on the control system at 0 s and it clearly shows that the eye will first start to move in the direction of the object and the neck will follow the eye. Near the end of the movement, the eye will move in the opposite direction of the neck movement since the neck movements will cause the eye to make compensatory movements. There is an overshoot in the visual error at the end of the movement. An extra mechanism like the vestibulo ocular reflex can be used to alleviate the problem (Shibata and Schaal, 2001a).

5.3 Oscillatory input

To create an oscillatory input, we used a tennis ball as a pendulum with an amplitude of approximately ±40 pixels. We let the system adapt its object models while actively tracking the object. The right side of Figure 6 shows that the eye immediately starts moving to reduce the visual error. The neck slowly follows the eye movement. Due to the feedback and visual processing delays we can see the visual error follows the oscillation of the swinging ball with the same frequency.

5.4 Face tracking

For the face tracking experiment, we used skin color as the initial model for the color cue. Note that the color cue alone is not sufficient for robust face tracking, if other objects with a similar color are present in the scene.

Figure 8 shows every 20th frame of an image sequence captured from the left camera while tracking faces. A blue cross marks the center of the image and a red cross the object position estimated by the DI tracking system. Video clips are available on our website[1].

One of the nice things about the DI tracking system is that it self-evaluates its cues and finds out how useful each of the cues is. Figure 7 shows that the system is able to automatically suppress the simple motion cue based on difference images, as soon as it becomes useless when the robot head starts to move.

6. Discussion

We have presented an adaptive, model-free tracking system implemented on an anthropomorphic robot head. Our goals were fourfold: First, the tracking system should be model-free, because we want to use it in the context of a learning system that develops object representations completely autonomously. Second, despite the lack of a detailed object model, the system should function robustly and handle fast movements over a wide range. Third, the system should operate in real-time, which in our case meant a frame rate of 30Hz. Fourth, the system should be biologically plausible because it will ultimately be a part of a larger embodied model for the development of object representations in the primate visual system. The system we have presented in here essentially meets all these requirements.

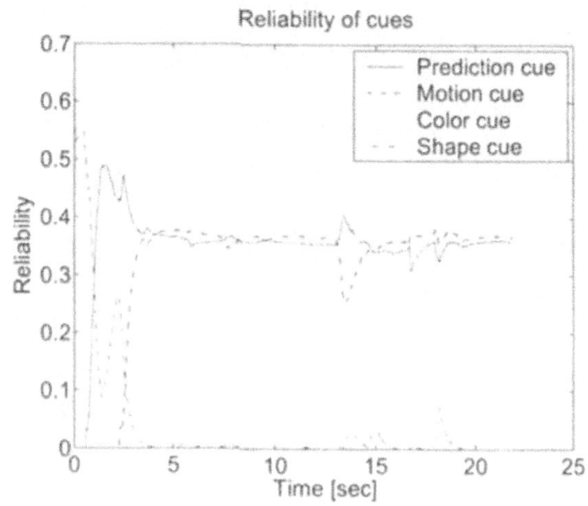

Figure 7: Reliability of cues as a function of time in face tracking experiment.

Our tracking system is based on the democratic integration idea (Triesch and von der Malsburg, 2001). The democratic cue integration used in the tracking system has the ability to bootstrap representations autonomously and does not require a detailed model of the object to be tracked (although some coarse prior knowledge is needed to specify which objects are "interesting"). It has been argued that this architecture mimics the robustness and flexibility known from biological vision systems. Furthermore, fast re-weighting effects as produced by the Democratic Integration mechanism can be readily observed in object tracking by humans (Triesch et al., 2001).

In the context of active tracking as demonstrated here, it is interesting to note that the system was fairly robust to the camera movements, despite its reliance on a simple motion detection cue based on difference images. This can be attributed to the democratic integration mechanism which automatically suppresses the motion detection cue during times of fast camera movements.

The idea of cue integration or sensor fusion has been very popular in diverse areas of computer vision. Helpful reviews, introductory papers and books published on the topic are available (Luo and Kay, 1989, Alomoinos and Shulman, 1989, Clark and Yuille, 1990, Abidi and Gonzalez, 1992). A number of different theoretical frameworks are being used for cue integration, most notably probability and information theoretic approaches using Bayesian inference (Pearl, 1993, Bayes, 1783, Manyika and Durrant-Whyte, 1994), Dempster-Shafer evidential reasoning (Shafer, 1976,

[1] http://csclab.ucsd.edu/publications/siab04_facetracking.mpg

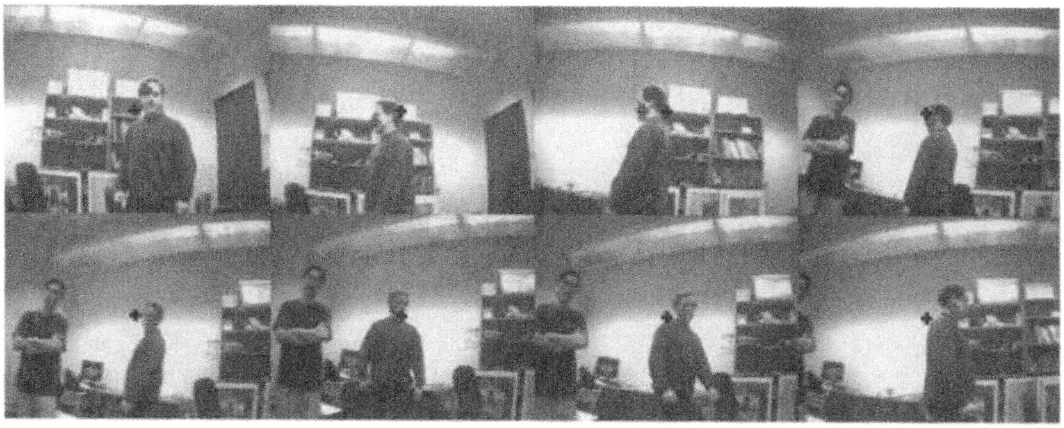

Figure 8: Sequence of images (top left to bottom right) of tracking a face. The red crosses indicate the estimated target position. The dark blue crosses mark the center of the image.

Hutchinson and Kak, 1992), and possibility theory (Zadeh, 1978, Dubois and Prade, 1992). In the areas of visual tracking, (Triesch and von der Malsburg, 2001) shows the robustness of the system with only weak visual cues. (Denzler et al., 2002) have combined DI-style cue integration with probabilistic fusion to work with multiple cameras for object tracking. While the Democratic Integration idea was originally introduced in the context of tracking, it is worth mentioning that it has also recently been successfully applied to the problem of model-free segmentation from video streams (Hayman and Eklundh, 2002). The potential benefits of such an approach for the purpose of autonomous learning of object representations is obvious. We will explore this issue in future work.

In order to exploit the complementary nature of the robot's redundant degrees of freedom (fast eye movements with only small kinematic range and slower neck movements with wide kinematic range) we used the idea of controller partitioning (Oh and Allen, 2001). This simple method works very effectively and is likely to be sufficient for our purposes. However, a number of improvements could easily be made if higher accuracy was required (e.g. for a highly foveated vision system). Most interesting to us seems the option of adding a feedback error learning component to the partitioned controller (Kawato, 1990). From the standpoint of biological plausibility, the partitioned control scheme is likely to be far simpler than the control mechanisms actually used by the primate visual system. At present, there seems to be little data about the coordination of neck and eye movements during tracking tasks. However, data from natural tasks where eye, head, and hand movements are recorded simultaneously suggests rather complicated forms of coordination between eyes, neck, and hands (Hayhoe et al., 1999).

Since tracking is only done with one "dominant" eye, we implemented a vergence control method to ensure coordination of both eyes. Our method is an extension to the one proposed in (Theimer and Mallot, 1994). It is reasonably accurate but currently needs several frames to make the eyes converge on the same point.

While none of the individual components of the system is entirely new, we consider their integration into our anthropomorphic robot head as a significant achievement. Future work will extend this system by adding the ability to autonomously learn and maintain representations for the objects in the robot's environment — bringing us one step closer to an embodied account of the development of the primate visual system.

Acknowledgments

This work was supported by the National Science Foundation under grants 0208451, 0233200, and 0329287, the UC Davis M.I.N.D. institute, and the National Alliance for Autism Research. We would like to thank the developers of the OpenCV libraries for making their software publicly available. We also thank Alan Robinson and Erik Murphy-Chutorian for comments on an earlier draft.

Appendix: Description of Individual Cues

This section describes the detailed specification of the cues. Figure 9 shows the saliency maps of the individual cues for a typical frame during tennis ball tracking. Most cues operate on grayscale images, only the color cue uses color images.

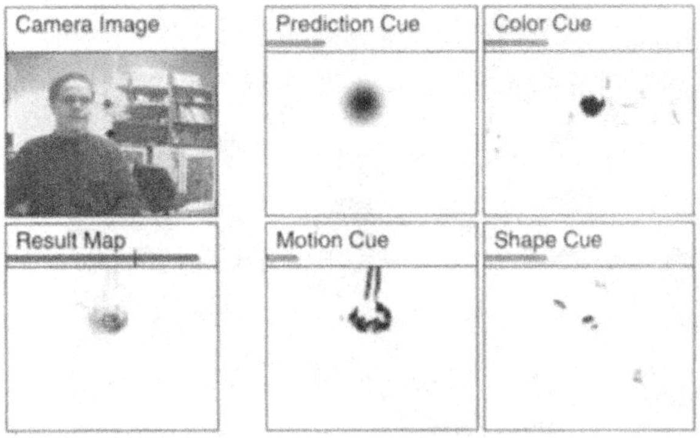

Figure 9: Saliency maps of the different cues during the tracking of swinging tennis ball. The length of the gray bars above the cue saliency maps represent the reliabilities of the cues. The bar above the result saliency map indicates the maximum saliency.

Motion Cue

The motion cue does not have any adapting object models and can be used for bootstrapping the other cues. It takes the current and the last input image and computes their difference image:

$$D(x,t) = |I(x,t) - I(x,t-1)|. \qquad (9)$$

All positions with differences over a threshold $T_{motion} = 15$ are set to one, all others to zero:

$$\tilde{A}(x,t) = 1 \text{ if } D(x,t) > T_{motion}, \text{ 0 otherwise.} \qquad (10)$$

This thresholded difference image is convolved with a gaussian filter to suppress camera noise and detect larger blobs of motion:

$$A(x,t) = \tilde{A}(x,t) * \mathcal{G}(x) \qquad (11)$$

The specific kernel size of the smoothing filter depends on the approximate size of the target. This motion cue is useful to detect moving targets as long as the cameras are stationary.

Color Cue

The color cue operates in YUV color space. It creates a two dimensional lookup table $L(u,v,t) \in [0,1]$ that contains a similarity value for every color in the discretized UV-color space.

This lookup table forms the object model of the color cue, along with a mean $\bar{Y}$ and deviation σ_Y of the brightness of the target object.

Every pixel x of the saliency map gets the similarity value that is assigned to the color found at the location in the source images, as long as its brightness is in the accepted range:

$$\tilde{A}(x,t) = \begin{cases} 0, \text{ if } \left|I(x,t) - \bar{Y}\right| > \sigma_Y \\ L\left(U(x,t), V(x,t), t\right) \text{ otherwise.} \end{cases} \qquad (12)$$

This saliency map is convolved with a gaussian filter to detect larger blobs of homogeneous color (11). The size depends on the expected size of the target object.

The first step of adapting the color object model is extracting the average color from a 3x3 image patch around the target position $\hat{x}(t)$.

This color value is converted to polar coordinates, resembling hue and saturation values $\bar{h}$ and $\bar{s}$. If the standard deviation of $\bar{h}$ in this image patch is smaller than a predefined threshold σ_h, the new object model is computed according to the following equation:

$$f(h,s,t) = \begin{cases} 1, \text{ if} \left|h - \bar{h}\right| < \sigma_h \wedge |s - \bar{s}| < \sigma_s \\ 0, \text{ otherwise} \end{cases} \qquad (13)$$

This range of similar colors has a rectangular shape in polar coordinates h and s, which is approximated with a pie slice polygon in the cartesian UV color space. This is plotted into the new lookup table $f(x,t)$, which is blended with the lookup table $p(x,t)$ according to (5).

Prediction Cue

The simple prediction cue tries to estimate the object location with a linear prediction based on the last two target positions:

$$\hat{X}(t) = \hat{x}(t-1) + (\hat{x}(t-1) - \hat{x}(t-2)). \qquad (14)$$

The saliency map is defined as a gaussian blob with $\sigma = 10$ at the estimated position:

$$A(x,t) = \exp\left(-\left(x - \hat{X}(t)\right)^2 / (2\sigma^2)\right). \qquad (15)$$

If the last target positions are unknown, or the predicted position is outside of the image boundaries, the whole saliency map is set to 0.1. This cue doesn't have a object model, the adaptation lies in the linear prediction.

Shape Cue

The shape cue computes the similarity of its shape model, a gray level template $P(x,t) = p_{shape}(x,t)$, to image patches of the same size around every pixel using a normed squared difference metric:

$$\tilde{A}(x,t) = \frac{\sum_{x'} (P(x',t) - I(x+x',t))^2}{\sum_{x'} (P(x',t)^2 \cdot I(x+x',t)^2)}. \qquad (16)$$

This metric provided by the OpenCV template matching function has the lowest values for the highest similarity. In order to get a saliency map with values between 0 and 1 we compute:

$$A_{min} = \min_x \tilde{A}(x,t), \text{ and } A_{max} = \max_x \tilde{A}(x,t)$$

$$A(x,t) = \max\left\{0,\ 1 - \frac{(\tilde{A}(x,t) - A_{min})}{K \cdot (A_{max} - A_{min})}\right\}, \quad (17)$$

with $K = 0.002$. A template adapted to a non-homogeneous surface leads to a very peaked response and is useful to precisely track a spot on the target object. For the adaptation of the prototype template the cue extracts an image patch centered at the target position in the input image. This image patch is blended to the current object model template as in (5).

References

Abidi, M. and Gonzalez, R., (Eds.) (1992). *Data Fusion in Robotics and Machine Intelligence.* Academic Press.

Alomoinos, Y. and Shulman, D., (Eds.) (1989). *Integration of Visual Modules.* Academic Press.

Bayes, T. (1783). An essay towards solving the problem in the doctrine of chances. *Philosophical Transactions of the Royal Society*, 53:370–418.

Becker, W. (1991). Saccades. *Vision & Visual Dysfunction Vol. 8:Eye movements*, pages 95–137.

Clark, J. J. and Yuille, A. L., (Eds.) (1990). *Data Fusion for Sensory Information Processing Systems.* Kluwer Academic Publishers.

Denzler, J., Zobel, M., and Triesch, J. (2002). Probabilistic integration of cues from multiple cameras. *Dynamic Perception, pp.309–314.*

Dubois, D. and Prade, H. (1992). *Data Fusion in Robotics and Machine Intelligence*, chapter Combination of fuzzy information in the framework of possibility theory, pages 481–505. Academic Press.

Hayhoe, M., Land, M., and Shrivastava, A. (1999). Coordination of eye and hand movements in a normal environment. *Invest. Opthalmol. & Vision Science*, 40.

Hayman, E. and Eklundh, J.-O. (2002). Probabilistic and voting approaches to cue integration for figure-ground segmentation. *ECCV (3)*, pages 469–486.

Hutchinson, S. and Kak, A. (1992). *Data Fusion in Robotics and Machine Intelligence*, chapter Multisensor strategies using dempster-shafer belief accumulation, pages 165–209. Academic Press.

Kawato, M. (1990). Feedback-error-learning neural network for supervised motor learning. *In Eckmiller R (Ed.) Advanced Neural Computers. Elsevier, North-Holland*, pages 365–372.

Kim, H., York, G., Burton, G., Murphy-Chutorian, E., and Triesch, J. (2004). Design of an anthropomorphic robot head for studying autonomous development and learning. *International Conference on Robotics and Automation (ICRA), accepted.*

Luo, R. C. and Kay, M. G. (1989). Multisensor integration and fusion in intelligent systems. *IEEE Trans. on Systems, Man, and Cybernetics*, 19(5):901–931.

Manyika, J. and Durrant-Whyte, H., (Eds.) (1994). *Data Fusion and Sensor Management: A Decentralized Information-Theoretic Approach.* Ellis Horwood, Chichester.

Oh, P. Y. and Allen, P. K. (2001). Visual servoing by partitioning degrees of freedom. *IEEE Trans. on Robotics and Automation*, 17(1).

Pearl, J., (Ed.) (1993). *Probabilistic Reasoning in Intelligent Systems.* Morgan Kaufmann Publishers, 2nd edition.

Shafer, G., (Ed.) (1976). *A Mathematical Theory of Evidence.* Princeton University Press.

Shibata, T. and Schaal, S. (2001a). Biomimetic gaze stabilization based on feedback-error learning with nonparametric regression networks. *Neural Networks*, 14(2):201–216.

Shibata, T. and Schaal, S. (2001b). Biomimetic smooth pursuit based on fast learning of the target dynamics. *IEEE Int. Conf. on Intelligent Robots and Systems.*

Theimer, W. M. and Mallot, H. A. (1994). Phase-based binocular vergence control and depth reconstruction using active vision. *CVGIP: Image Understanding, 60(3).*

Triesch, J., Ballard, D., and Jacobs, R. (2001). Fast dynamics of visual cue integration. *Perception*, 31:421–434.

Triesch, J. and von der Malsburg, C. (2001). Democratic integration:self-organized integration of adaptive cues. *Neural Computation*, 13(9):2049–2074.

Zadeh, L. (1978). Fuzzy sets as a basis for a theory of possibility. *Fuzzy Sets and Systems*, 1:3–28.

Morphology and Learning - A Case Study on Whiskers

Miriam Fend, Roland Abt*, Marco Diefenbacher*, Simon Bovet and Martin Krafft

Artificial Intelligence Laboratory
University of Zurich
Andreasstr. 15
8050 Zurich, Switzerland
fend@ifi.unizh.ch

* ETH Zurich
Department of Information Technology
and Electrical Engineering
8092 Zurich, Switzerland

Abstract

In many animal species, whiskers are an important sensory modality. Rats for example navigate, explore and recognize objects by actively whisking them, seals even use them to hunt in murky water. Whiskers also make an excellent model system for biologists due to their well defined anatomy. Recently, whiskers have also attracted an increasing interest by roboticists. We have built an artificial whisker system and mounted it on a robot. In this study, we investigate the role of whisker morphology on the behavior of a robot. For a robot to be able to navigate it has to react to obstacles appropriately yet in a flexible manner. For basic behaviors, associations between reflexes and stimuli can be learned using a model of classical conditioning, the distributed adaptive control (DAC) architecture. This model of associative learning is used to correlate whisker signals with collision signals from the robot's infrared sensors for whisker-based obstacle avoidance. The morphology of the whiskers is varied to test its influence on learning and behavior. We find that the performance of the robot is clearly influenced by its morphology. Furthermore, the weights learned under different morphologies vary according to the morphology. Finally, from the performance of the robot we can speculate about a suitable weight matrix for this task.

1. Introduction

Whiskers are a wide-spread touch sensor in biology. Although there are many animal species that have whiskers, the rat whisker system has been studied as a model system in biology for several reasons: first, rats can do amazing things with their whiskers. They can discriminate different surfaces by their textures (Carvell and Simons, 1990)(Guic-Robles et al., 1989), they can distinguish objects by their size and shape (Brecht et al., 1997) and they can of course use their whiskers for local navigation. Many other animal species also use whiskers for different purposes, for example cats judge the width of openings, seals use them for hunting in murky water (Dehnhardt et al., 1998) etc.. The whisker system is also anatomically a very well defined organ: the spatial arrangement of whiskers is highly conserved within each species. Each whisker lies on a precisely defined point in the grid of rows and arcs of the whisker pad, and the length of the whiskers always increases from the snout to the back of the animal's head (Brecht et al., 1997). Additionally, a region can be identified for each whisker along the processing pathway from brainstem via thalamus to the somatosensory cortex, which processes almost exclusively input from this one whisker only. Thus somatosensory stimulation can be well defined and the brain area affected by this stimulation is fairly easy to locate.

Figure 1: Picture of the AMouse with its whiskers and an omnidirectional camera.

While these reasons make the whisker system attractive to biologists, there are also good reasons for researchers in the field of artificial intelligence to be interested in the whisker system. Whiskers are a powerful sensory organ whose use for mobile robots has attracted an increasing number of scientists. There have been a number of publications on whiskers as binary touch (Zimmer, 1995) or as strain sensors (Will and Grossman, 1975) (Wang and Will, 1978) (Wijaya and Russell, 2002), most of them inspired by insect antennae. Often, two such passive antennae were mounted on a mobile robot, or a single, active antenna was moved across different shapes (Kaneko et al., 1998). Whiskers/antennae have proven suitable for fast obstacle avoidance (Jung and Zelinsky, 1996) (Cowan et al., 2003), but their more versatile use for texture recognition is just beginning to be recognized (Fend et al., 2003a) (Lungarella et al., 2002) (Hafner et al., 2003) (Russell and Wijaya, 2003).

In previous experiments (Fend et al., 2003b), we have investigated what the optimal morphology of whiskers on a robot could be. Since the most basic behavior that can be achieved with whiskers is locally avoiding obstacles, we started with this task for the robot. The robot was equipped with two whisker arrays of eight whiskers each with lengths of 4 to 6 cm. An arrangement not commonly found in nature performed best, where the long whiskers were in the middle of the robot, short whiskers on the side. While those experiments only focused on finding a morphology best suited for the task, the robot control was purely reactive involving only a minimum of computation. The question arose, whether the disadvantages of certain morphologies could be compensated for, if learning was included.

As a learning algorithm, we use distributed adaptive control (DAC), a form of associative learning inspired by classical conditioning. Since bumping into something usually hurts the animal, it is reasonable to assume that it has a set of reflexes associated with obstacles. When such a reflex and another sensory stimulation e.g. at the whiskers are consistently elicited simultaneously, associations can be formed and the whiskers can be used as an early alert system that something is getting very close to the animals face. A more detailed description of the algorithm and its motivation is given in the following section. Applying this learning algorithm to three different whisker morphologies we have found that avoiding collisions successfully is still highly dependent on the specific arrangement of the whiskers. We also find that the weight matrices learned on average depend strongly on the morphology. Finally we speculate how animals might deal with the disadvantage of the morphology, and how we as designers have to consider morphological issues.

1.1 Distributed Adaptive Control (DAC): Learning with Reflexes

A wide-spread and basic form of learning in natural systems is classical conditioning. It relies on a set of preprogrammed or unconditioned reflexes that can be triggered by an unconditioned stimulus, which is a certain sensory input. With Hebbian learning, other sensory inputs can form connections with the unconditioned stimulus and eventually trigger the unconditioned reflex without presence of the unconditioned stimulus. The DAC (DAC=Distributed Adaptive Control) architecture (Verschure et al., 1992) (Pfeifer and Verschure, 1992) is a control architecture for mobile robots inspired by a general model of conditioning (Verschure and Coolen, 1991). To test the influence of morphology in learning and behavior, we have set up a neural network architecture using DAC that closely models the learning just described. The network is schematically shown in figure 4. Here, input from the robot's infrared sensors signal collision with an object. In the case of collision, the preprogrammed reflex was to turn away from the encountered obstacle. The robot receives additional sensory input from the the whisker sensors. In the beginning of each experimental run, the neurons connected to the whiskers are connected to the neurons receiving signals from the IR sensors with zero weights. But with repeated exposure to obstacles, the weights of the whisker neurons j to the neurons i that elicit the avoidance reflex, grow stronger according to:

	R0	R1	R2	R3	R4	R5
motor 1	0.1	0.3	0.5	-1.0	-0.5	0.0
motor 2	0.0	-0.5	-1.0	0.5	0.3	0.1

Table 1: Preprogrammed weights connecting the neurons in the reflex layer (R0-R5) to the motor neurons.

$$\Delta w_{ij} = \eta \cdot a_j \cdot a_i \tag{1}$$

with the the learning rate $\eta = 0.05$ and the activities a_j and a_i of the whisker neurons and the reflex neurons respectively. The weights were kept between 0 and 1. To minimize random associations due to noise, the associated weights were diminished when a discrepancy according to the condition in equation 3 between whiskers and IR sensors was detected:

$$|a_j - a_{IR}| = 250 - \kappa \cdot 700 \cdot (w_{ij} - w_{iIR}) \tag{2}$$

$$\Delta w_{ij} = \kappa \cdot a_j \cdot a_{IR} \tag{3}$$

with the forgetting rate $\kappa = 0.1$, the activity a_{IR} and the weights to the reflex neurons w_{iIR} of the IR sensors. The activities of the whisker signals were normalized to values between 0 and 1000, the values in equation 3 have been determined heuristically.

2. The Artificial Whisker System

2.1 Hardware

All experiments described in this paper were done with the Artificial Mouse (AMouse) robot (figure 2). Designed in close cooperation with biologists and neuroscientists, this robot is based on a Khepera platform (Mondada et al., 1993) with two artificial whisker arrays. The artificial whisker sensor as described by (Lungarella et al., 2002) consists of a natural hair glued onto the membrane of a capacitor microphone (figure 3). The signals from the whiskers are dynamic and

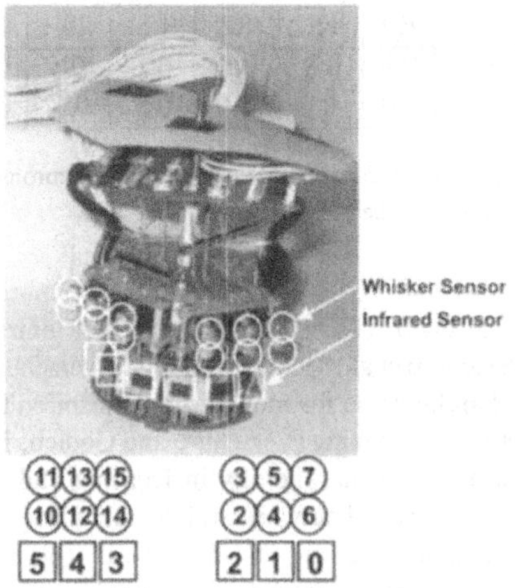

Figure 2: The robot as used in the experiments. On the bottom, the numbering of the individual sensors is given as they are read from the data acquisition hardware.

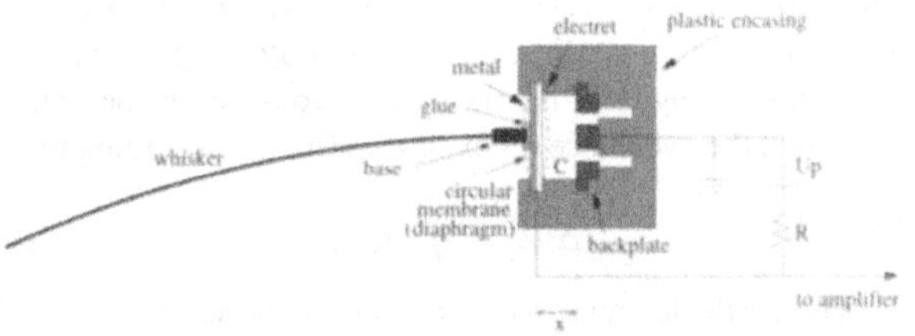

Figure 3: Schematic of the whisker sensor, taken from (Lungarella et al., 2002)

of a temporal nature, thus the sensor is capable of transducing high frequency information. To match the sampling of the infrared (IR) sensors, only a low sampling rate of 50 Hz per channel was used. The Khepera is equipped with eight infrared sensors, of which the six in the front were used as collision sensors for learning of obstacle avoidance. While whiskers are tactile sensors, IR sensors measure light, so they can be looked upon as simple visual input.

2.2 *Experimental Setup*

All experiments were conducted in a white arena of 96x73 cm. It contained several round or square obstacles of plastic and wood respectively. For some of the experiments, the training arena (figure 5(a)) was simplified in one corner to facilitate the formation of proper associations during learning. For each run, the robot started at the same position and with the same orientation. It explored the area for 2 min and during this learning period, the obstacle avoidance was based on the IR sensors and the forming connections with the whisker sensors. During the following testing period of maximal 3 min, the AMouse navigated only using its whiskers. If there was no stimulation of the whiskers, the robot moved forward by default.

The behavior of the robot was evaluated based on recordings with an overhead camera with two frames per second. After the experiment, the trajectories were generated by feature tracking based on the KLT library (Shi and Tomasi, 1994). The resulting trajectories of the exploration were pooled for each experimental condition, and the resulting cumulated trajectories for all runs of the same morphology are displayed in figure 7. For each morphology, a minimum of 11 runs was conducted.

3. Results

To evaluate the influence of the whisker morphology on behavior and learning, three different robot morphologies were tested (see figure 6). The first morphology used whiskers of an equal length of 7 cm. The second morphology had whiskers with a length of 5, 6 and 7 cm with the longest whiskers in the front and the shortest whiskers on the side of the robot. In a previous experiment with fixed controller (Fend et al., 2003b), the latter was the most successful morphology. The third morphology is the whisker arrangement also found in nature, namely long whiskers on the side and short whiskers in the front of the robot.

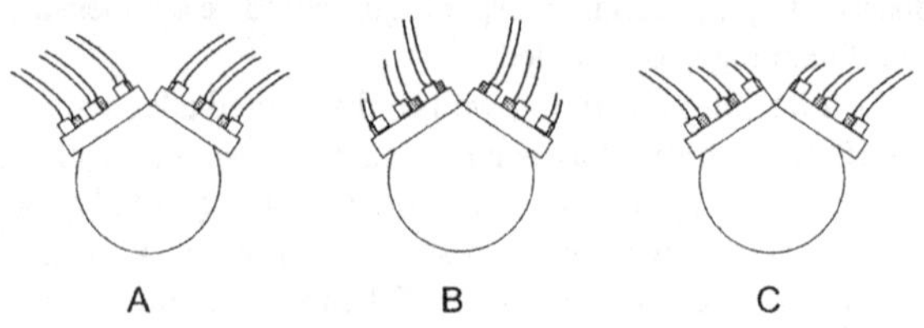

Figure 6: Robot morphologies. In (A), all whiskers have the same length, (B) long whiskers are in the front and (C) long whiskers on the side of the robot.

3.1 *Behavioral Results*

The performance of the robot at avoiding obstacles was evaluated during a 3 min exploration of the arena. The trajectories from all runs (figure 7) serve as a qualitative measure on how well the robot was able to explore the arena. Figure 7(d) shows the cumulated trajectories of 10 runs of obstacle avoidance solely based on the infrared sensors. This can be interpreted as the best possible behavior, since it provides the substrate for learning. The whisker-based behavior displayed in the plots 7(a), 7(b) and 7(c) shows clear differences depending on the morphologies. Only morphology A managed to drive around the round obstacle in the lower left and the upper right corner of the arena several times. In this area, morphology B performed worst.

To quantify the exploration of the arena, the area was divided into 10x10 pixel bins. For each bin, the probability p

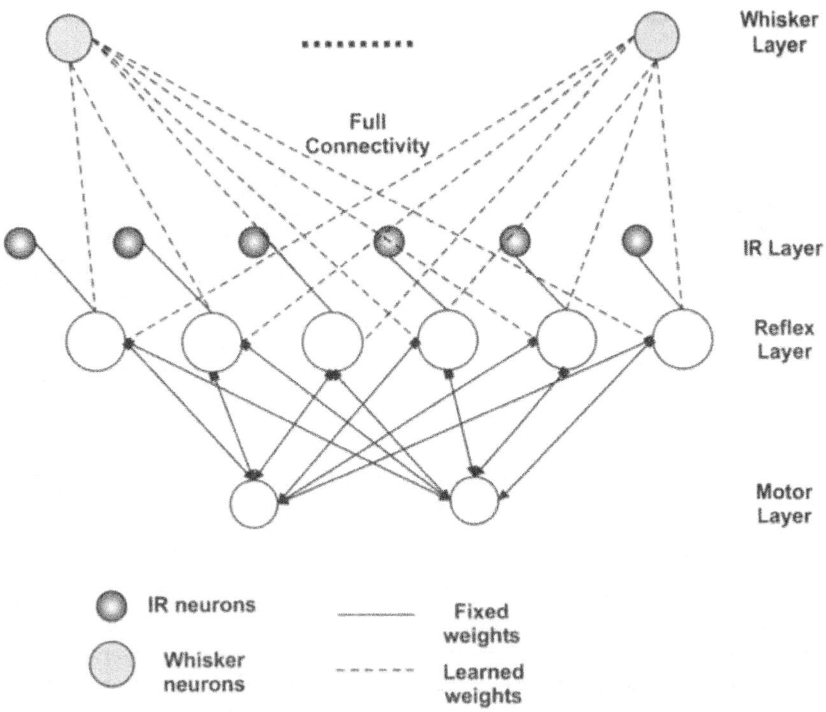

Figure 4: Schematic of the DAC architecture used in our experiments.

of the robot entering this bin was calculated from all trajectories (forming the set E). The entropy H over the set E was computed as:

$$H = -\sum_{a \in E} p(a) log_2 p(a) \text{ with } \sum_{a \in E} p(a) = 1 \quad (4)$$

A higher entropy corresponds to a more even exploration of the experimental area. For a uniform distribution, the entropy is maximal with H=11.59. Morphology A performs at H=9.67, B at H=8.87 and C attains H=9.14. Again, morphology A is closer to a uniform coverage of the area.

A different quantitative measure for the performance of the different morphologies is the mean time that the robot moved without colliding into an obstacle. As shown in figure 8, the morphology A with equally long whiskers performed best when evaluating the duration before collision.

3.2 *Analyzing Learning*

Another interesting perspective on the experiment is the analysis of the learned weights. Figure 9 shows the different mean weight matrices for the whisker neurons to the reflex layer neurons. Figure 9(d) displays a weight matrix we postulated to emerge, if the stimulation was evenly distributed over all sensors. It assumes that whisker neurons can form associations only with the reflex neurons that connect to IR sensors from the same side and which are thus spatially close. Due to the spatial arrangement on different sides of the robot and due to the limited range of the IR sensors, neighbouring sensors are the only ones that should be active simultaneously when encountering an obstacle. Comparing the weights in figure 7(c) as well as the other morphologies with these postulated weights, it becomes obvious that the natural morphology C develops weights that are most similar to the expected weight matrix.

To verify our hypothesis that these weights really are the ones learned for good performance, we analyzed the weight matrices of all morphologies with respect to the performance during testing (figure 10). Here, weight matrices are no longer averaged over a specific morphology, but pooled with weight matrices of similar performance. It can be seen that the weights that steered the robot longest without collision (figure 10(d)) are different from the expected weights in figure 9(d). While the postulated weights from the whisker neurons only connect to reflex neurons on the same side of the robot, the best-performing weights have connections to the contralateral reflex neurons as well. The weight pattern most similar to the postulated weights (figure 10(a)) actually performed worst.

4. Discussion

During the last years the importance of the physical instantiation of a behaving system has become widely recognized (Pfeifer and Scheier, 1999). In the context of embodied arti-

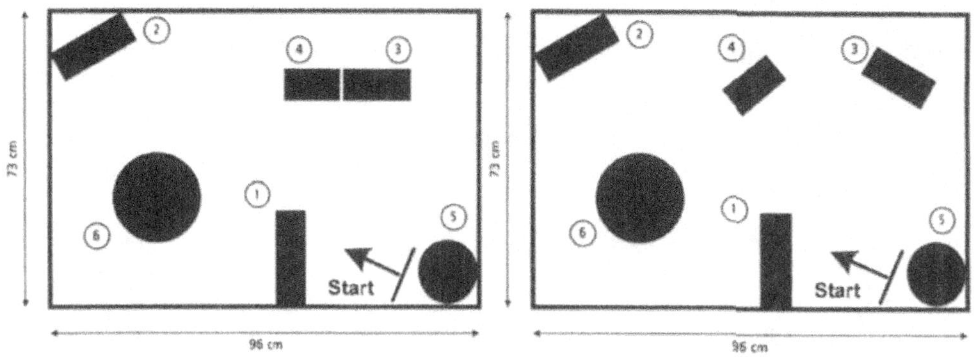

(a) (b)

Figure 5: Schematic of the experimental arena (a) Setup for training the robot (b) Test arena with one more tight passage in the upper right corner.

ficial intelligence, the morphology of an agent is also considered relevant for intelligence and the adaptivity of the agent's behavior. This study investigated how learning and behavior depend on the morphology of an agent in the specific case of an agent learning how to navigate with whiskers. As in our previous experiments, the performance of the agents varied with different morphologies and certain morphologies performed better than others. The morphology commonly found in nature performed worst, when measuring how long the robot explored the area without collision (figure 8). This morphology has long whiskers on the side and short whiskers in the front of the robot which poses the problem that in the main direction of movement, obstacles can only be recognized rather late. Possibly, there is not enough time left to turn completely when the obstacle is in an unfavorable position. The most successful morphology A employs equally long whiskers.

Another perspective on the performance of the robot takes into account, whether the robot managed to cover the whole experimental area. Again, morphology A showed the best performance regarding how evenly the arena was explored. Having long whiskers enables the robot to detect objects well in advance, but it could also cause a problem: If the robot has a sensory space that is far greater than the physical space it actually needs, having a longer sensing range might result in the misinterpretation of certain tight passages as being too small. This happened in the previous study for some morphologies. There the robot's sensory dimensions in certain morphologies exceeded the physical dimensions too much, thereby preventing the robot from entering narrow spots even though it would have been able to pass through.

In the experiments described in this paper, the robot learns to elicit a set of reflexes when certain whiskers are activated. The strength of each connection as well as the pattern of connections depends on what the robot learns. Although there is no reward for fitting through tight passages during testing or learning, simply associating neurons that are active at the same time balances the longer whiskers with the lower activity of the respective IR sensors. Thus, the robot morphology most successful at moving without collisions is also most successful at passing tight spots. It proves advantageous to have long whiskers everywhere as long as the robot can learn at what level of activation it has to turn away from an obstacle.

Our study has also another aspect, which analyzes the weights learned during training. We find clear differences in the associations formed between reflex neurons and whisker neurons depending on the whisker morphologies. In advance, we hypothesized that the most reasonable associations should correspond to the matrix displayed in figure 9(d). Here, the association between whisker neurons and spatially corresponding reflex neurons are highest, associations to neighboring reflex neurons are weaker, but there are no associations to contralateral reflex neurons. It turned out that the weights learned in the morphology performing best did not match this expectation. Therefore, we analyzed the weights of all runs sorted by the performance of that particular controller. Surprisingly, the weight matrix performing best had strong connections to the contralateral reflex neurons. To understand the effect of whisker stimulation on the motors in a specific case, we calculated the motor values for the case of an activity of 100 for whisker 11. For the badly performing matrix in figure 10(a) the resulting motor values are 11 and -7 for the right and left motor respectively. The same activity pattern yields 11 for the right and -20 for the left motor when processed by the matrix with the best behavioral performance. Stronger connections to the contralateral side thus elicit a stronger backwards motion on the contralateral motor, causing a smaller turn radius. This enables the robot to react faster and to successfully manoeuver in tight space. While this is an obvious advantage, it relies heavily on the

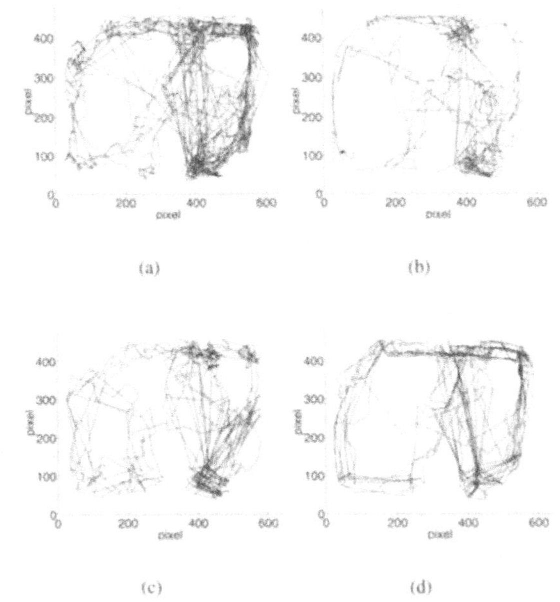

Figure 7: Cumulated trajectories for the three morphologies A(a), B(b) and C(c). (d) shows the trajectory for obstacle avoidance with IR sensors only.

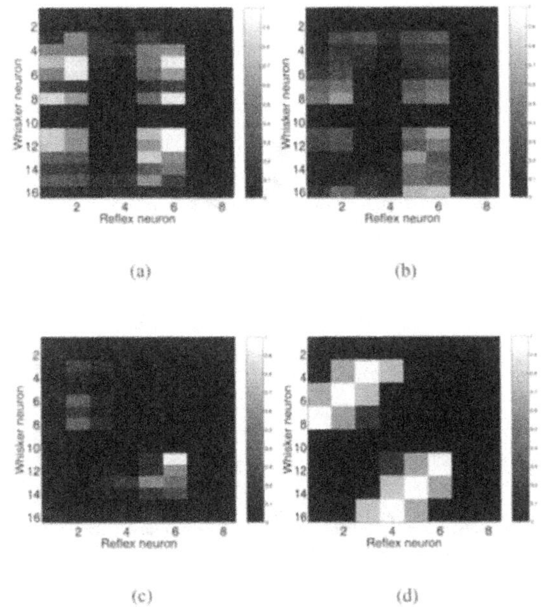

Figure 9: Learned weight matrices for the three different morphologies (A), (B) and (C) corresponding to figure 6. (D) Expected weight matrix

inbuilt connections of the reflex neurons to the motors. These preprogrammed reflexes already contained contralateral connections that were chosen to yield good performance when manoeuvering based on IR sensors.

The experiment placed a constraint on the time available for learning. This was done for two reasons: first we wanted to evaluate which morphology was able to learn to perform well in short time. Second, constraining the learning period also made the experiments more practicable. However, the training was too short to have perfect learning and we did some runs with significantly longer learning time that showed an increased performance. Still, the lesson from this restricted learning is that the morphology plays an important role for the acquisition of and the performance on obstacle avoidance.

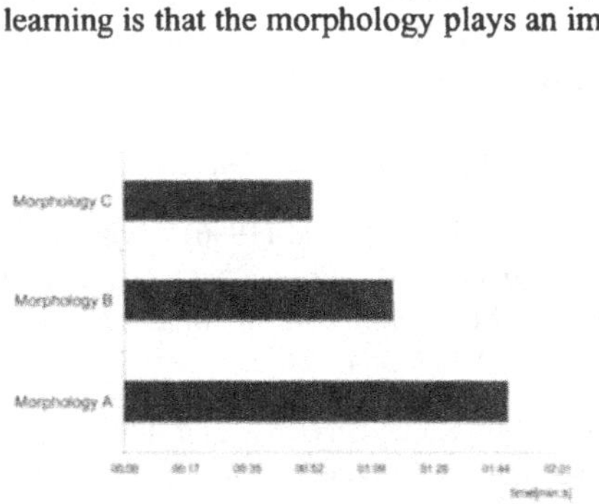

Figure 8: Mean time to collision for all three morphologies. Morphology A with all whiskers of equal length performed best.

But why is the natural whisker arrangement not the best for the obstacle avoidance, even though whiskers are used for this purpose by so many animal species? There are several aspects to be considered. Firstly, having long whiskers sticking out in front of the snout might be troublesome during eating. Secondly, the whisker pad of rodents for example is much denser than our robot model. Whiskers more to the back are still quite close to the snout, whereas in our robot, the whiskers are much further apart making their position more important. Most probably though, two other aspects are more prominent: as already described, whiskers have more functions than distance sensing, they are also a powerful touch organ. For tactile exploration, the natural morphology might be advantageous, because all the whisker tips lie in a plane. Objects can thus be whisked simultaneously by a maximal number of whiskers yielding rich information. Lastly, animals can move their whiskers actively and thus change their orientation dramatically. If necessary, a rat can point its whiskers actively to the front. We have not yet incorporated a flexibility in the morphology that is dependent on the behavioral state of the agent.

To conclude, the morphology of sensor arrangement is of paramount importance for the design of successful artificial agents. In this case study, it was not possible to compensate a slightly disadvantageous morphology with learning. Instead, the learning of the association of basic reflexes was strongly

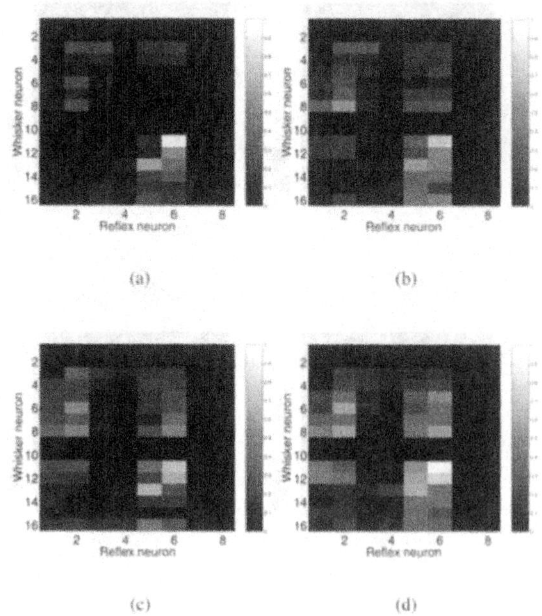

Figure 10: Weight matrices across all morphologies that were equally successful at avoiding collisions. (A) collision before 30 s (B) collision between 30 s and 1 min (C) collision between 1 and 2 min (D) collision after 2 min.

dependent on an appropriate morphology.

5. Conclusion and Future Work

In the experiments described, DAC proved to be a valid algorithm for learning whisker-based obstacle avoidance. In previous experiments, an unnatural whisker morphology turned out to be advantageous for obstacle avoidance. This time we tested whether this disadvantage of the natural arrangement would persist when the robot could learn. We found that using a restricted learning period, having equally long whiskers was advantageous. Furthermore and against our intuition, the most successful weight matrix for the task involved strong contralateral connections between whiskers and reflex neurons. In future work, we will investigate other functions of whiskers such as texture recognition and object exploration. For more advanced behaviors, vision will be integrated and the interaction between different sensory modalities will hopefully yield interesting behaviors not possible with a single sensory modality.

Acknowledgements

This research has been supported by the IST-2000-28127 European project (AMOUSE). The natural rat whiskers were kindly provided by Matthew Diamond at SISSA, Cognitive Neuroscience sector, Trieste, Italy. The preamplifier board was built by Dirk Naumann, ATM Computing. Regina Mudra from the Institute of Neuroinformatics, ETH Zurich gave valuable help and advice on tracking.

References

Brecht, M., Preilowski, B., and Merzenich, M. (1997). Functional architecture of the mystacial vibrissae. *Behavioral Brain Research*, 84(1-2):81–97.

Carvell, G. and Simons, D. (1990). Biometric analyses of vibrissal tactile discrimination in the rat. *Journal of Neuroscience*, 10(8):2638–2648.

Cowan, N. J., Ma, E. J., Cutkosky, M., and Full, R. J. (2003). A biologically inspired passive antenna for steering control of a running robot. In *International Symposium on Robotics Research*, Siena, Italy.

Dehnhardt, G., Mauck, B., and Bleckmann, H. (1998). Seal whiskers detect water movements. *Nature*, 394:235–236.

Fend, M., Bovet, S., Yokoi, H., and Pfeifer, R. (2003a). An active artificial whisker array for texture discrimination. In *IEEE/RSJ International Conference on Intelligent Robots and Systems (IROS)*.

Fend, M., Yokoi, H., and Pfeifer, R. (2003b). Optimal morphology of a biologically-inspired whisker array on an obstacle-avoiding robot. In Banzhaf, W., Christaller, T., Dittrich, P., Kim, J. T., and Ziegler, J., (Eds.), *Advances in Artificial Life - Proceedings of the 7th European Conference on Artificial Life (ECAL)*, volume 2801 of *Lecture Notes in Artificial Intelligence*. Springer Verlag Berlin, Heidelberg.

Guic-Robles, E., Valdivieso, C., and Guajardo, G. (1989). Rats can learn a roughness discrimination using only their vibrissal system. *Behavioural Brain Research*, 31(3):285–289.

Hafner, V., Fend, M., Lungarella, M., Pfeifer, R., Knig, P., and Krding, K. (2003). Optimal coding for naturally occurring whisker deflections. *Proceedings of the 10th International Conference on Neural Information Processing (ICONIP), Istanbul, June 2003.*

Jung, D. and Zelinsky, A. (1996). Whisker-based mobile robot navigation. *Proceedings of the IEEE/RSJ International Conference on Intelligent Robots and Systems (IROS)*, 2:497–504.

Kaneko, M., Kanayama, N., and Tsuji, T. (1998). Active antenna for contact sensing. *IEEE Transactions on robotics and automation*, 14(2):278–291.

Lungarella, M., Hafner, V., Pfeifer, R., and Yokoi, H. (2002). An artificial whisker sensor for robotics. *Proceedings of the IEEE/RSJ International Conference on*

Intelligent Robots and Systems (IROS), pages 2931–2936.

Mondada, F., Franzi, E., and Ienne, P. (1993). Mobile robot miniaturisation: A tool for investigation in control algorithms. *Proceedings of the Third International Symposium on Experimental Robotics*, pages 501–513.

Pfeifer, R. and Scheier, C. (1999). *Understanding Intelligence*. MIT Press, Cambridge, Mass., USA.

Pfeifer, R. and Verschure, P. (1992). Distributed adaptive control: A paradigm for designing autonomous agents. In Varela, F. and Bourgine, P., (Eds.), *Towards a practice of autonomous systems: Proceedings of the First European Conference on Artificial Life*, pages 21–30. Cambridge, MA: MIT Press.

Russell, R. A. and Wijaya, J. A. (2003). Object location and recognition using whisker sensors. *Australasian Conference on Robotics and Automation.*

Shi, J. and Tomasi, C. (1994). Good features to track. *IEEE Conference on Computer Vision and Pattern Recognition (CVPR'94)*, pages 593–600.

Verschure, P. and Coolen, A. (1991). Adaptive fields: Distributed representations of classically conditioned associations. *Network*, 2:189–206.

Verschure, P., Kröse, B., and Pfeifer, R. (1992). Distributed adaptive control: The self-organization of structured behavior. *Robotics and Auronomous Systems*, 9:181–196.

Wang, S. and Will, P. (1978). Sensors for computer controlled mechanical assembly. *The industrial robot*, March:9–18.

Wijaya, J. A. and Russell, R. A. (2002). Object exploration using whisker sensors. *Australasian Conference on Robotics and Automation.*

Will, P. and Grossman, D. (1975). An experimental system for computer controlled mechanical assembly. *IEEE Transactions on computers*, c-24(9):879–888.

Zimmer, U. (1995). Self–localization in dynamic environments. *IEEE/SOFT International Workshop BIES'95, Tokyo, Japan, May 30-31 1995.*

Simulating Whisker Sensors – on the Role of Material Properties for Morphology, Behavior and Evolution

Simon Bovet, Miriam Fend, and Rolf Pfeifer
Artificial Intelligence Laboratory
University of Zurich
Andreasstr. 15, 8050 Zurich, Switzerland
{bovet,fend,pfeifer}@ifi.unizh.ch

Abstract

Whiskers have recently received an increasing interest as touch sensors capable of providing rich information about the close environment of an agent. Many different animal species – especially rodents – heavily rely on them for tasks as varied as short-range navigation and object exploration. For roboticists, whiskers are not only interesting because of the unique physical process on which they are based (which is e.g. independent of illumination – a non trivial problem that vision-based modalities have to deal with), but also because of the potential overlap with other sensory modalities. Although there is a considerable body of literature about whiskers, some questions remain unanswered. For example: is there any appropriate morphological arrangement of the whiskers for a given task, since this arrangement is highly preserved within many animal species? What are the advantages of using flexible touch sensors rather than rigid ones? In this paper, we present a software that we have developed for the simulation and investigation of agents using whisker sensors. As an illustration of possible applications with this tool, we then present some results about the implications of the material properties of the whiskers on morphology, behavior and evolution.

1. Introduction

Whiskers are much more than simple touch sensors, and it is not surprising that they are widespread among many different animal species. Some insects have two advanced antennae that can compensate for the limitation of the sensing capability by their vision system (Kaneko et al., 1998). Seals can hunt in dim water by detecting with their whiskers the turbulence generated by the moving prey (Dehnhardt et al., 1998). Different mammals use their whiskers to navigate in the absence of visual cues without colliding with obstacles (Vincent, 1912, Beadle, 1977). Cats also use their whiskers to detect the contour of the prey's body to determine the head location (Leyhausen, 1979). Rats use active movements of their vibrissae (whisking) to extract information about the spatial properties of objects, including distance, size, shape and surface structure (Guic-Robles et al., 1989, Carvell and Simons, 1990, Brecht et al., 1997). The texture discrimination capabilities of rats can be compared to humans using their fingertips.

From an engineering point of view, whiskers offer several advantages. They are independent of the lighting conditions, they do not rely on the surface properties of the objects to be sensed and they do not involve strong contact with the environment. Moreover, whiskers do not require a complex electronic interface and can be considered as cheap and robust sensors (at least from a technical point of view).

Recognizing the importance of whiskers in the animal world, researchers have developed different robotics counterparts. Simple whisker sensors can be made from a flexible beam whose deflection at the base can be measured by a piezoelectric element or a simple switch (Will and Grossman, 1975, Wang and Will, 1978, Russell, 1984). Different kinds of long antenna-like whisker sensors have been mounted on mobile robots (Brooks, 1989, McKerrow, 1991, Zlajpah, 1999, Cowan et al., 2003) to provide warning of obstacles, in particular for high-speed wall-following (Jung and Zelinsky, 1996), or under the feet of a legged robot (Hirose et al., 1985, Schiebel et al., 1986) to detect ground proximity and obstacles. By measuring the torque at the base of the beam, it is possible to estimate the contact position along the whisker (Tsujimura and Yabuta, 1992, Kaneko et al., 1995). Active sensing (Russell, 1992, Kaneko et al., 1998, Ueno et al., 1998) and the use of whisker arrays (Russell and Wijaya, 2003) has also been reported, in particular to discriminate different surface textures (Hafner et al., 2003, Fend et al., 2003a).

To better study the interplay between sensor properties, agent controller and behavior, it is desirable to be able to vary the various parameters at will or even use artificial evolution to help us designing interesting systems. Since this is almost impossible in biology and quite demanding with robots, it is clear that simulation constitutes an interesting and useful solution. In this paper, we present a first attempt to simulate flexible whiskers that can be used by mobile agents in arbitrary environments. Due to the restricted amount of space of this paper, we only show a few demonstrations validating our model, and prefer to focus on some illustrations of how this simulation tool can be used.

1.1 Morphological Considerations

In many animal species – especially in rodents – the spatial arrangement of the whiskers on the animal's head is highly preserved within each species, and the length of the whiskers systematically increases from the anterior (rostral) whiskers to the posterior (caudal) whiskers. In addition, the neuroanatomy of rats shows that the prominent part of the somatosensory cortex devoted to the processing of whisker signals, the *barrel* cortex, has a beautifully preserved topology of the precisely arranged whiskers (Woolsey and der Loos, 1970).

These observations raise the question of how important the morphology of the whisker sensory system is. As a trivial example, cats can "measure" the width of a gap by help of a set of whiskers with different length arranged adequately, thus reducing the processing of the signals to a simple binary touch detection. Furthermore, animals use whiskers extensively as distance and collision sensors. Fast and easy evaluation of distances to objects is crucial when moving at high speed, e.g. when fleeing from predators or when hunting. Such evaluation of sensory information can be greatly facilitated by an appropriate morphology of the sensor distribution. In the simulation experiment presented below, we show that for a given task, a particular arrangement is found to be more suitable than others, in agreement with results found in a similar experiment conducted with a real robot.

1.2 Material Properties of Whiskers

For simple tasks – e.g. if whiskers are used as simple binary touch detectors – we could expect that flexible whiskers do not provide a significant advantage over rigid whiskers. However, we show that a change in the material properties of the whiskers leads not only to morphological modifications, but also to a substantial speed-up of the artificial evolution process, as well as a change in the obtained behavior.

2. The Simulation Software

Whereas some approaches focus on the detailed physical properties of real whiskers (Dechant et al., 2001, Neimark et al., 2003, Hartmann et al., 2003), the goal of the work presented here is to provide a more global framework, where the behavior of an agent situated in a relatively complex virtual environment can be investigated. This first step in simulating agents using flexible whisker sensors aims at assessing mainly macroscopic features (such as morphological arrangement of the sensors, or bending properties of the whiskers).

2.1 Simulating Flexible Whiskers

Simulation of flexible (or *deformable*) objects in virtual reality systems has been a challenging problem for years. A plethora of approaches can be found in the literature, and a description of how each model deals with the different trade-offs (accuracy vs. computation time; efficiency vs. stability; etc.) would be over the scope of this paper. For a survey, see (Gibson and Mirtich, 1997).

In order to fulfill our needs of having a model stable, fast, and simple enough to simulate agents with several flexible whiskers in a relative complex environment, we decided to start with the Open Dynamics Engine (Smith, 2003). This open-source library includes a fast and robust integrator and a set of collision detection algorithms which simulate the physics of 3D solid primitives connected by different types of joints. This provides the bases for simulating a robot moving in the virtual environment.

To simulate the flexible whiskers, we took a finite element method (FEM) approach. We extended the existing library by implementing the physics of a special type of joint, an *universal damped spring*: each whiskers is then divided into several elements (cylinders) connected pairwise by universal damped springs. A detailed description of the joint is given in the appendix.

2.1.1 Model Validation Examples

Figure 1 shows a fairly good quantitative agreement between simulation and theory in the case of a beam attached at one end and deflected by a force applied at the other end. The theoretical data are obtained by integrating the Euler-Bernoulli equation

$$\frac{d^2}{dx^2}\left(EI\frac{d^2w}{dx^2}\right) = p \qquad (1)$$

with the corresponding boundary conditions (w is the out-of-plane displacement of the beam, E its Young's modulus, I the moment of inertia of its cross-section and p the distributed loading).

Figure 2 shows qualitative data obtained for the case of a spider hair deformed by a flat surface. The ob-

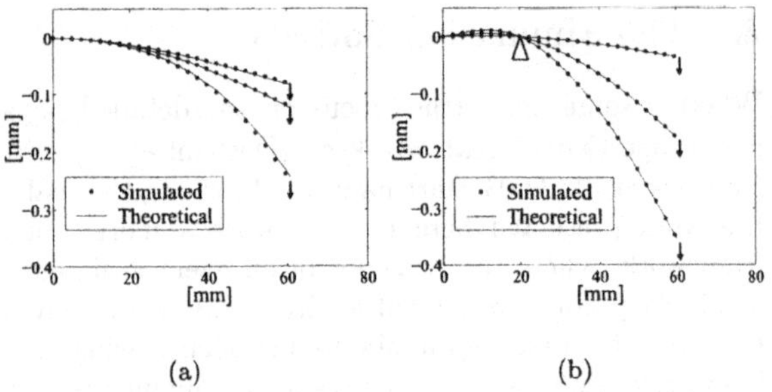

Figure 1: Comparison between simulated data and analytical solutions of deflected whiskers. (a) Perpendicular force (arrow) applied at whisker tip for different cross-section profiles. (b) The same with an extra support (triangle).

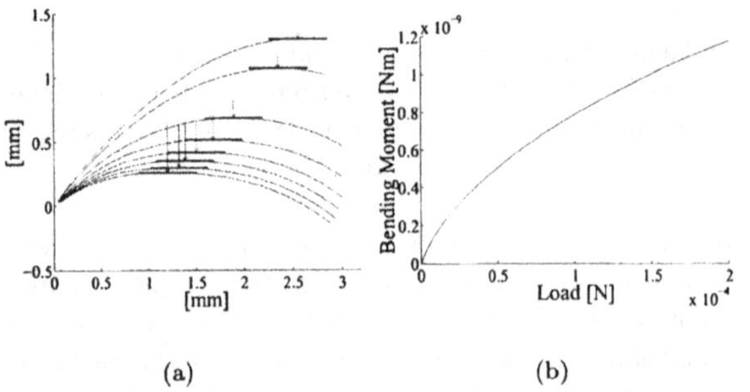

Figure 2: Deformation of a spider hair deflect from above by a flat surface. (a) With growing loads (arrow), the lever arm shortens. (b) Relation between the stimulating load an the resulting moment at the socket of the whisker.

tained results are comparable with measurements presented in (Dechant et al., 2001), where the values of the physical parameters of the whisker were taken from.

Figure 3 shows that the distance s between the fixed end of the whisker to the contact point can be determined from the angle θ and torque τ measured at the base of the whisker (Kaneko et al., 1998). In the case of a flexible beam with uniform cross-section, s is related to the compliance $C_\theta = \theta/\tau$ by

$$s = 3EI\frac{\theta}{\tau} \tag{2}$$

These few examples show that our model simulates satisfyingly the deformation of a whisker under different conditions.

2.2 Artificial Evolution

As an illustration of possible applications of the simulation presented so far, we show how artificial evolution reveals important aspects of material properties – even for

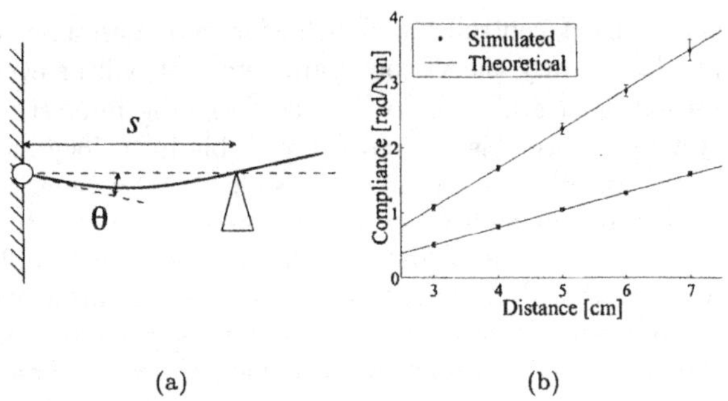

Figure 3: (a) Bending of a flexible beam with contact point. (b) Simulated and theoretical relation between contact distance s and compliance C_θ at the attachment point of a uniform whisker, for two values of Young's modulus.

Figure 4: Virtual environment. The agent is shown with its left whiskers in contact with an object it has to avoid. The floating sphere is the target the agent is seeking.

the simple case where sensors are used as binary touch sensors.

2.2.1 Agent and Environment

The agent consists of a wheeled robot (comparable to the Khepera platform) with eight whisker sensors distributed symmetrically on both side, as well as a sensor detecting the angle α_{target} to an arbitrary target. The position angle α_i and length l_i of whisker i is given by

$$\left.\begin{array}{rcl} \alpha_i &=& \frac{4-i}{3}\cdot\alpha_{\text{first}} + \frac{i-1}{3}\cdot\alpha_{\text{last}} \\ \alpha_{i+4} &=& -\alpha_i \\ l_i &=& \frac{4-i}{3}\cdot l_{\text{first}} + \frac{i-1}{3}\cdot l_{\text{last}} \\ l_{i+4} &=& l_i \end{array}\right\} i = 1\ldots 4 \tag{3}$$

The environment consists of a toroidal world (to avoid border effects) filled with cylindrical obstacles of different sizes placed randomly that the robot can detect with its whiskers (see figure 4). The task of the agent is to approach the target as quickly as possible and to avoid any contact with the obstacles. The target is initially placed at a random position, and each time the agent is closer to the target than a given threshold distance, the target is moved to another random position.

The agent is controlled by a subsumption-like architecture (Pfeifer and Scheier, 1999): when no object is detected by the whiskers, it behaves like a Braitenberg vehicle seeking for target. If an object is detected on one side, i.e. if one (or more) whisker(s) is deflected at its root more than a given threshold angle, the agent turns away for a given time period. If both sides detect some objects, the agent moves back for a given time period. The motor speeds m_{left} and m_{right} are set according to the following equations.

No whisker stimulation:

$$\begin{aligned} m_{\text{left}} &= v_0 + \gamma \tanh(\beta \cdot \alpha_{\text{light}}) \\ m_{\text{right}} &= v_0 - \gamma \tanh(\beta \cdot \alpha_{\text{light}}) \end{aligned} \quad (4)$$

Stimulation on the left/right:

$$\begin{aligned} m_{\text{left}} &= \pm v_1 + v_2 \\ m_{\text{right}} &= \mp v_1 + v_2 \end{aligned} \quad (5)$$

Stimulation on both sides:

$$\begin{aligned} m_{\text{left}} &= v_3 \\ m_{\text{right}} &= v_4 \end{aligned} \quad (6)$$

2.2.2 The Fitness Function

Each agent is evaluated for a fixed period of time. The fitness F of an agent is given by

$$F = a \cdot N_{\text{target}} - b \cdot N_{\text{collision}} + c \cdot N_{\text{move}} \quad (7)$$

where N_{target} and $N_{\text{collision}}$ are respectively the number of targets found and the number of collisions which occurred during the evaluation period. N_{move} counts the number of different positions (with respect to a discrete grid dividing the whole space) the agent moved to. This last term is added in order to avoid solutions where the agent would simply stay in place. a, b and c are positive constants with $a \approx b \gg c$.

2.2.3 Evolutionary Algorithm

The genotype of an agent consists of a set of 13 real values $\{w_1, \ldots, w_{13}\}$ with the phenotype mapping shown in table 1.

The first four genomes determine the morphology of the agent, the other genomes its behavior.

The evolution strategy used in this experiment is denoted as $(\mu + \lambda)$-ES. For an overview, the reader is referred to (Bäck et al., 1997). Each generation maintains

w_1	First Whisker Position (α_{first})
w_2	Last Whisker Position (α_{last})
w_3	First Whisker Length (l_{first})
w_4	Last Whisker Length (l_{last})
w_5	Forward Speed (v_0)
w_6	Turning Speed (v_1)
w_7	Turning Back Speed (v_2)
w_8	Left Backward Speed (v_3)
w_9	Right Backward Speed (v_4)
w_{10}	Turning Speed to Target (γ)
w_{11}	Target Sensitivity (β)
w_{12}	Turning State Delay
w_{13}	Backward State Delay

Table 1: Mapping between the genome values and their corresponding phenotypic features, see equations (3) to (6).

a population of $\mu = 10$ parents. For each new generation, the evolution strategy generates $\lambda = 20$ offspring by copying randomly selected parents and applying variation operators, such as mutation and cross-over. Note that mutation was not only applied to the offspring, but to all agents of the new population. This ensures that the best agents of a population have a finite lifetime (thus preventing from convergences to local solutions). For each run of artificial evolution, 500 generations were evaluated, starting each time from random initial genome values.

3. Results and Discussion

3.1 Optimal Morphology

Figure 5 shows the morphologies of the fittest agents found after 10 runs of 500 generations. Different observations can be made from this.

Firstly, all whiskers are systematically placed in the front of the agent rather than on its sides. This is not surprising since for the given task, the agent has to avoid obstacles he might bump into, i.e. the ones detected in front of the agent.

Secondly, the morphologies of the four fittest agents share a common trend: the longest whiskers are more often found in the front, with the shortest whiskers more on the side. These results agree with a similar experiment conducted on real robots (Fend et al., 2003b). In this latter experiment, different morphologies of a whisker array mounted on a Khepera robot were evaluated in an object avoidance task. The longest whiskers placed in the front were shown to be the most advantageous morphology.

Note that other arrangements of whiskers were found by the artificial evolution, such as morphologies 5 and 7 in figure 5, where the whisker tips lie almost in a plane.

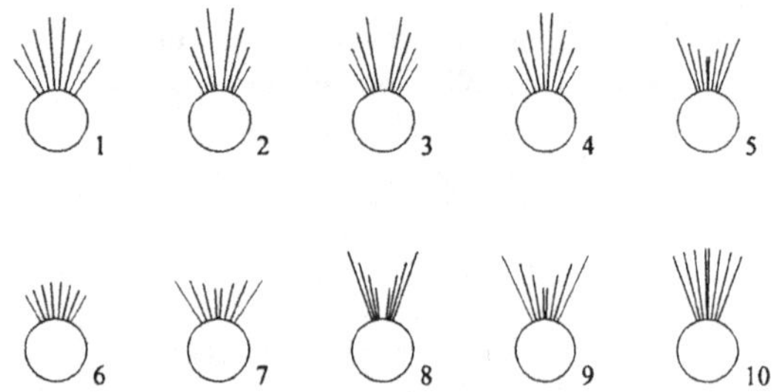

Figure 5: Morphologies of the fittest agents with flexible whiskers found by artificial evolution. The morphology of the fittest agent found after 500 generations is shown for each one of the 10 evolution runs. The agents are sorted by their fitness.

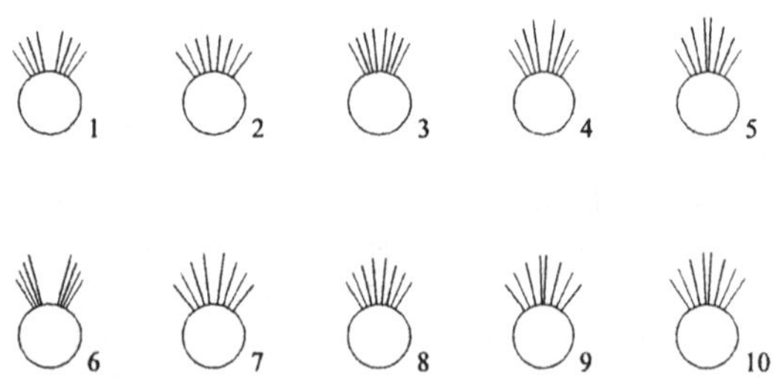

Figure 6: Same as figure 5, except that rigid whiskers were used instead of flexible whiskers.

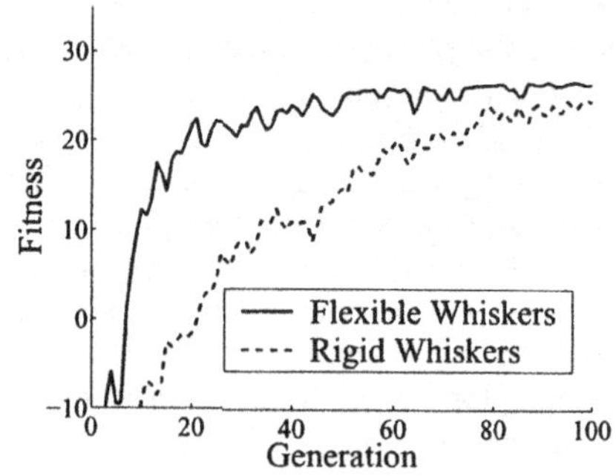

(a) Population Mean

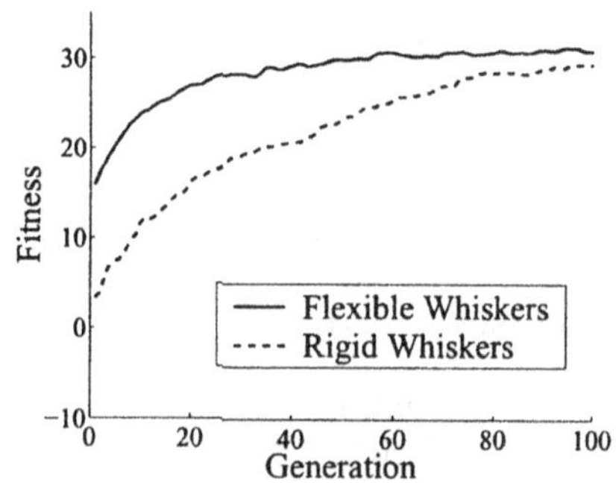

(b) Best Agent

Figure 7: Evolution of (a) the mean population fitness and (b) the fitness of the best performing agent in the population during the first 100 generations. The values for each curve are averaged over the 10 evolution runs.

3.2 Rigid vs. Flexible Whiskers

The same evolution runs as presented so far were performed again, this time replacing the flexible whiskers by rigid whiskers which could only bend at their root. Figure 6 shows the morphologies of the fittest agents obtained with rigid whiskers.

The comparison of figure 6 with figure 5 reveals two main differences. Firstly, rigid whiskers are found to be shorter than in the case of flexible whiskers. This difference can be explained by the fact that a small displacement of the tip of a rigid whisker produces the same deflection at the root as a larger displacement of the tip of a flexible whiskers. We also notice that artificial evolution did only find one type of morphology with rigid whiskers, namely longer whiskers in the front.

Another remarkable difference between rigid and flexible whiskers can be seen at the level of the artificial evolution. Figure 7 shows the average population fitness during the first 100 generations: the average population fitness increases more slowly in the case of rigid whiskers compared to the case of flexible whiskers. After the 100th generation, the fitness does not increase significantly anymore and stays at a comparable level in both case (the fitness in the case of rigid whiskers remains about 5% lower than in the case of flexible whiskers). One can speculate that the behavior of an agent using rigid whiskers is less robust against small changes in the parameters of the morphology and the controller: since the whiskers are less compliant, a small variation of the trajectory of the robot can produce a significant change in the sensory signals; the environment may therefore look quite different from the situated perspective of the agent, inducing possibly a drop of the agent's fitness. This could explain the discrepancy of the fitness landscape and why it is harder for the artificial evolution to find an adequate path in the genome parameter space.

An additional insight into these differences is provided by looking at the different controllers obtained by the evolution. Figure 8 shows the average and standard deviation of the phenotypic parameter values over the fittest agents of each run. The most significant difference between runs with flexible or rigid whiskers is found in the parameter corresponding to the backward speed in the turning state (v_2, see equation 5). The fittest agents using flexible whiskers move forward while turning away from an obstacle, whereas the fittest agents using rigid

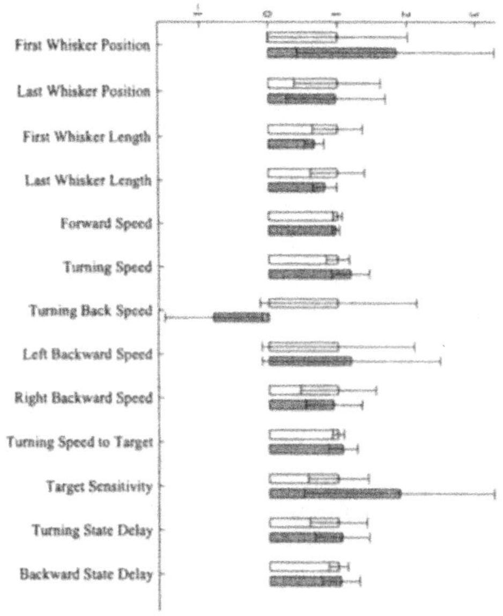

Figure 8: Comparison of the parameter values of the phenotypes for the best performing agents found after 500 generations. The mean value and standard deviation is shown both for the case of flexible (white bars) and rigid whiskers (dark bars). The values are normalized to the mean values for the case of flexible whiskers.

whiskers move backward. This indicates that with compliant whiskers, the agent does not have to avoid as much the obstacles as with rigid whiskers, allowing the first to move faster to the target (hence a higher fitness).

4. Conclusion and Future Work

In this paper, we presented the simulation software we have developed to study mobile agents using whiskers to perceive their proximal environment. We showed a few examples where a fairly good agreement is found between the simulated model of the whisker and the corresponding theoretical solutions.

To illustrate a possible application of this simulation software, we used artificial evolution to optimize the morphology and the controller of an agent using whiskers as touch sensors in a target seeking/object avoidance task. A particular morphology turned out to be more adequate, i.e. with longer whiskers in the front of the agent. This finding confirms results obtained by a similar experiment conducted with a real robot ??.

We then focused on the relation between the material properties of the whiskers, the agent's morphology and behavior, and artificial evolution. By only replacing flexible whiskers by rigid whiskers, different implications were observed: artificial evolution did not only find a different morphology, but also a different behavior, and the evolution speed was shown to be significantly slower when using rigid whiskers instead of flexible ones. This demonstrates the importance of considering – even for relatively simple tasks – the morphology and material properties when designing a situated agent interacting with its environment.

The limitations of the simulation software occur when trying to model more complex dynamical situations, e.g. real rat whiskers swept over different natural textures. For instance, it is impossible yet to obtain – in a reasonable amount of computation time – sensory signals comparable to data measured with real rat whiskers mounted on capacitor microphones, such as presented in (Fend et al., 2003a). More work needs to be done in order to understand and be able to simulate the complex dynamics of natural whiskers (non-linear damping of the whisker, friction with natural textures, etc.).

Nevertheless, several directions can be investigated with the simulation model at its current state, and constitute our future work. For instance: the use of whiskers for more demanding tasks, such as object recognition; the integration of different sensory modalities, such as vision; or the implementation of learning or discrimination algorithms.

Acknowledgment

This research has been supported by the IST-2000-28127 European project (AMOUSE).

Appendix

Model of the Universal Damped Spring Joint

This section describes in details the implementation of this joint. The interested reader is referred to (Smith, 2003) for a introduction to how implement joints in ODE.

Consider two bodies connected by a universal damped spring joint. Let p_i, $i = 1, 2$ be the positions of each body's center of mass, and R_i 3×3 rotation matrices corresponding to each body's orientation. The global position x of a point with position x'_i relative to the body i is:

$$x = p_i + R_i \, x'_i \tag{8}$$

In a rest position (see figure 9), let q_0 be the center of the joint, and $\{a_0, b_0, c_0\}$ an orthonormal basis with a_0 along the joint's main axis. We define q'_i as the "attachment" vector from p_i, and $\{a'_i, b'_i, c'_i\}$ the joint basis relative to body i:

$$\left.\begin{array}{lcl} q'_i & := & R_i^T \, (q_0 - p_i) \\ a'_i & := & R_i^T \, a_0 \\ b'_i & := & R_i^T \, b_0 \\ c'_i & := & R_i^T \, c_0 \end{array}\right\} \quad i = 1, 2 \tag{9}$$

As for a ball and socket joint, the first constraint is that

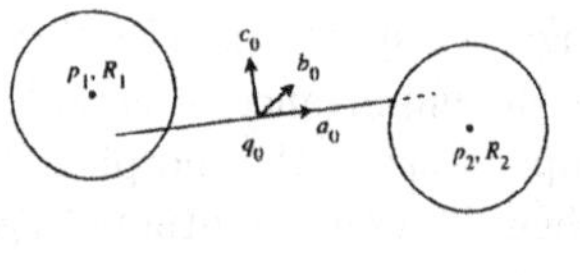

Figure 9: Two bodies connected by an universal damped spring joint.

both attachments coincide spatially:

$$p_1 + R_1\, q_1' = p_2 + R_2\, q_2' \qquad (10)$$

Another constraint of the joint is that both axes $b = R_1\, b_1'$ and $c = R_2\, b_2'$ must lie perpendicular to each other (if body 1 is twisted, body 2 will be twisted as well):

$$b \cdot c = 0 \qquad (11)$$

In other words, rotation of the two bodies about the direction perpendicular to the two axes should be equal:

$$p \cdot \omega_1 = p \cdot \omega_2, \quad p := \langle b \times c \rangle \qquad (12)$$

where ω_i is the angular velocity of body i, and $\langle \cdot \rangle$ the normalization operator

$$\langle x \rangle := \frac{x}{\|x\|}$$

If the joint is bent, i.e. if the axes $a_1 = R_1\, a_1'$ and $a_2 = R_2 a_2'$ are not parallel anymore, the angular damped spring has the effect of maintaining equal rotations of both bodies about an axis $q' = a_1 \times a_2$ perpendicular to a_1 and a_2. Since b and c might not be perpendicular due to numerical errors, q' is corrected to $q := \langle q' - (p \cdot q')p \rangle$ in order to be also perpendicular to p. We then have the additional constraint:

$$q \cdot \omega_1 = q \cdot \omega_2 \qquad (13)$$

The derivation of the constraints (10), (12) and (13) with respect to time leads to the following equation associated to the joint:

$$J_1 v_1 + \Omega_1 \omega_1 + J_2 v_2 + \Omega_2 \omega_2 = c + C\lambda \qquad (14)$$

where λ is the (unbounded) constraint force calculated during the semi-implicit integration of a time step. The 6×3 Jacobian matrices are given by:

$$\begin{aligned} J_1 &= \begin{pmatrix} 1_{3\times3} & 0_{3\times3} \end{pmatrix}^T & (15)\\ J_2 &= \begin{pmatrix} -1_{3\times3} & 0_{3\times3} \end{pmatrix}^T & (16)\\ \Omega_1 &= \begin{pmatrix} -\widehat{a_1} & p & q & r \end{pmatrix}^T & (17)\\ \Omega_2 &= \begin{pmatrix} \widehat{a_2} & -p & -q & -r \end{pmatrix}^T & (18) \end{aligned}$$

where $r := p \times q$, the operator $\widehat{\ }$ defined as

$$\widehat{a} := \begin{pmatrix} 0 & -a_z & a_y \\ a_z & 0 & -a_x \\ -a_y & a_x & 0 \end{pmatrix}$$

and

$$\begin{aligned} c &= \begin{pmatrix} \epsilon_1\, ((p_1 + R_1\, q_1') - (p_2 + R_2\, q_2')) \\ -\epsilon_2\, (b \cdot c) \\ \epsilon_3\, \|a_1 \times a_2\| \\ 0 \end{pmatrix} & (19)\\ C &= \operatorname{diag}(\eta_1, \eta_1, \eta_1, \eta_2, \eta_3, \eta_4) & (20) \end{aligned}$$

The error reduction parameters ϵ_i and the constraint force mixing factors η_i are set according the the corresponding spring constants k_i and damping factors d_i:

$$\begin{aligned} \epsilon_i &= \frac{k_i}{s\, k_i + d_i} & (21)\\ \eta_i &= \frac{1}{s\, k_i + d_i} & (22) \end{aligned}$$

s is the integration time step. The dissipative term η_4 is added to prevent a precession of the connected objects around the joint axis.

References

Bäck, T., Hammel, U., and Schwefel, H.-P. (1997). An overview of evolutionary algorithms for parameter optimization. *IEEE Transactions on Evolutionary Computation*, 1(1):3–17.

Beadle, M. (1977). *The Cat: History, Biology and Behavior.* Simon and Schuster, New York.

Brecht, M., Preilowski, B., and Merzenich, M. (1997). Functional architecture of the mystacial vibrissae. *Behavioral Brain Research*, 84(1-2):81–97.

Brooks, R. A. (1989). A robot that walks: Emergent behaviors from a carefully evolved network. Technical Report AI MEMO 1091, MIT.

Carvell, G. E. and Simons, D. J. (1990). Biometric analyses of vibrissal tactile discrimination in the rat. *Journal of Neuroscience*, 10(8):2638–2648.

Cowan, N. J., Ma, E. J., Cutkosky, M., and Full, R. J. (2003). A biologically inspired passive antenna for steering control of a running robot. In *International Symposium on Robotics Research*, Siena, Italy.

Dechant, H.-E., Rammerstorfer, F. G., and Barth, F. G. (2001). Arthropod touch reception: stimulus transformation and finite element model of spider tactile hairs. *J. Comp. Physiol. A*, 187:313–322.

Dehnhardt, G., Mauck, B., and Bleckmann, H. (1998). Seal whiskers detect water movements. *Nature*, 394:235–236.

Fend, M., Bovet, S., Yokoi, H., and Pfeifer, R. (2003a). An active artificial whisker array for texture discrimination. In *IEEE/RSJ International Conference on Intelligent Robots and Systems (IROS).*

Fend, M., Yokoi, H., and Pfeifer, R. (2003b). Optimal morphology of a biologically-inspired whisker array on an obstacle-avoiding robot. In Banzhaf, W., Christaller, T., Dittrich, P., Kim, J. T., and Ziegler, J., (Eds.), *Advances in Artificial Life - Proceedings of the 7th European Conference on Artificial Life (ECAL)*, volume 2801 of *Lecture Notes in Artificial Intelligence*. Springer Verlag Berlin, Heidelberg.

Gibson, S. and Mirtich, B. (1997). A survey of deformable modeling in computer graphics. Technical Report TR-97-19, Mitsubishi Electric Research Lab., Cambridge.

Guic-Robles, E., Valdivieso, C., and Guajardo, G. (1989). Rats can learn a roughness discrimination using only their vibrissal system. *Behavioural Brain Research*, 31(3):285–289.

Hafner, V., Fend, M., Lungarella, M., Pfeifer, R., König, P., and Körding, K. (2003). Optimal coding for naturally occurring whisker deflections. In *Proceedings of the 10th International Conference on Neural Information Processing (ICONIP)*, Istanbul.

Hartmann, M. J., Johnson, N. J., Towal, R. B., and Assad, C. (2003). Mechanical characteristics of rat vibrissae: resonant frequencies and damping in isolated whiskers and in the awake behaving animal. *Journal of Neuroscience*, 23(16):6510–6519.

Hirose, S., Masui, T., Kikuchi, H., Fukuda, Y., and Umetani, Y. (1985). Titan III: A quadruped walking vehicle. *Robotics Research - The Second International Symposium*, pages 325–331.

Jung, D. and Zelinsky, A. (1996). Whisker-based mobile robot navigation. *Proceedings of the IEEE/RSJ International Conference on Intelligent Robots and Systems (IROS)*, 2:497–504.

Kaneko, M., Kanayama, N., and Tsuji, T. (1995). 3-D active antenna for contact sensing. *Proceedings of the 1995 IEEE International Conference on Robotics and Automation*, pages 1113–1119.

Kaneko, M., Kanayama, N., and Tsuji, T. (1998). Active antenna for contact sensing. *IEEE Transactions on robotics and automation*, 14(2):278–291.

Leyhausen, P. (1979). *Cat Behavior*. Garland STMP Press, New York.

McKerrow, P. (1991). *Introduction to Robotics*. Addison-Wesley, Reading, Massachusetts.

Neimark, M. A., Andermann, M. L., Hopfield, J. J., and Moore, C. I. (2003). Vibrissa resonance as a transduction mechanism for tactile encoding. *Journal of Neuroscience*, 23(16):6499–6509.

Pfeifer, R. and Scheier, C. (1999). *Understanding Intelligence*. MIT Press, Cambridge, Mass., USA.

Russell, R. A. (1984). Closing the sensor-computer-robot control loop. *Robotics Age*, 6(4):15–20.

Russell, R. A. (1992). Using tactile whiskers to measure surface contours. In *Proceedings of the 1992 IEEE International Conference on Robotics and Automation*, pages 1295–1299.

Russell, R. A. and Wijaya, J. A. (2003). Object location and recognition using whisker sensors. *Australasian Conference on Robotics and Automation*.

Schiebel, E. N., Busby, H. R., and Waldron, K. J. (1986). Design of a mechanical proximity sensor. *Robotica*, 4:221–227.

Smith, R. (2003). ODE: Open dynamics engine. Available online: `http://q12.org/ode/`.

Tsujimura, T. and Yabuta, T. (1992). A tactile sensing method employing force/torque information through insensitive probes. *Proceedings of the 1992 IEEE International Conference on Robotics and Automation*, pages 1315–1320.

Ueno, N., Svinin, M. M., and Kaneko, M. (1998). Dynamic contact sensing by flexible beam. *IEEE/ASME Transactions on Mechatronics*, 3(4):254–264.

Vincent, S. (1912). The function of the vibrissae in the behavior of the white rat. *Behav Monographs*, 1(5).

Wang, S. S. M. and Will, P. (1978). Sensors for computer controlled mechanical assembly. *The industrial robot*, March:9–18.

Will, P. and Grossman, D. D. (1975). An experimental system for computer controlled mechanical assembly. *IEEE Transactions on computers*, c-24(9):879–888.

Woolsey, T. and der Loos, H. V. (1970). The structural organization of layer IV in the somatosensory region (SI) of mouse cerebral cortex. The description of a cortical field composed of discrete cytoachitectonic units. *Brain Research*, 17:205–242.

Zlajpah, L. (1999). On obstacle avoidance for redundant robots using tactile sensors. I.J. Rudas and L. Madar asz, editors.

Spatiotemporal processing of whisker input supports texture discrimination by a brain-based device

Anil K. Seth, Jeffrey L. McKinstry, Gerald M. Edelman, and Jeffrey L. Krichmar
The Neurosciences Institute, 10640 John Jay Hopkins Drive, San Diego, CA 92121, USA
seth@nsi.edu, mckinstry@nsi.edu, edelman@nsi.edu, krichmar@nsi.edu

Abstract

Sensory signals from whiskers are typically rich in both spatial and temporal structure, and are used by many animals to guide a variety of adaptive behaviors. To explore possible neural mechanisms underlying whisker-guided behavior, we constructed Darwin IX, a mobile physical device equipped with artificial whiskers and a neural simulation based on the rat somatosensory (whisker) system. We show that neuronal units with time-lagged response properties, together with the selective modulation of neural connection strengths, provide a plausible neural mechanism for the spatiotemporal transformations of sensory input necessary for both texture discrimination and selective conditioning to textures.

1. Introduction

Haptic sensory signals provided by vibrissae (whiskers) guide a variety of adaptive behaviors in many animals. For example, rodents use their whiskers to discriminate among different textures and to learn associations between specific textures and aversive or appetitive events (Carvell and Simons, 1990, Harvey et al., 2001, Prigg et al., 2002). Whisker signals typically have rich spatial and temporal structure as a consequence of interactions among the spatial arrangement of whiskers on the rat face, the spatial properties of the object being whisked, and the movement over time of the whiskers past the object; this movement can be generated by active sweeping of whiskers across object surfaces as well as by locomotion of the animal. Spatiotemporal processing is therefore fundamental to the neural operations underlying whisker-guided behavior.

The whisker system of the rat has been studied extensively as a model for spatiotemporal sensory processing (Moore et al., 1999, Ghazanfar and Nicolelis, 2001, Ahissar and Arieli, 2001). For example, neurons in layer V of primary somatosensory cortex (S1) have been shown to possess spatiotemporally defined receptive fields, i.e. individual neurons respond best to different whiskers as a function of post-stimulus time (Ghazanfar and Nicolelis, 1999). However, since most studies of the rat whisker system involve anesthetized or head-fixed animals, and many of these involve manipulating single whiskers, relatively little is known about the spatiotemporal operations of the complete system during natural whisker-guided behavior.

For example, the neural mechanisms underlying texture discrimination by the rat remain poorly understood. One possibility is that surface texture is encoded by intrinsic vibrational modes of the whiskers, so that different textures elicit different spectral profiles (Mehta and Kleinfeld, 2004). While this mechanism may be mechanically plausible (Fend et al., 2003), evidence is lacking for corresponding neural operations in the behaving rat. A second possibility, consistent with observations of spatiotemporal receptive fields, is that texture is encoded by patterns of convergent activity across multiple whiskers. However, direct neurobiological support for such a mechanism is again absent.

Experimental assessment of these possibilities requires specific hypotheses that incorporate neuronal mechanisms, phenotypes, and behavioral interactions with the environment. To test the idea that texture discrimination can be supported by convergent inputs from multiple whiskers, we constructed Darwin IX. Darwin IX is a mobile physical device equipped with artificial whiskers and a simulated nervous system incorporating neuroanatomical and neurophysiological features of the rat somatosensory (whisker) system, as well as neural areas involved in fear conditioning and learning. We investigate the hypothesis that neuronal units that respond to whisker input with unit-specific time delays (Saul and Humphrey, 1992), together with the selective modulation of neural connections strengths, provide a plausible neural mechanism for the spatiotemporal transformations of sensory input needed for both texture discrimination and selective conditioning to textures.

Darwin IX is the latest in a series of 'brain-based devices' (BBDs) that have been constructed over the last 12 years (Edelman et al., 1992, Almassy et al., 1998, Krichmar and Edelman, 2002, Seth et al., in press). All BBDs in the Darwin series have the following attributes: 1) An embodied morphology that allows for active exploration

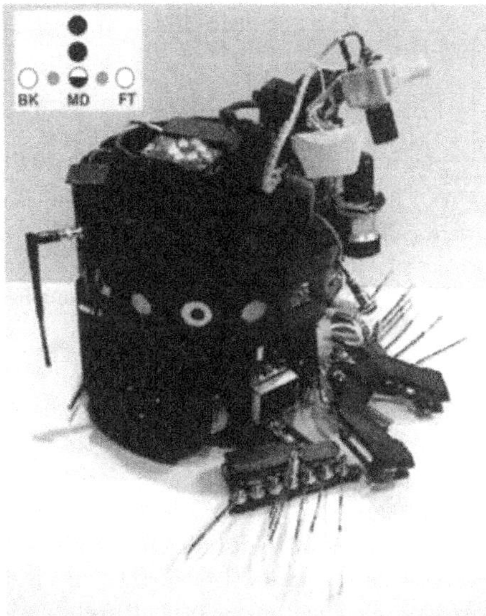

Figure 1: Darwin IX consists of a mobile base (Nomadic Technologies Inc.) equipped with several sensors and effectors including left and right whisker arrays, and a neural simulation running on a remote cluster of computer workstations. The arrangement of a whisker array is shown in the inset. Each array has 7 whiskers arranged in a row of 5 and a column of 3. Whiskers used for wall following are marked in white (BK,MD,FT). Whiskers that provide input to the neural simulation are marked in black. Note that one whisker (white/black) is used for both purposes. Whiskers marked in gray (small circles) are not used in the present model.

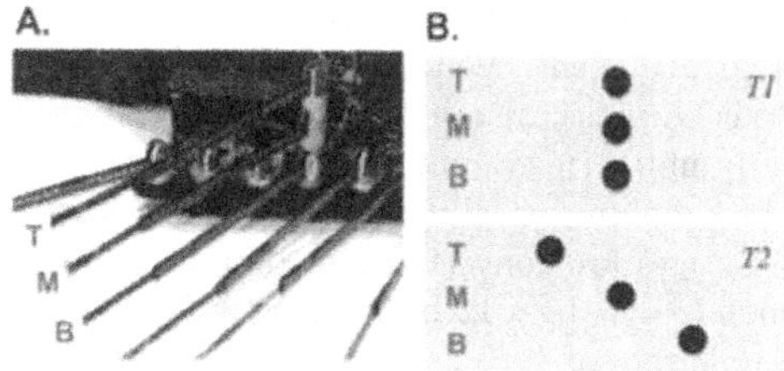

Figure 2: **A.** Detail of a whisker array: The top (T), middle (M), and bottom (B) whiskers in the column are labeled; these whiskers provide input to the neural simulation. **B.** Schematic of textures $T1$ and $T2$. Each texture consists of pegs embedded in a wall; pegs are aligned in rows corresponding to the whiskers in a column. Pegs in the top row deflect the top whisker (T), and similarly for pegs in the middle row (M) and the bottom row (B).

in a real-world environment, including sensors such as cameras, microphones, proximity sensors, and artificial whiskers. 2) A neural simulation to control the BBD's behavior, incorporating detailed neuroanatomy and neurophysiology based on vertebrate nervous systems. 3) A value system that signals the salience of environmental cues and that modulates plasticity in the nervous system resulting in modification of the device's behavior. A key feature of all BBDs is that they allow neuronal responses to be followed in detail at all levels of control, while the device is behaving. Such detailed correlations of behavioral and brain responses, which would not be possible in animal experiments, can provide insights into and form the basis for predictions regarding the neuronal basis of adaptive behavior.

In our experiments with Darwin IX, the device autonomously explored a walled environment containing two distinct textures each consisting of patterns of pegs embedded in the walls. It became conditioned to avoid one of the textures by association of this texture with an innately aversive simulated 'foot-shock'. Darwin IX demonstrated its conditioned behavior by freezing and then moving away from walls containing the texture corresponding to the aversive sources.

2. Methods

Darwin IX consists of a mobile robotic base containing infra-red (IR) proximity sensors, custom-built whisker arrays (Fig. 1), and a cluster of workstations that run the simulated nervous system and control the device's behavior. The cluster consists of six 1.4 GHz Pentium IV workstations running Message Passing Interface (MPI) parallel software under the Linux operating system. One workstation receives whisker and IR input information, and transmits motor commands, via a wireless modem. A microcontroller (PIC17C756A) onboard the base samples input and status from the sensors and controls RS-232 communication between the robotic base and the workstations.

A simulation cycle of Darwin IX consists of the following events: Sensory input is processed, the state of the neural simulation is updated, and motor output is generated. In our experiments, execution of each simulation cycle requires approximately 100ms of real time.

2.1 Artificial whisker design

Each whisker array consists of seven whiskers arranged in a column of three and a row of five (see Fig. 1, inset, and Fig. 2A). Whisker columns supply input to the simulated nervous system, while whiskers in the rows support innate avoidance and wall following behaviors (see *2.2. Innate behavior*). Each whisker consists of two 4cm by 0.63cm polyamide strips, adhered back-to-back, that are responsive to bend (Abrams/Gentile Entertainment). These strips are typically used as strain sensors in devices such as virtual reality gloves. Each strip has 20 resistive areas embedded regularly along its length, providing a resistance of $\sim$10KΩ when the strip is unbent and $\sim$50KΩ when the strip is maximally bent. Each strip detects bending in only one direction, hence the back-to-back arrangement. Voltage signals from each pair of strips (i.e. a whisker) are converted to a single signal ranging from 0V (maximum deflection in one direction) to 5V (maximum deflection in the opposite direction). These voltages are converted into digital signals, ranging from 0 to 255, by a 12-bit analog-to-digital converter at a sample rate of 40Hz.

During each simulation cycle of Darwin IX a 'packet' of 4 signals from each whisker is received by the neural simulation. For whisker w_k at simulation cycle t, the corresponding packet is $[w_{k1}(t), w_{k2}(t), w_{k3}(t), w_{k4}(t)]$, where $w_{k4}(t)$ is the most recent. The first signal in each packet, $w_{k1}(t)$, which we call the 'current whisker value', is used for guiding Darwin IX's wall following behavior; all four signals provide input to the neural simulation (see *2.4. Whisker input and lag cells*).

2.2 Innate behavior

Darwin IX is equipped with innate behavioral responses. The default behavior of Darwin IX is to move forward in a straight line at a speed of ~8cm/sec. If Darwin IX approaches a wall head-on to within a distance of ~4cm, an avoidance response is triggered by either of two IR sensors, one facing front-left, and the other facing front-right: The device stops, backs up (~10cm), and then turns $\sim\pi/6$ away from the wall before resuming it's default behavior.

Darwin IX has an innate freezing/escape response which is triggered by a simulated 'foot-shock'. A downward-facing IR sensor at the front of the device detects reflective construction paper ('foot-shock pads') on the floor of the arena; activation of this sensor evokes neural activity in area M_{ave} (see 2.3. *Neuroanatomy*) which triggers the behavioral response. This response consists of continued movement for ~5.5s, freezing for ~4s, then a turn away from the wall by an angle randomly chosen from the interval $[\pi/4, 3\pi/4]$.

Darwin IX also has an innate wall following capability such that, on encountering a wall, the device moves parallel to the wall at a distance suitable for the detection of embedded textures (see Fig. 2B). Wall following is based on signals from the backmost (BK), the 'middle' whisker, which is lowest of the vertical stack (MD), and the frontmost (FT) on each side (see Fig. 1, inset). For each of these whiskers, a running average of the current whisker value ($\overline{w}$) is maintained over 75 simulation cycles. This average is updated at every cycle except when the current whisker value differs by more than 10 from the corresponding average (signifying whisker deflection; recall that the range of the current whisker value is 0-255). For each whisker k on each side, a 'deflection' value wd is calculated as:

$$wd_k(t) = |w_{k1}(t) - \overline{w}_k(t)| \quad (1)$$

Side	Whisker	φ_w	$g_w(wd(t) > 0)$	$g_w(wd(t) \leq 0)$
Left	BK	89	0.05	0.075
	MD	33	0.1	0.2
Right	BK	110	0.075	0.15
	MD	28	0.15	0.30

Table 1: Wall following by Darwin IX. Scaling factors g_w and φ_w were calibrated according to the physical response characteristics of each whisker, in order to allow Darwin IX to follow walls at a distance appropriate for the detection of textures. Factor g_w varies according to the direction of the corresponding whisker deflection such that there is a bias in favor of turning towards a wall.

Wall following of a left(right) wall is triggered when any deflection value for the left(right) whiskers exceeds 15. During each cycle of wall following, adjustments are made to the speed of the wheel furthest from the wall (the contralateral wheel); the other (ipsilateral) wheel remains at the default speed of $W_{def} = 35$. Contralateral wheel speeds are set using:

$$W_{speed} = W_{def} - back - mid - front \quad (2)$$

where $front = 0.25(wd_{FT})$, $mid = g_w(wd_{MD} - \varphi_w)$, and $back = g_w(wd_{BK} - \varphi_w)$, and where mid and $back$ are bounded in the range ± 5, and g_w (a scaling factor) and φ_w (a threshold) were calibrated for each whisker according the physical response characteristics of the whiskers (see Table 1). Additionally, g_w varies according to the sign of the corresponding wd such that there is a bias in favor of turning towards a wall (see Table 1, columns 4 and 5).

2.3 Neuroanatomy

Darwin IX's simulated nervous system contains areas analogous to the somatosensory pathway in the rat brain, specifically the (ventromedial) nuclei of the thalamus, and primary and secondary somatosensory areas (in our model, $Th \rightarrow S1 \rightarrow S2$), as well as neural areas involved in fear conditioning and learning.

Areas $S1$ and Th are subdivided into left (L) and right (R) regions and further into 'top' (T), 'middle' (M) and 'bottom' (B) subregions, such that each subregion receives input from a single whisker in the column on the corresponding side. These subregions are analogous to so-called whisker 'barrels' (in rat S1) and 'barreloids' (in rat thalamus); as in Darwin IX, each whisker barrel/barreloid in the rat contains cells that respond preferentially to a specific whisker (Woolsey and Van der Loos, 1970).

A global diagram of Darwin IX's simulated nervous system is given in Fig. 3. It comprises 17 areas, 1101 neuronal units (see Table 2), and ~8400 synaptic connections (see Table 3). Neuronal units in area Th respond

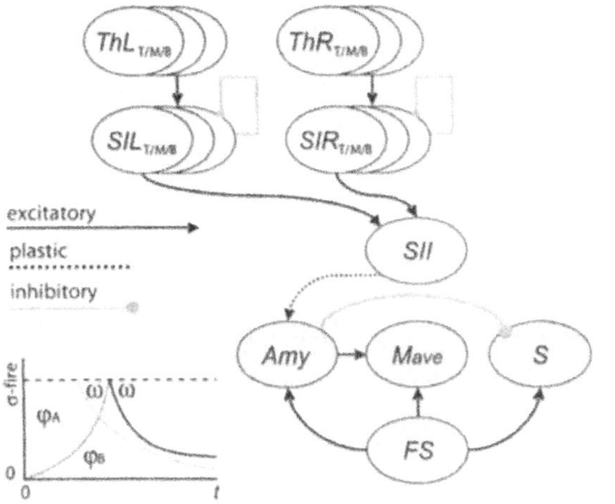

Figure 3: Global neuroanatomy of Darwin IX. See text for details. **Inset.** The operation of two idealized lag cells (A and B). The post-stimulus internal state of cell A (s_A^{in}) rises quickly (gray dashed line), at a rate determined by φ_A. The internal state of B (s_B^{in}) rises more slowly (black dashed line, φ_B). When the internal state of each cell reaches a threshold (σ^{fire}), output is generated (solid lines) which decays at a rate determined by ω.

Area	Size	σ^{fire}	ω	g
$Th(6)$	1x20	0.3	0.8	-
$S1(6)$	1x20	0.1	0.0	1.0
$S2$	30x30	0.2	0.8	1.0
S	1x1	-0.10	0.0	1.0
Amy	1x1	0.0	0.0	1.75
M_{ave}	3x6	0.0	0.0	1.0
FS	1x1	-	-	-

Table 2: Neuronal unit parameters for Darwin IX. Area Th consists of 6 separate 'barreloid areas' which take input directly from the 3 whiskers in each column on each side of Darwin IX. Activity in area FS is triggered by detection of 'foot-shock' pads by Darwin IX's downward-facing IR sensor. Size indicates the number of neuronal units in each area or sub-area. Neuronal units in each area apart from area FS have a specific firing threshold (σ^{fire}), a persistence parameter (ω) and a scaling factor (g).

to whisker input with unit-specific time delays (see *2.4. Whisker input and lag cells*). These units project topographically to the corresponding units of $S1$. Each barrel in $S1$ has local inhibitory connections which serve to increase the activity contrast among neuronal units. All barrels in $S1$ project to area $S2$ such that each neuronal unit in $S2$ takes input from 3 neuronal units, each of which is in a different barrel of either the left sub-area or the right sub-area of $S1$. This arrangement ensures that synaptic input to a neuronal unit in $S2$ is sparse and balanced. A deflection of a particular sequence of Darwin IX's whiskers leads to a spatiotemporal pattern of activity in $S2$. Such a dynamic sequence is comparable to that observed in the rat brain (Ghazanfar and Nicolelis, 1999).

Darwin IX's nervous system also contains areas supporting the acquisition of conditioned aversion (see *2.7. Aversive conditioning*). Area FS is activated by detection of a 'foot-shock' (see *2.8. Experimental environment and protocol*), and projects to areas Amy, M_{ave} and S. Area Amy is analogous to the amygdala, a neural area which has been widely implicated in the acquisition of conditioned fear (Maren and Fanselow, 1996). Area M_{ave} is analogous to a motor area, activity in which elicits an innate aversive freezing/escape response.

Activity in the simulated value system, area S, signals the occurrence of salient sensory events and this activity contributes to the modulation of plastic connection strengths in the pathway $S2 \rightarrow Amy$. Initially, S is activated by detection of simulated 'foot-shock'. Activity in S is analogous to that of ascending neuromodulatory systems in that it is triggered by salient events, influences large regions of the simulated nervous system (see *2.6. Synaptic plasticity*), and persists for several cycles (Schultz et al., 1997). After conditioning, S is activated by patterns of activity in $S2$ corresponding to the aversive texture (see *2.7. Aversive conditioning* and *2.8. Experimental environment and protocol*).

2.4 Whisker input and lag cells

Whisker input to the neural simulation is provided by transforming each whisker packet into a vector of 'difference values' according to:

$$\begin{aligned} \mathit{diff}_k = [&w_{k4}(t) - w_{k3}(t), w_{k3}(t) - w_{k2}(t), \\ &w_{k2}(t) - w_{k1}(t), w_{k1}(t) - w_{k4}(t-1)] \end{aligned} \tag{3}$$

for whisker w_k at simulation cycle t. These values from the whisker in each column provide input to the corresponding barreloids of area Th.

Each barreloid of area Th contains 20 'lag' cells; neuronal units which respond to input with cell-specific time delays (cells with similar properties have been found in the visual thalamus of the cat, see Saul and Humphrey, 1992). Each lag cell is characterized by an internal state (s_i^{in}), an output (s_i), and a cell-specific lag parameter set to be $\psi_i = \frac{0.2}{i}$, $i \in \{1, 2 \ldots 20\}$ for cell i in each barreloid. When triggered by a whisker deflection, the internal state s_i^{in} of cell i in the corresponding subregion increases at rate determined by the lag parameter ψ_i. When this internal state reaches a threshold, the cell begins to emit an output signal and s_i^{in} is reset to zero.

Because of differences in ψ_i among lag cells, cells in a barreloid fire with a range of delays, from 1 to 20 simulation cycles, following deflection of the corresponding whisker (see Fig. 3, inset).

Specifically, s_i^{in} in the barreloid corresponding to whisker w_k, is updated according to:

$$s_{ki}^{in}(t+1) = \begin{cases} 0.2 & s_{ki}^{in}(t) < 0.2, \overline{diff}_k(t) > 3.0 \\ 0 & s_{ki}^{in}(t) \geq \sigma_i^{fire} \\ (1+\psi_i)(s_{ki}^{in}(t)) & \text{otherwise} \end{cases} \quad (4)$$

where $\overline{diff}_k(t)$ is the average $diff_k(t)$ value (a value exceeding exceeding 3.0 signifies a whisker deflection), and σ_i^{fire} is a unit-specific firing threshold (Table 2).

The output s_{ki} is calculated using

$$s_{ki}(t+1) = \begin{cases} \tanh(10(\omega_i(s_{ki}(t)))) & s_{ki}^{in}(t) < \sigma_i^{fire} \\ \tanh(10(\omega_i(s_{ki}(t)) \\ \quad +(1-\omega_i)s_{ki}^{in}(t))) & \text{otherwise} \end{cases} \quad (5)$$

where ω_i determines the persistence of unit activity from one cycle to the next (Table 2). This value is fed as input into neuronal units in the corresponding barrel of area $S1$.

2.5 Neuronal dynamics

With the exception of the lag cells in area Th, the state of each neuronal unit in Darwin IX is determined by a mean firing rate variable s, which corresponds to the average activity of a group of roughly 100 real neurons over 100ms (s_{ki} of a lag cell is equivalent to the mean firing rate of neuronal units in other areas).

The mean firing rate (s) of each neuronal unit ranges from 0 (quiescent) to 1 (maximal firing). For all neuronal units, except those in areas Th and FS, the total input to unit i is given by

$$A_i(t) = \sum_{l=1}^{M}\sum_{j=1}^{Nl} c_{ij}s_j(t) \quad (6)$$

where M is the number of different anatomically defined connection types and Nl is the number of connections per type M projecting to unit i (see Table 3). The mean firing rate of unit i is given by

$$s_i(t+1) = \phi(\tanh(g_i(A_i(t) + \omega s_i(t)))) \quad (7)$$

where $\phi(x) = 0$ for $x < \sigma_i^{fire}$, otherwise $\phi(x) = x$, g_i is a scale factor, and ω determines the persistence of unit activity (Table 2).

Projection	Arbor	P	$c_{ij}(0)$
$Th \to S1$	[]0x0	1.0	13.0,15.0
$S1 \to S1(intra)$	[]2x8	1.0	-0.45,-0.6
$S1 \to S2$	special	-	0.25,0.25
$FS \to Amy$	non-topo	1.0	5.0,5.0
$Amy \to M_{ave}$	non-topo	1.0	40.0,40.0
$FS \to M_{ave}$	non-topo	1.0	50.0,50.0
$FS \to S$	non-topo	1.0	50.0,50.0
$Amy \to S$	non-topo	1.0	-50.0,-50.0
$S2 \to Amy$	non-topo	1.0	0.0001,0.0003

Table 3: Properties of anatomical projections in Darwin IX. Neuronal units connect with a probability (P) and projection shape (Arbor) which can be rectangular "[]" with a height and width (h x w), non-topographical "non-topo" where all pairs of neuronal units have an equal probability of being connected, or 'special' (see text). The initial connection strengths, $c_{ij}(0)$, are set randomly within the range (min, max). Connections marked with "intra" denote those within a sub-area.

2.6 Synaptic plasticity

Synaptic plasticity in Darwin IX acts to strengthen connections between simultaneously active neuronal units in areas $S2$ and Amy. This process is 'value dependent', i.e. the degree of change is modulated by activity in the simulated value system (area S) according to the following rule:

$$\Delta c_{ij}(t+1) = \eta s_j(t) BCM(s_i(t))(V(t) - 0.1) \quad (8)$$

where η is a fixed learning rate ($\eta = 1.4$), $s_i(t)$ and $s_j(t)$ are the activities of the post- and pre-synaptic units respectively and $V(t)$ is the mean activity in area S. The term $(V(t)-0.1)$ causes depression of plastic connections in the absence of value system activity.

The function $BCM()$ is based on the rule of Bienenstock et al. (1982) and is implemented as follows ($\rho = 6$, $\theta_1 = \theta_2 = 0.1$, $k1 = k2 = 0.45$):

$$BCM(x) = \begin{cases} 0 & x < \theta_1 \\ k_1(\theta_1 - x) & \theta_1 \leq \frac{\theta_1+\theta_2}{2} \\ k_1(x - \theta_2) & \frac{\theta_1+\theta_2}{2} \leq x < \theta_2 \\ \frac{k_2 \tanh(\rho(x-\theta_2))}{\rho} & \text{otherwise} \end{cases} \quad (9)$$

2.7 Aversive conditioning

Synaptic plasticity supports conditioned aversion to texture as follows. Area S maintains a baseline level of activity (0.1) in the absence of input ($\sigma_i^{fire} = -0.1$, see Table 2). Detection of a simulated foot-shock (see *2.8. Experimental environment and protocol*) causes neuronal units in area FS to produce a steady output of magnitude 1.0. This output pushes activity in area S above

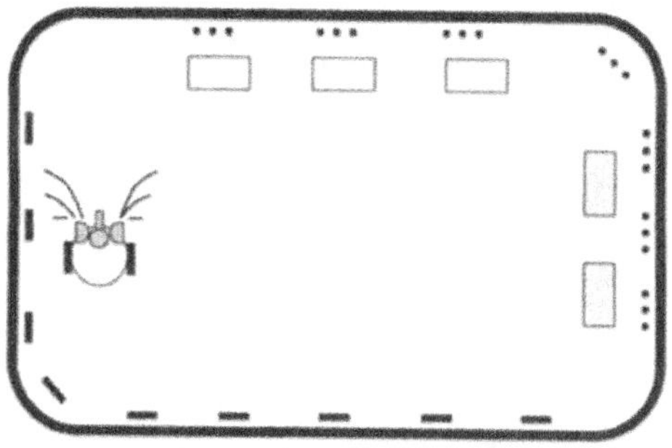

Figure 4: Experimental setup for Darwin IX. Darwin IX explored a walled enclosure (2.41m x 2.95m) with textures $T1$ and $T2$ on the walls. Instances of each texture were regularly spaced along the walls at intervals of ~30cm. Located on the floor adjacent to $T1$ patterns were 'foot-shock' pads made of reflective construction paper. Training and testing was repeated for each Darwin IX subject after exchanging the positions of $T1$ and $T2$.

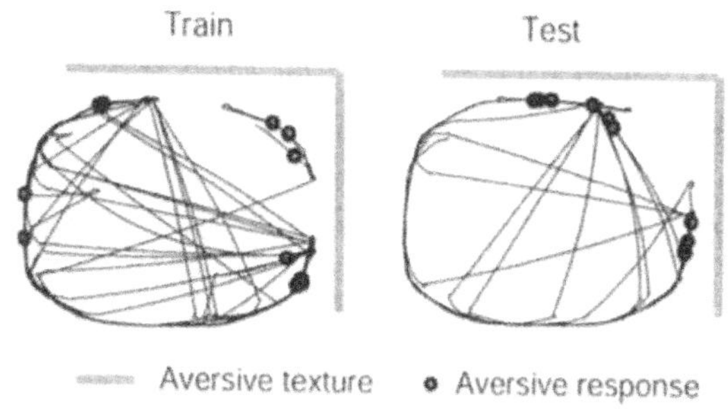

Figure 5: Conditioned aversion to textures by Darwin IX. The line tracings show the trajectory of a single Darwin IX subject during aversive conditioning to texture $T1$ (left), and during testing (right). Large black dots show locations of aversive responses, gray shading shows locations of $T1$. During testing all freezing/escape responses occurred in proximity to the aversive texture $T1$.

baseline, which causes potentiation of synapses onto neuronal units in area *Amy* from units in $S2$ corresponding to the currently present texture (see equations (8,9)). Freezing/escape responses are triggered by activity in M_{ave} exceeding a threshold (0.5) as a result of input from areas FS and/or *Amy* (see Fig. 3). This model also supports extinction of conditioned responses: If *Amy* is activated without any corresponding foot-shock, area S will be inhibited such that its firing rate will drop below baseline, and currently active synapses between $S2$ and *Amy* will be weakened (see equations (8,9)).

2.8 *Experimental environment and protocol*

Fig. 4 shows the overall arrangement of Darwin IX's environment. One texture ($T1$) consists of a vertically aligned column of pegs, the other ($T2$) consists of a vertically staggered set of pegs with offsets of ~6cm (see Fig. 2B). Two adjacent walls contained $T1$, the other two contained $T2$, and either $T1$ or $T2$ can be associated with a simulated aversive foot-shock. Note that Darwin IX typically travels in both clockwise and anti-clockwise directions around the environment such that the textures deflect both the left and right whisker arrays. Moreover, in the case of $T2$, the pattern of whisker deflection depends on the direction of travel.

Experiments were divided into training and testing stages. During training, either $T1$ or $T2$ was paired with foot-shock and Darwin IX autonomously explored its enclosure for 25,000 simulation cycles, corresponding to ~48 encounters with each wall and ~24 aversive responses to the simulated foot-shock. During testing, the foot-shock pads were removed and Darwin IX was allowed to explore its enclosure for 15,000 simulation cycles. Training and testing were repeated using three different Darwin IX "subjects" initialized with different random seeds, and pairing both $T1$ and $T2$ with foot-shocks (six training/testing episodes in total). During training and testing of each subject, responses of all neuronal units were recorded and saved for analysis. The position of Darwin IX was also continuously recorded by an overhead camera that detected an array of LEDs positioned on the top surface of the robotic device, the images from which were time-stamped for analysis.

3. Results

3.1 *Selective conditioning to textures*

Texture discrimination by Darwin IX subjects was assessed by monitoring trajectories during testing. Fig. 5 shows the behavior of a single Darwin IX subject during training in which foot-shock (the unconditioned stimulus, or UCS) was paired with $T1$ (the conditioned stimulus, or CS), and during testing, in which the foot-shock pads were removed. Most aversive responses, both unconditioned (UR, in response to the UCS), and conditioned (CR, in response to the CS), occurred in regions associated with the aversive texture. This response pattern is similar to that observed during conditioning of animals to aversive stimuli (Mackintosh, 1983).[1]

Taking into account data from all subjects reveals near perfect conditioned avoidance of aversive textures. During testing, Darwin IX subjects which were trained to avoid $T1$ made CRs on 96.6% (S.E. = 0.18%) of encounters with $T1$. When trained to avoid $T2$, these subjects made CRs on 97.9% (S.E. = 0.14%) of encounters with $T2$ during testing. Only 3.2% of all aversive responses

[1]Short movies showing Darwin IX's behavior during encounters with neutral and aversive textures can be viewed at www.nsi.edu/nomad/whisker/movies/.

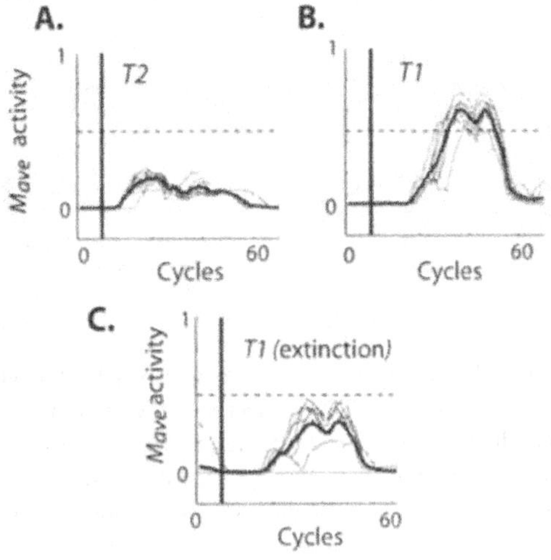

Figure 6: Neural activity in area M_{ave} during testing. Each panel shows mean M_{ave} activity over 60 time-steps for 10 encounters by the left whisker array with textures *T2* (A) or *T1* (B). The vertical line indicates the moment of deflection of the first whisker during each encounter, the thick line shows the grand mean M_{ave} response, and the dashed line shows the threshold for production of an aversive response. Panels A and B show encounters from early in testing (up to 10,000 time-steps). Panel C shows encounters from late in testing (after 20,000 time-steps).

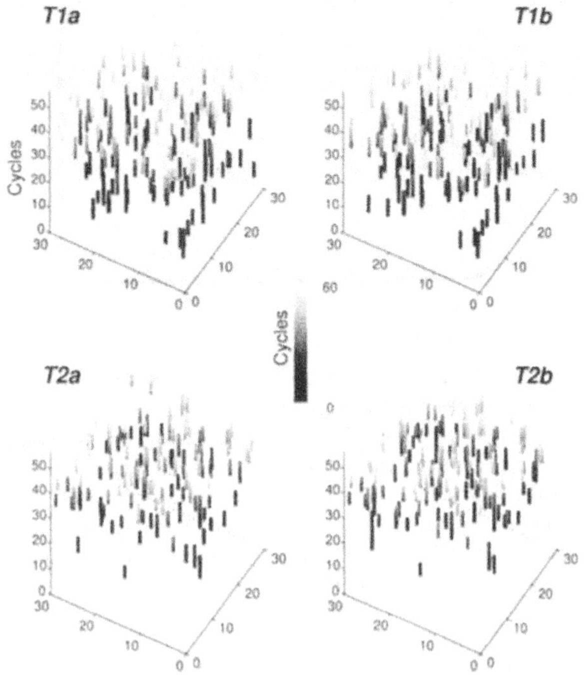

Figure 7: Spatiotemporal response properties of neuronal units in $S2$. The top panels show $S2$ activity during two separate encounters (*a*,*b*) with texture $T1$ by Darwin IX's left whisker column. The bottom panels show encounters of the same whiskers with texture $T2$. Each panel shows the activity of the 30x30 matrix of $S2$ neuronal units (x and y axes) over 60 post-stimulus cycles (indicated by the z axis as well as by the gray scale). Neuronal units are shown if their activity exceeded a threshold (0.25).

during testing occurred inappropriately, i.e. in response to whisker deflections by walls or by the texture not associated with foot-shock.

Fig. 6 illustrates the motor neuronal responses associated with the expression of conditioned aversion. Each panel shows mean activity in area M_{ave} from several encounters by the left whisker array with both neutral (*T2*) and aversive (*T1*, the CS) textures during testing. Fig. 6A shows that M_{ave} responses to the neutral texture (*T2*) are reliably weak, and hence do not trigger aversive responses. By contrast, M_{ave} responses to the CS (*T*1) are reliably strong (Fig. 6B) and do trigger aversive responses.

3.2 Extinction

Similar to behavioral data for aversive conditioning in animals (Mackintosh, 1983), Darwin IX's conditioned responses to aversive textures are subject to extinction. During testing, early encounters with the aversive textures evoke aversive responses. They also — as a result of inhibitory projections from Amy to S — suppress activity in the value system (area S). Inhibition of S activity below baseline leads to synaptic depression of active projections between $S2$ and Amy, and hence extinction of the conditioned response to the corresponding texture (see equations (8,9)). Note that during training, the inhibitory projections from Amy to S are counteracted by the excitatory projections from area FS to S.

The neural events underlying extinction in Darwin IX are illustrated in Fig. 6C, which shows M_{ave} responses late in testing (after ~20,000 time-steps). Peak M_{ave} responses to the CS have diminished below the threshold at which motor aversive responses are triggered.

3.3 Spatiotemporal neural activity

Darwin IX's ability to categorize textural stimuli is supported by spatiotemporal patterns of activity in neural area $S2$. Each texture deflects whiskers in a column in a specific temporal order. The lag cells in area Th and neural units downstream in $S1$ (see Fig. 3) present a pattern of activity with both a spatial component (i.e. the particular whisker) and a temporal component (i.e. the time since deflection). $S2$ responds to particular combinations of this $S1$ activity.

The population response of $S2$ to a texture was specific and repeatable. Fig. 7 shows representative $S2$

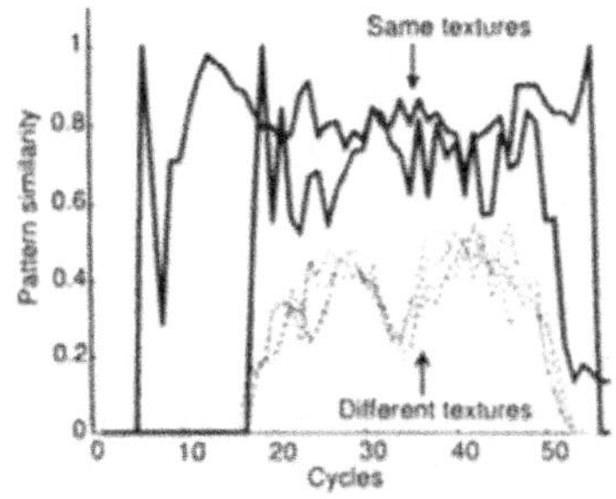

Figure 8: Similarity over time between pairwise combinations of activity patterns shown in Fig. 7, calculated as $sim(i,j) = v_i(t) \bullet v_j(t)$ where $v_i(t)$ is the normalized neuronal unit activity vector for pattern i at time t. Patterns representing the same texture are highly similar over time (solid lines; $sim(T1a, T1b)$, $sim(T2a, T2b)$), but patterns representing different textures are not similar (dashed lines; $sim(T1a, T2a)$, $sim(T1a, T2b)$, $sim(T2a, T1a)$, $sim(T2a, T1b)$).

activity patterns during testing of a Darwin IX subject. The top panels show $S2$ activity during encounters of the left whisker column with texture $T1$. The bottom panels show encounters of the same whiskers with texture $T2$. While all panels show complex spatiotemporal patterns of activity, the top panels are highly similar to each other, the bottom panels are also highly similar to each other, but the top and bottom panels are dissimilar. This observation is supported quantitatively by measures of pattern similarity over time for all possible pairs, calculated as the normalized vector product (see Fig. 8). As shown in this figure, there is high similarity between activity patterns representing the same texture, but not between activity patterns representing different textures.

4. Discussion

Darwin IX demonstrates the ability to categorize objects based on haptic data and to learn behavioral responses selective to specific textures. Observations of Darwin IX show that time-lagged neuronal responses to somatosensory input together with value-dependent synaptic plasticity provide a plausible mechanism for the spatiotemporal transformations of sensory input needed for texture discrimination. Darwin IX also shows that this mechanism can provide a basis for selective conditioned aversion to textures.

A key feature of Darwin IX is the incorporation of artificial whiskers that provide signals according to the degree of bend. Although the material properties of our artificial whiskers are clearly different those of rat whiskers, the rat somatosensory system appears capable of extracting similar signals. Szwed and colleagues have recently found neurons in the lower brainstem of the rat which are sensitive to pressure against objects, where pressure would correspond to degree of whisker bend (Szwed et al., 2003). Of course, rat whiskers may signal along many more dimensions than we capture in our design. For example, the intrinsic vibrational modes of whiskers may be significant for texture discrimination.

Distinguishing among textures using only the information provided by whiskers involves multiple spatiotemporal transformations. In our experiments, Darwin IX's self-generated movement transforms the spatial patterns of textures into temporal patterns of whisker deflections. These patterns are themselves transformed into spatiotemporal patterns of neural activity in area $S2$, via a combination of the intrinsic properties of lag cells in area Th and the arrangement of projections in the pathway $Th \rightarrow S1 \rightarrow S2$. As a result, neuronal units in $S2$ respond to specific combinations of whisker deflections with particular post-stimulus delays. Analysis of neural activity in $S2$ revealed the formation of spatiotemporal activity patterns corresponding to specific haptic perceptual categories (Fig. 7).

These observations are consistent with certain empirical data obtained from the rat. For example, Shuler and colleagues have shown that under bilateral whisker stimulation, the responses of rat S1 neurons to ipsilateral stimulation depended on both the spatial location and the relative timing of the bilateral stimuli (Shuler et al., 2001). Also, Ghazanfar and Nicolelis have identified cells in layer V of rat S1 that have spatiotemporal receptive fields, that is, individual neurons respond best to different whiskers as a function of post-stimulus time (Ghazanfar and Nicolelis, 1999). The spatiotemporal patterns of activity observed in area $S2$ of our model have similar properties: The sensitivity of a neuronal unit in $S2$ to a whisker deflection will vary over time as a function of the temporal pattern of preceding whisker deflections.

What neural circuitry might account for these sorts of spatiotemporal response properties? Our model shows the plausibility of a mechanism based on the convergence of inputs from different sources, each with different temporal dynamics. Empirical studies suggest that such circuits may exist in brains. For example, two classes of neurons — 'lagged' and 'non-lagged' — have been identified in the visual thalamus of the cat (Saul and Humphrey, 1992).

In the rat somatosensory system, the situation is less clear. While no direct equivalent to the 'lag' cells of our model have yet been identified, neurons have been found with latencies that vary as a function of input frequency during single whisker stimulation in restrained rats (Ahissar and Arieli, 2001). Also, neurons have been found which have specific latencies with respect to the onset of a whisking cycle during active whisking (Szwed et al., 2003). Both types of neuron may participate in convergent input mechanisms that transform temporal patterns into spatial patterns. Alternatively, Ghazanfar and Nicolelis (2001) suggest that the spatiotemporal properties of rat S1 neurons may depend on the convergence of asynchronous inputs from different brainstem nuclei, together with feedback from cortex. The resolution of these issues remains a challenge for neuroscientists.

A second question, which also remains open, is whether these spatiotemporal activity patterns are used to support texture discrimination. Our model shows that they can be, and it makes the prediction that disruption of the spatiotemporal patterning of activity in rat S1 should impair its ability to distinguish among textures. However, alternative mechanisms for texture discrimination can be envisaged. For example, texture may be encoded in terms of the intrinsic vibrational modes of whiskers (Mehta and Kleinfeld, 2004). In support of this idea, Fend and colleagues have shown that plucked rat whiskers appear to have a limited number of vibration frequencies, the relative amplitudes of which are modulated in specific ways by different grades of sandpaper (Fend et al., 2003). Of course, these mechanisms are not mutually exclusive and it is possible that the rat brain uses multiple pathways to achieve ethologically significant discriminations.

As well as a simulated somatosensory system, Darwin IX incorporates neural areas analogous to areas involved in fear conditioning and learning, and, as a result, reproduces several features of these phenomena.

Texture discrimination by Darwin IX was assessed behaviorally by a paradigm in which one texture (the CS) was paired with a simulated foot-shock (the UCS). In this paradigm, simulated foot-shock triggers a freezing/avoidance response (the unconditioned response, or UR) and activates a value system which modulates synaptic plasticity between $S2$ and Amy. Activity in Amy due to the CS evokes the same avoidance response (now called the conditioned response, or CR) via excitatory projections to M_{ave}. Repeated pairings of the CS and the US therefore result in a transference of the UCS-UR association to the CS (and CR). By these means, Darwin IX acquired a conditioned aversion to specific textures similar to that shown by animals in comparable situations (see Figs. 5 and 6) (Mackintosh, 1983, Harvey et al., 2001, Prigg et al., 2002).

Also consistent with animal studies (Mackintosh, 1983) is the extinction to conditioned aversion displayed by Darwin IX. During testing, repeated encounters with the CS in the absence of the US caused suppression of activity in the value system (area S) and a corresponding depression of active synapses between $S2$ and Amy. After many such encounters, the activity in Amy evoked by the CS became insufficient to drive M_{ave} above the threshold needed to evoke an avoidance response (Fig. 6).

Central to these behaviors is area Amy, which is analogous to the rodent amygdala. The amygdala has long been implicated in fear conditioning (Kluver and Bucy, 1937), and recent evidence has pointed to roles in the association of sensory information as well as in the execution of motor programs (Maren and Fanselow, 1996). Area Amy in the present model also sustains this dual role, taking input from area $S2$ as well as area FS, and projecting to M_{ave}.

The role of synaptic plasticity in the amygdala during fear conditioning remains a focus of research. Our model suggests a mechanism of plasticity between the amygdala and neurons carrying information about the CS (in our model, area $S2$). Consistent with this idea, synaptic potentiation in the amygdala has been induced *in vivo* by stimulation of afferent connections in putative CS pathways (Maren and Fanselow, 1995). However, establishing a direct link between such potentiation and learned behavior remains a challenge. Our model also suggests that CR extinction may involve synaptic depression in the same pathways that mediate acquisition; again, this hypothesis awaits experimental investigation.

Brain-based devices are not only useful for testing theories of the brain, they also provide a foundation for the development of intelligent machines that follow neurobiological rather than computational principles in their construction. Darwin IX extends previous BBDs in the series by incorporating a sense of touch, which allows texture discrimination and navigation in the dark.

Whisker-based systems have been the subject of increasing attention among engineers interested in building adaptive machines. However, previous work on whisker-based perception in robots has focused on texture discrimination or contact localization by fixed whiskers (Russell, 1992, Kaneko et al., 1998, Fend et al., 2003), or on obstacle avoidance by a mobile robot (Brooks, 1989, Jung and Zelinsky, 1996). Darwin IX extends these studies by showing texture discrimination using whisker arrays mounted on a behaving device, and by incorporating a neural simulation based on the rat somatosensory system.

5. Conclusion

Spatiotemporal transformations are fundamental to neural operations underlying adaptive behavior. Nowhere is this more evident than in the somatosensory system of the rat. Darwin IX illustrates a neural mechanism, based on the rat somatosensory system, which is capable of orchestrating spatiotemporal transformations of whisker signals to allow texture discrimination, and selective conditioning to textures, in a real world environment. Future developments in the Darwin series will explore the integration of somatosensory perception with other modalities, for example vision (Seth et al., in press), and other behavioral tasks, for example navigation.

Acknowledgment

This work was supported by the W.M. Keck Foundation and the Neurosciences Research Foundation. We thank Jim Snook, Donald Hutson, and Doug Moore for their contribution to the design of Darwin IX. We are also grateful to our anonymous reviewers for useful feeback.

Address correspondence to Anil K. Seth (*seth@nsi.edu*).

References

Ahissar, E. and Arieli, A. (2001). Figuring space by time. *Neuron*, 32:185–201.

Almassy, N., Edelman, G., and Sporns, O. (1998). Behavioral constraints in the development of neuronal properties: a cortical model embedded in a real-world device. *Cerebral Cortex*, 8:346–361.

Bienenstock, E., Cooper, L., and Munro, P. (1982). Theory for the development of neuron selectivity: orientation specificit and binocular interaction in the visual cortex. *Journal of Neuroscience*, 2(1):32–48.

Brooks, R. (1989). A robot that walks: Emergent behaviors from a carefully evolved network. *Neural Computation*, 1:153–162.

Carvell, G. and Simons, D. (1990). Biometric analyses of vibrissal tactile discrimination in the rat. *Journal of Neuroscience*, 10(8):2638–48.

Edelman, G., Reeke, G., Gall, W., Tononi, G., Williams, D., and Sporns, O. (1992). Synthetic neural modeling applied to a real-world artifact. *Proc. Nat. Acad. Sci. USA*, 89(15):7267–71.

Fend, M., Bovet, S., Yokoi, H., and Pfeifer, R. (2003). An active artificial whisker array for texture discrimination. In *Proceedings of the IEEE/RSJ International Conference on Intelligent Robots and Systems (IROS)*.

Ghazanfar, A. and Nicolelis, M. (1999). Spatiotemporal properties of layer V neurons of the rat primary somatosensory cortex. *Cerebral Cortex*, 9:348–361.

Ghazanfar, A. and Nicolelis, M. (2001). The structure and function of dynamic cortical and thalamic receptive fields. *Cerebral Cortex*, 11:183–193.

Harvey, M., Bermejo, R., and Zeigler, H. (2001). Discriminative whisking in the head-fixed rat: optoelectronic monitoring during tactile detection and discrimination tasks. *Somatosens. Mot. Res.*, 18(3):211–22.

Jung, D. and Zelinsky, A. (1996). Whisker-based mobile robot navigation. In *Proceedings of the IEEE/RSJ International Conference on Intelligent Robots and Systems (IROS)*, pages 497–504.

Kaneko, M., Kanayama, N., and Tsuji, T. (1998). Active antenna for contact sensing. *IEEE Transactions on Robotics and Automation*, 14(2):278–291.

Kluver, H. and Bucy, P. (1937). An analysis of certain effects of bilateral temporal lobectomy in the rhesus monkey, with special reference to 'psychic blindness'. *American Journal of Physiology*, 119:352–353.

Krichmar, J. and Edelman, G. (2002). Machine psychology: Autonomous behavior, perceptual categorization and conditioning in a brain-based device. *Cerebral Cortex*, 12(8):818–30.

Mackintosh, N. (1983). *Conditioning and associative learning.* Oxford University Press, Oxford.

Maren, S. and Fanselow, M. (1995). Synaptic plasticity in the basolateral amygdala induced by hippocampal stimulation in vivo. *Journal of Neuroscience*, 15:7548–7564.

Maren, S. and Fanselow, M. (1996). The amygdala and fear conditioning: has the nut been cracked? *Neuron*, 16(2):237–40.

Mehta, S. and Kleinfeld, D. (2004). Frisking the whiskers: Patterned sensory input in the rat vibrissa system. *Neuron*, 41:181–184.

Moore, C., Nelson, S., and Sur, M. (1999). Dynamics of neuronal processing in rat somatosensory cortex. *Trends in Neuroscience*, 22:513–520.

Prigg, T., Goldreich, D., Carvell, G., and Simons, D. (2002). Texture discrimination and unit recordings in the rat whisker/barrel system. *Physiol. Behav.*, 77.

Russell, R. (1992). Using tactile whiskers to measure surface contours. In *Proc. IEEE International Conference on Robotics and Automation*, pages 1295–1300.

Saul, A. and Humphrey, A. (1992). Evidence of input from lagged cells in the lateral geniculate nucleus to simple cells in cortical area 17 of the cat. *J. Neurophysiol.*, 68(4):1190–208.

Schultz, W., Dayan, P., and Montague, P. (1997). A neural substrate of prediction and reward. *Science*, 275(5306):1593–9.

Seth, A., McKinstry, J., Edelman, G., and Krichmar, J. (in press). Visual binding through reentrant connectivity and dynamic synchronization in a brain-based device. *Cerebral Cortex.*

Shuler, M., Krupa, D., and Nicolelis, M. (2001). Bilateral integration of whisker information in the primary somatosensory cortex of rats. *Journal of Neuroscience*, 21(14):5251–5261.

Szwed, M., Bagdasarian, K., and Ahissar, E. (2003). Encoding of vibrissal active touch. *Neuron*, 40:621–630.

Woolsey, T. and Van der Loos, H. (1970). The structural organization of layer IV in the somatosensory region SI of mouse cerebral cortex. The description of a cortical field composed of discrete cytoarchitectonic units. *Brain Research*, 17(2):205–42.

A biomimetic whisker for texture discrimination and distance estimation

DaeEun Kim* and Ralf Möller**

* Cognitive Robotics, Max Planck Institute for Human Cognitive and Brain Sciences Amalienstraße 33, Munich, 80799, Germany daeeun@cbs.mpg.de

** Computer Engineering Group, Faculty of Technology, Bielefeld University D-33501 Bielefeld, Germany moeller@techfak.uni-bielefeld.de

Abstract

Rodents use their whiskers (vibrissae) for tactile perception, specifically for texture discrimination and shape recognition. In this paper, we present an artificial whisker system biologically inspired by rodent whiskers. We first focus on the question, how the natural frequencies of the whisker bar affect the processing of tactile signals. For texture discrimination it is shown that separating the range of the tactile signals from the natural frequencies improves the classification quality. In contrast, natural frequencies prove to be a source of information for distance estimation. It is also demonstrated that the time course of deflection angles or deflection velocities can determine the distance of objects.

1. Introduction

Rats can use whiskers to discriminate surfaces and recognize objects even in darkness. The intriguing tactile perceptual abilities of rodents were investigated in behavioral experiments, and neurophysiological properties of the peripheral sensory apparatus and of the subsequent neural processing stages (Deschenes et al., 2003, Kleinfeld et al., 1999, Fanselow and Nicolelis, 1999, Fee et al., 1997).

Carvell and Simons (1995) studied texture discrimination in rats. They investigated two cases: a rough-rough texture discrimination between a narrow (1 mm) spacing gratings and a wider (down to 1.125 mm) spacing gratings, and a rough-smooth discrimination between 50 μm or larger spacing grooves and smooth surfaces. The experiments demonstrate that the rough-rough discrimination is only possible with multiple whiskers, while a single whisker is sufficient for the rough-smooth discrimination. Thus, the rough-rough discrimination appears to be based on intercolumnar processing within the barrel cortex. From this result, Neimark et al. (Neimark et al., 2003) argued that there are similarities between the whisker system and the cochlear structure (Geisler, 1998) in the auditory system of bats. Each whisker array consists of varying sizes of whiskers and a set of resonant frequencies can be defined depending on the size of whiskers. The set of resonant frequencies can be interpreted as a set of receptive fields for vibration signals. A recent physiological result showed that cortical neurons of rats can encode the product of frequency and amplitude, that is, the speed of vibration, which underlies texture discrimination (Arabzadeh et al., 2003).

There have been approaches to use mechanical probes or antenna to recognize objects. Russell (1992) and Kaneko et al. (1998) studied geometrical analysis when the probes touch an object with arbitrary shape. As was suggested by Ueno and Kaneko (1994), the contact point between a flexible beam and an object can be determined by observing the fundamental resonant frequency. Their model assumes a mass at the tip of the sweeping antenna, because more mass at the tip allows to determine the contact point only by the fundamental natural frequency. In fact, biological whiskers are conically shaped with large diameters at the base of the whiskers and small diameters at the tip rather than a large mass at the tip.

A biomimetic whisker system and its properties have been studied by Lungarella et al. (2002) and Fend et al. (2003). Their artificial whisker is based on a capacitor microphone with a real mouse whisker directly attached to the sensitive surface. Such a setup allows to directly explore the properties of natural whiskers. Recently texture discrimination experiments with artificial whiskers have been reported (Fend et al., 2003). Especially discriminating complex textures such as sandpapers proved to be a difficult task. Hipp and König (ETH Zurich, personal communication) reported a 50 % success rate for discriminating seven sandpapers with different grit numbers, using active sweeping of artificial whiskers.

In the biological whisker system, the whisker shaft is embedded in a follicle structure. Mechanoreceptors surrounding the shaft measure deflection in all directions, some of them with tonic, others with phasic response (Lichtenstein et al., 1990, Szwed et al., 2003,

Shoykhet et al., 2000, Hartings et al., 2000). Our artificial whisker system imitates this design in a very simplified way. In all our designs, the whisker is surrounded by sensors, and whisker deflection either affects these sensors without contact (magnetic sensors affected by magnets on the whisker shaft, or light sensors affected by a light source on the whisker shaft) or with contact (piezo sensors). Here we only present two of the sensor types, those based on magnetic and those based on piezo sensors. The artificial whiskers can detect the deflection angle and the deflection direction for both low frequencies (including static deflection) and high frequencies (texture-related signals).

In the following, we provide a mathematical analysis of the oscillations of the whisker bar, especially of its natural frequencies (section 2), we show that distance estimation can exploit changes in the natural frequencies (section 3.1), we report experiments on texture discrimination with a single whisker (section 3.2), and we present results on the distance estimation using active whisking (section 3.3).

2. Vibration analysis

The vibration of a biological whisker can be treated by mechanical beam analysis. In the following we describe three different experiments, how the natural frequencies can be determined for a given contact point. We assume that the whisker has a constant diameter from the base to the tip (in contrast to biological whiskers which usually have a conical shape where the diameter decreased towards the tip).

A dynamic movement of a beam and its vibration frequency can be analyzed by the Bernoulli-Euler equation (Weaver et al., 1990, Kelly, 2000) as given below:

$$\frac{\partial^2}{\partial x^2}\left[E\cdot I(x)\frac{\partial^2 y(x,t)}{\partial x^2}\right]+m(x)\frac{\partial^2 y(x,t)}{\partial t^2}=f(x,t)$$

This equation describes the time course of the transverse displacement y of a beam when a given force $f(x,t)$ is applied; see Figure 1. The transverse displacement at any point x and time t is denoted by $y(x,t)$ and the transverse force per unit length by $f(x,t)$. The system parameters are the mass per unit length $m(x)$ and the flexual rigidity $E\cdot I(x)$, where E is Young's modulus of elasticity and $I(x)$ the cross-sectional area moment of inertia passing through the center of the cross-sectional area. By simple beam theory, we assume that the rotation of an element is insignificant compared to the vertical translation, and the shear deformation is small in relation to the bending deformation.

The solution of the Euler-Bernoulli equation (Kelly, 2000, Weaver et al., 1990) is

$$\begin{aligned} y(x,t) &= [c_1\cos(\beta x)+c_2\sin(\beta x)+c_3\cosh(\beta x) \\ &\quad +c_4\sinh(\beta x)]\cdot u(\omega,t) \qquad (1)\end{aligned}$$

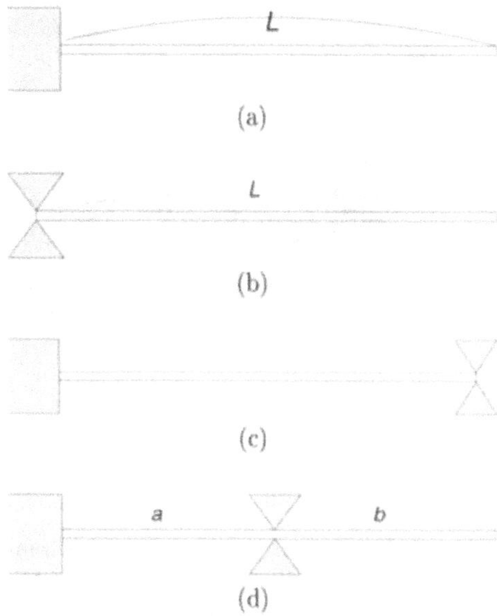

Figure 1: Boundary conditions of cylindrical beams (a) fixed-free (b) pinned-free (c) fixed-pinned (d) fixed-pinned-free

where $u(\omega,t)=\alpha_1\cos(\omega t)+\alpha_2\sin(\omega t)$ and $\beta^4=\frac{\omega^2 m}{EI}$.

Since the diameter of the whisker is constant, the cross-sectional area moment $I(x)$ of the uniform whisker bar will be constant. There are several boundary conditions of whisker beams depending on whether they are in touch with any object or how they are mounted as shown in Figure 1. For each situation, the distribution of natural frequencies varies.

For the fixed-free situation in Figure 1(a), one point of the cylindrical beam is fixed or clamped, and the other end point is free in space. The dynamic conditions are given as $y(0,t)=0$, $y'(0,t)=0$, $y''(L,t)=0$, $y'''(L,t)=0$ where L is the length of the beam and y' is the derivative of y with respect to x. These boundary conditions can be used for Equation (1). The solution will lead to

$$\cos(\beta L)\cosh(\beta L)=-1. \qquad (2)$$

From this formula, we can obtain a numerical solution of βL. We have $\omega=\beta^2\sqrt{\frac{EI}{m}}$ from Equation (1). If we set $\beta L=k$, then the natural frequencies can be estimated as

$$f=\frac{\omega}{2\pi}=\frac{k^2}{2\pi L^2}\sqrt{\frac{EI}{m}}=\frac{k^2}{2\pi L^2}\sqrt{\frac{EI}{\rho A}} \qquad (3)$$

where ρ is the material density, A is the cross-sectional area, L is the length of the whisker and the numerical solution of k is given as 1.875, 4.694, 7.855, 10.996, 14.137 ... by $\cos(k)\cosh(k)=-1$ in Equation (2).

For the pinned-free case shown in Figure 1(b), the boundary conditions are $y(0,t)=0$, $y''(0,t)=0$,

(a) magnetic sensors (b) piezo sensors

Figure 2: Artificial whiskers

$y''(L,t) = 0$, $y'''(L,t) = 0$. Its solution is $\tan(\beta L) - \tanh(\beta L) = 0$. For the fixed-pinned case shown in Figure 1(c), the boundary conditions are $y(0,t) = 0$, $y'(0,t) = 0$, $y''(L,t) = 0$, $y'''(L,t) = 0$. Its solution is $\tan(\beta L) - \tanh(\beta L) = 0$. The above three boundary conditions have been handled for biological whiskers (Neimark et al., 2003, Hartmann et al., 2003).

Now we develop a general form of boundary conditions, called fixed-pinned-free[1]. For the fixed-pinned-free case in Figure 1(d), we can divide the beam into two segments, a and b. One end of the whole beam is fixed and the other end is free. Also the natural frequencies of two segments should be the same. Then the boundary conditions are given as follows:

$$\begin{aligned} y_1(0,t) &= 0, \frac{\partial y_1(0,t)}{\partial x} = 0, \\ \frac{\partial^2 y_2(b,t)}{\partial x^2} &= 0, \frac{\partial^3 y_2(b,t)}{\partial x^3} = 0 \\ y_1(a,t) &= y_2(0,t) = 0, \\ \frac{\partial y_1(a,t)}{\partial x} &= \frac{\partial y_2(0,t)}{\partial x}, \quad \frac{\partial^2 y_1(a,t)}{\partial x^2} = \frac{\partial^2 y_2(0,t)}{\partial x^2} \end{aligned}$$

From the above conditions, we can obtain the solution as follows:

$$\begin{aligned} & \frac{2 - 2\cos(\beta a)\cosh(\beta a)}{\cos(\beta a)\sinh(\beta a) - \sin(\beta a)\cosh(\beta a)} \\ = & \frac{2 + 2\cos(\beta b)\cosh(\beta b)}{\cos(\beta b)\sinh(\beta b) - \sin(\beta b)\cosh(\beta b)} \end{aligned} \quad (4)$$

The fixed-free model is a special case of the fixed-pinned-free model with $a = 0$, $b = L$, which will lead to $\cos(\beta L)\cosh(\beta L) = -1$. Also the fixed-pinned model can be obtained from $a = L$, $b = 0$.

In the fixed-pinned-free model, the pinned position can be uniquely determined by investigating a series of natural frequencies. In the whisker sweeping process, the initial contact moment on an object can be modelled approximately with the fixed-pinned-free model. The contact position of whisker varies depending on where the edge of objects is. The above equation for the fixed-pinned-free model will be tested with an artificial whisker in the next section.

[1]The two segment approach has been first applied to a beam with a mass at the tip by Ueno and Kaneko (1994).

3. Experiments

Our experiments consist of three parts, (1) analysis of the natural frequencies (2) texture discrimination and (3) distance estimation. We use two types of artificial whiskers, based on hall-effect sensors and piezo sensors, which are displayed in Figure 2. The hall-effect magnetic sensors can measure the difference of magnetic flux at two sensors on opposite sides caused by the movement of whiskers to which a permanent magnet is attached. Directional sensitivity was achieved by using two pairs of sensors arranged orthogonally to each other. Thus, from each whisker, we obtain two signals describing deflection angle and deflection amplitude. Housing at the whisker base (small size 12 × 12 × 20 mm in Figure 2(a)) contains two sets of amplifiers to amplify the difference of sensor signals between a pair of magnetic sensors. The two amplifier output lines are shielded to prevent interference from external electromagnetic fields. The piezo-type whisker consists of four piezoelectric sensors and a rubber part surrounding the whisker. The vibration of whisker is transferred to the rubber and the piezoelectric sensors can sense the vibration. Piezo sensors have the advantage that the whisker base can be reduced to small size. It is a disadvantage of piezo sensors that no DC signal is delivered, so that an integrator circuit would be required to measure static deflections.

3.1 Natural frequencies

We first examine if the theoretical solution of natural frequencies can be observed in the experiments. For the experiment, we used a cylinder-shaped steel bar with 0.5 mm diameter. For this material and size, we obtain $E = 203.4$ GPa, $\rho = 7830$ kg/m^3, $I = \frac{1}{4}\pi r^4$, $A = \pi r^2$, $L = 25.9$ mm, $r = 0.25$ mm where r is the radius of the whisker bar. Then we estimate the natural frequencies as follows:

$$f_A = \{5.31,\ 33.30,\ 93.25,\ 182.75,\ 302.09,\ 451.28...\}.$$

In a similar way, we can obtain natural frequencies for various boundary conditions by numerical solution of Equation (4). In the fixed-pinned model with $a = L$ and $b = 0$, we have the natural frequencies

$$f_C = \{23.30,\ 75.52,\ 157.57,\ 269.46,\ 411.19,\ 582.75, ...\}.$$

If we have a pinned point along the whisker bar as an example of the fixed-pinned-free model (see Figure 1(d)), for instance, with the condition $\frac{a}{a+b} = \frac{a}{L} = 0.7$ for $L =$ 25.9 mm, its natural frequencies are given as

$$f_D = \{28.91,\ 66.30,\ 168.94,\ 291.68,\ 372.11,\ 582.65,\ ...\}.$$

To test the fixed-free model, the artificial whisker is clamped at the one end, and the tip of the whisker is free in open space. An impulse force was given at the tip of the whisker and the vibration signals were collected. For a given set of m data samples $d_1, d_2, ..., d_m$, we calculate the averaged amplitude spectrogram $\bar{S}$ as follows:

$$\bar{S} = \frac{1}{N} \sum_{n=1}^{N} \log |FFT(v_{a_n})|$$

$a_n = (n-1) * D + 1$ and v_{a_n} is a set of W samples such that $v_{a_n} = \{d_{a_n}, d_{a_n+1}, d_{a_n+2}, ..., d_{a_n+W-1}\}$. W is a window size for one FFT and this FFT calculation will be repeated N times where $N = \lfloor \frac{m-W+1}{D} \rfloor + 1$. In the experiments, we set $D = 100, W = 1000, m = 5,000$ (W-D (900) samples overlapped for FFT) and its sampling rate 6104 Hz, and the average amplitude spectrogram $\bar{S}$ will be evaluated over 5,000 data samples; this process of calculation of spectrogram for one set of data samples will be called one epoch. We collected five epoches of data as shown in Figure 3. There are dominant frequencies consistent with the theoretical resonant frequencies as expected. The measured natural frequencies $f = \{5.2,$ 32.3, 90.3, 179.4, 289.3, 416.3, ...$\}$ by experiments shown in Figure 3 slightly deviate from the theoretical frequencies f_A, but we observe a good correspondence for the first four resonant frequencies.

In a second experiment, we explored whether the contact position of an applied force can be determined by observing the natural frequencies. For this experiment, we use the drum setup described in section 3.2. The whisker was touching the drum at only one point and the drum (covered with a texture) was rotating underneath the whisker, in this way applying a time-varying force to the whisker at the given contact point.

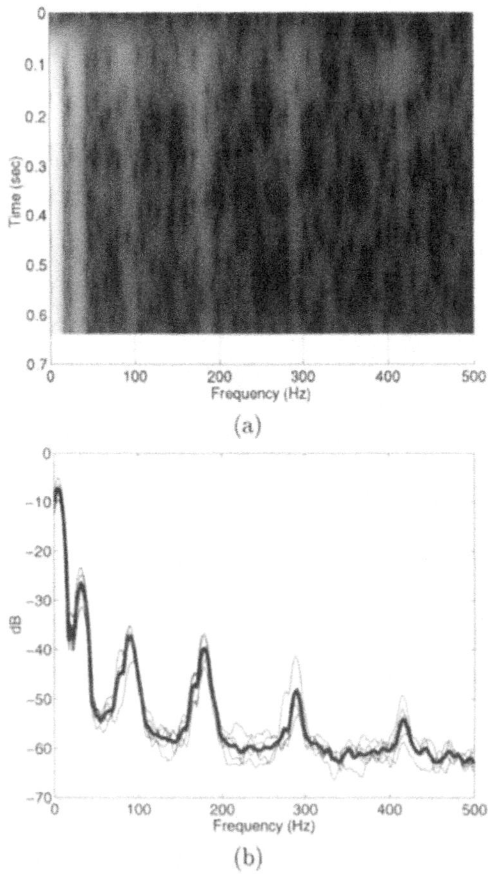

Figure 3: Frequency spectrum for a fixed-free whisker (a) example spectrogram for one epoch of data (bright colors mean high amplitudes (dB)) (b) averaged amplitude spectrogram over five epoches of data (thick: average, thin: each epoch)

The contact position of a beam while sweeping can be determined by observing the resonant frequencies (Ueno and Kaneko, 1994). The theoretical estimation of those resonant frequencies can be obtained from the fixed-pinned-free model in Equation (4). We tried to find the frequency values for the dominant peaks irrespective of surfaces when the contact position is changed. As expected, we can obtain a list of natural frequencies depending on contact position as shown in Figure 4. The experimental result for the first three resonant frequencies is close to the theoretical estimation. Larger discrepancies between theory and experiment only occur for higher natural frequencies. We suppose that this deviation can be explained by the fact that the theoretical solution does not consider damping which would have an impact on high frequencies.

3.2 Texture discrimination

To study texture discrimination, we use passive artificial whiskers in a drum setup (Figure 5). The diameter of

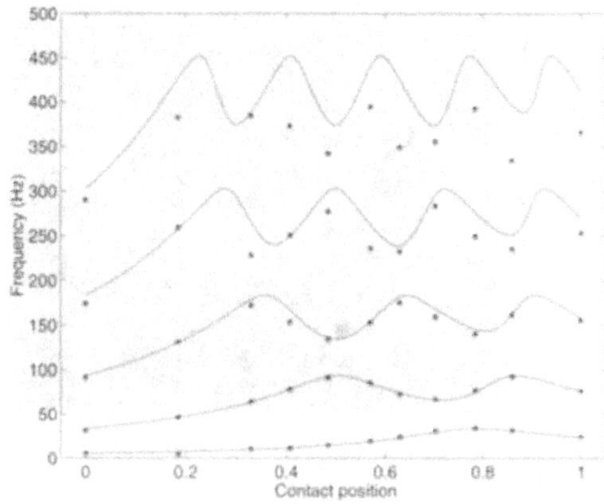

Figure 4: Natural frequencies varying with contact position (solid curves: analytical solution, dots(*): experimental results, the contact point is given relative to the whisker length)

the drum is 60 mm and a DC motor with encoder is driving the drum, and the rotation speed is stabilized by a feedback controller. In the experiments, the tip of the artificial whisker is placed on the rotating drum. Seven different sandpapers with grit numbers 40, 60, 80, 120, 150, 180 and 400 ($P40, ..., P400$) were wrapped around the drum.

From the signals of only one whisker channel (vertical) an averaged amplitude spectrum is determined as described in section 3.1. Based on this spectrum, we attempt to distinguish different textures. Normally the same dominant resonant frequencies (corresponding to the fixed-pinned model) always appear, regardless of surfaces. The differences between the textures are most pronounced in the region below 100 Hz with a given rotation speed of the drum. We also observe that the specific surface texture does not have an influence on the amplitude in regions close to the natural frequencies.

For the texture experiment, we set the rotation speed to a 1/4, 1/2, 3/4 of revolution per second, which will be called speed 1, speed 2, speed 3, respectively. The speed 1 corresponds to about 47 mm/sec. This speed is relatively small when compared with the sweeping speed of whiskers observed in rats, 200-2000 mm/sec (Carvell and Simons, 1995).

To avoid interference, the sweeping frequency and the natural frequencies of the whiskers should not be in the same range. This limits the sweeping frequencies in our active whisker setup. The steel whisker bar in use has the fundamental resonant frequency 5-60 Hz (artificial whisker length 80-270 mm), depending on the length of whiskers, so the corresponding sweeping speed values of the drum experiments lie in the appropriate range for our setup. Moreover, with lower rotation or sweeping speed a larger part of the amplitude spectrum is concentrated at lower frequencies, which avoids interference between the natural frequencies and the texture-related signals.

Figure 5: Drum setup

Initially we tested a hard paper texture with regular teeth shown in Figure 5 and it generated a high dominant peak at a low frequency 6 Hz with speed 2. We tested another periodic grating texture (40 pin cable, spatial frequency approximately 0.8/mm) and could be easily detectable by the averaged amplitude spectrogram. We observed a pronounced change in the dominant frequencies with varying rotation speed; as expected, higher rotation speed generated higher dominant frequencies.

A single artificial whisker can detect spatial frequencies of regular textures without difficulty, because a dominant peak at the corresponding frequency is constantly observed in the spectrogram as well as the average amplitude spectrogram. The discrimination of regular textures is thus a relatively easy task for an artificial whisker. As an example of a more "natural" texture we used sandpapers with different grit values in the texture discrimination experiment. They produce a broad band of frequencies in the spectrum rather than one specific frequency.

The average amplitude spectrogram from the sample data (W=500, D=50, m=10000, sampling rate 1523Hz) is normalized by dividing each logarithmic amplitude value by the total sum of the amplitudes. This normalization result will have a potential of being applied to the active sweeping case where the contact force may vary. Figure 6 shows examples of the normalized average spectrogram over ten epoches of data for each surface (smaller grit number means more rough surface). In Figure 6(a) rough surfaces (for instance, grit 40) have large amplitude in lower frequencies (0-40 Hz), while the smooth surface (grit 400) have large amplitude in higher frequencies (50-100 Hz). In Figure 6(d), a prominent natural frequency at approx. 50 Hz is visible. Texture-related signals are mostly found at frequencies below this natural frequency with speed 1.

To analyze how this prominent peak influences the quality of texture discrimination, we train a multilayer perceptron with the error back-propagation method for the classification of the different textures; the multilayer

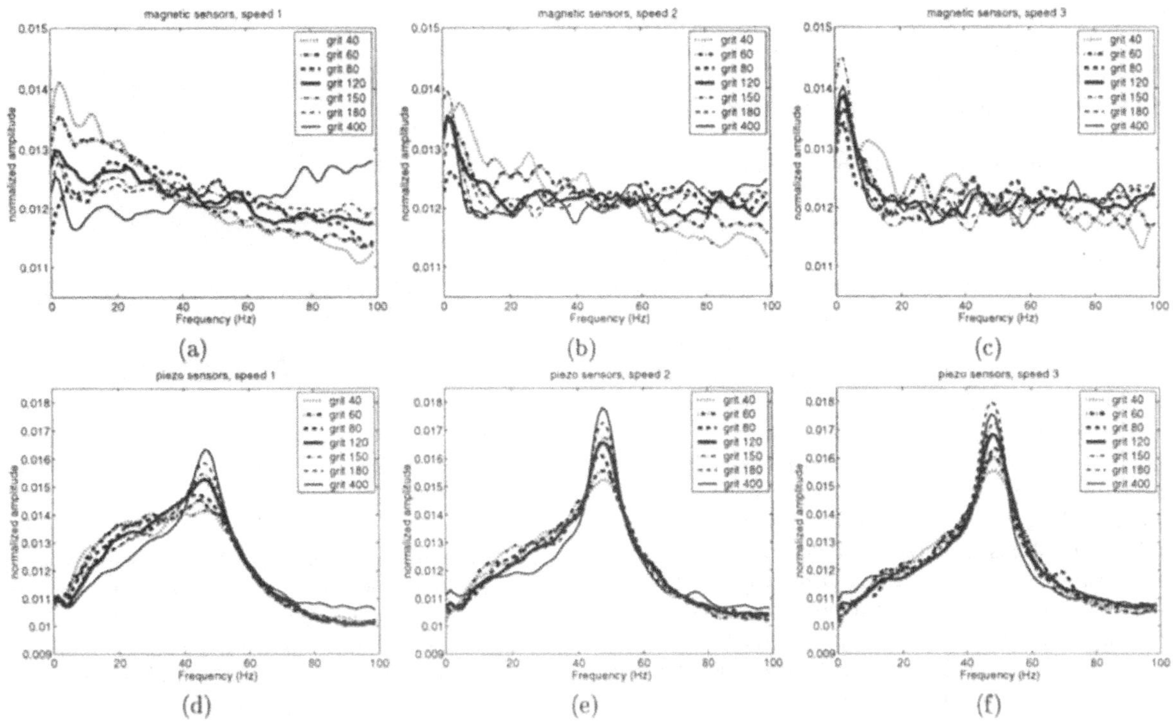

Figure 6: Examples of averaged amplitude spectrograms with magnetic sensors and piezo sensors on various surface textures (lines: average of ten epoch results) (a) speed 1 (b) speed 2 (c) speed 3 for magnetic sensors with 80 mm whisker length (d) speed 1 (e) speed 2 (f) speed 3 for piezo sensors with 175 mm whisker length

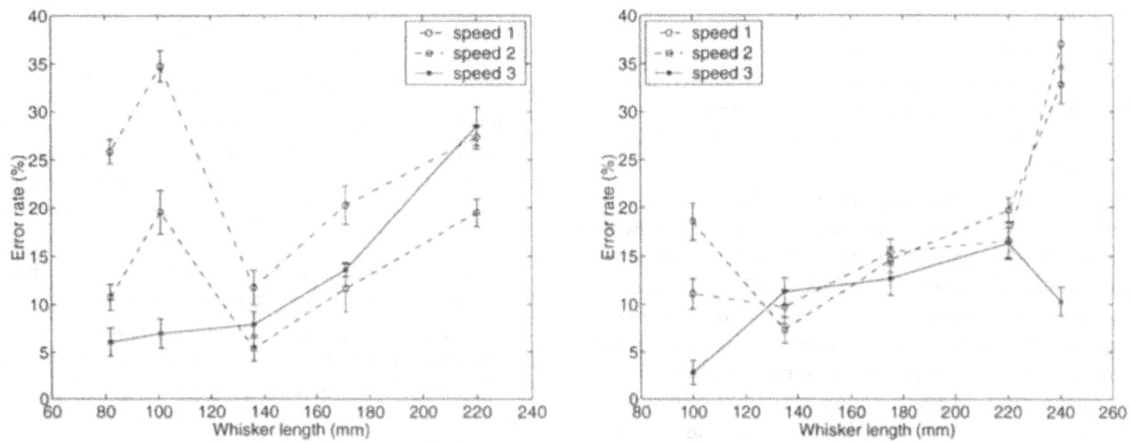

(a) magnetic sensors with 0.5 mm whisker diameter (b) piezo sensors with 0.5 mm whisker diameter

Figure 7: Classification error rates with varying sizes of artificial whiskers

perceptron has one hidden layer with 20 nodes and an output layer with 7 output nodes (7 categories). The number of nodes in the input layer depends on input frequency ranges; normalized amplitudes for frequencies 12-90 Hz are used for the input and they are sampled every 1.22 Hz, thus leading to 65 input nodes. We have seven different sandpaper textures and collected ten epoches of data for each texture. Among ten epoches of data, five epoches of amplitude spectrograms were used for training data and the other five sets were taken for test data. We measured the classification error rate for test data. The training of the neural network was repeated 25 times to calculate the average performance. The result with the average error rate is shown in Figure 7 with 95 % confidence intervals by assuming t-distribution. We used a variety of input ranges (12-25 Hz, 12-40 Hz, 12-50 Hz, 12-60 Hz, 12-90 Hz) for neural network training to recognize what frequency zone is effective for classification; several frequency zones starting from frequency 12 Hz were tested because a variation in the rotation speed of the drum changes the frequencies of texture signals. The number of input nodes will vary depending on the input ranges.

Figure 7 is displayed with the best performance among several input ranges for each size of whiskers. We can observe that long whiskers generally have higher error rates than short whiskers in the range of whisker length 120-240 mm. We believe that short whiskers can have more force on the surfaces and there seems to be a better transmission of the surface structure to the oscillation of the whisker bar. The natural frequencies interfere with the texture discrimination, since they occupy the same frequency range as the texture-related signals. The neural network fitting tends to utilize the activation of the natural frequency partly for their classification. For example, the piezo sensor whisker with length 220 mm and speed 1-3 has the best result for 12-40 Hz, and its first natural frequency is 32 Hz. Other results with long whiskers show that the neural network prefers to use the amplitude around the resonant frequency zone for correct classification. However, the classification performance only with a zone in immediate neighborhood of the resonant frequency is mostly worse than with other input ranges - e.g. 175mm length (40-60 Hz neighbor zone in Figure 6(d)) and 220 mm length (20-40 Hz neighbor zone). There is no clear separation among several different textures near the resonant frequency, because the texture signals easily trigger the natural frequency. Except for the above-mentioned cases, the overall result is that a whisker's natural frequencies interfere with the texture-related signals and thus make texture discrimination more difficult.

High rotation speed supports correct classification for small whiskers (80-100 mm), which means high sweeping speed leads to a clear separation on a variety of texture signals. We obtained up to 93-97 % success rate for classification with short whiskers and high sweeping speed. In the experiments piezo sensors have a better discrimination property for surface texture signals than magnetic sensors in many cases. We tested two different whisker diameters, 0.5 mm and 0.8 mm with magnetic sensors. The smaller diameter whiskers show significantly better results. It is believed that rodents prefer to use many small whiskers, called microvibrissae, which are positioned in the front, for texture detection (Hartmann, 2001, Brecht et al., 1997). To summarize, shorter whiskers with piezo-electric transducers are better suitable for texture recognition; these could be interpreted as a simple model of rodent microvibrissae.

3.3 Distance Estimation

The distance estimation experiments described in the following use active whisking, where a servo motor rotates a plate on which an array of whiskers is mounted. Its maximum rotation angle is 110 degrees and the whiskers on the plate share the same angular movement. Whisker sensors can be mounted in arbitrary position and orientation on the plate. An array of whiskers are mounted in the same horizontal plane with an angular spacing of neighbor whiskers, about 30 degrees. Since the whiskers are moved by the same servo motor, their protraction and retraction movements are synchronous. This closely corresponds to the whisking behavior of rodents — although each whisker is moved by a separate muscle (Sachdev et al., 2001), mostly all whiskers on one side move in synchrony (Wineski, 1985).

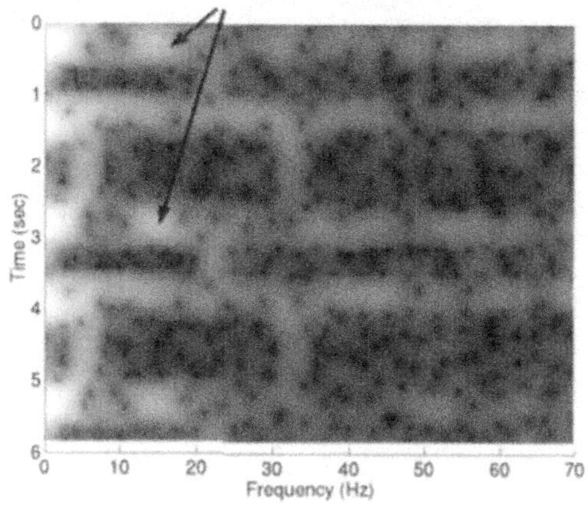

Figure 8: Detection of the contact point while continual sweeping (arrows: 15 Hz dominance)

Here, we investigate possible ways to estimate the distance between an object and the robot (the whisker base) with the artificial whiskers. First, we will test the aforementioned method for determining the contact point from the natural frequencies. In a second experiment we will measure the time course of deflection angles or deflection velocities to estimate the distance.

The contact position of the whisker can be determined by observing the natural frequencies as shown in Figure 8; 6 Hz and 33 Hz are dominant when the whisker is free (fixed-free model), and 15 Hz dominance appears when the middle of the whisker touches an object in a sweeping mode. It corresponds to one of natural frequencies for the pinned position or contact position in the fixed-pinned-free model. We can extract the distance information from the curves of natural frequencies shown in Figure 4. If the time course of the natural frequencies is available, then we can determine the lateral shape using the changing rate of the contact location in a sweeping mode. However, in a sweeping whisker, the signals are strongly dampened after contact, and thus it proves to be impossible to extract shape information directly.

A combination of distance information based on a variety of force directions by an active antenna has

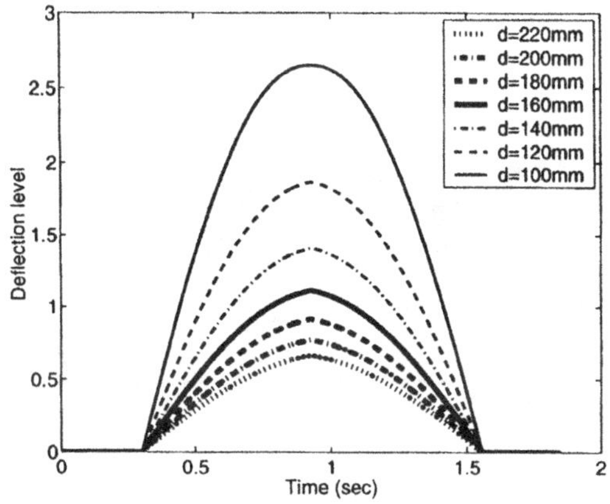

Figure 9: Theoretical deflection amplitudes depending on distance

been used to build the shape information of an object (Tsujimua and Yabuta, 1989). If the torque τ and the angular displacement α at the antenna base are given, the distance can be easily estimated by the equation $3EI\alpha/\tau$ (Kelly, 2000, Kaneko et al., 1998). In contrast, our artificial whisker system can record the deflection angle of a whisker, proportional to the contact force, but does not measure torque. Thus, we investigate if the distance can be obtained only from the time course of deflection angles.

For a given torque τ during contact by active whisking, we can calculate the displacement of the whisker (Kelly, 2000) as follows:

$$EI\tan\theta = \frac{\tau}{2d}x^2 - \tau x + \frac{1}{3}\tau d$$

where $\tan\theta$ is the slope at a position x along the whisker bar, and d is the distance of an object from the whisker base at the onset of contact. The magnetic sensors in the artificial whisker measure the deflection of whisker 10 mm in front of the clamped position (whisker base). Applying the sensor position $x = h = 10mm$ and the clamped position $x = 0$ to the above formula, we cancel out τ and obtain the following equation:

$$\tan\theta_1 = (\frac{3h^2}{2d^2} + \frac{3}{d} + 1)\cdot\tan\theta_0 \quad (5)$$

where h is the distance between the sensor position and the clamped position, and θ_0 and θ_1 are the angle of the clamped position and the sensor position, respectively ($\theta_0 = \theta_1 = 0$ at the onset time of contact). The angle θ_0 is the same as the angular movement of the whisker by a servo motor after contact with an object, and thus $|\theta_1 - \theta_0|$ is the deflection angle that the sensors measure. Figure 9 shows the theoretical estimation for deflection amplitude of $|\theta_1 - \theta_0|$ depending on the distance of an object (edge contact) in a period of protraction/retraction.

The voltage obtained from the horizontal channel of the whiskers increases with the deflection angle when the whisker is sweeping past an object. Figure 10(a) shows the voltage increase and decrease depending on the deflection angle in a protraction/retraction period; a hexahedron object is placed such that the whisker can contact the edge, and the sensor signal in the figure was pre-processed by a low-pass filter. One whisker touches an object at an earlier time than the other and its contact lasts longer. The anterior whisker (upper in the figure) thus has a larger convex curve than the posterior (lower in the figure). The deflection amplitudes from the sensors show a similar curve to the theoretical estimation shown in Figure 9, except the part of large deflection angles; for large deflection angles, the sensor signal of the whisker does not reflect correctly the corresponding deflection angle and it is caused by the current sensor structure. The level of the sensor signal at the peak is inversely proportional to the distance; larger deflection angle after contact with an object is observed in a closer distance.

We can estimate the distance by the deflection amplitude at a given angular position after contact. A distance mapping function $f(\nu, \delta, x)$ can be built for a given angular sweeping speed ν and $0 < \delta \leq \Delta$, where δ is the delay after the onset time t_i of contact with an object (during protraction of the whisker), Δ is the duration of contact, $t_i + \delta$ is a pick-up time to observe the deflection amplitude, and x is the level of the deflection signal at $t_i + \delta$. An alternative approach is to estimate the distance by the level of the peak point in the deflection velocity signal, as shown in Figure 10(b); the velocity is obtained by calculating the derivative of deflection amplitude signal. The peak time also gives the information of the angular position of the object in a sweeping area. This method with the deflection velocity does not need to consider the pick-up time for distance estimation. The distance estimation methods with deflection signals can produce the same result even with different sizes of the whiskers; Equation (5) shows that the deflection angle in our whisker system is not affected by the length or the material of the whiskers. Also different lateral shapes (e.g. cylinder and hexahedron) can be distinguished with deflection signals. Thus, observing the time course of deflection signals could be sufficient to determine the distance of objects from the whisker base. All the whiskers on a rotating plate have the same angular movement, and so a collection of the distance information from each whisker in an array can produce more correct distance estimation.

The onset time of contact or deflection can be crucial to determine the angular position of an object, when the sweeping speed and the sweeping angle are available. If the whiskers are stacked vertically, we can easily check the vertical shape of an object by observing the onset time of deflection. For example, a conically shaped object will produce a time-shift in whisker signals between

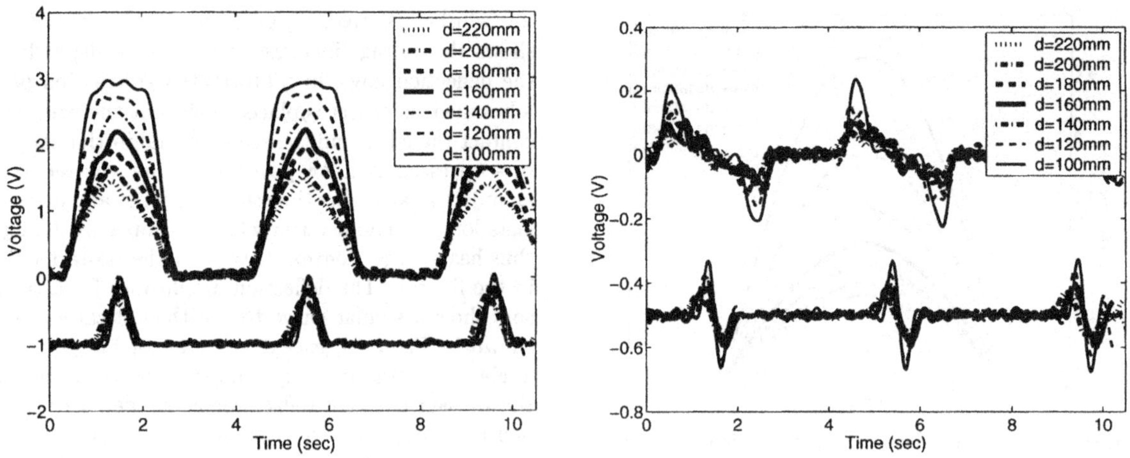

(a) deflection amplitude signal

(b) deflection velocity signal

Figure 10: Deflection amplitude and its velocity signals depending on distance

the upper whisker and the lower whisker. The time-shift can also be observed in the deflection velocity signals without difficulty.

In summary, the contact position of the whisker can be determined by observing the natural frequencies, the deflection amplitudes at a given delay time after contact, or the deflection velocity signals. Thus, a single whisker could be sufficient to estimate the distance.

4. Discussion

Active whisking generates a different amount of force at the tip of the whisker depending on how much the surface is tilted towards the whisker position. In this case, the texture discrimination result shown in section 3.2 may be changed. It will require an appropriate movement of the robot to a desirable position. We leave the study of head position for a good performance of texture discrimination to future work.

The sweeping frequency for active whisking is an important factor for texture discrimination. Rodents can use a sweeping frequency of 5-25 Hz (Berg and Kleinfeld, 2003, Carvell and Simons, 1995) They often set the sweeping frequency to 6-9 Hz for free whisking (corresponding to the fixed-free case). Once the rodent whisker is in contact with any object, the sweeping frequency of the whisker increases up to 25 Hz. In the behavioral experiments of rodents a higher sweeping frequency is associated with good performance of texture discrimination on the rough-rough task (Carvell and Simons, 1995). This fact agrees with our experimental results for short artificial whiskers; higher sweeping speed improves the texture discrimination performance.

It is still an open question how the brain system of rodents processes natural frequencies for texture discrimination. In our biomimetic whisker system, the natural frequencies easily interfere with the texture signals, even if they can be a source of distance estimation. Neimark et al. (2003) suggested a model of resonant frequencies in which small amplitude signals can be amplified by resonance of each whisker to facilitate detection of texture input stimuli. However, this model might not be applied to the texture discrimination task with our artificial whiskers. Currently the artificial whiskers have a low sweeping speed and large sizes ranging from 80 mm to 270 mm, when compared with rodent whiskers. It should be noted that our artificial whiskers are biologically inspired rather than model the real rodent whiskers.

We suggest that the time course of deflection signals or the deflection velocity signals during active whisking can estimate the distance and position of an object, which has been neglected so far for an active antenna system. This may partly support the neurophysiological results of rodents that cortex neurons encode the deflection velocity of rodent whiskers (Shoykhet et al., 2000). The deflection signals might be compared with the trigeminal ganglion neuronal responses during active whisking (Szwed et al., 2003).

5. Conclusion

In this paper, we present an artificial whisker system based on magnetic sensors or piezo sensors, which was developed to model the properties of its biological counterpart. We first showed the dynamic property of a whisker beam and analyzed the natural frequencies using the fixed-pinned-free model. The experiments with our artificial whiskers showed a close correspondence to the theoretical estimation of natural frequencies. It was also demonstrated that this property can be used to estimate the distance of objects from the whisker base.

For texture discrimination, we tested a variety of whisker lengths on sandpapers with seven different grit numbers. The natural frequencies easily interfere with the texture-related signals and decline the classification quality. Shorter whiskers with high sweeping speed tend to produce better results in the texture discrimination task. The deflection amplitude and deflection velocity signals by active whisking show the potential of estimating the distance of objects, which will be the basis of shape information.

From the results we derive the assumption that long whiskers (corresponding to macrovibrissae in rodents) are best serving as a distance detection array as Brecht et al. (1997) assumed, while short whiskers are primarily useful for texture discrimination (corresponding to microvibrissae). For future work, we will study the active whisking for texture discrimination, and apply the distance estimation methods to recognize an object with arbitrary shape.

6. Acknowledgements

The author would like to thank anonymous reviewers for useful comments and suggestions, Mathew E. Diamond (SISSA Trieste) for the discussion of biological whiskers and Jörg Hipp (ETH Zürich) for his information on their artificial whisker experiments. This work has been supported by the EU in the project AMOUSE (IST-2000-28127).

References

Arabzadeh, E., Petersen, R., and Diamond, M. (2003). Encoding of whisker vibration by rat barrel cortex neurons: Implications for texture discrimination. *Journal of Neuroscience*, Oct, 23(27):9146–9154.

Berg, R. and Kleinfeld, D. (2003). Rhythmic whisking by rat: Retraction as well as protraction of the vibrissae is under active muscular control. *Journal of Neurophysiology*, 89:104–117.

Brecht, M., Preilowski, B., and Merzenich, M. M. (1997). Functional architecture of the mystacial vibrissae. *Behavioural Brain Research*, 84:81–97.

Carvell, G. and Simons, D. (1995). Task- and subject-related differences in sensorimotor behavior during active touch. *Behavioural Brain Research*, 12(1):1–9.

Deschenes, M., Timofeeva, E., and Lavallee, P. (2003). The relay of high-frequency sensory signals in the whisker-to-barreloid pathway. *Journal of Neuroscience*, 23(17):6778–6787.

Fanselow, E. and Nicolelis, M. (1999). Behavioral modulation of tactile response in the rat somatosensory system. *Journal of Neuroscience*, 19(17):7603–7616.

Fee, M., Mitra, P., and Kleinfeld, D. (1997). Central versus peripheral determinants of patterned spike activity in rat vibrissae cortex during whisking. *Journal of Neurophysiology*, 78(2):1144–1149.

Fend, M., Bovet, S., Yokoi, H., and Pfeifer, R. (2003). An active artificial whisker array for texture discrimination. In *IEEE Int. Conf. on Int. Robots & Systems*. IEEE.

Geisler, C. (1998). *From sound to synapse*. Oxford University Press, New York.

Hartings, J., Temereanca, S., and Simons, D. (2000). High responsiveness and direction sensitivity of neurons in the rat thalamic reticular nucleus to vibrissa deflection. *Journal of Neurophysiology*, 83(5):2791–2801.

Hartmann, M. (2001). Active sensing capabilities of the rat whisker system. *Autonomous Robots*, 11:249–254.

Hartmann, M., Johnson, N., Towal, R., and Assad, C. (2003). Mechanical characteristics of rat vibrissae: Resonant frequencies and damping in isolated whiskers and in the awake behaving animal. *Journal of Neuroscience*, July, 23(16):6510–6519.

Kaneko, M., Kanayama, N., and Tsuji, T. (1998). Active antenna for contact sensing. *IEEE Transactions on Robotics and Automation*, 14(2):278–291.

Kelly, A. G. (2000). *Fundamentals of Mechanical Vibration*. Mc-Graw Hill, 2nd edition.

Kleinfeld, D., Berg, R., and O'Connor, S. (1999). Anatomical loops and their electrical dynamics in relation to whisking by rat. *Somatosensory and Motor Research*, 16(2):69–88.

Lichtenstein, S., Carvell, G., and Simons, D. (1990). Responses of rat trigeminal ganglion neurons to movement of vibrissae in different directions. *Somatosensory and Motor Research*, 7(1):47–65.

Neimark, M., Andermann, M., Hopfield, J., and Moore, C. (2003). Vibrissa resonance as a transduction mechanism for tactile encoding. *Journal of Neuroscience*, July, 23(16):6499–6509.

Sachdev, R., Sato, T., and Ebner, F. (2001). Divergent movement of adjacent whiskers. *Journal of Physiology*, 87:1440–1448.

Shoykhet, M., Doherty, D., and Simons, D. (2000). Coding of deflection velocity and amplitude by whisker primary afferent neurons: implications for higher level processing. *Somatosensory and Motor Research*, 17(2):171–180.

Szwed, M., Bagdasarian, K., and Ahissar, E. (2003). Encoding of vibrissal active touch. *Neuron*, 40:621–630.

Tsujimua, T. and Yabuta, T. (1989). Object detection by tactile sensing method employing force/torque information. *IEEE Trans. on Robotics and Automation*, 5(4):444–450.

Ueno, N. and Kaneko, M. (1994). Dynamic active antenna. In *Int. Conf. on Robotics and Automation*, pages 1784–1790. IEEE.

Weaver, W., Timoshenko, S., and Young, D. (1990). *Vibration problems in engineering*. Wiley, New York.

Wineski, L. (1985). Facial morphology and vibrissal movement in the golden hamster. *Journal of Morphology*, 183:199–217.

MOTOR CONTROL

A Simple, Adaptive Locomotion Toy-System

Jonas Buchli **Auke Jan Ijspeert**
Biologically Inspired Robotics Group (BIRG)
LSL, Swiss Federal Institute of Technology (EPFL), Lausanne
jonas.buchli@epfl.ch, auke.ijspeert@epfl.ch
http://birg.epfl.ch

Abstract

In order to successfully transfer biological principles to engineering problems, it is important to study the fundamental properties of biological systems. The goal is to arrive with useful abstractions that are (1) implementable, (2) testable by experiments and sample implementations, (3) retain the, for an engineering point of view, essential good properties of the biological systems. At BIRG, we are interested in the fundamentals of locomotion control and their possible applications to robotics. In this article, we present a simple, adaptive locomotion toy system that is of oscillatory nature. It is composed of two parts: an adaptive controller based on a nonlinear oscillator, and a mechanical system made of two blocks attached by an active and a passive spring. The controller is designed to be robust against perturbations, and to adapt its locomotion control to changing body kinematics or added external load. The tools to develop such a toy-system are a 2-scale nonlinear dynamical system, namely a Hopf oscillator with adaptive frequency, and a understanding of synchronization behavior of oscillators. A further central ingredient that will be discussed is the concept of asymmetric friction forces. We show that the system possesses several critical parameters. It is illustrated that the bifurcations connected with some of these parameters can be identified as non-smooth phase transitions and power law behavior. Links to biology and possible applications to robotics are discussed.

1. Introduction

Natural locomotion systems are of great interest for robotics as they present refined and robust solutions to a difficult engineering challenge. Namely, they manage to coordinate multiple degrees of freedom needed for locomotion, using signals of the right frequencies, phases and amplitudes (Ijspeert, 2003). However, any given natural locomotion system is of great complexity. In the case of a vertebrate locomotion system, it consists of an enormous number of muscle and neural cells and many other types of body parts such as bones, tendons, etc. And even the locomotion apparatus of single cell organisms turns out to be of astonishing complexity (Marx, 2003).

However, as many people have pointed out, simple copying of biological processes can not be what we have to strive for when trying to enhance engineering principles by biological inspiration. It is rather the identification of the principles at a given level of abstraction that makes adaptation of biological principles for engineering problems successful (Buchli and Ijspeert, 2004).

Therefore, we have to look for fundamental principles underlying locomotion. In order to identify these principles, a broad range of locomotion systems have to be looked at. This includes single cell movement, worm, snail, snake, insect and mammalian locomotion, different types of flying and swimming but also more abstract concepts like active Brownian particles and ratchet systems.

When we try to find a few abstract principles common to all types of locomotion the following key observations arise:

- Locomotion is the process of transforming energy (unordered movement of many parts, particles) into directed movement (coordinated movement of the particles).
- From the first point we can directly conclude that locomotion always implies an active system and consumes energy[1].
- Any form of locomotion implies asymmetric interaction forces with the environment, e.g. an asymmetric friction mechanism (it turns out that this mechanism is what renders locomotion into a energy consuming process).
- Biological locomotion is robust, i.e. works under a very broad range of external (temperature, external forces, textures, ...) and internal (body properties, fatigue, sicknesses, ...) conditions. There are adaptation processes inherent in the locomotion systems to ensure this robustness.

[1] The consumed energy can be potential energy (e.g. passive walkers) as well as energy from active mechanisms.

- The adaptation processes work on many, very different timescales. In other words, the time scales of, for instance, locomotion control, adaptation to fatigue, and adaptation to body development span over several orders of magnitude.

In order to make these fundamental principles accessible to theoretical and experimental investigation without the overhead of complicated experimental setups and difficult to control environments and subjects (all of which can easily hamper clear insight) we are looking for simple systems that, nevertheless, show all the properties listed above. Here we will report on a particular simple instance of a nonlinear dynamical system that possesses the described properties.

Modeling neuro-mechanical systems with oscillators During the last decades there have been growing numbers of attempts to model locomotion control systems with the help of oscillators (Buchli and Ijspeert, 2004). On the other hand bio-mechanical systems were modeled by spring mass systems or inverted pendulums - and are therefore of oscillatory nature (Blickhan, 1989, Full and Koditscheck, 1999). The logical consequence to model the complete neuro-mechanical systems by coupled oscillators has been used (Taga, 1995) and robotics controllers have been developed on the base of oscillators (Nakanishi et al., 2003). Oscillators are an important and widely used system to model phenomena in a broad range of fields. Often, simple oscillators with fixed parameters such as frequency and damping are used. Recently however, it was found that for modeling certain phenomena, oscillators with fixed frequencies are not sufficient. Therefore, oscillators with adaptive frequencies are investigated (Acebron and Spigler, 1998, Acebron and Spigler, 2000).

Exploiting adaptive frequency oscillators for locomotion control In this contribution we show that by using a Hopf oscillator with adaptive frequency we can devise a controller that adapts to the resonant frequencies of the mechanical system and therefore excites it. Together with asymmetric interaction forces with the environment – which turn out to be a fundamental property of any locomotion system – this leads to directed movement. There are 3 core ideas in our approach.

- The (bio)mechanical system is of oscillatory nature. I.e. it possesses resonant frequencies at which it can easily be excited.
- The controller is of oscillatory nature.
- The intrinsic frequency of the oscillator is adaptable and gets influenced by the mechanical system.

As we will see, by the proper choice of the mechanical system and the controller, a system that is self-exciting and adaptable can be devised. In order to convert the oscillatory movement of the mechanical system into directed movement a fourth fundamental property is necessary: The system needs some type of asymmetric interaction forces in our case asymmetric friction, i.e. the friction coefficients are lower for one direction compared to the other.

We use a simple mechanical spring-mass system to investigate the requirements needed for such a system to make directed movement. Furthermore, we are interested in a (simple) controller that is able to adapt to changing properties of the mechanical system (i.e. growth, mass changes, length changes).

2. A Simple, Adaptive Locomotion Toy-System

In order to study the aforementioned fundamental principles we devised a simple mechanical system controlled by an oscillator. We will describe the two main constituents of the system and their important properties along with how the two subsystems get connected in order to build the adaptive locomotion system.

Mechanical system The mechanical part of the system consists of two masses connected by two parallel springs S_a, S_p (cf. Fig. 1(a)), with different resting lengths l_a and l_p. For convenience, we will express them as $l_a = l_m + l_d$ and $l_p = l_m - l_d$, where l_m is the mean value and $2l_d$ the difference between both lengths. Both springs are linear springs with spring constant k_a, k_p. The system presented in this article is one-dimensional, in other words, the masses are assumed to move on, and remain in contact with, a horizontal plane and moving on a straight line. In the following, we present the mathematical description of such a system in terms of a system of differential equations. The system can be written as follows:

$$\dot{\mathbf{q}} = \mathbf{A}\mathbf{q} + F_c + F_r(\mathbf{q}) \tag{1}$$

where $\mathbf{q} = [x_1, v_1, x_2, v_2]^T$ are the state variables of the system (position and velocities of the two masses), $\mathbf{A}$ is the matrix describing the action of the spring on the masses (without any friction), F_r is the friction present in the system and F_c is constant force arising from the differences and mean value of the resting length of the springs. $\mathbf{A}$ takes the following form

$$\mathbf{A} = \begin{pmatrix} 0 & 1 & 0 & 0 \\ -\frac{k_a+k_p}{m} & 0 & \frac{k_a+k_p}{m} & 0 \\ 0 & 0 & 0 & 1 \\ \frac{k_a+k_p}{m} & 0 & -\frac{k_a+k_p}{m} & 0 \end{pmatrix} \tag{2}$$

where m are the masses attached at each end of the springs. And,

$$F_c = \frac{l_d(k_p - k_a) - l_m(k_a + k_p)}{m}[0, 1, 0, -1]^T \tag{3}$$

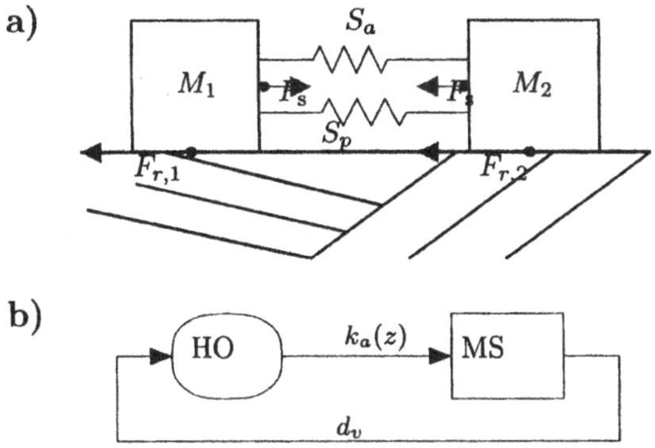

Figure 1: a) The mechanical sub-system of the adaptive locomotion toy system. It consist of two identical masses connected by a spring. From the mechanics of the system the forces acting on the body can be determined: F_s is the force that the springs exert on the blocks. It is of the same amplitude and opposite sign for the two masses. $F_{r,i}$ are the friction forces acting on the masses. With the help of this forces and Newton's law we are able to derive the system of differential equations that govern the mechanical system (Eq. 1). b) Schematic of the adaptive locomotion toy-system with its two main parts: The Hopf Oscillator (HO) and the spring mass system (MS). The Hopf Oscillator influences the mechanical system via the spring constant $k_a(z)$. The mechanical in turn disturbs the Hopf oscillator by coupling in the velocity difference d_v as an additive disturbance to the Hopf Oscillator equations.

With the choice of l_m and l_d the resting position of the masses can be influenced. The choice of the friction term F_r will be discussed below. Such a second order linear system possesses a resonant frequency, which can be calculated as:

$$\omega_m = \sqrt{\frac{2(k_a + k_p)}{m}} \tag{4}$$

In other words, this type of mechanical system can be interpreted as a band pass filter with the pass band around ω_m. For generating displacements of the system, we replace the spring S_a described above with an active spring whose spring constant k_a can be modulated by a controller (see next section). The active spring can be seen as an abstract muscle-like actuator that can both pull and push.

Hopf oscillator As controller/activator of the mechanical system, we use a Hopf oscillator:

$$\dot{z} = (\mu_h + i\omega_h)z - |z|^2 z + F(t), \quad z \in \mathbb{C} \tag{5}$$

which can be written in Cartesian coordinates:

$$\dot{x}_h = (\mu_h - r^2)x_h + \omega_h y_h + F_x(t) \tag{6}$$
$$\dot{y}_h = (\mu_h - r^2)y_h - \omega_h x_h + F_y(t) \tag{7}$$

where $x_h = Re(z), y_h = Im(z)$ and $r = |z| = \sqrt{x_h^2 + y_h^2}$. ω_h is the intrinsic frequency of the oscillator and $F_x(t) = Re(F(t))$ and $F_y(t) = Im(F(t))$ are perturbing forces. The solutions of this system will be discussed below. The Hopf oscillator is coupled to the mechanical system in the following way (cf. Fig. 1(b)): First, the spring constant k_a is made a function of z in the following form:

$$k_a = k_0 + a\frac{x_h}{r} \tag{8}$$

where k_0 is a constant for the offset of the spring constant and a is a coupling constant, where $a < k_0$ to ensure S_a remains a spring ($k_a > 0$). Since x_h will be an oscillating term, this choice of coupling leads to an undulatory modulation of k_a, where the frequency of the modulation will be ω_h. The division by radius r serves to normalize the maximum of the modulation term to 1.

Second, the Hopf oscillator has feedback from the mechanical system:

$$F(t) = c\,d_v \doteq -c(v_1 - v_2) \tag{9}$$

where c is a coupling constant. Here, the choice has been on the considerations that mainly the coordination between the two bodies is important. Therefore, the choice of a difference term[2]. The choice of the difference of velocities is chosen in order to have a more sensitive coupling in contrast to choosing the difference of positions[3].

Properties of the Hopf Oscillator The Hopf oscillator acts as a frequency selective amplifier (Eguíluz et al., 2000), i.e. frequency components of F(t) that are close to ω_h are amplified. Especially the setting $\mu_h = 0$ is special in the sense that the system undergoes a fundamental change at that point: For $\mu_h < 0$ the system exhibits a stable fixed point at $z = 0$, whereas for $\mu_h > 0$ a stable limit cycle occurs with radius $r = \sqrt{\mu_h}$. This phenomena is known as a Hopf bifurcation (Hopf, 1942). At $\mu_h = 0$, there is no signal oscillating at ω_h weak enough not to get amplified by the Hopf oscillator. Therefore, for that setting the Hopf oscillator can be considered an ideal amplifier[4]. This means that, if the resonant frequency ω_m of the mechanical system and the intrinsic frequency ω_h are close enough, any signal components close to the resonant frequency of the mechanical system get amplified. Thus, an excitation of the system can be expected.

In case of a biological setting, this would correspond to a case where the activation pattern of the muscles is well

[2]Such a difference is also more plausible from the practical and biological point of view, where no absolute values can play a role.

[3]It is expected, however, that with a proper choice of the coupling constant the qualitative behavior should not depend on this choice.

[4]We can not delve into the mathematical details of the Hopf oscillator in this contribution. For a more detailed treatment see e.g. (Kern and Stoop, 2003) and references therein.

adapted to the properties of the mechanical system, and, therefore, a very energy efficient mode of locomotion can be expected where only a small amount of energy has to be spent in order to maintain high locomotion velocity[5]. Animals possess adaptive processes which bring the locomotion system to (near) optimum mode of operation.

Adaptive frequency oscillator In our toy-system, adapting the oscillator to the mechanical systems means a proper choice or a tuning of ω_h. But, we can avoid manual tuning of ω_h by enhancing the dynamical system describing the Hopf oscillator. Instead of fixing ω_h we treat it as a variable with a corresponding differential equation describing the time evolution:

$$\dot{\omega}_h = f(\omega_h, q, t) \tag{10}$$

The choice of $f(\omega_h, q, t)$ that serves our purpose is chosen by understanding the synchronization behavior of two coupled oscillators. To get a good grasp on the effects of perturbations on a limit cycle system it is helpful to look at it in the phase plane representation (cf. (Strogatz, 1994)). In the phase plane all perturbations have a direction, i.e. they can be represented as a vector in that plane. It is known that a small perturbation of a limit cycle system can only affect that phase strongly if it perturbs the oscillator in direction tangential to the limit cycle, the perturbations perpendicular to the limit cycle are damped out[6]. Thus, depending on the state of the oscillator (the position of the phase point on the limit cycle) the perturbation accelerates the phase point or slows it down. If the perturbation is a periodic signal, with a frequency close to the intrinsic frequency of the oscillator, this results in an average acceleration or deceleration depending on the frequency difference and to synchronization. If we take this same effect to tune the frequency of the oscillator (on a slower time scale) the frequency should evolve toward the frequency of the perturbation. Therefore, the effect of $f(\omega_h, q, t)$ is the same as the effect leading to synchronization. Thus, (in average) driving ω_h toward ω_m. On that ground, we chose

$$\dot{\omega}_h = -\tau_h d_v \frac{y_h}{r} \tag{11}$$

where $\tau_h \ll 1$. However, the adaptation of ω_h happens on a slower time scale than the evolution of the rest of the system. This adaptation time scale is influenced by the choice of τ_h.

As explained, the output of the Hopf oscillator is coupled via setting $k_a = f(z)$ to the mechanical system. The mechanical system serves as a bandpass filter. Therefore, the Hopf oscillator, via the aforementioned feedback, slowly adapts its frequency to the mechanical system.

In order to measure the excitation of the system we can use the energy content of the mechanical part. There are two contributions to the energy of the mechanical system: The kinetic energy of the masses $E_k = \frac{1}{2}m(v_1^2+v_2^2)$ and the potential energy of the springs $E_p = \frac{1}{2}\left(k_a(x_2 - x_1 - l_m + l_d)^2 + k_p(x_2 - x_1 - l_m - l_d)^2\right)$. We will use this formalism to investigate the optimal excitation frequency of the system in the next section.

Friction Friction dissipates the energy of the system. Therefore, if no energy is pumped into the system the system will always come to standstill from any initial condition. We have tested the system with two different friction schemes: viscous friction and Coulomb friction. Both are introduced as asymmetric friction forces, i.e. the parameters are lower for one direction compared to the other. In viscous friction, the friction force is proportional to the velocity of movement:

$$F_r = [0, -\rho_1 v_1, 0, -\rho_2 v_2]^T \tag{12}$$

where

$$\rho_i = \begin{cases} \rho_+ & \text{if } v_i > 0 \\ \rho_- & \text{if } v_i < 0 \end{cases} \tag{13}$$

The second model that has been tested is the Coulomb friction model, in which the friction coefficients are constant:

$$F_r = [0, F_{c,1}, 0, F_{c,2}]^T \tag{14}$$

$$F_{c,i} = \begin{cases} -|F_s| & \text{if } F_s < \mu_S|F_N| \text{ and } v_i = 0 \\ -\frac{v_i}{|v_i|}\mu_{r,i}|F_N| & \text{otherwise} \end{cases} \tag{15}$$

where $F_N = gm$ is the normal force of the body on the ground, and F_s is the spring force acting on the body. Coulomb friction has a static mode (when $v_i = 0$) and a dynamic mode. In our case, the dynamic friction is made asymmetrical as follows

$$\mu_{r,i} = \begin{cases} \mu_+ & \text{if } v_i > 0 \\ \mu_- & \text{if } v_i < 0 \end{cases} \tag{16}$$

With the Coulomb friction scheme the system would be unstable since the increase in the input of energy is increasing faster then the dissipation of energy due to friction. In order to avoid this instability problem, the coupling from the Hopf oscillator is slightly changed. The coupling constant a is made dependent on the rate of change of distance of the two masses, i.e. the velocity difference:

$$a_{coulomb} = \begin{cases} a & \text{if } |v_1 - v_2| < v_{thr} \\ 0 & \text{if } |v_1 - v_2| > v_{thr} \end{cases} \tag{17}$$

In other words, the coupling is switched off, if, due to heavy excitation the masses separate too fast. Thus, the

[5] Everybody knows this effect from sitting on a swing and trying to get it to move. If one moves the legs at the right frequency, with a minimum effort the swing can be brought to breathtaking heights.

[6] This is, of course, a simplification of the real facts. In fact, perturbations perpendicular to the limit cycle can, in general, perturb the phase of the oscillator. However, for the case of phase oscillators, such as the Hopf oscillator, the above simplification comes very close to the real thing. To discuss the general case is beyond the scope of this article.

mechanical system runs passively for a short moment when reaching this maximum velocity until the spring forces and dissipation of energy bring it again below that threshold.

For this article the default friction model is the viscous model unless otherwise noted.

3. Simulation Results

The basic mode of locomotion is presented in Fig. 2(a). We show in vertical order consecutive snapshots of the mechanical system, at every 0.05 s. The locomotion of the system most closely resembles rectilinear-movement observed in some types of worms and snakes. The two masses exhibit an undulatory movement by which they are constantly stretching and compressing the springs between them. Due to the asymmetric friction forces one of the bodies is pushed (dragged respectively) toward one direction.

For the remainder of this section we would like to illustrate a few other important aspects of the presented locomotion toy-system. First, we list the parameters and initial conditions used in the simulations (unless otherwise noted):

Default parameter values			
parameter	value	parameter	value
μ_h	0	l_m [m]	12
τ_h	0.1	l_2 [m]	0.5
c [sm^{-1}]	0.1	ρ_+ [Nsm^{-1}]	0.2
a [Nm^{-1}]	10	ρ_- [Nsm^{-1}]	0.1
m [kg]	1	μ_s	0.3
k_0 [Nm^{-1}]	10	μ_+	0.21
k_p [Nm^{-1}]	100	μ_-	0.19
v_{thr} [ms^{-1}]	10		

Default initial conditions	
variable	initial value
z	0
ω_h [rads^{-1}]	12
x_1 [m]	0
v_1 [ms^{-1}]	0
x_2 [m]	11.595
v_2 [ms^{-1}]	0

In order to verify that the frequency calculated above is really the resonant frequency of the whole system (Hopf oscillator included) we make the following numerical experiment. We set $\tau_h = 0$, i.e. we go back to the fixed frequency oscillator. Then, we numerically integrate the system for different values of ω_h over a duration of 100 s. In Fig. 3, the resulting average energy content of the system is plotted against ω_h. The vertical lines depict the resonant frequency of the mechanical system and its 0.25, 0.5, 2,3 and 4 folds. As expected, the broadest peak is measured around the mechanical resonant frequency. Furthermore, there is a second strong peak (even higher amplitude but less broad) close to the harmonic $2\omega_m$. Apparently, the resonant effects at that point are very strong. Also around $0.25\omega_m$ and $0.5\omega_m$ resonant effects can be observed. They are however, much less strong. Furthermore, there seems to be a systematic shift of the resonant frequencies toward lower values for $\omega_h < \omega_m$ and toward higher values for $\omega_h > \omega_m$. In the lower panel the attained mean velocity of the center of mass is given. We see that that it follows the same pattern as the energy content of the mechanical system.

Henceforth, we introduce frequency adaptation into the Hopf oscillator by setting $\tau_h = 0.1$ and investigate

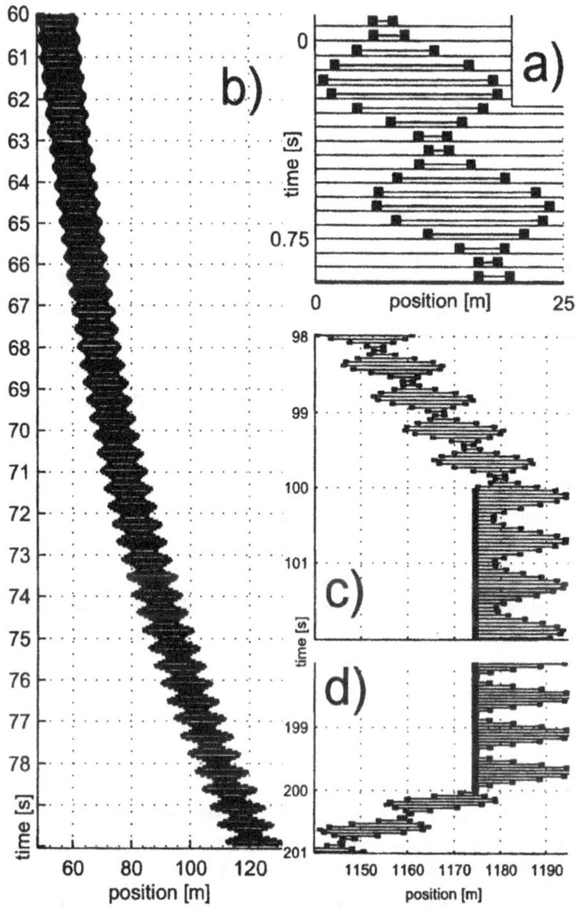

Figure 2: Illustration of the basic mode of locomotion. See text for description.

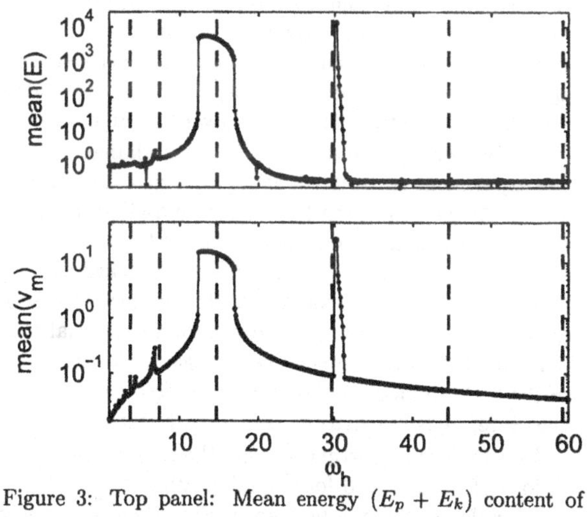

Figure 3: Top panel: Mean energy ($E_p + E_k$) content of the mechanical system for different $\omega_0 = [0, 60]$ at steady state. Bottom panel: Mean velocity. As clearly can be seen we have a broad peak around the resonant frequency of the mechanical system $\omega_m = 14.8324$.

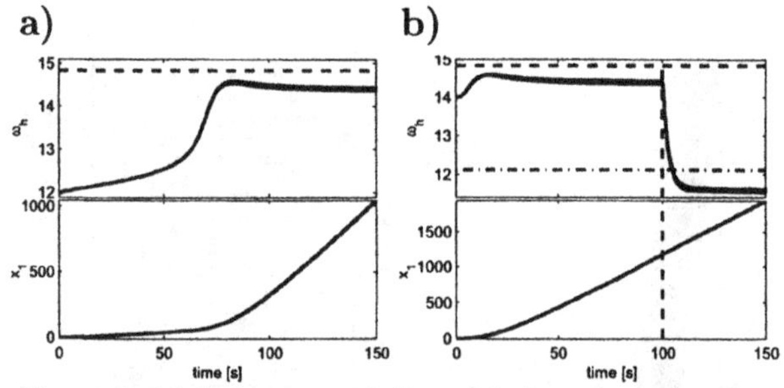

Figure 4: (a) This representative plot shows how the adaptation of ω_h increases the velocity of the locomotion system. ω_h starts at 12. The top panel shows the development of ω_h, the second panel shows the position x_1 of M_1. There is some activity in the system from the beginning, however the excitations gets much stronger the closer ω_h comes to to the resonant frequency of the mechanical system ω_m (vertical line). The steady state for ω_h is around 14.4, which is lower then the calculated resonant frequency of the passive mechanical system: $\omega_m = 14.8324$.
(b) To illustrate the robustness that is built into the locomotion system by its frequency adaptation we show a representative experiment in which the mass of the bodies is 1kg to 1.5kg at t=100s (vertical line). Immediately, the frequency of the oscillator ω_h starts to adapt to the new resonant frequency of the mechanical system (dash-dotted horizontal line).

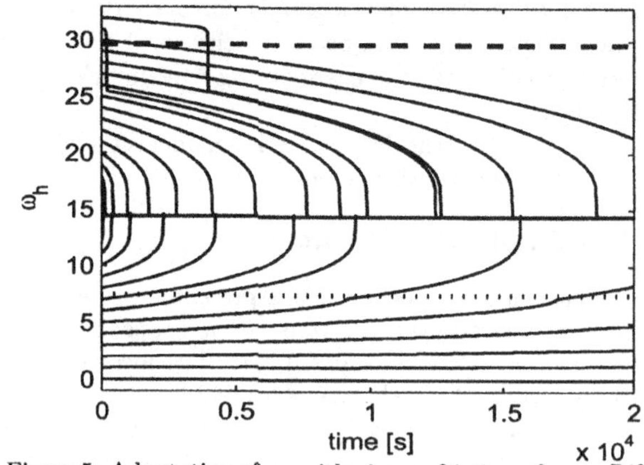

Figure 5: Adaptation of ω_h with viscous friction scheme. Different initial values for $\omega_h(0) = 0, 1, \ldots, 32$ have been chosen. Clearly, the further away the longer ω_h takes to reach its stable steady state. The horizontal lines depict $0.5\omega_m$ and $2\omega_m$, where from the energy diagram (Fig. 3) resonant effects have to be expected and their influence on ω_h clearly can be observed here. Refer to text for further description.

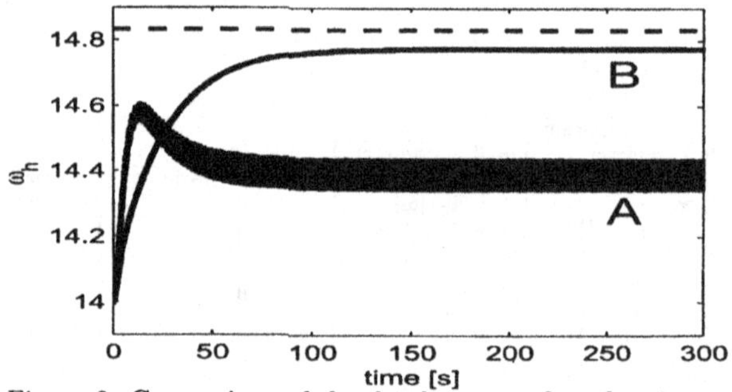

Figure 6: Comparison of the development of ω_h for viscous and Coulomb friction scheme. The Horizontal line indicates the resonant frequency of the passive mechanical system. (A) Viscous friction. (B) Coulomb friction. See text for further discussion.

the adaptation capability. We illustrate that the adaptation of ω_h leads to an increase of the forward velocity of the system. To get a first idea of how the system adapts ω_h and how this influences the speed of the mechanical system, we show a representative plot of ω_h and x_1 in Fig. 4(a) and the illustration of what happens with the masses from 60 to 80 seconds in Fig. 2(b). It can clearly be seen how the frequency adapts, first slowly then faster and finally reaches a steady state around $\omega_h = 14.4$. As can be seen the resonant effects start to excite the mechanical system heavily when ω_h passes at around 13 and the system starts to move forward at high speed.

In the next experiment, the mass is changed during the experiment at time $t = 100\,\mathrm{s}$ from $m = 1\,\mathrm{kg}$ to $m = 1.5\,\mathrm{kg}$. As we can see in Fig. 4(b) the oscillator quickly adapts its frequency to the new resonant frequency of the system and there is little change in the forward velocity. This corresponds for example to the biological case of growth of an animal or an addition of an external load. The adaptation to the massive change in the properties of the body immediately sets in, and after a few seconds the system has reached the new steady state.

As a further important aspect, the role of initial conditions of ω_h at $t = 0$ for the adaptation capability has been explored. As can be seen in Fig. 5, ω_h will always converge to the same value, but the initial conditions strongly affect the time ω_h needs to reach steady state. Furthermore, from the results in Fig. 3 we expect to see some clearly visible effects at $0.5\omega_m$ and $2\omega_m$ (the 1:2 and 2:1 resonant frequencies). At around $0.5\omega_m$ this leads to a slightly attractive area for ω_h (so called ghost points). Even more interesting is the effect at $2\omega_m$. From the energy diagram (Fig. 3) we know that the system reacts very strongly at that point. We can see that this violent reaction basically kicks the ω_h out of that region toward ω_m. So we see that even if the possible attained velocity is higher at $\omega_h = 2\omega_m$, this mode is not an attractor of the adaptation process.

Finally, in Fig. 6, a comparison between the viscous friction scheme and the Coulomb are presented. We see that the convergence for ω_h is slower in the case of the Coulomb scheme, and the final value is closer to the calculated resonant frequency of the passive mechanical system.

Influences of the Parameters Next, we want to shed light on the influence of the different parameters

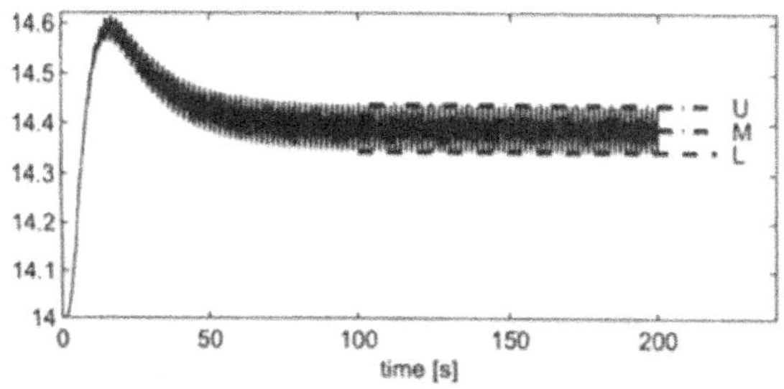

Figure 7: In order to quantify the steady state behavior of ω_h the mean m_{ω_h} and variance σ_{ω_h} have been measured after the system reached steady state (horizontal lines). Therefore, m_{ω_h} is a measure for (M) and σ_{ω_h} correlates with the distance (U)-(L). Note this figure is illustrative only – the actual values can change.

of the system. Note that the results in this section are of preliminary nature, since a lack of space and time does not allow to treat this topic in all detail in this article. There are a number of parameters which we can expect to influence the system fundamentally: First, the adaptation time scale of the Hopf oscillator τ_h directly influences the adaptation speed, but also the stability of the adaptation process. Second, the coupling constants between the Hopf oscillator and the mechanical system (a, c) can influence transient times as well as stability of the system. Third, the choice of the friction parameters $(\rho_+, \rho_-, \mu_r, \mu_s, \mu_d)$ obviously influences the locomotion speed of the system.

The mechanical parameters, such as masses, resting distances and spring constants, which seem to be important at first sight, turn out to be not so fundamental in a second look. They merely determine the time and length scales on which the solutions will be found but do not influence the type of solutions that are possible.

We will now present results of numerical simulations, highlighting the influences of the parameters listed above. First of all, the behavior of the attained steady state values for ω_h depending on the choice of the coupling parameter c will be illustrated. The parameter c influences how strongly the mechanical system perturbs the Hopf oscillator. In order to quantify the dependence of ω_h on this (and other) parameters, we use the mean (m_{ω_h}) and variance (v_{ω_h}) of ω_h after it reached the steady state (cf. Fig. 7). These values are measures for location and shape (width/depth) of the attractor for ω_h.

In Fig. 8(a), can be seen that there is a bifurcation[7] for a critical value of $c = C_c$. For $c < C_c$ there is a deviation of m_{ω_h} from the resonance frequency of the mechanical system ω_m. The deviation grows with an exponential increase in the variance when approaching C_c. At $c = C_c$ the mean and variance drop to 0, i.e. the system gets quenched. Such discontinuous behavior in the mean value and power law behavior of the variance are ubiquitous phenomena in phase transitions (bifurcations).

Next, we shed light on the influence of τ_h (cf. Fig. 8(b)) again we have a clear bifurcation behavior for a critical value of τ_h. For high values of $\tau_h > \tau_{h,c}$ it seems the system sets out onto a route into chaos (data not shown). The reason for that might be that for higher values of τ_h the timescales of the adaptation process and the other time scales of the system start to overlap, and therefore irregular behavior may occur. This conjecture however bases on visual inspection and is not confirmed yet. Again, we have a very clear power law behavior of σ_{ω_h}.

Now we turn our attention to the coupling from the Hopf oscillator to the mechanical system. The corresponding parameter is a. This parameter determines how strongly k_a is modulated by the activity of the Hopf oscillator (cf. Eq. 8). We measure the influence by calculating the mean value m_{v_1} of the attained velocity of M_1 at steady state. Here, we find a very clear linear dependence of m_{v_1} on a (cf. Fig 8(c)) for the explored range.

Finally, in Fig. 9, the influence of the friction asymmetry (ρ_+, ρ_n) is presented. After the transient phase the attained mean velocity of the center of mass is measured. As can be seen there is an exponential dependence of the velocity on the friction parameters.

Loss of feedback and disturbances It is also interesting to see how the system reacts when the feedback is cut from the mechanical system to the Hopf oscillator by setting $c = 0$. In Fig. 10(a), one can observe, that because of the setting of $\mu_h = 0$, the system keeps on moving, but at much reduced speed. If μ_h is set to slightly negative values $\mu_h = [-1, -10]$, the oscillations of the Hopf oscillator fade out and the system comes to a stand still (data not shown).

In a next step, we inhibit the frequency adaptation only by setting $\tau_h = 0$. In Fig. 10(b), can be seen how the system keeps on moving as normal, however it is not able to adapt to changing conditions anymore (data not shown).

In order to get an idea how the system behaves when it gets disturbed, we blocked one of the masses for 100 s. In Fig. 11, can be observed how the system reacts. Immediately the adaptation mechanism for ω_h starts to tune the system to the new resonant frequency $\omega_m = \sqrt{\frac{k_a + k_p}{m}}$. After releasing the blocks the system moves backward for a moment, before converging to the normal steady state again. What happens with the masses around $t = 100$ and $t = 200$ can also be seen in Fig. 2(c) and 2(d). Especially from Fig. 2(d) it can be understood where the

[7]Note that the names *bifurcation* and *phase transition* are basically two names for the same phenomena, even though with emphasis on different aspects. See (Haken, 1983) for a discussion.

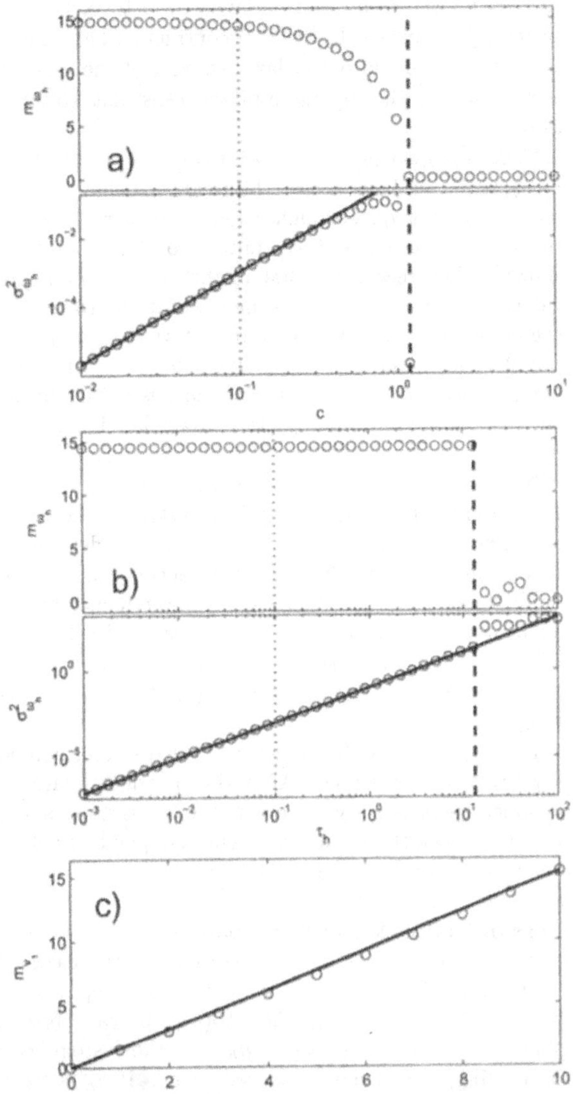

Figure 8: (a) Upper panel: the mean value attained for ω_h dependent on c (note the log scale). Lower panel: Variance for ω_h (note log-log scale). Note that for $c > C_c$ the variance drops to 0, therefore it is not drawn on the log scale. The vertical dotted line indicates the default setting of $c = 0.1$ and the dashed lines indicates the approximate value of C_c. The variance can be fitted with a power law $\sigma^2 \sim \tau^\alpha$. The exponent of the fit is $\alpha = 2.6959$. This behavior is typical for phase transitions.
(b) Dependence of mean and variance of ω_h on τ_h for steady state behavior. Again, the dotted line indicates the default setting of $\tau = 0.1$ and the dashed lines indicates the approximate value of $\tau_{h,c}$. The variance can be fitted with a power law $\sigma^2 \sim \tau^\alpha$. The exponent of the fit is $\alpha = 1.9903$. (c) The attained mean values m_{v_1} of v_1 after the system reached steady state depending on coupling parameter a. There is a very clear linear relationship $m_{v_1} = qa$. The fit is for $q = 1.5451$.

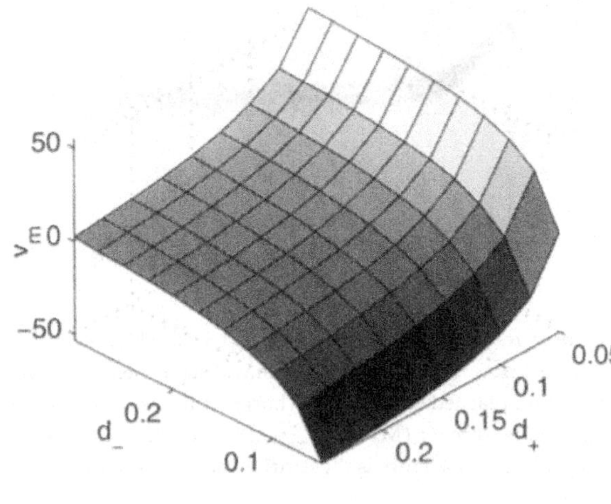

Figure 9: Influence of the friction asymmetry in viscous mode. The z axis is the position of m_1 after 200s for a given setting of ρ_- and ρ_+. White regions correspond therefore to fast forward locomotion, black to fast backward locomotion.

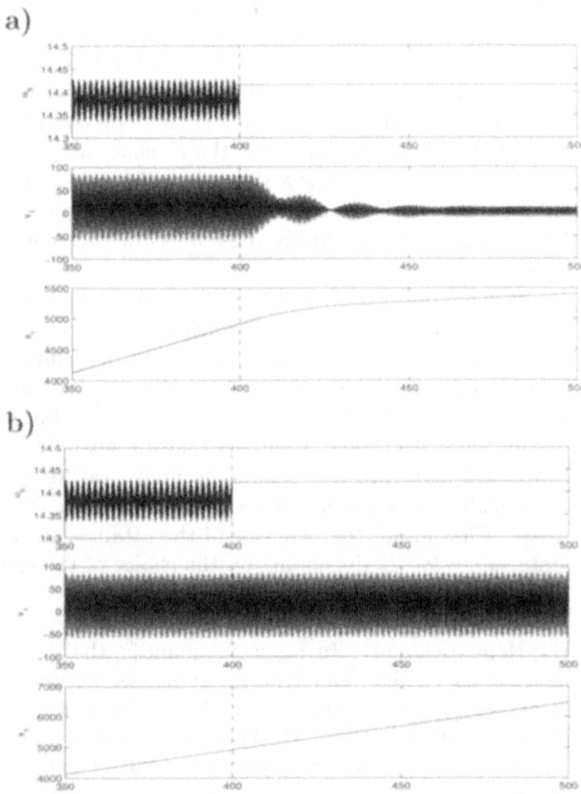

Figure 10: a) At $t = 400$ the feedback from the mechanical system to the Hopf Oscillator is cut by setting $c = 0$. This as well inhibits the frequency adaptation. b) At t=400 only the frequency adaptation is cut by setting $\tau = 0$. This means there is still feedback from the mechanical system to the Hopf oscillator. Therefore, the system keeps moving with the same speed. However, it can not react to changing mechanical properties anymore (not shown).

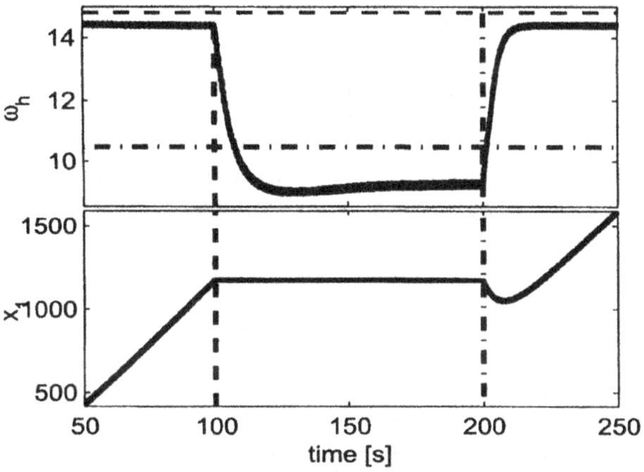

Figure 11: M_1 blocked at $t = 100$ (dashed vertical line) and released at $t = 200$ (dash dotted vertical line).

backward movement comes about. In the moment of the release of M_1, M_2 happens to move with high velocity toward M_1, which leads to acceleration backward in the moment of the release.

4. Discussion

In this article we have shown that, by using a Hopf oscillator with adaptive frequency, we can devise a controller that adapts to the resonant frequencies of a mechanical system and therefore excites it. We exploited the oscillatory nature of the systems and the understanding of synchronization behavior of oscillators. Together with asymmetric friction - which turns out to be a fundamental property of any locomotion system - this leads to directed movement. Asymmetric friction generalizes to asymmetric interaction forces for other types of locomotion, e.g. the elongated body of fishes have a low longitudinal and a large perpendicular drag coefficient, which by proper undulation of the body leads to forward movement. In the case of legged locomotion, the asymmetric interaction forces are implemented by having a stance phase with ground contact forces and a swing phase with zero interaction forces.

Two distinct time scales have been introduced into the dynamical system. Therefore, the system is a simple instantiation of a multi-scale[8] dynamical system. The fast time scale is what normally is associated with the control of locomotion. The slow time scale can be associated with learning or, more generally, with adaptive behavior. Thus, we see the descriptive power that is naturally built into such multi-scale dynamical systems. Especially, there is no distinction between learning algorithm and learning substrate. Moreover, it is worthwhile noting that for a full understanding of the behavior of such a system, the two time scales can not be treated completely separately.

[8]As a matter of fact it contains more than 2 time scales. Already a nonlinear oscillator possesses at least two time scales: One is the time scales of the oscillations and the other is the time scale at which it approaches its limit cycle.

Hitherto, multi-scale dynamical systems did not receive a lot of attention. One reason might be that there are some problems inherent in their investigation. One first problem that one encounters, is that, since there are many, possibly quite different, time scales in the system, the system has to be integrated in a way to encompass all time scales. In other words, the integration has to be exact enough (i.e. sampled fast enough) in order to calculate correctly the phenomena on the fastest time scale. On the other, hand the simulation has to run long enough in order to see the phenomena on the slowest time scale. The time scales are not fully separable since it is exactly the influence between them that is of interest. Furthermore, the interesting multi-scale systems usually exhibit strong nonlinear behavior and complicated bifurcation structure with many critical parameters, all of which makes it difficult the get a good understanding of the systems behavior. Recently, however, the interest in such systems is rising. One important reason is that cheap, fast workstations allow fast simulation of such systems and thus make their investigation more amenable.

Relation to biology We would like to stress the fact that the presented system is not meant to model detailed biological processes. It is rather a biologically inspired experimental system to investigate fundamental properties of locomotion (i.e. the generation of directed movement).

Nevertheless, and this is our hope, the treatment of such toy systems might lead to interesting questions for biologists. For instance, we have seen that our model reaches high velocities on higher order locking with the mechanical system. It would be interesting to see if the same phenomenon is observed in CPGs, and potentially, experimentally modify the CPG frequencies (e.g. through the application of excitatory neurotransmitters), and investigate whether it starts to work at these other resonant frequencies. In our case we saw a strong reaction, and it would be interesting to analyze the reaction of the natural system.

Furthermore, it is clear that the resonant frequencies of a locomotion system change very dramatically in an animal life time, and this at different time scales, e.g. due to the addition of external loads (e.g. carrying a prey) and to the growth of the animal. The natural locomotion systems have to cope with all such changes, and it would be interesting to explore the different types of adaptation mechanisms that play a role in keeping locomotion efficient. This is an area where the interplay between biology, engineering and mathematics can prove very fruitful.

Future work and outlook There are several avenues to build up on the presented results. First, there are many theoretical questions that may be asked. Even if the system seems to be a quite simple one, the influence of the parameters is far from trivial. It will be interesting to further investigate the bifurcation behavior of the system. A careful investigation of the influence of the parameters remains to be done. We know that the system contains several critical parameters as outline above. In the presented results, most of them have been chosen in order to be on the safe side, e.g. τ_h has been chosen small enough in order to have two well separated time scales in the system.

Second, there are several extensions that can be made to the systems, such as an extension to non-oscillatory systems, for instance. We see that by exploiting synchronization behavior we can devise a useful system. On the other hand, there exist generalizations of the notion of phase locking and synchronization for aperiodic and non-oscillatory systems (Kuramoto, 1984). This could indicate that the presented approach is an interesting path to follow also for investigation of other problems, such as cognition, pattern recognition, evolutionary models and the like. Other extensions include introducing other coupling schemes and different types of oscillators.

The control of speed and direction also needs to be addressed. More springs and oscillators can be introduced in order to get a system that can change its direction of locomotion (2/3D system). Furthermore, the system should have some control over speed. Currently there is only stop and full speed (disregarding the transients between the two modes). However, a control of speed could be introduced, e.g. with an intentional detuning of the oscillator and/or with a modulation of the amplitude of oscillatory control signals. There are various other possibilities, of which the respective advantages and disadvantages and their implications on the system have to be carefully studied.

Finally, it would be interesting to construct a real-world robotics implementation of this toy-system as well as to implement it as a realistic dynamical simulation in a physics-based simulator. The system (most probably in a slightly modified form) should prove interesting for several robotics applications such as snake robots, (ad-hoc) modular robotics, and legged robots.

Acknowledgments This research is funded by a Young Professorship Award to Auke Ijspeert from the Swiss National Science Foundation. We would like to thank Jun Nakanishi for constructive comments on an earlier version of this article.

References

Acebron, J. and Spigler, R. (1998). Adaptive frequency model for phase-frequency synchronization in large populations of globally coupled nonlinear oscillators. *Physical Review Letters*, 81:2229–2232.

Acebron, J. and Spigler, R. (2000). Uncertainty in phase-frequency synchronization of large populations of globally coupled nonlinear oscillators. *Physica D*, 141(1–4):65–79.

Blickhan, R. (1989). The spring-mass model for running and hopping. *J. Biomechanics*, 22(11–12):1217–1227.

Buchli, J. and Ijspeert, A. (2004). Distributed central pattern generator model for robotics application based on phase sensitivity analysis. In *Proceedings Bio-ADIT2004*, Lecture Notes in Computer Science. Springer Verlag Berlin Heidelberg. To appear, preprint at `http://birg.epfl.ch`.

Eguíluz, V., Ospeck, M., Choe, Y., Hudspeth, A., and Magnasco, M. (2000). Essential nonlinearities in hearing. *Phys Rev Lett*, 84(22):5232–5235.

Full, R. and Koditscheck, D. (1999). Templates and anchors: neuromechanical hypotheses of legged locomotion on land. *Journal of Experimental Biology*, 202:3325–3332.

Haken, H. (1983). *Synergetics. An introduction.* Springer Verlag Berlin Heidelberg, 3rd edition.

Hopf, E. (1942). Abzweigung einer periodischen Lösung von einer stationären Lösung eines Differentialsystems. *Ber. Math.-Phys., Sächs. Akad. d. Wissenschaften, Leipzig*, pages 1–22.

Ijspeert, A. (2003). Vertebrate locomotion. In Arbib, M., (Ed.), *The handbook of brain theory and neural networks*, pages 649–654. MIT Press.

Kern, A. and Stoop, R. (2003). Essential role of couplings between hearing nonlinearities. *Physical Review Letters*, 91.

Kuramoto, Y. (1984). *Chemical oscillations, Waves, and Turbulence.* Springer Verlag Berlin Heidelberg.

Marx, J. (2003). How cells step out. *Science*, 302(5643):214–216. Science News Focus. Cell Biology.

Nakanishi, J., Morimoto, J., Endo, G., Cheng, G., Schaal, S., and Kawato, M. (2003). Learning from demonstration and adaptation of locomotion with dynamical movement primitives. *Special issue of Robotics and Autonomous Systems.* To appear.

Strogatz, S. (1994). *Nonlinear Dynamics and Chaos.* Addison Wesley Publishing Company.

Taga, G. (1995). A model of the neuro-musculo-skeletal system for human locomotion. I. Emergence of basic gait. *Biological Cybernetics*, 73(2):97–111.

GasNets and other Evolvable Neural Networks applied to Bipedal Locomotion

Gary McHale* **Phil Husbands****
*Department of Informatics,
**Centre for Computational Neuroscience and Robotics
Centre for Research in Cognitive Science
University of Sussex, Falmer, Brighton BN1 9QH, UK
cglife@hotmail.com

Abstract

Evolutionary robotics relies upon techniques involving the evolution of artificial neural networks to synthesize sensorimotor control systems for actual or physically simulated robots. This paper is a comparative study of three principal types of artificial neural networks; the Continuous Time Recurrent Neural Network (CTRNN), the Plastic Neural Network (PNN) and the GasNet. An attempt is made to evolve networks capable of achieving locomotion with a physically simulated biped. Of the 14 distinct networks tested, GasNets were the only network to achieve cyclical locomotion, although CTRNNs were able to attain a higher level of average fitness.

1. Introduction

Evolutionary Robotics seeks to create real and simulated robotics that have economic value, and to gain insight into biological systems through suitable abstractions of these systems.

In Evolutionary Robotics, biologically inspired Artificial Neural Networks (ANN) are evolved using Genetic or Evolutionary Algorithms (GAs or EAs) to generate sensorimotor control systems. A significant challenge faced in this area is that the evolved neural network circuitry is rarely tractable to functional decomposition or deconstruction other than in the most trivial of cases (Beer et al., 1999). Whilst the characteristic equations associated with a specific network are a compact description of the network, we are as yet unable to predict from these equations, the dynamic characteristics of the network when it is embodied in an environmental agent.

In this paper empirical investigation is used to compare the performance of different network formulations in a common task; that of evolving bipedal locomotion in a physically simulated biped. It is hoped that through considering a number of networks applied to different but related tasks, we may discover certain network formulations that are of general or specific use. It is intended as a complementary study to work that sought to evolve locomotion in a physically simulated quadruped (McHale and Husbands, 2004). Whilst an in-depth analysis of the reasons why one network is superior to another is beyond the scope of this paper, a summary of relevant work is given in the latter stages of this report. A total of 14 different ANNs are assessed in this study.

2. Related Work

Continuous Time Recurrent Neural Network (CTRNNs), Plastic Neural Networks (PNN) and GasNets are all forms of Dynamic Recurrent Neural Networks (DRNNs). CTRNNs are the one of the simplest forms of DRNN. PNNs (Floreano and Mondada, 1996) seek to incorporate aspects of run-time learning through Hebbian adaptive network weights. GasNets are inspired by the action of Nitric Oxide as a neuromodulator (Husbands, 1998). GasNets and CTRNNs particularly have the potential to generate intricate dynamics on several different time scales, albeit by rather different means. This very likely, at least partially, accounts for their observed advantages over other less dynamic and unmodulated ANNs (Husbands et al., 1998, Smith et al., 2002).

This paper is the first study that includes GasNets, CTRNNs, PNNs and other forms of DRNN, in a performance comparison involving the generation of a suitably dynamic behavior - bipedal locomotion. This report extends work using a conventional DRNN applied to bipedal locomotion (Reil and Husbands, 2002). Other researchers have also addressed the issue of bipedal locomotion using evolutionary algorithms (Hase and Yamazaki, 1999) that incorporate neuromodulation (Ishiguro et al., 2003), variable morphology (Bongard and Paul, 2001), and use genetic programming (Ok et al., 2001). Other interesting work stresses the importance of reflexive responses in bipedal locomotion (Paul, 2004) in contrast to Central Pattern Generator (CPG) models.

3. Experimental Setup

Parameters were chosen as outlined by the original authors wherever possible (given the differences in genetic encoding). These original papers should be consulted for further experimental details if more information is required. Variations from their implementations are stated where relevant.

For the purposes of a fair comparison no assumption is made about the coupling of the underlying network nodes. The chosen task is to achieve bipedal locomotion *ab initio*, without predetermining that nodes are configured as coupled oscillators (Matsuoka, 1985).

3.1 Genetic Algorithm

The same distributed steady-state genetic algorithm is used on all networks. The population grid has dimensions of 10 by 10, to yield a total of 100 individuals. Competition is tournament based, with a tournament group comprising three individuals. A *principal* is selected at random from the grid. Two other population members are selected by a random walk originating at the principal's grid cell. The length of the random walk is an integer value in the range 1-4.

If the principal is the fittest then the weakest member is replaced by a mutated version of the principal. Otherwise the weakest member of the tournament set is replaced by a recombination of the two fittest individuals (using single point cross-over). This recombined genome is then mutated.

In these experiments the generation index is incremented after the evaluation of 100 individuals (comprising a pseudo-generation). Each neural network type was evaluated for 200 generations. This was carried out ten times with different random seeds for each network type.

Each fitness assessment starts with the biped in a stable standing position. The trial lasts a simulated 20 seconds. The fitness of the individual is taken to be the minimum of the distance traveled by either of the biped's feet, or hips. Each time-step for the neural network update lasts 0.025 seconds (or 1/40th of a second). Thus a 20 second trial corresponds to 800 neural network time-steps.

The trial was terminated if the biped fell below 50 percent of its original height and a fitness value assigned based on the distance traveled up to that point in time.

3.2 Physical Model

The computer code for the biped physical model was generated with the aid of a product called Autosim. Joints are simulated as torsional springs. Strictly speaking the motor output is actually a control signal. This signal is mapped to an angular displacement that corresponds to the rest position of a torsional spring. A change in the angular displacement of this rest position, will result in a torque applied to the lower limb attached to the joint (as the spring seeks to restore the joint to its new rest position).

The kinematic root has five degrees of freedom (two rotational and three translational). The biped is physically incapable of rotating in its roll axis. This prevents the biped from falling over on its side, although it is still free to fall forwards and backwards. Whilst this is not entirely physically realistic, it is sufficient for the purposes of this comparative study. This still remains a non-trivial problem; feet are modeled as point contact points, resulting in a dynamically unstable model after an initial displacement of the biped.

3.3 Sensorimotor System

Motor output signals are taken from the 1st 5th, 9th, 13th nodes in the network . These nodes were used to control the rest position of the torsional springs in the biped's left hip, right hip, left knee and right knee respectively. The output of these nodes was mapped to a range of -94 to 101 degrees angular hip displacement, and 0 to -89 degree angular knee displacement.

There are two sets of sensor configurations used in this experiment (nominally referred to as R and I configurations). The majority of simulations were carried out with a regular connection pattern (R configuration). Sensor input from the right foot contact sensor occurs at the 2nd, 6th, 10th and 14th nodes. Sensor input from the left foot contact sensor occurs at the 3rd, 7th, 11th and 15th nodes. Sensor input consisted of a binary 1.0 or 0.0 value depending upon whether or not the feet of the biped were in contact with the ground.

Additional simulations were carried out on GasNet networks with an irregular sensor connection pattern (I configuration). In this case sensor signals were input to the 8th, 9th and 10th nodes for the right foot, and the 11th, 12th, 13th and 14th nodes for the left foot. One of the reasons for this second sensor configuration was to see if there were any significant changes in the performance of circuits where sensor nodes were connected directly to motor nodes (i.e. in the case of the 9th and 13th nodes in the I configuration).

3.4 Genetic Encoding and Mutation

The genetic encoding strategy follows a similar approach for all networks. Network parameters are stored on a node or cell basis. Each gene comprises a list of real valued and integer parameters. Connection weights (where relevant) are also stored on a per node basis. The number of genetic parameters per node for each network type is shown in Table 1. Note that CTRNN's require us to store node weights on a per connection basis, in a fully connected network this means that we have N weights per node as evolvable parameters, where N is the total number of nodes in the network, plus the time constant and bias parameters, to yield a total of 18 parameters in this case.

Mutation takes place either after recombination, or after cloning of the principal tournament member (as described earlier). Mutation takes place at 20 percent of the nodes (rounded to 3 in a 16 cell network) selected at random. A single mutation event will result in the mutation of a single real or integer parameter in each of the randomly selected nodes. The magnitude of this mutation corresponds to 4 percent of the real valued parameters range with a probability of 0.2, and 1 percent of the parameters range with a probability of 0.8. In the case of integer parameters we follow a similar strategy of small mutations with a probability of 0.8 and large mutations with a probability of 0.2. These mutation parameters were chosen in preliminary experiments to avoid premature convergence and maintain a reasonable degree of phenotypic diversity across the different network varieties during evolution.

In addition to this, those networks where connection weights are under evolutionary control (such as conventional CTRNNs) undergo further mutation. Each randomly selected cell has all of its weights mutated (again by a factor of 4 percent with a 20 percent probability and 1 percent with an 80 percent probability).

Time constant initialization was devised to yield a wide range of of values. An exponent f was randomly selected from the set:

$$f\epsilon[-10,-8,-6,-4,-2,0,2,4,8,10]$$

A second random variable $r\epsilon[0.0,1.0]$ was then used to scale the value such that the time τ constant is calculated from:

$$\tau = 1.0 + r(10^f) \qquad (1)$$

The time constant mutation operator increments or decrements the exponent by 1 with a probability of 0.2, and generates a new value of $r\epsilon[0.0,1.0]$ with a probability of 0.8.

4. Network Details and Characteristic Equations

This section describes the network details of each of the varieties of networks tested. The focus is on the characteristic equations that govern the dynamics of each network variety.

A list of the network types is shown in Table 1. Each network type is assigned an index for reference purposes numbered between 1 and 14. The R letter denotes the regular sensor configuration, the I denotes the irregular sensor configuration.

Table 1: List of network types, and number of genetic parameters per node.

Type	Params.
1R Conventional CTRNN	18
2R Center-Crossing CTRNN	18
3R Basic PNN	3
4R Gas-Modulated PNN	17
5R CTRNN/PNN Hybrid	20
6R Conventional GasNet	15
7R Fully Recurrent GasNet	15
8R CTRNN/GasNet Hybrid no Gas	5
9R CTRNN/GasNet Hybrid with Gas	18
10I Conventional GasNet 16 Cell	15
11I Conventional GasNet 32 Cell	15
12I Fully Recurrent GasNet	15
13I CTRNN/GasNet Hybrid no Gas	5
14I CTRNN/GasNet Hybrid with Gas	18

It should be noted that Conventional GasNets comprise nodes that are spatially distributed. A parametric coding strategy is used where connections are determined for each node based on genetic parameters that define geometric arcs originating at each node. A node that falls within an excitatory or inhibitory arc are deemed to be electrically connected to the node at origin of the arc (Husbands, 1998). The consequence of this, is that conventional GasNets are only sparsely connected. This approach is used in the the following network types; 6R, 8R, 9R, 13I and 14I. All other network types are fully interconnected.

4.1 Conventional CTRNN - type 1R

This is a conventional CTRNN (Beer, 1995), fully recurrent, where node connection weights and biases are under evolutionary control.

$$y_i^{t+1} = y_i^t + \frac{T}{\tau_i}(-y_i^t + \sum_{j=1}^{N} \omega_{ji}\sigma(y_j^t + \theta_j) + I_i) \qquad (2)$$

$$i = 1, 2, ..., N$$

Where:

$y_i t + 1$ is the activation of the i'th node at time $t+1$.

y_i^t is the activation of the i'th node at time t.

τ_i is the time constant for the i'th node calculated according to equation equation 1.

I_i a sensor input to the i'th node where I is either 1 (in contact with the floor) or 0 (not in contact with the floor).

θ_j a bias term for the j'th node where $\theta\epsilon[-2,2]$.

T is the time slice (in this case T is set to 1).

ω_{ji} is the weight of the output from the j'th node to the i'th node where $\omega\epsilon[-4.0, 4.0]$.

σ is the logistic activation function.

$$\sigma(z) = \frac{1}{(1+e^{-z})} \tag{3}$$

4.2 Center-Crossing CTRNN - type 2R

The characteristic equation of the Center-Crossing CTRNN (Mathayomchan and Beer, 2002) is the same as that of the CTRNN (type 1). However initial biases are calculated such that:

$$\theta_i = \frac{-\sum_{j=1}^{N}\omega_{ji}}{2} \tag{4}$$

These authors suggest that populations seeded with center-crossing networks may be more likely to yield a wider range of dynamics than a population of random networks.

4.3 Basic PNN - type 3R

The key characteristic of PNN's (Urzelai and Floreano, 2000) is that connection weights vary over time based on Hebbian learning rules given by:

$$\omega_{ji}^{t} = \omega_{ji}^{t-1} + \eta\Delta\omega_{ji} \tag{5}$$

Where η is a learning rate ($0.0 < \eta < 1.0$) and ω_{ji} is the connection weight of the input to node i from node j. The adaptation rule $\Delta\omega_{ji}$ is genetically determined for each node. All inputs to a given node are subject to the same adaptation rule (referred to as node encoding by the original authors).

Where x is the activation of node j, which is an input to node i (which has an output activation of y), the adaptation rule is one of:

Plain Hebb Rule

$$\Delta\omega_{ji} = (1-\omega_{ji})x_j y_i \tag{6}$$

Post-Synamptic Rule

$$\Delta\omega_{ji} = \omega_{ji}(-1+x_j)y_i + (1-\omega_{ij})x_j y_i \tag{7}$$

Pre-Synaptic rule Rule

$$\Delta\omega_{ji} = \omega_{ji}(-1+y_i) + (1-\omega_{ji})x_j y_i \tag{8}$$

Covariance Rule

$$\Delta\omega_{ji} = \begin{cases} (1-\omega_{ji}) & \text{if } F(x_j, y_i) > 0 \\ (\omega_{ji})F(x_j, y_i) & \text{otherwise} \end{cases} \tag{9}$$

Where:

$$F(x_j, y_i) = \tanh(4(1-|x_i - y_j| - 2) \tag{10}$$

All nodes in the PNN are fully interconnected (with self-connections also supported). The rate of learning η can only assume one of four values (0.0, 0.3, 0.6, 0.9). The characteristic equation for the PNN is shown below:

$$y_i^{t+1} = \sigma(\sum_{j=1}^{N}\omega_{ji}^{t}(y_j^{t})) + I_i \quad i = 1, 2, ..., N \tag{11}$$

Where:

ω_{ji}^{t} is the adaptive weight for the j'th input to the i'th node.

σ is the standard logistic activation function.

I_i a sensor input to the i'th node where I is either 1 (in contact with the floor) or 0 (not in contact with the floor).

The term "basic" is used to differentiate it from other implementations an additional bias input is used for each node (Tuci and Quinn, 2003). The range of y_i is $[0, 2]$ for input neurons and $[0, 1]$ for hidden and output neurons, in line with other researchers (Blynel and Floreano, 2002).

4.4 Gas-Modulated PNN - type 4R

This is essentially the same as the basic PNN (type 3) with the exception that nodes whose weights are genetically determined to be modified by the Plain Hebb Rule, or Post-Synaptic Rule have their weights modified by diffused gases. For these two varieties of nodes, the weight modification rule becomes:

Gas Modified Plain Hebb Rule

$$\Delta\omega_{ji} = (\frac{c_{1i}^{t}}{c_{1i}^{t} + c_{2i}^{t}})(1-\omega_{ji})x_j y_i \tag{12}$$

Gas Modified Post-Synaptic Rule

$$\Delta\omega_{ji} = (\frac{c_{2i}^{t}}{c_{1i}^{t} + c_{2i}^{t}})(\omega_{ji}(-1+x_j)y_i + (1-\omega_{ij})x_j y_i) \tag{13}$$

Where:

c_{1i}^{t} is the concentration of gas 1 at the i'th node.

c_{2i}^{t} is the concentration of gas 2 at the i'th node.

The significance of these equations is that the rate of change in the weights of the inputs to these nodes will vary continuously with changes in the relative concentration of these two gases. When both gases have zero concentrations there is no change in weight.

4.5 CTRNN/PNN Hybrid - type 5R

This is a modification of the conventional PNN, with activation signals modified by a node based time constant under evolutionary control (in a similar fashion to conventional CTRNNs).

$$y_i^{t+1} = y_i^t + \frac{T}{\tau_i}(-y_i^t + \sum_{j=1}^{N} \omega_{ji}^t \sigma(y_j^t + \theta_j) + I_i) \tag{14}$$

$$i = 1, 2, ..., N$$

Where:

ω_{ji}^t is the adaptive weight for the j'th input to the i'th node.

4.6 Standard GasNet - type 6R

In GasNets (Husbands et al., 1998), node transfer functions can be modulated by local gas concentrations in the vicinity of the node. Nodes can also act as chemical emitters, under either gas or electrical stimulation. GasNet nodes exist in a geometric plane where internode distances determine gas concentrations and (in conjunction with additional genetic parameters) network connectivity. Under typical evolutionary parameters the GasNet connectivity rules result in a sparsely connected network.

$$y_i^{t+1} = \tanh[k_i^t(\sum_{j \epsilon C_i} \omega_{ji}\sigma(y_j^t + I_i)) + b_i] \tag{15}$$

Where:

k_i^t is a time-varying transfer function modulator. The value of k varies with gas concentrations at the i'th node.

C_i is the set of all nodes that have an input to the i'th node.

I_i a sensor input to the i'th node.

b_i a bias term for the i'th node.

The original GasNet diffusion model (upon which this implementation is based) is controlled by two genetically specified parameters, namely the radius of influence r and the rate of build up and decay s. Spatially, the gas concentration varies as an inverse exponential of the distance from the emitting node with a spread governed by r, with the concentration set to zero for all distances greater than r (Equation 16). The maximum concentration at the emitting node is 1.0 and the concentration builds up and decays from this value linearly as defined by Equations 17 and 18 at a rate determined by s.

$$C(d,t) = \begin{cases} e^{-2d/r} \times T(t) & d < r \\ 0 & \text{else} \end{cases} \tag{16}$$

$$T(t) = \begin{cases} H\left(\frac{t-t_e}{s}\right) & \text{emitting} \\ H\left[H\left(\frac{t_s-t_e}{s}\right) - H\left(\frac{t-t_s}{s}\right)\right] & \text{not emitting} \end{cases} \tag{17}$$

$$H(x) = \begin{cases} 0 & x \leq 0 \\ x & 0 < x < 1 \\ 1 & \text{else} \end{cases} \tag{18}$$

where C(d,t) is the concentration at a distance d from the emitting node at time t. t_e is the time at which emission was last turned on, t_s is the time at which emission was last turned off, and s (controlling the slope of the function T) is genetically determined for each node. The total concentration at a node is then determined by summing the contributions from all other emitting nodes (nodes are not affected by their own concentration, to avoid runaway positive feedback).

For mathematical convenience, in the basic GasNet there are two 'gases', one whose modulatory effect is to increase the transfer function gain parameter (k_i^t from equation 19) and one whose effect is to decrease it. It is genetically determined whether or not any given node will emit one of these two gases (gas 1 and gas 2), and under what circumstances emission will occur (either when the 'electrical' activation of the node exceeds a threshold, or the concentration of a genetically determined gas in the vicinity of the node exceeds a threshold. Note these emission processes provide a coupling between the 'electrical' and 'chemical' mechanisms. The concentration-dependent modulation is described by Equation 19, with transfer parameters updated on every time step as the network runs.

$$k_i^t = k_i^0 + \alpha C_1^t - \beta C_2^t \tag{19}$$

where k_i^0 is the genetically set default value for k_i, C_1^t and C_2^t are the concentrations of gas 1 and gas 2 respectively at node i at time t, and α and β are constants. Both gas concentrations lie in the range $[0, 1]$. Thus the gas does not alter the electrical activity in the network directly but rather acts by continuously changing the mapping between input and output for individual nodes, either directly or by stimulating the production of further virtual gas. The concentration dependent modulation can, for instance, change a node's output from being positive to being zero or negative even though the input remains constant. Any node that is exposed to a non zero gas concentration will be modulated. This set of interacting processes provides the potential for highly plastic systems with rich dynamics.

4.7 Fully Recurrent GasNet - type 7R

The fully recurrent GasNet uses the GasNet model of gas diffusion but adopts a fully recurrent connectivity model.

Whereas a conventional GasNet is sparsely connected, this network is fully connected.

$$y_i^{t+1} = \tanh[k_i^t(\sum_{j=1}^{N} \omega_{ji}\sigma(y_j^t + I_i)) + b_i] \quad (20)$$

$$i = 1, 2, ..., N$$

4.8 CTRNN/GasNet Hybrid no Gas - type 8R

This is a variation of the conventional CTRNN. In this case inter-node connectivity is determined by the approach used in GasNets. It uses a sigmoid transfer function, with bias and node time constants under evolutionary control.

$$y_i^{t+1} = y_i^t + \frac{T}{\tau_i}(-y_i^t + \tanh[K_i(\sum_{j\epsilon C_i} \omega_{ji}\sigma(y_j^t + I_i)) + b_i]) \quad (21)$$

Where:

K_i is a transfer function constant.

T is the time slice constant.

τ_i is the time constant for the i'th node.

4.9 CTRNN/GasNet Hybrid with Gas - type 9R

This is another variation of a conventional CTRNN, but in this case it is more extensively modified along the lines of GasNets. The network nodes transfer function is gas modulated, and network connectivity is based on the GasNet model. What remains of the the original CTRNN is the node time constant.

$$y_i^{t+1} = y_i^t + \frac{T}{\tau_i}(-y_i^t + \tanh[k_i^t(\sum_{j\epsilon C_i} \omega_{ji}\sigma(y_j^t + I_i)) + b_i]) \quad (22)$$

Where:

k_i^t is a time-varying transfer function modulator. The value of k varies with gas concentrations at the i'th node.

4.10 Conventional GasNet 16 Cell - type 10I

This network is based on the Conventional GasNet (type 6R) but uses the irregular sensor configuration. Another difference is that bias values are set such that $b_i\epsilon[-4, 4]$.

4.11 Conventional GasNet 32 Cell - type 11I

Other than the number of cells in this network, all other parameters are the same as those of the 16 cell gas net (type 10I). Note that of all the networks tested, this is the only one that is comprised of 32 cells or nodes.

Given the nature of the the GasNet connectivity algorithm 32 cell GasNets are likely to be more highly interconnected than 16 cell GasNets. The size of the plane remains constant, so a higher number of cells means a higher cell density. A larger number of cells will tend to fall within a given connection arc, thus resulting in a higher number of inter-cell connections per cell on average.

4.12 Fully Recurrent GasNet - type 12I

This network is based on the Fully Recurrent GasNet (type 7R) but uses the irregular sensor configuration. Another difference is that bias values are set such that $b_i\epsilon[-4, 4]$.

4.13 CTRNN/GasNet Hybrid no Gas - type 13I

This network is based on the CTRNN/GasNet Hybrid without Gas (type 9R) but uses the irregular sensor configuration. Another difference is that bias values are set such that $b_i\epsilon[-4, 4]$.

4.14 CTRNN/GasNet Hybrid with Gas - type 14I

This network based on the CTRNN/GasNet Hybrid with Gas (type 9R) but uses the irregular sensor configuration. Another difference is that bias values are set such that $b_i\epsilon[-4, 4]$.

5. Results

Table 2 shows the peak fitness of the best individual in the ten evolutionary runs for each network type, together with the standard deviation of the fitness of the fittest individual across all ten runs. Table 3 shows the average and median peak fitness value across all runs for each network type.

To put these fitness values in perspective 2.5 meters is attainable through a fast walk in about 10 seconds (attained by the Conventional GasNet type 10I), and a slow walk in about 20 seconds (attained by the Conventional GasNet type 6R). Bipeds that have traveled around 1 meter have typically taken around two steps. The worst performing Basic PNN typically only extended one leg, before slightly drawing the lagging leg forward.

Figure 1 shows three snapshots of the biped walking. This rendering is based on the results obtained from the 16 cell GasNet (type 10I).

5.1 Principal Results Summary

We can summarize the most important results as follows:

1. Conventional GasNets were the only networks to achieve cyclical bipedal locomotion.

2. Center-Crossing CTRNNs achieved the second highest peak fitness after the conventional GasNets.

3. Center-Crossing CTRNNs and CTRNNs attained the highest average fitness values.

4. Center-Crossing CTRNNs marginally outperformed conventional CTRNNs in peak and average fitness.

5. Basic PNNs without dynamic attributes performed the worst.

6. Both a CTRNN/PNN Hybrid and a Gas Modulated PNN improved on the performance of the Basic PNN.

7. The CTRNN/PNN Hybrid achieved comparable results to the Conventional CTRNN.

8. Fully Connected GasNets perform badly compared to Conventional GasNets. GasNet performance seems to decline with increased inter-connectivity (see type 11I vs 10I and 6R).

9. Conventional GasNets exhibit a higher variation in phenotype fitness than other network types. Conventional CTRNNs exhibit relatively low phenotypic variation.

10. Gas Modulated networks generally outperform their un-modulated counter-parts (see type R8 vs R9 and I13 vs I14).

11. GasNet performance is largely unaltered by minor sensorimotor configuration changes (see R vs I sensor configurations).

These results are generally consistent with those reported by GasNet researchers and the comparative studies cited earlier.

6. Discussion

An obvious question to ask when considering the relative performance of these networks is whether or not the relative performance differences are primarily a consequence of the number of degrees of freedom in parameter space. Table 1 lists the number of parameters per node. The basic PNN (3R) has the lowest number of parameters and displays the worst performance. However, the CTRNN/PNN Hybrid (5R) has the largest number of evolvable parameters, but it does not have the best performance. Whilst degrees of freedom in parameter space may be a contributing factor in the results described here, the internal dynamics of each network also play a significant role. The 16 cell GasNet (10I) has substantially superior peak fitness performance to the 32 cell GasNet (11I) and fully connected GasNet (12I),

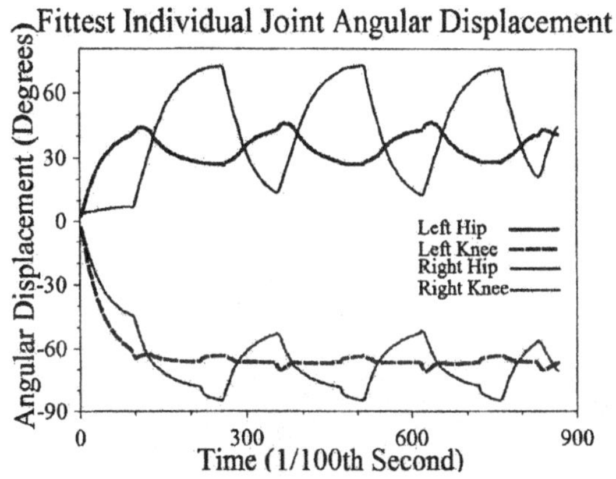

Figure 1: GasNet Biped Kinematic Data (Upper; Left Hip and Right Hip, Lower Right Knee and Left Knee).

Table 2: Peak fitness across all evolutionary runs, and Standard Deviation of the peak fitness (meters).

Network	Peak	SD.
6R Conventional GasNet	2.63	0.70
10I Conventional GasNet 16 Cell	2.62	0.66
2R Center-Crossing CTRNN	1.63	0.24
11I Conventional GasNet 32 Cell	1.31	0.29
5R CTRNN/PNN Hybrid	1.27	0.33
14I CTRNN/GasNet Hybrid with Gas	1.25	0.21
1R Conventional CTRNN	1.24	0.15
9R CTRNN/GasNet Hybrid with Gas	1.22	0.25
8R CTRNN/GasNet Hybrid no Gas	1.18	0.23
4R Gas-Modulated PNN	1.06	0.30
7R Fully Recurrent GasNet	0.98	0.09
13I CTRNN/GasNet Hybrid no Gas	0.72	0.13
12I Fully Recurrent GasNet	0.53	0.21
3R Basic PNN	0.24	0.01

Table 3: Average and median of distances traveled by the fittest individual at the end of each run (meters).

Network	Avg.	Med.
2R Center-Crossing CTRNN	1.13	1.19
1R Conventional CTRNN	1.11	1.18
6R Conventional GasNet	0.97	0.82
9R CTRNN/GasNet Hybrid with Gas	0.95	0.96
11I Conventional GasNet 32 Cell	0.94	1.08
14I CTRNN/GasNet Hybrid with Gas	0.92	0.96
10I Conventional GasNet 16 Cell	0.92	0.93
5R CTRNN/PNN Hybrid	0.90	1.00
8R CTRNN/GasNet Hybrid no Gas	0.90	0.95
13I CTRNN/GasNet Hybrid no Gas	0.61	0.66
4R Gas-Modulated PNN	0.57	0.72
12I Fully Recurrent GasNet	0.47	0.36
7R Fully Recurrent GasNet	0.35	0.38
3R Basic PNN	0.23	0.23

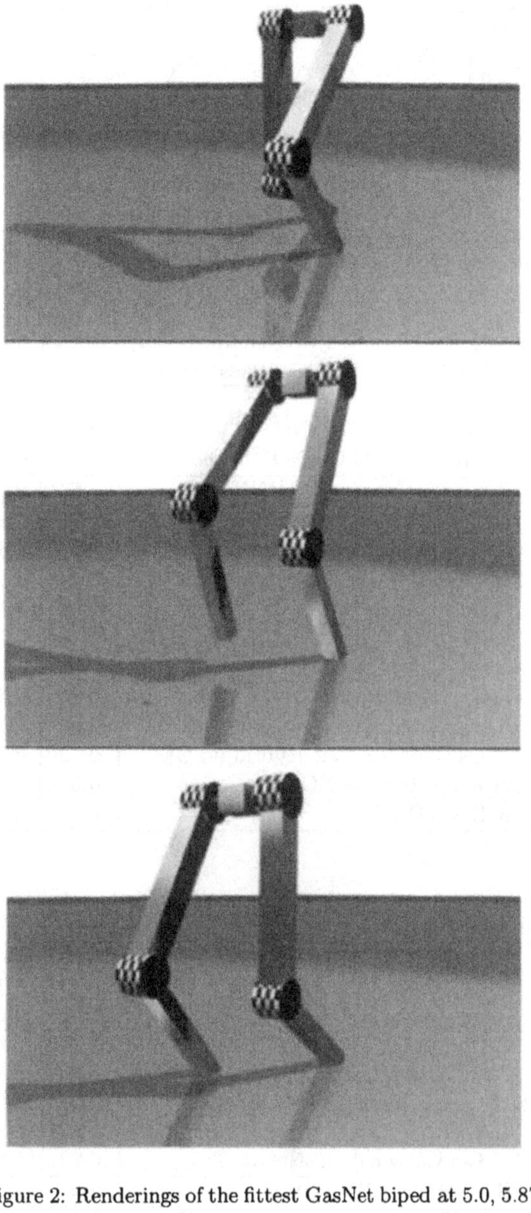

Figure 2: Renderings of the fittest GasNet biped at 5.0, 5.875 and 6.25 seconds.

despite having the same number of evolvable parameters. The number of parameters for a given network is a consequence of the network formulation, and whilst the low number of parameters used in the basic PNN may lead us to believe that it is under-specified in comparison with other networks, the key determinant is the saliency of a given parameter to the exploration of phenotype space. This is not something that can be easily determined without referring to the underlying dynamics of the alternative network formulations. As a consequence, the discussion in this section focusses largely on the underlying dynamics, rather than specifically address the issue of differences in parameter space.

6.1 *Temporal Adaptivity and Evolvability*

It is claimed that GasNets have a high evolvability due to their high capacity for *temporal adaptation* (Smith et al., 2002). A significant factor determining the activity of the network is the coupling of electrical and gas dynamics (Philippedes et al., 2002). In this work the authors showed that reducing the likelihood that a node pair are both chemically and electrically coupled, can improve the evolvability of the GasNet network even further. In particular they state;

> .. systems involving distinct yet coupled processes are highly evolvable when there is a bias towards loose coupling between the processes; this allows the possibility of 'tuning' one against the other without destructive interference.

The relatively poor performance of fully connected GasNets, where a high degree of coupling is forced into the network, tends to support this claim. Philippides et al. also refer to the multiple redundancies inherent in loosely coupled GasNets which potentially lead to increased numbers of routes through the evolutionary search space. These factors may help to account for the relatively high variance in GasNet fitness in comparison with CTRNNs, as well as the significantly higher peak fitness. The inherent dynamics are being shaped and explored in a very different way.

In the case of the one of the fittest GasNet bipeds (type I10) the reactive response of the right hip to the transition from non-contact to contact of the left foot appears to be primarily gas-mediated. In contrast the joint angular displacement dynamics are governed primarily by electrical activation signals. Whilst this may be an over-simplification (GasNets are integrated systems with co-dependencies between both gas and electrical signalling), it is easy to imagine circumstances in which there are independent phenotypic processes that have intrinsic time dynamics associated with them. The ability to explore these phenotypic temporal dynamics in parallel may be a significant factor in the evolvability of GasNets.

Concepts such as of temporal adaptivity and system coupling may be useful concepts in helping us to understanding the dynamics of such systems. However we still need to try and identify the specific characteristics of these networks that support temporal adaptivity. Two of the most obvious areas to consider are the frequency and phase characteristics of the GasNet.

6.2 Phase Space Exploration

Phase relationships between signals in an articulated agent are extremely important to achieve coordinated activity. One aspect of GasNets is that nodes are physically distributed in a virtual 2-dimensional space. A variation in the distance between nodes (through mutation for example) results in a phase lag or lead in the modulation of gas-coupled nodes. In this sense phase relationships are under direct evolutionary control. None of the other networks investigated in this study embody these characteristics. Simply put, in GasNets, node position mutations are operating directly in phenotype phase space (where nodes are chemically coupled).

In direct encoding, if a single parameter maps onto a specific phenotypic attribute that is largely independent of other attributes, this is likely to aid the efficient exploration of phenotypic space via mutation operators, since it may result in a smoother fitness landscape. In the case of articulated bipedal locomotion, we might imagine that the phase relationship between hip joints is a significant phenotypic attribute that affects overall fitness. A network model that can explore this phase relationship through mutation operators may exhibit greater temporal adaptation than one that cannot.

6.3 Frequency Space Exploration

Examination of the motor output signals of GasNets compared to other networks indicates that GasNets generate a motor control signals with a wide range of frequency components (particularly high frequency). It is easy to imagine that a network that exhibits signification oscillatory behavior over a wide bandwidths may have some advantages over networks in seeking to discover a solution suitable for articulated motion.

6.4 Temporal Dynamics and PNNs

It should be stated that there is no reason to expect that the Basic PNN should have performed particularly well in this problem domain. The environment remains constant over evolutionary time. As such there is no additional benefit to be gained by in-trial learning. Limited sensor stimulation may have resulted in a rapid decay in node activity in this implementation. However, the modification of the Basic PNN to incorporate richer time dynamics (e.g. type 4R and 5R networks) can improve the performance of PNN networks.

6.5 Complementary Study

A complementary study carried out by the authors of this paper (McHale and Husbands, 2004) investigated three of the network varieties described here (Conventional GasNet, Center-Crossing CTRNN and CTRNN/PNN Hybrid) in the evolution of locomotion in a physically simulated quadruped. Network morphologies were constrained to resemble coupled-oscillators. The initial population was comprised solely of symmetrical networks, with corresponding nodes connected by mutually inhibitory connections. The Open Dynamics Engine (ODE) physics simulation package was used, without any restrictions on the degrees of freedom of the quadrupeds kinematic root.

Whilst the GasNet network attained the highest fitness, the CTRNN/PNN Hybrid (introduced in this study) achieved comparable results, discovering several gaits between them. Whilst the Center-Crossing CTRNN also attained a high fitness level, it discovered only one gait, and average fitness was approximately half that of the other two networks. Qualitatively the gaits discovered by the CTRNN/PNN Hybrid appeared to be more reactive than those discovered by the GasNet, exhibiting some gaits that resembled ballistic walking. The evolved GasNets discovered more stable gaits (that continued if the trial period was extended), and generally appeared to rely more upon forced oscillations than the CTRNN/PNN Hybrid.

7. Acknowledgements

The original physical biped model was developed in collaboration with C.Buckley. Thanks should be given to the two anonymous referees for their constructive criticism and recommendations.

8. Conclusion

The results described here, taken in conjunction with those of complementary studies, suggest that GasNets offer a reliable solution in solving sensorimotor control problems for physically simulated agents. This paper also introduces new variants of conventional networks that may merit further study, such as the gas-modulated Hebbian network (4R) and CTRNN/PNN Hybrid (5R). The results of the complementary study described earlier remind us that we should be careful not to generalize from single experiments. The consideration of alternative dynamic models should be an ongoing part of the characterization of alternative network formulations. Future work will seek to carry out additional comparative studies for physically simulated hexapods and bipeds with unconstrained kinematic roots in a continuation of the work described here.

References

Beer, R. (1995). On the Dynamics of Small Continuous-Time Recurrent Neural Networks. *Adaptive Behavior*, 3(4):469–509.

Beer, R., Chiel, H., and J.C.Gallagher (1999). Evolution of CPGs for walking II. General principles and individual variability. *Journal of Computational Neuroscience*, 7(2):119–147.

Blynel, J. and Floreano, D. (2002). Levels of Dynamic and Adaptive Behaviorin Evolutionary Neural Controllers. In Hallam, B., Floreano, D., Hallam, J., Hayes, G., and Meyer, J.-A., (Eds.), *From Animals to Animats 7: Proceedings of the Seventh International Conference on Simulation of Adaptive Behavior (SAB 02)*, pages 272–281. MIT Press.

Bongard, J. and Paul, C. (2001). Making Evolution an Offer it Can't Refuse: Morphology and the Extradimensional Bypass. In Keleman, J. and Sosik, P., (Eds.), *Proceedings of the Sixth European Conference on Artificial Life (ECAL 2000)*, pages 401–412.

Floreano, D. and Mondada, F. (1996). Evolution of Plastic Neurocontrollers for Situated Agents. In Maes, P., Mataric, M., Meyer, J.-A., Pollack, J., and Wilson, S., (Eds.), *From Animals to Animats 4: Proceedings of the Fourth International Conference on Simulation of Adaptive Behavior (SAB 96)*. MIT Press.

Hase, K. and Yamazaki, N. (1999). Computational evolution of human bipedal walking by a neuro-musculo-skeletal model. *Artificial Life and Robotics*, 3:133–138.

Husbands, P. (1998). Evolving Robot Behaviours with Diffusing Gas Networks. In Husbands, P. and Meyer, J.-A., (Eds.), *EvoRobot98: Proceedings of 1st European Workshop on Evolutionary Robotics*, pages 71–86. Springer-Verlag LNCS.

Husbands, P., Smith, T., O'Shea, M., Jakobi, M., Anderson, J., and Phillipedes, A. (1998). Brains Gases and Robots. In Niklasson, N., (Ed.), *Proceedings of the 8th International Conference on Artificial Neural Networks (ICANN98)*, pages 51–64. Springer-Verlag.

Ishiguro, A., Fujii, A., and Hotz, P. E. (2003). Neuromodulated Control of Bipedal Locomotion Using a Polymorphic CPG Circuit. *Adaptive Behavior*, 11:7–17.

Mathayomchan, B. and Beer, R. (2002). Center-Crossing Recurrent Neural Networks for the Evolutionof Rhythmic Behavior. *Neural Computation*, 14:2043–2051.

Matsuoka, K. (1985). Sustained-Oscillations Generated by Mutually Inhibiting Neurons with Adaptation. *Biologocal Cybernetics*, 52:367–376.

McHale, G. and Husbands, P. (2004). Quadrupedal Locomotion : GasNets, CTRNNs and Hybrid CTRNN/PNNs Compared. Artificial Life 9: Proceedings of the Ninth International Conference on Artificial Life - Submitted.

Ok, S., Miyashita, K., and Hase, K. (2001). Evolving Bipedal Locomotion with Genetic Programming - Preliminary Report. In *Proceedings of the 2001 Congress on Evolutionary Computation (CEC2001)*, pages 1025–1032. IEEE Press.

Paul, C. (2004). Sensorimotor Control of Bipedal Locomotion Based on Contact Information. International Symposium on Intelligent Signal Processing and Robotics - Accepted.

Philippedes, A., Husbands, P., Smith, T., and O'Shea, M. (2002). Fast and Loose : Biologically Inpsired Couplings. In Standish, R., Bedau, M., and Abbass, H. A., (Eds.), *Artificial Life VIII: Proceedings of the Eight International Conference on Artificial Life*, pages 292–301. MIT Press.

Reil, T. and Husbands, P. (2002). Evolution of central pattern generators for bipedal walking in a real-time physics environment. *IEEE Transactions on Evolutionary Computation*, 6:159–168.

Smith, T., Husbands, P., Philippides, A., and O'Shea, M. (2002). M. neuronal plasticity and temporal adaptivity: Gasnet robot control networks. *Adaptive Behavior*, 10(3/4):161–184.

Tuci, E. and Quinn, M. (2003). Behavioural Plasticity in Autonomous Agents: A Comparison Between Two Types of Controller. In Cagnoni, S., Cardalda, J., Corne, D., Gottlieb, J., Guillot, A., Hart, E., Johnson, C., Marchiori, E., Middendorf, J.-A. M. M., and Raidl, G., (Eds.), *Applications of Evolutionary Computing: EvoWorkshops 2003*, pages 661–672, Berlin. Springer-Verlag.

Urzelai, J. and Floreano, D. (2000). Evolutionary Robots with Fast Adaptive Behavior in New Environments. In et al., A. T., (Ed.), *Third International Conference on EvolvableSystems: From Biology to Hardware (ICES2000)*. Springer Verlag.

Adaptive locomotion in a complex environment: simulation of stick insect gap crossing behaviour

Bettina Blaesing

Faculty of Biology, University of Bielefeld, Postbox 100131, D - 33501 Bielefeld, Germany
e-mail: bettina.blaesing@ uni-bielefeld.de

Abstract

In a complex three dimensional environment, the ability to climb across large gaps and obstacles is fundamental for a stick insect. The same ability is desired for a walking animat that operates in a similar habitat. In this study, the neural network simulation WALKNET that has been modelled on stick insect walking behaviour is used as a basis for the introduction of new behaviours that enable it to climb over large gaps. Gap crossing behaviour of stick insects has been studied to show which behaviours are most crucial for reaching the far edge. As slowing down of forward movement and searching movements of the front legs in the gap have been considered most important, these two behaviours have been analysed. Subsequently, new modules for velocity control, adaptation of swing amplitudes and generation of searching movements have been implemented into the WALKNET controller. With these innovations, the animat is able to climb across gaps of more than twice its normal step length, using similar strategies as the biological model. The new behaviours improve its adaptability and performance under challenging environmental conditions. Further results of experimental gap crossing studies are discussed with respect to their value for the walking simulation and robotic applications.

1. Introduction

Climbing across large gaps has been used as a paradigm to investigate how the walking behaviour of stick insects is adapted to the challenges of a complex environment. Gap crossing behaviour also represents an important ability in the behavioural repertoire of a walking machine, as large gaps and obstacles are likely to occur in a natural environment. Stick insect walking behaviour without or with slight disturbances has been thoroughly investigated (e.g. Cruse 1976 a, Cruse 1976 b, Cruse and Epstein 1982, Graham and Epstein 1985). In their natural environment, stick insects inhabit a three-dimensional maze of twigs and leaves which makes them a suitable biological model for locomotion on uneven terrain.

The aim of this work is to use the results of experimental studies of stick insect gap crossing behaviour to challenge the abilities of WALKNET (Cruse et al. 1998), an artificial neural network simulation that has been modelled on studies of stick insect walking behaviour. The WALKNET controller is fairly decentralized, it consists of six independent agents, one for each leg, that cooperate according to basic rules of leg coordination (Cruse 1990). The basic modules of each leg agent are the SWINGNET, a neural network that generates the swing movement, the STANCENET that controls the stance movement and a SELECTOR that decides which of the two has access to the motor output. Additionally, a TARGETNET provides the reference coordinates for the anterior extreme position (AEP) of the swing movement (Figure 1 A).

The gap crossing paradigm represents the most complex behaviour in stick insect locomotion that has been used to enhance the WALKNET simulation. So far, the WALKNET can cope with a variety of disturbances such as falling, stumbling, walking up and down stairs and even amputation of a leg (Kindermann 2002). Fitness for real life and physical environments has successfully been tested by implementation of the controller in different hexapod robots (Cruse et al. 1998). However, gaps of the size used here extend the abilities of the animat. Gaps provide a special difficulty as they have to be detected as absence of a tactile signal, not as a tactile signal like steps or obstacles. As no positive tactile signal is given by contact with the gap, an expectation of ground contact that is thwarted seems to be the key stimulus to detect a gap. In the stick insect, this "non-contact" signal can be received by the antenna or by the front leg (Blaesing and Cruse 2004 b).

In previous studies, it could be shown that gap crossing behaviour can be regarded as normal walking behaviour with small modifications. The main difference from undisturbed walking occurs during a phase of exploration that starts when the insect has detected the gap and ends when it has detected the far edge (Blaesing and Cruse 2004 b). Reaching the gap is directly followed by a decrease in walking velocity. After detecting the gap, the insect probes the space in front of it with its antennae and front legs with vigorous searching movements, which finally leads to tactile contact with the far edge of the gap. Tactile contact of an antenna or front leg with the far edge of the gap represents a "point of no return" after which the gap crossing behaviour is not terminated anymore. As visual input does not contribute to detecting the far side of the gap (Blaesing and Cruse 2004 b), it has not been regarded in this study. In *Drosophila*, locomotion in a stepping stone paradigm is also controlled by tactile cues but not by vision (Pick and Strauss 2002).

During the gap crossing sequence, parameters of single steps such as the posterior extreme position (PEP), the anterior extreme position (AEP) (Figure 1 A) and the trajectory of the swing movement are modified (Blaesing and Cruse 2004 a). Differences are so obvious that different step types have been defined as basic elements of gap crossing behaviour (Figure 1 B). In addition to normal walking steps and steps across the gap, short steps and tentative steps have been observed. Short steps are not displayed in Figure 1 B. Their swing trajectory is not different from the one of normal walking steps but their swing amplitude is shorter, approx. 30% of normal walking step amplitude (Blaesing and Cruse 2004 a, b). Tentative steps occur almost exclusively in the front legs. Both the tentative steps and the gap crossing steps contain a searching movement that is more complex than the normal swing trajectory (schematically displayed in Figure 1 B), but in the tentative steps the forward movement is reversed at some point and the leg is placed back on the first footbridge. This allows the front legs to apply a trial-and-error strategy by performing tentative steps until they can reach the far edge. The fictive AEP (fAEP) in the tentative step and gap crossing step marks the position where the tarsus crosses footbridge level.

In this study, I focus on two major modifications of the WALKNET controller that enable it to climb across large gaps in order to improve its adaptive properties:

1. Velocity of forward movement is slowed down after detecting the gap, in response to a non-contact signal that is given by thwarting of an expectation of ground contact. During slow forward movement, the target positions for the AEPs are shifted rearward so that the supporting legs perform short steps.
2. When stepping into a gap, the leg performs searching movements until it makes contact with the ground. Trajectories observed in searching movements are modelled based on gap crossing step trajectories of the front legs of stick insects. If no ground contact is made within a certain time, the direction of the swing trajectory is reversed and the front leg is placed back on the first footbridge close to its starting position (tentative step).

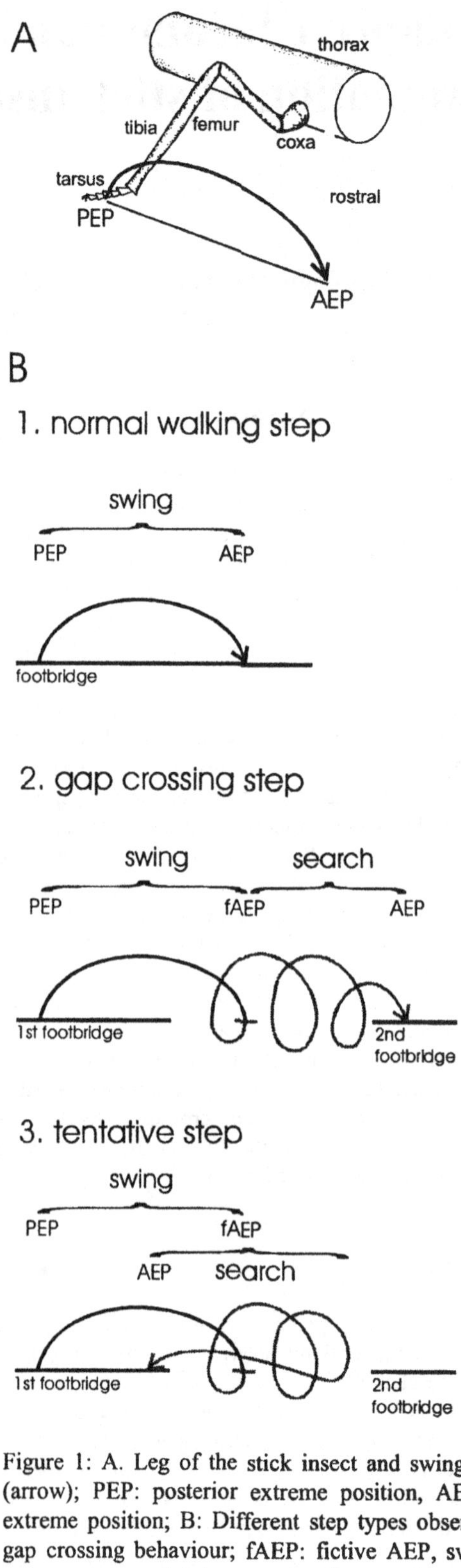

Figure 1: A. Leg of the stick insect and swing movement (arrow); PEP: posterior extreme position, AEP: anterior extreme position; B: Different step types observed during gap crossing behaviour; fAEP: fictive AEP, swing: initial swing movement into gap, search: searching movement, schematically

2. Experimental study

2.1 Material and method

Stick insects of the species *Aretaon asperrimus* were kept in mesh wire cages on bramble (*Rubus fruticosus)* and water *ad libitum* with an artificial day-night cycle of 12h : 12h. Average body length of the males used in this study is 51 ± 1.0 mm (mean ± SD), average step amplitude is approximately 17 mm (front legs: 17.1 ± 2.6 mm, n = 310, middle legs: 16.8 ± 2.7 mm, n = 326, hind legs: 17.6 ± 2.4 mm, n = 340). In all experiments, the animals were placed on a cardboard footbridge of 60 mm width, 300 mm length and 200 mm height, facing a second footbridge of the same type. The gap between the two footbridges was 30 mm wide. To exclude the influence of visual cues, all stick insects were blindfolded during the experiment. The insects were recorded walking along the first footbridge, climbing across the gap and walking on from a distance of 1.5 m by a overhead video camera (Sony EVI-D31). A simultaneous side view image of the walking animal was obtained via a mirror attached to the footbridges at a 45° angle. Video recordings (50 frames/s) were carried out in daylight with additional artificial lighting. From the video recordings, searching movements of the front legs in the gap were analysed in single frame modus, using customized software designed to read marked pixel coordinates as ASCII data. For the investigation of searching trajectories, the end of the tibia, femur tibia joint, coxa, head and end of thorax were marked in every frame throughout the gap crossing step (Figure 2). The angles of the three main leg joints were calculated from the data collected in screen coordinates using a customized software. For the velocity profiles, the front leg coxa was marked in the side view only in every tenth frame, i.e. with a frequency of 5 Hz in a short time window before and after reaching the gap. Sample sizes were N = 9 trials for searching movements and N = 14 for velocity profiles. Durations of single steps were measured for a larger number of trials (Table 1).

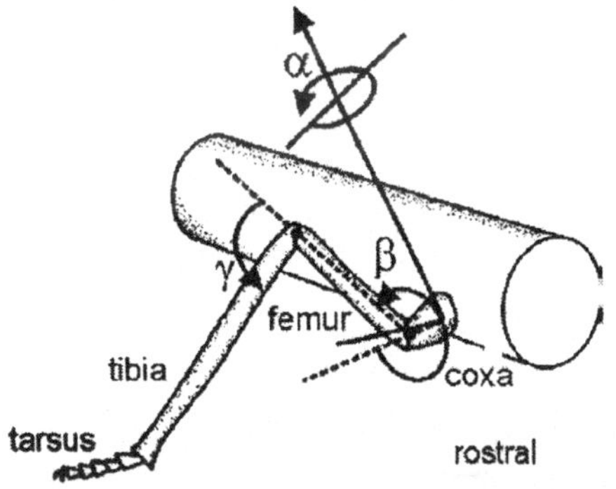

Figure 2: Leg of the stick insect with joint angles;
α subcoxal joint, β coxa trochanter joint, γ femur tibia joint

2.2 Results

When reaching a gap, the stick insect decreases its stance velocity almost to zero. Further forward movement consists of bouts of slow advance intermitted by short stops while the animal performs searching movements with its antennae and front legs. To determine which sensory signals the animal reacts to by slowing down, velocity profiles of forward movement have been triggered to two possible events. These events are 1. lowering the antenna into the gap and 2. stepping into the gap with a front leg. Figure 3 shows that slowing down occurs within the first 200 ms after initial non-contact of a front leg. The time relation between the first antennal non-contact and slowing down is likely to be an indirect effect of front leg non-contact that occurs shortly after antennal non-contact (385 ± 298 ms, n = 14).

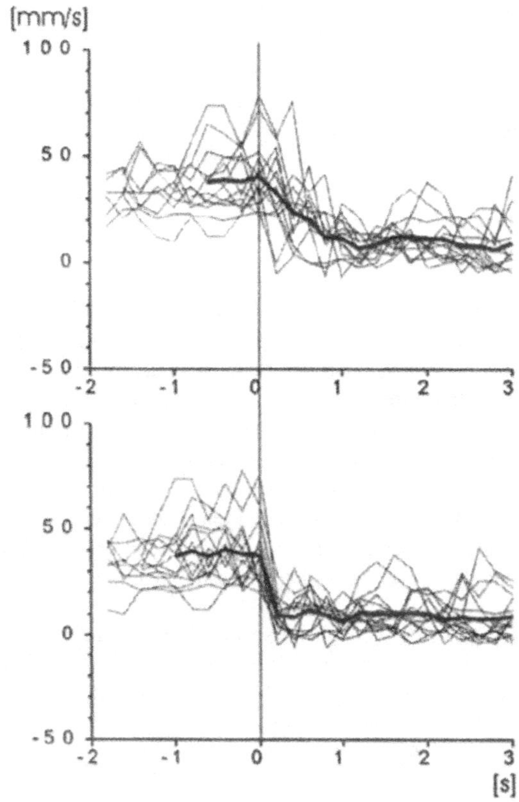

Figure 3: Velocity profiles of forward movement of the body over ground during a time window of 5 seconds at the beginning of the gap crossing sequence, n = 14 trials, bin width = 200 ms; in the two diagrams, the same data are triggered to different events; upper panel: t = 0 (vertical grey line) initial non-contact of the front leg (tarsus crosses footbridge level while swinging into the gap for the first time); lower panel: t = 0 (vertical grey line): initial non-contact of the antenna; heavy lines: arithmetic mean, thin lines: single trials

In the trajectory of gap crossing steps, an initial swing movement can be distinguished from a subsequent searching movement. The trajectory of the swing movement resembles a normal walking step whereas the searching trajectory is more irregular. A previous study has shown that the amplitude of the swing movement is slightly smaller than the swing amplitude of a normal walking step whereas the amplitude of the entire gap crossing step (swing movement and searching movement taken together) is larger than a normal walking step (Blaesing and Cruse 2004 b). Only in the second hind leg, gap crossing steps that consist of a single elongated swing movement without a searching movement occur frequently.

The time course of the joint angles during the searching movement shows different characteristics for the first and the second front leg that swings across the gap. In the first front leg gap crossing steps, joint angles show oscillations of about 2 Hz (α: 1.87 ± 0.63 Hz, n = 15; β: 2.04 ± 0.54 Hz, n = 29; γ: 1.89 ± 0.44 Hz, n = 25; period: α: 0.61 ± 0.26 s, β: 0.53 ± 0.16 s, γ: 0.56 ± 0.15 s). Searching movements in the γ joint show the largest amplitude, more than twice the amplitude of α and β joint movements (α: 34.6 ± 13.7°, β: 42.4 ± 14.6°, γ: 86.2 ± 27.7°). Five single examples of first gap crossing steps are displayed in Figure 4 in the upper five panels.

A characteristic that can be found in all examples is the negative phase coupling of oscillations in the β and γ joint (phase shifting of β relative to γ: 0.44 ± 0.08, n = 26). The β and γ joint movements are coupled in such a way that flexion occurs together with depression and extension occurs together with elevation. The signal of the α joint is not coupled to the other joints as strongly, however, in the beginning of the searching movement protraction occurs together with depression in the β joint. Later on during the searching movement, oscillations in the α joint become less clear and the leg is rather kept in a rostral position.

The gap crossing step of the second front leg needs less time and has a more stereotyped trajectory than the first front leg gap crossing step. In Figure 4, joint angles of the second front leg gap crossing step are displayed in the bottom panel together with their arithmetic mean.

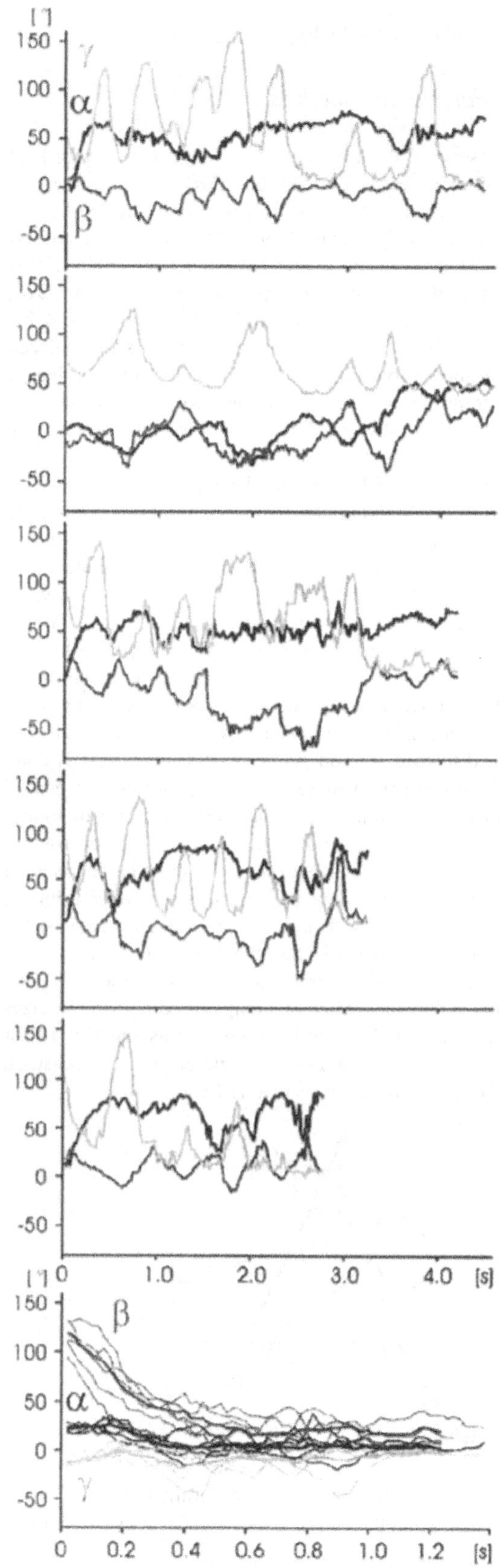

Figure 4: Time course of joint angles during gap crossing steps of the front legs; upper five panels: data of single trials of the first gap crossing step, bottom panel: data of seven trials of the second gap crossing step, thin lines: single trials; heavy lines: arithmetic mean; black: α joint, dark grey: β joint, light grey: γ joint; thin lines: single trials; heavy lines: arithmetic mean

For a larger number of trials, the duration of normal walking steps, short steps, gap crossing steps and tentative steps has been measured (Table 1). The gap crossing step of the first front leg takes longer than gap crossing steps of following legs. Especially the searching movement needs more time, which reflects the explorative character of the first gap crossing step. Searching movements of the other legs still need approximately three times longer than a normal walking step, only the second hind leg step that in most cases does not contain a searching movement needs less time.

step	FL duration [s]	n	ML duration [s]	n	HL duration [s]	n
WA	0.29 ± 0.09	22	0.27 ± 0.09	25	0.27 ± 0.07	23
SH	0.09 ± 0.60	94	0.09 ± 0.60	38	0.12 ± 0.90	34
GC1	1.33 ± 0.82	30	0.88 ± 0.50	30	0.87 ± 0.38	30
sw	0.25 ± 0.16	30	0.32 ± 0.33	30	0.27 ± 0.13	30
se	1.13 ± 0.85	29	0.69 ± 0.38	26	0.70 ± 0.37	27
GC2	0.88 ± 0.52	30	0.79 ± 0.52	30	0.45 ± 0.14	30
sw	0.22 ± 0.10	30	0.31 ± 0.13	30	0.40 ± 0.12	30
se	0.83 ± 0.49	26	0.63 ± 0.55	25	0.36 ± 0.18	5
TE	0.66 ± 0.25	38	-	-	-	-
sw	0.24 ± 0.08	38	-	-	-	-
se	0.42 ± 0.25	38	-	-	-	-

Table 1 : Mean swing duration of different step types; FL front leg, ML middle leg, HL hind leg, WA normal walking step (control), SH short steps, GC1 1st gap crossing step, GC2 2nd gap crossing step, TE tentative step, sw swing movement, se searching movement, n number of steps

The distribution of short steps of the different leg pairs during the gap crossing sequence is displayed in Table 2. The majority of all short steps is performed by the front legs after reaching the far edge of the gap (50%) and by the middle and hind legs during front leg exploration behaviour (27%). This indicates that short steps support slow forward movement and fine adjustment of body posture during gap crossing behaviour.

part of gap crossing sequence	FL	ML	HL
before front legs reach the far edge	11 7 %	27 16 %	19 11 %
after front legs reach the far edge	83 50 %	11 7 %	15 9 %

Table 2: Temporal distribution of short steps; numbers and percentages of short steps before and after the front legs reach the far edge of the gap, FL front legs, ML middle legs, HL hind legs, n = 166 steps

3. Simulation

A simple neuronal network that allows to generate swing trajectories has already been proposed by Cruse et al. (1995) as SWINGNET (see also Linder 2002). It can be described as a one layer perceptron that consists of three negative feedback controllers for the three joint angles (Duerr et al. 2003). Every neuron receives input from the actual joint angles α, β and γ that are measured by sensors. The reference values αt, βt and γt given by a preceding neural network represent the direction of the AEP at which the tarsus should be placed on the ground. This network has been called TARGETNET according to the targeting mechanism described in stick insects (Cruse 1979, Dean and Wendler 1983). The output values $\Delta\alpha$, $\Delta\beta$ and $\Delta\gamma$ of the SWINGNET are the angular velocities of the three joint angles for the next iteration.

In the SWINGNET, the three negative feedback controllers have been coupled via two weights on the β joint to generate a rise of the trajectory in z-direction (vertical axis in body coordinates), as otherwise the legs would not be lifted from the ground surface. In a later version named SWINGNET 2, this problem has been solved in a different way (Schumm and Cruse in prep.). The controllers of the three joint angles remain uncoupled, and a default target value β+ that makes the tarsus aim at a high position replaces the βt output of the TARGETNET. To decrease the output of the β controller again, a negative input that slowly increases over time is fed into the β neuron. This influence is gained from the swing signal given by the SELECTOR, a step function that is 1 in swing mode and 0 in stance mode, fed through a low pass filter. The time constant of this low pass filter basically determines the shape of the swing trajectory. SWINGNET 2 shows several advantages compared to the previous SWINGNET, it generates swing trajectories that resemble much more the swing trajectories observed in stick insects during straight and curve walking.

Two modifications have been added to the WALKNET controller in this study. The first modification concerns the neural network that generates the swing movement of the leg. SWINGNET 2 has been used as a basis for the SEARCHNET proposed here (Figure 5). The aim of this SEARCHNET is to maintain the abilities of SWINGNET 2 during undisturbed walking and to add the ability to perform searching behaviour similar to the one performed by stick insects in the case of stepping into a gap. Searching oscillations are generated in the β joint by a feedback loop (weight Q2). The oscillations are passed on to the α and γ joint (weights Q1, Q3) to trigger coupled searching movements in all angles. During undisturbed walking, the weights of the feedback loop are set to zero.

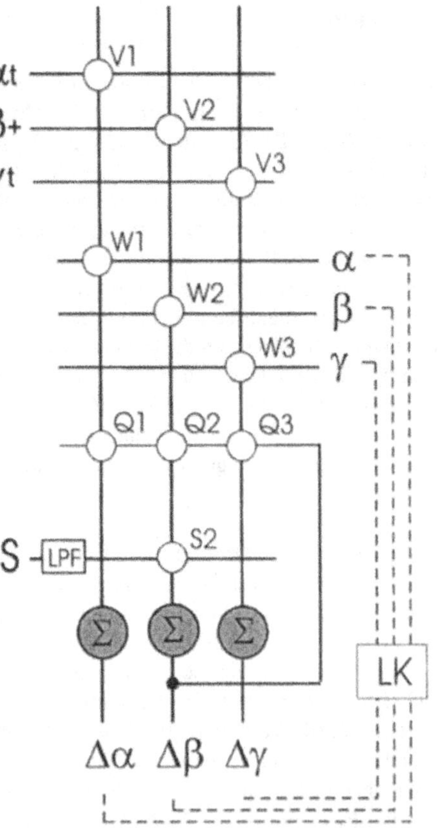

Figure 5: SEARCHNET, a neural network model that generates searching movements, based on SWINGNET 2; αt, γt: input from TARGETNET, β+: high default target value, α0: AEP for reversed swing movement, α, β, γ: sensory input, S: input from SELECTOR, Δα, Δβ, Δγ: output values, LPF low pass filter, LK leg kinematics, V1, V2, V3: positive weights on target input, W1, W2, W3 negative weights on sensory input, Q1, Q2, Q3 weights on feedback loop, S2 negative weight on input from SELECTOR

It has been shown that non-contact of a front leg when stepping into a gap represents a signal the animal reacts to by slowing down (Blaesing and Cruse 2004 b). The same non-contact signal is used by the SEARCHNET for "switching" from the initial swing trajectory to the searching trajectory. In the SEARCHNET, the non-contact signal is set if the z-coordinate of the tarsus relative to the coxa moves below a height threshold at which ground has been expected, slightly below ground level of the footbridge. During normal walking, the tarsus never moves below this threshold unless the leg is lowered into a gap or down a step. If the tarsus moves below the threshold, a variable *search* is set to 1 (Figure 6 A) which causes the following modifications: three weights Q1, Q2 and Q3 of the β feedback loop are set to non-zero values which causes oscillations in the β joint that are passed to the α and γ joints. Additionally, a slowly increasing additive influence on the target input αt shifts the tarsus forward during searching.

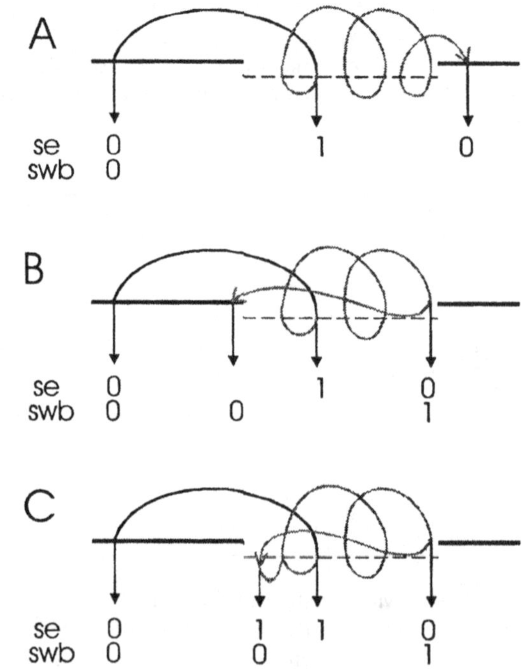

Figure 6: Variables *search* (se) and *swingback* (swb) act as switches during searching behaviour; A. when passing a height threshold, setting se = 1 initiates oscillatory searching movements; B. setting se = 0 and swb = 1 stops searching movements and initiates swinging backward; C. passing the height threshold again starts new forward searching movement (se = 1, swb = 0)

In the explorative phase of gap crossing behaviour, the front leg of the stick insect is repeatedly placed back onto the footbridge and steps into the gap again to continue searching behaviour before reaching the far side of the gap. These tentative steps (Figure 1 B) seem to be useful for adapting body posture and leg position during exploration and have therefore been included in the SEARCHNET. To generate tentative steps, a second switch named *swingback* has been implemented that is set to 1 after a certain number of time steps. Setting *swingback* to 1 causes the following modifications: *search* is set to 0 again and therefore no oscillations occur, the swing trajectory is generated in the same way as during walking (Figure 6 B). To direct the swing trajectory rearward, the input from the target net αt is replaced by the target value α0 that has been stored during the first iteration of the swing movement. After the reversed swing movement, the tarsus hits the ground and the leg switches into stance mode. It can start a new swing movement as soon as the rules of coordination allow (Cruse 1990). If the tarsus fails to reach the first footbridge after the reversed swing movement, it passes the height threshold again which sets *swingback* to 0 and *search* to 1 (Figure 6 C).

A second modification regards the control of walking velocity in the case of loss of ground contact. The velocity controller of the original WALKNET consists of a simple feedback loop: a predefined reference velocity vRef that can be tuned for slower or faster walking is compared to the actual stance velocity vSens. In the new module, vRef is set to zero if the z-coordinate of the tarsus relative to the coxa of any leg in swing mode moves below a threshold slightly below the surface level of the footbridge, the same threshold has already been applied in the SEARCHNET. Additionally, the αMin values of the legs in stance mode are set to zero. αMin is a small constant that determines the minimum angle velocity that has been introduced to the controller of stance movements to initiate walking and to maintain stability of walking in forward direction. Setting the global vRef value and the αMin values of the legs in stance mode to zero causes the animat to adjust its forward velocity to zero. When the tarsus is moved above the threshold again, vRef and αMin are set back to their default values. When stepping into a gap with a front leg, the animat therefore slows down its forward movement to zero in response to the front leg tarsus passing the threshold. New forward movement is initiated when the tarsus is moved up again by the SEARCHNET. During the oscillating searching movement generated by the SEARCHNET, the tarsus is moved above and below the threshold in an alternating way which results in bouts of forward movement alternating with stops (Figure 7, top panel).

During slow forward movement and while both legs of the same pair are in swing mode, the target position for the AEP is shifted rearward. The latter condition does not occur during undisturbed walking, but contralateral gap crossing steps overlap regularly. For the middle and hind legs, the target value depends on the position of the anterior leg, whereas the front leg has a fixed target position. During gap crossing, while vSens and vRef are below a threshold velocity, the x component of the target value is replaced by a smaller default value in all legs, which causes a backward shift of the AEP and therefore a shorter swing amplitude. The resulting short steps allow the animat to move closer towards the gap before stepping into it, so that a more appropriate body position can be adopted before starting a gap crossing step. This behaviour also prevents the front legs from "running away" by performing normal steps while the middle and hind legs cross the gap and thereby maintains a stable body posture.

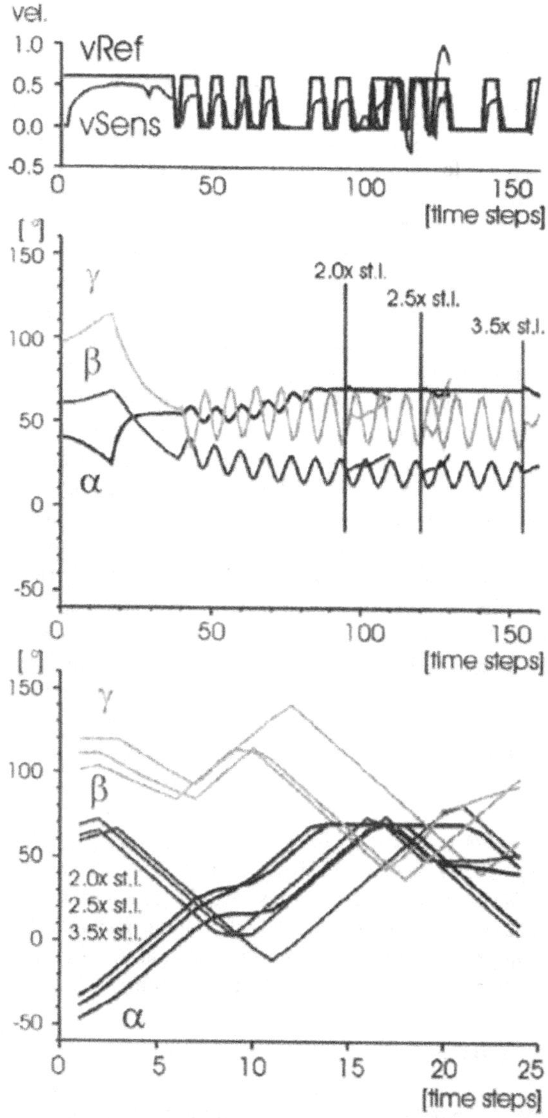

Figure 7: Simulated gap crossing steps of the front legs: time course of joint angles and velocity of forward movement over ground during; gap sizes: 2.0, 2.5 and 3.0 times normal step length (st.l.); top panel: velocity control during the first gap crossing step, black: reference velocity, dark grey: measured velocity; middle panel: data of the first gap crossing step (vertical lines mark end of 2.0, 2.5 and 3.0 step length trial as indicated); bottom panel: data of the second gap crossing step; black: α joint, dark grey: β joint, light grey: γ joint; time steps correspond to frames in the experimental data (50 Hz), 50 time steps correspond to 1 s

SEARCHNETs were implemented for all six legs, differences between the front, middle and hind legs consisted only in the weights as in the SWINGNET and in the additional ability of the front legs to perform tentative steps. After implementation of the SEARCHNETs, the

AEP shift for short steps and the enhanced velocity control, the animat is able to cross gaps of even larger width compared to its normal step amplitude than the animal in the experiment. In Figure 7 (middle and bottom panel), examples of gap crossing steps of the front legs are displayed for three trials with gap widths of 2.0, 2.5 and 3.0 times normal step amplitude. Searching trajectories generated by the SEARCHNET resemble the searching trajectories of the stick insect (Figure 4) in several of their main characteristics such as the main working space and the phase relationships of the three joint angles. Differences occur in the amplitude of oscillations especially in the γ joint in the first gap crossing step (Figure 7, middle panel). The amplitude is smaller in the simulation, however, increasing the amplitude of γ joint oscillations has been found to make the movement unstable. In the second gap crossing step (Figure 7, bottom panel), the amplitude of movements is larger which is caused by the fact that the PEP is more rearward than in the experimental data. The duration of simulated gap crossing steps is relatively shorter than the duration of stick insect gap crossing steps. This might be caused by the more stereotyped movement and by the forward shift of the α joint via αt that is terminated at the margin of the angle. In the three displayed trials, the far edge of the gap is reached after 95 time steps (gap width 2.0 times step amplitude), 120 time steps (gap width 2.5 times step amplitude) and 155 time steps (gap width 3.0 times step amplitude) which corresponds to 1.9 s, 2.4 s and 3.1 s. For comparison, a normal walking step takes 15 time steps (0.3 s, compare experimental data in Table 1). In trials with these three gap sizes, searching movements were successfully tested from different starting positions. Gaps of larger sizes could also be crossed, but success of the searching movement depended on the starting position.

4. Discussion

The locomotion of stick insects climbing over large gaps has proved to be a representative paradigm for adaptive walking behaviour in a complex environment (Blaesing and Cruse 2004 a, b). Adaptation of locomotor behaviour is not only important for insects but also for animats that travel in a complex environment. Especially walking robots that operate in uneven and probably unknown terrain have to be able to overcome large obstacles and gaps. The future perspective of this study is to improve adaptive locomotion of a TarryII robot that is used as a hardware platform for the WALKNET simulation (Frik and Guddat 2000). The same robot type is used for hardware simulation in a study of insect locomotion by Pick and Strauss (2002) who focus on the role of tactile control and effective foot structures.

A previous study has shown that gap crossing behaviour can be regarded as walking behaviour with small modifications with exception of the exploration part of the sequence. Most adaptations to different gap sizes and variations of behaviour occur during this phase that starts when the stick insect has reached the gap and ends when it has placed its front legs on the far side of the gap (Blaesing and Cruse 2004 b). During exploration behaviour, the animal slows down its walking velocity, moves slowly forward in short bouts and performs extensive searching behaviour with its front legs and supporting short steps with its middle and hind legs. Implementation of these behaviours into the WALKNET controller has been the main objective of this study.

It has been observed that non-contact of a front leg provides a signal that causes a change in the subsequent behaviour whereas antennal non-contact does not provide such a signal. This means that an expectation of ground contact has been thwarted in the case of the front leg but not in the case of the antenna. Slowing down occurs in response to stepping into the gap with a front leg. Apparently, thwarting of an expectation of ground contact at a certain height leads to this reaction. As the non-contact signal evokes behavioural changes in the insect, it has been used in the simulation as a signal for the velocity controller and for the beginning of the searching movement. A reaction to the initial front leg step into the gap has also been found regarding the fAEP of the swing movement into the gap: in the initial step into the gap in a trial, the fAEP is not different from the AEP of a normal walking step whereas it is shifted rearward in all following steps of the same trial. This result can be interpreted as the insect having gained "knowledge" about the gap (Blaesing and Cruse 2004 a).

The observation that antennae do not play a major role in detecting the gap allows to abandon the implementation of antennae as tactile probes at this state of the simulation. Stick insects with shortened antennae or other stick insect species with naturally short antennae use only their front legs for tactile exploration. However, antennae play a major role as tactile probes (Duerr et al. 2001). Touching the far edge of the gap with an antenna leads to behavioural changes during explorative behaviour, especially when crossing a larger gap than used in this study (Blaesing and Cruse 2004 a). When crossing a 30 mm gap, the first antennal contact with the far edge occurs before the gap crossing steps of the front legs, therefore no reaction is observed during the gap crossing steps analysed here. To exploit more of the results found for gap crossing behaviour and to enable the agent to look forward into its future, antennae longer than front legs would be helpful. Long antennae could help to literally plan ahead by focusing on a far goal while the legs step in different directions to adjust body posture.

In this study, searching movements of the front leg have been simulated as oscillations that roughly follow the same phase relationships as in the observed searching movements of the stick insect. For the other leg pairs, the same type of SEARCHNET has been used as for the front legs which has enabled the animat to perform successful searching and gap crossing behaviour. A searching reflex has also been proposed by Espenschied et al. (1996) as an important feature for a robot walking on rough terrain. The searching reflex has been implemented in the behaviour of Robot 2 (Beer et al. 1997). In this robot, the loss of foothold triggers a sequence of oscillatory movements of growing amplitude, biased in forward direction. The searching reflex enables the robot to traverse slatted ground with less than 50% ground surface provided.

A simulation of searching trajectories has been proposed by Duerr (2001). In this study, a neural network was trained on single trajectories performed by the middle leg of the stick insect *Carausius morosus* after stepping into a small ditch. The neural network used in this study generates cyclic searching movements as extension of the swing movement without any noticeable transition. Corresponding models were described for front leg and hind leg trajectories observed in *C. morosus*. (Duerr et al. 2002). For traversing a gap of the size used in the current study, however, different changes of behaviour have to be applied not only in the single leg but in the whole body. Velocity has to be controlled to provide a stable body posture during exploration which also requires a reduction in step length in the supporting legs. A signal that marks the transition from walking to exploration of the gap is useful in this case, especially when the complex behaviour observed in *A. asperrimus* in experiments with large gap sizes is regarded (Blaesing and Cruse 2004 a, b).

In the current study, searching movements of the legs are not the only feature that enables the animat to cross a large gap. In the case of large gaps relative to body size, supporting behaviour of the leg that are not actually crossing the gap becomes crucial. Like in stick insect locomotion, the simulated gap crossing behaviour is improved by short steps that allow the animat to move closer to the gap. The stick insect positions its body appropriately during exploration by performing short steps with the middle and hind legs, and more short steps are performed throughout the sequence by all leg pairs to adapt leg positions and body posture. During later parts of the sequence, short steps mainly occur in the front legs that still play an explorative role and support slow forward movement. In the simulation, short steps play an important role as they enable the animat to maintain a stable body posture while positioning its legs close to the gap. "Local stepping" has already been described as a means of maintaining stability and adjusting body posture in locusts walking on slippery surfaces or prior to climbing onto an obstacle (Pearson and Franklin 1984). Solutions for secure foot placement and adaptation of gait in robot locomotion have been proposed by Estrema and Gonzales (2002). By applying knowledge about its body (e.g. stability and kinematic margins) and the environment, their robot avoids stepping on forbidden zones that can also be interpreted as gaps in the ground or dangerous spots in a mine field. In this model, knowledge about the world has to be provided e. g. by a visual system, and based on this information a foothold planner determines which leg will be moved and where it will be placed. In contrast to the model proposed here, adaptive properties of decentralized control and exploration behaviour such as searching movements of single legs are not applied.

Tentative steps of the front legs can also be important for adapting body position in front of the gap. According to the rules of leg coordination (Cruse 1990), placing a front leg on the ground again might allow a different leg to go into swing mode and perform a short step. The application of tentative steps in the front leg SEARCHNETs improves the performance of the simulation with large gap sizes. Initiating a backward swing after 33 time steps which corresponds to the average duration of a tentative step in the biological model (Table 1), however, rather disturbs the searching behaviour of the animat. It has been observed that with increasing gap size, the number of tentative steps per trial and the mean duration of a tentative step also increases (Blaesing and Cruse 2004 a). In trials with a 50 mm gap, the mean duration of a tentative steps is 3.3 ± 6.5 s. Currently, quantitative simulation experiments are carried out testing the animat with different gap sizes and starting positions.

5. Conclusion

With the implementation of the main results from experimental studies, the WALKNET controller has been enabled to climb across gaps of more that twice its own step length. The simulated searching movements of the legs and controlled velocity of forward movement in front of the gap result in a behaviour that resembles the exploration behaviour of the stick insect. There are, however, more results from experimental studies using the gap crossing paradigm that could be included in the simulation. Stick insects utilize their long antenna for exploration. Rearward shifting of the posterior extreme positions and slight modifications of the rules of leg coordination increase the distance that can be reached with the front legs. Implementation of these additional modifications into the WALKNET controller will not only improve its performance in the gap crossing task. It will also improve its adaptability to other challenging walking conditions and contribute to understanding the neural basis of stick insect walking behaviour.

Acknowledgements

I am grateful to Prof. Dr. Holk Cruse for general support and for supervising my project, to Thorsten Roggendorf for WALKNET support and to Sebastian Stuebner for data collection. This work was supported by Grant no Cr 58/9-3 of the DFG and the graduate programme "Verhaltensstrategien und Verhaltensoptimierung" (DFG).

References

Blaesing B, Cruse H (2004 a) Mechanisms of stick insect locomotion in a gap crossing paradigm. *J Comp Physiol A* (in press)

Blaesing B, Cruse H (2004 b) Stick insect locomotion in a complex environment: climbing over large gaps. *J Exp Biol* (in press)

Beer RD, Quinn RD, Chiel HJ, Ritzmann RE (1997) Biologically inspired approaches to robotics. What can we learn from insects? *Communications of the ACM 40* (3), 31 – 38

Cruse H (1976 a) The function of the legs in the free walking stick insect, *Carausius morosus*. *J Comp Physiol A* 112, 235 – 262

Cruse H (1976 b) The control of the body position in the stick insect (*Carausius morosus*) when walking over uneven surfaces. *Biol Cybern* 24, 25 – 33

Cruse H (1979) The control of the anterior extreme position of the hind leg of a walking insect *Carausius morosus*. *Physiol Entom* 4, 121 - 124

Cruse H (1990) What mechanisms coordinate leg movement in walking arthropods? *Trends in Neurosciences* 13, 15 – 21

Cruse H, Brunn D, Bartling C., Dean J, Dreifert M, Kindermann T, Schmitz J (1995) Walking - a complex behavior controlled by simple networks. *Adaptive Behavior* 3, 385 – 418

Cruse H, Epstein S (1982) Peripheral influences on the movement of the legs in a walking insect *Carausius morosus*. *J Exp Biol* 101: 161 – 170

Cruse H, Kindermann T, Schumm M, Dean J, Schmitz J (1998) Walknet - a biologically inspired network to control six-legged walking. *Neural Networks*. 11, 1435 – 1447, R. Brooks, S. Grossberg (eds.)

Dean J, Wendler G (1983) Stick insect locomotion on a wheel: interleg coordination of leg position. *J Exp Biol* 103, 203 – 216

Duerr V (2001) Stereotypic leg searching-movements in the stick insect: Kinematic analysis, behavioural context and simulation. *J Exp Biol* 204 (9), 1589 – 1604

Duerr V, Krause A, Schmitz J, Cruse H (2003) Neuroethological concepts and their transfer to walking machines. *Int. J. Robotics Res.* 22 (3), 151 – 167 (invited paper)

Dürr V, König Y, Kittmann, R (2001) The antennal motor system of the stick insect *Carausius morosus*: Anatomy and antennal movements during walking. *J Comp Physiol A* 187 (2), 131-144

Espenschied KS, Quinn RD, Beer RD, Chiel HJ (1996) Biologically based distributed control and local reflexes improve rough terrain locomotion in a hexapod robot. Robotics Auton. Sysems 18, 59 – 64

Estremera J, Gonzalez de Santos P (2002) Free gaits for quadruped robots over irregular terrain. *The International Journal of Robotics Research* 21 (2), 115 – 130

Guddat M, Frik M (2000) Control of Walking Machines with Artificial Reflexes. In: *Advanced Motion of Animals and Machine. Second International Symposium on Impact and Friction of Solids, Structures and Intelligent Machines Montreal, Canada.*

Graham D, Epstein S (1985) Behaviour and motor output for an insect walking on a slippery surface. *J Exp Biol* 118, 287 – 296

Kindermann T (2002) Behavior and adaptability of a six-legged walking system with highly distributed control. *Adaptive Behavior* 9, 16 – 41

Linder C (2002) Self organisation ina simple task of motor control. In: *From Animals to Animats 7. Proceedings of the seventh International Conference on Simulation of Adaptive Behaviour.* B. Hallam, D. Floreano, J. Hallam, G. Hayes, J.-A. Meyer (eds.) MIT Press 185 – 195

Pearson KG, Franklin R (1984) Characteristics of leg movement and patterns of coordination in locusts walking on rough terrain. *The international Journal of robotic research* 3 (2), 101 – 112

Pick S, Strauss R (2002) Adaptive leg placement strategies in the fruit fly set an example for six-legged walking systems. In: *From Animals to Animats 7. Proceedings of the seventh International Conference on Simulation of Adaptive Behaviour.* B. Hallam, D. Floreano, J. Hallam, G. Hayes, J.-A. Meyer (eds.) MIT Press 70 – 71

Schumm M, Cruse H (in prep.) Control of swing movement: influences of differently shaped substrate

A simulation model of the locomotion controllers for the nematode *Caenorhabditis elegans*

John Bryden and Netta Cohen*
School of Computing
Leeds University
Leeds LS2 9JT
United Kingdom

Abstract

This paper presents a simple yet biologically-grounded model of the *C. elegans* neural circuit for forward locomotive control. The model considers a limited subset of the *C. elegans* nervous system, within a minimal two-dimensional environment. Despite its reductionist approach, this model is sufficiently rich to generate patterns of undulations that are reminiscent of the biological worm's behaviour and qualitatively similar to patterns which have been shown to generate locomotion in a model of a richer physical environment. Interestingly, and contrary to conventional wisdom about neural circuits for motor control, our results are consistent with the conjecture that the worm may be relying on feedback from the shape of its body to generate undulations that propel it forward or backward.

1. Introduction

There is an increasing body of work on the evolution of artificial neural circuits for motor control. These artificial circuits commonly share, or are inspired by, morphological and physiological features of biological model systems and are designed to mimic biological behaviour and functionality. Much of the work has been inspired by relatively complex motor control systems of vertebrates such as swimming in the lamprey (Ijspeert et al., 1999) and walking/swimming in the salamander (Ijspeert, 2001). Given partial information about the circuitry and by invoking a meaningful fitness function, artificial neural circuitry is evolved and its performance tested within the framework of a simulation model. The present work follows similar principles but is inspired by a much simpler, invertebrate nervous system, that of the nematode *C. elegans*.

The *C. elegans* nervous system's simplicity lends itself well to this approach. With one of the simplest known nervous systems, it has been called 'the hydrogen atom of systems neuroscience' (Ferrée, 2003). Its 302 neurons (White et al., 1986) represent a mere drop in the ocean as compared with the human's one hundred billion neurons. Considering its modest size, the nervous system is capable of rather sophisticated functions including traversing chemical and temperature gradients and even memory (Rose and Rankin, 2001).

One feature that makes the *C. elegans* nervous system an excellent candidate for study is the fact that, as an invertebrate, it is hard wired: Every individual has the same neural circuitry. In 1986, an extremely comprehensive map of the *C. elegans* nervous system was produced by White et al. (White et al., 1986). Every neuron and many of the connections have been identified. This, supplemented by the mapping of the *C. elegans* genome (The *C. elegans* Sequencing Consortium, 1998), provide a wealth of information, deeming *C. elegans* a modeller's paradise . While modelling an entire iguana (in the spirit of Daniel Dennett's 1978 call) may still be a daunting challenge for simulation modellers, a complete model of the *C. elegans* nervous system is slowly taking shape.

The worm's behaviour can be broken down into various components – forward and backward motion, turning, chemotaxis, thermotaxis – each of which can be studied independently due to the corresponding modularity of the nervous system. Previous work modelling the *C. elegans* nervous system has been done on chemotaxis (Ferrée and Lockery, 1999), the tap withdrawal reflex (Wicks et al., 1996) and locomotion control (Niebur and Erdös, 1993).

The modularity of the *C. elegans* nervous system is both structural and functional. One subsystem that stands out is the locomotion system, containing roughly 63 neurons and only sparsely connected to the rest of the nervous system. Within the locomotion nervous system, sub-modules can easily be identified. In particular, it appears that, to a good first approximation, the forward locomotion subsystem can be isolated from backward locomotion and turning, thus providing us with a relatively compact and well defined task. However, before delving

*Corresponding author: netta@comp.leeds.ac.uk

into the model itself, some additional background is provided about the *C. elegans* nervous system which led us to important assumptions in our model.

C. elegans is a small worm, generally growing to only 1mm in length (Wood, 1988). This simple organism, 959 cells in total (Wood, 1988), nonetheless has many of the features found in much larger and more complex organisms.

To move, *C. elegans* must lie on its side with contractions alternating dorso-ventrally (from top to bottom). Undulations propagate along the worm in the opposite direction to movement, propelling it forward or backward. Longitudinal muscles pull on the skin, an impermeable and elastic sealed high-pressure tube, to generate these undulations. Nerve processes that run along the body innervate the muscles. The worm also uses touch sensors in the head and tail to enable it to turn and reverse direction when it reaches an obstacle (Chalfie and White, 1988).

Like the locomotion of nematodes, the overwhelming majority of known motor activity in animals relies on the rhythmic contraction of muscles, which are controlled or regulated by neural networks. This rhythmic activity often consists of pairs of opposing muscles alternately contracting out of phase with each other. The same principle holds in *C. elegans*. In general, two different classes of mechanisms have been proposed to underlie such motor behaviour. One possibility is reflexes: the contraction of one muscle feeds back though the neural network to the opposing muscle, and vice versa. The second is that the neural network itself generates an endogenous rhythmic pattern to activate the muscles [see (Marder and Calabrese, 1996) for examples]. While historically surprising, such so-called *central pattern generating* (CPG) neural networks have been found throughout the animal kingdom and are generally believed to participate in almost all forms of motor behaviour.

Surprisingly, there is no known evidence of CPG circuits in *C. elegans*. For instance, many CPG circuits are dominated by inhibitory synapses. In contrast, an analysis of White's map of the *C. elegans* nervous system (White et al., 1986) established that the locomotion subsystem contains relatively sparse chemical synapses and, instead, abundant electrical (so-called *gap junction*) connections between neurons. These strong and fast bi-directional electrical connections typically serve to synchronise neurons. In a network with little inhibition and with abundant gap junctions, feedback-related mechanisms are more likely to underlie oscillations than centrally generated patterns.

An additional exciting difference between the *C. elegans* nervous system and most other nervous systems is the fact that its underlying neuronal mechanism is believed to use graded (otherwise known as electrotonic) potentials (Goodman et al., 1998). In contrast, most known neurons in the animal kingdom generate fast, high-amplitude electrical spikes called action potentials. These pulses have similar pulse-forms and are therefore referred to as all-or-none signals, to imply that they can be regarded as binary signals. In contrast, neurons with graded-potential states form analogue motor control circuits.

In this work we have generated a biologically-inspired toy model of the neural control of *C. elegans* forward locomotion that exhibits qualitatively similar behaviour to that shown by the biological system. Rather than using the standard model of a chain of central pattern generators, our model relies on feedback from the shape of the worm's body to the neurons to generate undulations that propagate along the body. The model considers a subset of the *C. elegans* nervous system, within a reduced environment. The model neurons have analogue states and the circuitry contains no synapses. These premises constitute a novel and compeletely unprecedented approach to modelling of motor control circuits.

This work makes some progress toward bridging the gap between our knowledge of the *C. elegans* neural circuitry and our models of its motor behaviour. Neural circuitry provides little insight on neural activity and even less on the resulting motor activity. Conversely, standalone models of motor behaviour that do not derive from the neural circuitry may or may not be consistent with it. Here, we build on our knowledge of the neural circuitry and neuronal properties, on the one hand, and on simulation models of the resulting motor patterns, on the other, to obtain results that are consistent with both. The model obtained and computer simulations thereof have led to specific testable predictions about the neural circuitry and its role in the control of forward locomotion in the worm.

The model of the neural circuit presented in this work relies to a large extent on a mechanical model for the locomotion of the worm by Ernst Niebur and Paul Erdös (Niebur and Erdös, 1991). However, Niebur and Erdös focused on mechanical aspects of an embodied worm, including detailed interactions with the physical environment. In contrast, the present work focuses on the neural circuitry and its ability to support robust motor control, while the physical model is kept to a bare minimum. This approach leads to a simple, almost toy model of the forward locomotion control. The model will allow us to investigate the minimal requirements for producing the desired control patterns.

A process of stages was used to generate our final circuit, each stage generating a model which was optimised according to a fitness function. In analogy with Ijspeert's lamprey work (Ijspeert et al., 1999), the worm was broken up into segments (despite the fact that worm has no vertebrae). Thus, in the first stage, an oscillatory neural circuit was obtained for modelling the local bending of a single segment. The second stage generated two coupled segments that would oscillate at a phase lag with feedback from the local bending. Finally eleven segments were linked together to complete the model of forward locomotion. The resulting model confirmed the feasibility of locomotion control with mechano-feedback.

The following section provides biological background on the nematode's nervous system and presents a simplified model of the locomotion based largely on the recent mechanical model of locomotion in *C. elegans* (Niebur and Erdös, 1991). A methodology section provides the equations used to model the neurons, their inputs from the shape of the body and their outputs to the muscles. A detailed description of the simplified circuitry for a single segment, a pair of segments and a complete (11-segment) model is provided. The results are ordered according to these stages and comprise a single solution of the problem and its analysis. The paper closes with a discussion of the work and its results.

2. Biological background

This section considers more detail on the models of *C. elegans* neurons, locomotion circuitry and a relevant existing model of locomotion in the nematode.

2.1 C. elegans neurons

Like many invertebrate neurons, *C. elegans* neurons do not have complicated structures. Most have only one axon and no dendrites. Furthermore, as mentioned above, it is unlikely that *C. elegans* neurons fire action potentials. In particular, research done on an individual, identified, *C. elegans* neuron (Goodman et al., 1998) found that it did not fire Na^+ action potentials and that it was unlikely that it fired Ca^{2+} action potentials. Similar experiments (Goodman et al., 1998) done on 40 other, unidentified neurons also showed similar results.

2.2 Forward locomotion

A mechanical model of the body of *C. elegans* was produced by Ernst Niebur and Paul Erdös (Niebur and Erdös, 1991). Their model worm was divided into segments, with each segment incorporating details of interior fluid pressure, elastic forces generated by the cuticle, muscular forces and external forces arising from movement within a groove of agar. Using computer simulations of the muscle excitation patterns, Niebur and Erdös demonstrated how locomotion could be generated in their model environment.

In a related publication (Niebur and Erdös, 1993), a number of different candidates for muscle excitation patterns were presented. One pattern which was sub-optimal but still capable of propelling the worm, stood out due to its simplicity and similarity with a model for the locomotion of snakes identified by Gray (Gray, 1950, Gray, 1953). This pattern (which they dubbed the *stretch receptor pattern*) generated contractions only in muscles located on the outer curve of the 'S' shape of the *C. elegans* body, toward the direction of the locomotion (see Figure 1). Niebur and Erdös demonstrated that the muscle excitation patterns in their model could in principle be generated even with very primitive models of motoneurons (Niebur and Erdös, 1991). Such minimal motoneurons would act as thresholding units and would require feedback from stretch receptors at specific locations along the worm. In particular, muscles in their model contract in response to bending of adjacent posterior segments. For undulations to occur, Niebur and Erdös' model requires one end of the body (the head in their case) to act as a pacemaker, or self-sustained oscillator. The mechanism for such oscillations is intentionally left unspecified.

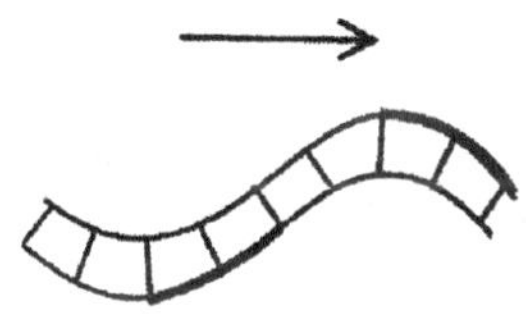

Figure 1: Diagram of the worm broken up into segments. The segments marked in heavy black indicate those that should be contracted [according to (Niebur and Erdös, 1991)] to move the worm in the indicated direction.

Niebur and Erdös based their model on the anatomy of the worm and information from White et al.'s map of the *C. elegans* nervous system (White et al., 1986). The map groups the 302 neurons into 118 classes. The locomotion neurons include ventral motoneurons, dorsal motoneurons and interneurons. The main neurons of interest for forward locomotion are 11 ventral motoneurons (class VB), 7 dorsal motoneurons (class DB) and two interneurons (left and right AVB). Laser-ablation experiments [in which behaviour is recorded before and after identified cells are ablated, (Chalfie et al., 1985)] have demonstrated that these three classes of neurons (VB, DB and AVB) are necessary for forward motion.

The asymmetry in the *C. elegans* circuitry (i.e., the different numbers of ventral and dorsal neurons) is striking. These 7 dorsal and 11 ventral neurons line up along the body and innervate four groups of longitudinally spaced muscles around them (approximately 2×24 dorsal and 2×24 ventral muscles connected to the 7 dorsal and 11 ventral neurons respectively). While the specific neuromuscular connectivity is, to our knowledge, unknown, it appears that motoneurons are generally connected to a number of nearby muscles.

To innervate the muscles in their lateral proximity, the axons of motoneurons of class VB and DB each extend along the corresponding part of the body. However, from this perspective, these motoneurons have axons that are longer than needed. They extend about one fifth of one body length further than the muscles they are thought to control; nonetheless, it appears that these axon extensions lack synapses, gap junctions or any other obvious function. It was therefore proposed (White et al., 1986) that the ends of these motor neurons contain *stretch receptors* that are sensitive to the posture of the body and are responsible for the feedback needed to generate the observed muscle excitation patterns. Interestingly, these axon extensions protrude in the direction of the tail, indicating that such stretch receptors could facilitate the propagation of a wave of neuronal excitation from the tail forward. Niebur and Erdös incorporate the assumption of stretch receptors into their locomotion model and showed that forward locomotion can be obtained under such conditions (Niebur and Erdös, 1991). (Surprisingly, however, their model assumes a mathematically equivalent formalism wherein oscillations initiate in the head of the worm and propagate backward.) We note that neither White et al. nor Niebur and Erdös give any indication for stretch receptors except at the far ends of these motoneurons.

The suggestion that mechanical feedback to the nervous system largely determines the neuronal activity appears consistent with other information about the neural circuitry. An analysis of the forward locomotion subcircuit shows that there are only electrical junctions between the main two classes of motoneurons (VB and DB). Furthermore, all VB and DB neurons are electrically coupled to the AVB neurons (which extend along the entire length of the body). This (and simple intuition about interneurons regulating the activity of motoneurons) suggests that perhaps the AVB interneurons function as a switch for forward locomotion. Since the two AVB neurons are electrically coupled and can be assumed to be acting in unison (White et al., 1986), they are approximated as a single unit (denoted here AVB).

3. Methods

A staged approach was chosen due to its ability to break up a complex problem, such as our need to train parameters for a complete circuit, into smaller components. The first stage was to see if it was possible to get one segment to oscillate and then generate a second adjacent segment that oscillates with a lagged phase from the first segment. Finally 11 segments would make a candidate model for the *C. elegans* forward locomotion control system.

3.1 Neuron Model

The graded response of a motoneuron of class VB or DB can be reduced to a simplified equation for the trans-membrane potential $V(t)$:

$$C\frac{dV}{dt} = -G(V-E) - I^{\mathrm{AVB}} - I^{\mathrm{coupling}} - I^{\mathrm{shape}}, \quad (1)$$

where C is the cell's membrane capacitance; E is the cell's resting potential; G is the total membrane conductance; $I^{\mathrm{AVB}} = G^{\mathrm{AVB}}(V^{\mathrm{AVB}} - V)$ is an input current due to gap junctional coupling with AVB (coupling strength $\mathrm{G}^{\mathrm{AVB}}$ and AVB voltage V^{AVB}); I^{coupling} is a similar input current due to coupling with other motoneurons, and $I^{\mathrm{shape}} = \sum_{j=1}^{n} \sigma_j^{\mathrm{stretch}}(\theta_j)$ is the stretch receptor feedback from the shape of the body, where θ_j is the bending angle of segment j and $\sigma_j^{\mathrm{stretch}}$ is a sigmoid response function of the stretch receptors to the local bending. Without loss of generality, we have set the capacitance C to 1 and the resting potential E to 0. Equation 1 assumes that the cell can be represented by a single compartment and that its dynamics can be reduced to a passive response (with a fixed conductance G). In addition, as an *ad hoc* simplification, the term I^{coupling} was omitted (as it was found to be of secondary significance in our model). The sigmoid function is given by

$$\sigma^{\mathrm{stretch}}(\theta) = \frac{A}{1+\exp\left(-k(\theta-\theta_0)\right)} \quad (2)$$

where the amplitude A, a steepness parameter k and the threshold θ_0 are constants.

3.2 Neural environment model

We used the simplest possible neural environment to reduce complexity in our simulations, especially where the physiological or anatomical details are not that well understood. The only direct inputs received by the locomotion neural network are through the AVB unit (which is connected to the rest of the nervous system). The sensory information from the head/tail was also ignored as its effect is only maintained through the state of the two neurons of class AVB.

The worm was split up into so-called *segments*: frictionless two-dimensional 'joints' between weightless rods. The bending angle of a segment θ was determined by a summation over contributions of motoneurons innervating the muscles of that segment. Neuromuscular junctions were modelled by sigmoid functions over the neuron's voltage state. The model of segment bending or stretching can be summarised as

$$\frac{d\theta}{dt} = \sum_{i=1}^{N_{\mathrm{VB}}} \sigma_i^{\mathrm{out}}(V) - \sum_{i=1}^{N_{\mathrm{DB}}} \sigma_i^{\mathrm{out}}(V). \quad (3)$$

where $\sigma^{\mathrm{out}}(V) = B/[1+\exp\left(-q(V-V_0)\right)]$ with constants B, q and V_0. This assumes that the muscles are effectively translating the neural output directly into segment stretch. Note that dorsal and ventral muscles contribute to bending in opposite directions.

3.3 Forward locomotion circuit

As a first simplification, the asymmetrical circuitry of the worm was reduced to a symmetrical circuit (with $N_{\mathrm{VB}} = N_{\mathrm{DB}}$). The circuit chosen for this paper used 11 segments (based on the number of VB neurons). Each segment was assumed to have one ventral neuron (N_{VB}=1), one dorsal neuron (N_{DB}=1), and one corresponding pair of muscles (VM and DM). The neurons were modelled as described in the above section. Figure 2 summarises this simplified symmetrical circuit.

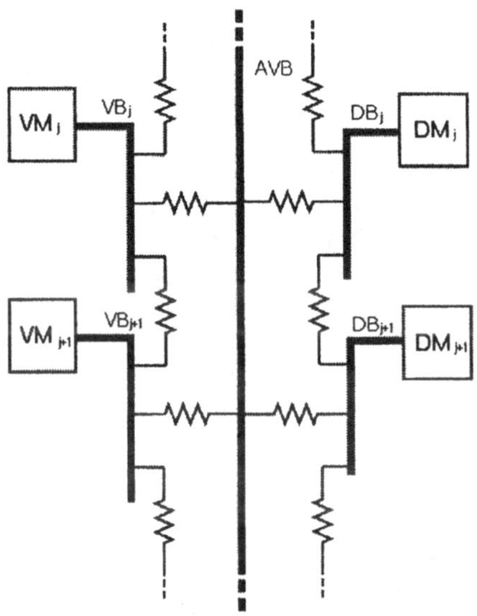

Figure 2: Schematic diagram of the reduced and symmetrized *C. elegans* forward locomotion neural circuitry. Electrical junctions (fast, bi-directional conduits through adjacent cell membranes) are shown as resistors. The electrical junctions between ventral and dorsal neurons (VB and DB, dashed lines) are shown for completeness, but are not actually included in the model. The Neuron somas are located at the anterior end of the neuron (not shown).

3.4 Self-oscillating segment

Our staged approach for building a model of the *C. elegans* locomotion controllers started with a single self-oscillating segment, representing the tail end of the worm. In this model, we have chosen to rely on the existence of stretch receptors, rather than requiring new assumptions about sophisticated neuronal properties. According to our model, the tail segment serves as a conditional pacemaker: It should oscillate whenever the AVB unit is on, and relies on feedback from its own bending angle to generate these oscillations. As mentioned before, the requirement that the oscillator be placed in the tail relies on the specific neuronal morphologies laid out by White et al. (White et al., 1986).

Figure 3 illustrates the single segment model. The

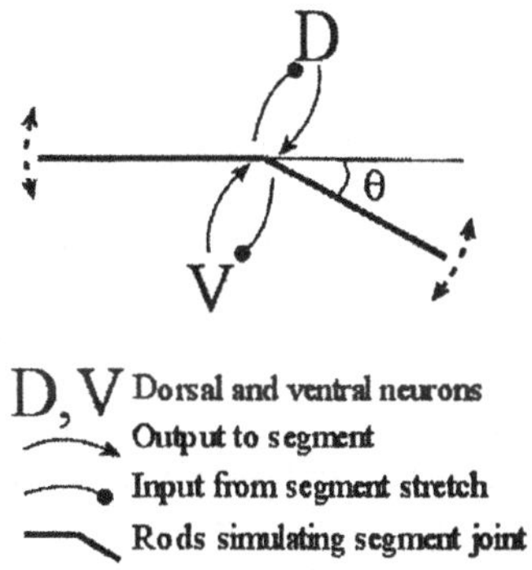

Figure 3: Diagram for the model of a self-oscillating segment.

imposed symmetry of our model distances any circuitry from biological realism, but provides a proof of concept that the neuronal and coupling mechanisms introduced here are sufficient to generate a locomotion-like pattern of activity. We note that despite the symmetry in this circuit, asymmetry is introduced through the neuronal parameters for ventral and dorsal neurons.

Desirable oscillations were identified by simulating the activity of the segment in time and testing for stable periods and stable amplitudes with minimal fluctuations or trends. For instance, the dynamics of an asymmetrical model such as this can include an unwanted downward trend in the segment angle θ.

Numerical analysis of the segment oscillations yielded a numerical value representing the fitness of a solution. The simulation was run three times, (i) with no input from AVB (I^{AVB}=0), (ii) with AVB input and (iii) with AVB input and white uniform noise in the segment stretch (θ). (The noise was added stochastically with a probability of 0.05 per time step and an amplitude range of 20% of the desired bending amplitude.

Two fitness functions were introduced: a main fitness function for assessing runs (ii) and (iii) and a control fitness function for run (i). The main fitness function F^{segment} was expressed as a product of various fitness components, as follows:

$$F^{segment} = f_\omega \cdot f_{\mathrm{std}(\omega)} \cdot f_\theta \cdot f_{\mathrm{stat}} \cdot f_{\mathrm{damping}} \cdot f_\epsilon \qquad (4)$$

where f_ω rewards any oscillation frequencies below some cut off; $f_{\mathrm{std}(\omega)}$ rewards low relative fluctuations in the undulation frequency; f_θ rewards bending angles of a prespecified amplitude; f_{stat} rewards stationary solutions; and finally f_ϵ and f_{damping} reward stable and undamped solutions, respectively. The control fitness function for run (i) rewarded low activity. The overall fitness of a solution was obtained by multiplying the fitness values of the three runs.

3.5 Phase-lagged segments

In order to generate a two-segment circuit, with phase-lagged oscillations, the solution obtained for a single self-oscillating segment (above) was used for both segments. One segment continued to serve as a pacemaker (tail segment, right) and the second segment served as a body segment (left). This body segment received mechano-feedback both from the tail and from its own body segment. The inclusion of local feedback is an important extension of the existing hypothesis, according to which feedback from posterior segments is sufficient. The reason for including local feedback will be explained in the Discussion below. Given the two-segment circuitry, it remained to search for fit mechano-feedback parameters connecting the tail and body segments $\sigma^{stretch}$. In addition, the conductance parameters for the body segment neurons G were re-trained.

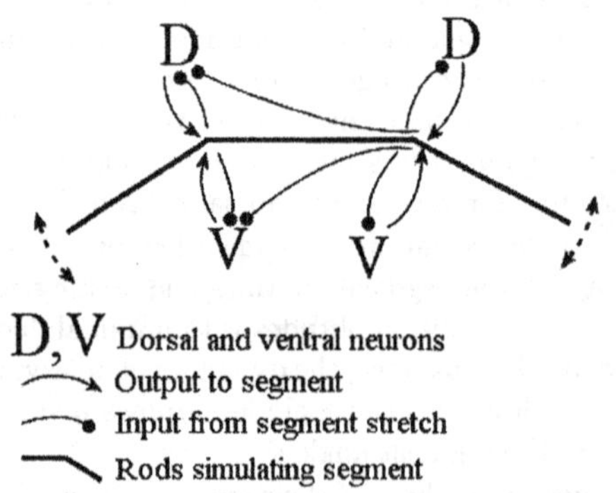

Figure 4: Diagram of the two-segment circuit. The phase-lagged (body) segment on the left should oscillate at a lagged phase relative to the self-oscillating (tail) segment on the right.

The fitness function used to identify desirable phase lags is given by:

$$F^{\text{lag}} = F^{\text{segment}} \cdot f_{\phi} \cdot f_{\Delta\omega} \tag{5}$$

where f_{ϕ} rewards oscillations at a predetermined phase-lag (see below) and $f_{\Delta\omega}$ rewards minimal discrepancies between the tail and body oscillation frequencies. To determine the optimal phase lag, we used the following argument. Assuming the length of the body includes 11 segments and spans a single wavelength of undulations λ, the phase lag is taken to be of the order of 0.1λ. However, since the neuronal excitation propagates forward from the tail and the physical undulations must propagate backward from the head, the phase lag is reversed to 0.9λ.

3.6 Eleven segment model

The body circuit consisted of a chain of ten phase-lagged (body) segments in sequence in front of a self-oscillating (tail) segment. In order to seek solutions for the entire body circuit, no fitness function was needed. In fact, it was sufficient to run simulations to confirm that the tail oscillations propagate along the body in the desired manner. A graphical tool was used to visualise the behaviour. The visualisation generating parallel copies of the model worm to endow it with a semblance of width.

3.7 Directed search algorithm

To generate fit solutions a directed search algorithm was developed and implemented on a parallel computer.[1] Parameters needed for each stage were generated using a modified random restart hill climber. ***Running solutions*** were assessed by applying fitness functions to simulation runs under a variety of conditions.

Once a working solution was obtained, mutations were stochastically applied to the solution. Mutated solutions that out perform the running solution would replace it. After 1000 mutations without a replacement, the algorithm would restart with a new random solution. The best solution overall was recorded. Improved search efficiency was obtained by mutating one in five parameters over its complete parameter domain rather than within a 1% boundary.

4. Results

Solutions found by the directed search algorithm are presented in this section, including a self-oscillating segment, a two-segment circuit using the single segment solution, and a full eleven-segment model of the forward locomotion circuit. The solutions obtained are shown to generate stable and robust oscillations. A visual comparison with a video of the biological worm is described. Finally, the consistency of the results with Niebur and Erdös' mechanical model is confirmed.

4.1 Self-oscillating segment

Approximately fifty searches of single self-oscillating segments (Figure 3) were performed (each generating over 300,000 solutions) with different random seeds. The solutions obtained gave rise to a wide variety of undulation waveforms, including square waves, sinusoidal-like waves, triangle waves, pulse waves and saw waves in the segment bending angle θ. Each solution consisted of 16 parameters: G, G^{AVB}, and the three σ^{stretch} parameters A, k and θ_0 [equations (1) and (2)] and the three σ^{out} parameters B, q and V_0 [equation (3)].

[1] Due to the high number of parameters (16), a systematic search of parameter space (with a simulation-based calculation of fitness for each parameter set) was deemed too lengthy a process.

While many of the solutions demonstrated neuronal oscillations leading to muscle pulse patterns, there was quite a lot of variation in the waveforms, frequencies and oscillation amplitudes. Many solutions consisted of symmetric or nearly symmetric circuits characterised by sinusoidal oscillations in the neurons, with corresponding anti-phase sinusoidal oscillations in the muscles. Other asymmetric solutions consisted of only one active muscle (with the other muscle static). Additional solutions consisted of hybrid circuits with one sinusoidally oscillating muscle and the other generating sharp pulses.

Most of the solutions obtained exhibited regulation by the AVB unit. Thus, neurons and muscles were relatively inactive when the voltage of AVB was set to zero. Further confirmation of this result was obtained in a related circuit model which exhibited frequency modulation by AVB voltage (Bryden, 2003).

Not all single segment solutions found were robust to random noise. Some altered their wavelengths once perturbed; in others the oscillations are lost altogether. Overall, the majority of solutions trained under noisy conditions demonstrated sustained stable activity, thus confirming their tolerance to random perturbation.

Figure 5 shows a particular solution with a square-like waveform. Square-waves constituted relatively rare solutions (only two patterns out of 50). In contrast, the majority of solutions found gave rise to either triangular or saw-shape waveforms.

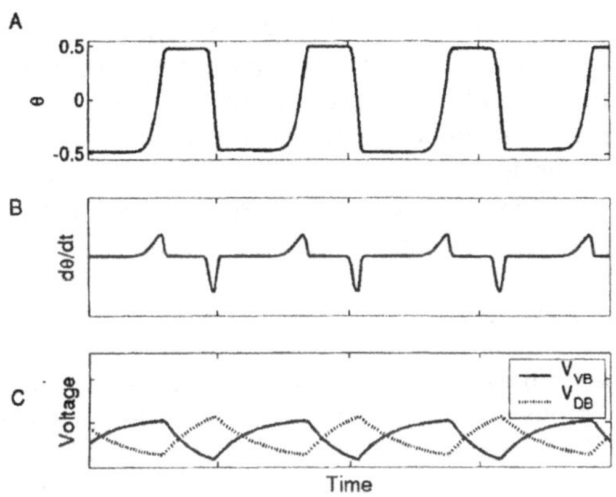

Figure 5: A fit solution of a self-oscillating segment. Graph A shows a trace of oscillations in segment bending angle (θ in radians); graph B shows the input to the muscles DM and VM (equal to $d\theta/dt$); and graph C shows the corresponding voltages of neurons DB and VB in time.

The figure presents time traces of the 'physical' bending of the segment (Graph A), the neuronal input to the muscles generating the bending (Graph B), and the underlying neuronal voltage oscillations (Graph C). Since the muscle activity is assumed to translate directly into bending, the trace of graph B is equivalent to $d\theta/dt$. As shown in the figure, in order to generate the square-like waveform, the input to the muscles should consist of two brief and opposite pulses (Graph B). These, in turn, are generated by anti-phase oscillations in the VB and DB neurons of the segment (Graph C).

Graphs B and C help us better understand the neuronal circuit mechanism. The initial condition is a strongly bent segment ($\theta \approx -0.5$). Consequently, the ventral neuron VB is receiving a positive feedback from the segment, whereas the dorsal neuron DB is receiving negative feedback. Accordingly, trace C begins with a build-up of the VB voltage and a similar but opposite decrease in DB voltage. However, the change in voltage has little effect on the muscles. As the VB voltage approaches its maximal amplitude, a threshold effect begins to activate the corresponding VM muscle. As the muscle bends the segment to the opposite configuration (positive θ), the mechanism is reversed. Now DB is excited and VB inhibited until a threshold in DB is reached and the activation of the dorsal DM muscle completes the cycle. Note that while dorsal and ventral activities are similar, the neuronal and muscle waveforms are not completely symmetric.

4.2 Phase-lagged segments

As described earlier (Figure 4), single segment solutions were combined to make phase-lagged segments. The single-segment solution used for this purpose is the one shown in Figure 5. Henceforth, we restrict our discussion to that solution.

The search algorithm was run fifty times with different seeds, testing around 20,000 solutions. Each solution consisted of 8 parameters: G was retrained, and the three $\sigma_{\text{tail}}^{\text{stretch}}$ parameters A, k and θ_0 were trained [equations (1) and (2)].

One solution is shown in Figure 6. As can be seen in graph A, the body segment oscillates slightly out of phase with the tail segment (appearing to precede it in time). The trace shows an initial transient period of approximately four cycles, during which the body segment gradually phase locks with the tail segment. Naturally, this transient behaviour is not characteristic of the biological worm and reflects the simplicity of the model used here. Furthermore, the directed search was not designed to penalise transient phases.

Graph B of the same figure illustrates how the output from the neurons is also phase-lagged, with the body-segment neurons displaying a similar pattern to the tail neurons. A comparison of this excitation pattern with that identified by Niebur and Erdös in (Niebur and Erdös, 1991) demonstrates that the two are qualitatively similar. The neurons only innervate muscles just before the segment bending is about to switch from one extremum to the other, i.e., when the segment lies on the outer curve of the 'S' shape of the *C. elegans* body, toward the direction of the locomotion (as shown in Figure 1).

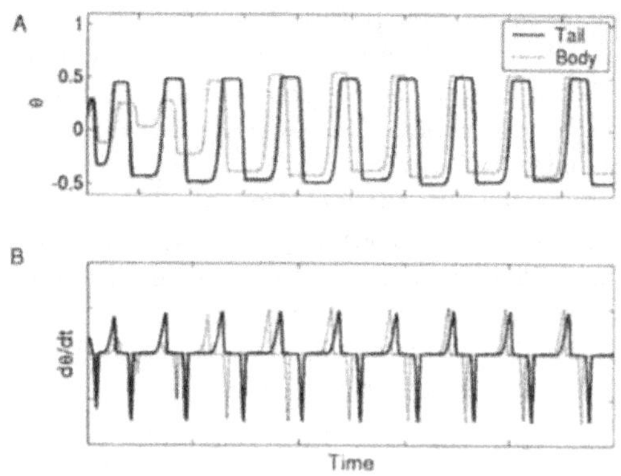

Figure 6: Two phase-lagged segments. Graph A shows the segment oscillations (θ is in radians); graph B shows the input to the segments from their muscles DM and VM (equal to $d\theta/dt$). The two segments oscillate approximately 9/10 of a cycle out of phase. An initial transient phase can be observed until the oscillations settle at a stable frequency and amplitude.

All fifty results from the search algorithm showed the same neural activity patterns. The only observed differences were in the amplitude ranges of the phase-lagged segment, the phase-lag and the durations of the transient phases at the starts of the oscillations. All solutions were characterised by a brief initial transient phase. The details of the transient phase depended on the initial condition of the simulation. We note that the steady-state periods of oscillation, amplitudes and phase-lags are independent of initial conditions.

4.3 Random perturbations

The addition of random perturbations demonstrated the robustness of the oscillation mechanism. Figure 7 presents a typical simulation result in which both segments were subjected to random perturbations. Note that the transient phase appears to be shorter, however this was just an artefact of that particular run. In general, perturbations did not significantly affect the duration of the initial transients.

4.4 Eleven phase lagged segments

To build up a complete model of the forward locomotion circuitry, ten phase-lag body segments (as shown in the previous section) were linked together in front of a self-oscillating tail segment. Figure 8 shows a sequence of four still frames from a simulation of an 11-segment circuit, juxtaposed with four frames of the biological worm. In each image, dots were placed alongside those segments of the simulation model whose muscles are active during that time frame. The biological worm moves on its side with the ventral and dorsal muscles generating lateral undulations. The simulated worm is also on its side with lateral undulations generated by the ventral and dorsal muscles.

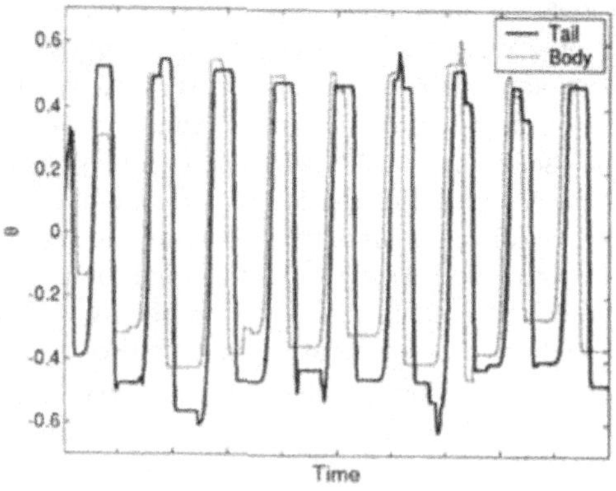

Figure 7: Two phase-lagged segments subjected to random perturbations. The segment bending angles θ are in radians.

The dots beside the worm are adjacent to segments where the neurons are excited enough to activate the muscles and move the segment. As can be seen they are in similar positions to the heavy black lines in Figure 1. This implies that the results found are consistent with the mechanical model developed by Niebur and Erdös (Niebur and Erdös, 1991). Videos of both the biological and the simulated worm moving can be viewed at `http://www.comp.leeds.ac.uk/johnb/celegans`.

5. Discussion

This paper presents a toy model of the forward locomotion controllers for *C. elegans*. The model uses mechano-feedback to generate oscillations at the tail and to propagate them along the body. We have shown that this mechanism can be used to generate undulations that propagate backward along the body thus sustaining forward locomotion in a simple simulation model. The model raises questions about the validity of typical motor control paradigms for the *C. elegans* circuit. In particular, the reliance on mechano-feedback in our circuit raises doubts about the applicability (to *C. elegans* motor control) of the often taken approach that the environment plays only a minor role in pattern generation.

The model incorporates knowledge about the neurobiology of the worm. Where experimental evidence is lacking, we followed biologically reasonable assumptions [in line with (Ferrée and Lockery, 1999,

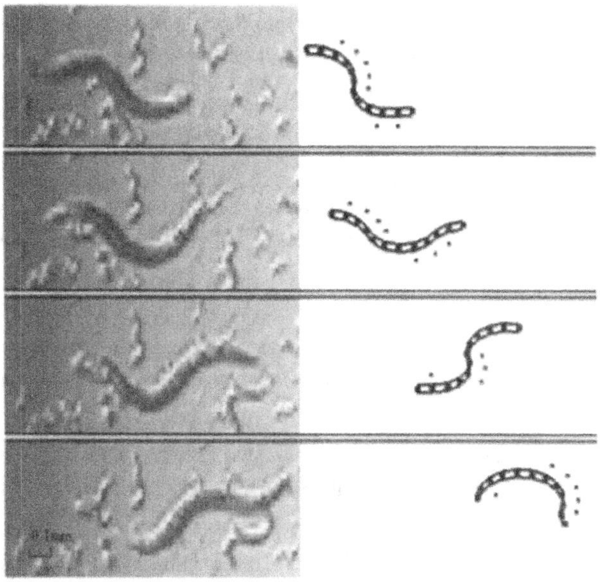

Figure 8: Comparison of biological worm with simulated worm. Left side shows four stills of an adult hermaphrodite *C. elegans* moving horizontally across agar, the worm is moving from left to right. Right side shows four stills from the output of the viewer application – the lateral movement of the worm has been introduced manually to make it easier to compare with the biological worm. Dots have been placed outside the worm adjacent to segments where $d\theta/dt$ is non-zero. Scales are approximate. Digital video was taken at Ian Hope's Laboratory, School of Biology, University of Leeds, LS2 9JT.

Ferrée and Lockery, 1998, Goodman et al., 1998, Niebur and Erdös, 1993, Niebur and Erdös, 1991)]. First, the neuronal activity was that of graded potential neurons. The models of the neurons were consistent with those identified by (Ferrée and Lockery, 1999, Ferrée and Lockery, 1998). The model also followed the general understanding that locomotion control circuitry was effectively isolated from the rest of the nervous system (White et al., 1986). In all these respects, the models generated are hoped to be biologically realistic.

In other respects, our model diverged from biological realism. Most importantly, the decision to partition the worm and its nervous system into structural segments was followed purely for convenience. In contrast, the biological worm has 11 ventral neurons and 7 dorsal neurons in its locomotion circuitry (White et al., 1986). There is no common factor which allows for such a circuit to be broken down into segments.

Secondly, the physical model for the worm was a simplistic one. This simple model demonstrates a neuronal mechanism through which waves of undulations are generated in a simulated worm. In the presence of friction, such undulations would have generated forward motion. However, this work stops short of building a mechanical model of the worm and its interactions with a physical environment.

The parsimony of our model has nevertheless allowed us to focus on the neural dynamics, without concern for the added complexity of mechanical effects. This approach enabled us to produce a detailed neuronal model of locomotion control as compared with (Niebur and Erdös, 1991). The results of the model were consistent with Niebur and Erdös' mechanical model. In particular the patterns of muscle activation generated by our model appear qualitatively similar to those found by Niebur and Erdös. Interestingly the search algorithm did find various other candidate patterns of muscle activation. An evaluation and comparison of these may prove fruitful.

The model chosen by Niebur and Erdös relied on stretch receptor feedback to generate propagation of undulations. This was our starting point as well. However, the neurons in Neibur and Erdös' model only have stretch receptors on the tips of the neurons, and therefore do not receive local feedback. Our simulations of such circuits all led to unstable solutions that are highly sensitive to perturbations (data not shown). Hence, the neurons in our model receive feedback from the local segment as well as feedback from the adjacent posterior segment. This indeed led to stable solutions that were robust to random perturbations both in the body and in the tail. Thus, one natural and testable prediction of our model is that the stretch receptors are located over the full length of the neuron bodies rather than at the tips alone.

Of course, it is possible that the mechanical dynamics of the biological worm are different to those demonstrated by the minimal environment chosen for our model. We hope that the consistency evident between our model and the mechanical model produced by Niebur and Erdös implies that mechanical effects extraneous to our model are compatible with it. If so, the addition of mechanical realism to the physical environment of our model might enable us to demonstrate the validity of our model further. Such extensions could shed light on the function of neurons and neural connections not explained by our model.

Proposed extensions to the model therefore include such additions of biological realism such as friction and elasticity. A performance-based fitness factor that takes account of how fast the worm was capable of moving would also be an interesting approach. An approach that used one stage, i.e., all parameters trained at the same time, may lead to fresh insights.

The simulation modelling techniques presented in this paper have enabled us to generate a feasible simplified model of the *C. elegans* forward locomotion controllers.

The approach of splitting the problem up into manageable smaller constituents proved successful in generating a solution that appears promising, both in terms of its biological realism and in terms of its functionality and robustness to noise. The application of a directed search algorithm led to satisfactory solutions at each stage of development. While a variety of solutions were found, in this paper we focused on one particular solution that appeared consistent with complementary models in the literature. The results in this paper reiterate the potential value in using a combined search/simulation modelling approach to the design (and indeed reverse engineering) of neural circuits for motor control in simple biological and bio-inspired systems.

6. Acknowledgements

Ian Hope has been very accommodating in allowing us access to *C. elegans* worms in his laboratory and has provided numerous suggestions. Special thanks to Jason Noble for many useful discussions and critical comments. David Gordon helped with the graphical software for visualisation of the simulation results.

References

Bryden, J. (2003). A simulation model of the locomotion controllers for the nematode *Caenorhabditis elegans*. Master's thesis, University of Leeds.

Chalfie, M., Sulston, J., White, J., Southgate, E., Thomson, J., and Brenner, S. (1985). The neural circuit for touch sensitivity in *Caenorhabditis elegans*. *Journal of Neuroscience*, 5(4):956–964.

Chalfie, M. and White, J. (1988). The nervous system. In Wood, W., (Ed.), *The Nematode Caenorhabditis elegans*. Cold Spring Harbor Laboratory, Cold Spring Harbor, NY.

Ferrée, T. C. (2003). Motivations for nemasys. `http://www.csi.uoregon.edu/projects/celegans/nemasys/Motivation.htm`.

Ferrée, T. C. and Lockery, S. R. (1998). Chemotaxis control by liner recurrent networks. *Journal of Computational Neuroscience: Trends in Research*, pages 373–377.

Ferrée, T. C. and Lockery, S. R. (1999). Computational rules for chemotaxis in the nematode *C. elegans*. *Journal of Computational Neuroscience*, 6:263–277.

Goodman, M., Hall, D., Avery, L., and Lockery, S. (1998). Active currents regulate sensitivity and dynamic range in *C. elegans* neurons. *Neuron*, 20:763–772.

Gray, L. (1950). The kinetics of locomotion of the grass snake. *Journal of Experimental Biology*, 26:354.

Gray, L. (1953). Undulatory propulsion. *Quarterly Journal of Microscopical Science*, 94:551.

Ijspeert, A. (2001). A connectionist central pattern generator for the aquatic and terrestrial gaits of a simulated salamander. *Biological Cybernetics*, 84(5):331–348.

Ijspeert, A., Hallam, J., and Willshaw, D. (1999). Evolving swimming controllers for a simulated lamprey with inspiration from neurobiology. *Adaptive Behavior*, 7(2):151–172.

Marder, E. and Calabrese, R. (1996). Principles of rhythmic motor pattern generation. *Physiological Reviews*, 76(3):687–717.

Niebur, E. and Erdös, P. (1991). Theory of the locomotion of nematodes. *Biophysical Journal*, 60:1132–1146.

Niebur, E. and Erdös, P. (1993). Modeling locomotion and its neural control in nematodes. *Comments on Theoretical Biology*, 3(2):109–139.

Rose, J. K. and Rankin, C. (2001). Analyses of habituation in *C. elegans*. *Learning and Memory*, 8:63–69.

The *C. elegans* Sequencing Consortium (1998). Genome sequence of the nematode caenorhabditis elegans. a platform for investigating biology. *Science*, 282:2012–2018.

White, J., Southgate, E., Thomson, J., and Brenner, S. (1986). The structure of the nervous system of the nematode c.elegans (the mind of a worm). *Philosophical Transactions of the Royal Society London. Series B, Biological Sciences*, 314(1165):1–34.

Wicks, S., Roehrig, C., and Rankin, C. (1996). A dynamic network simulation of the nematode tap withdrawal circuit: Predictions concerning synaptic function using behavioral criteria. *Journal of Neuroscience*, 16:4017–4031.

Wood, W. (1988). Introduction to *C. elegans* biology. In Wood, W., (Ed.), *The Nematode Caenorhabditis elegans*. Cold Spring Harbor Laboratory, Cold Spring Harbor, NY.

Planning the Sequencing of Movement Primitives

Marcelo Kallmann[1], Robert Bargmann[2*] and Maja Mataric'[3]

[1,3] Interaction Lab/Robotics Research Lab
Computer Science Department
University of Southern California
Los Angeles, California 90089-0781, USA
{kallmann,mataric}@usc.edu

[2] Computer Graphics Group
Max-Plank-Institut für Informatik
66123 Saarbrücken, Germany
bargmann@mpi-sb.mpg.de
(*Work done while at EPFL-VRlab)

Abstract

Neuroscience evidence supports the idea that biological adaptive behavior may utilize combination and sequences of movement primitives, allowing the motor system to reduce the dimensionality of movement control.

We present a framework, using sampling-based motion planning, that is able to automatically determine the sequencing of parametric movement primitives needed to execute a given motion task.

Our approach builds a search tree in which nodes are configurations reachable with one or more movement primitives, and edges represent valid paths connecting parent and child nodes. The paths are determined by a motion planner that operates in the parameter space of a single movement primitive. The search tree is expanded with A*-like best-first search using greedy problem-specific heuristics.

The benefits of our approach are twofold: 1) planning complex motions becomes more efficient in the reduced dimensionality of each movement primitive, and 2) the ability to plan entire motions containing heterogeneous types of constraints, such as collision-free, balanced, alternating support contacts, etc.

We present a general framework and several simulation results of statically stable biped walking motions among obstacles. The presented planning capabilities enable robots to better handle unpredicted situations, and can be used as a method of self-organization of higher-level primitives.

1. Introduction

The control of complex robot motions remains a key challenge in robotics. While the number of degrees of freedom (DOF) can characterize the complexity of a robot, the complexity of its motions is further influenced by the constraints they are subjected to.

Evidence from neuroscience supports the idea that complex and adaptive motor behavior might be obtained through the combination of motor primitives [TS00] [Ma02]. In the frog and rat, for instance, the presence of spinal force fields, when appropriately combined through supra-spinal inputs, results in the entire repertoire of observed movement [Gi+93]. Studies of human movement [TS00] also provide evidence towards an encoding of primitives.

In robotics, most work in this domain has focused on the design, learning, and combination of different kinds of motor primitives [SS98] [AM02] [Wi96]. In contrast, our work focuses on the problem of automatic adaptation and sequencing of movement primitives in order to satisfy a given motion task in an unknown environment.

We consider that a movement primitive affects the configuration of a robot through a proper parameterization, respecting a set of motion constraints. Our method is then able to plan motions traversing different configuration sub-spaces, each being covered by a single primitive. As a result, we are able to plan entire motions respecting heterogeneous types of constraints, such as: collision free, in balance, alternating support contacts, etc (see Figure 1).

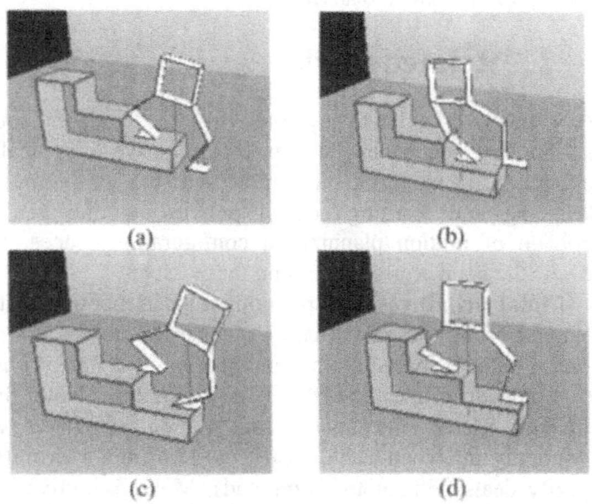

Figure 1: Climbing motion obtained with movement primitives designed to operate in the different support modes.

Our method finds the sequencing of primitives by means of a search tree in which nodes are configurations reachable by more than one movement primitive, and edges represent valid paths connecting parent and child nodes. A sampling-based motion planner operating in the parameter space of a single movement primitive determines each valid path. Therefore, edges represent primitive motions leading to nodes that serve as connection points allowing primitive change. The tree is expanded with A*-like best-first search using greedy problem-specific heuristics [RN95], until the desired task is satisfied.

The given vocabulary of movement primitives is responsible for reducing the planning complexity, which is related both to the number of DOF to be controlled and to the diversity of motion constraints to be satisfied. If the vocabulary of movement primitives is able to express the motions required to satisfy the task, the planner will easily find connection points between primitives, allowing the search tree to approach a solution. The chosen planner dictates the strategy of how movement primitives are adapted to overcome obstacles towards connection points.

Our method focuses on the generation of kinematic trajectories, assuming that a PD controller is available in order to transfer motions to the real robot. This choice is motivated by three main reasons:

- Kinematic plans are independent of motor commands, thus more suited to support the idea that complex motions are structured trough combination and sequencing of movement primitives.
- There is neuroscience evidence of representation of kinematic trajectory plans in the parietal cortex and inverse dynamics models in the cerebellum [SS98].
- We are able to plan motions respecting more difficult constraints, such as maintaining support contacts during locomotion or object manipulation.

We present a general framework, applicable to different problems and several simulation results of a statically stable biped walking among obstacles.

2. Related work

Very few works have attempted to plan complex robot motions by sequencing movement primitives. Typically, research has independently addressed the problems of designing and learning control policies, or the general problem of motion planning in configuration spaces. We review the main works addressing these issues.

Motor primitives Most previous work in robotics related to motor primitives has focused on the design and learning of control policies, based on the combination of different kinds of motor primitives [SS98] [AM02] [Wi96]. Note that, although motor, movement, and motion primitives are commonly interchangeable terms, the term motor primitive usually deals with motor commands. Motor primitives are often divided in two classes: oscillatory and postural (discrete). Being influenced by neuroscience, these works mainly concentrate on learning or skill acquisition, in dynamically simulated or real humanoid robots.

Motion planning Several algorithms are available for the general problem of robot motion planning [La91]. In particular, sampling-based methods provide general algorithms applicable to problems of arbitrary dimensionality of control. These methods can be divided into two main categories: multi-query and single-query.

Multi-query methods build a Probabilistic Roadmap (PRM) [Ka+96] that can be used for several different queries in a single static environment. The basic procedure consists of randomly sampling the configuration space, creating nodes when samples are valid, and connecting pairs of nodes each time the connection is tested to be valid and the nodes are considered to be close enough. Several variations to the basic PRM approach have been proposed [Be03] [Si+00] [BK00]. A good overview and comparison is given in [GO02].

Single-query methods are used when the environment is not static. Roadmaps are built specifically for each query, but, for better efficiency, trees are used instead of graphs. The Rapidly-exploring Random Tree (RRT) [KV00] [Va98] is a popular single-query method. Its basic idea is to expand nodes of the tree toward random samples until reaching the goal configuration. Another efficient method is based on Expansive Spaces Trees [Hs+99], where nodes in low-density locations are locally expanded. An efficient bi-directional version [SL01] incorporating lazy collision detection [BK00] is also available.

We make use of probabilistic roadmaps constructed in the valid portion of the configuration space covered by one movement primitive. Nodes of the roadmap are candidate configurations to serve as connection points to another movement primitive. Although the choice of a planner should take into account problem-specific issues, we use in this work an RRT [KV00] [Va98] approach to create roadmaps in the parametric space of movement primitives.

Legged locomotion Several different approaches have been presented for the control of legged locomotion. Genetic algorithms [RN95] have been used to adjust synapse weights of neural networks in central pattern generators [Ij01], and to evolve developmental programs based on pre-designed specific grammars [Fi+99]. Dynamic biped locomotion is usually achieved through the design of specific control policies [Zo+02], or based on the Zero Moment Point [Mi+95].

Our work is limited to statically balanced motions. Our approach can be seen as a complementary way to plan different patterns of motions required to overcome unanticipated situations.

Some works in the virtual reality and computer games domain have also addressed the locomotion problem of legged characters. In particular, motion planners were proposed for the multi-modal locomotion of a 2D character [KP01], and for planning footsteps locations, which can be connected with warped motion-captured sequences applied to virtual characters [Ch03].

The work presented by Kuffner et al. [Ku+01] [Ku+02] also relies on a search procedure in order to sequence pre-designed leg motions of a simulated humanoid robot. Their approach is more efficient than ours, however is limited to pre-designed motions. In our approach new types of motions are generated, for instance to avoid obstacles of any shape or use them as support.

3. Definitions and notations

Let d be the number of degrees of freedom (DOF) of a given robot and let C be the d-dimensional configuration space of the robot. A configuration in C is said to be valid if it satisfies problem-specific validity requirements. Examples of validity requirements are: collision-free, balanced, etc. The subset of all valid configurations in C is denoted as C_{free}.

Movement primitive As d might be too high and C might be too complex due to motion and environmental constraints, we are limited to manipulating the robot only by means of a finite set of movement primitives. Each primitive defines a control policy that locally alters a given configuration, according to a proper parameterization space. Therefore, each movement primitive $\mathcal{P}_i$, $i \in \{1, \ldots, n\}$, when instantiated at a given configuration $s \in C$, becomes a function of the type:

$$\mathcal{P}_i^s : S_i^s \rightarrow C \tag{1}$$

S_i^s is the parameter space of primitive i, instantiated at configuration s. Configuration s is also said to be the starting point of movement primitive $\mathcal{P}_i^s$. There are two main reasons for having the parameterization space dependable on s: first, parameters are often considered in relation to a local frame relative to s, and second, s might imply specific limits on the parameterization. We also allow that the dimension of the parameterization space change according to the instantiation.

Each primitive $\mathcal{P}_i$ has instantiation conditions to satisfy. If, for a given $s \in C$, the instantiation conditions of $\mathcal{P}_i$ cannot be verified, we say that $\mathcal{P}_i$ cannot start at configuration s. For example, a movement primitive for balanced knee flexion might be designed to start only when both feet are in contact with a proper support.

Each movement primitive has to be constructed in such a way as to allow the implementation of some required operators. These operators vary according to the motion planner selected to operate on the primitive parameter space. Typical motion planners considered in our framework require, for each instantiated primitive $\mathcal{P}_i^s$, the existence of a distance function and an interpolation function. The motion planner is responsible for building a roadmap graph connecting s to other configurations in the free portion of the configuration space covered by primitive $\mathcal{P}_i^s$.

Roadmap Let $C_i^s \subset C$ denote the image of movement primitive $\mathcal{P}_i^s$, i.e., for any $s \in C$, $C_i^s = \mathcal{P}_i^s(S_i^s)$ if $\mathcal{P}_i^s$ can be instantiated at s, and $C_i^s \equiv \varnothing$ otherwise. C_i^s represents the subspace of C covered by primitive $\mathcal{P}_i^s$.

We define the roadmap of a movement primitive instance as a single connected graph $R(\mathcal{P}_i^s)$. Nodes in the graph are valid configurations belonging to $C_i^s \cap C_{free}$, and edges in the graph represent valid paths in $C_i^s \cap C_{free}$ joining the edges endpoints. The starting node s of the primitive is required to be a node of $R(\mathcal{P}_i^s)$. For simplicity of notation, we may also refer to the nodes of a roadmap as the configurations of the roadmap.

Usually, for computational efficiency, roadmap edges are determined by checking the validity of the interpolation between the edge endpoints, and thus it is assumed that primitive $\mathcal{P}_i^s$ has a proper efficient interpolation function $interp_i^s$ of the type:

$$interp_i^s : S_i^s \times S_i^s \times [0,1] \rightarrow S_i^s \tag{2}$$

As usual, the interpolation function is parameterized over the interval $[0,1]$, so that $interp_i^s(p_1, p_2, 0) = p_1$, and $interp_i^s(p_1, p_2, 1) = p_2$, for a given pair of points $\{p_1, p_2\} \in S_i^s$. We say that the interpolation of a pair of points $\{p_1, p_2\}$ is valid if, for all $t \in [0,1]$, $\mathcal{P}_i^s(\ interp_i^s\ (p_1, p_2, t)\ )$ is a valid configuration, i.e., belongs to C_{free}.

Analogously, we say that there is a valid path joining $q_1 \in C_{free}$ and $q_2 \in C_{free}$ if there is an instantiated primitive $\mathcal{P}_i^s$ and two points $\{p_1, p_2\} \in S_i^s$, such that $q_1 = \mathcal{P}_i^s(p_1)$, $q_2 = \mathcal{P}_i^s(p_2)$, and the interpolation between p_1 and p_2 is valid.

Note that the interpolation function is defined in parameter space, meaning that paths are first generated in parameter space and then transformed into the configuration space for validity testing. With this formulation all the roadmap computation is done locally to the movement primitive, and does not require the computation of the inverse of the movement primitive function.

Besides the interpolation function, a distance metric is often required during the roadmap construction process. In Section 4 we describe an algorithm that constructs roadmaps following the RRT [KV00] [Va98] expansion strategy.

Roadmaps are used as a sampling strategy to transform the continuous parameterization of a movement primitive into a discrete set of configurations (the nodes of the roadmap), which are suitable for inclusion in a search tree.

Problem definition Let $\mathcal{P}_i$, $i \in \{1, \ldots, n\}$ be a given set of movement primitives manipulating a robot as defined above. Consider that the task to be accomplished is defined as a function that returns 1 if a given configuration q satisfies the task, and zero otherwise, i.e.:

$$task\,(q) : C \rightarrow \{0,1\} \tag{3}$$

The problem we want to solve is determining a sequence of configurations $(q_j)_{j=1,m}$, such that:

- The current robot configuration is equal to q_1
- $task\,(q_m) = 1$
- For each pair of configurations q_k, q_{k+1}, $1 \le k < m$, there is a valid path connecting q_k and q_{k+1}.

Note that the determination of each valid path requires the determination of the instantiated primitive that generates it (trough the primitive's interpolation function).

The solution $(q_j)_{j=1,m}$ implies the determination of a sequence of paths joining q_1 to q_m. For simplicity of notation, we also say that a valid path between two configurations exists when in fact there is a sequence of paths joining them.

When needed, we distinguish these two cases with the terms: composed path and direct path.

4. General method

Our approach is based on a search tree where nodes represent reachable valid configurations, and edges contain paths between parent and child nodes. At any point, a partial solution can be constructed by concatenating the paths in the unique sequence of edges joining the root node to a given node.

The algorithm starts by initializing the root of the tree with the current robot configuration, and then an expansion process adds nodes to the tree until the task function is satisfied. A cost is associated with each node and represents the cost of the path constructed so far. The root node is initialized with cost 0. A priority queue is used to efficiently store the leaves of the tree according to their priority of expansion.

Generally, an A* expansion [RN95] is followed, where the highest priority is given to the leaves with less heuristic cost. The simplest heuristic cost for a given node n sums the cost of n with an estimate of the distance to achieve the goal task from n. Several problem-specific heuristics can be added to this basic formula, and they are important in ensuring that the search tree grows toward the solution.

After initialization, the highest priority leaf q is removed from the priority queue (not the tree), and a roadmap is constructed from q. Nodes in the roadmap that allow primitive change are added to the priority queue as new leaves, and added to the search tree as children of q.

The following algorithm summarizes this expansion step:

```
expand ( tree, queue )
q = remove higher priority leaf from queue;
if task(q)==1, do:
   return composed path from root(tree) to q;
for i = 1 to n, do:
   if P_i can be instantiated at q, and
   P_i was not instantiated by q parent, do:
      R(P_i^q) = build_roadmap ( i, q );
      for all configurations q_r≠q in R(P_i^q), do:
         if q_r can be instantiated by a different
         movement primitive than P_i, do:
            edge = add_child ( tree, q, q_r );
            store in edge the path in R(P_i^q)
            joining (q,q_r);
            add leaf q_r to queue;
return null path;
```

```
build_roadmap ( i, s )
R(P_i^s) = init roadmap with node s
failures = 0
while failures<MaxTries, do:
   p_rand = random point in S_i^s;
   p_near = nearest node to p_rand in R(P_i^s);
   dist = distance ( p_near, p_rand );
   p_new = interp_i^s (p_near, p_rand, IncrementalStep );
   if ( p_near and p_new interpolation valid ), do:
      add node p_new and edge {p_near,p_new} to R(P_i^s);
      failures = 0;
   else
      failures++;
return R(P_i^s);
```

Procedure `expand` is called until a non-null path sequence (the solution) is returned. Each time a node is expanded, procedure `roadmap` returns a single connected roadmap graph. The nodes of the graph are candidate to become new leaves in the search tree.

Procedure `build_roadmap` implements a general roadmap construction method following the RRT [KV00] [Va98] expansion strategy. It requires the definition of two parameters. Parameter `MaxTries` specifies the number of failures that are needed in order to decide that the roadmap cannot grow any more. Parameter `IncrementalStep` provides control over the length of the edges in the graph. For more uniform resolution control, a length step measured in configuration space should be used whenever possible.

Several problem-specific issues can be addressed during roadmap construction. For instance, in the biped robot case presented in the next section, whenever a node in the roadmap is detected to be close to a configuration serving as a connection point, we move the node to that connection point.

Note that a random generator routine is used in procedure `roadmap` in order to generate the points in parameter space that guide the roadmap expansion. There is no problem if the free portion of the parametric space is much smaller than the whole parametric space. The RRT expansion strategy provides a suitable gradual exploration of the free space, and its implementation is simple and efficient. However, whenever possible, the choice of the roadmap construction strategy should be problem-(and primitive-)-specific.

As roadmaps represent a discretization of a continuous space, it is not possible to guarantee that the search tree will find the optimal solution. However, usually we are not interested in finding the optimal solution as the search tree easily becomes prohibitively large and greedy heuristics are always preferred. In addition, problem-specific optimizations are likely to be crucial for having acceptable running times for complex problems.

In the next section we demonstrate how our framework can be applied to generate motions for the control of a statically stable biped robot.

5. Biped robot example

We have implemented and tested the algorithm described in the previous section on planning statically stable walking motions for a biped robot moving in a planar environment containing polygonal obstacles. Obstacles are avoided during motion and are also used as support, allowing the generation of climbing sequences (see Figure 1).

The designed biped robot has a total of 9 DOF: the first two specify the position of the body center in the Cartesian plane. The remaining DOF are rotation angles: one to specify the orientation of the body, and three for the articulations of each leg (see Figure 2).

Each rotational DOF of the robot has specific lower and upper articulation limits. The two positional DOF have no limits, however when they are controlled by a movement primitive, they have limits imposed by the instantiated parameterization space of the primitive.

Configuration validity Let C be the 9-dimensional configuration space of the biped robot. We define a configuration $q \in C$ to be valid if q:

- Satisfies the articulation limits of the robot.
- Is collision-free, i.e., does not intersect with obstacles.
- Is in balance, i.e., its center of mass projects inside the support segment of the robot.

As the simulated robot is constructed in a planar environment, the validity tests have a straightforward implementation. However, some special care is required in geometric tests, such as deciding support contacts. In our implementation, all geometric tests are based on an epsilon precision distance.

When the two endpoints of a foot are close enough to an obstacle segment, but without crossing it, the foot is considered to be in contact with the segment and a support segment is defined. When both feet are in support, the support segment is increased to contain the support segment of the two feet.

For the computation of the center of mass we associate a mass value m_k, $k \in \{1, \ldots, 7\}$, to the center of each robot part P_k, each being a limb or the body. The center of mass position is then determined by the following weighted average sum:

$$(\sum center(P_k) m_k) / 7 \quad (4)$$

The robot is considered in balance if its center of mass vertically projects inside the support polygon. Collisions are detected whenever a limb segment crosses an obstacle segment or another limb segment.

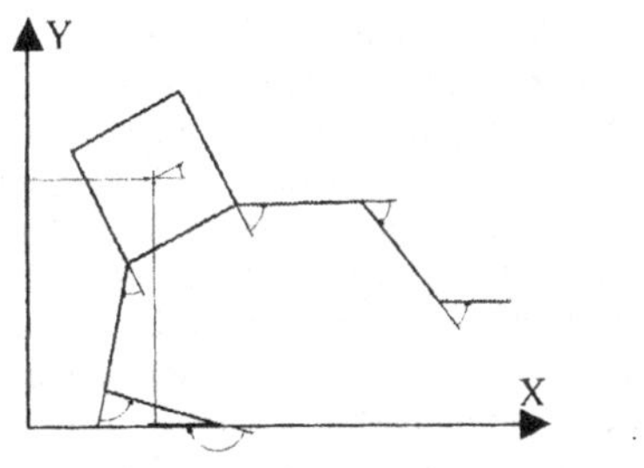

Figure 2: The planar biped robot and its nine DOF.

The defined task function checks if the center of mass of the robot is close enough to a desired location, according to a given precision.

Movement primitives Three movement primitives are defined, each specific to a support mode:

- Movement primitive $\mathcal{P}_L$ is used to move the right leg of the robot while balance is maintained only with the support of the left foot. Therefore the instantiation condition of this movement primitive requires support on the left foot.
- Movement primitive $\mathcal{P}_B$ was designed to move the body while the two legs remain attached to the floor. The instantiation condition is support on both feet, and its parameterization controls the position and orientation of the robot's body, keeping the feet fixed at their support location.
- Similarly to $\mathcal{P}_L$, $\mathcal{P}_R$ is designed to move the left leg while the robot keeps support on the right leg.

The parametric spaces of movement primitive $\mathcal{P}_L$ and $\mathcal{P}_R$ both have dimension 4. The affected rotational angles have the same range limits as those originally defined by the robot, except for body rotation, which is allowed only in the direction that favors the free leg to reach higher positions. Whenever body rotation is changed, the angles of the support leg have to be adjusted in order to maintain the support foot in the same place. This adjustment is done by employing a straightforward analytical Inverse Kinematics formulation for the foot to be fixed.

Movement primitive $\mathcal{P}_B$ defines specific range limits for the two translational DOF it controls. Let $s \in C_{free}$ be a configuration and assume that $\mathcal{P}_B$ can be instantiated at s, i.e., s has support in both feet. Let p be the body center point of the robot at s. The translational parameters of $\mathcal{P}_B^s$ are limited to be inside a rectangle of center p and sides with double the length of the body sizes. In this way we have a much smaller range of motion for the motion planner to explore the free portion of the parameterization space.

An Inverse Kinematics formulation is also applied to maintain both feet fixed at their original places while the translational and rotational DOF of primitive $\mathcal{P}_B$ are changed. Table 1 summarizes the main characteristics of the movement primitives.

Movement Primitive	Instantiation Condition	Type of Motion	Parametric Space Dimension
$\mathcal{P}_L$	support in left foot	moves right leg articulations and body rotation	4
$\mathcal{P}_B$	support in both feet	moves body, legs fixed with IK	3
$\mathcal{P}_R$	support in right foot	moves left leg articulations and body rotation	4

Table 1: Summary of used movement primitives.

Note that some configurations allow the instantiation of two different movement primitives. This happens when the configuration allows two different types of support, i.e., when the robot has support in both feet, and, at the same time, support in one foot alone. Such configurations serve as connection points between two different movement primitives. Figure 3 illustrates the different kinds of support modes.

Motion planner A single motion planner was implemented to operate in the parameter space of primitives $\mathcal{P}_L$, $\mathcal{P}_B$, and $\mathcal{P}_R$, and it closely follows the algorithm `build_roadmap` of Section 4.

The random function we used simply selects values inside the range defined for each parameter of each movement primitive. The interpolation function linearly interpolates the corresponding parameters of two given sets of parameters of a primitive.

As a distance function, instead of operating in the parameter space, we compute the mean of the Euclidian distances between the corresponding articulation points of two given configurations. The same distance function is used regardless of the movement primitive being considered.

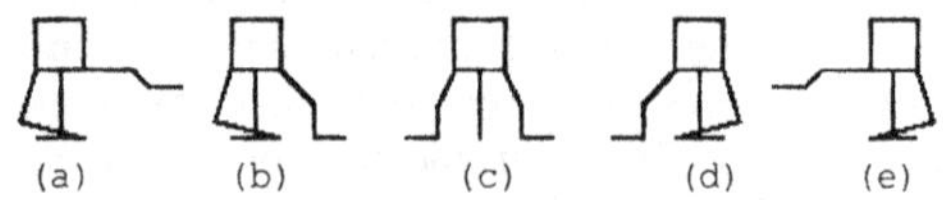

Figure 3: Example of configurations in different support modes: only in left support (a), simultaneous left and both feet support (b), only both feet support (c), simultaneous right and both feet support (d), and only in right support (e).

In order to promote the appearance of configurations in more than one type of support, we include a test during roadmap construction that detects configurations which are close to making contact with a new support. The test consists on measuring the distance between each foot of the configuration to its nearest obstacle. If one distance is smaller than the pre-specified snap distance, the configuration is adjusted with Inverse Kinematics in order to precisely place the foot in contact with the support. Such adjustment is critical, because the achievement of new supports is the only way to find configurations allowing primitive change. The specified snap distance trades the ability to find new supports with the ability to avoid obstacles, and should be set according to the environment.

Search Tree Heuristics After initialization, the search tree is expanded following the general algorithm given in Section 4. The heuristic cost of a leaf in the priority queue is:

$$hc(n) = \text{cost}(n) + \text{distance}(n \text{ mass center, target point}) \quad (5)$$

Term cost(n) is the cost of the path from the configuration at the root of the tree to n. The cost is defined as the length of the path, according to the same metric used for the roadmap construction: the average of the Euclidean lengths of the articulations paths. The target point is the same point considered by the task function, which tests if the center of mass is close enough to the target point.

The leaf with lowest heuristic cost has higher priority and will be removed first from the priority queue. Removed leaves are then expanded.

Figure 4 illustrates the expansion process. Figure 4a shows the roadmap constructed for the robot in a configuration with both feet in support. Each node in the roadmap represents a full configuration, but in this image, only the position of the body center is used to draw the roadmap. The roadmap in Figure 4b shows a mark (a cross) on each node allowing primitive change, which are the nodes becoming new leaves in the search tree. Figure 4c shows the robot configuration in the highest priority leaf, which is the one selected for node expansion, now with a single leg support. The roadmap in Figure 4d shows the coverage of the free foot in free space, and shows marks in two configurations again allowing a primitive change, and thus becoming new leaves in the next expansion of the tree. In all images, the circle identifies the root of the roadmaps.

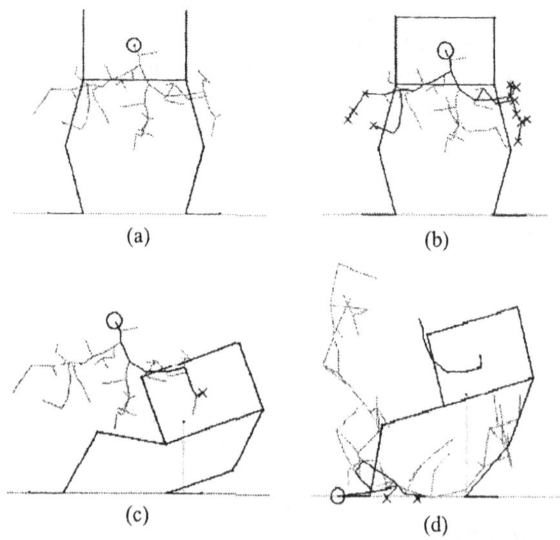

Figure 4: Node expansion example starting with both feet support.

Although the heuristic cost function is rather standard, three greedy optimizations were tested in order to reduce the branching factor and direct the search to the goal:

- The first optimization, and the most important, relies on the observation that some sequences of nodes in the search tree are not useful. This occurs when three adjacent nodes in the three have the support modes on: left foot, both feet, and left foot, respectively; or all have support simultaneously in both feet and in the left foot alone. The same occurs to the analogous sequences in relation to the right foot. Therefore, before inserting nodes in the search tree, we ensure that such sequences are not formed by checking the parent and grandparent nodes in the tree. More complex analysis in the search tree could be investigated, as, for instance, collapsing different nodes with the supporting feet located at the same place.
- Long locomotion sequences can easily demand prohibitively large time and memory consumption. The search tree is not suitable for planning long paths around a large amount of obstacles, but for planning leg motions for nearby obstacles. Therefore, whenever the current node n being expanded is sufficiently far away from the root node of the tree, the partial path until n can be stored, the whole tree deleted, and initialized again with the new root being n.
- Another optimization can be used to limit the creation of excessive nodes in the same location of the environment. A planar grid can be easily implemented

in order to control, for each cell, a maximum number of nodes occupying that cell. This procedure can speedup the search in some situations, but has some negative impact on the quality of the generated paths.

6. Discussion and results

Our implementation is able to find simple motions in a few minutes. Complex motions, however, require several minutes or more to be computed on a 2.8GHz Pentium 4. The required computation time is related both to the difficulty to overcome obstacles and to the chosen parameters of the planner and of the implemented heuristics. As example, the motion shown in Figure 5 took 50 minutes to be computed and generated a search tree of about 600K nodes.

The branching factor of the search tree is greatly influenced by the resolution step used in the motion planner. This step affects the number of nodes computed in each roadmap, and thus also the branching factor. With a small resolution step, the solutions are likely to be closer to an optimum, but large branching factors are not efficient and can quickly exhaust memory resources.

Parameter `MaxTries` (see Section 4) can be usually set to low values, allowing fast computation of roadmaps. However, in environments with many obstacles, larger values are required. In such cases, roadmaps only generate child nodes in the search tree after many random tries.

Due to the approximate nature of roadmaps, solutions are clearly not optimal. However, few optimal characteristics could be noted: large steps are generated towards the goal in the absence of obstacles and few finer steps are performed next to obstacles in order to determine the best position before overcoming them.

Figures 1, 5 and 6 show some of the results obtained. Note that the final result may exhibit some jerky motions, due to the exploration strategy of the roadmap computation. A simple and efficient smoothing process consists of trying to replace pair of points in the path by valid straight interpolations [Sc⁺02]. However, the presented results were not smoothed. Videos showing further results can be found at: `http://robotics.usc.edu/~kallmann/biped2d/`

Figure 5: Example motion over an obstacle.

Although the robot design used in this work is quite simple, we note that it is rather difficult to make it overcome obstacles by manually controlling the parameters of the movement primitives by means of a user interface. The validity constraints leave a small motion range in several situations, and precise positioning prior to obstacles is often required before overcoming them. Nevertheless, the planner successfully found several non-trivial solutions.

7. Conclusions

We have presented a search method for planning complex motions trough the concatenation of paths generated by biologically-inspired movement primitives. Our approach introduces a way to apply sampling-based motion planning techniques in order to compose motions controlled by heterogeneous types of parameterizations. We are thus able to respect different types of constraints, e.g., alternating support contacts. In addition, planning complex motions usually becomes more efficient in the reduced dimensionality of each movement primitive.

Our method is able to find motions for unanticipated situations. This adaptive behavior characteristic is of key importance and is supported by neuroscience evidence showing that complex motions in animals might be structured through the use of movement primitives.

The framework presented in this paper can be applied to other kinds of problems as well. Examples include mobile manipulators equipped with different primitives for locomotion and manipulation, and the automatic generation of higher-level primitives based on motions planned with simpler ones. For instance, portions of planned motions that are detected to be useful can be directly encoded and used as a new time-parameterized primitive.

In general, any control policy with a suitable parameterization can be treated as a movement primitive. However, the success of the best-search method greatly depends on the availability of efficient problem-specific heuristics; otherwise the branching factor of the search tree can easily become prohibitively large.

Acknowledgments

The first author is grateful to Dr. Anand Panangadan for fruitful discussions. This work was supported by a DARPA MARS 2020 Program grant NAG9-1444.

References

[AM02] R. Amit and M. J. Mataric', "Parametric Primitives for Motor Representation and Control", Proceedings of the IEEE International Conference on Robotics and Automation (ICRA-2002), Washington, DC, May 11-15, 2002, pp. 863-868.

[Be03] K. E. Bekris, B. Y. Chen, A. M. Ladd, E. Plakue, and L. E. Kavraki, "Multiple query probabilistic roadmap planning using single query planning primitives", Proc. of the International Conference on Intelligent Robots and Systems (IROS), Las Vegas, USA, 2003.

[BK00] R. Bohlin and L. Kavraki, "Path planning using lazy PRM", Proc. of the International Conference on Robotics and Automation (ICRA), San Francisco, USA, 2000, pp. 521-528.

[Ch03] M. G. Choi, J. Lee, S. Y. Shin, "Planning biped locomotion using motion capture data and probabilistic roadmaps", ACM Transactions on Graphics (TOG), 22(2), 2003, pp. 182-203.

[Fi+99] D. Filliat, J. Kodjabachian, and J.-A. Meyer, "Evolution of Neural Controllers of for locomotion and obstacle-avoidance in a 6-legged robot", Connection Science 11:223-240, 1999.

[Gi+93] S. F. Giszter, F. A. Mussa-Ivaldi, and E. Bizzi, "Convergent force fields organized in the frog's spinal cord", Journal of Neuroscience 13(2):467:491, 1993.

[GO02] R. Geraerts and M. Overmars, "A comparative study of probabilistic roadmap planners", Proc. of the International Workshop on Algorithmic Foundations of Robotics (WAFR), Nice, France, 2002.

[Hs+99] D. Hsu, J.-C. Latombe, and R. Motwani, "Path planning in expansive configuration spaces", International Journal of Computational Geometry and Applications, 9(4-5):495-512, 1999.

[Ij01] A. J. Ijspeert, "A connectionist central pattern generator for the aquatic and terrestrial gaits of a simulated salamander", Biological Cybernetics 84(5): 331-348, 2001.

[Ka+96] L. E. Kavraki, P. Svestka, J.-C. Latombe, and M. H. Overmars, "Probabilistic roadmaps for fast path planning in high-dimensional configuration spaces", IEEE Transactions on Robotics and Automation, 12(4):566-580, June 1996.

[KP01] M. Kalisiak and M. van de Panne, "A grasp-based motion planning algorithm for character animation", The Journal of Visualization and Computer Animation 12(3):117-129, 2001.

[Ku+01] J.J. Kuffner, K. Nishiwaki, S. Kagami, M. Inaba, and H. Inoue, "Footstep planning among obstacles for biped robots", Proc. of the International Conference on Intelligent Robots and Systems (IROS), Maui, USA, 2001.

[Ku+02] J.J. Kuffner, S. Kagami, M. Inaba, and H. Inoue, "Dynamically-stable Motion Planning for Humanoid Robots", Autonomous Robots, 12(1): 105-118, 2002.

[KV00] J. J. Kuffner and S. M. La Valle, "RRT-connect: an efficient approach to single-query path planning", Proc. of the International Conference on Robotics and Automation (ICRA), San Francisco, USA, 2000, pp. 995-1001.

[La91] J.-C. Latombe, "Robot motion planning", ISBN 0-7923-9206-X, Kluwer, Academic Publishers, 1991.

[Ma02] M. J. Mataric', "Visuo-motor primitives as a basis for learning by imitation", In K. Dautenhahn and C. Nehaniv, editors, Imitation in Animals and Artifacts, MIT, Press, 2002, pp. 391-422.

[Mi+95] K. Mitobe, N. Mori, K. Aida, and Y. Nasu, "Non-linear feedback control of a biped walking robot", IEEE International Conference. on Robotics and Automation (ICRA), 1995, pp. 2865-2870.

[RN95] S. J. Russell and P. Norvig, "Artificial intelligence: a modern approach", Prentice Hall, Englewood Cliffs, NJ, 1995.

[Sc+02] F. Schwarzer, M. Saha, and J.-C. Latombe, "Exact Collision Checking of Robot Paths", In Workshop on Algorithmic Foundations of Robotics, Nice, France, December 2002.

[Si+00] T. Simeon, J. P. Laumond, and C. Nissoux, "Visibility based probabilistic roadmaps for motion planning", Advanced Robotics Journal, 14(6), 2000.

[SL01] G. Sanchez and J.-C. Latombe, "A single-query bi-directional probabilistic roadmap planner with lazy collision checking", Proc. of the International Symposium on Robotics Research (ISRR), Nagoya, Japan, 2001.

[SS98] S. Schaal and D. Sternad, "Programmable pattern generators", In 3rd International Conference on Computational Intelligence in Neuroscience, Research Triangle Park, NC, 1998, pp 48–51.

[TS00] K. A. Thoroughman and R. Shadmehr, "Learning of action through combination of motor primitives", Nature 407:742–747, 2000.

[Va98] S. M. LaValle, "Rapidly-exploring random trees: a new tool for path planning", Technical Report TR 98-11, Computer Science Dept., Iowa State University, Oct. 1998.

[Wi96] M. W. Williamson, "Postural primitives: interactive bahavior for a humanoid robot arm", Proceedings of the 4th International conference on Simulation of Adaptive Behavior (SAB'96), MIT Press, 1996, pp. 124-131.

[Zo+02] F. Zonfrilli, G. Oriolo, and D. Nardi, "A biped locomotion strategy for the quadruped robot Sony ERS-210", IEEE International Conference. on Robotics and Automation (ICRA), 2002, pp. 2768-2774.

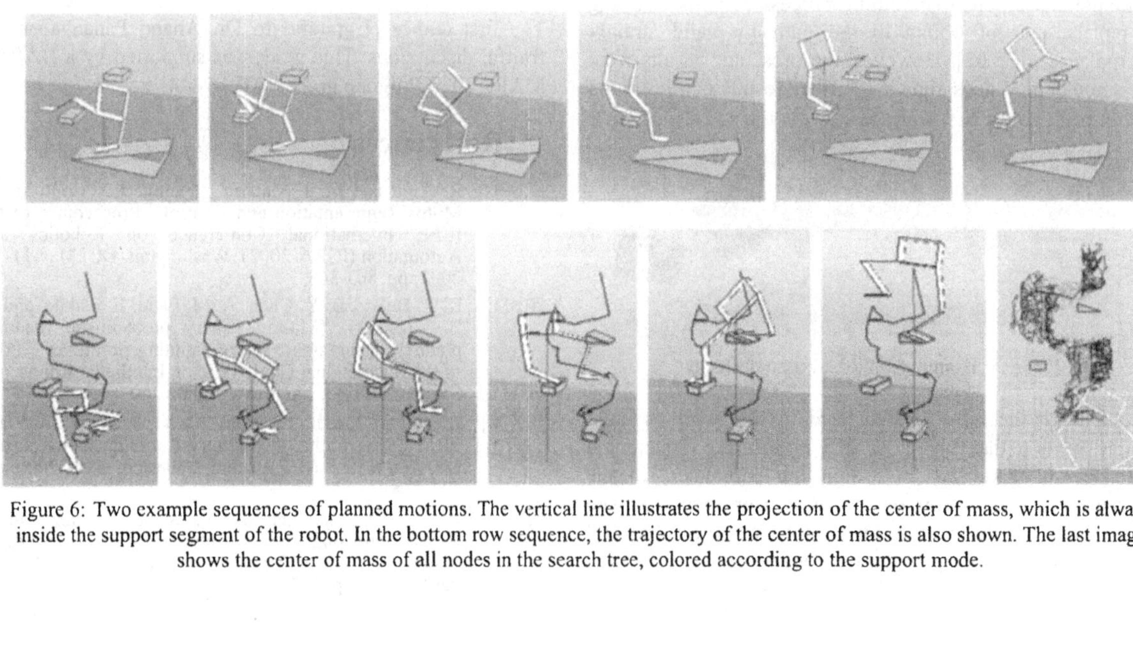

Figure 6: Two example sequences of planned motions. The vertical line illustrates the projection of the center of mass, which is always inside the support segment of the robot. In the bottom row sequence, the trajectory of the center of mass is also shown. The last image shows the center of mass of all nodes in the search tree, colored according to the support mode.

Biologically Inspired Adaptive Dynamic Walking of a Quadruped Robot

Hiroshi Kimura* **Yasuhiro Fukuoka***
*Graduate School of Information Systems
Univ. of Electro-Communications
1-5-1 Chofugaoka, Chofu
Tokyo 182-8585, Japan
{hiroshi, fukuoka}@kimura.is.uec.ac.jp

Avis H. Cohen**
**Dept. of Biology and
Institute for Systems Research
Univ. of Maryland, College Park
MD 20742, USA
avis@isr.umd.edu

Abstract

We have been trying to induce a quadruped robot to walk with medium walking speed on irregular terrain based on biological concepts. We propose the necessary conditions for stable dynamic walking on irregular terrain in general, and we design the mechanical system and the neural system by comparing biological concepts with those necessary conditions described in physical terms. PD-controller at joints constructs the virtual spring-damper system as the visco-elasticity model of a muscle. The neural system model consists of a CPG (central pattern generator), reflexes and responses. A CPG generates rhythmic motion for walking. A CPG receives sensory input and changes the period of its own active phase as responses. The virtual spring-damper system also receives sensory input and outputs torque as reflexes. The states in the virtual spring-damper system are switched based on the phase signal of the CPG. The physical oscillations such as the motion of the virtual spring-damper system of each leg and the rolling motion of the body are mutually entrained with the neural oscillations of CPGs. Consequently, the adaptive walking is generated by the interaction with environment. In this paper, we report our experimental results of dynamic walking on terrains of medium degrees of irregularity in order to verify the effectiveness of the designed neuro-mechanical system.

1. Introduction

Many previous studies of legged robots have been performed, including studies on running (Hodgins and Raibert 1991) and dynamic walking (Yamaguchi, Takanishi and Kato 1994; Kajita and Tani 1996; Chew, Pratt and Pratt 1999; Yoneda, Iiyama and Hirose 1994; Buehler et al. 1998) on irregular terrain. However, all of those studies assumed that the structure of terrain was known, even though the height of the step or the inclination of the slope was unknown.

On the other hand, studies of autonomous dynamic adaptation allowing a robot to walk over irregular terrain with less knowledge of it have been started only recently and by only a few research groups. One example is the recent achievement of high-speed mobility of a hexapod over irregular terrain, with appropriate mechanical compliance of the legs (Saranli, Buehler and Koditschek 2001; Cham et al. 2002). The purpose of this study is to realize high-speed mobility on irregular terrain using a mammal-like quadruped robot, the dynamic walking of which is less stable than that of hexapod robots, by referring to the marvelous abilities of animals to autonomously adapt to their environment.

As many biological studies of motion control progressed, it has become generally accepted that animals' walking is mainly generated at the spinal cord by a combination of a CPG (central pattern generator) and reflexes receiving adjustment signals from a cerebrum, cerebellum and brain stem (Grillner 1981; Cohen and Boothe 1999). A great deal of the previous research on this attempted to generate walking using a neural system model, including studies on dynamic walking in simulation (Taga, Yamaguchi and Shimizu 1991; Taga 1995; Miyakoshi et al. 1998; Ijspeert 1998, 2001), and real robots (Ilg et al. 1999; Tsujita, Tsuchiya et al. 2001; Lewis et al. 2003). But autonomously adaptive dynamic walking on irregular terrain was rarely realized in those earlier studies except for our studies (Kimura et al. 1999, 2001; Fukuoka et al. 2003). This paper reports on our progress in the past couple of years using a newly developed quadruped called "Tekken1&2," which contains a mechanism designed for walking in 3D space (pitch, roll and yaw planes) on irregular terrain. Especially, the self-contained (power autonomous) system was realized in Tekken2 (Fig.1) and we succeeded in outdoor experiments using it.

Table 1: Biological concepts of legged locomotion control.

	ZMP-based Control	Limit-Cycle-based Control	
		by Neural System (CPG and reflexes)	by Mechanism (spring and damper)
good for control of	posture and low speed walking	medium speed walking	high speed running
main controller	upper neural system acquired by learning	lower neural system (at spinal cord, brain stem, etc.)	musculoskeletal system through self stabilization

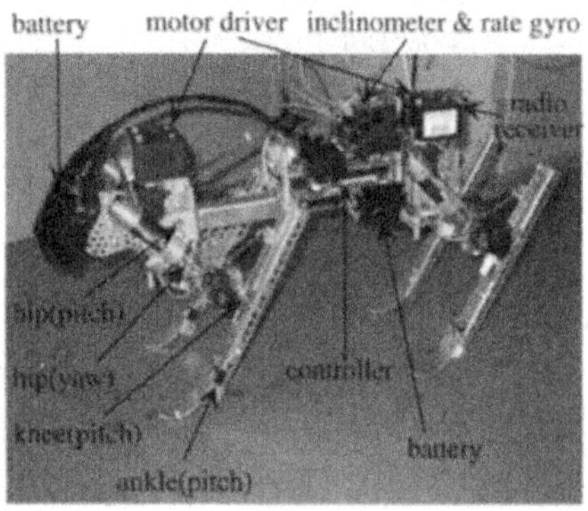

Figure 1: Self-contained quadruped robot: Tekken2. The length of the body and a leg in standing are 30 [cm] and 20 [cm]. The weight including batteries is 4.3 [kg]. The hip pitch joint, knee pitch joint and hip yaw joint are activated by DC motors of 23 [W], 23 [W] and 8 [W] through gear ratio of 20, 28 and 18, respectively.

2. Design concepts for adaptive walking

2.1 Legged locomotion control

Methods for legged locomotion control are classified into ZMP-based control and limit-cycle-based control (Table 1). ZMP (zero moment point) is the extension of the center of gravity considering inertia force and so on. It was shown that ZMP-based control is effective for controlling posture and low-speed walking of a biped (Takanishi et al. 1990) and a quadruped (Yoneda, Iiyama and Hirose 1994). However, ZMP-based control is not good for medium or high-speed walking from the standpoint of energy consumption, since a body with a large mass needs to be accelerated and decelerated by actuators in every step cycle.

In contrast, motion generated by the limit-cycle-based control has superior energy efficiency. But there exists the upper bound of the period of the walking cycle, in which stable dynamic walking can be realized (Kimura et al. 1990). It should be noted that control by a neural system consisting of CPGs and reflexes is dominant for various kinds of adjustments in medium-speed walking of animals (Grillner 1981). Full and Koditschek (1999) also pointed out that, in high-speed running, kinetic energy is dominant, and self-stabilization by a mechanism with a spring and a damper is more important than adjustments by the neural system. Our study is aimed at medium-speed walking controlled by neural system consisting of CPGs and reflexes (Table 1).

2.2 Necessary conditions for stable dynamic walking on irregular terrain

We propose the necessary conditions for stable dynamic walking on irregular terrain, which can be itemized in physical terms:

(a) the period of the walking cycle should be shorter enough than the upper bound of it, in which stable dynamic walking can be realized,

(b) the swinging legs should be free to move forward during the first period of the swing phase,

(c) the swinging legs should land reliably on the ground during the second period of the swing phase,

(d) the angular velocity of the supporting legs relative to the ground should be kept constant during their pitching motion or rolling motion around the contact points at the moment of landing or leaving,

(e) the phase difference between rolling motion of the body and pitching motion of legs should be maintained regardless of a disturbance from irregular terrain, and

(f) the phase differences between the legs should be maintained regardless of delay in the pitching motion of a leg receiving a disturbance from irregular terrain.

We design the neural system for these necessary conditions to be satisfied in order to realize adaptive walking.

2.3 Mutual entrainment between neural system and mechanical system

Motion generation based on biological concepts is illustrated as Fig.2, where a neural system and a mechanical system have their own non-linear dynamics. The characteristic of this method is that there is no adaptation through motion planning. These two dynamic systems are coupled to each other, generating motion by interacting with the environment emergingly and adaptively (Taga, Yamaguchi and Shimizu 1991; Taga 1995). We call this method "coupled-dynamics-based motion generation."

The coupled dynamic system can induce autonomous adaptation according to its own dynamics, under changes in the environment (e.g., adaptive walking on irregular terrain) and under adjustment of the neural system parameters by an upper-level controller (e.g., gait transition in change of walking speed). Therefore, we can avoid such serious problems in robotics as modeling of mechanical system and environment, autonomous planning, conflict between planned motion and actual motion and so on.

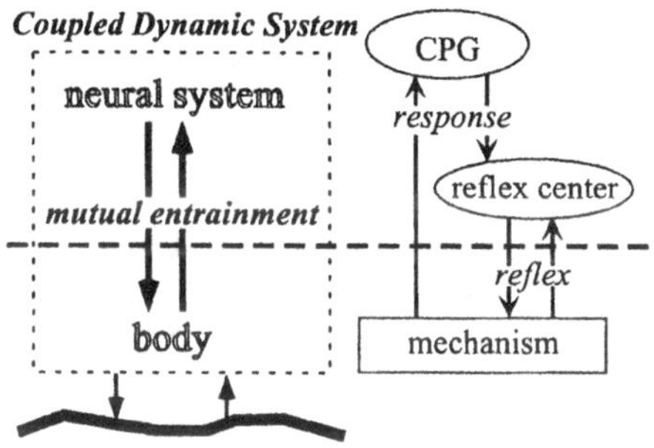

Figure 2: Motion generation in a biological system

2.4 Design of neural system

We construct the neural system centering a neural oscillator as a model of a CPG, since the exchange between the swing and stance phases in the short term and the quick adjustment of these phases on irregular terrain are essential in the dynamic walking of a quadruped where the unstable two-legged stance phase appears. On the other hand, Espenschied et al. (1996) constructed the gait pattern generator proposed by Cruse (1990) referring to a stick insect. They also employed the swaying, stepping, elevator and searching reflexes observed by Pearson et al. (1984) in a stick insect. Consequently, their neural system is more sensor dependent and more decentralized by using a non-oscillator type CPG.

The characteristics of our neural system in comparison with the one used by Espenschied et al. (1996) are:

(1) The cyclic period is mainly determined by the time constant of CPGs. This makes it easy for the necessary condition (a) described in Section 2.2 to be satisfied.

(2) The gait is mainly determined by the connecting weights of the CPGs network.

(3) As sensor feedback for adaptation on irregular terrain, a "response" directly and quickly modulating the CPG phase is employed in parallel with a "reflex" directly generating joint torque (Fig.2).

About the characteristic (3), it is well known in physiology that

- some sensory stimuli modify CPG activity and reflexive responses to sensory stimuli are phase dependent under CPG activity (Cohen and Boothe 1999).

Such interaction between CPG activity and a sensory stimulus is very important for adaptation, and corresponds to the necessary conditions described in physical terms in Section 2.2.

2.5 Design of mechanical system

Each leg of Tekken has a hip pitch joint, a hip yaw joint, a knee pitch joint, and an ankle pitch joint (Fig.1). The direction in which Tekken walks can be changed by using the hip yaw joints. Two rate gyro sensors and two inclinometers are mounted on the body in order to measure the body pitch and roll angles.

In order to obtain appropriate mutual entrainment between neural system and mechanical system, mechanical system should be well designed to have the good dynamic properties. In addition, performance of dynamic walking such as adaptability on irregular terrain, energy efficiency, maximum speed and so on highly depends on the mechanical design. The design concepts of Tekken are:

[a] high power actuators and small inertia moment of legs for quick motion and response,

[b] small gear reduction ratio for high backdrivability to increase passive compliance of joints,

[c] small mass of the lowest link of legs to decrease impact force at collision,

[d] small contacting area at toes to increase adaptability on irregular terrain,

[e] passive ankle joint mechanism to prevent a swinging leg from stumble on an obstacle quickly (Fig.3).

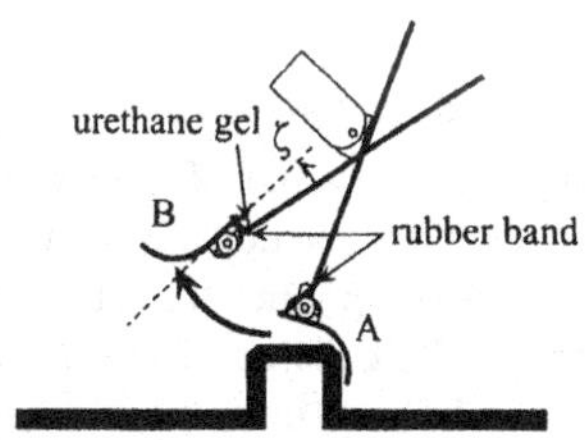

Figure 3: The ankle joint can be passively rotated:(A) if the toe contacts with an obstacle in a swing phase, and is locked:(B) while the leg is in a stance phase.

3. Implementation of neural system for adaptive walking

3.1 Rhythmic motion by CPG

Although actual neurons as a CPG in higher animals have not yet become well known, features of a CPG have been actively studied in biology, physiology, and so on. Several mathematical models were also proposed, and it was pointed out that a CPG has the capability to generate and modulate walking patterns and to be mutually entrained with a rhythmic joint motion (Grillner 1981; Cohen and Boothe 1999; Taga, Yamaguchi and Shimizu 1991; Taga 1995). As a model of a CPG, we used a neural oscillator proposed by Matsuoka (1987), and applied to the biped simulation by Taga et al. (1991; 1995). A single neural oscillator consists of two mutually inhibiting neurons (Fig.4-(a)). Each neuron in this model is represented by the following nonlinear differential equations:

$$\begin{aligned} \tau \dot{u}_{\{e,f\}i} &= -u_{\{e,f\}i} + w_{fe} y_{\{f,e\}i} - \beta v_{\{e,f\}i} \\ &\quad + u_0 + Feed_{\{e,f\}i} + \sum_{j=1}^{n} w_{ij} y_{\{e,f\}j} \\ y_{\{e,f\}i} &= \max\,(u_{\{e,f\}i}, 0) \\ \tau' \dot{v}_{\{e,f\}i} &= -v_{\{e,f\}i} + y_{\{e,f\}i} \end{aligned} \tag{1}$$

where the suffix e, f, and i mean an extensor neuron, a flexor neuron, and the i-th neural oscillator, respectively. $u_{\{e,f\}i}$ is u_{ei} or u_{fi}, that is, the inner state of an extensor neuron or a flexor neuron of the i-th neural oscillator; $v_{\{e,f\}i}$ is a variable representing the degree of the self-inhibition effect of the neuron; y_{ei} and y_{fi} are the output of extensor and flexor neurons; u_0 is an external input with a constant rate; $Feed_{\{e,f\}i}$ is a feedback signal from the robot, that is, a joint angle, angular velocity and so on; and β is a constant representing the degree of the self-inhibition influence on the inner state. The quantities τ and τ' are time constants of $u_{\{e,f\}i}$ and $v_{\{e,f\}i}$; w_{fe} is a connecting weight between flexor and extensor neurons; w_{ij} is a connecting weight between neurons of the i-th and j-th neural oscillator.

In Fig.4-(a), the output of a CPG is a phase signal: y_i

$$y_i = -y_{ei} + y_{fi} \tag{2}$$

The positive or negative value of y_i corresponds to activity of a flexor or extensor neuron, respectively.

We use the following hip joint angle feedback as a basic sensory input to a CPG called a "tonic stretch response" in all experiments of this study. This negative feedback makes a CPG be entrained with a rhythmic hip joint motion.

$$Feed_{e \cdot tsr} = k_{tsr}(\theta - \theta_0), \quad Feed_{f \cdot tsr} = -Feed_{e \cdot tsr} \tag{3}$$

$$Feed_{\{e,f\}} = Feed_{\{e,f\} \cdot tsr} \tag{4}$$

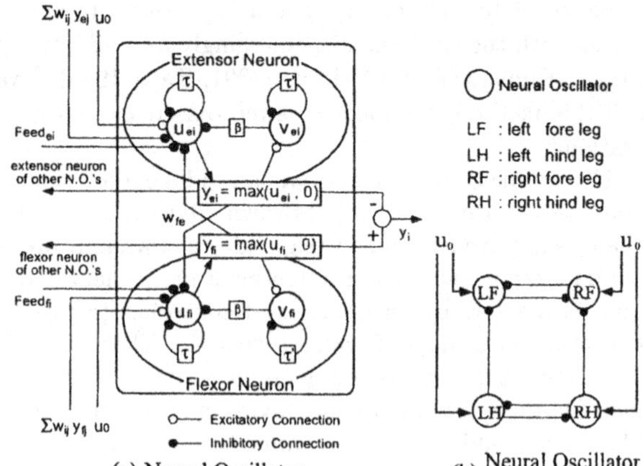

(a) Neural Oscillator (b) Neural Oscillator Network

Figure 4: Neural oscillator as a model of a CPG. The suffix $i, j = 1, 2, 3, 4$ corresponds to LF, LH, RF, RH. L, R, F or H means the left, right, fore or hind leg, respectively.

where θ is the measured hip joint angle, θ_0 is the origin of the hip joint angle in standing and k_{tsr} is the feedback gain. We eliminate the suffix i when we consider a single neural oscillator.

By connecting the CPG of each leg (Fig.4-(b)), CPGs are mutually entrained and oscillate in the same period and with a fixed phase difference. This mutual entrainment between the CPGs of the legs results in a gait. The gait is a walking pattern, and can be defined by phase differences between the legs during their pitching motion. The typical symmetric gaits are a trot and a pace. Diagonal legs and lateral legs are paired and move together in a trot gait and a pace gait, respectively. A walk gait is the transversal gait between the trot and pace gaits. We used a trot gait for most of experiments. But the autonomous gait transition in changing walking speed was discussed in our former study (Fukuoka et al. 2003).

3.2 Virtual spring-damper system

We employ the model of the muscle stiffness, which is generated by the stretch reflex and variable according to the stance/swing phases, adjusted by the neural system. The muscle stiffness is high in a stance phase for supporting a body against the gravity and low in a swing phase for compliance against the disturbance (Akazawa et al. 1982). All joints of Tekken are PD controlled to move to their desired angles in each of three states (A, B, C) in Fig.5 in order to generate each motion such as swinging up (A), swinging forward (B) and pulling down/back of a supporting leg (C). The timing for all joints of a leg to switch to the next state are:

$A \rightarrow B$: when the hip joint angle of the leg reaches the desired angle of the state (A)

$B \to C$: when the CPG extensor neuron of the leg becomes active ($y_i \leq 0$)

$C \to A$: when the CPG flexor neuron of the leg becomes active ($y_i > 0$)

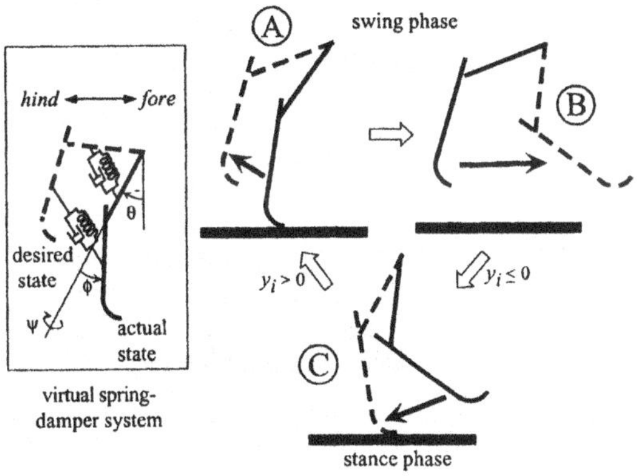

Figure 5: State transition in the virtual spring-damper system. The desired joint angles in each state are shown by the broken lines.

The desired angles and P-gain of each joint in each state are shown in Table 2, where constant values of the desired joint angles and constant P-gains were determined through experiments. Since Tekken has high backdrivability with small gear ratio in each joint, PD-controller can construct the virtual spring-damper system with relatively low stiffness coupled with the mechanical system. Such compliant joints of legs can improve the passive adaptability on irregular terrain.

3.3 CPGs and pitching motion of legs

The diagram of the pitching motion control consisting of CPGs and the virtual spring-damper system is shown in the middle part of Fig.6. Joint torque of all joints is determined by the PD controller, corresponding to a stretch reflex at an α motor neuron in animals. The desired angle and P-gain of each joint is switched based on the phase of the CPG output: y_i in Eq.(2) as described in Section 3.2. As a result of the switching of the virtual spring-damper system and the joint angle feedback signal to the CPG in Eq.(4), the CPG and the pitching motion of the leg are mutually entrained.

3.4 Reflexes and responses

It is known in physiology that the reflex to a stimulus on the paw dorsum in the walking of a cat depends on whether flexor or extensor muscles are active (as reviewed in Cohen and Boothe 1999). That is,

- when flexor muscles are active, the leg is flexed in order to escape from the stimulus,

Table 2: Desired value of the joint angles and P-gains at the joints used in the PD-controller for the virtual spring-damper system in each state shown in Fig.5. θ, ϕ and ψ are the hip pitch joint angle, the knee pitch joint angle and the hip yaw joint angle, respectively.

	P control	
angle in state	desired value[rad]	P-gain[Nm/rad]
θ in A	$1.2\theta_{C\to A}$	G_1
θ in B	-0.17	$G_2 v + G_3$
θ in C	$\theta_{stance}+$ body pitch angle	$-G_4 v + G_5$
ϕ in A & B	$*$	G_6
ϕ in C	0.61	G_7
ψ in all states	0	G_8

$\theta_{C\to A}$: the hip joint angle measured at the instance when the state changes from (C) to (A).
θ_{stance}: variable to change the walking speed.
body pitch angle: the measured pitching angle of the body used for the vestibulospinal reflex.
$*$ means that the desired angle is calculated on-line for the height from the toe to the hip joint to be constant.
v [m/s]: the measured walking speed of Tekken.

- when extensor muscles are active, the leg is strongly extended in order to prevent the cat from falling down.

We call these a "flexor reflex" and a "extensor reflex," respectively, and we assume that the phase signal from the CPG of the leg switches such reflexes.

The following biological concepts are also known (Ogawa et al. 1998):

- when the vestibule in a head detects an inclination in pitch or roll plane, a downward-inclined leg is extended while an upward-inclined leg is flexed (Fig.7).

We call this a "vestibulospinal reflex."

Referring to such biological knowledge, we employed the several reflexes and responses (Fig.6) to satisfy the necessary conditions (b)~(e) described in physical terms in Section 2.2 in addition to the stretch reflex and response described in Section 3.2 and 3.1. The necessary condition (f) can be satisfied by the mutual entrainment between CPGs and the pitching motion of legs, and the mutual entrainment among CPGs (Kimura et al. 2001).

3.4.1 Flexor and extensor reflexes

The flexor and extensor reflexes contribute to satisfy the conditions (b) and (c), respectively. In our former studies (Kimura et al. 1999, 2001), the stumble of a swinging

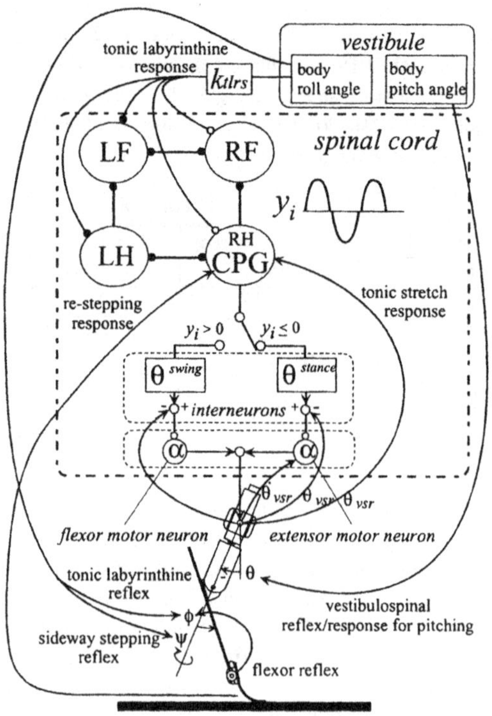

Figure 6: Control diagram for Tekken. PD-control at the hip yaw and knee pitch joints are eliminated in this figure.

leg on an obstacle was detected by force sensor, and the flexor or extensor reflex was activated afterwards. However, the problem was that the robot typically fell down due to the delayed flexing motion caused by the delay of sensing and large inertia of the leg while walking with short cyclic period. Therefore, we substitute the flexor reflex for the passive ankle joint mechanism (Fig.3) utilizing the fact that the collision with a forward obstacle occurs in the first half of a swing phase. The extensor reflex has not yet been implemented in Tekken.

3.4.2 Vestibulospinal reflex and response for pitching

We call the reflex/response for an inclination in the pitch plane a "vestibulospinal reflex/response for pitching," the role of which corresponds to the condition (d) and (f). In Tekken, hip joint torque in the stance phase is adjusted by the vestibulospinal reflex (Fig.7-(a)), since the body pitch angle is added to θ_{stance} in Table 2. For the vestibulospinal response, the following equations are used rather than Eq.(3), (4).

$$\begin{aligned} \theta_{vsr} &= \theta - (\text{body pitch angle}) \\ Feed_{\{e,f\}\cdot tsr\cdot vsr} &= \pm k_{tsr}(\theta_{vsr} - \theta_0). \end{aligned} \quad (5)$$

$$Feed_{\{e,f\}} = Feed_{\{e,f\}\cdot tsr\cdot vsr} \quad (6)$$

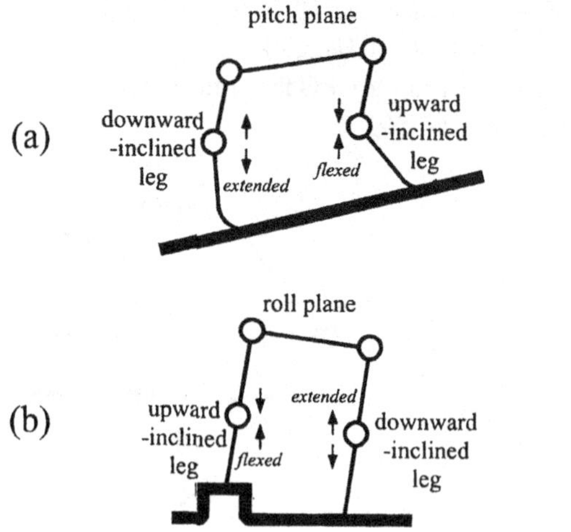

Figure 7: Downward-inclined and upward-inclined legs

3.4.3 Vestibulospinal reflex and response for rolling

We call the reflex/response for an inclination in the roll plane a "tonic labyrinthine reflex/response[1] ."

- *Tonic labyrinthine reflex*

The tonic labyrinthine reflex (TLRF) is employed as the adjustment of P-gain of the knee joint of the supporting legs (G_7 in Table 2) according to Eq.(7).

$$\tilde{G}_7 = \delta(leg)\ k_{tlrf} \times (\text{body roll angle}) + G_7 \quad (7)$$

$$\delta(leg) = \begin{cases} 1, & \text{if } leg \text{ is a right leg;} \\ -1, & \text{otherwise} \end{cases}$$

When an inclination of a body in roll plane (body roll angle) is detected, the knee joint P-gain of the downward-inclined legs is increased to extend those legs. In addition, the knee joint P-gain of the upward-inclined legs is decreased to flex those lges. As a result, the inclination of the body in roll plane is decreased (Fig.7-(b)).

- *Tonic labyrinthine response*

On the other hand, the tonic labyrinthine response (TLRS) is employed as rolling motion feedback to CPGs (upper left part of Fig.6). In Tekken, we made the body roll angle be inputted to the CPGs as a feedback signal expressed by Eq.(8), (9).

$$\begin{aligned} Feed_{e\cdot tlrs} &= \delta(leg)\ k_{tlrs} \times (\text{body roll angle}) \\ Feed_{f\cdot tlrs} &= -Feed_{e\cdot tlrs} \end{aligned} \quad (8)$$

$$\begin{aligned} Feed_e &= Feed_{e\cdot tsr\cdot vsr} + Feed_{e\cdot tlrs} \\ Feed_f &= Feed_{f\cdot tsr\cdot vsr} + Feed_{f\cdot tlrs} \end{aligned} \quad (9)$$

[1]The "tonic labyrinthine reflex" is defined in Ogawa et al. (1998). The same reflex is called "vestibular reflex" in Ghez (1991).

The rolling motion feedback to CPGs: Eq.(8), (9) contributes to an appropriate adjustment of the periods of the stance and swing phases while walking on irregular terrain (Fukuoka et al. 2003).

The rolling motion feedback to CPGs is important also in order to satisfy the condition (e). CPGs, the pitching motions of the legs, and the rolling motion of the body are mutually entrained through the rolling motion feedback to CPGs (Fig.8). This means that the rolling motion can be the standard oscillation for whole oscillations, in order to compensate for the weak connection between the fore and hind legs in the CPG network (Fig.4-(b)). As a result, the phase difference between the fore and hind legs is fixed, and the gait becomes stable.

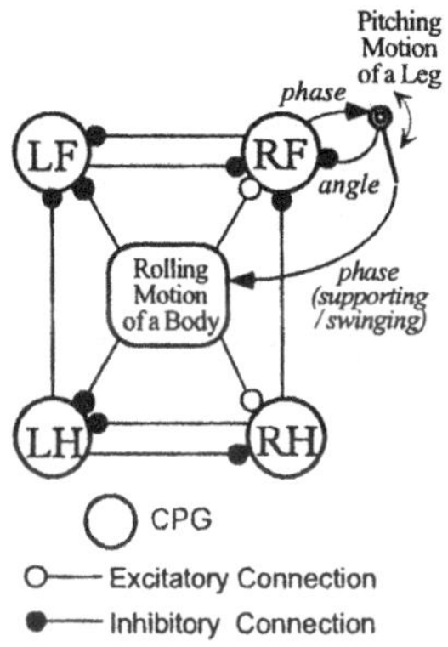

Figure 8: Relation among CPGs, pitching motion of a leg and rolling motion of a body

3.4.4 Stepping reflex to stabilize rolling motion

It is known that the adjustment of the sideway touch-down angle of a swinging leg is effective in stabilizing rolling motion against disturbances (Miura and Shimoyama 1994). We call this a "sideway stepping reflex," which helps to satisfy the condition (d) during rolling motion. The sideway stepping reflex is effective also in walking on a sideway inclined slope.

For examples, when Tekken walks on a right-inclined slope (Fig.9), Tekken continues to walk while keeping the phase differences between left and right lges with the help of the tonic labyrinthine response. But Tekken cannot walk straight and shifts its walking direction to the right due to the difference of the gravity load between left and right legs. In addition, Tekken typically falls down to the right for the perturbation from the left in the case of Fig.9-(a), since the wide stability margin: WSM [2] is small. The sideway stepping reflex helps to stabilize the walking direction and to prevent the robot from falling down while keeping WSM large on such sideway inclined slope (Fig.9-(b)).

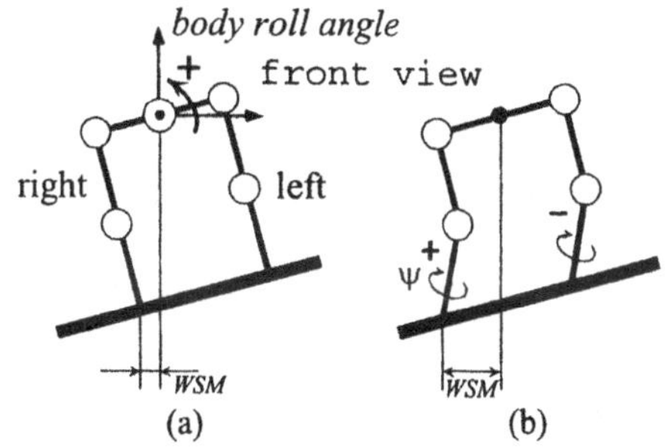

Figure 9: Walking on a sideway inclined slope. (a):Without a sideway stepping reflex, (b):With.

Since Tekken has no joint round the roll axis, the sideway stepping reflex is implemented as changing the desired angle of the hip yaw joint from 0 (Table 2) to ψ^* according to Eq.(10).

$$\psi^* = \delta(leg)\ k_{stpr} \times (\text{body roll angle}) \tag{10}$$

3.4.5 Stepping reflex to stabilize forward speed

In Table 2, P-gain of hip joints in the state (B) and (C) is adjusted using the measured walking speed: v. For examples, when v is increased in the state (B) as the swing phase, the swinging legs land on the forwarder position because P-gain is increased and the swinging legs are more strongly pulled forward. As a result, the increase of v is depressed since the forward motion of the inverted pendulum in the next stance phase is depressed by the gravity (Miura and Shimoyama 1994). On the other hand, when v is decreased in the state (C) as the stance phase while walking up a slope, the increase of P-gain generates additional torque against the gravity load at hip joints of the supporting legs. As a result, the decrease of v is depressed. These reflexes contribute to satisfy the condition (d).

3.4.6 Re-stepping reflex and response for walking down a step

When loss of ground contact is detected in a swing phase while walking over a ditch, a cat activates re-stepping to extend the swing phase and make the leg land on the forwarder position (Hiebert et al. 1994). We call this "re-stepping reflex/response," which is effective for the necessary condition (c) and (d) to be satisfied also in walking down a large step (Fig.10).

[2] the shortest distance from the projected point of the center of gravity to the edges of the polygon constructed by the projected points of legs independent of their stance or swing phases (Fukuoka et al. 2003)

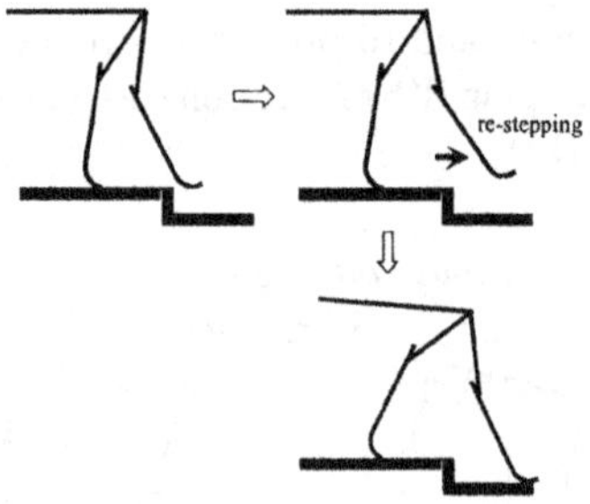

Figure 10: Re-stepping reflex and response

4. Experiments

4.1 Experiments under instantaneous disturbances

We made Tekken1 walk on several terrains of medium degree of irregularity in indoor environment (Fig.11) with all responses and five reflexes, those are, the stretch reflex, flexor reflex, vestibulospinal reflex for pitching, stepping reflex for forward speed and re-stepping reflex, among seven reflexes described in Section *3.4*. This means that Tekken can cope with the instantaneous disturbances while walking over irregular terrain consisting of series of steps and short slopes. Without a tonic labyrinthine response, the gait was greatly disturbed on irregular terrain, even if Tekken1 didn't fall down.

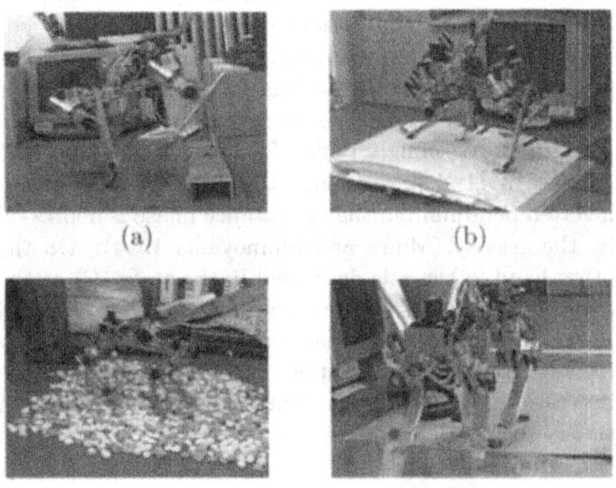

Figure 11: Walking over a step 4 [cm] in height: (a), walking up and down a slope of 10 [deg] in a forward direction: (b), walking over pebbles: (c) and walking over sideway slopes of 3 and 5 [deg]: (d)

Especially noteworthy, Tekken1 successfully walked down a large step with approx. 0.5 [m/s] speed using the re-stepping reflex/response (Fig.12). Without the re-stepping reflex/response, Tekken1 typically fell down forward because fore legs landed on the backwarder position excessively and could not depress the increased forward speed.

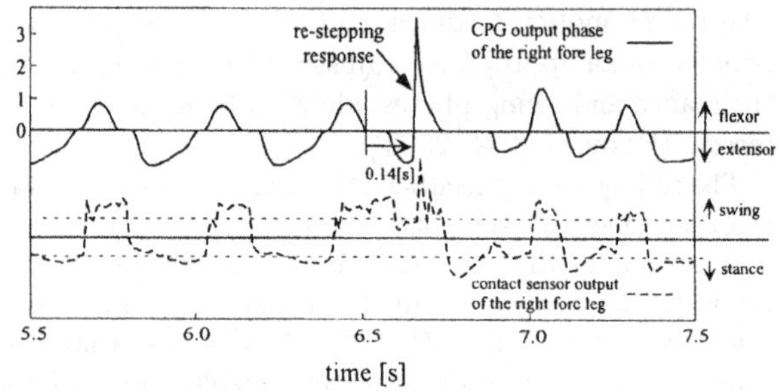

Figure 12: Walking down a step of 7 [cm] in height. A re-stepping response was activated when the contact of the right fore leg had not been detected for 0.14 [s] after the activity of the flexor neuron became zero.

4.2 Experiments under long-lasting disturbances

We made Tekken2 walk on a right-inclined slope of 4 [deg] (0.07 [rad]) in indoor environment with all responses and reflexes described in Section *3.4* in order to confirm the effectiveness of a tonic labyrinthine reflex and a sideway stepping reflex (Fig.9-(b)) under long-lasting disturbances. As a result of the experiment, the body roll angle and the hip yaw angle ψ of the right hindleg and left hindleg are shown in Fig.13. Tekken2 had walked on the right-inclined slope for 2 [sec], and walked on flat terrain afterwards.

In Fig.13, we can see that the body roll angle was positive (0.05~0.14 [rad]) while walking on the right-inclined slope. The hip yaw joint of the right hindleg moved to the outside of the body (right) by approx. 0.04 [rad] due to the sideway stepping reflex in the swing phase and moved to the inside of the body by approx. -0.04 [rad] due to the gravity load in the stance phase. On the other hand, the hip yaw joint of the left hindleg moved to the inside of the body (right) by approx. -0.04 [rad] due to the sideway stepping reflex in the swing phase and moved to the inside of the body furthermore by approx. -0.1 [rad] due to the gravity load in the stance phase. Consequently, Tekken2 succeeded in straight walking on the sideway inclined slope.

4.3 Outdoor experiments using Tekken2

Even on a paved road in outdoor environment, there exist a slope of 3 [deg] at most, bumps of 1 [cm] in height and small pebbles everywhere. With all responses and reflexes described in Section *3.4*, Tekken2 successfully maintained a stable gait on the paved road for 60 [sec]

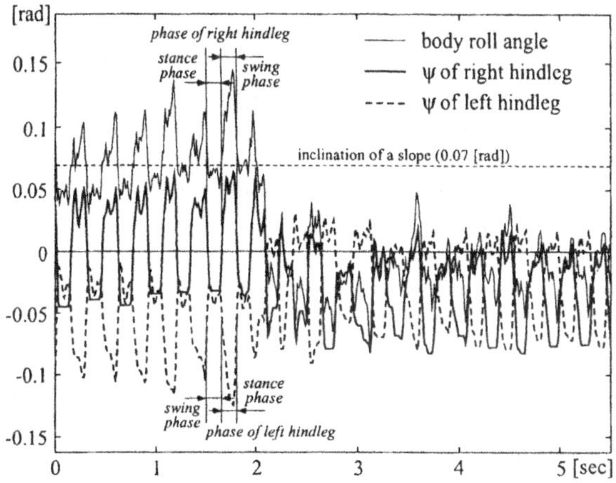

Figure 13: Walking on a right-inclined slope of 4 [deg]

Figure 14: Photos of walking in outdoor environment

with approx. 0.5 [m/s] speed while changing its walking speed and direction by receiving the operation commands from the radio controller (Fig.14).

5. Discussion : how to determine the values of parameters

Values of all parameters in the neural system including the virtual spring-damper system except for θ_{stance} were determined experimentally. But it should be noted that those values in each robot were fixed in all experiments independent of kinds of terrain. Although the size and weight of Tekken2 are different from those of Tekken1, the values of the parameters of CPGs used for Tekken2 were same with those used for Tekken1.

The most important subject in coupled-dynamics-based motion generation is to design and construct a neural system carefully, while taking into account the dynamics of a mechanical system and its interaction with the environment. In this study, we designed a neural system consisting of CPGs, responses and reflexes, while taking into account the characteristics of dynamic walking and utilizing knowledge and concepts in physics, biology, physiology and so on described in Section **2.** and **3.**. The relationship between parameters of CPGs and the mechanical system was previously analyzed to some extent (Kimura et al. 2001). Since the relationship between parameters of reflexes/responses and the mechanical system has not yet been investigated, we manually determined values of those parameters through experiments. The issue of how to construct a neural system suitable for a mechanical system corresponds to the issue of "embodiment" at the lowest level of sensorimotor coordination.

In addition, we pointed out the trade-off problem between the stability and the energy consumption in determining the cyclic period of walking on irregular terrain, and showed one example to solve this problem (Fukuoka et al. 2003).

6. Conclusion

In the neural system model proposed in this study, the relationships among CPGs, sensory input, reflexes and the mechanical system are simply defined, and motion generation and adaptation are emergingly induced by the coupled dynamics of a neural system and a mechanical system by interacting with the environment. To generate appropriate adaptation, it is necessary to design both the neural system and the mechanical system carefully. In this study, we designed the neural system consisting of CPGs, responses, and reflexes referring to biological concepts while taking the necessary conditions for adaptive walking into account.

We newly employed a tonic labyrinthine reflex, a sideway stepping reflex, and a re-stepping reflex/response to make the self-contained quadruped robot (Tekken2) walk in outdoor environment. In order to increase the degrees of terrain irregularity which Tekken2 can cope with, we should employ additonal reflexes and responses, and also navigation ability at the high level using vision.

MPEG footage of these experiments can be seen at: *http://www.kimura.is.uec.ac.jp.*

References

Akazawa, K., Aldridge, J. W., Steeves, J. D., and Stein, R.B. 1982. Modulation of stretch reflexes during locomotion in the mesencephalic cat. *J. of Physiol.* 329:553-567.

Buehler, M., Battaglia, R. , Cocosco, A. , Hawker, G., Sarkis, J., and Yamazaki, K. 1998. Scout: A simple quadruped that walks, climbs and runs. *Proc. of ICRA1998*, pp. 1707–1712.

Cham, J. G., Bailey, S. A., Clark J. E., Full, R. J., Cutkosky, M. R. 2002. Fast and Robust: Hexapedal Robots via Shape Deposition Manufacturing, *Int. J. Robotics Research*, 21(10-11):869–882.

Chew, C.M., Pratt, J., and Pratt, G. 1999. Blind walking of a planar bipedal robot on sloped terrain, *Proc. of ICRA99*, pp. 381–386.

Cohen, A. H., and Boothe, D. L. 1999. Sensorimotor interactions during locomotion: principles derived from biological systems. *Autonomous Robots* 7(3):239–245.

Cruse, H. 1990, What mechanism coordinate leg movement in walking arthropods?, *Trends in Neurosciences*, 12:15–21.

Espenschied, K.S., Quinn, R.D., Beer, R.D., and Chiel, H.J. 1996. Biologically based distributed control and local reflexes improve rough terrain locomotion in a hexapod robot. *Robotics and Autonomous Systems* 18:59–64.

Fukuoka, Y., Kimura, H., and Cohen, A. H. 2001. 2003. Adaptive Dynamic Walking of a Quadruped Robot on Irregular Terrain based on Biological Concepts, *Int. Journal of Robotics Research* 22(3-4):187–202.

Full, R. J., and Koditschek, D. E. 1999. Templates and anchors: neuromechanical hypotheses of legged locomotion on land. *J. Exp. Biol.* 202:3325–3332.

Ghez, C. 1991. Posture, *Principles of Neural Science* 3rd edn, Kandel, E.R., Schwartz, J.H, and Jessell, T.M., eds, Appleton & Lange, Norwalk, CT, p. 601.

Grillner, S. 1981. Control of locomotion in bipeds, tetrapods and fish. *Handbook of Physiology II* American Physiol. Society, Bethesda, MD, pp. 1179–1236.

Hiebert, G., Gorassini, M., Jiang, W., Prochazka, A., and Pearson, K. 1994. Corrective responses to loss of ground support during walking II, comparison of intact and chronic spinal cats, *J of Neurophys.* 71:611–622.

Hodgins, J. K., and Raibert, M. H. 1991. Adjusting step length for rough terrain locomotion. *IEEE trans. on Robotics and Automation* 7(3):289–298.

Ijspeert, A. J., Hallam, J., and Willshaw, D. 1998. From lampreys to salamanders: evolving neural controllers for swimming and walking. *Proc. of SAB1998*, pp. 390–399.

Ijspeert, A. J. 2001. A connectionist central pattern generator for the aquatic and terrestrial gaits of a simulated salamander. *Biolog. Cybern.* 84:331–348.

Ilg, W., Albiez, J., Jedele, H., Berns, K., and Dillmann, R. 1999. Adaptive periodic movement control for the four legged walking machine BISAM. *Proc. of ICRA1999*, pp. 2354–2359.

Kajita, S., and Tani, K. 1996. Adaptive gait control of a biped robot based on realtime sensing of the ground. *Proc. of ICRA1996*, pp. 570–577.

Kimura, H., Shimoyama, I., and Miura, H. 1990. Dynamics in the dynamic walk of a quadruped robot. *Advanced Robotics* 4(3):283–301.

Kimura, H., Akiyama, S., and Sakurama, K. 1999. Realization of dynamic walking and running of the quadruped using neural oscillator. *Autonomous Robots* 7(3):247–258.

Kimura, H., Fukuoka, Y., and Konaga, K. 2001. Adaptive dynamic walking of a quadruped robot using neural system model. *Advanced Robotics* 15(8):859–876.

Lewis, M. A., Etienne-Cummings, Hartmann, M. J., Xu, Z. R., and Cohen, A. H. 2003. An in silico central pattern generator: silicon oscillator, coupling, entrainment, and physical computation. *Biolog. Cybern.* 88:137–151.

Matsuoka, K. 1987. Mechanisms of frequency and pattern control in the neural rhythm generators. *Biolog. Cybern.* 56:345-353.

Miura, H., and Shimoyama, I. 1984. Dynamical walk of biped locomotion. *Int. J. Robotics Research* 3(2):60–74.

Miyakoshi, S., Taga, G., Kuniyoshi, Y., and Nagakubo, A. 1998. Three dimensional bipedal stepping motion using neural oscillators - towards humanoid motion in the real world. *Proc. of IRSO1998*, pp. 84–89.

Ogawa, T. et al., eds. 1998. *Nanzando's Medical Dictionary* 18th edn, Nanzando, Tokyo, p. 1211. (in Japanese).

Pearson, K.G., and Franklin, R. 1984, Characteristics of leg movements and patterns of coordination in locusts walking on rough teerain, Int. J. Robotics Research, 3:101–112.

Saranli, U., Buehler, M., and Koditschek, D. E. 2001. RHex: a simple and highly mobile hexapod robot. *Int. J. Robotics Research* 20(7):616–631.

Taga, G., Yamaguchi, Y., and Shimizu, H. 1991. Self-organized control of bipedal locomotion by neural oscillators. *Biolog. Cybern.* 65:147–159.

Taga, G. 1995. A model of the neuro-musculo-skeletal system for human locomotion II. - real-time adaptability under various constraints. *Biolog. Cybern.* 73:113–121.

Takanishi, A., Takeya, T., Karaki, H., and Kato, I. 1990. A control method for dynamic biped walking under unknown external force. *Proc. of IRSO1990*, pp. 795–801.

Tsujita, K., Tsuchiya, K., and Onat, A. 2001. Adaptive Gait Pattern Control of a QuadrupedLocomotion Robot, *Proc. of IROS2001*, pp. 2318–2325.

Yamaguchi, J., Takanishi, A., and Kato, I. 1994. Development of a biped walking robot adapting to a horizontally uneven surface. *Proc. of IRSO1994*, pp. 1156–1163.

Yoneda, K., Iiyama, H., and Hirose, S. 1994. Sky-Hook suspension control of a quadruped walking vehicle. *Proc. of ICRA1994*, pp. 999–1004.

ACTION SELECTION AND BEHAVIORAL SEQUENCES

Action Selection and Mental Transformation Based on a Chain of Forward Models

Heiko Hoffmann* **and Ralf Möller****

*Cognitive Robotics
Department of Psychology
Max Planck Institute for
Human Cognitive and Brain Sciences
80799 Munich, Germany
heiko.hoffmann@cbs.mpg.de

**Computer Engineering Group
Faculty of Technology
Bielefeld University
33594 Bielefeld, Germany
moeller@techfak.uni-bielefeld.de

Abstract

This study investigates whether goal-directed action selection and spatial mental transformations can be accomplished based on a chain of concatenated forward models. A forward model was trained with visual and motor data obtained from a mobile robot. A chain of identical forward models allowed predictions beyond one time step. We studied the accumulation of prediction errors theoretically and experimentally, and applied the chain model for two tasks. First, by applying an optimization procedure to the chain, a sequence of motor commands was determined that leads from an initial to a goal state. Second, we used the model for a mental transformation task driven by simulated motor commands.

1. Introduction

Forward models are internal sensorimotor models that predict how the sensory situation changes as a result of an agent's actions. The concept of forward models was originally introduced to motor control. However, recent results and new conceptual ideas broadly extended the range of cognitive capabilities in which internal sensorimotor models and specifically forward models are supposed to be involved. Wolpert et al. (2003) suggest that action observation, imitation, social interaction, and even 'theory of mind' could be based on these internal models. Hesslow (2002) presents a 'simulation theory of cognitive function', which 'postulates the existence of an associative mechanism that enables the preparatory stages of an action to elicit sensory activity that resembles the activity normally caused by the completed overt behavior'. Forward models could also play a role in spatial perception: understanding the behavioral meaning of an object may be an internal simulation of the sensory consequences of actions targeted at this object (Möller, 1999).

In this paper, we explore the capabilities of chaining forward models - concatenating input and output of two consecutive stages (see also Hesslow (2002), Jirenhed et al. (2001), and Tani (1996) for using a chain of forward models for prediction). Instead of trying to model humans or animals, we construct a simple model capable of performing goal-directed movements and mental transformation in a real environment. In this way, we hope to enlighten these 'human' capabilities. Further, we restrict our study by excluding any social ability like imitation or theory of mind - in that context, forward models were discussed by Wolpert et al. (2003) and Grush (2003), and implemented by Demiris and Hayes (2002).

Goal-directed planning usually requires not only the selection of a single motor command, but of a whole sequence. We suggest to understand the search for a motor sequence as an optimization in a chain of identical forward models. The input to the chain is the current sensory situation. The optimization criterion is that the sequence of actions results in an output of the last stage in the chain that matches the goal-situation. The free parameters are the motor commands for each stage. The idea is that the perceptual judgment leading to the selection of actions is not only based on visual data, but on visual data in the context of one's own behavior. We will show that the chain of forward models enables the selection of action sequences leading to a goal.

The chain can be further used for mental transformations. Here, the motor commands are given, and the forward model is used to predict the resulting sensory changes. However, the motor commands are not executed, but only drive the mental transformation of sensory situations. Forward models in this context were also discussed by Hesslow (2002) and Grush (2003). Here, we show that the knowledge of the sensorimotor relationships (incorporated in a trained forward model) can be the basis of perceptual judgment. Perceptual distortion is widespread in biological visual systems, for example due to inhomogeneous distribution of receptors or properties of the optical projection. Lines on a picture mostly do not look like lines on the retina or on the visual cortex. Thus, associating constant properties in the external world with a representation of the constancy may be difficult without a teacher who marks the changing image features as belonging to the same physical property. However, associating the visual changes with the movements causing these changes could be a means to accomplish perception of constancy. O'Regan and Noë (2001) suggested that the 'mastery of sensorimotor contingencies governing visual exploration' is the basis of visual perception. This seems similar to the above ideas. Different from our approach, O'Regan and Noë (2001) did not use forward models and robot experiments in their studies.

For our experiments, we use a mobile robot with omni-directional vision. We deliberately select a simple experimental setup where the robot is surrounded by a circle of obstacles. Full panoramic vision and the specific setup guarantee that all relevant visual information is available. Moreover, depth information can be directly extracted from a single image. Therefore, no memory is necessary, but the setup is restricted. Other robot studies working with a chain of forward models (Tani, 1996; Jirenhed et al., 2001) use recurrent neural networks with context layers. These allow more complex environments. But, omitting context units allows us to do predictions without the need to move the robot to initialize the context values. Omitting memory will further ease the theoretical investigation.

Two tasks are studied. The first task is an action selection task, where the robot's goal is to get close to an obstacle in a predefined direction. In the second task, mental transformation is used to decide if the robot is in the middle of the circle or not. This is based on the internal simulation of a rotation of the robot. The location of the omni-directional vision system is chosen such that it does not coincide with the robot's rotational axis. Otherwise, this would make the task trivial from the perspective of direct image processing.

Training and test are separated. Training data are collected by random exploration. The forward model is realized by a multilayer perceptron (MLP) (see Jordan and Rumelhart (1992) for MLPs used as forward models). The network associates a motor velocity vector (comprising the velocity of left and right side of wheels) with a change in the preprocessed and compressed visual information. A chain of concatenated MLPs enables the prediction of a change in sensory situation induced by a sequence of short movements with given velocity vectors. Since the accumulation of prediction errors can be critical, the increase of the error with increasing chain length is studied theoretically and compared with the performance of the MLP.

The remainder of this article is organized as follows: section 2 describes the methods used, the robot setup, the data collection, the image processing, the forward model network, the optimization methods for the goal-directed movements, and the mental transformation task. Section 3 shows the results of the robot experiments, which are discussed in section 4. Section 5 draws conclusions. Appendix A describes theoretically the accumulation of prediction errors.

2. Methods

2.1 Robot setup

We used a *Pioneer 2 AT* four-wheel robot from *ActivMedia Robotics* (see Fig. 1). It has differential steering and was equipped with a panoramic vision system based on an omni-directional hyperbolic mirror (middle size, wide view) from *Accowle* (Fig. 2)[1]. The camera's optical axis was positioned 12 cm in front of the robot's rotational axis. Images were grabbed at a resolution of 640 × 480 pixels. A circular shaped cover on top of the mirror prevented light entering directly into the lens without reflexion from the mirror. The illumination of the room was kept constant for all of the training and tests.

Figure 1: Pioneer robot with omni-directional camera surrounded by 15 red obstacles

2.2 Collection of training data

Training data were collected by random exploration. The goal was to record the changing images induced by a given motor command. The robot was put within a circle with an inner diameter of 180 ± 2 cm formed by red bricks (Fig. 1). A random velocity was chosen for the left and right side of wheels individually (v_L and v_R). The velocities ranged from −60 mm/sec to 60 mm/sec in steps of 20 mm/sec. The combination with both veloc-

[1] A *DFK 4303/P* camera and a *Pentax TS2V314A* lens were used.

Figure 2: Panoramic vision system

ities being zero was discarded. After a set of velocities was chosen, the robot maintained the given speed. Every two seconds, an image was recorded from the camera (Fig. 3) and stored. Recording started after guaranteeing the robot a one second acceleration phase. The series lasted up to a maximum of six shots (five 2 sec intervals) or until the robot got too close to one of the obstacles (this was determined with the help of the same kind of image preprocessing as described in section 2.4). In the first case, a new combination of velocities was chosen as above, and a new recording series started. In the latter case, the robot was either allowed to go only forward or backward, depending on if the obstacle was in the back or front, respectively. Forward movements were chosen randomly from a subset that fulfills $v_L v_R \geq 0$ and $v_L + v_R > 0$. Backward movements were chosen in an analogue way (here $v_L + v_R < 0$). This training scheme would result in more forward and backward movements vs. rotational movements (which fulfill $v_L v_R < 0$) because when the robot was next to the obstacles it was not allowed to do turns. Therefore, when rotation was possible, the rotational movements were chosen with a higher probability to adjust toward a balanced distribution of velocity combinations.

The actual wheel velocity was recorded during the 2 sec intervals. If it deviated by more than 10 mm/sec from the given value the series was stopped, and the interval's was data discarded. After that, a new series started, as above. The robot was able to pursue this kind of random exploration automatically without getting into physical contact with any obstacle.

Totally, 5466 intervals with 6808 images were recorded. The velocities were roughly evenly distributed (Fig. 4). There was a slight dominance for straight movements ($v_L = v_R$) and a lack for combinations like $v_L = 0$ and $v_R = 20$ mm/sec. The reason is that the front and back wheels are connected by a chain, making the robot occasionally stick to the floor during slow turns.

Figure 3: Image as seen through the mirror (part of the training set)

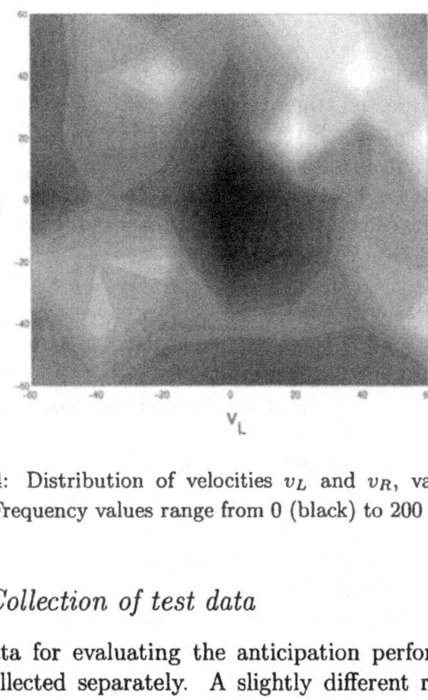

Figure 4: Distribution of velocities v_L and v_R, values in mm/s. Frequency values range from 0 (black) to 200 (white)

2.3 Collection of test data

Test data for evaluating the anticipation performance were collected separately. A slightly different random exploration scheme was used. The goal was to get random movement sequences instead of series with a constant motor command. For each interval, a new random velocity combination was used. A recording series consisted of eight 2 sec intervals starting from zero velocity. The first interval was discarded, thus leaving seven intervals each under identical conditions. The wheel velocity was monitored in the last 0.7 sec of each interval. If its mean value deviated by more then 10 mm/sec from the given value the whole recording series was discarded. Therefore, the limited acceleration forced us to make the random choice of velocities slightly dependent on the previous choice. Absolute velocity changes for each wheel by 80 mm/sec or more were not allowed.

The choice of velocities in each interval further depends on the encounters with obstacles. If the obstacle was in the front the robot moved backward, and if it was in the back the robot moved forward (same way as done for the training data). Additionally, the robot responded to an obstacle on the left or right side by turning the front of the robot toward the obstacles (choosing $|v_L| > |v_R|$, $v_L > 0$, and $v_R < 0$ for obstacles on the right side, and accordingly for the left side, changing roles of v_L and v_R). This behavior made the robot facing the obstacles with the front side, and thus, it was followed by a backward movement. Due to the location of the mirror in the frontal part of the robot (see Fig. 1), the bottom part of obstacles close to the rear part of the robot were occluded. Thus, this required to keep the back-side at a greater distance to the obstacles.

Totally, 138 series were recorded, with a total of 966 intervals and 1104 images.

2.4 Image processing

It proved to be impossible to use the original visual information in the training for the following reasons. First, the dimensionality is too high, and second, the pixels in the image change too rapidly. Therefore, the image was preprocessed to detect only a special class of objects and extract only a visual distance information in a few sectors. Image processing contained the following steps:

First, a contrast mechanism enhanced red objects ($R-(G+B)/2$). The result was smoothened with a binomial filter. Then, a threshold function was applied on all pixel values. In the resulting image, the red obstacles appear as white region on an otherwise black background (Fig. 5).

In 10 sectors (36° each), the distance from the center of the robot (within the image) to the closest object was determined (Fig. 5). These 10 distance values form the final representation of the sensory input to be processed by the network (Fig. 6).

The 'motor commands' v_L and v_R together with two corresponding image representations (as in Fig. 6) from two consecutive recordings (2 sec apart) form one training pattern. Each pattern is therefore a 22-dimensional vector. Before network training, the set of patterns is normalized to have zero mean and unit variance.

2.5 Forward model network

The basic ingredient of our network architecture is a forward model. It gets as input the sensory information

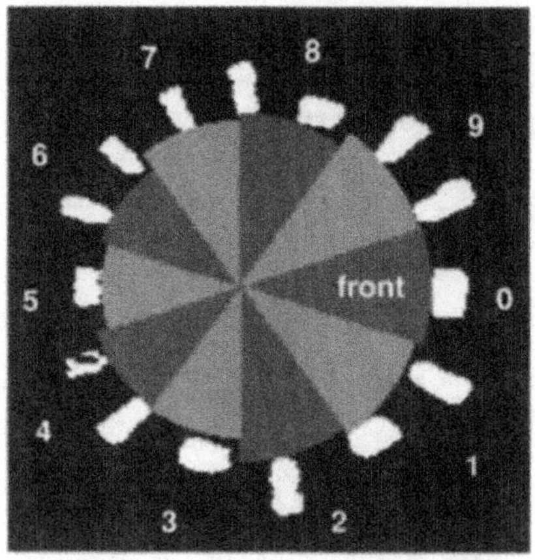

Figure 5: Distance information in 10 sectors derived from the camera image. Here, the data are taken from the image in Fig. 3

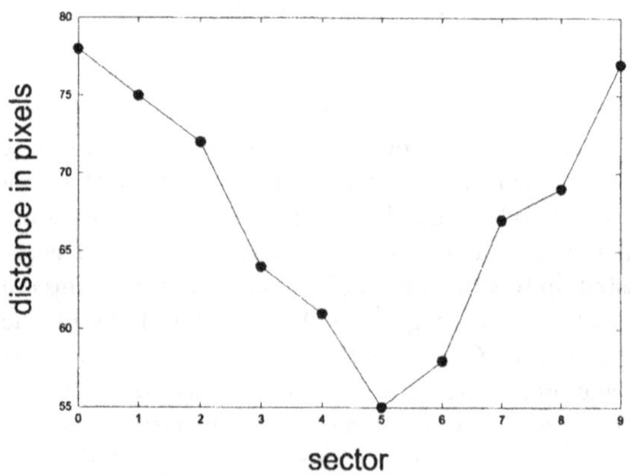

Figure 6: Visual distance in 10 sectors for the situation in Fig. 5

(Fig. 6) of one time step and the motor command consisting of the velocities v_L and v_R, and it predicts the sensory information of the next time step. Training data were collected as described in section 2.2.

To anticipate future sensory information beyond 2 sec, we feed the sensory output back into the sensory input (Fig. 7). This feedback completely overwrites the previous input. In each time step, the corresponding motor command M of the sequence is fed into the network. Thus, for illustration it seems more intuitive to replace the feedback by a chain of identical forward models (Fig. 8).

As a forward model we used an MLP with one hidden layer. The network's activation functions were the identity on input and output, and the sigmoidal function in the hidden layer. The MLP had 12 input neurons (two velocity values and the 10 sector values) and 10 output neurons (10 sector values). The hidden layer

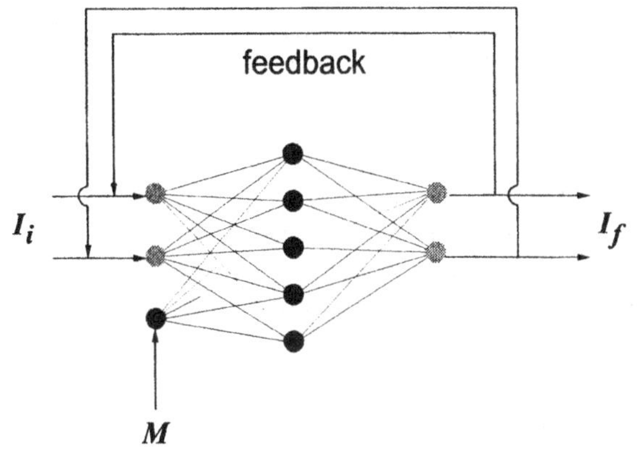

Figure 7: Forward model with feedback loop. The model maps the sensory information I_i to I_f in the context of the motor command M

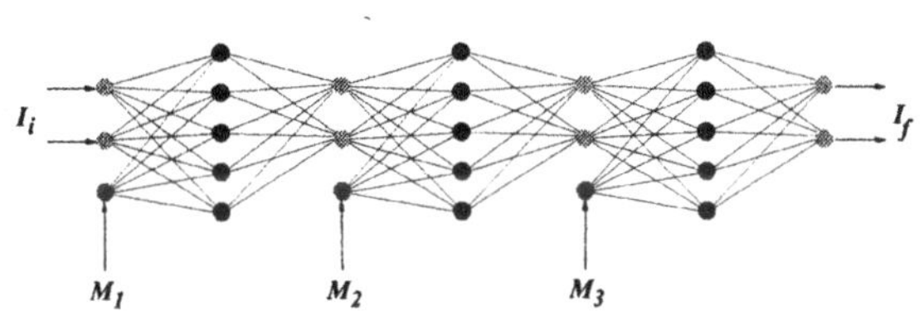

Figure 8: Concatenated chain of forward models. The sensory output of link t is the sensory input of link $t+1$

comprises 15 hidden units. This number seemed to be a good compromise between recall speed and accuracy. Higher numbers did not improve the performance noticeably. The network was trained on 5466 patterns with 3000 epochs of resilient propagation (RPROP) (Riedmiller and Braun, 1993).

The performance of anticipation was evaluated about the squared distance (squared error E^2) between the output $\mathbf{o}$ of a chain with n links and the real sensory information $\mathbf{r}$ after n 2 sec intervals,

$$E^2 = \sum_{i=1}^{10}(o_i - r_i)^2 \quad . \tag{1}$$

2.6 *Goal-directed movements*

The task in the planning of goal-directed movements is to find a series of motor commands (here velocity combinations) such that the final sensory information matches the desired value. Here, we treat this problem as an optimization task. The function to be optimized is the squared error between anticipated and desired goal. The free parameters are the velocities v_L and v_R for each time step (link in the chain).

We applied two different optimization methods, simulated annealing and Powell's method from the *Numerical Recipes* code (Press, 1992). The first is more suited to find a global minimum, whereas the second might be caught in a local minimum.

Simulated annealing is a stochastic method for searching the minimum value, occasionally allowing jumps to higher function values. The probability of these jumps is given by the Boltzmann distribution. The temperature parameter of the distribution is slowly reduced during simulation time according to an annealing scheme. Here, we used an implementation (Carter Jr., 1994) incorporating a scheme called 'Fast Simulated Annealing' (Szu and Hartley, 1987). Further, this implementation includes at the beginning an increase in the temperature to a point where a large jump in the function value occurs, and only then starts decreasing the temperature. We used the default parameters given in the implementation, except for the learning rate, which was set to 0.1, and at each temperature value the number of random steps, which was set to 20 times the number of free parameters. Random numbers were generated using the *ran1* code from the Numerical Recipes (Press, 1992).

Powell's method is based on conjugate directions, but does not need the evaluation of a gradient. We used the parameters as given in the Numerical Recipes (Press, 1992). The fractional tolerance of the function value was set to 10^{-4}.

Both optimization methods were initialized by setting all velocities to zero.

The treatment of the goal-directed movement as an optimization problem allows us to add penalty terms to the squared error to restrict the possible range of solutions. The choice of velocities beyond the range ± 60 mm/sec, used for training, was prohibited by punishing velocities outside this range with an additional term in the cost function. This was necessary because otherwise for goals out of the reach of one interval, the method could produce larger velocities as solutions for which no examples were available in the training set - there is no guarantee that the extrapolation of sensory predictions found by the network for these velocities are correct. Further, a penalty term was added to velocity series resulting in robot positions too close to the obstacles.

The above solution assumes a given number of chain links. Since the appropriate number of links - which corresponds to the duration of the movement sequence - is not known beforehand, we start with one link and increase the number of links in the optimization process. In each optimization step, we solve the optimization problem and test the optimization criterion. If the criterion is not yet met, the number of links is increased by one and the optimization is restarted (from zero velocities). This is repeated, until the anticipated sensory state matches the desired state (within 0.5 pixels - the resolution limit).

In our experiments, the goal state was not the complete sensory information, as in (1), but only the value

t_k in a predefined sector k. Thus, the squared error to be optimized is $E^2 = (o_k - t_k)^2$, with o_k equal to the predicted output in sector k.

2.7 Mental transformation

In the mental transformation task the robot has to assess whether it is standing in the center of the circle. In Fig. 3, the robot is roughly in the middle of the circle, but apparently this cannot be decided from the image representation (Fig. 6). The reason for this asymmetry is that the center of the robot differs from the optical axis of the camera.

Thus, instead of deciding the above question with pure image processing, we propose a different strategy. After observing the current image, the robot simulates a left and a right turn (around its rotational axis, i.e., $v_L = -v_R$), and anticipates the effect of these movements on the image representation. From the current position the robot simulated five rotational steps (2 sec each) to the left with the velocity $v_L = -40$ mm/sec and $v_R = 40$ mm/sec, and, also from the current position, five steps to the right at the opposite velocity. Five steps at this speed corresponded to a rotation of 72°. Since our environment always contained a full circle, it was not necessary to cover the entire 360° in the mental simulation. Then, the values of the frontal sector for the different representations were compared (altogether 11 values). If they had a variance of less than one pixel squared it was concluded that they were the same, and therefore, the position of the robot was in the center of the circle (which is the only point having same distance to the circle boundary in all directions).

3. Results

In this section, we first analyze the anticipation performance achieved by the chain of MLPs. Second, this chain is applied to goal-directed action-selection, and third it is used in a mental transformation task. All results presented in the following were obtained by using the same MLP as a forward model (see section 2.5).

3.1 Performance of anticipation

As we show in appendix A, the squared error is expected to increase linearly with the number of anticipation steps, under the assumption of randomly independent errors for each step. This is not likely for series with a constant motor command. Therefore, the test patterns were collected during random walks. Figure 9 shows the mean squared distance about the sensory values (distances in the 10 sectors) from the starting point of a sequence (see section 2.3). A pure random walk would result in a linear increase (same derivation as the one leading to (6), see appendix A). But, as the solid curve indicates, the squared distance increases stronger than linear. This is due to the slight dependence of velocities, as mentioned in section 2.3. The limited maximum range for random walks within the circle of obstacles reduces finally the increase in distance. For intermediate interval numbers, starting after 1 up to 6, the squared distance increases roughly linear (dashed line), as expected for random movement sequences. Therefore, we will only include the second to the sixth anticipation step for the test of the theoretic prediction of the linear increase of the squared error.

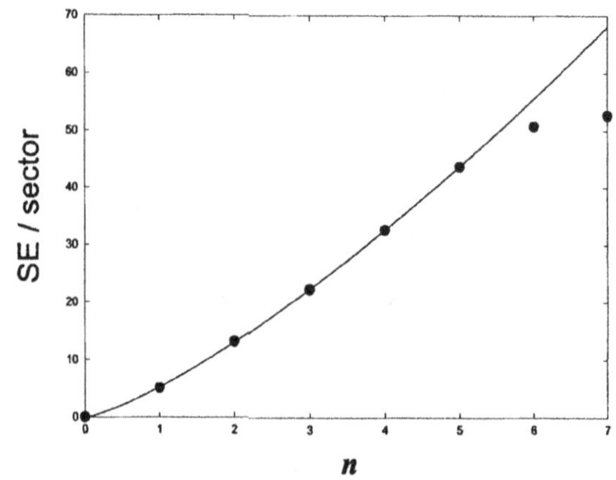

Figure 9: The mean squared distance (in pixels squared) of image sector values from the starting image representation for increasing length of the test series

The performance of the anticipation on the 138 test pattern series is shown in Fig. 10. The mean squared error was below 2.4 pixels squared for all tested chain lengths. This comes close to the image resolution. Starting from the second to the sixth step the squared error per sector increases linearly as predicted (a solid line is fitted to the data).

The MLP network was only trained on points lying on the manifold of valid sensory information embedded in a 10-dimensional space. This manifold is only two-dimensional (the robot has only two degrees of freedom, namely the distance to the center of the circle and the orientation of the robot). Therefore, the network's behavior on points slightly outside the manifold is not clear. Thus, we have a look at the effect of a small change in the input of the forward model. Let $\mathbf{f}(\mathbf{x})$ be the transformation the MLP does on the input $\mathbf{x}$. Figure 11 displays the mean change of the output, $\mathbf{f}(\mathbf{x}_0 + \mathbf{e}) - \mathbf{f}(\mathbf{x}_0)$, as a function of a small distortion $\mathbf{e}$. The sensory input to the network was taken from the collected test patterns. The velocities were chosen randomly within the range of the training values. For each test pattern, 100 $\mathbf{e}$ values were chosen randomly, ranging in magnitude from 0.02 to 2 pixels. The slope s in the diagram was evaluated with a linear fit, $s = 0.47$. It follows that the network

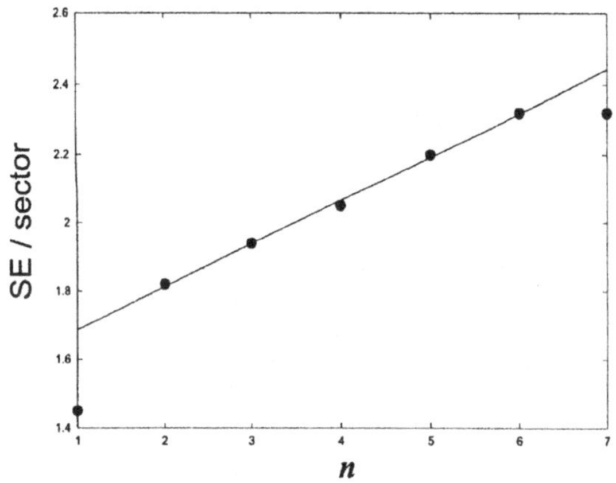

Figure 10: The anticipation performance on the 138 test series. The mean squared error (in pixels squared) per sector is shown as a function of the number n of chain links

reduces the magnitude of deviations from a valid input to about one half. See the discussion in section 4.

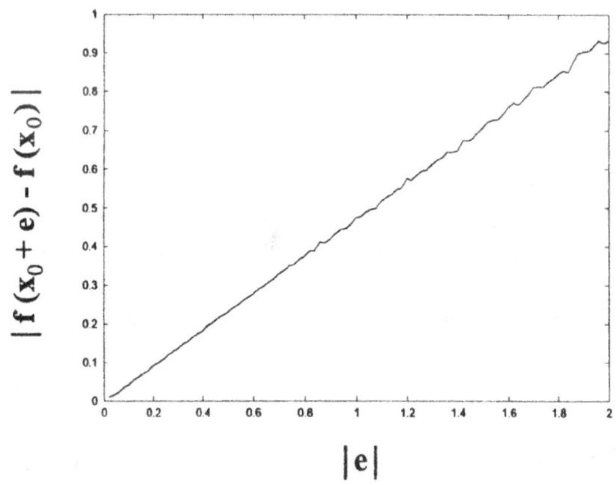

Figure 11: Stability of the forward model. The response to small deviations from a test pattern input is given. All values in pixels

3.2 *Goal-directed action selection*

The goal for the robot is to move in a way that the sensory value in a single given sector reaches a given value. Figure 12 illustrates the result of two typical movements. In example A, the goal was to make the sector in the back right (number 4 in Fig. 5) attain a low value (e.g. 50 pixels). The robot moved backward in a rightward curve. In example B, the front sector should attain a low value. Thus, the robot moved from the middle of the circle straight toward the obstacles.

To test the performance quantitatively, a random series of goals was chosen. A run consisted of choosing a goal and executing the resulting movement. The goal sector was chosen among the 10 sectors, and its value was chosen from the interval $[50, 65]$. With the given shape of the robot and the obstacles, it was physically possible for each sector to attain these values. At the beginning of each run, an image was taken, which was used as the starting point of the anticipation. At the end of a run, another image was taken for comparison with the desired goal. We did not change the position of the robot between two consecutive runs. The robot did two sequences including 50 runs. At the beginning of each sequence the robot was placed in the middle of the circle. This was done to increase the variety of movements, because during a sequence the robot happened to spend most of its time near the obstacles. Table 1 shows the result of these tests. Collisions with the obstacles during the test runs did not occur.

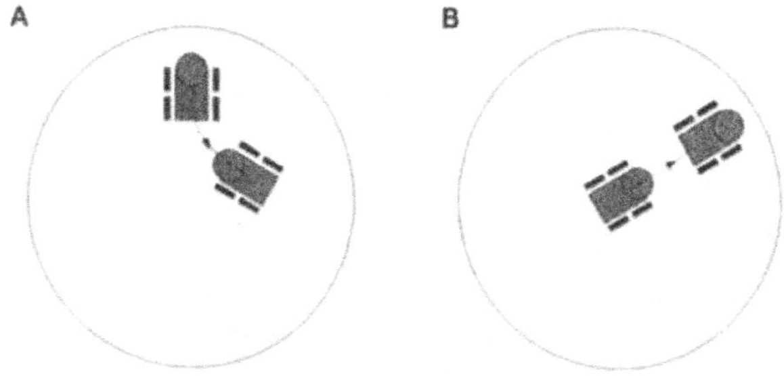

Figure 12: Typical goal-directed movements. In A, the goal is to have a short distance to an obstacle in the back right sector, and in B, in the front sector

Table 1: The result of 100 goal-directed movements for each optimization method is shown

Optimization method:	Sim. Anneal.	Powell
Found solution:	99%	96%
Exact hits:	18%	15%
Close within one pixel:	41%	46%
Closer to goal:	85%	83%
Right direction:	91%	91%
Mean squared error/link:	4.4	2.6
Mean number of links:	2.8	2.5

The two optimization methods gave very similar results. In almost all trials both optimization methods found a solution (96 to 99%). In 15 to 18% of the trials the final sector value matched exactly the desired value, and in almost half of the trials it was within one pixel of the desired value. In more than 80% of the trials, the final sector value was closer to the desired than to the initial value, and in more than 90% of trials, the final sector value was changed in the right direction (increasing or decreasing).

3.3 Mental transformation

The robot was placed in 20 random positions (concentrated in the center) with random orientation. For each position the robot had to decide, with the simulation strategy described in section 2.7, if it is standing in the middle of the circle. Figure 13 shows the result of this classification. Among the positions that were classified as center, the maximum distance to the center was 10 cm.

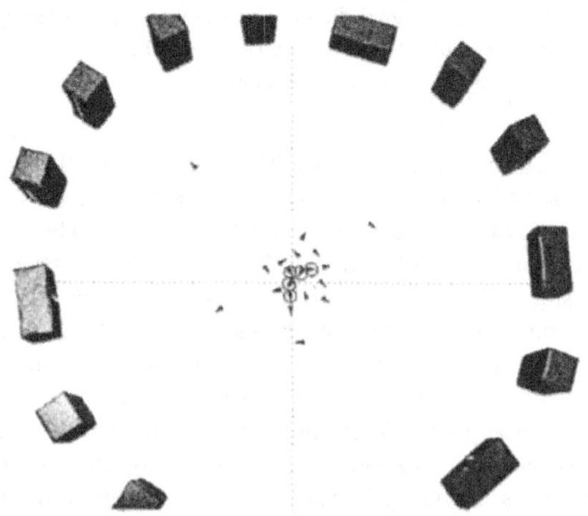

Figure 13: The performance based on mental transformation for detecting the center of the circle. Markers indicate position of the robot's rotational axis and direction the robot was facing. Markers surrounded by a circle represent trials in which the position was classified as center. The obstacle circle shown is the same as in Fig. 1

4. Discussion

Goal-directed motion planning requires a search in a high-dimensional motor space defined by a sequence of movements. Nothing is known about the structure of the optimization function defined over this space. The fact that Powell's method, which is a local minimization method, showed a similar performance than simulated annealing suggests that the presented task does not provide many local minima that are not global. Whether other environments have similar properties is not known.

Mental transformation could be used to judge the physical properties of the center position. No motor commands needed to be executed (this is not possible with the use of context layers (Tani, 1996)). No extra training, especially no teacher, was necessary to achieve a successful detection of the center of the circle. The maximum distance of a location, classified as center, to the real center (10 cm) is low compared to the circle diameter (180 cm), and to the length of the robot (40 cm). The remaining inaccuracy might be attributed to prediction errors, and to deviations from perfect symmetry in the circle of obstacles.

Distance comparisons in our mental transformation task were solely based on the compact image representation. Alternatively, given the demonstrated performance of the anticipation, the distance could be readily obtained about the time the robot needs to reach an obstacle with a constant straight forward movement. The time could be obtained with the same mental simulation as done for the rotational movements. This alternative was not chosen in the presented paper because of efficiency. First, the image representation was enough to tell if two distances are the same. Second, an additional simulation of a straight movement would add more errors. On the one side, if a high wheel velocity would have been chosen then the time resolution would have been poor because of the few 2 sec intervals. On the other side, a low wheel velocity results in many steps, and therefore, it leads to a higher prediction error (see the linear increase in squared error).

The overall performance was limited by mainly two technical deficiencies. First, the image resolution in the region showing the obstacles was low, about 140 pixels in diameter. This could be improved by using a lens with a smaller viewing angle, or a mirror with a viewing range more constrained to the neighborhood of the robot. Second, the robot did not very accurately reproduce the given wheel velocities. This problem was avoided in the collection of training and test sets by discarding inappropriate movements. However, we could not avoid it during the execution of the goal-directed movements. That explains the higher squared errors occurring there. An improvement would be a robot using only two driven wheels and caster wheels, which does turns on a floor more easily.

The presented tasks depend on the quality of prediction. Appendix A provides an estimation of the error accumulation. It relies on the assumption that a small deviation from a given point does not change the direction of the transformation. This seems plausible after comparing the size of the anticipation error - below 2.4 pixels squared (see Fig. 10) - with the range of sector values in Fig. 6. However, a test, as done for the MLP network (see Fig. 11) to reveal the effect of a small deviation in starting position, could not be done since the true transformation is unknown, and the training patterns do not lie dense enough to serve as a substitute.

In Fig. 10, the squared error grows at a smaller rate than expected with increasing chain length n, when compared to the squared error of a single step. According to (6) the slope should be the squared error of a single step. This discrepancy might be explained by considering that the 10-dimensional sensory data lie on a two-dimensional manifold. Thus, for a single anticipation step, errors could point in all 10 dimensions. But, since the anticipated states move only within the two-dimensional manifold, there is an increased chance for error compensation (as illustrated in Fig. 14). This works only if the network maps data outside its training domain onto the two-dimensional manifold (or close to) - which is not a guaranteed feature. But, luckily our network showed this behavior. The MLP decreases the distance to the manifold (see Fig. 11). To test the above idea, we computed the percentage of improvements in squared error during the anticipation series. The resulting value of 45% is much higher as expected for a random walk in 10 dimensions, and therefore supports our argument.

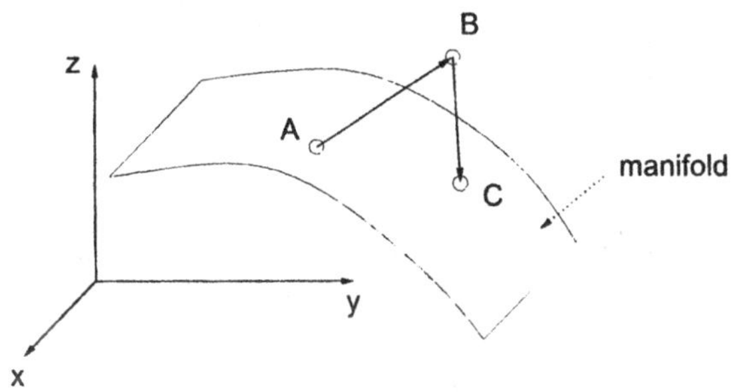

Figure 14: Compensation of the error in two prediction steps. The low dimensional data-manifold constraints the position of point C

The presented anticipation method relies on the complete current sensory information given. Therefore, the method could not cope with occluded vision. But, there is a solution to this problem. Given a partially occluded image, an association network model could be used to restore the complete sensory representation. In a previous study, we developed a model capable of this task (Hoffmann and Möller, 2003).

5. Conclusions and future work

Although the experimental setup used in our experiments is very simple and widely artificial, the results demonstrate that forward models can be used for both planning of goal-directed actions and mental transformation.

The agent learns from the interaction with the environment, no teacher is necessary. As a result, it obtains an internal sensorimotor model. This model is then applied to interpret sensory information and to select actions leading to a desired goal. An amalgamation of sensor processing and motor control might overcome the shortcomings of 'representationalist' approaches to visual perception, which, in contrast, restrict perceptual processes to the sensory domain (see Möller (1999)). We suggest that a sensorimotor approach as the one described in this work could explain spatial perceptual capabilities like the interpretation of the shape and the physical properties of objects or, as shown above, perceptual judgments of the spatial relation to objects.

We presented only a first simple example, but the approach offers extensions for future work:

- Obstacles arranged in a straight line could be detected based on mental simulating of straight movements.
- The robot could also learn a different environment like a triangle instead of a circle. It could be tested how well the model generalizes, for example, tackling the question if the robot can also operate in a square if it was only trained in circles and triangles.
- The above extensions will still work without the use of memory. But, we are also thinking of more complex mobile robot tasks, like detecting a dead-end based on internal simulation. Here, not all visual information will be available at any time.

6. Acknowledgments

The authors thank Wolfram Schenck for providing an implementation of the MLP network, Bruno Lara and Helmut Radrich for their effort in getting the robot running initially, Rainer Giedat for implementing a client-server interface for the optimization routines, Fiorello Banci for building a mounting for the camera, Karl-Heinz Honsberg for his assistance with electronic fixes, and the two anonymous reviewers for their comments.

A Estimate of error accumulation

We will show that the expectation value of the squared error of the anticipated sensor input only increases linearly with the number of chain links. Let $\mathbf{e}$ be the error of the feed-forward output of a single link. $\mathbf{e}$ is a vector with one component for each output neuron. We assume that the probability distribution of this error does not depend on the input of the network. Thus, all errors are independent of each other. Further, we assume that the error for each output neuron has zero mean and the same standard deviation σ.

On this basis, we compute the expectation value of the squared error. The total error of the chain output is the sum of the errors of the outputs of each link. To illustrate this, think of each correct transformation at one link as a line in a d-dimensional space, with d equal to the number of output neurons (Fig. 15).

A line connects an input point with an output point (of the transformation). The error at link i can be drawn as an arrow $\mathbf{e}_i$ at the end of a line (output point). This will result in a different starting point for the next line. If the error is small and the transformation function sufficiently smooth we can approximate that the displacement of the starting point does not change the direction

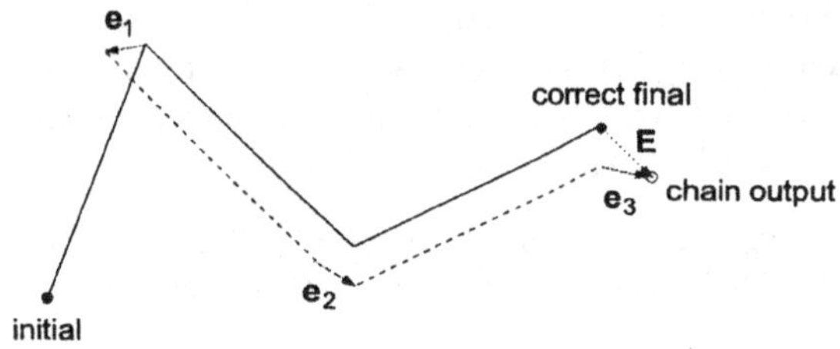

Figure 15: Error accumulation in a feed-forward chain. Each solid black line is the correct transformation for one link. The dashed black lines are the correct transformations for a slightly different starting point

and length of the next line, which would be then the correct transformation at the new starting point. Thus, the displacement $\mathbf{E}$ of the final point is the sum of the vectorial errors of each stage. Therefore, given n links, the total error E can be written as

$$E = |\sum_{i=1}^{n} \mathbf{e}_i| \quad . \tag{2}$$

We want to compute the expectation value of E^2,

$$< E^2 >=< (\sum_{i=1}^{n} \mathbf{e}_i)^2 > \quad . \tag{3}$$

Doing the square operation on the sum gives

$$< E^2 >=< \sum_i \mathbf{e}_i^T \mathbf{e}_i + \sum_{i,j \neq i} \mathbf{e}_i^T \mathbf{e}_j > \quad , \tag{4}$$

and using the linear property of the expectation value results in

$$\begin{aligned} < E^2 > &= \sum_i < \mathbf{e}_i^T \mathbf{e}_i > + \sum_{i,j \neq i} < \mathbf{e}_i^T \mathbf{e}_j > \\ &= \sum_i < \mathbf{e}_i^T \mathbf{e}_i > \quad . \end{aligned} \tag{5}$$

The last term vanishes because $\mathbf{e}_i$ and $\mathbf{e}_j$ are independent random variables, for $i \neq j$, and each variable has zero mean. The remainder is a sum over the variances for each link and dimension. Therefore,

$$< E^2 >= n\, d\, \sigma^2 \quad . \tag{6}$$

Thus, the expectation value of the squared error increases only linearly with the chain length.

References

Carter Jr., E. F. (1994). A general purpose simulated annealing class. http://www.taygeta.com/Classes.html.

Demiris, Y. and Hayes, G. (2002). Imitation as a dual-route process featuring predictive and learning components: a biologically-plausible computational model. In Dautenhahn, K. and Nehaniv, C., (Eds.), *Imitation in Animals and Artifacts*, pages 327–361. MIT Press.

Grush, R. (2003). The emulation theory of representation: motor control, imagery, and perception. *Behavioral and Brain Sciences.* in press.

Hesslow, G. (2002). Conscious thought as simulation of behaviour and perception. *Trends in Cognitive Sciences*, 6(6):242–247.

Hoffmann, H. and Möller, R. (2003). Unsupervised learning of a kinematic arm model. In Kaynaka, O., Alpaydin, E., Oja, E., and Xu, L., (Eds.), *Artificial Neural Networks and Neural Information Processing - ICANN/ICONIP 2003, LNCS*, volume 2714, pages 463–470. Springer.

Jirenhed, D.-A., Hesslow, G., and Ziemke, T. (2001). Exploring internal simulation of perception in mobile robots. *Lund University Cognitive Studies*, 86:107–113.

Jordan, M. I. and Rumelhart, D. E. (1992). Forward models: Supervised learning with a distal teacher. *Cognitive Science*, 16:307–354.

Möller, R. (1999). Perception through anticipation. a behavior-based approach to visual perception. In Riegler, A., Peschl, M., and von Stein, A., (Eds.), *Understanding Representation in the Cognitive Sciences*, pages 169–176. Plenum Academic / Kluwer Publishers.

O'Regan, J. K. and Noë, A. (2001). A sensorimotor account of vision and visual consciousness. *Behavioral and Brain Sciences*, 24:939–1031.

Press, W. (1992). *Numerical recipes in C: The art of scientific computing.* Cambrige University Press, Cambridge.

Riedmiller, M. and Braun, H. (1993). A direct adaptive method for faster backpropagation learning: The RPROP algorithm. In *Proceedings of the IEEE International Conference on Neural Networks*, pages 586–591.

Szu, H. and Hartley, R. (1987). Fast simulated annealing. *Physics Letters A*, 122(3,4):157–162.

Tani, J. (1996). Model-based learning for mobile robot navigation from the dynamical systems perspective. *IEEE Transactions on Systems, Man, and Cybernetics - Part B*, 26(3):421–436.

Wolpert, D. M., Doya, K., and Kawato, M. (2003). A unifying computational framework for motor control and social interaction. *Philosophical Transactions of the Royal Society of London. Series B*, 358:593–602.

Self-Growth of Basic Behaviors in an Action Selection Based Agent.

Olivier Buffet, Alain Dutech and François Charpillet
Loria - INRIA-Lorraine
Campus Scientifique - BP 239
54506 Vandœuvre-lès-Nancy CEDEX
France
dutech@loria.fr

Abstract

We investigate on designing agents facing multiple objectives simultaneously, that creates difficult situations, even if each objective is of low complexity. The present paper builds on an existing action selection process based on basic behaviors (resulting in a modular architecture) and proposes an algorithm for automatically selecting and learning the required basic behaviors through an incremental Reinforcement Learning approach. This leads to a very autonomous architecture, as the hand-coding is here reduced to its minimum.

1. Introduction

Action selection is the general high-level problem of choosing which action to perform when facing different parallel goals that can be heterogeneous and even conflicting. Although ethologists have studied this problem in depth, their models are not very easy to implement, as argued by (Tyrrell, 1993). A large number of implemented models rely on evaluating the quality of an action thanks to a kind of weighted sum of some quantities in order to evaluate the utility of that action related to the various high-level goals. To prevent the hard and tedious task of hand-coding and hand-tuning huge set of parameters associated with this action weighting and selection process, various *reinforcement learning* schemes have been studied (see (Lin., 1992), (Humphrys, 1996) for example).

Following (Humphrys, 1996) principles, we will use an action selection architecture where an agent knows different low-level behaviors (that we call *basic behaviors*) and must somehow choose an action based on the action's utility given by these basic behaviors. Previous works on this domain have focused on how to select this action, i.e. how to combine various quantities to select an action from a set of actions favored by the related basic behaviors or to suggest a new decision.

While in these works the basic behaviors are defined by hand and given to the agent, our aim is for the agent to define *by itself* the basic behaviors it will need in its action selection process so as to reach a complex satisfactory behavior. To do so, our suggestion is to give the agent the means to learn, from scratch, the basic behaviors it needs. This means, not only learning to achieve a given task, but also to *learn which tasks to learn.*

This kind of learning is made possible in the general framework of reinforcement learning thanks to a new mechanism of action selection (described more thoroughly in (Buffet et al., 2003)) that allows an agent to adaptively combine basic behaviors into a more efficient complex behavior. We show that the same mechanism can be used by the agent to detect if the new, slightly more complex, behavior learned is worth using as a *new basic behavior in future action selection processes.* Thus, starting from scratch, our agent incrementally grows its own *set* of basic behaviors from which, using action selection mechanisms, it can produce a complex behavior answering its needs and goals.

1.1 Motives

The motivations for an agent to learn by itself the basic behaviors needed to solve a complex task by action selection are many, some being highlighted below:

- It can be useful to remind that action selection should allow an agent to cope with complex tasks involving parallel, heterogeneous and sometimes concurrent goals.
- Learning new basic behaviors should bring better results than using intuitively hand-designed basic behaviors.
- Learning makes the design of an agent easier (no more long hand-tuning of parameters).
- Above all, the autonomy of an agent that can learn its own basic behaviors is greatly increased.

1.2 Our action selection framework

The basics of our action selection architecture must be clear:

- The agent does not know the model of the dynamics of its environment and can only partially perceive it. The goals of the agents are represented by a scalar reward signal that depends on the action and state of the agent.
- The agent has a set of basic behaviors. Each basic behavior should answer a simple goal/motivation.
- In a given situation (perception + internal drive), the agent uses information (mostly action utility) from its various basic behaviors to decide its global action.
- The action selection process is a "free-flow" as explained by (Tyrrell, 1993): the global action can be different from every action recommended by the basic behaviors. The agent is able to combine behaviors into a *new* action.
- The whole process gives the agent a *global complex behavior* that can cope with parallel, heterogeneous, conflicting needs and goals.

As said before, whereas in most previous works the set of basic behaviors is defined by the human designer[1], we aim for the agent to define and learn this set *by itself*, using an algorithm based on reinforcement learning.

1.3 Example

To illustrate our approach and to test our algorithms, we will use the well-known tile-world problem (see (Joslin et al., 1993), (Wooldridge et al., 1995)).

In the tile-world environment, as depicted on Fig. 1, the agent must push tiles into holes while avoiding to fall itself into holes. Intuitively this complex goal can be reached with at least two basic behaviors: pushing tiles in holes and avoiding holes.

Using our learning algorithm, our agent should learn by itself which basic behaviors are really useful for solving the complex task.

1.4 Articulation of the paper

To present our algorithm for automatically defining the basic behavior needed by an agent using our action selection architecture, we will first present our conception of basic behaviors (Section 2.). Then Section 3. explains how we combine basic behaviors in new behaviors, these new behaviors being potential new basic behaviors. The actual process of learning a set of basic behaviors is presented in Section 4., experimented and analysed in Section 5.. A discussion about our algorithm and some future works ends this paper.

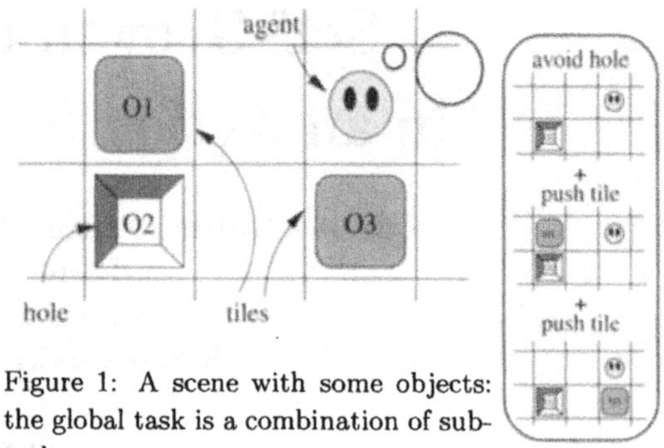

Figure 1: A scene with some objects: the global task is a combination of subtasks

2. Preliminaries

To go further in the presentation of our approach, we first need a precise formulation of what is called here a *basic behavior* (*bb*). As any behavior, it is guided by a *motivation*. This section thus describes what a motivation is, how to learn a behavior, what basic behaviors are, and finally how to use them.

2.1 Motivation Linked to a Behavior

A first aspect of a motivation is to know what are the good and bad results that the agent should try to reach or avoid. Having several possible elementary reward signals (two in the case of the tile-world: one for pushing tiles in holes and one for avoiding holes), a subset of these signals should be given.

Yet, the notion of motivation cannot be reduced to such a payoff. We need to also detail the situation through the types of objects taken into account (these types being instantiated once in a given world). If the agent only aims at pushing tiles in holes, it may be of interest for it to consider one hole and two tiles simultaneously, as one of the tiles could otherwise be an unseen obstacle.

In the present paper, a motivation is therefore described by 1- a subset of elementary reward functions and 2- a tuple of objects' types that are taken into account (what will be called a "type of configuration"). This will be represented as shown on Fig. 2.

Such a motivation is just a "reason" for acquiring an adapted behavior. Based on this, we now present a classical approach for learning this behavior.

2.2 Our Reinforcement Learning framework

The general Reinforcement Learning (RL) framework is the following (see (Sutton et al., 1998)). An agent per-

[1] when speaking of artificial agents

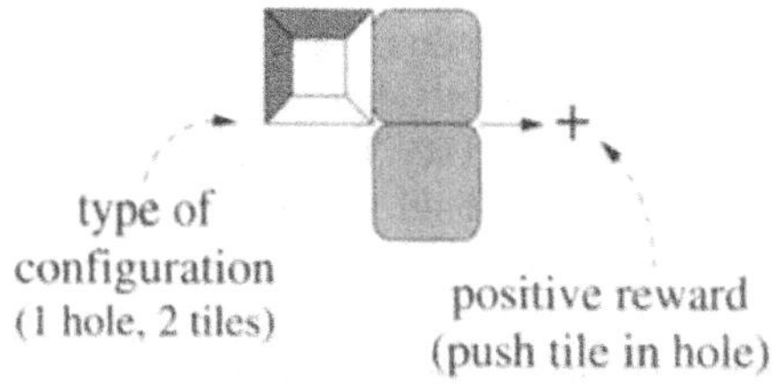

Figure 2: A schematic representation of a motivation for 1 hole, 2 tiles and a positive (+) reward when pushing a tile in a hole.

ceives its environment, decides of an action to alter this environment, and then receives a reinforcement signal. Using only this reinforcement signal (a scalar reward) and no information on the dynamics of the environment, various algorithms (like Q-Learning (Watkins, 1989), TD(λ) (Tesauro, 1992)) allow the agent to automatically learn a behavior (called a policy) to optimize the reward over time.

More formally, the RL framework can be expressed as Markov Decision Processes (MDP) (Puterman, 1994). Optimal solutions are deterministic policies π. Formally, a policy is a function that, for each *state* s of the system (agent + environment) gives an optimal action a. A utility function evaluating the "quality" of an action is often associated with a policy, and denoted $Q(s,a)$.

In our framework, we must face a crucial limitation to the ideal case of MDPs. As our agent has only a partial perception of the environment, it cannot know the full state of the system. The agent must then solve a non-Markovian process. The solution we have adopted is to use stochastic policies (as opposed to deterministic) that give a probability of choosing actions in a given state (see (Singh et al., 1994)). For a given and chosen basic behavior, such a policy can be learned using for example some gradient reinforcement algorithm (Baxter and Bartlett, 2001, Baxter et al., 2001).

2.3 Basic Behaviors

For a given motivation, any RL algorithm suited to partial observations can be used. It only requires: 1- to define the "guiding" reward function as the sum of selected elementary reward functions, and 2- to build the perception based on the objects' observations (the objects corresponding to the type of configuration). The first result that is obtained is a stochastic policy. Nevertheless, other information will need to be stored to make use of this "behavior": the type of configuration will serve to match the appropriate objects in the environment to the behavior, and the utility function Q will help weighting the importance of the behavior.

A **behavior** b is thus defined by a tuple $\langle C_b^T, P_b, Q_b\rangle$, where:

1. C_b^T is a type of configuration, that is, a tuple of types of objects involved in the behavior.

2. $P_b(a|o(c))$ is a stochastic decision policy. Given an appropriate configuration, it maps its observations to probability distributions over actions.

3. $Q_b(o(c),a)$ is the Q-table of this policy, giving the expected discounted reward of an observation.

Such a behavior answers to one given motivation. But our objective is to find a set of behaviors which can be combined together to obtain an efficient decision-making system. These selected behaviors are named **basic behaviors** (*bb*) and constitute a set noted $\mathcal{B}$. Whereas the first step should be the choice of this set, next section first presents the principle of our action selection process, which is based on the combination of basic behaviors.

2.4 Combining Basic Behaviors

For an agent confronted to a new observation, making a decision while following several basic behaviors is a two-step process. The agent first has to identify which sets of objects trigger some basic behaviors. This leads to several potential stochastic decisions to apply, which must be combined in a single one through a compromise.

We illustrate this process on Figure 1 with only two intuitive *bbs*: 1- b_p which considers one tile and one hole, and the reward for pushing a tile in a hole, and 2- b_a which considers one hole, and the reward for falling in a hole.

The scene analysis leads here to a decomposition in two reasons for using b_p: the "configurations" (tuples of objects) $cfg_1 = \langle O_1, O_2\rangle$ and $cfg_2 = \langle O_3, O_2\rangle$, and one reason for using b_a: the configuration $cfg_3 = \langle O_2\rangle$. To each pair (bb, cfg) is associated a probability distribution over action and the related Q-values, all this being used to compute a final probability distribution to apply, as shown on Figure 3.

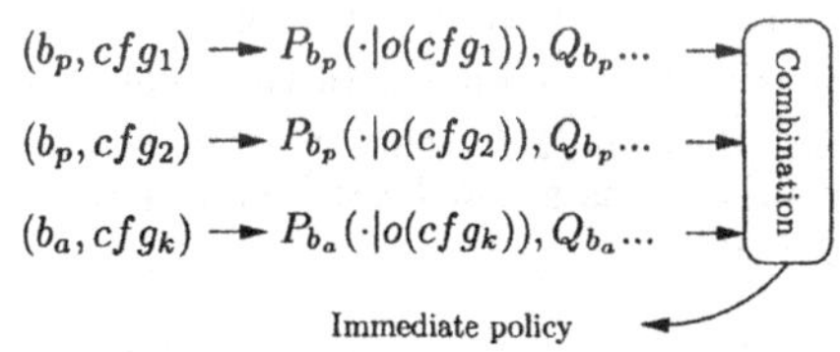

Figure 3: How to obtain a probability distribution over actions with the three identified (bb, cfg) pairs.

Note: Interested readers will find more details on the combination of basic behaviors in the appendix, Sec. A1.

3. How to detect a new *basic behavior*

In a few words, the principle of building a new basic behavior from an existing set of basic behaviors rely on three steps: learning the best combined behavior for a given motivation, trying to improve this behavior and deciding if it will then be used as a new basic behavior.

3.1 Learn a combination of basic behavior

As described in (Buffet et al., 2003), there is a learning algorithm that, given a set of basic behaviors $\mathcal{B} = \{bb_1, \cdots, bb_n\}$, can produce an efficient combination function of these behaviors. This combination is based on a parameterized function of the policies where the parameters are learned using a simulated annealing algorithm. Let's call the behavior resulting from this learning process a *combined behavior*: $bc = \mathcal{F}(\mathcal{B}, \Theta)$ where $\Theta = (\theta_1, \cdots, \theta_n)$ is a learned vector of parameter of the size of $\mathcal{B}$. Sec. A1 in the appendix gives more details on the combination algorithm.

3.2 Improvement

To improve the combined behavior thus obtained, our idea is to use an online gradient descent reinforcement learning (Baxter et al., 2001). The search done with such an optimization algorithm can help improving a policy, even in a non-Markovian reinforcement learning framework as ours. This search is initiated by using the combined policy of the combined behavior. With the new policy thus obtained, we can derive a new behavior (denoted *nb*) with its associated motivation and quality.

3.3 New basic behavior

The question now is to decide if this new behavior would improve the global behavior of the agent if used as a basic behavior in later action selection mechanism. This decision is based on a heuristic comparison between the quality of the combined behavior and of the new behavior. Our hypothesis is that a newly learned -and significantly better- behavior is likely to bring useful information to other behaviors that will be built on it.

If we call (V_{cb}) the quality of the combined behavior and (V_{nb}) the quality of the new behavior, the new behavior is chosen as a new basic behavior (and added to the set $\mathcal{B}$) if:

- $|V_{cb} - V_{nb}| > \sigma_s$ (to avoid errors due to close estimates) and,
- either $((V_{cb} < 0)$ and $(V_{nb} < \sigma_- * V_{cb}))$ (little improvement of a negative result)
- or $((V_{cb} \geq 0)$ and $(V_{nb} > \sigma_+ * V_{cb}))$ (major improvement of a positive result).

This criterion depends on three parameters that have a big influence on the overall algorithm, as detailed in Sec. 4.. In our experiment they have been tuned by hand. One important work would be to test their validity in other tasks.

4. More detailed algorithm

The question raised here is how to automate the choice of basic behaviors for the combination. As explained in Sec. A, it is of major interest since the intuitive selection of "minimalist" basic behaviors is not sufficient and since, moreover, this step is required for the agent to be truly autonomous.

4.1 Principle

Our proposition is to incrementally build a set of basic behaviors. To this end, starting with an empty set of basic behaviors, we progressively add more complex behaviors to this set, using already-selected basic behaviors to design the new ones. This leads to building a set of basic behaviors of increasing complexity.

The definition of a behavior's motivation (see Sec. 2.1) implicitly says that the "complexity" of a behavior comes from two sources: the size of the type of configuration and the combination of elementary reward functions.

To sum up the principle followed, a new type of configuration being chosen, the behaviors corresponding to the possible rewards' combinations are learned and tested to know whether they are of interest or not. Figure 4 illustrates the progression according to types of configuration: behaviors get more complex toward the leaves of the tree.

Using the process described in Sec. 3.3 for selecting a new basic behavior, next section explains the tree growth algorithm.

4.2 Tree-Growth

The idea is to grow a set of basic behaviors. We build an *exploration* tree of behaviors of increasing complexity in the number of objects. Each node of the tree is linked to a type of configuration and is in fact a short sub-tree of the possible reward combinations. In turns, using a breadth-first traversal of the tree, we compute the best behavior in each case, building upon the combination of earlier *bbs* (as explained in Sec. 4.1). Each new behavior is compared to the corresponding combination (criterion presented in Sec. 3.3) to determine if it must be added to the set of basic behaviors (the tree being then expanded at this node with new behaviors to explore).

Coming back to the example of Fig. 4, this tree's development follows more and more complex types of configurations and, for each of them, there is a short sub-tree of the possible reward combinations ("–" (respectively

"+") stands for the avoid-hole negative reward (resp. the push-tile-in-hole positive one), and "+−" combines both).

4.3 Algorithm

Before testing it, we present here a formal definition of our method in Algorithm 1. Among the data required by the algorithm are:

- d_{max}: As an ever-increasing tree could be possible, this parameter limits the maximum depth of the tree.
- *d*: Not only the immediate sons of a new basic behavior may be of interest for further exploration. Thus, this second parameter defines the depth of developed nodes. It can be compared to the *fringe depth* used in the U-tree algorithm presented in (McCallum, 1995).
- *Criterion*$(\cdot,\cdot)$: Already presented in Sec. 3.3, this is the criterion used to evaluate new behaviors.

Algorithm 1 A Growing Tree of Basic Behaviors

```
Require: Criterion(·,·): to evaluate new behaviors.
    d_max: maximum depth of the tree, and d: depth of
    developed sub-trees.
  T empty tree. B empty set.
  Add to T all behaviors with 0 to d types of object.
  for all Behavior b still to be visited (up to depth
  d_max) do
    Adapt the combination of current bbs from B:
    V_cb ← efficiency of this combined behavior.
    Learn b by a direct policy search (initialized by the
    combination):
    V_nb ← efficiency of this new behavior.
    if Criterion(V_cb, V_nb) = true then
      Add to T the sons of b with up to d more types
      of objects.
      Mark b as [basic]. Add it to B.
    end if
    Mark b as [visited].
  end for
Ensure: A set B of kept basic behaviors.
```

5. Experiments

5.1 Application to the Tile-World

Description of the Problem

The tile-world is a grid domain in which a cell may contain a hole, a tile or an agent. In the complete problem, the agent (only one is considered) has to push tiles in holes as often as possible, and avoids to go itself in these holes.

To detail the simulation, the agent can go freely in a hole (and also go out), but will get a negative reward doing so. Moreover, when a tile is pushed in a hole, both the tile and the hole disappear and reappear anywhere on the grid. Finally, to avoid some blocking situations, holes and tiles cannot be on cells of the grid's border.

In this complete complex problem, many tiles and holes must be handled. As it appears in previous examples (Fig. 1), a simple decomposition of the problem in basic behaviors can be made intuitively: 1- `[avoid hole]` (b_a, $\langle hole\rangle$) and 2- `[push tile in hole]` (b_p, $\langle tile,\ hole\rangle$). They will be used as a reference for our *bb*-tree generation (and noted as the set $\mathcal{B}_{ref}$).

Agent's Perceptions, Actions and Rewards

In these experiments, the agent has always all other objects of the environment in sight. The principle of locality is nevertheless respected through the imprecise perceptions. For any object O in the scene, the agent's perception of O gives: **near**(O) (tells if object O is in the 9-cells square centered on the agent (`true|false`)) and **direction**(O) (gives the object's direction (`N-NE-E-SE-S-SW-W-NW`)[2]).

The only actions available for an agent are to move one cell North, South, East or West (it cannot ask to stay on a cell). And to conclude, the reward given is +1 when a tile falls in a hole, −3 when the agent goes in a hole, and 0 otherwise.

Going back to Fig. 1, the agent's perceptions of the 3 objects are here:

O_1: $near(O_1) = false$, $direction(O_1) = W$
O_2: $near(O_2) = false$, $direction(O_2) = W$
O_3: $near(O_3) = true$, $direction(O_3) = S$

Parameters

The parameters appearing in the description of the tree-growth Algorithm 1 (including the criterion from Sec. 3.3) have been tuned by hand, and set as follows:

$$\sigma_s = 500 \qquad \sigma_- = 0.9 \qquad \sigma_+ = 2 \qquad d_{max} = 4 \qquad d = 1$$

Methodology

The experimentations carried out were twofold: 1- applying the tree-growth algorithm to produce a set of basic behaviors $\mathcal{B}_{tree}$ (in a 6×6 grid) and 2- testing its efficiency in various more complex situations (in an 8×8 grid).

[2] The direction is known even for far objects, but brings imprecise information.

The tests conducted in this second phase consisted in using the combination of resulting basic behaviors with various numbers of tiles and holes in the environment (up to five of each). It is compared to the two "intuitive" basic behaviors that serve as reference (as written in Sec.7 5.1). A combination's efficiency is measured through the total payoff received during 100, 000 simulation steps.

Note: due to a lack of space, only the major results are presented below. Nevertheless the discussion enters upon several interesting remarks on the complete experimentations.

5.2 *Analysis*

Tree-Growth

To first consider the immediate result of the tree-growth algorithm, Fig. 4 shows the developed tree of evaluated behaviors. The signs noted in parenthesis indicate that the behavior linked to this type of reward and the related type of configuration has been retained as a basic behavior.

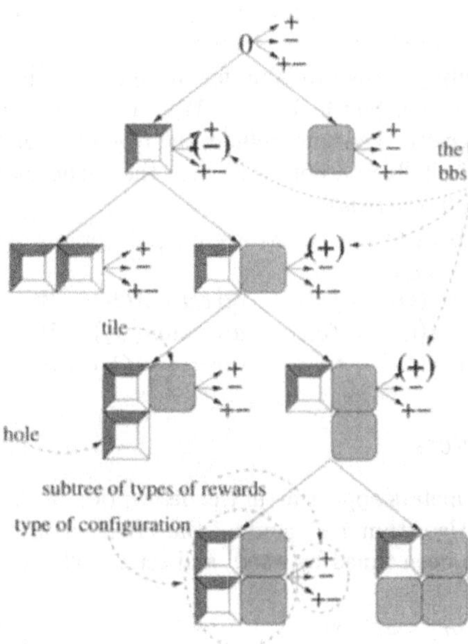

Figure 4: The tree of tested behaviors and (in parenthesis) the ones kept.

The three basic behaviors obtained (set $\mathcal{B}_{tree}$) correspond to the two intuitive ones (from $\mathcal{B}_{ref}$) along with a behavior (1-hole/2-tiles with the "+" reward) "specialized" for solving a blocking[3] (shown on Fig. 8 b)). The generated tree seems pretty satisfying as adding the "specialized" behavior answers a typical blocking situation.

[3]Not considering all objects regularly leads to such blockings.

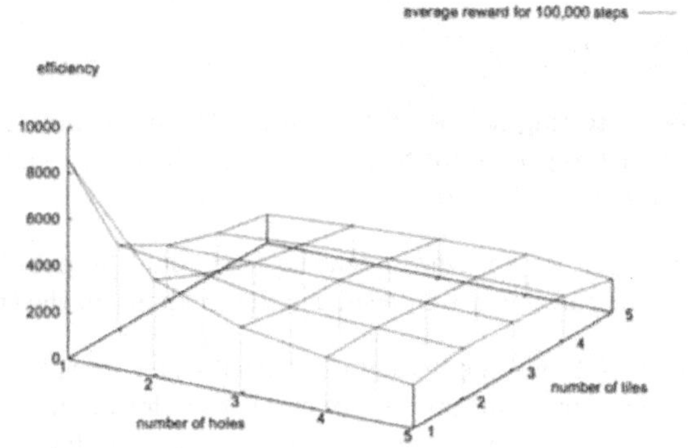

a) basic behaviors from $\mathcal{B}_{ref}$

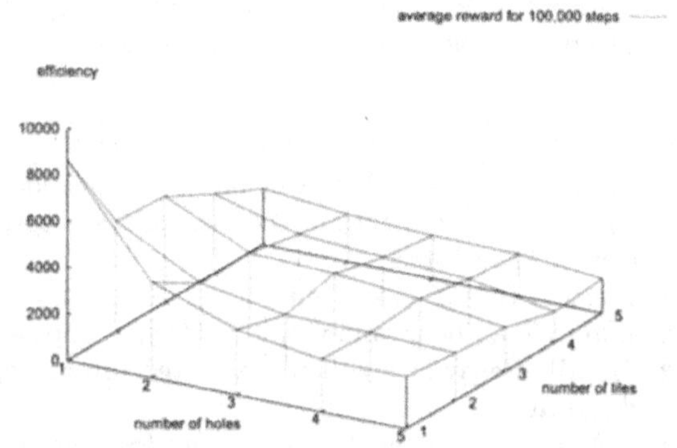

b) basic behaviors from $\mathcal{B}_{tree}$

Figure 5: Efficiency of the combination in more complex situations with different sets of basic behaviors.

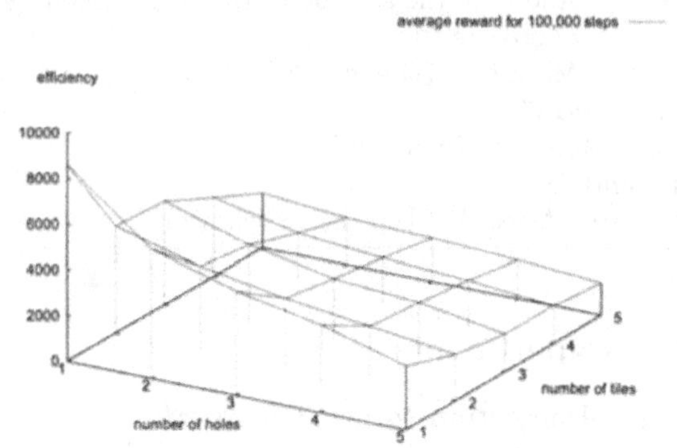

Figure 6: Identical tests as in Fig. 5 with *bbs* from $\mathcal{B}'_{tree}$: $\mathcal{B}_{tree}$ and the basic behavior considering 2 holes and 1 tile with only the "+" reward.

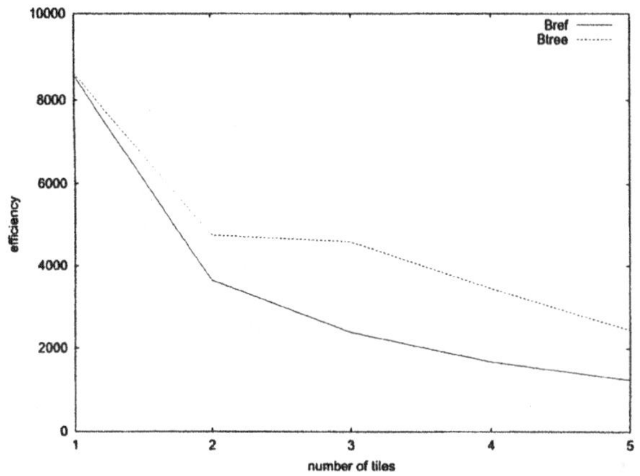

Figure 7: Comparison of $\mathcal{B}_{ref}$ and $\mathcal{B}_{tree}$'s efficiencies with 1 hole and several tiles.

However, the [2-holes/1-tile, "+" reward] behavior that solves another blocking (Fig. 8 a) in the appendix) is not added to the basic behaviors set. In fact, the increase in performance it brought does not fit the criterion ($V_{nb} \simeq 1.5 * V_{cb} < 2 * V_{cb}$) and, as a consequence, is left over. When we use a "weaker" criterion ($\sigma_+ = \mathbf{1.3}$), both specialized *bbs* are included in $\mathcal{B}$, along with 4 more complex ones. This brings difficulties, as the combination learning process is made more complex and more prone to local optima, even though individually, each of these behaviors answers an interesting specific situation. This problem comes rather from the combination algorithm itself than from the *bb* learning algorithm.

Tree's Efficiency

Figures 5 a) and b) present the efficiency of both sets of basic behaviors: $\mathcal{B}_{ref}$ and $\mathcal{B}_{tree}$, so as to compare the later (automatically generated through our algorithm) with the former (designed by hand). The x and y axes of the horizontal plane indicate the number of holes and tiles in the evaluated situations (the numerous objects to handle obliged to extend our grid to a size of 8×8). On the z axis is measured the efficiency in each situation (note: Table 1 also presents numerical values for some of these points).

At first glance, both surfaces are quite similar, with a common peak in the simple 1-tile/1-hole situation (to which a basic "push" behavior is dedicated). An important increase ($*1.3$) can be observed in the 2-tiles/1-hole case. It was in fact expected, as $\mathcal{B}_{ref}$ and $\mathcal{B}_{tree}$ only differ in the behavior managing the corresponding "push" motivation. Nevertheless, this increase appears to be persistent with a growing number of tiles to push (results are multiplied by 2) along the tiles-axis, as also shown on Fig. 7.

Considering a growing number of holes, the results from Fig. 5 b) are even closer to the reference surface. The remaining variations are partly due to the estimate of the efficiency, and we saw that a problem linked to the criterion prevented improvements in these cases with two holes and more.

Observation of a Weakest Criterion

Figure 6 shows the results obtained with the "weakest" criterion we already mentioned ($V_{nb} > \mathbf{1.3} * V_{cb}$) and a maximum tree-depth $d_{max} = 3$. Here, the behavior of the 2-holes/1-tile node (with the "+" reward) is now retained in set $\mathcal{B}$. It produces on the holes-axis a similar increase as the one observed previously on the tiles-axis. With this last set $\mathcal{B}'_{tree}$, the agent may even benefit from both improvements simultaneously, as shown on Table 1. The same table illustrates the fact that having useless *bbs* leads to a slight decrease in efficiency (see $\mathcal{B}_{tree}$ in cases 2/1 or 2/2, or $\mathcal{B}'_{tree}$ in case 1/2).

#holes/#tiles	$\mathcal{B}_{ref}$	$\mathcal{B}_{tree}$	$\mathcal{B}'_{tree}$
1 / 2	3648.3	**4747.5**	4680.5
2 / 1	4137.8	4089.6	**5660.8**
2 / 2	3142.3	2807.5	**3673.4**
2 / 3	2369.6	2873.9	**3356.4**
3 / 2	2468.3	2190.3	**3007.7**

Table 1: Comparison of the three sets' efficiencies. With more than five objects, the improvement gets harder to appreciate.

6. Discussion, similar work

If we come back to the motivations of our work, the main goal of action selection (i.e. being able to cope with multi-tasks problem) has been reached. Let us recall that classical Reinforcement Learning tools would not be able to manage more than one tile and one hole simultaneously (experiments are shown in (Buffet et al., 2002)). An agent using classical RL only finds out how to avoid falling in holes. Measuring their efficiency as done in Fig. 5 would produce a similar peak for one tile and one hole, the remaining surface lying with a null payoff. With our method, the agent is still able to behave correctly even with 5 holes and 5 tiles. Of course, there is a compromise to make between a reduced number of basic behaviors (fastest learning) and highest performances (in term of quality).

The outcome of our algorithm, and thus the global behavior of the agent, still depends on parameters and, more importantly, on a heuristic criterion. The parameters are linked to the exploration and growth of the tree of behaviors. As presented in Sec. 5., the influence of the criterion can be critical. If we keep with the same type of criterion, a more intelligent process would be to adapt the σ parameters to have a loose criterion at the start of the process and a more constraining criterion as the number of basic behaviors increases. That way, further behaviors would be added if they really bring something in term of performance. Of course, a more "intelligent" criterion and a way to automatically tune the parameters obviously need to be looked for.

It would be interesting to test and compare our work more thoroughly with similar projects. Testing are currently done on the "house environment" designed and used in (Humphrys, 1996), which is both easier and more difficult. It is easier as there is no need for several instances of the same basic behavior, but it is more difficult because the branching factor (due to many types of rewards and objects) in the exploration tree is high (8 for objects, 14 for rewards). Primary results show that the first basic behaviors learned are different from the basic behaviors hand-coded by Humphrys, but we have not the complete results to compare the global performances of the two approaches.

Closer to our work, (Digney, 1998) proposes a fully automated hierarchical Q-Learning algorithm that learns to create and structure "sub-modules" to solve a multi-task problem. One major difference with our work is that his agent has a perfect perception of its environment. The Q-modules learned using Digney's algorithm seem more strongly linked to the type of task at hand, and it is not clear if they could be re-used in very different environments. Furthermore, his approach to action selection is not scalable in the sense used in this paper: if many sub-tasks are similar, Digney must learn as many Q-modules whereas our algorithm would only need one basic behavior. On the other hand, Digney's work seems more automated (less dependent to parameters tuning) but still restricted to simpler tasks.

In fact, in our work, learning both basic behaviors and how to combine basic behaviors, did indeed increase the autonomous aspect of the agent. Of course, human interference is still needed, especially to tell about different kinds of objects and rewards in the environment. This is also the case in (Digney, 1998), and might seem inevitable from (Urzelai et al., 1998) point of view. In their work, (Urzelai et al., 1998) specifically aim at finding the right balance between human design and machine learning in efficiently designing autonomous agents; and they argue that human is more efficient for designing basic behaviors, even if they are crude and refined later on by the agent. It seems that our work, along with (Digney, 1998), and to some extends (Nehmzow et al., 1993), gives even less work to the human designer: he only has to provide rewards and perceptions types. Such approaches relate to the domain of shaping (Randløv and Alstrøm, 1998) and could be worth a comparison to cognitive development (Piaget, 1967, Weng, 2002). And we argue that it would be still interesting to go even further, but that is still one of the greatest challenge of AI: agents that auto-organize and represent their environment by themselves.

To keep on considering practical aspects, our approach, as presented here, requires working in a simulated environment, so as to be able to conduct controled experiments. In an application on a real-world problem, this would imply designing such a simulation, for example by learning a model of the environment. This leads to another important machine learning problem, but may appear more realistic than a crude trial-and-error approach in a real world.

7. Conclusion

7.1 Our Contribution

We have presented our work on designing complex behaviors for autonomous agents. This work builds on an algorithm set in the Action Selection framework where a complex behavior is a combination of basic ones. Our main contribution deals with automatically designing the set of basic behaviors that will be used by the combination algorithm. This task is crucial for building complex behaviors and is also a first step towards meta-learning: in some way, the agent is able to select the skills it needs.

Only reward signals and types of objects present in the environment are needed by the agent to incrementally define more and more complex behaviors. The added value of a new behavior is used to determine if it will be used to augment the set of basic behaviors of the agent, thus being used to further build more complex behaviors. This iterative process ends when no more complex behaviors can be added.

7.2 Validation

Our algorithm has been tested on the tile-world problem. It gave good results as it was able to propose a set of behaviors that is more complete than an intuitive one designed by hand. Furthermore, note that the complex problems it was confronted to could not be solved directly by classical RL.

7.3 Future Works

Still, several improvements and open problems are left. It is very important to work on a better criterion for selecting "pertinent" behaviors, as the current criterion is a crude heuristic. Testing our algorithm on other environments and problems is also necessary to assess more thoroughly its capacities and to better understand the influence of the few parameters of the algorithm that still need to be hand-tuned.

As the efficiency of the combination algorithm can be affected by the size of the basic behaviors' set, we think that the combination process could be studied and probably improved further.

Another complex domain we are interested in is the field of multi-agents systems. If we are confident in the capacity of our algorithm to deal with the added complexity in terms of number of objects, we need to investigate further on the influence of the local perception of the world and the dynamic induced by other agents.

A very interesting perspective would be for the agent to build itself *objects abstractions*. Currently, the types of objects present in the environment are given to the agent by a human designer, but we think that side effects of behavior selection could be used to automatically categorize objects in classes (like for example "obstacles", "goal", "to be moved", etc). This would be another step towards true meta-learning and greater autonomy in agents.

References

Baxter, J. and Bartlett, P. (2001). Infinite-horizon policy-gradient estimation. *Journal of Artificial Intelligence Research*, 15:319–350.

Baxter, J., Bartlett, P., and Weaver, L. (2001). Experiments with infinite-horizon, policy-gradient estimation. *Journal of Artificial Intelligence Research*, 15:351–381.

Buffet, O., Dutech, A., and Charpillet, F. (2002). Adaptive combination of behaviors in an agent. In *Proceedings of the 15th European Conference on Artificial Intelligence (ECAI'02)*.

Buffet, O., Dutech, A., and Charpillet, F. (2003). Automatic generation of an agent's basic behaviors. In *Proceedings of the 2nd International Joint Conference on Autonomous Agents and MultiAgent Systems (AAMAS'03)*.

Digney, B. (1998). Learning hierarchical control structure for multiple tasks and changing environments. In *Proceedings of the Fifth Conference on the Simulation of Adaptive Behavior (SAB'98)*.

Humphrys, M. (1996). Action selection methods using reinforcement learning. In *From Animals to Animats 4: 4th International Conference on Simulation of Adaptive Behavior (SAB-96)*.

Joslin, D., Nunes, A., and Pollack, M. E. (1993). Tileworld users' manual. Technical Report TR 93-12.

Lin., L.-J. (1992). Self-improving reactive agent based on reinforcement learning, planing and teaching. *Machine Learning*, 8:293–321.

McCallum, R. A. (1995). *Reinforcement Learning with Selective Perception and Hidden State*. PhD thesis, University of Rochester.

Nehmzow, U., Smithers, T., and McGonigle, B. (1993). Increasing behavioural repertoire in a mobile robot. In *From Animals to Animats: Proceedings of the Second Conference on the Simulation of Adaptive Behavior (SAB'93)*.

Piaget, J. (1967). *La Psychologie de l'Intelligence*. Armand Colin.

Puterman, M. L. (1994). *Markov Decision Processes–Discrete Stochastic Dynamic Programming*. John Wiley and Sons, Inc., New York, USA.

Randløv, J. and Alstrøm, P. (1998). Learning to drive a bicycle using reinforcement learning and shaping. In *Proceedings of the 15th International Conference on Machine Learning, (ICML-98)*, pages 463–471.

Singh, S., Jaakkola, T., and Jordan, M. (1994). Learning without state estimation in partially observable markovian decision processes. In *Proceedings of the 11th International Conference on Machine Learning (ICML'94)*.

Sutton, R., Precup, D., and Singh, S. (1998). Between MDPs and Semi-MDPs: Learning, planning, and representing knowledge at multiple temporal scales. Technical report, University of Massachusetts, Department of Computer and Information Sciences, Amherst, MA.

Tesauro, G. (1992). Practical issues in temporal difference learning. *Machine Learning*, (8):257–277.

Tyrrell, T. (1993). *Computational Mechanisms for Action Selection*. PhD thesis, University of Edinburgh.

Urzelai, J., Floreano, D., Dorigo, M., and Colombetti, M. (1998). Incremental robot shaping. *Connection Science Journal*, 10.

Watkins, C. (1989). *Learning from delayed rewards*. PhD thesis, King's College of Cambridge, UK.

Weng, J. (2002). A theory of mentally developing robots. In *Proceedings of the 2nd International Conference on Development and Learning (ICDL'02), MIT, Cambridge, MA*.

Wooldridge, M., Müller, J.-P., and Tambe, M. (1995). Agent theories, architectures and languages: A bibliography. In *Intelligent Agents II, IJCAI'95 Workshop*, pages 408–431.

A Appendix

We briefly present the action selection mechanism (learning and behavior combination) used in our framework. A detailed presentation of this process can be found in (Buffet et al., 2002).

A1 Scene Decomposition

The goal of the decomposition is to find which relevant basic behaviors are applicable for a given situation.

The environment of the agent is made of typed objects (such as holes and tiles in the example of Fig. 1). The agent's observation is a set of perceived objects of the environment. Each basic behavior is only concerned with some types of objects. Thus, they are "instantiated" by associating them with subsets of observable objects with appropriate types (such a set is called a **configuration** (cfg)). This leads to a decomposition of the scene in a set $\mathcal{BC}(o)$ of (*basic behavior*, *configuration*) pairs. Coming back to Fig. 1, three such pairs may here be extracted: $(b_a, \langle O_2\rangle)$, $(b_p, \langle O_1, O_2\rangle)$ and $(b_p, \langle O_3, O_2\rangle)$.

We have seen how each scene is decomposed in (bb, cfg) pairs. There is a stochastic decision policy for each such pair ($P_{bb}()$). And, as each configuration is associated with an observation $o(cfg)$, each pair implies a given probability distribution over actions $P_{bb}(a|o(cfg))$.

A1.1 Weighted Combination of basic behaviors

Each basic behavior bb of $\mathcal{BC}(o)$ identified in the scene decomposition step is associated with a policy of action $P_{bb}()$ and a quality $Q_{bb}()$. Based on this, the task of the combination step is to compute a new distribution of probabilities over the actions, thus determining the agent's immediate and global policy.

The combination can take various forms (see (Buffet et al., 2002)). The one we have used depends on a set $\Theta = (\theta_1, \cdots, \theta_n)$ of parameters and is expressed by the following formula (where $\mathcal{C}(b,o)$ is the set of observable configurations linked to a behavior b):

$$Pr(a|o) = \frac{1}{K} \cdot \frac{1}{k_{(o,a)}} \sum_{b \in \mathcal{B}} \underbrace{e^{\theta_b}}_{to\ learn} \underbrace{[\sum_{c \in \mathcal{C}(b,o)} |Q_b(o,c,a)| . P_b(o,c,a)]}_{already\ known}$$

$$(\text{with } k_{(o,a)} = \sum_{(b,c) \in \mathcal{BC}(o)} w_b(o,c,a) = \sum_{(b,c) \in \mathcal{BC}(o)} e^{\theta_b} . |Q_b(o,c,a)|)$$

A1.2 Learning the Weights

The global combined policy depends only on the set Θ of parameters and there are few of them as there is only **one parameter for each basic behavior** (i.e. for each different type of configuration).

As tuning these weights is an optimization problem (maximizing the expected average reward), these parameters can be learned by reinforcement using simulated annealing. The algorithm simply has to make a good estimation of the expected average reward for each set of parameters.

It is interesting to note that, to some extend, the whole action selection process is scalable as several instances of the same basic behavior can be used without having to learn new parameters.

A2 Open Problems

The algorithm that we briefly presented above was tested on the tile-world problem (described in Sec. 5.) by (Buffet et al., 2002) using only two "intuitive" hand-designed basic behaviors (`[push tile in hole]` and `[avoid hole]`). Despite "good results", optimality was not reached and several limitations were mentioned. Among them we want to mention the following ones:

- **Local optima.** Learning the parameters is difficult because of the many local optima simulated annealing can converge to. In fact, as the algorithm presented is an approximation scheme to solve a non-Markovian process, it is more than likely that no globally optimal solution exists. Nevertheless, a side effect of our method (see Sec. 4.1) could help finding an improved policy by refining the raw outcome of the behaviors' combination.
- **Choice of basic behaviors.** The set of available basic behaviors greatly influences the performances of the combination algorithm. Very general basic behaviors are required for learning and scalability, but for a particular task it would be great to adapt or propose new "ad hoc" basic behaviors more specialized. This would apply for example in the case of the two blocking situations on Fig. 8. This is exactly the purpose of the present article.

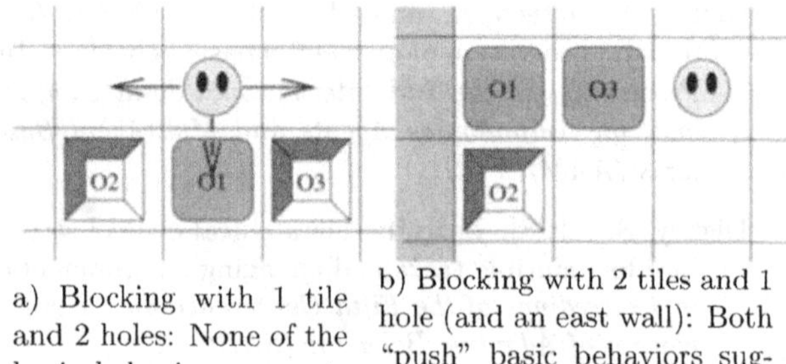

a) Blocking with 1 tile and 2 holes: None of the basic behaviors suggests to go south.

b) Blocking with 2 tiles and 1 hole (and an east wall): Both "push" basic behaviors suggest to go west, whereas it is not the right option. Going north would be better.

Figure 8: In both situations a) and b), none of the "intuitive" basic behaviors employed suggest an appropriate decision, whence an agent blocked for a long time.

Optimal and Non-Optimal Compromise Strategies in Action Selection

Frederick L. Crabbe
U.S. Naval Academy
Computer Science Department
572C Holloway Rd, Stop 9F
Annapolis, MD 21402
crabbe@usna.edu

Abstract

Devising a solution to the action selection problem rests on a definition of which actions are preferable over others. Among many properties suggested for action selection mechanisms, one prominent one is the ability to select compromise actions, i.e. actions that are not the best to satisfy any active goal in isolation, but rather compromise between the multiple goals. This paper performs a mathematical and experimental analysis of compromise behavior in an attempt to determine exactly how much compromise behavior aids an animat. It concludes that optimal compromise behavior is a small improvement over non-compromise behavior, but commonly proposed easy to compute compromise mechanisms are worse than non-compromise behavior. It further analyzes the properties of situations when compromise behavior is distinctly better than non-compromise behavior.

1. Introduction

Traditional Artificial Intelligence planning systems use search in order to fully characterize the space of actions a robotic agent (or animat) can select in a given situation. The animat considers the outcomes of possible actions into the future until it finds sequences of actions that achieve its goals. One feature of this approach is that given enough time, a planning system can determine the optimal actions for the animat. Of course, the issue of time is a fundamental problem for these planning systems: the animat may not have at its disposal the time needed in order to discover the optimal actions. In fact, often the amount of time required exceeds the age of the universe.

Behavior-based approaches to robotics and animats in general have been introduced to address these sorts of problems (Brooks, 1986, Arkin, 1998). These distributed reactive-style approaches are designed to generate "good enough" actions in a very small amount of time. Without optimality, there arises the important question of exactly what "good enough" means. In his now classic Ph.D. thesis, Tyrrell (Tyrrell, 1993) introduced a list of fourteen requirements for Action Selection Mechanisms. Of these, number twelve was "Compromise Candidates: the need to be able to choose actions that, while not the best choice for any one sub-problem alone, are best when all sub-problems are considered simultaneously." (p. 174) Tyrrell's list has had significant impact on the Action Selection field (e.g., Humphrys, 1996, Decugis and Ferber, 1998, Bryson, 2000, Girard et al., 2002), and a number of researchers have developed systems to meet the criteria he set out (e.g., Werner, 1994, Blumberg, 1994, Crabbe and Dyer, 1999). Meanwhile biologists have noted apparent compromise behavior among animals selecting between multiple targets (Morris et al., 1978, Latimer and Sippel, 1987, Bailey et al., 1990). The ability to consider compromise actions in an uncertain world makes great intuitive sense. When multiple goals interact, solving each optimally is not always optimal for the overall system. From the perspective of an animat in an environment where desired resources might go away at a moment's notice, a compromise movement in a direction in-between the resources might increase the animat's likelihood of obtaining one resource in the event that the other becomes unavailable.

Recent work has generated empirical results that seem to contradict the claim that the ability to consider compromise candidates is necessary (Bryson, 2000, Crabbe, 2002, Jones et al., 1999). Yet there have been few in-depth analyses of the nature of compromise actions and their effect on the overall success of an animat. This paper presents an extension of the work by Crabbe (Crabbe, 2002) and Hutchinson (Hutchinson, 1999) to investigate the nature of compromise actions, where an animat is faced with choosing between two destinations in the environment.

2. Problem Formulation

In this paper, we examine the situation where an animat has a goal to be co-located with one of two target locations in the environment. These could be, for instance, the locations of food, water, potential mates, or shelter. At any moment either or both of the targets can disappear from the environment. The animat must select an action that maximizes its chances of co-locating with a target before it disappears. This model is drawn from several scenarios in biology. For example, frogs or cricket males sometimes advertise for mates by emitting calls. The males may disappear with respect to the female through a cessation of signaling. This can occur either due to the actions of predators, the arrival of a competing female, or for internal reasons such as energy conservation. Another scenario in biology is that of a hunter such as a cat stalking prey such as birds in a flock, where an individual bird can fly at any moment (Hutchinson, 1999).

To approximate this type of scenario, we place an animat at the origin on a plane. Two targets are placed on the plane, one in the first and one in the second quadrant: T_a and T_b. The animat can sense the location of each target. Because the quality of the individual targets may vary, or the types of the targets may be different, the animat has two independent goals to be co-located with them. We call the strength of the goals G_a and G_b. The dynamism in the environment is represented with a probability p. This is the probability that any object in the environment will still exist after each time step. That is, any object will spontaneously disappear from the environment at each time step with probability $1-p$. Notationally, D_i^j is the distance from some location i to some location j, such that $D_{(0,0)}^{T_a}$ is the distance from the origin to the location of T_a. All distances are measured in the number of time steps it takes the animat to travel that distance.

When presented with this problem, the animat must select one action from the set of candidate actions. Because the task is a navigation task, all the candidate actions are movements to a location one time step away. Because the environment is continuous, not a discrete grid as in Tyrrell(Tyrrell, 1993), there are infinite candidate actions in our framework, each moving to a point on the unit circle centered at the animat's location. Comparing to Tyrrell's discrete world, our action choice may seem like some sort of "blending" of two choices. We argue that in fact, our formulation is identical to Tyrrell's except that the set of candidate actions our animats select from is much larger, i.e. infinite.

Once the animat has executed its action, it is again faced with the same action selection problem, unless one of the targets has disappeared and the best action is to move directly to the other target. As the animat repeatedly selects an action, the path it follows resembles a piece-wise linear approximation of a curved path (figure 1).

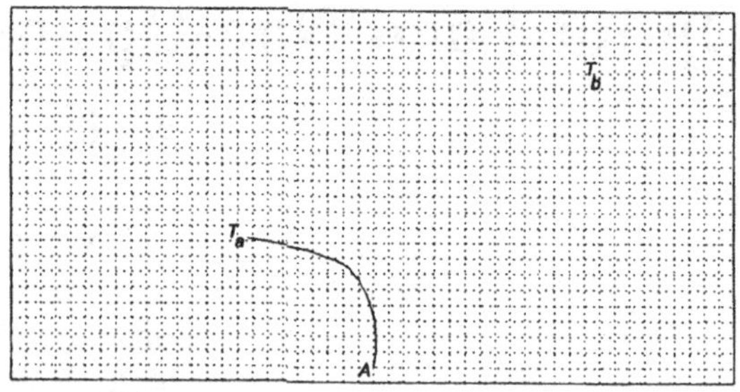

Figure 1: A possible path taken by an animat after repeated selection of a compromise candidate action. A marks the starting location of the animat, and T_a and T_b mark the locations of the targets.

It should be noted that the formulation used in this paper is one possible formulation. Another common formulation is where one of the goals is "proscriptive", that is, demands that certain actions not be performed. For example, a goal to avoid a hazard would proscribe against approaching the hazard. While Tyrrell explicitly used a proscriptive goal formulation as an example, he specifically hypothesized that "[m]ore generally, a *compromise candidate*, which might be beneficial to two or more systems to an intermediate degree, may be preferable to any of the candidates which are most beneficial for one system alone." (p. 170) (Tyrrell, 1993) This general hypothesis implies both the prescriptive version of the problem analyzed in this paper, as well as the proscriptive version[1]. Tyrrell's generalized hypothesis can be seen in the inclusion of prescriptive compromise behavior in many action selection mechanisms, such as Werner (Werner, 1994) and Montes-Gonzales et al. (Montes-Gonzales et al., 2000). Furthermore, outside the animat community, biologists have advocated in favor of the prescriptive version of compromise behavior for some time (Morris et al., 1978, Latimer and Sippel, 1987, Bailey et al., 1990).

3. Analytical Set-up

In order to investigate compromise candidates, we will analyze the initial configuration using Utility Theory (Howard, 1977). Utility Theory assigns a set of numerical values (utilities) to states of the world. These utilities represent the usefulness of that state to an animat. Expected Utility (EU) is a prediction of the eventual total utility an animat will receive if it takes a particular action in a particular state. The Expected Utility (EU) of taking an action A in a state S is the sum of the product

[1] By prescriptive, we mean that both goals are positive, enticing the animat to do something, rather than prohibiting actions.

of the probability of each outcome that could occur and the utility of that outcome:

$$EU(A|S) = \sum_{S_o \in Outcomes} P(S_o|A,S)U_h(S_o) \quad (1)$$

where $P(S_o|A,S)$ is the probability of outcome S_o occurring given that the animat takes action A in state S, and $U_h(S_o)$ is the historical utility of outcome S_o as defined below.

Assuming the animat is rational, the set of goals to consume objects will be order isomorphic[2] to the set of the animat's utilities of having consumed the objects. Therefore, EU calculated with utilities is order isomorphic with EU calculated with goals instead. For our purposes, we will assume that the goals and utilities are equivalent ($G_a = U(T_a)$).

Because a rational animat is expected to select the action with the largest EU, the historical utility of a state is the utility of the state plus future utility, or the max of the expected utility of the actions possible in each state:

$$U_h(S) = U(S) + \max_{A \in Actions} EU(A|S). \quad (2)$$

An animat can calculate EU using multiple actions in the future by recursively applying equations (1) and (2).

3.1 *Optimal Behavior*

We analyze compromise behavior by comparing a close approximation of optimal behavior with several non-optimal but easy to generate behaviors. We approximate the optimal behavior based on the technique used by Hutchinson (Hutchinson, 1999). This technique overlays a grid of points on top of the problem space and calculates the maximal expected utility of each location given optimal future actions. This is done recursively starting at the target locations and moving outward until stable values have been generated for all grid points.

The value we are trying to calculate is the expected utility of acting optimally at some location in a state where the two targets are still in the environment: $EU(Opt|T_a, T_b, (x,y))$. If θ is the angle of the optimal move for the agent at location (x,y) and (x',y') is 1 unit away from (x,y) in direction θ, then by equations 1 and 2 the expected utility of being at (x,y) is:

$$\begin{aligned} EU(Opt|T_a,T_b,(x,y)) =& p^2 EU(Opt|T_a,T_b,(x',y'))+ \\ & p(1-p)EU(Opt|T_a,(x',y'))+ \\ & p(1-p)EU(Opt|T_b,(x',y')), \end{aligned} \quad (3)$$

$$EU(Opt|T_a,(x',y')) = G_a p^{D^{T_a}_{(x',y')}}, \text{ and,} \quad (4)$$

$$EU(Opt|T_b,(x',y')) = G_b p^{D^{T_b}_{(x',y')}}. \quad (5)$$

the total expected utility is the expectation over four possible situations: both targets there, both targets gone, a there but b gone, and vice versa. When one of the targets disappears from the environment, the optimal action for the animat to take is to move directly to the other target, as shown in equations 4 and 5.

Using the above equations, the expected utility of each grid point in the environment can be calculated provided $EU(Opt|T_a,T_b,(x',y'))$ can be accurately determined and θ can be found. If (x',y') is a grid-point, this can be done easily, otherwise the local EU function must be interpolated from the expected utility values of the surrounding grid points. Using the interpolated surfaces for the local values of $EU(Opt|T_a,T_b,(x',y'))$, the value of θ can be determined by searching for the angle that maximizes the function described by equation 3. Once the expected utility is determined for a grid point, its value is then used to calculate the expected utility of its neighboring grid-points. This process is repeated until values are collected for all the grid points. Because the estimated utility value can change for a point when the values of its neighbors change, the values of all the points are repeatedly re-estimated until the values stabilize. The pseudocode for the algorithm is given in figure 2.

3.2 *Action Selection Mechanisms*

It is typically computationally prohibitive for an animat to calculate the optimal action using a technique similar to the one described in the previous section. Instead, most ethologists and AI researchers propose easy to compute action selection mechanisms that are intended to approximate the optimal behavior (Cannings and Orive, 1975, Fraenkel and Gunn, 1961, Römer, 1993). The mechanisms can be divided into two categories: those that select a single target and move directly toward it, and those that exhibit some sort of compromise behavior. In the former category, we will consider:

- Closest (C): select the closest target.
- Maximum Utility (MU): select the target with the higher utility (G_a or G_b).

[2] "Two totally ordered sets $(A, \leq)$ and $(B, \leq)$ are order isomorphic iff there is a bijection from A to B such that for all $a_1, a_2 \in A, a_1 \leq a_2$ iff $f(a_1) \leq f(a_2)$." (Weisstein, 2001)

```
Open is a list of grid-points that need to be updated.
Closed table of updated points.
N is the point currently being updated.
V_N is the current EU estimate at N.
repeat
    Open ← enqueue target locations
    Closed ← ∅
    repeat
        N ← dequeue from Open
        when N ∉ Closed
            EU(Opt|T_a, T_b, (x', y')) ← interpolated EU
                    function from neighboring points
            V_N ← max equation 3
            Open ← enqueue neighbors of N
            Closed ← add N
    until Open = ∅
until convergence
```

Figure 2: Algorithm for estimating the expected utility at all the grid points in the environment.

- Maximum Expected Utility (MEU): select the target with the higher expected utility if it were the only target in the environment ($p^{D^{T_a}_{(0,0)}} G_a$ vs. $p^{D^{T_b}_{(0,0)}} G_b$)

Of action selection mechanisms that exhibit compromise behavior, we will examine:

- Forces (F): the animat behaves as if it has two forces acting on it, where the strength of the force is proportional to the utility of the target divided by the square of the distance between the animat and the target location (e.g. $M = \frac{G_a}{D^{T_a}_{(0,0)}}$).

- Signal Gradient (SG): The animat behaves as if it is following a signal gradient. The targets emit a simulated "odor" that falls with the square of the distance from the target. The initial strength of the odor is proportional to the utility of the target. The animat moves to the neighboring location that has the strongest odor as the sum of the odor emanating from each of the two targets.

- Exponentially Weakening Forces (EWF): The pulling effects of the targets in the previous two mechanisms fall quadratically with distance, whereas the expected utility of targets falls exponentially with distance. Exponentially Weakening Forces tries to model this by behaving just as with the Forces mechanism, but the strength of the forces falls exponentially with distance, rather than quadratically. The magnitudes of the two vectors are $G_a p^{D^{T_a}_{(x,y)}}$ and $G_b p^{D^{T_b}_{(x,y)}}$.

The expected utility of each of these mechanisms can be calculated for any particular scenario by using equations 3, 4 and 5. Let n be the number of time steps it takes to come within a single move of a target using one of the mechanisms. Let T_e be the the eventual target that the animat arrives at, and ϵ be the distance between the animat after n steps and T_e. Finally, let $D^{T_a}_x$ be the distance from the location of the animat after the xth time step to the target T_a; let V be the current strategy, and $EU(V|x)$ be the expected utility after time step x. For the first n time steps, the expected utility at any time step x of using strategy V can be figured by:

$$\begin{aligned} EU(V|x) = & p^2 EU(V|x+1) + p(1-p)p^{D^{T_b}_{x+1}} G_b + \\ & p(1-p)p^{D^{T_a}_{x+1}} G_a, \end{aligned} \tag{6}$$

where the location of $x+1$ is determined by the strategy. The expected utility of the final step (when the animat is within ϵ of T_e) is:

$$EU(V|n) = (1-p^\epsilon)p^\epsilon p^{D^{T_f}_{T_e}} G_f + (1-p^\epsilon)p^\epsilon G_e + p^{2\epsilon} G_e. \tag{7}$$

These two equations can be combined to yield:

$$\begin{aligned} & EU(V|0) = \\ & \quad p^{2n} \left[(1-p^\epsilon)p^\epsilon p^{D^{T_f}_{T_e}} G_f + (1-p^\epsilon)p^\epsilon G_e + p^{2\epsilon} G_e \right] + \\ & \quad \sum_{i=1}^{n} p^{2i-1}(1-p) \left[p^{D^{T_b}_i} G_b + p^{D^{T_a}_i} G_a \right] \end{aligned} \tag{8}$$

Using this equation, we can evaluate each of the action selection mechanisms by generating a large number of scenarios and comparing the expected utility of each of the mechanisms with each other and with optimal behavior.

4. Experiments

In the first set of experiments we calculate and compare the expected utility of various strategies. In the second set, we determine some properties of scenarios that lead to good behavior by some of the strategies. For the experiments, we generated 50,000 compromise scenarios. In these scenarios, Target a (T_a) was randomly placed in the x-range of $(-100; 0)$ and the y-range of $(0; 100)$. Target b (T_b) was randomly placed in the x-range of $(0; 100)$ and the y-range of $(0; 100)$. The utility of each target is randomly generated in the range of $(0; 100)$. We also randomly generated p, the probability that a target would remain in the environment per time step, in the range $[0.9; 1)$[3].

4.1 Strategy Comparison Experiments

For each scenario, we calculated the expected utility of each of the action selection mechanisms described in

[3]Below 0.9, the probability that the animat would reach either of the targets approaches zero and is therefore less interesting.

	C	MU	MEU
% best	0.1471	0.5309	99.32
avg % EU over C	0.0	4.130	12.62
avg % EU over MU	9.353	0.0	15.31
avg % EU over MEU	-5.960	-8.491	0.0

Figure 3: Comparison of non-compromise strategies. For each strategy, the percentage of situations where that strategy works best is provided, along with the average expected utility improvement of each strategy compared to the other strategies.

	F	SG	EWF	Optimal
avg	-4.074	-2.791	-2.465	1.121
best	4.844	4.819	20.56	22.73

Figure 4: Comparison of compromise strategies to the non-compromise MEU. For each strategy, the average expected utility improvement over MEU and the single best expected utility improvement over MEU is provided.

the previous section: closest (C), maximal utility (MU), maximum expected utility (MEU), forces (F), signal gradient (SG), and exponentially weakening forces (EWF). We also estimated the expected utility of optimal behavior using the technique in section 3.. We compare the three non-compromise mechanisms (C, MU, and MEU) in the table in figure 3. In the table, we show the percentage of scenarios each strategy was the best (had highest expected utility). We also compare the average percentage improvement of expected utility using each strategy to using the other strategies. We see that the maximum utility strategy is clearly the best of the three as it almost always selects the better target and its overall expected utility is from 12% to 15% better than the other two strategies. It is interesting to note that Closest is a better strategy than Maximum Utility. This is likely because the expected utility falls exponentially with distance, so that closer targets are more useful than better targets.

The table in figure 4 compares the compromise based mechanisms with MEU. It shows both the average % improvement over MEU and the % improvement over MEU in the single best scenario. There are three interesting things to note in this table. The first is that the optimal strategy is only 1.1% better than the non-compromise based MEU. This is somewhat surprising, as one would expect that the optimal behavior would be much higher than a non-compromise approach. This result is consistent with the non-continuous space experiments of Crabbe (Crabbe, 2002) and the much smaller study in Hutchinson (Hutchinson, 1999). The result might also explain the results of researchers who find compromise behavior unnecessary. If the environment is sufficiently forgiving, the small improvement of optimal compromise behavior may go unnoticed.

The second thing to note is that *all* of the non-optimal compromise based strategies are *worse* than the MEU strategy. In order to double check these surprising results, we selected 50 initial configurations for which we calculated the expected utility using each of the strategies and we re-estimated the expected utility using Monte Carlo simulation. For each situation we ran 100,000 trials where an animat ran the selected strategy, and at each time step we randomly determined if each target remained in the environment. A trial ended when a target was reached or both targets disappeared. The total utility of all reached targets was divided by the number of trials to estimate the expected utility. We found that of the 50 scenarios selected, the Monte Carlo estimated expected utilities agreed with the directly calculated expected utilities to, on average, four decimal places, lending support the the accuracy of figure 4. These results may further explain why some researchers have found that compromise behavior is unnecessary: the commonly used compromise strategies perform worse than a non-compromise strategy. With respect to natural animals, this result implies that animals that exhibit compromise behavior are either: a) behaving non-optimally; b) using an as yet unproposed compromise based action selection strategy (or at least one of which we are unaware); c) behaving in that manner for reasons other than purely to compromise between two targets (Hutchinson (Hutchinson, 1999) suggests several other possible reasons for what appears to be two-target based compromise behavior).

The final property of figure 4 to note is that EWF is the best performing of the easy to compute compromise strategies that we tested. While it is certainly not conclusive, this may imply that the approach of decreasing the influence of farther targets exponentially is a good one for developing action selection strategies. If we examine the *best* scenario for EWF, we see that it is nearly as good as the best scenario for optimal. This brings up the questions of *when* are optimal and EWF most beneficial.

4.2 Analysis Experiments

While the conclusions of the previous section may be true as an average across all the initial scenarios generated, there may be types cases where the Optimal or EWF strategies are significantly better than the MEU. In this section, we use the idea of information gain from information theory to attempt to determine if this is so. For example, in previous work on a discrete two-choice simplification of the problem, Crabbe (Crabbe, 2002) claimed that a $p > 0.9$ was most important for there to be good compromise behavior, followed by the distances to the targets to be equal.

4.2.1 Information Theory

In order to analyze the properties of the situations in which there are good compromise candidates, we use the idea of information gain. This is the same technique used in the decision tree algorithms ID3 and C4.5 (Mitchell, 1997, Quinlan, 1983, Quinlan, 1993). The technique begins with a data set and a classification of the data elements. Our data will be the situations generated in previous experiments, and our classification will be whether or not the scenario has a compromise strategy with an EU better than MEU. The technique then considers a set of n-ary attributes which partition the data set. For each attribute, the technique considers the information gain of applying that attribute to partition the data set. Information gain is the decrease in entropy of the data set, where entropy is:

$$E(S) = \sum_{i \in Categories} -p_i \log_2 p_i \tag{9}$$

S is a data set, i is a category of the classification, and p_i is the proportion of S categorized as i. Entropy measures the inverse of the purity of a data set with respect to the classification. A set is most pure when all the data have the same classification, and least pure when the data is classifies evenly across all categories.

The information gain of dividing set S based on attribute A is defined as:

$$Gain(S, A) = E(S) - \sum_{v \in Values(A)} \frac{|S_v|}{|S|} E(S_v), \tag{10}$$

where v is a value of an attribute, $|S|$ is a size of set S, S_v is the subset of S for which attribute A takes value v, The attribute with the largest information gain is the attribute which best partitions the data set, and in our case is the property of a situation which most determines if there is a good compromise candidate.

The data describing a situation is continuous information, but attributes are n-ary classifiers. To cope with this, the attributes partition the values into ranges, such as every 0.05. The attributes we use are listed below:

- $D^{T_a}_{(0,0)}, D^{T_b}_{(0,0)}, D^{T_b}_{T_a}$. These three attributes are the distances between the key locations in the situation. The distances can be anywhere from 0 to 224, and the attributes break them up into ranges of 5 units.
- G_a, G_b. The goal values from 0 to 100, broken into ranges of 5 units.
- p. The value of p from 0 to 1, broken into ranges of 0.005.
- $R(D^{T_a}_{(0,0)}, D^{T_b}_{(0,0)})$. This encodes the relationship between $D^{T_a}_{(0,0)}$ and $D^{T_b}_{(0,0)}$, while keeping the values between 0 and 1 using the formula $\frac{D^{T_a}_{(0,0)}}{D^{T_a}_{(0,0)}+D^{T_b}_{(0,0)}}$. It is broken into ranges of 0.05.
- $R(D^{T_a}_{(0,0)}, D^{T_b}_{T_a})$. This encodes the relationship between $D^{T_a}_{(0,0)}$ and $D^{T_b}_{T_a}$, while keeping the values between 0 and 1 using the formula $\frac{D^{T_a}_{(0,0)}}{D^{T_a}_{(0,0)}+D^{T_b}_{T_a}}$. It is broken into ranges of 0.05.
- $R(D^{T_b}_{(0,0)}, D^{T_b}_{T_a})$ This encodes the relationship between $D^{T_b}_{(0,0)}$ and $D^{T_b}_{T_a}$, while keeping the values between 0 and 1 using the formula $\frac{D^{T_b}_{(0,0)}}{D^{T_b}_{(0,0)}+D^{T_b}_{T_a}}$. It is broken into ranges of 0.05.
- $R(G_a, G_b)$. This encodes the relationship between G_a and G_b, while keeping the values between 0 and 1 using the formula $\frac{G_a}{G_a+G_b}$. It is broken into ranges of 0.05.
- $R_l(G_a, D^{T_a}_{(0,0)}, G_b, D^{T_b}_{(0,0)})$ This encodes a *linear* relationship between $G_a, D^{T_a}_{(0,0)}, G_b$ and $D^{T_b}_{(0,0)}$, while keeping the values between 0 and 1 using the formula $\frac{G_a D^{T_a}_{(0,0)}}{G_a D^{T_a}_{(0,0)}+G_b D^{T_b}_{(0,0)}}$. It is broken into ranges of 0.05.
- $R_e(G_a, D^{T_a}_{(0,0)}, G_b, D^{T_b}_{(0,0)})$ This encodes a *exponential* relationship between $G_a, D^{T_a}_{(0,0)}, G_b$ and $D^{T_b}_{(0,0)}$, while keeping the values between 0 and 1 using the formula $\frac{G_a p^{D^{T_a}_{(0,0)}}}{G_a p^{D^{T_a}_{(0,0)}}+G_b p^{D^{T_b}_{(0,0)}}}$. It is broken into ranges of 0.05.

4.2.2 Optimal

We first classify the data based on whether the performance of the optimal behavior is a 1% or greater improvement over the performance of MEU for that scenario. Using equation 10, we find that attribute with the largest information gain is $R_e(G_a, D^{T_a}_{(0,0)}, G_b, D^{T_b}_{(0,0)})$ with a gain of 0.07243. Figure 5 shows a breakdown of the data as partitioned by the attribute. The x-axis is the range of values of the attribute, and the y-axis is the percentage of the data that falls in that range and is positively classified (the performance of the optimal behavior is a 1% or greater improvement over the performance of MEU). In the figure we see that the greatest likelihood of good compromise behavior is when the attribute is close to 0.5. This means that if $G_a p^{D^{T_a}_{(0,0)}}$ is close to $G_b p^{D^{T_b}_{(0,0)}}$ (when the expected utilities of the two targets are roughly equal), it is more likely that optimal behavior will confer a large advantage. This makes sense that the good compromise behavior would occur when the targets have the same expected utility.

The attribute with the next largest information gain is $R(D^{T_a}_{(0,0)}, D^{T_b}_{(0,0)})$ with a gain of 0.06917. Figure 6 shows a breakdown of the data as classified by the attribute. Once again, we see that the greatest likelihood of good

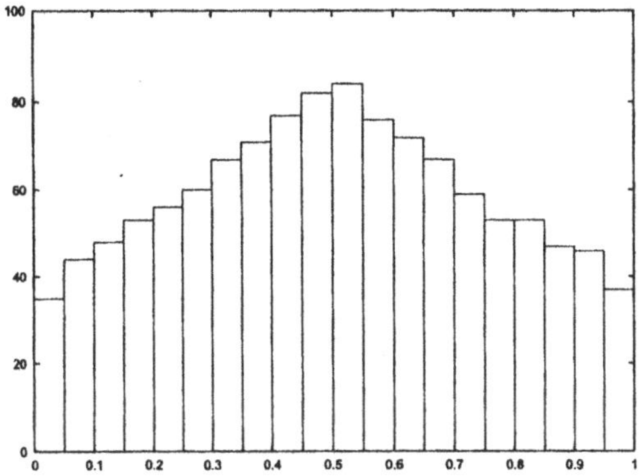

Figure 5: Breakdown of attribute $R_e(G_a, D^{T_a}_{(0,0)}, G_b, D^{T_b}_{(0,0)})$ when comparing optimal to MEU.

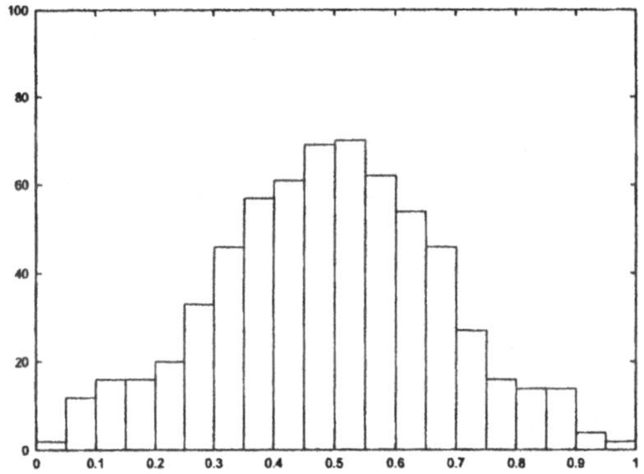

Figure 6: Breakdown of attribute $R(D^{T_a}_{(0,0)}, D^{T_b}_{(0,0)})$ when comparing optimal to MEU.

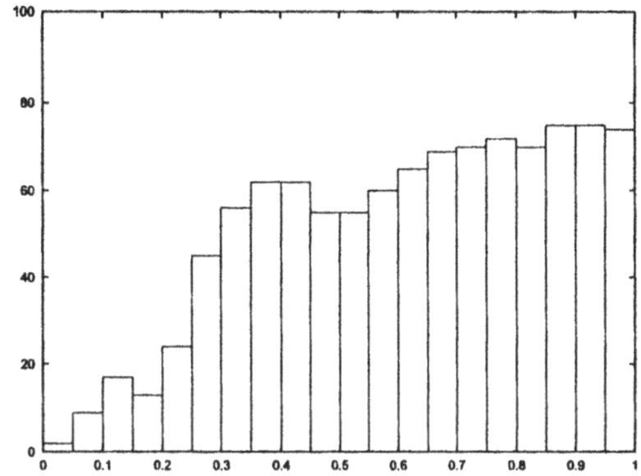

Figure 7: Breakdown of attribute $R(D^{T_a}_{(0,0)}, D^{T_b}_{T_a})$ when comparing optimal to MEU.

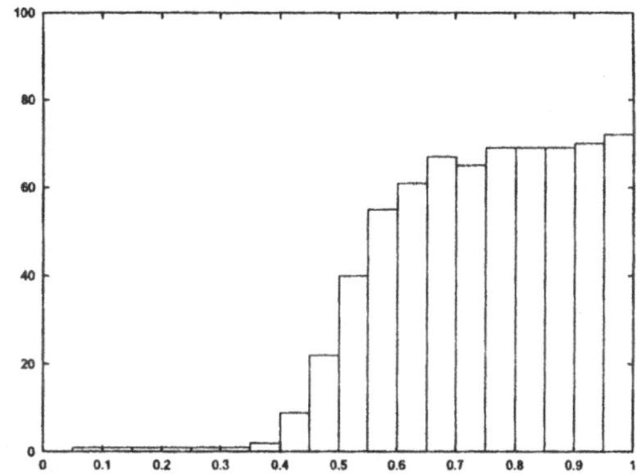

Figure 8: Breakdown of attribute $R(D^{T_a}_{(0,0)}, D^{T_b}_{T_a})$ when comparing EWF to MEU.

compromise behavior is when the attribute is close to 0.5, or in this case when $D^{T_a}_{(0,0)}$ and $D^{T_b}_{(0,0)}$ are roughly equal. We believe this is another manifestation of the need for a balance in expected utility for there to be good compromise behavior.

The next two attributes with high gain are $R(D^{T_a}_{(0,0)}, D^{T_b}_{T_a})$ and $R(D^{T_b}_{(0,0)}, D^{T_b}_{T_a})$, both with a gain of 0.027. Figure 6 shows a breakdown of the data as classified by $R(D^{T_a}_{(0,0)}, D^{T_b}_{T_a})$ (the plot of the other attribute is qualitatively identical). While the data in this figure is less distinct than in the previous figures, it does appear to show that there are *few* good compromises when the targets are relatively close to the animat but far away from each other.

4.2.3 Exponentially Weakening Forces

Although on average EWF performs worse than MEU, it is interesting to analyze the situations in which it performs better. If these situations could be recognized, then EWF could be applied only when they occur. Furthermore, it may lead to insight on why EWF outperformed the other non-optimal strategies. The data is classified based on whether or not EWF out performs MEU. The top two attributes in terms of information gain are $R(D^{T_a}_{(0,0)}, D^{T_b}_{T_a})$ and $R(D^{T_b}_{(0,0)}, D^{T_b}_{T_a})$, with gains of 0.1620 and 0.1569 respectively. Figure 8 shows a breakdown of the data as classified by $R(D^{T_a}_{(0,0)}, D^{T_b}_{T_a})$ and figure 9 shows a breakdown of the data as classified by $R(D^{T_b}_{(0,0)}, D^{T_b}_{T_a})$.

These graphs strongly suggest that good compromises using EWF occur when the targets are relatively far away from the origin and close to each other. This is backed up by examining the next highest attribute, $D^{T_b}_{T_a}$ (0.09223) in figure 10. Here we see that good compromises are more likely when $D^{T_b}_{T_a}$ is small, that is, the two targets are close together.

The last attribute we examine is p, with an information gain of 0.07903. Figure 11 shows its breakdown. It shows that good performance peaks at a p range of 0.91-

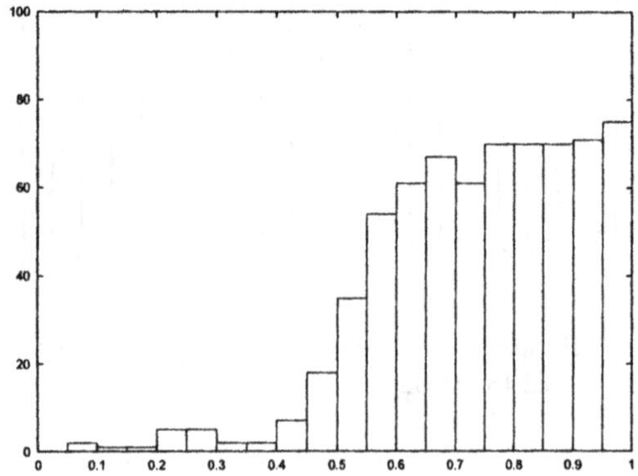

Figure 9: Breakdown of attribute $R(D^{T_b}_{(0,0)}, D^{T_b}_{T_a})$ when comparing EWF to MEU.

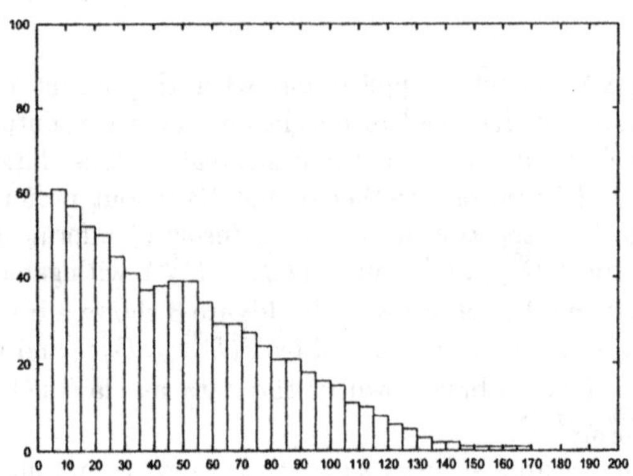

Figure 10: Breakdown of attribute $D^{T_b}_{T_a}$ when comparing EWF to MEU.

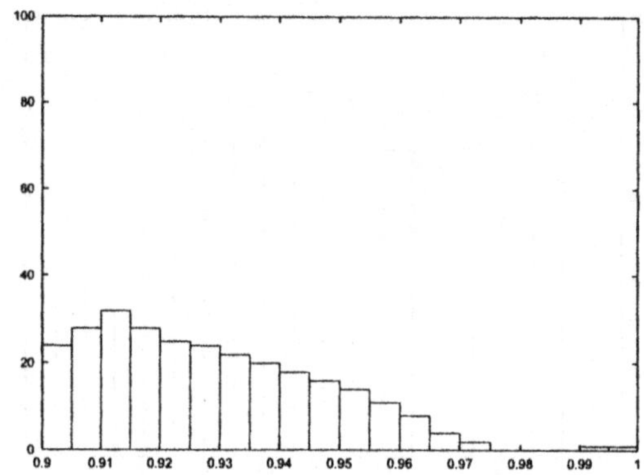

Figure 11: Breakdown of attribute p when comparing EWF to MEU.

0.915. This is consistent with Crabbe's previous results that good compromises occur with $p > 0.9$. That study only looked at p with a granularity of 0.1, and was thus unable to determine which values of $p > 0.9$ were best.

There is some overlap between the good attributes for the optimal behavior and for EWF, namely that the targets be relatively close together and far away from the animat. We have no evidence to explain why $R_e(G_a, D^{T_a}_{(0,0)}, G_b, D^{T_b}_{(0,0)})$ does not appear to be a factor for EWF, instead having a paltry information gain of 0.004855.

5. Conclusions and Future Work

In this paper we have analyzed the properties of action selection mechanisms in a scenario that has been of interest to both biologists and AI researchers in the past. We believe the paper makes several contributions. First, it derives a closed form equation for calculating the expected utility of any strategy in the multiple goal scenario (equation 8). This equation can also be trivially extended to any number of goals. As part of ongoing work, it has been modified to describe a proscriptive goal scenario was well. Next, the paper shows that among non-compromise strategies, closest is better than maximum utility, but the maximum expected utility is the best of the three. Maximum expected utility should be used if a non-compromise strategy is desired. When comparing compromise strategies, this paper concludes that optimal behavior performs only slightly better than maximum expected utility, and all other known compromise strategies perform worse than maximum expected utility. From this we conclude that animals that exhibit apparent compromise behavior are either using some unknown strategy or are doing so for some other reason. Of the non-optimal compromise strategies tested, exponentially weakening forces was the best, suggesting that if a better non-optimal compromise strategy exists, it may also involve forces that fall exponentially with distance. This is a clear area of potential future work. When analyzing when the optimal and exponentially weakening forces strategies are better, we concluded that good optimal strategies occur when the expected utility of the two targets are equal. We further concluded that good scenarios for the optimal strategy occur when the targets are far away from the origin and close together. For the exponentially weakening forces strategy, we also concluded that good scenarios occur when the targets are far away and close together. Explaining why equality of the expected utility of the targets does not appear to be a factor is another area of future work.

Acknowledgements

We would like to thank Chris Brown and Rebecca Hwa for many wonderful discussions. We would also like to thank the anonymous reviewers for several helpful comments. This work was sponsored in part by a grant from the Office of Naval Research, number N0001403WR20241.

References

Arkin, R. (1998). *Behavior-Based Robotics.* MIT Press, Cambridge, MA.

Bailey, W. J., Cunningham, R. J., and Lebel, L. (1990). Song power, spectral distribution and female phonotaxis in the bushcricket requena verticalis (tettigoniiddae: Orthoptera): active femail choice or passive attraction. *Animal Behavior.*

Blumberg, B. M. (1994). Action-selection in hamsterdam: Lessons from ethology. In *Proceedings of the Third International Conference on Simulation of Adaptive Behavior.*

Brooks, R. (1986). A robust layered control system for a mobile robot. *IEEE Journal of Robotics and Automation*, RA-2:14–23.

Bryson, J. (2000). Hierarchy and sequence vs. full parallelism in action selection. In *Proceedings of the Sixth International Conference on Simulation of Adaptive Behavior.*

Cannings, C. and Orive, L. M. C. (1975). On the adjustment of the sex ratio and the gregarious behavior of animal populations. *Journal of Theoretical Biology*, 55:115–136.

Crabbe, F. (2002). Compromise candidates in positive goal scenarios. In *Proceedings of the Seventh International Conference on Simulation of Adaptive Behavior.*

Crabbe, F. L. and Dyer, M. G. (1999). Second-order networks for wall-building agents. In *Proceedings of the International Joint Conference on Neural Networks.*

Decugis, V. and Ferber, J. (1998). An extension of maes' action selection mechanism for animats. In *Proceedings of the Fifth International Conference on Simulation of Adaptive Behavior.*

Fraenkel, G. S. and Gunn, D. L. (1961). *The Orientation of Animals.* Dover, New York.

Girard, B., Cuzin, V., Guillot, A., Gurney, K. N., and Prescott, T. J. (2002). Comparing a brain-inspired robot action selection mechanism with 'winner-takes-all'. In *Proceedings of the seventh international conference on simulation of adaptive behavior: From animals to animats*, pages 75–84.

Howard, R. (1977). Risk preference. In *Readings in Decision Analysis.* SRI International, Menlo Park, CA.

Humphrys, M. (1996). *Action Selection methods using Reinforcement Learning.* PhD thesis, University of Cambridge.

Hutchinson, J. (1999). Bet-hedging when targets may disappear: optimal mate-seeking or prey-catching trajectories and the stability of leks and herds. *Journal of Theoretical Biology*, 196:33–49.

Jones, R. M., Laird, J. E., Nielsen, P. E., Coulter, K. J., Kenny, P., and Koss, F. V. (1999). Automated intelligent pilots for combat flight simulation. *AI Magazine*, 20(1).

Latimer, W. and Sippel, M. (1987). Acoustic cues for female choice and male competition in the tettigonia cantans. *Animal Behavior.*

Mitchell, T. M. (1997). *Machine Learning.* McGraw-Hill, Boston.

Montes-Gonzales, F., Prescott, T. J., Gurney, K., Humphrys, M., and Redgrave, P. (2000). An embodied model of action selection mechanisms in the vertebrate brain. In *Proceedings of the Sixth International Conference on Simulation of Adaptive Behavior*, pages 157–166.

Morris, G. K., Kerr, G. E., and Fullard, J. H. (1978). Phonotactic preferences of female meadow katydids. *Canadian Journal of Zoololgy.*

Quinlan, J. R. (1983). Learning efficient classification proceedures and their application to chess and games. In Michalski, R., Carbonell, J., and Mitchell,

T. M., (Eds.), *Machine Learning: an artificial intelligence approach.* Morgan Kaufmann, San Mateo, CA.

Quinlan, J. R. (1993). *C4.5: Programs for Machine Learning.* Morgan Kaufmann, San Mateo, CA.

Römer, H. (1993). Environmental and biological constraints for the evolution of long-range signalling and hearing in acoustic insects. *Philosophical Transactions of the Royal Society B*, 340:179–185.

Tyrrell, T. (1993). *Computational Mechanism for Action Selection.* PhD thesis, University of Edinburgh.

Weisstein, E. W. (2001). *Eric Weisstein's World of Mathematics.* http://mathworld.wolfram.com/.

Werner, G. M. (1994). Using second order neural connection for motivation of behavioral choices. In *Proceedings of the Third International Conference on Simulation of Adaptive Behavior*, pages 154–161.

Using Hormonal Feedback to Modulate Action Selection in a Competitive Scenario

Orlando Avila-García and Lola Cañamero
Adaptive Systems Research Group
Department of Computer Science, University of Hertfordshire
College Lane, Hatfield, Herts AL10 9AB, UK
O.Avila-Garcia@herts.ac.uk, L.Canamero@herts.ac.uk

Abstract

In this paper we investigate the use of hormonal feedback as a mechanism to modulate a "motivation-based," homeostatic action selection mechanism (ASM) in a robot. We have framed our study in the context of a dynamic, multi-robot, competitive "two-resource" action selection problem. The introduction of competitors has important consequences for action selection. We first show how the interaction between robots introduces new forms of environmental complexity that affect their viability. Secondly, we propose a "hormone-like" mechanism that, modulating the input of the ASM, tackles these new sources of complexity.

1 Introduction

The choice of appropriate behavioral subsystems is one of the major problems of designing behavior selection architectures. While achieving flexible behavior adapted to complex environments requires a non-trivial set of subsystems, including too many leads to problems arising from overdesign. A balance must be found between the "machinery" used and the functionalities required from this underlying structure.

In biological systems, the bath of chemicals where neurons are placed, and that we will refer to here as neurohormones, is the major responsible for the plasticity of the nervous system. Neurohormones can act at multiple levels of the nervous system, from sensory to motor neurons, and can evoke a spectrum of different responses on different target neurons. For example, studies in invertebrates have shown that the same anatomical neural circuit can produce different motor patterns or behaviors in the presence of different neurohormones (see (Harris-Warrick *et al.* 1992) for a general overview). These modulatory inputs are thus seen as a set of instructions for building different functional networks from the same anatomically defined neural circuit (Kravitz 1988, Harris-Warrick *et al.* 1992).

Different mechanism inspired from the role of hormones have been proposed for behavioral learning and coordination (Brooks & Viola 1990, Grand *et al.* 1997), and emotion modelling (Cañamero 1997). In this paper, we take inspiration from the ability of neurohormones to rewire the same anatomically defined neural circuit; we explore a mechanism that modulates an action selection architecture to achieve different functionalities adapted to different environmental conditions. We have framed our study in the context of a dynamic "two-resource" action selection problem (TRP) extended with competitors, in which two robots competing for the same resources must choose between different activities in order to survive. We first show how the interaction between robots introduces various types of environmental complexity that affects the viability of the homeostatically controlled robots. Secondly, we propose a "hormonal feedback" that, acting at the sensory level, modulates the network of interrelations within the action selection architecture, biasing the behavior of the robot toward more adaptive responses to the environment—and giving rise to interesting emergent behavioral phenomena when both robots interact.

The remainder of the paper is as follows. Section 2 describes our action selection architecture, which is framed within the TRP with competitors (Section 3). Section 4 provides background theory necessary to understand the analysis of results. Section 5 reports the preliminary robotic experiments within the TRP, while in Section 6 we describe the implementation of a hormonal system to improve the performance of the robots, and the subsequent experiments. Finally, Section 7 draws some conclusions.

2 Behavior Selection Architecture

In previous work we defined different behavior (or action) selection architectures sharing the same general framework and components but implementing different Voting-Based (VB) and Winner-Take-All (WTA) policies. In this study, we have only used the WTA version of this architecture since our goal was to investigate action selection in a dynamic environment, and previous results (Avila-García & Cañamero 2002) showed that a WTA policy outperformed VB ones in dynamic environments[1]. Our architecture is neither strictly flat (parallel) nor hierarchical (structured), but a combination of both. It

[1]This does not imply that WTA outperforms VB in general, since in another study (Avila-García *et al.* 2003) a VB policy outperformed WTA in a static two-resource problem scenario.

consists of two layers—motivational and behavioral—linked through a synthetic physiology, leading to a two-step computation of intensity. This computation is parallel within each layer, but motivational intensity must be computed prior to the calculation of behavioral intensity, since the latter depends on the former. The motivational layer is made of motivational states that set the goals of the system—the tendency to satisfy bodily (physiological) or internal needs. The behavioral layer implements different ways in which those bodily needs can be satisfied. This distinction between motivations and behaviors is essential when implementing more than one behavior satisfying the same motivation (Toates 1986). Although for the simplified implementation of WTA that we use in this paper such a distinction is not necessary (see Figure 1), we have preferred to use the same model as in our previous studies to be able to compare our results, and also to allow scalability when other sources of dynamism are considered in future studies.

The Physiology consists of a number of survival-related, homeostatically controlled variables—abstractions representing the level of internal resources that the agent needs in order to survive. They must be kept within a range of values for the robot to stay "alive," thus defining a physiological space (Sibly and McFarland 1974) or viability zone (Ashby 1952, Meyer 1995) within which survival (continued existence) is guaranteed, whereas transgression of these boundaries leads to "death."

Motivational states are abstractions representing tendencies to behave in particular ways as a consequence of internal and external factors (Toates 1986). Internal factors are mainly (but not only) bodily or physiological deficits or needs (d_i), also traditionally known as "drives," that set urges to action to maintain the state of the controlled physiological variables within the viability zone. External factors are environmental stimuli or incentive cues (c_i) that allow to execute (consummatory) behaviors and hence to satisfy bodily needs. In our implementation, each motivation performs homeostatic control of one physiological variable. In that control, a feedback detector generates an error signal when the value of the controlled variable departs from its ideal value (set point). Thus, each motivation receives an intensity (activation level) proportional to the magnitude of the error. The motivational state is then the tendency to correct that error or deficit through the execution of behaviors. The homeostatic control of the physiological variable is also characterized by a limit or lethal boundary, the transgression of which leads to the robot "death." In addition to physiological deficits (d_i) and the presence of environmental cues (c_i), *other factors* may influence motivational states, such as the quality of the stimulus (e.g., palatability of food), abnormal bodily states (e.g., "illness"), hormonal levels (Cañamero 1997), etc. To account for these factors we have introduced a parameter (α) in the computation of the motivations' intensity (see Equation 1).

Motivation	Physiological Drive	Limit	Set point	Ext. Stim.
m_{cold}	$\downarrow d_{temperature}$	0	100	c_{heat}
$m_{fatigue}$	$\downarrow d_{energy}$	0	100	c_{food}

Table 1: Motivations used. Physiological drive represents the tendency of the motivation to decrement the physiological deficit until the correspondent variable reaches a set point of 100.

Behavior	Type	Stimulus	Effects on physiology (f_{ij})
b_{avoid}	*Reflex.*	Obstacle	$+0.2\,d_{temp}, +0.2\,d_{energy}$
b_{warmup}	*Consum.*	c_{heat}	$-1.0\,d_{temp}, +0.3\,d_{energy}$
b_{feed}	*Consum.*	c_{food}	$+0.3\,d_{temp}, -1.0\,d_{energy}$
b_{search}	*Appet.*	None	$+0.2\,d_{temp}, +0.2\,d_{energy}$

Table 2: Behaviors used by the WTA architecture.

Different combination rules for external and internal factors when computing motivational states have been discussed extensively by ethologists, e.g. (McFarland 1976). Following ideas proposed by Tyrrell (1993) and McFarland & Spier (1997), we have adopted a *Deficit* + *Cue* × *Deficit* model:

$$m_i = d_i + (d_i \times \alpha c_i) \tag{1}$$

Behaviors are coarse-grained subsystems (embedding simpler actions) that implement different competencies, similarly to those proposed in (Maes 1991, Cañamero 1997). Following the classical distinction in ethology (Hinde 1953) and more recent advances in neuroscience (Robbins & Everitt 1999), motivated behaviors can be consummatory (goal-achieving and needing the presence of an incentive stimulus to be executed) or appetitive (goal-directed search for a particular incentive stimulus). In addition to modifying the external environment, the execution of a behavior has an impact on (increases or decreases) the level of specific physiological variables. Therefore they are a mechanism to maintain the state of the physiological variables within the viability zone. In its Winner-Take-All form, our architecture will always execute the behavior that best satisfies the highest activated motivation.

3 The Competitive Two-Resource Problem

Our framework is an extension of the scenario known in the action selection literature as the two-resource problem (TRP). In the traditional TRP, a self-sufficient (biological or artificial) creature must continuously decide which of its two survival-related needs to satisfy by choosing between two resources available in the environment. The TRP presents the advantage of providing a standard that allows to compare results with those of other researchers, since it has been widely used to study action selection both in animals (see e.g., (McFarland 1974)) and in artificial creatures (see e.g., (Blumberg 1997, Spier & McFarland 1997, Girard *et al.* 2002)). Also, its simplicity (although not devoid of problems) favors a systematic analysis of results, and it has actually been characterized as the minimal scenario to test action selection mechanisms (Spier & McFarland 1997).

Our extension of this problem, the competitive two-resource problem (CTRP) consists in the introduction of two robots in the same environment simultaneously performing their own two-resource problem. The fact that they have to use the same resources to satisfy their needs introduces competition for those resources, as both robots might need access to the same resource at the same time.

3.1 Robotic Implementation of the TRP

Our competitive two-resource problem (CTRP) builds on a previous implementation of a typical static two-resource problem (TRP) (Avila-García *et al.* 2003), in which a single robot must choose between two types of static resources, i.e. the resources do not change location and are always equally accessible. Our particular implementation of the TRP, detailed in Tables 1 and 2, was as follows (see Figure 1 for a whole picture of the motivational system). Like any implementation of this problem, we used two environmental resources (incentive cues) that allow to satisfy two physiological needs, the deficits of which give rise to two motivational states. Motivations are satisfied by the execution of a consummatory behavior. The execution of each behavior reduces the deficit of one physiological variable, increasing the deficit of the other one.

In addition to consummatory behaviors (b_{feed} and b_{warmup}), our architecture also includes appetitive (b_{search}) and reflex (b_{avoid}) behaviors. It is important to point out that in this implementation only consummatory behaviors, but not appetitive or reflex, participate in the action selection process; the other two behaviors only are called upon to support their execution. Thus, b_{avoid} is a reflex behavior that is executed whenever an obstacle is present, and b_{search} is an appetitive behavior that is executed when the winner consummatory behavior cannot be executed because its related incentive stimulus is not present.

The particular robotic scenario we used to implement this TRP (see Figure 2) was as follows. Our robot, Tauron, was a modified version of the standard Lego Mindstorms "roverbot", with a controller programmed in C implementing the synthetic physiology and architecture described above (see Figure 1). We chose this simple design model to facilitate replication of our experiments. The robot was equipped with: two bumpers for obstacle avoidance, two frontal light sensors to detect brightness and darkness on the floor of the arena, and two motors providing differential steering to the wheels. The environment was a $1m \times 1m$ arena surrounded by a wall. The floor is made of 9 tiles (of $33cm \times 33cm$ each) of three types: "empty space" (uniform gray), two "heat sources" (white gradient), and two "food sources" (black gradient). The use of tiles does not mean that our environment is a "grid world" since our tiles have continuous physics.

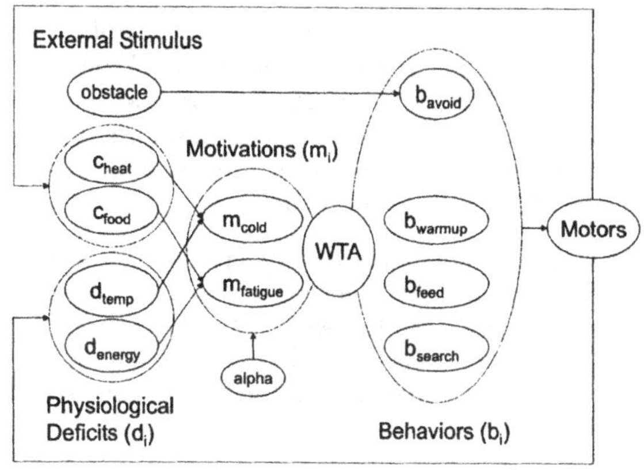

Figure 1: Motivational control system for our WTA architecture. The winner behavior modifies the physiology of the robot (negative feedback for the ongoing behavior) and sends commands to the motors, which can eventually change the perception of incentive cues (acting as a positive feedback).

Figure 2: Our robotic TRP scenario.

3.2 The CTRP and Environmental Complexity

In this study we extend our previous implementation of the TRP to create a dynamic environment in order to introduce new forms of environmental complexity. Dynamism is achieved by the addition of another robot in the same arena. This element adds complexity to the environment by introducing competition between both robots—hence we call this new problem the competitive two-resource problem (CTRP). We can identify at least two new types of environmental complexity, that we have studied in this paper:

(a) Due to the constraints of our implementation, two robots cannot have simultaneous access to the same resource, since a robot must be placed on top of a resource to consume it. This results in a continuous change in availability and accessibility of resources. For example, while one robot is consuming one resource, this type of resource will be less available for the other robot. Likewise, the particular resource that is being consumed will not be accessible to the other robot, which means that, from the point of view of the other robot, that resource has "disappeared" from its original position.

(b) When one robot is located on top of a resource—i.e. consuming it—the other robot might bump into it and push it out of the resource. This will result in the interruption of the ongoing consummatory activities, which for the robot that was executing them means that the resource has been extinguished.

4 Background Notions

4.1 Physiological Space

Physiological variables, controlled homeostatically, define a physiological space (McFarland 1974) or viability zone (Ashby 1952, Meyer 1995) within which survival (continued existence) is guaranteed, whereas transgression of these boundaries leads to death. In our TRP the physiological space has two dimensions, with each axis representing the deficit of each physiological variable (d_{temp}, d_{energy}). The origin of this space is the ideal state where both physiological needs are satisfied ($d_{temp} = 0$, $d_{energy} = 0$). Lethal boundaries are defined by $d_{temp} = 1$ and $d_{energy} = 1$ (Figure 3). We have used this physiological space to analyze the behavior and performance of our robots, with the help of the notions characterized in the remainder of this section.

4.2 Activity Cycles

Most work on action selection has focused on the choice of behavioral subsystems and individual behaviors resulting from different underlying selection mechanisms. However, McFarland and Spier (McFarland & Spier 1997) noted that a self-sufficient robot must perform a *basic cycle* of activities (and not only atomic behaviors) to maintain viability. The study of the physiological space of our robot should give us these cycles, which can then be used to analyze the behavior and performance of the robot.

Figure 3 (left) shows the basic cycle for our TRP implementation. During phase a, our robot executes the b_{search} behavior to find the resource c_{heat}. The phase labelled b corresponds to the execution of b_{warmup}, the consummatory behavior that decreases Temperature deficit (d_{temp}), also increasing the deficit of Energy (d_{energy}) to a lesser extent. When this need is satisfied, another episode of b_{search} is executed (label c), this time to obtain c_{food}. This is followed by the execution of b_{feed}, the consummatory behavior that helps to correct the Energy deficit (and also increments the deficit of Temperature slightly), in phase d. Upon satiation, the same basic cycle is repeated again.

Note that in our implementation the cycle is symmetric with respect to the diagonal of the physiological space; this is due to the fact that both aspects of our problem are symmetric in all respects (physiology, motivations, behaviors, etc.).

4.3 Satiation boundaries.

The point at which the robot stops consuming one resource—i.e., it finishes satisfying one motivation—can be termed *satiation boundary* (Figure 3, left). This limit defines the amplitude of the activity cycles, and can be obtained analytically.

Since our two problems are symmetric, we can assert that the point at which a motivation that is being satisfied (m_{cold}) loses the competition against the other ($m_{fatigue}$) is defined by $m_{cold} = m_{fatigue}$. At this moment, the robot is consuming a resource that leads to the satisfaction of m_{cold}, and hence the intensity of this incentive cue must be near the maximum ($c_{heat} \simeq 1$). In this situation, the incentive cue of the second motivation ($m_{fatigue}$) is disregarded ($c_{food} \simeq 0$). Therefore, using the *Deficit* + *Cue* × *Deficit* equation (1) that we have adopted to calculate motivational intensity, the final equation will be:

$$\begin{aligned} d_{ener} + (d_{ener} \times \alpha c_{food}) &= d_{temp} + (d_{temp} \times \alpha c_{heat}), \\ d_{energy} &\simeq (1+\alpha) d_{temp} \end{aligned}$$

The line determined by this equation is the satiation boundary of m_{cold} (Figure 3, left, SB_T), that is, the point at which there is no more tendency to decrease d_{temp}. Note that the parameter α, determining the influence of the incentive cue, determines also the slope of the satiation boundaries.

4.4 Opportunism and Persistence

Opportunism—taking advantage of the opportunities offered by the environment to satisfy a lower-priority need while searching to satisfy a higher-priority one—and persistence—the ability to keep working towards the satisfaction of the current goal—are two of the main features that a good action selection architecture must show (Maes 1991, Tyrrell 1993). These notions are also used in ethology to explain phenomena commonly observed in animal behavior related to decision-making, see e.g., (Slater 1999).

In our architecture, both phenomena are supported by the same positive feedback mechanism: The intensity of incentive cues. This is a well-known mechanism to avoid dithering (McFarland 1974). When the robot approaches a resource, the incentive cue (c_i) associated with it produces a positive feedback that increases the motivation (m_i) to consume that resource. This causes that under certain circumstances, a motivation wins the competition to be satisfied even when its related physiological deficit (d_i) is not the highest. In other words, in a system without such a positive feedback, a motivation would always win the competition when the physiological state is at one specific side of the diagonal of such space—that is, m_{cold} would always win whenever $d_{temp} > d_{energy}$. Figure 3 (center) depicts such a diagonal and the regions A and B where $m_{fatigue}$ and m_{cold} are

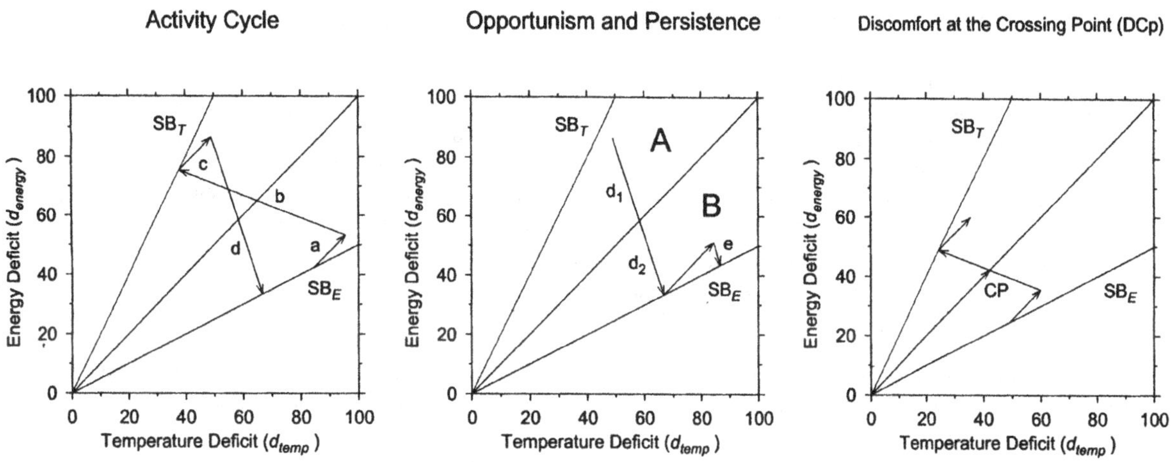

Figure 3: Left: Basic cycle of activities within the physiological space, with each axis representing the deficit of each physiological variable (d_{temp}, d_{energy}). Every activity cycle is depicted within the temperature and energy satiation boundaries (SB$_T$ and SB$_E$ respectively), and it consists of an appetitive phase (a) to look for c_{heat} (b_{search}); a consummatory phase (b) to decrease the Temperature deficit (b_{warmup}); followed by an appetitive phase (c) to look for c_{food} (b_{search}); and a consummatory phase (d) to decrease the Energy deficit (b_{feed}). Center: Situations of persistence (d_2) and opportunism (e) to satisfy $m_{fatigue}$—reducing Energy deficit. Right: Discomfort at the Crossing point (DCp) as the Euclidean distance to the point at which the cycle crosses the diagonal.

dominant, respectively—i.e., where they would be executed unconditionally. However, in our system, due to such a positive feedback, a motivation can be executed even when the physiological state is not within its dominance region—and that is always the case when its related incentive cue is higher than zero ($c_i > 0$).

Therefore, when the robot is next to a resource, the motivation to consume that resource increases, giving rise to two interesting phenomena: *opportunism* and *persistence*. We consider that our robot is being opportunistic when it starts satisfying a motivation—consuming its associated resource—when the physiological state is not in its dominance region. Figure 3 (center) depicts a typical opportunistic activity (e), when the robot decides to satisfy the motivation $m_{fatigue}$—reducing energy deficit d_{energy} while the physiological state is in B, the dominance region of m_{cold}.

We define persistence also as the act of satisfying one motivation out of its dominance region, but in this case the robot starts consuming such a resource in the "right" dominance region. Figure 3 (center) depicts a typical situation (d_2) of persistence. The robot starts satisfying $m_{fatigue}$ in A, but at some point the physiological state crosses to the other dominance region B. Due to the presence of the incentive cue, $m_{fatigue}$ continues being the winner until the physiological state reaches its satiation boundary (SB_E). Thus, there will be two typical phases in every consummatory process: a normal one, where there is no persistence (d_1), and the following phase with persistence (d_2), where the execution of the same behavior is only due to the presence of the incentive cue.

4.5 *Performance Indicators*

In previous studies we showed that, in order to successfully compare and understand the performance of action selection mechanisms in different tasks and environments, we must use more than one performance indicator (Cañamero *et al.* 2002, Avila-García & Cañamero 2002). We have extended that study developing the novel idea of understanding the action selection process in terms of activity cycles rather than single decisions (Avila-García *et al.* 2003). In that sense, a study of activity cycles must be included in any comparison between architectures. In this paper, we use four performance indicators, three based on the notion of viability and one on activity cycles, to asses quantitatively the performance of the robots.

4.5.1 *Viability Indicators*

We have used the notion of viability (Ashby 1952) as first criterion to assess performance. The behavior of the robot is said to be viable when it keeps the values of the essential variables within the boundaries of the physiological space. In our opinion, however, this notion of viability is too vague to provide a direct criterion to measure the goodness or performance of a behavior selection architecture, as it leaves several possibilities open. For example, the performance of a behavior selection architecture can be assessed in terms of the time it allows the robot to remain viable (survive) in a given environment. Longer life spans usually indicate better behavior selection performance but this correlation is not necessarily straightforward, since the "life quality" of the robot can be very different during its life span depending on how viability is preserved during that period. A robot can live a long life with poor "life quality" if the values of its essential variables are kept close to the critical zone (near the boundaries) of the physiological space for a long period. On the contrary, it can live a shorter life of "high quality" that ends due to accidental factors. "Life quality" can also have different interpretations. It can for example be measured in terms of global internal stability or "comfort" that takes into account the average level of satisfaction of all the essential variables simultaneously. It can also be seen in terms of how "balanced" the satisfaction of the different physiological needs is. Finally, the viability zone (physiological space) can be divided in different "regions", ranging from an "ideal zone" around the ideal state in which the physiological variables have their ideal values, to a "critical zone" near the lethal boundaries that lead to death if transgressed. These regions can indicate the "Risk of Death" (RoD) due to parameter overflow.

We have used three[2] indicators of viability in order to reflect various ways in which good viability can be obtained: Life span, Overall Comfort and Physiological Balance.

Life Span. It is the time that the robot survived (remained viable) during each run:

$$LS = t_{life}/t_{run} \tag{2}$$

where t_{life} is the number of steps that the robot lived and t_{run} is the maximum number of steps of one run.

Overall Comfort. It measures the average level of satisfaction of both physiological variables during a run:

$$OvC = \sum_{i=1}^{t_{life}} (1 - \overline{d_i})/t_{life} \tag{3}$$

where $\overline{d_i}$ is the mean of both physiological deficits.

Physiological Balance. It measures the homogeneity with which both physiological needs are satisfied during a run:

$$PhB = \sum_{i=1}^{t_{life}} (1 - \sigma^2(d_i))/t_{life} \tag{4}$$

where $\sigma^2(d_i)$ is the variance of both physiological deficits.

4.5.2 *Discomfort at the Crossing point (DCp)*

Our last performance indicator is related to activity cycles theory, and it exploits the fact that the cycles shift along the axis determined by the diagonal of the physiological space. Every cycle, in its way, crosses the diagonal of the physiological space in a certain point. The Euclidian distance from the origin to this point gives us a measure of how close the cycles get to the ideal state, and hence, how close to the goal the robot is. Our last indicator will thus be the average of Euclidean distances of the crossing points during the robot lifetime (Figure 3, right).

[2]The fourth viability indicator we have defined, Risk of Death, will be used to produce hormonal feedback in the experiments reported in Section 6.

5 Preliminary Experiments on the CTRP

In Section 3.2 we saw that the interaction between robots in our CTRP will produce at least two new forms of environmental complexity; in this preliminary experiments we show how they affect the performance of our robots. First, we report new experiments with two robots performing the CTRP, and then we compare the results with those obtained by a single robot in a static TRP, reported in a previous study (Avila-García *et al.* 2003). We discuss the results in terms of different performance indicators (described in the previous section) and behavioral phenomena. The arena and robot configuration used in these experiments are the same as those used in the static TRP (Section 3.1).

5.1 *Method*

We have tested the robots in a total of 16 runs of 1600 steps each. Each step represents a loop of the action selection mechanism, taking $260ms$[3]. This means that each run lasts about 7 minutes. For each run we initialized randomly physiological variables (to values within their viability range) and location of resources in the environment. The parameter α in the motivational intensity equation (Equation 1) is constant ($\alpha = 1$). We will compare our previous TRP's results with those obtained by the robot that performs worse in every run of the CTRP. Therefore, when we we speak of results obtained in the CTRP, we mean the results obtained by the robot that first died in every run.

5.2 *Results and Discussion*

Figure 4 shows the comparison of results obtained in the static two-resource problem (TRP) with those of the dynamic competitive two-resource problem (CTRP). In terms of Overall Comfort, Physiological Balance, and Discomfort at the Crossing point, performance within the CTRP is worse than within the TRP. This difference is statistically highly significant, and it shows that the two new forms of environmental complexity (see Section 3.2) highly affect the viability and cycles of the robots. In terms of Life Span, the results are not statistically significant; however, there is a tendency indicating that the robots live less when there are competitors.

To understand the results, let us analyze the new forms of environmental complexity (dynamism) introduced by the interaction between robots. Firstly, the availability and accessibility of resources causes unpredictability in appetitive phases—that is, when the robot starts to look for a resource the time that it spends to find it is more variable. This also causes that the robot, in the search process to find a specific resource, gets more resources of the other type, falling consecutively in an opportunism that eventually could drive it to death (over-opportunism).

In Figure 5 (left) we can see how the robot, looking for the resource c_{food} (to satisfy $m_{fatigue}$) finds consecutively

[3]Lego Mindstorms robots use a 16MHz microcontroller

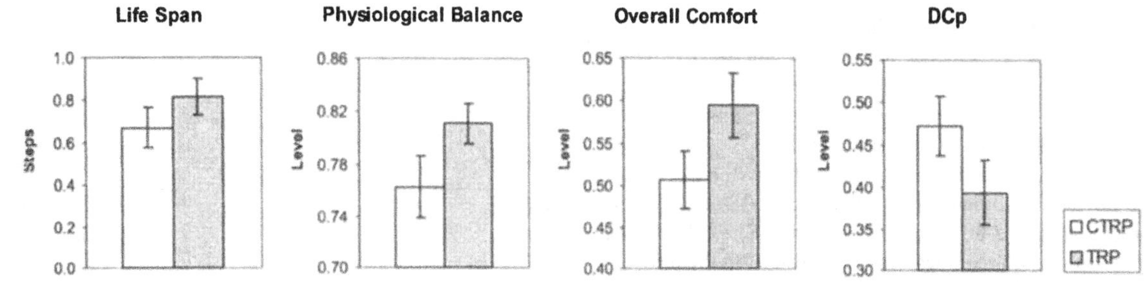

Figure 4: Average performance in terms of our four indicators. Standard error of the mean is shown within each bar.

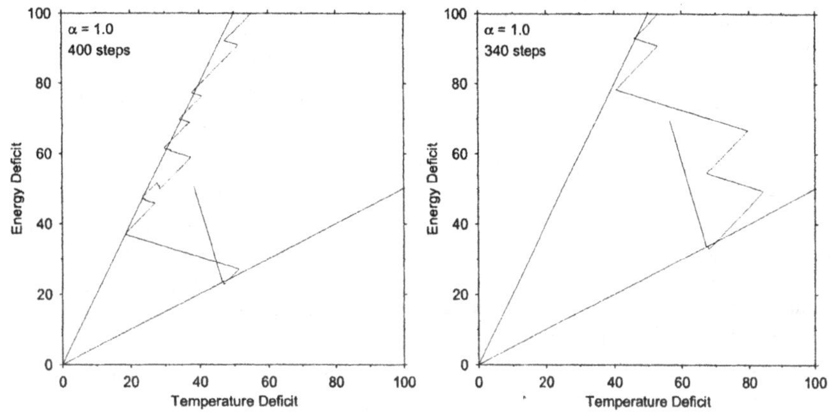

Figure 5: Left: During an appetitive phase looking for c_{food} to satisfy $m_{fatigue}$, Tauron finds consecutively several sources of c_{heat}. This makes it incur in opportunism in such a repetitive way that it finally dies (over-opportunism). Right: While consuming c_{heat}, Tauron is pushed out of the resource by the competitor, stopping its ongoing consummatory activity. Later, it gets again another c_{heat} source to continue reducing the Temperature deficit. It finalizes the activity cycles performing over-opportunism over c_{heat}.

the resources of the other type, c_{heat}, and this makes it incur in opportunism; however, it does so in such a way that the robot dies finally due to the execution of the sequence of opportunistic activities. Since the robot has no mechanism that allows it to detect when opportunism will drive it to death, it often performs it without any coherence.

The other source of complexity is the fact that consummatory activities can be suddenly interrupted. Figure 5 (right) reflects one of those situations; the first time that the robot is consuming c_{heat} (decreasing the level of Temperature deficit) it is interrupted by the other robot and has to look for the same resource again to carry on. The cycle of activities ends up with an opportunism that drives the robot to death.

The effect that both forms of dynamism have on the robot are: (a) the robot shows more bizarre activity cycles than in the static TRP; and (b) they reduce the viability of the cycles in every sense—distance to the origin, amplitude, etc. It seems now interesting to introduce some sort of mechanism that tackles those complexities to some extent, or that helps the robot to use them to compete advantageously. This is the purpose of our next set of experiments.

6 Experiments using Hormonal Feedback

To tackle the problems introduced by the new forms of dynamism described in the previous section, we investigated a mechanism to modulate action selection. As we saw in Section 4, the parameter α in the equation computing motivational intensity (Equation 1) determines the relevance given to external stimuli and also influences viability as reflected by the shape and position of the activity cycles in the physiological space (Avila-García *et al.* 2003). For this reason, we decided to act on (modulate) the perceptual level of the action selection mechanism in order to neutralize the problems produced by the introduction of dynamism. The idea that some phases of the action selection process can be performed at the sensory level is not new—see (Toates 1986, McFarland & Sibly 1975) for some references. In this paper we go beyond and propose that, by acting at that level, we can modulate an action selection architecture to adapt it to internal and/or external environmental circumstances.

Drawing on the architecture described in (Cañamero 1997) regarding hormonal ("emotional") modulation of action selection, we have implemented a "synthetic hormone" to modulate perception. Although in the original architecture different types of emotion-released hormones affected many different aspects of the architecture (e.g., synthetic physiology, arousal, attention, perception of the internal and external environment, etc), in this paper we only explore modulatory effects on the perception of external stimuli.

We therefore conducted new experiments in the CTRP scenario to test a modified version of the previous architecture upgraded with this hormonal mechanism. The robots have suffered minimum modifications with respect to the previous setting (see Section 3.1). The robot with the hormonal system, Uruk, is equipped with and extra light sensor to detect the presence of the competitor, a Tauron type robot. Likewise, Tauron is equipped with a sphere at its top that emits infrared light in all directions. Since Uruk's extra light sensor is aimed forwards, Uruk must be facing Tauron in order to detect the light coming from its infrared ball.

We have performed the same number of runs as in the preliminary experiments, but this time lasting 1200 steps, i.e, each run lasts about 5 minutes.

6.1 Hormonal mechanism

Neurohormones interact with the nervous system to modulate behavioral output (see e.g. (Kravitz 1988)), acting at different levels of the nervous system and evoking a spectrum of different responses on different target neurons. Based on these ideas, we have designed a "hormone-like" mechanism that modifies the perception of both incentive cues (c_{heat} and c_{food}) by acting on the parameter α of the motivational intensity equation (Equation 1). The same hormone also modifies the perception of the stimulus Obstacle that unconditionally fires the reflex behavior b_{avoid}. Its function is thus twofold:

(a) To avoid falling under excessive opportunism and persistence in situations where the Risk of Death (see Section 4.5) is high, therefore neutralizing the problems due to dynamism regarding accessibility and availability of resources. This is achieved by controlling the parameter α: The smaller α, the smaller the positive feedback (see Section 4.4). Therefore, variations in the hormonal level will modify the levels of opportunism and persistence.

(b) To potentiate the competition skills of the robot by enhancing its capacity to push the other robot out of the resources. This means, first, modifying accessibility and availability of resources—it will be easier to obtain a resource occupied by the competitor. Secondly, the complexity of the environment for the other robot increases. This is obtained by the cancellation of the perception of the stimulus Obstacle (bumpers), which produces the cancellation of the reflex behavior b_{avoid}. Therefore, hormonal levels determine whether the robot avoids obstacles or pushes them.

The question is now how to control that double functionality with only one hormone, and how to compute the level of such hormone. We do so by combining both external and

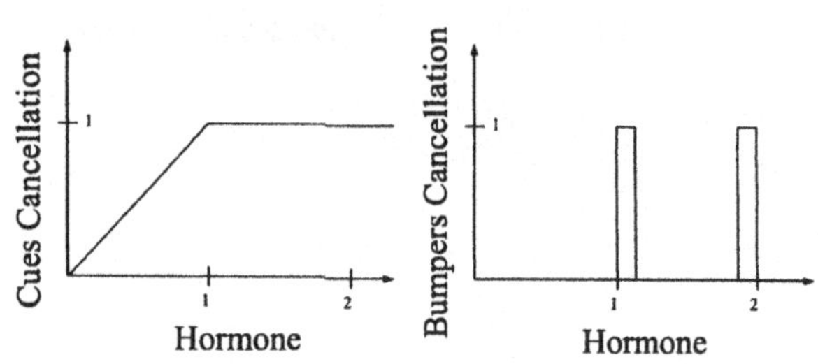

Figure 6: Cancellation functions of incentive cues (left) and bumpers (right) with respect to the hormonal level.

internal "appraisals"[4] of the risk of death. The internal assessment of Risk of Death (RoD) is given by the distance of the physiological state to the lethal boundaries, as follows:

1. If $d_{max} > 0.5$, then $RoD = d_{max} - 0.5/0.5$
2. Else, $RoD = 0$

where d_{max} is the maximum of both physiological deficits. This means that RoD is a number normalized between [0,1] that gives us a measure of the distance to the lethal boundary when the maximum deficit is higher than half that boundary.

The external risk of death is given by the perception of the competitor ($0 \leq Competitor \leq 1$). The hormonal level will be an additive function of internal Risk of Death (RoD) and the perception of the competitor:

$$Hormone = RoD + Competitor, \tag{5}$$

therefore $0 \leq Hormone \leq 2$.

Finally, we have to define the relation between the hormonal level and the cancellation of the perception of incentive cues and Obstacles (Figure 6). In order to achieve the first of the functionalities (a), the cancellation of α is directly proportional to the increment of the hormonal level. Therefore, when the RoD rises, α is reduced.

The second functionality (b) is obtained by cancelling the perception of the stimulus Obstacle when the stimulus Competitor is in front of the robot. There are two conditions that must be fulfilled to make a coherent pushing of the other robot. First, the robot must avoid getting engaged in fights when it has RoD. Second, it must only bump blindly into the other robot, not against the walls of the arena. To produce that effect the cancellation of the bumpers is at hormonal levels $Hormone \simeq 1$ and $Hormone \simeq 2$.

6.2 Results and Discussion

Figure 7 shows the comparison of results obtained by both robots, Uruk and Tauron, in the CTRP. In terms of Overall Comfort and Discomfort at the Crossing point results are alike. However, regarding Life Span and Physiological Balance the differences are statistically highly significant; Uruk outperforms Tauron.

[4]Although the term "appraisal" is often used to denote conscious evaluation, appraisal can also be unconscious and very automatic process—it is in this latter sense that we use it here.

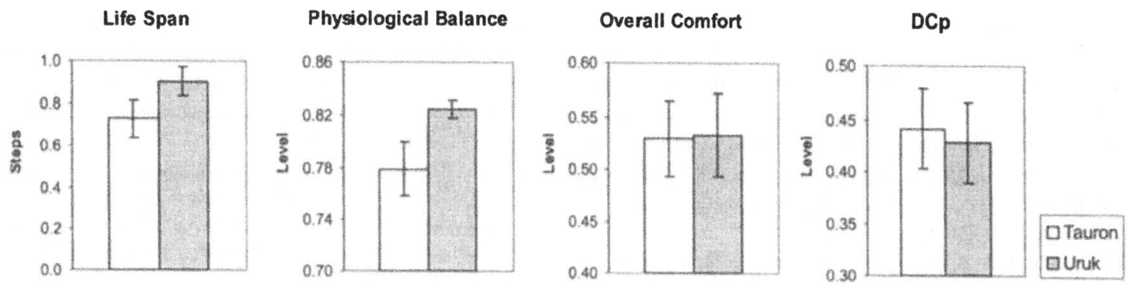

Figure 7: Average performance in term of our four indicators. Standard error of the mean is shown within each bar.

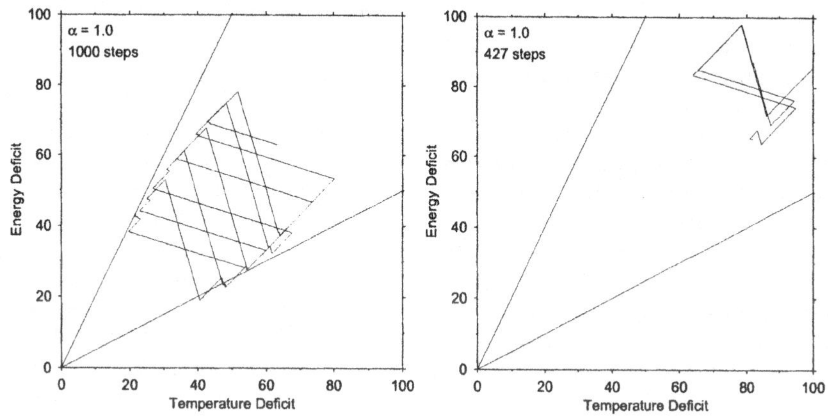

Figure 8: Activity cycles depicted by Uruk. Due to the hormonal system, satiation boundaries are reduced as the Risk of Death (RoD) increments. Both graphs depict satiation boundaries when there is not hormonal feedback, that is, with $\alpha = 1$. Left: Activity cycles with same amplitude due to the reduction of satiation boundaries—i.e., reduction of α. Right: Activity cycles evolving very close to the lethal boundaries through the previous reduction of α. Compare these cycles with those produced by Tauron in previous experiments (Figue 5).

Let us analyze the dynamics of the activity cycles in order to understand those quantitative results. Figure 8 (left) depicts cycles performed by Uruk in one of the runs. We can see how cycles keep the same amplitude when physiological deficits are higher than 50 ($d_{max} > 0.5$). This is because at that point the RoD starts to rise the hormonal level, which decreases the value of α. This results in a reduction of the satiation boundaries, and hence the persistence. In Figure 8 (right) we can see the drastic reduction of persistence at high levels of RoD. This process also counteracts the tendency to perform opportunism under risk of death. This is the main reason for such good results in terms of Physiological Balance.

In terms of Life span, the results are also the demonstration that the second functionality of the hormonal system improves the performance of the robot. This functionality leads it to push the other robot out of the resources, hence improving its competition skills.

This double functionality of the hormonal system gives rise to some emergent behaviors. The first one is stopping opportunism and persistence when Uruk is consuming a resource and it detects its competitor approaching. This could be interpreted by an external observer as abandonment of a situation (waiting for the other robot at the resource) that is disadvantageous to compete. Instead, the robot will leave the resource and go straightforward towards the competitor until it reaches it; at that moment, if there is some level of RoD, the bumpers of Uruk will not be cancelled and it will avoid Tauron—showing a behavior that an observer could interpret as "fear" after evaluating the competitor. On the contrary, if there is not RoD, the hormonal system will cancel the bumpers and Uruk will push Tauron unconditionally—as if it showed some sort of "aggression" against it. If we study the whole picture as external observers, the previous behavioral phenomena could well be interpreted as some sort of "protection of resources".

7 Conclusion

Neurohormones have the ability of rewiring the same anatomically defined neural circuit to build the appropriate network for the behavior needed at any moment. Inspired in this functionality, we have investigated the role of an "hormone-like" mechanism to produce adaptive behavior in an action selection architecture. We have carried this study in two steps.

In a first step, we have discussed the extension of a robotic implementation of a static two-resource problem (TRP) into a dynamic competitive two-resource problem (CTRP). In this extension, with only the introduction of a second robot—a competitor—new forms of environmental complexity are generated due to the interaction between robots. These forms of complexity effectively reduce the viability of the robots. Using activity cycles analysis, we have also shown that the action selection mechanism used within the TRP presented incoherences within the CTRP, performing pathological opportunism and persistence.

In a second step, we have proposed a first rudimentary "hormonal mechanism" to tackle these new forms of environmental complexity. Hormonal feedback modulates the action selection mechanism (ASM)—the interconnections between the sensory elements and other elements of the ASM—adapting its output to different environmental conditions. This modulation gives rise to adaptive behavioral patterns to the changing environmental conditions, as well as interesting emergent behavioral phenomena. This mechanism can be seen in relation to autopoietic systems (Maturana & Varela 1980), which change their internal structure in response to environmental circumstances. A modulatory hormonal system improves the *structural coupling* between an agent and its environment in a way that is different from other mechanism such as learning or evolution, for which "past solutions" are "overwritten" by new ones. Hormonal modulation extracts different functionalities from the same architectural "skeleton" to produce adaptive behavior to any of the environmental conditions of the agent's niche.

Acknowledgments

We are grateful to René te Boekhorst for his helpful comments on an earlier version of this paper. Orlando Avila-García is supported by a research scholarship of the University of Hertfordshire.

References

Ashby, W.R. (1952). *Design for a Brain*. Chapman & Hall.

Avila-García, O. and Cañamero, L. (2002). A Comparison of Behavior Selection Architectures Using Viability Indicators. In *Proc. International Workshop on Biologically-Inspired Robotics: The Legacy of W. Grey Walter*, pp. 86–93. 14–16 August 2002, Bristol HP Labs, UK.

Avila-García, O., Cañamero, L., and Boekhorst, R. (2003). Analyzing the Performance of "Winner-Take-All" and "Voting-Base" Action Selection Policies within the Two-Resource-Problem. In *Proc. Seventh Eutopean Conference in Artificial Life* (ECAL03), pp. 733–742. Springer-Verlag.

Blumberg, B. (1997). Old Tricks, New Dogs: Ethology and Interactive Creatures. *PhD. Thesis*, MIT Media Laboratory, Learning and Common Sense Section.

Brooks, R. A. & Viola, P.A. (1990). Network Based Autonomous Robot Motor Control: from Hormones to Learning. In R. Eckmiller (Editor), *Advanced Neural Computers*, pp. 341–348. Elsevier Science Publishers.

Cañamero, L.D. (1997). Modeling Motivations and Emotions as a Basis for Intelligent Behavior. In W.L. Johnson, ed., *Proc. First Intl. Conf. Autonomous Agents*, 148–155. ACM Press.

Cañamero, L., Avila-García, O., Hafner, E. 2002. First Experiments Relating Behavior Selection Architectures to Environmental Complexity. In *Proc. 2002 IEEE/RSJ International Conference on Intelligent Robots and Systems* (IROS 2002), 3024–3029. IEEE Press.

Girard, B., Cuzin, V., Guillot, A., Gurney, K.N., Prescott, T.J. (2002). Comparing a Brain-Inspired Robot Action Selection Mechanism with "Winner-Takes-All". In *Proc. Seventh Intl. Conf. on Simulation of Adaptive Behavior*. MIT Press.

Grand, S., Cliff, D., Malhotra, A. (1997). Creatures: Artificial Life Autonomous Software Agents for Home Entertainment. *Proc. First Intl. Conf. on Autonomous Agents*, 22–29. ACM Press.

Harris-Warrick, R.M., Marder, E., Selverston, A.I., Moulins, M., eds. (1992). *Dynamic Biological Networks*. The MIT Press.

Hinde, R.A. (1953). Appetitive Behaviour, Consummatory Act, and the Hierarchical Organization of Behaviour –with special reference to the great tit (Parus Major). *Behaviour*, 5, 189–224.

Kravitz, E.A. (1988). Hormonal Control of Behavior: Amines and the Biasing of Behavioral Output in Lobsters. *Science 241*, September 30, 1175-1781.

Maes, P. (1991). A Bottom-Up Mechanism for Behavior Selection in an Artificial Creature. In J.A. Meyer and S.W. Wilson, eds. *Proc. First Intl. Conf. on Simulation of Adaptive Behavior*, 238–246. Cambridge, MA: MIT Press.

Maturana, H.R., Varela F.J. (1980). *Autopoiesis and Cognition. The realization of the Living.* Dordrecht, Holland: D. Reidel.

McFarland, D. (1974). Experimental Investigation of Motivational State. In *Motivational Control Systems Analysis*, D. McFarland, ed., 251–282. London: Academic Press.

McFarland, D. (1976). Form and function in the temporal organization of behavior. *Growing Points in Ethology*, Bateson P., Hinde R. ed. Cambridge University Press.

McFarland, D. and Sibly, R. (1975). The behavioural final common path. *Phil. Trans. Royal Society (Series B)*, 270, 265-293.

McFarland, D. and Spier, E. (1997). Basic Cycles, Utility and Opportunism in Self-Sufficient Robots. *Robotics and Autonomous Systems*, 20, 179-190.

Meyer, J.A. (1995) Artificial Life and the Animat Approach to Artificial Intelligence. In M. Boden, ed., *Artificial Intelligence*. Academic Press.

Robbins, T.W. and Everitt, B. J. (1999). Motivation and Reward. In M. J. Zigmond, F. E. Bloom, S. C. Landis, J. L. Roberts, L. R. Squire, ed., *Fundamental Neuroscience*, 1245–1260. Academic Press.

Sibly, R.M. and McFarland, D. (1974). A State-Space Approach to Motivation. In D. McFarland, ed., *Motivational Control Systems Analysis*, 213–250. London: Academic Press.

Slater, P.J.D. (1999). Motivation. In *Essentials of Animal Behaviour*, 57–83. Cambridge University Press.

Spier, E. and McFarland, D. (1997). Possibly Optimal Decision Making under Self-Sufficiency and Autonomy. *J. Theor. Biol.*, 189, 317-331.

Toates, F. (1986). *Motivational Systems*. Cambridge Univ. Press.

Tyrrell, T. (1993). Computational Mechanism for Action Selection. *PhD. Thesis*, Centre for Cognitive Science, Univ. of Edinburgh.

NAVIGATION

A Spiking Neuron Model of Head-Direction Cells for Robot Orientation

Thomas Degris* **Loïc Lachèze*** **Christian Boucheny**** **Angelo Arleo****

*AnimatLab - LIP6
8 rue du Capitaine Scott
75015 PARIS
thomas.degris@lip6.fr
loic.lacheze@lip6.fr

**LPPA - Collège de France - CNRS
11 place M. Berthelot
75005 PARIS
angelo.arleo@college-de-france.fr
christian.boucheny@college-de-france.fr

Abstract

This paper proposes a bio-mimetic model of head-direction (HD) cells implemented on a real robot. The model is based on spiking neurons to study the temporal aspects of state transitions of the HD cell activity following reorienting visual stimuli. The short transient latencies observed experimentally are reproduced by the model. We focus on the integration of angular velocity inertial signals provided by accelerometers. This integration is realized by a continuous attractor network modeling the interaction between the lateral mammillary nucleus (LMN) and the dorsal tegmental nucleus (DTN), two structures belonging to the HD cell anatomical circuit. Relevant parameters defining the connections between LMN and DTN are determined by a genetic algorithm.

1. Introduction

Self-orienting agents, that is navigating systems able to estimate their own orientation autonomously and on-line, might rely upon an internal direction representation. Extracellular recordings from freely-moving rats show the presence of head-direction (HD) cells, limbic neurons whose firing activity is correlated with the current direction of the head of the animal (Taube, 1998). A HD cell n has a unique preferred direction θ_n for which it discharges maximally. The directional coding of the HD cells is independent of the animal location, and provides, therefore, an allocentric directional representation.

The properties of the HD cell system resemble those of a compass, but the allocentric coding of HD cells does not depend on geomagnetic fields. Rather, the preferred directions are anchored to visual cues in the environment. For instance, rotating a dominant visual cue by an angle $\Delta\theta$ causes a shift $\Delta\theta$ of all the preferred directions of the HD cells (Taube, 1998, Zugaro et al., 2001). Besides the control of visual cues upon HD cells, experiments show that inertial self-motion signals (e.g. vestibular) strongly influence the dynamics of the HD cell system. For instance, when the environmental light is turned off, the directional coding is maintained (Taube, 1998). In addition, when the head of the animal rotates, vestibular signals (coding for the head angular velocity ω) are integrated to update the HD representation over time.

HD cells have been observed in several regions centered on the brain limbic system, in particular the postsubiculum (PSC), the anterodorsal thalamic nucleus (ADN), and the lateral mammillary nucleus (LMN) (Taube, 1998). Inertial self-motion signals are likely to converge onto the HD system via the dorsal tegmental nucleus (DTN) (Bassett and Taube, 2001) which projects inhibitory connections directly to LMN. Visual inputs enter the HD system via the PSC which receives afferents from the visual cortex (Taube, 1998). Fig. 1 shows the anatomical interconnections between PSC, ADN, LMN and DTN.

Maintaining a sense of direction autonomously is crucial for both biological and artificial systems involved in navigation. For instance, path integration allows a navigating agent to infer its location and its heading relative to a departure point (Mittelstaedt and Mittelstaedt, 1980). This implies estimating the rotational as well as the translational components of motion continuously. In addition, lesions to the HD cell system disrupt the allocentric location representation of hippocampal place cells (Knierim et al., 1995), neurons whose firing activity is correlated with the location of the animal. Consequently, the sense of direction is necessary for establishing place cell representations.

This paper presents a neuro-mimetic model of the HD system and focuses on its robotic implementation. To reflect the properties of the rat HD cell system, the model must provide a stable and persistent HD representation robust to noise and must permit the integration of an angular velocity signal.

The model is based upon a continuous attractor/integrator neural network modeling the DTN-LMN interaction. An attractor model of HD cells is a neural network in which units encoding neighboring directions are recursively connected by strong excitatory synapses, whereas units having distant pre-

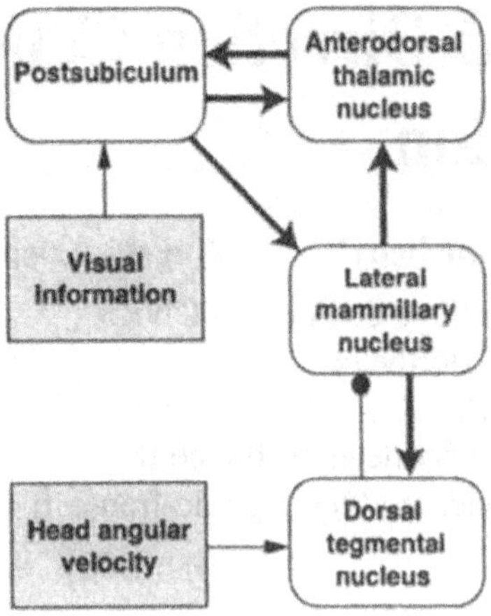

Figure 1: Anatomical connections between the postsubiculum (PSC), the anterodorsal thalamic nucleus (ADN), the lateral mammillary nucleus (LMN) and the dorsal tegmental nucleus (DTN). Arrows and circles indicate excitatory and inhibitory connections, respectively.

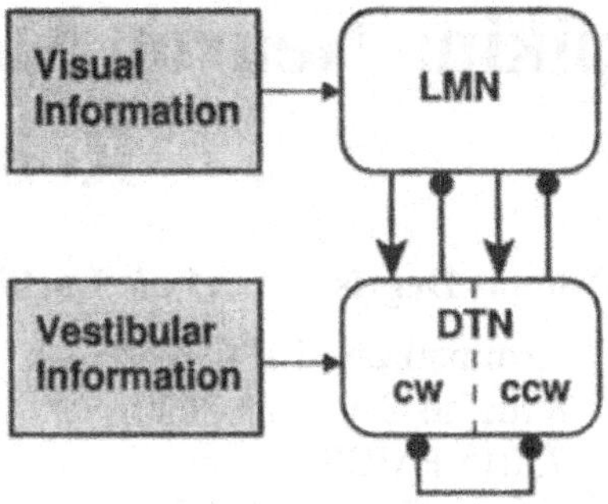

Figure 2: A view of the DTN-LMN interaction. An attractor network is formed by the excitatory connections from LMN to DTN (arrows) and the inhibitory connections from DTN to LMN (circles). Also, inhibitory connections within DTN are employed to integrate angular velocity signals.

ferred directions strongly inhibit each other. The interaction between the excitation and the inhibition in the neural network determines the intrinsic dynamics of the attractor. This allows the system to reproduce properties exhibited by rat HD cells. For instance, the system is able to encode a stable HD representation where a subpopulation of HD cells encoding similar directions are active while others remain silent. This HD representation can be updated by extrinsic input signals (e.g. vestibular angular velocity) utilized to shift the ensemble hill of activity of the HD cells over the directional space.

The integration of angular velocity realized by the model is not perfect in a mathematical sense. Consequently, the model needs a calibration method to avoid a cumulative tracking error. We simulate a vision-based calibration by generating an external stimulus applied to the HD cell system. Electrophysiological studies by (Zugaro et al., 2003) have focused on the temporal aspects of the preferred direction update in ADN. The quantitative results show a short update latency of approximately $80 \pm 10\ ms$. Our model uses spiking neurons and allows us to study the temporal aspects of the attractor state transitions. Particularly, the model reproduces the fast update transient exhibited by the rat HD cells observed by (Zugaro et al., 2003).

Several models of HD cells have been previously proposed (Skaggs et al., 1995, Zhang, 1996, Redish et al., 1996, Goodridge and Touretzky, 2000). These models, however, have not been tested by means of robotic implementations. The HD model by (Arleo and Gerstner, 2000) has been validated on a real robot but it uses rate coding computational units which make it impossible to study the dynamics of the system for short time windows. A robotic compass system has been proposed by (Gourichon et al., 2003). However, the authors focus on static visual information only, and do not integrate angular velocity over time. The model presented here is similar to our previous work (Degris et al., 2004) that puts forth an attractor neural network of spiking neurons to model the DTN-LMN circuit. In contrast to that model, the attractor system presented here does not employ recurrent excitatory projections within the LMN network, consistent with anatomical findings (Allen and Hopkins, 1989). Furthermore, our previous model has been validated by means of numerical simulations only. Here we stress the importance of endowing a mobile robot with a HD system using spiking neurons, and we cope with implementation constraints such as real time performance.

2. Methods

We model HD cells by means of a modular artificial neural network. Fig. 2 shows a view of the model. The architecture consists of an attractor-integrator network composed by a population LMN of $N_E = 1000$ excitatory directional units and two populations DTN$_{cw}$ and DTN$_{ccw}$ of $N_I = 500$ inhibitory directional units each. The HD units within each network have evenly distributed preferred directions.

The dynamics of the HD system is primarily controlled by idiothetic signals (i.e. self-motion inertial inputs) which determine the directional selectivity property of the formal HD cells. Allothetic information (e.g. visual input) can be used to occasionally modify the system's dynamics and calibrate the HD cell activity.

2.1 Robotic Platform

In order to validate the model experimentally, we have implemented it on a mobile Pekee[1] robot (Fig. 3). The behavior of the robot is monitored by an infrared camera above the arena tracking two LEDs positioned on the robot. The relative positions of the two LEDs are sampled at about 60 Hz, which

[1]The Pekee robot is commercialized by Wany Robotics, http://www.wanyrobotics.com

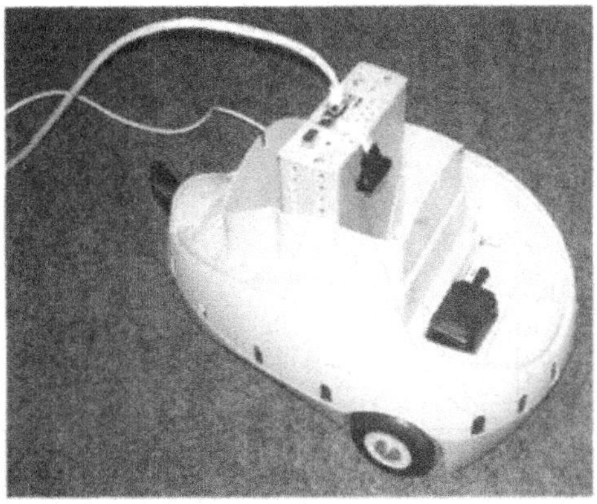

Figure 3: The mobile Pekee robot with the MT9 device (the black box in front of the robot) for sensing inertial self-motion signals.

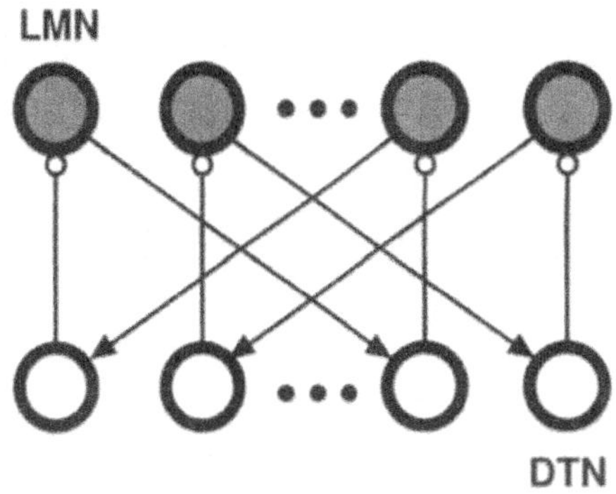

Figure 4: The attractor network is based on excitatory connections from LMN to DTN (arrows) and inhibitory connections from DTN to LMN (circles).

allows us to measure the actual orientation $\theta(t)$ of the robot over time.

The robot's sensory system consists of an odometer and a Xsens MT9[2] device for sensing inertial self-motion signals. The MT9 device, composed of gyroscopes, accelerometers and magnetometers, provides a real time absolute orientation resulting of the fusion of the inputs of these sensors.

The low-level controller of the robot runs on the on-board computer, whereas the HD model and the high-level controller run on a remote PC Pentium IV.

2.2 Attractor network

In order to generate and maintain a HD signal $\bar{\theta}$, we consider a continuous attractor network based on the interconnections between LMN and DTN (Blair et al., 1998). Fig. 4 shows the connections between formal neurons in LMN and DTN. Note that, because there is no experimental evidence for lateral connections in LMN (Allen and Hopkins, 1989), there is no recurrent excitation in the LMN of the model (Song and Wang, 2002).

LMN projects excitatory connections to DTN. The weight w_{kn} of these connections is such that a neuron $n \in$ LMN with a preferred direction θ_n projects to a cell $k \in$ DTN with a preferred direction θ_k according to a Gaussian weight distribution

$$w_{kn} = W_e^{max} \cdot \exp\Big(- \frac{(\theta_n - \theta_k + \pi + \delta_e)^2}{2\sigma_e^2} \Big) \qquad (1)$$

where $W_e^{max} = 63.3$ is the maximum weight of excitatory connections and $\sigma_e = 67.2°$ is the width of the Gaussian profile. The values of these parameters have been determined by means of a genetic algorithm (Sec. 2.6). The term δ_e is the angular offset of the intermodule connections with $\delta_e = 50°$ if $k \in \text{DTN}_{cw}$ and $\delta_e = -50°$ if $k \in \text{DTN}_{ccw}$.

[2]The MT9 device is commercialized by Xsens Motion Technologies, http://www.xsens.com

Global inhibition, necessary to implement the center-surround attractor scheme, is provided by the connections from DTN to LMN. The weight w_{nk} of these connections is distributed according to

$$w_{nk} = W_i^{max} \cdot \exp\Big(- \frac{(\theta_k - \theta_n)^2}{2\sigma_i^2} \Big) \qquad (2)$$

where $W_i^{max} = 59.5$ is the maximum weight of inhibitory connections, and $\sigma_i = 147.3°$ is the width of the Gaussian weight distribution (also these two parameters have been set by means of the genetic algorithm described in Sec. 2.6).

Because of the real-time constraint, only the connections with a weight $w > 0.75 * W^{max}$ are actually implemented in the network. This avoids the propagation of spikes along synapses with small weight and drastically reduces the computational complexity of the model. The intrinsic dynamics of the DTN-LMN attractor network settles the system down to stable (self-sustained) attractor states, in which subpopulations of HD cells with similar preferred directions are active while the others remain silent (Zhang, 1996, Blair et al., 1998).

2.3 Integrator

To integrate non-zero angular velocities (i.e. to shift the stable state over the continuous attractor space according to the angular velocity ω), an unbalanced cell activity between DTN_{cw} and DTN_{ccw} is generated. The neuronal responses of the DTN_{cw} and DTN_{ccw} cells are correlated with both the head direction $\theta(t)$ and the angular velocity $\omega(t)$. In particular, the activity of directional cells in DTN_{cw} and DTN_{ccw} is modulated proportionally to $|\omega(t)|$ during clockwise and counterclockwise turning, respectively. Neurons encoding the animal's head angular velocity have been observed in the parietal, somatosensory, and visual cortices (McNaughton et al., 1991, Blair and Sharp, 1995).

Fig. 5 shows the lateral connections between DTN_{cw} and DTN_{ccw}. These connections are distributed such that a neuron i with a preferred direction θ_i projects to a cell l with a

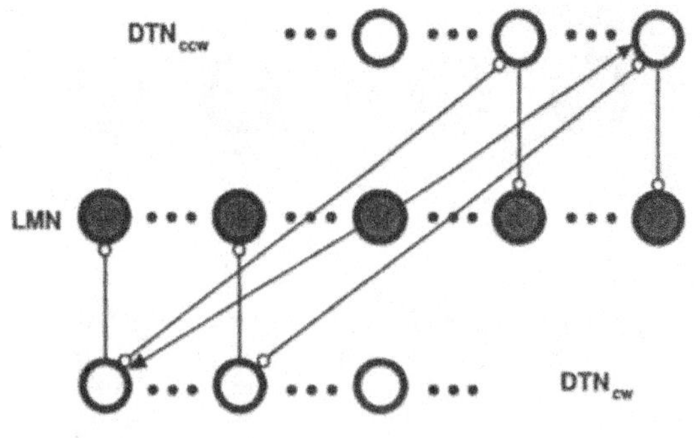

Figure 5: The integrator is based on inhibitory lateral connections (circles) between DTN_{cw} and DTN_{ccw}.

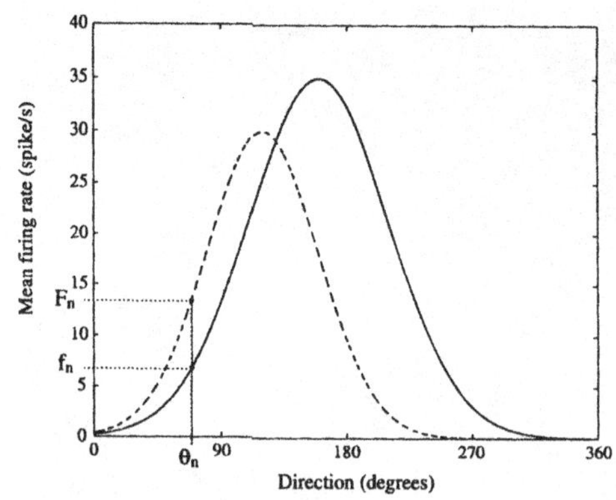

Figure 6: Evaluation of a set of parameters at time t. For each neuron n, the desired Gaussian profile F_n (dashed line) is compared to the generated Gaussian profile f_n (solid line).

preferred direction θ_l according to

$$w_{li} = W_{li}^{max} \cdot \exp\left(- \frac{(\theta_i - \theta_l + \delta_{li})^2}{2\sigma_{li}^2} \right) \quad (3)$$

where the maximum weight $W_{li}^{max} = 25.6$ and the width $\sigma_{li} = 35.1°$ have been determined by the genetic algorithm (Sec. 2.6). The term δ_{li} is the angular offset of the inter-module connections with $\delta_{li} = -30°$ if $i \in \text{DTN}_{cw}$ and $\delta_{li} = 30°$ if $i \in \text{DTN}_{ccw}$. Only the connections with a weight $w_{li} > 0.90 * W_{li}^{max}$ are actually implemented in the network.

During clockwise head turns, for instance, DTN_{cw} cells inhibit the left side of the LMN hill of activity encoding the current direction θ (i.e. introduce an asymmetry within the recurrent coupling between LMN and DTN). The lateral shifted connections between DTN_{cw} and DTN_{ccw} reinforce the asymmetry in DTN. During clockwise head turns, DTN_{cw} cells inhibit the left side of the DTN_{ccw} hill. Thus, the inhibition on the right side of the LMN hill is decreased when the inhibition by cells in DTN_{cw} on the left side is increased. This yields a clockwise shift $\Delta\theta$ of the LMN activity profile. The speed of this shift is proportional to $|\omega|$.

2.4 *Interpreting the directional output*

At each time t, the ensemble activity of a population of neurons provides the estimate $\bar{\theta}(t)$ of the allocentric heading $\theta(t)$ of the robot. In order to compute $\bar{\theta}(t)$, we apply a population vector scheme (Georgopoulos et al., 1986) to decode the ensemble HD cell activity:

$$\theta(t) = \arctan\left(\frac{\sum_n^N sin(\theta_n)\delta(t - t_n)}{\sum_n^N cos(\theta_n)\delta(t - t_n)} \right) \quad (4)$$

where $\delta(t - t_i)$ is a Dirac function equal to 1 if the neuron n fires at time t, 0 otherwise.

2.5 *Neuron and synapse model*

Both excitatory and inhibitory formal units are leaky integrate-and-fire neurons. Let $V_l = -70\ mV$, $V_t = -50\ mV$, and $V_r = -55\ mV$ denote the resting membrane potential, the firing threshold, and the reset potential of a formal neuron, respectively. Let $I(t)$ be the total synaptic drive received by a cell from the afferent units. The dynamics of the membrane potential $V(t)$ is given by

$$C \cdot \frac{dV(t)}{dt} = -g \cdot (V(t) - V_l) + I(t) + I_\epsilon(t) \quad (5)$$

where C and g are, respectively, the membrane capacitance and leak conductance ($\tau = C/g$ is the membrane time constant) with $C = 0.5\ \mu F$ for an excitatory cell and $C = 0.2\ \mu F$ for an inhibitory cell; $g_m = 25\ nS$ for an excitatory cell and $g_m = 20\ nS$ for an inhibitory cell. The refractory period is 2 ms and 1 ms for an excitatory neuron and an inhibitory neuron, respectively.

$I_\epsilon(t)$ is the background input activity of a neuron and is defined as

$$I_\epsilon(t) = rnd_{0,1} \cdot K(t) \quad (6)$$

where $rnd_{0,1}$ is a random number uniformly drawn from $[0, 1]$. The factor $K(t)$ is constant for excitatory neurons and equal to 1200. For inhibitory neurons

$$K(t) = K_i + K_\omega \cdot \omega(t) \quad (7)$$

where $K_i = 100$, $K_\omega = 0.44$, and $\omega(t)$ is the angular velocity at time t. The value of K_ω has been determined by the genetic algorithm described in Sec. 2.6.

The synaptic input to a neuron n is taken as

$$I(t) = D \cdot s(t) \quad (8)$$

where $D = 0.01$ is a constant factor and $s(t)$ is a gating variable whose dynamics is given by

$$s(t + \Delta t) = (1 - D) \cdot s(t) + \sum_{k=0}^{N} \delta(t - t_k) \cdot w_{nk} + S(n) \quad (9)$$

where $\Delta t = 1\ ms$ is the time step, $\sum_{k=0}^{N} \delta(t - t_k) \cdot w_{nk}$ is a weighted sum over the spikes emitted by the presynaptic neurons k and w_{nk} is the strength of the connection from a presynaptic unit k to the postsynaptic neuron n, with $w_{nk} > 0$ for

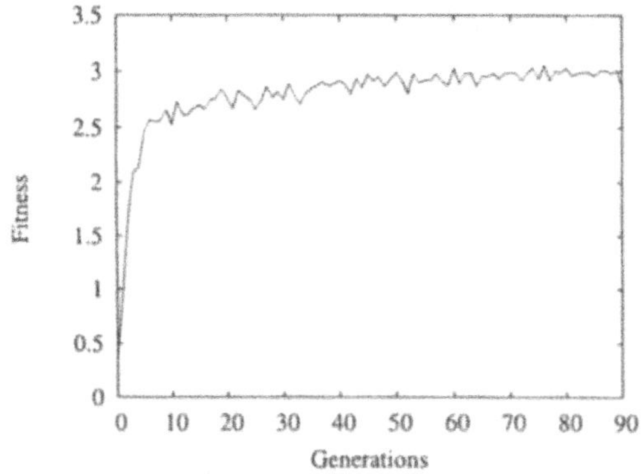

Figure 7: Evolution of the average fitness function over generations.

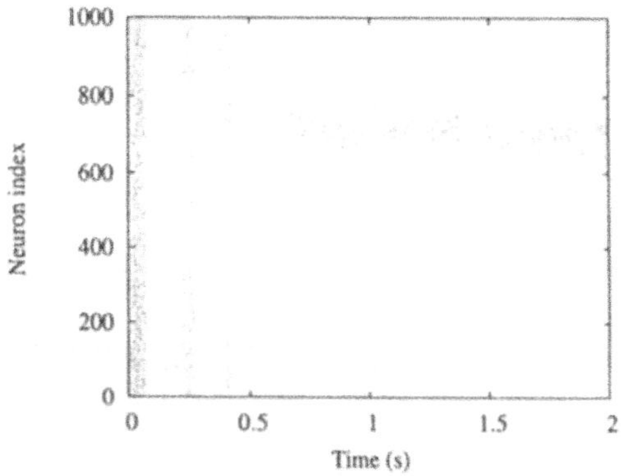

Figure 8: Emergence of a stable state in LMN in the absence of any polarizing stimulus. The hill of activity is centered at a random direction $\bar{\theta}_e$. Spikes are represented by dots.

excitatory connections and $w_{nk} < 0$ for inhibitory connections. The function $S(n)$ represents a stimulus that can be applied to the excitatory LMN neurons, and defines the number of spikes conveyed by the stimulus to a neuron n at time t. $S(n)$ is distributed over the neurons $n \in$ LMN according to a gaussian profile with amplitude $W_S = 540$ and width $\sigma_S = 90°$.

2.6 *Genetic Algorithm*

Three sets of Gaussian connections are used in the model: LMN to DTN, DTN to LMN, and the lateral connections between DTN_{cw} and DTN_{ccw}. As seen in Eqs. 1, 2 and 3, two parameters are relevant for determining the weight of a connection: the maximum weight W^{max} and the width of the Gaussian σ. All the six weight parameters as well as the factor K_ω used in Eq. 7 have been determined by using a genetic algorithm in simulation. This learning procedure allowed us to find a set of parameters that provides the model with appropriate stability and integration properties.

The fitness function measures the difference between the desired Gaussian activity profile $F_n(t)$ in LMN and the profile $f_n(t)$ generated by the current set of parameters (Fig. 6). The desired profile function $F_n(t)$ at time t is determined by:

$$F_n(t) = A \cdot \exp^{\frac{-(\theta_n(t)-\theta_\mu(t))^2}{2\sigma^2}} \tag{10}$$

where $A = 40\ Hz$, $\sigma = 40°$ and $\theta_\mu(t)$ is the current robot direction in the simulation. The fitness function $\phi(t)$ is given by:

$$\phi(t) = \frac{\sum_{n=1}^{N_E} \exp^{\frac{|f_n(t)-F_n(t)|}{k}}}{N} \tag{11}$$

where $f_n(t)$ is the frequency of the neuron n at time t and $k = 17.4$ a constant.

To improve the selection of appropriate sets of parameters, generations are tested through four different steps M:

1. stability of the model: $\theta_\mu(t) = \theta_{initial}$ is constant and the angular velocity ω is null, $M = 0$,
2. integration of a low angular velocity: $\theta_\mu(t + \Delta t) = \theta_{initial} + \omega \cdot t$ where $\omega \simeq 15°/s$, $M = 1$,
3. integration of a medium angular velocity: $\theta_\mu(t + \Delta t) = \theta_{initial} + \omega \cdot t$ where $\omega \simeq 45°/s$, $M = 2$,
4. integration of a high angular velocity: $\theta_\mu(t + \Delta t) = \theta_{initial} + \omega \cdot t$ where $\omega \simeq 110°/s$, $M = 3$.

Only those neural networks with a fitness $\phi \geq 0.75$ for a given step M are selected and tested in the phase $M + 1$. The final fitness ϕ_P for a given set of parameters P is defined by

$$\phi_P = M + \sum_{t=0}^{T} \frac{\phi(t)}{T} \tag{12}$$

where T is the duration of the experiment. The evolution of the fitness over generations is shown Fig. 7.

3. Results

First, we study the emergence and stability of a hill of activity to assess the attractor property of the HD model. Second, we study the update of the HD signal based on the integration of an ideal angular velocity input. Third, we apply to the model a real angular velocity signal provided by the MT9 device. Finally, we simulate a visual-based calibration of the system and focus on the latency of the update of the HD cell preferred directions.

3.1 *Emergence and stability of an attractor state*

The attractor dynamics of the model yields the emergence of a stable state from random noise. Thus, in the absence of external stimuli, the HD population activity profile settles down to a self-sustained state in which only a subpopulation of cells with similar preferred directions discharges tonically. The blob of activity encodes a constant direction $\bar{\theta}_e$ (as shown in Fig. 8). The attractor state is established after a transitory

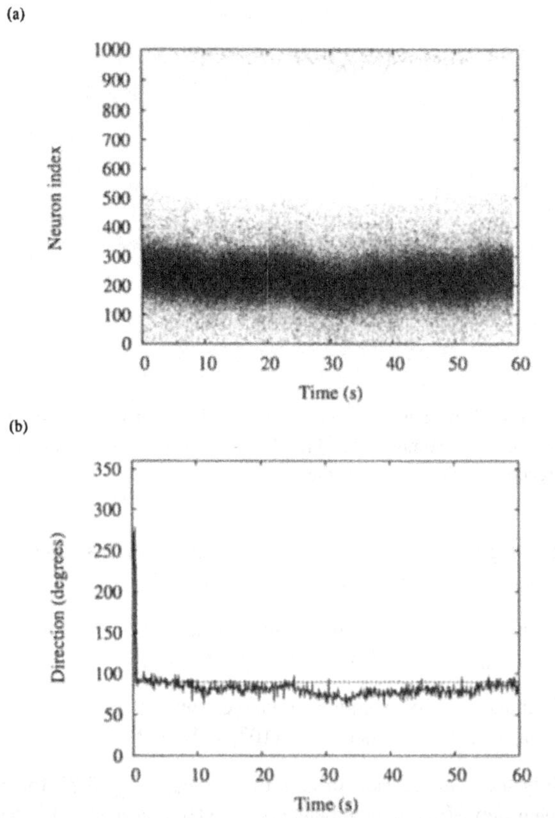

Figure 9: (a) Rastergram of the activity of the HD cells in LMN over 60 s. A polarizing external stimulus (applied at $\theta_S = 90°$ during the first 10 ms) generates a stable attractor state. After stimulus removal ($t > 10$ ms) the self-sustained state persists over time. (b) Stability of the HD representation over time. The dashed line represents the initial direction θ_S set by the stimulus during the first 10 ms. The continuous line is the direction computed by population vector coding (Eq. 4).

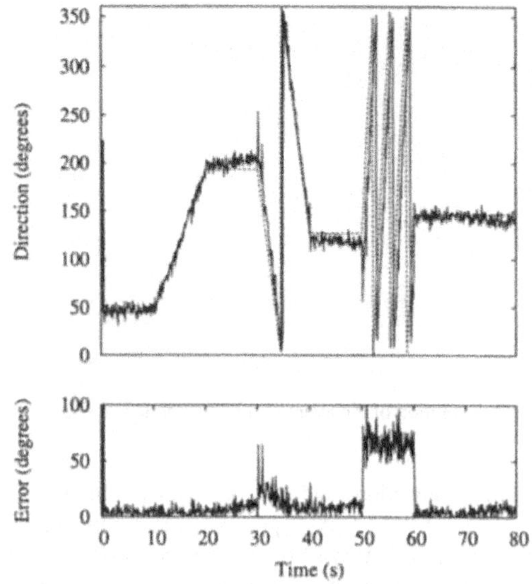

Figure 10: Integration of ideal angular velocities by the model. Top: a velocity $\omega_1 = 15°/s$ is applied for $10\ s < t < 20\ s$; a velocity $\omega_2 = 45°/s$ is applied for $30\ s < t < 40\ s$; and a velocity $\omega_3 = 110°/s$ is applied for $50\ s < t < 60\ s$. The dashed line represents the ideal integration of these angular velocities. The continuous line represents the direction signal of the HD cells of the model. Bottom: the integration error of the HD system over time.

period (of approximately 700 ms) characterized by an oscillatory activity pattern. The relative weight between inhibition and excitation determines the equilibrium reached after the transient period.

The model can also be initialized to a specific direction θ_s by applying an external stimulus S to LMN. The rastergram in Fig. 9 (a) shows the spike activity of the N_E cells in LMN when a polarizing input centered at $\theta_S = 90°$ is applied during the first 10 ms. After stimulus removal ($t > 10$ ms), the attractor state persists over the time. This corresponds to the situation in which the head of the animal is immobile (the head angular velocity ω is zero) and oriented in a given direction θ_S. The mean peak spike frequency and the width of the hill of activity, averaged over $T = 60$ s, are about 30 Hz and 200°, respectively.

The center of mass of the ensemble firing pattern, computed according to Eq. 4 and averaging over $T = 60$ s, is about $\bar{\theta} = 86°$. Fig. 9 (b) shows the stability of the directional representation over 60 s. The mean error $e = \sum_t |\bar{\theta}(t) - \theta_S| / T$ is about 9°.

3.2 Integration

When the head of the rat rotates (i.e. $\omega \neq 0$), the HD cell system integrates the angular velocity in order to update the HD representation over time.

We first investigate the integration done by the model in the case of three ideal angular velocity signals, $\omega_1 = 15°/s$, $\omega_2 = 45°/s$ and $\omega_3 = 110°/s$. Each angular velocity input is applied during a 10 s period. Fig. 10 shows the response of the HD cell system to these three velocity profiles as well as the integration error over time. The average error of the estimation of the direction by the model is about 12°. Qualitatively, the diagram shows that the larger the angular velocity, the larger the error.

The above simulation does not reflect the noise present in real experimental conditions. For instance, when sending a motor command to the robot, the motion generated by the actuators is noisy. Fig. 11 compares an ideal angular velocity profile (motor command) to the real angular velocity of the robot. The robot is asked to turn on the spot at different random speeds for different random time periods. The angular velocity of the robot is computed from the absolute direction given by the video tracking system. The comparison shows

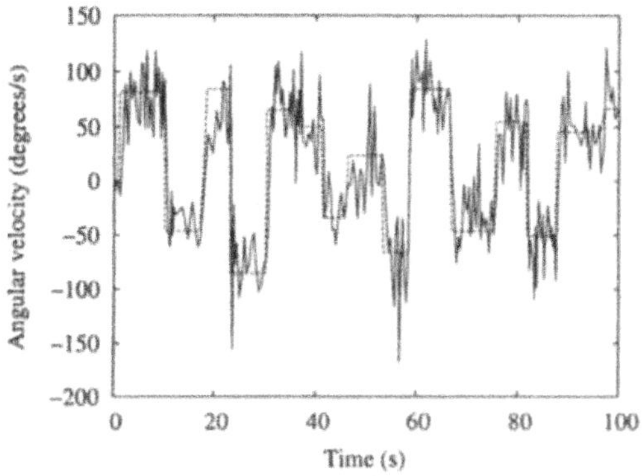

Figure 11: Example of noisy angular velocity profile. The dashed line represents the command sent to the robot. The continuous line indicates the angular velocity computed by the tracking system.

that, given a constant angular velocity command, the real angular velocity of the robot is unstable.

Since the navigating robot has to be able to estimate its current heading autonomously, it must integrate the angular velocity signals coming from of one of the on-board sensory systems. The robot is equipped with odometers measuring locomotor signals and accelerometers (the MT9 device) providing inertial signals.

Fig. 12 compares the real direction of the robot (provided by the video tracking system) and the directions computed by the MT9 device and by the odometers. The experimental protocol is the same as the one used previously. Let $e = \sum_t |\theta'(t) - \theta(t)|/T$ denote the mean error, averaged over $T = 100$ s, between the estimation $\theta'(t)$ and the actual direction $\theta(t)$ of the robot. We average e over 26 trials and find an error of $e_{MT9} = 64°$ for the directional signal from the MT9 device and an error of $e_{odo} = 71°$ for the odometers.

Note that, given the experimental protocol (a robot rotating on the spot), non-systematic errors, like wheel slippage or irregularities on rough surface, are unlikely to occur. Despite the small difference between e_{MT9} and e_{odo}, the angular velocity input provided by the MT9 device is more reliable than the signal from the odometers. For instance, if the robot was pulled up and displaced passively by the experimenter, the odometry would be seriously impaired whereas the MT9 device would be able to sense the movement. Therefore, we decided to use the MT9 device signal as input for our HD model.

At time $t = 0$, the HD system is initialized with respect to an absolute direction θ_{MT9}. As the robot starts rotating, the signals from the MT9 device are integrated by the model which starts shifting its internal representation to track the angular displacements of the robot. Fig. 13 shows an example of the update of the HD signal $\bar{\theta}(t)$ encoding the current robot heading $\theta(t)$ as well as the tracking error over time. The mean error of the model, over 26 trials of $T = 100$ s each, is 33° relative to the input of the MT9 device and about 73° relative

(a)

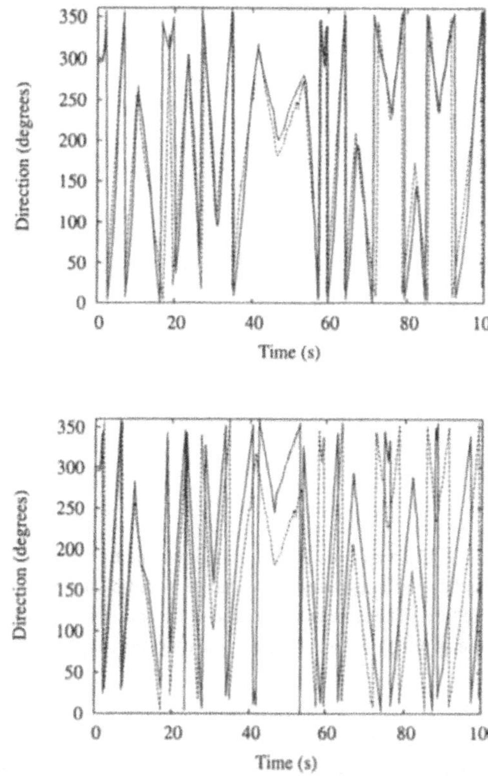

(b)

Figure 12: Tracking of the robot actual direction (represented by the dashed line) by the MT9 device (a) and by the integration of the input from the odometers (b).

to the output of the video tracking system.

As shown in Fig. 13, the integration realized by the model is not perfect in a mathematical sense. Thus, the system needs to be calibrated to avoid cumulative errors over time. In this paper, we do not describe any calibration mechanism (Arleo and Gerstner, 2000). Rather, we focus on the temporal aspects of the update of the preferred directions of HD cells by a reorienting visual stimulus.

3.3 *Short update latencies following reorienting stimuli*

Electrophysiological findings suggest that HD cells are controlled by salient visual cues (Taube, 1998). Recently, (Zugaro et al., 2003) have measured the time necessary to update the preferred directions of rat HD cells by a reorienting visual landmark. The experimental setup consisted of a black cylinder with a white card attached to the inner wall used as a visual cue. The preferred directions of the HD cells were first measured in light conditions. Then, in the dark, the cue card was rotated by 90°. When, the light was turned back on the HD cells updated their preferred directions to reflect the

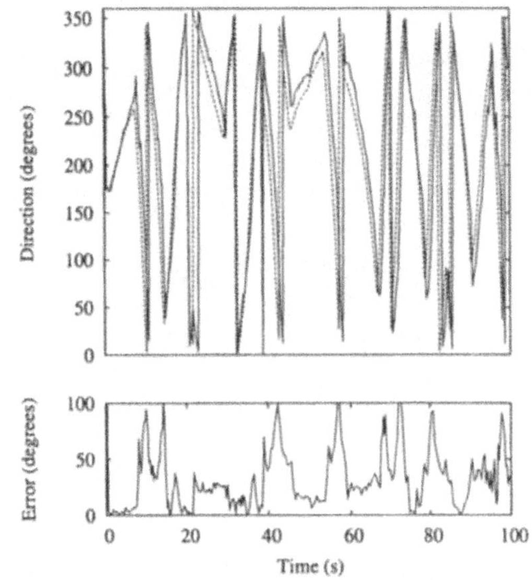

Figure 13: Update of the HD directional coding by the model (continuous line) by integrating the signal provided by the MT9 device (dashed line). The error of integration of the model compared to the direction given by MT9 is shown in the diagram below.

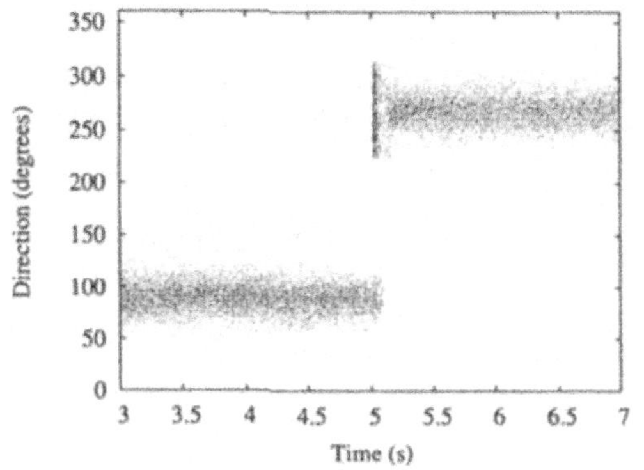

Figure 14: Raster plot showing the response of the HD system to a 180° reorienting visual stimulus applied at time $t_{S'} = 5\ s$. A new blob of activity is established at $\theta_{S'} = 270°$ whereas the previous hill of activity is extinguished.

new visual configuration. A mean latency of $80 \pm 10\ ms$ was observed for the establishment of a new blob following the reorientation event. On the other hand, $140 \pm 10\ ms$ were necessary for the extinction of the hill of activity existing prior the reorientation event.

To reproduce these results, the model is first initialized with an external stimulus S centered at a specific direction $\theta_S = 90°$. Second, at time $t_{S'} = 5\ s$, a reorienting stimulus S' is applied at $\theta_{S'} = 270°$ (simulating the transition from dark to light conditions). Fig. 14 shows the response of the system to the reorienting visual stimulus which triggers a 180° update of all preferred directions. The HD system is reoriented according to the directional reference frame anchored to S'. As a consequence, the attractor network settles to a new stable state abruptly.

The resulting update latency for the establishment of the new blob of activity is about $30\ ms$, whereas the transient delay for the extinction of the previous attractor state is about $110\ ms$. The attractor dynamics is such that inhibition occurs after the increase in the overall activity of the network. This causes a longer update time for the extinction of the previous hill of activity. These results are consistent with those reported by (Zugaro et al., 2003). Note that in the model the transmission delay necessary for the visual signals to reach HD cells is not taken into account. There are no experimental data reporting this transmission delay. However, (Galambos et al., 2000) show that a visual stimulus takes already about $30\ ms$ to go from the retina to the primary visual cortex.

4. Discussion

The main goal of this work was to endow a robot with an allocentric direction representation. We adopted a neuro-mimetic approach inspired by the properties of head direction (HD) cells, neurons observed in the rat's limbic system.

In order to obtain a stable and persistent HD signal, the model relies on an attractor neural network modeling the neural circuit formed by the lateral mammillary nucleus (LMN) and the dorsal tegmental nucleus (DTN). These two anatomical structures are relevant for generating and maintaining the internal direction representation of rat HD cells. The model reproduces the anatomical interconnectivity between LMN and DTN. The attractor network has no lateral excitatory connections to reflect the absence of experimental evidence for recurrent excitation in the DTN-LMN circuit. Similar to the rat HD cell system, the model can integrate angular velocity signals to update the direction representation over time. Lateral inhibitory connections in the formal DTN are used for this purpose.

In contrast to earlier HD models, we employ spiking neurons to focus on the temporal aspects of the dynamics of the attractor network. For instance, electrophysiological data suggest a very rapid reorientation of the HD representation following changes in the visual scene. The model is able to reproduce these short update latencies.

We stress the importance of validating the model in real experimental conditions in order to make it useful for autonomous navigating systems. We assess the ability of the model to integrate an angular velocity signal $\omega(t)$ in real-time. We employ an inertial sensor to estimate the angular displacements of the robot. The integration of $\omega(t)$ is affected by a drift of the internal direction representation over time. Therefore, a calibration mechanism, based for instance on vision, will be further developed. Experimental findings show that visual information is conveyed to the HD system via the postsubiculum (PSC). The model may be extended by adding a PSC module performing the vision-based calibration.

The HD model presented here provides a subcomponent for a larger autonomous navigating system. This work is part of Psikharpax, a project that aims at developing a bio-mimetic navigation and action selection model embedded on a real robot.

Acknowledgments

This work was supported by the ROBEA program of the CNRS and by the ACI Neuroscience du Ministère de la Recherche. The authors thank S. Gourichon, S. I. Wiener, N. Brunel, O. Sigaud and J. A. Meyer for useful discussions.

References

Allen, G. V. and Hopkins, D. A. (1989). Mamillary body in the rat: topography and synaptology of projections from the subicular complex, prefrontal cortex, and midbrain tegmentum. *Journal of Comparative Neurology*, 286:311–336.

Arleo, A. and Gerstner, W. (2000). Modeling rodent head-direction cells and place cells for spatial learning in bio-mimetic robotics. In Meyer, J.-A., Berthoz, A., Floreano, D., Roitblat, H., and Wilson, S. W., (Eds.), *From Animals to Animats VI*, pages 236–245, Cambridge MA. MIT Press.

Bassett, J. P. and Taube, J. S. (2001). Neural correlates for angular head velocity in the rat dorsal tegmental nucleus. *The Journal of Neuroscience*, 21(15):5740–5751.

Blair, H. T., Cho, J., and Sharp, P. E. (1998). Role of the lateral mammillary nucleus in the rat head direction circuit: A combined single unit recording and lesion study. *Neuron*, 21:1387–1397.

Blair, H. T. and Sharp, P. E. (1995). Anticipatory head direction signals in anterior thalamus: Evidence for a thalamocortical circuit that integrates angular head motion to compute head direction. *The Journal of Neuroscience*, 15(9):6260–6270.

Degris, T., Brunel, N., Sigaud, O., and Arleo, A. (2004). Rapid response of head direction cells to reorienting visual cues: A computational model. *Neurocomputing (in press)*.

Galambos, R., Szabo-Salfay, O., s, P. B., Palhalmi, J., Szilagyi, N., and Juhasz, G. (2000). Temporal distribution of the ganglion cell volleys in the normal rat optic nerve. *Proceedings of the National Academy of Sciences USA*, 97(24):13454–13459.

Georgopoulos, A. P., Schwartz, A., and Kettner, R. E. (1986). Neuronal population coding of movement direction. *Science*, 233:1416–1419.

Goodridge, J. P. and Touretzky, D. S. (2000). Modeling attractor deformation in the rodent head-direction system. *Journal of Neurophysiology*, 83(6):3402–3410.

Gourichon, S., Meyer, J.-A., Ieng, S., Smadja, L., and Benosman, R. (2003). Estimating ego-motion using a panoramic sensor: Comparison between a bio-inspired and a camera-calibrated method. In Holstein, H. and Labrosse, F., (Eds.), *AISB03 Symposium on Biologically Inspired Vision, Theory and Application*, pages 91–101.

Knierim, J. J., Kudrimoti, H. S., and McNaughton, B. L. (1995). Place cells, head direction cells, and the learning of landmark stability. *The Journal of Neuroscience*, 15:1648–1659.

McNaughton, B. L., Chen, L. L., and Markus, E. J. (1991). Dead reckoning, landmark learning, and the sense of direction: A neurophysiological and computational hypothesis. *Journal of Cognitive Neuroscience*, 3:190.

Mittelstaedt, M. L. and Mittelstaedt, H. (1980). Homing by path integration in a mammal. *Naturwissenschaften*, 67:566–567.

Redish, A. D., Elga, A. N., and Touretzky, D. S. (1996). A coupled attractor model of the rodent head direction system. *Network*, 7(4):671–685.

Skaggs, W. E., Knierim, J. J., Kudrimoti, H. S., and McNaughton, B. L. (1995). A model of the neural basis of the rat's sense of direction. In Tesauro, G., Touretzky, D. S., and Leen, T. K., (Eds.), *Advances in Neural Information Processing Systems 7*, pages 173–180, Cambridge, MA. MIT Press.

Song, P. and Wang, X.-J. (2002). An attractor network model without recurrent synaptic excitation for head direction neurons in rodents. *Society for Neuroscience Abstracts*, page 584.8.

Taube, J. S. (1998). Head direction cells and the neurophysiological basis for a sense of direction. *Progress in Neurobiology*, 55:225–256.

Zhang, K. (1996). Representation of spatial orientation by the intrinsic dynamics of the head-direction cell ensemble: A theory. *The Journal of Neuroscience*, 16(6):2112–2126.

Zugaro, M. B., Arleo, A., Berthoz, A., and Wiener, S. I. (2003). Rapid spatial reorientation and head direction cells. *The Journal of Neuroscience*, 23(8):3478–3482.

Zugaro, M. B., Berthoz, A., and Wiener, S. I. (2001). Background, but not foreground, spatial cues are taken as references for head direction responses by rat anterodorsal thalamus neurons. *The Journal of Neuroscience*, 21(RC154):1–5.

Navigation and Cognitive Map Formation Using Aperiodic Neurodynamics

Derek Harter **Robert Kozma**
Department of Computer Science
University of Memphis
Memphis, TN 38152 USA
{dharter|rkozma}@memphis.edu

Abstract

Biological brains are saturated with complex dynamics. Artificial neural network models abstract much of this complexity away and represent the computational process of neuronal groups in terms of simple point, and sometimes periodic attractors. But is this abstraction justified? Aperiodic dynamics are known to be essential in the formation of perceptual mechanisms and representations in biological organisms. Advances in neuroscience and computational neurodynamics are helping us to understand the properties of nonlinear systems that are fundamental in the self-organization of stable, complex patterns for perceptual, memory and other cognitive mechanisms in biological brains.

Much of this new understanding of the principles of self-organization in biological brains has yet to be used to improve the performance of animats and other biologically inspired models of behavior generation. In this paper we review some of the findings of how biological brains may use aperiodic dynamics in the formation of perceptual mechanisms. We discuss some models of this formation of chaotic attractors for perceptual categorization. And finally we present some work using these models to develop cognitive maps and navigation behavior in an autonomous agent.

1. Introduction

The study of nonlinear dynamics has blossomed in all areas of science in the past decades for many reasons. Nonlinear dynamics provide new conceptual and theoretical tools that allow us to understand and examine complex phenomena that we have never been able to tackle before. Nonlinear dynamics seem to show up everywhere, in physical systems like electrical circuits, lasers, optical and chemical systems. But such dynamics are especially ubiquitous in the biological world, from fractal growth patterns in biological development and city formation to the self-organizing characteristics of population models, and the importance in regulating healthy biological rhythms such as the beating of the heart.

Nonlinear systems in critical states have many interesting properties. Phenomenon such as stochastic and chaotic resonance (Kozma and Freeman, 2001) are known which enable such systems to actually detect the presence of signals much better in noisy environments than nonlinear systems are capable of doing. Their greatest interest lies however in their fundamental relationship to self-organization and emergence of complex patterns and behaviors in complex environments. Aperiodic dynamics are both an indication of and a mechanism for the emergence of such self-organizing properties.

Insights in nonlinear systems theory are beginning to be applied to understanding the dynamics of the brain, and how such processes produce cognition (Freeman, 1999, Tsuda, 2001, Freeman, 2003). Aperiodic dynamics are know to play a fundamental role in the mechanisms for the self-organization of meaning in mammalian perceptual systems (Skarda and Freeman, 1987, Freeman, 1991). Neurological evidence has shown that perceptual meanings (of recognized smells) are created through the formation and dissolution of chaotic attractors in the olfactory bulb. We will discuss this example of the self-organization of a perceptual pattern of meaning. We use this type of organization in aperiodic systems to model the formation of place cells in the simulated hippocampus of an autonomous agent. And we also show how these mechanisms may be used to produce goal-directed navigation in an autonomous agent.

2. K-Sets: A Neurodynamical Population Model of Brain Dynamics

2.1 *Aperiodic Dynamics in Olfactory Systems*

In their influential paper, Skarda and Freeman (Skarda and Freeman, 1987) argued that chaos, as an emergent property of intrinsically unstable neural masses, is very important to brain dynamics. In ex-

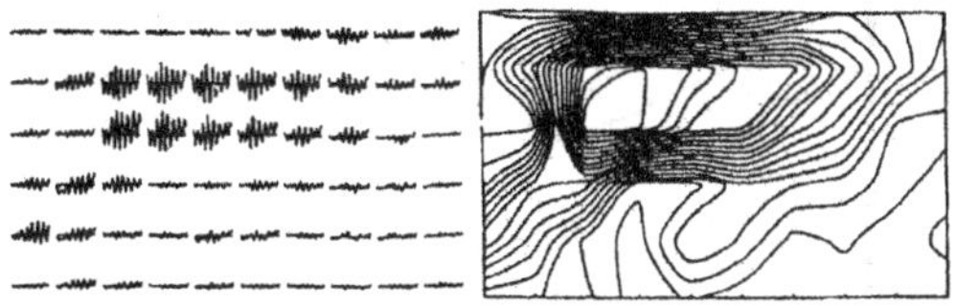

Figure 1: EEG carrier wave patterns (left) and contour map (right) of olfactory cortex activity in response to a recognized smell stimulus (from Freeman, 1991, p. 80)

periments carried out on the olfactory system of trained rabbits, Freeman was able to demonstrate the presence of chaotic dynamics in EEG recordings and mathematical models. In these experiments, Freeman and his associates conditioned rabbits to recognize smells, and to respond with particular behaviors for particular smells (e.g. to lick or chew). They captured EEG recordings of the activity in the olfactory bulb, before and after training for the smells.

The EEG recordings revealed that in fact, chaotic dynamics (as shown by the observed strange attractors) represented the normal state when the animal was attentive, in the absence of a stimulus. These patterns underwent a dramatic (nonlinear) transition when a familiar stimulus was presented and the animal displayed recognition of a previously stored memory (through a behavioral response). The pattern of activity changed, very rapidly, in response to the stimulus in both space and time. The new dynamical pattern was much more regular and ordered (very much like a limit cycle, though still chaotic of a low dimensional order). The spatial pattern of this activity represented a well defined structure that was unique for each type of odor that was perceptually significant to the animal (e.g. conditioned to recognize). Figure 1 shows an example of such a recorded pattern, after recognition of a stimuli, in the EEG recordings and the associated contour map. In this figure after recognition, all of the EEG waves are firing in phase, with a common frequency (which Freeman called the carrier wave). The pattern of recognition is encoded in the heights (amplitude modulations) of the individual areas. The amplitude patterns, though regular, are not exact limit cycles and exhibit low dimensional chaos. In other words, different learned stimuli were stored as a spatio-temporal pattern of neural activity, and the strange attractor characteristic of the attention state (before recognition) was replace by a new, more ordered attractor related to the recognition process. Each (strange) attractor was thus shown to be linked to the behavior the system settles into when it is under the influence of a particular familiar input odorant.

Freeman suggests that "an act of perception consists of an explosive leap of the dynamic system from the basin of one (high dimensional, in the attentive state) chaotic attractor to another (low dimensional state of recognition) (Freeman, 1991). These results suggest that the brain maintains many chaotic attractors, one for each odorant an animal or human being can discriminate. Freeman and Skarda speculate on many reasons why these chaotic dynamics may be advantageous for perceptual categorization. For one, chaotic activity continually produces novel activity patterns which can provide a source of flexibility in the individual. But since chaos is a ordered state, such flexibility is under control. As Kelso (Kelso, 1995) remarks, such fluctuations continuously probe the system, allowing it to feel its stability and providing opportunities to discover new patterns. Another advantage of chaos is that it allows for very rapid switching between attractors, which random activity is not able to do. Freeman also proposed that such patterns are crucial to the development of nerve cell assemblies and for other biologically motivated reasons.

2.2 K-Set Model of Aperiodic Dynamics

The K-set hierarchy, developed by Freeman and associates (Freeman, 1975, Freeman, 1999, Skarda and Freeman, 1987, Freeman, 1991), is both a model of neural population dynamics and a description of the architectures used by biological brains for various functional purposes. The original purpose of the K-set was to model the dynamics observed in the olfactory perceptual system. The lowest level of the hierarchy, the K0 set, provides a basic unit that models the dynamics of a local population of tens of thousands of neurons. The dynamics of the K0 set are described by a second order ordinary differential equation feeding into an asymmetric sigmoid function:

$$ab\frac{d^2x(t)}{d^2t} + (a+b)\frac{dx(t)}{dt} + x(t) = f(t) \qquad (1)$$

This equation was determined by measuring the electrical responses of isolated neural populations to stimulation and other conditions. The a and b parameters are time constants that were determined through such physiological experiments. $x(t)$ is the pulse density of the modeled neural population, in other words the average number of neurons that are pulsing in the population at any given point in time. $f(t)$ is a nonlinear asymmetric sigmoid function describing the influence of incoming activation, and is given in equation 2.

$$f(t) = k[1 - exp(-\frac{e^{v-1}}{k})] \qquad (2)$$

A K0 unit models the dynamics of an isolated neural population. From the basic K0 unit can be built up architectures that capture the observed dynamics of increasingly larger functional brain areas. The KI models excitatory-inhibitory feedback populations. KII models interacting excitatory-inhibitory populations and correspond to organized brain regions such as the olfactory bulb (OB) or the prepyriform cortex (PC). KIII combine 3 or more KII populations to model functional brain areas such as perceptual cortex or hippocampus, and are capable of aperiodic dynamics of the type observed in these regions to, for example, derive meaning from perceptual senses. In the simulations presented in this paper, we use a discretized version of the K-model (described in (Harter and Kozma, 2003, Harter and Kozma, 2002)) developed for use in large-scale autonomous agent simulations.

In the original K model, the purpose of the KIII set was to model the chaotic dynamics observed in rat and rabbit olfactory systems (Freeman, 1987, Shimoide et al., 1993, Freeman and Shimoide, 1994). KII are capable of oscillatory behavior, as described above. When three or more oscillating systems (KII) of different frequencies are connected through positive and negative feedback, the incommensurate frequencies can result in aperiodic dynamics. The dynamics of the KIII are produced in just this manner, by connecting three or more KII units of differing frequencies together. The KIII set was not only capable of producing time series similar to those observed in the olfactory systems under varying conditions of stimulation and arousal, but also of replicating power spectrum distributions characteristics of biological and natural systems in critical states (Solé and Goodwin, 2000, Bak et al., 1987).

The power spectrum is a measure of the power of a particular signal (or time series as for example that obtained from an EEG recording of a biological brain) at varying frequencies. The typical power spectrum of a rat EEG (see Figure 2, top) shows a central peak in the 20-80 Hz range, and a $1/f^{\alpha}$ form of the slope. The measured slope of the power spectrum varies around $\alpha = -2.0$. $1/f^{\alpha}$ type power spectra are abundant in nature and are characteristic of critical states, between order and randomness, at which chaotic processes operate. Power spectra of biological brains have been observed to vary from $\alpha = -1.0$ to $\alpha = -3.0$. The atypical part of the experimental EEG spectra is the central peak, indicating stronger oscillatory behavior in the γ frequencies. This central peak in the 20-80 Hz range is known as the γ frequency band, and is associated with cognitive processes in biological brains. The K-models are capable of replicating the power spectra of biological EEG signals, as shown in Figure 2, bottom (Harter and Kozma, 2003, Freeman, 1991).

The KIII sets are capable of organizing perceptual categories in the fashion observed in biological perceptual systems. The KIII used as such a pattern classifier is very robust and compares well with more standard methods of pattern classification (Kozma and Freeman, 2001). In this paper we use a discretized version of the K-sets developed for use in autonomous agents, called KA sets (Harter and Kozma, 2003, Harter and Kozma, 2004).

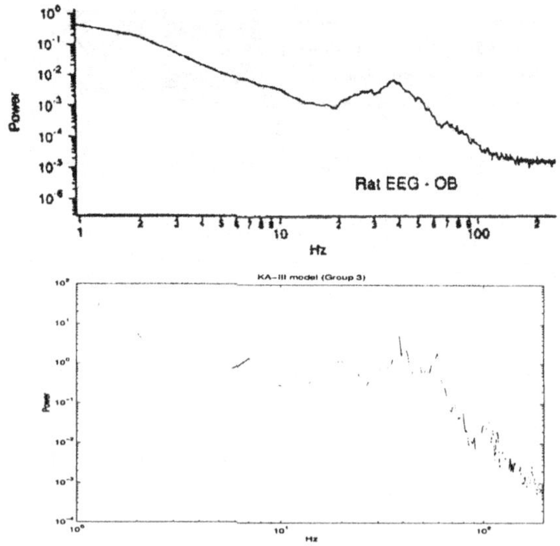

Figure 2: The power spectrum of a rat Olfactory Bulb EEG is simulated with the KA-III model. The calculated "1/f" slope of the EEG and model is approximately -2.0. Rat OB data from (Kay et al., 1995), KA power spectrum from (Harter and Kozma, 2003)

3. Place Cell Formation using Aperiodic Neurodynamics

3.1 Experimental Architecture and Environment

Perceptual meanings are formed through aperiodic attractors in the spatio-temporal activation of neuronal groups in the perceptual cortex. The same basic mechanisms of aperiodic dynamic in perception are also used by the biological brain in other areas to form memory and behavior producing structures (Kozma et al., 2003). In this section we use the basic KA-III architecture to demonstrate the formation of place cells in a simulated hippocampus of an autonomous agent.

In this experiment, we used the Khepera virtual environment simulator (Michel, 1996). Figure 3 (bottom left) shows the morphology of the Khepera agent. The Khepera robot is a simple agent that contains 8 infra-red and 8 light sensors. It has two independently controlled wheels that allow it to move forward, backward, and turn left and right in place. The environment for this experiment is shown in figure 3. In the environment we place 8 light sources, which will be used as salient environmental locations (i.e. they can be thought of as food sources for the agent in the environment). The light sources are detectable to the agent at a distance, and the range where

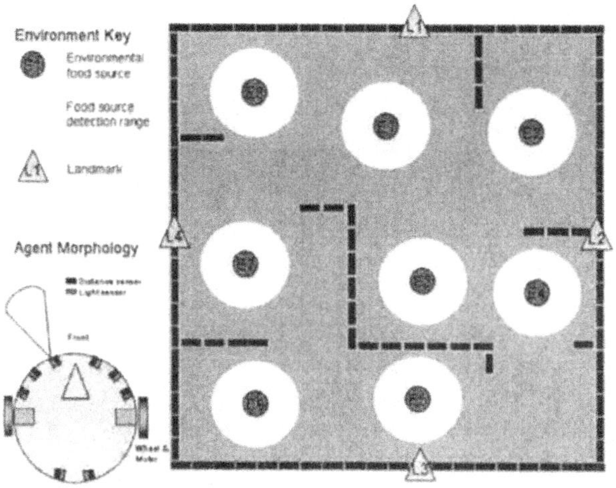

Figure 3: Agent morphology (bottom left) and environmental setup for hippocampal simulations. The environment contains landmarks, used as allocentric reference points by the agent, and salient environmental locations, such as food sources. The agent is only able to detect the presence of a food source when it is within a particular range of it.

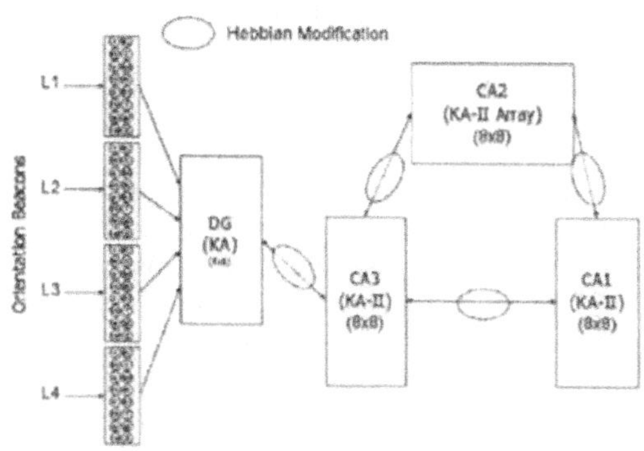

Figure 4: Architecture of KA-III hippocampal simulations

the light source is detectable is indicated in Figure 3. In addition to the 8 salient environmental locations, there are 4 landmarks. The landmarks are always detectable to the agent, and it knows the distance and direction to each of the 4 landmarks as part of its sensory information.

The architecture of the simulated hippocampus is shown in Figure 4. The portions of the architecture that form the cognitive map of the environment are simulated by a KA-III. These are the CA1, CA2 and CA3 areas, and are based on biological evidence of the structure of the hippocampus. Each of the CA areas contains an 8x8 array of KA-II units (for a total of 64 units in each CA region). The KA-II units in a CA region are recurrently connected to the 8 nearest neighbors of that unit in the grid. Each CA region is connected to the other 2. That is to say, each of the KA-II units in a CA region has projections to KA-II units in other regions. The fan-out ratio is 1 to 8 in the projections between regions. The interconnection of these 3 CA regions via inhibitory and excitatory feedback forms a KA-III. The connections between CA regions were changed using Hebbian modification.

Orientation beacons are fed into the hippocampal simulation through the DG region (Figure 4, left). The DG contains an 8x8 matrix of KA-0 units. Orientation signals from the 4 landmarks are fed into the DG units. Each of the 4 landmarks has 8 units associated with the direction to the landmark, and 8 units associated with the distance. Directions are broken into 8 cardinal units, North, NorthEast, East, SouthEast, South, SouthWest, West and NorthWest. Units are sensitive to the direction of a particular landmark, though we use a graded response with a normal distribution which allows more than 1 unit to be active. Similarly there are 8 cardinal distance values VeryClose, Close, MediumClose, Medium, MediumFar, Far, VeryFar, Distant. Again a graded response with normal distribution is applied to the units. The DG area connects with the CA3 area, and the connections between these areas are also subject to Hebbian modification.

3.2 Method

We use two types of learning in the simulation, Hebbian modification and habituation. Hebbian modification only occurs when the robot is within a certain range of a light source. Therefore the light sources provide a certain valence signal that acts as a stimulus to learn environmentally salient locations. When the robot is not within proximity to a light source, no reinforcement signal is produced. During these times habituation of the stimulus occurs. This has the effect of lessening the response of the simulated hippocampus to unimportant regions in the environment (Kozma and Freeman, 2001).

The expected effect of this stimulation is to form 2 distinct types of dynamical patterns in the CA regions. When the agent is out of range of an environmentally salient location, the dynamics should be in the high-dimensional chaotic state, receptive to input but not indicative of recognizing a salient event. When in range of a light source, the system should transition to a low dimensional attractor, indicative of recognition of the important location. Further, the spatial amplitude modulation patterns in the CA regions upon such recognition should form 8 unique patterns, one for each of the recognized regions.

The agent is allowed to roam in the environment, using a low level mechanisms to produce efficient, but random wandering. The agent roams for a set time period (1000 seconds) during which sensory data is perceived and

Hebbian modification occurs as described previously.

3.3 Results

We first look at the amplitude modulation (AM) patterns produced by the hippocampal simulation characteristic of the 8 salient regions in the agents environment. In Figure 5 we show examples of the amplitude of the activity of the 8x8 units in the CA3, displayed as contour maps. We next explain these contour maps and how they were produced.

After the agent was allowed to wander in its environment, and Hebbian modification was applied to the simulated hippocampus, we tested the response of the simulated hippocampus in the following manner. We randomly chose 4 points close to each of the 8 environmentally salient regions. We placed the agent at these points in the environment for 1 second where it received the appropriate perceptual inputs (e.g. distance and direction information to the 4 environmental landmarks). We captured the responses of the 8x8 units in the CA3 region during this 1 second of activity. We then calculated the standard deviation of the 8x8 CA3 units during the 1 second of activity to come up with a single matrix of 8x8 values characterizing the amplitude of the activity of each of the units. We plotted these amplitude patterns as contour maps and displayed the results in Figure 5.

The results in Figure 5 show visually the response of the simulated hippocampus after learning. The CA3 responds with similar AM patterns for points close to a particular location; for example test points 1,2,3 and 4 at location 8 show 2 valleys to the left of the 8x8 CA3 and 3 or so peaks near the bottom of the CA3 region. However, the AM patterns formed at the different locations seem to be different and distinct from one another.

Although the contour maps give evidence of the formation of unique AM patterns in response to environmental locations, we can more clearly see the results by measuring the similarity of the AM patterns to one another. In Table 1 we show the results of measuring the Euclidean distance between the 4 points tested at the 8 environmental locations. We show each of the target location/test patterns and which of the other AM patterns (excluding itself) the target was closest to. The AM pattern is treated as a 64 dimensional vector and Euclidean distance is used to determine closeness of one AM pattern to another. As can be seen, the AM patterns formed in the CA3 region of the KA-III hippocampal simulation form fairly coherent patterns. Patterns around an environmentally salient location tend to be most similar to one another. The performance is not perfect, however. For example, location 2 test points 3 and 4 ended up being closest to AM patterns formed at location 7. The fairly unique formation of AM patterns around the environmental locations can be interpreted as the formation of "place cell" like patterns in the simulated hippocampus.

Table 1: Results of similarity (distance) measure of CA3 hippocampal AM patterns

Target		Closest		Target		Closest	
Loc	Test	Loc	Test	Loc	Test	Loc	Test
1	1	1	3	5	1	5	2
1	2	1	1	5	2	5	1
1	3	1	1	5	3	5	4
1	4	1	2	5	4	5	3
2	1	2	3	6	1	6	2
2	2	2	3	6	2	6	3
2	3	7	2	6	3	6	2
2	4	7	1	6	4	1	1
3	1	3	4	7	1	7	2
3	2	3	3	7	2	7	1
3	3	3	2	7	3	6	2
3	4	3	1	7	4	7	3
4	1	4	2	8	1	8	4
4	2	4	1	8	2	8	4
4	3	4	1	8	3	8	4
4	4	4	1	8	4	8	2

As another example of the aperiodic attractors formed by the hippocampal simulation, we show a state space plot of the activity of a single unit in the CA3 region. Figure 6 shows a state space representation of the activity of a unit (in column 3, row7) in the CA3 region. The state space plots represent the activity of the unit plotted against itself with a 12ms time delay. Again we show the activity of the unit in response to 4 tests (rows in the figure) at the 8 environmental locations (columns in the figure).

This figure shows an example of how the KA-III may produce subsymbolic type representations of the input patterns. In this case, the unit seems to have formed at least 2 distinct aperiodic attractors. The unit activity at locations 1, 5 and 8 seem to be similar, and also attractors for locations 2,3,4,6, and 7 though there may be some difference in the dynamics among the second group. Although the attractors produced are not perfectly consistent, they may be capturing some kind of similarity in a portion of the perceptual input between these 2 groups of locations. They therefore are dividing the environment into 2 subsymbolic patterns. Each of the 64 units in the CA3 region, when you observe the state space plots, performs a similar but different partitioning of the input stimuli. The resulting attractors in the 8x8 CA3 region produce the unique AM patterns observed at each of the locations in the environment.

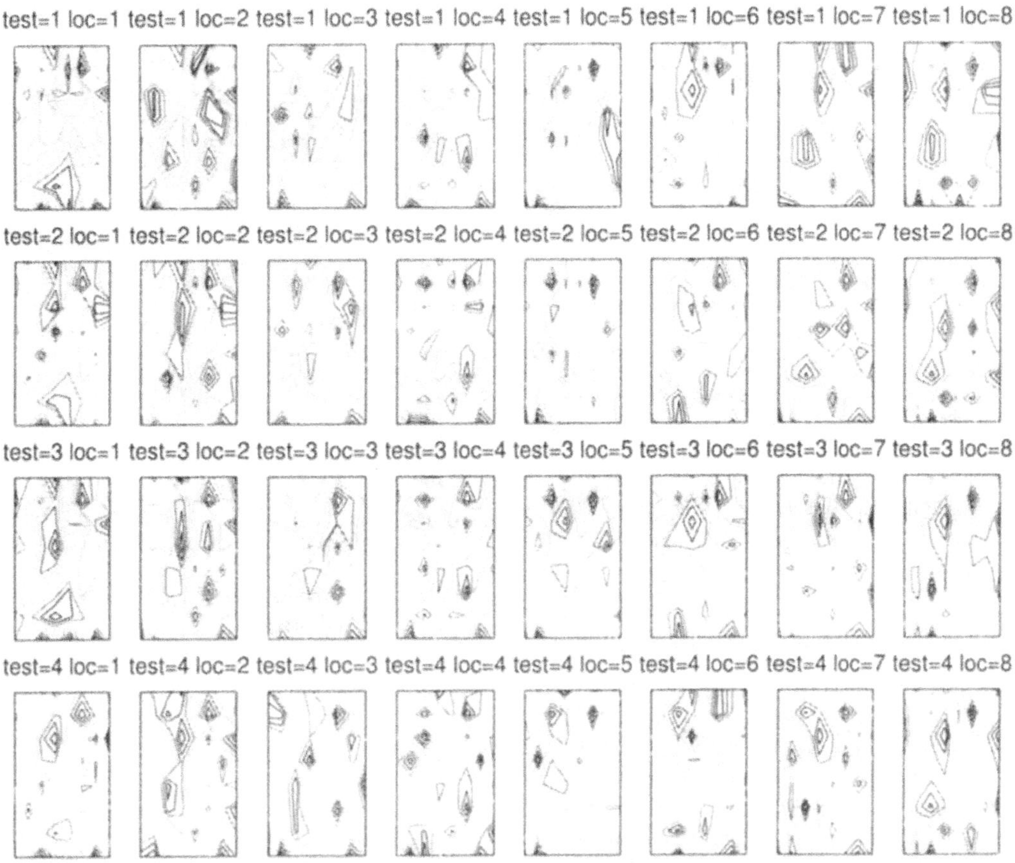

Figure 5: Example of AM patterns formed in the CA3 hippocampal region. In this figure we show contour maps of amplitudes of the activity in the 8x8 units in the CA3 region. We show the AM patterns for 4 test cases around each of the 8 salient locations in the environment.

4. Goal-Directed Navigation

4.1 Experimental Architecture and Environment

In this section we use the KA-III model to perform goal-oriented navigation in a simulated Martian-like debris field that might be encountered by a Martian planetary rover (Huntsberger, 2001, Tunstel, 2001). We use the Webots simulation environment (Michel, 2003), which is a physics based 3-d graphical simulation environment of real robotic platforms (see Figure 7).

The architecture of the agent for this simulation is shown in Figure 8. In this experiment the agent is hard-coded to instinctually move and avoid objects, using the agents touch and distance sensors (Figure 8, right). A goal location is selected in the environment, which emits a simulated gradient field that can be detected by the agents sensors. We think of the gradient field as a type of chemical gradient, such as might be followed by an animals olfactory system to locate a target. The gradient is invariant with respect to the obstacles in the environment, therefore the agent must combine local information about obstacles with the global information provided by the gradient to successfully navigate to the target.

4.2 Method

The simulated olfactory system receives sensory input on the strength of the smell signal from sensors positioned around the robots body. The task of the simulated olfactory system is to turn the agent towards the strongest direction of the gradient in order to follow the gradient to its source. The agent is trained by receiving positive reinforcement signals when it moves in a direction that reduces the distance from the agent to the goal. These signals are used to guide the Hebbian modification of weights among units between the simulated perceptual system and from the perceptual to the motor systems. Reinforcement, however, is not received if the olfactory system is interfering with the object avoidance signals

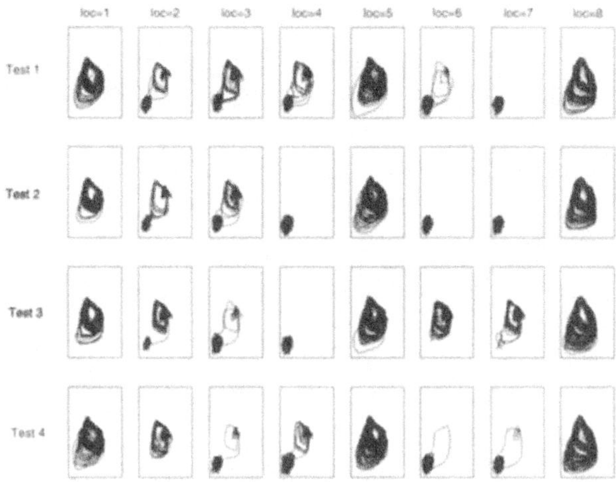

Figure 6: State space plot of activity of unit (column 3, row 7) in CA3 region during testing of KA-III hippocampal simulation.

and causing the agent to run into an obstacle.

4.3 Results

In Figure 9 we show one result of the agents behavior after training. The agent starts in the lower left corner of the debris field and successfully finds a path to the goal location. In this example, the agent combines local information to perform object-avoidance, with global information in order to follow the environmental gradient.

The previous figure showed one successful navigation of the agent to the goal location. It is necessary, however, to note that with this simple perception/action system the agent can easily get lost or fall into unproductive, repetitive behavior, depending on the starting/goal positions and obstacle placement. For example, sometimes the agent follows the gradient towards the goal only to find that it is blocked by obstacles. The object-avoidance behaviors can cause the agent to turn away and go back the way it came for a time. However the gradient-following may then kick back in and cause the agent to turn back in the direction it has already explored. This repetitive behavior represents a failure of the agent to remember its environment (or become bored), and will be discussed next.

5. Discussion

The two experiments discussed in this article represent isolated pieces of a complete organism, namely a perceptual/motor system for goal-directed navigation, and a hippocampus for place cell formation. We are currently working on bringing such pieces of an organism to form a complete intentional agent.

As described previously, the purpose of the KIII set

Figure 7: The Webots environment used for the goal-directed navigation demonstration.

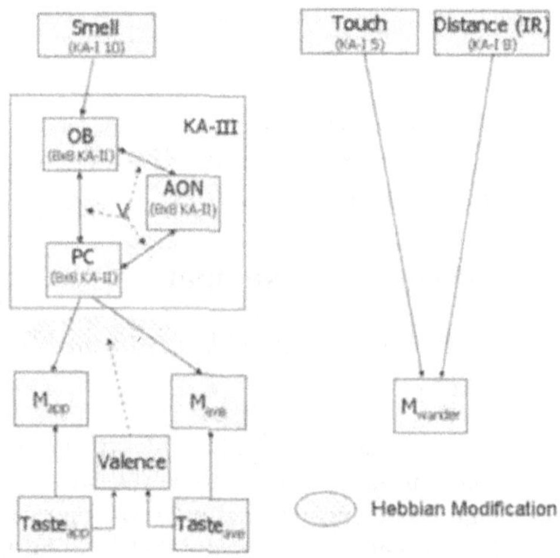

Figure 8: Architecture of the neural model used for the goal-directed navigation demonstration. The agent is hardcoded to avoid obstacles and wander in its environment using touch and distance sensors (right side of figure). The agent receives sensory information from a simulated smell gradient produced at the goal location (left of figure). The agent is given reinforcement when it moves closer to its goal, which allows for Hebbian modification to take place.

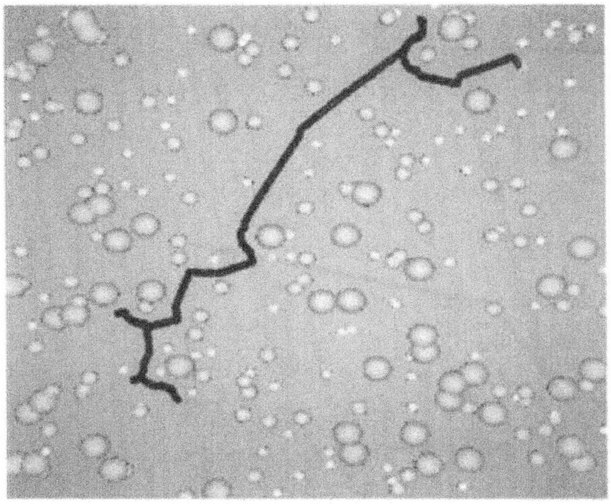

Figure 9: Results of a simulated run of the agent in navigating to its goal location (upper right, green rock). The agent avoids local obstacles while following the environmental gradient to the goal. The environment is a Mars-like simulation of a debris field encountered in typical Martian rover landing sites.

is to model the aperiodic dynamics observed in the sensory systems of biological brains, and to begin to understand how such dynamics may take part in the formation of meaning for the organism. While perception is an important component in the production of intelligent behavior, it is only a small part of the whole. One insight of the embodiment movement in cognitive science is that studying pieces of the cognitive puzzle (perception, memory, etc.) may in many cases miss important points on how behavior emerges from the pieces working together as a whole in a complete autonomous agent (Freeman et al., 2003).

The KIV architecture is a model of what biologists believe may be the simplest neurological structure capable of basic intentional behavior, the limbic system (Kozma et al., 2003, Kozma and Freeman, 2003). The limbic system is composed of four basic functional areas: sensory/perceptual areas, memory and orientation, value system (needs and goals) and motor systems. These four areas can roughly be described as the "What" "Where" "Why" and "How" functions respectively. Figure 10 is a schematic representation of the KIV model of the limbic system. The limbic appeared phylogenetically earliest among simple reptiles and amphibians, and is present in more advanced forms in all mammals.

The hypothesis captured in the KIV model is that the same types of aperiodic dynamics that have been shown to be crucial to the formation of meaning in perceptual systems are also necessary for the formation of memory and motor maps, as well as the hierarchical organization of competing goals. Therefore, at the heart of each of the sensory, memory and valence systems lies a KIII component, which is capable of producing aperiodic dynamics of the type observed in biological perceptual systems. We will describe each of the four areas in detail next.

What (Sensory/Perception) Aperiodic dynamics have been observed in rabbit sensory systems, and are believed to play an important part in the formation of meanings. In the KIV architecture, external signals (exteroception) arrive from the environment (Figure 10 top). Each sensory channel is mapped into meanings of interest to the organism, through a process of the formation of chaotic attractors due to learning. In the figure, the OB (olfactory bulb), PC (prepyriform cortex) and AON (anterior olfactory nucleus) are three groups of KII sets that form a sensory KIII. In a full model, each sense would be handled by one or more KIII groups of its own (not shown in figure). The formation and recognition of salient meanings in the environment provide the animal with a sense of "What" important things are in its immediate environment that can or should be dealt with. The navigation example described in this paper used a sensory KIII to develop gradient following behaviors.

Where (Orientation, Memory) A primitive hippocampus is the center of more long term memory and orientation functions in simple limbic systems. In the KIV architecture, the formation of cognitive maps of the environment, and the determination of the orientation of the organism in its environment (both locally and globally) is taken care of in the hippocampus. The orientation function of the hippocampus is depicted in the KIV model (Figure 10 left) as receiving orientation signals from the environment. The three CA regions (CA1, CA2 and CA3) form a KIII set that is responsible for the formation of memories of the environment of the organism. The formation of cognitive maps and so called place cells are performed in the hippocampal KIII region. A hippocampus using the KIII architecture can take advantage of the flexibility of aperiodic dynamics to form such representations. The hippocampal simulation discussed in this paper showed one example of how place cells may form using these types of aperiodic attractors.

Why (Goals, Drives, Value Systems) Figure 10 on the right comprises the value system of the KIV architecture, and mediates the production of behavior to guide the organism in completing goals and tasks to satisfy its needs. This system keeps track of the reasons "Why" the organism is doing what it is doing. In the valence system, internal signals that monitor basic needs (such as food, or avoiding damage) are fed into the system (Figure 10 right). Another KIII forms the heart of the system for forming and balancing a goal landscape of the organism (HT, DB and septum).

How (Motor Actions) The motor system (Figure 10 bottom) is responsible for directing actual effectors for "How" the organism will carry out behaviors in pursuit

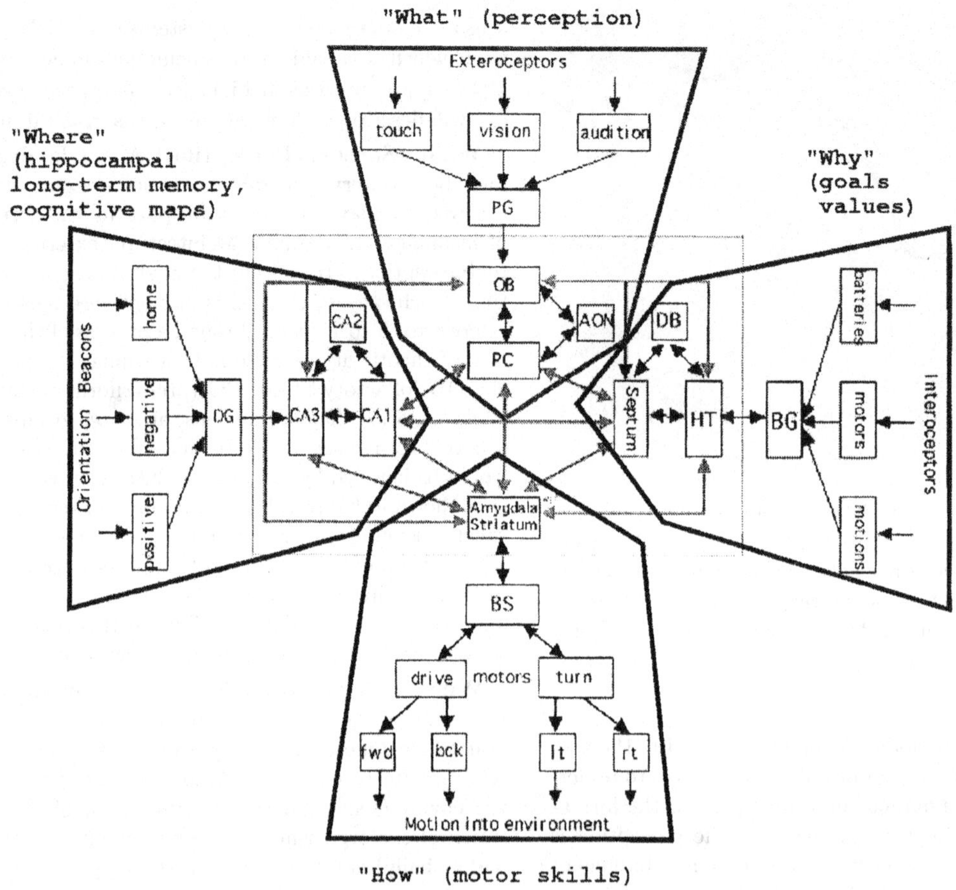

Figure 10: The KIV architecture. Based on figure from (Kozma and Freeman, 2003). Notations: PG - periglomerular; OB - olfactory bulb; AON - anterior olfactory nucleus; PC - prepyriform cortex; HF - hippocampal formation; DG - dentate gyrus; CA1, CA2, CA3 - curnu ammonis sections of the hippocampus; MF - midline forebrain; BG - basal ganglia; HT - hypothalamus; DB - diagonal band; SP - septum

of its goals. The amygdala provides the goal-oriented direction for the motor system that is superimposed on local tactile and other protective reflexes.

As mentioned previously, a simple reactive perceptual/motor system is insufficient to detect and prevent unproductive, repetitive behavior. What is needed is something like the limbic system. The limbic system, with its perceptual, long-term memory and goal-oriented areas, is capable of organizing and keeping track of multiple and sometimes conflicting goals, as well as remembering features of the environment and forming cognitive maps. We are working towards building models of a complete limbic system for an autonomous agent, through the realization and implementation of the KIV architecture.

6. Conclusion

The self-organization of spatio-temporal patterns in nonlinear systems are essential to cognitive mechanisms in biological brains. We need to better understand how such mechanisms operate in order to build better models of cognition and smarter autonomous agents. This paper has demonstrated how the formation of perceptual AM patterns may also account for the formation of place cells patterns in hippocampal cognitive maps. We have also shown how the KA-III perceptual mechanism may be used to implement simple tropic behaviors to follow gradients in an environment in order to perform goal-oriented navigation. We are working towards realizing a complete intentional agent by combining the perceptual, memory, goal and motor pieces of our agents into a complete KIV model of the basic limbic system.

Acknowledgment

This work was supported by NASA Intelligent Systems Research Grant NCC-2-1244.

References

Bak, P., Tang, C., and Wiesenfeld, K. (1987). Self-organized criticality: An explanation of 1/f noise. *Physical Review Letters*, 59(4):381–384.

Freeman, W. J. (1975). *Mass Action in the Nervous System*. Academic Press, New York, NY.

Freeman, W. J. (1987). Simulation of chaotic EEG patterns with a dynamic model of the olfactory system. *Biological Cybernetics*, 56:139–150.

Freeman, W. J. (1991). The physiology of perception. *Scientific American*, 264(2):78–85.

Freeman, W. J. (1999). *How Brains Make Up Their Minds*. Weidenfeld & Nicolson, London.

Freeman, W. J. (2003). The wave packet: An action potential for the 21st century. *Journal of Integrative Neuroscience*. in press.

Freeman, W. J., Burke, B. C., and Holmes, M. D. (2003). Aperiodic phase re-setting in scalp EEG of beta-gamma oscillations by state transitions at alpha-theta rates. *Human Brain Mapping*, 19:248–272.

Freeman, W. J. and Shimoide, K. (1994). New approaches to nonlinear concepts in neural information processing: Parameter optimization in a large-scale, biologically plausible corticle network. In Zornetzer, (Ed.), *An Introduction to Neural and Electronic Networks*, chapter 7, pages 119–137. Academic Press.

Harter, D. and Kozma, R. (2002). Simulating the principles of chaotic neurodynamics. In *Proceedings of the 6th World Multi-Conference on Systemics, Cybernetics and Informatics (SCI 2002)*, volume XIII, pages 598–603, Orlando, FL.

Harter, D. and Kozma, R. (2003). Chaotic neurodynamics for autonomous agents. *IEEE Transactions on Neural Networks*. submitted.

Harter, D. and Kozma, R. (2004). Aperiodic neurodynamics using a simplified k-set neural population model. In *Proceedings of 2004 International Joint Conference on Neural Networks (IJCNN 2004)*, Budapest, Hungary. submitted.

Huntsberger, T. (2001). Biologically inspired autonomous rover control. *Autonomous Robots*, 11:341–346.

Kay, L., Shimoide, K., and Freeman, W. J. (1995). Comparison of eeg time series from rat olfactory system with model composed of nonlinear coupled oscillators. *International Journal of Bifurcation and Chaos*, 5(3):849–858.

Kelso, J. A. S. (1995). *Dynamic Patterns: The Self-organization of Brain and Behavior*. The MIT Press, Cambridge, MA.

Kozma, R. and Freeman, W. J. (2001). Chaotic resonance - methods and applications for robust classification of noisy and variable patterns. *International Journal of Bifurcation and Chaos*, 11(6):1607–1629.

Kozma, R. and Freeman, W. J. (2003). Basic principles of the KIV model and its application to the navigation problem. *Journal of Integrative Neuroscience*, 2(1):125–145.

Kozma, R., Freeman, W. J., and Erdi, P. (2003). The KIV model - nonlinear spatio-temporal dynamics of the primordial vertebrate forebrain. *Neurocomputing*, 52-54:819–826.

Michel, O. (1996). Khepera simulator package version 2.0. Downloaded from WWW at http://wwwi3s.unice.fr/ om/khep-sim.html. Freeware mobile robot simulator written at the University of Nice Sophia-Antipolis.

Michel, O. (2003). Webots v4.0 3-d physics based mobile robot simulator. www.cyberbotics.com.

Shimoide, K., Greenspon, M. C., and Freeman, W. J. (1993). Modeling of chaotic dynamics in the olfactory system and application to pattern recognition. In Eeckman, F. H., (Ed.), *Neural Systems Analysis and Modeling*, pages 365–372. Kluwer, Boston.

Skarda, C. A. and Freeman, W. J. (1987). How brains make chaos in order to make sense of the world. *Behavioral and Brain Sciences*, 10:161–195.

Solé, R. and Goodwin, B. (2000). *Signs of Life: How Complexity Pervades Biology*. Basic Books, New York, NY.

Tsuda, I. (2001). Towards an interpretation of dynamic neural activity in terms of chaotic dynamical systems. *Behavioral and Brain Sciences*, 24(4).

Tunstel, E. (2001). Ethology as an inspiration for adaptive behavior synthesis in autonomous planetary rovers. *Autonomous Robots*, 11:333–339.

Adaptive Motor Primitive and Sequence Formation in a Hierarchical Recurrent Neural Network

Rainer W. Paine **Jun Tani**

RIKEN Brain Science Institute
Laboratory for Behavior and Dynamic Cognition
2-1 Hirosawa, Wako-shi, Saitama, 351-0198 JAPAN
rpaine@brain.riken.jp; tani@brain.riken.jp

Abstract

This study describes how complex goal-directed behavior can be obtained through adaptation in a hierarchically organized recurrent neural network using a genetic algorithm. Robot simulations showed that different types of dynamic structures self-organize in the lower and higher levels of the network for the purpose of achieving complex navigation tasks. Behavior primitives are switched in a top-down way through lower level parametric bifurcation structures. In the higher level, a topologically ordered mapping of initial cell activation states to motor-primitive sequences self-organizes by utilizing the initial sensitivity characteristics of nonlinear dynamical systems. The biological plausibility of the model's essential principles is discussed.

1. Introduction

It is widely believed that behavior systems develop certain hierarchical or level structures for achieving goal-directed complex behaviors, and that these structures should self-organize through interactions with the environment. It is reasonable to assume that an abstract event sequence is represented in a higher level, while its detailed motor program is generated in a lower level. Arbib (1981) proposed the idea of movement primitives (also referred to as perceptual-motor primitives, motor schemas, or motor programs), which are a compact representation of action sequences for generalized movements that accomplish a goal. Evidence of such primitives in animals has been found (Giszter et al., 1993; Mussa-Ivaldi et al., 1994), and human studies also indicate their role in complex movement generation (Thoroughman & Shadmehr, 2000). Once such motor primitives develop early in the life of an organism, diverse behaviors can emerge by learning to combine them in a multitude of complex sequences.

A movement primitive can be formalized as a "control policy", encoded using a few parameters in the form of a parameterized motor controller, for achieving a particular task (Schaal, 1999). The motor primitives are routine motor programs that repeatedly appear in the sequences of the motor patterns. Once such motor primitives develop early in the life of an organism, diverse behaviors can emerge by learning to combine them in a multitude of complex sequences.

Mataric (2002) showed examples of behavior generation through primitive combination in the context of imitation learning using a virtual humanoid robot. In her computational architecture, a cluster of motor primitives was organized in a lower level. Then, a Hidden Markov model in the higher level learned how the primitives were combined into sequences in order to recognize and regenerate the human instructor's behavior patterns. In related work, Amit and Mataric (2002) used Self-Organizing Maps (SOMs) to hierarchically control postural and oscillatory movement primitives in a simulated robotic arm.

Hochreiter and Schmidhuber (1997) proposed a so-called long-term and short-term memory connectionist model for hierarchical sequence learning which focuses on the problem of sequence segmentation. The essential idea in this study was to learn long and complex sequences by dividing them into chunks of sub-sequences.

Tani and Nolfi (1999) extended the idea of the Mixture of Experts (Jacobs et al., 1991; Jordan & Jacobs, 1994) by introducing level structures. In their experiments with a simulated mobile robot, the robot learned to perceive sensory-motor flow as hierarchically articulated. In contrast to this local representation scheme utilizing expert modules for representing primitives, Tani (2003) proposed a distributed representation scheme where multiple primitives can be embedded in a single recurrent neural network (RNN) in terms of a "forward model" (Kawato et al., 1987). Each primitive can be accessed by a control parameter called the "parametric bias" (PB). The higher level RNN combines the lower level primitives in sequences by learning and sending the corresponding PB sequences to the lower level RNN.

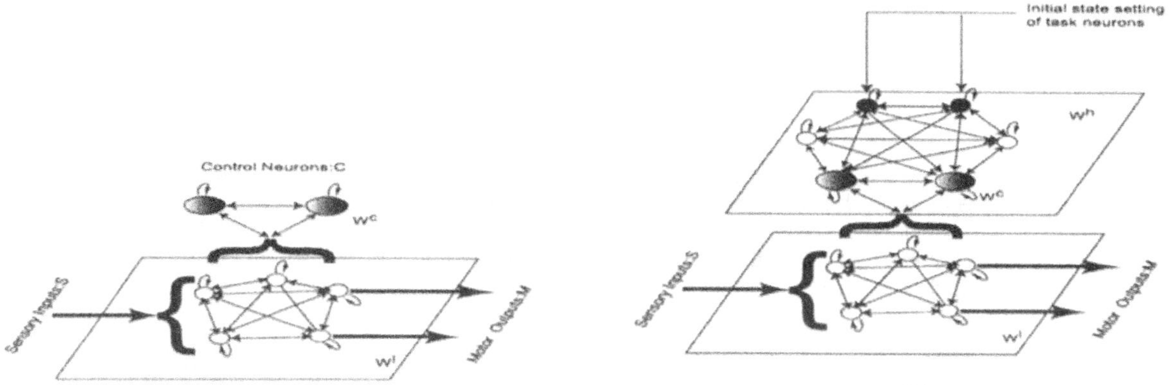

(a) (b)

Figure 1: Conceptual diagram of network architecture.

However, these neural network schemes seem to have some potential drawbacks. One major problem is that the network cannot be adapted dynamically through trial and error. This is due to the fact that the above mentioned neural network approaches depend heavily on a supervised learning scheme using teaching signals. In contrast, reinforcement learning schemes using "macro-actions", or sequences of primitives, can learn to effectively combine primitives to solve sequential tasks simply through environment-mediated rewards. However, inappropriate use of macro-actions may retard learning. Appropriate macro-actions must generally be tuned manually for specific environments, although work is being done to automate the process (McGovern et al, 1997, 2001).

Another problem is the setting of time constant parameters for the network dynamics at each level. The time constant in the lower level has to be determined based on the shifting frequency of the lower primitives. The higher level time constant must further be determined based on those of the lower level primitives as well as task features. Such parameter setting is usually done manually by experimenters, which requires certain a priori knowledge about the task environment. In order to overcome these problems, this paper introduces a novel scheme using an evolutionary adaptation mechanism through a genetic algorithm (GA). The evolutionary robotics community has shown that the GA scheme allows for the self-organization of dynamic adaptive behaviors in sensory-motor systems (Tucci et al., 2002; Nolfi & Floreano, 2000).

In this paper, we study the dynamic adaptation process of multiple levels of continuous time recurrent neural networks (CTRNNs) applied to a navigation task using a simulated robot. Through the experiments, we will demonstrate that a hierarchically organized network can perform well in adapting to complex tasks through combination with a GA. We will focus on how motor primitives are self-organized in the lower level, and how they are manipulated in the higher level.

2. Methods

2.1 General Model

The neural network model utilized in the current paper consists of two levels of fully connected CTRNN (Yamauchi & Beer, 1994; Blynel & Floreano, 2002). The lower level network, as shown in Figure 1, receives sensory inputs and generates motor commands as outputs. This network is supposed to encode multiple sensory-motor primitives, such as moving straight down a corridor, and turning left or right at intersections or to avoid obstacles in the navigation task adopted in this study.

A set of external neural units, called the "control neurons", are bidirectionally connected to all neurons in the lower level network. The control neurons influence lower level network functions and favor the generation of particular motor primitives. Through evolution of both the lower level internal synaptic weights (W_l) and the interface weights (W_c) between the control neurons and the lower level neurons, a mapping between the control neurons' activities and the sensory-motor primitives stored in the lower level network is self-organized. Modulation of the control neurons' activities causes shifts between generating one primitive and another. The scheme is analogous to the idea of the parametric bias in Tani (2003) and the command neuron concept (Aharonov-Barki et al., 1999; Edwards et al., 1999; Teyke et al., 1990). How might more complex tasks, such as navigation in an environment, be generated? Such tasks require generating sequences of motor primitives. We propose that a higher level network may modulate the activities of the control neurons through time to generate sequences of lower level movement primitives (Figure 1b). The higher level network evolves to encode abstract behavior sequences utilizing the control neurons.

It is assumed that the desired sequences will be generated if adequate nonlinear dynamics can be self-organized in the higher level network. As will be described in detail later, the robot becomes able to navigate to multiple goal positions when starting from the same initial position in the maze environment. Therefore, the higher level network has to encode multiple sequence patterns, which have to be retrieved for the specified goal.

We utilize the initial sensitivity characteristics of nonlinear dynamic systems in order to initiate different sequences. When the robot is placed at the initial position in the environment, the internal values of all the higher level neurons are set to 0.0, except for two neurons called the task neurons (Figure 1b). The initial values of the task neurons are set corresponding to the specified goal positions. These goal-specific initial task neuron activities were evolved through the same genetic algorithm that yielded the network's synaptic weights.

We assume that an appropriate sequence pattern that enables the robot to navigate to the kth goal can be generated by setting adequate initial activities (γ_k) for the task neurons when the higher level internal connective weights (W_h) are adequately generated through evolution. This idea of utilizing the initial sensitivity of the network dynamics is similar to the studies of Nishimoto & Tani (2003) and Blynel (2003). Further, Tanji & Shima (1994) showed in the monkey that the pre-Supplementary Motor Area neuronal activities during the motor preparation period might encode abstract behavior sequences through the initial state values of the network dynamics. Nishimoto & Tani (2003) showed that a Recurrent Neural Network (RNN) learns to generate various action sequences by setting different initial context unit activities using the back-propagation learning method. Blynel (2003) found that goal positions can be "remembered" through the activations of hidden neurons in a reinforcement learning task when Genetic Algorithm evolution is applied to a single level CTRNN.

Evolution of the two-level network used here goes through two phases. In the first phase (Experiment 1), the network shown in Figure 1a evolves to perform collision-free left and right turns in a T maze environment. Both the lower level internal synaptic weights (W_l) and the control neuron interface synaptic weights (W_c) are evolved while adequate activation values of the control neurons are determined for the left and right turns. At this stage, there is no higher level network with task neurons. Instead, the neuronal activation bias (θ, Equation 2) of the two control neurons is free to evolve differently for the left and right tasks. The synaptic weights are identical for the two turn directions.

In the second phase (Experiment 2), the evolved lower level network is extended by adding and evolving the higher level network (Figure 1b). The task of the robot is to find ways to reach multiple goals from the same starting position. In this phase, only the higher level internal connective weights (W_h) and goal-specific task neuron initial activities (γ_0) are evolved. The bottom level internal weights and control neuron interface synaptic weights (W_c) are kept constant from the first phase's T maze task. Thus, collision free left or right turning does not need to be re-evolved. Instead, the second learning phase focuses solely on goal finding through the appropriate turn sequence generation.

2.2 Continuous-Time Recurrent Neural Networks

All neurons in the simulations presented here used the following equations and parameters for a CTRNN, based on those of Blynel & Floreano (2002). In Equation 1, γ_i is the internal activation state (cell potential) of the ith neuron. τ is the time constant of the neuron. It affects the rate of neuronal activation in response to the kth external sensory neuronal activation, I_k, and signals from the jth presynaptic neuron with activity A_j. The signal from the presynaptic neuron is weighted by weights w_{ij}, and the sensory input is weighted by w_{ik}. N is the number of neurons in the network, and S is the number of sensory receptors which send input signals to the network.

$$\frac{d\gamma_i}{dt} = \frac{1}{\tau_i}\left(-\gamma_i + \sum_{j=1}^{N} w_{ij}A_j + \sum_{k=1}^{S} w_{ik}I_k\right) \quad (1)$$

The presynaptic neuronal activity (A_j) is defined in Equations 2 and 3. θ is a bias term and σ is the standard logistic function, defined in Equation 3.

$$A_j = \sigma(\gamma_j - \theta_j) \quad (2)$$

$$\sigma(x) = 1/(1+e^{-x}) \quad (3)$$

Numerical integration was carried out using the Forward Euler method. The update rule of the neuronal activation state γ_i for each integration time step is given by Equation 4. n is the iteration step number and Δt is the time step interval, defined as 0.2.

$$\gamma_i(n+1) = \gamma_i(n) + \frac{\Delta t}{\tau_i}\left(-\gamma_i(n) + \sum_{j=1}^{N} w_{ij}A_j(n) + \sum_{k=1}^{S} w_{ik}I_k\right) \quad (4)$$

Except for the previously mentioned task neurons, whose initial activation is dependent on the movement goal, the neuronal activation, γ_i, is initialized to 0 at the start of integration, $\gamma_i(0) = 0$.

2.3 Genetic Encoding

The following parameters are genetically encoded within the following ranges. τ is the time constant (Equation 1). θ is the activation bias (Equation 2). w is the synaptic weight, and $\gamma_{task}(0)$ is the initial activation of the task-dependent neurons of the higher level network.

$$\tau \in [1,70]; \theta \in [-1,1]; w \in [-5,5]; \gamma_{task}(0) \in [-10,10]$$

Each parameter is encoded using 5 bits and the following encoding rule to generate analog parameter values. Thus, each parameter value is generated by a linear scaling of the analog value within a given range to the range of 0 to 31 in binary code. The encoded parameter value (P_e) from the 5 bit string is given in Equation 5. The analog parameter value (P_a) is given by Equation 6 for an analog parameter value in the range $P_a \in [\min, \max]$.

$$P_e = \sum_{i=1}^{5} bit_i \cdot 2^{5-i} \quad (5)$$

$$P_a = \min + \frac{P_e \cdot (\max - \min)}{2^5 - 1} \quad (6)$$

In the T maze experiment, the network consists of eight inputs, five bottom level units, and two control neurons, for a total of 515 bits encoding τ, θ, and w. In the Eight-Goal-Maze experiment, the network is as above, except that it also includes 4 higher level neurons. The task neurons' initial activity ($\gamma_{task}(0)$) is also encoded, yielding a genome length of 715 bits.

2.4 Genetic Algorithm

A standard genetic algorithm (GA) with mutation but without crossover was employed to evolve the weights and parameters (τ, θ, w, $\gamma_{task}(0)$) of the network (Mitchell, 1998; Goldberg, 2002). The mutation rate per bit was set at 2% for all simulations reported here. (Higher mutation rates of up to 10% were tried, but the fitness of the resulting populations became progressively more unstable without improvements in the best robot performance). The population consisted of 80 robots. The twenty robots with the best fitness reproduced each generation. Each of the reproducing robots made 4 copies of itself with mutation. Of the best robot's offspring, one was an exact copy of the parent without mutation. Simulations were generally run for 50 to 200 generations, and the performance of the best robots was analyzed.

2.5 Network Architecture

In Experiment 1, the network depicted in Figure 1a consists of 8 sensory inputs, scaled to a range of 0 to 1, which are sent to the bottom level of 5 neurons. The bottom network receives from and sends signals to the two control neurons. In Experiment 2, the control neurons are further connected to the higher level of 4 neurons (Figure 1b). Two of the higher level neurons are the "task neurons", whose initial activities determine the particular turn sequence which will be generated to reach a goal.

The outputs of the first two bottom level neurons are taken as motor command signals to the simulated robot wheels. The actual neuronal outputs are analog signals, but the robot controller can use only integer speed commands. The analog signals are therefore rounded to the nearest integer. The simulated wheel speed is in the range of 0 to 4, corresponding to speeds of 0 to 0.32cm per second. Note that the wheels are allowed to turn only in a forward direction in these experiments for simplicity.

2.6 Experimental Setup

All experiments reported here were executed using a simulated Khepera II robot in the Webots 3 robot simulator. Simulations were run on a Plathome computer with a 2.0 GHz processor. Simulations ran at about 40 times real Khepera speed. Data were analyzed using Matlab Release 13.

Inputs to the neural network consisted of the signals from seven infra-red proximity sensors (2 left, 2 front, 2 right, and 1 rear) and one downward facing ground sensor (illustrated by rays protruding from robot in Figure 2). The input was modified by randomly adding or subtracting 5% of its value as noise. All sensor inputs to the network were scaled to a range of 0 to 1. In all experiments, the robot was repositioned to the starting point and the trial was ended if the robot either collided with a wall or reached a goal area. Goal areas were defined by differently colored floors. The floor sensor was used by the controller to detect when the robot found a goal, triggering the appropriate fitness reward, ending of the trial, and repositioning of the robot.

3. Experiments

3.1 Experiment 1: T-Maze Task

Experiment 1 is designed to evolve a bottom level network which contains movement primitives of left and right turning behavior at intersections as well as collision-free straight movement in corridors. The same lower level and control neuron weights are used for both right and left turns. The only difference between the left and right turn controllers is in the bias values (θ) of the two control neurons. Intuitively, this might correspond to different sets of cortical "control" neurons becoming associated with each of the lower level movement primitives. Parallel connections from the bottom level to the control neurons might develop, yielding the same weights to both sets of control neurons. Intrinsic differences in the control neurons' responses (as through the different θ bias values used here) to the lower level signals would determine with which motor primitives each set of control neurons became associated.

The evolutionary runs consisted of up to 200 generations with 2 epochs and 3 trials per robot of the 80 robot population. Each trial was run for 500 time steps, starting at the same position at the bottom of the T maze. Different bias values (θ) evolved in the control nodes for the left and right turning tasks in epochs 1 and 2, respectively. All other parameters were identical in the left and right turning tasks. In epoch 1, fitness was awarded to robots that turned to the left at the intersection based on the following fitness rule. In epoch 2, fitness was awarded to robots that turned to the right. Each robot ran 3 trials per epoch.

Experiment 1 uses a two-component fitness rule (Equation 7). The first component (F_{OA}) consists of a reward for straight, fast movements with obstacle avoidance. The fitness rule of Floreano & Mondada (1994) was adopted for this purpose and is shown in Equation 8. V is the wheel speed scaled to a range of 0 to 1. $\overline{V}$ is the average speed of the two wheels. ΔV is the absolute value of the difference in speeds between the two wheels, and is used to reward straight movements. S_{max} is the maximum robot sensor value, scaled to a range of 0 to 1, and is used to reward obstacle avoidance.

$$F = F_{OA} + F_{goal} \tag{7}$$

$$F_{OA} = \overline{V} \cdot (1 - \sqrt{\Delta V}) \cdot (1 - S_{\max}) \tag{8}$$

The second component of the fitness rule, F_{goal}, rewards the robot for finding a goal. The goal is located to the left of the intersection for epoch 1, and to the right for epoch 2. The robot is linearly rewarded, based on its position, for approaching and reaching the goal. Greater reward per time step is received linearly as the robot approaches the goal, starting at the middle of the top of the T maze.

At the start of each trial, the robot was placed at the same starting position at the bottom of the T maze. Three different starting orientations (facing 135°, 90°, and 45°; that is, left, straight, and right, respectively) were employed, one for each of the three trials per epoch. Further, motor noise was added to the wheel speed commands sent to the robot. The integer speed command was increased or decreased by one with a probability of 20% on each simulation time step. The varying starting orientations and motor noise were used to ensure that the robot would experience wall collisions early during evolution, so that controllers with obstacle avoidance would be more likely to evolve. Both the bottom level and control neurons were free to evolve in this experiment.

3.2 Experiment 2: Eight-Goal Task

The Eight-Goal maze depicted in Figure 2 is a combination of T maze-like environments with eight different goals at the ends of each T maze component. Thus, combinations of the same turn primitives evolved in experiment 1 should allow the robot to reach the different goals. In experiment 1, different sets of control neurons with differing internal dynamics (due to differing θ bias values) became associated with particular motor primitives, as will be described later. In experiment 2, it is shown how the activity of a *single* set of control neurons can be modulated over time to generate a *sequence* of motor primitives. Further, it is shown how varying only the initial activation of the task neurons in the higher level network can lead to the generation of different network activation time courses by which multiple primitive sequences are generated. Thus, the routes to multiple goals can effectively be stored in a single network, with a single set of synaptic weights and corresponding initial activation values of the task neurons.

The bottom level genome, including the weights for the connections to the control neurons, from experiment 1 was used in this experiment and held constant. Only a single set of synaptic weights and parameters in the higher level network, and multiple sets of the initial task neuron activities, were free to evolve. The experiment consisted of up to 200 generations, with 12 epochs per generation and 2 trials per epoch. Each of the 12 epochs evaluated a different set of higher level task neuron initial activities, using the fitness rule described below. Further, each task neuron set was run for two trials, in order to evaluate the stability of the robot's goal-finding ability. Robots which found multiple goals repeatedly were rewarded more than those which found them only intermittently. Further, robots which found some stable goals tended to be rewarded more

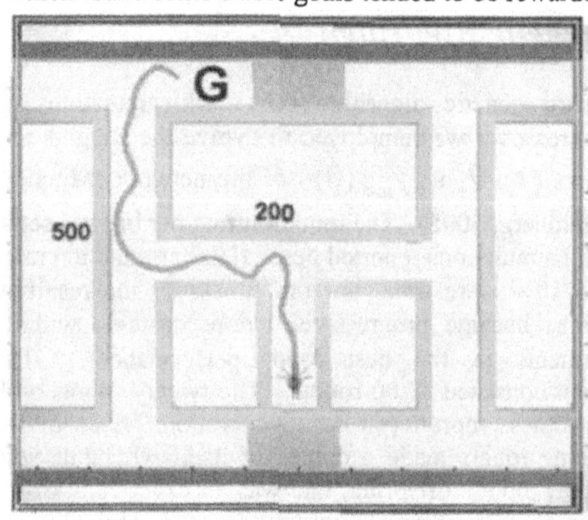

Figure 2: Eight-Goal Maze environment, showing trajectory for a Left, Right, Right sequence. G= Goal location. Turns occur at steps 200 and 500 in the neuronal activity traces for this sequence in Figure 4.

than robots that found more goals haphazardly. As in experiment 1, a trial was ended when the robot either found a goal or collided with a wall. The robot was then repositioned to the same starting point. Since the bottom level genome from experiment 1 was used here, obstacle avoidance and turn primitives were present in the first generation. Therefore, only one starting orientation was used (90°, i.e., straight up) and no motor noise was added to the speed commands sent to the robot wheels (but sensor noise was still present).

In experiment 2, the fitness rule (*F*) consists solely of a reward for finding goals consistently (Equation 9).

$$F = \sum_{g=1}^{Ngoals} \max_{i=1:Ntasks} \Big(\sum_{trial=1}^{2} R_{gi} \Big) \qquad (9)$$

Here, a fixed reward (*R*) per new goal (*g*) found is given to the robot. In experiment 2, Ngoals = 8, and Ntasks = 12.

Each robot has 12 sets of task neuron initial activities (*i*) which are evaluated. Each set has two trials in which to find a goal. A robot which finds a goal on both trials receives twice the reward of finding the goal on only one trial. If a different set of task neuron initial activities leads to the same goal, then the reward is the maximum of the reward given to the two different task neuron sets. Thus, a robot with multiple task neuron sets that find a goal on only one of the two trials will receive less reward than a robot with one task neuron set that finds the goal on both trials.

4. Results/Analysis

4.1 Experiment 1: T-Maze Task

Collision avoidance and left and right turning behavior emerged within 63 generations. Left turns were generated for one set of control neuron bias values, and right turns were generated for another set. Although both left and right turns could be generated, the robot exhibited oscillatory movements after collision avoidance. That is, it turned away from one wall too much and headed towards the opposite wall instead of straightening its path through the lower part of the T maze. (As mentioned previously, the robot started from three different orientations, leftward, straight up, and rightward, requiring it to avoid wall collisions early during each trial.) The controller was therefore allowed to evolve further. By generation 189, fewer fluctuations occurred after collision avoidance and the lower level genome was used for experiment 2.

Control neuron 0 appears to be critical in determining whether a left or right turn is generated. Turns occur at approximately time step 200, at which time the control neuron activity is 0.2 for the left turn and 0.6 for the right turn. The relation between turn behavior and the activities of control neurons 0 and 1 is further quantified in the phase plot of Figure 3. 441 trials with different sets of the two control neurons' activities, held constant for each trial, (A=[0:1], step size = 0.05) were run and the resulting turn directions recorded. Note that this figure also applies to the activity of the control neurons evolved in experiment 2, since the same bottom level genome, including the synaptic weights between the bottom level and control neurons, was used in both experiments 1 and 2. The turn phase plot of Figure 3 shows a clear bifurcation, or appearance of new movement behavior with the control neuron activity change, with two distinct regions of stability for left (black) and right (white) turns, for the corresponding combinations of control neuron 0 and neuron 1 activities. The gray squares indicate unstable regions in which wall collisions occur. In both experiments 1 and 2, the evolved control neuron weights tend to suppress the activity of control neuron 1. The phase plot shows that the smallest region of instability between left and right turns occurs for such small control neuron 1 activities. Further, the weights from control neuron 0 to the two motor output nodes (w_{jk}= 5.0, -3.4) of the bottom level have a greater magnitude than the weights

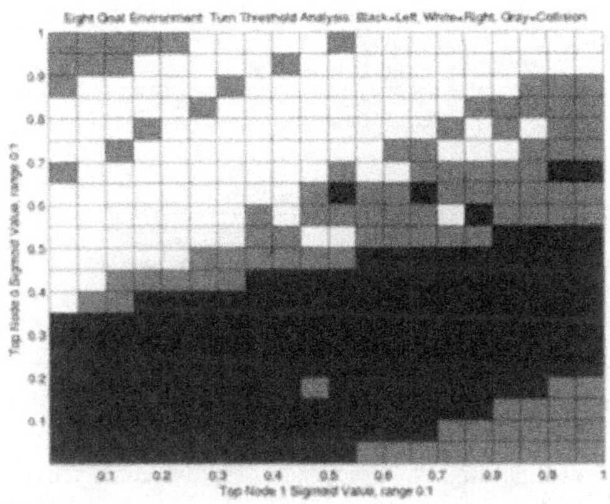

Figure 3: Phase analysis of turn direction as a function of control neuron activation. X and Y axes: neuronal activities of control neurons 1 and 0, respectively. Black = Left turn, White = Right turn, Grey = Collision with wall.

from control neuron 1 (w_{jk}=-1.1, 2.0). Thus, control neuron 0 has the dominant effect on the turn direction of the lower level's output neurons. It therefore appears that the lower level, which receives and processes all sensory inputs, has a dominant role in collision avoidance. Collision avoidance competes with the control neurons' "turn" signals. In the unstable regions of the phase plot (grey in Figure 3), the control neurons' "turn" commands inappropriately override the lower level's collision avoidance, triggering collisions when the robot turns toward and collides with a wall.

4.2 Experiment 2: Eight-Goal Task

The best robot became able to reach up to 7 different goals stably within 24 generations by evolving the higher level network and control neurons. The turn sequence for each goal was determined by a particular set of initial task neuron activities ($\gamma_{task}(0)$). Since 12 different sets of $\gamma_{task}(0)$ were evaluated per robot, multiple sets would sometimes lead to the same goal. Each $\gamma_{task}(0)$ set was evaluated for the stability of its goal. Some values of $\gamma_{task}(0)$ led to repeatable goal-finding performance, while others were

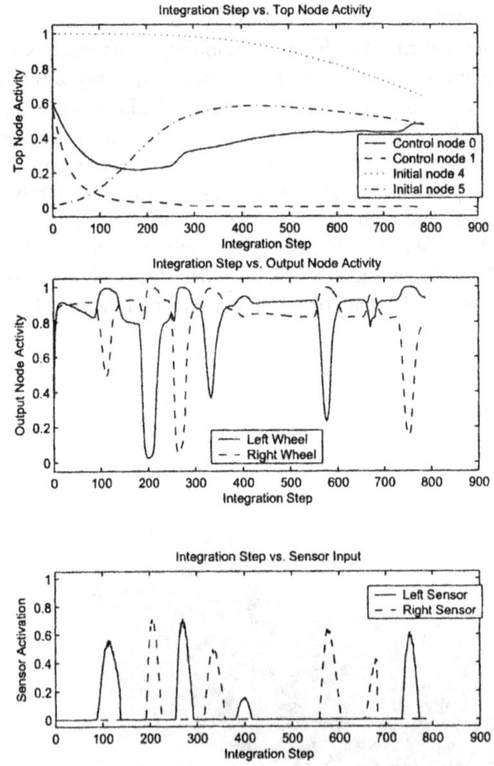

Figure 4: Left-Right-Right turn sequence. Top: Neuronal activity of control and task (initial node) neurons; Middle: Lower level motor output node activity; Bottom: Left and right sensor activities.

unstable, leading to different goals on different trials, or even to wall collision. Although evolution led to controllers which could find all 8 goals, the best controller could only find 7 goals stably. The definition of stability used here is reaching a particular goal, with the corresponding evolved $\gamma_{task}(0)$ values, on at least 70% of trials.

Different $\gamma_{task}(0)$ activities led to differently fluctuating patterns in the control neuron activities (Figure 4). As in the results of experiment 1 reported above, control neuron 0 had the greatest influence on turn direction, and exhibited the greatest fluctuations during various turn sequences. Left turns were generated when its activity was below a threshold of approximately 0.35, and right turns were generated when the activity was above the threshold. The activity of control neuron 1 tended to be suppressed, leading to a smaller region of instability at the transition from left to right turns in the phase plot of Figure 3. Figures 2 and 4 show the trajectory and neuronal activities, respectively, for a left-right-right turn sequence. The results shown here are for a controller which learned to reach six goals stably in 25 generations. They show large amplitude fluctuations of the control neurons due to a small evolved value of the time constant τ in Equation 1.

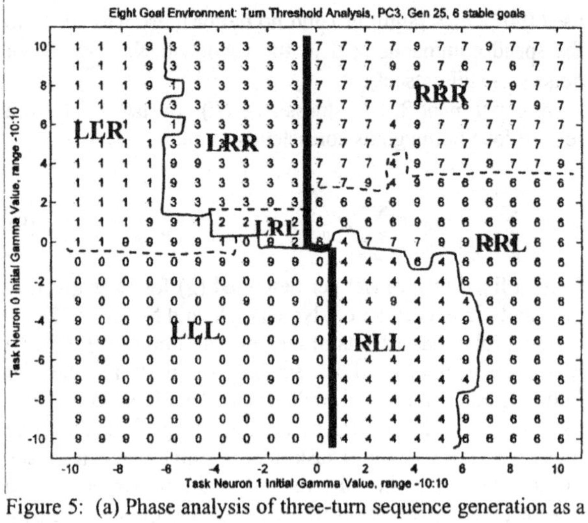

Figure 5: (a) Phase analysis of three-turn sequence generation as a function of task neuron initial activity, $\gamma_{task}(0)$. X and Y axes: Initial activities of task neurons 1 and 0, respectively. Plotted numbers correspond to sequences (L=Left, R=Right) as in Table 1.

The amplitude of the control neurons' fluctuations is also significant because larger amplitude fluctuations may render the controller more robust to noise. Although 5% sensor noise was used throughout these experiments, motor noise (increasing or decreasing the wheel speed command by 1 unit with 20% probability) was used only in experiment 1 in order to facilitate the development of obstacle avoidance. Motor noise was also tested during separate evolutionary runs in experiment 2, and found to decrease the number of stable goals found from a maximum of 7 (without motor noise) to 5 (with motor noise). Note that the total number of goals found was eight, with or without motor noise. Thus, additional noise added to the wheel motor commands increases the instability of the turn sequences.

Goal instability was most often seen in the final turn direction of the three-turn sequences learned. This final turn instability can be appreciated by noting that the control neuron activity tends to hover around the turn threshold near the end of the trial. As seen in the phase plot of Figure 3, this region is unstable.

Figure 5 shows an analysis of the movement sequences generated for the range of task neuron initial activities, $\gamma_{task}(0) \in [-10,10]$, in the evolved controller of experiment 2. 441 sets of initial task neuron activities were tested and the resulting turn sequences recorded. The numbers in the figure correspond to movement sequences as labeled in the figure, e.g., LRL for left, right, and left turns. The sequence patterns self-organize into well-defined, topologically ordered clusters in the $\gamma_{task}(0)$ space. First, the $\gamma_{task}(0)$ space is grossly clustered based on the first turn direction, left or right, of the movement sequence, as shown by a thick solid line in Figure 5. Each of these two clusters is then further divided into sub-clusters, depending

SEQUENCE	BINARY CODE	INTEGER VALUE
LLL	000	0
LLR	001	1
LRL	010	2
LRR	011	3
RLL	100	4
RLR	101	5
RRL	110	6
RRR	111	7
Collision	---	9

Table 1: Movement Sequence representation used in Figure 5. L=Left Turn; R=Right Turn; 0=Left Turn; 1=Right Turn.

on the second turn direction of the movement sequence, as shown by a solid line. These sub-clusters are still further divided into smaller clusters, depending on the third turn as shown by the dashed lines.

Thus, turn sequences are hierarchically ordered into progressively smaller regions of the initial task neuron activity space, as additional turns are added. In other words, as the complexity of the movement sequence increases, so too does the initial sensitivity to the task neuron activities.

This mapping of initial task neuron activity to particular sequences is an emergent property of the evolved controller. Different evolutionary runs yield different cluster patterns, but the general trend of distinct, topologically ordered sequence regions remains. This self-organized, topologically ordered mapping of sequences, with increasing initial sensitivity to task node activities as movement sequence complexity increases, is notable in such a small network, and is reminiscent of the fractal distribution of sequences mapped in the parameter space of Nishimoto & Tani (2003). Indeed, it would be interesting to see if fractal structure could be found in controllers branching out to larger numbers of goals.

5. Discussion/Conclusion

The work presented here describes a novel hierarchical model of behavioral sequence memory and generation. It recalls in general terms the hierarchical organization of movements in the primate spinal cord, brainstem, and cortical regions. Different types of dynamic structures self-organize in the lower and higher levels of the network. A parametric bifurcation in the control neurons' interaction with the lower level allows top-down behavioral switching of the primitives embedded in the lower level. Utilizing the initial sensitivity characteristics of nonlinear dynamic systems (Fan et al., 1996), a topologically ordered mapping of initial task neuron activity to particular behavior sequences self-organizes throughout the development of the network. The interplay of task-specific top-down and bottom-up processes allows the execution of complex navigation tasks.

One unique feature of the current model is the hierarchical organization of the network and its training. The bottom level network represents movement primitives, such as collision avoidance and turning at intersections. Since it must directly deal with quickly changing environmental stimuli, its time constants have become small through adaptation so that the neuronal activity of the output neurons ($\tau_0 = 1$, $\tau_1 = 1$ in Equation 1) can change rapidly to drive the robot's movement in real time. In contrast, the higher level represents sequences of the lower level primitives over longer time spans. Accordingly, the task neuron time constants have adapted to be large ($\tau_{task0} = 70$, $\tau_{task1} = 52$ in Equation 1) so that neuronal activity changes much more gradually and is less affected by short-term sensory changes.

The neurons of the higher level receive no direct sensory inputs, but are gradually influenced by them through the control neurons, which are fully connected to the input-receiving bottom level. This system is reminiscent of the organization of sequence generation in primates, as is elucidated by the studies of Tanji & Shima (1994) and Ninokura et al. (2003). In the former study, cellular activity in monkeys' supplementary motor area (SMA) was found to be selective for the sequential order of forthcoming movements, much as the task neurons' initial activities determine future movement order in the current model. In the latter study, distinct groups of cells in the lateral prefrontal cortices (LPFC) of monkeys were found to integrate the physical and temporal properties of sequentially reached objects, in a manner analogous to integration of higher level sequential information and lower level sensory input by the control neurons in the present model.

Although other models of sequence generation have been trained in a modular fashion because it was felt necessary to achieve the task (Yamauchi & Beer, 1994), the current work begins by explicitly evolving simple movement primitives, such as straight movements, collision avoidance, and turning at corners. The next level of the hierarchy subsequently develops to utilize the lower level primitives in complex movement sequences. One can envision further levels of complexity, with higher levels representing sequences of sequences for different sets of tasks, in a manner analogous to the "chunking" phenomenon observed in human memory of data sequences (Sakai et al., 2003). The beauty of this system is that the synaptic connections need not grow without bound as the number and complexity of sequences increases. As shown here, a *single* network can represent *multiple* complex movements through modulation of the activities of a small number of "task" neurons.

One may argue that hierarchical structure has been imposed on our system, and that no such structure is absolutely needed to complete the task. Indeed, Tucci et al. (2002) showed that the sequence generation task, which

Yamauchi & Beer (1994) felt required a modular approach, could indeed be generated with a single, non-modular, network. Further, Siegelmann & Sonntag (1995) showed that first and higher order recursive networks are computationally equivalent. However, the theoretical possibility that one giant first order network can carry out the same tasks as modular, hierarchically structured systems implies nothing about the relative ease with which either system can be generated artificially or biologically.

Although the initial sensitivity of the movement sequences generated to task neuron activations was an emergent feature of the system found by self-organization of network parameters through a genetic algorithm, the model architecture was predetermined, and the details of the network training influenced the specific functions that were assumed by different components of the architecture. Given that the current network architecture is loosely based upon the primate motor system's hierarchical design, one might expect it to perform better than a less biologically plausible giant first-order network that encompasses both simple movement primitives as well as their combination into complex sequences. This assumption will be tested in future work.

One may further question the need for hierarchical training of the current architecture. Given that humans show a clear progression of movement learning, from simple to complex movement patterns as they develop (Needlman, 2003), one might assume that there is an advantage, either in learning rate or the final skill level attained, to such an incremental, hierarchical learning organization.

Future work will test whether such complex movements can be learned "from scratch", without an externally imposed succession of increasingly difficult tasks. Further, it will be of great interest to see whether a similarly intricate dynamic structure self-organizes in the task-dependent neuronal activity of a network without such hierarchical movement learning.

Acknowledgments

The authors wish to thank Dario Floreano and Jesper Blynel, at EPFL in Switzerland, for their help and hospitality during the early stages of this work. Thanks are also due to Miki Sagara for her organizational skills and help in the preparation of this manuscript. This work would also not have been possible without the technical assistance of Yuuya Sugita and Ryunosuke Nishimoto.

References

Aharonov-Barki, R., Beker, T., Ruppin, E. (1999)
Spontaneous Evolution of Command Neurons, Place Cells and Memory Mechanisms in Autonomous Agents
In: *Advances in Artificial Life*, ECAL '99, vol. 1674, Lecture Notes in Artificial Intelligence, Springer Verlag.

Amit, R., Mataric, M. J. (2002)
Parametric Primitives for Motor Representation and Control
Int. Conf. on Robotics and Automation (ICRA), Washington DC, May 11-15, 2002.

Arbib, M. A. (1981)
Perceptual structures and distributed motor control
In: Brooks, V. B. (Ed.), *Handbook of Physiology*, Section 2: The Nervous System (Vol. II, Motor Control, Part 1), pp. 1449-1480, American Physiological Society.

Blynel, J. (2003)
Evolving Reinforcement Learning-Like Abilities for Robots
In: Tyrrell, A, Haddow, P.C., and Torresen, J (Eds.), *Evolvable Systems: From Biology to Hardware: 5th International Conference*, ICES 2003.

Blynel, J., Floreano, D. (2002)
Levels of dynamics and adaptive behavior in evolutionary neural controllers
In: Hallam, B., Floreano, D., Hallam, J., Hayes, G., Meyer, J.A. (Eds.), *From Animals to Animats 7: Proceedings of the Seventh International Conference on Simulation of Adaptive Behavior*. MIT Press, Bradford Books, Cambridge, MA.

Edwards, D. H., Heitler, W. J., Krasne, F. B. (1999)
Fifty years of a command neuron: the neurobiology of escape behavior in the crayfish
Trends in Neurosciences, 22(4), 153-161.

Fan, J., Yao, Q., Tong, H. (1996)
Estimation of Densities and Sensitivity Measures in Nonlinear Dynamical Systems
Biometrika, 83, 1, 189-206.

Floreano, D., Mondada, F. (1994)
Automatic creation of an autonomous agent: genetic evolution of a neural-network driven robot
In: Cliff, D., Husbands, P., Meyer, J., Wilson, S.W. (Eds.), *From Animals to Animats 3: Proceedings of the Third Conference on Simulation of Adaptive Behavior*. MIT Press, Bradford Books, Cambridge, MA.

Giszter, S. F., Mussa-Ivaldi, F. A., Bizzi, E. (1993)
Convergent force fields organized in the frog's spinal cord
Journal of Neuroscience, 13(2): 467-491.

Goldberg, D.E. (2002)
The Design of Innovation: Lessons from and for Competent Genetic Algorithms
Kluwer Academic Publishers, Boston, MA.

Hochreiter, S., Schmidhuber, J. (1997)
Long short-term memory
Neural Computation, 9(8), 1735-1780

Jacobs, R., Jordan, M., Nowlan, S., Hinton, G. (1991)

Adaptive mixtures of local experts
Neural Computation, 3(1), 79-87.

Jordan, M., Jacobs, R. (1994)
Hierarchical mixtures of experts and the EM algorithm
Neural Computation, 6(2), 181-214.

Kawato, M., Furukawa, K., Suzuki, R. (1987)
A hierarchical neural network model for the control and learning of voluntary movement
Biological Cybernetics, 57, 169-185.

Mataric, M. J. (2002)
Sensory-motor primitives as a basis for imitation: Linking perception to action and biology to robotics
In: Dautenhahn, K., Nehaniv, C. L. (Eds.), *Imitation in Animals and Artifacts*, pp. 391-422, MIT Press.

McGovern, A., Barto, A. G. (2001)
Accelerating Reinforcement Learning through the Discovery of Useful Subgoals
Proceedings of the 6th International Symposium on Artificial Intelligence, Robotics, and Automation in Space: i-SAIRAS 2001

McGovern, A., Sutton, R. S., Fagg, A. H. (1997)
Roles of macro-actions in accelerating reinforcement learning
Proceedings of the 1997 Grace Hopper Celebration of Women in Computing, 13-18.

Mitchell, M. (1998)
An Introduction to Genetic Algorithms
MIT Press, Cambridge, MA.

Mussa-Ivaldi, F. A., Giszter, S. F., Bizzi, E. (1994)
Linear combination of primitives in vertebrate motor control
Proceedings of the National Academy of Sciences, USA, 91: 7535-7538.

Needlman, R.D. (2003)
Growth and Development
In: Behrman, R. E., Kliegman, R. M., Jenson, H. B. (Eds.)
Nelson Textbook of Pediatrics, 17th Ed., Part II, Chapter 10, p. 33. W. B. Saunders Publishing

Ninokura, Y., Mushiake, H., Tanji, J. (2003)
Integration of Temporal Order and Object Information in the Monkey Lateral Prefrontal Cortex
Journal of Neurophysiology, 10, 1152.

Nishimoto, R., Tani, J. (2003).
Learning to Generate Combinatorial Action Sequences Utilizing the Initial Sensitivity of Deterministic Dynamical Systems
Proc. of the 7th International Work-Conference on Artificial and Natural Neural Networks (IWANN'03). Springer, pp. 422-429.

Nolfi, S., Floreano, D. (2000)
Evolutionary Robotics: The Biology, Intelligence, and Technology of Self-Organizing Machines
MIT Press, Bradford Books, Cambridge, MA.

Sakai, K., Kitaguchi, K., Hikosaka, O. (2003)
Chunking during human visuomotor sequence learning
Experimental Brain Research, 152(2), 229-242.

Schaal, S. (1999)
Is imitation learning the route to humanoid robots?
Trends in Cognitive Sciences, 3, 233-242.

Siegelmann, H.T., Sontag, E.D.(1995)
On the computational power of neural nets
Journal of Computer and System Sciences, 50(1), 132-150.

Tani, J. (2003)
Learning to generate articulated behavior through the bottom-up and the top-down interaction processes
Neural Networks, 16, 1, 11-23.

Tani, J., Nolfi, S. (1999)
Learning to perceive the world as articulated: An approach for hierarchical learning in sensory-motor systems
Neural Networks, 12: 1131-1141

Tanji, J., Shima, K. (1994)
Role for supplementary motor area cells in planning several movements ahead
Nature, 371, 413-416.

Teyke, T., Weiss, K. R., Kupfermann, I. (1990)
An identified neuron (CPR) evokes neuronal responses reflecting food arousal in Aplysia
Science, 247, 85-87.

Thoroughman, K. A., Shadmehr, R. (2000)
Learning of action through combination of motor primitives
Nature, 407, 742-747

Tucci, E., Quinn, M., Harvey, I. (2002)
An Evolutionary Ecological Approach to the Study of Learning Behavior Using a Robot-Based Model
Adaptive Behavior, 10(3/4), 201-222.

Yamauchi, B., Beer, R.D. (1994)
Sequential behavior and learning in evolved dynamical neural networks
Adaptive Behavior, 2(3), 219-246.

LEARNING AND ADAPTATION

Competitive-Cooperative-Concurrent Reinforcement Learning with Importance Sampling

Eiji Uchibe[*,**] **Kenji Doya**[*,**]
[*]Dept. of Computational Neurobiology, ATR Computational Neuroscience Labs.
[**]CREST, Japan Science and Technology Agency
2-2-2 Hikaridai, "Keihanna Science City," Kyoto 619-0288 Japan
uchibe@atr.jp doya@atr.jp

Abstract

The speed and performance of learning depend on the complexity of the learner. A simple learner with few parameters and no internal states can quickly obtain a reactive policy, but its performance is limited. A learner with many parameters and internal states may finally achieve high performance, but it may take enormous time for learning. Therefore, it is difficult to decide in advance which architecture and algorithm should be used for a new task. In this paper, we propose a new framework for selecting an appropriate policy out of a set of heterogeneous reinforcement learning modules and for correctly improving the policies of all learning modules including those not selected, using the method of importance sampling. In this framework, multiple heterogeneous learning modules sharing the same sensory-motor system can compete to act and cooperate to learn, allowing the overall learning system to obtain a good performance faster. We show in a simulation of partially-observable pole balancing task and robotic experiments of battery-pack foraging and partially observable T-maze tasks that a complex learning module trained with the proposed method can actually learn faster than when it is trained alone, by exploiting task-relevant episodes generated by suboptimal, but fast-learning modules.

1. Introduction

Reinforcement learning (RL) (Sutton and Barto, 1998) is an attractive learning framework with a wide range of possible application areas. Yet there are critical unsolved problems in real-world application of RL. However, it often takes a prohibitively long time to obtain a good policy, and the learning algorithm and parameters have to be tuned for each task. Then, RL with multiple modules is a focus of active research because it can potentially be useful in scaling reinforcement learning to complex problems. In his popular book "Society of Mind," Minsky suggested that intelligent behavior is a result of cooperation in a society of primitive agents (Minsky, 1986). Neuroscience research revealed heterogeneous learning modules working in parallel for fast and robust learning of skilled behaviors (Hikosaka et al., 1999, Nakahara et al., 2001).

Singh (Singh, 1992) has proposed compositional Q learning in which an agent learns multiple sequential decision tasks with multiple learning modules. In Singh's compositional Q learning method, learning modules are switched on the basis of TD-error (Singh, 1992). In Multiple Model-based Reinforcement Learning (MMRL), modules are selected on the basis of the accuracy of prediction model of each module (Doya et al., 2002). In these previous approaches, each module has the same structure and algorithm, and only the module that is currently selected for action learns from the experience. In this paper, we consider a method for concurrently training a set of heterogeneous learning agents that share the same sensory-motor system, which resembles the parallel architecture of the brain.

A critical issue in applying reinforcement learning to a novel task is how to select the right state representation and the learning algorithm. If sensory inputs are coarsely discretized, or a small number of basis functions are used in function approximation, the ultimate performance is limited. On the other hand, if discretization is too fine or there are too may parameters in function approximation, learning can be slow. Furthermore, an agent with memory bits or recurrent units may solve partially observable problems, but learning can be slower and prone to be trapped in suboptimal solutions with many internal states.

For the algorithm, we have a choice between two major types: Action Value function based Reinforcement Learning (AVRL) and Policy Gradient based Reinforcement Learning (PGRL). AVRL, such as Q learning (Watkins and Dayan, 1992) and SARSA (Rummery and Niranjan, 1994), has been applied to many tasks with remarkable success (Asada et al., 1996,

Asada et al., 1999, Gaskett et al., 2000), but it can perform badly if the environment is not Markovian. PGRL (Baird and Moore, 1999, Kimura et al., 1995, Peshkin et al., 1999, Meuleau et al., 1999, Sutton et al., 2000) can be applied to continuous and hidden-state problems, but the learning speed of PGRL is slower than that of AVRL for a Markovian task. In practice, experimenters try different combinations of representations and algorithms and then select the best combination. In order to eliminate the need for such human hand-tuning, we explore a learning framework in which an agent has multiple heterogeneous learning modules and selects an appropriate module according to the task and the experience. More specifically, we consider how a single agent can best utilize multiple learning modules that have different state representation and learning algorithms but share the same experience.

In this paper, we propose a novel framework named *Cooperative-Competitive-Concurrent Learning with Importance Sampling* (CLIS) for selecting an appropriate module for action and correctly improving the policies of all learning modules, including those not selected, based on the method of importance sampling (Rubinstein, 1981). The agent has multiple learning modules with different numbers of parameters and learning algorithms and improves their multiple policies simultaneously to accomplish the given task. Because the modules that are not selected can learn from the experience obtained by the actions of another module, it is possible to obtain purposive behaviors efficiently and rapidly. We apply the proposed method to simulation of a partially-observables pole balancing task and to two robotic experiments of a battery-pack foraging task and partially observable T-maze task. The result shows that a complex learning module trained with the proposed method can actually learn faster than when it is trained alone, by exploiting task-relevant episodes generated by suboptimal, but fast-learning modules.

2. Basics of Reinforcement Learning

Reinforcement learning is a computational paradigm for an agent to learn a policy, state-to-action mapping, based on a scalar reward feedback. When the agent observes environmental state s_t and outputs an action a_t, the agent makes a transition to state s_{t+1} and receives a reward r_t. The aim of the agent is to learn a policy $\pi = \Pr(a|s)$ in order to receive maximal cumulative reward.

The value of a state s under policy π is given by

$$V^\pi(s) = \mathrm{E}_\pi\{R(s)\} = \mathrm{E}_\pi\left\{\sum_{k=0}^{\infty}\gamma^k r_{t+k}|s_t = s\right\},$$

where E_π denotes the expected value given that the agent follows policy π. γ is a parameter, $0 \le \gamma \le 1$,

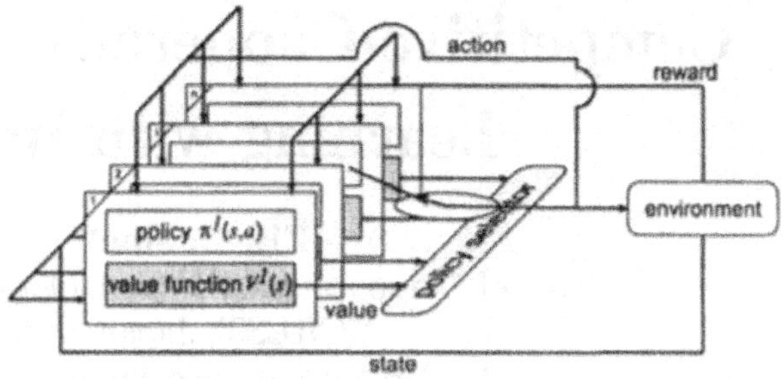

Figure 1: A multi-module learning system.

called a discount rate. $R(s)$ is the weighted sum of rewards starting from s, called return. Similarly, the value of taking action a in state s under policy π is given by

$$Q^\pi(s,a) = \mathrm{E}_\pi\left\{\sum_{k=0}^{\infty}\gamma^k r_{t+k}|s_t = s, a_t = a\right\}.$$

V^π and Q^π are called the state-value and the action-value functions, respectively.

In AVRL, policy π is derived from the action value function. In PGRL, policy π is updated with regard to value function V^π. In either case, an important issue is how to gather episodes, sequences of state, action, and reward, in order to estimate the value function rapidly and accurately. In general, the value function of the policy of an agent is estimated in either an *on-policy* or *off-policy* manner.

- On-policy learning: estimate the value function V^π or Q^π while the agent follows policy π.
- Off-Policy learning: estimate V^π or Q^π while the agent follows another policy π'.

We use off-policy learning for training multiple modules with different policies from episodes obtained by a single policy.

3. Concurrent Reinforcement Learning with Importance Sampling

3.1 Competitive Policy Selection

Figure 1 shows the basic architecture of the proposed modular learning method. The agent has N learning modules with a state value function V^i and a policy π^i $(i = 1, \dots, N)$. At the beginning of each episode, the agent selects the policy of one module following the probability

$$\Pr(i|s_0) = \frac{\exp(\beta_{\mathrm{Sel}} V^i(s_0))}{\sum_{j=1}^{N}\exp(\beta_{\mathrm{Sel}} V^j(s_0))}, \tag{1}$$

where s_0 is an initial state and β_{Sel} is a positive parameter called the inverse temperature. Low β_{Sel} causes (nearly) equi-probable selection of all modules, while high β_{Sel} causes selection of the module with the highest value with probability close to one.

Let π^{Beh} and π^i be the *behavior policy* of the agent and the *target policy* of the i-th module, respectively. During each episode, $\pi^{\text{Beh}} = \pi'$ only for the module that was selected by equation (1). While the agent collects the experiences by using the behavior policy π^{Beh}, all of the modules improve their policies π^i and state value functions V^i from the experience obtained by π^{Beh}, using the importance sampling method below.

3.2 Cooperative Learning by Importance Sampling

Let us consider how to train all modules $i = 1, \ldots, N$ using an episode obtained by a selected policy π^{Beh}. Studies of "off-policy" reinforcement learning suggest that it is possible and sometimes better to estimate the value function of a policy π by following a behavior policy $\pi^{\text{Beh}} \neq \pi$. The choice of distribution obviously makes a difference to the efficiency of the method. Importance sampling (Rubinstein, 1981) is a method to compensate the mismatch between the target policy π and behavior policy π^{Beh}.

Let h_m be the sequence of states, actions, and rewards of the m-th episode:

$$h_m = \{s_0, a_0, r_1, s_1, a_1, r_2, \cdots, s_{t-1}, a_{t-1}, r_t, s_t\}.$$

After M episodes are experienced, the Monte Carlo estimate of the state value function is given by (Sutton and Barto, 1998)

$$V^{\pi}(s) = \frac{1}{M} \sum_{m=1}^{M} R_m(s) \frac{\Pr(h_m|\pi)}{\Pr(h_m|\pi^{\text{Beh}})}. \tag{2}$$

Here, R_m is the actual return in the m-th episode:

$$R_m(s) = r_{t_m(s)} + \gamma r_{t_m(s)+1} + \cdots + \gamma^{T_m - t_m(s)-1} r_{T_m - 1},$$

where $t_m(s)$ is the first time that state s is visited and T_m is the time of termination of the m-th episode. $\Pr(h_m|\pi)$ and $\Pr(h_m|\pi^{\text{Beh}})$ are the probabilities that the episode h_m occurs by following π and π^{Beh}, respectively. Since $\Pr(h_m|\pi)$ is represented by

$$\Pr(h_m|\pi) = \Pr(s_0) \prod_{t=0}^{T_m-1} \pi(s_t, a_t) \Pr(s_{t+1}|s_t, a_t),$$

$\Pr(h_m|\pi) / \Pr(h_m|\pi^{\text{Beh}})$ is given by

$$\frac{\Pr(h_m|\pi)}{\Pr(h_m|\pi^{\text{Beh}})} = \prod_{t=0}^{T_m-1} \frac{\pi(s_t, a_t)}{\pi^{\text{Beh}}(s_t, a_t)} = \prod_{t=0}^{T_m-1} \rho_t, \tag{3}$$

where $\rho_t = \pi(s_t, a_t)/\pi^{\text{Beh}}(s_t, a_t)$ is the correction factor to deal with the differences between the behavior and the target policies. Here, the knowledge of the environmental dynamics $\Pr(s_{t+1}|s_t, a_t)$ is not needed to compute $\Pr(h_m|\pi)/\Pr(h_m|\pi^{\text{Beh}})$. We only need the ratio ρ_t of the action selection probabilities. Note that the behavior policy should satisfy $\pi^{\text{Beh}}(s, a) > 0$ if $\pi(s, a) > 0$.

$\prod_t^{T_m-1} \rho_t$ will often increase or decay rapidly over time, especially if the behavior policy π^{Beh} and the target policy π are very different. In this case, importance sampling may be numerically unstable. Therefore, we introduce the maximum value of $\prod_t^{T_m-1} \rho_t$ to avoid divergence. That is, equation (3) is modified as follows:

$$\frac{\Pr(h_m|\pi)}{\Pr(h_m|\pi^{\text{Beh}})} = \max\{ \prod_{t=0}^{T_m-1} \rho_t, \;\; C\}, \tag{4}$$

where C is a positive constant.

3.3 Implementation with Importance Sampling

3.3.1 Action Value Function based Reinforcement Learning

In this section, we explain the importance sampling method for AVRL. AVRL algorithms such as Q learning (Watkins and Dayan, 1992) and SARSA (Rummery and Niranjan, 1994) define the policy π^{AV} according to the current estimate of the action value function Q^{AV}. SARSA is an on-policy algorithm, while Q learning is an off-policy algorithm. If the environment is an MDP (Markov Decision Process), Q learning is guaranteed to obtain the optimal value function (Watkins and Dayan, 1992). Furthermore, SARSA often obtains good memoryless stochastic policies for POMDP problems by setting an appropriate eligibility trace λ (Loch and Singh, 1998).

Here, we formulate the learning rules for MDP, where the state is perfectly observable. We adopt the importance sampling method proposed by (Precup et al., 2000, Precup et al., 2001) for our implementation. The action value function can be represented as

$$Q^{\text{AV}}(s, a) = \frac{1}{M} \sum_{m=1}^{M} R_m(s, a) \prod_{t=0}^{T_m-1} \rho_t, \tag{5}$$

where $R_m(s, a)$ is the actual return following (s, a) of the m-th episode

$$R_m(s, a) = r_{t_m(s,a)} + \gamma r_{t_m(s,a)+1} + \cdots \gamma^{T_m - t_m(s,a)-1} r_{T_m-1}.$$

Here, $t_m(s, a)$ is the first time that state s and action a are visited and T_m is the time of termination of the m-th episode. If we consider function approximations based on a set of feature vectors $\{\phi_k(s, a)\}$, the action value function can be represented by

$$Q^{\rm AV}(s,a)=\sum_{k=1}^{n} w^{\rm AV}(k)\phi_{s,a}(k),$$

where $w^{\rm AV}$ and N denote a learned parameter and the number of features, respectively. The update rules of the action values are given by

$$\begin{cases} c_t &= \max\{\rho_t c_{t-1},\ C\}, \\ \delta_t &= r_t+\gamma\rho_{t+1}Q_t^{\rm AV}(s_{t+1},a_{t+1})-Q_t^{\rm AV}(s_t,a_t), \\ T_t^{\rm AV}(k) &= \lambda\gamma\rho_t T_{t-1}^{\rm AV}(k)+c_t\phi_{s_t,a_t}(k), \\ w_t^{\rm AV}(k) &= w_{t-1}^{\rm AV}(k)+\alpha_{\rm AV}\delta_t T_t^{\rm AV}(k), \end{cases} \tag{6}$$

Here, $T^{\rm AV}$ is an eligibility trace, λ is an eligibility parameter ($0\le\lambda\le 1$), and $\alpha_{\rm AV}$ is a learning rate. If the target policy is identical to the behavior policy ($\rho_t=1$), these update rules become the standard update rules of SARSA.

The policy must satisfy $\pi^{\rm AV}(s,a)>0$ for all s and a in order to compute equation (4). A stochastic policy for AVRL can be represented by

$$\pi^{\rm AV}(s_t,a_t)=\frac{\exp(\beta_{\rm AV}Q^{\rm AV}(s_t,a_t))}{\sum_{a'}\exp(\beta_{\rm AV}Q^{\rm AV}(s_t,a'))}, \tag{7}$$

where $\beta_{\rm AV}$ is the inverse temperature.

3.3.2 Policy Gradient based Reinforcement Learning

In PGRL, a policy π is represented by a parameter vector $\boldsymbol{w}$, which is updated to the direction of the gradient of the state value function V^π. Glynn showed that the gradient could be estimated for an finite-horizon MDP (Glynn, 1986). PGRL algorithms like REINFORCE (Williams, 1992) have recently been re-evaluated because of their robust performance with non-Markovian tasks (Baird and Moore, 1999, Kimura et al., 1995, Meuleau et al., 1999, Sutton et al., 2000).

REINFORCE is basically an on-policy method (Sutton and Barto, 1998) because it estimates the gradient at a given point in the policy space by acting precisely in the manner of the corresponding policy during learning trials. Suppose that policy $\pi^{\rm PG}$ is parametrized by a vector $\boldsymbol{w}=\{w_1,\cdots,w_n\}$. In PGRL, a single average expected return $V^{\rm PG}(\boldsymbol{w})$ is employed as an objective function

$$V^{\rm PG}(\boldsymbol{w})={\rm E}_{\pi^{\rm PG}}\left[V^{\pi^{\rm PG}}(s)\right].$$

If the derivative of $V^{\rm PG}(\boldsymbol{w})$ can be calculated for each $w(k)$, it is possible to achieve exact gradient ascent by making updates

$$w_t(k)=w_{t-1}(k)+\alpha_{\rm PG}\frac{\partial}{\partial w(k)}V^{\rm PG}(\boldsymbol{w}),$$

where $\alpha_{\rm PG}$ is the step-size parameter.

The application of the importance sampling to Williams' REINFORCE was discussed by (Meuleau et al., 2001). The state value function in consideration of importance sampling is given by

$$V^{\rm PG}(\boldsymbol{w})=\frac{1}{M}\sum_{m=1}^{M}R_m\frac{\Pr(h_m|\boldsymbol{w})}{\Pr(h_m|\pi^{\rm Beh})}\Pr(h_m|\pi^{\rm Beh}),$$

where $\Pr(h_m|\boldsymbol{w})$ is the probability in which episode h_m occurs by using the policy parametrized by the vector $\boldsymbol{w}$. The gradient can be computed with importance sampling by

$$\frac{\partial}{\partial w_k}V(\boldsymbol{w})=\sum_{t=0}^{T_m-1}\gamma^t r_t\Pr(h_m|\pi^{\rm Beh})\frac{\partial}{\partial w_k}\frac{\Pr(h|\boldsymbol{w})}{\Pr(h_m|\pi^{\rm Beh})}.$$

From equation (3), $\Pr(h_m|\boldsymbol{w})/\Pr(h_m|\pi^{\rm Beh})$ can be easily computed by the products of the ratio ρ_t of the two policies. Consequently, the update rules of PGRL with importance sampling are given by

$$\begin{cases} c_t &= \max\{\rho_t c_{t-1},\ C\}, \\ T_t(k) &= T_{t-1}(k)+\frac{\partial}{\partial w_k}\ln\Pr(a_{t-1}|s_{t-1}), \\ w_t(k) &= w_{t-1}(k)+\alpha_{\rm PG}\gamma^t r_t c_t T_t(k), \end{cases} \tag{8}$$

where $T_t(k)$ is an eligibility trace for the policy gradient method. If the target policy is identical to the behavior policy, i.e. $\rho_t=1$, equation (8) is the same as the policy gradient learning rule proposed by (Baird and Moore, 1999) and (Kimura et al., 1995).

In PGRL, the stochastic policy is represented by a parameter vector. In the simulation below, we use softmax action selection rules by assigning the policy parameter as $\boldsymbol{w}=\{q_{s,a}\}$, $(s=1,\dots,n_s; a=1,\dots,n_a)$

$$\pi^{\rm PG}(s_t,a_t)=\frac{\exp(\beta_{\rm PG}q_{s_t,a_t})}{\sum_{a'}\exp(\beta_{\rm PG}q_{s_t,a'})}, \tag{9}$$

where $\beta_{\rm PG}$ is the inverse temperature. Note that q_{s_t,a_t} is a policy parameter and need not be the action value function. Although it is possible to learn $\beta_{\rm PG}$ because it is also a policy parameter, we fix $\beta_{\rm PG}$ to a constant value for simplicity. By using the above equations, PGRL learns the parameter vector $\boldsymbol{w}$ online. Although this implementation does not require a state value function $V^{\rm PG}$ explicitly, we compute $V^{\rm PG}$ by using equation (2) for the sake of module selection by equation (1).

3.3.3 Temporal Difference Learning with Hand-Made Controller

It is not realistic that the robot executes various actions randomly. Suppose that a mobile robot navigates in a crowded environment. This mobile robot finally learns

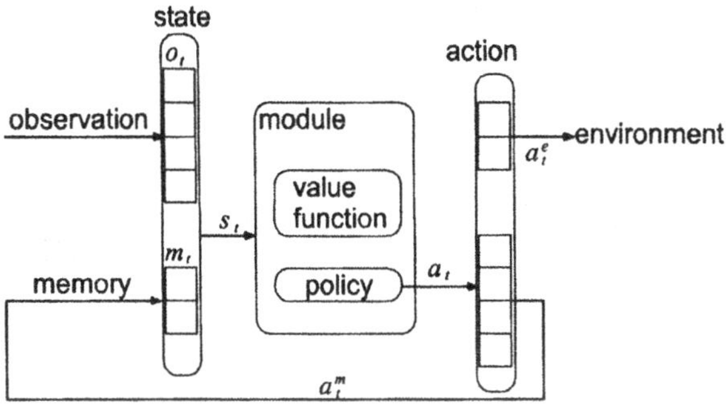

Figure 2: Learning module with "external" memory bits.

collision avoidance behavior after it makes collisions with the obstacles. This is not a problem in computer simulation. However, the collisions with obstacles in the real environment will cause serious hardware troubles in many cases. In order to realize robot learning in the real environment, it is important to introduce a priori knowledge. If the designer has some *a priori* domain-specific knowledge, CLIS enables incorporating it in a handmade controller, which improves initial performance and learning of other modules.

Implementation is quite simple. Suppose that the state value function is represented by

$$V^{\mathrm{HM}}(s) = \sum_{k=1}^{n} w^{\mathrm{HM}}(k)\phi_s(k),$$

where w^{HM} is a learned parameters. w^{HM} is updated by using the rule similar to equation (6):

$$\left\{\begin{array}{rcl} c_t & = & \max\{\rho_t c_{t-1},\ C\}, \\ \delta_t & = & r_t + \gamma\rho_{t+1}V_t^{\mathrm{HM}}(s_{t+1}) - V_t^{\mathrm{HM}}(s_t), \\ T_t^{\mathrm{HM}}(k) & = & \lambda\gamma\rho_t T_{t-1}^{\mathrm{HM}}(k) + c_t\phi_{s_t}(k), \\ w_t^{\mathrm{HM}}(k) & = & w_{t-1}^{\mathrm{HM}}(k) + \alpha_{\mathrm{HM}}\delta_t T_t^{\mathrm{HM}}(k). \end{array}\right. \tag{10}$$

3.4 Learning Modules

Figure 2 shows the general structure of a learning module. In order to deal with POMDP (Partially Observable Markov Decision Process) problems, we adopt an architecture with "external" memory bits (Peshkin et al., 1999). This method has the following features.

1. The state s_t of the agent is given by the sensor information o_t and the memory information m_t with l-bits; $s_t = (o_t, m_t)$.
2. Its output a_t consists of the external actions a_t^e that make state transitions in the environment and augmented actions a_t^m that change the state of the memory.
3. For each memory bit, there are two actions: One action sets the memory bit to 1, and the other sets it to 0. Thus for memory bits of length l, the number of actions is $2l$.

If the length of the memory bits is sufficient to accomplish the given task, the agent is able to obtain the optimal policy.

In order to reduce the size of the action space, the agent only selects either a_t^e or a_t^m at each decision step. This approach introduces a bias, since we have to lose at least one time-step each time we want to change the content of the memory. However, it can be adjusted in most algorithms by assuming that memory-setting actions finish immediately (Peshkin et al., 1999).

4. Partially-observable Pole Balancing

4.1 Task setting

In order to verify our method, we ran a number of experiments with the cart pole problem. The objective here is to apply forces to a cart on a track in order to keep a pole hinged to the cart from falling over. Let us denote by F the applied control force, by θ the pole's angle, by x the cart's position on the track. The dynamics of motion is modeled by the following nonlinear ordinary differential equations:

$$\begin{array}{rcl} \ddot{\theta} & = & \dfrac{g\sin\theta + \frac{\cos\theta(\mu_c \mathrm{sign}(\dot{x}) - F - m_p l\dot{\theta}^2\sin\theta)}{m_c+m_p} - \frac{\mu_p\dot{\theta}}{m_p l}}{l\left(\frac{4}{3} - \frac{m\cos^2\theta}{m_c+m_p}\right)}, \\ \ddot{x} & = & \dfrac{F + m_p l(\dot{\theta}^2\sin\theta - \ddot{\theta}\cos\theta) - \mu_c \mathrm{sign}(\dot{x})}{m_c + m_p}. \end{array}$$

The specifications for the simulated pole and cart are the same as used in previous work (Barto et al., 1983): length of the track: $L = 4.8$ [m]; mass of the cart: $m_c = 1.0$ [kg]; mass of the pole: $m_p = 0.1$ [kg]; half length of the pole: $l = 0.5$ [m]; gravity $g = 9.8$ [m/s^2]; coefficient of friction of cart on track: $\mu_c = 5.0 \times 10^{-4}$; coefficient of friction of pole on cart: $\mu_p = 2.0 \times 10^{-6}$. For all sensor readings in the simulated environment 10% of noise is added. Five actions $F = \pm 10, \pm 5, 0$ [N] were prepared for this task. Control actions were applied at the rate of 50 [Hz]. At the beginning of each episode, θ was randomly initialized between -12 and 12 [deg], and $x = \dot{x} = \dot{\theta} = 0$. If the pole falls past a given angle from vertical ($\theta < -12$ or $\theta > 12$) or if the cart reaches the end of the track, an episode is terminated on failure with a negative reward -100. In addition, each step costs $0.001|F|$ as a penalty. One episode lasts for at most 20 [s].

This is an easy control task when all the state variables (x, $\dot{x}$, θ, and $\dot{\theta}$) are observable. But here we consider a case when only x and θ are observable. We discretized

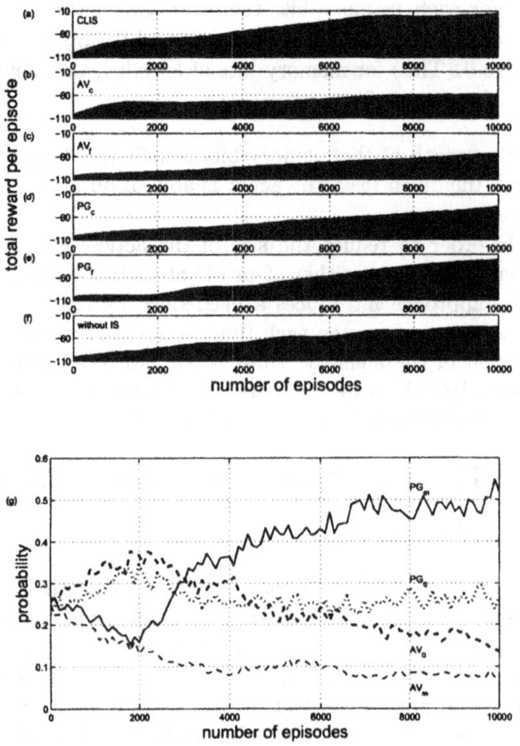

Figure 3: Experimental results of partially-observable pole balancing task. (a)-(f) total reward per episode, averaged over 10 simulation runs. (g) the probability of module selection.

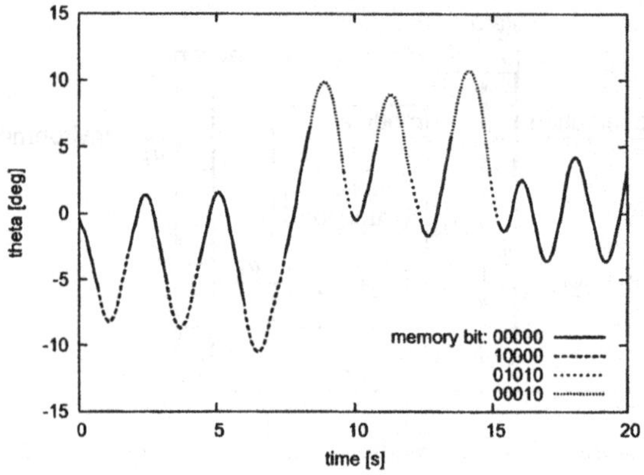

Figure 4: Typical trajectory of the pole angle θ in POMDP setting.

x and θ into eight bins. Thus, the number of observations o_t was 8^2. We prepared four learning modules (AV_0: AVRL without memory bit, AV_m: AVRL with memory bit, PG_0 PGRL without memory bit, and PG_m: PGRL with memory bit). The memory length was $l = 5$. In order to compare the performance of our method, we also tested four learning modules (AV_0, AV_m, PG_0, and PG_m) independently. The learning meta-parameters were $\gamma = 0.95$, $\lambda = 0.2$, $\alpha_{AV} = 0.2$, $\alpha_{PG} = 0.01$ and $C = 10$, which were tuned in preliminary experiments. We conducted ten simulation runs for each experiment.

4.2 Experimental results

Figure 3 (a) compares the performance of the proposed method with that of each learning module trained alone. In the early stage of learning, AV_0 quickly learned a policy which can hold the pole till the time-out if the initial condition is favorable. The reason why AV_0 could learn in POMDP environment would be due to the effects of eligibility trace (Loch and Singh, 1998). While the performance of AV_0 saturated after about 1000 episodes, both PG_0 and PG_m gradually improved the performance. The performance of the proposed method was almost the same as that of AV_0 in the early stage of learning. This is not surprising since AV_0 is selected most frequently in early stage, as shown in Figure 3 (b). However, what is remarkable is that the proposed method achieved better performance faster than the best-performing single module, in this case PG_m.

Next, we investigated how the proposed method used the memory bit in this task. Figure 4 shows a typical trajectory obtained by PG_m in the proposed method. In this figure, four different lines were used according to the state of the memory bit. For example, when all memory bits were set to 0s, the solid line was selected to draw the trajectory. In POMDP setting, since the agent had no information about $\dot{x}$, and $\dot{\theta}$, θ oscillated constantly even though it learned sufficiently. It was found that the agent usually clear all the bits with θ in $[-5, 5]$. When θ made a transition to the area with $|\theta| > 5$, the agent set one memory bit.

The reason why the proposed method could learn faster than PG_m alone was that the PG_m module in the proposed method explored more relevant part of the state space, thanks to the policy of AV_0 in the early stage. The performance of the method without importance sampling was worse than that of PG_m as shown in Figure 3 (a). This shows that direct use of the experience without importance sampling can have a bad influence on learning.

5. Real Mobile Robot Learning

5.1 Cyber Rodent

Then, we performed experiments with a rodent like robot, Cyber Rodent (Elfwing et al., 2003, Eriksson et al., 2003). The goal of the Cyber Rodent project is to understand neural mechanisms necessary for the artificial agents that have the same fundamental

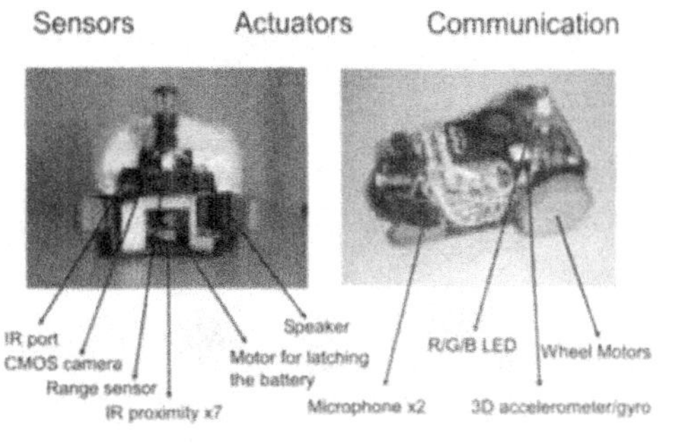

Figure 5: Cyber Rodent

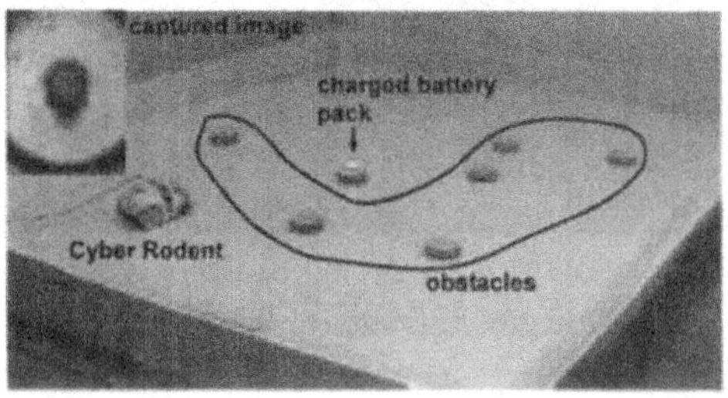

Figure 6: The first task: Battery capturing task.

constraints as biological agents: self-preservation and self-reproduction. Cyber Rodent is a two-wheel driven mobile robot with an omni-directional vision system, eight distance sensors, color LEDs for visual signaling, an audio speaker and two microphones, and an infrared port for copying programs, shown in Figure 5.

5.2 *Battery Capturing Task*

5.2.1 *Task Settings*

The task for Cyber Rodent is to capture a charged battery pack in an environment as shown in Figure 6. The environment consists of one charged battery pack and six dummy battery packs as obstacles, and walls around the experimental field. Although this task is simple, it is difficult to obtain appropriate behaviors from scratch in the real environment.

When Cyber Rodent captured the charged battery pack, the episode terminates and Cyber Rodent receives a reward value 1. We assign a negative reward -0.1 for collisions with obstacles. In addition, when pre-specified time interval (30 [s]) expires, the episode terminates. Each charged battery pack is equipped with a red LED, which Cyber Rodent can by its visual system. The visual is not sensitive to obstacles and walls. Each proximity sensor measures the distance up to 30 [cm], but it is not possible to distinguish the battery pack, obstacles, and walls.

Cyber Rodent has seven actions; go straight fast/slowly, turn left/right, rotate to left/right, and go backward. Control actions were applied at the rate of 5 [Hz]. An RBF network is used to approximate the action value functions of AVRL and the policies of PGRL. The feature vector $\{\phi_k(s)\}$ are modeled as Gaussian functions with center $\boldsymbol{\mu}_k$ and spread σ:

$$\phi_s(k) = \exp\left\{-\|(\boldsymbol{x} - \boldsymbol{\mu})/\sigma\|^2\right\},$$

where $\boldsymbol{x}$ is a sensor vector, which consists of the center position of the battery pack in the image plane and readings of the five proximity sensors.

For this experiment, we prepared five learning modules (HM: simple hand-made controller, $\mathrm{AV_c}$: AVRL with coarse representation, $\mathrm{AV_f}$: AVRL with fine representation, $\mathrm{PG_c}$: PGRL with coarse representation, and $\mathrm{PG_f}$: PGRL with fine representation). The policy of HM is designed based on the position of the battery pack in the image. Cyber Rodent can efficiently capture the battery pack if there is no obstacles between them. However, it often collides with obstacles since the proximity sensors are not considered at all. Feature vectors used in four RL modules ($\mathrm{AV_c}$, $\mathrm{AV_f}$, $\mathrm{PG_c}$, and $\mathrm{PG_f}$) are computed from the center position of the charged battery pack in the image plane and the values of the proximity sensors. While 30 features are used for coarse representation, 60 are used for fine representation. The center positions μ_k are determined by the designer. The learning meta-parameters were $\gamma = 0.95$, $\lambda = 0.2$, $\alpha_{\mathrm{HM}} = 0.05$, $\alpha_{\mathrm{AV}} = 0.05$, $\alpha_{\mathrm{PG}} = 0.05$ and $C = 10$, which tuned in preliminary experiments.

5.2.2 *Experimental Results*

To compare the performance, we also tested five learning modules separately. We conducted three experiments for each setting. Figure 7 (a)-(f) shows the learning curves of the number of captured battery packs in 10 episodes. Since the policy of HM did not change, the average performance of HM was also almost constant (Figure 7(f)). After about 40 episodes, CLIS came to capture the battery pack (Figure 7(a)). Eventually, CLIS achieved the best performance after 300 episodes. Although $\mathrm{AV_c}$ could obtain the battery pack a little earlier than CLIS, the final performance was worse than that of CLIS (Figure 7(b)). $\mathrm{AV_f}$ needed more than 100 episodes to accomplish the task (Figure 7(c)). However, the performance was gradually improved, and the $\mathrm{AV_f}$ could obtain the battery pack at about 80 percent after 300 episodes. The performances of $\mathrm{PG_c}$ and $\mathrm{PG_f}$ slightly exceeded that of HM after about 130 episodes. Since $\mathrm{PG_f}$ has the same state space as $\mathrm{AV_f}$, it is expected to obtain the same performance as that of $\mathrm{AV_f}$. In this experiment, however, $\mathrm{PG_f}$ was not able to learn the task within the maximum number of episodes we set.

Figure 7 (g) shows the selected learning module. In the early stage of learning, HM was frequently selected by

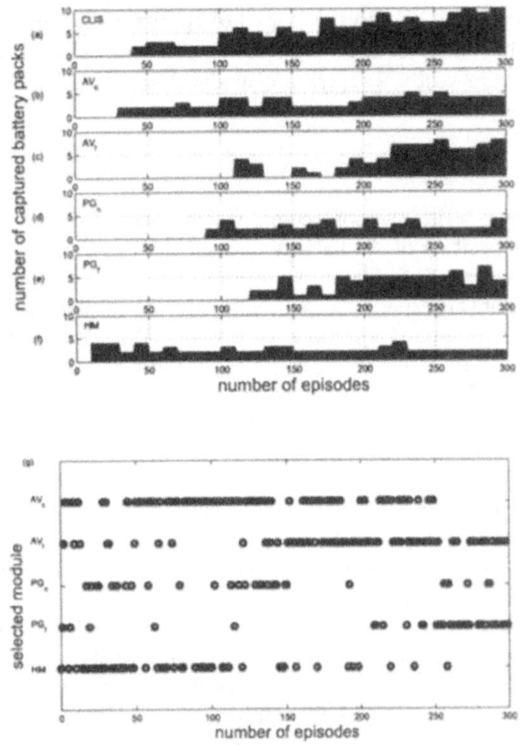

Figure 7: Experimental results in capturing task in the real environment. (a)-(f) the number of captured battery packs in 10 episodes for each learning module. (g) the selected learning module.

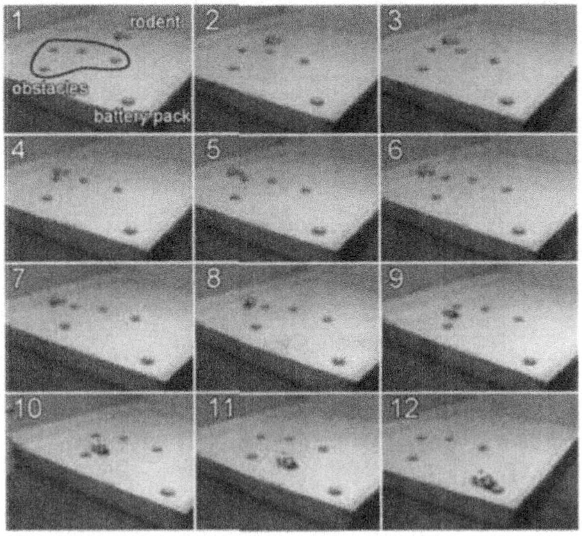

Figure 8: Obtained behaviors

Cyber Rodent. CLIS efficiently collected the experiences by using HM and trained the other four learning modules simultaneously. After 30 episodes, the probability for selecting AV_c, which was trained with the experience of HM, then increased. After 150 episodes, CLIS frequently chose AV_f. These indicate that CLIS collected meaningful experiences by switching HM, AV_c, and AV_f according to the progress of learning.

Finally, a typical behavior obtained by CLIS is shown in Figure 8. Since the battery pack was outside its visual range at the beginning of the episode, Cyber Rodent moved forward until proximity sensors reacted to the obstacles and walls to find the battery pack.

5.3 Partially Observable T-Maze

5.3.1 Task Settings

The second task we consider is a partially observable T-maze navigation task shown in Figure 9. Cyber Rodent has to move from the fixed starting position to a goal position. The goal position can be either to the left or right of the T-junction. Although the battery pack is placed at the correct goal position, the battery pack can not be observed. Instead, there is a "road sign" near the start position that the robot can perceive with its vision system. If the road sign is visible, the goal is on the right. This task requires Cyber Rodent to memorize a particular bit of information, although it is not clear before learning.

In this task, Cyber Rodent receives a positive reward: $r = 1$ when it reaches the goal position. If, instead, it reaches the opposite position, it receives a negative reward $r = -0.1$. In addition, when pre-specified time interval (15 [s]) expires, the episode terminates. The goal position is randomly switched by the experimenter.

We prepared six learning modules (LF: left wall follower (following the left wall and turning left if possible), RF: right wall follower, AV_0: AVRL without memory bit, PG_0: PGRL without memory bit, AV_m: PGRL with one memory bit), and PG_m: PGRL with one memory bit). LF and RF are hand-coded using the proximity sensor inputs. The state space for AV_0 and PG_0 is constructed based on the readings of the five proximity sensors, and one bit for the road sign (visible or not). In addition, PG_m has one memory bit. RBF network with 5 units is used to code the reading of the proximity sensors. The same action set described in 5.2.1 is used.

5.3.2 Experimental Results

Figure 10 shows the experimental results. Since the goal position was randomly determined, the prepared controllers RF and LF could reach the correct goal with 50 % probability (Figure 10(f) and (g)). Although this task seems easy to accomplish, each learning module trained alone failed to reach either goal, often resulting in success

Figure 9: The second task: Partially observable T-maze.

rate less than 50 % (Figure 10 (b)-(e)).

CLIS acquired good performance after 180 episodes on average. However, CLIS sometimes failed to navigate to the correct goal position after 300 episodes because of accidental reset of the memory bit. Figure 10 (h) shows the selection of learning modules by CLIS. Two hand made controllers (LF and RF) were frequently selected until about 200 episodes, and finally AV_m and PG_m were frequently used.

6. Conclusion

This paper proposed a new framework for selecting an appropriate module and correctly improving the policies of all learning modules including those not selected, based on the method of importance sampling. We have applied our proposed method to the partially-observable cart pole problem and real robot experiments. One of the interesting points is that a complex RL system can learn faster with the help of a simpler RL system that can not obtain the best performance.

We have also tested rejection sampling method to compensate for the mismatch between the behavior policy and target policy. However, the actual learning time was not so shortened because rejection sampling method can be wildly inefficient if the behavior policy approximates the target policy poorly. In fact, correction by using equation (4) introduces bias in general. However, we should make the best possible use of experience because real world experience is very expensive.

One promising application of CLIS is behavior learning in evolutionary robotics (Nolfi and Floreano, 2000). In general, a genetic algorithm or genetic programming of learning agents requires a long time for training and evaluation of many individuals (controllers) sequentially by using a limited number of robot hardware units. CLIS can greatly shorten the computational time by training and evaluating multiple individuals concurrently by using a single hardware unit.

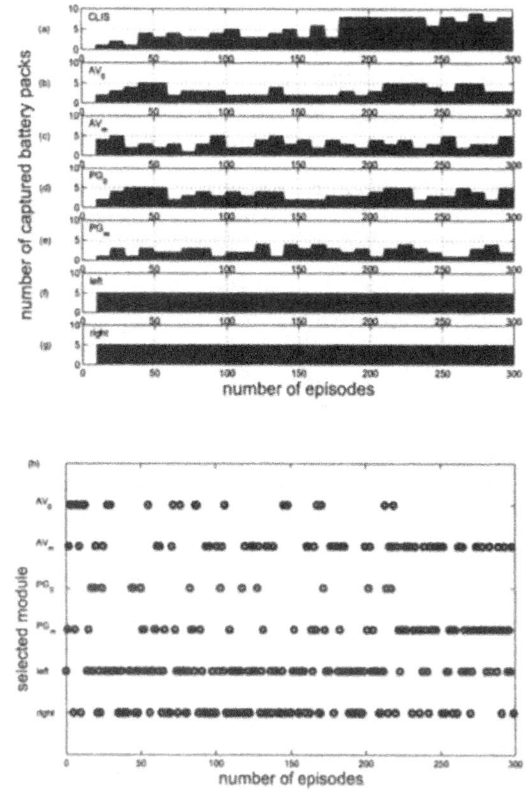

Figure 10: Experimental results in capturing task in the real environment. (a)-(g) the number of captured battery packs in 10 episodes for each learning module. (h) the selected learning module.

Acknowledgments

The research reported here was supported in part by a contract with the Telecommunications Advancement Organization of Japan entitled, 'Research on Human Communication.'

References

Asada, M., Noda, S., Tawaratsumida, S., and Hosoda, K. (1996). Purposive behavior acquisition for a real robot by vision-based reinforcement learning. *Machine Learning*, 23:279–303.

Asada, M., Uchibe, E., and Hosoda, K. (1999). Cooperative behavior acquisition for mobile robots in dynamically changing real worlds via vision-based reinforcement learning and development. *Artificial Intelligence*, 110:275–292.

Baird, L. C. and Moore, A. W. (1999). Gradient descent for general reinforcement learning. In *Ad-*

vances in Neural Information Processing Systems 11. MIT Press.

Barto, A. G., Sutton, R. S., and Anderson, C. W. (1983). Neuronlike elements that can solve difficult learning control problems. *IEEE Transactions on Systems, Man, and Cybernetics*, 13:835–846.

Doya, K., Samejima, K., Katagiri, K., and Kawato, M. (2002). Multiple model-based reinforcement learning. *Neural Computation*, 14:1347–1369.

Elfwing, S., Uchibe, E., and Doya, K. (2003). An evolutionary approach to automatic construction of the structure in hierarchical reinforcement learning. In *Proc. of the Genetic and Evolutionary Computation Conference*, pages 507–509.

Eriksson, A., Capi, G., , and Doya, K. (2003). Evolution of the meta-parameters in reinforcement learning. In *Proc. of the IEEE/RSJ International Conference on Intelligent Robots and Systems.*

Gaskett, C., Fletcher, L., and Zelinsky, A. (2000). Reinforcement learning for a vision based mobile robot. In *Proc. of the IEEE/RSJ International Conference on Intelligent Robots and Systems.*

Glynn, P. W. (1986). Stochastic approximation for Monte-Carlo optimization. In *Proc. of the 1986 Winter Simulation Conference*, pages 356–365.

Hikosaka, O., Nakahara, H., Rand, M. K., Sakai, K., Lu, X., Nakamura, K., Miyachi, S., and Doya, K. (1999). Parallel neural networks for learning sequential procedures. *Trends in Neuroscience*, 22(10):464–471.

Kimura, H., Yamamura, M., and Kobayashi, S. (1995). Reinforcement learning by stochastic hill climbing on discounted reward. In *Proc. of the Twelfth International Conference on Machine Learning*, pages 295–303.

Loch, J. and Singh, S. (1998). Using eligibility traces to find the best memoryless policy in partially observable markov decision processes. In *Proc. of the Fifteenth International Conference on Machine Learning.*

Meuleau, N., Kim, K.-E., Kaelbling, L. P., and Cassandra, A. R. (1999). Solving POMDPs by searching the space of finite policies. In *Proceedings of the Fifteenth Conference on Uncertainty in Artificial Intelligence*, pages 417–426, San Francisco, CA. Morgan Kaufmann.

Meuleau, N., Peshkin, L., and Kim, K.-E. (2001). Exploration in gradient-based reinforcement learning. Technical report, AI Memo 2001-003, MIT, April 2001.

Minsky, M. (1986). *The Society of Mind.* New York: Simon and Schuster.

Nakahara, H., Doya, K., and Hikosaka, O. (2001). Parallel Cortico-Basal Ganglia Mechanisms for Acquisition and Execution of Visuo-Motor Sequences – A Computational Approach. *Journal of Cognitive Neuroscience*, 13:626–647.

Nolfi, S. and Floreano, D. (2000). *Evolutionary Robotics, The Biology, Intelligence, and Technology of Self-Organizing Machines.* MIT Press.

Peshkin, L., Meuleau, N., and Kaelbling, L. P. (1999). Learning policies with external memory. In *Proc. of the Sixteenth International Conference on Machine Learning*, pages 307–314.

Precup, D., Sutton, R. S., and Dasgupta, S. (2001). Off-policy temporal-difference learning with function approximation. In *Proc. of the Eighteenth International Conference on Machine Learning.*

Precup, D., Sutton, R. S., and Singh, S. (2000). Eligibility traces for off-policy policy evaluation. In *Proc. of the Seventeenth International Conference on Machine Learning*, pages 759–766. Morgan Kaufmann.

Rubinstein, R. (1981). *Simulation and the Monte Carlo method.* New York: Wiley.

Rummery, G. and Niranjan, M. (1994). On-line Q-learning using connectionist systems. Technical report, Technical Report CUED/F-INFENG/TR 166, Engineering Department Cambridge University.

Singh, S. P. (1992). Transfer of learning by composing solution of elemental sequential tasks. *Machine Learning*, 8:323–340.

Sutton, R. S. and Barto, A. G. (1998). *Reinforcement Learning.* MIT Press/Bradford Books.

Sutton, R. S., McAllester, D., Singh, S., and Mansour, Y. (2000). Policy gradient methods for reinforcement learning with function approximation. In *Advances in Neural Information Processing Systems 12*, pages 1057–1063. MIT Press.

Watkins, C. J. C. H. and Dayan, P. (1992). Q-learning. *Machine Learning*, pages 279–292.

Williams, R. J. (1992). Simple statistical gradient-following algorithms for connectionist reinforcement learning. *Machine Learning*, 8:229–256.

Estimating Future Reward in Reinforcement Learning Animats using Associative Learning

George D. Konidaris **Gillian M. Hayes**
Institute of Perception, Action and Behaviour
School of Informatics, University of Edinburgh
James Clerk Maxwell Building, King's Buildings
Mayfield Road, Edinburgh EH9 3JZ
Scotland, UK

Abstract

We introduce a model of animat reinforcement learning where an associative learning element is placed between the animat's reinforcement learning component and its internal reinforcement functions. This element forms an impression of the sensory stimuli present near a goal and uses it to make an initial estimate of the value of newly discovered state-action pairs in tasks where reward is necessarily delayed.

We then describe the implementation of Peaches 'n Cream, a simulated robot that implements one version of the model where behaviour-based reinforcement learning is used in a puck foraging task. The results suggest that once given an initial simple task to learn from, the resulting associations significantly speed up learning in a later, more complex task, and provide further evidence that other learning methods may be used in conjunction with reinforcement learning to make it feasible for situated agents.

1. Introduction

Reinforcement learning provides an intuitively appealing animat learning model because it is well suited to the problems animats face (Sutton, 1990), and because it provides a principled way to build animats that function according to a set of internal drives.

However, when such drives are expressed as reward functions, they often lead to situations where a long sequence of unrewarded actions culminates in a single reward obtained by the satisfaction of one of the drives. Learning when reward is thus delayed is difficult both in terms of finding optimal solutions and in terms of finding any solution at all, because the reward signal is not sufficiently informative to guide the agent to a solution. This creates a problem for animats in domains where the agent does not have sufficient time to fully explore the problem space in search of a solution (e.g., where the animat's energy levels are low and it needs to find a battery) and is therefore (at least initially) concerned with quickly finding any solution at all.

One way to alleviate this problem is to reduce the state space by making it as task-relevant as possible (e.g., Mahadevan and Connell (1992)). However, some learning problems are intrinsically difficult, and require a more informative reward signal to guide the initial search for a first solution, after which rapid propagation of reward through the state space (either using traces or some form of offline learning) can guide the agent in optimising it.

In this paper we introduce the use of an associative learning element that acts between the animat's internal drives and its reinforcement learning component. This element learns to associate environmental stimuli with estimated reward when performing an initial, relatively easy task. In later tasks, when a new state is encountered, its value can then be estimated based on its similarity to prior reward states, and the animat has effectively learned a heuristic that it can use to guide search. Animats built with this model are therefore first exposed to a task where they can learn the correct associations, and thereafter may be exposed to more difficult tasks where the associative element can provide more information than a single reinforcement signal alone. The model is thus aimed at increasing the autonomy of animats that use reinforcement learning by freeing the designer from having to specify more explicit reinforcement functions.

We then present the implementation of a control system for a simulated robot that performs reinforcement learning in a simulated arena using one implementation of the associative element described here. By comparing the results obtained by Peaches 'n Cream in a complex environment after having learned in a more simple one to those obtained without the use of an associative learning mechanism, we show that such a mechanism is capable of speeding up learning in complex tasks.

2. Background

Exploration in reinforcement learning is usually performed using ϵ-greedy methods, where the animat makes a random move some fraction (ϵ) of the time. ϵ-greedy methods guarantee asymptotic coverage of the state space, and are usually combined with the assignment of optimistic initial values to states to encourage exploration. The agent thus starts with no knowledge of the state space, and is drawn to the unexplored parts of the space by their higher values (Sutton and Barto, 1998), making random choices every now and then. The optimistic estimates are gradually replaced as the animat learns the actual value of each state.

However, Konidaris and Hayes (2004) make the claim that ϵ-greedy methods (along with all other methods that emphasise asymptotic coverage) are not appropriate for situated agents that must act as best they can given the knowledge, time, and resources they have available. A hungry animat should not make random decisions every now and then, in the hope of stumbling across food. Rather, it should find the food it knows where to find, and explore when it isn't hungry. We therefore separate exploration in situated reinforcement learning into two separate concerns: first, how to find some initial solution path, and second, how to cover the state space in order to find better solutions than the first. This paper deals with the former problem; Konidaris and Hayes (2004) describe an exploration drive that deals with the latter.

When an animat is trying to find any solution at all, it is essentially performing something akin to a search through the state space where it may only keep one state "open" at a time. The use of optimistic initial values (the most common solution method to this problem) is thus essentially equivalent to a blind or depth-first search because all states estimations are equivalently optimistic.

Matarić (1994) suggests the use of "progress indicators" or some other form of more informative reinforcement signals to help guide search. The major drawback to this is that it requires extra engineering effort, and does not map well to the analogy of reinforcement functions as internal drives. The balance to be struck here is between a priori knowledge making learning easier and principled design making learning harder. However, there is a middle ground: the animat may be allowed to learn its own richer reward signal from experience without having to have it specified in advance.

3. Estimating Initial Values using Associative Learning

We propose that instead of having a very informative but difficult to engineer reinforcement signal, animats should be able to learn which sensory signals indicate the nearby presence of a reward state from experience. This sensory signature can then be used as a heuristic measure, with states that have similar signatures to states that have in the past contained a reward being given higher exploration values than states that do not.

An agent using this mechanism would start off without any knowledge, and therefore perform as badly as agents without any associations, but would learn them over its lifetime in perhaps many different related problem instances. Such a system would be an instance of layered learning (Stone and Veloso, 2000), where associative learning is used to make reinforcement learning more efficient.

A good example here is the use of smell by some agents to find food. It is not necessarily true that this knowledge is built in – humans did not evolve specifically to like the smell of coffee, for example, but have learned to associate it with the act of drinking coffee. Once this has been learned, office workers are more likely to expect to find a coffee dispensing machine in corridors that smell of coffee than in those that do not. This knowledge does not have to be obtained through inference; simple association will do.

Learning would thus proceed as follows. First, the agent is placed in an environment that is either relatively simple, or where it has sufficient time to properly explore. Whenever the agent encounters a goal state, it makes a note of the sensory stimuli that are especially strong there and thus learns to associate the presence of those stimuli with reward. Later, when encountering new states, the relevant stimuli levels are used to determine how similar the new state is to the expected goal state characteristics, and this is in turn used as a heuristic estimate of the value of the state.

There are three specific issues to be dealt with here: determining which stimuli are relevant, determining some form of relative weighting for them (since some are more important then others), and then using this weighting to estimate initial reward. The exact mechanism used to perform the associative learning will in general depend on the characteristics of the relevant stimuli and the properties of the task. In the following section, we describe an example task and outline the implementation of an associative learning mechanism that is appropriate to it.

4. Puck Foraging in a Simulated Arena

In order to test the model proposed above, we developed a working system where a plain reinforcement learning animat could be compared against one that made use of associative learning to estimate rewards.

4.1 The Task

The experimental task chosen was a simulated version (in Webots 4 (Webots, 2003)) of the puck-foraging task used to validate the behaviour-based reinforcement learning model introduced in Konidaris and Hayes (2004).

A mobile robot was placed in an arena constructed from orthogonal white walls, and was driven by three internal needs: the need to find food (represented by a green cylindrical puck), the need to explore, and the need to return to its home area. These needs were governed by a circadian cycle that started the robot off as hungry, then switched to exploration mode when food was found, and then switched to homing mode at the end of a predetermined time (the cycle length). If the robot failed to find food by the end of the cycle length, it proceeded directly to the homing phase without exploring. Upon returning home, the robot started the cycle again, and ran for ten cycles in total.

The robot was required to perform its task by following the walls of the arena. At each wall (either a left or a right wall – we omit corridors here for simplicity) the robot could at each end go straight or turn against the wall after it has ended, giving four possible actions for each wall. In addition, walls that were connected to other orthogonal walls to form internal corners did not permit turns. The robot thus also had to discover which turns were not permitted and avoid them.

The robot's aim was to maximize the reward corresponding to its currently active drive. For the homing and feeding phases, it received a reward of 1200 points for returning home and finding food, respectively, and lost 1 point every half second it failed to. In the exploration phase, the robot received a reward of between −1200 and 1200 points for each landmark-action pair, depending on the ratio of the number of times that pair had been taken previously against the average number of pair experiences, with rarer transitions receiving higher rewards.

Figure 1: The First Arena

The two models were first tested in a relatively simple environment, shown in Figure 1, which aimed to provide a fairly simple training case for the associative learner. Peaches 'n Cream is in the bottom left corner, which serves as its home area, and the puck can be seen in the bottom right corner. Although this arena does not look imposing, it was nearly as complex (at 20 distinct walls) as the most complex arena used in Konidaris and Hayes (2004). However, in this work we allowed the robot fifteen minutes of simulated time per cycle as opposed to three minutes of real time[1].

Figure 2: The Second Arena

The models were then tested in a much more difficult environment, shown in Figure 2, with 44 distinct walls. This environment was intended to provide a difficult problem instance where the associations learned in the first environment could guide the robot's exploration. In addition, the cycle length was reduced to ten simulated minutes, which significantly increased the difficulty of the problem since the most direct route from the robot's home area to the puck took approximately eight simulated minutes. Peaches 'n Cream started out in the bottom right corner of the second arena (within the box that serves as its home area) with the food puck in the bottom left.

4.2 The Robot

The robot used in the tests, Peaches 'n Cream, was a simulated standard khepera robot with a colour vision turret. Peaches used a ported version of the reinforcement learning control software written for Dangerous Beans, and detailed in Konidaris (2003) and Konidaris and Hayes (2004). We therefore provide only a high level outline of it here, along with a description of the associative learning modules added to it.

The architecture of the control system can be viewed as having three distinct control layers. The first layer

[1]Note that the simulated time and real time measures are not directly comparable. In our experience the model we used in simulation was approximately half as fast as the real robot described in Konidaris and Hayes (2004).

performs obstacle avoidance and wall-following using a standard set of behaviors. The second performs landmark detection and distributed map building. The distributed topological map is a graph where a *place behavior* or node is allocated to each distinct landmark the animat has encountered so far, and with connections between place nodes indicating that the animat can move directly between them with some simple (recorded) sequence of behaviors. Each place behavior is able to detect when it should be active, and the place behavior that is active following it, as well as the decisions made to move there, and form connections to them. Another behavior is responsible for merging place behaviors that represent the same state and creating new ones when a wall is encountered for which no suitable behavior exists. This layer is equivalent to the distributed map-building system developed by Matarić and Brooks (1990). An example map visualisation taken from the first arena is shown in Figure 3.

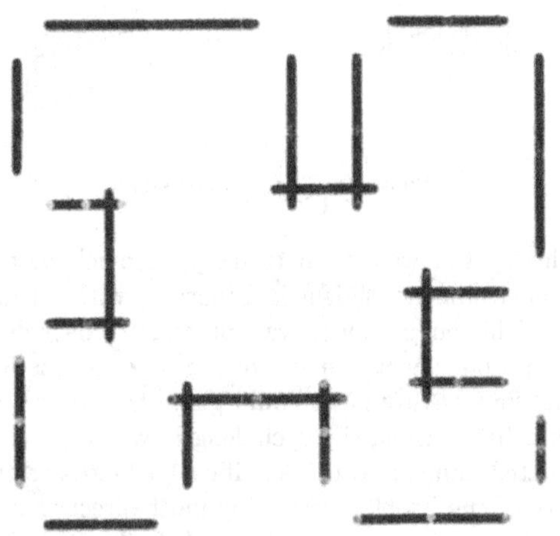

Figure 3: A Sample Topological Map of the First Arena

The final control layer performs reinforcement learning over the topological map, with nodes in the map acting as states and edges as actions in the reinforcement learning problem. Each place behaviour maintains a set of action values, and updates them using the TD(0) update rule and a full backup (Sutton and Barto, 1998). Updates occur all the time, at all behaviors, but the currently active place behavior is able to request the action with the highest learned action value for execution. We refer to the resulting structure as a *reinforcement learning complex* rather than a reinforcement value table because of its dynamic structure. Note that there are actually three distinct complexes (one for each drive) embedded in the same map.

In the standard reinforcement learning model, actions that have not yet been executed are assigned an estimated value of 300 for the homing and reward drives, and an exact value of 1200 for the exploration drive. This optimistic initial value encourages the agent to try these transitions, but it does so uniformly, resulting in an unguided search of the state space until a reward point has been found.

The version of Peaches 'n Cream with the associative learning model was given a set of five extra sensors, each returning a real number between 0 and 1, and each subject to 10% uniformly distributed noise. The first sensor returned a randomly generated value, while the third and fifth returned a constant and cycling value, respectively. The second and fourth sensors returned a measure that represented the straight line distance to the homing and puck points, respectively. The sensor readings were scaled so that they read as 1 at the goal point (the middle of the homing area or the puck), and 0 at a distance of about 700 units away (with the first and second arenas measuring 500 and 1100 units in both directions, respectively). They thus acted as smell sensors with a signal emanating from the goal but decreasing with distance, although they could be interpreted as any abstract distance heuristic.

The associative learner collected samples of each sensor reading every half second, and every time a puck or homing reward was obtained. These samples were used to estimate the mean sensor reading and variance for each sensor in puck reward states, home reward states, and over the arena as a whole, with the intention of using sensors that show a markedly different distribution near goal states as indicators. In order to determine which sensors were markedly different, a t-test was performed, measuring the difference between the reward state distributions and the overall arena distribution. Using the resulting t_i value for each sensor s_i, a weight w_i was calculated as follows:

$$w_i = \begin{cases} 0 & \text{if } t_i < 2.5 \\ \frac{t_i - 2.5}{t_{tot}} & \text{otherwise} \end{cases}$$

where t_{tot} was the sum of all of the t_i values of at least 2.5. The weights thus sum to 1, and weight each factor in proportion to its difference from the overall arena distribution. The cutoff value of $t = 2.5$ was chosen as a cutoff that gives a low probability of mistaken significance ($p < 0.05$) for a relatively small sample size. The similarity factor k for a particular state was then estimated as:

$$k = \sum_i w_i d(s_i, e_i)$$

where s_i is the reading of the ith sensor and e_i is its expected reward mean. The function $d(s_i, e_i)$ simply returns the inverse of the fraction of e_i that s_i is off by – for example, $d(s_i, e_i)$ returns 0.75 if e_i was 0.8 and s_i was either 0.6 or 1.0. The initial transition value was then estimated by the reinforcement learning system as:

$$E_R = 300 + 500k$$

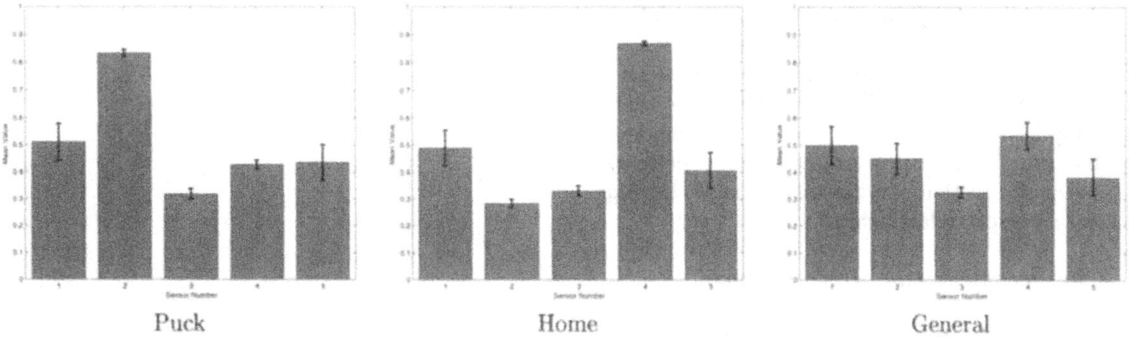

Figure 4: Averaged Sensor Means Learned from the First Arena

This guarantees every transition an optimistic initial value of at least 300 (because k is always positive) but allows some (those that are similar to goal states) to receive estimates that are proportionally higher, although never quite as high as an actual reward state[2].

5. Results

In this section we present the results obtained in the task described above. We first cover the associations learned by the associative learning version of Peaches 'n Cream in the first arena, and then compare the performance of associative robots using these learned values to that obtained by reinforcement learning robots without association.

5.1 The First Arena: Associations

The means of the sensor readings obtained by ten robots at the puck, the homing area, and the first arena generally are shown in Figure 4. In the puck goal case the sensor corresponding to the puck distance sensor (sensor 2) was consistently very high, and similarly in the homing goal case the homing distance sensor was consistently very high. The other sensors do not differ significantly from their values in the rest of the arena.

Figure 5 shows the mean t-values for the puck goal. Here, only the puck distance sensor has a sufficiently high t value to be above the cutoff, although in some of the individual robots the homing distance sensor with a low mean value was just above the cutoff of 2.5. Figure 6 shows the t values for the homing case, where both the homing distance sensor and the puck distance sensor are above the threshold and thus considered significant, although again in some individual robots the puck distance sensor fell just below the significance threshold.

Both of these extra sensor significances make sense since the goals were fixed relative to each other and it is

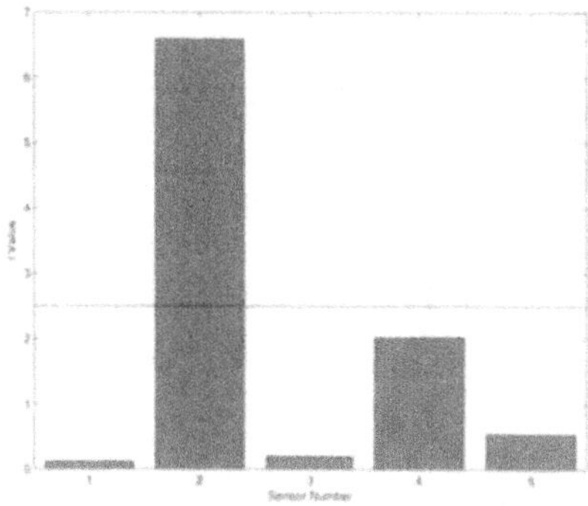

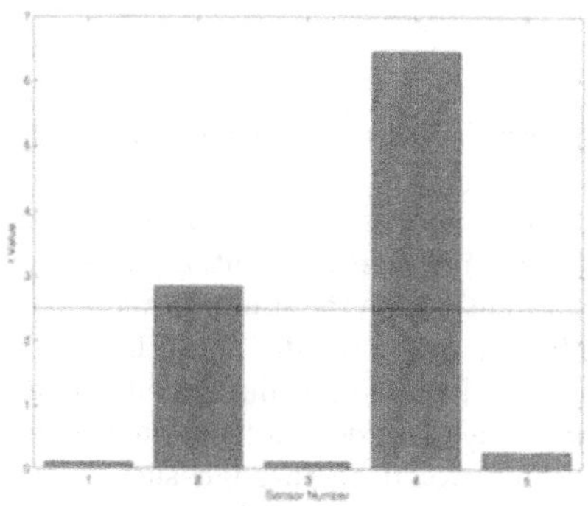

Figure 6: Home t Values

therefore reasonable for the robots to associate fixed (in this case low) values from these sensors with reaching the goal. Note that the puck goal association with the homing distance sensor is correctly lower than that of the homing goal association with the puck distance sensor because the two places where a puck reward could be obtained were further apart than the two places where the robot could enter the home arena, and were therefore likely to vary more widely.

Because each weight is proportional to the sensor's t value's height above the threshold, the alternate sen-

[2]Note that all of the constants presented in this section were determined by trial and error, guided by the desire to have a puck reward just too far for the robot to reach in time to have a value of zero.

sors were given low weightings by the associative learner and thus did not adversely affect performance. We can therefore conclude that the associative learning element was able to correctly and reliably learn the associative relationship between the two distance sensors and their goals, but that in general several learning experiences may be required so that the animat can avoid learning coincidental sensory relationships.

5.2 The Second Arena: Performance

Figure 7 shows the average puck reward against cycle number for both models over ten test runs in Arena 2. The associative learning robots were given the means and variances outlined in the previous section as a priori knowledge.

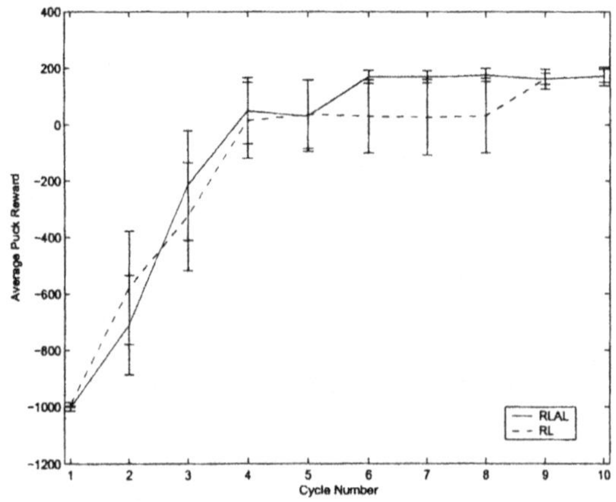

Figure 7: Average Puck Reward over Time

In Figure 7 the associative learning animats appear to be able to find the puck initially slightly faster and more consistently than the standard reinforcement learning agents. However, the difference here is hard to discern because often the standard reinforcement learning robots would explore deep into the maze, and when returning home would go through such long paths that they stumbled across the puck on the way. This is the way that most of the pucks through the runs were discovered, and it skews the puck reward data.

This is illustrated by Figures 8 and 9, which show two representative robot paths, with the position of the robot starting off black and fading over time. Figure 8 shows the route an associative learning robot took back home once it had reached the end of a cycle length, but before it had discovered a robust way home. Note that it starts out in the bottom corner near the puck and proceeds to slowly but definitely move toward the home area in the bottom right corner.

In Figure 9, however, it is clear that the robot is essentially making random choices without repetitions until it arrives back at the home arena, thus covering virtually

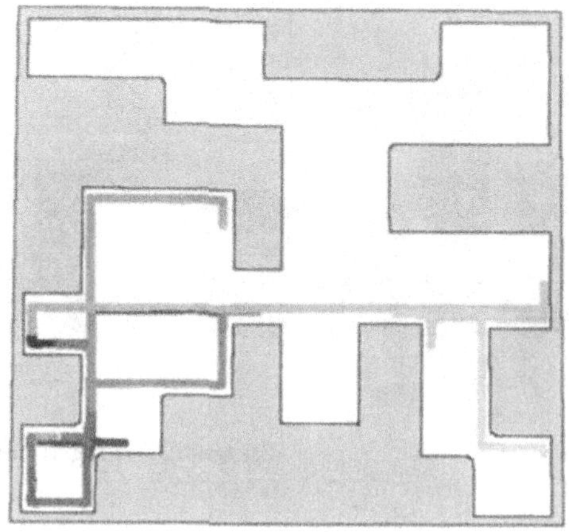

Figure 8: An Associative Robot Returning Home

the entire arena and discovering the puck. In particular, it comes very close to returning home fairly early on, but then turns around and goes the other way, before eventually wandering back home.

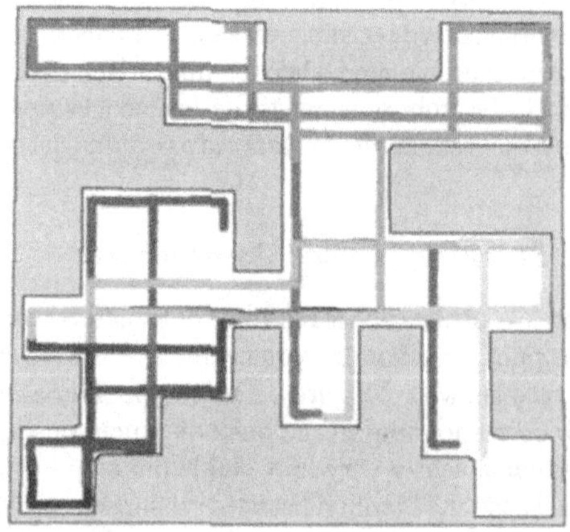

Figure 9: A Non-Associative Robot Returning Home

Figure 10, showing the average home reward against cycle number over the same ten runs, shows the difference between the two models clearly. The associative learners are able to return home much more quickly than the standard reinforcement learners, at least initially. The associative learners have higher average reward values and smaller error bars up until the 4th cycle, when the two graphs level out. Thus the associative element was capable of speeding up learning initially even though eventually the systematic but unguided standard reinforcement learning agent caught up.

Although the differences between the mean home reward values for each cycle as shown in Figure 10 are fairly large, they cannot be used directly to determine

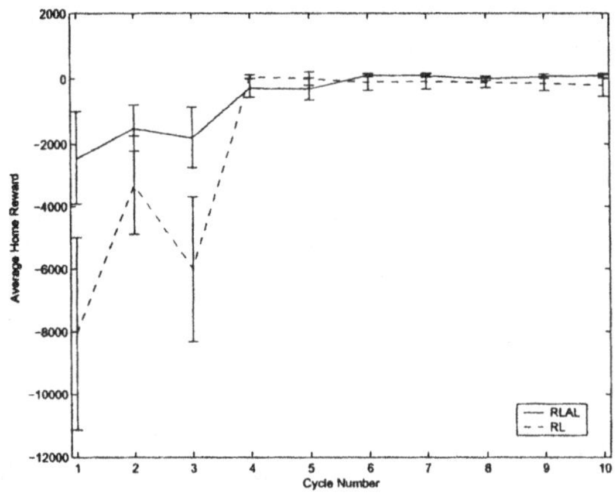

Figure 10: Average Homing Reward over Time

the significance of the results because the structure of the task means that the animat's performance in one cycle is heavily biased by its performance in the cycles before it. Instead, we use the average total homing rewards over time for comparison, which are shown in Figure 11. A t-test indicates that this difference is statistically significant ($p < 0.05$), and this, combined with the significant qualitative differences in search behavior shown in Figures 8 and 9, demonstrates clearly that the associative learning model performs better than the standard reinforcement learning model.

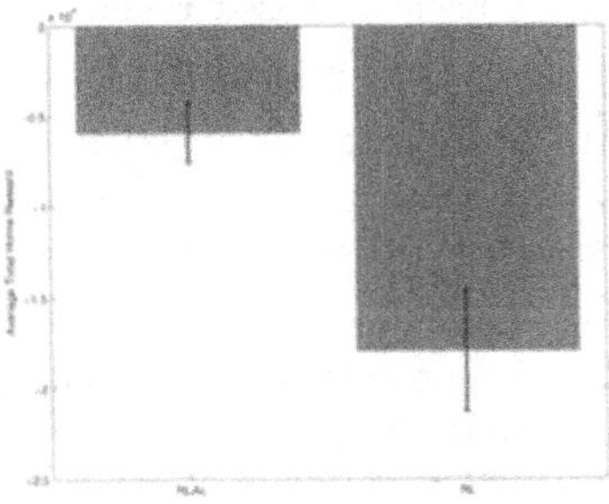

Figure 11: Average Total Home Reward

6. Discussion

6.1 Significance

The results and model presented in this paper are significant for two reasons. The first is that they suggest that reinforcement learning animats may be able to use their own experiences to learn their own, more informative, initial reinforcement learning estimates without having to have them built in. This has the potential to increase the autonomy of animats built using reinforcement learning because it removes the need for highly informative but hand-crafted reinforcement signals.

This is superior to simply adding the signal to the reinforcement learning state space for two reasons. First, even an impoverished representation of such a signal would significantly increase the size of the state space – the addition of only one bit of extra information would double it – and thereby impede the learning capabilities the agent. Second, such a signal simply does not belong in the state space, because it is not a degree of freedom in the state space – rather, its presence at certain states is highly correlated with a high return at those states. As such, many combinations of state and signal simply could never be experienced by the agent, and the ones that could would carry no more information as states than those without the signal. In other words, the signal adds no information to the state space, but it instead adds information to the reward landscape embedded in the state space. Thus, the system preserves the advantages of using reinforcement learning and provides a principled way to build value driven agents without losing the ability to use extra sensory cues to improve the agent's performance.

The second significant aspect of the research presented here is that it provides evidence for the claim made in Konidaris and Hayes (2004) that additional learning models may be added to an already layered reinforcement learning system in order to speed learning. Although there are several ways this could be done, the existence of one such model and the results given here prove that it is feasible in at least one interesting case, and strengthen the argument that layered learning can bring substantial benefits to situated agents.

6.2 Limitations

There are two primary limitations with the work presented here. The first is that it reports the results of the use of the model given here for only one task (puck foraging and homing in a simulated arena) and in only one trial instance (the second arena). In particular, a task more suited to highlighting the difference between an associative and standard reinforcement learning agent should provide clearer and more decisive results. There are also several aspects of the work presented here that could be expanded upon or improved. For example, the arena used does not contain a local maximum "trap" for the associative learning agent. Such a trap would likely be completely explored by the agent before it moved on, thus slowing it down. This problem should not occur in real smell sensors because they work through air flow rather than distance, but in general no heuristic method will be without flaws.

Another potentially useful aspect of the model that has not been tested here is the notion that it should work with a discrete sensor or one that is not strictly monotonic with respect to distance or gradient based, provided the sensor is relatively consistent and has some unusual value near the goal state.

However, although we cannot therefore claim that in general this methodology will always work, these results do show that it is capable of speeding learning in a fairly difficult problem and that the approach is promising.

The second primary limitation is that the task was carried out in a simulator, rather than on a real robot. In this case, the simulator allowed for the use of virtual sensors for association, and rapid experiments in larger arenas than would have been feasible had the arenas required physical construction. We feel that this justifies its use given the fact that the reinforcement learning model and task have been previously validated on a real robot (Konidaris and Hayes, 2004), and because smell sensors similar to those that we require already exist (e.g., Duckett *et al.* (2001)).

6.3 Implications

The results given here imply that informative and carefully structured reinforcement signals may not be necessary when designing animats to handle complex tasks, provided those animats are allowed to go through a training stage first. This is loosely analogous to the process of shaping (Dorigo and Colombetti, 1998).

One of the negative implications of this is that if such an animat first encounters a poorly or even maliciously constructed environment, it may learn incorrect associations which could hinder its ability to solve later tasks. This may not be too serious since the animat may be able to correct its associations over time, but in some cases it may perform badly in subsequent tasks.

The use of this model also provides further evidence for the suggestion that the form and function of representation is tied to its use in learning models, and that the layering of learning models may create complex representational structures. In the model given here, the associative weights are meaningless outside of the associative learner, and similarly, the associative learner's role is difficult to explain outside of the context of the reinforcement learning complex.

7. Conclusion

This paper has introduced a model of animat reinforcement learning that includes the use of an associative learning element. The associative element learns to estimate the initial value of a state based on its similarity to the sensory characteristics experienced at previous goal states, and thereby provides a heuristic which guides the animat's search for an initial solution. The results of an experimental scenario outlined here show that this model is capable of speeding up reinforcement learning in a complex task, once the animat has been exposed to a simpler training task. These results go some way toward relieving animat designers of the difficult task of hand-crafted informative reinforcement signals for problems in which reward is necessarily delayed.

Acknowledgments

We would like to thank the Institute for Perception, Action and Behaviour at the University of Edinburgh for the use of its resources, and two anonymous referees for their useful comments. George Konidaris is employed under the HYDRA project (EU grant IST 2001 33060).

References

Dorigo, M. and Colombetti, M. (1998). *Robot Shaping: An Experiment in Behavior Engineering*. MIT Press/Bradford Books.

Duckett, T., Axelsson, M., and Saffiotti, A. (2001). Learning to locate an odour source with a mobile robot. In *Proceedings of the IEEE International Conference on Robotics and Automation (ICRA 2001)*, Seoul, Korea.

Konidaris, G. (2003). *Behaviour-Based Reinforcement Learning*. Master's thesis, School of Informatics, University of Edinburgh.

Konidaris, G. and Hayes, G. (2004). Behavior-Based Reinforcement Learning. Submitted to *Adaptive Behavior.*

Mahadevan, S. and Connell, J. (1992). Automatic programming of behavior-based robots using reinforcement learning. *Artificial Intelligence*, **55**(2–3).

Matarić, M. (1994). Reward functions for accelerated learning. In *Proceedings of the Eleventh International Conference on Machine Learning*, pages 181–189.

Matarić, M. and Brooks, R. (1990). Learning a distributed map representation based on navigation behaviors. In R. Brooks, editor, *Cambrian Intelligence: the Early History of the New AI*. The MIT Press, Cambridge, Massachusetts.

Stone, P. and Veloso, M. (2000). Layered learning. In *Proceedings of the 11th European Conference on Machine Learning*, pages 369–381, Barcelona, Spain. Springer, Berlin.

Sutton, R. (1990). Reinforcement learning architectures for animats. In J. Meyer and S. Wilson, editors, *From Animals to Animats: Proceedings of the International Conference on Simulation of Adaptive Behavior*. MIT Press.

Sutton, R. and Barto, A. (1998). *Reinforcement Learning: An Introduction*. MIT Press, Cambridge, MA.

Webots (2003). http://www.cyberbotics.com. Commercial Mobile Robot Simulation Software.

Hebbian reinforcement learning in a modular dynamic network

Emmanuel Daucé
UMR Movement and Perception
Faculty of sport sciences
University of the Mediterrannean
163 avenue de Luminy, CP 910
13288 Marseille cedex 9
France
dauce@esm2.imt-mrs.fr

Abstract

We present a multi-population dynamic neural network model with binary activation and a random interaction pattern. The weights parameters have been specified in order to distinguish excitatory populations from inhibitory populations. Under specific parameters, we design functional modules composed of two populations, one of excitatory neurons, one of inhibitory neurons. Such modules are found to display a weak chaotic activity, and to react toward incoming stimulations with increasing synchronization. We also present the design of topologically structured neural maps. We then combine such modules for the design of a perception/action network composed of one sensory module and two concurrent motor modules. Such a network is coupled with the dynamics of an inverted pendulum, showing spontaneous phase transitions toward various attractors, each attractor corresponding to a particular structural coupling within the agent/environment system. This spontaneous versatility is then exploited in a reinforcement learning paradigm where two separate reinforcement path are defined, one for the positive rewards, the other for the negative rewards. The learning experiment shows a fast adaptation to the constraints, followed by a slower phase where the behavior is improved. No degradation of the behavior is found for continuing learning, i.e. the learned behavior is preserved for long lasting time.

1. Introduction

Brain structures at different scales are in majority composed of self-feeding connections, and the connections between different structures/areas are always reciprocal (see for instance (Hupé et al., 1998) for the visual cortex). This massive recurrence suggests a major role of self-feeding dynamics in the processes of perceiving, acting, and learning, and more generally in maintaining the organism alive. Lot of simulation and theoretical work has been done about the spontaneous dynamic properties of massively connected recurrent neural networks (Sompolinsky et al., 1988, Doyon et al., 1993, Cessac, 1995, Ben-Yishai et al., 1995, Hansel and Sompolinsky, 1996, Camperi and Wang, 1998, Wang, 1999, Compte et al., 2000, Moynot and Samuelides, 2002, Izhikevich, 2003, Daucé, 2004), whereas lot remains to be done on the question of the *learning* processes taking place in such networks (Hertz and Prugel-Bennett, 1996, Daucé et al., 1998, Daucé et al., 2002).

Reinforcement learning offers a relevant framework for the study of such learning processes in large recurrent networks. Reinforcement learning is a well-known semi-supervised learning paradigm, where the environment occasionally sends to the agent reinforcement signals, which may be positive (reward) or negative (penalty). The aim of a reinforcement learning algorithm is to modify the agent's behavior in order to maximise the reward (and/or to minimize the penalty). This paradigm is of course well adapted in case an agent has to explore an unknown environment, where foes and traps have to be avoided and feeding sources and strategies have to be found. The agent is supposed to adapt by its own means, i.e. one can not select agents over generations, like in genetic algorithms. Every reinforcement learning method needs an *exploratory* process in order to test and experiment various structural couplings and a *selection* process in order to retain the most appropriate ones. For instance, in statistical reinforcement methods(Sutton, 1988, Watkins and Dayan, 1992), the transitions probabilities are initially random and uniform in order to explore the largest scope of interactions, so that learning tends to reduce the uncertainty, through the selection of the most appropriate transitions.

The aim of this paper is to use some of the native dynamical properties of large networks, such as chaos, attractor switching and synchronization, as the core of the exploratory process. We also want to bring reinforcement methods one step closer to biological plausibility.

The paper is organized as follows. We first present the formalism of the multi-populations neural model, composed of several populations, which can either be purely excitatory or purely inhibitory (section 2.). We then present in section 3. some simulations showing the spontaneous dynamics of different "modules" composed of excitatory and inhibitory populations. Section 4. presents the controller architecture. This controller is composed of 3 functional modules : one sensory module an two motor modules. The last section presents our Hebbian reinforcement method, through different protocols (closed loop and on-line protocols), and simulations giving the convergence properties under various parameter settings.

2. A dynamic network model

The full model formalism is given here. We tried by the way to make this paper self-explanatory and testable[1].

2.1 Model setting

Our model is a discrete-time dynamical system with parallel update, where the state of the system at time t both depends on the previous state of the system $\mathbf{x}(t-1)$ and on the input $\mathbf{u}(t-1)$, i.e

$$\mathbf{x}(t) = f(\mathbf{x}(t-1), \mathbf{u}(t-1)) \tag{1}$$

where $\mathbf{x}$ is a state array, $\mathbf{u}$ is an input array and f is a global (nonlinear) operator. One can notice that our system is deterministic as soon as the input signal is set according to a deterministic process.

A network is defined as a pool of P interacting populations of neurons, of respective sizes $N^{(1)}, ..., N^{(P)}$. The global number of neurons is $N = \sum_{p=1}^{P} N^{(p)}$. The synaptic weights from population q toward population p are stored in a matrix $\mathbf{J}^{(pq)}$ of size $N^{(p)} \times N^{(q)}$ (possibly sparse). The state vector of population p at time t is $\mathbf{x}^{(p)}(t)$, of size $N^{(p)}$. The initial conditions $x_i^{(p)}(0)$ are set according to a random draw in $\{0,1\}$.

At each time step $t \geq 1$, $\forall (p,q) \in \{1,..,P\}^2$,

$$\mathbf{h}^{(pq)}(t) = \mathbf{J}^{(pq)}\mathbf{x}^{(q)}(t-1) \tag{2}$$

is the *local field* array of population q toward population p.

We also consider spatio-temporal input signals $\mathbf{u}^{(p)} = \{\mathbf{u}^{(p)}(t)\}_{t=1..+\infty}$, where $\mathbf{u}^{(p)}(t)$ is an input array of size $N^{(p)}$. The input $\mathbf{u}^{(p)}(t)$ acts like a bias on each neuron[2]. Then, the global equation of the dynamics is :

$$\forall t \geq 1, \forall p \in \{1, ..., P\}$$
$$\mathbf{x}^{(p)}(t) = H\left(-\theta^{(p)} + \mathbf{u}^{(p)}(t-1) + \sum_{q=1}^{P} \mathbf{h}^{(pq)}(t)\right) \tag{3}$$

The activation potential, which corresponds to a linear combination of afferent local fields and input minus activation threshold $\theta^{(p)}$, is a real valued array. The activation function H is the Heaviside function so that the neuron state is a binary array, which takes its values in $\{0,1\}$.

2.2 Weights setting

2.2.1 Random Recurrent Neural Networks

Our network belongs to the category of Random Recurrent Neural Networks (RRNNs), so that the weights obey to a random draw. The principal consequence of this setting is the "almost sure" non-symmetry of the connectivity pattern, so that one can not ensure the convergence of the dynamics toward a fixed point attractor. Autonomous RRNN's (i.e. $\forall t, \mathbf{u}(t) = 0$ in eq.(1)) are discrete time dynamical systems, that can for instance display a generic quasi-periodicity route to chaos while progressively increasing the gain of a continuous transfer function (Doyon et al., 1993).

Each family of weights $\mathbf{J}^{(pq)}$ (weights from population q toward population p) are set according to a uniform distribution. The main parameters are $\bar{J}^{(pq)}$ (weights mean), $\sigma_J^{(pq)}$ (weights standard deviation) and $\rho^{(pq)}$ (weights sparsity), such that the expectation of the weights is $E\left(J_{ij}^{(pq)}\right) \simeq \frac{\bar{J}^{(pq)}}{N^{(q)}}$ and the variance is $\text{var}\left(J_{ij}^{(pq)}\right) \simeq \frac{\left(\sigma_J^{(pq)}\right)^2}{N^{(q)}}$. The precise weights settings are given in Annexe A.

The activation threshold $\theta^{(p)}$ have scalar values (they are identical for every neuron of a given population).

2.2.2 Biological constraints

In this paper, in order to remain coherent with elementary biological requirements, we rule the weights of a given population $q \in \{1, ..., P\}$ to be either purely excitatory or purely inhibitory. This constraint implies additional dependencies between $\bar{J}^{(pq)}$, $\sigma_J^{(pq)}$, $N^{(q)}$ and $\rho^{(pq)}$. The parameter $\rho^{(pq)}$ is derived from $\bar{J}^{(pq)}$ and $\sigma_J^{(pq)}$ such that :

- When $\bar{J}^{(pq)} > 0$, the weights lower bound is 0, such that every weight is positive or null.

[1] Matlab code can be obtained by simple demand at dauce@esm2.imt-mrs.fr.

[2] On the contrary to Hopfield system (Hopfield, 1982), the input doesn't correspond to the initial state $\mathbf{x}^{(p)}(0)$ of the network.

- When $\bar{J}^{(pq)} < 0$, the weights upper bound is 0, such that every weight is negative or null.

Details are in Annexe A.

3. Excitatory-inhibitory interactions

Most of the cortical an sub-cortical layers are composed of interacting populations of excitatory and inhibitory neurons. We present in this section a large scale structure of interaction that we further call a "module". In order to take into account some basic physiological features, we suppose that our two-populations module grossly models a cortical column, such that :

- local dynamics dominate incoming signals.
- excitatory neurons can receive and send signals, inhibitory neurons only act locally.
- excitatory neurons represent 70-90% of the total population.

The parameters setting globally defines the way the two populations interact, independently of the population sizes (i.e. the dynamic properties are invariant with the sizes, provided the sizes are larger enough, i.e. >50). Two different parameter settings are given in the following sections.

3.1 *Simple modules*

The excitatory population is population 1. The inhibitory population is population 2. Inputs are only displayed on the excitatory layer, i.e. $\forall i, \forall t, u_i^{(2)}(t) = 0$ (see eq.(3)). We take binary inputs such that $\forall i, \forall t, u_i^{(1)}(t) \in \{0, 1\}$. We set the population sizes to $N^{(1)} = 1000$, $N^{(2)} = 200$.

We define two global parameters k and d which help to define the interaction pattern. The way the two populations interact are characterized by the *asymmetry* k between excitatory and inhibitory influences. Parameter k acts on the mean value of weights distributions. The *eccentricity* d mostly acts on the scattering/standard deviation of the weights such that :

$$\bar{J} = \begin{pmatrix} \frac{1}{2} & -\frac{k}{2} \\ \frac{k}{2} & -\frac{k}{2} \end{pmatrix} \quad \sigma_J = \begin{pmatrix} \frac{1}{2d} & \frac{\sqrt{k}}{2d} \\ \frac{\sqrt{k}}{2d} & \frac{\sqrt{k}}{2d} \end{pmatrix}$$

The activation threshold relates to the individual excitabilities. The thresholds are small, so that a small excitation can initiate the neuron to spike. The thresholds are of the order of 1/10 of the expanse of the weights interval, i.e. $\theta^{(1)} = 0.1$, $\theta^{(2)} = 0.1k$.

We present on figure 1 the reaction of a network toward input presentation. Once the system defined, the dynamical system (2-3) is initialized with a binary random vector taking values in $\{0, 1\}$. The spontaneous activity (without input) is displayed from $t = 1$ to $t = 100$. This activity is weak, irregular, with a slight synchronization (synchronous bursts appear as small peaks on the mean activity). The synchronization is stronger on inhibitory population. Then, an input is sent on 15 excitatory neurons. This input appears as a black band on first population activity. The network then tends to become more synchronous with almost periodic bursts of activity. This synchrony appears to be stronger in the inhibitory population. When the input is removed, the network turns back to its initial spontaneous weak activity.

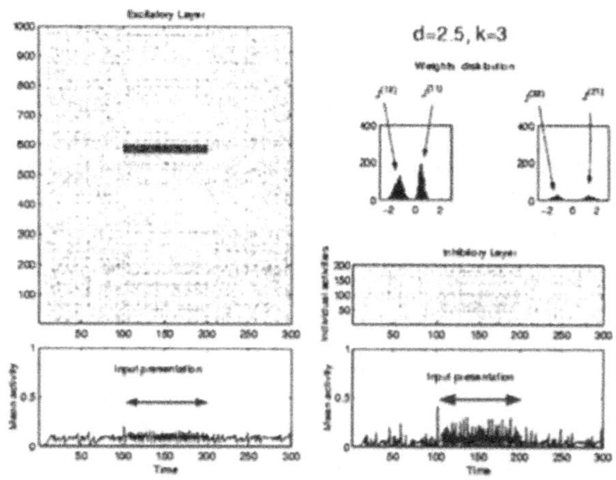

Figure 1: **Input presentation**. The network is composed of $N^{(1)} = 1000$ neurons and $N^{(2)} = 200$ neurons. The effective sparsity are $\rho^{(11)} = 0.8\%$, $\rho^{(12)} = 11\%$, $\rho^{(21)} = 2\%$, $\rho^{(22)} = 11\%$. An input is sent at time 100-200 on neurons 585-600. **Upper left** : global activity of the excitatory layer. **Lower left** : mean activity on excitatory layer. **Middle right** : global activity of the inhibitory layer. **Lower right** : mean activity on inhibitory layer. **Upper right insets** : distribution of the sum of incoming weights from excitatory and inhibitory populations, for 1000 excitatory neurons (left) and 200 inhibitory neurons (right).

The dynamical transition toward input presentation can be compared to a phase transition, i.e. the crossing of a bifurcation point in the parametric space. This kind of phase transition during input presentation has been observed for long time in neurobiology, see for instance (Skarda and Freeman, 1987), and also (Mac Leod and Laurent, 1996). More generally, synchronizing behaviors in unitary delays networks depend on the asymmetry k, i.e. inhibition has to dominate excitation for the network to produce synchrony. This point has still be noted in simulation works, see for instance (Bush and Sejnowski, 1996, Izhikevich, 2003).

3.2 Topologically organized modules

Topologically structured dynamical systems are defined by a *space* (or a *map*, which defines a distance), a *field*, possibly representing the states of uniformly distributed units, and a *process* by which those units interact. The nature of this interaction process is supposed to depend on the distance. Topologically structured dynamical systems have been introduced in neural modeling with the Neural Field of Amari (Amari, 1977).

Our maps are derived from Amari's neural map. We thus introduce a topology in the weight structure: We define a parameter $r^{(pq)}$ which represents the neighborhood density from population q toward population p (also called neighborhood radius). The weights are thus adapted according to the (normalized) distance between neuron i and neuron j. The smaller is $r^{(pq)}$, the tighter is the neighborhood. Close links are strongly enhanced (according to the initial random draw) whereas distant links fade to zero. One can notice that introducing a neighborhood factor tends to increase the sparsity of the weights. In concrete terms, every link such that $d_{ij}^{(pq)} > \pi r^{(pq)}$ is suppressed, i.e. set to zero, such that for $r^{(pq)} \leq 1$, the effective sparsity is $\rho^{(pq)'} = r^{(pq)}\rho^{(pq)}$. We take a simple $1D$ closed ring geometry. Map settings are in Annexe B.

Topological maps are used as perceptual modules in the next section. More details on random recurrent neural maps (as models of short term memory) can be found in (Daucé, 2004).

4. Perception-action network

The 2-population networks we have described in previous sections are now combined in order to build a perception/action network. A functional module is composed of 2 populations of neurons : one excitatory population and one inhibitory population. As previously said, a module can be seen as a rough approximation of a cortical column. A module can own a topology or not :

- topologically structured modules are associated with the perception processes;
- unstructured modules are associated with the action processes.

Designing a perception-action network also means to specify the environment through which the system interacts, and also to specify a task. As a first lookahead, we made the choice to minimize the environment complexity and apply our system on a very well known and documented task : the control of an inverted pendulum.

The environment is thus composed of 2 variables : the angular position ϑ and the angular velocity $\dot{\vartheta}$, and its dynamics is given by :

$$\begin{cases} \frac{d\vartheta}{dt} = \dot{\vartheta} \\ \frac{d\dot{\vartheta}}{dt} = g\sin(\vartheta) - 2\dot{\vartheta} + F \end{cases}$$

where $g = 9.81$ is the earth attraction and F is an external force (coming by the control system). We suppose that a measure of angular position ϑ can be done every 5 ms. The force F is also updated every 5 ms. The system stable fixed point is $\vartheta = \pi$. The task consists in maintaining the angular position within given bounds, say $[-\pi/15, \pi/15]$, whereas the natural pendulum tendency is to fall toward $\vartheta = \pi$.

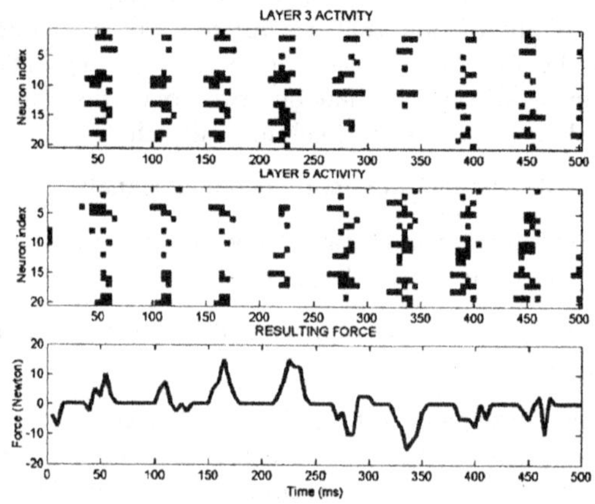

Figure 2: **Motor output production.** The upper figure gives an example of layer 3 activity, in case layer 3 is composed of 20 neurons. The middle figure presents layer 5 activity in the same conditions. In that simulation, the two layers are synchronized. The lower figure presents the resulting force, which comes from the difference of population activities. Time is in ms.

The control system is composed of 3 modules, each module owning two populations. In order to allow fast simulations, we limited the module sizes to 260 neurons (i.e. 200 excitatory neurons and 60 inhibitory neurons).

- The sensory module "S1" corresponds to populations 1 (excitatory) and 2 (inhibitory). Its topology is defined by $r^{(11)} = 0.2$, $r^{(12)} = 0.6$. The input is sent on layer 1. It activates 2% of the neuron around the reference position $\left\lfloor N^{(1)} \times \left(\frac{15\vartheta(t)}{2\pi} + \frac{1}{2}\right)\right\rfloor$.
- Two motor modules are defined. Module "M1" corresponds to populations 3 (exc.) and 4 (inh.), and module "M2" corresponds to populations 5 (exc.) and 6 (inh.). The force F is defined as the difference between the mean activities of populations 3 and 5, i.e.

$$F(t) = 50\left(m^{(3)}(t) - m^{(5)}(t)\right)$$

where $m^{(p)}(t)$ is the mean activity on layer p ($m^{(p)}(t) \in [0,1]$). Modules M1 and M2 act concurrently, one leading to the left, the other leading to the right. An example of network output is given on figure 2.

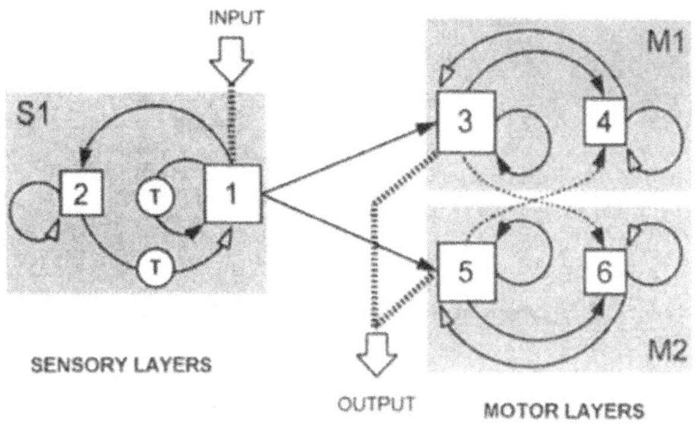

Figure 3: **Interconnection pattern for the perception-action network** The network is composed of 6 populations. Filled arrows represent excitatory connections. Unfilled arrows represent inhibitory connections. A functional module is composed of two strongly interconnected populations, one excitatory and one inhibitory. Sensory module **S1** is composed of populations 1 and 2, motor module **M1** of populations 3 and 4, motor module **M2** of populations 5 and 6. Module S1 owns a topology in its interconnection pattern (topological links are mentioned with symbol "T"). The 3 modules are interconnected through excitatory links (the inhibitory neurons only act locally). Module S1 sends (excitatory) signals toward modules M1 and M2. Modules M1 and M2 can inhibit each other through the excitation of their neighbor inhibitory layer (dotted links). Those lateral interactions are initially set to zero.

The global interconnection pattern is given on figure 3. The weights and threshold settings are comparable with the ones of sections 3.. Weights parameters are $k = 3$ and $d = 6$ with :

$$\bar{J} = \begin{pmatrix} \frac{1}{2} & -\frac{k}{2} & 0 & 0 & 0 & 0 \\ \frac{k}{2} & -\frac{k}{2} & 0 & 0 & 0 & 0 \\ \frac{1}{2} & 0 & \frac{1}{2} & -\frac{k}{2} & 0 & 0 \\ 0 & 0 & \frac{k}{2} & -\frac{k}{2} & 0 & 0 \\ \frac{1}{2} & 0 & 0 & 0 & \frac{1}{2} & -\frac{k}{2} \\ 0 & 0 & 0 & 0 & \frac{k}{2} & -\frac{k}{2} \end{pmatrix}$$

$$\sigma_J = \begin{pmatrix} \frac{1}{2d} & \frac{\sqrt{k}}{2d} & 0 & 0 & 0 & 0 \\ \frac{\sqrt{k}}{2d} & \frac{\sqrt{k}}{2d} & 0 & 0 & 0 & 0 \\ \frac{1}{2d} & 0 & \frac{1}{d} & \frac{\sqrt{k}}{d} & 0 & 0 \\ 0 & 0 & \frac{\sqrt{k}}{d} & \frac{\sqrt{k}}{d} & 0 & 0 \\ \frac{1}{2d} & 0 & 0 & 0 & \frac{1}{d} & \frac{\sqrt{k}}{d} \\ 0 & 0 & 0 & 0 & \frac{\sqrt{k}}{d} & \frac{\sqrt{k}}{d} \end{pmatrix}$$

We present on figure 4 the spontaneous behavior of the network/pendulum system. The initial pendulum position is randomly set in interval $[-\frac{\pi}{30}, \frac{\pi}{30}]$ rad, and the velocity is randomly set in interval $[-0.2, 0.2]$ rad/s. Some remarks can be made :

- Layer 1 activity directly adapts to the input, and develops a local "bundle" attractor in the vicinity of the stimulation. This attractor is driven by the input. Its radius of this is of the order of 10% of the map.

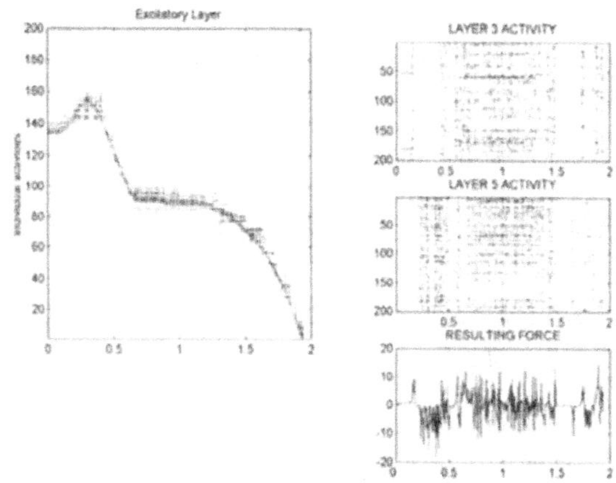

Figure 4: **Spontaneous dynamics of the network/pendulum system.** Left : layer 1 activity. The input (red trace) corresponds to the pendulum position which is measured every 5 ms (i.e. one time step corresponds to 5 ms). The pendulum overtakes the perceptual field around $t = 2$s. Upper right : layer 3 activity. Lower right : layer 5 activity. Middle right : network motor output, which is the difference between layer 3 mean activity and layer 5 mean activity. Time is in s.

- The activity of layers 3 and 5 organize under layer 1 influence. As in figure 1, They tend to display a synchronized activity when they get stimulated. The property of synchrony allows to reach a significant global activity, which defines the force to be sent to the output. The mean force amplitude is of the order of 5 N. The maximum force amplitude is of the order of 20 N.
- Even synchronized, the activity of layers 3 and 5 remains chaotic, and their pattern of activity is both driven by layer 1 activity *and* local reentrant activity (the activity of layer 5 between 0.2 and 0.5 s is thus mostly a self-feeding activity). This strong *versatility* in motor activity allows the system to explore and experiment various motor responses.
- The activities of layers 3 and 5 are balanced, i.e. the probability of a left command is almost the same as the probability of a right command. There may be only one layer to react to a given perceptual configuration, which can cause fast angular variations (for instance between 0.2 and 0.5 s).

More generally, the network/pendulum system is found to display a complex activity profile, where phase transition from one chaotic attractor to the other can be observed within the pattern of activity (i.e. it is not caused by an external parametric change). Such dynamics are often called *itinerant chaotic dynamics*, see for instance (Tsuda, 1991). This property is of course of great interest in the perspective of a sensory-motor exploration where every attractor may correspond to a *structural coupling* between the agent and the environment (Varela et al., 1991).

5. A Hebbian reinforcement learning paradigm

5.1 Hebbian reinforcement in dynamic neural networks

Our *exploratory* process relies on the self-generated chaotic activity. We thus have to define a *selection* process through which the better configurations will be stabilized (see introduction). We use for this purpose the principle of the *Hebbian trace*. In an active network, a Hebbian reinforcement factor can be defined according to the pre-synaptic and post-synaptic activities, i.e. $\forall (p,q) \in \{1,...,p\}^2$

$$\mathcal{H}_J^{(pq)}(t) = \frac{\alpha^{(pq)}}{N_{\text{aff}}^{(pq)}} \mathbf{x}^{(p)}(t)\mathbf{x}^{(q)}(t-1)^T$$

where $\mathbf{x}^{(q)}(t-1)^T$ is the transpose of $\mathbf{x}^{(q)}(t-1)$. This Hebbian term represents the correlated activities from population q toward population p. In order to refine the weight selection, we add a local factor called the *cooperative limitation factor*. This factor is based on the local field of the target neuron, i.e. $\forall (p,q) \in \{1,...,p\}^2$

$$\mathcal{H}_J^{(pq)}(t) = \frac{\alpha^{(pq)}}{N_{\text{aff}}^{(pq)}} \left[\left(1 - H(\mathbf{h}^{(pq)} - \theta^{(p)})\right) .\mathbf{x}^{(p)}(t) \right] \mathbf{x}^{(q)}(t-1)^T \quad (4)$$

where the "." operator represents a term to term product. It namely means that a weight adaptation will be allowed if (and only if) the activation of the target neuron relies on the cooperation between several local fields, i.e. for instance a cooperation between local recurrent stimulation and distant stimulation. If the source neuron is the only cause of target neuron activation, then no adaptation is done (for it is not necessary to reinforce a still strong connection). This factor reduces by 80-90% the number of selected links.

Those Hebbian terms are calculated every time step, and stored in a Hebbian trace with a slow decay, i.e.

$$\mathcal{T}_J^{(pq)}(t) = \beta \mathcal{T}_J^{(pq)}(t-1) + \mathcal{H}_J^{(pq)}(t)$$

with $\beta = 0.95$, so that the "half-life" of a Hebbian trace is 20 time steps, i.e. 100 ms. The Hebbian trace thus memorizes the most recent correlated activities from layer q toward layer p.

The Hebbian trace being stored, the most delicate aspect of Hebbian reinforcement is the choice of the effective weights reinforcement according to positive or negative reward. In case of positive reward, the rule has to simulate and stabilize the current attractor. In case of negative reward, the rule has to favor the transition toward a different pattern of activity, i.e. to "inhibit" the current attractor. We thus defined two reinforcement path, one devoted to positive rewards, the other devoted to negative rewards (see figure 5) :

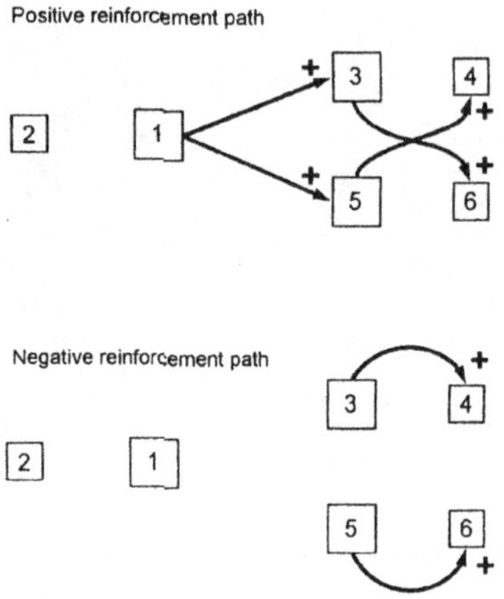

Figure 5: **The reinforcement path.**

- In case of positive reward, we both reinforce the excitatory path between the sensory module and the motor modules (visuo-motor path), and also the lateral connection toward the neighbor module inhibitory population, in order to reinforce the most active module in case of asymmetric activity (motor lateral path).
- In case of negative reward, we only reinforce the path corresponding to the motor module self inhibition, in order to weaken the most active module in case of asymmetric activity (local path).

Those reinforcements *only take place on excitatory links*, and the learning rule is such that the selected links can only be strengthened. The learning parameters are small enough for the weights adaptation to remain weak according to the weights initial values. We concretely set $\alpha^{(31)} = 0.1$, $\alpha^{(51)} = 0.1$, $\alpha^{(63)} = 0.15$, $\alpha^{(45)} = 0.15$ for the positive reinforcement path, and $\alpha^{(43)} = -0.15$, $\alpha^{(65)} = -0.15$ for the negative reinforcement path. The other $\alpha^{(pq)}$'s are equal to zero.

5.2 Closed loop protocol

The closed-loop learning protocol is the following. The angular position and velocities are randomly set according in intervals $[-\frac{\pi}{30}, \frac{\pi}{30}]$ and $[-0.2, 0.2]$. Then, the network/pendulum dynamics is iterated until a reinforcement signal is generated. The reinforcement signal mainly relies on the pendulum velocity. The aim is to maintain the velocity as small as possible i.e. :

- $R(t) = 1$ if $t > 300$ ms and $|\dot{\vartheta}(t)| < 0.05$ rad/s.
- $R(t) = -1$ if $|\dot{\vartheta}(t)| > 0.5$ rad/s or $|\vartheta(t)| > \frac{\pi}{15}$

The weights are then modified according to the Hebbian trace :

$$\text{If } \left(R(t)\mathcal{T}_J^{(pq)}(t) > 0\right)$$
$$\mathbf{J}^{(pq)} \leftarrow \mathbf{J}^{(pq)} + R(t)\mathcal{T}_J^{(pq)}(t)$$

Due to the decay term, only the most recent correlations get reinforced through the current Hebbian trace. The system is resetted after every trial. This series of operations is repeated for numerous trials, while the the weights get progressively modified according to the rewards.

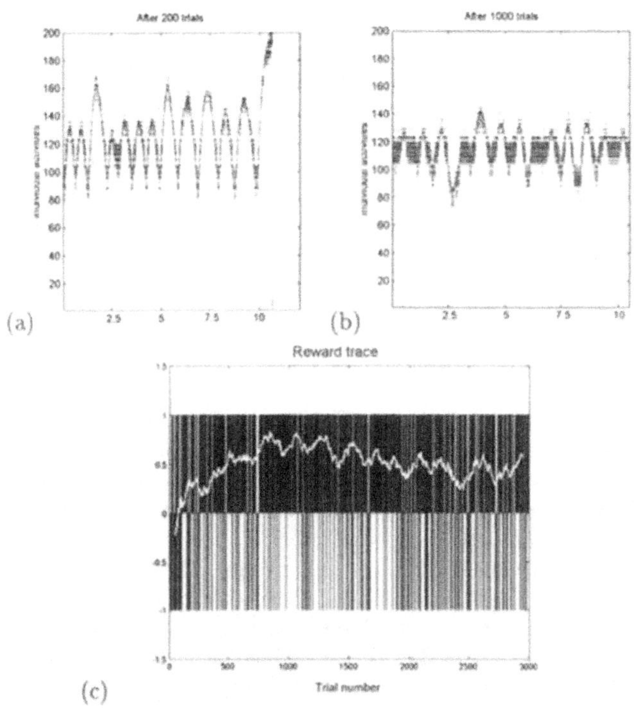

Figure 6: **Control adaptation during a learning experiment.** (a) Layer 1 activity displaying the pendulum control after 200 trials. Time is in s on the x-axis. (b) Layer 1 activity displaying the pendulum control after 1000 trials. (c) Rewards evolution during 3000 trials with their mean value on a sliding window of 100 trials (white line).

The result of a typical learning simulation is presented on figure 6. The experiment is repeated for 3000 trials. The successive reward values are shown on figure 6(c). The main observations are :

- The mean reward rapidly increases during the first steps of the experiment. Figure 6(a) shows that the network is already controlling the pendulum after 200 trials, which is a rather fast result, even if the task complexity is not too strong. The pendulum trajectory owns a periodicity, with various amplitudes. The network has learned to associate a negative force with the upper map activity (i.e. layer 5 dominates layer 3), and a positive force with the lower map activity (i.e. layer 3 dominates layer 5), whereas the two layers activities remain balanced in the middle of the map. The pendulum amplitude is found to increase so that it reaches the bound after 10 s.
- Then the mean reward slowly increases and seems to reach a plateau between 1000 and 1500 trials. This period corresponds to a control improvement. The resulting control is more accurate and tight, as shown on figure 6(b). Most of the activity takes place on the center of the map, where the two motor modules balance each other, with small fluctuations on one side or the other, so that the pendulum remains in the vicinity of $\pi = 0$ for unbounded time.
- The control can be said to be achieved after 1000 trials. However, we extended the number of trials in order to test the preservation of the learned behavior. Despite the slow reward decrease, we observed no significant alteration of the pendulum control accuracy.

This experiment illustrates the feasibility of a reinforcement learning protocol on a dynamic neural network using two separate Hebbian path which tend to stabilize favorable interaction patterns. The learning of the pendulum control is found to be very reproducible and robust, and the learning protocol, although relying on a simple Hebbian process, is found to be fast, giving significant results in 100-200 trials, and although capable of a slow improvement.

5.3 On-line protocol

For the on-line protocol, we allow several rewards during one trial, and there is no difference between exploration and exploitation processes. The only reset condition is the crossing of the angular bounds, so that the trial duration is equal to the control duration. In order to experiment various situations, we however limited the trial duration to 1000 time steps (i.e. 5s), so that our on-line protocol is not a "pure" on-line protocol.

Three improvements are moreover introduced in order to avoid weight drift :

- The rewards are adaptive, i.e. the reward amplitude depends on its probability of appearance (***habituation*** principle). When most of the rewards are negative, a positive reward is more significant. On the contrary, when most of the rewards are positive, a negative reward is more significant. We thus define a mean reward trace r, initially equal to 0. Then,

$r' = 0.9r + 0.1$ in case of positive reinforcement, and $r' = 0.9r - 0.1$ in case of negative reinforcement, such that $R(t) = (1 - r')/(r' + 1)$ in case of positive reinforcement, and $R(t) = (1 + r')/(r' - 1)$ in case of negative reinforcement. Then, r is updated with the value of r', and the process goes on until the next reinforcement. A typical series of reinforcement values is given on figure 7.

- A slight forgetting term is added on the Hebbian rule, i.e.

$$\text{If } \left(R(t)T_J^{(pq)}(t) > 0\right)$$
$$\Delta \mathbf{J}^{(pq)} \leftarrow \left(1 - \frac{R(t)}{1000}\right) \times \Delta \mathbf{J}^{(pq)} + R(t)T_J^{(pq)}(t)$$
$$\mathbf{J}^{(pq)} \leftarrow \mathbf{J}_0^{(pq)} + \Delta \mathbf{J}^{(pq)}$$

with $\Delta \mathbf{J}^{(pq)}$ initially set to 0 and $\mathbf{J}_0^{(pq)}$ is the initial weight matrix.

- The minimal interval between 2 different rewards is 20 time steps i.e. 100 ms.

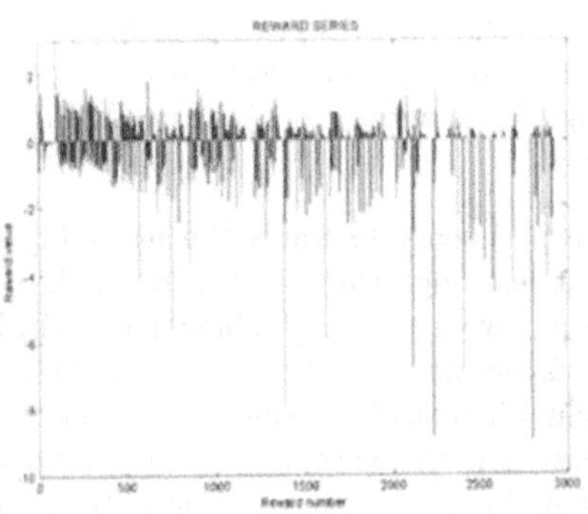

Figure 7: **Series of reward values in a typical on-line learning session.**

With those improvements, we empirically tested the convergence properties of our method, under three condition. Every condition has been tested on 20 networks. The first condition corresponds to the reinforcement path given in figure 5. In the second condition, the positive path is limited to the visuo-motor path (no lateral reinforcement). In the third condition, the positive path is limited to the motor lateral path (no visuo-motor reinforcement). Under those three conditions, we tested the median control duration over the 20 networks, for increasing trial numbers (i.e. ongoing learning process). the median control duration is measured over 20 networks, in a 10 trials width window (i.e 100th control duration over 200 control duration values around trial number +/-5). Note that every simulation was stopped after 5s of control, so that the mean would not be representative of the real control duration. Results are given on figure 8.

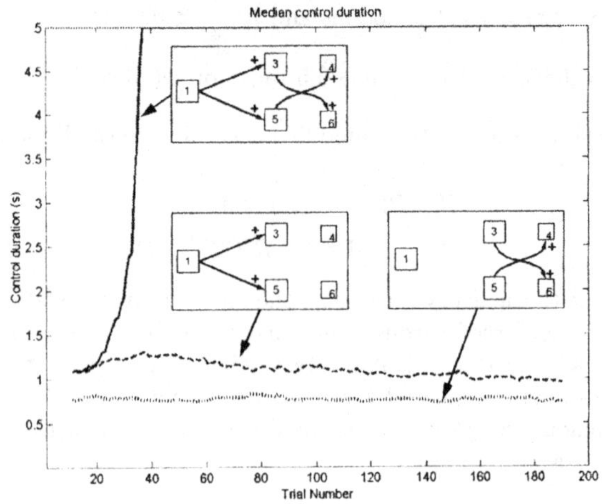

Figure 8: **Median control duration under three reinforcement conditions.** Trial number is on the x-axis, median control duration over 20 networks is on the y-axis (in seconds). The plain line corresponds to the full positive reinforcement path condition. The dashed line corresponds to the single visuo-motor path. The dotted line corresponds to the single motor lateral path.

This figure is self explanatory. First of all, the dual learning path (visuo-motor/motor-motor) is robust and operant over every network we tested. 40 to 50 trials are necessary for displaying a robust pendulum control (often less!). Second, single local path reinforcement or single visuo-motor path reinforcement do not display any significant increase of the control duration. A tight cooperation between the two processes is thus found to be necessary for the process to converge. The activation of the two path thus appears as a necessary condition for the learning process to be effective.

The nature of the process taking place during learning is not fully highlighted at present time. We can however give some clues :

- The negative reinforcement path doesn't seem to take much part in the reinforcement process. This mostly relies on limitations induced by the cooperative factor (eq. 4). Indeed, the local excitatory population is initially the only source of activation for the inhibitory population, until the learning process allows lateral inhibition to be activated. In the current settings of the network, the negative reinforcement path is not fully operant in the first steps of the learning process.

- The selection of a proper motor command relies on an exploration/selection process, where the exploration grounds on the versatility of spontaneous chaotic activity. The positive rewards often occur when one of the two motor layers overtakes the other, so that a light inbalance tends to be twice consolidated by the process : the stronger layer receives a supplementary activation from the sensory layer through the visuo-motor path, and simultaneously weakens the activity of its neighbor throuh the motor lateral path.

- There are however some severe limitation in the competition process which avoid the learning process to diverge. First, the cooperative limitation factor prevents the motor layers to be too strongly driven by the sensory layer (the sensory layer activation can not overtake local activation). Second, the indirect neighbor inhibition prevents to extinguish the neighbor activity. It must indeed be known that local excitatory-inhibitory interactions tend to produce synchrony. An indirect effect of the lateral inhibition reinforcement is to increase the synchrony (and thus the cooperation) between the two motor layers.
- Why is one path such necessary to the other for the full process to collapse in case one is missing? One can just conjecturate that the two path may balance each other, as the visuo-motor path feeds the motor excitatory layers and the lateral path feeds the inhibitory layers, so that the global level of activity may remain approximately equal. When one of the two path is missing, the network activity may diverge, either with an excess of excitation or inhibition.

6. Outlook

So the main results of our simulation experiment is to demonstrate the feasibility of reinforcement learning using the intrinsic dynamical properties of large random recurrent networks, and a biologically compatible Hebbian learning rule. The reinforcement experiment we presented can be seen as a first lookahead toward more complex and realistic learning tasks, possibly taking place on real robotic agents. There is a need of intensive simulation toward various constraints, environments and reinforcement signals in order to stabilize the parameters, learning protocol and operational/generalization capabilities. One can also ask whether several concurrent behaviors may be learned in the same network. One can for instance imagine a robotic agent with several sensory modules exciting several concurrent motor modules, each one devoted to the control of a particular muscle flexion or extension.

More generally, this article aimed to demonstrate the relevance of a dynamic system framework in neural modeling and control, which may provide a unified approach at different scales, including the neural level, the structural level and the agent/environment interaction level (Guillot and Daucé, 2002). Such an approach could help to shed new lights on biological functions and structures, and also to give new protocols and methodologies for the design of artificial life-like systems. This work is thus to be completed.

Annexe A : Weights settings

For two given populations p and q, we set $\rho_0 = \frac{\left(\bar{J}^{(pq)}\right)^2}{3\left(\sigma_J^{(pq)}\right)^2 N^{(q)}}$ and $\rho^{(pq)} = \frac{4\rho_0}{1+3\rho_0}$, which is the sparsity. We also set $\sigma^* = \frac{\sigma_J^{(pr)}}{\sqrt{4-3\rho^{(pq)}}}$, which is the effective weights deviation and $N_{\text{aff}}^{(pq)} = \rho^{(pq)} N^{(q)}$ represents the expectation of the number of afferent weights arriving from population q. $N_{\text{aff}}^{(pq)}$ is such that $\left|\frac{\bar{J}^{(pq)}}{N_{\text{aff}}^{(pq)}}\right| - \sqrt{\frac{3}{N_{\text{aff}}^{(pq)}}}\sigma^* = 0$, i.e. $N_{\text{aff}}^{(pq)} = \frac{\left(\bar{J}^{(pq)}\right)^2}{3(\sigma^*)^2}$.

A given weight $J_{ij}^{(pq)}$ is then equal to 0 with probability $1 - \rho^{(pq)}$, and takes its value randomly in interval $\left[\frac{\bar{J}^{(pq)}}{N_{\text{aff}}^{(pq)}} - \sqrt{\frac{3}{N_{\text{aff}}^{(pq)}}}\sigma^*, \frac{\bar{J}^{(pq)}}{N_{\text{aff}}^{(pq)}} + \sqrt{\frac{3}{N_{\text{aff}}^{(pq)}}}\sigma^*\right]$, with probability $\rho^{(pq)}$, i.e.

$$J_{ij}^{(pq)} = \begin{cases} \frac{\bar{J}^{(pq)}}{N_{\text{aff}}^{(pq)}} + \frac{\sigma^*}{\sqrt{N_{\text{aff}}^{(pq)}}} b & \text{with probability } \rho^{(pq)} \\ 0 & \text{with probability } (1 - \rho^{(pq)}) \end{cases} \tag{5}$$

where b is set according to $\mathcal{U}(0,1)$, which is a uniform distribution in $[-\sqrt{3}, \sqrt{3}]$.

Annexe B : Map settings

$\forall i \in \{1, ..., N^{(p)}\}, \forall j \in \{1, ..., N^{(q)}\}$, we calculate the distance

$$\delta_{ij}^{(pq)} = 2\pi \times \min\left(\left|i/N^{(p)} - j/N^{(q)}\right|, 1 - \left|i/N^{(p)} - j/N^{(q)}\right|\right)$$

so that $\delta_{ij} \in [0, \pi]$, and then we calculate a gaussian normalized neighborhood factor

$$\nu^{(pq)}(\delta) = \frac{\sqrt{2\pi}}{r^{(pq)}} \exp\left(-\frac{1}{2}\left(\delta/r^{(pq)}\right)^2\right)$$

We finally modify the weights according to this neighborhood factor

$$\tilde{J}_{ij}^{(pq)} = J_{ij}^{(pq)} \times \nu^{(pq)}\left(\delta_{ij}^{(pq)}\right)$$

Knowing that

- $\int_0^\infty \nu^{(pq)}(\delta)d\delta = \pi$, and, for small $r^{(pq)}$, $\int_0^\pi \nu^{(pq)}(\delta)d\delta \simeq \pi$.
- The repartition of $\delta_{ij}^{(pq)}$'s is uniform in $[0, \pi]$,

For $r^{(pq)} < 1$, weights enhancements and weights decays are globally balanced, so that the expectation of the weights remains unchanged by this transformation. On the contrary, the weights standard deviation is increased by a factor of the order of $1/r^{(pq)}$. In order to avoid too large weights distortion, we define a adaptation factor which is designed in order to approach $1/r^{(pq)}$ for $r^{(pq)} < 1$, and 1 for $r^{(pq)} \geq 1$, i.e. $\kappa^{(pq)} = 1 + \frac{\exp\left(-\left(r^{(pq)}\right)^2\right)}{r^{(pq)}}$ and the weights standard deviation is set to $\sigma^{(pq)\prime} = \frac{\sigma^{(pq)}}{\sqrt{\kappa^{(pq)}}}$

Acknowledgments

This work is supported by ACI "neurosciences intégratives et computationnelles", thème "temps et cerveau" and by the ACI "Robea".

References

Amari, S. (1977). Dynamics of pattern formation in lateral-inhibition type neural fields. *Biological Cybernetics*, 27:77–87.

Ben-Yishai, R., Lev Bar-Or, R., and Sompolinsky, H. (1995). Theory of orientation tuning in visual cortex. *Proc. Nat. Acad. Sci. USA*, 92:3844–3848.

Bush, P. and Sejnowski, T. (1996). Inhibition synchronizes sparsely connected cortical neurons within and between columns in realistic network models. *J. Comput. Neurosci.*, 3(2):91–110.

Camperi, M. and Wang, X.-J. (1998). A model of visuospatial working memory in prefrontal cortex: Recurrent network and cellular bistability. *J. of Computational Neuroscience*, 5:383–405.

Cessac, B. (1995). Increase in complexity in random neural networks. *Journal de Physique I*, 5:409–432.

Compte, A., Brunel, N., Goldman-Rakic, P. S., and Wang, X.-J. (2000). Synaptic mechanisms and network dynamics underlying spatial working memory in a cortical network model. *Cerebral Cortex*, 10:910–923.

Daucé, E. (2004). Short term memory in recurrent networks of spiking neurons. *Natural Computing.* in press.

Daucé, E., Quoy, M., Cessac, B., Doyon, B., and Samuelides, M. (1998). Self-organization and dynamics reduction in recurrent networks: stimulus presentation and learning. *Neural Networks*, 11:521–533.

Daucé, E., Quoy, M., and Doyon, B. (2002). Resonant spatiotemporal learning in large random recurrent networks. *Biological Cybernetics*, 87:185–198.

Doyon, B., Cessac, B., Quoy, M., and Samuelides, M. (1993). Control of the transition to chaos in neural networks with random connectivity. *Int. J. of Bif. and Chaos*, 3(2):279–291.

Guillot, A. and Daucé, E. (2002). *Approche dynamique de la cognition artificielle.* Lavoisier.

Hansel, D. and Sompolinsky, H. (1996). Chaos and synchrony in a model of a hypercolumn in visual cortex. *J. Comp. Neurosc.*, 3:7–34.

Hertz, J. and Prugel-Bennett, A. (1996). Learning synfire chains: turning noise into signal. *Int. J. Neural Systems*, 7:445–450.

Hopfield, J. (1982). Neural networks and physical systems with emergent collective computational abilities. *Proc. Nat. Acad. Sci.*, 79:2554–2558.

Hupé, J.-M., James, A., Payne, B., Lomber, S., Girard, P., and Bullier, J. (1998). Cortical feedback improves discrimination between figure and background by v1, v2 and v3 neurons. *Nature*, 394:784–787.

Izhikevich, E. M. (2003). Simple model of spiking neurons. *IEEE trans. on neural networks*, 14(6):1569–1572.

Mac Leod, K. and Laurent, G. (1996). Distinct mechanisms for synchronization and temporal patterning of odor-encoding cell assemblies. *Science*, 274:976–979.

Moynot, O. and Samuelides, M. (2002). Large deviations and mean-field theory for asymmetric random recurrent neural networks. *PTRF*, (123-1):41–75.

Skarda, C. and Freeman, W. (1987). How brains make chaos in order to make sense of the world. *Behav. Brain Sci.*, 10:161–195.

Sompolinsky, H., Crisanti, A., and Sommers, H. (1988). Chaos in random neural networks. *Phys. Rev. Lett.*, 61:259–262.

Sutton, R. (1988). Learning to predict by the method of temporal differences. *Machine learning*, 3:9–44.

Tsuda, I. (1991). Chaotic itinerancy as a dynamical basis of hermeneutics in brain and mind. *World Futures*, 32:167–184.

Varela, F., Thompson, E., and Rosch, E. (1991). *The Embodied Mind.* MIT Press.

Wang, X.-J. (1999). Synaptic basis of cortical persistent activity: the importance of nmda receptors to working memory. *The Journal of Neuroscience*, 19(21):9587–9603.

Watkins, C. and Dayan, P. (1992). Q-learning. *Machine learning*, 8:279–292.

Learning human-like Movement Behavior for Computer Games

Christian Thurau
Applied Computer Science
Bielefeld University
P.O.Box 100131, 33502 Bielefeld, Germany
cthurau@techfak.uni-bielefeld.de

Christian Bauckhage
Applied Computer Science
Bielefeld University
P.O.Box 100131, 33502 Bielefeld, Germany
cbauckha@techfak.uni-bielefeld.de

Gerhard Sagerer Applied Computer Science
Bielefeld University
P.O.Box 100131, 33502 Bielefeld, Germany
sagerer@techfak.uni-bielefeld.de

Abstract

Modern Computer Game AI still relies on rule-based approaches, so far failing to develop a convincing, human-like opponent. Towards the development of more human-like computer game agents, we propose an approach for learning strategies by observation of human players, principally viewing the design of a computer game agent as a problem of pattern recognition. First we introduce a Neural Gas based grid learning of an internal representation of the 3D game-world. Then we discuss the use of a learning potential fields approach, establishing a mapping from world state vectors to corresponding potential field forces for agent guidance. The training data being used is acquired by observation of human players acting in the 3D game environment of a commercial computer game. Finally, some experiments are presented.

1. Introduction

Over the last years computer games changed their appearance from simple card or reaction games to complex virtual world simulations. Although the graphical advances are impressive, modern games most often ruin realistic world impressions by poor implementation of their virtual inhabitants, artificial agents (game term:*bots*). Up to now, actions and behaviors of game agents are scripted or rely on (at times very complex) finite state machines (Cass, 2002). It is at least questionable, if finite state machines will lead to convincing, human like, game agents.

Laird outlined, that computer games are an excellent experimental environment, when it comes to the investigation of human level AI (Laird and v. Lent, 2000), and Isla (Isla and Blumberg, 2002) suggested some of the possible research directions in Game AI. Nevertheless, up to now there has been little effort in applying techniques from pattern recognition or machine learning. While learning movement primitives for humanoid robots (Schaal et al., 2004) or implementations of autonomous behavior for simulated worlds (Terzopoulus et al., 1994)(Tomlinson and Blumberg, 2002) are quite common, so far no one seems to use computer games as an application for learning human behavior in virtual worlds. A closer look at these games features will reveal some of the opportunities.

On the one hand there are the gaming worlds, which are inspired by the real world. That does not only take into account the graphical representation or game physics, but also the tasks a human player has to fulfill, which are a lot less abstract than for example in a game of chess. Especially the genre of MMORPG's (Massively Multiplayer Online Role Playing Games) focuses on a diversity of aspects, the social interaction with other human players taking the most important role, since these games want to be understood as a world simulation, giving the player the opportunity to be whatever he wanted to. Although these games can be seen as a simulation, they are not from a scientist point of view:"..[computer games are] real products and real environments on their own that millions of humans vigorously interact with and become immersed in.."(Laird and v. Lent, 2000).

On the other hand, artificial agents in a computer game can "embody" a human player. They can be restricted to a human's world perception and reaction spectrum, which makes the adaption of training samples for classification approaches less of a problem (in many games a human player and a game agent share the same interface to the game world via the network protocol). Further, just like in a simulation, there is no need for low-level operations, such as object recognition or basic motor-control. Most importantly, although it is a game, human players show a lot of different aspects of human act-

Figure 1: A game bot's (foreground character) possible perception (left picture) and reaction (right picture) spectrum.

ing and thinking, providing interesting challenges for learning approaches.

Consider for instance the well known genre of first persons shooters (FPS) games on which we will concentrate in our experiments. They provide a complex, feature rich 3D game world, in which humans compete each other over the Internet. Every player taking part has control over a single character and therefor has only a limited local view onto the game world, Figure 1 shows a game-bot and his possible perception and reaction spectrum. The main goal is to survive as long as you can, by picking up different types of items, and shoot as much enemy players as you can, by using different types of weapons. The game physics stretch what can be commonly understood as realistic physics, nevertheless the players are bound to it (we are currently focusing on more comic like FPS games, instead of "realistic" FPS games, since they do not only add realism to the physics but also to the violence). Although these games share a violent background, players are more interested in the sport aspects of the game, team based modifications are highly popular. Which at first seems like a mindless shooter game, turns out to get very complex, when the aspect of a sport competition comes into play, no wonder there are numerous Internet sites which deal with all kinds of different tactics on how to play the game in a successful manner, simple rules do not necessarily lead to a simple game.

What was said for computer games in general, holds for FPS games as well: Usually, finite state machines are used for agent control. These agents offer a challenge rather by perfect aiming than by a superior game play, and can be easily distinguished from a human player.

The FPS games fast pace does not leave much time for thinking. In order to play the game successfully, the human player has to develop a set of *reactive behaviors* gathered from his own gaming experiences. Thus a direct mapping of the k last world states $\vec{s}_t, \vec{s}_{t-1}, \ldots, \vec{s}_{t-k}$ onto a bot's reaction $\vec{a}_t$ seems to be possible, it is exactly what a human player is doing most of the time. These games popularity offers the opportunity to gain access to large training data sets. Not only by observation of an ongoing match, but also by downloading so called *demo files*. These demos are basically records of the network code of an ongoing match and can fully reproduce a single match from the perspective of one player. A lot of sites on the Internet give access to thousands of those demo files, each of which containing hundreds of training samples. The basic idea is to learn human-like behavior through analysis of these freely available training sets.

Although the overall problem of learning human-like behavior for an FPS game agent can be reduced to building up a set of reactive behaviors, it can be understood as a common task dependent problem solving. Doing so, we can divide the player's actions into global strategies and more localized tactics and basic reactive behaviors. Global strategies are, in our domain, mostly targeted at controlling the 3D gaming world, by gaining access to important items and holding off opposing players from certain map areas. Where tactics usually include enemy movement pattern anticipation and appropriate weapon selection and usage. Behaviors are very similar to the described reactive behaviors and build up movement/action primitives. Dividing human reactions into the described categories simplifies the overall problem and allows for an in depth analysis of each separate layer of a human's thinking reaction process. Of course, some kind of merging function, selecting a layers activation level has to be developed.

We already showed, that purely reactive behaviors can be learned by observation of a human player (Thurau et al., 2003) by means of a combination of Self-Organizing-Maps and Multi-Layer-Perceptrons. In this paper we concentrate on methods for learning a *game strategy*. Although there are other important behaviors to learn, this one is crucial, since it often makes up for the impression on how smart a human or a game bot is. A good aim would never add to a human like artificial agent's impression (the opposite is the case, human players are often thought of using an "aiming-bot", when there aim is too good), where a smart movement in the gaming world, a plan behind certain movements, would certainly create a convincing, realistic impression.

Strategic decisions in a FPS game are mostly based on decision processes on picking up certain items and guarding areas. Since *Artificial Potential Fields* (Arkin, 1998) provide an computationally inexpensive and well examined approach for navigational, goal oriented movement control, and allow for an overlay of concurrent goals, we chose them for pathfinding. Because of the strong coupling between a player's decision at timestep t and the world state $\vec{s}_t$, we want to learn the mapping between world states and corresponding potential field forces, section 2.2 deals with this subject. To overcome some common potential field problems, we are using an *Avoiding The Past Pheromone Trail*, which is described in section 2.3. In section 3., we present some first experiments and their evaluation and finally end with a small conclusion and an outlook on our future work in section 5. First, however, the next section describes our approach for learning an internal grid based representation of the 3D gaming-world, using a *Neural Gas* algorithm.

2. Learning plans

2.1 Neural Gas for Waypoint Learning

For a representation of the virtual world we are learning a grid based *waypoint map* using a Neural Gas (Martinez et al., 1993) algorithm, an improved k-means algorithm showing good performance when it comes to topology learning (Martinez and Schulten, 1991). The training data used for Neural Gas learning consists of all positions $\vec{p} = \{x, y, z\}$ a human player held during various plan executions, thus staying very close to the actual human movement paths by only taking visited places into account.

Applying a Neural Gas algorithm to the player positions results in a set of position prototypes, places a player visited frequently. The number of prototypes dependes obviously on the 3D map size. In our experiments a number of 1100 prototypes showed good results where 19000 player positions were used for training. In a second step neighbouring prototypes are interconnected. The edges connecting the nodes are not computed using the classical hebbian topology learning (Martinez and Schulten, 1991). Drawing edges based on concurrent activity would certainly lead to problems on one-way paths. Therefor we only connected two nodes, if there was an observabel player movemement from one node to the other. This finally results in a topological map representation or waypoint map. The edges are stored in a $n \times n$ matrix $C_{i,j} \in N^+$ where n denotes the number of units/neurons. If $C_{i,j} = 0$ unit i is not connected to unit j, if $C_{i,j} = 1$ unit i is connected to unit j, Figure 2 shows a resulting waypoint map. The node distances can be computed in advance and are stored in a $n \times n$ matrix D, where $D_{i,j} \in \mathcal{R}^+$ denotes the distance from node i to node j.

Using a nearest neighbour criterion, every position on the 3D map can be assigned to a single node. A movement from one node to another is only possible if a connection between the two exists, and can be easily translated into "real" movements in the 3D map.

When considering potential fields which will guide an agent through the 3D world, the waypoint representation has several advantages over directly moving in 3D. Potential fields placed on certain nodes won't influence the agent's behavior through walls or one-way paths. Besides, we can be sure that when our agent is moving along the waypoint nodes, he won't fall of ledges or run into walls.

2.2 Learning Potential Fields

A world state vector is denoted as $\vec{s} = \{s_1, s_2, \cdots, s_n\}$, where $s_1, s_2, \cdots, s_3$ describes certain world aspects or internal states of an agent. In our experiments we used the player's inventory (which weapons or items did he already picked up) as well as his current armor and health value as state vector variables.

Similar encountered situations will most always lead to similar strategic actions or movement patterns. Therefor we are grouping similar observed player states using a Neural Gas algorithm, resulting in a number of state vector prototypes, representing more common game situations. In a second step each observed player position is assigned to one state prototype, effectively forming a number of situation dependent movement behaviors. Each movement behavior begins when entering a new state space cluster, and ends when a state change occurs. Most state changes -in the state space we consider- occur when items are picked up. Therefor almost all movement patterns end with an item pickup, implicitly defining item positions as the goals to attain.

The state space clustering finally leads to a number of movement patterns which were executed, when the human player encountered a certain situation. A single movement behavior can be denoted as:

$$m_i = [m_{i,1}, m_{i,2}, m_{i,3}, \cdots, m_{i,k}]$$

where m_i denotes a complete movement pattern and $m_{i,k} \in N_1^n$ denote a single node, where n denotes the length off the overall movement pattern m_i.

To rebuild the human movement behavior, we decided to use potential fields for agent guidance. Potential fields originating in a node in the waypoint map produce certain potential on all nodes, which decrease with the distance to the potential field source. Since any node is a possible potential field source, the potential f_j at node j is denoted as:

$$f_j = \sum_{i=1}^{n} \frac{F_i}{d_{i,j}^2} \tag{1}$$

where $d_{i,j}^2$ denotes the distance from node i to node j and F_i denotes the potential field strength at node i.

For every given state space cluster we have to find a potential field force configuration, which will lead to the observed movement patterns. Since the movement in a potential field

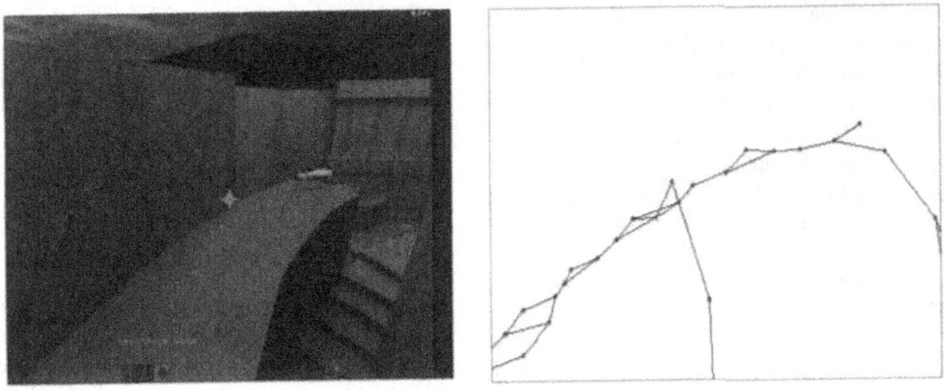

Figure 2: A screenshot of a small part of the map and the corresponding waypoint map

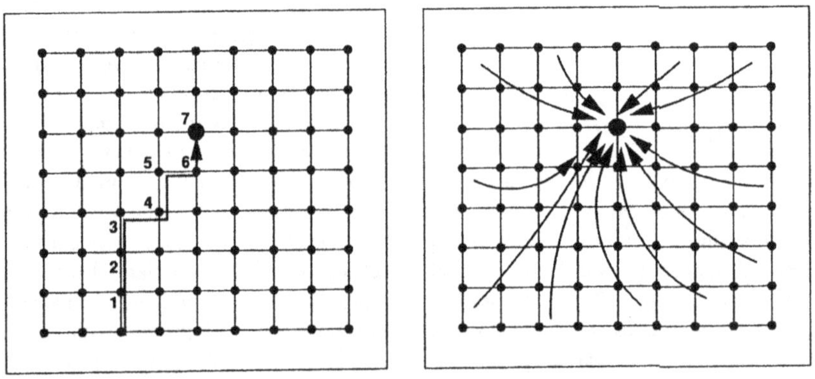

Figure 3: The left figure shows the observed movement path and the assigned discrete potentials, the right the learned potential field forces

does always follow the steepest gradient, there has to be an increase in the potentials f_i found on every node along a specific movement pattern. Consequently we construct a set of possible potentials for every node along a certain movement pattern, by assigning ascending discrete values to the node sequence m_i.

$$m_i = [(m_{i,1}, f_1 = 1), (m_{i,2}, f_2 = 2), \cdots, (m_{i,k} = k, f_k = k)]$$

where f_n denotes the assigned potential for a node n. This is done with all observed movement paths belonging to a certain state space cluster, giving more weight to more common plans by summing up all observed potential distributions. As a side effect arbitrary movements which do not approach the final goal node $m_{i,k}$ are always assigned lower potentials than their successor nodes, effectively favoring goal oriented nodes and still producing goal oriented potential distribution as Figure 4 illustrates.

In the next step a potential field strength F_i for every node i which leads to the constructed potential distribution, needs to be computed, the following error function has to be minimized:

$$E = \frac{1}{2} \sum_{i=1}^{n} (f_{i,desired} - f_i)^2 \qquad (2)$$

where $f_{i,desired}$ denotes the sum over all assigned potentials of a node belonging to a state prototype. A gradient descent leads to the update rule for the potential field force F_i of a node i:

$$F_i + 1 = F_i - \alpha \sum_{k=1}^{n} \left(f_{i,desired} - \sum_{l=1}^{n} \frac{F_l}{{d_{k,l}}^2} \right) \frac{1}{{d_{k,i}}^2} \qquad (3)$$

where α denotes a learning rate which is decreased during the training process.

Applying the learning rule, we end up with a maximum of active potential fields which is equal to the number of passed nodes assigned to a state prototype. Figure 3 illustrates the relation between the observed movement path and resulting potential field forces, given a simple goal oriented movement pattern.

On the first sight this methods seems a bit ineffective, a shortest path algorithm like Dijkstra might provide the same

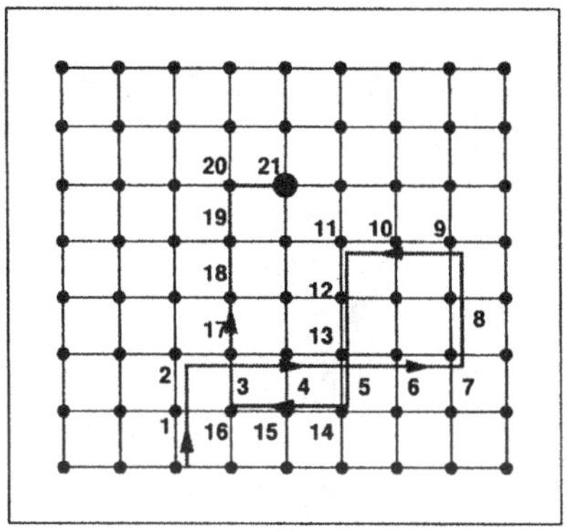

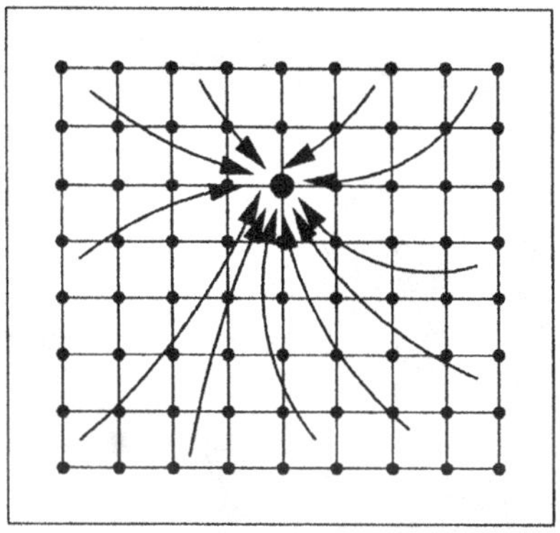

Figure 4: An observed movement pattern with some arbitrary, non goal oriented movement, does still produce goal oriented potential fields.

results. But the problem we are facing is different from finding the shortest path to a specific map location. Learning a strategic movement behavior implies, that we are not searching for the shortest path, but for the strategically most appealing. Which could also mean to guide the agent on a longer path, if it provides strategic advantages (a better path could be in darker areas, avoiding open terrain and so on), Figure 5 illustrates how a longer way would be favored over the shorter one.

If the number of training samples is limited, the potential field learning effectively results in a shortest path way finding to the most attractive area, when approaching from a direction which doesn't provide any local guidance. With more training samples available more information on how to approach certain important nodes is provided, resulting in "smarter" movements and a better local guidance.

While the overall approach is similar to learning a value function known from dynamic programming or reinforcemeant learning (Sutton and Barto, 1998), here it is merely a representation of specific movement patterns. A further improvement of certain strategies by means of reinforcement learning, though desirable, is not intended at the moment. Besides, in its current form our approach is efficient and computationally inexpensive, thus well suited for fast paced real time computer games.

However, the usage of potential fields comes along with some problems. On the one hand we might encounter local maxima, stucking the agent at unimportant nodes, on the other hand we have a relatively poor guidance when the distance to originating potential field nodes increases. The next section deals with both problems.

2.3 *Dealing with Potential Field flaws*

Avoiding the past potential fields (T.Balch and R.Arkin, 1993) serve as a local spatial memory, to avoid already visited nodes. At fixed time intervals a repulsing potential force at the current node is deposited. To reduce interference with goal based potential fields, the avoid-past potential field forces are reduced over time, thus effectively resulting in a pheromone trail, known from ant-algorithms. In addition the area of effect is reduced compared to the goal based potential field forces.

These pheromone trails have two purposes. First, local maxima can be avoided by pushing the agent to nearby nodes, hopefully back to the field of influence of goal orientated potential fields. Second, areas which provide poor guidance because of the distance to goal oriented potential fields can be escaped easily, by reinforcing a chosen direction.

The overall potential distribution on waypoint map node m_j, at timestep t is now computed as follows:

$$f_j(t) = \sum_{i=1}^{n} \frac{\mathcal{F}_i}{{d_{i,j}}^2} - \sum_{i=1}^{n} \frac{\mathcal{P}_i(t)}{{d_{i,j}}^3} \quad (4)$$

where $P_i(t)$ denotes the Pheromone concentration at node i, at timestep t, and with $P_i(t+1) = P_i(t) - \varrho P_{evap}$, where ϱ is a diffusion parameter and P_{evap} is the fixed evaporation parameter for an occupied node.

Both parameters P_{evap} and ϱ have to be carefully chosen, to provide the expected effects and to not interfere too much with the goal oriented potential field forces. If chosen properly to fit the premise of very local influence on the goal oriented potential fields, a recomputation of the goal oriented potential field forces is not necessary. However, our experiments were not influenced by moderate changes to the parameters, after all the field of influence of a pheromone is more localized than a potential field.

If goal oriented potential fields are not available, the pheromone trail generates an "explore map" behavior, which, though not intended, adds to a more human-like agent impression. A proper guidance in any given world state is to favor over time consuming non goal oriented movements, evolved from the pheromone trail, just relying on the pheromone trail would not produce convincing results.

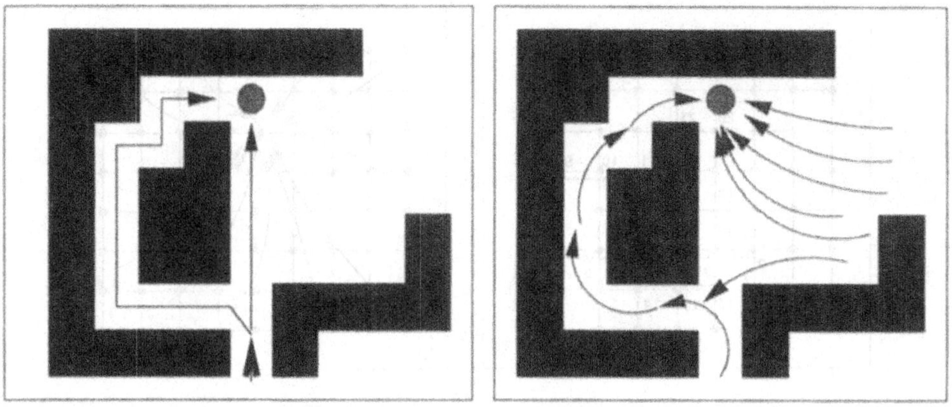

Figure 5: The left figure shows an observed movement path, where the green arrows denotes the actual observation, the red one the shortest path to the goal. The right figure shows the learned potential field forces, which would guide the agent along the longer path, when approaching the goal from the initially observed direction.

3. Experiments

3.1 Simple Tasks

The experimental environment consists of the commercial computer game Quake II and a Quake II Matlab client. To focus on the presented approach human players were presented a certain task, which had to be executed and was afterwards used for training.

The agent control mechanism is quite simple. With the knowledge about the potential field forces, the potentials for all nodes, connected to the currently occupied node, can be computed very fast. From the list of possible successor nodes the one with the highest potential is selected. Environmental influences or item pickups change the agents state vector $\vec{s}$ and result in a change to the overall potential field forces, thereby leading the way to other nodes.

The training data used consisted of the observation of 20 different tasks on one map, mainly consisting of item pickup sequences. For each task a separate bot was trained. To estimate the game-bot performance, we evaluated four different starting locations (game term: "spawn point") from which it had to find its way according to the human behavior in the training samples. Performance can be measured over the number of goals the bot reached, and the time it took him to do so. Table 1 shows some of the results. Generally our approach showed very good results, failing to reach the goals in the predefined order only once. In that particular failed experiment, the game-bot got stuck at some part of the map and didn't manage to reproduce the desired behavior in less than two minutes, however, he might have reached the goals later on, since the pheromone trail would have driven him towards an (even weak) attractor.

Visual/graphical comparison of the human and artificial player's movement paths gives further insights in how well the potential fields are able to learn human movement patterns. Figure 7 shows the observed game agent's movement path in comparison to the human's movement path. While Figure 6 illustrates the agent's behavior when starting from a different spawn point, then the one observed in the training samples.

Task	SP 1	SP 2	SP 3	SP 4
Set 1	13.89	17.60	16.51	8.31
Set 2	33.30	25.71	24.50	31.42
Set 3	8.42	9.00	11.43	3.02
Set 4	22.49	19.30	23.62	14.89
Set 5	11.64	9.21	13.61	14.01
Set 6	16.82	16.52	22.70	16.42
Set 7	33.30	35.03	36.40	31.54
Set 8	28.01	–	–	–
Set 9	41.50	41.24	41.80	46.33
Set 10	20.02	14.71	24.82	15.71
Set 15	39.19	46.48	41.20	36.21
Set 20	40.2	55.10	32.41	50.91

Table 1: The bot had to complete every learned task when starting from four different locations SP 1-4. To give an indication about the agent's performance, we measured the time in sec. it took him to reach all goals in the predefined order, thereby reproducing the human movement behavior.

3.2 More Complex Behaviors

Since strategic movements do not depend on the execution of a single, isolated movement pattern, but on a combination of situative adequate movement paths, we decided to test the described approach with more complex behavior learning. Hereby we instructed a human player to play the game Quake II, as if an enemy was present (again we simplified the situa-

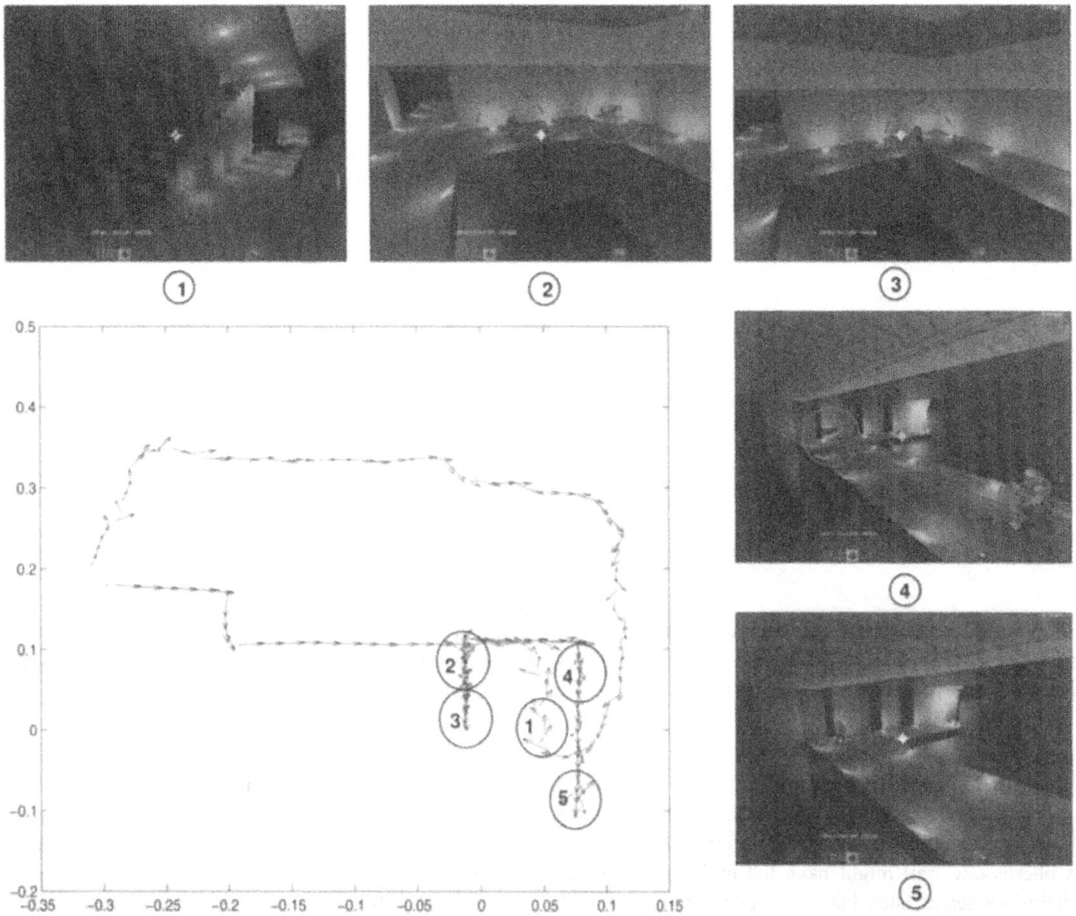

Figure 6: The bot started from a map location, other than the one used during the learning process. The red arrows show all bot positions and its velocity, while the blue arrows show the positions and velocity of the observed human player's movement path. Picture 1 shows, where the bot moves up a staircase and comes close to the observed movement pattern. In Picture 2 he is targeting the first item and picks it up at 3, reproducing the observed human behavior. A state change occurs, switching the potential field forces. The bot moves on to the next goal in 4 and finally reaches it at 5.

tion in order to focus on the strategical aspects). This resulted in a set of training samples, where the player collected various items in different orders, dependent on the initial starting position, and after that securing important areas by cycling different types of armor (a quite common behavior shown by many players).

This experiment setup, too, yielded good results, indicating that the approach is also working reasonably well for a larger scale. The artificial agent moved around the 3D gameworld in a convincing manner and managed to reproduce the observed behavior. He picked up several different types of items, before he mostly stayed at one area, usually located around an important armor item, and only moved away when driven by its pheromone trail or when there were sudden changes in its state vector (for example a decrease in its armor value due to an attack by another player). Usually the agent reached a state where he owned all weapons and filled up his armor value to the maximum.

In comparison to the first experimental setup, no fixed movement behavior sequence was reproduced. Instead, seemingly intelligent behavior emerged, by evolving a mixture of appropiate observed movement patterns.

4. Evaluation

As Table 1 indicates, the movement patterns were successfully learned for almost all tasks, and the agent managed to reproduce the observed behavior. Also local maxima or low potentials on certain nodes could be avoided using the described avoiding the past pheromone trail. Only task number 8 showed problems. The bot got stuck in three out of four different runs. The main problem in that experiment was that the first goal was located very close to the last goal, thus the agent accidentally picked up the last item at the beginning of

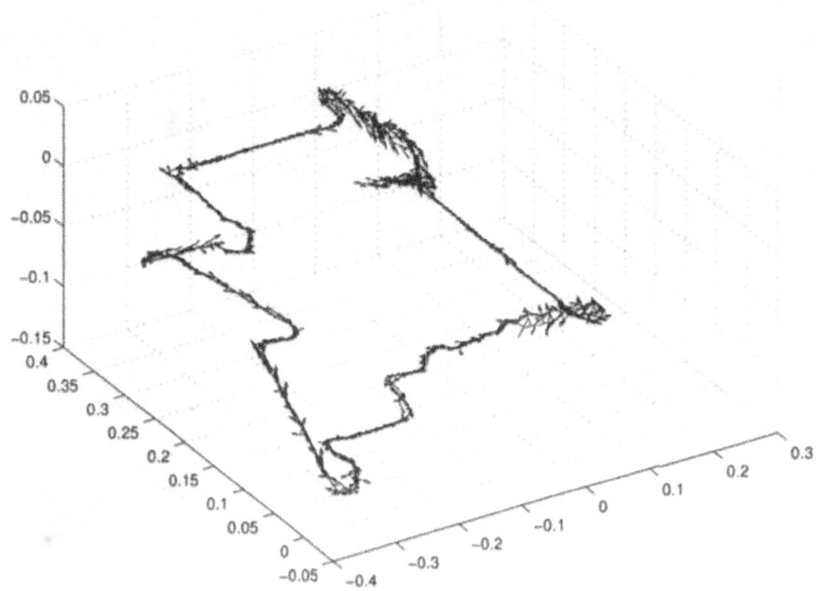

Figure 7: Visualization of a human and a game agents movement path in x, y, z. The blue arrows indicate the game bot's current positions and its velocity, the red arrows show the observed human player's position and velocity, which was used for learning the potential field forces. The bot started from the same positions as the human player and managed to fully imitate the human movement behavior.

its run, which somehow trapped it in a internal state which was associated with only very weak potential field forces and therefor poor guidance throughout the map. However, the agents pheromone trail might have led him to pick up the other items sooner or later, but we skipped the experiments after a waiting time of two minutes.

The usage of more training samples generally avoids that problem, since all states should have a sufficient number of potential field forces associated with it. So far we couldn't watch the problem of entering an almost undefined state, when using larger sets of trainings samples for learning more complex behaviors. The second experiment indicates that common complex strategic behaviors can indeed be learned using our described approach. The bot behaved in a very convincing manner, showing combinations of all observed movement patterns, while still following a more global strategy, presented in all training demos. Sudden state changes lead to understandable behavior switching and added to the overall good impression.

5. Conclusion

In order to learn a functional mapping of a computer game agents internal state on to strategic decisions, we investigated human movements in a virtual environment, the computer game Quake II.

To find a suitable way of representing the 3D game world, we learned a grid based topology of the 3D game world from the observation of a human player's movements, applying a Neural Gas algorithm. Then we learned potential field forces for agent guidance, to create human-like situation dependent movement patterns. At first we clustered the player's state space into several state prototypes, to which observed corresponding movement patterns were assigned. In a second step potential field forces, leading to the observed movement patterns, were computed using a simple error minimization. To deal with local maxima and occasionally poor potential field guidance, we were using an avoiding the past pheromone trail, evaporated from the agent's position.

And indeed, our experiments show very good imitation of the observed human behavior. Besides reaching the only implicitly encoded aims, the approach paid attention to strategic decisions of the human player on how to reach certain goals - without any knowledge of the terrain structure. Thereby staying close to the observed movement patterns.

The second experiment finally showed that larger numbers of strategic movement patterns can be decomposed and learned using our approach, covering a greater spectrum of situation dependent strategic movements and leading to more realistic sequences of movement patterns.

6. Future Work

Since not all items are available all the time, we have to investigate mechanisms of learning a time dependent potential field.

So far, except for the agents own pheromone trail, the potential fields were centered on fixed map positions. While this provides a goal oriented movement, localized potential fields attached to moving objects, other players or projectiles could certainly lead to some interesting behavior, although this might be more suited for learning tactics, since the overall strategy is mostly determined through fixed item/ area locations.

In order to use real demo files downloaded from the Internet to model a specific player's style of play, we need to find a way to extract the strategy when there are concurring tactical and reactive behaviors. A simple smoothing of the overall movement path might reveal the long term goals.

Despite the good results, we can not be sure if the world state variables, expressing the state - potential field dependency are indeed the "really" important state variables. Datamining methods applied to real demos could provide useful insights on the relevance of certain state variables and their further influence on the learned potential field forces.

References

Arkin, R. C. (1998). *Behavior-Based Robotics*. MIT Press.

Cass, S. (2002). Mind games. *IEEE Spectrum*, pages 40–44.

Isla, D. and Blumberg, B. (2002). New challenges for character-based ai for games. In *Proceedings of the AAAI Spring Symposium on AI and Interactive Entertainment*.

Laird, J. E. and v. Lent, M. (2000). Interactice Computer Games: Human-Level AI's Killer Application. In *Proc. AAAI*, pages 1171–1178.

Martinez, T., Berkovich, S., and Schulten, K. (1993). Neural gas network for vector quantization and its application to time-series prediction. In *IEEE Transactions on Neural Networks*, pages 558–569.

Martinez, T. and Schulten, K. (1991). A neural gas network learns topologies. In *Artificial Neural Networks*. Elseviers Science Publishers B.V.

Schaal, S., Peters, J., Nakanishi, J., and Ijspeert, A. (2004). Learning movement primitives. In *International Symposium on Robotics Research (ISRR 2003)*. Springer.

Sutton, R. S. and Barto, A. G. (1998). *Introduction to Reinforcement Learning*. MIT Press.

T.Balch and R.Arkin (1993). Avoiding the past: A simple but effective strategy for reactive navigation. In *IEEE International Conference on Robotics and Automation*, pages 678–685.

Terzopoulus, D., Tu, X., and Grzeszczuk, R. (1994). Artificial fishes: Autonomous locomotion, perception, behavior, and learning in a simulated physical world. In *Artificial Life*, pages 327–351.

Thurau, C., Bauckhage, C., and Sagerer, G. (2003). Combining Self Organizing Maps and Multilayer Perceptrons to Learn Bot-Behavior for a Commercial Computer Game. In *Proc. GAME-ON*, pages 119–123.

Tomlinson, B. and Blumberg, B. (2002). Alphawolf: Social learning, emotion and development in autonomous virtual agents. First GSFC/JPL Workshop on Radical Agent Concepts.

On Ashby's homeostat: A formal model of adaptive regulation

J. Michael Herrmann
Institut für Nichtlineare Dynamik
Universität Göttingen
37073 Göttingen, Germany
michael@chaos.gwdg.de

Michael Holicki and Ralf Der
Institut für Informatik
Universität Leipzig
04109 Leipzig, Germany
holicki,der@informatik.uni-leipzig.de

Abstract

Ashby's homeostat performs a random search in the parameter space of a linear system until a stable behavior is retrieved following e.g. structural perturbations. We consider a continuous adaptive version of the homeostat where the parameters are determined by a gradient dynamics with respect to a given goal state. In this way much shorter transients are achieved by a parameter dynamics that runs on the same time scale as the state evolution. The model allows for analytical solution and forms in so far an interesting toy model for nonlinear adaptive systems with implications for control mechanisms in natural or artificial embedded agents.

1. Introduction

Ashby's homeostat (Ashby 1948, 1952) was invented and actually built half a century ago. It was intended as a working illustration of the principle of homeostasis, a term coined by Cannon (1932) in the 1920s. Presently, we are witnessing a revived interest in the concept of homeostasis in various fields ranging from synaptic plasticity (e.g. Turrigiano, 1998) to the control of autonomous robots (Di Paolo, 2000).

Originally homeostasis was conceived as an approach to complex control mechanisms in living beings. Blood pressure, e.g., depends on many local variables which are in turn affected by many hormonal, nervous and chemical processes, but it is stabilized globally without a central control unit. Ashby studied properties that underly such mechanisms in living beings from a theoretical point of view and tried to transfer this principle to the design of control architectures in machines. He found it natural in this context to consider dynamical systems involving switching processes, intermittency, memory and dynamic relations with the environment. In particular the notion of ultrastability appeared to him of central importance for a theory of the mechanisms underlying life.

Ashby claimed that in complex dynamical systems the notion of linear stability is not sufficient. Such systems should be able to cope with large deviations from a target state or even to produce them in order to achieve structural reorganization in their course. Ultrastability refers to systems which in addition to their states also possess dynamical parameters which are supposed to modulate the system dynamics such that eventually a target state is approached even in the presence of structural deformations.

In the cybernetical era the notion of homeostasis has served as a kind of metaphorical explanation of various phenomena ranging from physiological control to social relations. Relevant work has been performed by Zemanek (1958) and his student Hauenschild (1956), who constructed a replica of Ashby's homeostat, including some display devices such as a face with movable parts driven by the output of the homeostat and a glass screen where two spots of light were controlled each by two units of the homeostat. According to Wilkins (1968) they considered also coupling to an external environment, but perhaps this refers only to these output devices.

In the early 1960s Haroules and Haire (1960) constructed a homeostat with 16 units (cf. also Williams, 1961) called Jenny, it is however not clear whether any results have been obtained. Finally, Wilkins (1968) has started an even larger project on homeostats, which unfortunately could not be continued. Only recently (Di Paolo, 2000, 2002, Balaam, 2001, Ducker, 2001) the idea of homeostasis experienced a revived interest. Schwefel (1994) has noted that the function of Ashby's original homeostat bears some similarity with evolutionary algorithms.

The present contribution tries to relate more directly to the early works for several reasons, while leaving aside the peculiarities of the hardware implementation of the homeostat. We will see that the main idea of the homeostat can be formulated in terms of continuous adaptive dynamical systems and treated analytically, where explicit solutions can be obtained at least in a number of special cases. Practical applications of the homeostatic principle to robot control (Der et. al., 1999, 2002) are beyond the scope of the present paper, but have given a starting point for the present work. Further, the learning homeostat may turn out to provide a standard model with implications for various problems in the theory of adaptive behavior, autonomous robotics, neural learning dynamics, and nonlinear control. We will first present a formal account of homeostatic dynamical systems. Later we will introduce an adaptive dynamics which adjusts the parameters of the system, provide some analytical considerations and numerical results. Finally, we will discuss presently relevant implications of homeostatic dynamics.

2. Ashby's homeostat

Ashby has built a machine consisting of four rotatable magnets whose deviations from their target orientation gave rise to currents in a number of coils that in turn influenced the orientation of the magnets. The strengths of the interactions among the magnets are subject to switching processes realized by pseudo-random resistor values produced by multi-selectors. The dynamics of the system can be described mathematically (Ashby, 1948, 1952, cf. also the detailed presentation in Hauenschild, 1956) by a seemingly linear system in $n = 4$ dimensions.

$$\dot{x}_i = \sum_{j=1}^{n} a_{ij} x_j \qquad (1)$$

The self-couplings a_{ii} are adjusted by hand to negative values initially, such that the system has a bias towards stability. The coefficients a_{ij} for $i \neq j$ are subject to a switching dynamics and are automatically chosen from a set of 9 possible values, which are fractions of 0.48, 0.73, 0.89 and 1.0 of some maximal value, the corresponding negative values or zero. Switching occurs if a variable x_j reaches a threshold $|x_j| \geq \theta = \frac{\pi}{4}$ with the target being at the origin. For each of the four vectors (a_{21}, a_{31}, a_{41}), (a_{12}, a_{32}, a_{42}), (a_{13}, a_{23}, a_{42}), and (a_{14}, a_{24}, a_{34}) there are 25 combinations of values stored out of the $9^3 = 729$ possible ones such that the system (1) includes up to $25^4 = 390625$ different dynamical behaviors.

The system state x is allowed to evolve for some time. The system will either approach some attractor inside the critical surface or eventually reach a critical surface. If $|x_i| \geq \frac{\pi}{4}$ a new set of random variables a_{ij} for all off-diagonal couplings towards unit i is drawn such that the system is changed each time a potential instability is encountered. Since in the hardware implementation the critical surface is simply given by a mechanical constraint, the state is assumed not to evolve further once the critical surface is reached even if the newly chosen couplings do not reintroduce stability into the system. The switching process stops if the coefficients of (1) are such that all eigenvalues have negative real parts.

In the stable region of the combined parameter and state space, the system tolerates small noise. But by an external impact that is sufficiently strong, e.g. the loss of a physical connection between node i and j or the reverse of the sign of an a_{ij}, the system returns via a large deviation in state space to a possibly different parameter setting with the desired dynamical properties. For random couplings, negativity of all eigenvalues is realized only with probability 2^{-n}, i.e. homeostats with large n will converge only after very long transients.

Further, oscillatory behavior or slow dynamics may occur inside the critical region such that the time between switching events may be long. If the eigenvalues are of small negative real part the system will be prone to noise. The fact that the system dynamics stops when x_i reaches the critical surfaces implies a separation of time scales which is not clear for the envisaged applications of homeostasis. When the critical surface is specified by soft thresholds, state may be controlled to avoid the boundaries and to stay near the target state by the exploitation of gradients of the threshold functions.

3. The learning homeostat

While in Ashby's approach stabilizing weights are found by random search, we suggest an incremental weight update, because it is reminiscent to the modification of synaptic efficacies in neural networks and will turn out to show a directed convergence.

The parameters a_{ij} in (1) are to be selected such that the system is stable at the fixed point $x = 0$. We will now discuss a learning scheme for the parameters a_{ij} which in a sense realizes Ashby's original ideas in a supposedly straightforward way. We consider the more general case that the system is to be stabilized at an arbitrarily given target state $\tilde{x}$. Because $x^* = 0$ is the only fixed point of (1), an adaptable inhomogeneity is required which can shift the fixed point towards $\tilde{x}$.

$$\dot{x}_i = \sum_{j=1}^{n} a_{ij} x_j + b_i \tag{2}$$

The aim is realized if the matrix $A = (a_{ij})$ is invertible and has eigenvalues with negative real parts, such that the inhomogeneity $b = (b_i)$ can be obtained by $b = A^{-1}\tilde{x}$.

In order to derive an adaptation rule for the parameter matrix A we introduce the target state $\tilde{x}$ which is to be approached by the system state x. Hence, the goal is to minimize

$$E = \frac{1}{2} \|x - \tilde{x}\|^2 . \tag{3}$$

Note that only future states $x(t+\tau)$, $\tau > 0$, can be influenced by present parameter changes. More precisely, we have

$$E = \frac{1}{2} \|x(t+\tau) - \tilde{x}\|^2 , \tag{4}$$

instead of (3), where $x(t+\tau)$ can be approximated for small τ by $x(t+\tau) \approx x(t) + \tau\dot{x}(t)$. Specifically for the system (2) we have

$$x(t+\tau) \approx x(t) + \tau\left(A(t)x(t) + b(t)\right).$$

The objective function (4) becomes

$$\begin{aligned} E = & \tfrac{1}{2} \|x + \tau(Ax+b) - \tilde{x}\|^2 \\ & \approx \textstyle\sum_i (x_i - \tilde{x}_i)^2 + \tau(x_i - \tilde{x}_i)\left(\sum_j a_{ij}x_j - b_i\right) + O(\tau^2) \end{aligned}$$

and suitable parameter changes may be determined by gradient descent

$$\begin{aligned} \Delta a_{ij} &= -\varepsilon \tfrac{\partial E}{\partial a_{ij}} = -\varepsilon\tau\,(x_i - \tilde{x}_i)\, x_j \\ \Delta b_i &= -\varepsilon \tfrac{\partial E}{\partial b_i} = -\varepsilon\tau\,(x_i - \tilde{x}_i) \end{aligned} \tag{5}$$

In matrix form this reads

$$\begin{aligned} \Delta A &= -\varepsilon\tau\,(x - \tilde{x})\, x^T \\ \Delta b &= -\varepsilon\tau\,(x - \tilde{x}) \end{aligned} \tag{6}$$

We assume that the parameter update is effective after a time delay τ_1 such that $A(t+\tau_1) = A(t) + \Delta A$ with $0 < \tau_1 \ll \tau$. For $\tau \to 0$, $\tau/\tau_1 = \text{const}$, we consider

$$\frac{A(t+\tau_1) - A(t)}{\tau_1} = -\varepsilon\frac{\tau}{\tau_1}(x - \tilde{x})\, x^T$$

and obtain from the analogous relation for b and (2) the dynamical system (with ε rescaled by τ/τ_1)

$$\begin{aligned} \dot{x} &= Ax + b \\ \dot{A} &= -\varepsilon\,(x - \tilde{x})\, x^T \\ \dot{b} &= \varepsilon\,(x - \tilde{x}) \end{aligned} \tag{7}$$

The same learning dynamics is obtained when aiming directly at a decrease of the distance from the target state $\tilde{x}$. This task can be expressed as the minimization of the function

$$\begin{aligned} E &= \frac{1}{2}\frac{d}{dt}\|x - \tilde{x}\|^2 = \frac{1}{2}\sum_i \frac{d}{dt}(x_i - \tilde{x}_i)^2 \\ &= \sum_i (x_i - \tilde{x}_i)\,\dot{x}_i = \dot{x}^T(x - \tilde{x}) \end{aligned}$$

A differential error function, namely,

$$E = \frac{1}{2}\|\dot{x}\|^2$$

gives rise to the dynamical system

$$\begin{aligned} \dot{x} &= Ax + b \\ \dot{A} &= -\varepsilon(Ax + b)x^T \\ \dot{b} &= -\varepsilon(Ax + b) \end{aligned} \tag{8}$$

Here stationary states are given by $Ax + b = 0$, such that $x^* = -A^{-1}b$. Since no explicit information on the target state x^* is present in the adaptation rules there is a multitude of solutions.

4. Analysis of simple homeostats

We will first consider a homeostat with a one dimensional state space and with $b = 0$. The system (7) reduces to

$$\begin{aligned} \dot{x} &= ax \\ \dot{a} &= -\varepsilon x^2 \end{aligned} \tag{9}$$

which describes the dynamics in the two dimensional (x, a)-space with target $x_0 = 0$. The dynamics of (9) can be easily understood qualitatively. If $a(0) < 0$, the state x will quickly run into the fixed point $x^* = 0$, while $a\,(t)$ tends to become more negative. a is strictly decreasing also for $a(0) > 0$, but now the magnitude of x grows as long as a is above zero. At $a = 0$ the state x reaches its maximal deviation from x_0 such that a will further decrease below zero, such that now x returns towards x_0. Eventually a saturates, and x further converges exponentially to zero with a nearly fixed rate, cf. Fig. 1.

The final value of a can be identified by exploiting the existence of a first integral of (9). By calculating the derivative $\frac{d}{dt}F(x,a)$ it is easily checked that the quantity

$$F(x,a) = \varepsilon x^2 + a^2 = \text{const} \tag{10}$$

is conserved by the dynamics (9). This means that due to $x(\infty) = 0$ the value $a(\infty)$ is given by

$$a(\infty) = -\sqrt{a^2(0) + \varepsilon x^2(0)},$$

where the sign is determined by the above qualitative consideration. Thus, for $a(0) > 0$ and $\varepsilon x^2(0) \ll a(0)$ the parameter a simply perform a sign switch, i.e. the parameter dynamics turns out to perform a directed, smooth analog of Ashby's original switching mechanism.

The conservation law (10) can be used for an explicite solution of (9), because it allows to eliminate one dimension. Introducing polar coordinates

$$\begin{aligned} x &= \tfrac{1}{\sqrt{\varepsilon}} r \sin\phi \\ a &= r\cos\phi \end{aligned} \tag{11}$$

we have $r^2 = \text{const}$ instead of (10). Inserting (11) into (9) we find a closed equation for ϕ

$$\dot{\phi} = r\sin\phi \tag{12}$$

which can be solved exactly as

$$\phi(t) = \pm 2\arccos\left(\pm\left(1 \pm \exp(2rt)\tan^2\left(\frac{\phi(0)}{2}\right)\right)^{-\frac{1}{2}}\right)$$

where the various signs reflect symmetry properties of (12). $\phi^* = 0$ is an unstable fixed point of (12) such that for $\phi(0) > 0$ ($\phi(0) < 0$) the dynamics converges towards $\phi = \pi$ ($\phi = -\pi$) i.e. to $x = 0$ and $a = -r$. A special solution of (9) is shown in Figure 1.

In case of a non-zero bias b we consider

$$\begin{aligned} \dot{x} &= ax + b \\ \dot{a} &= -\varepsilon x^2 \\ \dot{b} &= -\varepsilon x \end{aligned} \tag{13}$$

If x is close to zero, its evolution will be essentially determined by b and assume the same sign as b, which results eventually in a decay of b towards zero. If x and b have the same sign, the dynamics is similar to the case without a bias. Analogously to (10), for (13) it can be shown that

$$\frac{d}{dt}\left(\varepsilon x^2 + a^2 + b^2\right) = 0, \tag{14}$$

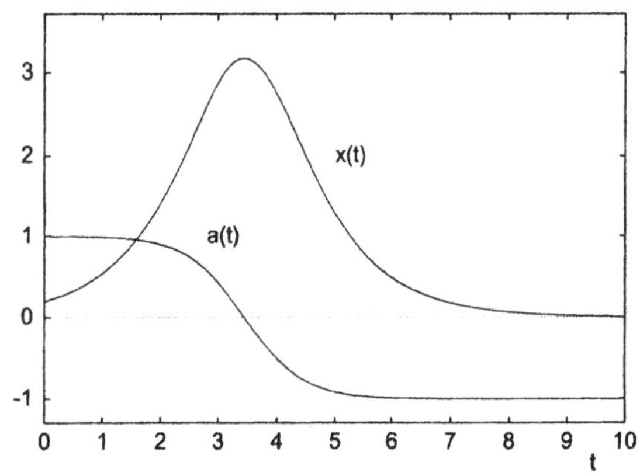

Figure 1: Time course of state x and parameter a for the one dimensional homeostat. Initial values are $a(0) = 1$ and $x(0) = 0.2$ and the learning rate is $\varepsilon = 0.1$. The excursion of x is reversed as soon as a changes sign. When x finally tends to zero, a settles near $-a(0) = -1$.

In order to find an explicite solution, equation (13) can be transformed into spheric coordinates

$$\begin{aligned} \sqrt{\varepsilon}x &= r\cos\phi\sin\theta \\ a &= r\sin\phi\sin\theta \\ b &= r\cos\theta \end{aligned}$$

such that r is invariant with respect to the dynamics and we are left with the reduced system

$$\begin{aligned} \dot{\phi} &= -r\cos\phi\sin\theta + \sqrt{\varepsilon}\sin\phi\cot\theta \\ \dot{\theta} &= -\sqrt{\varepsilon}\cos\phi \end{aligned} \tag{15}$$

The second equation has fixed points at $\phi = \frac{\pi}{2} + k\pi$, such that the first equation implies $\theta = \frac{\pi}{2} + l\pi$. The linearization of (15) yields eigenvalues $\frac{1}{2}\left(\pm r - \sqrt{r^2 - 4\varepsilon}\right)$ and $\frac{1}{2}\left(\pm r + \sqrt{r^2 - 4\varepsilon}\right)$ with negative real part if $k + l$ is an odd number. The eigenvalues are real if $4\varepsilon < r^2$, i.e. for large ε the solution is fluctuating before it settles to a stable fixed point.

5. Multidimensional homeostat

5.1 Homogenous case

Consider now the multidimensional case of (9)

$$\begin{aligned} \dot{x} &= Ax \\ \dot{A} &= -\varepsilon x x^T \end{aligned} \tag{16}$$

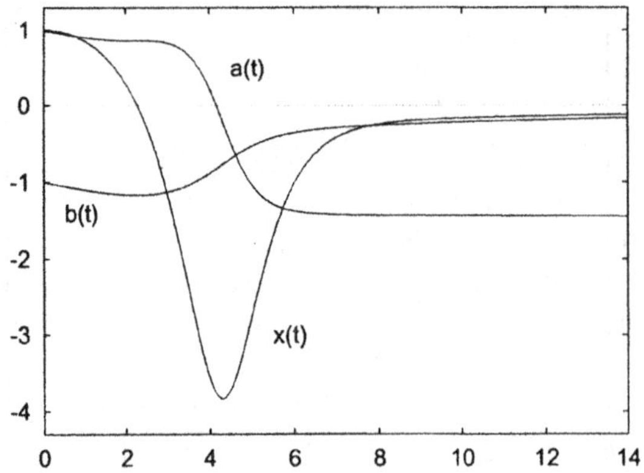

Figure 2: Time course of state x and parameters a and b for the one dimensional homeostat with bias. Initial values are $x(0) = 1$, $a(0) = 1$ $b(0) = -1$, and $\varepsilon = 0.1$. For this set of initial conditions the excursion of x undergoes now a sign change which is due to the initially large negative bias. As soon as b and x have the same sign the bias is no longer interfering the stabilizing dynamics and the further evolution is similar as shown in Figure 1, but with a smaller terminal value of a.

where $x \in \mathbf{R}^n$, $A \in \mathbf{R}^{n\times n}$. $n = 4$ corresponds to Ashby's homeostat. The expression

$$F(x, A) = \varepsilon x x^T + AA^T \tag{17}$$

is a first integral of (16), i.e. the dynamics of the full system is restricted to the set

$$\{(x, A)|F(A, x) = V = \text{const}\}$$

with V being a semidefinite symmetric matrix.

For a symmetric parameter matrix the convergence of the state dynamics can be shown easily, thus we will not assume that A is symmetric. We consider the dynamics of $r^2 = x^T x$ which obeys

$$\dot{r^2} = \dot{x}^T x + x^T \dot{x} = x^T \left(A^T + A\right) x$$

for (16). The norm of x is thus governed only by the symmetric part of A. r^2 tends to zero if $B = A^T + A$ has only strictly negative eigenvalues. Let b_i an eigenvector of B and β_i the corresponding eigenvalue and assume that A has a full set of n eigenvectors. Then, taking the time derivative of the eigenvalue equation we obtain

$$\dot{\beta}_i b_i + \beta_i \dot{b}_i = \dot{B} b_i + B \dot{b}_i$$

or, using $\dot{B} = -2\varepsilon x x^T$ and multiplying by b^T from the left,

$$\dot{\beta}_i b_i^T b_i = -2\varepsilon b_i^T x x^T b_i + b_i^T \left(B - \beta_i\right) \dot{b}_i$$

where the last term vanishes, because b_i^T is an eigenvector of $B^T = B$. Thus, we have

$$\dot{\beta}_i = -2\varepsilon \left\|x^T b_i\right\|^2 / \left\|b_i\right\|^2 \leq 0, \tag{18}$$

i.e. all eigenvalues λ_i with $x^T b_i \neq 0$ are strictly decreasing. Because we can assume generic initial conditions, i.e. $x^T(0)\, b_i(0) > \eta > 0$, and that $x^T b_i$ even grows for positive β_i all eigenvalues will eventually become negative.

Assume $A(0)$ and $A(\infty)$ being symmetric with real entries and of full rank. If $\|x(0)\| \ll 1$ and $x(\infty) = 0$, then both $A(0)$ and $A(\infty)$ are generalized roots of V and we can write approximately

$$A(\infty) = QA(0)$$

where $Q^{-1} = Q^T$ is an orthogonal matrix. We further have that

$$A(\infty) = A^T(\infty) \Rightarrow QA(0) = A(0)Q^T.$$

Thus

$$Q = A(0)Q^T A(0)^{-1}$$

which implies that Q and Q^T have the same eigenvalues, i.e. $\lambda = \bar{\lambda}$, thus the eigenvalues are real and, because Q is orthogonal, they are ± 1. In this way we have obtained an analogon of the sign-switch property from the one dimensional case.

5.2 Global analysis

Eq. (17) implies that

$$\varepsilon x x^T + AA^T = V$$

with $V = V^T$. By diagonalizing V via $V = TDT^T$, where T is an orthogonal matrix and D is a diagonal one with non-negative entries, we define

$$y = \sqrt{\varepsilon} T x, \quad G = TAT^T. \tag{19}$$

Using (19) the system (16) transforms into

$$\begin{array}{rcl} \dot{y} & = & Gy \\ \dot{G} & = & -yy^T \end{array} \tag{20}$$

where the first integral (17) becomes now

$$D = yy^T + GG^T = \tilde{F}(G, y) = \text{const.} \tag{21}$$

which can be written as

$$\tilde{C}\tilde{C}^T = \tilde{D} \tag{22}$$

where

$$\tilde{C} = \begin{pmatrix} 0 & 0 \\ y & G \end{pmatrix} \in \mathbf{R}^{(n+1)\times(n+1)}$$

$$D = \begin{pmatrix} 0 & 0 \\ 0 & D \end{pmatrix} \in \mathbf{R}^{(n+1)\times(n+1)}$$

The matrix $\tilde{C}$ in Eq. (22) can by termed a generalized root of $\tilde{D}$. It can be proven that the set

$$M_D = \{(G,y) \in \mathbf{R}^{n\times n} \times \mathbf{R}^n : F(G,y) = D\} \tag{23}$$

is an $\frac{n(n+1)}{2}$-dimensional compact submanifold of $\mathbf{R}^{n\times n} \times \mathbf{R}^n$, if the matrix D has full rank, which is the case for almost every initial condition.

This allow to characterize the manifold of fixed points of (16). (21) allows to eliminate the y equation in (20) and to consider the reduced dynamics

$$\dot{G} = GG^T - D. \tag{24}$$

The set of fixed points of (24) is determined by $GG^T = D$, hence any stationary G can be written as $G = \sqrt{D}O$, where $O^T = O^{-1}$ is an arbitrary orthogonal matrix. We can express the set of solutions by $\sqrt{D}O(n)$ where $O(n)$ denotes the set of orthogonal matrices. Because every solution G has full rank, at the fixed point of the joint dynamics we have $y = 0$. Consequently, the fixed-point manifold $\sqrt{D}O(n) \times \{0\} \subset M_D$ is of the same dimension as $O(n)$, i.e. $\frac{n(n-1)}{2}$. Recalling that D is fixed by the initial values, we arrive at the reduced form

$$\dot{y} = \sqrt{D}$$

The manifold which combines the dynamics of the full set of variables of (16) can also be characterized using (21). Since $\det D \neq 0$, the set M_D (23) is diffeomorphous with M_I, where I is the $n \times n$ unit matrix. The corresponding diffeomorphism is given by $G \to \sqrt{D}G \quad y \to \sqrt{D}y$. Instead of (21) we can restrict ourself thus to the special case

$$yy^T + GG^T = I$$

For any y we define an orthogonal $n \times n$ matrix R_y with $\det R_y = 1$ and a positive semi-definite diagonal matrix $D_{|y|}$, such that

$$\begin{aligned} R_y GG^T R_y^T &= R_y(I - yy^T)R_y^T \\ &= \begin{pmatrix} 1-|y|^2 & 0 \\ 0 & I_{(n-1)\times(n-1)} \end{pmatrix} = D_{|y|}^2 \end{aligned}$$

Further, we define an orthogonal $n \times n$ matrix V_y with $\det V_y = -1$ which satisfies

$$V_y = R_y \begin{pmatrix} -1 & 0 \\ 0 & I_{(n-1)\times(n-1)} \end{pmatrix} R_y^T.$$

Then, each matrix in the set $U_1 = M_I \cap \{(G,y) : \det G > 0, |y| < 1\}$ can be represented by the form $(R_y D_{|y|} R_y^T T, y)$ and each one from $U_2 = M_I \cap \{(G,y) : \det G < 0, |y| < 1\}$ by $(R_y D_{|y|} R_y^T V_y T, y)$, resp., where T is orthogonal with $\det T = 1$. Both U_1 and U_2 are diffeomorphous to $\mathcal{SO}(n) \times E^n$, where $SO(n)$ is the set (group) of orthogonal matrices with positive determinant and E^n is the open unit sphere $M_I = \bar{U}_1 \cup \bar{U}_2$

$$\lim_{|y|\to 1} R_y D_{|y|} R_y^T T = \lim_{|y|\to 1} R_y D_{|y|} V_y R_y^T$$

$$= T_y \begin{pmatrix} 0 & 0 \\ 0 & I_{(n-1)\times(n-1)} \end{pmatrix} T_y^T T = g(T,y)$$

$$g(T_1,y) = g(T_2,y),\ T_1,T_2 \in \mathcal{SO}(n) \ \Leftrightarrow T_1 = T_2$$

The two parts of the set of solutions can be joined by identifying the margins of $\bar{E}^n$ (E.g. for $n = 2$ the margins are one dimensional circles surrounding two dimensional disks, which are glued together yielding an object which is topologically equivalent to a sphere). The joint set of solutions M_I is thus homeomorphous to $SO(n) \times S^n$, with S^n being the n dimensional sphere. Formally this process is denoted as

$$\begin{aligned} & \left(\mathcal{SO}(n) \times \bar{E}^n\right) \cup_{\mathrm{id}_{\mathcal{SO}(n)\times S^{n-1}}} \left(\mathcal{SO}(n) \times \bar{E}^n\right) \\ = \;& \mathcal{SO}(n) \times S^n \end{aligned}$$

Since M_D is homeomorphous to M_I if $\det D \neq 0$, we have also that the set M_D is homeomorphous to $SO(n) \times S^n$. Remember that D is given by the initial conditions of (16).

By construction it is clear that with respect to the above homeomorphism the fixed-point manifold is

$$\mathcal{SO}(n) \times \{\text{"north pole"}\} \cup \mathcal{SO}(n) \times \{\text{"south pole"}\},$$

although here the somewhat artificial nature of this representation becomes obvious: the state vector x always converges to the same fixed point $x = 0$, but the parameter matrix develops in two qualitatively different ways. In order to prove the stability of a fixed point of (16) we need to know the tangents of the manifold M_D in that fixed point. Expansion of the equation $GG^T + yy^T = D$ around the fixed point $(\sqrt{D}T, 0)$ yields

$$\delta G\sqrt{D}T + T^T\sqrt{D}\delta G^T = 0.$$

δG is thus tangential with respect to the fixed point manifold $\sqrt{D}\mathcal{O}(n)$ or is zero, while δy remains arbitrary. The eigenvalues in the direction of the tangents of the fixed-point manifold are all zero. Perturbations with $\delta G = 0$ behave within a linear approximation as

$$\dot{\delta y} = \sqrt{D}T\delta y,$$

such that the stability of a fixed point is determined by the eigenvalues of $\sqrt{D}T$.
Consider for example $n = 2$ and $D = \mathrm{diag}(\alpha^2, \beta^2)$. An two-dimensional orthogonal matrix X has the form

$$\begin{pmatrix} \cos\phi & \sin\phi \\ -\sin\phi & \cos\phi \end{pmatrix}$$

if $\det X = 1$. In this case the eigenvalues

$$\frac{1}{2}\left(\cos(\phi)(\alpha+\beta) \pm \sqrt{\cos(\phi)^2(\alpha+\beta)^2 - 4\alpha\beta}\right)$$

are obtained. If $\det X = -1$, i.e. for matrices of the form

$$\begin{pmatrix} \cos\phi & \sin\phi \\ \sin\phi & -\cos\phi \end{pmatrix}$$

we find analogously

$$\frac{1}{2}\left(\cos(\phi)(\alpha-\beta) \pm \sqrt{\cos(\phi)^2(\alpha-\beta)^2 + 4\alpha\beta}\right)$$

5.3 Multidimensional with bias

Here the simple homeostat with a bias term (7) is formulated for an n dimensional state space, i.e. we consider the system

$$\dot{x} = Ax + b.$$

The parameters A and b are modified such that x approaches the target state $\tilde{x}$, which is expressed as the gradient dynamics (6)

$$\begin{aligned} \dot{A} &= -\varepsilon\,(x-\tilde{x})\,x^T \\ \dot{b} &= -\varepsilon\,(x-\tilde{x}) \end{aligned}$$

which as derived from (4). We find a conservation law

$$0 = \frac{d}{dt}\left(\varepsilon(x-\tilde{x})(x-\tilde{x})^T + AA^T + bb^T\right).$$

as a multidimensional analog of (14). The considerations on dynamics and stability are basically similar as in the homogeneous case, but more involved due to the additional dynamical bias vector as will be seen in the following section.

5.4 "Concomitant" learning

In the dynamical systems like (7) or (8) the parameter and the state dynamics are coupled via the learning rate ϵ. If ϵ is not infinitely small, parameter and state dynamics severely influence each other. In particular there is no seperation into a learning and performance phase so that we call this scenario concomitant learning. Tis interaction is particularly strong if the bias is included. This is seen most promoinently when comparing the dynamics $\dot{x} = Ax + b$ with the similar dynamics $\dot{x} = A\,(x + b)$, the difference being only the different definitions of the bias. Using the error function $E = \|x\,(t) + \tau\dot{x}\,(t) - \tilde{x}\|^2$ the combined system dynamics is

$$\begin{aligned} \dot{x} &= A\,(x+b) \\ \dot{A} &= -\varepsilon\tau(x-\tilde{x})(x+b)^T \\ \dot{b} &= -\varepsilon A^T(x-\tilde{x}) \end{aligned}$$

There is again a conservation law

$$\frac{d}{dt}\left(\varepsilon(x-\tilde{x})(x-\tilde{x})^T + AA^T\right) = 0$$

For a one dimensional state space the above system can be formulated in spheric coordinates (cf. (15)) which leads to the reduced dynamics

$$\begin{aligned} \dot{\phi} &= \frac{2\sqrt{\varepsilon}\cos\theta\left(\sqrt{\varepsilon}\cos\theta+\cos\phi\sin\theta\right)(\tilde{x}\cos\phi - r\cos(2\phi)\sin\theta)}{\cos(2\phi)+\cos(2\theta)} \\ \dot{\theta} &= \frac{2\sqrt{\varepsilon}\sin\phi\left(\sqrt{\varepsilon}\cos\theta+\cos\phi\sin\theta\right)(\tilde{x}\sin\theta + r\cos(2\theta)\cos\phi)}{\cos(2\phi)+\cos(2\theta)} \end{aligned} \tag{25}$$

MATHEMATICA produces a rather complex solution of (25) for the case $\tilde{x} = 0$. The system (25) has many fixed points. Of particular interest are the ones given by $\sqrt{\varepsilon}\cos\theta + \cos\phi\sin\theta = 0$, which form a one dimensional manifold of fixed points in (ϕ, θ)-space. The stability properties along the manifold vary due to sign switches of the other terms. Further fixed points of (25) are at (for $\tilde{x} = 0$), $\phi = \pm\frac{\pi}{2}$, $\theta = \pm\frac{\pi}{2}$ with all combinations of signs and zero eigenvalues of the Jacobian. The stationary states will thus be on the appropriate parts of the $\sqrt{\varepsilon}\cos\theta + \cos\phi\sin\theta = 0$ manifold.

6. Dynamics with noise

We consider both the case of the states being subject to Gaussian white noise

$$\begin{aligned} \dot{x} &= Ax + \sigma\xi \\ \dot{A} &= -\varepsilon x x^T \end{aligned} \tag{26}$$

as well as also the parameters are affected by noise

$$\begin{aligned} \dot{x} &= Ax + \sigma\xi \\ \dot{A} &= -\varepsilon x x^T + \rho\zeta \end{aligned} \tag{27}$$

both ξ and ζ being of zero mean and unit variance. It is obvious that for state noise (26) the general behavior of the homeostat is not much affected, expect in the sense that the eigenvalues of A diverge to $-\infty$ (with a rate proportional to $\sqrt{t}$) because $\dot{A}$ is continuously receiving negative contributions by the noisy fluctuations of x. From the numerical studies we observe also that the off-diagonal elements of A decay to zero if the noise is uncorrelated among different dimensions. Therefore the negative contributions to A directly change the eigenvalues.

For the noise affecting both states and variables the off-diagonal elements of A do not vanish, but rather perform a random walk with increasing amplitude. The real parts of the eigenvalues of A may thus return to the zero, but are pushed back as soon as a some deviation of x has been caused by the momentary instability, cf. Figure 3.

In particular if $\sigma \ll \varepsilon$ the state dynamics will stay very close to the fixed point of the deterministic dynamics, such that A performs essentially an unbiased random walk which is, however, reflected from the x-axis if an eigenvalue of A reaches zero from below. The dynamics of x looks spiky, i.e. every time when the system becomes unstable x performs a substantial deviation from zero which, however, quickly is reset by the response of the A dynamics, cf. Figure 4.

For weak parameter noise temporary instabilities are not excluded, and for larger noise they continue to occur, although at a lower rate.

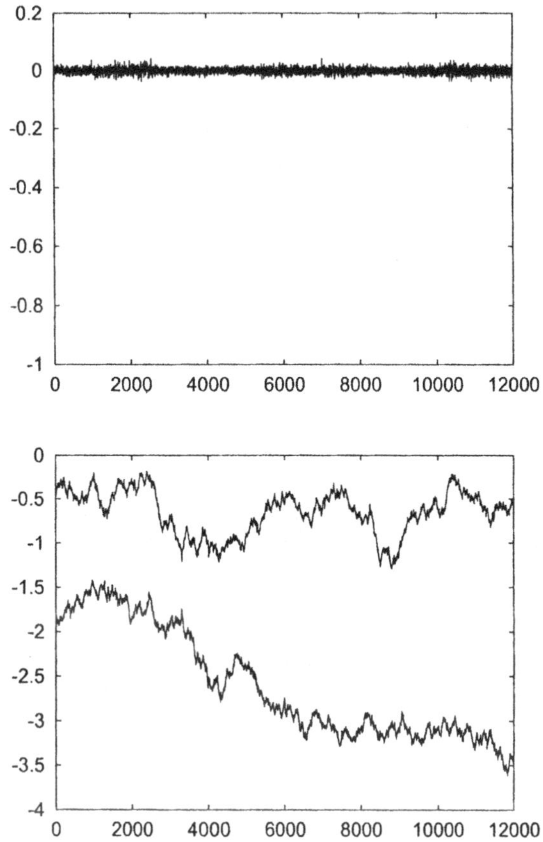

Figure 3: Dynamics with state noise in a two-unit homeostat. $\sigma = \rho = 0.1$. The top image presents one coordinate of the state, the bottom one the two eigenvalues of the parameter matrix.

7. Discussion

Parameter changes in Ashby's homeostat are performed by switching, i.e. whenever a state variable reaches a certain threshold, a subset of the parameters is replaced by a random configuration. Disregarding the specific architecture of the homeostat we may consider the parameter changes in the homeostat to occur on a much faster time scale than the state dynamics. In this way it is not possible to guide the parameters into suitable directions.

Adaptive systems, on the other hand, are mostly based on the opposite assumption, that parameter changes are much slower than the state dynamics. If the states are in a simple attractor, the information available from the states is limited and will not be extendable to other regions of the state space. The consequence is usually an increase of stability at the cost of an insensitivity to changes. This is in other words an instance of the stability-plasticity dilemma. Therefore, the state dynamics has to be controlled in some way in order to enable learning. If, on the other hand, the state dynamics is unstable, a slow parameter adaptation may be unable to prevent the system from divergence. Although there are ways to cope with such problems, such as resetting, shuffling or weighting the states as inputs to the learning algorithm, on-line learning in living systems or in machines as envisioned by Ashby requires parameter

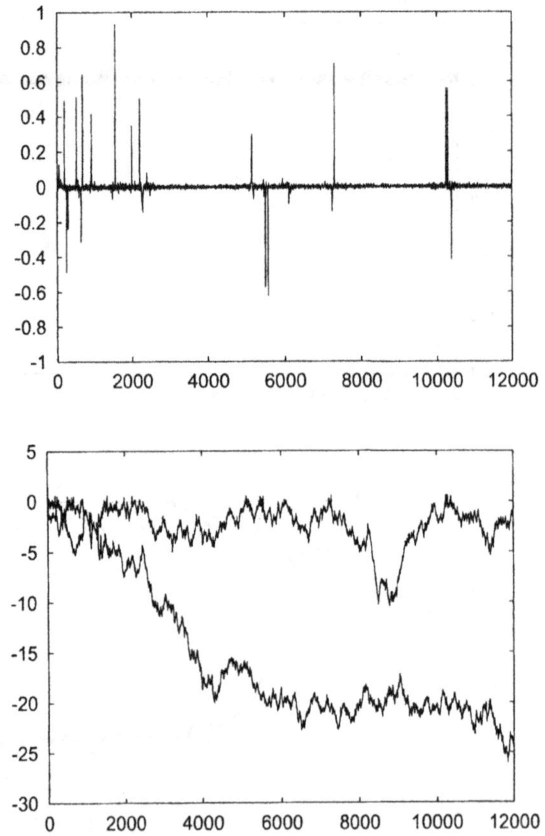

Figure 4: Dynamics with parameter noise in a two-unit homeostat. $\sigma = 0.1$ and $\rho = 1.0$.. The top image presents one coordinate of the state, the bottom one the two eigenvalues of the parameter matrix.

dynamics on a time scale which is comparable with that of the state dynamics. In this way states and parameters are dynamically equivalent and form a joint dynamical system with inseparable time scales, although generally different dynamical behaviors. Because then all variables are subject to intrinsic control, the agent can be improved only by structural modifications, such as performed by genetic algorithms.

If under certain conditions the parameter dynamics is converged, the corresponding behavior is governed by the state dynamics, i.e. on small scales ultrastability implies stability. If the external conditions change or some internal failure occurs, then the stability of the state dynamics may be lost and the parameter dynamics is required to reinstall stability of the target region. For larger excursions from the locally stable state, ultrastability requires thus the existence of possibly complex returning trajectories. Globally stable systems such as stable linear systems are in this sense not ultrastable, because ultrastability does not refer just to the dynamical variables (states) of the system, but also to the parameters, which can no longer be considered as fixed. The joint dynamics of parameters and states will thus be nonlinear and give rise to all kinds of nonlinear phenomena such as bifurcations, self-organization etc.

Although ultrastable system can cope with large scale changes, more complex systems than the one considered here should possess capabilities beyond thus, such as active learning and memory, e.g. in order to be able to predict long excursion or to initiate them in order to approach other goal states without an essential reorganization of the system properties. One step in this direction relates to active destabilization or, ideally, criticalization of the state dynamics. In such adaptation schemes the parameter dynamics tends to create center manifolds of the state dynamics (instead of stable fixed points of limit cycles) which allows for low-risk explorative behavior near the center and fast stabilization in the presence of unexpected external perturbations.

Hauenschild (1956) claims that the long transients in large homeostats can be avoided if the connectivity matrix is a block matrix, i.e. the described function separates into subfunctions which are controlled by low-dimensional servo-systems. More interesting is the possibility of considering blocks of different dynamical behaviors to subspaces of the state space. This allow to model systems with partial feedback arising in incomplete observable problems or in models that include an autonomous external or internal dynamics. Similarly, different learning rates or objective function may be applied to different blocks of the connectivity matrix, such as to achieve optimal exploration in certain directions while stabilizing other ones.

Ashby's homeostat is an attempt to mimic natural systems which are both resistive and reactive to the external world and are able to survive even structural disruptions. Prominent to Ashby's ideas is the switching of system parameters subject to some internal criteria (like the state reaching the boundary). In the present paper we introduce instead a parameter dynamics which rests on gradient descending an objective function. We have shown that in simple cases the adaptive system created in the way shares prominent features with Ashby's homeostat. In particular we observe a parameter switching dynamics as an emergent feature of the system to maintain stability.

To conclude, we have considered a particularly simple example of a nonlinear system which allows for analytical considerations some of which have been presented here. Applications of the present formal model will pose further questions such as the behavior in the presence of colored or correlated noise, the effect of time delays, the capabilities for tracking control and explorative dynamics.

References

W. R. Ashby (1948) The homeostat. *Electronic Engineering* **20**, 380ff.

W. R. Ashby (1952) *Design for a brain: The origin of adaptive behaviour.* London: Chapman & Hall. (second printing 1954, second edition 1960).

W. R. Ashby (1956) An introduction to cybernetics. Chapman and Hall: London.

W. R. Ashby (1981) *Mechanisms of Intelligence: Writings of Ross Ashby.* Intersystems Publications, Salinas, CA.

A. J. Balaam (2001) Analysing homeostatic adaptation to sensorimotor disruptions. MSc thesis. University of Sussex at Brighton.

W. B. Cannon (1932) *The wisdom of the body.* New York: Norton.

R. Der, U. Steinmetz, F. Pasemann (1999) Homeokinesis – A new principle to back up evolution with learning. In: M. Mohammadian (ed.) *Computational Intelligence for Modelling, Control, and Automation.* IOS Press, 43-47.

R. Der, J. M. Herrmann, R. Liebscher (2002) Homeokinetic approach to autonomous learning in mobile robots. In: R. Dillmann, R. D. Schraft, H. Wörn: *Robotik 2002. VDI-Berichte* **1679**, 301-306.

M. Ducker (2001) Homeostatic mechanisms in spiking neurons and spike-time dependent plasticity. MSc Thesis, University of Sussex at Brighton.

A. Hauenschild (1956) *Der Homeostat: Ein erweitertes Servosystem zur Nachahmung von Rückkapplungsvorgängen im Lebewesen.* Diplomarbeit. Technische Hochschule Wien, Institut für Schwachstromtechnik (NF).

E. A. Di Paolo (2000) Homeostatic adaptation to inversion in the visual field and other sensorimotor disruptions. In: J.-A. Meyer, A. Berthoz, D. Floreano, H. L. Roitblat, S. W. Wilson (eds.) From Animals to Animats 6. Cambridge, Mass, MIT Press, 440-449.

E. A. Di Paolo (2002) Fast homeostatic oscillators induce radical robustness in robot performance. Proc. of SAB'2002, MIT Press.

G. G. Haroules, P. F. Haire (1960) Jenny: An improved homeostat. *Air Force Cambridge Research Center, AFCRC-TN-60-379.*

H.-P. Schwefel (1994) On the evolution of evolutionary computation. In: J. M. Zurada and R. J. Marks II, C. J. Robinson (eds.) *Computational Intelligence - Imitating Life.* IEEE Press, Piscataway NJ, 116-124.

J. G. Taylor (1962) The behavioral basis of perception. New Haven: Yale University Press.

G. G. Turrigiano, K. R. Leslie, N. S. Desai, L. C. Rutherford, S. B. Nelson (1998) Activity-dependent scaling of quantal amplitude in neocortical neurons. *Nature* **391**, 892 - 896.

C. Wilkins (1968) A new homeostat. *BCL Report* 8.3, University of Illinois, Urbana-Champain.

R. E. Williams (1961) Static and dynamic responses of the homeostat Jenny. *Air Force Cambridge Research Laboratories, AFCRL-505.*

H. Zemanek (1958) La tortue de Vienne et les autres travaux cybernetics. *Proc. 1st Int. Cong. Cybernetics*, Namur, Belgium.

Toward a Dynamical Systems Analysis of Neuromodulation

Chris Buckley, Seth Bullock and Netta Cohen
Biosystems Group, School of Computing
University of Leeds, UK
christb|seth|netta@comp.leeds.ac.uk

Abstract

This work presents some first steps toward a more thorough understanding of the control systems employed in evolutionary robotics. In order to choose an appropriate architecture or to construct an effective novel control system we need insights into what makes control systems successful, robust, evolvable, etc. Here we present analysis intended to shed light on this type of question as it applies to a novel class of artificial neural networks that include a neuromodulatory mechanism: GasNets.

We begin by instantiating a particular GasNet subcircuit responsible for tuneable pattern generation and thought to underpin the attractive property of "temporal adaptivity". Rather than work within the GasNet formalism, we develop an extension of the well-known FitzHugh-Nagumo equations. The continuous nature of our model allows us to conduct a thorough dynamical systems analysis and to draw parallels between this subcircuit and beating/bursting phenomena reported in the neuroscience literature.

We then proceed to explore the effects of different types of parameter modulation on the system dynamics. We conclude that while there are key differences between the gain modulation used in the GasNet and alternative schemes (including threshold modulation of more traditional synaptic input), both approaches are able to produce tuneable pattern generation. While it appears, at least in this study, that the GasNet's gain modulation may not be crucial to pattern generation , we go on to suggest some possible advantages it could confer.

1. Introduction

Evolutionary roboticists employ a range of control architectures as varied as the research questions they address. Particular control systems are adopted for many different reasons. Some researchers wish to model a particular biological phenomenon and place emphasis on incorporating this particular aspect (Alexander & Sporns, 2002). Other researchers, (Beer, 2000; Tuci, Quinn, & Harvey, 2002) choose controllers for their perspicuity, attempting to incorporate as few *a priori* assumptions as possible. From an engineering perspective, however, we wish to employ highly evolvable control architectures—ones that facilitate fast and effective evolution of high-quality solutions to the kinds of task we are typically faced with. Unfortunately, as of yet there are few guiding principles to help researchers choose an appropriate class of control system.

Recent successes of one novel type of artificial neural network control architecture, the GasNet (Smith, Husbands, & O'Shea, 2001), have led researchers to consider the evolvability of this architecture, its causes and generality. Unlike most artificial neural networks, GasNets include processes inspired by biological *neuromodulation* in addition to those modelled after the more traditionally understood process of *neurotransmission*. Although GasNets have only been tested on a small range of tasks to date, their success suggests that neuromodulation may increase the evolvability of this class of control system across a range of real-world tasks (Smith, Husbands, Philippides, & O'Shea, 2002).

In order to assess this hypothesis, it is important to try to identify which aspect or set of aspects of the GasNet architecture might be responsible for this increased evolvability. In doing so, we hope to be able to reveal design principles underlying successful control systems. Here we report the initial steps in a dynamical systems study of the GasNet system. Our goal is to gain a deeper insight into the possible root causes of the increase in evolvability exhibited by the GasNet. We will start by briefly outlining the idea of neuromodulation and suggesting a working definition to be used in the rest of this study.

2. Neuromodulation in Neuroscience

The traditional bias of modern neuroscience is to assume that fast synaptic transmission constitutes the primary means of neuronal interaction. Neuronal communication is taken to be a composite of three dominating ideas (Katz, 1999):

1. Fast—pulses or on-off responses act on the 10 millisecond time scale.
2. Point-to-point—a neuron's neighbourhood is completely specified by the incoming synaptic connections and the outgoing neuronal branches of its dendritic tree.
3. Inhibitory/excitatory—synaptic connections either increase or decrease the activation of a target neuron.

The transmission of synaptic pulses is confined to the synaptic cleft and is mediated by neurotransmitters. Hence the amalgamation of these notions is commonly known as *neurotransmission*, an idea with historical roots in a 1954 paper by Hodgkin and Huxley (HH) on the squid giant axon (Hodgkin & Huxley, 1954). This seminal paper is still regarded as the progenitor of all modern neuronal models. Hodgkin and Huxley were particularly interested in the electrical signalling or firing of nerve cells, primarily because of its ubiquity throughout the nervous system, but also because the idea of electrical circuits were very much a dominating paradigm of the era. Consequently, but perhaps not intentionally, studies of the synapse has dominated subsequent research. Although it has often been recognised that synaptic pulses do not tell the whole story and that many more subtle processes take place, these processes have been largely ignored. Recently, this shortcoming is being comprehensively challenged. Many neuroscientists have voiced dissent against the exclusiveness of neurotransmission (Harris-Warrick, Nagy, & Nusbaum, 1992; Abbot, LeMasson, Siegel, & Marder, 2003; Turrigiano, 1999). The "patch clamp" is an experimental technique which has generated an avalanche of very detailed data on neuronal activation (Poggio & Glaser, 1993, Ch 1). Researchers discovered that many phenomena do not fit easily into the picture provided by neurotransmission. For example, Harris-Warrick remarks that "it is no longer possible to discuss sensory processing or motor coordination without considering the role that non-traditional forms of neuronal activity and communication play" (Harris-Warrick et al., 1992).

These non-traditional processes have been collectively grouped under the term *neuromodulation*. Although the word has been used for over 20 years, the ubiquity of such processes has only just been acknowledged. Throughout this paper we will employ a working definition of neuromodulation provided by Katz that casts neuromodulation as the antithesis or complement of neurotransmission:

> "Any communication between neurons, caused by the release of a chemical that is either not fast, or not point-to-point or not simply excitation or inhibition" (Katz, 1999, p.3).

Neuromodulators act over a range of spatial and temporal scales. Neurotrophines and hormones, for instance, are large macromolecular chemicals that are transported via the circulatory system. They can change many properties of the neural tissue that they come into contact with, affecting the rates of synaptic growth and intrinsic properties of neurons (Turrigiano, 1999). Although they act over large volumes of tissue, they are relatively slow and diffuse. In comparison amphipathic neurotransmitters are small molecules that can pass through lipid tissue. Consequently they act over small volumes but are relatively fast (Changeux, 1993). They have been implicated in short term changes to synaptic strengths and the process of learning. Crudely, whereas neurotransmission has been conceived of as analogous to the operation of an electrical circuit, one can visualize neuromodulation as waves of gases and liquids diffusing from neurons or perhaps neuronal modules. They affect volumes of neural tissue and change the functionality of the neurons and synapses within it. In contrast to the dominating paradigm of electrical circuitry, a colourful term sometimes used to convey this alternative idea is "the liquid brain" (Changeux, 1993; Husbands, Philippides, Smith, & O'Shea, 2001).

3. Neuromodulation in Robotics, GasNets

Generally, dynamic artificial neural networks are represented as the simplest extension of the feedforward McCulloch-Pitts perceptron (Arbib, 1998, p.4-11). They were first popularised by Hopfield in the 1970's (Haykin, 1999; Hopfield, 1982), and are described by equations of the following form:

$$u_i(t+1) = F\left(\sum_{ij} \omega_{ij} u_j(t) + \theta_i\right) + I_i \tag{1}$$

The neural activation, u_i, of a network node, i, represents the mean firing rate of either a single neuron or an average over an ensemble of neurons. Each such node possesses a threshold, θ_i, and receives synaptic input from any neighbour, j, weighted by a synaptic link, w_{ij}, as well as external input, I_i. The function F is known as the transfer function and is generally of sigmoidal form, i.e. approximately linear in its mid-range with saturating limits. Note, the only dynamic variables are the neural activations, every other parameter is fixed before the network operates.

In the GasNet formalism this network of nodes is embedded in a 2-dimensional continuous space (Husbands et al., 2001). The neural activation of a node is now described by an equation of the following form:

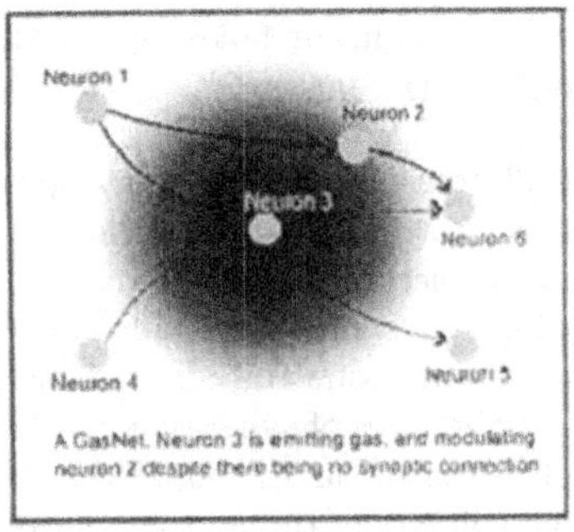

Figure 1: Gas diffuses across a control network embedded in a 2-d place, from (Husbands et al., 2001)

$$u_i(t+1) = F(\kappa_i[\sum_{ij} \omega_{ij} u_j(t)] + \theta_i) + I_i \qquad (2)$$

This differs from equation (1) only by the presence of a prefactor, κ_i, which is proportional to the concentration of "gas" in the vicinity of the node. Furthermore, each node may possess the ability to emit slowly diffusing gas under certain conditions—e.g., when either gas concentration at the node's location, or the node's neural activation, exceeds some fixed, node-specific threshold. Thus, as gas ebbs and flows across the plane in which the node's are located, it modulates the sum of the inputs in a multiplicative fashion, deforming the network's weight space (see Figure 1).

This gas mechanism exhibits all three qualities of neuromodulation as defined above. It is not point-to-point but diffuse in nature, it acts on a slower timescale than the electrical signals, and finally it neither simply excites nor inhibits neurons, but rather *modulates* the gain, or the steepness, of their transfer functions. Given this triumvirate of novel functionality, plus the embedding of the network in a continuous space, it is unclear exactly which element or combination of elements in the design of a GasNet is responsible for the increased evolvability that they exhibit.

Smith et al. (2002) made the first attempt at analysis of the GasNet architecture. In particular, they identified a subcircuit of the GasNet control system that had been evolved to solve a robot control problem, and showed that it contributed to the "temporal adaptivity" of the evolved solution.[1] This subcircuit was capable of tuneable pattern generation, in that it could produce electrical oscillations that were slowly modulated by gas such that rhythmic behaviours were readily achieved. However the discrete nature of the GasNet system hindered the completion of a more comprehensive dynamical systems analysis.

Here we will attempt to extend and generalise this analysis to some degree by developing and analysing a simple, idealised, continuous neuromodulatory network. Our model is based on the well-known FitzHugh-Nagumo equation (Murray, 1989, p.161-166), but inherits some important properties from the original GasNet formalism. We will explore the extent to which behaviours exhibited by GasNet sub-systems are characteristic of neuromodulatory systems more generally. In doing so, we can take advantage of both the continuous nature of the FitzHugh-Nagumo equation, which readily admits of dynamical systems analysis, and the body of existing analysis of the FitzHugh-Nagumo system. Moreover, the ubiquity of the FitzHugh-Nagumo model in neuroscience ensures that we have some chance of relating our findings on neuromodulation both to the adaptive behaviour and neuroscience communities.

4. Neuromodulation in the FitzHugh-Nagumo Equations

4.1 The FitzHugh-Nagumo Equations

While the (HH) equations were extraordinarily successful at reproducing key experimental data, a deep understanding of the underlying dynamics was hindered by their inherent complexity. In the early 1960's, the FitzHugh-Nagumo model (FHN), a simple, analytically tractable, yet non-trivial reduction of HH, was developed (FitzHugh, 1961; Nagumo & Yoshizawa, 1962). Electrical pulses are derived from the differential permeability of the neural tissue to chemical ions. The dominant ionic species in nerve membranes are potassium and sodium, but in general there are many ionic species acting over many timescales. Since the timescales of the ionic species in the HH model are not of the same order, simplification could be achieved by assuming that sufficiently fast variables would settle to their steady state values almost instantaneously. This allowed certain variables to be eliminated from the Hodgkin-Huxley equation. The FHN model is given in equations (3) and (4).

$$F(v,w) \equiv \frac{dv}{dt} = f(v) - w + I_\alpha, \ f(v) = v(a-v)(v-1) \quad (3)$$

$$G(v,w) \equiv \frac{dw}{dt} = \frac{bv - \gamma w}{\tau} \quad (4)$$

Here, v is the membrane potential, while w plays the role of the ionic currents. The remaining terms I_α, b, γ and τ are all positive constants and $0 < a < 1$. This model has been extraordinarily successful and displays

[1] The term "temporal adaptivity" refers to the ability of an agent to respond to changes on a wide range of timescales. Biological agents are faced with an environment that exhibits regularities at scales ranging from seconds and minutes to diurnal and annual rhythms. They must adapt their internal structures to mirror and exploit these regularities. The ability to do this has also been suggested as a vital aspect of artificial control systems.

many of the key phenomena discovered in the original Hodgkin-Huxley model. For example FHN readily exhibits excitable and oscillatory behaviour (Murray, 1989, p. 164). Consequently FHN seems a natural choice for investigating some of the issues raised by GasNet research.

Although the FHN equation involves only two free variables (alongside time) we cannot solve it directly. Instead progress can be made by investigating the stability of equilibrium states of the model. Equilibrium behaviour is defined as the state, or sequence of states, that the system settles to after some finite transient. The equilibrium points correspond to points in the phase space where all the derivatives with respect to time of the system are equal to zero. For FHN, the equilibrium points can be found by setting the derivatives of (3) and (4) to zero, giving equations (5) and (6).

$$w = f(v) + I \tag{5}$$

$$w = \frac{b}{\gamma} v \tag{6}$$

If we plot these equations we obtain the nullclines of the system (see Figure 2). For the whole system to be in equilibrium, derivatives of both v and w must equal zero, this point is represented by the intersection of the nullclines.

These equilibria can be stable (if perturbed from this value the system quickly returns) or unstable (if perturbed from this point the system may not return). With the aid of linear stability analysis it is possible to derive a condition for the local stability of any given equilibrium point. For details of such analysis the reader is referred to (Strogatz, 1994; Murray, 1989). For a very good introduction to dynamical systems theory for robot control systems see (Beer, 1995). In the case of FHN, stability is determined by a single quantity known as the *trace*. For this system the condition for stability is given below. Where v^* is the value of v at equilibrium.

$$Tr(A) = \frac{\partial f(v^*)}{\partial v} - \frac{\gamma}{\tau} < 0. \tag{7}$$

The FHN equations exhibit three classes of behaviour which are determined by the number and stability of the equilibrium points. Figures 2a and 2b have a single intersection which is locally and globally stable. All initial conditions of the equations will relax to this point. Such configurations represent excitable systems—perturbations generate short-lived spiking followed by a return to quiescence. Alternatively if the nullcline cross at their centres they can exhibit fixed point or cyclic behaviour,Figure 2c and Figure 2d respectively.

The empty circles drawn on all plots represent the bifurcation points of the system. They delimit the region, for a particular set of parameters, within which

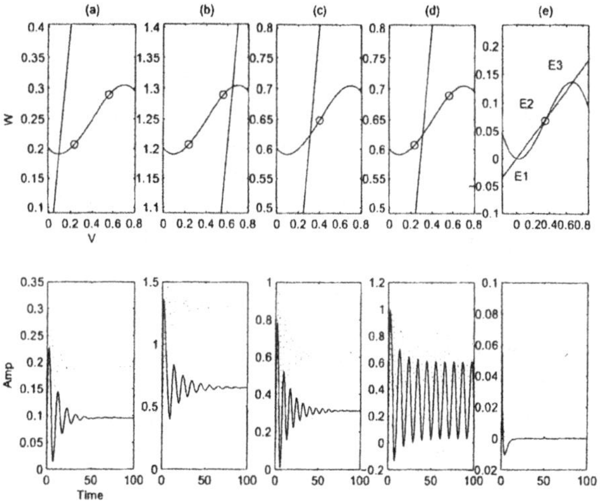

Figure 2: Modes of FHN behaviour: (a) and (b) represent non-oscillatory fixed-point behaviour but are excitable under perturbation. If the nullclines cross at their centres then the system can exhibit a fixed point (c) or cyclic behaviour (d). (e) shows a configuration with multiple equilibrium points. In general E1 and E3 are stable, while E2 is unstable. The circles drawn on all plots represent the bifurcation points of the system. They indicate the region within which the value of v^* is such that the first term of equation (7) makes the *trace* positive and hence unstable (which gives oscillations in this system). Outside this region the *trace* is negative and stable.

the value of v^* is such that the first term of in equation (7) makes the *trace* positive and hence the system is unstable (which gives oscillations in this configuration). Outside this region the *trace* is negative and the system is stable. Note the size of this region will be altered by the right hand term of equation (7). We will come back to this effect later in this work. Figure 2e presents a multistable configuration of the nullclines. It exhibits three equilibrium points with two stable (E1 and E3) and one unstable (E2). It can be shown that cyclic behaviour is not possible in this configuration.

Our work will focus on the case depicted in Figure 2d, the only case in this system that supports oscillation and is thus capable of rhythmic pattern generation. This configuration of nullclines has been implicated in the production of effective robot control. For example (Mathayomchan & Beer, 2002) evolved recurrent neural networks to perform a simple ball catching task. It was discovered that seeding the initial evolutionary population with "centre crossing" networks increased the speed and reliability of the evolutionary process.

Furthermore the ability to alternate between oscillatory and non-oscillatory behaviour has been identified in a host of neural systems. A parameter change that causes a qualitative change in a system's dynamics (e.g., a transition from non-oscillatory to oscillatory behaviour) is known as a bifurcation. The bifurcation in this particular case is known as the Hopf bifurcation and has been the focus of a great deal of investigation, both in its own right (Guckenheimer, Gueron, & Harris-Warrick, 1993) and as a model for biological systems (Rinzel & Ermentrout, 1989). In general it is thought that systems existing at or near bifurcations in their dynamics have greater potential for exhibiting interesting behaviour—they are more computationally rich (Poggio & Glaser, 1993, Ch 3). Bifurcation points have also been directly implicated in modern research on neuromodulation. For example, researchers argue that it might be advantageous for systems to dwell near bifurcation boundaries, thus making them sensitive to neuromodulatory input (Fellous & Linster, 1998).

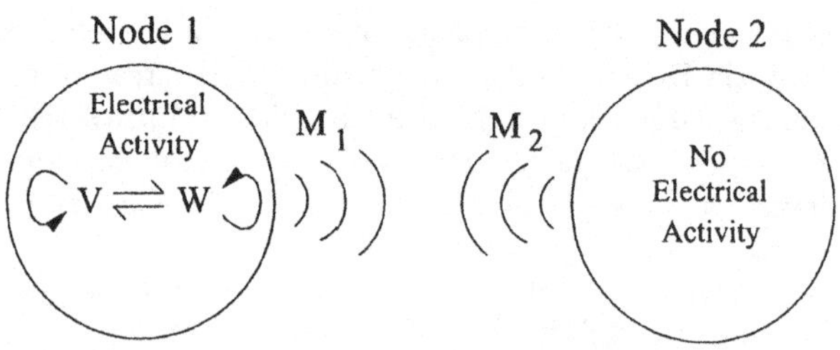

Figure 3: A dynamical pattern generator circuit: The electrical behaviour of node 1 is described by the FHN equation. It releases a modulator, M_1, when its electrical potential, v, rises above its electrical threshold, T_v. Node 2 is triggered to release a second modulator, M_2 when the concentration of M_1 rises above its modulator threshold T_C. Reciprocally, the concentration of this second modulator affects the parameters of node 1's FHN equation.

4.2 A Dynamical Pattern Generator

A GasNet "discrete dynamical pattern generator" was outlined by (Smith et al., 2002). It is a simple system of coupled neurons that utilizes a gas dynamic to modulate electrical oscillations such that tuneable rhythmic patterns of activity are generated. This key behaviour is obtained by slowly switching the system from an unstable to a stable dynamic. In order to explore a similar mechanism using the FHN equation, we added to it a modulatory mechanism analogous to the gas employed in the GasNet.

Our simple system comprises one node governed by the FitzHugh-Nagumo equation that emits a modulator, M_1, when its electrical activation rises above a fixed threshold, T_v. A second node is modelled in a much more simple fashion. Should the concentration of M_1 at this node rise above a fixed threshold, T_C, it emits a second modulator, M_2. The first node is sensitive to M_2, in that the concentration of this modulator affects the parameters of the node's FHN equation. Note that no electrical activity is modelled for the second node, which is merely a source of modulator that can be switched on and off (see figure 3). Our equations for modulator growth and decay at both nodes are given in (8)-(11).

$$\frac{dC_1}{dt} = H_1(v)G_1 + (H_1(v) - 1)D_1 \tag{8}$$

$$H_1(v) = \begin{cases} 1, & \text{if } v > T_v \\ 0, & \text{otherwise} \end{cases} \tag{9}$$

$$\frac{dC_2}{dt} = H_2(C_1)G_2 + (H_2(C_1) - 1)D_2, \tag{10}$$

$$H_2(C_1) = \begin{cases} 1, & \text{if } C_1 > T_C \\ 0, & \text{otherwise} \end{cases} \tag{11}$$

The concentration of each modulator is represented by C_i, with their specific growth and decay rates denoted G_i and D_i, respectively. Each node's heavyside function returns unity when it is emitting modulator, zero otherwise.

The concentration of M_2 affects the behaviour of node 1 in the following manner. Increasing concentration of M_2 decreases the γ parameter of node 1 in the manner described by equation (12), where C_{2max} represents a ceiling concentration value for M_2, and $[\gamma_{min}, \gamma_{max}]$ describes a legal range of values for γ. This γ modulation affects the node's nullclines in a fashion analogous to the gain modulation described in section 3. That is, γ could be considered to play a similar role to that played by κ in the GasNet equation (2).

$$\gamma = \gamma_{max} - \frac{C_2}{C_{2max}}(\gamma_{max} - \gamma_{min}) \tag{12}$$

The system is initialised in an oscillatory configuration. The initial nullclines of the system are depicted by the solid lines and crosses (the bifurcation points) in Figure 4d. Figures 4a and 4b represent the v and w time traces of the system with the fixed electrical threshold T_v superimposed over v. Figure 4c shows the build-up and decay of modulators M_1 and M_2, while figure 4d displays how the nullclines change as a result of modulation. The w nullcline oscillates between the solid and dashed lines, altering the system's equilibrium *position.* This change in the equilibrium position alters the first term of equation (7) from a negative to positive value and hence from an unstable oscillating solution (inside the crosses) to a stable fixed point (outside the crosses). Furthermore, because γ also affects the *trace* through the second term in equation (7), the modulation also alters the *range* of the oscillatory region. As modulation moves the w nullcline from the solid to the dashed line, the portion of the v nullcline associated with oscillatory behaviour grows from the section spanned by the two crosses to that spanned by the two circles.

The general effect of this modulation is to produce a beating/bursting system, with fast oscillation of the v and w variables within a low-frequency, modulated packet. Initially, as the system oscillates, M_1 builds, stimulating the emission of M_2. As the concentration of M_2 rises, it decreases γ such that node 1's equilibrium position lies outside the oscillatory region. The delay between the build up of M_1 and M_2 produces the low frequency packet. There is also a smoothing effect on M_2, since while M_1 displays small amplitude, high-frequency oscillations as it builds, these are not present in the dynamic of M_2.

This behaviour strongly resembles that of the discrete dynamical pattern generator identified within an evolved GasNet solution (Smith et al., 2002). The amplitude and frequency of the slow packet can easily be tuned by altering the speed and maximum concentration of the modulators. This shows that important aspects of the GasNet functionality can be instantiated and understood in a wider class of continuous systems which offer an advantage in terms of tractable analysis. Hence, we have demonstrated that dynamical pattern generation of the kind described for the GasNet is not particular to the specifics of that formulation of neuromodulation.

Furthermore the continuous nature of this system allows us to draw parallels between its behaviour and bursting behaviours referred to in the neuroscience literature (Rinzel & Ermentrout, 1989). The extent to which the behaviour of our simple model can be used to explore bursting remains an open question. However, in light of this observation, it may be interesting to use bursting ideas from neuroscience as a tool for understanding and designing robot control systems.

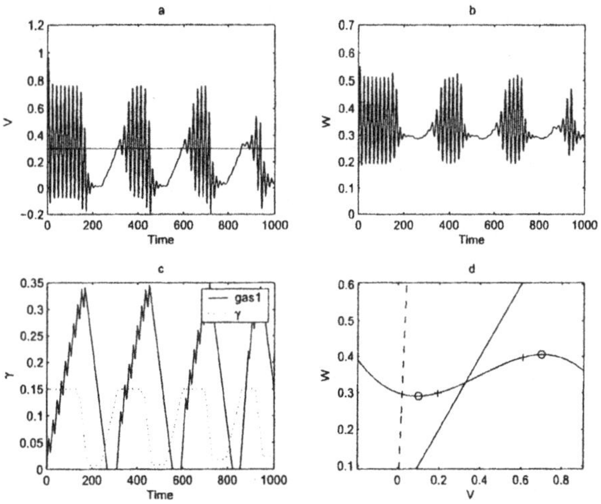

Figure 4: The output of the dynamical pattern generator: The figure shows the output of the system under γ modulation. Panels a and b show v and w, respectively. Figure c depicts the growth and decay of modulators M_1 and M_2. Figure d displays the movement of the nullclines and bifurcation points—the solid lines and crosses depict their positions in the absence of M_2, while the dashed line and circles denote their locations in the presence of a maximum concentration of M_2.

5. Not Inhibitory or Excitatory

We will now take a closer look at the movements of the nullclines during modulation of several kinds. In particular we wish to understand whether multiplicative gain modulation is necessary to effect dynamical pattern generation.

One of the key tenets of neuromodulation is that it "is not simply excitatory or inhibitory" but modulatory—i.e. it alters behavioural parameters, rather than merely activation levels. Given the tractability of our model, we are in a position to explore this assertion more closely. First of all let us assess what we mean by "not simply excitatory or inhibitory". In the GasNet formulation this idea is quite clear—the gain parameter acts as a *multiplicative* factor in the update function in contrast to the merely additive character of synaptic inputs,see equation (2).

The idea of multiplicative connections is not new. Indeed, Pollack (1987) argued that greater computational power would not come about without their use. Typically the output of a particular node is calculated as a function of the sum of the synaptic inputs. Pollack thought that connections that multiplied the sum of the synaptic inputs were equally as important. Furthermore he believed that full Turing-complete computability could not be realized without such connections. This was later shown not to be the case (Siegelmann & Sontag, 1995). Nevertheless, Pollack goes on to state that "multiplicative connections remain a critical and under appreciated component of neurally inspired computing" (Pollack, 1999). Although the addition of multiplicative connections may not confer new functionality on simple additive networks, they may alter what behaviour is easily obtained, and hence network *evolvability*. The next section will look closely at the different types of parameter modulation possible in our augmented FHN model.

5.1 Modulation in the FHN Equation

Figure 5 shows the change in the nullclines under various parameter modulations of equations (3) and (4). The circles mark the oscillatory region on the v nullcline.

Figure 5a shows the effect of modulating γ. The *trace* is affected in two ways: because γ affects the gradient of the w nullcline, it changes the equilibrium position, v^*, and hence the first term of equation (7). Furthermore, it also changes the second term of equation (7), which scales the oscillatory region (note the difference between the locations of crosses and circles).

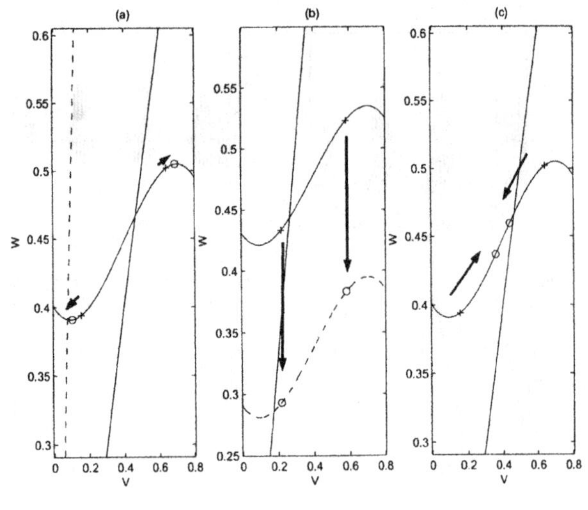

Figure 5: A transition from oscillatory to non-oscillatory behaviour caused by three types of modulation: (a) γ modulation, (b) I_γ modulation, and (c) τ modulation. Each panel displays the movement of nullclines and bifurcation points—solid lines and crosses depict their positions in the absence of modulation, while dashed lines and circles denote their locations in the presence of modulation. See text for details.

In contrast the size of the oscillatory region does not change under I_γ modulation, which merely translates one nullcline (see figure 5b). The only change to the *trace* is due to the first term of equation (7). This type modulation is closely analogous to synaptic input in equation (1). Perturbing I_γ is analogous to raising or lowering a GasNet node's electrical threshold, i.e. increasing or decreasing its level of activation. Hence, this type of change is not traditionally associated with neuromodulation since, at root, it is merely "excitatory or inhibitory".

Perhaps the most interesting type of modulation considered here is shown in Figure 5c, for the parameter τ. Perturbing this parameter produces no change in the nullclines and hence no change in the position of the equilibrium point, v^*. Nevertheless, this kind of modulation affects the second term of the *trace* equation, and hence alters the range of the portion of the v nullcline associated with oscillatory behaviour. The effect of manipulating τ is to scale the w axis while preserving the nullcline configuration.

In general there are two ways the nullclines can change, first the equilibrium point can be *translated*, which corresponds to a change in the first term of equation (7). Second, altering certain parameters can change the *size* of the oscillatory region, which corresponds to a change in the second term of equation (7). While γ (gain) modulation achieves a mixture of both effects, I_γ (threshold) modulation produces pure translation, and τ (time constant) modulation achieves pure scaling of the oscillatory region, leaving the equilibrium position of the system unchanged.

5.2 Comparing Different Modulation Types

If the neuromodulatory mechanisms in GasNets are responsible for enhanced evolvability (perhaps through facilitating the construction and tuneability of dynamic pattern generators of the kind explored above) it is important to understand how the different aspects of neuromodulation contribute. Is the fact that GasNets employ gain modulation critical? Our definition of neuromodulation (not fast, not point-to-point, not inhibition/excitation) suggests that it might be since alternative styles of modulation (e.g., threshold modulation) are merely equivalent to excitation/inhibition. Here we explore the ability of different types of modulation to effect dynamic pattern generation in an attempt to discover whether the particular kinds of "multiplicative" modulation feted in the literature are in some sense behaviourally superior. Might simple threshold modulation do the job (despite not counting as neuromodulation)?

Figure 6 shows our FHN model system under I_γ modulation (i.e. equation (12) is modified such that the γ, γ_{min} and γ_{max} terms are replaced by equivalent I_γ terms). It successfully produces beating behaviour analogous to that seen under γ modulation. Variation in M_2 causes a vertical translation of the v nullcline such that the equilibrium point, v^*, lies at times inside, and at other times outside, the region associated with oscillatory behaviour. The size of the oscillatory region remains unchanged.

Figure 7 shows the system under τ modulation. Again, it successfully exhibits beating behaviour. However, this is not achieved by translation of the nullclines, but rather by scaling the region of the v nullcline associated with oscillatory behaviour. In the absence of M_2, oscillatory behaviour is associated with the portion of the v nullcline spanned by the two crosses. As the concentration of M_2 increases, these points move together, reducing the size of the oscillatory region, until they collide at a point indicated by the open circle. At or above this level of M_2 concentration, no oscillatory behaviour is possible.

In both cases (as well as the case of γ modulation described earlier), slowly modulating a particular system parameter achieves beating by allowing the system to alternate between non-oscillatory and oscillatory modes of behaviour. The manner in which this alternation is achieved is all that varies. Thus, even though we have discovered some fundamental differences between the different forms of modulation, each remains able to support a dynamical pattern generator circuit. In particular, we have shown that threshold modulation (that is merely inhibitory/excitatory) is sufficient in this regard, despite not satisfying our definition of neuromodulation.

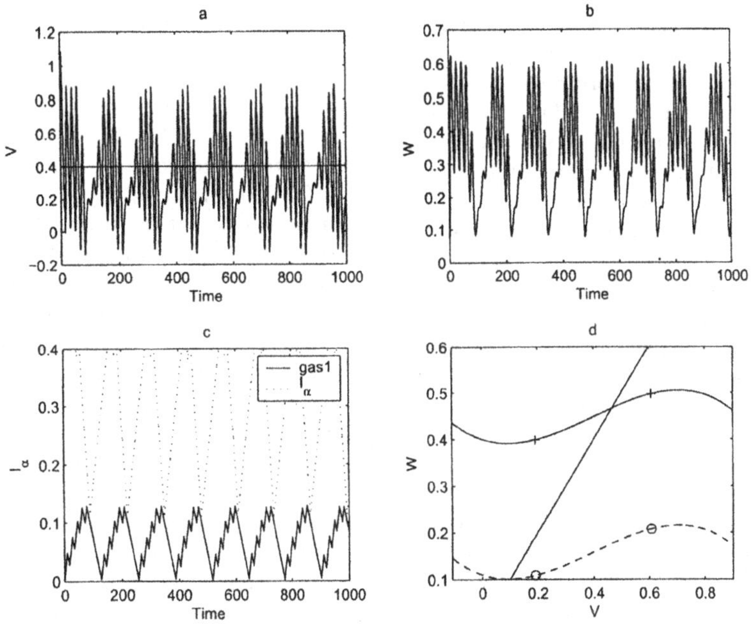

Figure 6: Output of the dynamical pattern generator under I_γ modulation.

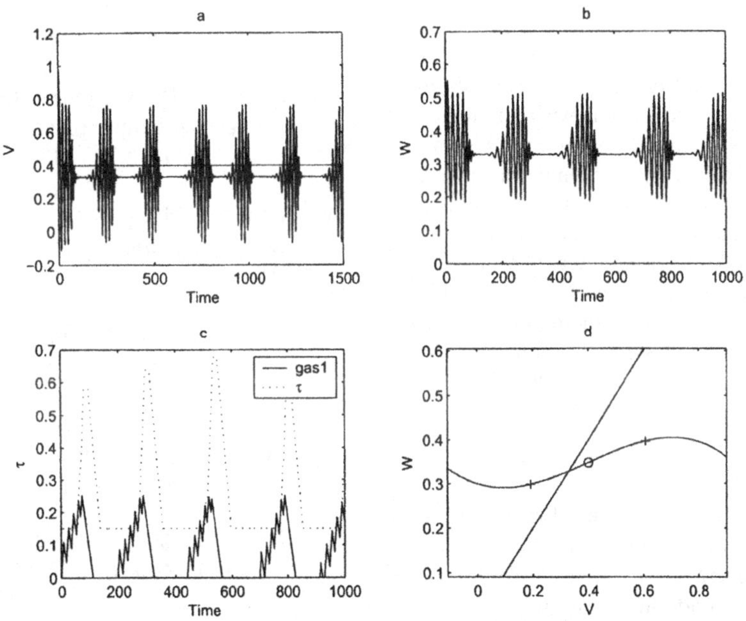

Figure 7: Output of the dynamical pattern generator under τ modulation.

However, while we have seen that it is possible for pattern generation to be achieved via a number of modulatory mechanisms, our current work suggests that the differences between gain modulation and threshold modulation are significant in some respects. Specifically, imagine an arbitrarily configured system of the kind described here that happens to be quiescent. Our analysis shows that gain modulation will be capable of exciting oscillatory behaviour in *any* such system. By contrast, mere threshold modulation will in general be unable to do so. Future papers will seek to characterise this and other differences between the behavioural and evolutionary potential of different modulatory mechanisms.

6. Conclusion

We have constructed a simple model of neuromodulation based on the FHN equation. Using this model we were able to obtain very similar behaviour to that produced by dynamical pattern generators observed in evolved GasNets. This lends weight to the idea that important GasNet behaviours are not specific to the particular idiosyncrasies of that formalism, but are more generic features of neuromodulation. We also noted that the observed behaviours were reminiscent of beating/bursting system studies in the neuroscience literature (Rinzel & Ermentrout, 1989).

We gave a working definition of neuromodulation suggested by (Katz, 1999). The GasNet contains all three characteristics of this definition. In particular gain modulation in the GasNet is concordant with the idea that neuromodulation is neither excitatory nor inhibitory, but rather modulatory. We investigated what this meant for our model system, observing how the system's nullclines changed under different kinds of parameter modulation. We concluded that different parameter modulations exhibited key differences in their mechanics. While I_γ modulation, which we argued was analogous to modulation of electrical threshold or synaptic input, could only *translate* the nullclines, γ modulation, which is analogous to gain modulation, could also affect the *size* of the oscillatory region (our third mechanism, τ modulation, was only able to affect this aspect).

Given these differences we explored their effect on the ability of the system to reproduce the behaviour of a dynamical pattern generator subcircuit. We discovered that while each type of modulation was able to effect the dynamic pattern generation behaviour, each achieved this in a different manner. Initially this leads us to believe that the slow dynamic of the modulation (its growth and decay) is perhaps more important than its additive or multiplicative nature. However, our most recent work suggests that multiplicative interactions may nevertheless be important for system evolvability.

7. Future Work

The analysis and results presented here are only a first step toward a complete answer to the question: "what is the source of GasNet evolvability"? One obvious next step is to consider the role of spatial embedding in GasNet evolvability. To what extent is the low-dimensionality of the space within which GasNet nodes are embedded implicated in their ability to solve robot control tasks. Second, we wish to extend our simple model to consider small assemblies of multiple neurons, and the role of neuromodulation in their self-organisation and self-regulation.

We believe that in order to answer these questions, we need to confront some of the deeper issues concerning neuromodulation, not only with respect to behaviour-based robotics but also in the wider context of the neuroscience community. In particular, given growing evidence that multiple parallel processes at different spatial and temporal scales co-operate (operate together) across neural systems, we would like to know, as Eve Marder puts it, "what factors stabilise network operation in the face of so many competing influences?" (Poggio & Glaser, 1993, chapter 2)

References

Abbot, L. F., LeMasson, G., Siegel, M., & Marder, E. (2003). Acvtivity-dependent modification of intrinic neuronal properties., 151–166.

Alexander, W. H., & Sporns, O. (2002). An embodied model of learning, plasticity, and reward. *Adaptive Behavior, 10*(3/4), 143–159.

Arbib, M. A. (Ed.). (1998). *The Handbook of Brain Theory and Neural Networks.* MIT Press, Cambridge, MA.

Beer, R. D. (2000). Dynamical approaches to cognitive science. *Trends in Cognitive Sciences, 4*(3), 91–99.

Beer, R. (1995). On the dynamics of small continuous-time recurrent neural networks. *Adaptive Behavior, 3*(4), 471–511.

Changeux, J. (1993). Chemical signaling in the brain. *Scientific American, 8*, 58.

Fellous, J.-M., & Linster, C. (1998). Computational models of neuromodulation. *Neural Computation, 10*(4), 771–805.

FitzHugh, R. (1961). Impulses and physiological states in theoretical models of the nerve membrane. *Journal of Biopyhsics, 1*, 445–466.

Guckenheimer, J., Gueron, S., & Harris-Warrick, R. (1993). Mapping the dynamics of a bursting neuron. *'Philosophical Transactions of the Royal Society of London, Series B, 341.*

Harris-Warrick, R. M., Nagy, F., & Nusbaum, M. P. (1992). Neuromodualtion of the stomatogastric networks by identified neurons and transmitters. In Harris-Warrick, R. M., Marder, E., Selverston, A. I., & Moulins, M. (Eds.), *Dynamic Biological Networks: The Stomatogastric Nervous System*, pp. 251–291. MIT Press, Cambridge, MA.

Haykin, S. (1999). *Neural Networks: A Comprehensive Foundation.* Prentice Hall, New Jersey.

Hodgkin, A., & Huxley, A. F. (1954). A quantitative description of membrane current and its application to excitation and conduction in nerve. *Journal of Physiology (London), 117*, 500–544.

Hopfield, J. J. (1982). Neural networks and physical systems with emergent collective computational abillities. *Proceedings of the National Academy of Sciences, 79*, 2554–2558.

Husbands, P., Philippides, A., Smith, T. M. C., & O'Shea, M. (2001). The shifting network: Volume signalling in real and robot nervous systems. In Kelemen, J., & Petr, S. (Eds.), *Fifth European Conference on Artificial Life: ECAL2001*, pp. 23–36. Springer, Heidelberg.

Katz, P. S. (Ed.). (1999). *Beyond Neurotransmission: Neuromodulation and its Importance for Information Processing.* Oxford University Press, Oxford.

Mathayomchan, B., & Beer, R. (2002). Center-crossing recurrent neural networks for the evolution of rhythmic behavior. *Neural Computation, 14*, 2043–2051.

Murray, J. D. (Ed.). (1989). *Mathematical Biology.* Springer, Heidelberg.

Nagumo, J. S. Arimoto, S., & Yoshizawa, S. (1962). An active pulse transmission line simulating nerve axon. *Proceedings of the Institute of Radio Engineers, 50*, 2061–2071.

Poggio, T. A., & Glaser, D. A. (Eds.). (1993). *Exploring Brain Functions: Models in Neuroscience.* John Wiley and Sons, New York.

Pollack, J. B. (1987). On connectionist models of natural language processing. PhD Dissertation in Computer Science, University of Illinois.

Pollack, J. B. (1999). Connectionism: Past, present, and future. *Artificial Intelligence Review, 3*, 3–20.

Rinzel, J., & Ermentrout, B. (1989). Analysis of neural excitabillity and oscillations. In Koch, C., & Sergev, I. (Eds.), *Methods in Neuronal Modeling*, pp. 251–291. MIT Press, Cambridge, MA.

Siegelmann, H. T., & Sontag, E. D. (1995). On the computational power of neural nets. *Journal of Computer and System Sciences, 50*(1), 132–150.

Smith, T., Husbands, P., & O'Shea, M. (2001). *Not* measuring evolvability: Initial exploration of an evolutionary robotics search space. In *Congress on Evolutionary Computation: CEC2001*, pp. 9–16. IEEE Press.

Smith, T., Husbands, P., Philippides, A., & O'Shea, M. (2002). Neuronal plasticity and temporal adaptivity: GasNet robot control networks. *Adaptive Behavior, 10*(3/4), 161–184.

Strogatz, S. H. (1994). *Nonlinear Dynamics and Chaos.* Addison-Wesley, Reading, MA.

Tuci, E., Quinn, M., & Harvey, I. (2002). Evolving fixed-weight networks for learning robots. In *Congress on Evolutionary Computation: CEC2002*, pp. 1970–1975. IEEE Press.

Turrigiano, G. G. (1999). Homeostatic plasticity in neuronal networks: The more things change, the more they stay the same. *Trends in Neuroscience, 22*, 221–227.

Homeostatic Plasticity in Recurrent Neural Networks

Hywel Williams
Biosystems, Informatics Research Institute, University of Leeds, Leeds LS2 9JT, UK
hywelw@comp.leeds.ac.uk

Abstract

Recent work in the neurosciences has highlighted the existence of homeostatic plasticity mechanisms that regulate the activity in neuronal networks to a constant level. Functionally similar mechanisms are applied to continuous-time recurrent neural networks. Homeostatic plasticity is found to increase the sensitivity of neural networks to changes in input, giving richer dynamics and robustness to severe changes in the mean level of input. Simulated robots controlled by continuous-time recurrent neural networks using homeostatic plasticity display more complex behaviours than those controlled by non-plastic networks. Synaptic scaling in randomly parameterised networks is observed to generate phototactic behaviour without further adaptation.

1. Introduction

When we are cold, we shiver. When we are too warm, we start to sweat. Both of these mechanisms help us to maintain our body temperature at a constant level. Alternatively, we may move closer to the fire or shed a layer of clothing, or on a longer timescale we may learn new behaviours; jumping up and down to keep warm, a dip in the pool to cool off. Homeostasis is maintained by a combination of physiological response, behaviour and learning.

Recent work in the neurosciences has identified a number of plastic mechanisms by which homeostasis of neural activity is maintained. This paper draws inspiration from these mechanisms to explore the effect of homeostatic plasticity in dynamic recurrent neural networks, such as those that are often used in evolutionary robotics. The evolutionary optimisation of these networks for dynamic control tasks is a significant challenge, and networks using homeostatic plasticity may provide an evolutionary substrate offering improved evolvability and greater robustness of evolved solutions.

Section 2 gives some background to the paper, describing biological neural homeostasis and relevant work in evolutionary robotics. Section 3 covers the experimental method used, the results from which are presented in section 4. Sections 5 and 6 attempt to interpret the results and offer some thoughts on their implications.

2. Homeostatic Adaptation

2.1 Ashby's Homeostat

The connection between homeostasis of internal variables and adaptive behaviour was first recognised by Ashby in the 1950s (Ashby, 1952). Ashby talked about two kinds of homeostatic adaptation: adaptive behaviour that maintains homeostasis of internal variables, and adaptation by which such homeostasis-maintaining behaviour is created. Ashby created an analogue electrical device called the Homeostat to demonstrate the idea of ***ultrastability***; plastic change triggered by loss of homeostasis and continuing until homeostasis is regained.

2.2 Neural homeostasis

It is now accepted in neuroscience that many neural circuits adapt to changing patterns of activity and appear to dynamically maintain their overall level of activity around some set point (Turrigiano, 1999, Turrigiano and Nelson, 2000). While the precise feature of neural activity that is regulated is not known (it may be mean firing rate, mean calcium concentration or some other feature) it is clear that neural activity tends towards a constant level in the long term. It is also clear that there are a variety of mechanisms by which this homeostasis is accomplished, amongst which are mechanisms affecting the strength of synaptic connections and the intrinsic excitability of individual neurons.

2.2.1 Synaptic scaling

One way of maintaining a relatively constant level of activity in the face of a prolonged increase or decrease in firing rates is by altering the strengths of excitatory synaptic connections. If the overall level of firing of a neuron falls too low, an increase in the strength of all its excitatory connections should help to raise it; similarly, if the level of firing gets too high a decrease in the strength of excitatory connections should help to reduce it. This form of homeostatic plasticity is known as ***synaptic scaling*** and specific mechanisms that enable it have been identified (Turrigiano et al., 1998, O'Brien et al., 1998). It is thought that the trigger signal is some function of post-synaptic firing rate (Leslie et al., 2001), which

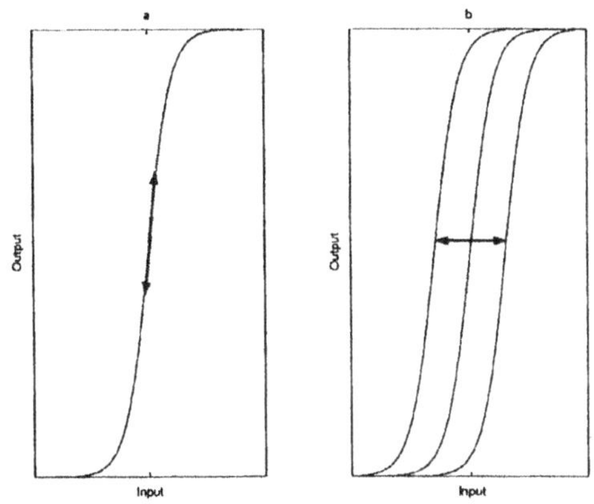

Figure 1: Homeostatic plasticity: (a) synaptic scaling increases/decreases the effect of input, (b) intrinsic plasticity translates the activation curve.

causes a multiplicative scaling of the strengths of all the neuron's afferent synaptic connections. Contrast this with the synapse-specific changes associated with Hebbian correlation-based plastic mechanisms such as long-term potentiation or depression. Synaptic scaling affects the function of a neuron by altering its rate of firing for a given amount of stimulus; in effect it moves the neuron along its input/output curve (see figure 1).

2.2.2 *Intrinsic plasticity*

Another way of regulating activity around a constant level is alteration of the intrinsic excitability of neurons. If firing rate is consistently too high, a reduction in the excitability of the neuron should reduce it; if firing rate is consistently too low, increased excitability should raise it. Alteration of excitability assumes some intrinsic plasticity of neurons; such plasticity has been reported (Turrigiano et al., 1994). The effect of this plasticity on intrinsic excitability has also been demonstrated in cultured pyramidal cortical neurons, where externally imposed damping of activity led to lowered spiking thresholds (Desai et al., 1999). Intrinsic plasticity of the excitability of neurons affects the neuron's function by changing the firing rate response to a given input; effectively this translates the neuron's input/output curve along the output axis (see figure 1).

2.3 *Homeostatic adaptation in neural networks for robot control*

To the best of the author's knowledge there is only one previously published study of homeostatic adaptation in artificial neural networks, which is the work presented by Di Paolo (2000) . Using ideas from Ashby and Turrigiano, Di Paolo developed a system that incorporated a form of homeostatic adaptation into a continuous-time recurrent neural network used to control a phototactic robot. In Di Paolo's scheme, each synapse in the network had a Hebbian learning rule associated with it. The learning rules were only activated when the firing rate of the post-synaptic neuron passed out of a prescribed range, becoming either too high or too low; when post-synaptic firing was inside the range there was no plasticity.

Di Paolo's work was an exploration of the utility of Ashby's concept of homeostatic adaptation for providing robustness to sensorimotor disruptions. Networks were evolved using a genetic algorithm to give phototactic behaviour in a simulated robot, with fitness awarded for phototaxis and for maintenance of neural activation within the target range. The successful controllers were then tested for their robustness to inversion of their visual field, performed by swapping the positions of the two light sensors. Di Paolo found that around half of the evolved controllers that gave long-term stable phototaxis were able to adapt to this disruption.

While Di Paolo's work produced some very interesting results, the types of homeostatic plastic mechanism implemented in his experiment were qualitatively different to those observed in neuroscience. Di Paolo's use of plasticity triggered when neural firing is too high or too low fits in with the neuroscientific data, but the form of the plasticity does not. Di Paolo used Hebbian learning, which acts on the strengths of individual synapses. In contrast, the synaptic scaling reported by Turrigiano acts upon groups of synapses afferent to a node. Intrinsic plasticity does not operate on synaptic strengths at all but on the excitability of the neuron. In addition, the homeostatic plasticity observed in biological neural networks gives negative feedback, whereas Hebbian rules are seen by neuroscientists as likely to provide positive feedback on both activity levels and synaptic strengths (Turrigiano and Nelson, 2000).

3. Method

While this study is clearly influenced by Di Paolo's work and makes use of a similar experimental set-up for studying robot behaviour, its aims and objectives are significantly different. The intention here is to apply the kinds of homeostatic plasticity observed in neuroscience to continuous-time recurrent neural networks (CTRNNs), of the type commonly described in the evolutionary robotics literature (Beer and Gallagher, 1992, Beer, 1996, Harvey et al., 1997, Slocum et al., 2000, Nolfi and Floreano, 2000), and to examine the effect of homeostatic plasticity on network dynamics and on the behaviour of simulated robots controlled by these networks.

This section describes the networks used, the way in which the homeostatic plasticity mechanisms identified by Turrigiano are implemented, and the robot simulation. Networks using the two types of homeostatic plasticity (synaptic scaling and intrinsic plasticity) will be compared with non-plastic networks throughout.

3.1 CTRNN

The CTRNNs used here are made up of artificial neurons defined by the equations 1 and 2 given below. Neuron state is governed by equation 1, in which y is the neuron potential, τ is a constant governing the decay of potential with time, w_i and z_i are the connecting synapse strength and firing rate respectively of the ith afferent neuron, and I is any external input the neuron receives.

$$\tau \dot{y} = -y + \sum_{i=1}^{N} w_i z_i + I \tag{1}$$

The firing rate z of a neuron is given by equation 2, in which b is the bias term for the neuron. Firing rate is thus given by a sigmoid function of potential returning values in the range $[0, 1]$ as shown in figure 2(a).

$$z = \frac{1}{1 + e^{-(y+b)}} \tag{2}$$

The networks used are all fully connected, including self-connections. Synapses can be inhibitory or excitatory depending on their sign, and may have varying strengths; this is represented by an associated signed real weight value w. Values are initialised from the following ranges: $\tau \in [0.4, 4.00], b \in [-3.00, 3.00], w \in [-8.00, 8.00]$. Networks are updated using Euler integration with a time step of 0.2. Certain nodes are designated as input or output (sensor or motor) nodes as appropriate and are the only nodes that directly interact with the external environment; all other nodes are hidden nodes and only interact with the external environment via input/output nodes.

3.2 Homeostatic plasticity mechanisms

Homeostatic plasticity is incorporated into the CTRNN by defining a permitted range for the firing rates of neurons, corresponding to Turrigiano's postulated set level of activity about which homeostasis is maintained. Plasticity is triggered whenever the firing rate of a neuron is too high or too low and is proportional to the amount by which the firing rate has exceeded its range. This notion is captured by the use of a plastic facilitation function that varies with firing rate (Di Paolo, 2000); plastic facilitation ρ is zero when the firing rate is within the target range and rises or falls to ± 1 outside this range, as shown in figure 2(b). The target range is arbitrarily set to $[0.3, 0.7]$.

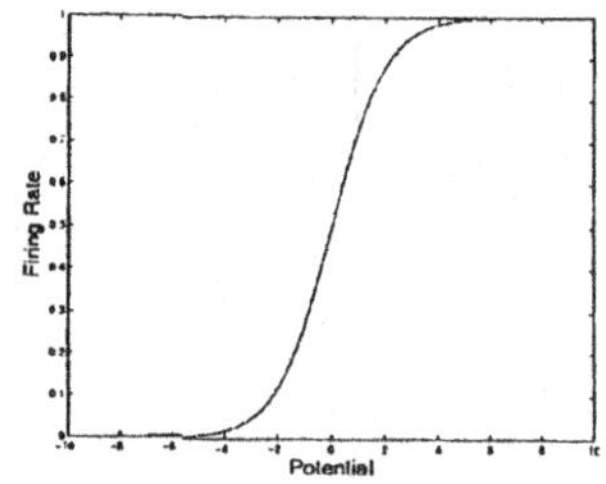

(a) Sigmoid activation curve showing upper and lower bounds of target range

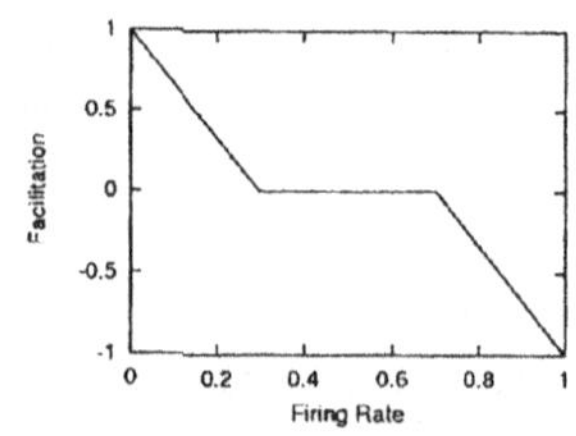

(b) Plastic facilitation as a function of firing rate

Figure 2: Plastic facilitation depends on firing rate

3.2.1 Synaptic scaling

When the firing rate of a neuron goes outside the prescribed range all afferent synapses to that neuron are multiplicatively scaled. The scaling is directional; it acts so that weights are changed in the direction most likely to bring the neuron firing rate back into bounds. Scaling is applied to both inhibitory (negative weight) and excitatory (positive weight) synapses. If the firing rate is too high, then excitatory synapse strengths are scaled down and inhibitory synapse strengths are scaled up. If the firing rate is too low, excitatory inputs are scaled up and inhibitory inputs are scaled down. Scaling up or down here refers to the absolute value of the synaptic weight, so that scaling down a negative weight makes it less negative. The size of the change is determined by the plastic facilitation ρ, by a node learning rate parameter μ, and by the current magnitude of the weight. The plasticity rule for synaptic scaling is therefore expressed by equation 3, with $\mu \in [0.00, 0.10]$. Weight values are restricted to the range $[-8.00, 8.00]$ and are clipped where they go outside this range.

$$\Delta w = \rho \mu |w| \tag{3}$$

3.2.2 Adaptive bias

This mechanism is an implementation of the intrinsic plasticity described by Turrigiano (Turrigiano, 1999, Turrigiano and Nelson, 2000). When a neuron's firing rate goes outside the prescribed range, the bias term of the neuron is shifted to make the neuron more or less likely to fire depending on what is required to bring the firing rate back into bounds. If firing rate is too low the bias is increased, effectively translating the sigmoid activation function so that the neuron is more excitable and hence more likely to fire. If firing rate is too high the bias is decreased, making the neuron less excitable. The size of the change depends on the plastic facilitation ρ and the node learning rate μ. The plasticity rule for intrinsic plasticity is therefore given by equation 4, with $\mu \in [0.0, 0.10]$

$$\Delta b = \rho\mu \tag{4}$$

3.3 Robot simulator

The behavioural effect of homeostatic plasticity at an agent level is studied by incorporating the plastic mechanisms into CTRNNs used to control a simple simulated robot. The robot is placed in a featureless infinite plane along with a single light source. The robot's light sensors and control network are then sequentially updated, with network output being used to calculate the motion of the robot relative to the light source.

3.3.1 Simulated robot

The robot is modelled as a circular body with two motors mounted at either ends of an axle along its diameter and two light sensors mounted at angles of $PI/3$ radians from the forward direction, as shown in diagram 3.

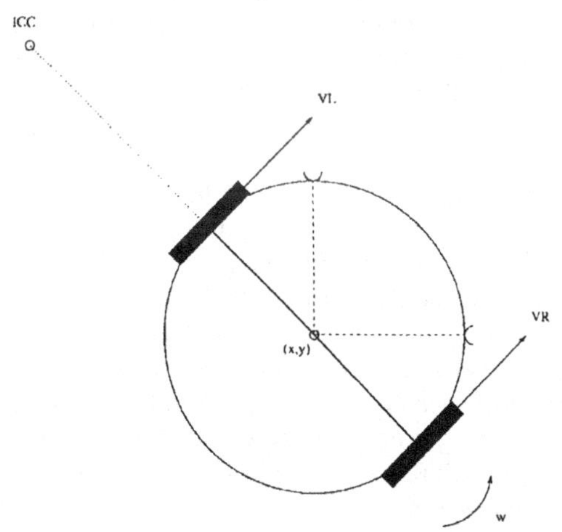

Figure 3: Diagram of photosensitive robot

3.3.2 Robot kinematics

The simulated robot kinematics are adapted from Dudek and Jenkin (2000) . Differential drive steering is accomplished by the two motors, which may give thrust both forwards and backwards, allowing the robot to spin while stationary or to move forwards or backwards with any instantaneous angular velocity. Robots are assumed to have negligible mass, so that the motor output can be taken as the tangential velocity of the robot at the motor mount point. The instantaneous angular velocity ω of the robot is calculated using equation 5, and this is used to calculate the motion of the robot using the system of equations 6, where V_L and V_R are the instantaneous velocities of the left and right motors respectively, d is the length of the axle, ICC denotes the instantaneous centre of curvature (the imaginary point about which the robot orbits at any given instant), β is the current heading of the robot, and (x, y) is the robot's position. Details of how this system is derived can be found in Dudek and Jenkin (2000). The equations are updated using time step $\delta t = 0.2$.

$$\omega = (V_R - V_L)/d \tag{5}$$

$$\begin{aligned} x' &= (x - ICC_x)cos(\omega\delta t) - (y - ICC_y)sin(\omega\delta t) + ICC_x \\ y' &= (x - ICC_x)sin(\omega\delta t) + (y - ICC_y)cos(\omega\delta t) + ICC_y \\ \beta' &= \beta + \omega\delta t \end{aligned} \tag{6}$$

3.3.3 Light sensors

Light sources are of fixed intensity 1000 unless otherwise stated. The light sensors return a signal value proportional to source intensity and the inverse square of the distance between the source and the sensor. Sensors are assumed to be mounted on top of the agent, so that they are never in the shadow of the robot's body.

3.3.4 Control network

The robot is controlled by a fully connected 6-node CTRNN of the type described above. Two nodes are designated sensor nodes and each receive the signal from one of the light sensors modified by an agent-symmetrical gain parameter (drawn from the range $[0, 10.00]$) as external input. Two other nodes are designated motor nodes, and the output from these nodes is mapped to the range $[-1.00, 1.00]$ and modified by an agent-symmetrical gain (again drawn from the range $[0, 10.00]$) to give the instantaneous velocity created by that motor. Both sensory input and motor output have noise added before the gain is applied, drawn from a uniform distribution in the range $[-0.25, 0.25]$.

4. Results

4.1 Network dynamics

The first experiments were performed on stand-alone CTRNNs, that is, without embedding the networks in the robot simulation environment. These tests were to get a feel of how network dynamics varied with the different forms of homeostatic plasticity. Random sets of network parameters (weights, bias terms, decay constants, learning rates) were generated and networks formed from them. Network behaviour was observed with random input fluctuating around a mean level of zero and with random input fluctuating around a steadily increasing base level. The results reported are for fully-connected 3-node networks with a single input node and two hidden nodes.

4.1.1 Random input with zero mean

Randomly generated networks were updated for 400 timesteps (2000 Euler integration steps). The input signal to the input node was reset every 20 time steps to a value randomly drawn from a uniform distribution in the range $[-1.00, 1.00]$ and remained constant between each change. Samples of network behaviour are given in figures 4, 5 and 6, which show the dynamics of three typical examples of each network type. As can be seen from these plots, the behaviour of the plastic and non-plastic networks was significantly different. The plots show the firing rates of the the three nodes over time, as well as plastic changes where appropriate. The dashed horizontal lines on the plots mark the upper and lower boundaries of the target range; above and below these lines plasticity is triggered.

Non-plastic CTRNNs typically settle to a steady equilibrium with each neuron giving a constant output. The input node is sometimes affected by changes in input signal, but the other nodes typically do not change their activity in response to a change in input to the network; even when the firing rate of the input node changes in response to a change in input signal these changes are rarely transmitted to the other nodes. This is because in a randomly generated fixed-weight CTRNN there is a strong chance that the nodes will reach an equilibrium with firing rate close to the maximum or minimum level, which because of the sigmoid shape of the activation function means that changes in input have little effect on output.

The dynamics of networks with both forms of homeostatic plasticity are more complex than those of non-plastic networks. The homeostatic plasticity functions to keep the firing rates of all the nodes within the target range, as can be seen by inspection of the changes in synaptic weights or biases which occur at times when the firing rate of a node goes outside the target range. This restriction means that the inputs to each node are fluctuating around the region of the sigmoid activation curve where the gradient is greatest; at this point the nodes are most sensitive to changes in input. All nodes are sensitive to changes in each other's firing rate, and the input node is additionally sensitive to changes in the input signal. Thus a change in the input signal is almost always transmitted via the input node to the hidden nodes.

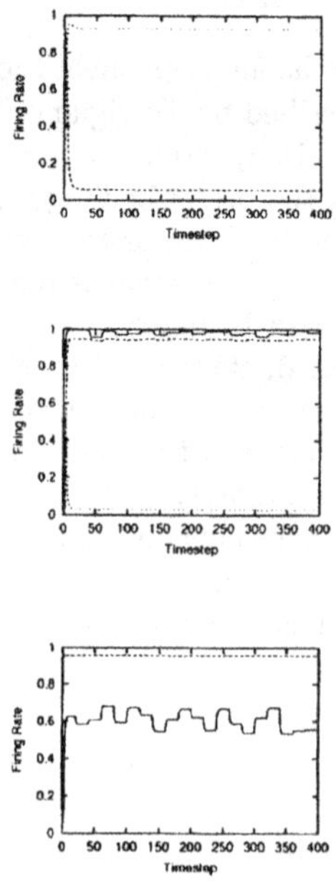

Figure 4: Behaviour of three typical non-plastic 3-node CTRNNs. Input signal changed every 20 timesteps, randomly drawn from the range $[-1.00, 1.00]$. There is a tendency towards saturated/constant firing rates, and changes in the firing rate of the input node do not affect the firing rates of the hidden nodes.

The increased sensitivity of the homeostatic networks is shown in figure 7, which shows the mean change in the firing rate of each node in the network caused by a change in the input signal. This statistic is calculated by measuring the change in the firing of each node following each change in the input signal. Readings are taken just before the next change in input signal so that the network has time to settle on a steady activation level following the previous change. The mean change in firing rate was calculated for 1000 randomly generated networks of each type to give an overall mean value plotted in figure 7. The plots show that randomly generated non-plastic CTRNNs typically have a very low sensitivity to changes in external input, whereas the CTRNNs with homeostatic plasticity have much greater sensitivity.

CTRNNs with synaptic scaling showed higher sensitiv-

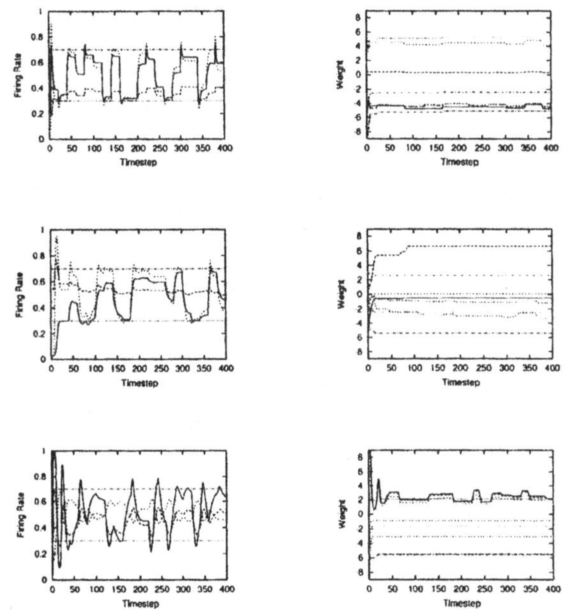

Figure 5: Behaviour of three typical 3-node CTRNNs with synaptic scaling. Input signal changed every 20 timesteps, randomly drawn from the range $[-1.00, 1.00]$. Plots in the left-hand column show node firing rates, plots in the right-hand column show the connection weights for the network. Horizontal dotted lines in the plots of firing rates show the upper and lower bounds of the target range. Firing rates are kept within the target range by the synaptic scaling; changes in connection weights are correlated with periods when firing rates leave the target range. Dynamics are richer than those of non-plastic CTRNNs.

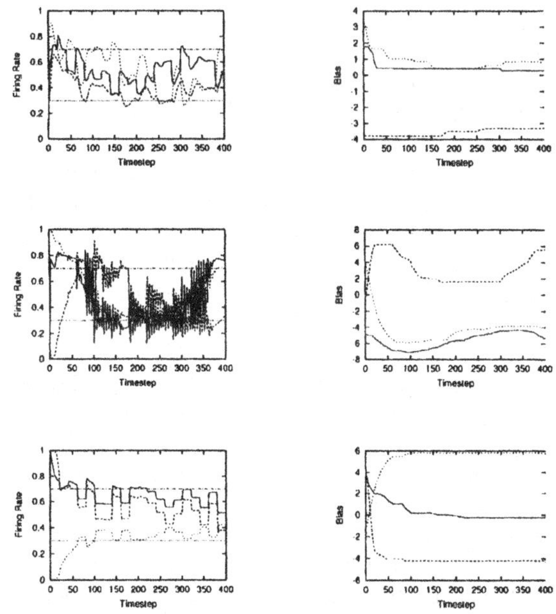

Figure 6: Behaviour of three typical 3-node CTRNNs with adaptive bias. Input signal changed every 20 timesteps, randomly drawn from the range $[-1.00, 1.00]$. Plots in the left-hand column show node firing rates, plots in the right-hand column show the bias terms for each node. Horizontal dotted lines in the plots of firing rates show the upper and lower bounds of the target range. Firing rates are kept within the target range by the adaptive bias; changes in node bias terms are correlated with periods when firing rates leave the target area. Dynamics are richer than those of non-plastic CTRNNs.

ity to input fluctuations than the non-plastic CTRNNs on all nodes. The input node is more sensitive than the hidden nodes, which is explained by its direct exposure to the changing input signal; the hidden nodes are only affected by the changing input indirectly through their interaction with the input node. CTRNNs with adaptive bias are the most sensitive of all the network types, with all the nodes showing relatively large changes in firing rate in response to fluctuating network input. This is because the adaptive bias translates the sigmoid activation curve for each node so that its point of steepest gradient is positioned roughly at the mean level of input to the node, giving the largest possible change in firing rate in response to a given change in input. The increased sensitivity of the CTRNNs with adaptive bias compared to those with synaptic scaling perhaps suggests that adaptive bias is more effective in adjusting the nodal response than synaptic scaling.

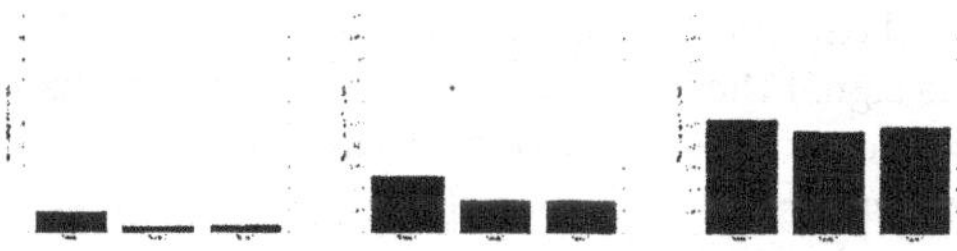

(a) Non-plastic CTRNN (b) CTRNN with synaptic scaling (c) CTRNN with adaptive bias

Figure 7: Mean change in node firing rates caused by a change in external input signal. Input signal randomly drawn from range $[-1.00, 1.00]$, applied to node 1 (left-most column in each graph). Homeostatic plasticity increased network sensitivity. Sensitivity of networks using synaptic scaling was greater than that of non-plastic networks, while networks using adaptive bias were most sensitive of all.

4.1.2 Random input with increasing mean

Random networks were generated as before and tested for their sensitivity to fluctuations in input around an incrementally increasing base input level. Every run started with input base level $base = -20$. During the run the base level was incremented through the series $\{-20, -19.5, -19, ..., 19, 19.5, 20\}$. After each base level increment, network sensitivity was tested as before, but in this case with input drawn from a uniform distribution with range $[base - 1.00, base + 1.00]$. Base level was incremented every 100 time steps and input signal re-assigned every 10 timesteps. Figure 8 shows mean change in node firing rates against input base level, measured across 1000 randomly generated networks of each type.

The non-plastic CTRNNs were relatively insensitive to changes in input signal; they were only sensitive to changes in input signal for input base levels close to zero, and for more extreme levels of input they were completely insensitive. The explanation is that at very high or very low levels of external input the input node of the non-plastic CTRNNs gives a saturated response and the small fluctuations in the input signal have no effect on its firing rate, so the change in input signal is not communicated to the hidden nodes. Even when the input node is not saturated, these networks are relatively insensitive for the reasons outlined above.

CTRNNs with synaptic scaling were more sensitive than the non-plastic CTRNNs for the whole range of input base levels, but were still significantly less sensitive at extreme levels of network input. At very high or low levels of input the input node is actually less sensitive than the other nodes to input fluctuations; this initially seems counter-intuitive, but can be explained by the saturation of its response. The hidden nodes are not directly affected by the external input and so do not saturate, retaining their sensitivity. Synaptic scaling acts to make them more sensitive, so that even a small change in the signal they receive from the input node can cause a significant change in their activation.

The filtering of the input signal through the input node helps the hidden nodes to remain sensitive when the base level is extreme, but also stops them from becoming as sensitive as the input node when the input signal is moderate. The skew of the peak in figure 8(b) to the left of the origin is hard to account for (it might be expected to be centred on the origin). This may be a feature of the way the synaptic scaling was implemented. Possible explanations involving the direction of the incremental change in input base level (from low to high) can be ruled out as the same skew appears when input base level is changed in the other direction.

CTRNNs with adaptive bias remained highly sensitive across the whole range of input base level. This is because the bias term can directly cancel out the external input, as can be seen from equations 1 and 2; the external input translates the activation curve in one direction and the adaptive bias term translates it back again. Thus the input node does not saturate and the network retains its sensitivity. The dip in sensitivity at the leading edge of the plots in figure 8(c) is accounted for by the time taken for the plastic mechanism to adapt to the initial low level of input.

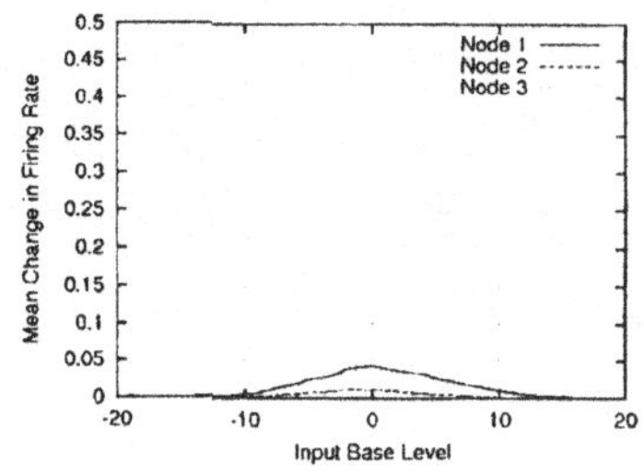

(a) Non-plastic CTRNN

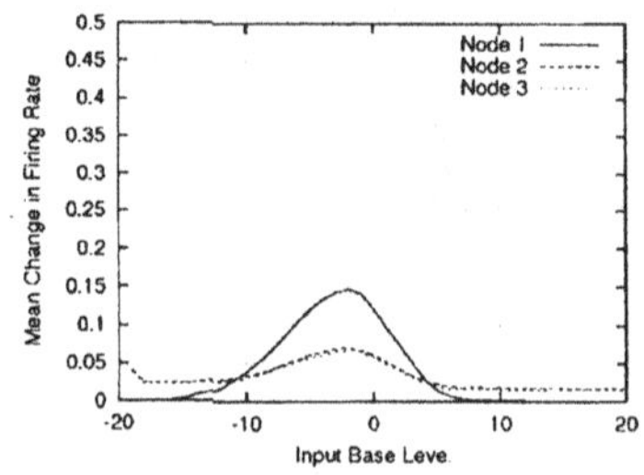

(b) CTRNN with synaptic scaling

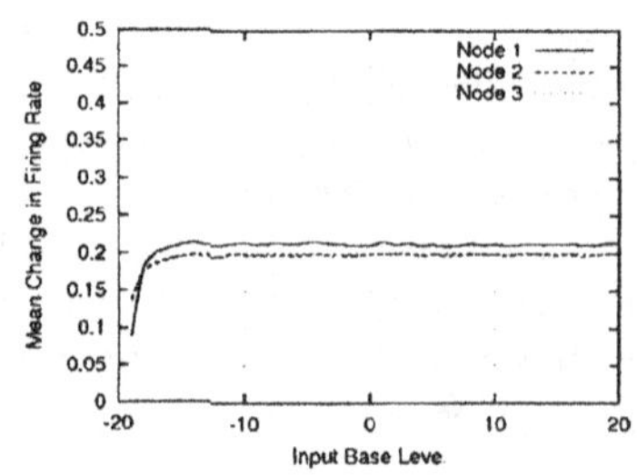

(c) CTRNN with adaptive bias

Figure 8: Mean change in node firing rates caused by a change in external input signal. Input signal randomly drawn from range $[base - 1.00, base + 1.00]$ and applied to node 1, for $base$ incremented from -20 to 20 (step size 0.5). Compared to non-plastic networks, networks with synaptic scaling showed greater sensitivity across the whole range of base input level. Networks with adaptive bias showed the greatest sensitivity, with a high level of sensitivity consistent across the whole range (the early lower level is due to the time needed for adaptation).

4.2 *Robot behaviours*

The next set of experiments were performed with the simulated robot described previously. Parameter sets for 6-node CTRNNs and sensor/motor gains were randomly generated and instantiated as robot controllers. It is hard to quantify robot behaviour; instead this section will try and give a descriptive account of the typical behaviours of robots controlled by each kind of CTRNN. Figures 9, 10, 11 and 12 show examples of the robot behaviours described; these plots show motion for 4000 time steps (20000 Euler integration steps) where the light source was randomly repositioned every 800 time steps. The symbol $\times$ marks the position of light sources and the symbol $+$ marks the start position of the robot. Note the different scales of the plots.

Robots controlled by randomly generated non-plastic CTRNNs almost always rotated on the spot and ignored the light source. Sometimes they moved in circles, but their motion was consistently a steady rotation. This is explained by the observations of the behaviours of fixed random CTRNNs; they are not generally sensitive to external input and tend to reach an equilibrium state with each neuron firing at a constant rate. If the motor neurons fire at a constant rate the differential drive kinematics will produce a constant angular velocity for the robot, which gives the types of rotational movement observed. An example of this is shown in figure 9. The rotation cannot be seen in the plot as the robot was spinning about its axis almost without any movement relative to the light source. The net distance travelled by the robot is very small, and is due to gradual drift as the robots rotation is very slightly influenced by the different light intensities its sensors experience as it rotates.

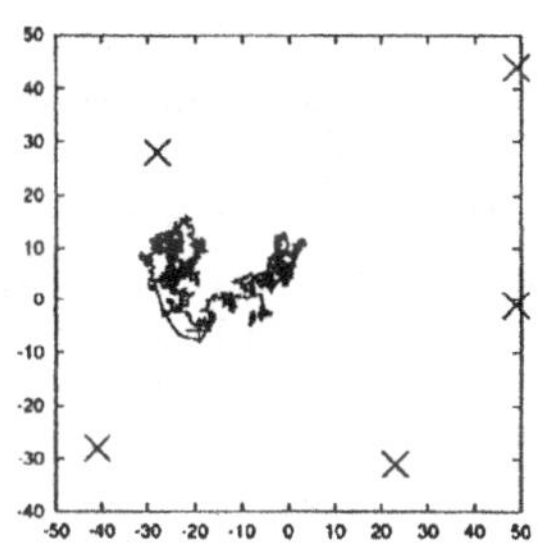

Figure 9: Behaviour of robot controlled by random non-plastic CTRNN.

In contrast to the non-plastic CTRNNs, robots controlled by the CTRNNs with synaptic scaling displayed quite complex behaviours. Since the synaptic scaling ensured that they were sensitive to external input (light intensity being at a low enough level not to force input node saturation except when very close to the source), these agents were strongly influenced by the light source. Their motion, as seen in figure 10, is complex; the easiest description is to liken it to that of a moth near a flame. The robot approaches the source, then veers away, then approaches, then spends some time moving erratically away from the light source. Periods of phototactic behaviour occur, but then cease, as do periods of photo-aversion. The difference in behaviour for different randomly generated networks is considerable, but the plot shown is representative of the general class of behaviours observed.

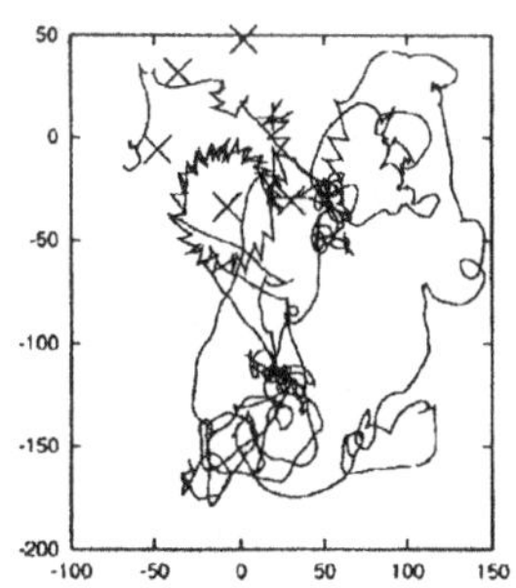

Figure 10: Behaviour of robot controlled by random CTRNN with synaptic scaling.

Interestingly, while many of these agents showed occasional periods of phototaxis, some (roughly 10% by visual inspection) of the agents controlled by CTRNNs with synaptic scaling displayed stable phototaxis. In these cases the typical pattern was an early period of erratic behaviour, while the synaptic scaling configured the network to be sensitive to external input, followed by consistent phototaxis. An example of this is shown in figure 11, where the robot approaches and then circles the light sources.

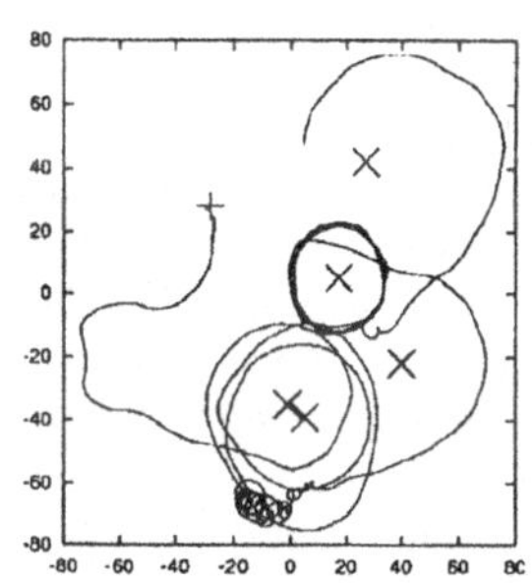

Figure 11: Phototaxis by robot controlled by random CTRNN with synaptic scaling.

Robots controlled by CTRNNs with adaptive bias also displayed interesting behaviours, which were generally some form of cycloidal motion (see figure 12). The cycloidal motion is caused by the interaction of the adaptive bias with the random set of fixed weights. Without the adaptive bias term the network is non-plastic and would tend to produce a rotational movement as described above, but with the adaptive bias the rotation is modulated by the changing bias term to become cycloidal. In most cases the bias term actually becomes a part of the network dynamics, oscillating in phase with the oscillations of neural firing rates. The cycloidal motion means that the robot moves relative to the light source, in contrast to the near-stationary rotation of the non-plastic CTRNN controllers. The cycloidal motion displayed was generally the same irrespective of how far the agent was from the light source (figure 12), which can be explained by reference to the dynamics of the CTRNNs with adaptive bias described earlier. In a similar manner to how it was able to negate the large changes in external input base level in the previous section, the adaptive bias negates the change in the base level of light intensity so that the motion caused by the fixed set of weights remains unchanged.

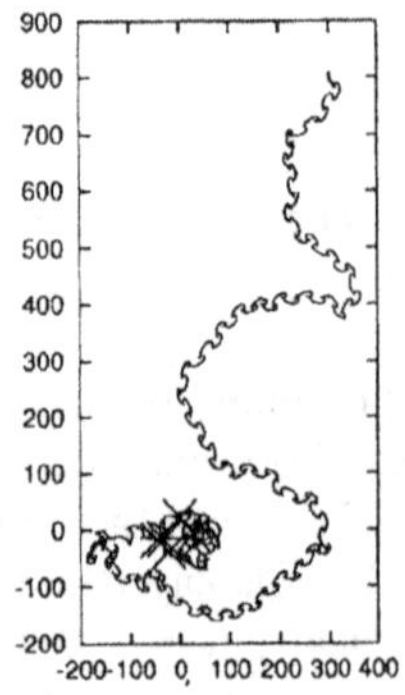

Figure 12: Behaviour of robot controlled by random CTRNN with adaptive bias.

5. Discussion

We have seen that homeostatic plasticity in CTRNNs makes them much more sensitive to changes in their input signal. Synaptic scaling significantly increases the sensitivity of all nodes. Plasticity of intrinsic excitability (implemented here as an adaptive bias term) makes a still bigger increase in sensitivity while also allowing the network to cope with extreme changes in the level of external input received. When these networks are randomly parameterised and used as controllers for a photosensitive robot, they produce more interesting behaviours than randomly parameterised non-plastic CTRNNs. Random non-plastic networks make robots rotate, random networks with adaptive bias cause robots to display cycloidal motion, and random networks with synaptic scaling give complex behaviours including consistent phototaxis.

It is tempting to think that the networks with synaptic scaling would be a good substrate for the evolution of robot controllers using genetic algorithms, since randomly generated networks of this kind produce interesting behaviours before any evolutionary optimisation has taken place. The networks with adaptive bias might also be seen as a good evolutionary substrate because of their sensitivity, especially given the improved evolvability of centre-crossing recurrent neural networks for the generation of rhythmic behaviours (Mathayomchan and Beer, 2002). The adaptive bias serves to make the CTRNN a dynamically centre-crossing network. However, a cautionary note is required.

In some preliminary work that there is not space to report on fully here, CTRNNs were successfully evolved for phototaxis in the same simulated robot scenario used above, achieving good results in few (< 10) generations with a population size of 50. It was found that the networks with homeostatic plasticity were actually less evolvable than the networks without plasticity, giving consistently lower fitness scores. There are several possible reasons for this, many of which are concerned with the design of fitness functions that allow a fair comparison of evolvability, which is not straightforward. Plastic networks have more parameters than non-plastic networks, and require longer trials to allow the plasticity to configure the network, amongst other problems.

A more general reason why homeostatic plastic networks may not necessarily be more evolvable is to do with the utility of sensitivity. Intuitively, it seems that more sensitive networks should be more use; networks must change their behaviour in response to changing stimuli if they are to perform any non-trivial function. However, the sensitivity given by homeostatic plasticity as implemented here comes at a cost; it precludes the networks from making use of the beneficial effects of node saturation. An example of such a beneficial effect is given by the non-plastic CTRNNs evolved to perform phototaxis, which all made use of a simple strategy: have one motor node constantly saturated (thus giving a fixed contribution to the angular velocity of the agent) and the other motor node sensitive to changes in the activation of one of the sensor nodes. Put another way, use one motor node to give forward motion and the other to steer. This strategy gives the robot a cycloidal motion that slowly spirals towards the light source, the angular velocity increasing as it draws near until it rotates on the spot when close to the source. [1] This cheap and robust phototaxis strategy is not available to the networks with homeostatic plasticity, since their plasticity acts to stop any of their nodes becoming saturated.

Another point of interest is that the evolved controllers incorporating homeostatic plasticity were more robust

[1] As an aside, it was found that agents with only one sensor could reliably perform phototaxis by this method. The agents that had two sensors only made use of one of them, meaning that sensor inversion was a trivial disruption (since either sensor could supply the required stimulus).

than the non-plastic controllers. The networks with adaptive bias were able to cope with dramatic changes in the intensity of the light sources, whereas the other networks were much less able to do so. The adaptive bias networks continued to perform phototaxis when light source intensity was scaled up or down by as much as 2 orders of magnitude, whereas the synaptic scaling and non-plastic networks were able to cope only with changes up to around ±50% in most cases. This fits in with the results presented above showing the ability of the adaptive bias networks to cope with large changes in the level of external input. Another example of the robustness of the homeostatic plastic networks is that the synaptic scaling networks were often able to recover phototaxis after sensor inversion, despite having phototaxis strategies that relied on the use of both sensors. This is reminiscent of the results achieved by Di Paolo (2000).

One significant area that demands further exploration is that of the timescales of the plasticity mechanisms. In the experiments reported upon here the plastic mechanisms operate at timescales that are much slower than the timescales of neural activation, but they are still very fast compared to the biological mechanisms that are their inspiration. The implications of different timescales for network dynamics will be examined in future work.

6. Conclusion

The results described in this paper are modest, but are enough to suggest that homeostatic plasticity is a worthwhile avenue of exploration. Both synaptic scaling and adaptive bias terms make CTRNNs more sensitive to changes in input, and adaptive bias also gives robustness to large changes in the mean level of input. Simulated robotic agents controlled by CTRNNs with homeostatic plasticity are more sensitive to their environment and display more complex behaviours than those controlled by non-plastic CTRNNs. Randomly generated CTRNN controllers using synaptic scaling to maintain homeostasis of neural firing were repeatedly observed to perform stable phototaxis after an initial period of adaptation. Also, evolved solutions based on homeostatic plastic CTRNNs display greater robustness to some sorts of perturbation.

Acknowledgements

This paper has benefited greatly from discussions with Jason Noble, Chris Buckley, and the rest of the Biosystems group at the University of Leeds.

References

Ashby, W. (1952). *Design for a Brain.* Chapman and Hall, London.

Balaam, A. (2001). *Analysing homeostatic adaptation to sensorimotor disruptions.* Unpublished MSc thesis, University of Sussex, UK.

Beer, R. (1996). Toward the evolution of dynamical neural networks for minimally cognitive behavior. In Maes, P., Mataric, M., Meyer, J., Pollack, J., and Wilson, S., (Eds.), *From Animals to Animats 4: Proceedings of the 4th Intl. Conf. on Simulation of Adaptive Behavior*, pages 421–429, Cambridge, MA. MIT Press.

Beer, R. and Gallagher, J. (1992). Evolving dynamic neural networks for adaptive behavior. *Adaptive Behavior*, 1:91–122.

Desai, N., Rutherford, L., and Turrigiano, G. (1999). Plasticity in the intrinsic excitability of neocortical pyramidal neurons. *Nature Neuroscience*, 2:515–520.

Di Paolo, E. (2000). Homeostatic adaptation to inversion in the visual field and other sensorimotor disruptions. In Meyer, J., Berthoz, A., Floreano, D., Roitblat, H., and Wilson, S., (Eds.), *From Animals to Animats 6: Proceedings of the 6th Intl. Conf. on Simulation of Adaptive Behavior*, pages 440–449, Cambridge, MA. MIT Press.

Di Paolo, E. (2002). Evolving robust robots using homeostatic oscillators. Technical Report CSRP 548, School of Cognitive and Computing Sciences, University of Sussex, UK.

Di Paolo, E. (2003). Organismically-inspired robotics: Homeostatic adaptation and natural teleology beyond the closed sensorimotor loop. In Murase, K. and Asakura, T., (Eds.), *Dynamical Systems Approach to Embodiment and Sociality*, pages 19–42, Adelaide. Advanced Knowledge International.

Dudek, G. and Jenkin, M. (2000). *Computational Principles of Mobile Robotics.* Cambridge University Press, Cambridge.

Floreano, D. and Mondada, F. (1996). Evolution of plastic neurocontrollers for situated agents. In Maes, P., Mataric, M., Meyer, J., Pollack, J., and Wilson, S., (Eds.), *From Animals to Animats 4: Proceedings of the 4th Intl. Conf. on Simulation of Adaptive Behavior*, Cambridge, MA. MIT Press.

Harvey, I., Husbands, P., Cliff, D., Thompson, A., and Jakobi, N. (1997). Evolutionary robotics: the sussex approach. *Robotics and Autonomous Systems*, 20(2-4):205–224.

Leslie, K., Nelson, S., and Turrigiano, G. (2001). Postsynaptic depolarization scales quantal amplitude in neocortical pyramidal neurons. *J. Neurosci.*, 21:RC170.

Mathayomchan, B. and Beer, R. D. (2002). Center-crossing recurrent neural networks for the evolution of rhythmic behavior. *Neural Computation*, 14:2043–2051.

Nolfi, S. and Floreano, D. (2000). *Evolutionary Robotics: The Biology, Intelligence, and Technology of Self-Organizing Machines.* MIT Press, Cambridge, MA.

O'Brien, R., Kamboj, S., Ehlers, M., Rosen, K., Fischbach, G., and Huganir, R. (1998). Activity-dependent modulation of synaptic ampa receptor accumulation. *Neuron*, 21(5):933–935.

Slocum, A., Downey, D., and Beer, R. (2000). Further experiments in the evolution of minimally cognitive behavior: From perceiving affordances to selective attention. In Meyer, J., Berthoz, A., Floreano, D., Roitblat, H., and Wilson, S., (Eds.), *From Animals to Animats 6: Proceedings of the 6th Intl. Conf. on Simulation of Adaptive Behavior*, Cambridge, MA. MIT Press.

Turrigiano, G. (1999). Homeostatic plasticity in neuronal networks: the more things change, the more they stay the same. *Trends in Neuroscience*, 22:221–228.

Turrigiano, G., Abbott, L., and Marder, E. (1994). Activity-dependent changes in the intrinsic properties of cultured neurons. *Science*, 264:974–977.

Turrigiano, G., Leslie, K., Desai, N., Rutherford, L., and Nelson, S. (1998). Activity-dependent scaling of quantal amplitude in neocortical pyamidal neurons. *Nature*, 391:892–895.

Turrigiano, G. and Nelson, S. (2000). Hebb and homeostasis in neuronal plasticity. *Current Opinion in Neurobiology*, 10:358–364.

EVOLUTION

Evolving Walking: The Anatomy of an Evolutionary Search

Chad W. Seys
Dept. of Electrical Engineering and Computer Science
Case Western Reserve University
Cleveland, OH 44106
chad.seys@cwru.edu

Randall D. Beer
Dept. of Electrical Engineering and Computer Science
Dept. of Biology
Case Western Reserve University
Cleveland, OH 44106
beer@eecs.cwru.edu

Abstract

The evolution of a continuous time recurrent neural network central pattern generation for walking is characterized and found to proceed in two phases. The first phase spans the beginning of the search through the generation at which a "breakthrough" individual is discovered. The second phase proceeds from that generation forward. The first phase is most quickly completed if each succeeding population is as random as possible. Hence GA searches performed at lower mutation variances require more generations to discover a breakthrough individual than higher mutation variances searches. In the second phase the best fitness in the population most rapidly increases at low mutation variances. The role of parameter space structure in these trends will be examined.

1. Introduction

Evolutionary algorithms (EVAs) have long played an important role in adaptive behavior research (Beer & Gallagher, 1992; Cliff et al., 1993; Nolfi & Floreano, 2000). From an engineering perspective, EVAs promise an opaque design methodology whereby solutions can be produced to complex problems without necessarily understanding them. Scientifically, EVAs promise a means of concretely exploring theoretical ideas with a minimum of a priori assumptions. Both activities would greatly benefit from an improved understanding of the behavior of evolutionary search.

Within the EVA community, a great deal of work has been done on the analysis of evolutionary algorithms (van Nimwegen et al., 1997; Gao et al., 1998). However, these studies often make unrealistic assumptions (e.g., infinite population size), or are carried out with highly simplified fitness functions. It is not at all clear how results obtained from such studies apply to the evolution of controllers for situated, embodied, and dynamical agents, where the fitness spaces can be high-dimensional, discontinuous, and extremely irregular. Furthermore, an agent's fitness often depends in very indirect ways on the parameters being evolved. In a neurally-controlled agent, for example, the parameters determine neural dynamics, which interact with body dynamics and environment dynamics to produce behavior, which is then evaluated by a behavioral fitness function.

In this paper, we examine in some detail the behavior of a simple evolutionary algorithm on a walking task. A significant advantage of this task is that the neural model, body model and fitness function are well-understood (Chiel, et al., 1999; Beer et al., 1999). First, we describe the general features of evolutionary searches on this task. Then we divide the evolutionary search into two distinct phases, showing that the EVA behavior is rather different in the two phases and explaining the basis of these differences. Finally, we show how our results can be used to both predict features of the evolutionary search and to improve the performance of the EVA on this task. The paper concludes with a discussion of the broader applicability of our results.

2. Methods

The agent used in the experiments is the same as used in Beer and Gallagher (1992) and consists of a body with a single leg (of nominal length 15) with a foot, and effectors to move them. The leg is connected to the body by a joint which allows effectors to apply clockwise and counterclockwise torques which sum to produce a resultant torque which may act to move the leg through an angle of $\pm\pi/6$. If the foot is raised the opposing torques are allowed to have a maximum value of 1/40. When the foot is lowered, the resultant torque applies translational force on the body with a maximum value of 1/20. If the body has no support, the translational velocity is 0, while if the body is supported the acceleration is equal to the translational force. The leg is allowed to stretch past its nominal length (and angle) if the foot is lowered, but no forces can be applied. Finally, if the distance between the body's center of mass and foot is too great (greater than 20 along the axis in the direction of travel), the leg snaps back to nominal length and maximum clockwise angle, the body loses support, and translational velocity is 0. In these experiments an agent is always created with the leg in the fully forward (counterclockwise) position.

The effectors of the agent are controlled by a three neuron continuous time recurrent neural network (CTRNN). Each neuron i has the state equation

$$\frac{dy_i}{dt} = \frac{1}{\tau_i}\left(-y_i + \sum_{j=1}^{N} w_{ji}\sigma(y_j+\theta_j)\right)$$

where y_i is the state of each neuron, τ_i is the time constant, w_{ji} is the weight of the connection between the jth and i^{th} neuron, θ_j is bias term, and $\sigma(x)=1/(1+e^{-x})$ is the standard logistic activation function. The neurons are named after the function of the effectors they control: Forward Swing (FS), Backward Swing (BS), and FooT (FT). The foot is down when FT's output is above 0.5 and up otherwise. When the foot is up, the output of BS and FS scale the torques swinging the leg, when down these outputs scale the translational force applied to the body.

A real valued genetic algorithm (GA) (Bäck, 1996) was used to evolve the CTRNN parameters. Each 3 neuron neural circuit was represented as 15 real values. A generation of GA search contains 100 individuals described by a point in 15 dimensions restricted to [-1,1]. The first generation of individuals is created by choosing points randomly from this space. After initialization, succeeding generations are created from the proceeding generation by copying the highest fitness individual (a.k.a. "the elite") and then by mutating parents chosen for reproduction by a linear rank based method. A parent vector is mutated into a child vector by adding a displacement vector created by randomly choosing a direction in the fifteen dimensional space and then randomly choosing a magnitude from a Gaussian distribution with a mean 0 and variance of σ^2 (also called "mutation variance" throughout this paper). If the point resulting from the addition of the parent vector and displacement vector lies outside [-1,1] then those dimensions are rounded ("clipped") to within the range.

To evaluate the fitness of an individual, the neural circuit is instantiated and the coupled neural circuit/body simulated. A neural circuit is instantiated by linearly mapping the point representing the individual in search space to CTRNN parameter space (with ranges [-16,16] for weights and biases, [0.5,10] for time constants). The motor neuron outputs are initialized to zero. The coupled neural network body are numerically integrated over 220 time steps by the forward Euler method at a step size of 0.1 and the individual's fitness is the distance covered in this time. (The actual fitness used is the average velocity, but because the amount of time is fixed it is equivalent, and easier, to think of the fitness in units of single leg insect steps.) The distance of one full step gives the agent a fitness of approximately 0.125, two steps $2*0.125=0.25$ and so on. An optimal agent (Beer et al., 1999) has time to go the distance of a little over five steps in the time allowed.

3. General Characteristics

All GA searches on this task have a common structure when best fitness per generation is plotted. Initially all searches have a best fitness of approximately 0.125. No improvement to the best fitness is made for a varying number of generations until an individual is found with a fitness greater than approximately 0.125. After this individual is discovered, there is a steady ratcheting upwards in the best fitness which eventually levels off to different levels in each search as the theoretical maximum of 0.627 (Beer et al., 1999) is approached. We call the discovery of a greater than 0.126 fitness individual "breakthrough". (As will be discussed later, 0.125 individuals are qualitatively different than 0.126 individuals.) The period of the search which occurs before breakthrough will be referred to as "phase I" and the period after breakthrough "phase II".

Figure 1 shows three searches performed nominal mutation variance (0.5) for this problem. One can see that all three searches begin with a best fitness of 0.125 and remain at this fitness for a varying number of generations until breakthrough occurs. In these searches, breakthrough occurs at generations 6, 42, and 116 as marked by the small arrows.

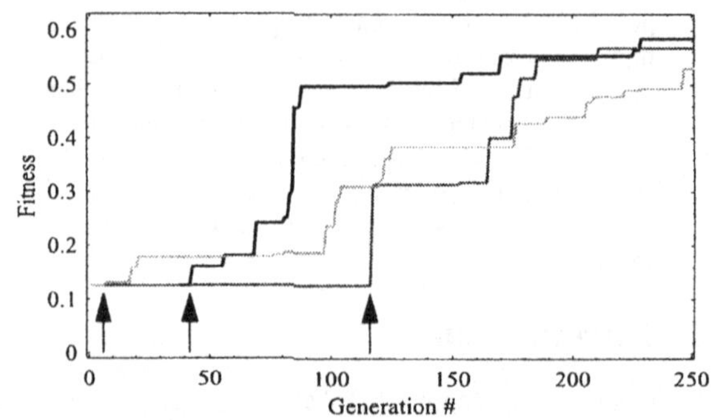

Figure 1: Best fitness by generation of three searches Small arrows indicate "breakthrough", which separates phase I from phase II.

There are parallels here to work on neutral networks. The pattern of constant fitness while a population moves over parameter space is related to the concept of neutrality, phase I is similar to to an epoch, and breakthrough is analogous to the discovery of a portal. Information on neutral networks can be found in van Nimwegen (1999).

4. Phase I

Phase I lasts from the beginning of search until the population contains at least one individual whose fitness is greater than 0.126. The breakthrough generation distribution is very non-normal, with a strong skew toward lower generations. Figure 2 is a plot of the fraction of GA searches which have broken through at the generation

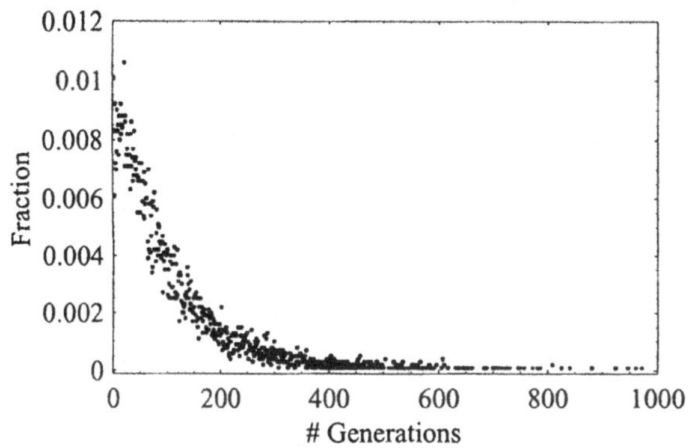

Figure 2: Fraction of breakthroughs per generation for 10,000 GA searches at nominal mutation variance.

indicated. After 250 generations slightly greater than 90% of searches have broken through, while after 1000 generations more than 99% have completed phase I.

This distribution can be fit with an exponential function $N(t) = N_0 e^{-pt}$. The exponential function is commonly used to fit random processes such as radioactive decay which occur with a constant probability per unit time. The data are converted into the appropriate form by calculating the fraction of searches yet to breakthrough at each generation. Then N(t) gives the fraction of breakthroughs at generation t, N_0 the initial fraction not broken through (should be around 1), and p the probability of breakthrough per generation. Ignoring the last 10% of searches in the tail of the distribution, the best fit equation has $N_0 = 0.984$, p = 0.00962, and a r^2 value of 0.998. The median[1] (i.e. generation at which half of all breakthroughs have taken place) is predicted to be at generation 72, which is very near the observed median of 71.

Interestingly, random sampling in "generations" of 100 individuals discovers a breakthrough individual with a median of 73 "generations". Randomly generating individuals within parameter space and checking "generations" of 100 individuals for at least one breakthrough individual essentially simulates a GA search in which one generation (of 100 individuals) is completely unrelated to the next generation. Thus, the GA search at nominal mutation variance is behaving like random sampling at the level of breakthroughs per generation.

Not only does the breakthrough event occur with random-like equal probability per generation, but the population's fitness change per generation also has analogies to random processes. To plot how the population's fitness is changing, a line is drawn from the fitness of each parent in one generation to the fitness of all that parent's children in the next generation. Figure 3 is a lineage web of the first 60 generations of search A above. One can see that each generation parents which have a

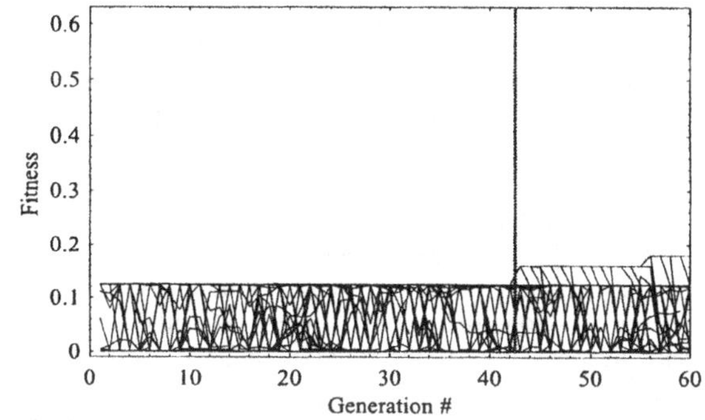

Figure 3: Phase I lineage web of search A from Figure 1. "X"s result from parents in one fitness band having children in the other fitness band – a very common occurrence. Phase I ends at generation 43 when a ≥ 0.126 individual is discovered (gray vertical line).

fitness of 0.125 have 0 fitness children as well as 0.125 fitness. Similarly 0 fitness parents often have children which have a 0.125 fitness as well as the same fitness. Graphically this appears as the two bands at 0 and 0.125 fitness connected by "X"s. By the lack of lines changing direction between 0 and 0.125 fitness it can be concluded that most individuals are either 0 or 0.125 fitness and very few individuals appear at intermediate fitness values. A plot (Figure 4) of individuals' fitness distribution in every generation of ten searches up until breakthrough confirms that a vast majority of individuals are either of 0 or 0.125 fitness. This pattern suggests a dynamic equilibrium between two relatively large fitness plateaus with relatively few intermediary fitness regions.

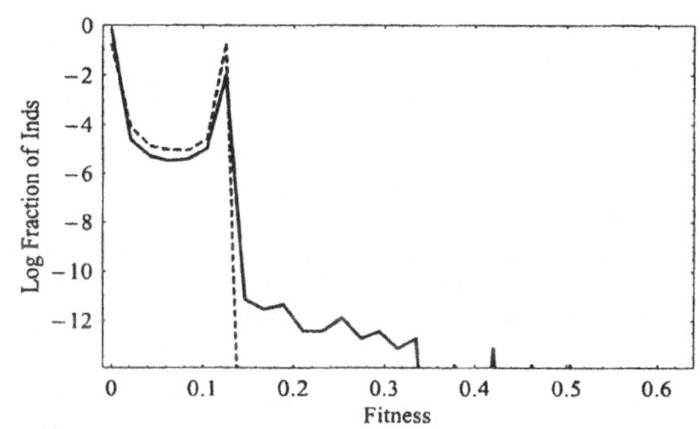

Figure 4: Distribution of fitnesses for A) GA search in phase I (dashed line) and for B) a random sample of parameter space (solid line). Random sampling and GA search at nominal mutation variance result in a similar distribution of individuals less than 0.125 fitness. Adapted from Ames (2003)

Non-oscillating neural networks are all that is required to achieve the fitnesses observed in phase I. Zero fitness is achieved simply by not moving the agent's body at all: behaviorally this can be done either by never putting the foot down to make contact with the ground or by leaving the agent's leg in the initial fully forward position. A fitness of 0.125 is given to individuals which move the distance equivalent to a full swing of the leg. Monotonically changing neural network outputs are all that is required for both of these fitnesses. Intermediary fitnesses are less

[1] The median is more stable than the mean in describing data with this distribution: the median breakthrough generation is 71 whether the GA search is carried out to 250 or 1000 generations. Setting those searches which have not broken through equal to the maximum generation, the mean is 93 for 250 generations 105 for 1000 generations.

common because the sigmoid activation function makes it easier for neurons to be "on" or "off" rather than an output resulting in a step midway between all or nothing.

Random sampling of individuals from parameter space have a fitness distribution (Figure 4) very similar to the GA search in phase I. (The exception is individuals greater than 0.125 fitness: there are no individuals greater than 0.125 in phase I.) The close match between the fitness distributions is most easily explained by GA search populations covering the same regions of parameter space as random sampling. This suggests that at the nominal mutation variance GA search is emulating random sampling.

Up until now it has been argued that GA search *at nominal mutation variance* has been analogous to random sampling. At lower mutation variances this analogy breaks down immediately. Figure 5 shows the median generation of breakthrough for a wide range of mutation variances compared to the median number of random individuals in "generations" of 100 needed to discover at least one breakthrough individual. At mutation variances greater than approximately 0.3 the GA search does at least as well as random sampling while at mutation variances less than approximately 0.3 the median number of generations to finding a breakthrough individual is larger than random sampling.

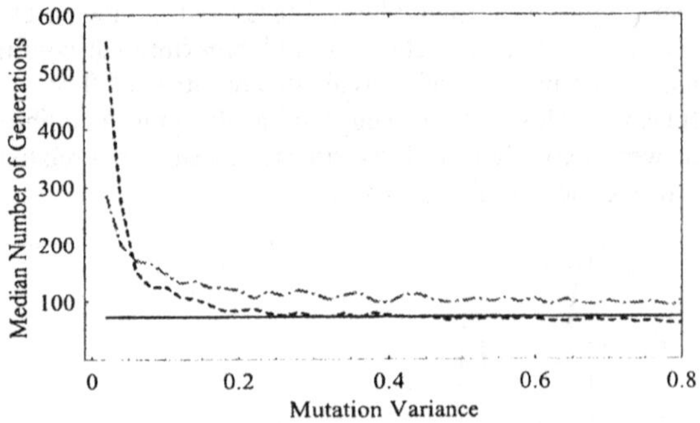

Figure 5: Median number of generations to discovery of a ≥ 0.126 fitness individual. A) Evenly dashed line is GA search, B) syncopated line is diffusion search, and C) solid line is random samples of 100 individuals. At higher mutation variances (>2) all lines converge.

Selection has a varying role over the range of mutation variances. This can be determined by running "diffusive searches" at the same mutation variances as the GA searches and comparing their median generation to breakthrough. Diffusive search is performed by turning off elitism and having each parent generate exactly one child using the same procedure as GA search.

At very low mutation variances, less than 0.05, one can see that the GA search performs more poorly than diffusive search. This is most likely due to premature convergence of the GA search.

At higher mutation variances, between 0.05 and 2, GA search does better than diffusion search. The reason for this is currently unknown: the slim majority parents of breakthrough individuals in diffusion search are nearer 0 in fitness than 0.125, eliminating the possibility that the 0.125 areas are "nearer" breakthrough areas in general. Turning off elitism and turning down the selective advantage given to higher ranked individuals eliminates this gap.

At very high mutation variances (greater than 2) GA, diffusion, and random sampling all have nearly the same median generation of breakthrough. This is not surprising because as mutation variance is increased each generation becomes less related to the previous and approximates a random sample.

The pattern of performance of GA search in phase I search and random sampling of individuals lead to some conclusions about the structure of parameter/fitness space. There are vast plateaus of fitness in parameter space at 0 and 0.125. Above 0.125 fitness parameter space areas are at least 5 orders of magnitude less common. Because of a lack of fitness gradients in this parameter/fitness terrain, search in phase I does well at high mutation variances by emulating random sampling.

5. Phase II

Phase II begins when a "breakthrough" individual greater than or equal to 0.126 fitness is found. A breakthrough individual must have a neural network whose outputs are not monotonicly changing because a fitness of greater than or equal to 0.126 is achievable only by covering a distance equivalent to more than one full swing of the leg. In other words the agent must repeat some movements such as move the leg both backward and forward and both raise and lower it's foot (at the appropriate time).

Phase II has a well defined beginning, but not a well defined end because it is very difficult to find an individual with the theoretical maximum fitness by GA search. A lower, more easily achieved fitness which qualitatively remains a good walker is desirable for purposes of measuring the elapsed number of generations between breakthrough and "end of phase II". Individuals with a fitness of 0.5 show no obvious behavioral deficits (i.e. seem to walk without difficulty) and are not too difficult for the GA to find.

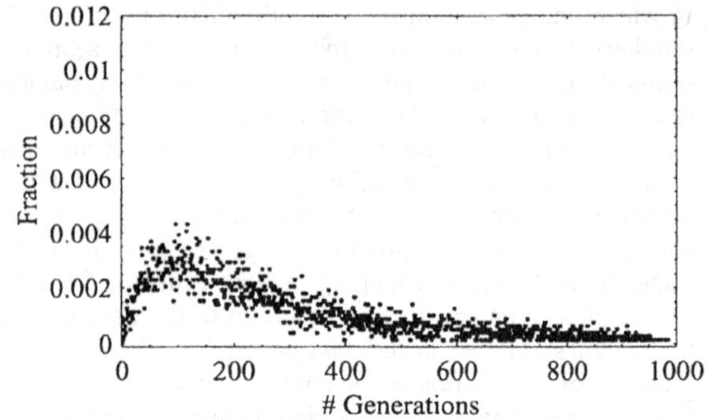

Figure 6: Fraction of 10,000 GA searches which have an elapsed number of generations between breakthrough and end of phase II.

Using these criteria for the beginning and end of phase II a plot (Figure 6) can be made of the elapsed number of generations. In contrast to the breakthrough distribution in the previous section, this distribution initially rises to peak around 100 generations elapsed and a descending phase for the remainder of the distribution Approximately 55% of searches are able to complete phase II when allowed 250 generations after breakthrough and slightly less than 80% if allowed 1000 generations.

The exponential function does not fit this distribution as nicely as the phase I breakthrough distribution. Following the same procedure as outlined in phase I, the best fit equation has $N_0 = 1.12$, $p = 0.00385$, and a r^2 value of 0.999. The median is predicted to be at generation 180, which is not near the observed median of 213.

Lineage webs of phase II (e.g. Figure 7) allow us to examine patterns in the population fitness change between breakthrough and 0.5 fitness. The majority of the population continues to exist throughout the search in a band of fitness from 0 to 0.125. A very small fraction of the population has a fitness of greater than 0.125.

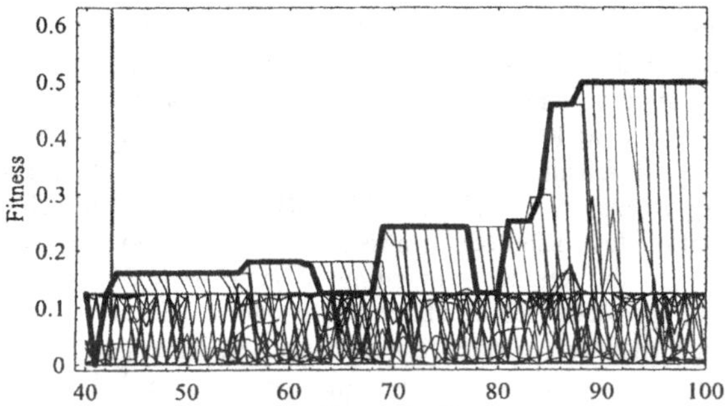

Figure 7: Phase II lineage web of search A from Figure 1. The elite individual increases in fitness every generation. The lineage which leads to the best individual in the last generation loses and gains fitness. Phase II begins at breakthrough (vertical gray line).

Of the individuals with a fitness greater than 0.125, most of them are descendants of the first breakthrough individual: of the occurances of the elite individual being replaced by a higher fitness individual, less than 4% of the time the individual is unrelated to the elite. This is not too surprising given that during the course of each search the fitness of the elite individual progressively increases above breakthrough fitness, making its replacement more and more difficult.

The descendants of the elite quickly diffuse away from the high fitness regions. Most children quickly join the 0 to 0.125 fitness band of the population. Less than than 6% of the elite's descendants ever achieve a fitness higher than the elite. Of those, the more closely related the descendant is, the more likely it is to replace the elite. Of those descendants that do replace the elite, 70% of the time it is a child, 13% a grandchild, 6% a great-grandchild, 3% a great^2-grandchild, 2% a great^3-grandchild, and less than 1% for all others. In Figure 7 the thicker line traces the lineage which produces the elite individual in the last generation of the search. Along this lineage one can see instances of both the elite's child (generation 56) and of great-grandchildren of varying degrees (generation 68, 81) replacing the elite.

A smaller mutation variance more effectively explores the region of parameter space which contains the above breakthrough fitness individuals. A survey of GA searches at various mutation variance was performed to characterize the median elapsed number of generations between the discovery of a breakthrough individual and the discovery of an individual with a fitness greater than or equal to 0.5 (Figure 8). The median elapsed number of generations increases with increasing mutation variance. The best mutation variance for completion of phase II is approximately 0.05, $^1/_{10}$ the magnitude of the nominal mutation variance taking roughly $^3/_4$ as many generations. The sensitivity of elapsed number of generations is not as great as in phase I. This may be because the distribution of displacements between parent and child are Gaussian even at high mutation variances most of the parent-child displacements remain small.

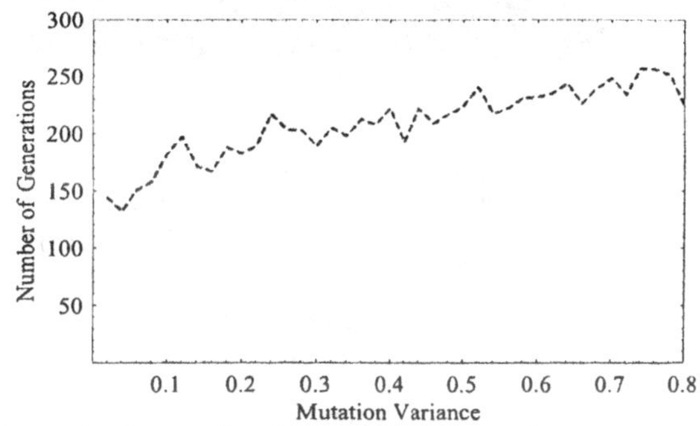

Figure 8: Mean number of generations between discovering a ≥ 0.126 and then a ≥ 0.5 fitness individual. The elapsed number of generations slowly increases with increasing mutation variance.

The improvement at low mutation variances is due to more distantly related descendants replacing their elite parents. For 0.05 mutation variance around 8% of the elite's descendants eventually attain a higher fitness than the elite compared to less than 6% for 0.5 mutation variance. Of these, fewer children replace the elite than at the nominal mutation variance (57% versus 66%), while more distantly related descendants replace the elite at a higher rate. The low mutation variance appears to allow descendants to jostle around longer in the vicinity of the high fitness region.

Taken together, phase II is best characterized as a GA search, not randomly searching large, flat fitness plateaus, but instead more like hill-climbing a smaller parameter/fitness region with a gradient. The rise and fall of the generations elapsed from breakthrough to 0.5 reflects the variation in distance between the breakthrough individual and the greater than 0.5 region over searches. Finally, at lower mutation variances the elite's descendants are given more of a chance to find higher fitness regions rather than diffuse off into the low fitness plateaus that occupy the majority of parameter space.

The above description of the structure of parameter space can be shown to exist for lower than 15 dimensional slices through parameter/fitness space. A two dimensional slice of parameter space is made by instantiating individuals

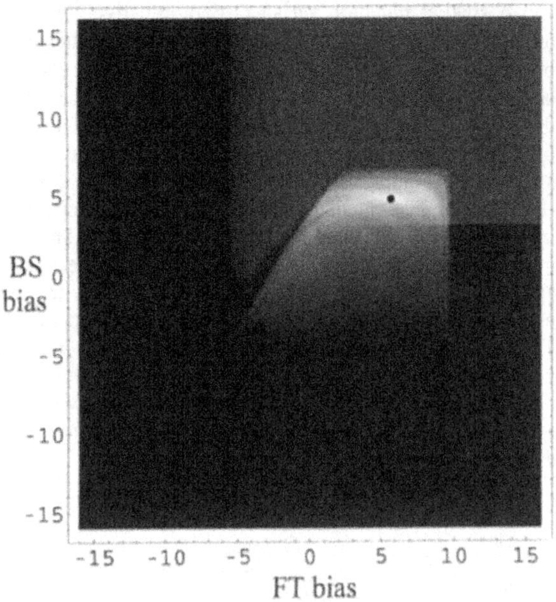

Figure 9: A two dimensional slice through fifteen dimensional parameter space. Dot is individual found by GA search. Gray scale is log of fitness. Zero fitness – black, 0.125 fitness – large dark gray area in upper right, lighter is higher fitness.

all of whose parameters are fixed but two. The resulting fitness of the individuals is plotted in gray scale at the coordinate containing the values of the two varying parameters. Most randomly chosen slices through parameter space will show fitnesses no greater than 0.125. If one centers a slice on a high fitness individual one gets a view of parameter space such as in Figure 9. The dot is a high fitness individual (found by GA search) from which the slice was extended by varying the biases of the foot and back swing motor neurons. One can see large flat expanses of zero fitness (black) and 0.125 fitness (gray area in upper right). It is these regions that the population occupies during phase I. Phase II begins when an individual discovers a greater than 0.125 point in centrally located "ridged" region of parameter space. (Note that a less than 0.125 individual on the ridged region will not have a reproductive advantage due to individual in the higher fitness 0.125 region.) The descendants of this breakthrough individual slowly ratchet (when using elitism) or climb their way up to the highest fitness areas.

6. Applications

Typically single leg stepper searches are run at a single mutation variance. In these searches, mutation variance is a balance between a high mutation variance and a low mutation variance. The high mutation variance does well in the first part of the search when the population is initially randomly scattered on the low fitness plateaus which make up the majority of parameter space. The low mutation variance does well exploring the smaller regions of higher fitness parameter space.

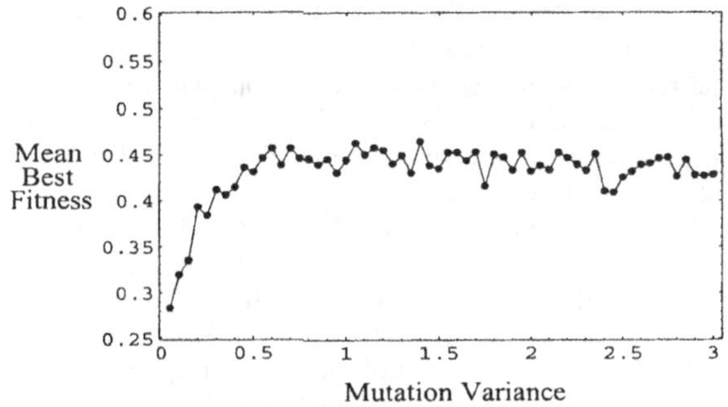

Figure 10: Averaged best (last generation) fitnesses over a range of mutation variances.

The current results can be applied towards explaining the findings of Mathayomchan (2002) whose examination of mean best fitness over a range of mutation variances was used to determined the nominal mutation variance for the single leg stepper problem. In these experiments the best fitness of each 250 generation search was recorded and averaged. The searches were performed on a variety of mutation variances. The mean best fitness was found to be very low at low mutation variances and increases steeply with increasing mutation variance until it levels off at approximately 0.6 (Figure 10). The current results show that the low mean best fitness at small mutation variances is due to the search terminating at generation 250 without having discovered an above 0.126 individual. Note that at high mutation variances the mean best fitness apparently remains unchanged. This may be explainable by the relatively more robust performance of phase II with respect to mutation variance.

If one is not restricted to only one mutation variance one can decrease the median number of generations from the beginning of search to end of search. With a single mutation variance of 0.5, the median generation for a complete search is approximately 400 generations. If two mutation variances are used with a switch from one to the other at breakthrough, then one can do better. Choosing 0.5 mutation variance until breakthrough and then 0.05 thereafter results in a median of approximately 300 generations for a savings of 100 generations.

Decreasing the mutation variance over time is reminscent to the strategies of simulated annealing in which a "temperature" analogous to mutation variance is changed using an annealing schedule. The annealing schedules usually decrease over the long term, letting solutions settle to local minima, often with punctuation with increases in temperature whose goal is to break the solution out of local minima. (Kirkpatrick et al. 1983, Salamon et al. 2002) In the case of the single leg stepper problem, the switch to lower mutation variance is made at a specific point in the search using problem specific knowledge.

It may be possible reduce the computing time required for the search by using the observation that after breakthrough occurs, most of the population is not needed for the continued improvement in the fitness of the individuals. Specifically, descendants of the breakthrough individual have a much higher probability of finding a new high fitness than a random individual and unrelated individuals may be safely ignored. All of the relatives of the breakthrough individual may not even be needed: If a child does not become the elite, it's lineage diffuses via mutation away from the region. With passing generations the children may eventually become equivalent to a unrelated individual. If this happens they can also be eliminated from the population. Thus it may be possible to speed the search by reducing the number of individuals in the population during phase II.

7. Conclusions

The layout of parameter/fitness space for the single leg stepper problem is such that one mutation variance does not explore all features equally well. A larger mutation variance does best in the initial search because high fitness regions are such a small proportion of parameter space that a random search is needed rather than a gradient following search. Once an above breakthrough fitness region is found a lower mutation variance does better because children are more likely to be created on nearby above breakthrough fitness regions.

These findings apply to many systems with complex dependencies among parameters. In these cases the intersection of parameter values which "work" is often very small. Instead, parameter space may be occupied by "background" fitnesses which are easily achievable by chance and supply no information as to the location of the higher fitness regions of interest. Under these conditions random search along with problem specific knowledge might be used to discover the higher fitness regions of interest followed by GA search to tune the initial solution.

Acknowledgments

This research has been supported in part by grant EIA-0130773 from the NSF and a NSF IGERT grant in neuromechanics at CWRU.

Thanks to the Beer Nuts for giving helpful suggestions on the revisions of this paper.

References

Ames, J.C. (2003) *Design methods for pattern generation circuits*, Master thesis, Department of Electrical Engineering and Computer Science, Case Western Reserve University.

Bäck, T. (1996) *Evolutionary Algorithms in Theory and Practice*, Oxford University Press

Beer, R.D. and Gallagher, J.C. (1992) Evolving dynamical neural networks for adaptive behavior. *Adaptive Behavior* **1**(1):91-122.

Beer, R.D. (1995) On the dynamics of small continuous-time recurrent neural networks. *Adaptive Behavior* **3** (4):471-511.

Beer, R.D., Chiel, H.J. and Gallagher, J.C. (1999) Evolution and analysis of model CPGs for walking II. General principles and individual variability. *J. Computational Neuroscience* **7**(2):119-147.

Chiel, H.J., Beer, R.D. and Gallagher, J.C. (1999) Evolution and analysis of model CPGs for walking I. Dynamical modules. *J. Computational Neuroscience* **7**(2):99-118.

Cliff, D., Harvey, I., and Husbands, P. (1993) Explorations in evolutionary robotics. *Adaptive Behavior* **2**(1):73-110.

Kirkpatrick, S., Gelatt, Jr., C. D., Vecchi, M.P. (1983) Optimization by Simulated Annealing, *Science* 220 (4598): 671-680.

van Nimwegen, E., Crutchfield, J.P., and Mitchell, M. (1997) Finite Populations Induce Metastability in Evolutionary Search. *Physics Letters A* **229**(2), 144-150.

van Nimwegen, E. (1999) *The Statistical Dynamics of Epochal Evolution*. PhD thesis, Bioinformatics Group, University of Utrecht.

Gao, Y., Qi, X., and Palmieri, F. (1998) Comments on theoretical analysis of evolutionary algorithms with an infinite population size in continuous space I. basic properties of selection and mutation. *IEEE Trans. Neural Netw.* **9**(2), 341-343.

Mathayomchan, B. and Beer, R.D. (2002) Center-crossing recurrent neural networks for the evolution of rhythmic behavior. *Neural Computation* **14**:2043-2051.

Nolfi, S., Floreano, D., (2000) *Evolutionary Robotics*, MIT Press.

Salamon, P., Sibani, P., Frost, R., (2002) *Facts, Conjectures, and Improvements for Simulated Annealing*, Society for Industrial & Applied Mathematics.

The Force Model: Concept, Behavior, Interpretation

Ralf Salomon*
* Department of Electrical Engineering and Information Technology
University of Rostock
18051 Rostock, Germany
ralf.salomon@etechnik.uni-rostock.de

Abstract

Most experiments in research on autonomous agents and mobile robots are performed either in simulation or on robots with static physical properties; evolvable hardware is hardly ever used. One of the very rare exceptions is the *eyebot* on which Lichtensteiger and Eggenberger have evolved simplified insect eyes (L. Lichtensteiger and P. Eggenberger, 1999). Even though substantially improved, the evolutionary models currently applied still lack both scalability and noise-resistance. To tackle these problems, this paper proposes a biologically-inspired force model for this class of real-world applications. The simulation results clearly indicate that this model provides a significant improvement over existing limitations. Furthermore, this paper argues that the force model is of more general utility.

1. Introduction

In its long history, (classical) artificial intelligence has developed its strengths in areas, such as playing games, developing game theory, automatic theorem provers, etc. Most of this research focuses on algorithmic questions that are more or less bound to a formal framework. Since the beginning of the 90ies, however, a new research area has emerged, for which Brooks has coined the term new artificial intelligence (new AI). New AI aims at understanding (natural) intelligence and its underlying mechanisms by building systems that exhibit "intelligent" behavior ??? (R. A. Brooks, 1991a, 1991b, R. Pfeifer and C. Scheier, 1999). These systems are often realized as mobile robots, which are supposed to operate in dynamically changing, partially unknown environments without any human control (they are thus attributed *autonomous*).

New AI prefers a synthetic approach, i.e., understanding by building. In order to reach its research goals, new AI draws a significant amount of inspiration from natural systems. It therefore often investigates (biological) hypotheses and aims at validating them in simulation or on particularly-designed robots. For obvious reasons, most validations are done in simulation (see, for example, the conference series *Simulation of Adaptive Behavior*), but some are indeed done on physical robots (L. Lichtensteiger and P. Eggenberger, 1999, R. Salomon, 1996, R. Salomon and L. Lichtensteiger, 2000). In most of these experiments, the robot's morphology, the positions of its sensors, etc., are predetermined, and adaptation concerns mostly the robot's controller, which is given in software in virtually all cases; *evolvable hardware* is still used only in exceptional cases (cf. (D. Keymeulen, M. Iwata, Y. Kuniyoshi, and T. Higuchi, 1999)).

Section 2. discusses the *eyebot*, which has been published in the recent literature (L. Lichtensteiger and P. Eggenberger, 1999) and is a nice example of a robot with evolvable hardware components. Lichtensteiger and Eggenberger used evolutionary algorithms to evolve the sensor distribution of an insect eye. As has been mentioned earlier, autonomous agents are supposed to freely move around and should not collide with obstacles. Therefore, Section 2. also explains how a suitable sensor distribution can be used to estimate the lateral distance to an obstacle by means of a very simple neural network, in which all connections have *equivalent* weights.

Unfortunately, evolution in hardware suffers from immense time requirements. On the eyebot, for example, one single fitness evaluation takes about one minute. Thus, if an evolutionary algorithm would consider a population of about 60 individuals, which are mostly considered not many, the execution of one single generation would already take an hour of experimentation time. In this setup, experimentation time is a very limited resource. Consequently, as Section 3. summarizes, subsequent research accelerated this application by developing different coding schemes (R. Salomon and L. Lichtensteiger, 2000). Despite the achievable performance improvements, the scalability remained strongly limited, which allows for the evolution of insect eyes with only a few receptors for practical reasons. Furthermore, the consideration of noise, which is omnipresent in such hardware setups, imposes severe problems; in many cases, the evolutionary algorithms converge at poor solutions.

In order to tackle the problems discussed above, Section 4. proposes a new, biologically-inspired coding approach, called the *force model*. In essence, this model can be considered a reformulation of the original task. Due to its very nature, this force model is not limited to the evolution of an inset eye, but may be transferred to other real-world applications. Section 5. summarizes the experimental setup, Section 6. then

Figure 1: The "eyebot" consists of a chassis, an on-board controller, and sixteen independently-controllable facet units (see Fig. 2), which are all mounted on a common vertical axis.

presents some results, which indeed indicate a significant reduction of the problem's complexity, which offers the evolution of more complex systems. Section 7. tries to analyses the mechanisms responsible for the obtained performance improvements. Section 8. finally concludes with a brief discussion.

2. Background: The Eyebot

This section summarizes previous research and includes the description of the robot, its eye, the neural network controller, as well as the motion parallax phenomenon.

2.1 *The Eyebot*

Inspired by biological evidence (T. S. Collett, 1978, N. Franceschini, J. Pichon, and C. Blanes, 1992, G. A. Horridge, 1978), Lichtensteiger and Eggenberger (1999) constructed the *eyebot* (Fig. 1) to model the eye of an insect, such as the house fly. It consists of a chassis, an on-board controller, and sixteen independently-controllable facet units, which are all mounted on a common vertical axis. A facet unit (Fig. 2) basically consists of the sensor, a thin tube, two cog-wheels, a motor, and a potentiometer. By means of the cog-wheels, the motor can position the facet within a range of about 200 degrees, and the potentiometer provides feedback about its actual position, i.e., its angle α_i. The thin opaque tube is used to reduce the sensor's aperture to about two degrees.

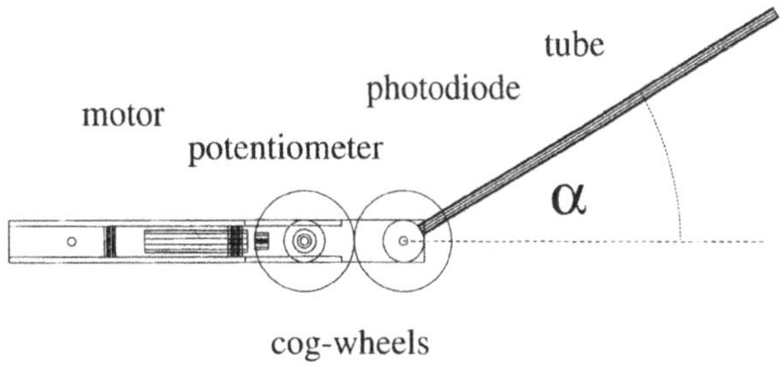

Figure 2: The robot's facet basically consists of the sensor (the photo diode), a thin tube, two cog-wheels, a motor, and a potentiometer. By means of the cog-wheels, the motor can position the facet within a range of about 200 degrees, and the potentiometer provides feedback about its actual position.

These tubes are the primitive equivalent to the biological ommatidia (T. S. Collett, 1978, N. Franceschini, J. Pichon, and C. Blanes, 1992, G. A. Horridge, 1978). It should be noted that such a low-cost construction is subject to several imprecisions and tolerances, which might be sensed as noise during operation.

2.2 *Motion Parallax: A Mathematical Description*

Figure 3 sketches how a phenomenon, also known as motion parallax, can be utilized to avoid obstacles. Let us assume that the compound eye, presented as the observer R, is at a fixed position. If the obstacle X is moving at constant speed v, the observer views the obstacle under different angles α_1, α_2, and α_3 at different time steps t_1, t_2, and t_3. Let r denote the vector from observer R to obstacle X and let v_t denote the component of v that is perpendicular to vector r. For the distance d of closest approach, the following relation holds

$$d = r \sin \alpha \,. \tag{1}$$

Since $v_t = v \sin \alpha$, the angular velocity $\omega = \dot{\alpha}$ with which the image of X moves through the visual field of the observer is

$$\omega = \frac{v_t}{r} = \frac{v \sin \alpha}{r} \,. \tag{2}$$

If the agent can estimate the velocity v and can measure both α and ω, it can calculate its distance r to the obstacle at any time. Solving eq. (2) for r and substituting into eq. (1) leads to

$$d = \frac{v}{\omega} \sin^2 \alpha \,. \tag{3}$$

Let us assume that the agent uses some sensors each of which can detect the obstacle if it appears under a particular angle α. The agent can then estimate the angular velocity $\omega = d\alpha/dt \approx \Delta\alpha/\Delta t$ by the change of α per time interval:

$$d = \frac{v}{(d\alpha/dt)} \sin^2 \alpha \approx v \frac{\Delta t}{\Delta \alpha} \sin^2 \alpha \,. \tag{4}$$

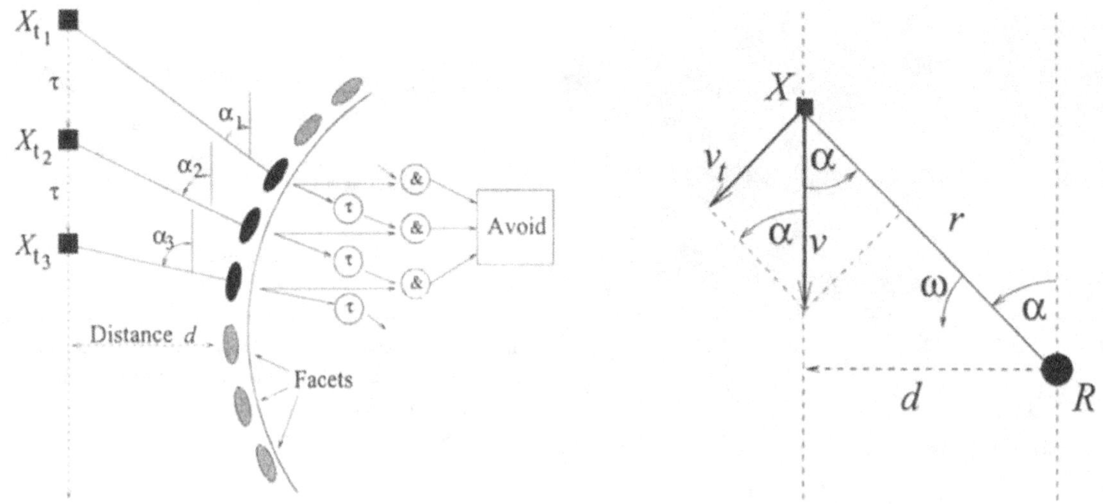

Figure 3: Left: due to their small aperture of about two degrees, the insect eye's facets recognize a moving object at different times $t_1, \ldots t_3$. With some units that maintain their activity during a time period τ after stimulation and output units that perform an and-operation over neighboring input pairs, a neural network is able to detect too small a distance of closest approach to an obstacle. Right: the angular velocity ω under which an obstacle X is seen equals $\omega = v_t/r = v \sin\alpha/r$ and depends on the actual angle α leading to a distance of closest approach $d = (v/\omega)\sin^2\alpha$.

Similar to biological systems, an agent with s facets can estimate $\omega \approx \Delta\alpha/\Delta t$ by utilizing a simple neural network, which consists of s input units u^I and s-1 output units u^O. As can be seen in Fig. 3, each output unit u_i^O is connected to two input units u_i^I and u_{i+1}^I each of which is in turn connected to one facet with all connections being topology preserving, i.e., neighboring facets signal to neighboring inputs, which in turn connect to the same output unit. Furthermore, each input unit u_i^I has an associated time constant τ_i during which the unit remains active after it has been triggered by an appropriate input.

Each triple u_i^I, u_{i+1}^I, and u_i^O constitutes a motion sensor. An input unit u_i^I is activated by the appearance of a sufficiently-high "dark-to-bright" stimulus. Then, this unit remains active during the decay time τ_i. If also the neighboring input unit u_{i+1}^I becomes active during this time interval, the output unit u_i^O is triggered (due to its "and" operation). If however, the stimulus moves too slowly, the first input neuron u_i^I becomes inactive and the output unit u_i^O is not triggered. Depending on its relative velocity v, the agent may trigger an appropriate avoidance action if the sum of active output units exceeds a critical threshold, which can be, for example, dependent on v (see also eq. (4)). For an agent to be successful, it is essential to avoid obstacles only if necessary, since an agent that constantly avoids obstacles is rather useless. Therefore the parameter τ_i determines a critical speed between two neighboring sensors.

An agent is in principle able to calculate the distance d of closest approach by utilizing only two facets. However, this calculation would be limited to a particular angle of its visual field, which would be too restrictive for real-world mobile robots. By using a compound eye, the agent is able to anticipate potential collisions regardless of the angle under which the object is seen. That is, two compound eyes would provide an agent with an almost 360° view.

2.3 Evolving A Suitable Sensor Distribution

Equation (4) for calculating the latera distance d is highly non-linear in α. If all network delays τ are assumed equivalent, then the time it takes the obstacle to be recognized by two neighboring facets must be τ as well (see also Fig. 3). This immediately implies that the distance between two intersections of the facets and the moving path must be equivalent as well. This, in turn, suggests the following fitness function

$$f = \sum_{i=1}^{s-1} (t_i - \tau)^2 . \tag{5}$$

As already mentioned in the introduction, this paper focuses on the development of a potentially non-uniform sensor distribution (i.e., the α_i's) (L. Lichtensteiger and P. Eggenberger, 1999), which constitutes the eye's morphology, for a simple, prespecified network, rather than developing a complicated neural network for a – perhaps inadequate – sensory system. From Eq. (4), it can be easily derived that the target distribution is $d\alpha/dt \sim \sin^{-1}\alpha$. It should be furthermore noted that all benchmark tests are done in simulation but that particular results are validated by physical robot experiments.

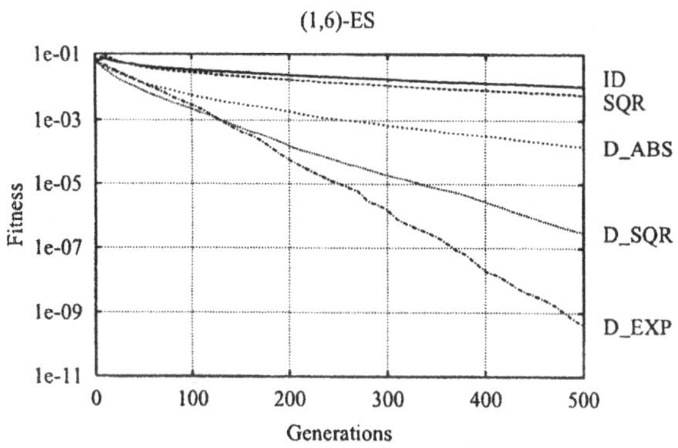

Figure 4: The performance of the (1,6)-evolution strategy significantly depends on the chosen coding scheme.

3. Previous Research and Problem Description

Most classical optimization procedures would try to optimize the given objective (Eq. (5)) by directly applying mathematical operators to the problem-specific variables, which are the angles α_i in the problem at hand. Examples for such operators are the gradient or even the Hessian-matrix.

Rather than applying mathematical (optimization) operators to only one object, most evolutionary algorithms consider a population of different objects and apply simple (random) variation operators, such as mutation and recombination/crossover. It is widely believed that this population-based approach has certain advantages, such as resistance to noise, not getting trapped in local optima, not being biased by the designer's preferences, and so forth.

It might be interested to note that also most evolutionary algorithms apply their variation operators directly to the individuals' problem-specific variables. However, this is *not mandatory*, because the evolutionary context distinguishes between genotype (the parameters x_i to be considered) and the object's phenotype (i.e., the angles α_i in the problem at hand). Evolutionary algorithms therefore employ am appropriate mapping function; but surprisingly many research attempts consider only the identity function.

Even though successful, preliminary experiments (L. Lichtensteiger and P. Eggenberger, 1999) (as mentioned above) indicated slow convergence when using the identity mapping functions. Some other previous research (R. Salomon and L. Lichtensteiger, 2000) investigated the efficacy of the following mapping functions (coding schemes): ID: $\alpha_i = x_i$; SQR: $\alpha_i = x_i^2$; D_ABS: $\alpha_{i+1} = \alpha_i + |x_i|$; D_SQR: $\alpha_{i+1} = \alpha_i + x_i^2$; and D_EXP: and $\alpha_{i+1} = \alpha_i + c * \exp(x_i)$, with c denoting an arbitrary scaling constant, which was set to $c = 1.8$ for historical reasons. All coding schemes assume that the angles α_i are numbered from front to side (see Fig. 3).

Figure 4 shows that the choice of the coding scheme significantly influences the efficacy of a simple (1,6)-ES (please, see Section 5. for the experimental setup). Further results indicate that evolution strategies converge significantly faster than genetic algorithms on this particular problem (R. Salomon and L. Lichtensteiger, 2000).

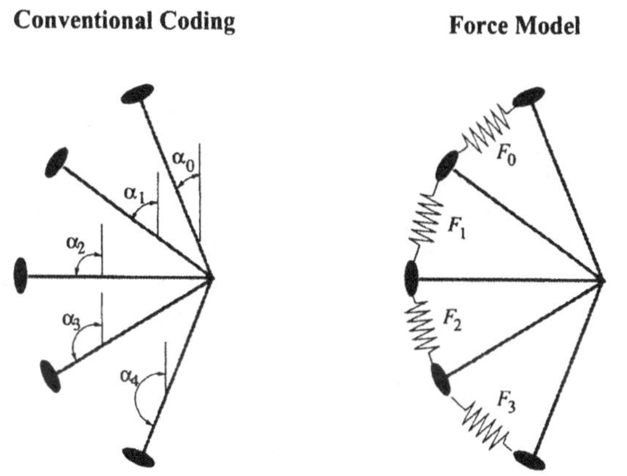

Figure 5: Both figures illustrate an artificial insect eye with five facets. A conventional model (left-hand-side) would directly or indirectly focus on the various angles α_i. By contrast, the force model (right-hand-side) interprets the objective variables as forces F_i between two neighboring facets, and derives the various angles as $\alpha_{i+1} = \alpha_i + F_i(\alpha_n - \alpha_0)/(\sum_{j=0}^{n-1} F_j)$.

Despite all achieved performance improvements, the following two problems remain:

1. The scalability is still unsatisfactory for an increased number of facets s.
2. All coding schemes reacted very sensitive to a noisy fitness evaluation, which is the normal case in a real-world setup.

The next section proposes a new model, called the *force model*, to tackle these problems by new different approach.

4. The Force Model

The various coding schemes discussed so far lack one common problem. A modification of one of the very first angles moves all other facets by that amount. This can result in drastic fitness changes under certain circumstances; unfortunately, these models do not provide a gradually decreasing shift. From a theoretical point of view, it might be better, if a shift of x-degrees of one facet would be distributed in small portions $x/(n-1)$ over the remaining $n-1$ facets. This idea has led to the following force model:

- The eye consists of $n+1$ facets from which the first and last are kept in fixed positions, i.e., α_0 and α_n are fixed, by the agents body.
- An individual has n objective variables which are directly or indirectly considered as forces F_i between the two

facets i and $i+1$. Due to these n forces F_i, the remaining $n-2$ facets evolve their final positions.

- The individual angels α_i are not directly coded. Instead, the model *virtually simulates* springs between neighboring facets. These springs expand and/or shrink depending on some internal forces. Since all facets' tubes are of equal length, the angle α_i is directly proportional to the spring's extension. The force model adjusts the angle between two neighboring facets in proportion to the spring's internal force (and thus to its extension). The angular opening between two neighboring facets is then $\alpha_{i+1} - \alpha_i = F_i / \sum_j F_j$. Since the model assumes equal material constants, it is sufficient to only consider forces (instead of spring extensions). The force model derives all angles as follows:

$$\alpha_{i+1} = \alpha_i + F_i \frac{\alpha_n - \alpha_0}{\sum_{j=0}^{n-1} F_j} \tag{6}$$

To repeat shortly: The force model employs $n+1$ facets from which two are kept at fixed positions. Between those $n+1$ facets, the model considers n forces with which the remaining $n-1$ facets evolve their final positions. In other words, this model has one more genotype parameter than phenotype attributes. In the remainder of this paper, the number of free facets is denoted by $s = n - 1$. (Please remember that the model employs n+1 facets from which two are hold at fixed positions.)

The force model draws its inspiration from some biological observations. Consider a bunch of cells. The insertion of another cell (or a small number of cells) at some particular place would have rather "strong" effect its vicinity but a rather small global one. The force model behaves similarly. A particular force F_i twice as strong as before would almost double the angle between facets i and $i+1$, and would approximately decrease all other angles by the same amount but divided by n.

5. Methods

The term evolutionary algorithms refers to a class of heuristic population-based search procedures that incorporate random variation and selection, and provide a framework that mainly consists of genetic algorithms (D. E. Goldberg, 1989), evolutionary programming (L. J. Fogel, 1962, D. B. Fogel, 1995), and evolution strategies (I. Rechenberg, 1973, 1994, H.-P. Schwefel, 1995).

Even though all evolutionary algorithm have their own peculiarities, they share a lot of common features. All evolutionary algorithms maintain a population of μ individuals, also called parents. In each generation g, an evolutionary algorithm generates λ offspring by copying randomly selected parents and applying variation operators, such as mutation and recombination. It then assigns a fitness value (defined by a fitness or objective function) to each offspring. Depending on their fitness, each offspring is given a specific survival probability. For a good overview of these algorithms, the interested reader is referred to (T. Bäck, U. Hammel, and H.-P. Schwefel, 1997).

Since in the compound eye all angles α_i are real-valued parameters, this paper employs evolution strategies (I. Rechenberg, 1973, 1994, H.-P. Schwefel, 1995) and the breeder genetic algorithm (H. Mühlenbein and D. Schlierkamp-Voosen, 1993).

5.1 Evolution Strategies

In their simplest form, evolution strategies maintain one global step size σ for each individual, and they typically apply mutations to all n parameters $x_i, 1 \le i \le n$, i.e., $p_m = 1$, as follows

$$x_i \leftarrow x_i + \sigma N(0,1) \;, \tag{7}$$

with $N(0,1)$ denoting normally-distributed random numbers with an expectation value of 0 and a standard deviation of 1. Each offspring inherits the step size from its parent, and prior to mutation, the inherited step size is modified by multiplication with a lognormally-distributed random number[1] $\exp(N(0,1))$. In all experiments, the step size was initially set to $\sigma = 0.01$. This simple evolution strategy is denoted as (μ,λ)-ES, or (μ+λ)-ES. The first selection scheme indicates that the algorithm chooses the parents for the next generation *only* from the offspring, whereas the latter selection scheme selects from the union of the offspring *and* previous parents, i.e., μ-fold elitism. In addition, some evolution strategies also feature various recombination operators (see (T. Bäck, U. Hammel, and H.-P. Schwefel, 1997) for further details), such as discrete and intermediate recombination. More elaborate evolution strategies feature individual step sizes σ_i, one for each parameter x_i. Since these forms require relatively large population sizes of $\lambda \ge 200$ to work properly (H.-P. Schwefel, 1997), and since experimentation time is a severe constraint for the final real-world experiments, these forms are not considered here.

5.2 The Breeder Genetic Algorithm

The breeder genetic algorithm (H. Mühlenbein and D. Schlierkamp-Voosen, 1993), denoted as (μ,λ)-BGA for short, is a genetic algorithm variant that is especially tailored to continuous parameter optimization. The breeder genetic algorithm also encodes all parameters x_i as floating-point numbers, and implements mutation by adding or subtracting small random numbers. It normally applies a mutation to each parameter with probability $p_m = 1/n$. In addition, it features different crossover operators, such as discrete recombination, extended intermediate recombination, and extended line recombination (see (H. Mühlenbein and D. Schlierkamp-Voosen, 1993) for further details). It is recommended (H. Mühlenbein and D. Schlierkamp-Voosen, 1993) to use dis-

[1] Constant factors, such as 1.5, 1.0, and 1/1.5, might work as well; see also (I. Rechenberg, 1973).

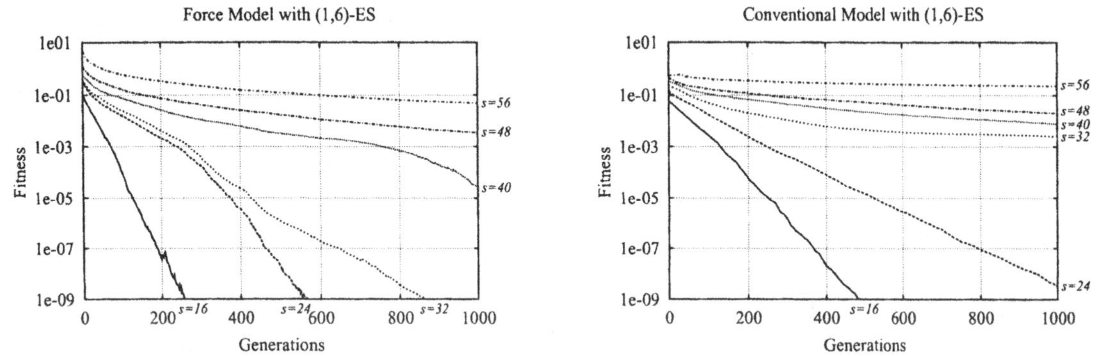

Figure 6: Performance of the (1,6)-ES for various numbers of facets s.

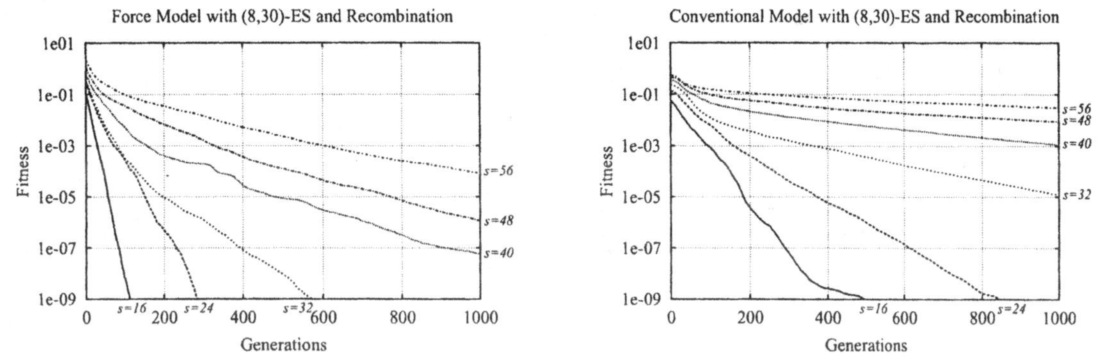

Figure 7: Performance of the (8,30)-ES and recombination for various numbers of facets s.

crete recombination with $p_r = 0.5$. The current breeder genetic algorithm's mutation operator (D. Schlierkamp-Voosen and H. Mühlenbein, 1994) is typically defined as

$$x_i \leftarrow x_i \pm A2^{-ku}\,, \quad u \in [0,1)\,, \tag{8}$$

with "+" and "-" being selected with equal probability, A denoting the mutation range, k denoting a precision constant, and u being a uniformly-distributed random number. Unless otherwise stated, $A = 0.1$ and $k = 16$ were used in all experiments. Previously, discrete mutations were used (H. Mühlenbein and D. Schlierkamp-Voosen, 1993). Furthermore, the breeder genetic algorithm features simple elitism by preserving the best parent from the previous generation in case all offspring have worse fitness.

5.3 *The Fitness Function*

The fitness of a particular sensor distribution is given by the sum of the squared deviations to the time constant τ (τ is equal for all neurons in order to obtain the most simple network):

$$f = \sum_{i=0}^{s} (t_i - \tau)^2\,. \tag{9}$$

When moving with speed v, the time interval t_i is given by the lateral distance between two neighboring sensors

$$t_i = \frac{d}{v}\left(\frac{1}{\tan(\alpha_i)} - \frac{1}{\tan(\alpha_{i+1})}\right)\,. \tag{10}$$

To allow the consideration of noisy fitness evaluations, this paper also considers the following fitness definition

$$f = \sum_{i=0}^{s} ((1 - xN(0,1))t_i - \tau)^2\,, \tag{11}$$

with x denoting the Gaussian-distributed noise level.

Unless otherwise stated, $d/v = 1$ and $\tau = 0.15$ have been used in all experiments. With $\alpha_0 = 20$ degrees and s free facets, the system has s remaining free parameters $\alpha_1 \ldots \alpha_s$.

6. Results

Figures 6 to 9 compare some representative performance aspects of the two models under consideration. The figures on the left-hand-side alway refer to the force model, whereas the figures on the right-and-side always refer to the rather conventional α_i-coding scheme. All shown graphs are averages over 20 independent runs.

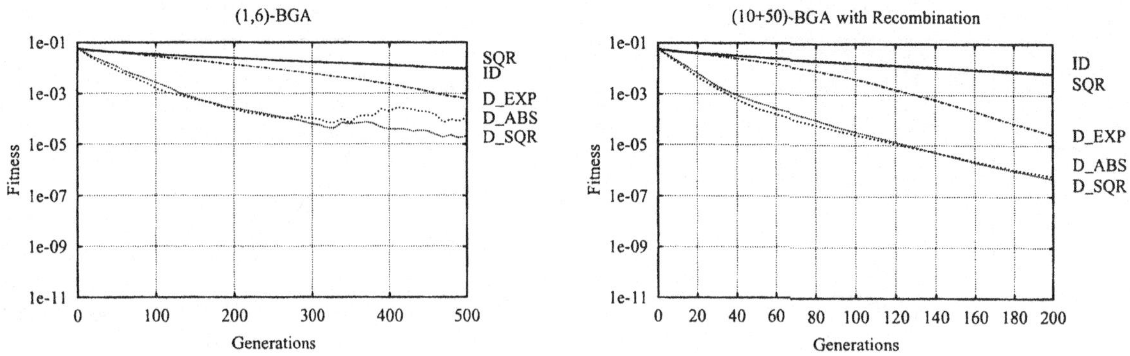

Figure 8: Performance of the (1,6)-BGA and the (10+50)-BGA for only s=16 facets and different coding schemes. Compare also to Fig. 4.

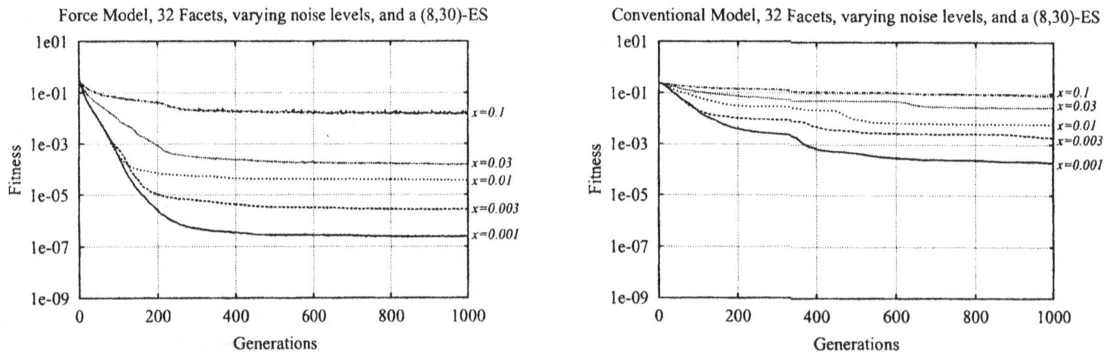

Figure 9: Performance of the (1,20)-ES for various noise levels x and s = 32 facets.

Figure 6 indicates that with the force model, the considered evolution strategies find precise solutions significantly faster; especially for larger numbers of facets s, the difference increases up to an order of magnitude. This gain is practically relevant, since it reduces the experimentation time of real-world experiments from an entire day to approximately three hours. As Fig. 7 shows, evolution strategies yield a similar behavior when changing their parametrization, i.e., μ and λ.

The results obtained by using the BGA are not very satisfying. As Fig. 8 shows, the results (in comparison to Fig. 4) for using only s=16 facets are already very poor. Since the performance progressively degrades when using significantly more facets, the presentation of additional results seem not worth it.

In addition to the presentation of some practically-relevant performance gains, this paper wants to emphasize on the force model's advantage in the case of a noisy fitness evaluation. Figure 9 clearly shows that the force model exhibits both a significantly higher rate of descent and a final accuracy orders of magnitude better than the conventional α_i-coding model. For example, with a noise level of 0.1% ($x = 0.001$), the force model achieves a final fitness value of about 2.4 10^{-7}, a value never reached by the conventional model; by contrast, the conventional model takes twice as long to end up more than three orders of magnitude worse.

At this point, two short notes should be made. First, the resulting sensor distribution is not affected by the proposed coding scheme, since the fitness functions (eq. (11)) has not been changed. Second, a generalization to other problems in this domain is straight forward; the particular design step is to account for give real-world physical constraints.

7. Interpretation

This section addresses the question of why the force model significantly improves the evolutionary process. The obtainable speed up is – in a sense – mainly due to the normalization given in eq. (6). The significant effect caused by this little change can be understood as follows: Let us assume, for the sake of simplicity, that the fitness function has only to parameters. In the conventional model, the evolutionary process has to evolve to a pair of very specific values. In a three-dimensional surface landscape, the optimum is represented by the lowest point of a two-dimensional parabola-like function.

Figure 10 depicts how the optimization process might be finding the *only* global optimum. It is evident that the successful optimization steps necessarily decrease as the opti-

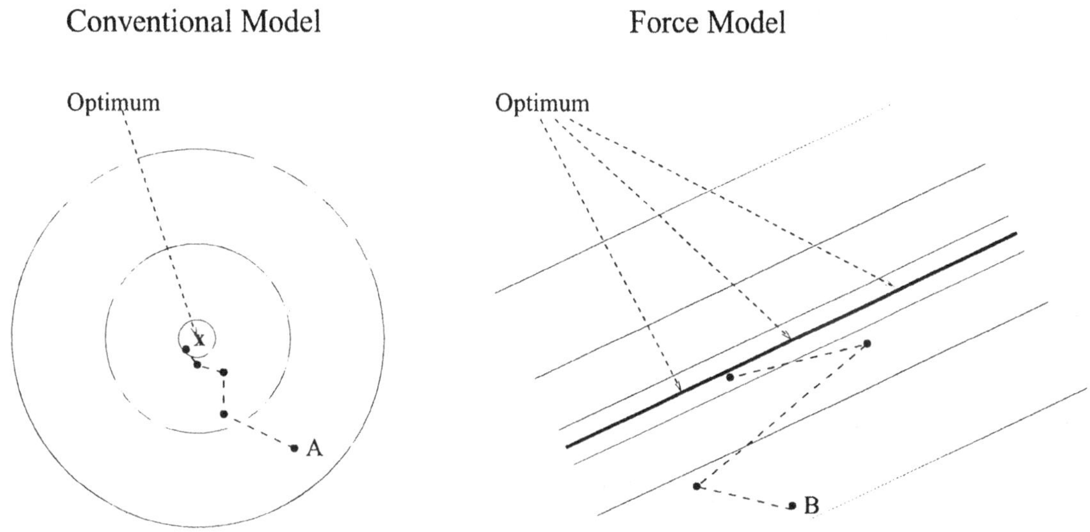

Figure 10: The figure on the left-hand-side shows the rather conventional model that has exactly one optimum. As this figure shows, the possible steps that yield a fitness improvement are constantly decreasing as the optimization process progresses. The figure on the right-hand-side illustrates the situation of the force model. Rather than having *only* one optimum, the force model has infinitely many that are all along a particular line. As can be seen, the optimization process has many more possibilities to achieve an improvement, and consequently, the optimization process might be much faster. "A" and "B" denote some comparable starting points.

mization process progresses. Hence, the optimization process requires an increasing amount of time.

The situation changes in case of the force model. Due to the normalization step, the evolutionary process does consider *only* the quotient of these two parameters. That is, instead of searching for two specific values x_1^o and x_2^o, the force model aims at finding $x_1/x_2 =$ opt. In this case, however, as Fig. 10 shows, the optimum changes from a single point to an entire line (the base of a valley) with infinitely many optimal relations $x_1/x_2 =$ opt. Hence, the force model's normalization step at least reduces the complexity. Apparently, this modification also makes the step size adaptation easier, and thus accelerates the optimization process. It furthermore seems as if the existence of infinitely many optima also positively affects the presence of noise.

8. Discussion

This paper has proposed the force model for the evolution of a simplified insect eye directly in hardware. Despite its acceleration, the force model behaves very resistant to external disturbances (noise), which are omnipresent in real-world experimental setups. Furthermore, the force model applies a normalization which reduces the number of parameters by one.

The achieved improvements are probably mainly due to the model's normalization. This normalization changes the fitness landscape's structure. If, for example, a two-dimensional instance of the conventional model contains a unique optimum at α_1^o, α_2^o, the normalized model contains infinite many optima at $\alpha_1^o = c \times \alpha_2^o$. Future research will be devoted to a thorough theoretical analysis.

References

T. Bäck, U. Hammel, and H.-P. Schwefel (1997). Evolutionary Computation: Comments on the History and Current State. *IEEE Transactions on Evolutionary Computation,* 1(1):3-17.

R. A. Brooks (1991a). Intelligence Without Reason. *Proceedings of the 12th Intl. Conference on Artificial Intelligence (IJCAI-91),* Morgan Kaufmann, San Mateo, CA, 569-595.

R. A. Brooks (1991b). Intelligence without representation. *Artificial Intelligence,* 47:139-159.

T. S. Collett (1978). Peering – a locust behavior pattern for obtaining motion parallax. *Journal of Experimental Biology,* 76:237-241.

L. J. Fogel (1962). Autonomous Automata. *Industrial Research* 4:14-19.

D. B. Fogel (1995). *Evolutionary Computation: Toward a New Philosophy of Machine Learning Intelligence.*

N. Franceschini, J. Pichon, and C. Blancs (1992). From insect vision to robot vision. *Philosophical Transactions of the Royal Society of London B,* 337:283-294.

D. E. Goldberg (1989). *Genetic Algorithms in Search, Optimization and Machine Learning.* Addison-Wesley, Reading, MA.

G. A. Horridge (1978). Insects which turn and look. *Endeavour* 1:7-17.

L. Lichtensteiger and P. Eggenberger (1999). Evolving the Morphology of a Compound Eye on a Robot. *Proceedings of the Third European Workshop on Advanced Mobile Robots (Eurobot '99).* IEEE Piscataway, NJ, pp. 127-134.

D. Keymeulen, M. Iwata, Y. Kuniyoshi, and T. Higuchi (1999). Online Evolution for a Self-Adapting Robotic Navigation System Using Evolvable Hardware. *Artificial Life Journal,* 4(4), 359-393.

H. M¨uhlenbein and D. Schlierkamp-Voosen (1993). Predictive Models for the Breeder Genetic Algorithm I. *Evolutionary Computation.* 1(1):25-50.

R. Pfeifer and C. Scheier (1999). *Understanding Intelligence* MIT Press, Cambridge, MA.

I. Rechenberg (1973). *Evolutionsstrategie.* Frommann-Holzboog, Stuttgart. Also printed in (Rechenberg, 1994).

I. Rechenberg (1994). *Evolutionsstrategie.* Frommann-Holzboog, Stuttgart.

R. Salomon (1996). Increasing Adaptivity through Evolution Strategies. *From Animals to Animats 4: Proceedings of the Fourth International Conference on Simulation of Adaptive Behavior.* 411-420.

R. Salomon and L. Lichtensteiger (2000). Exploring different Coding Schemes for the Evolution of an Artifi cial Insect Eye. *Proceedings of The First IEEE Symposium on Combinations of Evolutionary Computation and Neural Networks.* 10-16.

D. Schlierkamp-Voosen and H. M¨uhlenbein (1994). Strategy Adaptation by Competing Subpopulations. *Parallel Problem Solving from Nature (PPSN III).* 199-208.

H.-P. Schwefel (1995). *Evolution and Optimum Seeking.* John Wiley and Sons, NY.

H.-P. Schwefel (1997). Evolutionary Computation — A Study on Collective Learning. *Proceedings of the World Multiconference on Systemics, Cybernetics and Informatics, vol. 2.* 198-205.

Evolving a neurocontroller through a process of embryogeny

Diego Federici
Norwegian University of Science and Technology
Department of computer and information science
N-7491 Trondheim, Norway
federici@idi.ntnu.no

Abstract

We introduce a model of cellular growth that generates neurocontrollers capable of guiding simple simulated agents in a harvesting task.

The morphogenesis of the neurocontroller is itself controlled by an evolved artificial neural network. The neural network operates only on local variables and chemical concentrations and is thought as a flexible model of a gene regulatory system and cell metabolism.

The model is designed in order to increase the evolvability of the growth mechanism, which constitutes a serious issue in artificial embryogeny.

Also, to increase the flexibility of development, organisms are grown in embryonal stages, which allows an incremental refinement of development.

Neurocontrollers are organized in horizontal layers, with vertical input and output pathways. Within the same layer, neurons can have only local connections. On one side this limits the information needed for connectivity and on the other makes the system easy to implement in hardware.

Results show that the system is capable of developing appropriate neurocontrollers in most of the evolutionary runs.

1. Introduction

The single most serious problem in the field of evolutionary computation (EC) is the combinatorial explosion of the search space as target designs grow in size.

Since the dimensionality of the search space in EC is given by the genotypic space (G-space) while designs reside in the phenotypic space (P-space), one possible conclusion is to reduce the first without loosing the possibility to produce the best solutions in the latter.

Concerning the evolution of artificial neural networks (ANN), the problem is that the number of weights typically grow quadratically with the size of the layers, while the possible connections among layers grows combinatorially.

Examples taken from biological organisms suggest that the solution may lay in a not trivial mapping from genotype to phenotype. When we look to the development of the human brain we can see that the ontogenetic and epigenetic processes decompress the estimated 10^4 human genes to 10^{14} synaptic connections.

Inspired by such considerations, several indirect encoding schemes have been proposed. By allowing multiple reuses of the same genes, indirect encodings decompress a smaller genotype in the final phenotype through a process of development.

The term 'Artificial Embryogeny' (AE) has been proposed to describe "the subdiscipline of evolutionary computation (EC) in which phenotypes undergo a developmental phase" (Stanley and Miikulainen, 2003).

In AE, it is possible to distinguish two major evolutionary approaches to development of ANN's. The first one, is aimed at the evolution of a grammar which is repeatedly applied to the phenotype. Examples include the Matrix Rewriting scheme (Kitano, 1990), the Cellular Encoding (Gruau, 1994), Edge Encoding (Luke and Spector, 1996) and the GenRe system (Hornby and Pollack, 2001).

A second approach evolves the regulatory system of a cell with its metabolism and ability to duplicate. Cells are usually capable of releasing chemicals which diffuse in a simulated 2D environment, grow selective connections to neighboring cells and sometimes move following some chemical gradient.

There are two different ways to model the cell's regulatory system. In biological organisms, DNA RNA and proteins interact in complex patterns which are still matter of debate. Still, it has been suggested that this gene regulatory system could be modeled by Boolean Networks (Kauffman, 1993). In (Dellaert and Beer, 1996) the authors have have developed a Gene Regulatory Network (GRN) capable of evolving a neurocontroller. Other models of GNRs have been proposed in (Jacobi, 1995) and (Bentley, 2003).

Inspired by cellullar automata (CA), another approach is to regulate the development with a set of rules that, upon matching the particular cellular state, activate specific responses. The model presented in this paper belongs to this category.

Also inspired by CAs, Bentley and Kumar proposed a model which does not yet develop ANNs but tiling patterns (Bentley and Kumar, 1999). Cells can only be of a single type and the aim is to develop perfect tessellating patterns. Rule preconditions are composed of strings matching the local North-West-South-East neighborhood (NWSE) of each cell and their absolute position. *Don't care* symbols are also allowed. Results showed that the systems performed and scaled better than a direct encoding method. On the other side, the best solutions developed had very regular phenotypes. This left open the question whether the system could be used with targets of higher complexity.

Miller extended Bentley's model and developed more complex patterns (Miller, 2003). He allowed 4 different cell types (colors) and a chemical undergoing diffusion. Cells based their growth program on their type and the types and chemical concentrations from the 8 neighboring cells. The growth program itself was a boolean network that was evolved with the Cartesian Genetic Programming. Results showed evolved phenotypes resembling the target with only very few misplaced cells.

Additionally, Miller analyzed the behaviour of the evolved phenotype after the developmental step in which the fitness was computed and when subjected to severe mutilations. Phenotypes were shown to regrow the missing parts regaining a striking resemblance to the target. The self repair feature is very interesting since it was not selected for during evolution. A similar resistance to damage was also observed by this author in (Federici, 2004).

In this paper we extend this model of cellular development. Seeking a proof of the viability of these systems, we impose a functional target to the evolutionary search.

This, on one side imposes a bigger search space, more than 50 times bigger than previously tried, and on the other offers the opportunity to test an encoding strategy for the development of a given functionality.

In the genotype, instead of using lookup tables or boolean functions to control development, we adopted a recurrent neural network with a single hidden layer (neuromorpher). On one side, recurrent networks are an accepted model of genetic regulatory systems and on the other, by allowing continuous changes of their outputs, they allow a finer tuning of the program and smoother fitness landscapes.

The mature phenotypes constitute themselves neurocontrollers for situated agents. The evolved neurocontrollers are composed of spiking neurons with only local connections. Compared to standard ANNs, the neurocontroller is more biologically plausible and offers a better hardware implementability, resistance to damage and scalability.

The following of the paper is organized as follows: **section 2** contains the agent's task description, **section 3** the developmental system, **section 4** the neurocontroller structure, **section 5** details of the evolutionary model, **section 6** results, **section 7** conclusions.

2. Agent's task and selection rules

Agent's are selected for their ability in a simple harvesting task. On a 60x60 toroidal map 240 agents compete for the collection of 320 food resources. While they search for food, agents must also avoid to crash against each other and avoid to collect poison resources (also 320). The fitness is computed as the sum of collected reinforcements over 100 activation steps. Reinforcements are calculated as follows:

+0.8	if	visiting a tile containing food
-0.8	if	visiting a tile containing poison
-0.3	if	colliding with another agent
-0.1	if	the agent did not move

Consumed resources are regenerated in random positions. At a given iteration, a tile can only contain either food or poison. Also, only one agent can occupy a given position. In case of a collision, the moving agent is unable to reach its destination.

The best 25% of the agent survive and reproduce. 10% of the new agents are produced by crossover and all the offspring undergoes mutation.

3. The development model

Phenotypes are developed starting from a single egg (zygote) placed in the center of a 8x8 2D grid. Cells (see figure 1) are characterized by internal and external variables. Internal variables define the cell state and move with it, while external ones (chemicals) belong to the environment and follow a simple diffusion law.

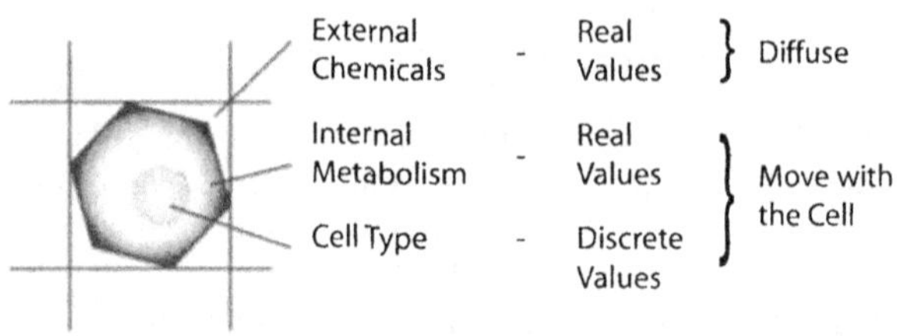

Figure 1: Description of the variables used for development

At each development step, any existing cell can release chemicals, change its own type and internal metabolism and produce new cells in the cardinal directions North, West, South and East. If necessary, existing cells are pushed sideways to create space for the new ones (see figure 2). When a cell is pushed outside the boundaries of the grid it is permanently lost.

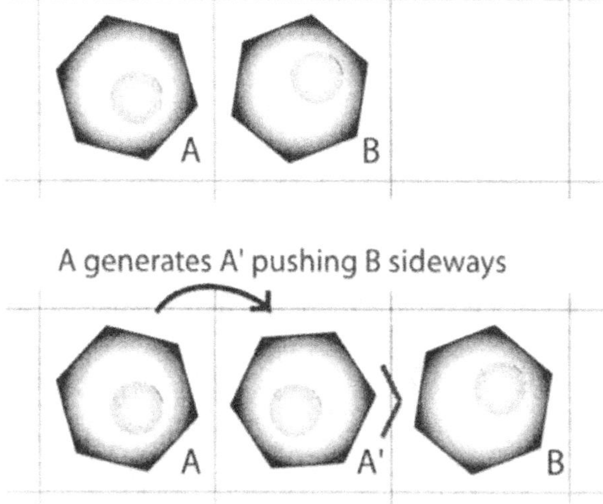

Figure 2: Placement of new cells

Morphogenesis is governed by the program, the neuromorpher, defined in the genotype. The neuromorpher (figure 3) receives in input the current cell internal and external variables, and the cell types of the neighboring cells in the four cardinal directions. Its output determines the new internal and external variables of the cell and, in case, the internal variables of the newly generated cells. An additional local variable, the cell age, is set to 1 at birth and decays exponentially to 0.

The neuromorpher is a model of a gene regulatory system, and it is implemented by a standard feedforward ANN[1], and the genotpye is a direct gray-coded representation of the ANN. The use of a neural network to control the development process allows, we believe, a greater degree of adaptivity for the definition of the appropriate growth rules. We have also tested the same mechanism in the development of various 2D patterns (Federici, 2004).

4. Agent's neural model

At every stage of development, a phenotype defines a neural network. For simplicity, in this paper the phenotype is first brought to maturation with a given number of development steps, and then used to control the movement of an agent.

The evolved neurocontroller (figure 4) is composed of a single layer. Each neuron can have only local connections to itself and to neighbors in the cardinal directions. Also it can specify a single input and a single output.

[1]even if, for an existing cell, the internal variables provide an immediate source of feedback, so that the system is actually recurrent

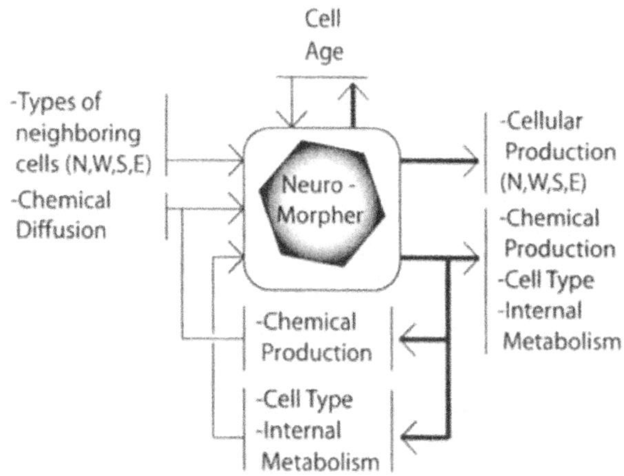

Figure 3: Inputs and outputs of the growth program

Neurons are spiking units. A spike is generated if the current net input exceeds the neuron's threshold.

The sensory input activates the hidden layer during the first activation step only. The output of the neurocontroller is computed after a given number of activation steps.

Outputs are normalized in $[-1, 1]$. An output of 1 (-1) is possible if all excitatory (inhibitory) units connected to that output are spiking in the last step of activation of the neurocontroller. Otherwise the output assumes a value in the range, proportional to the neural activity among the connected neurons.

While intra-layer weights and thresholds take values in $[-2, 2]$, input and output weights can only have integer values in $\{-1, 0, 1\}$.

For the task presented herein, neurocontrollers have 9 inputs and 2 outputs. There are two possible types of sensors, one which detects food and poison (F-P) and the second the presence of an agent (A) on a particular tile.

The F-P sensor is set to 1 if food is present in the target tile, to -1 if poison is present, and 0 otherwise. The A sensor is set to 1 if an agent is detected, -1 otherwise.

In each of the cardinal directions, agents possess F-P and A sensors, while only a F-P sensor for the tile currently occupied (figure 5).

The two output nodes are used to select the agent's direction of movement. The direction taken is the one with strongest absolute output level. If both output levels are within a $[-.15, .15]$ region, no action is taken and the agent does not move.

4.1 construction of the neurocontroller

The position, internal and external variables of each cell select for their functionality in the organism:

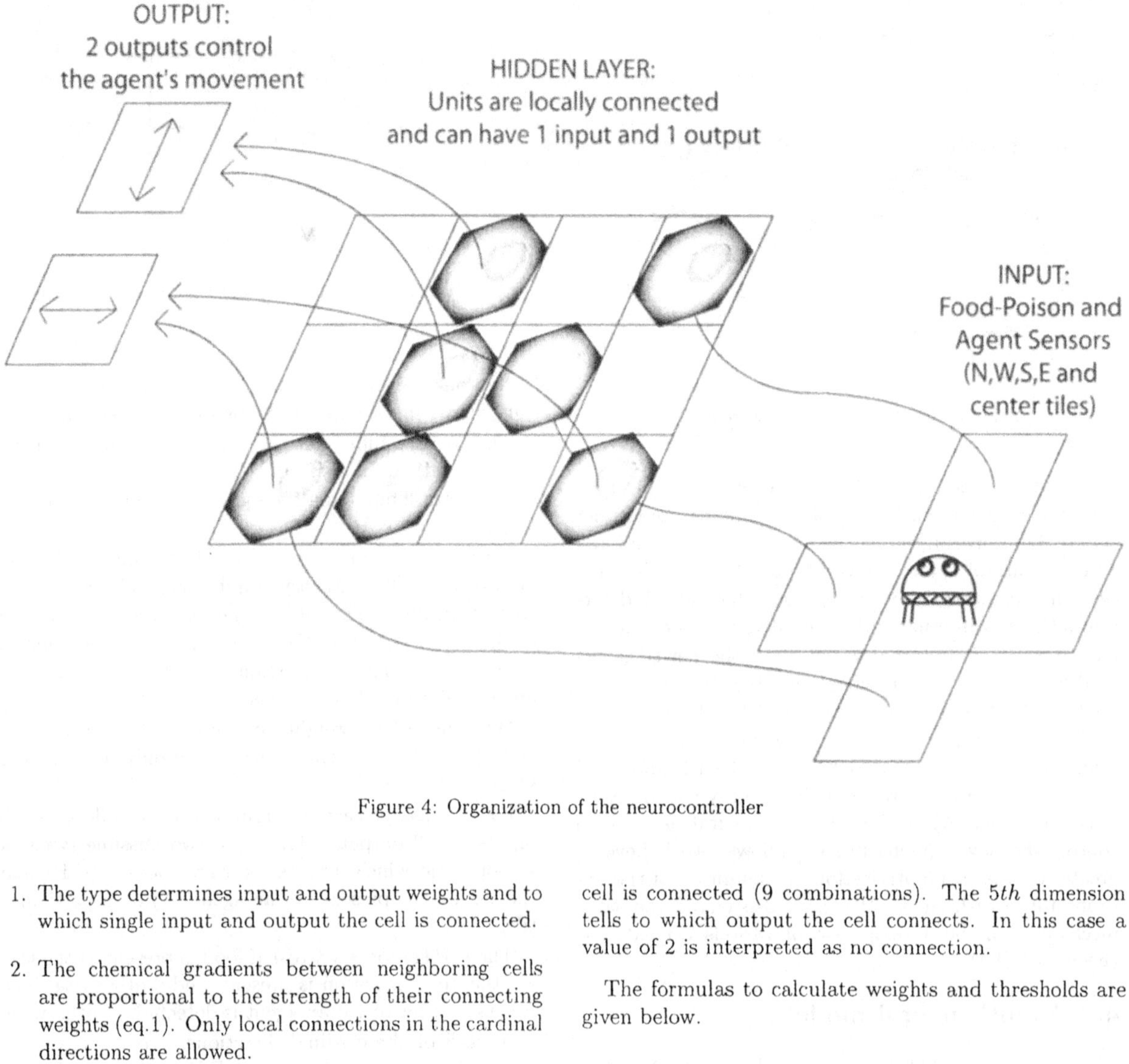

Figure 4: Organization of the neurocontroller

1. The type determines input and output weights and to which single input and output the cell is connected.
2. The chemical gradients between neighboring cells are proportional to the strength of their connecting weights (eq.1). Only local connections in the cardinal directions are allowed.
3. The internal metabolism specifies the recurrent weight (eq.2) and the cell's threshold (eq.3).

The cell type is represented by a ternary vector of 5 dimensions. This gives rise to $3^5 = 243$ different cell types. The first and second dimensions contain the input and output weights, where a value of 0 is read as no connection, a value of 1 gives an excitatory connection (+1), and a 2 an inhibitory connection (-1). The following 2 dimensions contain the number of the input to which the cell is connected (9 combinations). The $5th$ dimension tells to which output the cell connects. In this case a value of 2 is interpreted as no connection.

The formulas to calculate weights and thresholds are given below.

$$W_{i,j} = Chemical_i^{Ext,1} - Chemical_j^{Ext,2} \quad (1)$$

$$W_{i,i} = Chemical_i^{Int,1} - Chemical_i^{Int,2} \quad (2)$$

$$\Theta_i = 2\left(Chemical_i^{Int,3} - \frac{1}{2}\right) \quad (3)$$

Where $Chemical_i^{Ext,k}$ and $Chemical_i^{Int,k}$ stand for the concentrations of the kth external or internal chemical for cell i.

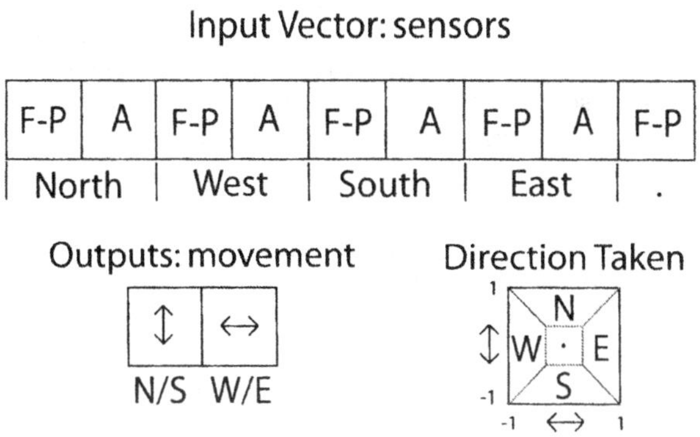

Figure 5: neurocontroller input and output. F-P stands for Food-Poison and A for Agent sensors. The direction marked with a '.' denoted the cell currently occupied by the agent. The N/S and W/E outputs specify the agent's direction of movement

4.2 genotype structure and operators

The neuromorpher is a feedforward ANN with 31 inputs, a single hidden layer with 8 units, and 47 outputs. The adopted transfer function is the hyperbolic tangent. The genotype is a direct gray-code representation of the weight matrices.

Mutation operates by selecting each weight with a .05 probability and shifting it by adding a value extracted by a normal distribution of zero mean and .01 variance. Crossover generates two new individuals shuffling the parent's hidden units.

5. The evolutionary model

*F*amily Grouping

One of the characteristics of AE systems is that evolutionary improvements have saltatory characteristics. Under these conditions a positive innovation can increase the reproductive chance of a particular strain reducing the chance of survival of all others. This often produce premature convergence to local optima.

To reduce these over-exploitative effects, we have adopted a family based grouping mechanism. Since small differences at genotype level can have a great impact on the development of the organisms, grouping is done at the phenotypic level.

Every generation, individuals are taken in random order and grouped by their phenotypes. For this task only the cell types and their positions are taken in consideration.

In order to belong to a family, individuals must have less than 4 phenotypic differences from the family prototype. If an individual does not fit in any family it will generate a new one, and its phenotype will be taken as new family prototype.

Within the same families, organisms are ranked by their fitness. In case of a tie, younger individuals are given a higher rank. The preference for younger individuals reduces the chances of stagnation (Miller, 2003). All those individuals which have a rank lower than the fourth have their fitness reduced:

$$fitness\ used\ for\ selection \quad = \quad fitness \cdot 0.9^{rank-3}$$

In figure 6, we show an example of how ranking counteracts the population convergence following an innovation.

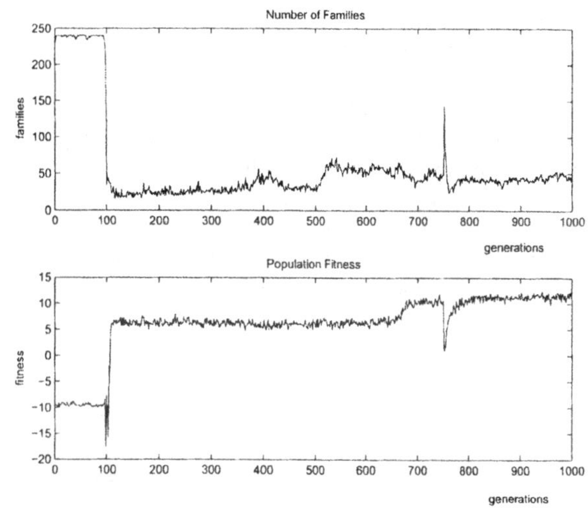

Figure 6: number of families and fitness. Around generation 100 an innovation favors a restricted group of genotypes. The number of families drops as these individuals are selected for reproduction. Ranking does not allow a single strain to take over the entire population and avoids stagnation.

*E*mbryonal Stages

Another problem with the evolution of AE systems is that, because of gene reuse, occasional mutations may cause huge phenotypic changes. This effect reduces the correlation between G-space and P-space, which has well known negative effects on system evolvability.

A recent study conducted on the Matrix Rewriting model (Kitano, 1990) shows that this effect increases with the complexity of the phenotypes (Lehre and Haddow, 2003). Since organisms of high complexity are probable targets of evolutionary methods, these considerations pose serious questions on the viability of AE methodologies.

In biological organisms it is possible to observe a relationships between ontogeny and philogeny:

> It is generally observed that if a structure is evolutionary older than another, then it also appears

> earlier than the other in the embryo. Species which are evolutionary related typically share the early stages of embryonal development and differ in later stages. [...] If a structure was lost in an evolutionary sequence, then it is often observed that said structure is first created in the embryo, only to be discarded or modified in a later embryonal stage. (Wikipedia,)

Motivated by these considerations[2], the author has developed a method consisting of incremental embryonal stages (Federici, 2004).

Every embryonal stage is characterized by a single growth program. At the beginning of the evolutionary search, phenotypes are developed with a single embryonal stage. When a certain generation or fitness score are reached, a new stage is added. For this new stage, the novel growth program is initialized as a copy of the one from the previous stage. From now on, development will proceed in two successive phases.

At first it executes the growth program belonging to the first stage, while, after a predetermined developmental step, development switches to the second[3]. Additional stages are added likewise (see figure 7).

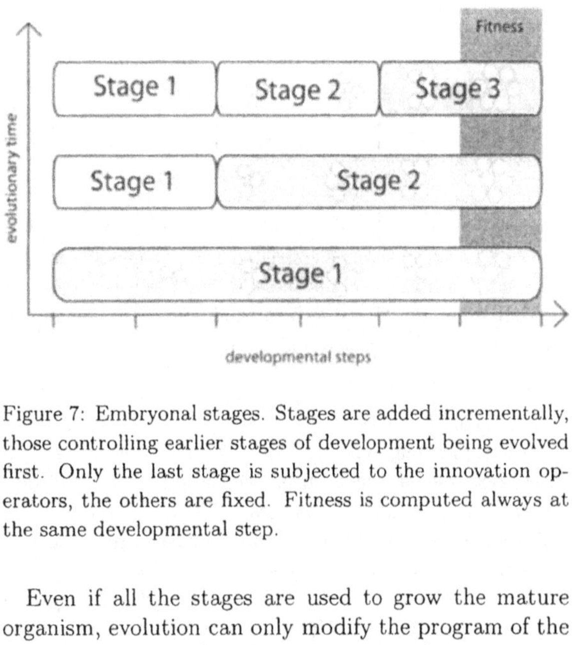

Figure 7: Embryonal stages. Stages are added incrementally, those controlling earlier stages of development being evolved first. Only the last stage is subjected to the innovation operators, the others are fixed. Fitness is computed always at the same developmental step.

Even if all the stages are used to grow the mature organism, evolution can only modify the program of the newest one, while the others are preserved intact.

This allows an incremental refinement of embryogenesis.

Even if the genotype size is increased linearly with the number of stages, as the search space is extended and explored incrementally, the evolvability is actually increased. This is because, 1) additional stages are created as a copy of the previous ones so that the evolutionary search will proceed without discontinuities, 2) the search is restricted to the refinement of the latest stage and therefor of fewer developmental steps, simplifying the task.

This effect should be particularly useful for high complexity phenotypes, since the change of the genotype of late embryonal stages will not affect early development causing catastrophic changes in development.

For the simulations presented in this paper, three embryonal stages have been used. The second stage is introduced at the 500th generation, and the third at the 750th. When all the stages are activated, the first grows the phenotype until the 5th developmental step, the second until the 7th and the third completes the maturation with step 8.

6. Results

We have evolved 20 populations with 240 agents. The average fitness of each population computed over the last 5 generations ranges from a minimum of 0.5 to a maximum of 13.4, with an overall average of 6.8.

To achieve a good fitness, individuals have to develop 3 different abilities.

1. Collect Food: when an agents is next to a tile containing food it must approach it.

2. Avoid Agents: agents must avoid to collide with other agents.

3. Avoid Poison: agents must not enter tiles containing poison.

Being that these are reactive agents, the optimal strategy would be to simply proceed in a preferred direction and steer only if confronted by an other bot, a tile containing poison or because detecting food in an other direction.

After evolution, we have taken the best individuals of each population and monitored their behavior in controlled enviromental conditions, such as with the presence of food, poison or other bots within the vision range.

Even if these controlled behavioral tests can check the agent reactions only on a subset of all the possible situations, they provide an objective description of the evolved control strategies. We were able to classify the collecting behavior in 7 categories of increasing quality (figure 8).

6.1 collecting behavior

No agent developed the ability to collect food in the direction opposite to the preferred direction of movement. On the other side, the advantage for the development of

[2]which should not be confused with the discredited Recapitulation theory

[3]notice that, as the two programs are identical, the introduction of new stages does not alter the development process.

	Description	grade	Nr
1	Always move in a single direction	low	3
2	As type 1 but also approach food in an additional direction	low	3
3	As type 1 but additionally avoid poison ahead	ok	3
4	As type 3 but also approach food in an additional direction	ok	2
5	As type 3 but also avoid agents ahead	good	4
6	As type 5 and collect food in an additional direction	good	4
7	As type 4 but approach food in three directions	good	1

Figure 8: the description of the 7 evolved behavioral strategies, a qualitative grade given to them, and the number of populations that developed them.

this ability is quite limited. By proceeding in a single direction, the tile left behind will be most often empty.

Also, the 60x60 toroidal world is quite crowded since there are 240 agents. In these conditions collecting food in all direction (strategy 7) can lead to globally unstable strategies. If an agent is surrounded by too many greedy individuals, it may be unable to collect food and score a low fitness score. Population-wise it might be as good to approach food only in two (strategy 6 and 4) or one (strategy 5 and 3) preferred directions, leaving not collected resources to individuals of the same specie.

6.2 *agent avoidance*

Similar considerations may also explain why only 4 populations (strategy 5) developed the ability to avoid agents in the preferred direction of movement. Since a homogeneous population will share the same preferred direction of movement, encounters with other agents will be very limited.

6.3 *poison avoidance*

Six populations (strategy 1 and 2) did not develop the ability to steer away from tiles containing poison, and therefore classified with a low grade. Still, the average population fitness of the two strategies is quite high, being 5.1 and 8.0 respectively. The reasons for this, probably found at the level of population behavior.

6.4 *analysis of an evolved neurocontroller*

To provide an example of an evolved neurocontroller, we have selected the best individual from a population of type 6.

Selected steps of the growth process are displayed in figure 10. Rectangles represent neurons, while synapses are plotted as cones of various dimensions (see figure 9). The bigger a cone the stronger the connection. White (black) cones stand for inhibitory (excitatory) connections.

A small circle at the right top of each neurons gives the amplitude of the recurrent weight.

Threshold values are represented by small vertical bars inside the neuron rectangle. Bars extending upwards have positive values.

Inside the neuron-rectangle two strings can be found. The topmost string indicates how and which input is connected to the neuron:

symbol	input to the unit
+AX	+1 only if agent is present
-AX	+1 only if agent is not present
PX	+1 only if poison is present
FX	+1 only if food i spresent

where X is either {N,W,E,S,.}.

The second string indicates towards which direction an activated neuron tries to move the agent.

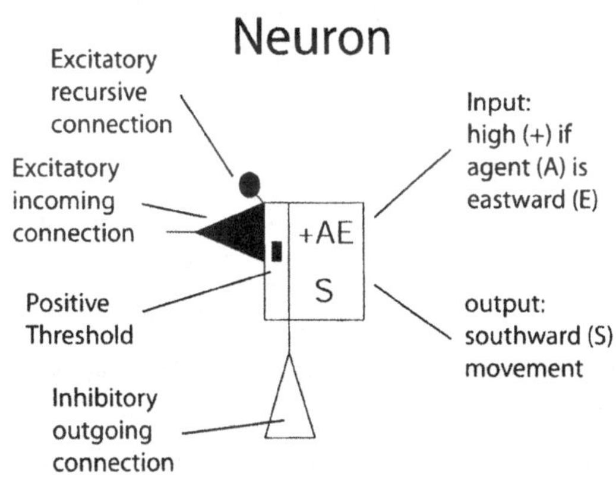

Figure 9: Explanation of the symbols used in the neurocontroller plots

The neurocontroller final activation in four important behavioral tasks is shown in figure 11.

Figure 11A contains the neural activity during the exploration task. In this case the agent proceeds in the default direction (east). The task is accomplished with the activation of 2 neurons for both the move south and move east actions. Since there are more connections to the north-south output, scaling favors the eastward direction.

In figure 11B, a single neuron is responsible for steering the agents towards the food. The extra activity sent to the north-south output suffice for the direction change.

In figure 11C, several neurons are activated by the presence of poison to the east. Only one though, is connected to the output and responsible for the avoidance behavior.

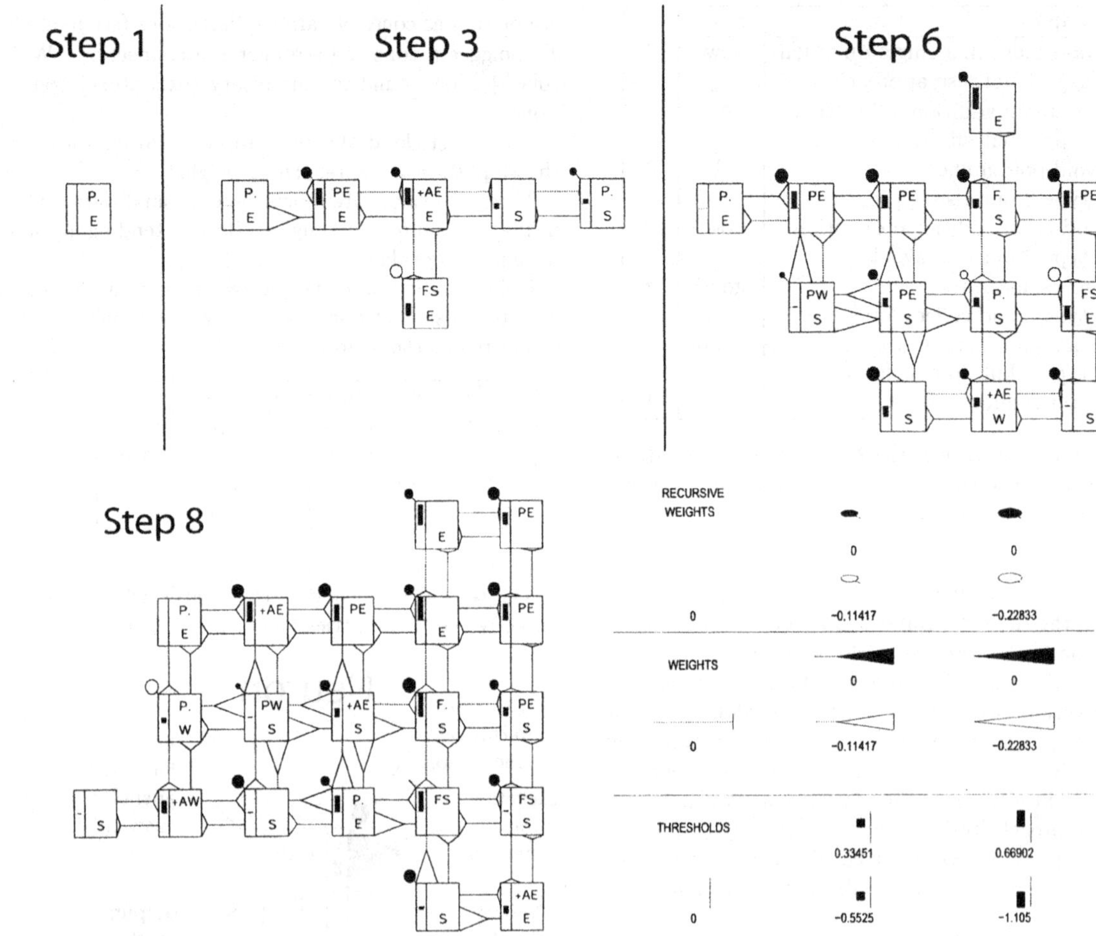

Figure 10: Development of the neurocontroller. The single zygote of step 0 grows into the functional organism by step 8.

In figure 11D, the agent must decide if it should enter a tile containing both food and an agent. In this case an agent detector correctly steers the bot away.

It is interesting to notice, that in this case, the only functional lateral weight goes from the neuron to the top left to the unit immediately below it. Without that connection the agent would start exploring southward loosing its ability to avoid agents and poison.

7. Conclusions

This paper contains an explorative work in the development of an Artificial Embryogeny model based on cell chemistry.

We have presented a novel model capable of developing a functional neurocontroller for a simple harvesting task of situated agents.

The optimal control policy is based on three behaviours: collecting a food resource, avoiding a poison resource and avoid other agents.

The quality of the results allows a cautious optimism towards the presented approach. Over the 20 simulations, 14 produced individuals with good control strategies. Among the other 6, 5 where still able to score acceptable average fitness scores. On the other side, behavioral tests showed their inability in the poison avoidance task.

Compared to similar previous work, these simulations have extended considerably the search space. Here there are 243 different types of cells and 5 internal and external chemicals.

The increased system's evolvability is given by three basic features: the ANN used to encode the growth program, the family based grouping, and the embryonal stages.

- Compared to other encodings, ANNs allows a smooth refinement of the morphogenesis process.

- Because of the saltatory characteristics of evolution in AE systems, grouping based on phenotype metrics plays a fundamental role avoiding the takeover of a single genotypic strain.

- Adding several embryonal stages reduces the catastrophic effects of mutations affecting early developmental steps. This allows an incremental refinement of the growth program and increases the correlation between genotype and phenotype, expecially, we believe, for phenotypes of higher complexity.

The neuromorpher and neurocontroller introduced in this paper are organized in a bidimensional layer and present only local lateral connections. On one side, local connections decrease dramatically the routing information and therefore the search space. On the other, these characteristics make the system feasible for an hardware implementation such as the one presented in (Tufte and Haddow, 2003). This is true both for the evolutionary system and for the final evolved organisms.

From the analysis of the best scoring individuals, it appears that chemical gradients are not a good way to encode synaptic weights.

Since external chemicals diffuse in the environment, neighboring cells tend to have similar concentrations. This causes the connections to have mostly the same signs and values. Under such conditions, processing through lateral connections is very limited. Internal chemical concentrations seem more appropriate.

8. Further work

Probably the biggest limitation of the model presented is the way that input and output connections are encoded. The schema is very inefficient requiring a cell type for each possible input-output combination.

One possibility is to augment the encoding using the unique spatial disposition of each neuron. Also, this technique could permit the stacking of several neural layers one over the other.

A direction that is surely worth considering, is to activate and test the neurocontrollers also during development. The interaction with the environment would allow the possible integration of epigenetic processes.

Acknowledgements
I wish to thank Julien Miller for the useful discussions that inspired the work presented in this paper, and Keith Downing and Gunnar Tufte for the valuable suggestions.

References

Bentley, P. (2003). Evolving fractal gene regulatory networks for robot control. *Proceeding of ECAL 2003, 753–762.*

Bentley, P. and Kumar, S. (1999). Three ways to grow designs: A comparison of embryogenies for an evolutionary design problem. *Proceeding of the Genetic and Evolutionary Computation Conference 1999, 35-43.*

Dellaert, F. and Beer, R. (1996). A developmental model for the evolution of complete autonomous agents. In *proceedings the fourth international conference on Simulation of Adaptive Behavior*, pages 393–401.

Federici, D. (2004). Increasing evolvability for developmental programs. *Under submission to the Genetic and Evolutionary Computation Conference 2004.*

Gruau, F., (Ed.) (1994). *Neural Network Synthesis using Cellular Encoding and the Genetic Algorithm.* PhD Thesis, Ecole Normale Sup'erieure de Lyon.

Hornby, G. S. and Pollack, J. B. (2001). The advantages of generative grammatical encodings for physical design. In *Proceedings of the 2001 Congress on Evolutionary Computation*, pages 600–607. IEEE Press.

Jacobi, N. (1995). Harnessing morphogenesis. *Proceeding of information processing in cells and tissues, 29-41.*

Kauffman, S., (Ed.) (1993). *The origins of order.* New York: Oxford University Press.

Kitano, H. (1990). Designing neural networks using genetic algorithms with graph generation system. *Complex Systems, 4(4):461–476.*

Lehre, P. and Haddow, P. C. (2003). Developmental mappings and phenotypic complexity. *Proceedings of the 2003 Congress on Evolutionary Computation.*

Luke, S. and Spector, L. (1996). Evolving graphs and networks with edge encoding: Preliminary report. In Koza, J. R., (Ed.), *Late Breaking Papers at the Genetic Programming 1996 Conference*, pages 117–124.

Miller, J. (2003). Evolving developmental programs for adaptation, morphogenesys, and self-repair. *Proceeding of European Conference on Artificial Life 2003, 256–265.*

Stanley, K. and Miikulainen, R. (2003). A taxonomy for artificial embryogeny. *Artificial Life 9(2):93–130.*

Tufte, G. and Haddow, P. C. (2003). Insertion of functionality into development on an sblock platform. *Proceedings of the 2003 Congress on Evolutionary Computation.*

Wikipedia. Ontogeny and phylogeny. *http:// en2.wikipedia.org /wiki /Ontogeny_and_phylogeny.*

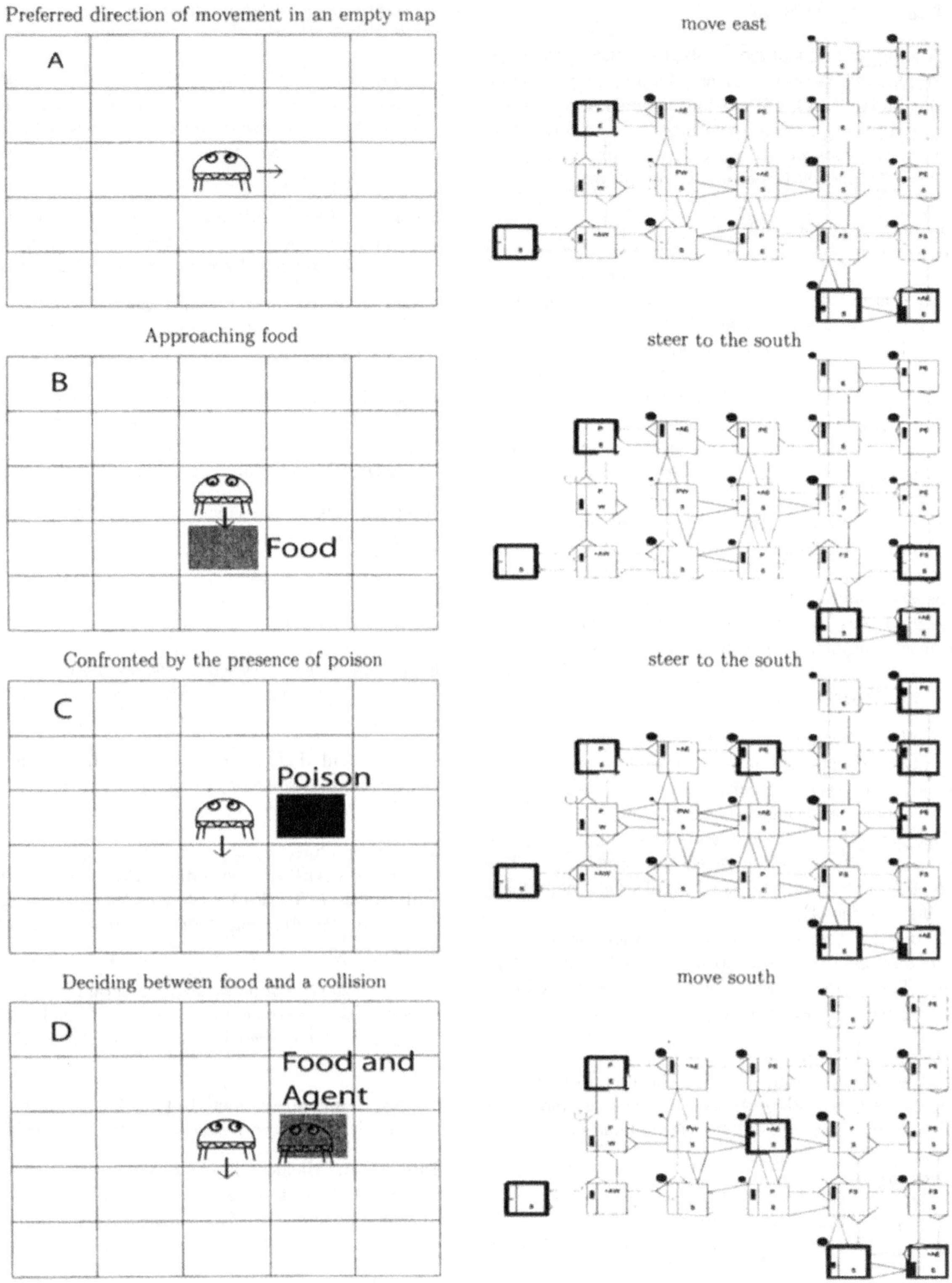

Figure 11: neural activity in various situations. A) without anything in sight, B) when spotting food, C) when the preferred direction of movement contains poison, D) when tempted by some food. Highlighted rectangles show activated neurons.

COLLECTIVE AND SOCIAL BEHAVIORS

Coordination and Behaviour Integration in Cooperating Simulated Robots

Gianluca Baldassarre **Domenico Parisi** **Stefano Nolfi**
Institute of Cognitive Sciences and Technologies,
National Research Council,
Viale Marx 15, 00137 Rome, Italy
baldassarre@ip.rm.cnr.it parisi@ip.rm.cnr.it nolfi@ip.rm.cnr.it

Abstract

This paper shows how a group of evolved physically-linked robots are able to display a variety of highly coordinated basic behaviours (coordinated motion, coordinated obstacle avoidance, coordinated light approaching) and to integrate such behaviours into a single coherent behaviour. In this way the group is capable of searching and approaching a light target in an environment scattered with obstacles, furrows, and holes and of dynamically changing its shape in order to pass through narrow passages. The paper analyses in detail the emerged basic behaviours and shows how the coordination of the group relies upon robust self-organising principles based on a traction sensor that allows the single robots to perceive the "average" direction of motion of the rest of the group.

1. Introduction

Consider the scenario shown in Figure 1 and Figure 8. A group of robots is placed in a maze with obstacles, furrows and holes. The mission of the robots is to explore the maze and search for a light target. Some robots form a *swarm-bot* (i.e. a group of physically linked robots with a particular topological structure) in order to pass over furrows and holes in which they would fall by moving alone. The challenges proposed by this scenario are several: how can the assembled robots move in the same direction? How can they avoid obstacles? How can they approach the light once it is in sight? To face these challenges, the robots should be able to display coordinated behaviours (e.g. coordinated movement, coordinated obstacle avoidance, and coordinated light approaching) and to integrate these behavioural capabilities into a single coherent behaviour. This paper will show how these problems can be solved (a) by suitably designing the hardware of the robots and by providing them with a traction sensor that allows them to feel the direction and the intensity of the traction caused by the group, and (b) by developing the controllers of the robots through an evolutionary method that allows evolving robots to exploit behavioural properties that emerge from the fine-grained interactions between the robots and between the robots and the environment.

Figure 1: A group of eight assembled robots searches a light target (white sphere at the top left of the picture) in a maze with obstacles, furrows, and holes. Notice how single robots, as the three robots right behind the group, get stuck in holes and furrows.

Collective robotics is a growing research area within the broader field of robotics (Bonabeau et al., 1999; Grabowski et al., 2003). Part of this research area focuses on *distributed coordination* of groups of cooperating robots that have to accomplish common tasks (Baldassarre et al., 2002; Ijspeert et al., 2001). Distributed coordination implies that the characteristics of the collective behaviour exhibited by the group, such as the basic behaviours through which the task is accomplished, the roles played by the different robots, the synchronisation problems raised by their interactions, and so on, are not managed centrally by one or few "leader robots" but are the result of a self-organization process. This self-organization process is the result of the interactions between simple behavioural rules based on local sensory information, followed by the single robots. Depending on the specific tasks to be accomplished, distributed coordination might have some advantages compared to central coordination in terms of (a) resistance to failure of some robots, (b) reduced communication needs, and (c) simplicity of single robots in terms of sensors, actuators, and computational capabilities.

Coordination and behaviour integration is also a central topic in the study of collective behaviour in animals. In social insects, in particular, the interaction between simple behavioural rules followed by each individual might lead to rather complex collective behaviours such as chaining and formation of complex 3D structures (Anderson et al., 2002), collective nest building (Camazine et al., 2001), collective food retrieval (Kube and Bonabeau, 1998). These behaviours tend to rely upon self-organisation mechanisms (Camazine et al., 2001) such as positive feedbacks (e.g.: "do what the majority does") or negative feedbacks (e.g. "perform a given behaviour only while a certain stimulus is above a certain threshold"). Self-organization and behaviour coordination is also important in more complex animals such as non-human primates. For example baboons may use "move grunts" to coordinate the group's transition from resting to moving (Seyfarth and Cheney, 2003). Even humans, though endowed with sophisticated capacities such as language and planning to manage coordination, still heavily exploit distributed coordination. In these cases single individuals might even benefit from distributed coordination while being unaware of it (e.g. consider the coordination of individuals' actions through markets' prices: Von Hayek, 1945). Research studies such as those presented in this paper should produce insights on the general mechanisms of distributed coordination that might underlie all these types of behaviours from different realms.

As mentioned above, in our research evolutionary techniques are used to evolve the behaviours of the robots in simulation. Unfortunately, the time required to run evolutionary experiments can be prohibitively long, especially when detailed simulations of real robots and environments are used as it has been done here. To overcome this problem evolving robots were selected for their ability to solve critical aspects of the problem and then the final task was solved on the basis of the fact that they were able to generalize their abilities to new circumstances. This practical necessity, however, allowed shedding some light on what might be an important general property of nature, that is, searching for simple "building blocks" that can be re-arranged to generate new and more complex behaviours.

Section 2 presents the experimental setup. Section 3 illustrates the results in terms of overall behaviour. Section 4 and 5 respectively analyse the functioning of the basic behaviours and their integration. Section 6 draws conclusions.

2. Experimental setup

The scenario described in the introduction includes some of the problems that are being faced within a research project funded by the European Union ("SWARM-BOTS") which is developing *swarm-bots*, that is groups of fully autonomous robots able to physically connect/disconnect to form a single robotic system that can move and assume different physical shapes in order to solve problems that cannot be solved by single robots. Each robot is provided with its own neural controller and can only have access to local information, so connected or disconnected robots forming swarm-bots should coordinate in order to display coherent behaviours. This research focussed on the problem of how producing coordinated movements in a team of robots that are already assembled. Other research carried out within the project is tackling problems of how robots can self-assemble/disassemble.

Each robot (Mondada et al., in press) has a cylindrical body with a diameter of 116 mm and consists of a mobile base ("chassis") with two drive mechanisms each controlling a track and a teethed wheel (Figure 2), and a main body ("turret") provided with two grippers, one rigid and one flexible, that allow the robots to self-assemble and to grasp objects. Only the first gripper was used in the experiments presented below. The turret has a motor with which it can rotate with respect to the chassis. Each robot is provided with a number of different sensors (Mondada et al., in press), but only the traction sensor described below has been used in the experiments reported in this paper.

Figure 2: The hardware prototype of an individual robot.

Given that the robots of the project are still under production, a simulator software based on the SDK Vortex™ toolkit (Critical Mass Labs, Canada), which allows programming realistic simulations of dynamics and collision of rigid bodies in 3D, was designed. Given the high computation costs of simulations, the evolutionary experiments were speeded up as follows: (a) only few relevant characteristics of the sensors, actuators and body of the robot were simulated; (b) the size of the robots and the gravitational acceleration coefficient were reduced to have the possibility of increasing the simulation time step without having instabilities; (c) the controller of the robots was evolved in a simplified environment (see Figure 3) and then tested in the more complex environment shown in Figure 1 (for a comparison of the behaviour of simulated and real robots, and for more details on the simulator, see Mondada et al., in press).

The motor system of a robot was modelled by four wheels: two lateral motorised wheels that model the external wheels of the real robot and two spherical passive wheels placed at the front and at the back that stabilise the robot. The four wheels were connected to the chassis, which underpins the rotating turret modelled as a cylinder (Figure 3). The turret was endowed with a simplified gripper which was modelled by creating a physical joint between two robots when needed (Figure 3). The active and passive wheels had a radius of 1.15 and 0.575 cm, respectively. The turret had a radius of 2.9 cm and a height of 2.3 cm.

During evolution, spherical collision models were used for all the wheels and for the chassis, as they required less computations, but equivalent results were obtained by testing evolved controllers with the collision models shown in Figure 3. The gravitational acceleration coefficient was set at 9.8 cm/s^2. This low value, that caused a low friction of the wheels on the ground, was compensated for by setting the maximum torque of the motors at a low value, 70 dynes-centimetre. The friction coefficient was set at 0.6 (Vortex simulates friction according to the Coulomb friction model). The desired speed of the wheels was allowed to vary within the range [–5, +5] radians per second. The desired speed applied to the turret-chassis motor was equal to the difference between the desired speed of the left and right wheel times 0.26. The effect of this was that when the chassis turned the turret turned in the opposite direction, so that its orientation did not change with respect to the environment. This greatly helped the robots to turn the chassis when attached to other robots. The state of the sensors and the motors, and the differential equations used by Vortex to simulate the bodies' dynamics, were updated every 100 ms.

Figure 3: Four simulated robots linked up to form a linear structure. For each robot, the cylinders and the spheres respectively represent the motorised and the passive wheels. The cylindrical structure represents the turret. The arrow indicates the orientation of the turret. The black segment between two robots represents the physical link between them. The white line above each robot, that goes from the robot's centre towards a point on its perimeter, indicates the direction of traction and, with its size, the intensity of traction. The environment consists of a flat terrain and a light source (large white sphere at the top left corner of the picture).

Each robot was provided with a "traction sensor", placed at the turret-chassis junction (Figure 4). This sensor returned the direction (angle with respect to the chassis' orientation) and the intensity of the force of traction (henceforth called "traction") that the turret exerted on the chassis. Traction was caused by the movements of both the connected robots and the robot's own chassis. Notice that, by being assembled, the turrets of the robots *physically integrated* the forces produced by all robots. As a consequence, the traction sensor measured the mismatch between the directions toward which the entire group and the robot's chassis were trying to move. The intensity of the traction measured the size of this mismatch. As it will be shown below, given that robots are physically connected the traction sensor informs robots on what the other robots are doing and thus implements a sort of implicit communication. Traction, seen as a vector, was affected by a 2D noise with a size of ±5% of its maximum length.

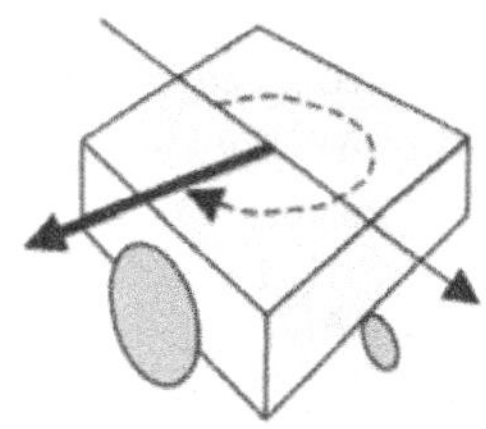

Figure 4: Traction force detected by the robot traction sensor. The parallelepiped represents the chassis. The turret has not been drawn for clarity. The large and small gray circles respectively represents the right motorised wheel and the front passive wheel. The thin arrow indicates the orientation of the chassis, the bold arrow indicates the vector of the traction force that the turret exerts on the chassis, the dotted arrow indicates the angle of the traction.

Each robot was also endowed with four light sensors, positioned on the perimeter of the turret, simulated by using a sampling procedure (see Miglino et al., 1995). A noise of ±5% of the maximum intensity was added to the sensors. Shadows were simulated by computing the geometrical projections of obstacles. In order to provide information about the light gradient with respect to the orientation of the chassis (this greatly eased control since the wheels were connected to the chassis) four virtual light sensors were simulated. These virtual sensors were activated on the basis of the weighted average of the activation of the two closest light sensors, with weights proportional to the angular distance of them from the virtual sensor (Figure 5).

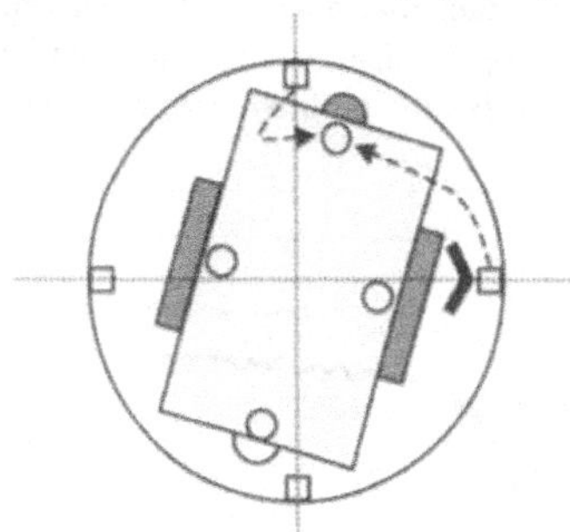

Figure 5: The four empty squares represent the light sensors placed on the turret. The four empty circles represent the virtual light sensors located on the chassis. As an example, the dotted arrows indicate the sensors that were used to compute the activation of the virtual sensor located on the top side of the picture.

Each robot's controller (Figure 6) consisted of a neural network with nine sensory neurons directly connected to two motor neurons. The first four sensory neurons encoded the intensity of the traction from four different preferential orientations with respect to the chassis (rear, left, front and right). Each sensory neuron had an activation proportional to the cosine of the angle between the sensor's preferential orientation and the traction direction when the angle was within [–90, +90] degrees, and zero otherwise. This activation was then scaled by the traction intensity. The next four sensory neurons were activated by the four virtual light sensors described above. The last sensor was the bias unit that was always activated with 1.0. The activation state of the two motor neurons was normalized within [–5, +5] rad/s and was used to set the desired speed of the two corresponding wheels and of the turret-chassis motor.

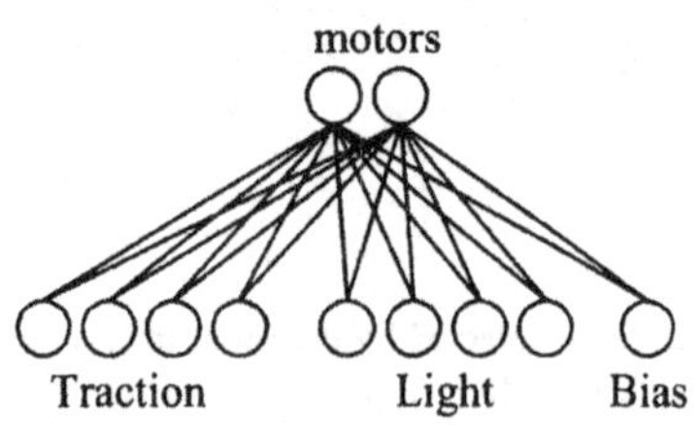

Figure 6: The neural controller of each robot consisted of a two-layer neural network with nine sensory neurons directly connected to two motor neurons.

The connection weights of the neural controllers were evolved (Nolfi & Floreano, 2000). The initial population consisted of 100 randomly generated genotypes that encoded the connection weights of 100 corresponding neural controllers. Each connection weight was represented in the genotype by eight bits that were transformed into a number in the interval [–10, +10]. Each genotype encoded the connection weights of four identical neural controllers, that were used to control the four robots linked up to form the swarm-bot shown in Figure 3. The case of groups of robots with different neural controllers in which different robots might specialize by assuming different roles was not studied here. Although investigating this aspect would have been interesting, the fact that near-optimal performance was obtained without specialization indicates that specialization is not needed to solve the problem discussed here.

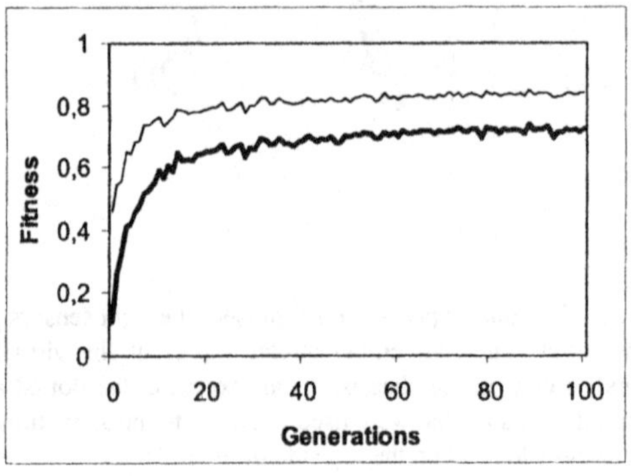

Figure 7: Fitness throughout 100 generations. Thin line: fitness of the best swarm-bot of each generation. Thick line: average fitness of the population. Each line indicates the average performance over 20 replications of evolution.

Each swarm-bot was tested in six "epochs", each lasting 150 time steps of 100 ms each. The light source was present only in three of the six epochs. At the beginning of each epoch the orientations of the chassis of the four robots were randomly assigned and, when present, the light source was placed at a random selected position at a distance of 100 cm from the swarm-bot. The 20 best genotypes of each generation were allowed to reproduce by generating five copies each, with 3% of their bits replaced by a new randomly selected value. The evolutionary process lasted 100 generations. The evolution was replicated 20 times by starting with different initial random genotype populations.

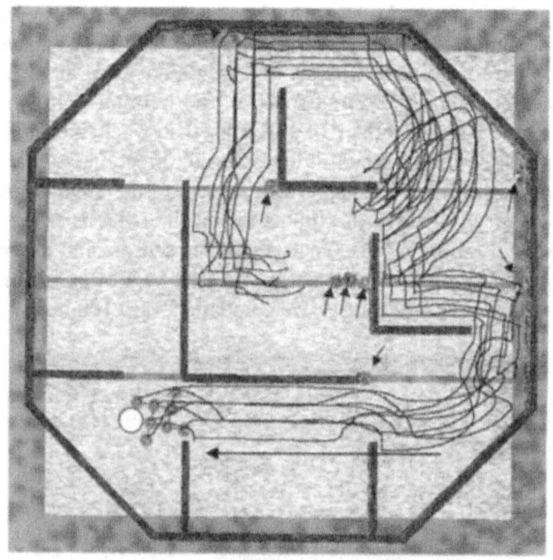

Figure 8: The trajectory produced by a star-shaped swarm-bot in the environment shown in Figure 1. The shape of the swarm-bot is depicted in its final position near the light target represented by the white sphere. The black irregular lines indicate the trajectories followed by the eight robots forming the swarm-bot. While the eight single robots (indicated by short arrows) get stuck in furrows, the swarm-bot passed over them, succeeds to free its robots that fall in the holes near the walls, and searches and finds the light that was not visible from the starting position (centre of graph). As soon as the light is in sight, the swarm-bot reaches it by following a quite direct path (long arrow).

To develop swarm-bots able to explore the environment and to approach the light target when it was in sight, the swarm-bots were selected for the ability to move as fast and as straight as possible when the light was off, and to move toward the light when the light was on. More specifically, the fitness of a swarm-bot was computed by summing the Euclidean distance between the centre of mass of the swarm-bot at the beginning and at the end of the epoch in the three epochs in which the light was off, and the Euclidean distance travelled by the swarm-bot toward the light in the three epochs in which the light was on. To normalize the value of the fitness between [0.0, 1.0], the total fitness of one swarm-bot was computed by dividing the average distance travelled during one epoch by the maximum distance travelled by a single robot moving straight at maximum speed for 150 steps. Notice how the few and short *epochs focussed* only on the two critical aspects of the final task, namely coordinated motion and light approaching, since an evolution directly tackling the final task (see Figure 8) would have required a prohibitive amount of time.

3. The evolved behaviour

Figure 7 shows how the fitness of the best swarm-bot and the average fitness of the population increase throughout the generations. By testing for 100 epochs the swarm-bots of the last generations of the 20 replications of the evolution, it is find that the performance of the best swarm-bot is 0.87 while the average performance is 0.78 with a standard error of 0.07. This means that all evolved robots are able to coordinate so as to move effectively toward the light when the light is on and to move straight when the light is off.

By testing evolved swarm-bots in new conditions (the controllers used are *identical* to those evolved with groups of four robots forming a linear structure) it can be seen that they are able to generalize their coordinated motion and coordinated light approaching ability in rather different circumstances and also to display new interesting behaviours (cf. Baldassarre et al., 2003). In particular, by placing the evolved swarm-bots in the environment shown in Figure 1, it can be seen that they are able to explore the environment avoiding walls and to display a coordinated light approaching behaviour as soon as the light is not shadowed by walls.

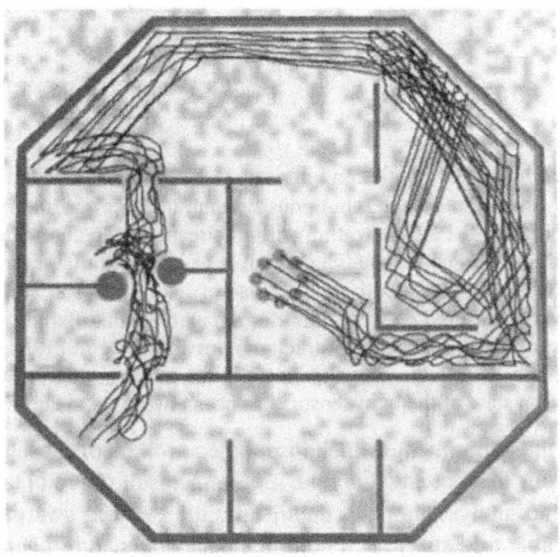

Figure 9: A swarm-bot with a circular shape formed by robots connected through flexible links. The irregular lines, that indicate the trajectories of the single robots, indirectly provide an indication of how the shape of swarm-bot changes while it is moving. With respect to the environment shown in Figure 8, this environment includes additional obstacles and a narrow passage formed by two close cylindrical objects.

Figure 8 shows how the same neural controllers (i.e. the controllers evolved in a team of four robots forming a linear structure) are able to generalize their ability to swarm-bots made up by a *larger number* of robots (eight robots instead of the four used during the evolutionary process) and assembled to form a *different shape* (a star shape instead of a linear shape). This graph and direct observation of behaviour indicate that the robots forming the swarm-bot, whose chassis initially have different orientations, quickly negotiate a common direction of motion and move along such direction by compensating further mismatches arising during the movement. Later on, when one or more of the robots forming the swarm-bot hit a wall, the swarm-bot changes direction by displaying a very effective and coordinated obstacle avoidance behaviour. After avoiding one wall, the swarm-bot keeps moving straight in the arena until it hits another obstacle. The combination of coordinated motion and collective obstacle avoidance behaviour allows the swarm-bot to explore the environment. When finally the swarm-bot sees the light, it reaches it following a straight path.

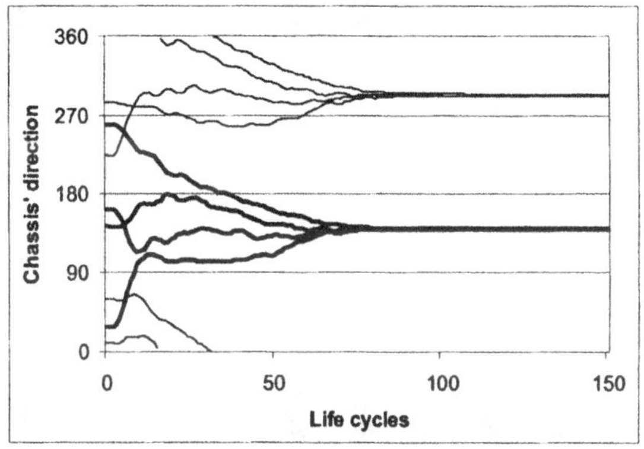

Figure 10: The orientation of the chassis of four robots assembled to form a linear structure placed in the simple environment shown in Figure 3 with the light turned off. The thick and thin lines represent data obtained in two independent tests in which the robots' chassis are initially assigned different random orientations.

Figure 9 shows how the same neural controllers (i.e. the controllers evolved with a group of four robots forming a linear structure through rigid links) are able to generalize their ability to swarm-bots having a circular shape and in which robots are connected through *flexible links* (i.e. links that, at their centre, have a hinge joint with a passive degree of freedom around a vertical axis). Swarm-bots formed by robots connected through flexible links might modify their shape while moving. Interestingly, as shown in Figure 9, the evolved neural controllers display an ability to dynamically adapt the shape of the swarm-bot to the local environmental configuration in order, for example, to pass through narrow passages. As explained in detail below, this result can be explained by considering that the robots that see the light move toward it and, by doing so, produce a traction in that direction that is felt by the other robots that react accordingly. The deformation of the shape of the swarm-bot that allows the swarm-bot itself to pass the narrow passage can be explained by considering the fact that flexible links can bend thus allowing the swarm-bot to change its shape as a result of the movements of the different robots and of collisions. The generalization ability with respect to the shape of the swarm-bot, that in turn depends on how the turrets of the robots are assembled and on the position of the links (in the case flexible links), can be explained by considering that the control system receives only sensory information with respect to the orientation of the chassis.

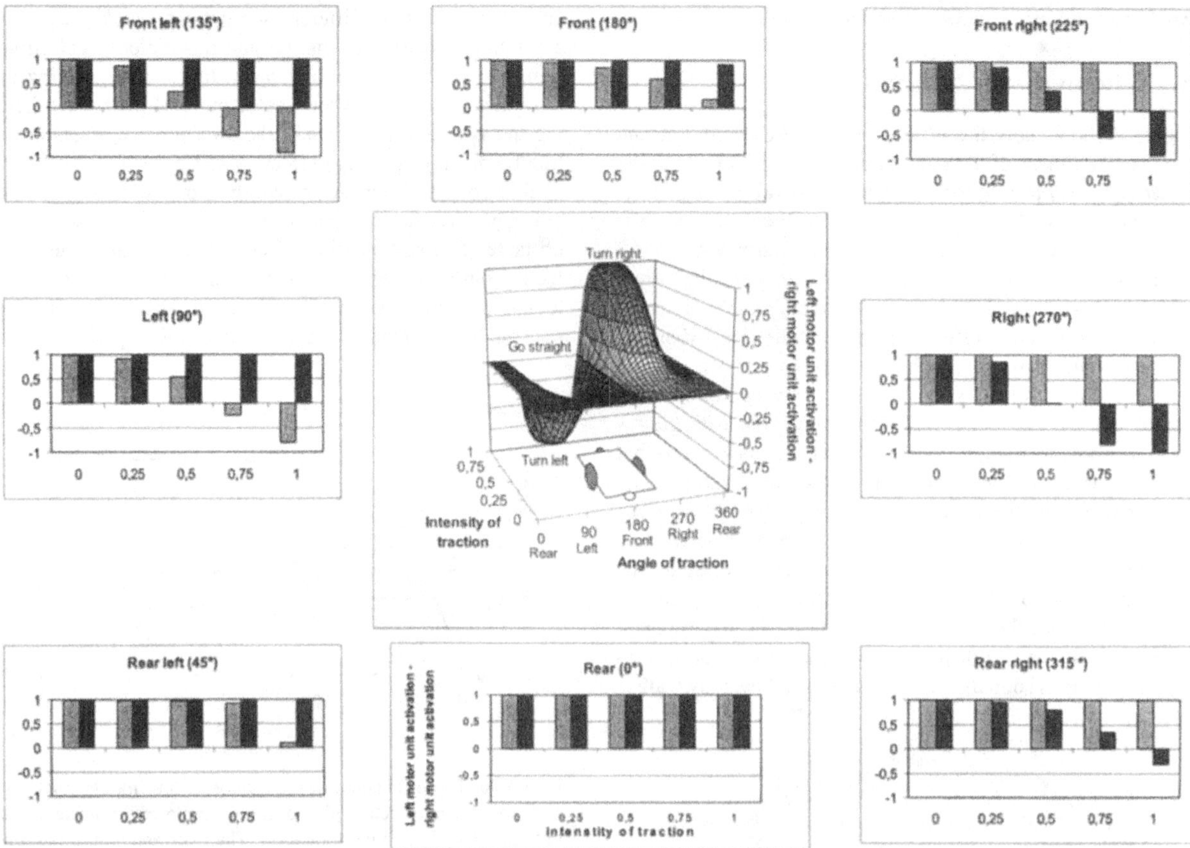

Figure 11: The picture in the centre shows how a robot reacts to a traction with different angles and intensities. The vertical axis indicates the difference between the left and the right motor neurons that, in turn, indicates whether the robot goes straight, turns left or turns right, with different speed. The schematic little picture represent a chassis and should aid the "visualization" of the direction of traction with respect to the chassis itself: the white little wheel represents the rear of the chassis and corresponds to an angle of traction of 0°, measured anticlockwise. The histograms show the desired speed of the left and right wheel as indicated by the activation state of the left and right motor neurons (respectively light grey and dark grey columns of the histograms, measured on the y-axis) for a traction with different angles (see the title of each histogram) and intensities (histograms' x-axis shows four different intensities of traction normalized in [0.0, 1.0]). Data of histograms were obtained with a typical evolved neural controller by manually setting the activation state of the sensory neurons and by measuring the corresponding activation state of the motor neurons (the activation state of the light sensors has been always set to zero).

Above it has been mentioned that, for efficiency reasons, evolution was focussed on some critical aspects (coordinated motion and light approaching) of the final task. When the controllers obtained in *different replications* of the evolution are tested in the final task (see Figure 8 and Figure 9) they display different behaviours. For example, some swarm-bots are more stubborn than others in passing through narrow corridors, some explore the arena very efficiently by doing a sort of collective wall following, some sometimes get stuck in narrows passages when the light is in sight, some other rotate on the spot with some particular initial chassis' orientations (i.e. they reach a stable local minimum in terms of the minimisation of the traction intensity). Such a variety of behaviours produces the possibility for the experimenter to choose the controllers of those replications that better satisfy the requirements of the final task.

To analyse more in detail how the evolved swarm-bots are able to display such complex behaviours and to generalize their ability to such different circumstances the next two sections will describe how different basic behaviours are produced and how they are integrated.

4. Analysis of the basic behaviours

First, it is analysed how evolved swarm-bots perform coordinated motion when the light is off or in shadow. In this condition the robots (a) start to pull/push in different directions, (b) orient their chassis in the direction where the majority of the other robots are pulling, and (c) move straight along the direction that emerges from the initial negotiation by compensating successive mismatches that arise while moving. As shown in Figure 10, the absolute direction that emerges from the robots' negotiation changes in different tests depending on the initial orientation of the robots, but the robots always converge toward a single direction.

By analysing how evolved robots react to traction with different angles and intensities (Figure 11) and by observing the behaviour of the corresponding swarm-bots, the control strategy of a robot might be described as follows:

1) When the traction comes from the front (about 180°), the robot is oriented toward a direction that is close to the mean direction of the other robots. In this situation, when the intensity of the traction is low the robot moves straight. When the intensity is higher (above 0.25, see Figure 11), the robot tends to turn left. However, this latter condition tends to not take place given that when the robots are oriented toward similar directions, the intensity of the traction tends to be low.
2) When the traction comes from the left or the right side (i.e. the angle of the traction significantly differs from 0° and 180°) there is a significant mismatch between the orientation of the robot and the mean orientation of the other robots. In this condition the robot turns toward the direction of traction, that is towards the mean direction of the other robots, by turning left when the traction comes from the left side and right otherwise. The speed of turn is proportional to the intensity of the traction.
3) When the traction comes from the rear (about 0°) the robot goes straight at maximum speed independently of the intensity of the traction and even if its orientation might significantly differ from the mean orientation of the other robots.

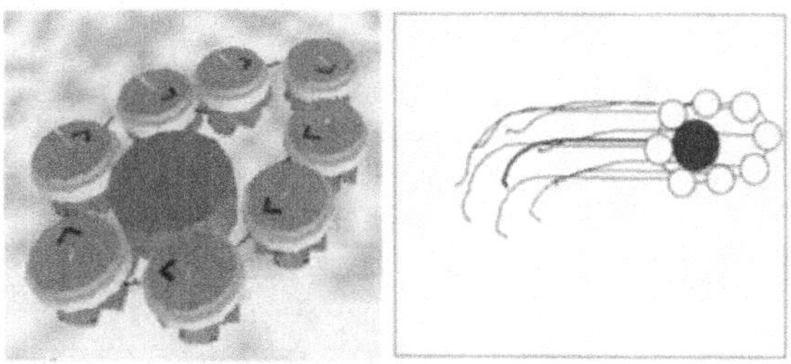

Figure 12: *Left*: Eight robots connected around an object that can move freely. *Right*: the irregular thin and bold lines respectively represent the trajectories left by the robots and the object. The white and black circles respectively represent the final position of the robots and of the object.

Summarising, the ability to display coordinated motion is the result of a *conformist tendency*, that is, a tendency to follow the direction of the traction that provides an indication of the average direction of the other robots (see previous points 1 and 2). Coordinated motion is also a result of a *stubborn tendency*, that is, a tendency to persevere in one direction independently of the intensity of the traction. The stubborn tendency is due to the following factors: (a) the fact that turning toward the direction of the traction takes time (i.e. the mismatch cannot be compensated for instantaneously), (b) the tendency to go straight when the intensity of the traction is low, independently of its direction, and (c) the tendency to go straight when the traction comes from the rear (i.e. from around 0°, see point 3 above). The stubborn tendency is likely to play an important role in the ability of the robots to keep the equilibrium state once a common direction of movement has emerged from the negotiation and to avoid never-ending negotiation phases in which robots keep changing orientation in order to eliminate small mismatches (in this case the swarm-bot would enter into a limit cycle dynamics without ever reaching a stable state).

Another effect of the stubborn tendency, with particular reference to the tendency to move straight when the traction come from the rear, is that evolved robots spontaneously display coordinated object pushing/pulling behaviours when linked to or around an external object. Figure 12 shows how, after coordinating toward a single direction, robots start to push/pull the object in the direction that emerged from the negotiation notwithstanding the fact that the friction with the ground of the object produces a traction in the opposite direction with respect to the direction of the group's motion.

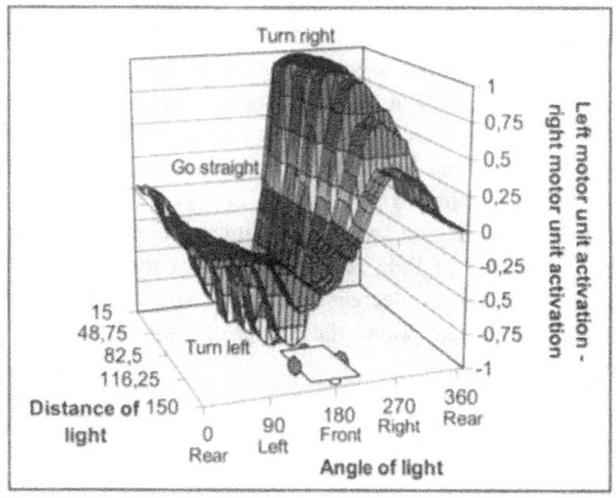

Figure 13: This graph shows how a robot reacts to different orientations and distances with respect to the light. The vertical axis indicates the difference between the left and the right motor neurons that, in turn, indicates whether the robot goes straight, turns left or turns right with different speed. Data have been obtained for a typical evolved neural controller by placing a single robot at different orientations and distances with respect to the light and by recording the corresponding activation state of the motor neurons. The activation state of the traction sensors was always set to zero.

Now a description of how evolved robots spontaneously display an obstacle avoidance behaviour and how swarm-bots consisting of several assembled robots display a coordinated obstacle avoidance behaviour will be presented. When a single robot hits an obstacle, the collision produces a traction with a direction that points away from the obstacle. The robot follows this traction and so avoids the obstacle. When a swarm-bot consisting of several assembled robots hits an obstacle, only a single or few robots collide with it. However, the resistance of the obstacle will propagate through the turrets of these robots to the turrets of the other robots via the links. As a consequence, all the robots will start to turn away from the obstacle, eventually tending to select slightly different directions. Traction will immediately average between this tendencies and will guarantee a synchronous well coordinated turning of the whole swarm-bot. Indeed, direct observation of behaviour shows how the single robots perform a surprisingly highly synchronised turning when the swarm-bot hits an obstacle.

Finally, how evolved swarm-bots display a collective light approaching behaviour is now explained. The fact that individual robots display such behaviour can be explained by analysing how evolved neural controllers react to light sources located at different angles and distances with respect to the robot. As shown in Figure 13, when the traction is null, evolved robots follow the light gradient by turning left or right when the light is respectively located at the left or at the right side of the robot's chassis.

The ability to display a coordinated light approaching behaviour despite the fact that different individual robots forming a swarm-bot have different sensory information (because of their different relative positions and because of shadows) can be explained by considering, once more, the ability to coordinate through the traction sensor and the effects of the motor behaviour of each individual robot on the other assembled robots. Indeed, as soon as a single robot starts to perceive and follow a light gradient turning in the direction of the light, it creates a traction force that is felt by the other robots which, as a consequence, will also turn in the direction of the light. To confirm this, Figure 14 shows an experiment where a single robot provided with light sensors is able to drive a swarm-bot composed of up to twelve robots deprived of light sensors towards a light target. Although performance decreases with the size of the swarm-bot, the experiment shows the power of traction in supporting coordination.

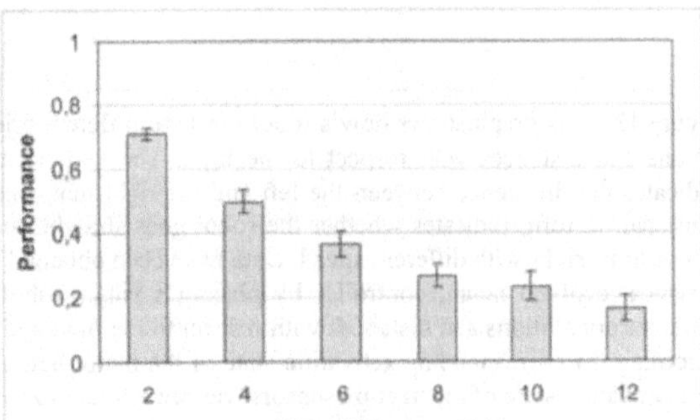

Figure 14: Performance of swarm-bots composed of two to twelve robots forming a linear structure in the simple environment shown in Figure 3 with the light always on. Only one robot forming the swarm-bots is provided with the light sensors while the light sensors of the other robots are always set to zero. Data have been obtained by using the usual neural controller evolved with four robots forming a linear swarm-bot. Columns and bars respectively indicate averages and standard errors over 60 epochs.

This section can be concluded by observing that the coordinated motion behaviour, by relying upon the conformist tendency, is a nice example of coordination based on a mechanism of self-organisation named "positive feedback". This mechanism allows groups of individuals to converge on the same selection, among some possible alternatives, without a centralised decision maker (cf. Camazine et al., 2001).

5. Integration of behaviours

The ability to integrate the different behaviours described in the previous section into a single coherent behaviour results from the combination of three mechanisms.

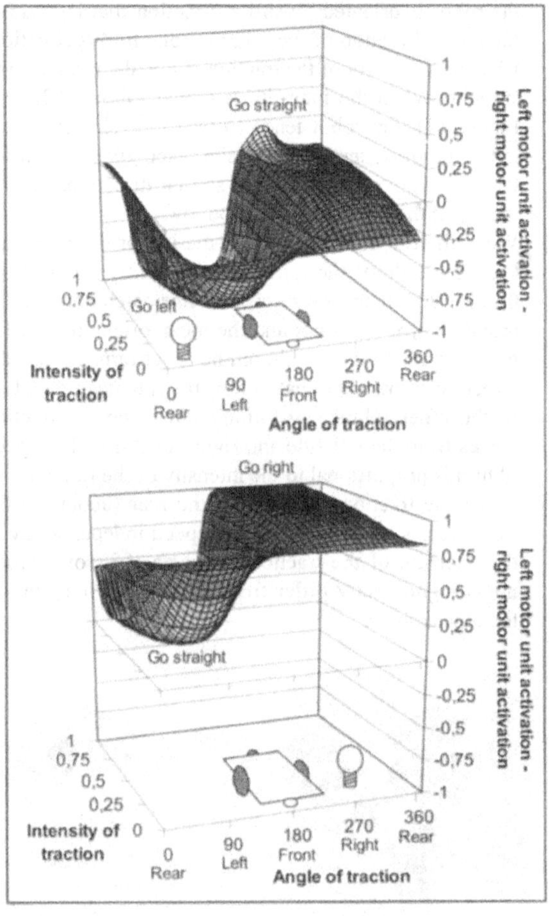

Figure 15: The pictures show how a robot reacts to traction with different direction and intensities when it also perceives a light at a distance of 100 cm coming from the left (90°) or from the right (270°) side (top and bottom pictures, respectively). The y-axis indicates the difference between the left and the right motor neuron that, in turn, indicates whether the robot goes straight or turns left or right with different speed.

As the reader might have already realised from the analyses presented above, one important mechanism is the ability to coordinate through the traction sensor and through the effects that arise from the fact that the robots are physically linked. In fact, as it has been shown previously, the ability to coordinate through the traction sensor and the fact that robots are physically assembled do not only allow swarm-bots to display coordinated motion but also play a crucial role in the ability of the swarm-bot to display coordinated obstacle avoidance and coordinated light approaching behaviours. In fact, the tendency to coordinate through the traction sensors and links between the robots assure that the swarm-bot produces a coherent and coordinated behaviour even when the individual robots have different, incomplete or noisy perceptions. In other words, the ability to coordinate through traction constitutes an important behavioural building block also for the ability to display other more complex behaviours.

The second mechanism, already mentioned above, consists in the fact that the control strategies responsible for the ability to display the different basic behaviours described in the previous section are independent of the particular shape of the swarm-bot. This is due to the fact that the direction of the traction returned by the traction sensor corresponds to the direction of the force that the turret exerts on the chassis *independently* of the particular *turret's orientation.* This characteristic is important not only because it allows the swarm-bot to behave robustly independently of the number of robots, of the way in which robots are assembled, and of possible damages that might affect some individual robots, but also because it allows the swarm-bot to behave robustly even when its shape changes dynamically by adapting to the current situation of the environment.

A third important aspect consists in the fact that the ability to coordinate through the traction sensor and the ability to approach the light and to avoid obstacles integrate without interfering with each other. This can be seen by examining, for example, how the evolved robots react when the information coming from the traction and the light sensors are consistent or inconsistent.

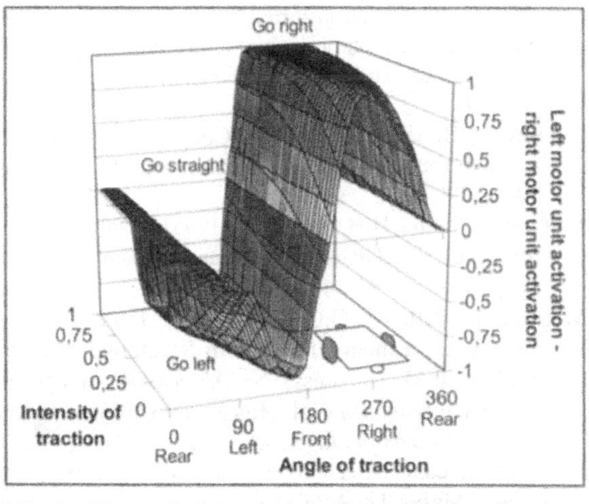

Figure 16: The graph shows the behaviour of a robot when it perceives a traction with different angles and intensities and a light with the same angle as the traction and a fixed distance of one metre.

Figure 15 shows how an evolved robot reacts to traction with different directions and intensities when it also perceives a light at a fixed distance of 100 cm coming from a fixed position. When the light comes from the left side (Figure 15, top graph) the robot turns mostly left by varying the speed of turning on the basis of the intensity of the traction. On the contrary, when the light comes from the right side (Figure 15, bottom graph) the robot turns mostly right by varying the speed of turning on the basis of the intensity of the traction. The combination of these two tendencies allows the robots to approach the light and to maintain the group's coherence at the same time. Figure 15, however, only shows how the control mechanisms responsible for the two behaviours interact from a static point of view. The interaction from a dynamical perspective is even more interesting.

One first important aspect to notice is that the three basic behaviours described above are different forms of *taxis* or *anti-taxis* (i.e. behaviours through which a robot should approach or avoid a stimulation source; notice how the surface graphs of Figure 11 and Figure 13 have similar shapes), and as a consequence in some conditions they can suitably integrate and sum up. A second important aspect is that the behaviour based on traction is a special type of taxis in that it is not related to a stimulus anchored to the environment, as the light pursuing behaviour, but it is related to the average motion direction of the group. This direction tends to change dynamically on the horizontal plane (due to the mechanisms described above) so as to eventually coincide with the direction of the light. When the two directions match completely, the tendency to move toward the average direction of the group and toward the light gradient tend to sum up and to amplify each other as shown in Figure 16.

The possibility of integrating the coordination behaviour based on traction with other behaviours eliminates the need to use behavioural selection mechanisms such as the subsumption architectures (Brooks, 1986) in which only one control mechanism among many is activated at any one time. The possibility of avoiding behaviour selection mechanisms, in turn, eliminates the need of evolving arbitration mechanisms able to select the right behaviour at the right time (e.g. coordinated motion or obstacle avoidance or light approaching). Moreover, it allows evolution to exploit the synergies that might emerge from the interplay between different basic behaviours.

It should be noted that the fact that the control mechanisms which are responsible for the three different basic behaviours cooperate and never interfere with each other is not only due to the characteristics of the basic behaviours themselves but also to the way in which sensory information is encoded in the sensors, to the organization of the robots' body, and to the interaction of these characteristics with the environment. For instance, the possibility of integrating obstacle avoidance and coordinated motion behaviours is due to the fact that robots collide with obstacles with the turret (since the turret is bigger than the chassis) and this generates a traction with a direction opposite with respect to the collision. A second example refers to the obstacle avoidance and the light approaching behaviours. The fact that obstacles shadow the light has the effect that the anti-taxis behaviour related to obstacles never interferes with the taxis behaviour related to light. This means that the possibility of finding simple and robust solutions, from the point of view of the control system of the robots, crucially depends on the way in which sensory and motor information are encoded, on the structure of the robot's body, and on the relation of these with the environment's properties.

6. Conclusions and future work

The paper has described how an evolved group of physically assembled robots can solve problems that could not be solved by single robots, on the basis of simple control mechanisms, a suitable integration of them, and robust coordination mechanisms. In particular, the analysis of the evolved robots indicates that the ability to coordinate through a "traction sensor", that in turn allows exploiting the physical interaction between assembled robots, not only is at the basis of the ability of the robots to display coordinated motion but it also constitutes an important building block for the ability to display other more complex behavioural capabilities.

The possibility of achieving these results can be ascribed to (a) the use an evolutionary technique, and (b) a careful design of the robots' hardware structure. The importance of the robots' hardware structure can be explained by considering that the particular shape of the robots' body, the type and positions of the sensors, and the way in which sensory information is encoded in the neural controllers might crucially affect the complexity of sensory-motor mapping that should be produced by the control systems of the robots and the possibility of integrating different behavioural abilities. These characteristics have been carefully designed and often re-designed on the basis of the results obtained by exploratory experiments performed in simulation. The importance of the latter point can be explained by considering that, as discussed elsewhere (Nolfi and Floreano, 2000; Baldassarre, Nolfi, and Parisi, 2003), artificial evolution allows evolved robots to find solutions that exploit useful behavioural characteristics emerging from the interaction between the control systems of the robots, the structure of their bodies, and the external environment, including the *social* environment made up by other robots.

This work has focussed on the coordination problems faced by some robots that are manually assembled by the experimenter before being tested. Future work will focus on the interesting issue of self-assembling robots, eventually triggered by the challenges of tasks that can be suitably tackled only by assembled robots, for example passing over furrows, passing over high obstacles, linking up around a heavy objects in order to drag them, forming a supporting structure for other robots.

Acknowledgements

This research has been supported by the "SWARM-BOTS" project founded by the Future and Emerging Technologies program (IST-FET) of the European Commission under grant IST-2000-31010.

References

Anderson C., Theraulaz G., Deneubourg J.-L. (2002). Self-assemblages in insect societies. *Insectes Sociaux*, vol. 49, pp. 1-12.

Baldassarre G., Nolfi S., Parisi D. (2002). Evolving mobile robots able to display collective behaviour. In Hemelrijk C. K. (ed.), *International Workshop on Self-Organisation and Evolution of Social Behaviour*, pp. 11-22. Zurich: Swiss Federal Institute of Technology.

Baldassarre G., Nolfi S., Parisi D. (2003). Evolution of collective behaviour in a group of physically linked robots. In Raidl G., Guillot A., Meyer J.-A. (eds.), *Applications of Evolutionary Computing - Proceedings of the Second European Workshop on Evolutionary Robotics*, pp. 581-592. Berlin: Springer Verlag.

Bonabeau E., Dorigo M., Theraulaz G. (1999). *Swarm intelligence*. Oxford: Oxford University Press.

Brooks R.A. (1986). A robust layered control system for a mobile robot. *Journal of Robotics and Automation*, vol. 2, p. 14-23.

Camazine S., Deneubourg J.L., Franks N.R., Sneyd J., Theraulaz G., Bonabeau E. (2001). *Self-organization in biological systems*. Princeton, NJ.: Princeton University Press.

Dorigo M., Trianni V., Sahin E., Labella T.H., Gross R., Baldassarre G., Nolfi S., Deneubourg J.-L., Mondada F., Floreano D., Gambardella L.M. (submitted). Evolving self-organising behaviour for a swarm-bot. *Autonomous Robots (Special issue on Swarm intelligence)*.

Grabowski R., Navarro-Serment L.E., Khosla P.K. (2003). An army of small robots. *Scientific American*, November, pp. 42-47.

Ijspeert A.J., Martinoli A., Billard A., Gambardella L.M. (2001). Collaboration through the exploitation of local interactions in autonomous collective robotics: The stick pulling experiment. *Autonomous Robots*, vol. 11, pp. 149-171.

Kube R.C., Bonabeau E. (1998). Cooperative transport by ants and robots. *Robotics and autonomous systems*, vol. 30, pp. 85-101.

Miglino O., Lund H.H., Nolfi S. (1995). Evolving mobile robots in simulated and real environments. *Artificial Life*, vol. 4, pp. 417-434.

Mondada F., Pettinaro G.C., Guignard A., Ivo K.W., Floreano D., Deneubourg J.-L., Nolfi S., Dorigo M., Gambardella L.M. (in press). SWARM-BOT: A new distributed robotic concept. *Autonomous Robots (Special issue on Swarm intelligence)*.

Nolfi S., Floreano D. (2000). *Evolutionary Robotics*. Cambridge, Mass.: The MIT Press.

Seyfarth R.M., Cheney D.L. (2003). Signallers and receivers in animal communication. *Annual Review of Psychology*, vol. 54, pp. 145-173.

Von Hayek F. (1945). *American Economic Review*, vol. 35, pp. 519-30.

Effects of the topology of social networks on information transmission

Jason Noble, Simon Davy, and Daniel W. Franks
School of Computing, University of Leeds, Leeds LS2 9JT, UK
[jasonn|wavy|dwfranks]@comp.leeds.ac.uk

Abstract

Social behaviours cannot be fully understood without considering the network structures that underlie them. Developments in network theory provide us with relevant modelling tools. The topology of social networks may be due to selection for information transmission. To investigate this, we generated network topologies with varying proportions of random connections and degrees of preferential attachment. We simulated two social tasks on these networks: a spreading-innovation model and a simple market. Results indicated that non-zero levels of random connections and low levels of preferential attachment led to more efficient information transmission. Theoretical and practical implications are discussed.

1. Introduction

The origin of social behaviour was a major transition in evolutionary history. Sociality opens up a wide range of new strategies for doing well in the world, and many of these strategies involve sharing information. Social animals have evolved elaborate communication systems, such as the dance used by a honeybee to tell others about the location of a food source (von Frisch, 1967). Social behaviour combined with the ability to learn means that animals are not limited to individual trial-and-error, but can learn from the behaviour of their conspecifics. Rats, for example, can learn from other rats about which foods are safe to eat (Galef, 1996); chimpanzees can socially learn new ways of exploiting a food source, such as fishing for termites by probing termite mounds with thin sticks (Tomasello, 1996). In our own species, of course, communication and social learning have reached great levels of sophistication and have made possible our technological progress.

The study of communication and social learning in animal species is the province of behavioural ecology and of ethology. In the human case, linguists are interested in understanding how language is used to communicate, how it is learned, and how it evolved. In the artificial sciences, researchers attempting to build useful multi-agent systems (in either robotic or software form) have taken inspiration from both biology and linguistics. Across all of these domains of inquiry, progress has been made by focusing on dyadic interactions, such as a young animal learning from its parent, or two people having a conversation. However, the bigger picture—who talks to whom, who learns from whom, etc.—has generally been neglected. Social behaviours such as communication or learning cannot be fully understood without considering the social network structures over which information transfer occurs. Indeed, social behaviour both shapes and is shaped by such networks. This paper represents a first step in taking social networks more seriously: the aim is to simulate some simple social tasks over a range of possible network configurations in order to see whether particular topologies of the social network promote more efficient information transmission.

The neglect of social networks in much of the existing work on social behaviour is not down to simple laziness on the part of the investigators. Given the mathematical and computational formalisms that have typically been used to express theories and models, rich representations of network structure were simply not possible. The individuals in a population are assumed to constitute a fully connected network, and an instance of learning or communication takes place between two randomly chosen individuals. (Consider classic adaptive behaviour work on communication such as Steels, 1996). But it is clear that populations of social organisms are not randomly structured in this way. Depending on the species, geographical ranges, family groupings, dominance hierarchies, cliques, dialects, and many other factors could all influence the likelihood of any two individuals interacting. Given advances in computing power and individual-based modelling techniques, it is now possible to construct models of social behaviour in which complex social networks can be directly represented. It is an open research question as to the extent to which current theories will need to be revised once the simplifying assumption of random social interactions has been dropped.

Recent simulation results have indicated that considering social network structure can certainly affect the conclusions drawn from a model. In ongoing work on the iterated learning model of language transmission, Smith and Hurford (2003) have looked at what happens when a language learner receives input from varying numbers of both parents and siblings. Smith and Hurford have found that incorporating the social network in this way places severe restrictions on the conditions for optimal language transmission: contact needs to be limited to a small number of parents with minimal input from siblings. Similarly, in a model of social learning, Noble and Franks (2002) found that the details of who learns from whom affected the efficiency of different social learning strategies. Imitation worked best if directed towards a parent, whereas emulation (a less sophisticated strategy involving attention to positive outcomes experienced by other agents without the ability to copy their specific actions) was most effective given input from randomly encountered conspecifics.

Given that social network structure makes a difference to social behaviour, it makes sense to try to capture it in our models of such behaviours. But how should we do so? How exactly are human and animal social networks typically structured? These are big questions. Fortunately, the recent explosion of interest in network theory provides us with the theoretical tools needed to start answering them. Network theorists have given us classification schemes and generative models of network structure, as well as various statistical measures that can be used to characterize networks. Space precludes a complete discussion of network theory here, although see Newman (2003) for an excellent introduction. However, we will briefly consider several basic types of networks in order to illustrate the thinking behind the model presented in the next section.

Perhaps the simplest type of network is a random graph (Erdös & Rényi, 1959), in which every node is equally likely to be connected to every other node. The only parameter needed to describe such a network is the average number of connections per node (also known as the average node degree). Random graphs have been a useful mathematical formalism but it is widely recognized that they are not a good model for any real-world network. Characteristic properties of random graphs include a short average path length and very low transitivity. The average path length of a network is simply the mean of the minimum distances between every pair of nodes, where paths are measured along connections. Transitivity refers to the likelihood that a node's neighbours will themselves be connected to each other (usually measured in terms of a clustering coefficient C, being the average of the proportions of the $n(n-1)/2$ possible inter-neighbour connections that exist for each node, with C being defined as 1 for nodes with less than two neighbours).

Small-world networks (Watts & Strogatz, 1998) have received a great deal of attention lately. The idea is to start with a network that is connected in some regular way, such as a large ring of nodes in which each one is connected to its two nearest neighbours on the left and its two nearest neighbours on the right. Such a network is highly transitive ($C = 0.5$ for the setup given) but the average path length will be long as many hops are needed to connect a node on one side of the ring with a node on the other. However, if a small proportion of the connections are randomly rewired, this effectively creates short cuts across the ring, without destroying the overall clustered structure of the network. The result is a small world network, with the notable combination of a high degree of transitivity and a short average path length. Anecdotal evidence suggests that human social networks may share these characteristics: many of our friends know each other (transitivity), and we can often find surprisingly short chains linking ourselves and an arbitrary stranger (short path lengths).

In the small-world network example given, the node degree is constant at 4 connections per node. In many real world networks, the distribution of node degrees is not flat, but highly variable. In particular, Barabási and Albert (1999) noted that node degrees are often found in a scale-free or exponential distribution. That is, there are many nodes with only a few connections, while a few nodes have many connections. (Another way of putting this is to say that the distribution of node degrees is long-tailed or heavily skewed to the right.) Consider for example the internet, where most machines are connected to only one or two other machines, whereas major hubs may have many thousands of connections. Barabási and Albert suggested the *preferential attachment* algorithm for generating such networks: nodes are added to the network one at a time, and upon its arrival each node will make one or more connections to existing nodes. However, these connections are not made at random but are allocated preferentially to existing nodes that already have a higher-than-average number of connections. Thus, nodes rich in node degree get even richer as the algorithm continues. Barabási and Albert explored the properties of scale-free networks generated using preferential attachment and found that, despite the absence of a regular substrate, they share the critical properties of small world networks: high transitivity and a short average path length. They also have a negative node degree correlation, which means simply that a connection from a high-degree node is likely to lead to a low-degree node and vice versa.

Network theorists employing the concepts reviewed above have looked at social networks as one example among many. This work has been descriptive in nature, and has focused on idiosyncratic human cases for which data is readily available (e.g., scientific collaboration networks), but it has indicated that social networks have distinct structures in comparison to other real-world examples such as computer networks. Newman and Park (2003) found two distinctive features of social networks. The first was that transitivity is unusually high. The second feature was that social networks show positive node degree correlation. This can be illustrated by the idea that well-connected people tend to know lots of other well-connected people, and loners tend to know (if anyone) other loners. Positive node degree correlation stands in contrast to the negative correlation seen in scale-free networks such as the internet, where hubs connect to single machines and vice versa.

Newman and Park's work is not definitive—there may be other distinctive features of human social networks not yet enumerated—and it is not at all clear that these properties will also apply to the social networks of other species. However, the fact that social networks can be shown to have distinctive structures raises what is for us an exciting theoretical question. Social networks in the natural world must be a product of the social behaviours of the agents involved. These behaviours must have been subject to selective pressure of some kind; either natural selection in the case of instinctive social behaviours, or some sort of cultural selection in the case of learned behaviours. It seems likely that the quality of the information an agent receives via its social network was a factor in this selective process. It therefore follows that we might expect some naturally occurring social networks to have been optimized for their efficiency in transmitting information between members of a population. If this is so, what sort of structure would we expect the networks to have? Could the properties observed by Newman and Park, for instance, be the product of selection for individuals whose social behaviours lead them to form networks with just these characteristics? One goal of the work presented here is to begin to answer this question by comparing the effectiveness of different network topologies in supporting the transfer of information. Of course, it is also possible that the structures of social networks are spandrels; that they have not been directly selected for but are the by-products of selective pressure on other aspects of social behaviour. This conclusion would be supported by showing that networks optimized for information transfer look nothing like real social networks.

Another goal of our work is entirely practical. In building a multi-agent system to perform some function, we can take inspiration from biology but there is no requirement for biological fidelity. So what sort of network topology should be imposed in order to facilitate information sharing between agents? The multi-agent system we have in mind is a form of market-based control for the coallocation of jobs and resources in Grid computing (Davy, Djemame, & Noble, 2003). Other authors (Byde, Salle, & Bartolini, 2003) are interested in a similar market-based control system for resource allocation in large scale ($\sim 50,000$ node) computing clusters. In both systems, processors and jobs are treated as economic agents in a marketplace, with jobs bidding for suitable processors, and processors offering CPU cycles for sale. When dealing with large-scale systems, centralized allocation approaches and all-to-all connection schemes cannot work due to congestion. Clearly what is needed is some type of network linking the agents, over which bids and offers can be transmitted. However, nothing is known about what sort of network topology will best support the operation of such a market. In the simple model presented below, we have chosen one of two social-information-sharing tasks with this problem in mind. More generally, we suspect that a better understanding of social networks will have payoffs for multi-agent system design, in that designers will have better guidelines for deciding who or what an artificial agent should communicate with, who it should trust, who it might want to imitate, etc.

2. The model

Each node in our model represents an individual, such as a person, an animal, or an artificial agent. We began with isolated nodes positioned on a regular spatial substrate, i.e., a grid. Our networks consisted of 2,500 nodes arranged in a 50×50 array. Of course, real social networks are not formed simply through every individual linking to their four immediate neighbours in space. However, we wanted to capture the idea that social networks have the physical world as their ultimate substrate, and, generally speaking, increased physical proximity between two individuals means a higher probability of a social connection existing. The various network layouts that we explored deviated from this underlying spatial grid to greater and lesser degrees.

Our aim was to generate a wide range of network topologies. In order to do this we devised two axes on which the layouts of our networks could vary (not to be confused with the two dimensions of the underlying spatial grid). For the first axis of variation we took inspiration from work on small-world networks, in which network properties change dramatically as a few random, long-range connections are added. We therefore varied the probability with which a connection would be either local or random; we called this parameter R. When $R = 0$, all links are local (see Figure 1). When $R = 1$, all links are random (Figure 2). A major finding with respect to small-world networks is that small proportions of random connections can make a big difference. We therefore investigated the following values for R: $0, \frac{1}{256}, \frac{1}{128}, \frac{1}{64}, \frac{1}{32}, \frac{1}{16}, \frac{1}{8}, \frac{1}{4}, \frac{1}{2}$, and 1.

In order to instantiate local links, we needed some definition of a local neighbourhood. The neighbourhood for our nodes consisted of the 24 other nodes contained in a 5×5 square centred on the focal node. (Nodes close

Figure 1: A sample network with $R = 0$ and $P = 0$, i.e., local and egalitarian connections. This could become a small world network with the addition of a few long-range "short-cuts", i.e., with $R \approx 0.05$. Note that this and subsequent sample networks have only 64 nodes and 128 connections (i.e., 4 per node); 2500 nodes and 10000 connections were used in the simulation runs.

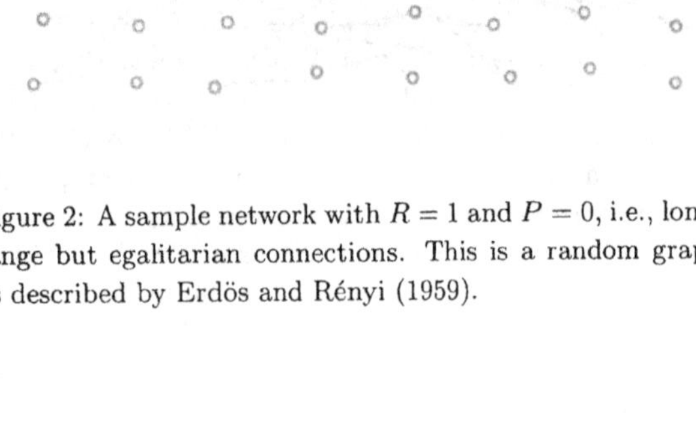

Figure 2: A sample network with $R = 1$ and $P = 0$, i.e., long-range but egalitarian connections. This is a random graph as described by Erdös and Rényi (1959).

Figure 3: A sample network with $R = 0$ and $P = 1$, i.e., local and preferential connections. Note the formation of local "hubs".

to the edges and corners of the 50×50 grid had a correspondingly smaller neighbourhood.) A local connection meant a connection to a randomly chosen node within the local neighbourhood. A random connection meant a connection to a randomly chosen node from anywhere else on the network. Self-connections and multiple connections were not allowed in either case.

The second axis of variation was based on the preferential attachment idea, and we called the associated parameter P. When $P = 0$, connections were egalitarian, and no node was more likely to be selected for a connection than any other (see Figures 1 and 2). That is, all 24 local neighbours were equally eligible to receive a local connection, and all of the other 2499 nodes in the network were equally eligible to receive a random connection. As the value of P increased, connections were made in an increasingly preferential way. The value $(n+\delta)^P$ was calculated for each eligible node, where n is the node degree and δ is a small positive value (0.1 in the runs described below). The node that actually received a connection was chosen by roulette-wheel selection over these values. The parameter P thus acts as a preferential exponent, governing the strength of the bias towards well-connected nodes. The point of the value δ was to avoid stalemates in network generation by ensuring that even a node with no connections had a non-zero probability of selection. Figures 3 and 4 illustrate networks with $P = 1$. The following values of P were investigated: 0, 0.2, 0.4, 0.6, 0.8, 1.0, and 1.2.

One variable that affects information transmission across a network in an obvious way is connectivity, i.e., the average number of connections per node. If a network has low connectivity, then whole sections of it may be isolated from each other, for example, and information transmission between sections would be impossible. In order to investigate the effects of topology rather than brute connectivity, we held the number of links constant

Figure 4: A sample network with $R = 1$ and $P = 1$, i.e., long-range preferential connections. Prominent hubs can be seen, and the distribution of node degree is heavily right-skewed.

at 10,000 in all of our experiments. Connections were bidirectional, so this represented an average node degree of 8. This level of connectivity guaranteed that all or nearly all nodes in the network would be grouped into a "giant component" or connected section.

To actually construct sample networks, we began by sorting the nodes into a random order and giving each node at least one initial connection, selected according to the current values of R and P. This procedure ensured that there would be no singleton nodes. Then the remainder of the 10,000 connections were allocated. Note that in this phase the prevailing values of R and P were used in the selection of both source and destination nodes for each connection: if $R = 0$ and $P = 1$, for example, a random neighbourhood would be selected, one node within that neighbourhood preferentially selected as the source, and then a destination node preferentially selected from within the 5×5 square centred on the source.

Once the layout was established, we ran two simple social tasks across the network in order to find out what sorts of topology best support social information transfer. The two tasks were quite different in character, but each task had clearly defined success measures to ensure that our results would be unambiguous. For each task, we also looked at how well the network coped with a sudden localized change. This work is exploratory, so we were not strongly committed to particular hypotheses about what we would find. However, the network theory literature suggested some promising questions. Might there be a "sweet spot" for the classic small-world network, at $R \approx \frac{1}{32}$? Does node degree distribution make a difference? Could it be that the highly connected nodes in a skewed distribution function as information centres, increasing the overall efficiency of the network?

The first task was a basic spreading-innovation or behavioural-contagion model of social learning. We were mindful of Kawamura's (1959) research on Japanese macaques, in which one animal came up with a behavioural innovation (i.e., washing potatoes in a stream before eating them) and the novel behaviour could be seen to spread gradually throughout the social network. In our model, a random node was chosen as the source of the behavioural innovation. Every time-step an individual would observe one of its social partners, chosen at random from among the nodes it was connected to, and would adopt the novel behaviour if the chosen partner had already done so. If the chosen partner had not yet adopted the new behaviour, nothing happened. The measure of effective information transmission was simply the time taken for the novel behaviour to reach 99% prevalence in the population.[1]

The second task was more complex: each node represented an agent in a market. Every agent can both buy and sell, but only to its social partners, i.e., the nodes it is connected to. The market trades in a single indistinguishable commodity, let's say bicycles. Each agent can only possess one bicycle at a time, and a randomly chosen 50% of the agents begin the simulation owning a bicycle. Each agent has a personal valuation of a bicycle that is known only to itself; this is drawn from a uniform random distribution between \$1 and \$1000. Each agent also has access to a medium of exchange—we can think of this as a bank account with an infinite overdraft.

The agents operate according to very simple financial principles set out in a paper by Gode and Sunder (1993); they are "zero-intelligence" (ZI) traders. (We chose the ZI trading architecture not because we think it is a sensible way to trade but because it is probably the simplest way to set up a working market.) Each turn, an agent will make either an offer or a bid to a random partner, depending on whether the agent has or does not have a bicycle respectively. Would-be buyers will bid at a price randomly chosen from between the minimum price (\$1) and their private valuation. This is the core logic of the ZI trader: it makes no sense to bid more than you think a bicycle is worth, and by bidding less, you may pick up a bargain. Similarly, would-be sellers will make a random offer somewhere between the maximum price (\$1000) and their private valuation. Offers received by agents who already have a bicycle are ignored, as are bids received by agents who have no bicycle to sell. However, if the neighbour has a bike and receives a bid higher than their valuation, or has no bike and receives an offer under their valuation, a trade will take place. The market continues to operate in this way until a complete time-step passes without any successful trades being made.

[1]In fact there were no strict time-steps in either task; random nodes were chosen and allowed to act. After 2,500 random choices, i.e., one per node in the network, a nominal time-step was assumed to have passed. However there was no guarantee that a particular node would be chosen exactly once during each time-step.

Compared to the spreading innovation model, the economic task is not as obviously related to the spread of information. However, the bids and offers do transmit information about private valuations, and the net result of trading is to transfer bicycles to agents that valued them highly, and cash to agents that did not. We can therefore measure the efficiency of the market by looking at the mean increase in utility across the population of agents. We assume that an agent's utility is equal to its private valuation of the bicycle, if it ends up owning one, plus the amount of money it has in the bank, which may be either positive or negative. Note that the initial average utility will be around \$250, as half of the agents have bicycles and the average valuation is \$500. After a period of trading we can calculate the change in this score and use that as an index of market efficiency. We measured the change in utility at two points: after ten time-steps had elapsed, and after the market had stabilized.

For both tasks we wanted to let the network reach an initial equilibrium and then see how quickly it would recover from a sudden local disturbance. In both cases we disrupted the function of 25% of the nodes, corresponding to the northwest quadrant of the underlying spatial grid. For the spreading innovation task, we assumed that all individuals in the northwest quadrant had suffered an abrupt memory failure and lost the ability to perform the novel behaviour. We then measured how long it took for the novel behaviour to regain 99% prevalence in the population. For the economic task, we abruptly took the bicycles away from any individuals in the northwest quadrant who possessed one. We then recalculated the average utility, allowed the market to run for ten time-steps, and recorded the improvement in average utility that was achieved within that time.

3. Results

For each combination of 10 values of R and 7 values of P, we generated 20 sample networks and recorded the results on the spreading-innovation and economic tasks. Thus our primary results, shown in Figures 5 and 6, are measures of average success on each task across the space of the P and R values investigated.

Our network generation algorithm was successful in producing valid networks, and most of the time a connected graph resulted (overall mean number of nodes outside the giant component was only 0.48 out of 2,500). In order to be sure that our network topologies had the expected properties, we measured several basic network statistics. These were average path length (sampled over 5,000 random pairs of nodes rather than being calculated exhaustively), clustering coefficient (defined earlier), node degree correlation, and skewness of the degree distribution. This last was calculated using the 1st percentile, the median, and the 99th percentile of the node degree distribution, combined to give a standard measure of skewness as follows:

$$\frac{(n_{99} - n_{50}) - (n_{50} - n_1)}{n_{99} - n_1}.$$

A value of zero on this skewness measure indicates a symmetrical distribution, while positive values, up to a maximum of 1.0, indicate increasingly long tails to the right.

Beyond simply measuring success on each task for particular values of R and P, we also constructed multivariate analyses of variance for the success score on each task against the four network statistics listed above. Although R and P would themselves be useful predictors in a multivariate model, we did not include them. The idea was to put ourselves in the position of investigators looking at a real-world network, where the relevant network statistics could be measured with some degree of error but "behind the scenes" variables such as R and P were not available. We hoped that these multivariate models would help us to identify which network properties were associated with successful outcomes.

Results for the spreading-innovation task are shown in Figure 5. Note that this graph shows time-to-convergence data, so lower values are better. The results are very clear: the novel behaviour reaches 99% prevalence most rapidly (in around 15 time-steps) when the probability of random connections (R) is equal to 1, and the preferential exponent (P) is low. This means that the most successful network for the rapid spread of an innovative behaviour would be a classical random graph. For lower values of R, i.e., as the networks become more locally connected, the time taken to reach 99% prevalence greatly increases. Although R has the strongest effect on the outcome, it is also clear that increasing values of P, i.e., more preferentially connected networks with more heavily skewed degree distributions, are associated with slightly longer convergence times.

A multivariate analysis of the data was highly significant ($F_{4,1395} = 758.9$, $p < 2.2 \times 10^{-16}$) and explained 68% of the variance in the time to reach 99% prevalence. Average path length, clustering coefficient, and skewness of degree distribution were all significant predictors, and were all positively associated with longer times. The outstanding predictor was average path length, however, which, fitted alone, explained 65% of the variance in convergence times. The rapid spread of an innovation around a social network appears to depend chiefly on that network having short average path lengths.

When the network was disrupted by resetting the behaviour of the 625 agents in the northwest quadrant, the story was very similar. The fastest recovery of 99% prevalence of the novel behaviour was found in networks with high R and low P.

Results for the economic task were very different, and are shown in Figure 6. This graph shows the mean in-

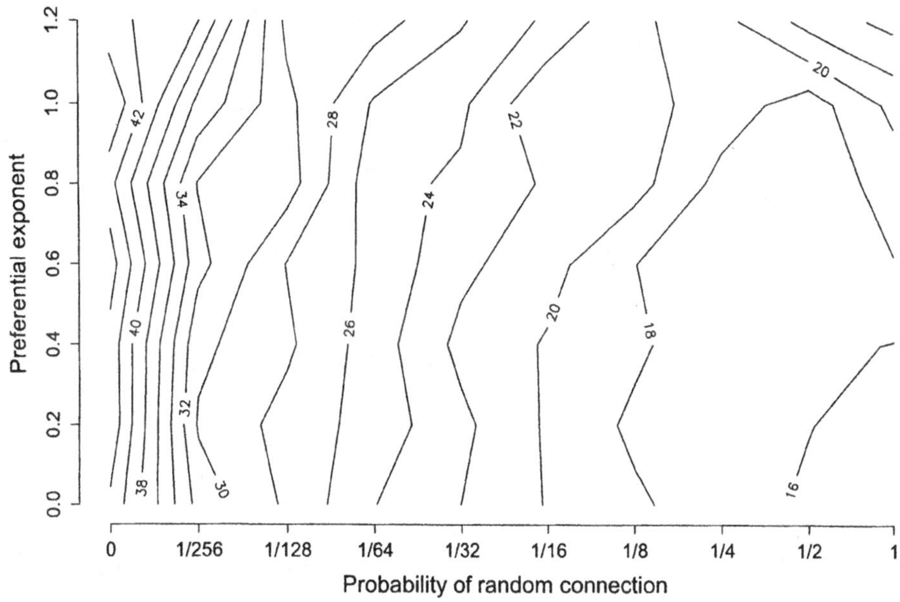

Figure 5: For the spreading-innovation task, the graph shows the number of time-steps until the novel behaviour reached 99% prevalence in the population for various network architectures. Each data point is averaged over 20 runs.

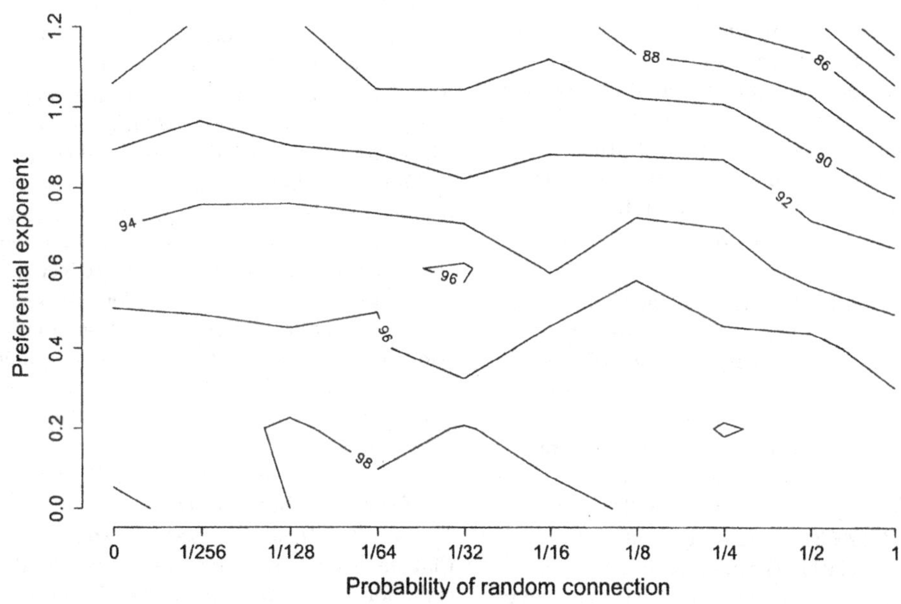

Figure 6: Change in average utility per agent after ten time-steps of trading in the economic task, for various network architectures. Each data point is averaged over 20 runs.

crease in utility per agent, which is a measure of market efficiency, and thus, unlike Figure 5, higher values are better. We measured the change in average utility both after 10 time-steps, and after the market had stabilized. The pattern of results was similar in both cases, and Figure 6 shows the situation after 10 time-steps. The first point to note is that high values of the preferential exponent P are associated with lower increases in average utility. The most impressive increases in utility are found when $P = 0$ and $R = \frac{1}{64}, \frac{1}{32}$ and $\frac{1}{16}$. In other words, the most efficient markets are found when the network has an egalitarian distribution of node degrees, and a small proportion of random connections. This corresponds to Watts and Strogatz's (1998) original small-world model.

The multivariate analysis of the data was again highly significant ($F_{4,1395} = 462.9$, $p < 2.2 \times 10^{-16}$) and in this case explained 57% of the variance in added utility. The clustering coefficient and the skewness of the node degree distribution were both significant predictors, and both had a negative association with added utility. The clustering coefficient fitted alone explained 14% of the variance in utility, but the skewness of the node degree distribution explained 54% of the variance when fitted alone. Thus the efficiency of a market appears to depend primarily on an egalitarian distribution of connections, but also on transitivity being low.

When the market was disrupted by removing all bicycles from the northwest quadrant of the grid, however, the results were more in line with the spreading-innovation task: the fastest recovery of utility was found when the network had many random connections. The value of P was not especially important, although it was associated with slightly poorer recovery of utility. However, high values of R led to much greater improvements in utility, e.g., a mean improvement after 10 time-steps of 16.7 when $R = 1$ compared to only 3.55 when $R = \frac{1}{32}$ ($P = 0$ in each case).

One point of concern in our results was that we only observed very moderate levels of positive degree correlation in any of the networks in our space of possible topologies. The highest values for node degree correlation (only around 0.025) were seen for low levels of random connections and egalitarian node degree distributions; as preferential attachment increased, the node degree correlation dropped to around −0.23. However, Newman and Park (2003) argue that significantly positive node degree correlation (popular people know other popular people, and loners know other loners) is one of the defining characteristics of human social networks. We therefore put together an alternative network generation algorithm that was designed to produce positive node degree correlations: the probability of a connection going from a source to a target node was proportional to how closely the target node's degree matched that of the source node. The parameter R applied to these networks in the usual fashion, but the parameter P was no longer relevant. This algorithm succeeded in producing positive node degree correlations, with an overall average correlation of 0.23 across the different values of R. We re-ran both of the social tasks on these networks but in all cases the results were not substantially different from those for the equivalent network with egalitarian connections (i.e., $P = 0$).

4. Conclusions

The higher the proportion of random connections, the faster an innovative behaviour spread throughout the social network. On the other hand, a modest proportion of random connections led to the most efficient market performance. In both cases, an egalitarian distribution of node degrees was more effective: the heavily skewed distributions generated by preferential attachment just seem to slow things down.

With the benefit of hindsight it seems obvious that high levels of random connections will be advantageous on the spreading-innovation task, as random connections ensure a short path length between any two nodes which in turn allows the most rapid spread of the novel behaviour. Similarly, in both the spreading-innovation and the economic tasks, the fastest recovery from the disruption of one quadrant of the underlying grid was seen in networks with many random connections. We can draw a parallel here with recovery from a disaster in human society: if your town has been flooded, for instance, you will do well to get help and temporary accommodation from friends who live some distance away (i.e., random connections) rather than seeking help from neighbours whose houses are also flooded (local connections).

We did not have concrete hypotheses concerning our results, but there were suspicions that small-world networks might do well. This was borne out with respect to the economic task, with the highest increases in average utility seen in networks with a few percent random connections and a fairly flat distribution of node degrees. In addition, if we return to the spreading-innovation task (Figure 5) and look at the performance of networks with $R = \frac{1}{16}$ and $P = 0$, for example, we find that they reach 99% prevalence in 19.87 time-steps. This is not as impressive as 15.78 time-steps for the networks with $R = 1$, but far better than the figure of 41.60 time-steps when $R = 0$ ($P = 0$ in all cases). It would be wrong to say that small-world networks were the optimal design for social information transfer—indeed, an important message from the results is that different topologies favour different social tasks. However, small-world networks look like a reasonable compromise given the tasks and topologies explored so far.

The suspicion that preferential attachment and skewed node degree distributions might promote information sharing turned out to be completely wrong. For the economic task one might imagine that the high-degree nodes could help things along by acting as "dealers", being more likely to buy and sell bicycles from their many trading partners several times before the market stabilized. However, the other side of the coin with a skewed node degree distribution is that many nodes have relatively few connections, and in the marketplace task this means that they will have a lower chance of being connected to an appropriate partner to make a trade, i.e., to someone who has a bike and a lower valuation than one's own, or to someone who has no bike but a higher valuation than yours.

One of our theoretical goals was to show what we would expect social networks to look like if they had been optimized for social information transfer. Although we would be far more confident if we had covered a wider range of social tasks, based on what we have done so far we can predict that optimized networks should show a non-zero level of random connections and a near-symmetrical node degree distribution. The validity of the first prediction is hard to assess and possibly needs to wait for both more modelling work and more empirical measurements of average path length and clustering coefficients in real social networks. Still, it is clear that in a population of animals, for example, having many long-range or random connections would entail costs (e.g., the energy costs of travelling around to meet all of these distant connections). So even if a high proportion of random connections had selective value for an individual due to information-gathering payoffs, this would need to be balanced against the cost of maintaining such links.

The second prediction, that node degree distribution should be symmetrical, is clearly not seen in nature. We therefore speculate that the skewed node degree distributions found in natural social networks are not the result of selection for efficient information transmission but are produced by some other factor, e.g., competition for status.

A practical goal of the work was to find out what sorts of network topologies would best support a marketplace populated by self-interested agents. On this point we can be clear: our results suggest that the overlay networks for trading in the systems currently being designed by Davy et al. (2003) and Byde et al. (2003) should be classic small-world networks, with about 1 in 32 connections being random and a flat distribution of node degrees.

We admit that in some of these conclusions we are perhaps pushing the results of an early, exploratory model a little too hard. Clearly a wider range of social tasks needs to be investigated. The two tasks that we did consider could be examined in more detail as well: it is not clear how the spreading-innovation model would respond to the addition of factors such as noisy perception, multiple competing innovations, and a tendency to copy the majority behaviour among one's neighbours. The economic task dealt with about the simplest market imaginable: something as simple as allowing the agents to possess more than one bicycle at a time might alter the results significantly.

In future models we also want to look at what happens when the network topology (or, more specifically, the mechanisms and behaviours that generate that topology) is under evolutionary pressure. Currently the network architecture is static and pre-specified, but this is not the case with real social networks. Indeed, your social network is not imposed on your communicative and learning behaviours, it is in fact shaped by those very behaviours: if you talk to someone and receive useful information from them, you are likely to talk to them again in future and thus your social network has altered. Capturing such reciprocal effects will be critical to future models. This would leave the way open for tackling much bigger questions about what sorts of social network formation and maintenance strategies—in people, animals, robots, or software agents—are likely to win out in the struggle for selection.

Acknowledgements

Our thanks to Dave Cliff of HP Labs for useful discussions.

References

Barabási, A.-L., & Albert, R. (1999). Emergence of scaling in random networks. *Science, 286*, 509–512.

Byde, A., Salle, M., & Bartolini, C. (2003). Market-based resource allocation for utility data centers. Technical report HPL-2003-188, Hewlett Packard Laboratories, Bristol, UK.

Davy, S., Djemame, K., & Noble, J. (2003). The application of bioinspired engineering principles to Grid resource allocation. In *Proceedings of the Nineteenth Annual UK Performance Engineering Workshop.*

Erdös, P., & Rényi, A. (1959). On random graphs. *Publicationes Mathematicae, 6*, 290–297.

Galef, Jr., B. G. (1996). Social enhancement of food preferences in Norway rats: A brief review. In Heyes & Galef (1996), pp. 49–64.

Gode, D. K., & Sunder, S. (1993). Allocative efficiency of markets with zero-intelligence traders: Market as a partial substitute for individual rationality. *Journal of Political Economy, 101*, 119–137.

Heyes, C. M., & Galef, Jr., B. G. (Eds.). (1996). *Social Learning in Animals: The Roots of Culture.* Academic Press, San Diego, CA.

Kawamura, S. (1959). The process of sub-culture propagation among Japanese macaques. *Primates*, *2*, 43–60.

Newman, M. E. J. (2003). The structure and function of complex networks. *SIAM Review*, *45*, 167–256.

Newman, M. E. J., & Park, J. (2003). Why social networks are different from other types of networks. *Physical Review E*, *68*(036122).

Noble, J., & Franks, D. W. (2002). Social learning mechanisms compared in a simple environment. In Standish, R. K., Bedau, M. A., & Abbass, H. A. (Eds.), *Artificial Life VIII: Proceedings of the Eighth International Conference on Artificial Life*, pp. 379–385. MIT Press, Cambridge, MA.

Smith, K., & Hurford, J. R. (2003). Language evolution in populations: Extending the iterated learning model. In Banzhaf, W., Christaller, T., Dittrich, P., Kim, J. T., & Ziegler, J. (Eds.), *Advances in Artificial Life: Seventh European Conference on Artificial Life (ECAL'03)*, Lecture Notes in Artificial Intelligence, pp. 507–516. Springer, Berlin.

Steels, L. (1996). Emergent adaptive lexicons. In Maes, P., Matarić, M., Meyer, J.-A., Pollack, J., & Wilson, S. W. (Eds.), *From Animals to Animats 4: Proceedings of the Fourth International Conference on Simulation of Adaptive Behavior*, pp. 562–567. MIT Press / Bradford Books, Cambridge, MA.

Tomasello, M. (1996). Do apes ape?. In Heyes & Galef (1996), pp. 319–346.

von Frisch, K. (1967). *The Dance Language and Orientation of Bees*. Belknap Press / Harvard University Press, Cambridge, MA.

Watts, D. J., & Strogatz, S. H. (1998). Collective dynamics of 'small-world' networks. *Nature*, *393*, 440–442.

Evolving Functional Self-Assembling in a Swarm of Autonomous Robots

Vito Trianni **Elio Tuci** **Marco Dorigo**
IRIDIA - Université Libre de Bruxelles
50, avenue F. Roosevelt - CP 194/6
1050 Bruxelles
{vtrianni,etuci,mdorigo}@ulb.ac.be

Abstract

The goal of this study is the design of controllers for robots capable of physically connecting to each other, any time environmental contingencies prevent a single robot to achieve its goal. This phenomenon is referred to as *functional self-assembling.* Despite its relevance as an adaptive response, functional self-assembling has been rarely investigated within the context of collective robotics. Our task requires the robots to navigate within a rectangular corridor in order to approach light bulbs positioned on the opposite end of the corridor with respect to their starting positions. Aggregation and assembling are required in order to traverse a low temperature area, within which assembled robots navigate more effectively than a group of disconnected agents. The results of our empirical work demonstrate that controllers for a group of homogeneous robots capable of functional self-assembling can be successfully designed by using artificial neural networks shaped by evolutionary algorithms.

1. Introduction

This paper addresses the problem of synthesizing controllers for groups of autonomous robots—referred to as *s-bots*—capable of adaptively connecting to each other, forming an assembled structure—referred to as *swarm-bot.* The *swarm-bot* is a self-assembling, self-organizing artifact that exploits the cooperation among its simple components, the *s-bots*, in order to solve problems a single individual cannot cope with. The focus of this paper is on *functional self-assembling*, that is, the self-organized creation of a physically connected structure, which should be *functional* to the accomplishment of a particular task. In other words, our goal is the design of controllers for *s-bots* capable of connecting to each other (i.e., forming a *swarm-bot*) any time environmental contingencies prevent the single *s-bot* to achieve its goal.

Self-assembling occurs in a wide range of natural systems. In particular, it characterizes the behaviour of many social insects (see Anderson et al., 2002, for a review). A striking example of functional self-assembling can be observed in ants of the species *Œcophilla longinoda* (see Lioni et al., 2001). These ants are able to build chains connecting one to the other, creating bridges that facilitate the passage of other ants (see Figure 1, top).

Functional self-assembling is a problem of fundamental importance within the SWARM-BOTS project,[1] wherein this research is conducted. By drawing inspiration from the behaviour of social insects (see Bonabeau et al., 1999), functional self-assembling can be viewed as an adaptive response of a group of autonomous agents to cope with environmental conditions which prevent them from carrying out their task. For example, the navigation of a group of robots can be hindered by the presence of a trough larger than the size of the body of a single robot, which can not be avoided and can be passed only by forming an assembled structure capable of bridging the gap (see Figure 1, bottom).

Despite its relevance as an adaptive response, functional self-assembling has been rarely investigated within the context of collective robotics. Several works in the literature have focused on the study of collective behaviour of autonomous agents, which due to the limits of their morphological structure, are not capable of self-assembling (see Cao et al., 1997; Liu and Wu, 2001). Other studies focus on the development of assembled structures of non-autonomous robotic units, that can be differently connected to each others by the experimenter and in some cases are able to self-reconfigure their shape (see Castano et al., 2000; Yim et al., 2000). With respect to this literature, the agents developed by the SWARM-BOTS project represent a step further. The *s-bots* are morphologically designed to autonomously carry out most of the tasks generally explored in collective robotics studies (e.g., foraging, clustering and sorting,

[1]A project funded by the Future and Emerging Technologies Programme (IST-FET) of the European Community, under grant IST-2000-31010.

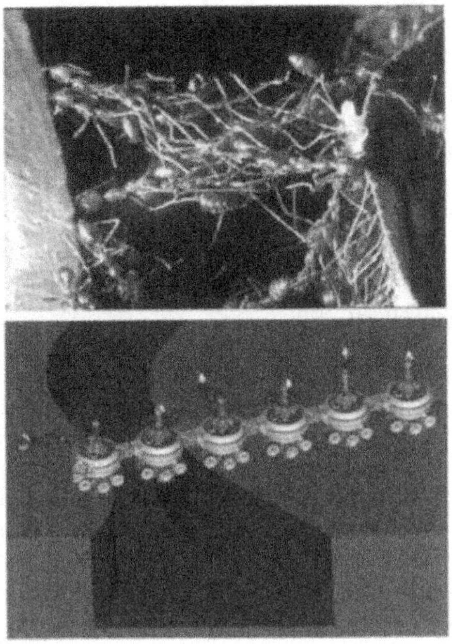

Figure 1: Examples of functional self-assembling. Top: ants of the species *Ecophilla longinoda* forming a chain (This image has been kindly provided by Dr. Arnaud Lioni). Bottom: a graphical representation of a *swarm-bot* structured in a chain while passing over a trough.

navigation and surveillance), as well as to perform functional self-assembling. However, despite the facilities provided by the mechanics and electronics of our agents,[2] the design of control systems for autonomous robots capable of functional self-assembling is not a trivial task.

The complexity of functional self-assembling resides in the nature of the individual mechanisms required to bring forth the coordinated movements that lead firstly to the formation of the assembled structure, and subsequently to the collective motion of the assembled structure. Moreover, functional self-assembling is never *per se* the goal of the agents, but it generally requires to be integrated within the behavioural repertoire of agents capable of performing several different adaptive responses. In particular, within the SWARM-BOTS project, we are interested in investigating scenarios in which the *s-bots* should be capable of:

1. Independently performing a specific task. For example, if assembling is not required, *s-bots* should be capable of individually performing phototaxis.

2. Aggregating in order to allow subsequent assembling. That is, if assembling is required by particular environmental contingencies, the *s-bots* should be capable of bringing forth the conditions which facilitate functional self-assembling. Aggregation is one of the first steps in order to form an assembled structure—i.e., a *swarm-bot*.

3. Moving coordinately in order to assemble. That is, each *s-bot* should find the correct position with respect to another *s-bot* in order to be able to establish a connection.

4. Moving coordinately in order to improve the effectiveness of the behaviour of the assembled structure. For example, in order to pass over a trough, the assembled *s-bots* should negotiate a common direction of movement which brings the *swarm-bot* on the other side of the trough.

5. Disconnecting. That is, once the environmental contingencies do not require any longer the assembled structure, the *s-bots* should disconnect and carry out their goal independently.

This set of capabilities is commonly required by many complex tasks a *swarm-bot* should be able to carry out. In addition to the already mentioned example of passing over a trough larger than the body of a single *s-bot*, we can imagine a situation in which the *s-bots* have to climb a steep slope, or the case in which *s-bots* have to navigate through a very rough terrain. In this paper, we consider a simplification of these situations, that allows us to study the basic mechanisms that underpin functional self-assembling.

The task studied in this paper requires the *s-bots* to approach a light source located at the end of a corridor. Assembling is required to navigate in a "low temperature" area in which a *swarm-bot* can navigate more effectively than a group of disconnected *s-bots*. When located in the low temperature area, the aggregation of the *s-bots* should facilitate the subsequent assembling through their gripper element (see Section 2.1 for more details). Notice that, in our experiments, the "temperature" can be perceived by a single binary sensor which returns 1 if the *s-bots* are in a high temperature area, and 0 otherwise. This is a strong simplification with respect to other scenarios, in which the *s-bots* might be required to employ more complex sensory-motor skills in order to perceive those environmental contingencies that require assembling—e.g., the detection of a trough or of the increased roughness of the terrain. Moreover, the *s-bots* should perform individual and collective obstacle avoidance and coordinated motion. Obviously, these actions might require the use of more complex visual and proprioceptive sensors than the ones our simulated *s-bots* can rely on, in particular if the *s-bots* are located in an environment more complex than an empty corridor.

If compared to tasks that require *s-bots* to self-assemble in order to pass over a trough or to navigate

[2] Details regarding the hardware and simulation of the *swarm-bot* can be found in Mondada et al. (2004) and in the project web-site (`www.swarm-bots.org`).

on rough terrain, our scenario looks rather simple. It also relies on implementation details—such as the simulation of the gripper element, and the concept of energy, see Section 2. for further details—which might hinder the porting of the evolved controllers from simulated to real *s-bots*. However, we believe that these simplifications do not trivialise the significance of our experimental setup. They instead are required in order to allow us to begin our exploration within the domain of functional self-assembling. Another issue that induced us to take a simple or minimal approach concerns the methodology, as illustrated below.

In this study, we employed the evolutionary robotics (ER) methodology to develop integrated (i.e., single) control systems, which are capable of coordinating the actions of autonomous agents required to perform a broad repertoire of individual and collective responses. ER represents a way to automate the design of control systems for autonomous robots, using algorithms based on Darwinian evolution. The controllers are usually artificial neural networks, whose parameters are set by the evolutionary algorithm (see Harvey et al., 1992; Meyer et al., 1998; Nolfi and Floreano, 2000, for reviews of the field). Recently, the ER approach has been successfully employed to evolve controllers for groups of homogeneous robots required to act cooperatively in order to achieve simple goals. Homogeneous groups are those in which the controllers of all the members of the group are derived from the same genotype—they are clones. For example, Quinn et al. (2003) evolved robots to perform a coordinated motion task, obtaining a situated specialization in distinct functional roles. Trianni et al. (2004) studied coordinated motion in a group of assembled robots that had to navigate in an arena with holes and open borders without falling down. Groß and Dorigo (2004) evolved cooperative transport behaviours for a group of two robots.

From the method's point of view, our task looks more demanding than those mentioned above. Moreover, at the best of our knowledge, we are not aware of any other study in which a similar methodology has been employed to design controllers for robots capable of displaying both individual and collective behaviours which require the coordination of such a rich sensory and motor apparatus, as in our task. In fact, the latter requires controllers—i.e., artificial neural networks—capable of bringing forth a variety of individual and collective responses triggered by visual stimuli (from the light sensors), acoustic stimuli (from the sound sensors), and proprioceptive stimuli (from the traction sensors), to appropriately control the state of various different actuators (loudspeakers, wheels, gripper, see Section 2.2 for more details). In consideration of this, the study of a minimal and simplified—rather than a more complex and more appealing—scenario looks more suitable to begin our exploration on functional self-assembling. Obviously, our intention is to use this study as a stepping stone toward the development of controllers for *s-bots* capable of displaying functional self-assembling in more complex environmental circumstances.

We believe that our empirical work, although preliminary, brings significant contributions: (i) at the best of our knowledge, our experiments represent one of the first works in which controllers for a group of homogeneous robots capable of functional self-assembling are described; (ii) our results prove that, despite the complexity of the task, artificial neural networks, shaped by evolutionary algorithms, can cope with complex scenarios in which a single robot's controller is required to bring forth different individual and collective responses based on the appropriate control of the state of various actuators (i.e., wheels, loudspeakers, gripper element) triggered by the local information coming from various sensors (i.e., traction sensors, microphones, ambient-light sensors).

1.1 Structure of the paper

In what follows, we first describe, in Section 2., the methodology used. In Section 2.1, we give a detailed description of the task. In Section 2.2, we describe the simulation model of the *s-bots* employed in this work. In Section 2.3, we illustrate the type of artificial neural network used to control the behaviour of the *s-bots* and the genetic algorithm used to shape the network. Section 2.4 describes in details the evaluation function. In Section 3., we show experimental results. In Section 4., some conclusions are drawn.

2. Methods

In this section, we describe the methodology defined for the evolution of controllers for *s-bots* capable of displaying individual and collective coordinated motion, phototaxis, obstacle avoidance, aggregation and self-assembling. In the following subsections, we describe in detail the task, we illustrate the morphology of the *s-bots*, the methodology employed to develop their controllers and the evaluation function used.

2.1 Description of the task

The task requires navigation within a rectangular corridor in order to approach light bulbs positioned on the opposite end with respect to the *s-bots*' starting positions (see Figure 2). The corridor (4 meters long, 1 meter wide) is divided in an area of high temperature and an area of low temperature (respectively, light and dark gray in Figure 2). Aggregation and assembling are required in order to traverse a low temperature area, within which a *swarm-bot* (i.e., assembled *s-bots*) navigates more effectively than a group of disconnected *s-bots*. The effectiveness of the navigational strategies is correlated with the amount of "energy" required by the *s-bot* to explore the corridor. In the area of high temperature, each *s-bots* saves more of its energy by navigating disconnected, while in the area of low temperature, each *s-bot* saves more energy by navigating assembled—i.e., by forming a *swarm-bot*. If, while navigating, an *s-bot* exhausts its energy, it is not able to move any more. The *s-bots* do not have any information concerning their energy level. However, the *s-bots* can reach the light bulbs before running out of energy if they properly react to the characteristics of the environment. In particular, an optimal strategy requires the *s-bots* (i) to individually move toward the light bulbs as long as the temperature remains high; (ii) to aggregate by exploiting the sound signaling system they are provided with as soon as the temperature drops; (iii) to continue their phototactic behaviour in an assembled structure (i.e., by forming a *swarm-bot*) throughout the low temperature area.

It is worth mentioning that the "energy level" of an *s-bot* does not refer to any physical property of the *s-bots*. Moreover, it does not imply the use of unrealistic sensors, which cannot be instantiated on the real *s-bots*. In fact, the *s-bots* do not have access to their own energy level. The energy level has been mainly introduced to define a simple task that requires self-assembling. In fact, the energy level plays an important role in the evaluation procedure, as described in Section 2.4.

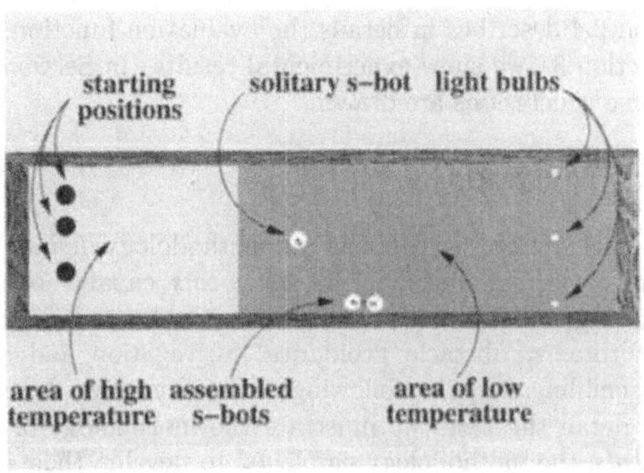

Figure 2: A graphical representation of the task. See text for details.

At the beginning of each trial, three *s-bots* are randomly positioned and oriented at one end of the corridor, in the area of high temperature. The light bulbs, located at the opposite end of the corridor, can be perceived by the *s-bots* from anywhere within the corridor. The intensity of the light which impinges upon the *s-bots* light sensors decreases quadratically with the distance from the light sources. Each *s-bot* s has an initial amount of energy $e_s = 1$, which must be higher than a certain threshold $\epsilon = 0.01$ for the *s-bot* to be able to move. The energy level of each *s-bot* can increase or decrease depending on:

1. The temperature of the area in which the *s-bot* is currently located. The temperature is 1 if the *s-bot* is in a high temperature area, 0 if it is in a low temperature area.
2. The state of the *s-bot*'s loudspeaker. An *s-bot* emits a tone to signal its position to other *s-bots*. This signaling behaviour can facilitate the aggregation of the group, which is a prerequisite for the assembling.
3. Whether the *s-bot* is assembled or disassembled.

More precisely, when *s-bot* s is assembled in a *swarm-bot* formation, it loses its energy in the area of high temperature and it increases its energy in the area of low temperature, as described by the following equation:

$$e_s(t+1) = e_s(t) + \tau \cdot ((1 - \Gamma(t)) - e_s(t)), \quad (1)$$

where $e_s(t)$ is the energy level of the sth *s-bot* at cycle t, $\tau = 0.2$ is a time constant governing the speed of the energy variation and $\Gamma(t)$ is the temperature sensed by the sth *s-bot* at cycle t in its current position. When an *s-bot* is not connected but it emits a sound signal, it loses energy in both areas. In the areas of low temperature its energy converges to a low but non-null value. This is described by the following equation:

$$e_s(t+1) = e_s(t) + \tau \cdot (k(1 - \Gamma(t)) - e_s(t)), \quad (2)$$

where $k = 0.1$ is a constant. In all the other situations, the *s-bot* increases its energy in areas of high temperature and loses it in areas of low temperature:

$$e_s(t+1) = e_s(t) + \tau \cdot (\Gamma(t) - e_s(t)). \quad (3)$$

The time constant τ guarantees that the *s-bots*' energy level varies smoothly according to the state of the system as described above. This smooth variation gives time to the control system of each *s-bot* to react to the new environmental situation in order to perform appropriate actions, before its energy level drops under the threshold $\epsilon = 0.01$.

The simulation is deliberately noisy, with noise added to all sensors (see Section 2.2 for more details). This is also extended to the environmental parameters: at the beginning of each trial, the point in which the temperature changes from high to low is redefined randomly within certain limits.

2.2 The simulation model

We developed a simulation software based on Vortex^{TM}, a 3D rigid body dynamics simulator that provides primitives for the implementation of detailed and realistic physics-based simulations (see also Mondada et al., 2004, for more details about the simulator). We have defined a simple *s-bot* model that at the same time allows fast simulations and preserves those features of the real *s-bot* that were important for the experiments (see Figure 3). The *s-bot* has a differential drive motion provided by a traction system composed of four wheels: two lateral, motorized wheels and two spherical, passive wheels placed in the front and in the back, which serve as support. The four wheels are fixed to the chassis, which also holds the cylindrical rotating turret. The turret can rotate around its axis, and it holds many sensory systems. Connections among *s-bots* can be made using a virtual gripper, which is modeled by dynamically creating a joint between two *s-bots* when needed. The position of the virtual gripper is represented by an arrow painted on the turret. Finally, the turret also holds a loudspeaker that can be controlled to produce a tone perceived by the other *s-bots*.

Figure 3: The simulation model of the *s-bot*.

Each *s-bot* is provided with a *traction sensor*, which detects the forces that are applied to the junction between the chassis and the rotating turret. This particular sensor proved to be important to evolve coordinated motion strategies in a *swarm-bot* and can also work as a bumper for individual and collective obstacle avoidance (Baldassarre et al., 2003; Trianni et al., 2004). Four variables encode the traction force information from four different preferential orientations with respect to the chassis (front, right, back and left, see Baldassarre et al., 2003, for more details). Each *s-bot* is also equipped with three directional microphones, used to detect the tone emitted by other *s-bots*, and with a temperature sensor. Finally, the ambient light is sensed using eight light sensors mounted on the rotating turret. We have defined two virtual light sensors positioned on the front and on the back of the chassis, whose readings are computed using the two closest light sensors of the turret. The information provided by the virtual light sensor does not vary significantly with the rotation of the turret with respect to the chassis, and it facilitates the evolution of phototactic behaviours. Noise is simulated for all sensors, adding a random value uniformly distributed within the 5% of the sensors saturation value.

Concerning the actuators, *s-bots* can control the two wheels, independently setting their speed in the range $[-6.5, 6.5]$ rad/s. The loudspeaker can be switched on, simulating the emission of a continuous tone, or it can be turned off. The virtual gripper can be closed or open, in order to connect to another *s-bot*. However, we force the gripper to stay open if a connection fails. Finally, the motor controlling the rotation of the turret is used, even though it is not directly controlled by the evolved neural network. When *s-bots* are not connected, this motor ensures the alignment between the turret and the chassis. On the contrary, when an *s-bot* is connected to other *s-bots* to form a *swarm-bot*, the turret can rotate freely.

2.3 The controller and the genetic algorithm

Homogeneous groups of *s-bots* are controlled by artificial neural networks, whose parameters are set by an evolutionary algorithm. A single genotype is used to create a group of individuals with an identical control structure. Each *s-bot* is controlled by a fully connected, 14 neuron continuous time recurrent neural network (CTRNN, see also Beer, 1995, for details). We chose to employ a 14 neurons network because it corresponds to the smallest network which has at least one neuron associated with each input signal and one neuron for each output signal (see Figure 4). Each neuron, indexed by i, is governed by the following state equation:

$$\frac{dy_i}{dt} = \frac{1}{\tau_i}\left(-y_i + \sum_{j=1}^{14} \omega_{ji} z_j + gI_i\right) \qquad (4)$$

$$\text{where} \qquad z_j = \frac{1}{1 + e^{-(y_j+\beta_j)}}$$

Here, by analogy with real neurons, y_i is the cell potential, τ_i the decay constant, β_j the bias term, z_j the firing rate, ω_{ji} is the strength of synaptic connections from the jth neuron to the ith neuron, I_i the intensity of the sensory perturbation on sensory neuron i. The neurons either receive direct sensor input or are used to set the state of an *s-bot*'s actuators (i.e., the wheels speed, the state of the gripper and the loudspeaker). There are no internal neurons. All but four of the neurons receive direct input from the robot sensors (for the remaining four, $gI_i = 0$). Each input neuron is associated with a single sensor, receiving a real value in the range $[0.0, 1.0]$, which is a simple linear scaling of the reading taken from its associated sensor.[3]

The four remaining neurons are used to control the *s-bot*'s actuators, after mapping their cell potential y_i

[3] Specifically, neurons N_1 to N_4 take input from the 4 variables encoding the traction force, neurons N_5 to N_7 take input from the sound sensors (i.e., the directional microphones), N_8 and N_9 from the virtual light sensors, and N_{10} from the temperature sensor.

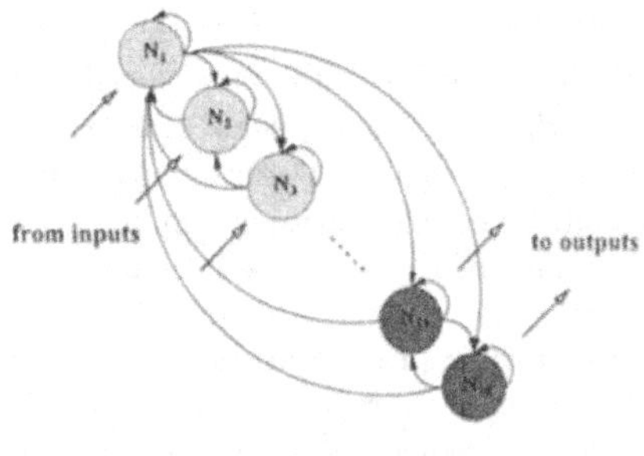

Figure 4: A graphical representation of the artificial neural network employed to control the *s-bots*. The nodes in gray represent those which receive input from the *s-bots* sensors. The nodes in black represent those whose activation values are used to set the *s-bots* actuators.

onto the range [0.0, 1.0] by a sigmoid function. Two of them are used to set the *s-bot*'s wheels speed, linearly scaling the output into [−6.5, 6.5]. The third motor neuron is used to set the state of the loudspeaker, which is turned on if the neuron output is higher than 0.5, and off otherwise. The last motor neuron controls the gripper actuator, trying to set up a connection if the neuron output is higher than 0.5, and keeping the gripper open otherwise.

The strengths of the synaptic connections, the decay constants, bias terms and the gain factor are all genetically encoded parameters. Cell potentials are set to 0 each time a network is initialised or reset. State equations are integrated using the forward Euler method with an integration step-size of 0.1.

In order to set the parameters of the *s-bots*' controllers, a simple generational genetic algorithm (GA) is employed (Goldberg, 1989). Initially, a random population of 100 genotypes is generated. Each genotype is a vector of 1800 binary values—8 bits for each of the 225 parameters, that is 196 connections, 14 decay constants, 14 bias terms, and 1 gain factor. Subsequent generations were produced by a combination of selection with elitism and mutation. Recombination is not used. At every generation, the best 20 genotypes are selected for reproduction, and each generates 4 offspring. The genotype of the selected parents is copied in the subsequent generation; the genotype of their 4 offspring is mutated with a 5% probability of flipping each bit. One evolutionary run lasts 1000 generations.

The binary values of a genotype were mapped to produce CTRNN parameters with the following ranges:

- connection weights: $\omega_{ji} \in [-6, 6]$.
- biases: $\beta_j \in [-2, 2]$.
- gain factor: $g \in [1, 13]$.

Concerning the decay constants, the genetically encoded parameters were firstly mapped onto the range [−1, 1] and then exponentially mapped onto $\tau_i \in [10^{-1}, 10]$.

2.4 The evaluation function

During the evolution, a genotype is mapped into a control structure that is cloned and downloaded in all the *s-bots* taking part to the experiment (i.e., homogeneous group of *s-bots*). Groups of $n = 3$ *s-bots* are evaluated 5 times—i.e., 5 trials. Each trial differs from the others in the initialisation of the random number generator, which influences mainly the *s-bots* starting positions and the point beyond which the temperature drops from 1 to 0. In each trial θ, the lifetime of an *s-bot* is limited to 600 simulation cycles, corresponding to 60 seconds of real time. The behaviour of the *s-bots* is evaluated according to an evaluation function F_θ that takes into account the individual contribution of each *s-bot* s:

$$F_\theta = \frac{1}{n^3} \cdot \left(\sum_{s=1}^{n} d_s \cdot \sum_{s=1}^{n} e_s \cdot c \right), \quad (5)$$

where the factors d_s, e_s and c are explained below.

- d_s rewards *s-bots* that perform phototaxis; this fitness component is computed as follows:

$$d_s = \begin{cases} 0.1 \cdot \dfrac{x_{f,s} - x_{i,s}}{x_\Gamma - x_{i,s}} & \text{if } x_{f,s} \le x_\Gamma \\ 0.1 + 0.9 \cdot \dfrac{x_{f,s} - x_\Gamma}{x_M - x_\Gamma} & \text{if } x_\Gamma < x_{f,s} \le x_M \\ 1 & \text{otherwise} \end{cases} \quad (6)$$

  where $x_{i,s}$ and $x_{f,s}$ are respectively the initial and final x coordinate of the sth *s-bot* position, x_Γ is the x coordinate in which the temperature drops from 1 to 0, and x_M is the x coordinate of the light bulbs position.[4]

- e_s is the final energy possessed by the sth *s-bot*, at cycle $t = 600$. The variation of the energy $e_s(t)$ of the sth *s-bot* at cycle t is regulated by (1), (2) and (3) as discussed in Section 2.1.

  This fitness component rewards *s-bots* that end their lifetime with a high amount of energy. For example, if we compare groups of *s-bots* that managed to reach the end of the corridor close to the light bulbs, those which proved to be capable of assembling early in response to the decrease in the environmental temperature will get a higher fitness score than those which did not perform such collective response.

[4] The coordinate system used has the x and y axes parallel respectively to the long and short wall of the corridor. The origin of the axes is positioned at the bottom left corner of the corridor.

- c is the maximum size of a *swarm-bot* observed at the end of the trial, ranging from 1 (no connections among *s-bots*) to n (all *s-bots* connected in a single *swarm-bot*). This fitness component rewards *s-bots* that reach the end of the corridor assembled in a *swarm-bot* formation. Recall that, due to the characteristics of the environment—an initial area of high temperature is followed by an area of low temperature at the end of which the light bulbs are located—successful *s-bots* should terminate the trial in *swarm-bot* formation close to the opposite end of the corridor with respect to their starting position.

The average performance of the group F is computed averaging the evaluations F_θ performed in each trial θ. This value corresponds to the fitness of the genotype: it is used to select which genotypes will reproduce in the current generation, but is not in any sense a reinforcement directly available to the *s-bots*.

3. Results

Ten evolutionary runs, each using different random seeds, were run for 1000 generations each. Two runs out of ten ended up successfully by producing controllers capable of displaying functional self-assembling. Figure 5 shows the fitness of the best group of *s-bots* and the average fitness of the population for each generation. Two prototypic runs are shown: a particularly successful one (top) and an unsuccessful one (bottom). Note that, the fitness curve does not increase monotonically because each genotype is evaluated every generation. Therefore, even in the absence of mutations, genotypes with relatively high fitness—e.g., those corresponding to the elite— might not obtain the same fitness in the following generation.

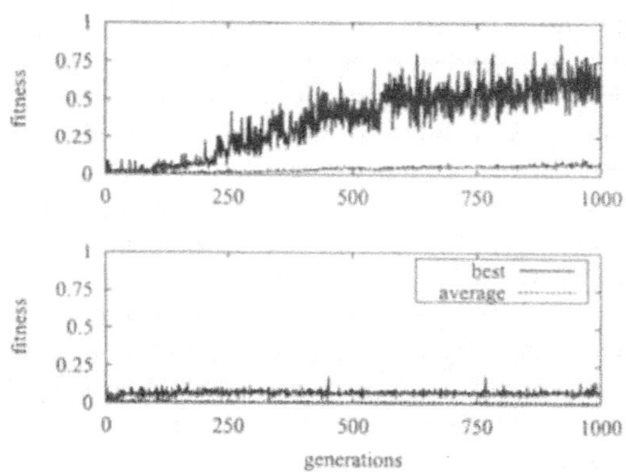

Figure 5: The graphs show the fitness of the best group of *s-bots* (thick line) and the normalized average fitness of the population (thin line), for each generation, for a successful run (top graph) and an unsuccessful one (bottom graph).

We examined the behaviour of the best evolved group

Table 1: Average fitness and standard deviation of the best evolved controller of the last generation of each run.

seed	1	2	3	4	5
avg.	0.505	0.572	0.052	0.031	0.028
std.	0.218	0.289	0.028	0.003	0.002
seed	6	7	8	9	10
avg.	0.006	0.030	0.292	0.031	0.058
std.	0.007	0.002	0.325	0.002	0.029

of *s-bots* of the last generation for each run, in order to establish whether it evolved functional self-assembling. The fitness of each of these groups was re-evaluated 250 times by using the equations illustrated in Section 2.4. The corresponding results are shown in Table 1. The first two runs are the most successful ones, showing an average fitness higher than all the others, which did not result in a satisfying performance.

An analysis of the controllers produced by the unsuccessful runs revealed that these groups of *s-bots* were able to solve the task only in part. We observed that, while in these runs the *s-bots* were capable of phototaxis and obstacle avoidance, only in few runs they were able to properly react to the decrease in temperature, and only in the run n. 8 the *s-bots* were occasionally capable of functional self-assembling. On the contrary, in the two successful runs, the groups of *s-bots* showed the complete repertoire of behaviours required by the task.

3.1 Behavioural analysis: self-assembling

In this section, we illustrate the results of post-evaluation tests performed on one of the successful evolutionary run. Similar results can be observed in the post-evaluation tests of the other successful run, with a small difference as far as it concerns the time required for the formation of the self-assembled structure (i.e., the *swarm-bot*).

Figure 6 shows how the covered distance[5] and the energy level of each *s-bot* vary over time. Looking at these graphs, it is possible to distinguish four different phases: individual phototaxis, aggregation, self-assembling and collective phototaxis. At the beginning of the trial—from cycle 0 to the time indicated by the empty circle—the three *s-bots*, located in the high temperature area and with full energy, perform individual phototaxis, as shown by the continuous line.

The second phase starts when the *s-bots* enter the low temperature area. Three phenomena can be observed in Figure 6: aggregation, decrease in the energy level and signaling behaviour. Aggregation is indicated by the covered distances of the three *s-bots* (see continuous lines), which reach similar values before the end of the

[5]The covered distance refers to the distance between the current position of the *s-bot* and the starting position, along the x axis.

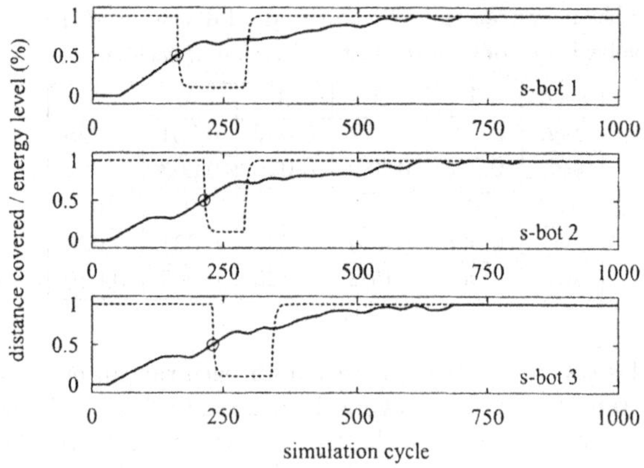

Figure 6: The graphs refer to a post-evaluation of the best evolved group of three *s-bots*. In particular, each graph shows how the covered distance along the corridor (continuous line) and the energy level (dashed line) of an *s-bot* vary during a post-evaluation which lasts 1000 simulation cycle. The empty circles indicate the time when an *s-bot* enters the low temperature area.

phase. The decrease in the energy level, according to equations (2) and (3), indicates that the *s-bots* move independently. Since the energy level converges, for each *s-bot*, to the value $e_s = k$ (see equation (2)), we can deduce that the *s-bots* react to the temperature decrease by switching on their loudspeaker, signaling their position to the other *s-bots*. This should in principle facilitate the aggregation. However, we observed that the *s-bots* tend also to exploit other "affordances", such as the walls of the corridor, in order to get close to each other (Gibson, 1977). A complete understanding of this process would require further analysis, that will be carried out in the future.

The third phase corresponds to self-assembling. In Figure 6, this phase is indicated by an increase in the energy level (dashed line), caused by the *s-bots* connecting to each other when located in the low temperature area (see equation (1)). In this particular case, *s-bots* 1 and 2 self-assemble first, while *s-bot* 3 joins the *swarm-bot* later. Collective phototaxis is performed during the last phase. Here, *s-bots* move assembled in a *swarm-bot* that approaches the light bulbs, as indicated in Figure 6 by the synchronous increase of the covered distance (see continuous lines).

3.2 Disassembling of the swarm-bot

In an additional series of post-evaluation tests, we looked at the capability of the *swarm-bot* to disassemble as a reaction to an increase of the environmental temperature. Recall that this circumstance has never been encountered by the *s-bots* during the evolutionary phase.

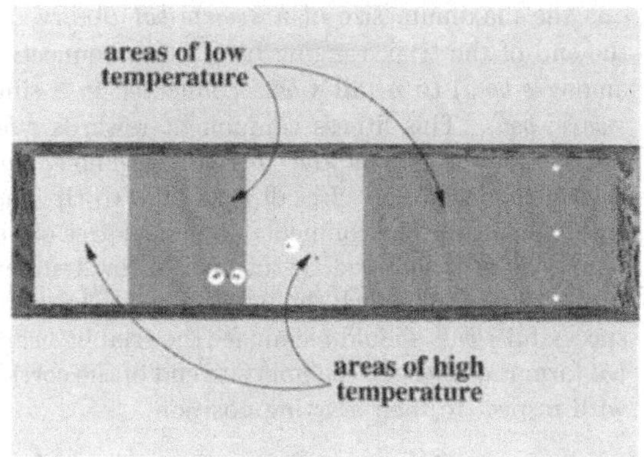

Figure 7: A graphical representation of the environment with two high temperature and two low temperature areas. This environment has been used for post-evaluation to check whether the *s-bots* capable of assembling were also capable of disassembling in response to an increase in the environmental temperature.

Therefore, disassembling should be considered an additional capability of the evolved controllers, which confers robustness to the system. We placed the *s-bots* in a corridor with four temperature areas: two high temperature and two low temperature areas (see Figure 7).

The graphs in Figure 8 show how the covered distance and the energy level of each *s-bot* vary in time while the *s-bots* move down the corridor toward the light bulbs. In this case, we focus our attention on how the *s-bots* react to the transition from low to high temperature areas. In fact, the transitions from high to low temperature areas result in a variation of the covered distance and of the energy levels similar to what was observed and discussed for Figure 6.

The transition of the *s-bots* from low to high temperature areas is indicated in the graphs by a filled circle. Roughly speaking, this transition is characterized by two different phases. Initially, a decrease in the energy level is observed, according to equation (1), when an *s-bot*, still assembled in a *swarm-bot* formation, perceives the new environmental condition (high temperature). Subsequently, the *s-bots* progressively disconnect from each other, which results in a gain in the energy level. In the particular case illustrated in Figure 8, *s-bot* 1 is the first to perceive the high temperature area and consequently to disassemble from the *swarm-bot*. It is possible to notice that *s-bot* 1, after disconnecting, moves back and forth, experiencing twice the low-to-high temperature transition. Similarly, *s-bot* 2 disconnects from *s-bot* 3 as soon as it ends up in the high temperature area. Consequently, *s-bot* 3 finds itself alone in the area of low temperature. It is possible to notice that its energy level drops, according to equation (2), due to the fact that the *s-bot* has the loudspeaker turned on. Nevertheless, the *s-bot* still has enough energy to perform individual phototaxis and to approach the high temperature area. Once in the high temperature area, its energy level increases again, indicating that the *s-bot* has switched off the loudspeaker. Its covered distance indicates that the *s-bot* approaches faster the light bulbs, reaching and finally connecting to the other 2 *s-bots*.

4. Conclusions

In this paper, we have described a set of experiments aimed at the design of controllers for a homogeneous group of robots—the *s-bots*. Our goal was the evolution of a single neural network capable of integrating a repertoire of individual and collective behaviours. In particular, we focused on functional self-assembling, that is, on the ability of the *s-bots* to react to particular environmental contingencies forming a physically connected structure—the *swarm-bot*.

Our results show that the evolutionary robotic methodology is promising: the evolved controllers are capable of displaying individual and collective obstacle avoidance, individual and collective phototaxis, aggregation and self-assembling. To the best of our knowledge, our experiments represent one of the first works in which (i) functional self-assembling in a homogeneous group of robots has been achieved and (ii) evolved neural controllers successfully cope with such a complex scenario, producing different individual and collective responses based on the appropriate control of the state of various actuators triggered by the local information coming from various sensors.

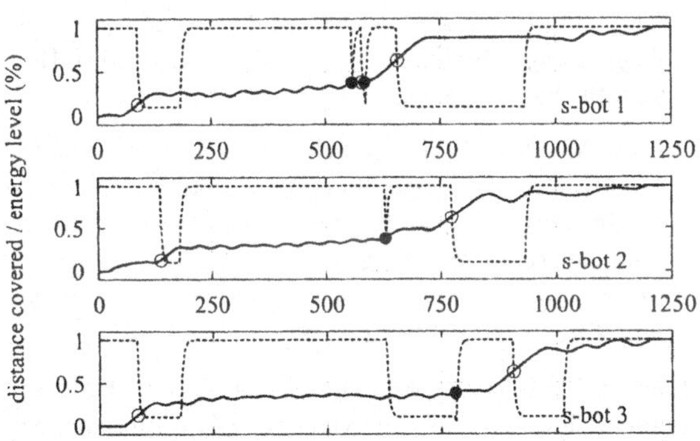

Figure 8: Each graph shows how the distance to the light bulbs (continuous line) and the energy level (dashed line) of an *s-bot* vary during a post-evaluation which lasts 1250 simulation cycle. The empty circles indicate the time when an *s-bot* enters a low temperature area. The filled circles indicate the time when an *s-bot* enters a high temperature area.

We approached the study of functional self-assembling by identifying possible scenarios in which the *s-bots* could require the formation of a *swarm-bot*. In Section 1., we provided some examples of these scenarios, such as passing over a trough or navigation on a rough terrain. Instead of directly facing these challenges, we have chosen to simplify the task, in order to focus our attention on the basic mechanisms that underpin functional self-assembling. The task studied in this paper is the result of a careful analysis of the common features observed in those scenarios that require self-assembling. If compared to tasks that require *s-bots* to self-assemble in order to pass over a trough or to navigate on rough terrain, our scenario looks simpler. However, we believe that this is a important initial step to be taken before proceeding further.

The obtained results are satisfying, though only two evolutionary runs out of ten were successful. We believe that, on the one hand, it is important to improve our methodology in order to consistently obtain successful behaviours. On the other hand, these results are very promising, as they suggest that solutions to complex tasks can be obtained using the evolutionary robotics methodology.

In the future, we aim at the improvement of the results discussed in this paper. First of all, we would like to consistently evolve successful controllers, by improving the evaluation function and by exploring the parameter space of the neural controllers and of the genetic algorithm we used. Furthermore, we would like to synthesize more efficient behaviours, in particular concerning the use of the sound signaling system for aggregation and self-assembling.

Another interesting issue is given by the disassembling capabilities of the *swarm-bot*. Disassembling can be improved in many different ways. In particular, we would like the *s-bots* to be capable of using the sound signalling system both to favour aggregation and self-assembling (switching on the loudspeaker) and to trigger the disassembling of connected *s-bots* (turning it off). In this way, disassembling would not be exclusively under the control of the *s-bot* that makes a connection, but it could also be influenced by the *s-bot* that "receives" a connection.

Finally, we intend to study functional self-assembling triggered by a temporal integration of the information available to an *s-bot*. That is, self-assembling should arise when no other possible solution has been found while the *s-bots* are experiencing the environment. This implies that an *s-bot* should be capable of integrating some information obtained through its sensors during time, before trying to self-assemble with other *s-bots*. For example, when trying to reach a location that is barred by the presence of a trough, an *s-bot* should first try to search the surroundings for a feasible passage. When such a passage does not exist, the iterated experience of the presence of the trough should trigger the self-assembling phase.

Acknowledgments

This work was supported by the "SWARM-BOTS" project, funded by the Future and Emerging Technologies programme (IST-FET) of the European Commission, under grant IST-2000-31010. Marco Dorigo acknowledges support from the Belgian FNRS, of which he is a Senior Research Associate, through the grant "Virtual Swarm-bots", contract no. 9.4515.03, and from the "ANTS" project, an "Action de Recherche Concertée" funded by Scientific Research Directorate of the French Community of Belgium. The information provided is the sole responsibility of the authors and does not reflect the Community's opinion. The Community is not responsible for any use that might be made of data appearing in this publication.

References

Anderson, C., Theraulaz, G., and Deneubourg, J.-L. (2002). Self-assemblage in insects societies. *Insectes Sociaux*, 49:99–110.

Baldassarre, G., Nolfi, S., and Parisi, D. (2003). Evolution of collective behavior in a team of physically linked robots. In Gunther, R., Guillot, A., and Meyer, J.-A., (Eds.), *Proceedings of the Second European Workshop on Evolutionary Robotics*, pages 581–592. Springer Verlag, Berlin, Germany.

Beer, R. D. (1995). A dynamical systems perspective on agent-environment interaction. *Artificial Intelligence*, 72:173–215.

Bonabeau, E., Dorigo, M., and Theraulaz, G. (1999). *Swarm Intelligence: From Natural to Artificial Systems.* Oxford University Press, New York, NY.

Cao, Y. U., Fukunaga, A. S., and Kahng, A. B. (1997). Cooperative mobile robotics: Antecedents and directions. *Autonomous Robots*, 4:1–23.

Castano, A., Shen, W., and Will, P. (2000). CONRO: Towards deployable robots with inter-robot metamorphic capabilities. *Autonomous Robots*, 8:309–324.

Gibson, J. J. (1977). The theory of affordances. In Shaw, R. and Bransford, J., (Eds.), *Perceiving, Acting and Knowing. Toward an Ecological Psychology*, chapter 3, pages 67–82. Lawrencw Erlbaum Associates, Hillsdale, NJ.

Goldberg, D. E. (1989). *Genetic Algorithms in Search, Optimization and Machine Learning.* Addison-Wesley, Reading, MA.

Groß, R. and Dorigo, M. (2004). Evolving a cooperative transport behavior for two simple robots. In Liardet, P., Collet, P., Fonlupt, C., Lutton, E., and Schoenauer, M., (Eds.), *Artificial Evolution - 6th International Conference, Evolution Artificielle, EA 2003, Marseille, France, October 2003*, volume 2936 of *Lecture Notes in Computer Science*, pages 305–317. Springer Verlag, Berlin, Germany.

Harvey, I., Husbands, P., and Cliff, D. (1992). Issues in evolutionary robotics. In Meyer, J.-A., Roitblat, H., and Wilson, S., (Eds.), *From Animals to Animats II: Proceedings of the 2nd International Conference on Simulation of Adaptive Behavior*, pages 364–373. MIT Press/Bradford Books, Cambridge, MA.

Lioni, A., Sauwens, C., Theraulaz, G., and Deneubourg, J.-L. (2001). Chain formation in *Œcophilla longinoda. Journal of Insect Behaviour*, 15:679–696.

Liu, J. and Wu, J. (2001). *Multiagent Robotic Systems*, volume 21 of *International Series on Computational Intelligence.* CRC Press, Boca Raton, FL.

Meyer, J.-A., Husbands, P., and Harvey, I. (1998). Evolutionary robotics: A survey of applications and problems. In Husbands, P. and Meyer, J.-A., (Eds.), *Proceedings of the First European Workshop on Evolutionary Robotics.* Springer Verlag, Berlin, Germany.

Mondada, F., Pettinaro, G. C., Guignard, A., Kwee, I. V., Floreano, D., Deneubourg, J.-L., Nolfi, S., Gambardella, L. M., and Dorigo, M. (2004). SWARM-BOT: A new distributed robotic concept. *Autonomous Robots, Special Issue on Swarm Robotics.* To appear.

Nolfi, S. and Floreano, D. (2000). *Evolutionary Robotics: The Biology, Intelligence, and Technology of Self-Organizing Machines.* MIT Press/Bradford Books, Cambridge, MA.

Quinn, M., Smith, L., Mayley, G., and Husbands, P. (2003). Evolving controllers for a homogeneous system of physical robots: Structured cooperation with minimal sensors. *Philosophical Transactions of the Royal Society of London, Series A: Mathematical, Physical and Engineering Sciences*, 361:2321–2344.

Trianni, V., Nolfi, S., and Dorigo, M. (2004). Hole avoidance: Experiments in coordinated motion on rough terrain. In *Proceedings of the 8th International Conference on Intelligent Autonomous Systems (IAS-8)*, Amsterdam, The Netherlands. To appear.

Yim, M., Duff, D., and Roufas, K. (2000). PolyBot: A modular reconfigurable robot. In *Proceedings of the 2000 IEEE/RAS International Conference on Robotics and Automation*, volume 1, pages 514–520. IEEE Robotics and Automation Society, Piscataway, NJ.

How can a life-like agent evoke its emotions for users? — A method for inducing users' natural behaviors to establish "mutual adaptation"

Takanori Komatsu
Department of Media Architecture, Future University-Hakodate
116-2 Kamedanakano
Hakodate 041-8655, JAPAN
komatsu@fun.ac.jp

Abstract

In terms of creating a smooth communication between a human and an interactive agent, some researchers have argued that they must form a mutual adaptation process; that is, both should learn to respond and adapt appropriately to each other's behaviors. Moreover, to create a mutual adaptation process between them, an agent is required to induce its partner's adaptations or natural behaviors. Some studies have already reported that giving a certain physical body to such an agent is reasonable; however, doing so for all kinds of interactive agents is unrealistic. To achieve an interactive agent without a physical body, I assumed that this agent should express certain information that can evoke its specific emotions for a user, because it is expected that we humans will empathize with certain entity which seems to have some emotions. I then carried out experiments to clarify which kinds of information the agent expressed (its synthesized sounds and its unilateral actions) could evoke the specific emotions for humans. The results of these experiments could be explained by the following hypothesis: when the temporal transitions in expressed information regardless of any modalities conform to the law of conservation of energy, this information evokes an information sender's (an agent) positive emotions for an information receiver (a human); On the other hand, when the specific temporal transitions deviate from the law of conservation of energy, this information evokes a sender's negative emotions. In near future, this hypothesis would contribute to a basic methodology to create a mutual adaptation for various applications that are required to communicate with humans, even for these applications without having a physical body.

1. Introduction

Recently, various interactive agents have been applied to intelligent interface systems in our ordinary life, such as those for home information appliances and wireless communication terminals (Addlesee *et al.*, 2001; Brumitt *et al.*, 2000; Kidd *et al.*, 1999; Weiser, 2001). Obviously, we have to create a smooth means of communicating with these agents to obtain the maximum benefits of their abilities and capacities. Some researchers have argued that a human and an interactive agent must engage in a **mutual adaptation process** to establish a smooth communication between them (Komatsu *et al.*, 2002; Komatsu *et al.*, 2003; Yamada and Yamaguchi, 2002). This mutual adaptation process consists of repeating the following actions: a user adapts to an agent, next the agent adapts to the user by inducing and then using this user's induced adaptation, i.e., both learn to respond and adapt appropriately to each other's behaviors by inducing and then using this induced partner's adaptation. It can be said that such a process is a kind of adaptive behaviors and that this would commonly be observed in pairs who can communicate smoothly, e.g., parents and their children (Snow, 1997; Fernald and McRoberts, 1996), dogs and their owners (Katcher, 1983). Therefore, this mutual adaptation process should also be achieved in a relationship between a user and a desired interactive agent. Here, the one of the important requirements for creating a mutual adaptation is that both (a human and an agent) must be able to induce a partner's adaptations or natural behaviors. For instance, imagine that an agent with only a learning function is interacting with a human. In this case, I assumed that they cannot create a mutual adaptation, because this agent is just unilaterally receiving information from a human. Therefore, an interactive agent must be equipped both with a function of learning about a partner and one for inducing the partner's adaptations or natural behaviors.

Up to now, several HCI (human-computer interaction)

and HRI (human-robot interaction) studies have focused on how to induce a human's adaptations or natural behaviors (Imai *et al.*, 2002; Ishiguro *et al.*, 2003; Ogata *et al.*, 2003). For example, Ono *et al.*,(2003) argued the "body entrainment theory" that holds that an interactive agent should have a physical body, and its body configuration must be similar to humans. This is because this study found that when two humans are communicating smoothly, they usually exhibit gestures that demonstrate that both are entrained to each other's gestures. Otherwise, if one of the pair is prohibited to express any gestures, this prohibition eventually disrupted their smooth communication. Some studies in social psychology also reported similar results with the above HCI and HRI studies; Harada (2002) reported that we humans generally hesitate to speak naturally to an agent that does not have an actual physical body, such as a life-like agent in a computer display; while we can speak rather naturally to an agent that does have an actual physical body, e.g., NEC PaPeRo robot (NEC Personal Robot Center, 2001) that Harada actually used in her experiment. Therefore, one can say that giving some physical body to an agent that must communicate smoothly with humans is reasonable. However, giving an actual physical body to every interactive agent, for example, a speech interface for a car navigation system or a small communication terminal such as a PDA, is unrealistic. Therefore, considering how an artificial agent without a physical body induces a user's adaptations or natural behaviors is worthwhile.

To realize such an agent, I assumed that this agent express certain information that evokes the agent's specific emotions for a human, even though this does not have a physical body and actual emotions. Because it is expected that we humans will empathize with the certain entity which seems to have some emotions. Specifically, I conducted two experiments to observe and analyze which kinds of information could evoke an agent's specific emotions for a human. In the first experiment, I observed how humans estimate the agent's emotions from its artificial synthesized sounds. In the second experiment, I observed how humans estimate an agent's emotions from its unilateral actions appearing on a computer display. Then, if I could find the specific information that can evoke an agent's specific emotions for a human, I believe that this information will be able to induce users' adaptive behaviors and this would eventually contribute to the basic methodology to create a mutual adaptation for various application that are required to communicate with humans, even for these applications without having a physical body.

Moreover, if the specific information that can evoke the agent's specific emotions obtained in two experiments had common features, a universal mechanism that humans could use to detect an agent's emotion regardless of the modalities of the expressed information would be found. In that case, it can be said that this finding has an impact on the various emotion detecting studies, not only in HCI and HRI but also in cognitive science or social psychology.

2. Experiment 1: Can we estimate an agent's emotions from its artificial synthesized sounds?

2.1 Purpose and Settings

Nagasaki and I performed an experiment to observe how humans estimate an agent's (speaker's) emotions from its artificial synthesized sounds when the agent's sounds were presented to humans [1]. Specifically, we focused on two kinds of prosodic features in the agent's sounds: F0 contour and duration. In this experiment, we prepared the 44 different types of artificial synthesized sounds that had 11 different F0 contours and four different durations. Specifically, the four durations were 189 ms, 418 ms, 639 ms, and 819 ms. And 11 F0 contours had the following design (see Figure 1):

1. The transition range of F0 values between the onset and endpoint in the sound stimuli were 0 Hz, 25 Hz, 50 Hz, 75 Hz, 100 Hz, or 125 Hz.
2. The F0 contours were downward or upward.
3. The F0 contour transition rate was consistent and linear.

Participants were 23 Japanese (17 men and 6 women; 20 - 42 years old). Hearing tests showed that no participants had any hearing problems. As the experimental procedure, an experimenter gave the participants the following experimental instructions.

1. The synthesized sounds were presented to participants as sound stimuli, and they were asked to assume that these sound stimuli were answering speeches in a telephone conversation (actually this was not true).
2. After each sound stimulus was presented, they were asked to estimate the emotional state of this speaker by choosing one of the following three emotional states:
    - Disagreement, e.g., surprising, doubting.
    - Hesitation, e.g., being lost for words.
    - Agreement, e.g., acceptance, back-channel feedback.

[1] This experiment was conducted as the joint study with Ms. Yasuko Nagasaki. Actually she introduced the acquired results in this experiment for the other speech processing study. For a detailed description of this experiment, see Komatsu and Nagasaki (2004), Nagasaki and Komatsu (2004a, 2004b).

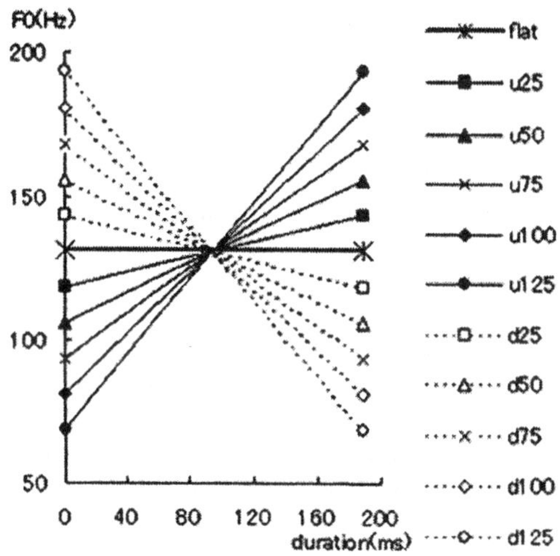

Figure 1: Eleven different F0 contours (duration: 189 ms). For example, "u25" indicates that the F0 transition range was 25 Hz with upward slope (increasing intonation).

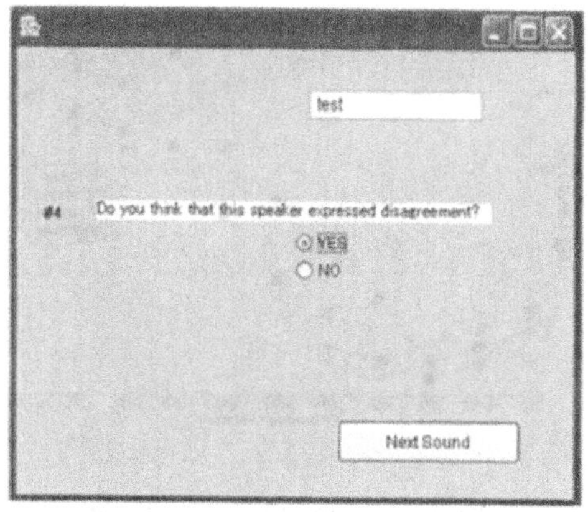

Figure 2: View of software developed for this experiment.

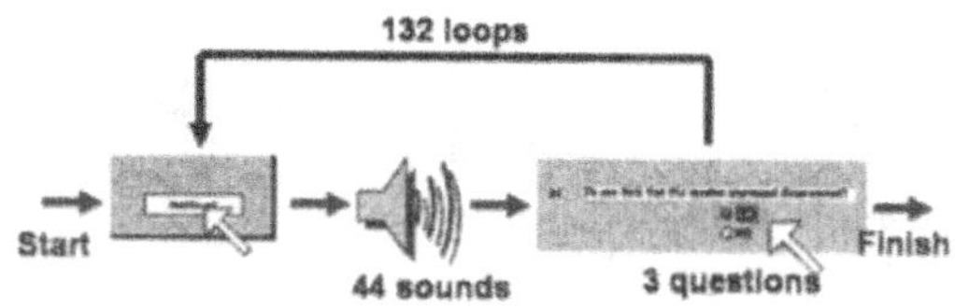

Figure 3: Specific procedure for Experiment 1.

Actually, these three emotional states correspond to valence evaluations such as "good or bad"; the emotional state "disagreement" corresponds to the negative valence value ("bad"), "agreement" to the positive value ("good"), and hesitation to a zero value. Reeves and Nass (1996) reported that *every response that people have, regardless of their talents and training in expression, begins with an extremely simple evaluation of whether people, places or objects are good or bad, likeable or not likeable, worthy of approach or something to avoid, and that the psychology literature about this valence evaluation "good or bad" is huge and diverse, encompassing hundreds of studies with similar results.* They also mentioned that *our old brains are hard-wired to distinguish "good" versus "bad" as naturally and efficiently as possible.* Therefore, we can say that the three emotional states that we focused on in this study were equivalent with the "primitive emotions of a speaker (an agent)."

As an experimental tool, we developed software that can both present sound stimuli and record participants' answers (Figure 2). In the actual experiment among 44 synthesized sounds, a randomly selected sound stimulus was presented to participants. Afterward, one of three emotional states was randomly selected and presented to them in a question on the window's label, e.g., "Do you think that this speaker expressed DISAGREEMENT?" Specifically, 132 trials (44 synthesized sounds x 3 emotional states) were conducted (Figure 3).

2.2 Results

We analyzed which kinds of artificial synthesized sounds evoked specific emotional states for the participants; Figure 4 shows which sounds evoked the emotional state "disagreement," Figure 5 the sounds rated as "hesitation," and Figure 6 the ones for "agreement." In these figures, the horizontal axis lists the types of F0 contours and the labels on this axis, such as "u25" or "d100," are the same ones used to distinguish the contours in Figure 1. The vertical axis shows the interpretation rates that give the proportion of participants answering "yes" in the corresponding trials.

2.2.1 Case 1: Disagreement

At first, we analyzed the interpretation rates in terms of durations of sound stimuli. The disagreement interpretation rate, for the stimuli of 189-ms durations (11 sounds out of 44 ones) was 52.6%, for 418 ms (11 sounds) 48.2%, 639 ms 52.6%, and 819 ms 50.6%. So we found no significant differences in interpretation rates in terms of duration ($F(3,66) = .70$, n.s.).

Next, we analyzed the interpretation rates in terms of upward and downward F0 contours. The interpretation rate of upward slopes (20 sounds out of 44 ones) was 89.1%, and for downward (20 sounds) 15.4%. This result clearly indicated that the upward slopes evoked "disagreement" ($F(1,22) = 219.66$, $p<.01$).

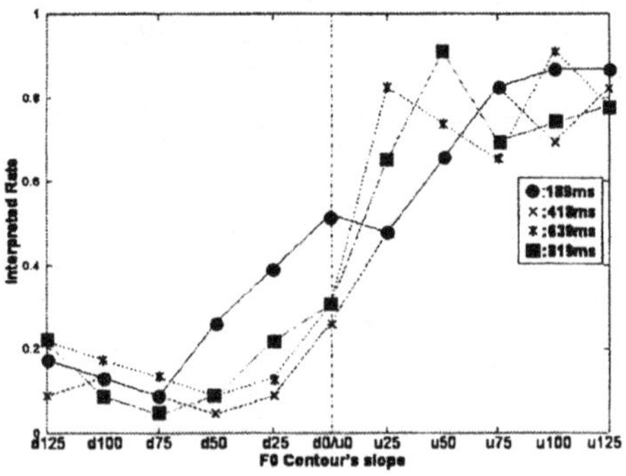

Figure 4: Relationship between F0 contours and rate of "disagreement" interpretation for each duration.

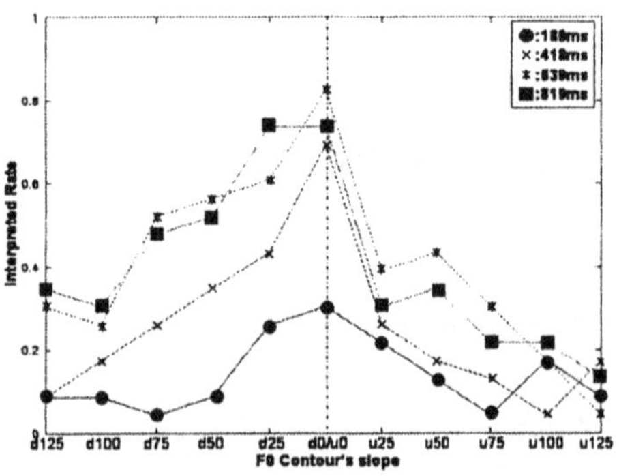

Figure 5: Relationship between F0 contours and rate of "hesitation" interpretation for each duration.

Then, we analyzed the effects of transition ranges of F0 contour. For upward slopes, 25 Hz (u25, 4 sounds) had a 73.9% disagreement interpretation rate, 50 Hz (u50, 4 sounds) was rated at 90.2%, 75 Hz (u75) 93.5%, 100 Hz (u100) 92.3%, and 125 Hz (u125) 95.6%. This result showed that significant differences in these five types of upward slopes ($F(4,88) = 9.8$, $p<.01$), and the Newman-Keuls test revealed significant differences between u25 and the range u50 - 125. Therefore, we concluded that most upward slopes evoked "disagreement." For downward slopes, 25 Hz (d25, 4 sounds) was interpreted as disagreement at an interpretation rate of 20.7%, for 50 Hz (d50, 4 sounds) it was 8.7%, for 75 Hz (d75) 11.9%, 100 Hz (d100) 16.3%, and 125 Hz (d125) 19.6%. No significant differences were found for these stimuli ($F(4,88) = 1.68$, n.s.).

In sum, this result indicated that **sound stimuli with upward slopes (increasing intonation) regardless of its duration evoked the emotional state "disagreement."** Conversely, the sound stimuli with downward slopes (decreasing intonation) did not evoke "disagreement."

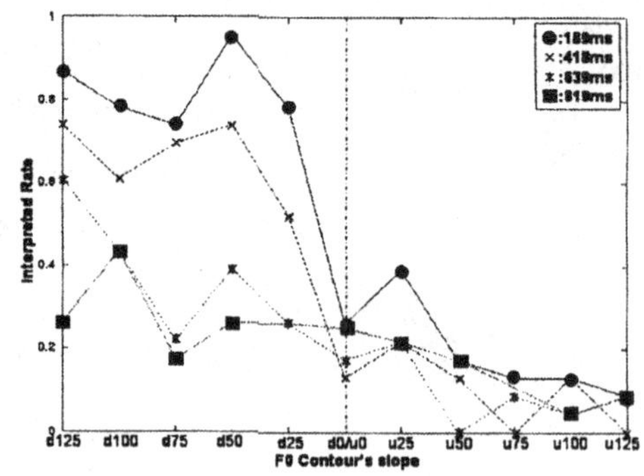

Figure 6: Relationship between F0 contours and rate of "agreement" interpretation for each duration.

2.2.2 *Case 2: Hesitation*

We analyzed the interpretation rates in terms of durations of sound stimuli. The stimuli of 189-ms durations (11 sounds) had hesitation interpretation rate 14.6%, for 418 ms it was 29.2%, for 639 ms 45.1%, and 819 ms 44.7%. This result demonstrated significant differences for these four durations ($F(3,66) = 18.1$, $p<.01$), and the Newman-Keuls test showed that there were significant differences between (1) 189 ms and the range 418 - 819 ms, and (2) 418 ms and the range 639 - 819 ms. This demonstrates that the longer durations had higher "hesitation" interpretation rates.

Next, we analyzed the hesitation interpretation rates in terms of upward and downward F0 contours. The interpretation rate of hesitation for upward slopes (20 sounds) was 22.8%, while that for downward slopes was 37.0%. This result indicated significant differences between upward and downward slopes ($F(1,22) = 11.64$, $p<.01$), and we concluded that the downward slopes showed higher hesitation interpretation rates.

Then, we analyzed the effects of the transition ranges of F0 contour. Of the upward slopes, u25 (4 sounds) had a hesitation interpretation rate of 36.9%, for u50 it was 30.4%, for u75 19.6%, u100 15.2%, and u125 12.0%. From this result, we found significant differences among them ($F(4,88) = 8.89$, $p<.01$), and the Newman-Keuls test revealed significant differences between (1) u25 and the range u75 - 125, and (2) u50 and the range u75 - 125. This result demonstrates that the slower upward slopes (such as u25 and u50) had higher interpretation rates as hesitation. Additionally, among downward slopes, d25 (4 sounds) had a hesitation interpretation rate of 57.6%, for d50 it was 42.4%, for d75 38.0%, d100 26.1%, and d125 20.6%. As was true of the upward slopes, this result showed significant differences among them (F(4,88)

= 11.23, $p<.01$), and the Newman-Keuls test revealed significant differences between (1) d25 and the range d50 - 125, (2) d50 and the range d100 - 125, and (3) d75 and d125. This means that the slower downward slopes had higher interpretation rates.

To sum up, the result indicated that **the slower slopes with the longer duration evoked "hesitation."** Moreover, most upward slopes did not evoke "hesitation," even though the longer duration did so.

2.2.3 *Case 3: Agreement*

We analyzed the interpretation rate of agreement in terms of durations of sound stimuli. The interpretation of agreement for the stimuli of 189-ms durations was 53.4%, for 418 ms it was 43.5%, for 639 ms 24.9%, and 819 ms 29.3%. This result showed significant differences ($F(3,66) = 19.42$, $p<.01$), and the Newman-Keuls test revealed significant differences between (1) 189 ms and the range 418 - 819 ms, and (2) 418 ms and the range 639 - 819 ms. This indicated that the shorter durations received the higher interpretation rates.

Next, we analyzed the interpretation rate in terms of upward and downward F0 contours. The agreement interpretation rate of upward slopes was 12.4%, while for downward it was 62.4%. Differences between upward and downward slopes were significant ($F(1,22) = 67.5$, $p<.01$), and this result indicated that the downward slopes evoked "agreement."

Next, we analyzed the effects of the transition ranges of F0 contours. For upward slopes, the interpretation rate for u25 was 23.9%, for u50 it was 13.0%, for u75 5.4%, u100 9.8%, and u125 9.8%. This result showed significant differences among them ($F(4,88) = 5.48$, $p<.01$), and the Newman-Keuls test revealed significant differences between u25 and the range u50 - 125. This demonstrated that most upward slopes did not evoke "agreement." For downward slopes, the agreement interpretation rate for d25 was 53.3%, for d50 it was 65.2%, for d75 56.4%, d100 66.3%, and d125 70.7%. This result demonstrated significant differences among them ($F(4,88) = 3.12$, $p<.05$); however, the Newman-Keuls test only showed significant differences between d25 and d125. Therefore, most downward slopes evoked "agreement."

In sum, the result indicated that **the downward slopes with shorter duration evoked "agreement,"** while the upward slopes with longer duration did not do so.

2.3 *Discussion*

Our results showed that synthesized sounds as sound stimuli with specific F0 contours and durations evoked specific emotions for the participants as described below.

1. Sounds with upward slopes (increasing intonation) regardless of their duration evoked the emotional state of "disagreement" for participants.
2. Slower downward slopes with longer duration evoked the emotional state "hesitation."
3. Downward slopes (decreasing intonation) with shorter duration evoked the emotional state "agreement."

Here, I can state that Nagasaki and I observed the specific temporal transitions of two properties in synthesized sound (F0 contour and duration) evoked the agent's specific emotions for participants. Finally, I summarized this observed phenomenon as the following preliminary hypothesis;

The specific temporal transitions in expressed information evoke an information sender's specific emotions for an information receiver.

To generalize the above preliminary hypothesis regardless of the modalities of the expressed information, I looked closely into the meanings of these temporal transitions of F0 contours and durations **in terms of the laws of conservation of energy**, especially as they relate to potential energy in uttered sounds. In general, natural speech sounds exhibit a "natural damping phenomenon" so that the F0 and power values are gradually decreased. Moreover, the shorter speech sounds consume less energy for utterance. Therefore, I argue that sounds with downward slopes of F0 contour and shorter duration conform to the laws of conservation of energy; while, the sounds with upward slopes and longer duration deviate from those laws. Actually, Sugito *et al.*,(2000) reported that when certain speech sounds deviate from the natural damping phenomenon, the listener tends to focus on this deviated part of the speech and then sense attention. To sum up, I can state the following hypothesis to explain the results of this experiment based on the above preliminary one in terms of the laws of conservation of energy.

When temporal transitions in expressed information regardless of any modalities conform to the laws of conservation of energy, they will evoke an information sender's positive emotions (e.g., the emotional state "agreement") for an information receiver. Conversely, when temporal transitions in expressed information deviate from the laws of conservation of energy, they evoke an information sender's negative emotions (e.g., the emotional state "disagreement").

To test this hypothesis, I needed to observe whether the specific temporal transitions in other modalities evoke an information sender's specific emotions for an information receiver and how they do so. Therefore, I conducted another experiment to observe whether expressed information in other modalities can satisfy the above hypothesis. Specifically, in the second experiment, I focused on the agent's behaviors appearing on a computer screen as the visual information, rather than the agent's synthesized sound as the audio information used in this Experiment 1.

3. Experiment 2: Can we estimate the agent's emotion from its unilateral actions?

3.1 Purpose and Settings

I conducted a second experiment to observe how humans estimate an agent's emotions from its unilateral actions. Specifically, simple animations of a white rectangle moving from left to right on a black background were prepared and presented as an agent's unilateral actions (Figure 8). An experimenter asked participants to assume that these rectangle's motions were a part of the trajectory of a sign language speaker's finger (actually this was not true). In this experiment, I prepared 36 different animations that had nine different velocity slopes and four different durations. Specifically, the four durations were 250 ms, 500 ms, 750 ms, and 1000 ms. The nine velocity slopes were designed like the F0 contours in Experiment 1 (see Figure 7).

Participants were 11 Japanese (10 men and 1 woman; 20 - 24 years old) and they were asked to answer the question of which of three emotional states this agent had, the same as in Experiment 1, disagreement, hesitation and agreement.

I used the same software system as in Experiment 1; however, animation files replaced with the sound wav files. As an experimental procedure in Experiment 2, a randomly selected animation among 36 animations was presented to the participants. Afterward, one of three emotional states was randomly selected and presented to them in a question on the window's label, for example, "Do you think that this actor expressed HESITATION?" Specifically, 108 trials (36 animations x 3 emotional states) were conducted (Figure 9). Each animation window was 640 x 480 pixels and its physical size on the 11-inch computer display used in this experiment was about 13 x 10 cm.

3.2 Results

I analyzed which kinds of the rectangle's actions evoked which specific emotional states; Figure 10 shows those actions that evoked "disagreement," Figure 11 shows those which evoked "hesitation" and Figure 12 shows those for "agreement."

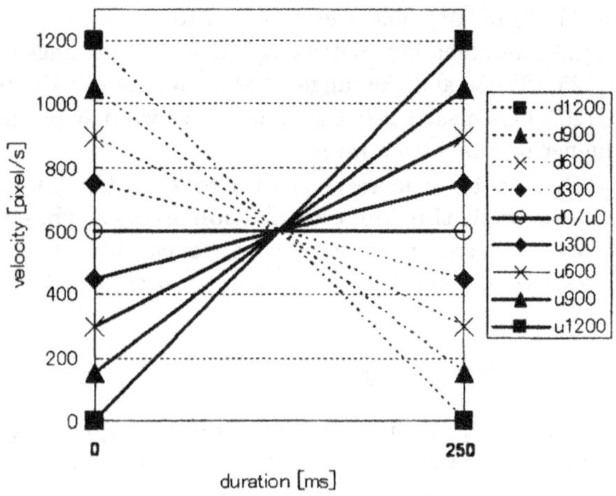

Figure 7: Nine different velocity slopes (duration: 250 ms). For example, "u1200" indicates the velocity transition range of 1200 [pixel/s] with upward slope (increasing velocity).

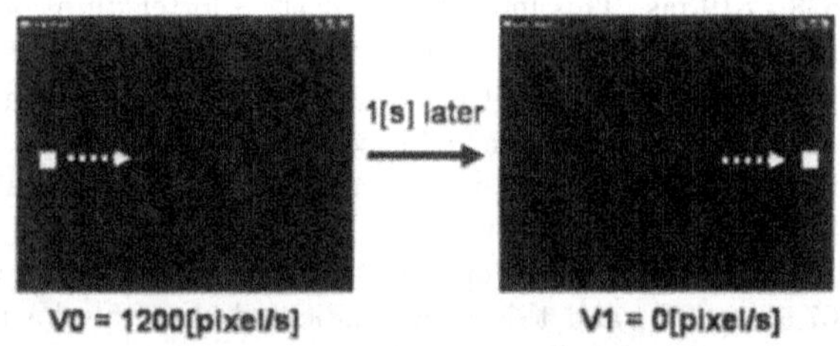

Figure 8: Animation stimulus for "d1200" velocity slope and 1000 ms duration.

3.2.1 Case 1: Disagreement

First, I analyzed the interpretation rates in terms of durations of animations as visual stimuli. The disagreement interpretation rate of the 250-ms durations stimuli (9 animations out of 36 ones) was 68.7%, that of the 500-ms duration (9 animations) was 38.4%, for 750 ms it was 23.2%, and for 1000 ms 27.3%. This result showed significant differences among these four durations ($F(3,30) = 11.1$, $p<.01$); however the Newman-Keuls test only revealed significant differences between 250 ms and the range 500 - 1000 ms, so this result demonstrated no significant differences except for the value 250 ms which was probably too short for participants to distinguish whether this agent expressed "disagreement."

Next, I analyzed the interpretation rates in terms of upward and downward velocity slopes. The rate of disagreement interpretation for upward slopes (16 animations) was 50.0%, while that of downward slopes (16 animations) was 31.8%. This result clearly indicated that the upward slopes evoked "disagreement" ($F(1,10) = 12.8$, $p<.01$).

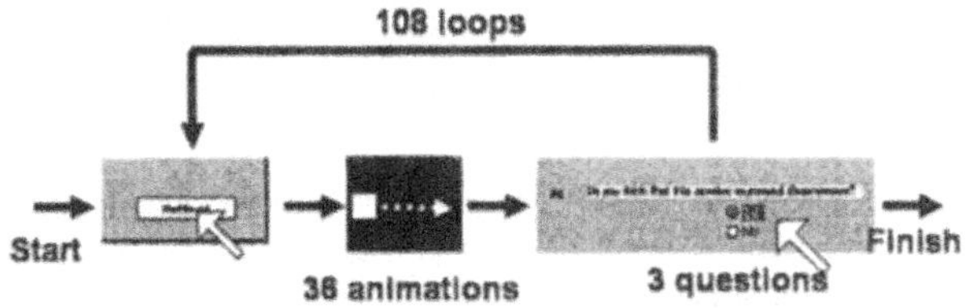

Figure 9: Specific procedure for Experiment 2.

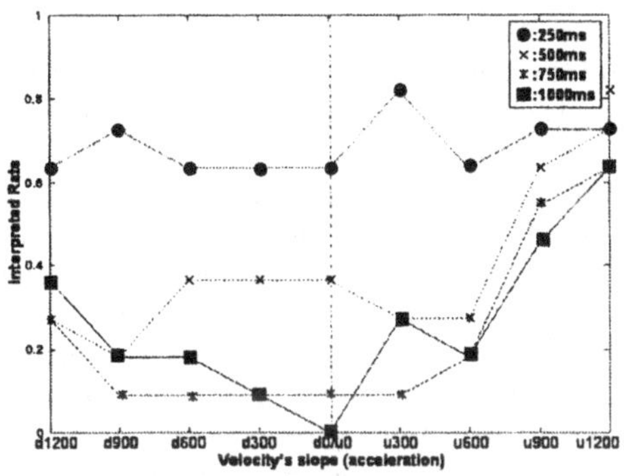

Figure 10: Relationship between velocity slope and "disagreement" interpretation rate for each duration.

Then, I analyzed the effects of the velocity transition ranges. Of the upward slopes, the disagreement interpretation rate for u300 (4 animations) was 50.0%, for u600 (4 animations) it was 47.2%, for u900 63.3%, and u1200 70.5%. This result revealed significant differences for these four types of upward slopes (F(3,30) = 5.9, p<.01), and the Newman-Keuls test revealed significant differences between (1) u300 and the range of u900 - 1200, and (2) u600 and the range u900 - 1200. Therefore, I concluded that the steeper upward slopes (faster increasing velocity) evoked "disagreement." For the downward slopes, the interpretation rate for d300 (4 animations) was 43.2%, for d600 (4 animations) it was 45.5%, for d900 45.5%, and d1200 54.5%. No significant differences were found for these stimuli (F(3,30) = 1.43, n.s.).

To sum up, these results indicated that **the visual stimuli with the steeper upward slopes (faster increasing velocity) evoked the emotional state "disagreement."** Conversely, the visual stimuli with downward slopes (decreasing velocity) did not evoke "disagreement."

3.2.2 Case 2: Hesitation

I analyzed the interpretation rates in terms of durations of visual stimuli. The hesitation interpretation rate for stimuli of 250-ms durations (9 animations) was 13.1%, that for 500 ms was 23.2%, for 750 ms it was 43.4%, and for 1000 ms 58.6%. This result revealed significant differences in these four durations (F(3,30) = 9.3, p<.01), and the Newman-Keuls test showed significant differences between (1) 250 ms and the range 750 - 1000 ms, and (2) 500 ms and the range 750 - 1000 ms. This means that the longer durations received higher interpretation rates for "hesitation."

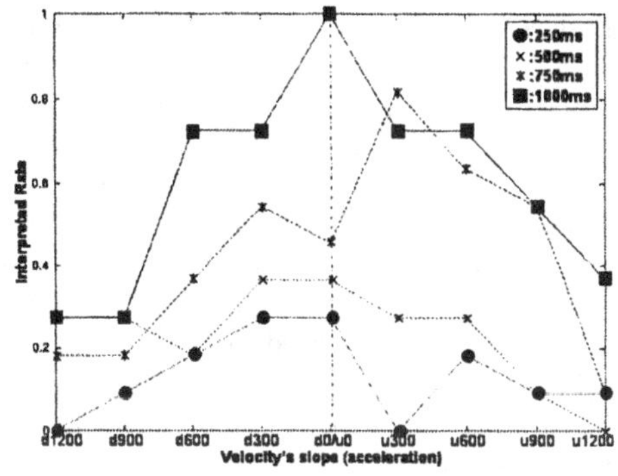

Figure 11: Relationship between velocity slope and "hesitation" interpretation rate for each duration.

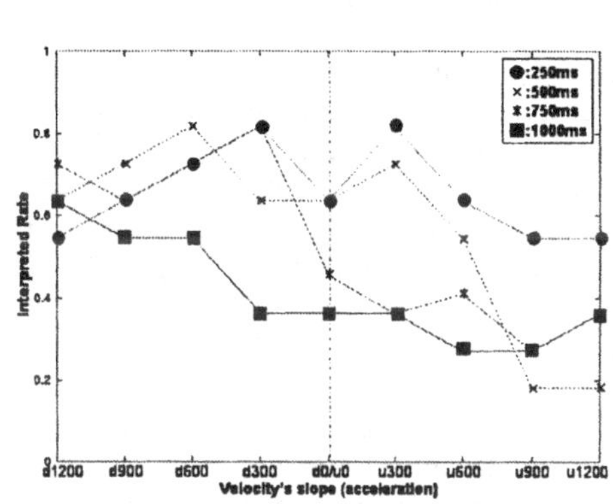

Figure 12: Relationship between velocity slope and "agreement" interpretation rate for each duration.

Next, I analyzed the interpretation rates in terms of upward and downward velocity slopes. The interpretation rate of upward slopes (16 animations) was 34.1%, while downward was 30.7%. This result indicated that there were no significant differences between upward and downward slopes (F(1,10) = .25, n.s.).

Then, I analyzed the effects of the velocity transition ranges. For upward slopes, u300 (4 animations) received a 40.9% interpretation rate, for u600 it was 36.4%, u900 29.6%, and u1200 13.6%. This result showed that there were significant differences among them (F(3,30) = 8.15, p<.01), and the Newman-Keuls test revealed significant differences between (1) u300 and the range u900 - 1200, and (2) u600 and the range u900 - 1200. This means that the slower upward slopes (such as u300 and u600) had the higher interpretation rates. For downward slopes, d300 (4 animations) was rated 34.1%, for d600 it was 29.6%, for d900 13.6%, and d1200 13.6%. As with the upward slopes, this result showed significant differences among them ($F(3,30) = 5.71$, $p<.01$), and the Newman-Keuls test showed significant differences between (1) d300 and the range d900 - 1200, and (2) d600 and the range d900 - 1200. In short, the slower downward slopes received the higher interpretation rates.

In sum, the result indicated that **the slower slopes with the longer duration evoked "hesitation."**

3.2.3 Case 3: Agreement

I analyzed the interpretation rates for agreement in terms of durations of visual stimuli. The agreement interpretation rate for stimulus of 250-ms durations was 69.7%, for 500 ms it was 57.6%, for 750 ms 55.6%, and 1000 ms 41.4%. This result revealed no significant differences among them ($F(3,30) = 2.72$, $p<.075$); however, a significant tendency between 250 ms and 1000 ms was found. This result indicated that shorter durations received the higher interpretation rates of agreement.

Next, I analyzed the interpretation rate in terms of upward and downward velocity slopes. The agreement interpretation rate for upward slopes was 45.5%, and for downward slopes 66.5%. Significant differences between upward and downward slopes were found ($F(1,10) = 9.83$, $p<.01$), and these indicated that the downward slopes evoked "agreement."

Then, I analyzed the effects of the velocity transition ranges. For upward slopes, the interpretation rate of u300 was 38.6%, for u600 it was 45.5%, for u900 34.1%, and u1200 36.7%. Thus, no significant differences were found among them ($F(3,30) = 1.46$, n.s.). For downward slopes, the agreement interpretation rate for d300 was 50.0%, for d600 it was 52.3%, for d900 50.0%, and d1200 54.6%. This result also showed no significant differences among them ($F(3,30) = .25$, n.s.).

To sum up, the result indicated that **the downward slopes with the shorter duration evoked "agreement,"** while the upward slopes with the longer duration did not.

3.3 Discussion

As a result, I confirmed that the animation as visual stimuli with specific velocity slopes and durations also evoked the agent's specific emotional states as follows.

1. Actions with upward slopes (increasing velocity/positive acceleration) regardless of their duration evoked the emotional states "disagreement."
2. Slower slopes (less acceleration) regardless of their slope types with longer duration evoked the emotional states "hesitation."
3. Downward slopes (decreasing velocity/negative acceleration) with shorter duration evoked the emotional states "agreement."

Thus, I could observe that specific temporal transitions of two properties in a rectangle's unilateral actions (velocity slope/acceleration and duration) evoked the agent's specific emotions for participants.

The results of this experiment were then analyzed in terms of the laws of conservation of energy, including the effects of friction in the agent's unilateral actions. Specifically, the agent's actions with upward slopes (positive acceleration) deviate from the laws of conservation of energy and the effects of friction, and these actions actually evoked negative emotions such as "disagreement" for participants. On the other hand, the actions with downward slopes (negative acceleration) conform to the laws of conservation of energy and the effects of friction, and these actions evoked positive emotions such as "agreement." Therefore, these results supported the hypothesis mentioned in the last section as follows.

When the temporal transitions in expressed information regardless of any modalities conform to the laws of conservation of energy, they evoke an information sender's positive emotions (e.g., emotional states such as "agreement") for an information receiver. On the other hand, when the temporal transitions in expressed information deviate from the laws of conservation of energy, they evoke an information sender's negative emotions (e.g., such as the emotional state "disagreement").

4. General Discussion

4.1 Are humans sensitive to the laws of conservation of energy?

The results of the above two experiments supported my hypothesis: *an agent's expressed information that deviates from the laws of conservation of energy evokes negative emotion, while that which conforms to the laws of conservation of energy evokes positive emotions.* Actually, some studies in developmental psychology reported the similar phenomena to that of my hypothesis, i.e., infants are already sensitive to the physically impossible events that deviate from the laws of conservation of energy. Specifically, infants increase their attentions toward events that violate their understanding of the physical world, or that are physically impossible (Spelke, 1985; Baillargeon, 1986; Baillargeon and DeVos, 1991). Thus, it can be said that humans (even infants) are generally sensitive to the events that deviate from the laws of conservation of energy, including the effects of friction or a natural damping phenomenon. Therefore, to say that humans tend to feel some negative valence values from the specific information that deviates from the laws of conservation of energy is reasonable, and it can be said that this humans' general tendency accords with my hypothesis. Thus, this evidence also would suggest that humans' emotion detecting mechanism can be explained in terms of the laws of conservation of energy in expressed information, and show some possibilities that the results could have an impact on the various emotion-detecting studies, not only in HCI and HRI but also in cognitive science or social psychology.

4.2 Conclusion

In terms of creating a smooth communication between a human and an artificial agent, some researchers have argued that they should form a mutual adaptation process: both must learn to respond and adapt appropriately to each other's behaviors. Moreover, to create mutual adaptation processes between them, the agent must induce the partner's adaptation or natural behaviors. Some studies have already reported that giving some physical body to such agent is reasonable. However, to do so for all kinds of interactive agents is unrealistic. To achieve an interactive agent that does not have physical body, I assumed that this agent should express certain information that can evoke its emotions for a user, because it is expected that we humans will empathize with certain entity which seems to have some emotions. In this study, I conducted two experiments to clarify which kinds of information the agent expressed (its synthesized sounds and its unilateral actions) evoke specific emotions for humans. In the first experiment, I observed how humans estimate the agent's emotions from its artificial synthesized sounds. In the second experiment, I observed how humans estimate the agent's emotions from its simple unilateral actions appearing on a computer display. I then summarized the results of these experiments as follows:

When the temporal transitions in expressed information regardless of any modalities conform to the laws of conservation of energy, they evoke an information sender's positive emotions (e.g., emotional states such as "agreement") for an information receiver. On the other hand, when the temporal transitions in expressed information deviate from the laws of conservation of energy, they evoke an information sender's negative emotions (e.g., emotional state such as "disagreement").

In addition, this also would provide some suggestions that humans' emotion detecting mechanism can be explained in terms of the laws of conservation of energy in expressed information, and could have an impact on the various emotion-detecting studies, not only in HCI and HRI but also in cognitive science or social psychology.

In these experiments, the participants just passively received the presented stimuli, such as synthesized sounds and a rectangle's unilateral movements. In actual communication, however, humans are not only receiving some information from a communication partner but also behaving positively toward this partner. Therefore, as a subsequent study, I am planning to conduct an experiment in which participants will receive some stimuli from the agent as responses to participants' active behaviors toward this agent. If the results acquired from this subsequent experiment conform to the results acquired in this study, these new results will provide further support for my hypothesis.

And also, up to now, it is not yet clear whether the agent's evoking the specific emotions for humans eventually induce these humans' adaptation or natural behavior in actual human-agent (robot/computer) interaction. So, in a consecutive study, I must examine whether my hypothesis could facilitate the smooth mutual adaptation and human-agent interaction. If I could succeeded in revealing this issue, this result would be utilized to induce users' adaptive behaviors and also would contribute to a basic theory to create a mutual adaptation for various applications that are required to communicate with humans, even for these applications without having a physical body.

References

Addlesee, M., Curwen, R., Hodges, S., Newman, J., Steggles, P., Ward, A. and Hopper, A.: Implementing a satient computing system, *IEEE Computer*, 34(8), 50–56 (2001).

Baillargeon, R.: Representing the existence and the location of hidden objects: Object permanence in 6- and 8-month-old infants, *Cognition*, 23, 21–41 (1986).

Baillargeon, R., and DeVos, J.: 1991 Object Performance in young infant: Further evidence. *Child Development*, 62, 1227–1246 (1991).

Brumitt, B., Mayers, B., Krumm, J., Kern, A. and Shafer, S.: EasilyLiving: Technologies for Intelligent Environments, In *Proceedings of the 2nd International Symposium on Handheld and Ubiquitous Computing (HUC2000)*, pp.12–29 (2000).

Fernald, A., and McRoberts, G. W.: Prosodic bootstrapping: A critical analysis of the argument and the evidence, In J. L. Morgan., and L. Denuth (Eds.), *Signal to syntax: Bootstrapping from speech to syntax in early acquisition*, pp. 365–388, Lawrence Erlbaum Associates (1996).

Harada, T. E.: Effects of agency and social contexts faced to verbal interface system; In *Proceedings of the 19th Japanese Cognitive Science Society Annual Meeting (In Japanese)*, pp. 14–15 (2002).

Imai, M., Kanda, T., Ono, T., Ishiguro, H., and Mase, K.: Robot Mediated Round Table: Analysis of the Effect of Robot's Gaze, In *Proceedings of 11th IEEE International Workshop on Robot–Human Interaction (RO-MAN2002)*, pp. 411–416 (2002).

Ishiguro, H., Ono, T., Imai, M., and Kanda, T.: Development of an interactive humanoid robot "Robovie" —An interdisciplinary approach, In R. A. Jarvis., and A. Zelinsky (Eds.), *Robotics Research*, pp. 179–191, Springer (2003).

Katcher, A. H., and Beck, A. B.: *New Perspectives on Our Lives with Companion Animals*, the University of Pennsylvania Press (1983).

Kidd, C. D., Orr, R., Abowd, G. D., Atkeson, C. G., Essa, I. A., MacIntyre, B., Mynatt, E., Starner, T. E. and Newstetter, W.: The Aware Home: A living laboratory for ubiquitous computing research, In *Proceedings of the 2nd international workshop in cooperative buildings (CoBuild99)*, pp.190-197 (1999).

Komatsu, T., and Nagasaki, Y.: Can estimate the speaker's emotional state from her/his prosodic features? — Effects of F0 contour's slope and duration on perceiving disagreement, hesitation, agreement and attention, In *Proceedings of the 18th International Congress on Acoustics (ICA2004)*, (to appear).

Komatsu, T., Suzuki, K., Ueda, K., Hiraki, K. and Oka, N.: Mutual Adaptive Meaning Acquisition by Paralanguage Information: Experimental Analysis of Communication Establishing Process, In *Proceedings of the 24th Annual Meeting of the Cognitive Science Society (CogSci2002)*, pp. 548–553 (2002).

Komatsu, T., Utsunomiya, A., Suzuki, K., Ueda, K., Hiraki, K., and Oka, N.: Toward a Mutual Adaptive Interface: An interface induces a user's adaptation and utilize this induced adaptation, and vice versa. In *Proceedings of the 25th Annual Meeting of the Cognitive Science Society (CogSci2003)*, (CD-ROM) (2003).

Nagasaki, Y., and Komatsu, T.: Can people perceive different emotions from a non-emotional voice by modifying its F0 and duration?, In *Proceedings of Speech Prosody 2004 (SP2004)*, (2004a, to appear).

Nagasaki, Y., and Komatsu, T.: Fundamental frequency as a cue to estimate speaker's emotional state, In *Proceedings of the 18th International Congress on Acoustics (ICA2004)*, (2004b, to appear).

NEC Personal Robot Center.: http://www.incx.nec.co.jp/

Ogata, T., Masago, N., Sugeno, S., and Tani, J.: Collaboration Development through Interactive Learning between Human and Robot, In *Proceedings of the 3rd International Workshop on Epigenetic Robotics*, pp. 99–106 (2003).

Ono, T., Imai, M., Kanda, T., and Ishiguro, H.: Embodied communications between humans and robots emerging from entrained gestures, In *Proceedings 2003 IEEE International Symposium on Computational Intelligence in Robotics and Automation*, pp. 558–563 (2003).

Reeves, B., and Nass, C.: *The Media Equation*, CSLI publications (1996).

Snow, C. E.: Mother's Speech Research; From Input to Interaction, In C. E. Snow., and C. A. Ferguson (Eds.), *Talking to Children: Language Input and Acquisition*, Cambridge University Press (1977).

Spelke, E. S.: Preferential looking method as a tool for the study of cognition of infancy, In G. Gottelieb., and N. Krasnegor (Eds.), *Measurement of audition and vision in the first year of postnatal life*, Norwood (1985).

Sugito, M., Inukai, T., and Sadanobu., T.: Sentence structure and prosody, In Spoken Language Working Group (Eds.), *Speech and Grammar (in Japanese)*, pp.3–20, Kuroshio Publications (1997).

Yamada, S., and Yamaguchi, T.: Mutual Adaptation to Mind Mapping in Human-Agent Interaction; In *Proceedings of the 11th IEEE International Workshop on Robot–Human Interaction (ROMAN-2002)*, pp.105–110 (2002)

Weiser, M.: Some computer science issue in ubiquitous computing, *Communication of the ACM*, 36(7), 75–84 (1993).

Hamelin: A model for collective adaptation based on internal stimuli

Vincent Thomas* **Christine Bourjot*** **Vincent Chevrier*** **Didier Desor****

*LORIA - MAIA Project
Campus Scientifique - BP 239
54506 Vandoeuvre-Les-Nancy CEDEX
France
{vthomas, bourjot, chevrier}@loria.fr

** Laboratoire de neurosciences comportementales
Université Henri Poincare Nancy 1 - BP 239
54506 Vandoeuvre-Les-Nancy CEDEX
France
Didier.Desor@scbiol.uhp-nancy.fr

Abstract

Groups of rats confronted to an increasing difficulty to reach food organize themselves. The emergent organizational structure is characterized by a distribution of two different behavioral profiles among the rats. This specialization is stable, robust and presents adaptive properties.

Hamelin, the simulation system we propose is based on the coupling of two existing models: adaptive response thresholds and dominance relationships. It manages to reproduce the organization observed by biologists and presents some interesting adaptive properties. The originality of this model is that local internal needs (hunger) are distributed among the collectivity without an explicit global representation. The collectivity profits from individual adaptation abilities.

1. Introduction

Animal societies manage to organize themselves to face problems in their environment. The way they ensure a global coherence of the collectivity on the basis of individual decisions has attracted a lot of attention. A lot of studies have focused on collective behaviors (Bonabeau and Theraulaz, 1999) (Camazine et al., 2001) observed in several species: ant colonies (Deneubourg et al., 1990) (Corbara et al., 1993), social spiders (Dury et al., 2001) (Bourjot and Chevrier, 2001), fish schools (Reynolds, 1987) (Kunz and Hemelrijk, 2003), termites (Miramontes and DeSouza, 1996) or primates (Hemelrijk, 1996). Some principles at the root of some of those phenomena have been extracted, like for example the stigmergy principles (Grasse, 1959) which can explain various collective phenomena from the nest building of wasps or termites (Bonabeau and Theraulaz, 1999) to the collective foraging behavior of ant colonies (Deneubourg et al., 1990).

An interesting collective phenomenon that can be observed in various species is the specialization behavior. Specialization is a way to assign different tasks to the animals belonging to a colony. These phenomena have been observed in various insects colonies like ants (Bonabeau and Theraulaz, 1999) or wasps (Theraulaz et al., 1990). Specialization have some specific assets. A specialized system in which tasks have already been assigned is supposed to be more efficient, because there is no need to reassign the tasks to agents, which can cost time and need a lot of communication. Moreover, specialized individuals can collect experience from the tasks they perform and thus, the overall performance of the system can be higher.

This paper focuses on task allocation and specialization observed in groups of rats when they have to face an increasing difficulty to reach food (Desor et al., 1991). Under certain conditions, a differentiation between individuals with same abilities can be observed and leads to a division of labor: some rats fetch food whereas other reach food by stealing it. The phenomenon of task allocation in insect societies has been modeled by a mechanism called response thresholds model (Bonabeau and Theraulaz, 1999) (Theraulaz et al., 1998). This model has been extended to explain the genesis of specialization and task allocation to a population of homogeneous individuals. However, it is based on an external stimulus which can be perceived by various individuals, whereas the phenomenon we have focused on is not characterized by external stimulus but by local internal stimuli: the hunger of the rats put in the system. We introduced in task allocation a second mechanism which can be observed in nature and which can explain how the differentiation can appear on the basis of these local stimuli.

This mechanism is a second model: self-organizing hierarchies. This kind of model which can be found in (Hemelrijk, 1996) (Hogeweg and Hesper, 1986) (Bonabeau et al., 1996) is wide-spread and has been used to explain some phenomena like coalition formation observed in primates group (Hemelrijk, 1996) which was at first attributed to high cognitive abilities.

The model we present in this paper consists in the coupling of response thresholds model with self-organizing hierarchies. It manages to reproduce the global phenomenon by organizing the society of rats according to a hierarchical structure. The task allocation is then made on the basis of this organization between individuals. A coupling of these two models has already been seen in literature (Theraulaz et al., 1990) but it relies on external stimuli. The originality of the coupling we propose is that its goal is to distribute internal stimuli among agents.

This paper presents in the first part the biological experiments that have been conducted in the behavioral neuroscience laboratory of Henri Poincare University of Nancy and explains why it can be described as an adaptive self-organized system. Then, it depicts the model we have developed and validates its conformity with respect to biological observations. The next part dwells on the properties such a complex system can exhibit and analyzes them. To conclude, we will propose an interpretation and few hints we are planning to follow in order to reuse mechanisms of this model for collective problems solving.

2. Biological phenomenon

2.1 *Specialization experiment*

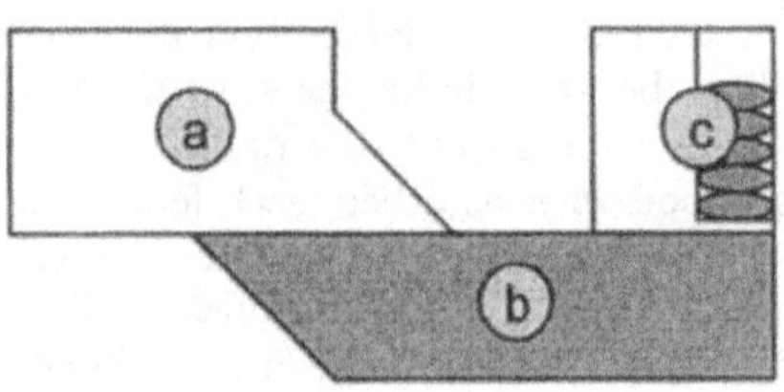

Figure 1: Environmental setup composed of a cage (a), an immersed corridor (b) and a feeder (c).

The diving-for-food situation can be considered as representative of an adaptive process for environmental exploitation by a group. It is a complex social task in which, for a group of 6 rats introduced in a specific environmental setup (Figure 1), the food accessibility is made difficult by the progressive immersion of the corridor, the only path of access to the feeder. This experimental schedule leads in a few days to the emergence of a differentiation characterized by two behavioral profiles: (a) a carrier profile characterizing rats which dive into water and bring the food back to the cage, and (b) a non-carrier profile characterizing rats which never dive, but get food by stealing it from the carriers (Desor et al., 1991) (Desor and Toniolo, 1992). The social differentiation regularly occurs with the same respective proportions of carriers and non-carriers of about 50%. It remains stable for several months and has been observed in mice and rats, including Long-Evans, Wistar and Sprague-Dawley rat strains.

2.2 *Other experiments*

Other investigations have been undertaken to determine the characteristics of this process.

In these pharmacological experiments, a drug, the Diazepan, has been administered to 50% of the 6 rats introduced in the environmental setup (Schroeder et al., 1998). This drug has for direct consequence the decrease of the anxiety of drugged rats. Specialization still occurs in the system. The same profiles can be observed with qualitatively the same proportions. However, a more precise study highlights an over-representation of carrier rats in the drugged subgroup (80% of the rats), whereas at the same time, an over-representation of non-carrier rats (80%) in the non-drugged sub-group can be observed.

This lead biologists to two conclusions. The first one is that this experiment stressed upon the importance of anxiety toward water, which would lead the less anxious rats to dive first, and consequently be aggressed first by their more anxious conspecifics. The second one, is that it is the social environment of rats which determines the adopted profiles. It explains why, whereas the non-drugged rats have not suffered from any alteration, an over-representation of a specific profile (the non-carrier rat profile) is nevertheless present.

Finally, the introduction of rats that have already been specialized in previous experiments, lead to a re-differentiation. The social differentiation (profiles and sizes of groups) that occurs in this case is the same as the differentiation observed when the rats are undifferentiated at the beginning of the experiment.

2.3 *Adaptive self-organized system*

The profile adopted by a rat depends on its anxiety, its social environment and the interactions that occurred in the system. Other experiments have also shown that a single rat dropped in the environmental setup exhibits a carrier profile. It implies that each rat has the physical ability to cross the immersed corridor.

All these hints lead biologists to assume that the system is ruled by a robust underlying mechanism that allow the system to self-organize. This mechanism consists in two kinds of adaptation: the first one is an individual adaptation: each rat of the system manage to sustain its needs in a specific way (by stealing or by diving). The second is a collective one: some rats don't dive and the system exhibits a distribution of roles among the individuals. This distribution has been interpreted by biologists as a limitation of the number of carrier rats appearing in the system. The next step we have followed to explain the social differentiation consisted in simulating this biological system.

3. The Hamelin Simulation

The problem we face is to study the apparition of a global pattern: the specialization between individuals who may have no explicit representation of the whole collectivity. To verify the connection between individual local behaviours and the observed collective phenomenon, an agent based model called Hamelin has been built. Hamelin has been designed following a parsimony principle that searches to reproduce the phenomenon with the simplest assumptions.

In this section, we will stress upon the different components of the agent-based model we have designed: the agents, the environment and the possible interactions in the system.

3.1 Agents

Each agent is the computer implementation of a rat. Agents are reactive agents: they have no planning abilities nor social representation and are ruled by stimulus-response rules which make them act according to their instantaneous partial perceptions.

The internal state of an agent is characterized by 4 variables which have been proven to have importance during biological experiments. These variables are:

- Its hunger h which represents the need for food and constitutes the motivation for the agent.
- The possessed amount of food *Food* implemented as the size of the owned pellet.
- The strength of the agent s, which corresponds to its ability to win when it is involved in a fight (to catch a pellet or to defend itself).
- Its anxiety toward water θ corresponding to its reluctance to dive into water.

Due to the parsimony principle we have followed, these variables are not directly linked. For example, the strength of an agent is not directly related to the value of its food and hunger variables but such relationships between variables (which are not required to observe specialization) could be investigated in the future.

The activity of the agent is a combination of three behavioral items that will be depicted further in the paper:

- a diving item
- an attacking item
- an eating item

An agent perceives locally its environment. It knows if it owns a pellet and can detect if other rats in the cage possess pellets. Finally, it must be noticed that the perceptions of agents are still very simple: a rat cannot identify and thus can not recognize the other rats present in the system.

3.2 Environment

The environment represents the environmental setup. It is characterized by the length of the immersed corridor and by the size of the pellets in the feeder. The characteristics are implemented through two variables: τ the time needed to consume entirely a pellet and η the energy absorbed during a time step of eating a pellet. The energy contained in a new pellet is thus $\tau\eta$.

3.3 Behavioural items

Each of the behavioral items (diving, attacking and eating) is stochastically triggered or stochastically carried out. The associated probability is computed according to the internal state of the rat and biological observations. When the item is triggered, the associated action is effectively performed and a reinforcement alters the internal state of the agents allowing them to learn and modify their behaviors according to their past actions.

3.3.1 Diving

This action is considered by an agent when it does not possess a pellet.

Probability This behavioral item is computed according to adaptive response thresholds (Bonabeau and Theraulaz, 1999) based on the hunger of the agent and its anxiety toward water. The hunger of an agent and the probability for it to dive are positively correlated. On the contrary, the higher its anxiety toward water, the lower the same probability. These tendencies are computed according to the following formula:

$$P_{diving} = \frac{h^2}{h^2 + \theta^2}$$

Results of the action When decision is made, the action is automatically and instantaneously succeeded. The agent dives into water, manages to take a pellet from the feeder and comes back to the cage. The agent possesses then a new pellet

$$Food \leftarrow \tau$$

Reinforcement Moreover, when the action is performed, the anxiety of the rat is reduced according to adaptive response thresholds models.

$$\theta \leftarrow \theta.\delta_\theta$$

$\delta_\theta \in [0, 1]$ is a global parameter. Thus, the agent learns to react more quickly to the same hunger. This reinforcement is responsible for the individual adaptation and the genesis of task allocation between agents. The more an agent has dived, the less reluctant it will be to dive. Finally, agents tend to forget their past experience and their anxiety increases gradually ($\delta_f \in [0, 1]$ also a global parameter of the system is responsible for the increase and 1000 the highest possible anxiety).

$$\theta \leftarrow \theta + (1000 - \theta).\delta_f$$

3.3.2 *Attacking*

This action is considered when a rat has no pellet in its paws and perceives pellets in the cage.

Probability This behavioral item is systematically triggered but the results are stochastically determined according to the relative strength of the two individuals involved in the fight. The probability for the aggressor (agent A) to steal the pellet from the victim (agent B) is computed according to dominance relationships.

$$P(winning_A) = \frac{s_A}{s_A + s_B}$$

Results When the action is successful, the aggressor A manages to steal the pellet from the victim B. The size of the received pellet depends thus on how much of the pellet the loser has already eaten.

$$Food_A \leftarrow Food_B, Food_B \leftarrow 0$$

Otherwise, nothing happens.

Reinforcement Whether the action is successful or not, the strength of the winner (A or B) is reinforced whereas the strength of the looser is decreased ("winner and looser effect"). Alterations of strength are computed according to dominance formula presented in (Hemelrijk, 1996):

$$s_A = s_A + (Winning_A - \frac{s_A}{s_A + s_B}) * \delta_s$$

$$s_B = s_B - (Winning_A - \frac{s_A}{s_A + s_B}) * \delta_s$$

δ_s is also a global parameter of the simulation and $Winning_A$ is 1 if A won the fight, or 0 otherwise.

3.3.3 *Eating*

This action is automatically performed. If the agent owns a pellet, it has for results the decrease of its hunger and the decrease of the size of its pellet.

$$h \leftarrow h - \eta, Food \leftarrow Food - 1$$

Otherwise, the hunger of the agent increases.

$$h \leftarrow h + 1$$

4. Validation of adaptation

Computer experiments have been conducted to confront the results of the model with the results observed in biological experiments.

4.1 *Individual adaptation*

When a single agent is confronted to a difficulty of accessing food, there is no possibility of interacting with other agents and its behavior is only ruled by a response threshold model. The rat learns to dive and to reduce its hunger, according to adaptive response model properties. There is individual behavior reinforcement and individual adaptation to the situation.

The next part will focus on the adaptation of the system when some rats are put together in the environment.

4.2 *Collective adaptation: Specialization*

4.2.1 *Parameters of conducted simulations*

number of rats	δ_θ	δ_s	δ_f
6	0.95	0.15	0.00001
τ	η	$\theta_{initial}$	$s_{initial}$
10	2	600	10

Table 1: Parameters of conducted simulations

We have studied the model for some fixed initial parameters. These parameters are summed up in the Table 1. Some of these parameters are global parameters (δ_θ, δ_s and δ_f). They have been empirically tuned to effectively observe the specialization phenomenon among initially homogeneous agents. These parameters will stay unchanged in all conducted experiments. The other parameters (the number of rats, η, τ, $\theta_{initial}$ and s_{intial}) represent the conditions under which a specific kind of experiment has been undertaken. They will be altered in the next experiments presented in the paper.

4.2.2 *Observation of specialization*

The specialization observed with parameters from Table 1 divides the group of agents into two different sub-

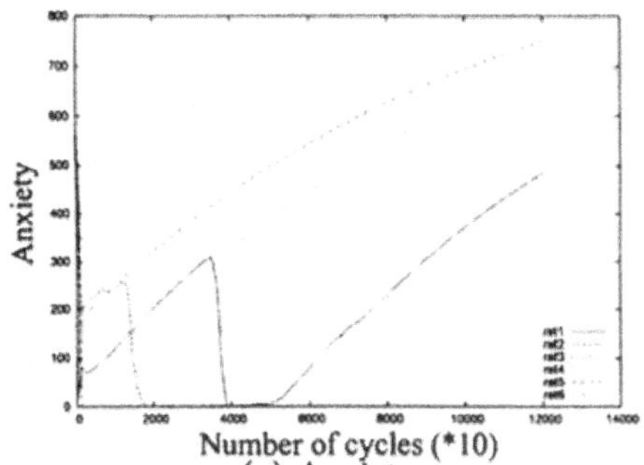

(a) Anxiety

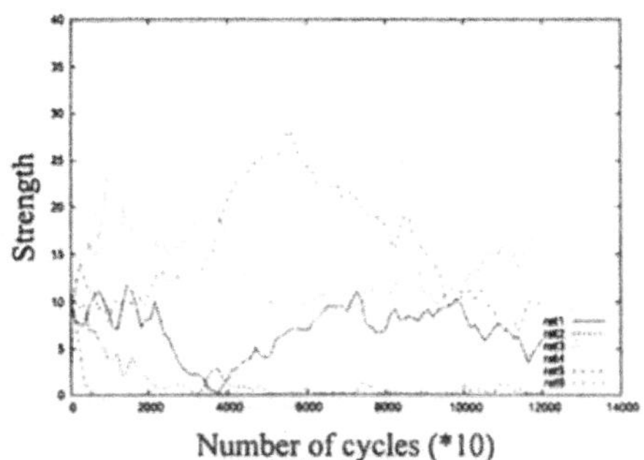

(b) Strength

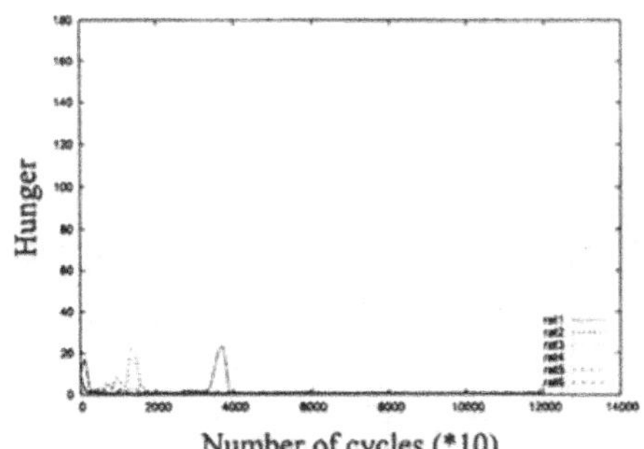

(c) Hunger

Figure 2: Computer-simulated specialization

groups. The first one contains agents with low anxiety and low strength. These agents are prone to dive, loose their stealing attempt against agents belonging to the second subgroup and correspond to the carrier rats. The second subgroup contains agents with high anxiety and high strength. These agents never dive, manage to fulfill their needs (hunger) by stealing the pellets from agents of the first subgroup and thus correspond to non-carrier rats.

The figure 2 shows results obtained by the execution of a single simulation. 4 agents (number 1, 3, 4 and 6) are non-carrier agents whose anxiety rises (Fig 2(a)) and who have a high rank in the hierarchical structure (Fig 2(b)). 2 agents are diving agents. The hungers of all the rats are controlled and are close to 0, except for some events (profile inversions) we will discuss in the future.

For the initial values presented in table 1, the size of the subgroups of the various experiments are almost constant. We conducted 100 experiments and two global states have been observed: one state with 3 carrier agents and 3 non-carrier agents which occurs in about 60% of the experiments that have been made and a second state with 4 non-carrier agents and 2 carrier agents which appears in about 40% of conducted experiments. Moreover, in all the experiments, the hungers of the agents stay bounded and relatively low.

The system manages to reproduce collective adaptation: some agents manage to avoid facing water element, the whole need of the society is satisfied and self-maintenance of the system is reproduced.

4.2.3 Stability

During these experiments, it has been stated that the profiles associated to agents evolve. However, whenever a profile change occurs, it is compensated by another change of profile. It ensures the constant size of the profile subgroups. For the conducted experiments, we have observed an average number of profile inversions by experiment of 2.1 (standard deviation of 1.22). This behavior is the consequence of the function we have chosen at first to decide the results of a stealing attempt between agents.

This function equals to the ratio of the strength of the involved agents. This function leads to a unstable hierarchical organization. Dominance ranks (determined according to relative strength) evolve in this hierarchy. An agent whose rank was high and whose rank has decreased must now dive to fetch food and simultaneously, the agent whose rank has increased can now access pellets by stealing them from less strong agent. This change of profile is accompanied by a temporary increase of hunger because the rat whose rank has decreased need to re-adapt itself to this new situation.

Other experiments (consisting in computing the probability of succeeding in stealing attempt as the ratio of the square of strengths) have shown that it is possible to make these profile inversions disappear to obtain a system with stable organization.

4.3 Re-differentiation

Specialization still occurs if we alter the initial values of the agents introduced in the system, simulating the fact that agents have already been differentiated in previous experiments.

When the system contains initially only carrier agents (low initial anxiety $\theta_{initial} \leftarrow 50$ and low initial strength $s_{initial} \leftarrow 1$), specialization can still be observed. The same division in two groups, each one characterized by a profile, occurs and the sizes of these groups are globally the same as the sizes observed in previous experiments. 3 global states have been detected: one constituted by 4 non-carrier and 2 carrier agents which appears in about 30% situations, another constituted by 3 non carrier and 3 carrier agents which appears in about 60% of situations and a last constituted by 2 non-carrier and 4 carrier agents which appears in about 10% of situations for 100 experiments. The number of inversion of profiles also rises (an average number of 4.8).

This experiment shows that the collective adaptation leads almost to the same global states independently of the history of the agents introduced in the system.

4.4 Hamelin for the biology

The simulation managed to reproduce the specialization that has been observed in real experiments. The reproduced computer specialization is characterized by the same profiles, the same global proportions of profiles, the ability to re-differentiate a system composed of agents that have already been differentiated. Finally, because the hungers of the agents are close to 0, the system is assumed to reproduce the self-maintenance of the biological system.

Hamelin has shown that no social cognition nor individual recognition is necessary to explain specialization observed in groups of rats confronting environmental difficulties to reach food. Hamelin shows also that reinforcement rules that have already been observed in other animal collectivities can explain such a phenomenon. Furthermore, Hamelin manages to reproduce the two kinds of adaptation observed: individual adaptation when a single rat is put in the environmental setup and collective adaptation leading to specialized agents.

Hamelin constitutes a first step to apprehend the biological mechanisms responsible of social differentiation in these diving-for-food situations.

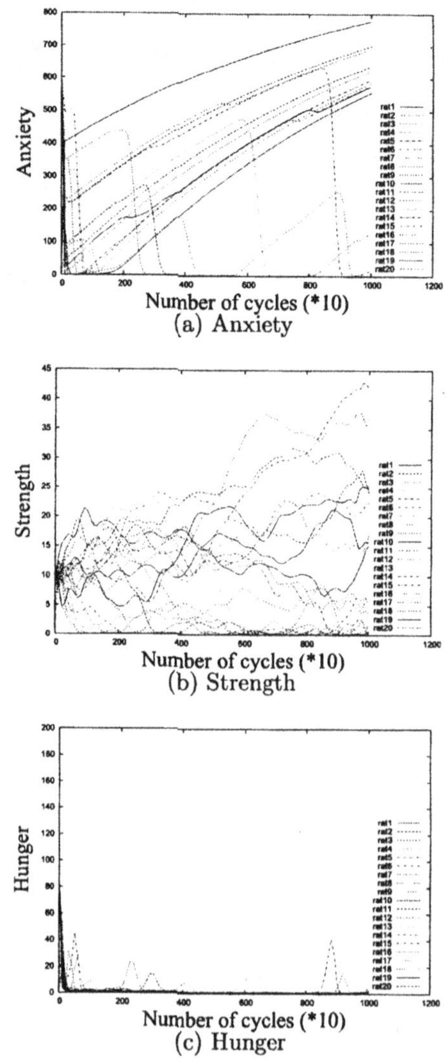

Figure 3: Adaptation to the number of agents

5. Analysis of the system

Hamelin system can also be seen as a multi-agent system which generates global pattern (specialization) on the basis of individual local rules (the behavioral items of the agents). In order to have a better assessment of the adaptation abilities of the computer system and to have a better understanding of the principles involved, other experiments that have not been conducted with real rats have been undertaken with the Hamelin system. These experiments are presented in the following.

5.1 Adaptation to the number of agents

We have focused on the influence of the number of agents in the differentiation. We have conducted experiments with 20 agents (see Figure 3) and parameters from Table 1. The agents have still no social cognition and cannot recognize other individuals. Nevertheless, differentiation is still observed and the global hunger of the system is still very low except for profile inversions which are more frequent than with less agents.

The global state depends on the conducted experiment, but the number of carrier agents is comprised between 12 and 14. It shows that the system manages to adapt itself to social environment of the rats without needing any representation of the agents of the system.

Moreover, adding or taking out an agent during execution don't hinder the system which manages to re-adapt itself to the new situations.

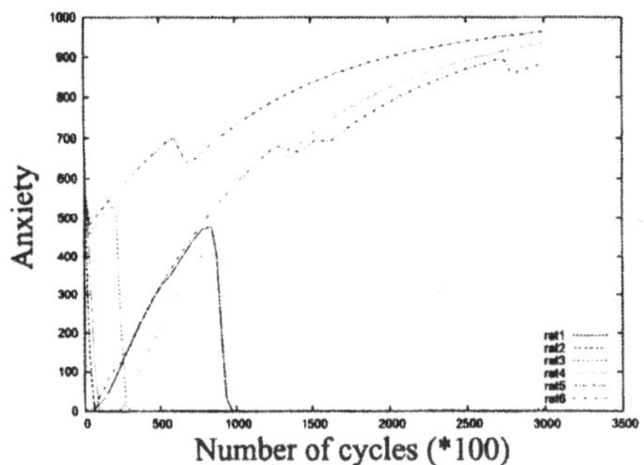

(a) Anxiety

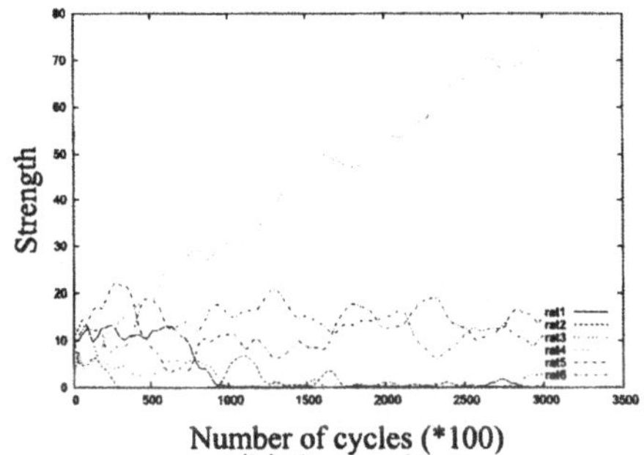

(b) Strength

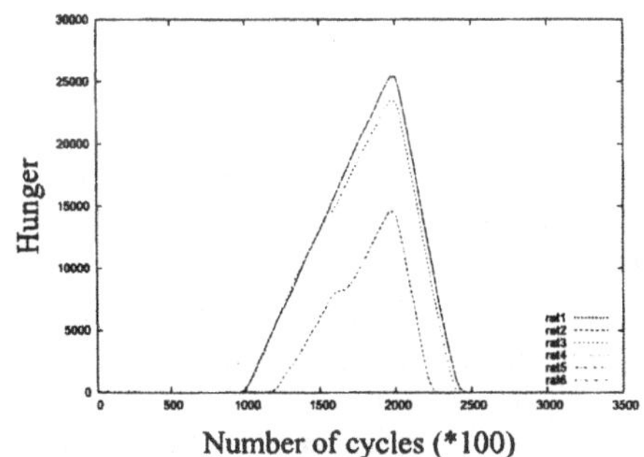

(c) Hunger

Figure 4: Adaptation to external condition during runtime(At 100000 step, τ is set to 2, and at 20000 step, it is reset to 10)

5.2 *Adaptation to external conditions*

We have also studied the way the system reacts when data describing the environment change during runtime. Simulations based on the parameters of the table 1 have been conducted. The first 100000 time steps, these parameters stay unchanged. After the 100000 step, the τ value is modified from 10 to 2. This new value corresponds to a strong constraint on the system. When an agent reaches the feeder and takes a pellet, the pellet is small from now on and contains less energy. In consequence, to fulfill the need of the collectivity, more agents have to adopt a carrier profile.

In the conducted computer experiments, we have observed a re-differentiation as shown in the figure 4. An agent (the first rat) which was a non-carrier agent become a carrier agent. Its anxiety drops, meaning that he dives into water and learns to do so. Moreover, this change of profile is not compensated by the appearance of a new non-carrier agent in the system, the global state of the system has thus changed to response to the increase of difficulty presented by the environment. However, this re-differentiation is not sufficient to provide enough food to the collectivity because of the harshness of the new situation: the hunger of agents increases linearly. After 200000 steps, the τ is re-affected to its initial value 10, agents manage to fulfill the needs and hungers decrease to 0.

The system manage to readapt itself to changing environmental conditions, but if the constraints are too important, some agents still don't manage to eat.

5.3 *Modification of organization process*

In some cases, the hunger of agents increases linearly and questions the viability of the collectivity. This is due to the absence of transmission of information on the system. In these cases, carrier agents manage to fetch enough food for reducing the hunger of the non-carrier agents but not enough for them. Because of the partial perceptions of the agents, this kind of situation cannot be detected by the non-carrier agents which are not incited to change their behavior. One way to transmit this information is to introduce a feedback by making the stealing attempts dependent on the hunger of the aggressed rats.

To do so, we have modified the probability of winning a fight according to the hunger of the potential victim. If this hunger is above a specific threshold, the stealing attempt is always a failure.

When this alteration of interaction is implemented in the system (with a threshold of 100), the problem of hungers increase disappears and the system readapts itself in a more efficient way. The figure 5 presents the graphics for the same scenario as the one of the figure 4. It can be noticed, that the system readapts itself at the 100000 step. At this moment, the difficulty to fetch food is very important, and all the rats except one become carrier rats. The number of rats is then sufficient for the fulfillment of the needs of the group. When, after 200000 steps, the difficulty is decreased the system readapts itself to a state close to its former global state.

5.4 *Hamelin properties*

The Hamelin simulation can be considered as a complex system inspired by biological observations. This system manages to organize behaviors of agents during runtime according to interactions occurring in the system. This organization appearing in the system is the consequence

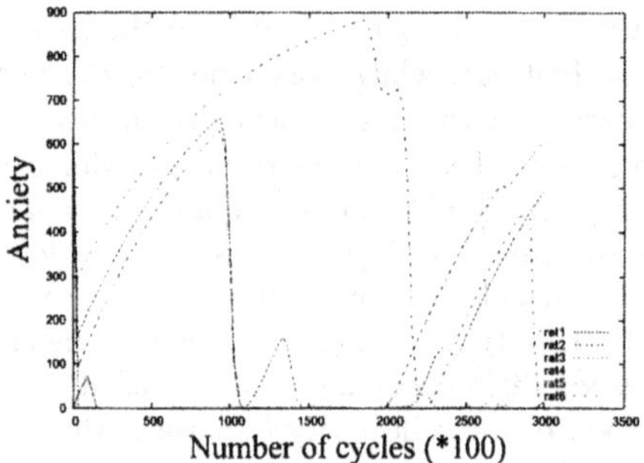

(a) Anxiety

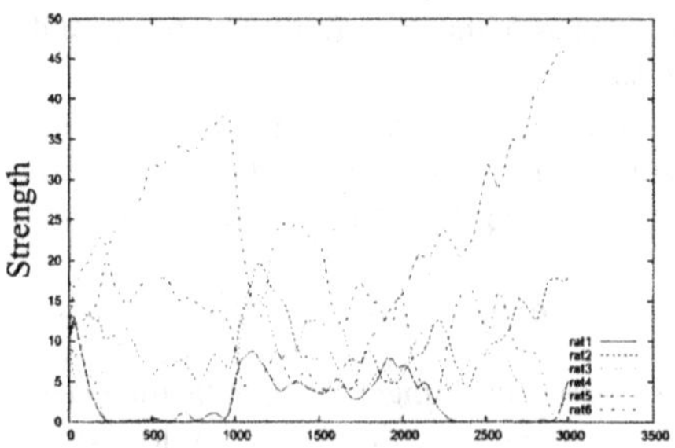

(b) Strength

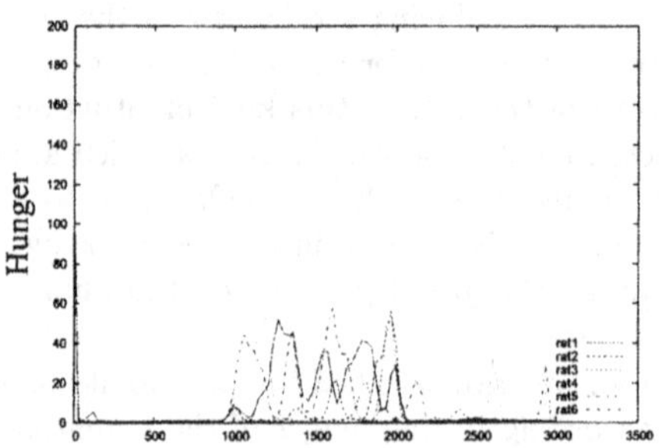

(c) Hunger

Figure 5: Specialization observed after adding a feedback in fighting interactions

of the partial observations of agents and of simple reinforcement rules. This allows the system to readapt itself to various conditions without needing an explicit representation of these changes.

The system presents interesting adaptive properties: it manages to readapt itself to external conditions, to the number of agents put in the system and to different initial conditions as it has been presented in re-differentiation experiments.

6. Discussion

The last part of the paper is a discussion about this system and the underlying mechanisms. The main goal of this part is to have a better point of view of the system in order to generalize the interesting principle involved.

6.1 Interpretation of Hamelin

The system is characterized locally by a need to fulfill (the hunger of each rat) and by a difficulty to reach the resource needed to fulfill it. Each agent has its own need and can accede to the resource by two different ways: it can steal the pellet from other agents or dive and take a pellet from the feeder.

Initially individuals are undifferentiated, they have no preferred way of acquiring the resource. The collective dynamics of the system leads to an organization of these agents where specialization concerning the way of acquiring the resource emerges: some agents prefer to accede directly to the resource by diving into water, other prefer to obtain it indirectly by interacting with agents.

This differentiation process is the consequence of a coupling of two reinforcement based models: the first model, adaptive response thresholds, rules the direct acquiring way and reinforces the diving decision each time an agent reaches food directly. The second, dominance relationships rules the indirect acquiring way and the associated behavioral item is reinforced each time an agent manages to possess food by stealing it. The coupling of these two mechanisms is responsible for the dynamics of the system and the appearance of specialized profiles.

A more detailed analysis of the first profile inversions of figure 2 shows that those two models have not the same functionality.

Dominance relationships allow the distribution of pellets in the system. Agents manage to steal pellets according to their dominance value. Dominance relationship is then a way to transfer the pellets among agents (from the weakest agents to the strongest). Because the pellet is the resource needed by individuals, the transfer of pellets can be interpreted as a transfer of needs among the agents from the strongest to the weakest.

Response thresholds model allows the weakest agents (which concentrate the need of the collectivity) to adapt themselves to their hunger due to the environmental constraints and to the social pressure generated by hierarchical organization. This adaptation is an individual one: agents learn to dive more early in order to reduce their own hunger.

Hamelin can thus be seen as a coupling between an organizational process responsible for the transfer of internal stimuli through the system and a local individual adaptation process enable the fulfillment of individual needs. The transfer of local information and the possibility to solve locally a part of the problem by the more experienced agent (the one whose anxiety is the lowest) lead to the adaptation of the collectivity without an explicit representation of the global needs of the system.

6.2 *Hamelin as a problem solving process*

It must be noticed that the Hamelin model is close to the satisfaction-altruism model of (Simonin and Ferber, 2000). In this model, agents have to explore an environment. The method proposed to achieve cooperation is to couple a model of selection of actions and a model of interaction between agents. The interaction model is based on satisfaction signals and social pressure: if the satisfaction of an agent is very low, it can make the other agents move to free the path.

The interaction model and the social pressure of satisfaction-altruism model can be compared with the dominance relationship model of Hamelin and the action selection model with the response threshold model. The satisfaction-altruism and Hamelin model are based on the same abstract principle of coupling an interaction model to transfer local information through collectivity and a model of selection of action. This comparison let us think that this general principle can be a general mean to achieve cooperation between agents for collective problem solving.

7. Conclusion

In this paper, we have presented a simulation of a phenomenon of specialization observed in groups of rats when they face a diving-for-food situation. This simulation, Hamelin, is a self-organized process based on local interactions between agents.

It manages to reproduce the differentiation phenomenon and the collective adaptation of the group characterized by the viability of the collectivity while limiting the number of diving agents. The computer simulated specialization leads to the same profiles as those observed in biological experiments and reproduces the re-differentiation of individuals. It is based on reactive agents without any social abilities and proves thus that developed cognitive abilities are not required to explain the specialization occurring in this situation.

Furthermore, the simulation can be analyzed independently of biological considerations and turns out to have broader adaptive properties: it can adapt and readapt itself to changes of external conditions and to the number of agents. The core of the model is a coupling between an organization process which is in charge of distributing needs in the collectivity and an individual adaptation process.

Two perspectives are considered for this work. One consists in associating costs to the performed actions in order to assess the collective efficiency of the process we have described and to reuse the same dynamics to solve collectively problems of accessing resources.

The second consists in keeping in mind the main principle of the Hamelin system: the use of an organizational process to profit from individual adaptation abilities for collective task solving. It requires to express this organization process in a formal framework (Decentralized Markov Decision Processes for example) where a collective problem can be described and where a local evaluation of the associated costs can guide the organizational process. Future works consist in expressing an organization principle similar to dominance relationship that can be guided by the task to be achieved.

References

Bonabeau, E. and Theraulaz, G. (1999). *Swarm intelligence.* Oxford university press.

Bonabeau, E., Theraulaz, G., and Deneubourg, J. (1996). Mathematical models of self-organizing hierarchies in animal societies. *Bulletin of Mathematical Biology*, pages 661–717.

Bourjot, C. and Chevrier, V. (2001). Multi-agent simulation in biology: application to social spiders case. In *Agent Based Simulation, Passau, Germany*, pages 18–23.

Camazine, C., Deneubourg, J., Franks, N., Sneyd, J., Theraulaz, G., and Bonabeau, E. (2001). *Self-Organization in Biological Systems.* Princetown university press.

Corbara, B., Drogoul, A., Fresneau, D., and Lalande, S. (1993). Simulating the sociogenesis process in ant colonies with manta. In *Toward A Practice of Autonomous Systems: Proceedings of the First European Conference on Artificial Life*, pages 224–235.

Deneubourg, J., Aron, S., Goss, S., and Pasteels, J. (1990). The self-organizing exploratory pattern of the argentine ant. *insect behavior*, pages 159–168.

Desor, D., Krafft, B., Toniolo, A., and Dickes, P. (1991). Social cognition in rats : incentive behaviour related to food supply. In *Proc XXIInd Int. Ethologica conference.*

Desor, D. and Toniolo, A. (1992). Incentive behaviour in structure groups of rats : about the possible occurence of socio-cognitive processes. In *Comparative approach in sciences cognitives, Aix en Provence.*

Dury, A., Vakanas, G., Bourjot, C., Chevrier, V., and Krafft, B. (2001). Multi-Agent Simulation to Test a Coordination Model of the Prey Capture in Social Spiders. In *13th European Simulation Symposium - ESS'01, Marseille, France*, pages 831–833.

Grasse, P. P. (1959). La reconstruction du nid et les coordinations interindividuelles chez bellicositermes natalensis et cubitermes sp., la théorie de la stigmergie : essais d'interprétation du comportement des termites constructeurs. *Ins. Soc.*, 6:41–84.

Hemelrijk, C. K. (1996). Dominance interactions, spatial dynamics and emergent reciprocity in a virtual world. In Press, M. T. M., (Ed.), *Proceedings of the fourth international conference on simulation of adaptive behavior*, volume vol. 4, pages pp. 545–552, Cambridge.

Hogeweg, P. and Hesper, B. (1986). Knowledge seeking in variable structure models. *Modelling and Simulation Methodology in the Artificial Intelligence Era.*

Kunz, H. and Hemelrijk, C. (2003). Artificial fish schools: collective effects of school size, body size, and body form. *Artificial Life*, pages 237–253.

Miramontes, O. and DeSouza, O. (1996). The nonlinear dynamics of survival and social facilitation in termites. In *Journal of Theoretical Biology*, pages 373–380.

Reynolds, C. W. (1987). Flocks, herds, and schools: A distributed behavioral model. *Computer Graphics*, 21(4):25–34.

Schroeder, H., Toniolo, A., Nehlig, A., and Desor, D. (1998). Long-term effects of early diazepan exposure on social differentiation in adult male rats subjected to the diving-for-food situation. *Behavioural Neuroscience*, pages 1209–1217.

Simonin, O. and Ferber, J. (2000). Modeling self satisfaction and altruism to handle action selection and reactive cooperation. In *6th conference of simulation and adaptive behaviour*, pages 314–323.

Theraulaz, G., Bonabeau, E., and Deneubourg, J. (1998). Response threshold reinforcement and division of labour in insect societies. *Proc. Roy. Soc. London B 265*, pages 327–332.

Theraulaz, G., Goss, S., Gervet, J., and Deneubourg, J.-L. (1990). Task differentiation in polistes wasp colonies: a model for self-organizing groups of robots. In *First International Conference on Simulation of Adaptive Behavior*, pages 346–355.

BASC, a bottom-up approach to automated design of spatial coordination

Fabien Flacher[*,**] **Olivier Sigaud**[**]

[*]Dassault Aviation, DGT/DPR/ESA
78, Quai Marcel Dassault, 92552 St-Cloud Cedex

[**]AnimatLab-LIP6, 8, rue du capitaine Scott, 75015 PARIS
fabien.flacher@lip6.fr olivier.sigaud@lip6.fr

Abstract

The design of spatial coordination mechanisms for a group of agents is a challenging problem to which the top-down approach of traditional Artificial Intelligence often fails to find convincing solutions. Bottom-up approaches have given more promising results, but they often rely on a tedious manual hand-crafting that raises scalability issues. In this paper, we show that combining a bottom-up spatial coordination mechanisms with specially designed evolutionary methods to search the space of solutions is an efficient approach to such problems. More precisely, we show the benefits of our platform, BASC, through a quantitative comparison with previous work published by Balch and Hybinette and we conclude on the methodological issues raised by our work.

1. Introduction

Imagine you are a gnu in the wilderness. To avoid feeding one of the local predators, you need to move adequately. For example, you must not go too far away from your mates, staying in the pack of gnus, the shape of which will globally reduce the number of individuals exposed to danger. You can also complicate the task of your predators by making difficult an attack by surprise. For that purpose, you can move so as to avoid obstructing the line of sight of your mates, so that each one can concentrate his attention on a piece of savannah, while the others cover the rest. In order to do this, at each instant, you must observe your neighbors, locate and orient yourself according to their movements, in order to satisfy the collective objective of the group: this is an example of spatial coordination problem.

The scope of this problem is very large. It concerns as well animation in movies, educational, entertainment, insect colony and military simulations (Van Panurak et al., 2000, Collins et al., 2000) and collective robotics (Luke and Spector, 1996, Nolfi and Floreano, 1998). Thus a lot of researchers in different fields try to understand or to reproduce natural mechanisms in order to realize spatial coordination.

This problem raises specific difficulties: the agents should coordinate themselves in an environment often partially observable and should consider their goals, constraints and the goals and constraints of all other agents in order to be correctly located and oriented. This generates a strong interdependence between the movements of each agent. These properties participate in the complexity of the design of the spatial coordination mechanisms. The partial observability problem prevents the use of classical motion planning algorithms which generally need a complete knowledge of the motion problem and the strong interdependence between movements makes ineffective most of the top-down approaches originating from traditional Artificial Intelligence. Such approaches consist in predicting, from the problem specifications, all the possible configurations and associating *a priori* a solution to each relevant configuration. But, predicting all possible configurations of a spatial coordination problem is all the more tedious that the complexity of the interdependence between movements increases with the number of agents.

From these observations, some researchers have tried to apply techniques drawn from Situated Artificial Intelligence or the Animat approach. They rely on a bottom-up method decomposed into two parts: the generation of a variety of potential solutions and the auto-organization or the automatic selection of solutions corresponding to the specifications of the problem.

Important results in this domain stem from Reynolds' *steering behaviors* (Reynolds, 1987). He defined for every agent a combination of speed vectors built from local information and making it possible to avoid inter-agent collision, to match velocity or to gather flock. He demonstrated the generation of successful coordinated group behavior (the so-called *boids*) from the auto-organization of a small set of local behavioral modules. But an important limitation of his approach results from the manual definition and tuning of all speed vectors.

Comparable results were obtained by collective robotics researchers such as Matarić (Matarić, 1994, Matarić, 1995). She exhibited a set of basic behaviors[1] which, once carefully tuned, suffice to solve complex spa-

[1]exploration, aggregation, dispersion and homing

tial coordination problems involving robots, such as encircling or flocking. However, as Reynolds, she had to pre-define these basic behaviors manually.

Arkin and colleagues also investigated spatial coordination mechanisms (Arkin, 1989, Arkin, 1992). They built a theory which decomposes a behavior into *schemas*, sorts of behavioral atoms, represented as attractive or repulsive vector fields. Schemas are combined by summing attractive fields for goals and repulsive fields for obstacles. The collective behavior emerges from the combination of schemas. In some extensions of this theory (Balch and Arkin, 1998, Balch and Hybinette, 2000), Balch did show that the generation and maintenance of formations can be synthesized by combining schemas. But, despite of Balch's innovations, their approach suffers from the same drawbacks as those of Reynolds or Matarić, concerning the manual pre-definition of a set of basic schemas.

Globally, most bottom-up approaches handling the spatial coordination problem were developed according to a similar principle and share analogous limitations. They manually define a set of basic solutions which is consistently restricted and strongly depending on the specific expertise of the designers. From the combination and auto-organization of these solutions, a spatial coordination mechanism emerges, which is again manually selected by designers. Thus, if the overall mechanism does not fit, designers must exploit their expertise to modify or to tune some parameters in the solution set or in their combination. It is clear that such a process is often tedious and would collapse in front of large scale problems. Papers from Arkin and colleagues regularly exposing new schemas to extend the scope of their theory underline the difficulty to find and to manually tune an ideal set of basic schemas[2].

Arisen from this observation, our approach, BASC, is an extension of these bottom-up approaches to the spatial coordination problem, in particular that of Arkin and Balch, with evolutionary mechanisms. Our goal is to get collective behaviors without any manual tuning. For that purpose, we define in section 2. a formal model which reproduces the properties of Arkin and Balch's theory and is completely customizable by a search algorithm. Thus, we have no pre-tuned schemas, but rather a generic definition of any schema. An exploration and automatic selection algorithm manipulates this definition and extracts a specific solution to a given problem.

Since we extend Arkin and Balch's framework, we reproduce in section 3. one of their experiments, consisting in maintaining a group of vehicles in formation while they move to a specified goal and avoid obstacles. In section 4., we verify Balch's results and then we show that we automatically obtain a solution comparable to that of Balch with much less implication from the designer.

[2]this ideal set is also sought by Matarić

We then compare our approach to other frameworks coupling the evolutionary tuning of parameters to bottom-up spatial coordination mechanisms, and we show that BASC both looks for a solution in a larger search space and finds them more efficiently in that space. We conclude on the methodological issues raised by this work.

2. A Bottom-up Approach to Spatial Coordination

2.1 About BASC

Our bottom-up approach for the generation of spatial coordination properties consist of into two modules.

A first module allows to synthesize spatial coordination mechanisms resuming and extending principles from the bottom-up approaches mentioned in the introduction. As in the schema theory, this module represents the control of the movement of agents as force fields. But we extend this framework by giving a more generic and parametrized definition of these forces.

Most spatial coordination problems being fundamentally multi-objective, a second module is devoted to the exploration and the automatic selection of controllers by approximating the Pareto Front[3] of the problem. This module is implemented as a Genetic Algorithm (GA) specialized in this type of optimization.

2.2 Spatial coordination synthesis

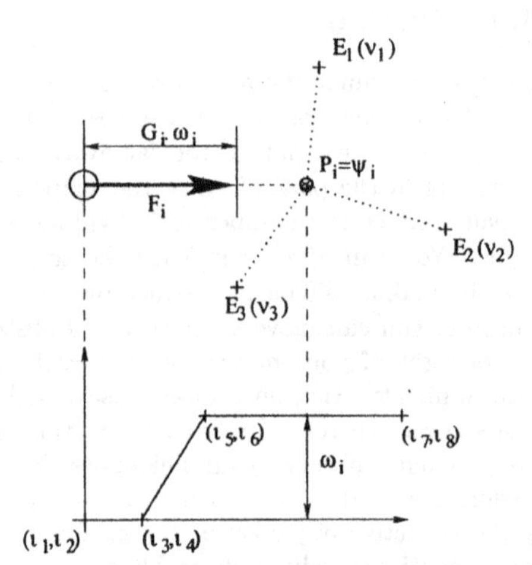

Figure 1: *Example of the construction of one force F_i. The point of interest P_i, targeted by F_i, is built as the barycenter of the perceived elements E_1, E_2, E_3 respectively weighted by ν_1, ν_2 and ν_3. The function ω_i, parametrized by $\iota_1, \cdots, \iota_8$ returns the partial intensity of F_i, finally modulated by the gain G_i.*

[3]the set of *non-dominated* solutions on all the criteria, see section 2.3.

A spatial coordination mechanism needs to correctly position and orient the considered group of η mobile agents. In order to compute this overall displacement, each agent moves by following a movement law built from a combination of forces. Thus, the future movement $\overrightarrow{M_i}$ of an agent A_i is modelled by a set of δ forces $\overrightarrow{F_k}$. Each of these forces is a vector originating from the agent and built from a direction and an intensity.

$$\overrightarrow{M_i} = \sum_{k=1}^{\delta} \overrightarrow{F_k} \tag{1}$$

The direction of a force $\overrightarrow{F_k}$ is obtained by connecting the agent A to a particular point of the environment, named *point of interest.* We define the *points of interest* P_k as barycenters of a some potentially relevant elements of the environment. To determine this barycenter at each time step, the agent A_i establishes a list of elements E_j perceived in its immediate surrounding. The barycenter P_k is then built by associating a weight to every object E_j perceived and referenced. Thus, we define for each force a linear combination ψ_k, parametrized by α reals, taking α elements E_i of the environment $\mathcal{E}$ as entry and producing the point P_k :

$$\psi_k^{\nu_1,\cdots,\nu_\alpha}(E_1,\cdots,E_\alpha) = P_k = \sum_{j=1}^{\alpha} \nu_j.E_j, \quad \sum_{j=0}^{\alpha} \nu_j = 1 \tag{2}$$

The intensity of a force $\overrightarrow{F_k}$ is obtained via the product between a gain G_k and a normalized function ω_k according to the distance of the agent A to the corresponding point of interest P_k. In practice, we define this function of intensity ω_k as a simple piece wise linear function, parametrized by β reals, taking the distance $\parallel \overrightarrow{AP_k} \parallel$ as entry and producing the intensity of the force $\overrightarrow{F_k}$:

$$\omega_k^{\iota_1,\cdots,\iota_\beta} = \begin{cases} \frac{\iota_4-\iota_2}{\iota_3-\iota_1}(x-\iota_1)+\iota_2, & x\in[\iota_1,\iota_3] \\ \vdots & \\ \frac{\iota_{i+3}-\iota_{i+1}}{\iota_{i+2}-\iota_i}(x-\iota_i)+\iota_{i+1}, & x\in[\iota_i,\iota_{i+2}] \\ \vdots & \\ \frac{\iota_\beta-\iota_{\beta-2}}{\iota_{\beta-1}-\iota_{\beta-3}}(x-\iota_{\beta-3})+\iota_{\beta-2}, & x\in[\iota_{\beta-3},\iota_{\beta-1}] \end{cases} \tag{3}$$

The gain G_k represents the relative importance of $\overrightarrow{F_k}$ w.r.t. the other forces exercized on the agent.

The movement function $\overrightarrow{M_i}$ of the agent A_i is given by equation (4).

$$\overrightarrow{M_i} = \sum_{k=1}^{\delta} G_k \times \omega_k(\parallel \overrightarrow{A_iP_k} \parallel) \times (\sum_{j=1}^{\alpha} \nu_j.\overrightarrow{A_iE_j}) \tag{4}$$

Finally, with our formalism, we are able to synthesize a spatial coordination mechanism from η movement functions $\overrightarrow{M_i}$, $i \in [1,\ \eta]$, one for each agent A_i. Thanks to the parametric aspect of our model, we can represent this mechanism by a set of ρ reals (cf. equation (5)).

$$\rho = \eta.\rho_h, \quad \rho_h = \delta.(1+\alpha_M+\beta) \tag{5}$$

2.3 *Automatic exploration and selection*

Given a multi-objective optimization problem, a solution must be simultaneously optimized on several criteria. Such a solution is an optimum in the partially ordered space of criteria, and several optimal solutions can be equivalent w.r.t. the criteria they optimize[4]. Such solutions are said *non dominated.* For a given multi-objective problem, the set of *non dominated* solutions is called the Pareto Front (Pareto, 1896).

Since spatial coordination problems can often be formalized as multi-objective optimization problems, we use a GA freely inspired from the one Goldberg applied to the approximation of Pareto Fronts (Goldberg, 1989). This algorithm is efficient enough, flexible and simple to implement, even if several more powerful algorithms have been published recently (Deb, 2001). As any such algorithm, it can be divided into three sub-parts: the genetic encoding representing the parameter setting of a solution, the genetic operators allowing the random modification of solutions and the selection methods allowing the convergence toward the final solution set. These three sub-parts are described in the following sections.

2.3.1 *Genetic encoding*

In our framework, since the number of agents can increase, we choose to look for an homogeneous spatial coordination mechanism. This means that the same movement function $\overrightarrow{M_i}$ is bred and applied to all agents. Thus, as indicated in equation (5), the number ρ_h of parameters coding this function is independent of the number of agents η.

Moreover, in our GA, one genome codes a movement function for one agent. As previously shown, this function is decomposed into δ attraction-repulsion forces. As a consequence, the genome consists of δ independent chromosomes. Every chromosome represents the gain, the direction function and the intensity function determining one force with $(1+\alpha+\beta)$ real numbers.

The real numbers encoding a genome are chosen in [0, 1]. Any manipulation of these values realized by the genetic operators keep them in [0, 1]. During the transformation of the genome into the movement function, these genetic parameters are mapped to the adequate intervals which depend on the specific application.

[4]For instance, if we try to maximize two criteria c_1 and c_2, a solution can be maximal on c_1 and not on c_2. On the contrary, another solution can be maximal on c_2 and not on c_1. In fact, none of these solutions is globally optimal, they do not dominate each other.

2.3.2 Genetic operators

We use two genetic operators: mutation and cross-over. The mutation operates by random modifications and results in a local exploration in the neighborhood of a solution. The cross-over exchanges genetic materials between two solutions and investigates the search space between different solutions.

The mutation operator is applied by adding a random value depending on a normal law to each parameter of the genome. This addition is made for every value with a probability p_M. The cross-over is applied with a probability p_C in two steps. First, the chromosomes of two solutions are gathered into a set of $(n_{C1} + n_{C2})$ chromosomes. Then, n_{C3} chromosomes are randomly chosen in this set and copied into a new individual.

2.3.3 Selection algorithm

The selection algorithm is designed to retain the most promising genomes and to apply the genetic operators to them so as to generate a new population of solutions. The Pareto dominance principle is applied to perform the selection process. It consists in using the dominance relation to rank each genome from its performances and to derive its selection probability. We used a ranking method based on the *non dominance strata*. The first stratum is the Pareto Front and every genome in this stratum is assigned the rank 1. The second stratum consists of the *non dominated* solutions remaining when the first stratum is removed. The corresponding genomes are assigned the rank 2, etc. The selection probability is proportionally distributed according to the rank of the genomes. Then, as many genomes as needed to create a new population are randomly selected from a roulette wheel algorithm (Goldberg, 1989) on this probability mapping.

3. Experimental set-up

3.1 Simulation environment

The experiments presented here are inspired from the work of Balch and Hybinette on a formation maintenance problem (Balch and Hybinette, 2000). On this problem, Balch applied the schema theory as a spatial coordination model. He mapped his expertise of this problem into the parameter setting of a set of schemas. We chose to reproduce this experiment because of the similarity between BASC and this schema theory. Indeed, our formalism includes the one defined by Balch and Arkin and extends it toward a completely automated design process. Our concept of point of interest is equivalent to that of point aimed by the schema, our intensity function is equivalent to Balch and Arkin's function of magnitude and we both modulate it by a gain.

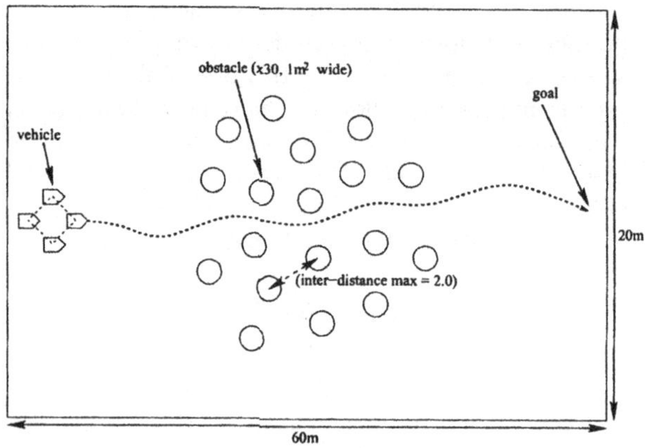

Figure 2: *Formation maintenance experiment: a group of vehicles must reach a goal while staying in formation and avoiding obstacles.*

At the origin, this experiment was developed on the *Team Bots platform*[5] by Balch to prototype control architectures for unmanned ground military vehicles. Our simulation resumes their specifications. As indicated in figure 2, it implies four holonome vehicles moving across a field as quickly as possible while maintaining a geometric formation and avoiding collisions with obstacles and other vehicles. The field is 20m x 60m wide. 30 obstacles, each $1m^2$ wide, are distributed around the center of the field in a way that prevents the group of vehicles from crossing the field of obstacles without deforming its formation[6]. We consider the geometric formation indicated on the same figure, the four vehicles moving in diamond inscribed into a circle of radius $r_d = 1.5$ meter and following a leader vehicle.

3.2 Performance criteria

Because of the conflict between maintaining formation and avoiding the obstacles, we need to use explicitly different criteria to evaluate the collective behavior of our agents. Thus, we defined three separate criteria (*formation maintenance*, *obstacle avoidance* and *goal reach*), as follows :

- A punishment is determined at each time step from the position error to the ideal formation configuration. The final reward on the run duration t_M is the average over the run. Let P_1^t, P_2^t, P_3^t and P_4^t be the correct position in the formation at instant t. We have the following fitness f_1 for the *formation maintenance* criterion:

[5] avaible online at `http://www.cs.cmu.edu/coral/minnow`.

[6] A maximum inter-distance of 2.0 between each obstacles is imposed.

$$f_1(a) = \frac{1}{t_M} \sum_{t=1}^{t_M} r[e_f(a)],\ a \in \mathcal{A} \quad (6)$$

where e_f is a function giving the formation error of a given group of agents, r a reward function and $\mathcal{A}$ the group of agents:

$$e_f(a) = \sum_{k=1}^{4} |\overrightarrow{A_k P_k}|, r(x) = \begin{cases} 1.0 & \text{if } x < 0.05 \\ \frac{0.5-x}{0.45} & \text{if } x \in [0.05, 0.5] \\ 0.0 & \text{if } x > 0.5 \end{cases} \quad (7)$$

- The number of collisions between vehicles and obstacles during the run is memorized. The difference between the number of collisions q and a maximum number of authorized collisions C_M give the fitness f_2 for the *obstacle avoidance* criterion. Formally, we have:

$$f_2(a) = \frac{c_M - q(a)}{c_M},\ a \in \mathcal{A} \quad (8)$$

- When the assigned time for the task resolution is over, a reward is determined from the average distance of all vehicles to the goal. This reward is the fitness f_3 for the *goal reach* criterion. Formally, if G is the goal and e_g the function giving the cumulated position error of a group of agents w.r.t. the goal at the assigned time for the resolution, we have:

$$f_3(a) = r[e_g(a)],\ e_g(a) = \sum_{k=1}^{4} |\overrightarrow{A_k G}|, a \in \mathcal{A} \quad (9)$$

3.3 Balch's schema-based controller

In order to make a comparative study between our model and Balch's, we first reproduced a schema-based controller according to (Balch and Hybinette, 2000). This controller consists of six schemas: `avoid-static-obstacles`, `avoid-robots`, `move-to-goal`, `move-to-unit-center`, `maintain-formation` and `noise`. At each time step, each schema generates a vector representing a component of the movement of each vehicle. Each vector is multiplied by a gain indicating the relative importance of its associated schema. The six resulting vectors are added to obtain the global direction and intensity of the movement.

The six output vectors of schemas are determined from two functions which return the magnitude of the vector according to the distance to the point aimed by the schema. These functions are detailed in figure 3. The first function is decreasing and is parametrized by the real numbers M and S. Associated to a negative gain, this function realizes an repulsion which increases if the agent get closer to the target, an obstacle for the `avoid-static-obstacles` schema or a robot for the `avoid-robots` schema. The second function is increasing and is parametrized by the real numbers C and D. Associated to a positive gain, this function realizes an attraction which increases if the agent goes away from the target, a goal for the `move-to-goal` schema, the barycenter of the group for the `move-to-unit-center` schema or the ideal position in the formation for the `maintain-formation` schema. This ideal position is the closest attachment site of the vehicle. Attachment sites are built from the position of the others robots, as indicated in figure 4. The output vector of the last schema `noise` is built from a random direction in $[O, 2\pi]$. It is parametrized by P, the number of iterations between any change of direction.

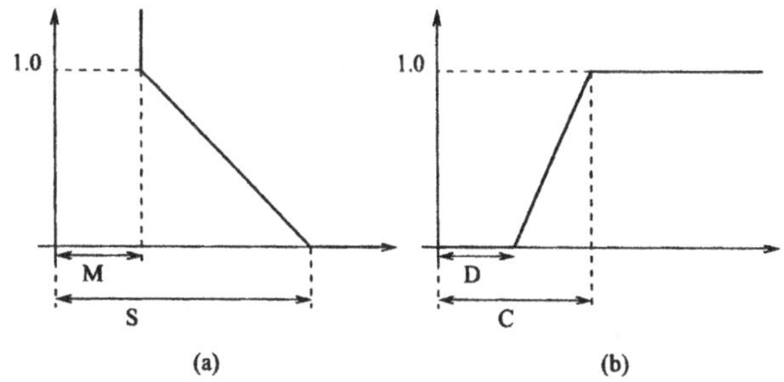

Figure 3: *Intensity functions used by `avoid-static-obstacles` and `avoid-robots` schemas (a), and `move-to-goal`, `move-to-unit-center` and `maintain-formation` schemas (b)*

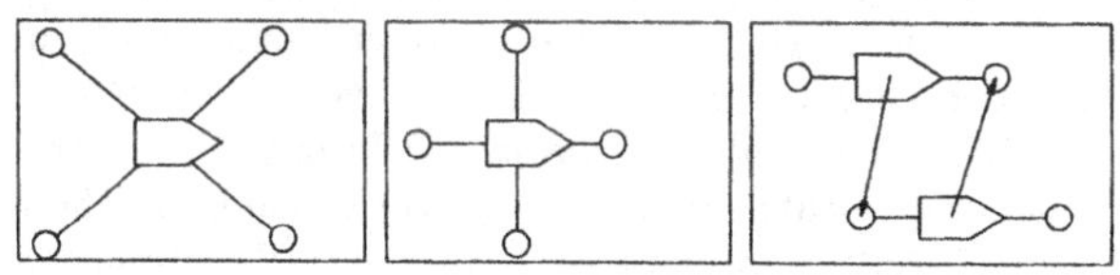

Figure 4: *From the point of view of each vehicle, every other vehicle has several attachement sites, shown with circles, which are the target of the `maintain-formation` schema.*

The functions parameters and gains of the various schemas developed by Balch and rebuilt here are summarized in table 1. The corresponding solution will be referred to as S_{balch} in the following.

3.4 Controller generated by BASC

We submitted the same task to BASC in order to obtain a controller for this specific spatial coordination problem. The evolution tuned the gains, directions and

Schemas	Gains
`avoid-static-obstacles` $M = 0.664$ $S = 2.564$	-1.1
`avoid-robots` $M = 0.4$ $S = 2.3$	-1.1
`move-to-goal` $D = 0.0$ $C = 0.0$	0.7
`move-to-unit-center` $D = 2.0$ $C = 3.0$	0.6
`maintain-formation` $D = 0.0$ $C = 1.0$	1.3
`noise` $P = 5.0s$	0.1

Table 1: *Balch's schema-based controller parameters.*

$\delta = 6$ chromosomes / genome		
type	parameters per chromosome	parameters per genome
gain G_k	1	6
direction function $\nu_{k,1}, \ldots, \nu_{k,\alpha}$	$\alpha_M = 6$	36
intensity function $\iota_{k,1}, \ldots, \iota_{k,\beta}$	$\beta = 10$	60

Table 2: *Genome encoding of a controller.*

intensity functions parameters (cf. section 2.3.1) during 5 days[7]. 102 parameters on the whole describe the global movement function of each agent as indicated in table 2. These parameters are optimized via 5 evolutions, launched with a population of $n_P = 100$ individuals and a probability of cross-over $p_C = 40\%$, and mutation $p_M = 5\%$. The maximum time assigned to a simulation is $t_M = 500$, the maximum number of authorized collisions is $C_M = 30$ and every evaluation of a controller is realized from $n_E = 25$ initial positions. The average value of the criteria is calculated on these various trials to evaluate the controllers. We present in the following section a comparison of the solution S_{balch} and the best solution obtained with BASC, S_{basc}.

4. Results

4.1 Performance comparison

The results obtained with S_{balch} correspond to those presented in (Balch and Hybinette, 2000). However, this equivalence is essentially qualitative because Balch described only a single quantitative evaluation of his solution: the robustness of task duration when the number of agents increases. As we can notice in figure 5, S_{balch} is scalable on this criterion, since the performance remains close to the average value 623.36 (with a standard deviation of 7.2% of the average duration). Balch concludes that his solution his scalable because the reported average duration standard deviation represents only 10% of this average, which seems to indicate that this performance is stable in spite of the increasing number of agents.

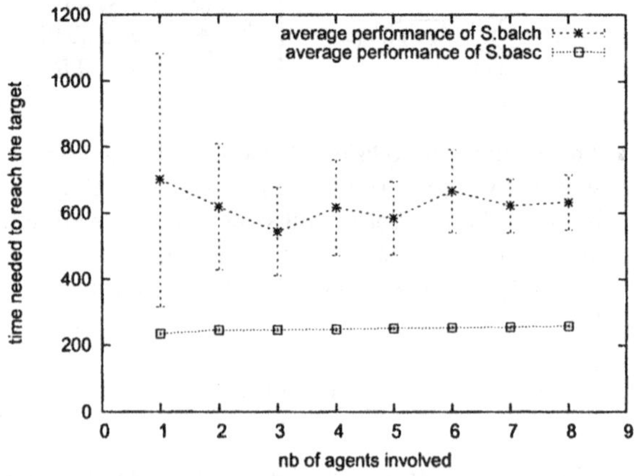

Figure 5: *Task duration scalability. Curves show the average duration over 100 runs, starting with random initial obstacles and vehicle positions. As in Balch's results, S_{Balch} is scalable though we additionally show here with error-bar that is not very robust. Moreover, S_{basc} is also scalable, perfectly robust (the error-bars are shown but reduced to points), and globally faster.*

Table 3.a reports the average performances of S_{balch} and figure 6.1 illustrates the average progress of these performances on 100 runs. From these data, we can observe that S_{balch} generates a correct spatial coordination mechanism which is only disrupted during the crossing of the obstacle field. As shown in figure 6.1, as soon as the group enters the obstacle zone (stage (1.b)), it slows down and forgets the formation until it leaves the zone (stage (1.c)). Then the group resumes a satisfactory speed and a correct formation. This restoring capability of the nominal behavior of the group illustrates the stability and robustness of the generated spatial coordination mechanism. However, the performance of 89.3% to the *goal reach* criterion does not only result from the time lags. Indeed, in 10% of cases, the group fails to exit the cluttered zone. This gap justifies the important standard deviation (14.9%) and error-bars (see figure 5) measured on this performance. This phenomenon results from the fact that Balch's solution adds to each agent an `avoid-static-obstacles` schema for every detected obstacle. When the configuration of the obstacles field is such that the cumulated sum of all `avoid-static-obstacles` schemas begin to be superior

[7]BASC is a Java 1.4.1 code source executed on a 2.4 GHz Pentium IV operating under linux.

solutions	obstacles avoidance	formation maintenance	goal reach
a: solutions tested during 500 iterations			
S_{balch}	90.2 (13.9)	53.1 (10.9)	89.3 (14.9)
S_{basc}	98.5 (3.1)	29.1 (4.8)	100.0 (0.0)
b: solutions tested during 190 iterations			
S_{basc}	98.5 (3.1)	67.6 (8.0)	89.8 (0.8)
$S_{basc} - \vec{F_4}$	99.8 (1.2)	62.4 (6.9)	88.3 (1.2)
c: confidence* of difference between $S_{basc(190)}$ and			
S_{balch}	> 99.9%	> 99.9%	> 95.0%
$S_{basc} - \vec{F_4}$	> 95.0%	> 95.0%	> 95.0%

* *using standard distribution-free Mann & Whitney U test*

Table 3: *Performances in % of the different solutions on the 3 criteria. Results are averaged over 100 runs (standard deviation indicated into brackets). Part c of the table gives the statistical significance of difference between the solutions* $S_{basc(190)}$, S_{balch} *and* $S_{basc} - F_4$.

to the cumulated sum from all others schemas, the group stops moving toward the goal, keeping stuck in a local minimum. The more dense the obstacle field is, the more likely it is to encounter such configurations.

Quite as for S_{balch}, the task realization duration from S_{basc} is scalable w.r.t. the number of agents (cf. figure 5). Furthermore, S_{basc} is globally faster than S_{balch} since it needs only 248.91 steps on average (with a standard deviation of 2.7%) to cross the field. More exactly, we notice in table 3.a that the solution S_{basc} perfectly manages to reach the goal (100 % with null standard deviation) and is thus more robust than S_{balch} on this criterion. Nevertheless, S_{balch} is clearly better concerning the *formation maintenance* while both *obstacle avoidance* performances are equivalent. These solutions do not dominate each other in Pareto dominance terms because they are each specialized on one of the aspects of the problem. In fact, a detailed study of the average performances of S_{basc} illustrated in figure 6 shows that the *formation maintenance* performance of S_{basc} is continuously decreasing though the vehicles always succeed to reach the goal (stage (2.c)). Actually, once the goal is reached, the vehicles cannot correctly maintain the formation because they do not stop moving and oscillate around the goal. Since the *formation maintenance* evaluation is a cumulated reward (see equation (6)), S_{basc} is finally weaker than S_{balch} on that criterion. But as shown on table 3.b, when we test S_{basc} during 190 iterations[8] only, the *goal reach* performance of the two solutions become equivalent while S_{basc} starts performing better than S_{balch} on *formation maintenance*. In fact, in that case, a Mann & Whitney statistical test on the average difference significance (see table 3.c), shows that S_{basc} is significantly better than S_{balch} on all criteria. Thus, at a given time w.r.t. the realization of the task, S_{basc} dominates S_{balch} in Pareto dominance terms.

[8] 190 iterations is the average duration needed by S_{basc} to obtain a similar performance on *goal reach* criterion than that obtained by S_{balch} in 500 iterations

force	gain	barycenter P_k expression	function
$\vec{F_1}$	-38.95	$\overrightarrow{AP_1} = \overrightarrow{AT}$	ω_1
$\vec{F_2}$	-16.50	$\overrightarrow{AP_2} = \overrightarrow{AV_1}$	ω_2
$\vec{F_3}$	-90.10	$\overrightarrow{AP_3} = \overrightarrow{AO_1}$	ω_3
$\vec{F_4}$	-6.11	$\overrightarrow{AP_4} = 0.86.\overrightarrow{AT} + 0.14.\overrightarrow{AO_2}$	ω_4

Table 4: *The four forces used by* S_{basc}. *G is the goal,* V_1 *is the closest vehicle and* O_i *is the* i^{th} *closest detected obstacle.*

4.2 S_{basc} analysis

Solutions obtained via BASC can be difficult to analyze. The large number of parameters in our model gives rise to a large variety of possible solutions. Even so, the solution selected by the GA in this vast set is often complex since the generated behavior results from the interaction of several dynamical interferences. However, with some simplifications, we made this analysis and partially understood how S_{basc} operates. Its genome codes for 4 forces summarized in table 4 and figure 7.

The role of the first three forces is rather obvious:

- $\vec{F_1}$ attracts every vehicle towards the goal,
- $\vec{F_2}$ tends to place vehicles on a circle of radius r_0. Indeed, from the point of view of each vehicle, this force is directed towards the closest other vehicle. Moreover, by looking at the intensity function ω_2, we see that it crosses the x axis at $x_0 = 2.26$. As a consequence, if the vehicle influenced by $\vec{F_2}$ is less than x_0 meters away from the closest vehicle, ω_2 returns a negative value, generating a repulsive force: the concerned vehicle is then repelled by the closest vehicle. Otherwise, it is attracted. Thus $\vec{F_2}$ generates a stabilization at distance x_0 to the closest vehicle. As all the vehicles move in the same way, they place themselves on a circle of radius $r_0 = \frac{x_0}{\sqrt{2}} \approx 1.60$ and as far from each other as possible.
- $\vec{F_3}$ repulses every vehicle from the closest obstacle.

Each of these three forces handles one of the three aspects of the problem. Nevertheless as we can notice in table 3.c, the *formation maintenance* is significantly less efficient if we remove the force $\vec{F_4}$ as confirmed by a Mann & Whitney statistical test. $\vec{F_4}$ is oriented towards a point which is always very close to the goal and seems

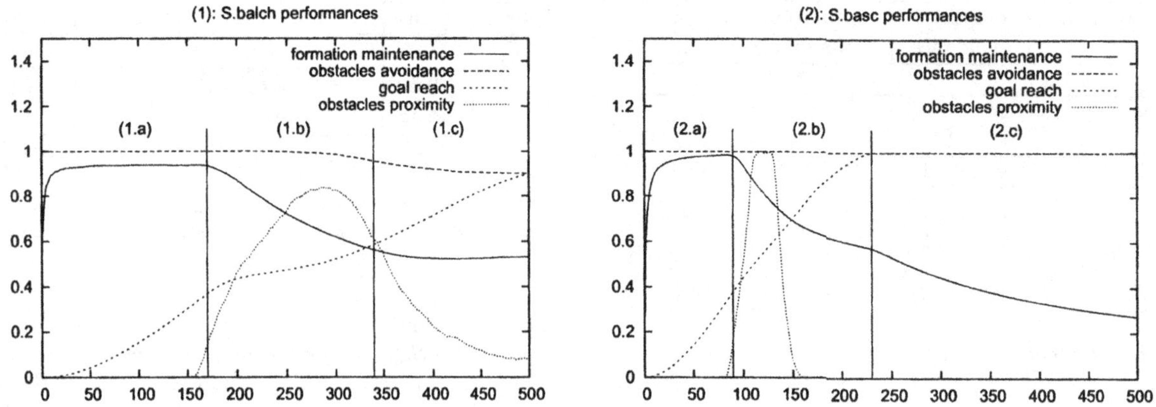

Figure 6: *Performances of S_{balch} (1) and S_{basc} (2). Each curve shows the average progress overs 100 runs on each criteria. The obstacles proximity curves show an evaluation of the virtual collisions number between vehicles and obstacles (the amplitude of this curve is divided by 20 for the sake of clarity). The average behavior of S_{balch} is decomposed into three stages: a stage (1.a) during which vehicles successfully maintain formation and get closer to the goal; a stage (1.b) during which vehicles slow down (constant part of the goal reach curve) and deform their formation (drop of the formation maintenance curve); and a last stage (1.c) during which vehicles resume their previous coordination by getting closer to the goal and maintaining a correct formation. The average behavior of S_{basc} is decomposed into the same stages, but at the last stage vehicles do not stop moving and oscillate around the goal.*

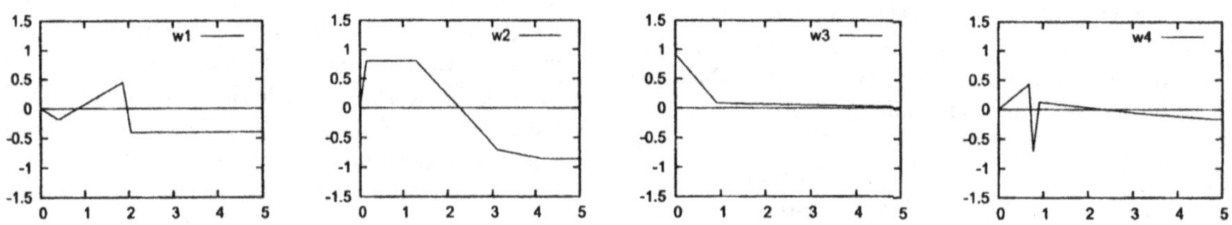

Figure 7: *Evoldved intensity functions ω_1, ω_2, ω_3, ω_4 modulating the intensity of the forces F_1 to F_4 in S_{basc}.*

to play a role in the orientation of the formation in presence of obstacles. But a more detailed geometrical analysis is still necessary before we completely understand its mode of operation.

5. Discussion

Our experiments have demonstrated the possibility to automatically generate solutions to a difficult spatial coordination problem. In particular, the automatic exploration and selection properties provided by our bottom-up approach bring methodological advantages: they significantly reduce the implication of the designer and can even find solutions that they would not succeed to design.

Indeed, it is clear that Balch's schema-based solution that we manually rebuilt tackles the obstacles field crossing problem and is also scalable w.r.t. the number of agents. Admittedly, the simplicity of the schema theory makes it possible to exploit human expertise to develop new schemas. But we have shown that, in front of the variability of this complex problem, the schema-based solution supplied by Balch does not always realize a robust spatial coordination mechanism: it fails in front of too dense fields of obstacles. An expert could possibly correct the lack of robustness, by finding the flaws and manually tuning schemas to generate a new solution. But, as soon as the scale of the problem increases, new interactions problems between agents will arise, increasing the number of possible configurations. As the difficulty to prevent other flaws in the new built solutions increases, experts must re-iterate this endless manual trial-and-error design process, making it more and more tedious. The alternative bottom-up approach developed in BASC is exempt from this drawback. The trial-and-error design process is automated by using a GA. As a result, the effort necessary to define each basic schema is not required anymore and the system will find solutions adapted to the particular problem at hand.

Relying on evolutionary generate-and-test methods,

the system can find original solutions, eventually exploiting counterintuitive relations between agents and their environment, as it was the case in the experiment presented in section 4.2. Indeed, where an expert would probably propose forces such as $\overrightarrow{F_1}$ to $\overrightarrow{F_3}$, BASC has added a force $\overrightarrow{F_4}$ which results in improved performance and requires careful analysis to be understood. Eventually, original solutions proposed by BASC can be reused by the expert in the context of more complex problems.

Furthermore, we have shown that the solution obtained via our bottom-up approach builds a spatial coordination mechanism which dominates the mechanism manually built by Balch. This fact speaks in favor of BASC, but must be considered carefully. Indeed, GAs have a tendency to over-specialize the solutions they generate, *i.e.* to optimize them for the particular conditions under which they are tested. Here, for example, the solutions are tested over 25 different runs during the evolution and on 100 different runs in the final evaluation. The solution discovered via BASC is more successful on these 100 tests and, though each one starts with different initial positions, these tests share a certain number of properties, specific to this experiment (strong density of obstacles concentrated in the center of the zone of operation, for instance). These shared properties can be exploited by the GA to specialize the solutions for these particular cases. They will be more successful in one type of experiment (the one tested), but globally less successful under very different conditions.

From this remarks, we must admit that a comparison with S_{balch} would have been more fair if the solution had been specifically tuned with a GA under the same conditions. Actually, evolutionary algorithms have been applied as optimization processes or extensions of several bottom-up approaches to spatial coordination problems. In particular, works extending on the initial ideas of Reynolds proposed a reduced version of the *steering behaviors* automatically tuned by genetic programming algorithm (Reynolds, 1992, Reynolds, 1994). In this extended version, the basic behaviors can represent in theory any navigation mechanism. Unfortunately, Reynolds has only applied these ideas to mono-agent navigation or predator-prey problems and did not investigate the capacity of these approaches to generate spatial coordination mechanisms. Similarly, Arkin and colleagues did extend the theory of schemas with evolutionary approaches. For example, results exposed in (Ram et al., 1994) show that some model parameters such as the range of influence of some schemas (equivalent to the parameters C D of `maintain-formation` or `move-to-goal` schemas) or the gains representing their relative importance in the final combination can be optimized by a GA. In our framework, this would correspond to setting all the parameters δ_i, α_i to an *ad hoc* value and letting the others ($G_{k,i}$, $\beta_{k,i}$) evolve thanks to a basic GA. But, as it is the case for Reynolds, these works were only applied to the automated design of schemas in a mono-agent navigation case.

The GA dilemma is well-known: either the algorithm searches through the complete parameter space, and the algorithm may be too slow to converge, or the search space is restricted by setting some parameters manually, and the quality of the solutions found by the GA heavily relies on this manual definition. (Reynolds, 1994) and (Ram et al., 1994) have chosen the second solution, but they did so in an implicit way without considering that the parameters they have set manually could have been evolved too. Our approach rather consists in making all the parameters explicitly customizable by a GA, and eventually setting some of them by hand if the search space needs to be reduced. We have tested the possibility of setting some parameters manually in the context of a complex industrial project not described here.

Furthermore, since the search space is potentially very large, our formalism and the corresponding evolutionary algorithm have been specialized to efficiently find good solutions to spatial coordination problems. Indeed, whereas Ram (Ram et al., 1994) and Reynolds (Reynolds, 1994) just evolve a set of parameters with a basic GA without taking into account the semantics of the parameters, our framework has been defined so that the mutation and cross-over operators maximize the chance to discover better solutions out of promising ones. For instance, there are two important *building blocks* (Goldberg, 1989) in our formalism: the point of interest corresponding to a barycenter of relevant objects in the environment, and the definition of a force binding the agent to such a point of interest. Our cross-over operator has been defined so that new solutions can incorporate as such some forces used by the previous solutions. As a consequence, the chromosomes defining useful forces are kept and spread naturally into the population. We have shown in section 4.2 that this feature results in the efficient identification of a set of meaningful forces adequately solving the spatial coordination problem.

Moreover, our system explicitly takes into account the fact that a lot of spatial coordination problems are multi-objective. As a consequence, we use a dedicated Pareto optimization algorithm while all the works discussed so far rely on a single-objective standard GA.

However some authors also take into account the multi-objective aspect of spatial coordination problems. (Pirjanian and Mataric, 2000) define a set of behaviors as objective functions mapping the decisions variables towards $[0, 1]$. Thanks to their model, the agent can foresee the consequences of its actions and choose Pareto-optimal actions according to multicriteria decision theory. On the same line, (Panatier et al., 2000) use potential fields to automatically build for every agent a model of the behavior of the others. Thus, each agent can foresee the short term future situation and select the best action. Nevertheless, both models impose strong *a priori* constraints on the definition of the basic behaviors, as in the case of Reynolds or Mataric.

Finally, though being dedicated to spatial coordination problems, our framework is not specialized towards any particular subclass of such problems, and has already been applied successfully to several different applications, from a multi-agent instance of the flock-herding problem (Flacher and Sigaud, 2002) to industrial military simulations. Being focused exclusively on simulation applications, we have not tried to address the issue of applying our framework to robotics, but Arkin and Balch did so in the case where the robots accurately know each other's positions, which shows the feasibility of such a transposition.

6. Conclusion and Future Work

As a general conclusion, we have shown that using a bottom-up approach combining a parametric model of spatial coordination with a special purpose GA is a powerful methodology which requires a small design effort and is more efficient in terms of performance than other bottom-up approaches relying either on manual design or on weaker evolutionary methods.

Adopting our approach implies several methodological issues. First, the problem consisting in designing directly a spatial coordination mechanism has been changed into finding the evaluation criteria that match the requirements of the applications, which is not a trivial task. Second, the result of the evolutionary process may eventually be hard to analyze and rely on counterintuitive mechanisms, but this drawback can also be an asset if the analysis reveals an innovative solution to the problem. Third, depending on the complexity of the application, the designer must decide which parameters should be set manually and which should be submitted to the evolutionary process. Here, there is a tradeoff between the designer and the CPU available. In order to make this tradeoff less critical to the usability of our approach, our agenda of future research consist in incorporating in our framework efficient methods coming from the Learning Classifier Systems paradigm, which should hopefully result in a much faster convergence.

Acknowledgments

The authors thanks Dassault Aviation for the financial support that made this work possible.

References

Arkin, R. C. (1989). Motor Schema-Based Mobile Robot Navigation. *The International Journal of Robotics Research*, 8(4):92–112.

Arkin, R. C. (1992). Behavior-based robot navigation for extended domains. *Adaptive Behavior*, 2(1):201–225.

Balch, T. and Arkin, R. C. (1998). Behavior-based Formation Control for Multi-robot Teams. *IEEE Transactions on Robotics and Automation*, 14(6):1–15.

Balch, T. and Hybinette, M. (2000). Social Potentials for Scalable Multi-Robots Formations. In *IEEE International Conference on Robotics and Automation (ICRA-2000)*, pages 73–80.

Collins, D. J., Agah, A., Wu, A. S., and Schultz, A. C. (2000). The effects of team size on the evolution of distributed micro air vehicles. In *Proceedings of the Genetic and Evolutionary Computation Conference (GECCO-2000)*, pages 949–956, Las Vegas, Nevada, USA. Morgan Kaufmann.

Deb, K. (2001). *Multi-Objective Optimization using Evolutionary Algorithms*. Wiley.

Flacher, F. and Sigaud, O. (2002). Spatial Coordination through Social Potential Fields and Genetic Algorithms. In *From Animals to Animats 7: Proceedings of the Seventh International Conference on Simulation of Adaptive Behavior*, pages 389–390, Cambridge, MA. MIT Press.

Goldberg, D. E. (1989). *Genetic Algorithms in Search, Optimization, and Machine Learning*. Addison Wesley, Reading MA.

Luke, S. and Spector, L. (1996). Evolving teamwork and coordination with genetic programming. In *Genetic Programming 1996: Proceedings of the First Annual Conference*, pages 150–156, Stanford University, CA, USA. MIT Press.

Matarić, M. J. (1994). *Interaction and Intelligent Behavior*. PhD thesis, MIT AI Mobot Lab.

Matarić, M. J. (1995). Designing and Understanding Adaptive Group Behavior. *Adaptive Behavior*, 4(1):51–80.

Nolfi, S. and Floreano, D. (1998). Co-evolving predator and prey robots: Do arm races arise in artificial evolution? *Artificial Life*, 4:311–335.

Panatier, C., Sanza, C., and Duthen, Y. (2000). Adaptive Entity thanks to Behavioral Prediction. In *From Animals to Animats 6: Proceedings of the Sixth International Conference on Simulation of Adaptive Behavior (supplement book)*, pages 295–303, Cambridge, MA. MIT Press.

Pareto, V. (1896). *Cours d'Economie Politique*. Rouge, Lausanne.

Pirjanian, P. and Mataric, M. (2000). Multi-robot target acquisition using multiple objective behavior coordination. In *Proceedings of the 2000 IEEE International Conference on Robotics and Automation*.

Ram, A.and Arkin, R., Boone, G., and Pearce, M. (1994). Using genetic algorithms to learn reactive control parameters for autonomous robotic navigation. *Adpative Behavior*, 2(3):277–304.

Reynolds, C. (1987). Flocks, Herds and Schools: a Distributed Behavioral Model. In *Computer Graphics*, volume 21, pages 25–34.

Reynolds, C. (1992). An evolved, vision-based behavioral model of coordinated group motion. In *From Animals to Animats 2: Proceedings of the Second International Conference on Simulation of Adaptive Behavior*, pages 384–392.

Reynolds, C. (1994). Evolution of corridor following behavior in a noisy world. In *From Animals to Animats 3: Proceedings of the Third International Conference on Simulation of Adaptive Behavior*, volume 3, pages 402–410.

Van Panurak, H., Brueckner, S., Sauter, J., and Matthews, R. (2000). Distinguishing Environmental and Agent Dynamics: A Case Study in Abstraction and Alternate Modeling Technologies. In *ESAW Workshop at ECAI'00*.

The More Radical, the Better: Investigating the Utility of Aggression in the Competition among Different Agent Kinds

Matthias Scheutz **Paul Schermerhorn**
Artificial Intelligence and Robotics Laboratory
Department of Computer Science and Engineering
University of Notre Dame
Notre Dame, IN 46556
{mscheutz,pscherm1}@nd.edu

Abstract

Aggression is wide-spread in nature and seems to serve, among others, an important role in the interspecies competition for resources. In this paper, we argue that displaying aggression as a means to signal action tendencies (in particular, the probability to continue an encounter) is beneficial for social groups and show that discriminating between "own" and "other" is more beneficial than treating "other" the same as "own". In particular, we demonstrate that aggression plays a crucial role in strategies applied to "other". To test the theoretical discussion, we define a seven basic agent types which give rise to 42 different *discriminating agents*, i.e., agents with different strategies for "own" and "other". In extensive simulation studies we show that discriminating agents, which assume an aggressive attitude towards others, while playing a strategy that distributes resources fairly among "own", are ultimately the most successful ones. We discuss the implications of these results for natural and artificial agents and conclude with a brief outlook on further studies.

1. Introduction

Competition for resources is wide-spread in nature and has been studied extensively by ethologists. Typically, conflicts or contests among individuals are considered games (in a game-theoretic sense, e.g., (Maynard Smith, 1974, Maynard-Smith, 1982), etc.), where all players want to maximize their payoff and the winner is determined after a number of rounds (e.g., after one round in games like the "Prisoner's Dilemma", the "'Dove-Hawk Game", or the "Game of Chicken"). Some of these games can be repeated many times (e.g., the "Iterated Prisoner's Dilemma" (Axelrod and Hamilton, 1981) , or the "Sequential Assessment Game" (Enquist and Leimar, 1983)) in order to study the change in strategies based on previous outcomes, while others might end terminally after one round (e.g., when both players accept the dare in a "Game of Chicken"). Given the payoff matrices for these games (i.e., the payoff for winning and losing), different strategies for playing the game can be studied. In particular, it is often possible to determine whether games have one or more *Nash equilibria* (or *evolutionary stable strategies*), which determine the best response to a best reply.

Many real life incarnations of these above games are special iterated versions, called *stopping games* ((Shmaya et al., 2003, Touzi and Vieille, 2002) where a (possibly varying) cost is associated with playing the game. Different from standard game-theoretic approaches, players typically have to pay in every round (e.g., the basic bet in poker games or the cost of fighting for a period of time). Furthermore, the payoff is not determined after each round (as is typically the case in repeated versions of the above mentioned games), but only when one player quits, which often happens when one player lacks the resources to continue the game. In the competition for food in the animal kingdom, for example, it may mean that the animal runs out of energy and dies. Consequently, it may not be possible for an individual to play what would have otherwise been an optimal strategy.

To account for the influence of the context, in which competitions for resources are embedded in nature, we view the "competitions games" as part of the larger "survival game", in which agents have to survive for long enough to be able to procreate.[1] Specifically, we study and compare different strategies for resolving conflicts among individuals in the context of a simple discrimi-

[1] Elsewhere we have called the games played by individuals in the context of larger games "embedded games" (Scheutz and Schermerhorn, 2004).

nation between "own" (i.e., members of one's own kind) and "other" (i.e., members of another kind). The role of aggression as a means to signal action tendencies in order to resolve conflicts better and faster will be of particular interest, given that aggressive behavior is widespread in nature and that ethologists and anthropologists have repeatedly emphasized the utility of aggression (e.g., see (Lorenz, 1963, Lorenz, 1977, Brain, 1979, Eibl-Eibelsfeld, 1974)).

The paper is organized as follows: we start with a description of a very general conflict scenario that has many instances in natural and artificial domains and consider different strategies for resolving conflicts. We argue that displaying aggression as a means to signal action tendencies (in particular, the probability to continue an encounter) is beneficial for social groups. We then argue that discriminating between "own" and "other" is more beneficial than treating "other" the same as "own" and that aggression plays a crucial role in strategies applied to "other". To test the theoretical discussion, we define a seven basic agent types which give rise to 42 different *discriminating agents*, i.e., agents with different strategies for "own" and "other". In extensive simulation studies we show that discriminating agents, which assume an aggressive attitude towards others, while playing a strategy that distributes resources fairly among "own", are ultimately the most successful ones. We discuss the implications of these results for natural and artificial agents and conclude with a brief outlook on further studies.

2. Conflicts, Aggression, and Survival

Conflicts arise in nature because resources such as food, mates, territory, etc. are often scarce. Typically, there are more individuals interested in a kind of resource than is available in their environment, so a conflict ensues in which individuals compete for the resources they desire or need. The strategies used in such a competition can take various forms, ranging from brute force fighting (disregarding any possible signals of opponents about their behavioral disposition) in which the stronger individual is likely to win, to careful negotiations in which a decision is reached based on comparing advantages and disadvantages of the conflict situation with respect to their expected utility. It can also be carried out at different levels of social organization, from conflicts among two individuals, to conflicts among small groups of individuals (e.g., among families), to conflict between nations. In its most general form, the conflict will end in one of three ways: (1) some agents win, the others retreat, (2) nobody gets the resource (everybody loses), or (3) the resource can be shared (everybody gets a part, but not the full resource). In natural settings, individuals need to survive long enough to be able to have offspring, to which they can pass on their genes. This, in turn, will also guarantee the survival of their kind (i.e., the group to which they belong, e.g., as determined by their genetic features).

In the context of survival, aggression levels, which depend on various external, but also on internal, individual factors (e.g., (Barlow et al., 1986)), seem to play an important role. And while aggression may serve several regulatory purposes for organisms (e.g., (Brain, 1979, Lorenz, 1977, Albert et al., 1993)), its role in social conflicts that relate to *interspecies competition* (as opposed to *intraspecies competition*) for resources is off particular importance for the present investigation.

When animals fight for resources (food, mates, territory, etc.), the contests typically involve various displays of aggression or prowess (e.g., (Adamo and Hanlon, 1996, DeCourcy and Jenssen, 1994, Hofmann and Schildberger, 2001)). These expressions of aggression, whether facial expressions, gestures, or other demonstrations of strength or determination, can be construed as signals that communicate the probability with which an animal will continue to fight (if put in a contest): roughly speaking, the stronger the display, the higher the likelihood that the animal will keep fighting after a given time. Several questions arise: (1) is displaying this information beneficial, (2) what strategy should animals play, (3) is there an optimal strategy, and (4) is cheating (i.e., displaying wrong action tendencies) beneficial (at the individual and/or the group level)?

Previous work with agents that display their action tendencies–whether to continue an encounter or whether to abort it–has shown that taking other agents' displayed action tendencies into account leads to better group outcomes (Scheutz and Schermerhorn, 2004). For example, if it is obvious that an opponent is very likely to continue to fight over the resource (i.e., that it has a high action tendency to fight) and ultimately win the encounter, then it is not in an agent's best interest to enter the fight when it is less likely to continue to fight (i.e., it has a lower action tendency to fight) and win the encounter, thereby wasting resources fighting while gaining no benefit. Retreating immediately may also be costly, but compared to the cost of prolonged fighting, it is in the agent's best interest to retreat. Furthermore, it is in the more aggressive agent's best interest for its opponent to leave early, because prolonged fights reduce the net benefit of the resource being contested.

3. Conflict resolution in the competition for resources

In this study, we consider a particular kind of embedded two-player stopping game: the competition of individuals for food in the larger context of foraging for food in order for them to survive long enough to be able to procreate. Each agent has an internal energy store, from which energy is subtracted for movements, fighting, and even just existing. In other words, there is a cost associated with each action which is measured in terms of energy, and ingesting food is the only way to replenish the internal energy store. During each agent's search for food, there may be occasions in which two or more agents are trying to obtain the same food source. These encounters then lead to the embedded stopping games, where agents can decide to quit and leave the food for the other agent(s), or to fight for food. Costs (e.g., for fighting) are assessed at every step of the game. When participants run out of resources, they are removed from the (larger) game, and hence cannot have offspring. In the evolutionary context, this means that their genes are not passed on.

There is ample evidence that in nature the differences between opponents with respect to their capacity to inflict harm or impose cost on each other matters in making decision about whether to fight for a resource or whether to retreat (and leave the resource to the opponent) (e.g., see (Maynard-Smith, 1982, Parker, 1974)). In addition, the value of the resources plays a crucial role (a factor we will ignore in the experiments to simplify the agent model and the theoretical analysis). Hence, a very general model will need to include some measure of the kind of harm that could be done and the likelihood that it will be done as well as the measure for the value of the resources over which the conflict is carried out.

3.1 The "embedded" competition game

We will use the following abbreviations for constants that determine the benefits and costs involved in the competition for a resource: B_W denotes the benefit of winning an encounter (i.e., the expected utility of the resource), B_L denotes the benefit of losing an encounter (i.e., the expected utility of obtaining other resources), C_F denotes the cost of fighting to the individual, and C_R denotes the cost of retreating. All costs are measured in terms of energy. In general, we can assume that the utility of winning outweighs the utility of losing, i.e., $B_W + C_F > B_L + C_R$.

We model an agent's decision making process as a random variable that assumes values in {FIGHT, RETREAT}. We denote the probability (in $[0, 1]$) that an agent will fight by P_S and that it will retreat by $1 - P_S$ (for opponents we will use the P_A and $1 - P_A$, respectively).

Each competition can then be described by a sequence of interactions (or "rounds") which continue as long as both player decide to fight and end as soon as one player decides to retreat. The utility function $U(1)$ determining the expected outcome for a one round interaction is then given by:

$$\begin{aligned} U(1) \quad = \quad & P_S \cdot P_A \cdot C_F + P_S \cdot (1 - P_A) \cdot (B_W + C_F) + \\ & (1 - P_S) \cdot P_A \cdot (B_L + C_R) + \\ & (1 - P_S) \cdot (1 - P_A) \cdot (B_L + C_R) \end{aligned}$$

The above equation can be further transformed and simplified into

$$U(1) = P_S \cdot ((1 - P_A) \cdot B_W + C_F) + (1 - P_S) \cdot (B_L + C_R)$$

In general, the utility for an $(n + 1)$-round game can be determined in terms of the utility of a 1-round game and the cost of the "fight-fight" outcome, denoted by C_F (note that for an $n + 1$ round-game, there must have been n preceeding rounds with "fight-fight" outcomes):

$$U(n + 1) = n\dot{C}_F + U(1)$$

The utility function $U(n)$ is strictly decreasing. Consequently, it is clear that long games only incur costs and that, therefore, the best game (other things being equal) is a one-round game. $U(1)$ can be maximized by player S depending on player A's choice in the following way: suppose P_S is fixed, then player A should play $P_A = 1$ if $(1 - P_S) \cdot B_W + C_F > B_L + C_R$.[2] Conversely, if $(1 - P_S) \cdot B_W + C_F < B_L + C_R$, then player A should play $P_A = 0$. For $(1 - P_S) \cdot B_W + C_F = B_L + C_R$ player A will be indifferent between $P_A = 0$ and $P_A = 1$ (any value in $[0, 1]$ will pay $B_L + C_R$). Similarly, player S will be indifferent if player A plays P_A such that $(1 - P_S) \cdot B_W + C_F = B_L + C_R$. Therefore, the pair (P_A, P_S) such that $(1 - P_S) \cdot B_W + C_F = (1 - P_A) \cdot B_W + C_F = B_L + C_R$ is a Nash equilibrium.[3]

It may be tempting for player S to play $P_S = 1$ as this promises the highest payoff. However, in the absence of information about P_A this can lead to prolonged games, e.g., if both S and A decide to play "fight" (i.e., $P_S = P_A = 1$), where the "payoff" per round is the only the cost C_F with no benefits to either contestant (and reduced expected utility for both players in the next round).

If information about the likely behavior of the other participant is available, however, then better decisions can be reached that will ensure short and for some strategies one-round games. Contestants can, for example, indicate their behavioral disposition in the upcoming conflict, i.e., their respective apriori probability to choose "fight" or "retreat" in the encounter. In nature, one such indicator displayed by animals in terms of facial expressions and gestures (and used in this investigation) is the individual's level of aggression, which is causally linked to intensity and duration of a fight. Based on

[2] In general, if an expression α (e.g., $(1-P_S) \cdot B_W + C_F$ is greater than an expression β (e.g., $B_L + C_R$), then $x \cdot \alpha + (1 - x) \cdot \beta$ has a maximum for $x = 1$ for all $x \in [0, 1]$.

[3] Whether there are other equilibria will depend on the particular choice of the involved parameters.

these apriori probabilities, contestants can gauge their opponents' willingness to fight and figure it into their decisions.

For example, an agent might reduce its propensity to fight if it notices that the opponents' aggression level, and thus likelihood to engage in long encounters, is higher than the agent's own aggression level. Such an adjustment pays off, because the likelihood of long encounters is reduced and the more likely outcome–that the agent with the higher aggression level will win–is reinforced.

A one-shot rule to end games after the first round based on such encounters is what we called the "rational rule" (Schermerhorn and Scheutz, 2003, Scheutz and Schermerhorn, 2004): agent S plays 0 if $P_S <= P_A$ and 1 otherwise. This rule is based on the assumption that contestants do not know the actual value of B_W, B_L, C_F, and C_R, hence they cannot compute whether $(1 - P_A) \cdot B_W + C_F > B_L + C_R$. Furthermore, the rule is intended to guarantee one-round games, which are not guaranteed if any of the Nash equilibria are played: suppose that both agents indicate probabilities p such that $(1-p) \cdot B_W + C_F$, then they both should play "fight", which will lock them into a prolonged game with only costs. The rational rule, on the other hand, prevents this by forcing the contestant with the lower probability to "give up" right away and still get at least $B_L + C_R$ (instead of the winner's $(1-p) \cdot B_W + C_F$).

Of course, this rule not is *fair* in that repeated encounters between the same two individuals will lead to the same outcome, i.e., the same individual will win over and over again. In a group that means that the top individual (with the highest level of aggression) will have $n \cdot ((1 - p) \cdot B_W + C_F)$ payoff after n encounters and the bottom one (with the lowest level of aggression) will have $n \cdot (B_L + C_R)$, the others being in-between depending on their encounters and aggression levels. From an evolutionary perspective, this will lead to an "arms race" of aggression levels (as the more aggressive is bound to win), which eventually will lead to agents that always fight and only "give up" when they die (Schermerhorn and Scheutz, 2003). These extended conflicts are very costly, and as a consequence a population of such agents will dwindle or become extinct altogether.

A fair way to distribute resources equally among members in a group is to take turns (e.g., (Neill, 2003)). In a two-agent environment, players would alternate between getting $B_W + C_F$ and $B_L + C_R$, thus getting the average payoff $(B_W + C_F + B_L + C_R)/2$ every turn. For environments with more agents, turn-taking is not as straightforward as it is possible that both agents in a conflict took their turn at losing in the previous encounter, so both will expect to win in the current encounter. There are a number of ways to implement a fair turn-taking rule for multi-agent environments. An agent could, for example, keep track of the other agents it interacted with and figure out whose turn it is to win next based on the interaction history. Such an approach, however, is not only computationally expensive, but also requires the ability to track other agents and their encounters over time (which, for example, presumes mechanisms for reidentify agents). We have proposed a simple tallying mechanism that keeps track of the number of encounters the agent won or lost by increasing or decreasing its aggression base aggression level. This adaptation process leads to higher aggression levels if the agent has not been able to obtain the desired resources and to lower aggression levels if the desired resource has been obtained. Formally, the rule can be defined as follows:

Definition [The Turn-Taking Rule] Let r be the basic aggression level of an agent A and let m be the current aggression level ($r = m$ if the rule has never been applied). Then the turn-taking rule $TT(m)^+$ is defined (for losses) as follows: if $m \geq r$, then $TT^+(m) = m + (1-m)/2$; if $m \leq r/2$, then $TT^+(m) = 2 * m$; else $TT^+(m) = r + (2m-r)(1-r)/2r$ (this maps values in the interval $(r/2, r)$ into $(r, (1-r)/2)$). Similarly, $TT(m)^-$ is defined (for wins) as follows: if $m \geq r + (1-r)/2$, then $TT^-(m) = m - (1-m)$; if $m \leq r$, then $TT^-(m) = m/2$; else $TT^-(m) = r/2 + r(m-r)/(1-r)$ (this maps $(r, (1-r)/2)$ into $(r/2, r)$).

This rule effectively keeps track of how often an individual was able to obtain a resource by increasing or decreasing the level of aggression relative to the rest level and the current level of aggression in a way that the computation is straightforward and inexpensive (in fact, the rule can easily be implemented by a few recurrent perceptrons).

We have shown (Schermerhorn and Scheutz, 2003) that this rule, if combined with the "rational rule" described above, is *fair* in a clearly specified sense: the absolute difference between the number of wins and losses of agents in a group in conflicts is bound by a constant c for all agents: $|wins - losses| < c$, where c depends on the size of the group. This means that at the worst an individual member of the group will have c more losses than wins.[4]

4. Strategies and Agent Models

The above discussion suggests that signaling action tendencies through displays of aggression is beneficial for the survivability of individuals in an embedded game. In particular, making "rational decisions" while taking

[4]It is important to note that this is a theoretical limit. Our previous simulations show that in practice, this number is almost never reached (especially not for large populations), hence the group practically is even more balanced than one would expect from the upper bound c.

turns using the above suggested mechanism to adapt aggression levels has proved superior to all other strategies in previous simulation studies with agents that do not discriminate between "own" or "other". The question to be answered here is whether this is also true of agents that are capable of discrimination and if not, what the best strategy would be for such agents. We will in the following first define seven different kinds of agents and then report the performance ordering of new agents created from combinations of two different out of the seven strategies. These agents apply one strategy to agents from their own kind and one to agents from other kinds, respectively.

4.1 7 Non-Discriminating Agent Types

We distinguish two different classes of agents, call them *social* and *asocial* agents, depending on whether or not they use perceptions of opponents' aggression in their decision to fight or retreat. All asocial agents decide whether they will enter and stay in a conflict situation based only on their own aggression level in a probabilistic fashion: the more aggressive they are, the more likely they are to stay. In the experiments reported here the mapping between aggression level and action tendency is the identity function, but other functions are possible. Social agents also decide their actions in the same probabilistic fashion but include perceptions of other agents' aggression levels, possibly to a point where they reduce their own decisions to non-probabilistic certain actions (i.e., either "fight" or "retreat").

While we assume that all agents display their aggression for the following experiments, it is worth mentioning that the display for asocial agents has no added benefit as they do not use perceived aggression levels in their decision making. Hence, the strategies for asocial agents would work exactly the same for non-signaling agents.[5]

In the class of asocial agents, we distinguish three kinds: the (prototypical) *asocial* agents, which play their aggression level (i.e., a value in $(0, 1)$), *timid agents*, which always play 0, and *aggressive agents*, which always play 1.[6]

In the class of social agents, we distinguish two kinds: the (prototypical) *social* agents, which increase or lower their action tendencies based on both their own and the others' aggression levels, and the *rational* agents, which implement the "rational rule" from above. Prototypical social agents use the following rule: if $P_A > P_S$, then they play $P_S - (P_A/P_S) \cdot (P_A - P_S)$, if $P_A < P_S$, they play $P_S + (1 - P_S/P_A) \cdot (P_S - P_A)$, and is both levels are equal they simply play P_S.[7]

Finally, we consider a third category, that of *adaptive agents*, which implement the "turn-taking rule" and thus change their aggression levels based on past encounters (contrary to the above described social and asocial agents, which modify only their behavioral disposition based on their perceptions, but keep their aggression levels constant throughout their lifetime). In particular, we consider (prototypical) *asocial adaptive* agents and *rational adaptive* agents.

4.2 42 Discriminating Agents Types

Based on the above non-discriminating agent types, we can define 42 different discriminating agent types by endowing agents with the ability to distinguish perceptually between "own" and "other" and applying different strategies depending on the membership of the opponent.[8] We will denote these agents by the name of the strategy used against others followed by the name used against their own (e.g., an "aggressive asocial agent" is aggressive towards others and asocial toward its own without any modification of aggression, although the action tendencies towards other is to always "fight").

Of the 42 discriminating agents we will focus on the six asocial discriminating agents, i.e., *the timid*, the *aggressive*, the *social*, the *the rational*, the *the adaptive asocial*, and the *the adaptive rational* asocial agents. The choice for the class of asocial agents is based on the previous studies that show that asocial agents are worse than social agents (both when placed in direct competition with them and when among their own). The question is whether discriminating among "own" and "other" can boost the performance of this agent kind above the performance of the others. Given that we compare 6 discriminating agents to 7 non-discriminating ones, we will consider 42 pairings of discriminating and non-discriminating agents altogether.

5. Simulation Experiments

All simulations described in this section were carried out in the simulated world agent grid experimentation system SWAGES under development in our lab.[9]

[5] It is also possible to augment the strategy of social agents so as to work with non-signaling agents by simply making assumptions about their level of aggression. E.g., in the case of the rational strategy described above an agent could start by ascribing no aggression to the other agent and then systematically increasing the ascribed aggression level such that it will leave a conflict in time before the accumulated cost of fighting outweighs the potential benefit of winning.

[6] Note that we use "asocial agent" to denote a class of agents as well as an agent kind for ease of terminology as the intended interpretation will always be clear from the context.

[7] For a more detailed discussion of this rule, see Scheutz and Schermerhorn, forthcoming. We are currently exploring the relationship between a similar rule used for the analysis of dominance interactions by (Hemelrijk, 1996).

[8] We exclude those 7 discriminating kinds that apply the same strategy to their own kinds and to other kinds, which effectively amounts to no discrimination and thus reduces them to the respective non-discriminating agents.

[9] SWAGES is an agent-based artificial life simulation built on top of the SimAgent toolkit, both of which are freely available at http://www.nd.edu/~airolab/swages and

For the simulation reported here, we configured the simulation environment for a "one-resource foraging task". Instances of the resource, call them "food", contain 800 energy units, pop up at random locations within a predetermined rectangular 1800 x 1800 food area–agents have a size of 9 in comparison–at a predefined frequency (1 resource per cycle) and stay indefinitely until consumed by an agent.

Agents are in constant need of energy and, hence, constantly engaged in foraging for resources. They have a visual system that computes directional force vectors that are scaled by the square of the distance to each resource within their sensory range (of 300 distance units). The vectors are summed and the agent moves in direction of the resultant vector at a speed of at most 4 units per cycle (for details of the architecture, see (Scheutz, 2001, Scheutz and Schermerhorn, 2002)). The energy expenditure for movement is the square of their speed. In addition, each agent consumes one unit of energy per cycle for processing.

Agents also have touch sensors, which they use to detect whether they can ingest a resource or whether they are about to collide with another agent, which means that they enter a competition (typically, this happens when food is close by, but does not have to involve a food source). In the former case, an ingestion mechanism will be triggered to ingest the food item (which takes one cycle), while in the latter a self-preservation collision detection mechanisms will be triggered unless suppressed by the agent. This mechanism is coupled with a reflex that will the agent from crashing into other agents (which would be fatal) by moving them away from the potential collision site at a speed of 7. This reflex is also triggered when agents lose a conflict, hence its duration determines C_R, which on average is 330 energy units (simply because of movement at a higher than usual speed). $C_F = 50$, in contrast, is charged to agents that suppress this mechanism and fight for the resource.

Agents also have an "energy alarm" for self-preservation, which automatically limits their overall speed to 1 if their energy level drops below $Energy_{crit} = 400$ (the speed will remain at 1 until the energy level is raised above the critical level again).

After a certain age α (measured in terms of simulation cycles), agents reach maturity and can procreate asexually, if their energy levels are above the minimum necessary for procreation (set to 2200). The energy necessary for creating the offspring (2000) is subtracted from the parent, and a new agent will appear in the vicinity of the parent in the subsequent simulation cycle. Parents pass all the parameters of all of their strategies to their offspring without modification (e.g., mutation).

http://www.cs.bham.ac.uk/research/poplog/newkit.tar.gz. It consists of a continuous, potentially unlimited two-dimensional surface populated with various kinds of spatially extended objects, in particular, different kinds of agents and resources they need as well as various kinds of static and moving obstacles.

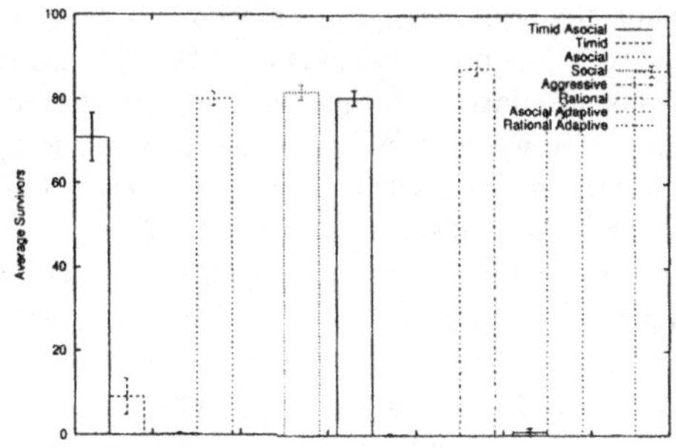

Figure 1: The average number of survivors in the experiment sets comparing timid-asocial discriminating agents to all seven non-discriminating agent kinds.

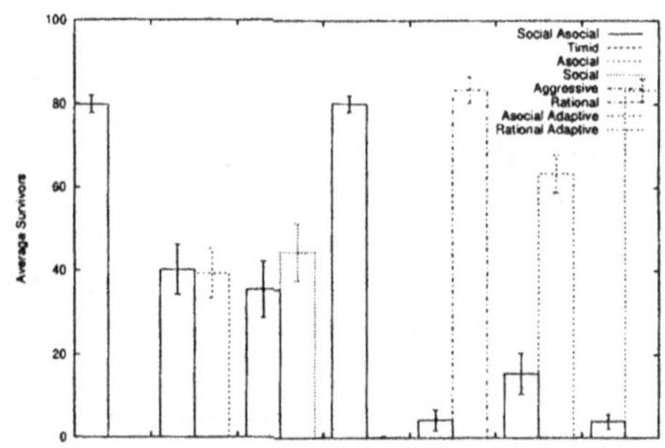

Figure 2: The average number of survivors in the experiment sets comparing social-asocial discriminating agents to all seven non-discriminating agent kinds.

5.1 Experiments and Results

We conducted 42 sets of experiments, each consisting of 40 different runs for 10000 cycles each with different random initial conditions (the same 40 different initial conditions were used in all 42 sets to guarantee a fair comparison). In all runs, 25 discriminating and 25 non-discriminating agents were placed at random locations within the 1800 x 1800 food area in the environment together with 50 randomly placed food items. The initial energy of all agents was set to 2000 and their initial aggression levels were distributed following a Gaussian distribution with spread 0.125 around 0.5.

Previous results from simulations of multi-species environments (Scheutz and Schermerhorn, 2003, 2004) with non-discriminating agents show that it is better to be "timid" than to be "aggressive", and better to be "social" than "asocial", and even better to be "rational" or to take turns on top of being rational. In short, merely aggressive agents showed the worst performance,

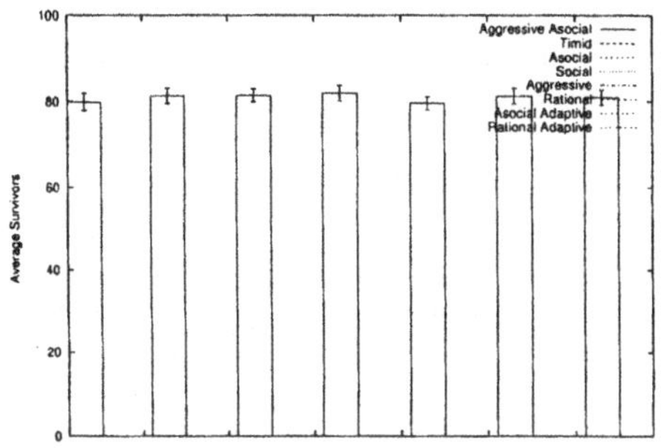

Figure 3: The average number of survivors in the experiment sets comparing aggressive-asocial discriminating agents to all seven non-discriminating agent kinds.

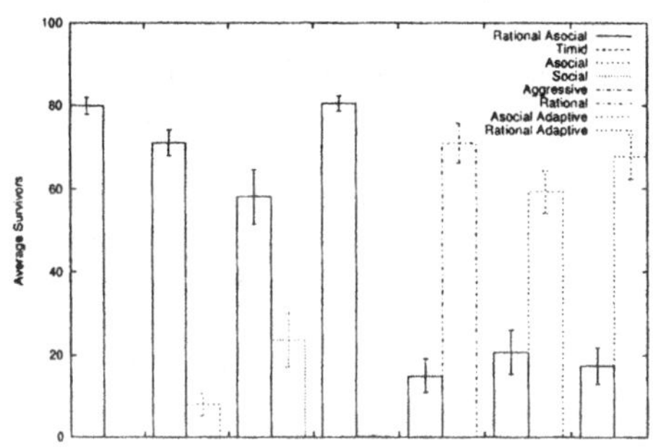

Figure 4: The average number of survivors in the experiment sets comparing rational-asocial discriminating agents to all seven non-discriminating agent kinds.

and all asocial agents performed worse than all social or adaptive agents.

Adding discrimination, however, changes the picture and we see that applying two standards (in this case, strategies for interactions), one for "own" and a different one for "other", pays off, even if the two employed strategies used in isolation are not very good at all. Figures 1 through 6 show the average number of survivors at the end of the simulation run for each discriminating agent kind compared to those of all non-discriminating kinds (the error bars depict the 95% confidence intervals).

The results show that asocial agents with the highest level of aggression against others (i.e., an "always fight" strategy) outperform all other combinations of asocial discriminating agents in addition to outperforming all non-discriminating agents (see Figure 3 compared to the others). In fact, there is no other combination that always leads to the extinction of all non-discriminating agent kinds within only a few generations.

Discriminating timid asocial agents only perform well against other timid and aggressive agents, where they

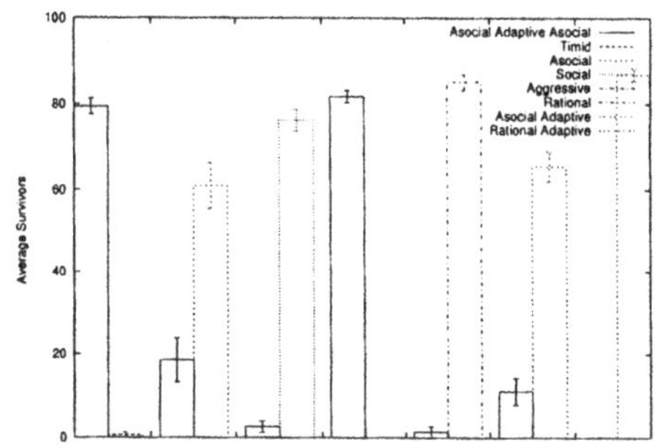

Figure 5: The average number of survivors in the experiment sets comparing asocial adaptive-asocial discriminating agents to all seven non-discriminating agent kinds.

can reap the benefits from their intragroup conflicts. They lose against all other kinds, similar to the timid non-discriminating agents.

Social asocial discriminating agents perform better than timid and aggressive, on a par with prototypical asocial and social agents, but worse than non-discriminating rational agents or adaptive agents (i.e., the benefit of adaptation is still higher than that of discrimination). Note that they manage to survive against the non-discriminating rational adaptive agents, which is possible because they will occasionally win encounters against them (when their adaptation has lowered their aggression level to a point where multiple losses in a row are possible, which significantly increases the likelihood for the social strategy to win).

The discriminating rational asocial agents show an expected high performance against asocial and prototypical social agents given that they play the best social strategy against others (the second best performance after the discriminating aggressive asocial agents). Compared to non-discriminating rational agents, however, their intraspecies asocial conflicts reduce their overall performance.

Finally, the discriminating adaptive agents suffer from the same interspecies conflicts that eventually reduce their fitness. Adaptation caused by conflicts with others also seems to work against them (compared to the discriminating rational asocial agents, for example). Especially, if the majority of opponents are "others", discriminating adaptive asocial agents will have raised their aggression to very high levels, which in turn are likely to lead to harmful "fight-fight" encounters in intragroup competitions.

5.2 *Analysis*

The results of the simulation studies may seem somewhat surprising at first glance. However, we believe that they should not come completely unexpected at

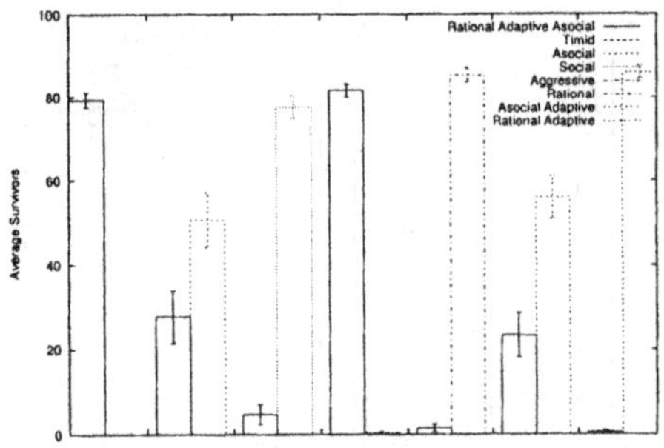

Figure 6: The average number of survivors in the experiment sets comparing rational adaptive-asocial discriminating agents to all seven non-discriminating agent kinds.

second thought, if the goals of discriminating vs. non-discriminating (i.e., intraspecies vs. interspecies) competitions are taken to be different. Roughly, the goal in non-discriminating competitions is to maximize the individual's payoff, while in discriminating competitions, as we would like to suggest, it is to minimize the relative payoff of all competing members in other groups compared to competing members in the own group. In other words, a strategy for intergroup competition is good if it inflicts more damage on members of the other group than on members of the own group, while a strategy for intragroup competitions is good if it gives more benefit to the individual than to other contestants. It is important to note that these two conditions are not symmetric. The discriminating rational asocial agents, for example, attempt to maximize their own utility while playing against others, rather than harming the others as in the case of the discriminating aggressive asocial agents. Consequently, despite their generally high performance against any other non-discriminating agent types (see Figure 4), they do not even come close to the performance of the discriminating aggressive asocial agents. We will illustrate this asymmetry with a brief example.

Consider an environment with two non-discriminating agents of one kind (call it A) and two discriminating agents of another kind (call it B). There are 6 possible mixed two-way interactions of these four agents, i.e., where at least one A-agent interacts with at least one B agent: $a1 - b1$, $a2 - b2$, $a1 - b2$, $a2 - b1$, where the remaining two agents do not interact, and $a1 - b1/a2 - b2$, and $a1 - b2/a2 - b1$. Suppose the B agents are aggressive asocial agents with aggression level 0.3 and 0.5, respectively, and the A-agents are rational agents with aggression level 0.6. and 0.75, respectively. The total expected utility for A summed over the six encounters is $6 \cdot (C_R + B_L)$, while it is $6 \cdot (C_F + B_W)$ for the B agents. Now we consider the intraspecies competition, where the A-agents get a combined utility $C_F + B_W + C_R + B_L$, while the B-agents will at most get that much, given that with probability 0.15 their interaction is prolonged. If we consider only one-round encounters for simplicity sake, then B-agents get $0.35(C_F + B_W) + 0.15 \cdot (C_F + B_W) + 0.35 \cdot (C_R + B_L) + 0.15 \cdot C_F$. Combining these two utilities we can see that the discriminating agents do much better on average than the non-discriminating agents given that the discriminating agents play the harshest possible strategy against "others". We can also see that it is important for the discriminating agents to be confrontational , as playing a non-confrontational strategy like "always retreat" would not have worked (rather it would have resulted in higher average payoffs for the non-discriminating agents, as exemplified in the results shown in Figure 1).

6. Discussion

One way to understand why the difference between discriminating and non-discriminating agents gives rise to such a significant increase in performance is to appreciate that discriminating agents have the option to choose how to distribute resources to non-group members, whereas non-discriminating agents do not have this option (a change in strategy affects non-members and members alike). The simulation studies confirm our analysis that once this choice becomes an option, then the best strategy is not one that favors others, but rather represses them. It is almost ironic that discriminating aggressive agents do better when other agents are around, as they gain more from others' losses (because of their aggression) than they gain from interactions among themselves. Yet, their strategy is so "successful" that they quickly cause the extinction of non-discriminating agents (in all the above simulation the total number of competitors of discriminating aggressive asocial agents never exceeded 200, while their numbers were in the thousands).

Given that the worst strategy, being aggressive and being inclined to fight all the time, for non-discriminating agents turns out to be the best for discriminating ones (as far as "others" are concerned), one would expect aggression to evolve if discrimination is possible, and one would expect discrimination to evolve, if (perceptual) categorization is possible (as discrimination, the way we modeled it here, is based on perceptual difference, which requires a categorization mechanism that is capable of distinguishing between "own" and "other").

The outcomes of these investigations may, thus, help answer the question why there are aggression mechanisms in biological organisms in the first place, when it seems that aggression is not beneficial in the non-discriminating case. One answer given by Lorenz, for example, is that intra-species aggression serves to force animals to distribute better over a given territory (as it prevents animals from sticking together, thus having to compete for scarce resources in a small area, while resources are plentiful elsewhere (Lorenz, 1963)).

A different answer based on our results, which would, of course, require further investigation, might be this: suppose species developed an aggression mechanism because it was beneficial to be aggressive towards others in the competition for resources, thus furthering (or maybe even ensuring) the survival of their kind. Once such a mechanism is in place, there are two possibilities for its activation: either through innate categories of "other kinds" or through an adaptation phase, where the difference between own and other is learned and an association between other and the aggression mechanism is formed. In that case, the association between own and other is not fixed as in the innate case, but amenable to change and thus subject to modifications during an individual's' lifetime. While this adds flexibility and thus higher adaptivity to individuals (as they may be able to survive in environments with other competitors than their current ones as well), it also opens up the possibility of forming associations between "own" and the aggression mechanism. It is then possible, that in the absence of other species, the discrimination is applied to the own kind and a new division between other and own is formed such that the aggression mechanism is associated with new category of "other", which really is a subgroup of "own".

Of course, this consideration assumes that there is a (possibly innate) mechanism that is disposed to divide individuals in the environment into own and other. The categorization could be based on many factors ranging from difference in physical appearance to mental attitudes. The important point here is that there is great evolutionary utility to having such a mechanism that will elicit aggressive behavior towards others, i.e., a mechanism that causes the individual to look out for differences among other individuals, categorize other individuals in terms of these differences into "own" and "other", and elicit aggressive behavior towards individuals that are categorized not "own", but "other". In that case, the problem would not be with the aggression mechanism *per se*, but rather with the lack of target individuals against which the aggression is to be employed.[10]

Note that there are good evolutionary reasons to make this mechanism innate, i.e., to prevent individuals from failing to develop such a categorization, as the failure to develop such a discrimination might not serve the group as a whole, and consequently (future) individuals. Specifically, it may give rise to "free-riders" benefit from their conspecifics' aggression towards others (e.g., an example in the context of a war between nations might be soldiers that do not put their lives at risk in combat.[11]

[10]It seems that this consideration is consistent with what it commonly believed about the utility or role of sports in modern societies.

[11]It seems that main reason why people fight is to not let their "buddies" down, see (Little, 1964).

Finally, we would like to mention that the above proposed agent architectures fits within the "frustration-aggression theory of "aggression" (e.g., see (Berkowitz, 1989)). Specifically, the turn-taking mechanism can be taken to feed into the frustration level, so that whenever a resource is not obtained frustration grows according to the turn-taking rule, which then feeds into aggression, which is subsequently displayed. In this setup, agents get more aggressive as their frustration increases (which increase because they do not obtain their resources based on the decision mechanism built into the turn-taking rule). Eventually they will get the desired resource and aggression levels will subsequently drop (as frustration levels drop caused by goal accomplishments).

7. Conclusion

In this paper, we have addressed the question whether aggression is a useful control mechanism in the competition of agents for resources. Specifically, we demonstrated that aggression, while not beneficial by itself, is highly beneficial in combination with discrimination between groups (i.e., between agents from one's own group and all other agents). Extensive simulation experiments showed that other combinations of strategies for the seven basic agent types considered in the paper do not fare as well. This should be an encouraging result for those who would like to ground a theory of aggression in evolutionary pressures. Although we have only been able to sketch a possible account, it should be clear how our conjecture could, at least in principle, be tested in further, much more elaborate simulation experiments with agents that have the respective architectural capacities (such as categorization, learning mechanisms, etc.).

The above results also suggest another interesting direction for future research, which is feasible in the short term and which we intend to pursue. While in principle discriminating agents derive their increase in performance from the harm they inflict on others, it is not clear to what extent this actually beneficial for them and how, for example, the above investigated aggressive asocial agents would fare against other discriminating agents (that also played an aggressive strategy against "others"). In particular, it would be interesting to determine the relative productive success of the discriminating agents that had more encounters with agents from another kind (i.e., which mostly played the aggressive strategy) compared to their conspecifics, which mostly interacted with their own kind. This should shed light on the dynamics of the interactions between agents of different kinds and help answer questions about why some biological species seem to be intrinsically more aggressive than others.

Finally, we would like to point out that the results are likely to have implications beyond the biological domain, as the general game-theoretic principles do not only apply to the competition for resources that are immediately relevant and a constitutive part of the embedded survival game. Rather, they are likely to apply to many social situations and groups of different social structure. If this is correct, then more detailed studies of the kinds of embedded games we have considered in this paper might contribute to our understanding of the implications of aggression to the social dynamics as they manifest themselves in national and international conflicts.

References

Adamo, S. A. and Hanlon, R. T. (1996). Do cuttlefish (cephalopoda) signal their intentions to conspecifics during agonistic encounters? *Animal Behavior*, (52):73–81.

Albert, D., Walsh, M., and Jonik, R. (1993). Aggression in humans: what is its biological foundations? *Neurosci Behav Rev*, (17):405–425.

Axelrod, R. and Hamilton, W. D. (1981). The evolution of cooperation. *Science*, 211:1390–1396.

Barlow, G., Rogers, W., and Fraley, N. (1986). Do midas cichlids win through prowess or daring? it depends. *Behav Ecol Sociobiol*, (19):1–8.

Berkowitz, L. (1989). Frustration-aggression hypothesis: examination and reformulation. *Psychological Bulletin*, (106):59–73.

Brain, P. (1979). Classical ethology and human aggression. In Brain, P. and Benton, D., (Eds.), *The biology of aggression*, pages 603–623.

DeCourcy, K. R. and Jenssen, T. A. (1994). Structure and use of male territorial headbob signals by the lizard *Anolis carolinensis*. *Animal Behavior*, (47):251–262.

Eibl-Eibelsfeld, I. (1974). Aggression in the iko-bushmen. *Aggression*, (52):1–20.

Enquist, M. and Leimar, O. (1983). Evolution of fighting behavior: decision rules and assessment of relative strength. *Journal of Theoretical Biology*, (102):387–410.

Hemelrijk, C. K. (1996). Dominance interactions, spatial dynamics, and emergent reciprocity in a virtual world. In *Proceedings of the Fourth International Conference on the Simulation of Adaptive Behavior*, pages 545–552.

Hofmann, H. A. and Schildberger, K. (2001). Assessment of strength and willingness to fight during aggressive encounters in crickets. *Animal Behavior*, (62):337–348.

Little, R. (1964). Buddy relations and combat performance. In Janowitz, M., (Ed.), *The new military*.

Lorenz, K. (1963). Das sogenannte bse. *Naturgeschichte der Aggression*.

Lorenz, K. (1977). Aggressivitt: arterhaltende eigenschaft oder pathologische erscheinung? *Aggression und Toleranz*.

Maynard Smith, J. (1974). The theory of games and the evolution of animal conflicts. *Journal of Theoretical Biology*, pages 209–221.

Maynard-Smith, J. (1982). *Evolution and the Theory of Games*. Cambridge University Press, Cambridge.

Neill, D. (2003). Cooperation and coordination in the turn-taking dilemma. In *14th International Conference on Game Theory at Stony Brook*.

Parker, G. (1974). Assessment strategy and the evolution of fighting behaviour. *Journal of Theoretical Biology*, pages 223–243.

Schermerhorn, P. and Scheutz, M. (2003). Implicit cooperation in conflict resolution for simple agents. In *Agent 2003*.

Scheutz, M. (2001). The evolution of simple affective states in multi-agent environments. In Cañamero, D., (Ed.), *Proceedings of AAAI Fall Symposium*, pages 123–128, Falmouth, MA. AAAI Press.

Scheutz, M. and Schermerhorn, P. (2002). Steps towards a theory of possible trajectories from reactive to deliberative control systems. In Standish, R., (Ed.), *Proceedings of the 8th Conference of Artificial Life*. MIT Press.

Scheutz, M. and Schermerhorn, P. (2004). The role of signaling action tendencies in conflict resolution. *Journal of Artificial Societies and Social Simulation*, 1(7).

Shmaya, E., Solan, E., and Vieille, N. (2003). An application of ramsey theorem to stopping games. *Games and Economic Behavior*, 2(42):300–306.

Touzi, N. and Vieille, N. (2002). Continuous-time dynkin games with mixed strategies. *SIAM J. Control Optim*, 4(41):1073–1088.

Imitation in embodied agents results in self-organization of behavior

Bart Jansen[1] Tom ten Thij[2] Tony Belpaeme[1] Bart De Vylder[1]
Bart de Boer[2]
[1]Artificial Intelligence Lab
Vrije Universiteit Brussel
Pleinlaan 2, B-1050 Brussels, Belgium
http://arti.vub.ac.be

[2]Artificial Intelligence
Rijksuniversiteit Groningen
Grote Kruisstraat 2/1, 9712 TS Groningen, The Netherlands

Abstract

This paper addresses the construction and learning of behavior through imitation. We explore imitation as a framework to create and sustain behavior in a population of agents. Two experiments are presented, in simulation and on actual robots. These demonstrate how the behavior self-organizes through the interplay between the imitative interactions, the environment and the embodiment of the agents.

1. Introduction

Imitation in robotic agents has received considerable attention in the last decade (e.g. Kuniyoshi et al., 1994; Gaussier et al., 1998; Schaal, 1999; Alissandrakis et al., 2001; Billard and Matarić, 2001; Breazeal and Scassellati, 2002; Demiris, 2002). Imitation manifests itself in its different guises throughout the natural world (for a review see Whiten and Ham, 1992). In particular higher-order primates and humans seem to benefit profoundly from imitative learning (Nadel and Butterworth, 1999). Because of this, imitation receives quite some attention from robotics researchers (Dautenhahn and Nehaniv, 2002), especially as imitation is a powerful means to transfer skills between agents in a way that comes natural to humans. Most robotic studies emphasize on constructing artifacts that implement a particular form of learning by demonstration. Typically, in this form of learning a pre-defined repertoire of behaviors is transferred from one agent (a teacher) to another (the student). The repertoires of behaviors and the roles of the imitating agents are fixed. Hence the focus of this kind of work lies on machine learning methodologies for imitation. Our work, on the other hand, takes a different point of view. Instead of constructing an imitation application, we study imitation as an adaptive and flexible way to acquire new behavior using a robotic multi-agent system.

Through imitation a learner acquires skills and behaviors from peers or parents by observing and mimicking those skills and behaviors. While imitating, the imitator needs to observe behavior and map it onto its motor system. This is called the correspondence problem. What is remarkable is that the bodies of the parties involved in imitation can be largely dissimilar. Children, for example, have smaller bodies than the parents that they imitate; still this poses little problem. Meltzoff and Moore (1977) for example report how neonates can imitate adult facial expressions, thereby showing that they have the innate capability to solve the correspondence problem for dissimilar bodies[1]. In the experiment reported here, we likewise endow our agents with the capacity to solve the correspondence problem: all agents are capable of mapping observed actions on their own motor system.

We are particularly interested in the constraints posed by the embodiment of the agents. Imitation cannot be studied without considering the embodiment of the agents. Their perception and their physical implementation poses limits to what they can observe and what actions they can perform. This principally influences the behavior that can be successfully imitated, and determines the collective behavior that will be sustained in a population of imitating agents. This is well illustrated by linguistic communication; imitation forms one of the cornerstones of the acquisition of language and is, for example, crucial for the learning of speech sounds (Kuhl

[1]Incidentally, this might be related to mirror neurons (Rizzolatti et al., 1996). These neurons have been found in monkeys and are active both when performing a certain action and when observing that action.

and Meltzoff, 1996). Speech is learned by imitating others, but the nature of speech sounds depends largely on the perception and production of speech, i.e. on the embodiment of our auditory and acoustic apparatus.

2. The imitation game

The imitation between agents is implemented as an *imitation game*. The imitation game[2] was introduced in the context of research into speech (de Boer, 2000). In this context it is logical to use imitation as a learning mechanism, as children learn to speak through imitation, but imitation was also used as a means to put pressure on agents to develop sounds that are as distinctive as possible. In reality such pressure would be caused by the function of the sounds for distinguishing meaning: as different speech sounds are associated with different meanings, the need to differentiate between two different meanings would push apart these speech sounds. However, implementing these dynamics would have made the simulations needlessly complex. The aim of the imitation game was to develop a repertoire that contains as large a number of clearly distinctive speech sounds as possible. At the same time, these sounds had to be distinguishable and learnable under the physical and cognitive constraints of production, perception and learning.

It turns out that the imitation game is also very useful when developing a repertoire of actions. As will be explained below, some of the peculiarities of the imitation game can be better understood when one takes its background in the learning of speech sounds into account.

The generic imitation game (whether for speech or actions) is based on a population of agents that each have an open repertoire of categories. In principle, meaningful imitation games can be played in a population of just two agents, but the population size can be much larger. In a larger population, convergence towards a shared repertoire of categories will be slower, but it will happen nevertheless. The basic dynamics of the game, however, can be demonstrated in a population of just two agents.

Crucial is that the agents in the population do not have fixed roles, nor do they start with pre-defined categories in their repertoires. They start out empty and no agent can therefore be the teacher from which the other agents learn their categories. A shared repertoire emerges through repeated interactions between the agents. Although in many cases it is possible to identify teachers and students in learning of real language, there are cases, especially in the formation of pidgin- and creole languages, where such teacher-student roles cannot be identified, and a shared repertoire of speech sounds still emerges.

Each agent develops a repertoire of categories. It is important that agents have true categorical perception (Harnad, 1987). This means that when they perceive an input from a continuous domain (either a speech sound or an action) they perceive it as one of the discrete categories in their repertoire, never as something in between. This is certainly the case in speech sounds (e.g. Cooper et al., 1976) and also appears to work well when working with actions.

In any imitation game, two agents are randomly selected from the population. One agent is randomly assigned the role of *initiator*; the other gets the role of *imitator*. The initiator is the agent who initiates the imitation game. The imitator imitates and the initiator evaluates how well the imitator managed to imitate. It is important to notice that the only things the agents can observe are the expressions produced by either the other agents or themselves. This means that there is no such thing as telepathy; the agents do not have direct access to the other agents' categories.

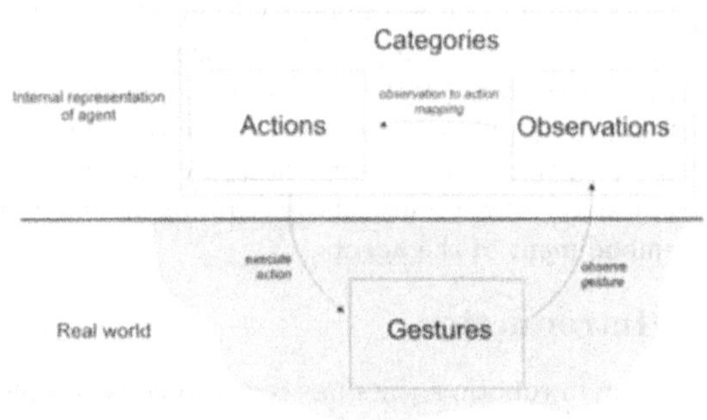

Figure 1: Agents have categories, which are made up of actions and observations. Actions can be expressed as gestures in the real world, and observing these gestures results in observations. And observation in turn can be mapped onto categories using inverse kinematics.

To start the imitation game, the initiator selects one random category from its repertoire (unless its repertoire is empty, in this case it starts with creating a random category). It then expresses this category (by uttering speech, or in this paper, by making a gesture with a robot arm). Usually, the expression will not be perfect, such that when an agent expresses the exact same category twice, the results will not be identical. The imitator perceives this expression and finds the nearest category. This effectively implements categorical perception: each expression is always perceived as a category; details of the expression are lost in categorization (nevertheless, a copy of the exact expression is retained for the duration of one game, as it might be needed to improve the categories). The imitator then expresses the category it found. This expression, too, is imperfect. It is

[2]Of course, Alan Turing in 1950 already used the term *imitation game* in the context of artificial intelligence for the experiment now commonly known as the Turing Test. This is not related to our use of *imitation game*.

perceived by the initiator, who, like the imitator, finds its closest category. If this closest category is the same as the category it initially expressed, the game is successful. Apparently both agents' categories are equally similar and sufficiently distinguishable such that twice-repeated categorical perception of imperfect expressions does not change the category. However, when the category the initiator perceives is not the same as the category it initially selected, the game is a failure. Apparently, the agents' categories are not sufficiently similar, or perhaps they are not sufficiently distinguishable. In each case, the initiator gives feedback to the imitator. This feedback consists of a single bit of information saying whether the game succeeded or failed. Now both agents can update their repertoires in order to improve expected future performance at the imitation game. In the implementation game as implemented here, the imitator makes the most important changes.

Table 1 and 2 provide pseudo code for the imitation game, while figure 1 illustrates the relation between the internal representation of an agent and its actions in the world. Agents have a repertoire C of categories, each category c has an action *c.action* and an observation *c.observation* associated with it. Actions are a tuple containing start and end configuration of the robot arm[3], while observations are a series of 3D-coordinates.

initiator	imitator
if $C = \emptyset$ **new-category**(C) $c \leftarrow$ random from C $c.use \leftarrow c.use + 1$ **execute** $c.action$	
	observe O_1 **if** $C = \emptyset$ **new-category**(C) **else** $c_{rec} \leftarrow$ category from C so $c_{rec}.observation$ closest to O_1 **execute** $c_{rec}.action$.
observe O_2 $c'_{rec} \leftarrow$ category from C so $c'_{rec}.observation$ closest to O_2 **if** $c = c'_{rec}$ $c.success \leftarrow c.success + 1$ send feedback "success" **else** send feedback "failure"	
	update($c_{rec}, O_1, feedback, C$)
do-other-updates()	**do-other-updates**()

Table 1: Pseudo code of the imitation game

3. Monitoring the performance

It is important to define good measures for monitoring the performance of the agents. Four measures were defined: the imitative success, the number of categories, the average category variance and the information flow.

3.1 Imitative success

The *imitative success* is simply the average number of successful imitation games. Imitative success is calculated as a running average over the last $n = 100$ games. Due to the nature of the game, the imitative success is already high at the start of the game, when agents have only few –not even very similar– categories. This is best explained by a small example: if the initiator has only one category, and it expresses that category as an action, anything that the imitator does is matched to that category. So the initiator judges the imitation to be successful, resulting in a high success score. This is why at the beginning of the experiment, the imitative success is already quite high. What is interesting, is looking if the imitative success stays high as agents continue to create and learn new categories.

[3]Nothing prevents the action from being a series of motor commands, instead of just the start and end configuration of the motors. In the experiments reported here, for practical considerations, actions are only a start and end position of the arm.

3.2 Number of categories

It is informative to monitor the number of categories that agents possess, as this tells us something about the noise in the environment. The number of categories that agents are able to acquire depends on what they can perceive and express. Hence, the embodiment of the agents is the limiting factor on acquiring categories.

3.3 Average category variance

Consistently successful imitation games are not possible without shared categories, i.e. categories that are similar between agents. However, it is useful to define an explicit measure of similarity of the repertoires of all agents. First we introduce a method for comparing the repertoires of two agents. *Category Variance (CV)* is the *scaled sum of minimum distances* applied to the categories of both agents and is given in equation 1. A and B are agents, a and b are categories, $a \in A$ means that a is an category in the repertoire of agent A and $|A|$ denotes the number of categories of agent A. The distance function $d(a, b)$ on categories of agents used here is Dynamic Time Warping on the observations associated to the categories.

Procedures

update($c_{rec}, O, feedback, C$)
$c_{rec}.use \leftarrow c_{rec}.use + 1$
if feedback = "success"
  $c_{rec}.success \leftarrow c_{rec}.success + 1$
  shift-closer(c_{rec}, O)
else
  if $c_{rec}.success / c_{rec}.use >$ *treshold*
    $C \leftarrow C \cup$ **construct-category**(O)
  else
    shift-closer(c_{rec}, O)

shift-closer(c, O) $c.action \leftarrow c.action +$ *shift* .(**find-action**(O) $- c.action$) $c.observation \leftarrow$ **execute** ($c.action$)
find-action(O) find action that produces O using inverse kinematics
new-category(C) $c.action \leftarrow$ random $((x_1, y_1, z_1), (x_2, y_2, z_2))$ $c.observation \leftarrow$ **execute** ($c.action$) $C \leftarrow C \cup c$
construct-category(O) $c.action \leftarrow$ **find-action**(O) $c.observation \leftarrow$ **execute** ($c.action$)

do-other-updates(C)
$\forall c \in C$ **do**
  if $c.success/c.use <$ *throwawaythreshold* **and**
    $c.use >$ *minimumuses*
    $C \leftarrow C \setminus c$
with probability *addprobability* **do new-category**(C)
merge-categories(C)

merge-categories(C)
for all $c_1, c_2 \in C,\ c_1 \neq c_2$
if $\lVert c_1 - c_2 \rVert <$ *merge-threshold*
  $c_{new} \leftarrow (c_1 + c_2)/2$
  $c_{new}.success \leftarrow c_1.success + c_2.success$
  $c_{new}.use \leftarrow c_1.use + c_2.use$
  $C \leftarrow C \setminus c_1 \setminus c_2$
  $C \leftarrow C \cup c_{new}$

Table 2: Important procedures used in the imitation game

$$CV(A,B) = \frac{\sum_{a \in A} \min_{b \in B} d(a,b) + \sum_{b \in B} \min_{a \in A} d(a,b)}{|A| + |B|} \quad (1)$$

The category variance allows us to calculate the *Average Category Variance* ($\overline{CV}$), which is given in equation 1, of the population $P = \{A_i\}_{i=1\ldots N}$ consisting of N agents, indicating how similar the categories of all agents are on average.

$$\overline{CV}(P) = \frac{2}{N(N-1)} \sum_{i=1}^{N-1} \sum_{j=i+1}^{N} CV(A_i, A_j) \quad (2)$$

3.4 Information flow

When taking an information theoretical perspective on the imitation game, one could view the initiator as the source or sender of information, the physical system consisting of the robot arm and the stereo head as the noisy channel and the imitator as the destination or receiver. The source can send different symbols, corresponding to the different action categories of the initiator. The communication can succeed or fail, indicated by the success or failure of the imitation game: we assume that a symbol was correctly transmitted if its repetition by the receiver over a similar channel yields the original symbol according to the sender. If the source emits symbols randomly and independently chosen according to a fixed distribution, the information content (in bits) or entropy H of a symbol depends on its frequency p and is given by:

$$H = \log_2(\frac{1}{p}). \quad (3)$$

The initiator selects actions with equal chance out of his repertoire, which makes the frequency of each action inverse proportional to the number of categories. This results in the following measure of information flow:

$$I = \begin{cases} \log_2(|A|) & \text{if the game succeeds} \\ 0 & \text{if the game fails} \end{cases} \quad (4)$$

with $|A|$ the number of categories of the initiator. If $|A|$ is a power of 2, I gives exactly the number of bits necessary to distinguish between these categories.

The creation of a new category by an agent has a twofold effect on the information flow. As this new category needs time to propagate through the population and get adopted by other agents, in the short term the imitative success and hence the information flow will decrease. In the long term, however, if the imitative success is able to recover, the information flow will increase as the number of categories increases.

4. Experimental setup

The imitation experiments, both in simulation and on physical systems, are implemented as embodied agents, consisting of a robot arm and a stereo head. At the moment one such setup is present at our laboratory and different agents use the same physical system during an experiment (figure 2).

As an agent must be able to observe other agents' actions as well as its own, the stereo head is shared between two agents playing an imitation game. This causes all agents to have the same point of view on an action, in contrast with the situation where two agents sit opposite each other and thus have different points of view on the action. Although this circumvents a number of interesting issues, such as the recognition of self and other, and the use of deictic coordinates, it presents a necessary simplification to test the principles of the imitation game. Another simplification that is used, is that the actions an agent can perform are limited to motion trajectories of the robot arm from one point to another.

Within the limitations of this simplified setup, similar results have been achieved with simulations and the real setup, as will be shown in section 5. These show that both setups are viable and that results achieved in simulation are valid for the robotic setup.

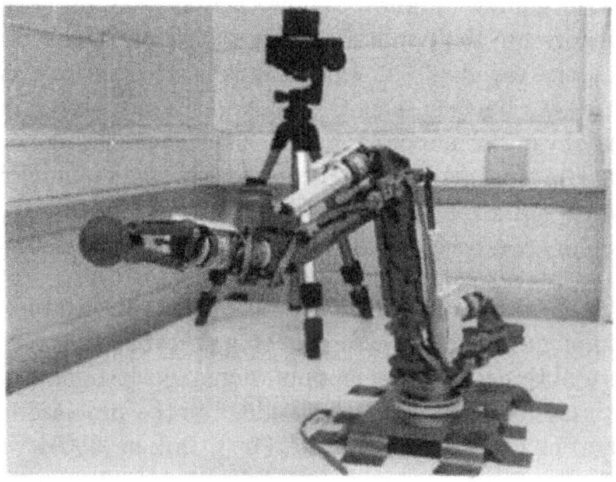

Figure 2: The setup consists of a 6 DOF arm and a stereo camera for respectively performing and perceiving actions.

The robot arm, stereo head and stereo processing software are commercially available. The arm we use is the teach-robot[4]. It has six degrees of freedom and is equipped with a gripper. The arm is position-driven; one can send goal motor positions to it but not change the motors' speed nor intervene during the execution of a command. The forward and inverse kinematics are known (De Vylder, 2002). In order to facilitate image processing and stereo vision, a colored ball is attached to the gripper of the arm. Currently only the position of this ball is focused on when observing an action. The vision system contains a MEGA-D stereo head, which delivers a series of simultaneously taken left and right images. The Small Vision System (SVS) (Konolige, 1997) is used for camera calibration and stereo vision. When an agent observes an action (both its own and that of another agent), it tracks the ball and builds a series of three-dimensional coordinates. We will describe this process in some detail.

For each pair of image frames, the left image is searched for possible ball segments using color templates, the right image is left unprocessed. A simple tracking mechanism then selects the most probable ball segment. Of all the pixels of this segment, the 3D-coordinates are calculated using SVS. Finally, the current ball position is estimated using expectation maximization (Bilmes, 1998). If the SVS calculation fails or is not reliable, the right image is searched for the most probable ball segment as well. The center of the ball segments in the right and left image are then assumed to correspond, from which we can estimate the ball's 3D-coordinates. This process runs at 10 frames per second.

The observation of an action by an agent results in a time series of 3D points. The interval between successive points is determined by the frame rate of the vision system and is independent of the duration of an action. As this duration varies over different actions, the length of such time-series varies as well. We assume that an agent knows when to observe its own actions and it is informed about the starting and stopping of actions of others.

Despite the fact that agents all use the same physical setup, it is important to notice that action execution and observation are subject to considerable amounts of noise. Among other disturbing factors are the inaccurate robot control (up to 2 cm inaccuracy in some workspace regions) and the varying illumination conditions. This noise poses limitations on the number and kinds of actions the agents can discriminate, but is inherent in working with embodied agents acting in the real world.

5. Results

5.1 Simulation results

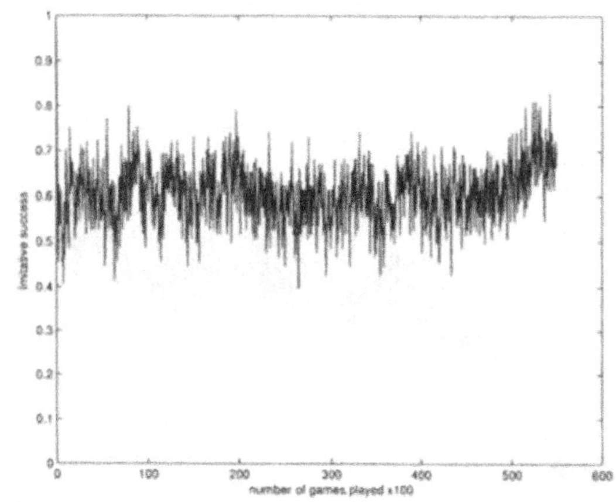

Figure 3: Average imitative success for the simulation.

Before performing experiments on real robots, we have implemented the imitation game in simulation, to verify the functionality of the imitation game. In the simulations a model of the kinematics of our physical robot was used to calculate simulated observations for actions. To approximate uncertainties found in the real system, noise is added to articulations of the robot arm and to

[4]Microelectronic Kalms

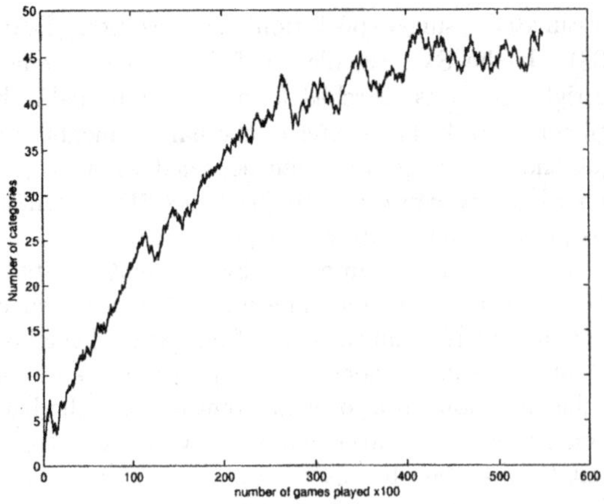

Figure 4: Average number of categories for the simulation

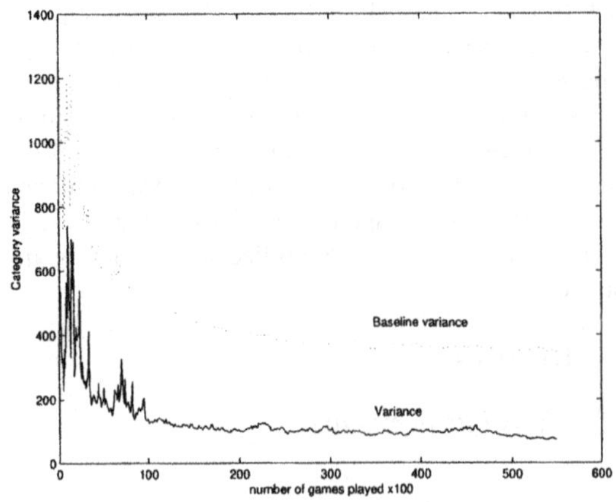

Figure 5: Average category variance for the simulation.

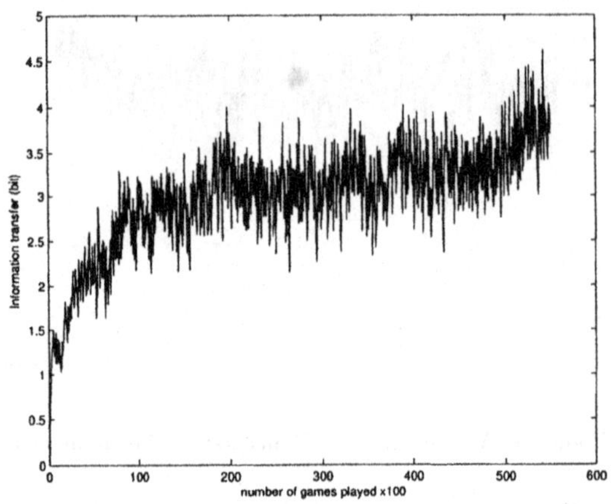

Figure 6: Information transfer for the simulation.

the perception. The model consists of a basic geometrical model of the robot arm, without simulating electrical and mechanical properties of the arm. The model takes as input 6 joint angles, and returns the position of all joints and of the end-effector (De Vylder, 2002). Noise is added to the simulation by adding a normal distributed term to these positions.

The simulation has 10 agents, and runs 55000 imitation games. Figure 3 shows the evolution of imitative success over all 55000 games, note how the success stays stable. The agents perform better than baseline imitative success of 0.53, which was determined by letting the agents first create random categories and then play imitation games. Figure 4 shows the average number of categories in the population. Figure 5 shows the average category variance $\overline{CV}$, which tells us how similar categories are between agents, $\overline{CV}$ is markedly better than random performance. Figure 6 shows how the information transfer increases as the agents continue playing imitation games, at 55000 games the agents have learned enough categories to transfer up to 4 bits of information.

These results illustrate that the agents succeed in developing a shared repertoire of categories. Similar results were obtained in larger populations and in simulations with higher amounts of noise added to the simulated observations, see (Jansen et al., 2003; Jansen, 2003). The work described here is distinguished from previous work by the fact that the imitation game now performs -and performs well- on a real-world system. This shows that the game implements dynamics that converge reliably under various conditions, both in simulation and on actual systems.

5.2 Robot experiment results

As experiments on the actual robotic setup run slower, the experiment uses only two agents. 13000 imitation games are played to obtain the results shown here.

Figure 7 through 10 show the imitative success, the number of categories, the average category variance and the information transfer. The success is again stable and slightly higher than in the simulation (due to the fact that only two agents are learning from each other, which is easier than having 10 agents learn from each other). The number of categories has not yet started to settle, meaning that both agents have not yet reached their limit of distinguishing and producing actions. The average category variance is considerably lower than baseline performance, showing that the categories resemble each other quite well. Figure 11 again illustrates this by plotting all actions of both agents at game 13000. And finally the information transfer is close to 6 bits, showing that both agents have acquired a coherent enough repertoire of categories to express 2^6 successfully distinguishable actions.

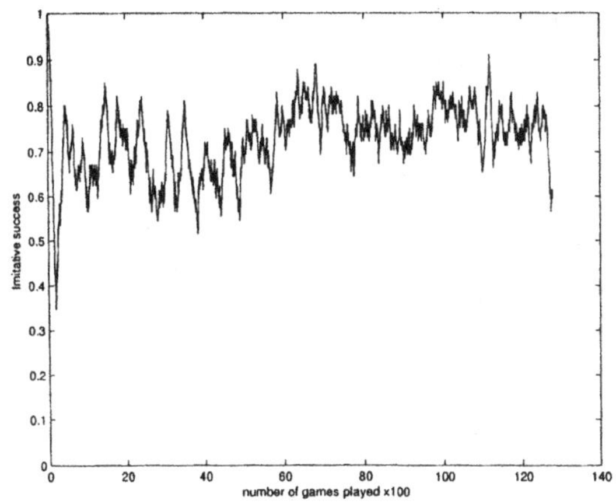

Figure 7: Average imitative success for the robotic setup.

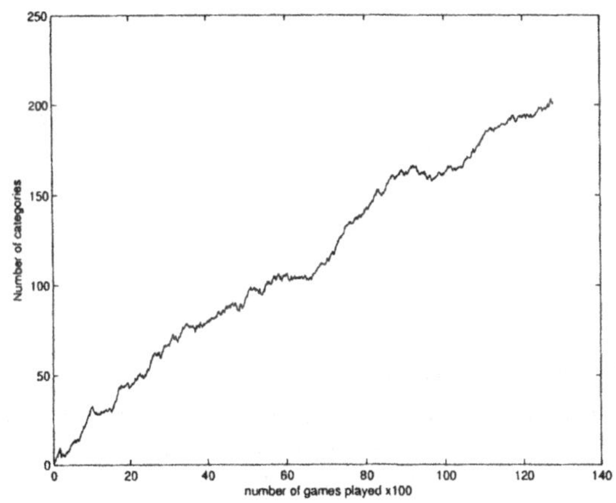

Figure 8: Average number of categories for the robotic setup

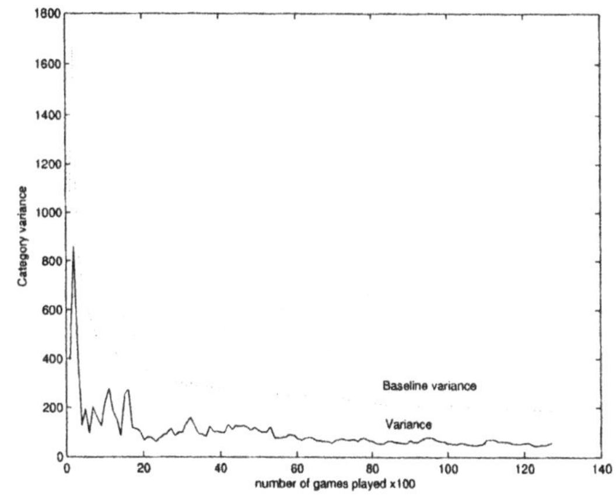

Figure 9: Average category variance for the robotic setup.

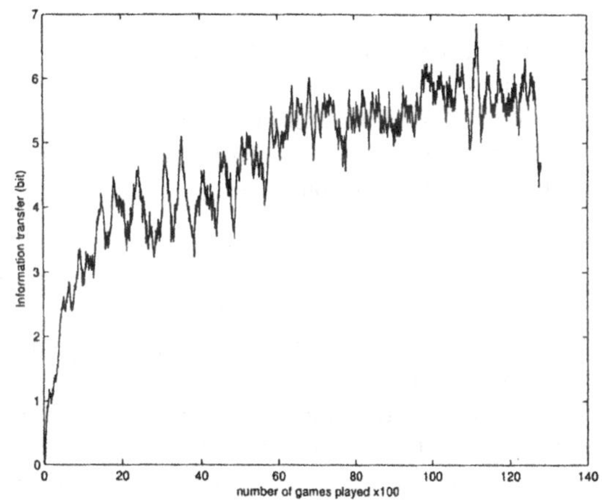

Figure 10: Information transfer for the robotic setup.

6. Conclusion

We demonstrated how actions can emerge and be sustained in a population of agents through simple imitative interactions. These interactions, called imitation games, are basic turn taking interactions between two agents, where one agent demonstrates an action and the other imitates. Successful imitation reinforces actions, while unsuccessful actions will disappear over time from the agent's repertoire. The imitation between the agents forms a self-organizing system. As with any self-organizing system it has two components: random fluctuation, implemented by the agents creating random actions, and selective feedback, implemented by the successfulness of imitations having an influence on which actions are retained in the agent's repertoire. This feedback provides pressure on the actions to be imitable, and is the driving factor for attaining a coherent and shared action repertoire in the population.

This is related to the theories of Steels (Belpaeme et al., 1998; Steels, 2000), who views human language and cognition as a self-organising system. In this view, human conceptualisation and language are the result of social and linguistic interactions between individuals in a population. Language and conceptualisation are thus open-ended systems in constant flux and evolve both synchronically and diachronically under influence of its users.

The take home message of this paper is that imitation serves to tune internal representations (in this case the internal representations are action categories, but these could also be speech sounds or sign language) of agents without the need to explicitly transfer these representations between agents. The agents' internal representations become shared, without resorting to some kind of "telepathic" transfer of representations. Moreover, the representations become shared in a robust manner, as imitation weeds out representations that are poorly observable or reproducible. These representations are also

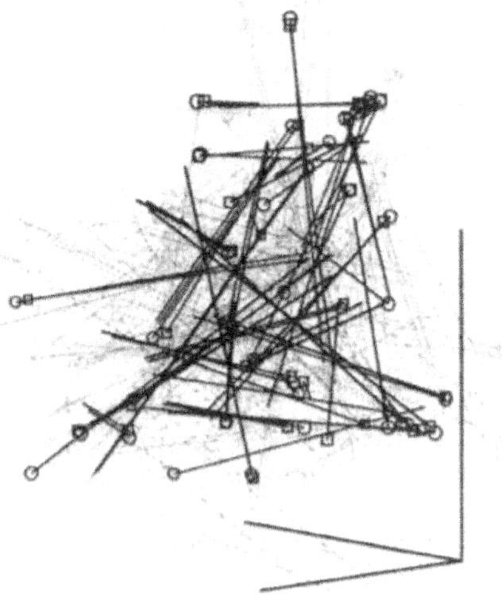

Figure 11: The actions of two agents plotted in 3D space, to demonstrate how actions resemble each other. Squares and circles denote the starting point of the action for agent 1 and 2 respectively. The most successful actions have been plotted in black, less successful actions in grey.

adaptive, as the dynamics of the imitation game serve to discard any representations that are unsuccessful and gracefully accepts changes to the parameters, embodiment or the environment.

In the experiments reported here, the agents acquire repertoires of action categories that are not situated. The actions do not serve to communicate, as they for example would in sign language. Nor do they serve to manipulate objects or perform tasks, such as stacking blocks. A logical extension to the imitation game described in this paper, would be to make the actions that agent learn "useful". For this however, the agents would need intentionality. They would need to understand what it is that the other is trying to demonstrate. Is the other agent just waving its arm? Or is the action aimed at pushing a block across the table?

To build agents that actually imitate the use of an action, we would need to construct a basic form of Theory of Mind for agents. This would enable agents to deduce the intention behind an action, and would lead agents to imitate the intent of an action, instead of the action itself. We hope to learn from developmental psychology how intentionality can bootstrap itself, without being programmed explicitly into the agents. One route, which might be promising, looks at how infants learn to coordinate their body movements; new born babies learn to coordinate their limbs and explore the impact of their actions on the environment through play (Vygostky, 1967). Perhaps creating a challenging and stimulating environment for the agent and letting them explore the impact of their actions on the environment, might bootstrap the self-organization of intentional behavior, which could then later be used to acquire new skills through imitation.

Acknowledgements

Bart Jansen is sponsored by a grant from the Institute for the Promotion of Innovation by Science and Technology in Flanders (IWT). Tony Belpaeme is a postdoctoral fellow of the Flemish Fund for Scientific Research (F.W.O. Vlaanderen).

References

Alissandrakis, A., Nehaniv, C. L., and Dautenhahn, K. (2001). Through the looking-glass with ALICE – trying to imitate using correspondences. In Balkenius, C., Zlatev, J., Kozima, H., Dautenhahn, K., and Breazeal, C., (Eds.), *Proc. First International Workshop on Epigenetic Robotics: Modeling Cognitive Development in Robotic Systems*, volume 85. Lund University Cognitive Studies. September 17-18, 2001, Lund, Sweden.

Belpaeme, T., Steels, L., and van Looveren, J. (1998). The construction and acquisition of visual categories. In Birk, A. and Demiris, Y., (Eds.), *Proceedings of the 6th European Workshop on Learning Robots*, Lecture Notes on Artificial Intelligence, Berlin. Springer.

Billard, A. and Matarić, M. J. (2001). Learning human arm movements by imitation: Evaluation of a biologically inspired connectionist architecture. *Robotics and Autonomous Systems*, 37(2-3):145–160.

Bilmes, J. A. (1998). A gentle tutorial of the EM algorithm and its application to parameter estimation for gaussian mixture and Hidden Markov Models. Technical report, U.C. Berkeley technical report TR-97-021.

Breazeal, C. and Scassellati, B. (2002). Challenges in building robots that imitate people. In Dautenhahn, K. and Nehaniv, C. L., (Eds.), *Imitation in Animals and Artifacts*, pages 363–390. The MIT Press, Cambridge, MA.

Cooper, F. S., Delattre, P. C., Liberman, A. M., Borst, J. M., and Gerstman, L. J. (1976). Some experiments on the perception of synthetic speech sounds. In Fry, D. B., (Ed.), *Acoustic Phonetics*, pages 258–283. Cambridge University Press, Cambridge, UK.

Dautenhahn, K. and Nehaniv, C. L., (Eds.) (2002). *Imitation in Animals and Artifacts.* The MIT Press, Cambridge, MA.

de Boer, B. (2000). Self-organisation in vowel systems. *Journal of Phonetics*, 28(4):441–465.

De Vylder, B. (2002). Forward and inverse kinematics of the Teach-Robot. Technical report, AI-MEMO 02-02, Artificial Intelligence Laboratory – Vrije Universiteit Brussel.

Demiris, Y. (2002). Imitation as a dual-route process featuring predictive and learning components: a biologically-plausible computational model. In Dautenhahn, K. and Nehaniv, C. L., (Eds.), *Imitation in Animals and Artifacts*, pages 327–361. The MIT Press, Cambridge, MA.

Gaussier, P., Moga, S., Quoy, M., and Banquet, J.-P. (1998). From perception-action loops to imitation processes: A bottom-up approach of learning by imitation. *Applied Artificial Intelligence*, 12(7-8):701–727.

Harnad, S., (Ed.) (1987). *Categorical Perception*. Cambridge University Press, Cambridge.

Jansen, B. (2003). An imitation game for emerging action categories. In Banzhaf, W., Christaller, T., Dittrich, P., Kim, J. T., and Ziegler, J., (Eds.), *Advances in Artificial Life, 7th European Conference on Artificial Life (ECAL 2003), Dortmund, Germany*, volume 2801 of *Lecture Notes in Computer Science*, pages 800–809. Springer.

Jansen, B., De Vylder, B., Belpaeme, T., and de Boer, B. (2003). Emerging shared action categories in robotic agents through imitation. In Dautenhahn, K. and Nehaniv, C. L., (Eds.), *Second International Symposium on Imitation in Animals and Artifacts 2003*, pages 145–152. The Society for the Study of Artificial Intelligence and the Simulation of Behavior.

Konolige, K. (1997). Small vision system: Hardware and implementation. In *Proceedings of the Eighth International Symposium on Robotics Research*. http://www.ai.sri.com/konolige/svs/Papers.

Kuhl, P. K. and Meltzoff, A. N. (1996). Infant vocalizations in response to speech: Vocal imitation and developmental change. *The Journal of the Acoustical Society of America*, 100(4):2425–2438.

Kuniyoshi, Y., Inaba, M., and Inoue, H. (1994). Learning by watching: Extracting reusable task knowledge from visual observation of human performance. *IEEE Transactions on Robotics and Automation*, 10(6):799–822.

Meltzoff, A. and Moore, M. K. (1977). Imitation of facial and manual gestures by human neonates. *Science*, 198:75–78.

Nadel, J. and Butterworth, G., (Eds.) (1999). *Imitation in infancy*. Cambridge studies in cognitive and perceptual development. Cambridge University Press, Cambridge, UK.

Rizzolatti, G., Fadiga, L., Gallese, V., and Fogassi, L. (1996). Premotor cortex and the recognition of motor actions. *Cognitive Brain Research*, 3:131–141.

Schaal, S. (1999). Is imitation learning the route to humanoid robots? *Trends in Cognitive Sciences*, 3(6):233–242.

Steels, L. (2000). Language as a complex adaptive system. In Schoenauer, M., (Ed.), *Proceedings of Parallel Problem Solving from Nature VI*, Lecture Notes in Computer Science. Springer, Berlin, Germany.

Vygostky, L. S. (1967). Play and its role in the mental development of the child. *Soviet Psychology*, 5(3):6–18.

Whiten, A. and Ham, R. (1992). On the nature and evolution of imitation in the animal kingdom: Reappraisal of a century of research. In Slater, P., Rosenblatt, J., Beer, C., and Milinski, M., (Eds.), *Advances in the study of behavior*, volume 21, pages 239–283. Academic Press, San Diego, CA.

ADAPTIVE BEHAVIOR IN LANGUAGE AND COMMUNICATION

A Connectionist Approach to Learn Association between Sentences and Behavioral Patterns of a Robot

Yuuya Sugita
BSI, RIKEN
Hirosawa 2-1, Wako-shi
Saitama 3510198 JAPAN
sugita@bdc.brain.riken.go.jp

Jun Tani
BSI, RIKEN
Hirosawa 2-1, Wako-shi
Saitama 3510198 JAPAN
tani@bdc.brain.riken.go.jp

Abstract

We present a novel connectionist model for acquiring the semantics of a simple language through the behavioral experiences of a real robot. We focus on the "compositionality" of semantics, a fundamental characteristic of human language, which is the ability to understand the meaning of a sentence as a combination of the meanings of words. We also pay much attention to the "embodiment" of a robot, which means that the robot should acquire semantics which matches its body, or sensory-motor system. The essential claim is that an embodied compositional semantic representation can be self-organized from generalized correspondences between sentences and behavioral patterns. This claim is examined and confirmed through simple experiments in which a robot generates corresponding behaviors from unlearned sentences by analogy with the correspondences between learned sentences and behaviors.

1. Introduction

Implementing language acquisition systems is one of the most difficult problems, since not only the complexity of the syntactical structure, but also the diversity in the domain of meaning make this problem complicated and intractable. In particular, how linguistic meaning can be represented in the system is crucial. This problem has been investigated for many years.

In this paper, we introduce a connectionist model that acquires the semantics of a simple finite language with respect to correspondences between sentences and the behavioral patterns of a real robot. An essential question is how compositional semantics can be acquired in the proposed connectionist model without providing any representations of the meaning of a word or behavior routines *a priori*. By "compositionality", we refer to the fundamental human ability to understand a sentence from (1) the meanings of its constituents, and (2) the way in which they are put together. It is possible for a language acquisition system that acquires compositional semantics to derive the meaning of an unknown sentence from the meanings of known sentences. Consider the unknown sentence: "John likes birds." It could be understood by learning these three sentences: "John likes cats."; "Mary likes birds."; and "Mary likes cats." That is to say, generalization of meaning can be achieved through compositional semantics.

From the point of view of compositionality, the symbolic representation of word meaning is advantageous for processing the linguistic meaning of sentences. In general, AI-based models employ semantic symbols to represent word meanings and have much affinity with compositionality in terms of the meanings of sentences (e.g., (Thompson and Mooney, 1998)). Thus, this approach assumes that the meanings of words (i.e., lexicon) is independent of the usages of words in sentences (i.e., syntax).

According to this observation, various learning models have been proposed to acquire the embodied semantics of language. For example, some models learn semantics in the form of correspondences between sentences and non-linguistic objects, i.e., visual images (Roy, 2002), sequences of video images (Siskind, 2001), or the sensory-motor patterns of a robot (Iwahashi, 2003, Steels, 2000). In these models, the meanings of words are labeled by the words themselves (a.k.a., semantic symbols). Consequently, a significant part of semantic learning can be reduced to the learning of the syntactic structure of the sentence. This means that the acquisition of meanings of sentences can be translated into two relatively separate steps, the acquisition of word meanings and the acquisition of syntax.

Although this separated learning approach seems to be plausible from the requirements of compositionality, it causes inevitable difficulties in representing the grounded meaning of a sentence. A priori separation of lexicon and syntax requires a pre-defined manner of combining word meanings to compose the meaning of a sentence. As a result, these approaches require relatively heavy manual pre-programming to realize compositional semantic representations. Not all of these models aim for self-organization of grounded compositional semantic representations. We nevertheless point out possible problems with their approaches within the framework of acquiring grounded compositional semantics.

In Iwahashi's model, the class of a word is assumed to be given prior to learning its meaning because different acquisition algorithms are required for nouns and verbs (c.f., (Siskind, 2001)). Roy's model does not require a priori knowledge of word classes, but requires the strong assumption, that the meaning of a word can be assigned to some pre-defined attributes of non-linguistic objects. This assumption is not realistic in more complex cases, such as when the meaning of a word needs to be extracted from non-linguistic spatio-temporal patterns, as in case of learning verbs.

Recently, several connectionist models for acquiring embodied language have been proposed (Billard, 2002, Sugita and Tani, 2002). These models don't require separate treatment of words and syntax. However, they can demonstrate only few compositional characteristics. Also, it cannot be said that embodied semantics is self-organized fully from scratch, since the models assume behavior primitives *a priori*.

In this paper, we discuss an essential mechanism for self-organizing embodied compositional semantic representations. Our model implements compositional semantics by utilizing the generalization capability of a recurrent neural network (RNN), where the meaning of each word cannot exist independently, but emerges from the relations with others (c.f., reverse compositionality, (Fodor, 1999)). In this situation, a sort of generalization can be expected, such that the meanings of novel sentences can be inferred by analogy with learned ones.

The experiments were conducted using a real mobile robot with an arm and with various sensors, including a vision system. A finite set of two-word sentences consisting of a verb followed by a noun was considered. We assume that a sentence is represented as a sequence of pre-defined words, however, employ neither semantic symbols nor composition rules *a priori*. Our analysis will clarify what sorts of internal neural structures should be self-organized for achieving compositional semantics grounded to a robot's behavioral experiences. Although our experimental design is limited, the current study will suggest an essential mechanism for acquiring grounded compositional semantics, with the minimal combinatorial structure of this finite language (Evans, 1981).

2. Task Design

The aim of our experimental task is to understand an essential mechanism for self-organizing compositional semantics grounded to the behavior of a robot. In the training phase, our neural network model learns the relationships between sentences and the corresponding behavioral sensory-motor sequences of a robot in a supervised manner. It is then tested to generate behavioral sequences from a given sentence. We regard compositional semantics as being acquired if appropriate behavioral sequences can be generated from unlearned sentences by analogy with learned data.

Our mobile robot has three actuators, with two wheels and a joint on the arm; a colored vision sensor; and two torque sensors, on the wheel and the arm (Figure 1a). The robot operates in an environment where three colored objects (red, blue, and green) are placed on the floor (Figure 1b). The positions of these objects can be varied so long as the robot sees the red object (R) on the left side of its field of view, the blue object in the middle (B), and the green object (G) on the right at the start of every trial of behavioral sequences.

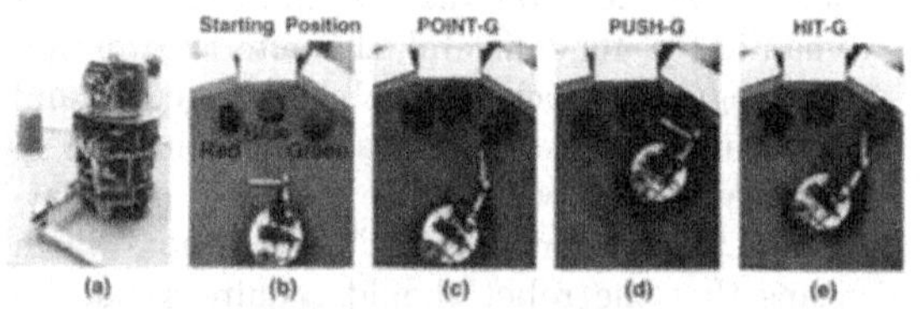

Figure 1: The mobile robot (a) starts from a fixed position in the environment and (b) ends each behavior by pointing at (c), pushing (d), or hitting (e) either the red, blue, or green object.

The robot learns nine categories of behavioral patterns, consisting of pointing at, pushing, and hitting each of the three objects, in a supervised manner. These categories are denoted as POINT-R, POINT-B, POINT-G, PUSH-R, PUSH-B, PUSH-G, HIT-R, HIT-B, and HIT-G (Figure 1c-e). It should be noted that the robot learns these behaviors as sensory-motor time sequences, in which there are no obvious relationships among behavioral categories. For example, one can not generate a sensory-motor sequence belonging to POINT-R as a combination of POINT-B and PUSH-R.

The robot also learns sentences which consist of one of three verbs (**`point`**, **`push`**, **`hit`**) followed by one of six nouns (**`red`**, **`left`**, **`blue`**, **`center`**, **`green`**, **`right`**). The meanings of these 18 possible sentences are given in terms of fixed correspondences with the 9 behavioral categories (Figure 2). For example, "**`point red`**" and "**`point left`**" correspond to POINT-R, "**`point blue`**" and "**`point center`**" to POINT-B, and so on.

In these correspondences, "**`left`**," "**`center`**," and "**`right`**" have exactly the same meaning as "**`red`**," "**`blue`**," and "**`green`**" respectively. These synonyms are introduced to observe how the behavioral similarity affects the acquired linguistic semantic structure.

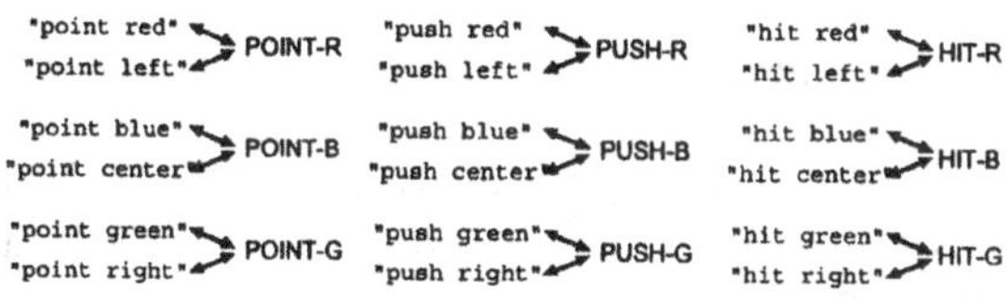

Figure 2: The correspondences between sentences and behavioral categories. For each behavioral category, there are two corresponding sentences.

3. Proposed Model

Our model employs two RNNs with parametric bias nodes (RNNPBs) (Tani, 2003, Tani and Ito, 2003, Ito and Tani, 2003) in order to implement a linguistic module and a behavioral module (Figure 3). The RNNPB, like the conventional Jordan-type RNN (Jordan and Rumelhart, 1992), is a connectionist model for learning time sequences. The linguistic module learns the above sentences represented as time sequences of words (Elman, 1990), while the behavioral module learns the behavioral sensory-motor time sequences of the robot. To acquire the correspondences between the sentences and behavioral sequences, these two modules are connected to each other by using the parametric bias binding method. Before discussing this binding method in detail, we introduce the overall architecture of RNNPB and both modules.

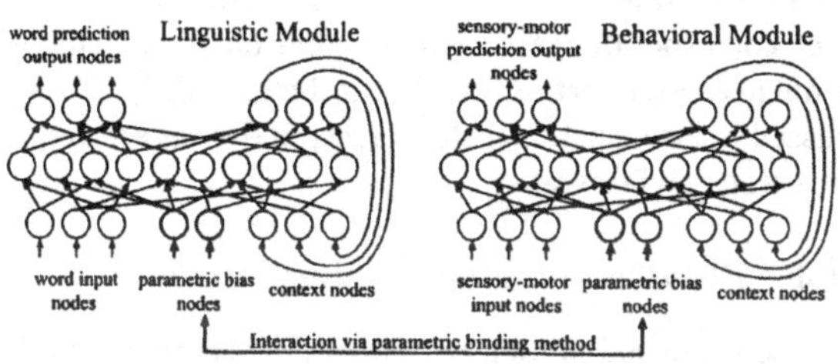

Figure 3: Our model is composed of two RNNs with parametric bias nodes (RNNPBs), one for a linguistic module and the other for a behavioral module. Both modules interact with each other during the learning process via the parametric bias method introduced in the text.

3.1 RNNPB

The RNNPB has the same neural architecture as the Jordan-type RNN except for the PB nodes in the input layer (c.f., each module of Figure 3). Unlike the other input nodes, these PB nodes take a specific constant vector throughout each time sequence, and are employed to implement a mapping between fixed-length vectors and time sequences.

Like the conventional Jordan-type RNN, the RNNPB learns time sequences in a supervised manner. The difference is that in the RNNPB, the vectors that encode the time sequences are self-organized in PB nodes during the learning process. The common structural properties of all the training time sequences are acquired as connection weight values by using the back-propagation through time (BPTT) algorithm, as used also in the conventional RNN (Jordan and Rumelhart, 1992, Rumelhart et al., 1986). Meanwhile, the specific properties of each individual time sequence are simultaneously encoded as PB vectors (c.f., (Miikkulainen, 1993)). As a result, the RNNPB self-organizes a mapping between the PB vectors and the time sequences.

The learning algorithm for the PB vectors is a variant of the BPTT algorithm. For each time sequence, the back-propagated errors with respect to the PB nodes are accumulated for all time steps to update the PB vectors. Formally, the update rule for the PB vector p_{x_i} encoding the i-th time sequence x_i is given as follows:

$$\delta^2 p_{x_i} = \frac{1}{l_i}\sum_{t=0}^{l_i-1} error_{p_{x_i}}(t) \qquad (1)$$

$$\delta p_{x_i} = \epsilon \cdot \delta^2 p_{x_i} + \eta \cdot \delta p_{x_i}^{old} \qquad (2)$$

$$p_{x_i} = p_{x_i}^{old} + \delta p_{x_i}. \qquad (3)$$

In equation (1), the update of PB vector $\delta^2 p_{x_i}$ is obtained from the average back-propagated error with respect to a PB node $error_{p_{x_i}}(t)$ through all time steps from $t = 0$ to $l_i - 1$, where l_i is the length of x_i. In equation (2), this update is low-pass filtered to inhibit frequent rapid changes in the PB vectors.

Here, we introduce an abstracted operational notation for the RNNPB to facilitate a later explanation of our proposed method of binding language and behavior. After successfully learning the time sequences $x_1, \cdots, x_N$, the RNNPB can generate a time sequence x_i from its corresponding parametric bias p_{x_i}. By applying an operator $RNNPB$, the generation of x_i is described as follows:

$$RNNPB(p_{x_i}) \rightarrow x_i, \quad i = 1, \cdots, N. \qquad (4)$$

We note here that the actual generation process of a time sequence is implemented by iteratively utilizing the RNNPB with input vectors for each time step. Both the environmental sensory information and the internal prediction of the RNNPB are employed as input vectors depending on the required functionality of the module.

Furthermore, the RNNPB can be used not only for sequence generation processes but also for recognition processes. For a given sequence x_i, the corresponding PB vector p_{x_i} can be obtained by using the update rules for the PB vectors (equations (1) to (3)), without updating the connection weight values. This inverse operation for generation is regarded as recognition, and is hence denoted as follows:

$$RNNPB^{-1}(x_i) \rightarrow p_{x_i}, \quad i = 1, \cdots, N. \qquad (5)$$

The other important characteristic nature of the RNNPB is that the relational structure among the training time sequences can be acquired in the PB space through the learning process. This generalization capability of RNNPB can be employed to generate and recognize unseen time sequences without any additional learning. For instance, by learning several cyclic time sequences of different frequency, novel time sequences of intermediate frequency can be generated (Ito and Tani, 2003).

3.2 Linguistic Module

The linguistic module learns the sentences shown in section 2to acquire the underlying syntactic structure. The sentences are represented as a time sequence of words, which starts with a fixed starting symbol. Each word is locally represented, such that, each input node of the module corresponds to a specific word and only one input node takes 1.0 while the others take 0.0. The module has 10 input nodes for each of 9 words (`point`, `push`, `hit`, `red`, `left`, `blue`, `center`, `green`, and `right`) and a starting symbol. The module also has 6 parametric bias nodes, 4 context nodes, 50 hidden nodes, and 10 prediction output nodes. This representation is almost the same as in Elman's previous work (Elman, 1990). However, the acquired dynamics shows a different characteristic nature.

Elman's model predicts a probabilistic distribution of the next words. Therefore, the prediction output can be a probabilistic superposition of word vectors. In contrast, our linguistic module can deterministically predict the next word by utilizing a given PB vector. Recall that a PB vector encodes a specific sentence by means of equation (4). Thus, RNNPB acquires properties of specific sentences in addition to common syntactic properties of sentences. We'll see the role of this characteristic nature in treating the meaning of a sentence in a later section.

3.3 Behavioral Module

The behavioral module learns the time sequences of sensory-motor vectors involving the robot's behaviors presented in section 2.After successfully learning them, the module generates motor commands and a prediction of the sensory image at the next time step from the current sensory-motor vector (Tani, 1996).

The robot can generate behavior by iteratively utilizing the module with changing sensory inputs to generate motor commands every one-third of a second. A behavioral sequence is thus created by sampling three sensory-motor vectors per second during a trial of the robot's behavior. Typical behavioral sequences are about 5 to 25 seconds long, and therefore have about 15 to 75 sensory-motor vectors.

A sensory-motor vector is a real-numbered 26-dimensional vector consisting of 3 motor values (for angular velocities of 2 wheels and an angle of the arm joint), 2 values from torque sensors (of the wheels and the arm), and 21 values encoding the visual image. The visual field is divided vertically into 7 regions, and each region is represented by (1) the fraction of the region covered by the object, (2) dominant hue of the object in the region and (3) the bottom border of the object in the region, which is proportional to the distance of the object from the camera[1]. We note here that the visual information is not the most important for the acquisition of semantics. It occupies 21 of 26 dimensions in the sensory-motor vector only due to the characteristic nature of visual information. All the sensory-motor information is complementary and interdependent.

The module has 26 input nodes for the sensory-motor vector, 6 parametric nodes, 4 context nodes, 70 hidden nodes, 6 prediction output nodes which correspond to 3 motor values, 2 required torque values, and a hue value for the center region of the visual field. In the actual behavior generation process, the predicted motor values are used as the actual motor values at the next time step. The rest of the values of the sensory vector are not predicted to reduce the learning time. The module can enable the robot to generate behavior appropriately without predicting the entire sensory-motor vector.

It should be emphasized that the behavioral sequences are not separable. Hence they can not be decomposed into plausible primitives, unlike the sentences which can be broken down into words. This implies that no direct correspondences between some parts of behavioral sequences and certain words can be established. As discussed in later sections, the correspondences are acquired in a non-trivial way.

3.4 Parametric Bias Binding Method

In the proposed model, corresponding sentences and behavioral sequences are constrained to have the same PB vectors in both modules. Under this condition, corresponding behavioral sequences can be generated naturally from sentences.

When a sentence $\boldsymbol{s}_i$ and its corresponding behavioral sequence $\boldsymbol{b}_i$ have the same PB vector, we can obtain $\boldsymbol{b}_i$ from $\boldsymbol{s}_i$ as follows:

$$RNNPB_B(RNNPB_L^{-1}(\boldsymbol{s}_i)) \rightarrow \boldsymbol{b}_i \qquad (6)$$

where $RNNPB_L$ and $RNNPB_B$ are abstracted operators for the linguistic module and the behavioral module, respectively.

The PB vector $p_{\boldsymbol{s}_i}$ is obtained by recognizing the sentence $\boldsymbol{s}_i$. Because of the constraint that corresponding sentences and behavioral sequences must have the same

[1] For the region in which there is no colored area, the hue takes a pre-defined constant value 1.0, and the bottom border position takes 0.0, which designates very far.

PB vectors, p_{b_i} is equal to p_{s_i}. Therefore, we can obtain the corresponding behavioral sequence $\boldsymbol{b}_i$ by utilizing the behavioral module with p_{b_i}. In the same way we can also obtain the $\boldsymbol{s}_i$ from $\boldsymbol{b}_i$. Thus, sentences and behavioral sequences are connected bi-directionally as observed in the mirror system (Rizzolatti et al., 1996).

The constraint is implemented by introducing an interaction term into part of the update rule for the PB vectors (equation (3)).

$$p_{s_i} = p_{s_i}^{old} + \delta p_{s_i} + \gamma_L \cdot (p_{b_i}^{old} - p_{s_i}^{old}) \quad (7)$$

$$p_{b_i} = p_{b_i}^{old} + \delta p_{b_i} + \gamma_B \cdot (p_{s_i}^{old} - p_{b_i}^{old}) \quad (8)$$

where γ_L and γ_B are positive coefficients that determine the strength of the binding. Equations (7) and (8) are the constrained update rules for the linguistic module and the behavior module, respectively. Under these rules, the PB vectors of a corresponding sentence $\boldsymbol{s}_i$ and behavioral sequence $\boldsymbol{b}_i$ attract each other.

Actually, the corresponding PB vectors p_{s_i} and p_{b_i} need not be completely equalized to acquire a correspondence. The epsilon errors of the PB vectors can be neglected because of the continuity of PB spaces.

3.5 Generalization of Correspondences

The compositional semantics of a simple language self-organized in our model is now explained. Originally, "compositionality" referred to the characteristic nature of semantics in which the meaning of a sentence can be represented as a combination of the meanings of the words. However, for compositional semantics, a substantial requirement is the meaning of an unlearned sentence can be derived from the meanings of known sentences.

Our model implements compositional semantics without introducing any explicit representations of the meanings of words. We instead regard the model as acquiring compositional semantics when it can generate appropriate behavioral sequences from all sentences without learning all correspondences.

To achieve this, an unlearned sentence and its corresponding behavioral sequences must have the same PB vector. Nevertheless, the PB binding method only equalizes the PB vectors for given corresponding sentences and behavioral sequences (e.g., equation (7) and (8)).

Implicit binding, or in other words, inter-module generalization of correspondences, is achieved by dynamic coordination between the PB binding method and the intra-module generalization of each module. The local effect of the PB binding method spreads over the whole PB space, because each individual PB vector depends on the others in order to self-organize PB structures reflecting the relationships among training data. Consider the situation in which the linguistic PB vector of "`point red`" is perturbed via the PB binding method. To keep the relationship among sentences, the perturbation of "`point red`" is propagated to "`point *`" and "`* red`", and then spreads over the whole linguistic PB space. A similar process occurs in the behavioral module.

Thus, the PB structures of both modules closely interact via the PB binding methods. Finally, both PB structures converge into a common PB structure, in which the structures of both sentences and behavioral sequences are unified. Under the condition that both modules share a common PB structure, all corresponding sentences and behavioral sequences then share the same PB vectors automatically.

4. Experiments

To observe self-organization of the compositional semantics of language based on the behavioral experiences of a robot, we designed three experiments. Here, we briefly explain each experiment.

In experiment I, only the linguistic module was employed to investigate the acquisition of the pure syntactic structure of language. The module acquired the complete syntax by learning 14 of the 18 possible sentences.

In experiment II, only the behavioral module was employed to investigate the acquisition of the pure embodied structure of the behaviors. The module learned sensory-motor sequences of all nine behavioral categories in a supervised manner.

To generate behaviors robustly, the behavioral module needs to acquire not the behavioral trajectories themselves but the functionality to generate motor commands coupled with the environment. Thus, 10 different sensory-motor sequences were given for each behavioral category, and a total of 90 training sequences were generated through human guidance. To differentiate these sequences, the positions of the objects in the environment were slightly varied for each generated sequence. The variation was at most 20 percent of the distance between the starting position of the robot and the original position of each object in every direction.

Finally, in experiment III, both modules were employed to investigate the acquisition of the compositional semantics. As in the previous experiment, the linguistic module learned 14 of the 18 sentences and the behavioral module learned the behavioral sequences of the nine categories. For each sentence, five different pairs of sentences and the corresponding behavioral sequences were learned. Thus, the linguistic module learned 70 sentences, and the behavioral module learned 70 behavioral sequences with binding and 20 without binding. In addition, the behavioral module learned the same 90 behavioral sequences without binding.

5. Results and Analysis

The results of the last experiment showed that the compositional semantics of a simple language could be acquired by generalizing the correspondences between sentences and behavioral sequences in the proposed model. By employing the acquired semantics, unlearned sentences could be recognized and understood in order to generate appropriate behaviors.

This generalization was achieved by sharing a common PB structure between modules. In the following analysis, we show that this common structure possesses (1) the combinatorial properties of the pure linguistic structure and (2) a metric based on the similarities of behavioral sequences from the pure behavioral structure.

5.1 Linguistic Module: Syntactic Structure

In this section we analyze the results of experiment I, in which only the linguistic module learned. The final averaged output error of each node was 0.0060 after 50000-step learning. Analysis of the linguistic module reveals two important properties: (1) acquisition of syntax through the process of generalizing sentences, and (2) the structure of PB space, which reflects the combinatorial structure of the sentences.

In this experiment, 14 of the 18 two-word sentences were employed as training sentences. The four remaining sentences, "`point green`", "`point right`", "`push red`", and "`push left`" were used to evaluate the intra-module generalization.

The linguistic module acquired the underlying syntax from the given sentences. This was confirmed by the fact that the module could correctly generate only grammatical sentences. In other words, the minimized regeneration error of grammatical sentences (verb-noun) is much smaller than of ungrammatical 2-word sequences (verb-verb, noun-verb, noun-noun). For a given 2-word sequence s, we define the minimized regeneration error as the total RMS error between s and the regenerated sequence $\hat{s}$, where

$$\hat{s} \leftarrow RNNPB_L(RNNPB_L^{-1}(s)). \tag{9}$$

Here, it should be noted that the most likely PB vector encoding s can be obtained by the recognition process. Therefore, $\hat{s}$ is the most likely estimation of s in the generable sequences of the module.

Next, we show the acquired PB space representing the combinatorial structure of the sentences. Figure 4a is a plot of the PB vectors obtained by recognizing all the sentences including the unlearned ones. The PB space is 6-dimensional, so only two selected parametric nodes are plotted here[2]. We can find three congruent substructures for each verb (Figure 4b), and six congruent sub-structures for each noun (Figure 4c)[3].

The congruency of the sub-structures for verbs and nouns represents the combinatorial structure of the sentences. This means that the words included in a sentence determine the PB vector of the sentence. For example, the PB vector of "`push green`" must be on the cross-over point between sub-structures for `push` and `green`. In addition, the PB vectors of the unlearned sequences take part in these sub-structures, and this supports intra-module generalization.

5.2 Behavioral Module: Behavioral Similarity

In this section, we analyze the results of experiment II, in which only the behavioral module learned. The final averaged output error of each output node was 0.012 after 50000-step learning.

After successful learning, the robot can generate the specified behavior regardless of the arrangement of the objects so long as the red, blue, and green objects are placed from the left to the right (Figure 5). The behavioral module generalizes behavioral sequences as functions which generate motor commands depending on the current sensory images, and therefore the robot can generate appropriate behavioral trajectories even in a noisy environment.

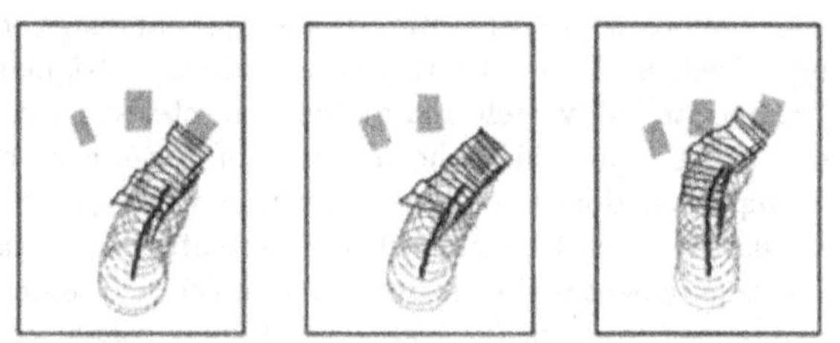

(a) (b) (c)

Figure 5: These three trajectories (a, b, and c) show that the robot could regenerate PUSH-G behavior regardless of the position of the objects. The robot could regenerate behavioral sequences appropriately so long as the red, blue, and green objects were placed in order from the left to the right, and the robot could see the target object at the starting position. This means that the behavioral module learned the given behavioral patterns by acquiring appropriate sensory-motor functions. The thick black lines represent trajectories of the center of the robot, and the thin black lines represent trajectories of the tip of the arm.

Although no combinatorial structure among the behavioral categories could be seen in the acquired structure of the behavioral module, we did find a metric reflecting the similarity of behavioral sequences (Figure 6). In this figure, the 6-dimensional PB vectors acquired at

[2] Unlike in figure 6 and figure 7, we did not employ the conventional principal component analysis (PCA) method to determine the axes. The plot projected on the surface obtained by PCA is not suitable for seeing congruent sub-structures because each sub-structure is plotted on a line.

[3] You can find similar diagrams in "Cours de linguistique générale (de Saussure, 1966)."

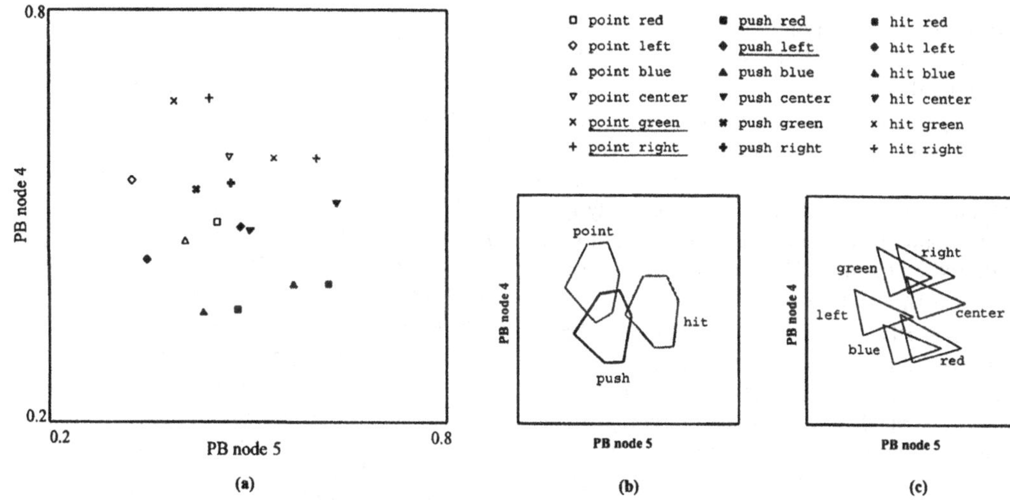

Figure 4: A plot of the linguistic PB space acquired without binding. The underlined sentences were unlearned. Two of six nodes are selected for ease of observing the combinatorial structure (a). The plot has three congruent sub-structures for each verb (b), and six congruent sub-structures for each noun (c).

the learning phase was projected onto a surface which maximizes the deviation of the plots using the conventional principle component analysis (PCA) method.

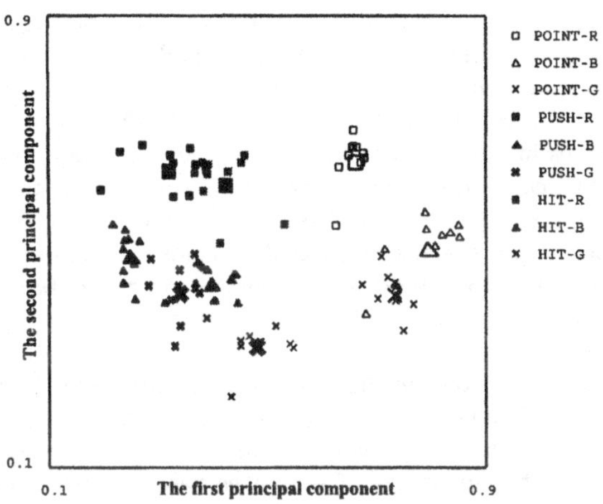

Figure 6: A plot of the behavioral sequences in the PB space acquired without binding. The large plots show the averaged PB vectors for each behavioral category. The 6-dimensional PB space was projected onto a surface that maximizes the deviation of the plots by using the PCA method. The accumulated contribution ratio of the two axes is 70%.

As nine clusters were observed, corresponding to the nine behavioral categories, we can conclude a metric based on the behavioral categories is internally represented in the PB space. This was confirmed by an additional experiment, in which it was shown that the robot could robustly regenerate the behavioral trajectory of each behavioral category using a PB vector value from the corresponding cluster.

We could not find a combinatorial arrangement of the behavioral clusters differing from the linguistic PB space shown in section 5.1. In the plot of the averaged PB vector over each behavioral category, this non-combinatorial arrangement is clearly observed (see large plots in figure 6). For example, it is not possible to estimate the PB vector of PUSH-G from the relationship among the PB vectors of PUSH-B, HIT-G and HIT-B. This confirms that the PB structure reflecting the behavioral sensory-motor sequences has no combinatorial property.

5.3 *Unified Structure*

In this section, we analyze the results of experiment III, in which we combined the linguistic and behavioral modules by using the PB binding method to learn the sentences and behavioral sequences simultaneously. The final averaged output error of each node was 0.0091 for the linguistic module, and 0.025 for the behavioral module after 50000-step learning. The analysis reveals that the PB binding method could fill an essential role in self-organizing the compositional semantics of language through the behavioral experiences of the robot.

As mentioned in section 4,the training data for this experiment did not include all the correspondences. Four sentences (“**`point green`**”, “**`point right`**”, “**`push red`**”, and “**`push left`**”) were not included in the training data for the linguistic module. As a result, although the behavioral module was trained with the behavioral sequences of all behavioral categories, those in two of the categories, whose corresponding sentences were not in the linguistic training set (POINT-G and PUSH-R), could not be bound (c.f., Figure 2).

The most important result was that these dangling behavioral sequences could be bound with appropriate sentences. That is to say, the resulting semantics could recognize all four unlearned sentences and properly generate the corresponding behaviors. This means that both modules acquired a common PB structure by generalizing the given correspondences.

Comparing the PB spaces of both modules shows that they indeed shared a common structure as a result of binding. The linguistic PB vectors are computed by recognizing all the possible 18 sentences including 4 unseen ones (Figure 7a), and the behavioral PB vectors are computed at the learning phase for all the corresponding 90 behavioral sequences in the training data (Figure 7b). The acquired correspondences between sentences and behavioral sequences can be examined according to equation (6). In particular, the coincidence of the four unlearned correspondences ("`point green`"↔POINT-G, "`point right`"↔POINT-G, "`push red`"↔PUSH-R, and "`push left`"↔PUSH-R) demonstrates acquisition of the underlying semantics, or the generalized correspondences.

The acquired common structure has two striking characteristics: (1) the combinatorial structure originated from the linguistic module, and (2) the metric based on the similarity of behavioral sequences originated from the behavioral module. The interaction between modules enabled the linguistic PB space (Figure 7a) to simultaneously self-organize not only the combinatorial structure but also the embodied structure. We can see this embodied structure introduced into the linguistic PB space as the similarity of the PB vectors of sentences that correspond to the same behavioral category. For example, the two sentences corresponding to POINT-R ("`point red`" and "`point left`") are encoded in similar PB vectors. Recall here that such a metric nature was not observed in experiment I (Figure 4a). All nouns were plotted symmetrically in the PB space by means of the syntactical constraints.

At the same time, the combinatorial structure was introduced into the behavioral module (Figure 7b). In contrast to the behavioral PB space acquired in experiment II (Figure 6), we can find geometric regularity in the relationships among behavioral categories.

The PB vectors of the unlearned sentences and the corresponding behavioral sequences successfully coincided without binding because of the common structure shared by both modules. As explained in section 3.5, the structural unification of both PB spaces is realized by the local interaction of PB vectors. When a corresponding sentence and behavioral pattern pair is bound according to equations (7) and (8), the original PB structural regularity of both modules recedes. For example, the displacement of the linguistic PB vector of "`point red`" breaks the congruency reflecting the syntactic relationships among sentences (c.f., Figure 4). In the subsequent learning process of the linguistic module, all linguistic PB vectors are updated to recover regularity among sentences according to equations (1) to (3). Thus a local perturbation introduced by the PB binding spreads over the whole PB space. We especially note that the learning process affects the PB vectors for the unlearned sentences owing to the intra-module generalization. Similar processes are also observed in the behavioral module. As a result of the iterative interaction between the modules, they share a common PB structure.

The above observation thus confirms that the embodied compositional semantics was self-organized through the unification of both modules, which was implemented by the local PB binding method. We also made experiments with different test sentences, and confirmed that similar results could be obtained.

6. Discussion and Summary

Our simple experiments showed that the minimal grounded compositional semantics of our language can be acquired by generalizing the correspondences between sentences and the behavioral sensory-motor sequences of a robot. Our experiments could not examine strong systematicity (Hadley, 1994), but could address the combinatorial characteristic nature of sentences. That is to say, the robot could understand relatively simple sentences in a systematic way, and could understand novel sentences. Our 2-word language is very similar to the language $\mathcal{L}_0$ in (Evans, 1981), and the argument about the compositional semantics of $\mathcal{L}_0$ holds true for our language as well. Therefore, our results can elucidate some important issues about the compositional semantic representation.

In our experiments, compositionality was implemented in a non-trivial way. The embodied meaning of a word was implicitly realized in terms of the relationships among the meanings of sentences based on behavioral experiences. That is to say, the robot could understand a sentence by means of a generated behavior as if the meaning of the sentence were composed of the meanings of the words included in it. Our model does not require any pre-programming of syntactic information, such as symbolic representation of word meaning, a predefined combinatorial structure in the semantic domain, or behavior routines. Instead, the essential structures accounting for both compositionality and generalization are fully self-organized in the iterative dynamics of the RNN, through the structural interactions between language and behavior using the PB binding method.

We claim that the acquisition of word meaning and syntax can not be separated from the standpoint of the

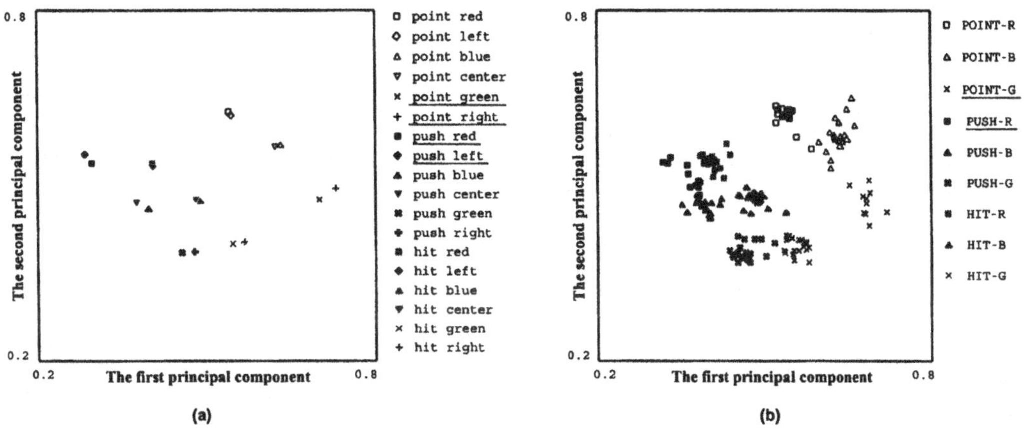

Figure 7: PB plots of the bound linguistic module (a) and the bound behavioral module (b). Both plots are projections of the PB spaces onto the same surface determined by the PCA method. Here, the accumulated contribution rate is about 73%. Unlearned sentences and their corresponding behavioral categories are underlined.

symbol grounding problem (Harnad, 1990). The meanings of words depend on each other to compose the meanings of sentences (Winograd, 1972). Consider the meaning of the word "`red`." The meaning of "`red`" must be something which combines with the meaning of "`point`", "`push`" or "`hit`" to form the grounded meanings of sentences. Therefore, a priori definition of the meaning of "`red`" substantially affects the organization of the other parts of the system, and often results in further preprogramming. This means that it is inevitably difficult to explicitly extract the meaning of a word from the meaning of a sentence.

Our model avoids this difficulty by implementing the grounded meaning of a word implicitly, in terms of the relationships among the meanings of sentences based on behavioral experiences. Thus, the robot can understand "`red`" through its behavioral interactions in the designed tasks in a bottom-up way (Tani, 1996). A similar argument holds true for verbs. For example, the robot understands "`point`" through pointing at red, blue, and green objects. Moreover, it should be noted that the meanings of nouns and verbs also depend on each other. One can not understand that a verb takes a noun as its object prior to the acquisition of semantics.

Next, we focus on generalization in the behavioral module and its effects on the common semantic structure. We assume that the meaning of a sentence is represented as a corresponding category of behavioral patterns. Although clarifying to what extent this assumption is appropriate is an important future goal, this assumption plays an essential role in the self-organization of a common semantic structure. For example, behavioral generalization (or categorization) enables the linguistic module to learn the semantic similarity of "`point red`" and "`point left`", which can not be learned from syntactic properties.

In a separated learning approach, the meaning of a sentence is represented as a combination of the meanings of words, which are labeled by the words themselves. This assumption allows the acquisition of sentence meanings to be translated into two relatively separate steps, the acquisition of word meanings and the acquisition of syntax (Kirby and Hurford, 2001). However, it is difficult to find the correspondence of a word meaning in the spatio-temporal behavioral patterns. This difficulty could be avoided by introducing an adequate set of behavioral routines. However it is often a non-trivial task and tends to result in a significant amount of preprogramming.

In contrast, our model acquires not just a mere mapping between sentences and behavioral patterns. The PB spaces of both modules (a.k.a., domains of the acquired mapping) acquire the internal structures suitable for realizing a structural mapping through mutual interaction between both modules. As mentioned in section 5.2, the behavioral module learned multiple behavioral patterns in a generalized manner, and categorized them with respect to the relationships between sensory images and motor commands. Therefore, the semantic domain is self-organized as some sort of a functional space, in which each function outputs motor commands from various input sensory information to generate behavioral patterns in a specific category. This behavioral generalization is important not only to the self-organization of the common semantic space, but also to realizing robust generation of behavioral patterns regardless of the perturbations in the environment.

To the summary, the current study has shown the importance of generalization of the correspondences between sentences and behavioral patterns in the acquisition of an embodied language. In future studies, we plan to apply our model to larger language sets. In the current experiment, the training set consists of a large fraction of the legal input space, when compared with related works. Such a large training set is needed because our model has no a priori knowledge of syntax and composition rules. However, we think that our model requires relatively fewer fraction of sentences to learn a larger language set, for a given degree of syntactic complexity.

References

Billard, A. (2002). Imitation: a means to enhance learning of a synthetic proto-language in an autonomous robot. In Dautenhahn, K. and Nehaniv, C. L., (Eds.), *Imitation in Animals and Artifacts.* MIT Press.

de Saussure, F. (1966). *Course in General Linguistics.* McGraw Hill.

Elman, J. L. (1990). Finding structure in time. *Cognitive Science*, 14:179–211.

Evans, G. (1981). Semantic Theory and Tacit Knowledge. In Holzman, S. and Leich, C., (Eds.), *Wittgenstein: To Follow a Rule.* London: Routledge and Kegan Paul.

Fodor, J. (1999). Why Compositionality Won't Go Away: Reflections on Horwich's 'Deflationary' Theory. Technical Report 46, Rutgers University.

Hadley, R. F. (1994). Systematicity revisited: reply to Christiansen and Chater and Niklasson and van Gelder. *Mind and Language*, 9:431–444.

Harnad, S. (1990). The symbol grounding problem. *Physica D*, 42:335–346.

Ito, M. and Tani, J. (2003). Generalization and Diversity in Dynamic Pattern Learning and Generation by Distributed Representation Architecture . Technical Report 2003-3, Lab. for BDC, Brain Science Institute, RIKEN.

Iwahashi, N. (2003). Language acquisition by robots – Towards New Paradigm of Language Processing –. *Journal of Japanese Society for Artificial Intelligence*, 18(1):49–58.

Jordan, M. and Rumelhart, D. (1992). Forward models: supervised learning with a distal teacher. *Cognitive Science*, 16:307–354.

Kirby, S. and Hurford, J. R. (2001). The Emergence of Linguistic Structure: an Overview of the Iterated Learning Model. In Parisi, D. and Cangelosi, A., (Eds.), *Computational Approaches to the Evolution of Language and Communication.* Springer Verlag.

Miikkulainen, R. (1993). *Subsymbolic Natural Language Processing: An Integrated Model of Scripts, Lexicon, and Memory.* MIT Press.

Rizzolatti, G., Fadiga, L., Galles, B., and Fogassi, L. (1996). Promotor cortex and the recognition of motor actions. *Cognitive Brain Research*, 3:131–141.

Roy, D. K. (2002). Learning visually grounded words and syntax for a scene description task. *Computer Speech and Language*, 16.

Rumelhart, D. E., Hinton, G. E., and Williams, R. J. (1986). Learning internal representations by error propagation. In Rumelhart, D. E. and Mclelland, J. L., (Eds.), *Parallel Distributed Processing.* Cambridge, MA: MIT Press.

Siskind, J. M. (2001). Grounding the Lexical Semantics of Verbs in Visual Perception using Force Dynamics and Event Logic. *Artificial Intelligence Research*, 15:31–90.

Steels, L. (2000). The Emergence of Grammar in Communicating Autonomous Robotic Agents. In Horn, W., (Ed.), *Proceedings of European Conference of Artificial Intelligence*, pages 764–769. IOS Press.

Sugita, Y. and Tani, J. (2002). A Connectionist Model which Unifies the Behavioral and the Linguistic Processes: Results from Robot Learning Experiments. In Stamenov, M. I. and Gallese, V., (Eds.), *Mirror Neurons and the Evolution of Brain and Language.* John Benjamins.

Tani, J. (1996). Model-Based Learning for Mobile Robot Navigation from the Dynamical Systems Perspective. *IEEE Trans. on SMC (B)*, 26(3):421–436.

Tani, J. (2003). Learning to generate articulated behavior through the bottom-up and the top-down interaction process. *Neural Networks*, 16:11–23.

Tani, J. and Ito, M. (2003). Self-Organization of Behavioral Primitives as Multiple Attractor Dynamics: A Robot Experiment. *IEEE Trans. SMC-A*, 33(4):481–488.

Thompson, C. A. and Mooney, R. J. (1998). Semantic lexicon acquisition for learning natural language interfaces. Technical Report AI98-273, The University of Texas at Austin.

Winograd, T. (1972). Understanding natural language. *Cognitive Psychology*, 3(1):1–191.

Imitation is not enough for lexicon learning

Jason Fleischer
The Neurosciences Institute
10640 John Jay Hopkins Drive
San Diego, CA 92121 USA
fleischer@nsi.edu

Jonathan Shapiro
Department of Computer Science
University of Manchester
Manchester M13 9PL UK
jls@cs.man.ac.uk

Abstract

Lexicon learning systems need to be concerned with more than just producing symbol usage agreement between agents, which is easy to acquire through imitation. Lexicon learners should also explicitly attempt to increase the mutual information between their symbol usages (a measure of the usefulness of the symbols for transferring information between agents). This paper argues that, although many lexicon learning algorithms presented in the literature do attempt to create highly informative symbol usages implicitly, there are good reasons to make the mutual information of symbol usages an explicit goal of the lexicon learning system. Some first steps in this direction are provided in this paper. It presents lexicon learning experiments using both purely imitative and explicitly information maximizing algorithms. The results of these experiments are used to support the thesis of this paper, that lexicon learning algorithms should explicitly attempt to produce high mutual information symbol usages.

1. Introduction

Lexicon learning is a task where a group of agents learn to use symbols to label meanings. The goal of the task is to enable agents to transmit meanings to each other using the symbols contained in the lexicon. Therefore, a lexicon could not be called successful, in any sense, unless a signal-receiving agent could reliably come up with the same meaning that the signal-transmitting agent intended.

This interpretation of lexicon learning has significantly influenced the way it has been approached in the literature. The primary focus of recent research has been on how the agents can use symbols to denote meanings *coherently* (i.e., that meanings can be reliably transmitted). However, there has been little focus on another important aspect of lexicon learning: that the goal of communication is to transmit information. As such, it is little use for the agents to have a coherent lexicon if they cannot distinguish between meanings that are useful for task-accomplishing behavior or survival. For instance, a degenerate case of enforcing coherence in the lexicon is simply to label every meaning with the same symbol; such a lexicon would be of no use in trying to transfer information from one agent to another.

Much of the original work in lexicon learning asked the question, "Under what circumstances will agents evolve an explicit signaling system?" See, for instance, the work of (Werner and Dyer, 1992; MacLennan, 1992). More recently, a number of approaches have looked at how lexicon formation might occur as a result of agent's imitating each other's symbol usages (e.g., Steels, 1998; Batali, 1998). These approaches have focused on the contribution of imitation to the learning process, and the dynamics of lexicon formation. However, they do not directly address the problem of learning an informative lexicon, but instead typically have "hidden" features that help create informative symbol usages. These safeguards are neither theoretically motivated nor do they have any performance guarantees.

In contrast, we are interested in the use of lexicon learning among robots to enable them to communicate with each other about a task (Fleischer, 2004). In the course of this work, an imitative lexicon learning algorithm was implemented and found to be susceptible to the degenerate case of labeling every meaning with the same symbol. The reason for this problem was two-fold: (1) the robot data is noisy and does not lend itself to easy extraction of meanings, and (2) most imitative lexicon learning algorithms contain some hidden features that ensure the creation of an informative symbol usage, whereas the algorithm implemented above did not. This result exposed an aspect of imitative symbol learning that is not explicitly addressed in the literature: imitation cannot guarantee an informative lexicon in the absence of other mechanisms.

This paper addresses the problem by calling for research into lexicon learning methods that explicitly address the problem of learning informative symbol usages, and by investigating one such algorithm.

Section two of this paper argues that there are benefits to be had by making informative symbol usages an explicit goal of the lexicon learning system. First, a definition of the symbol usage informativeness is presented, along reasons why an informative symbol usage is desirable for a set of communicating agents. Next, two of the more known examples of imitation-based lexicon learning algorithms from the literature are presented, and it's illustrates how they contain implicit or hidden safeguards that cause these algorithms to create informative symbol usage.

In order to show the potential problems of imitation learning, section three presents an imitation-based symbol learning algorithm that lacks the implicit safeguards of the approaches in the literature. Experiments demonstrate how it performs in both simulated lexicon learning tasks and also in a location-naming task on a mobile robot. The simulation results give an indication of both ideal performance and what types of meaning confusion will most adversely affect the information of the learned symbol usages. The robot lexicon learning task is an example of what can go wrong with imitation learning when there is a lot of perceptual noise.

Section four makes a first stab at a learning system that explicitly seeks more informative symbol usages. The learning system is based on negotiation; when agents disagree about how to label a meaning, they enter negotiations where each proposes a labeling that will increase its estimate of symbol usage information entropy. Agents only accept a proposal that does not decrease their own estimates of information entropy. Simulation results show that this learning system can, in the right circumstances, produce more informative symbol usages than the imitation algorithm. In terms of absolute performance, however, imitation can always be tuned to produce more informative symbol usages. An analysis of this problem is offered.

The conclusions section ties together the strands of argument about why information maximization should be an explicit goal of lexicon learning. Finally, future directions for producing information maximizing algorithms are discussed.

2. Background

2.1 Measuring information

Information theory provides tools useful both for evaluating how informative a set of symbol usages is, and also for providing a measure which can be maximized explicitly during learning to ensure that the symbol usages are informative. The concepts of information theory were introduced by Shannon (1948), and a standard textbook in the area is (Cover and Thomas, 1991).

Formally, an information channel is a random variable X, that can take a discrete and finite set of values (or signals) $\mathcal{X} = \{x_1, x_2, \ldots x_n\}$ that occur with probabilities $p(x_1), \ldots, p(x_n)$, such that $p(x_i) \geq 0 \ \forall i \in \mathcal{X}$ and $\sum_{i=1}^{n} p(X_i) = 1$, forming the discrete distribution $P(X)$. Then the entropy, H, of the channel is defined as

$$H(X) = -\sum_{i=1}^{n} p(x_i) \log_2 p(x_i).$$

which is, in fact, the average information content of the channel. Note that the maximum of this function occurs when all the signals are equiprobable. That is, the maximum information content of communication will take place when all the symbols are equally likely to occur. Also, information entropy will be zero when the probability of a single symbol is one, and all the others are zero. This is the degenerate case where every meaning is labeled by the same symbol.

Mutual information (also called information gain) is the amount of information that we gain about one random variable if we know the value of another. It can be defined as,

$$H(X;Y) = H(X) + H(Y) - H(X,Y).$$

Mutual information is a useful measure of the information content of the symbol usage between two agents. In a test set of meanings, each agent generates a set of symbols to describe those meanings. Each agent's description of the test set is a random variable, which we'll call S_a for the symbols used by agent a, and S_b for agent b. The mutual information between the symbol usages is then $H(S_a; S_b)$. Note that $H(S_a; S_b)$ can never be greater than $\min(H(S_a), H(S_b))$. Also, $H(S_a; S_b) = 0$ only if the two variables are independent, and is equal to the maximum only if the two variables are maximally correlated. So this measure has a maximum and minimum, and progresses in a fashion that is natural to the quality we want to describe.

The mutual information between symbol usages in a test set of meanings is used in this paper as a means to evaluate how informative the symbol usages learned by the agents are. However, this quantity requires knowledge of the output of all the agents on the entire set of inputs. When an agent is employing a learning algorithm that explicitly tries to maximize information it will be unlikely to have access to such global performance knowledge. Instead, the agent can assume that by increasing the information entropy of its own symbol usages, something which is much easier to estimate, that it will increase the mutual information between symbol usages. This method is discussed further in section 4.1, which presents the explicit algorithm.

2.2 Why maximize information?

The research in lexicon learning to date has predominantly focused on the methods by which agents learn and the properties of learning such lexicons. Much work has also taken the approach of trying to understand how natural language evolved and how humans create meaning. If the purpose of investigating lexicon learning is to better understand lexicon formation and language evolution in humans, then explicit information maximization algorithms should be of interest. It seems unlikely that imitation is the only mechanism involved in lexicon acquisition and language evolution; there could be specific features of these systems whose function it is to maximize the informativeness of the symbols used.

In contrast, one also might be interested in having agents learn a lexicon in order to *accomplish a task*. Agents that work together in a group can increase their task performance by using some form of communication (Balch and Arkin, 1994). One can obviously design and implement a communication system for such agents, but there are possible benefits to a communication system that the agents adapt for their own uses. These are the same arguments that are normally made for adaptive behavior in general: more robustness, the ability to react to changing circumstances, the ability to tackle problems that the agents were not explicitly designed to solve.

If the purpose of lexicon learning is to enable agents to communicate about objects in the world (e.g., robots describing navigation routes to each other by naming landmarks along the route), then it becomes very important that the symbols used should be useful for the task. Defining the usefulness of a symbol system for a task is a difficult thing, that perhaps can only be assayed on a case-by-case basis. However, it is impossible for a symbol system to be useful if it is not informative. Take the example of a lexicon that names landmarks experienced by mobile robots. If almost every landmark is called by the same name, then a route description generated using that lexicon would be of very little navigational use to a robot trying to follow the route.

When task accomplishment is important, then it seems obvious that informative symbol usages are an important goal. When mechanisms for maximizing information in the lexicon are implicit, such as in the methods presented in the next section, then it can be difficult to ensure that the results will be as desired. The best way to make sure that symbol usages remain informative is to make it an explicit goal of the learning system.

2.3 Literature

This section presents two well-known examples of imitative lexicon learners that do not explicitly try to maintain the information content of symbol usages. However, since both are able to produce lexicons with high mutual information they must have some implicit mechanisms that keep them from devolving to the solution of naming all meanings with the same symbol. The following paragraphs illustrate how these hidden mechanisms work.

Steels (1996) introduced naming games as a method of lexicon learning. A naming game is a game in that sense that agents try to "win" by successfully communicating with each other. If both agents understand a symbol to represent the same meaning then a naming game is successful. Losses, i.e., communication failures where the agents do not use a symbol to describe the same meaning, give the agents data they use to do better at the game in the future. Steels and colleagues have introduced a variety of "language games" that perform lexicon learning[1]. Although the language games have different algorithms, they can all be described by the same general framework:

> Two agents are selected at random from the population. One agent is randomly selected as the speaker, and the other as the listener. Both agents are presented with a context containing several objects. The speaker selects a topic object from the context, and produces an utterance to describe it. The speaker chooses the most "successful" symbol in its repertoire to represent the topic object, or makes one up if it doesn't already know one for that object. Both agents increment counters of symbol usage and naming game success, whose values they use to determine which symbol to generate when speaking.

Note that the naming game is imitative because the agents use their estimations of symbol occurrence and success rate to decide which symbols to use. That is, by observing the usage and communicative success of the rest of the population, they determine their own symbol usage.

One particularly important element of the language games, for the purposes of this discussion, is that the algorithm makes up a new word (i.e., symbol) randomly to describe any new object. The likelihood of the population as a whole using the same symbol to describe multiple meanings (and thereby loosing information in its symbol usages) is directly related to how an agent randomly chooses a new word to describe an object. Imagine two agents, each encountering different objects for the first time. They now (separately) make up words for those objects and use them in naming games with other agents. If both agents are fairly likely to make up the same word, then its possible for the population as a whole to adopt that word to represent two different, and perfectly distinguishable, objects. Unfortunately, the mechanism used in the naming games to make up new words is never completely explained in the literature, so it the true probability of this circumstance occurring is not known.

[1]The naming game, the observational game, the guessing game, the selfish game. These are explained and reviewed in (Vogt, 2000) and (Steels, 1998).

Note that if selection is equiprobable and the number of words in the possible set is orders of magnitude greater than the number of objects multiplied by the number of agents, then there is very little chance of symbol usages having low information. This can be seen from the following argument. (1) Since agents learn from the symbol production of the rest of the group, it is not possible for the final symbol assignment for a meaning to be anything other than one of the symbols made up by an agent first encountering that meaning. (2) It is therefore unlikely that one symbol will be used for multiple meanings if there are many more symbols than possible occurrences of an agent first experiencing a meaning. Yet some of the naming games reported in the literature do have symbol usages where more than one meaning generates the same symbol (see the "specificity" measurements in Vogt, 2000). This naturally raises the question of how tunable this information loss is, and whether it can effectively be prevented if it is not wanted.

Another well-known approach to imitative lexicon learning is that presented by Batali (1998). In this simulation, the agents learn a lexicon where the symbols are composed of morphological components, in the same way that words are built out of letters. The agents themselves are recurrent neural networks that, when presented with a meaning vector as input, produce strings of symbols as outputs. The strings are truncated at 20 characters (even if the recurrent network would produce more), and each character can be one of four letters.

Learning is directly imitative. An agent acting as the hearer trains its network to produce the same meaning encoded by the speaker. The same network that interprets symbols to meanings is also used to send symbols when the agent acts as speaker, so training as the hearer affects production as speaker.

Note that the set of potential strings (the functional equivalent of symbols) is of size 4^{20}. Since there are only 100 possible meanings we are once again confronted with a case where it is very unlikely that random initializations would produce the same symbol being assigned to different meanings. However, it is more complex than the case presented by the naming games of Steels. The associations between meaning and symbol are produced by a recurrent network. Therefore, the relationship between input and output is much more complex than the case of simply counting occurrence and success rates. Over the course of training, it seems inevitable that the network weights will become modified so that the agent will produce symbols that it neither originally produced after initialization nor was trained to produce. In this case, the question of how to control the information content of symbol usage has no obvious solution.

3. Imitation learning

This section demonstrates that an imitation based algorithm without the hidden safeguards can develop very uninformative symbol usages. This provides a proof of the dangers of relying only on imitation, which motivates the development of mechanisms that explicitly maximize information in the lexicon.

3.1 Algorithm

At this point it is important to distinguish between a meaning and a perception. In this paper, a meaning is a category or object that has an identity independent of agent perceptions. That is, it is a real-world location or it is an abstract category in a simulation that the agents are trying to communicate about. In some of the experiments, agents will have different perceptions of these meanings. For instance, they may be unable to differentiate between two meanings. In the robot experiments the agents self-organize categorizations, thereby creating arbitrary meanings from unlabeled data that will be different from agent to agent.

The lexicon learning system is based on each agent maintaining estimates of the conditional probability that it should generate a particular symbol, given that it has perceived a particular meaning. When an agent is required to generate a symbol to represent a perception, it generates the maximum likelihood symbol (the symbol with the highest conditional probability given the perception experienced).

The set of possible perceptions and symbols is fixed before lexicon learning occurs. For instance, in a simulation there may be ten meanings to be learned. These meanings will not change identity or grow in number during the simulation. Likewise, the agents might be given five arbitrary symbols to label these meanings at the beginning of lexicon learning, and the symbols will not change identity or grow in number during the learning process.

The conditional probabilities are stored in a single-layer neural network, with no bias input, using the softmax activation function. The input to the neural network is a 1-of-n encoded vector of the current meaning (as perceived by the agent — recall that agents may have different perceptual abilities). The activations of the output nodes are the conditional probabilities given the perception on the input nodes. Note that a single-layer network is sufficiently powerful to encode the probabilities because the inputs are mutually exclusive and therefore orthogonal in the input space. The weights of every agent's neural network are initialized using a zero-mean normal distribution.

The agents then proceed to learn a lexicon by imitating each other's usage. The total training data is partitioned into training, validation, and test sets of equal

no error	disagreements	0 (0)
	$H(C_a; C_b)$	3.29 (0.01)
	$H(S_b; S_b)$	2.55 (0.05)
	unique symbols	6.57 (0.20)
systematic error	disagreements	28.2 (1.8)
	$H(C_a; C_b)$	3.26 (0.01)
	$H(S_a; S_b)$	2.21 (0.05)
	unique symbols	6.53 (0.18)
random error	disagreements	18.2 (0.7)
	$H(C_a; C_b)$	2.53 (0.04)
	$H(S_a; S_b)$	1.93 (0.04)
	unique symbols	6.40 (0.15)

Table 1: The mean (std. err.) performance over 30 trials for various perceptual error models in the simulation experiments. See section 3.1 for an explanation of the models and the measurements. The results are over a test set of 200 randomly drawn meanings.

sizes (300 each for simulation, 640 each for the robot data). The training set is re-ordered randomly at the beginning of every epoch. Training occurs on-line according to the following method:

> Each agent perceives a meaning in the training set according to its abilities, and encodes it as an input vector for it's neural network. Each agent then generates the symbol that has the maximum likelihood on the outputs, and transmits it to the other agent. When an agent receives the other one's output it turns the symbol into a 1-of-n encoded vector which is used as a training pattern for its own neural network. The network is trained using on-line gradient descent on the cross-entropy error function, with a learning rate of 0.01 and no momentum.

At the end of an epoch of training the agents are evaluated on a validation set. An error is counted for every meaning in the validation set which the agents generate two different symbols to describe. Training continues until a maximum number of epochs is reached (1000 for simulation, 10000 for robot experiments), validation error is zero, or validation error has not decreased for a certain number of epochs (500 in both cases).

3.2 Simulation experiment

In this simulation the agents were given ten symbols to label ten meanings. These ten meanings occur with equal frequency, but sometimes one or both of the agents may systematically or randomly confuse meanings with each other. In other words, the agents may not perceive the same meaning, and therefore the imitative symbol learning they perform will be subject to noise in the target values.

There are two ways the agents can perceive the ten meanings:

No error Both agents perceive perfectly.

Systematic error Agent a perceives meaning #1 as meaning #2 50% of the time and agent b perceives meaning #3 as meaning #2 50% of the time.

Random error Agent a confuses all meanings with $p = 0.1$. When confused it picks any other of the nine meanings with uniform probability. Agent b perceives perfectly.

In both systematic and random error models roughly 10% of meanings will be confused by one or another of the agents. Thirty simulations were run of each model, with different randomly initialized agents in each simulation.

The results of these simulations are summarized in table 1. Four different measures of agent performance are presented here, giving information on how much agreement there is on symbol usage, on how much information it is possible to transmit, how much information was transmitted, and how many symbols were used. The number of disagreements is the number of times the agents produced different symbols for the 300 randomly drawn meanings (w/ uniform probability) in the test set. $H(C_a; C_b)$ is the mutual information between the agent's *perceptions* of the meanings, i.e., the meanings with the perceptual error model applied. This is the maximum possible value that the symbol usage mutual information, $H(S_a; S_b)$ could ever hope to achieve. Finally, the number of unique symbols in the test set is also recorded, telling us how many of the ten possible symbols were used in the test set.

The empirical results show us that the imitative learning algorithm will produce a symbol usage that contains on average 77% of the information contained in the meanings when the number of symbols equals the number of meaning. It should be noted this loss of information is a "best-case" scenario. If the meanings are not equiprobable then it can only become worse.

Looking at the effect of the perceptual error shows firstly that both sorts of errors significantly reduce the amount of information in the learned symbol usages. What is interesting to note, however, is that in the case of the random error, the loss in $H(S_a; S_b)$ probably comes directly from the loss in $H(C_a; C_b)$ from the perceptual errors. In the random error model the symbol usage mutual information is 76% of the perception mutual information, a close match to the 77% found with perfect perceptions. However, systematic errors seems to cause a loss of symbol mutual information that is not accounted for by the corresponding loss of perception mutual information. In the systematic error experiments the symbol usage obtains only 68% of the mutual information of the perceptions.

measurement	mean	std err
disagreement rate	26.3%	0.4%
$H(C_a; C_b)$	2.20	0.02
$H(S_b; S_b)$	1.38	0.41
unique symbols	3.13	0.12

Table 2: Test set results for the robot landmark lexicon learning experiment (30 trials, 20 symbol lexicon), where each agent uses its own 20 node SOM to categorize landmarks. See section 3.3 for an explanation the setup.

These results tell us three important things. First, we shouldn't expect an imitative system without safeguards to learn symbol usages with as much mutual information as exist in the agent's perceptions. Secondly, any sort of perceptual confusion will only aggravate this problem. Thirdly, for a given level of loss in perceptual ability, a systematic error would seem to cause a greater loss of mutual information as a percentage of the possible information in the meanings. Other experiments that are not reported here with different numbers of meanings, symbols, and noise levels all support these conclusions.

3.3 Robot experiment

This section presents an experiment using the imitative symbol learning algorithm to try and generate a lexicon for use in a robotic navigation task. The lexicon consists of names for landmark types (in a human sense, something akin to "T-junction" rather than a name for one specific location like "T-junction #42"). The interested reader can find further details on both landmark lexicon learning and route description and navigation for this task in (Fleischer, 2004).

In fact, this experiment is originally responsible for our investigation of the topic of explicit information maximization in lexicon learning. The low information symbol usages learnt by the imitation algorithm spurred the need to develop a better method of lexicon learning.

In this experiment, two agents with different random initializations are trained on the perceptions of a single physical robot. Therefore, there is never any doubt about what the "meaning" is — both agents have the same raw sensor input. However, the sensor input is not a meaning in and of itself, but rather a high-dimensional, noisy, representation of the local area around the robot[2]. Each agent produces a fixed set of meanings for itself by training a Self-Organizing Map (SOM) (Kohonen, 1993) on a batch of representative landmark perceptions. The agents then categorize raw sensor data during lexicon learning by using the identity of the highest activation node of the map for a given sensory input as the agent's perception of that landmark. In other words, each agent will have similar but not identical perceptions of landmarks. Thus, we can expect both systematic and random perceptual error during lexicon learning.

The agents trained 20 node (4x5 toroidal topology) SOM networks on robot data representing 200 meters of travel in three different office corridors. They then used a a different 110 meters of data, taken from a subset of the same corridors, to train their lexicons. The lexicon learning data was randomly partitioned into training (1/2), test (1/4), and validation (1/4) sets. Training continued to a maximum of 10,000 epochs, but otherwise learning parameters are identical to those of the simulation. The agents were given a 20 symbol lexicon to learn.

Lexicon learning was repeated according to this setup 30 times with different agent initializations and different training data splits each time. The results can be seen in table 2, and they are quite disappointing. The agents produced a symbol usage with only 62% of the possible mutual information[3]. In light of the simulation results presented in section 3.2, it is clear that the agents must be suffering from extensive perceptual noise. This is, in fact, the case. By looking at any given pair of SOMs and sensory data sets, clear cases of systematic errors can easily be found. Also, when the two agents are given identical SOMs (i.e., they now have identical perceptions of the landmarks), they can learn to produce the normal (for the case of equal numbers of symbols and meanings) 76% of the possible mutual information.

These disappointing results highlight the importance of maintaining mutual information in the symbol usages, if the goal of lexicon learning is to produce a communication system that is useful for transmitting information between agents.

4. Negotiation learning

This section describes experiments using an algorithm that attempts to explicitly increase the mutual information of symbol usages. The agents do this by negotiating with each other about what symbol they should use for a particular meaning.

The motivation of using negotiation comes from the desire to produce explicit maximization of symbol informativeness without trying to compute things the agent will not typically have access to, such as the probability of a particular perception occurring (difficult in different environments) or symbol usage probabilities of all agents in the communication group (difficult when there are many agents). Although the scheme is not motivated by a particular cognitive or psychological model, it seems that explicit negotiation does take place when

[2] In actuality, a 20x20 occupancy grid updated with the data of the last 10 inches of travel, covering an area of 100x100 inches around the robot.

[3] Note that information entropy is a logarithmic scale, so seemingly small percentages of change can be very important in terms of system operation. In fact, the robots were essentially unable to navigate using this type of landmark description lexicon.

adult humans play games requiring the development of coordinated descriptions of spatial events (Garrod and Doherty, 1994), although in their case negotiation seems to function as a method of selecting which of several possible alternative description schemes to use in a particular game, rather than which scheme to use "for all time" as we're investigating here.

4.1 Algorithm

The negotiating agents are substantially similar to the imitation agents described in section 3.1. They generate symbols using the same type of neural network, and are trained using the same parameters.

However, a negotiating agent also maintains an estimate of the occurrence rate of symbols. It does this by counting the number of times it uses each symbol during a single training epoch. Before training begins each agent runs through the training set of meanings by itself in order to generate an initial estimate of the occurrence rates. At the end of an epoch, the agent replaces its current estimated symbol probabilities with the occurrence rates from the last epoch. Let us denote agent a's current estimate of the occurrence rate of symbol s as $P_a(s)$. Note that

These probability estimates are used by the agent to try and maximize the information entropy of its own symbol usage, which in turn makes it possible for the mutual information of symbol usages to become higher. Entropy is at a maximum when the symbols are equiprobable; equiprobable symbol use is therefore the direct goal of the negotiating agent.

The negotiation protocol is as follows:

1. Each agent, $\{a, b\}$, receives the meaning, perceives it according to its ability, and produces a symbol, $\{s_a, s_b\}$ using the method described in section 3.1.

2. If $s_a = s_b$ then train the networks using that symbol and return to step 1.

3. Otherwise each agent makes an ordered list of all symbols that occur equally or less often (according to its estimation) than the symbol it produced in step 1. The least used symbols are first in the list.

4. Each agent proposes a symbol (agent a proposes s_a^{prop}) from the top of its list and sends this symbol to the other agent.

5. Each agent calculates its willingness to accept the other's proposal. For agent a this value is

$$r_a = log \frac{P_a(s_a)}{P_a(s_b^{prop})}$$

6. If at least one of the signals is non-negative, then the agent with the largest willingness (r) accepts the other agent's proposal. In case of a tie, a winner is chosen randomly. Each agent trains its network using the winning proposal as the target. Return to step 1.

7. Otherwise, each agent removes the symbol from the top of its list. If both agent's lists are now empty they return to step 1. If at least one agent still has proposals left to make they return to step 4.

Obviously, coherent symbol assignment (where the agents agree on the symbols to use for all meanings) is a fixed point of this protocol. It's also the case that it can be shown that this protocol will always produce a coherent symbol assignment provided that (1) the agents perceive the meanings in exactly the same way (no disagreement from perceptual confusion of meanings), and (2) that the learning system, in this case the neural network, is capable of making the assignments proposed by the protocol (no disagreement because the meaning-symbol mapping dictated by the algorithm cannot be learned).

4.1.1 Proof of convergence

Theorem: Given a set of input patterns, and a pair of agents with different symbol assignments for those patterns, there is at least one pattern for which a symbol assignment is accepted and agreed upon by the agents.

Proof by contradiction: Assume that there are two different symbol assignments such that *no* symbol proposal made by one of the agents is accepted by the other. Consider the set of symbols that agent a uses more than any other symbols,

$$S_a^0 = \{s \mid P_a(s) \geq P_a(s') \forall s' \in S\},$$

where S is the set of all symbols. For any pattern which agent a assigns to a symbol in this set, any proposed symbol would be accepted by a. Thus, since we assume no proposal is accepted, the two agents must already agree on the patterns assigned to these symbols. Now consider the symbols used by agent a which are used more than any other symbols except those in S_a^0,

$$S_a^1 = \{s \in S \setminus S_a^0 \mid P_a(s) \geq P_a(s') \forall s' \in S \setminus S_a^0\}.$$

Note that agent b cannot propose a symbol in S_a^0 for any pattern that a assigns to a symbol in S_a^1 because a and b agree on those symbols and therefore must have the same estimate of occurrence rate. So a will accept any symbol proposed by b, and since we are assuming that no proposal will be accepted, it must be the case that for all patterns which a assigns to a symbol in S_a^1, b must assign the same pattern to the same symbol. Likewise, we build sets $S_a^2, S_a^3, \ldots S_a^i$ (where i is the number of such sub-sets in S) and show that the two agents must agree on the symbol-pattern mappings in those sets. The same argument can be used from agent b's standpoint. But then we have shown that the agents agree on symbols for all patterns, and have contradicted the assumption that they have not.

Corollary: If the learning systems can learn to assign any symbol to any pattern without affecting the mappings from the other patterns, this protocol will converge to a coherent set of symbol assignments.

The single-layer neural network in question will produce symbol usage coherence because the input patterns are orthogonal and there is no bias term in the outputs to link the symbol mappings of the patterns.

However, this algorithm does not converge to the optimal (maximum possible entropy) solution. This happens for two reasons. First of all, agreement is a fixed point in the system. So as soon as the agents agree on how to label a particular meaning, they will no longer be able to change that, even if some other assignment they make later in learning would mean that mutual information would be increased by using a different symbol. Secondly, an agent is not actually increasing the information entropy of its symbol usage, but is employing a heuristic approximation to direct maximization. The acceptance criteria will only increase information of symbol usage if the change in symbol occurrence rates after the new assignment is small and of equal size for all symbols. There are many times when this is might not be the case, e.g. if the agent assigns a new symbol to a meaning that occurs 70% of the time.

4.2 Simulation experiment

The simulations presented here are identical in procedure to the no noise simulations presented in section 3.2. They demonstrate that the negotiation algorithm *can* outperform the imitation algorithm it was designed to replace in certain circumstances, but that the imitation algorithm can be tuned to produce a more informative lexicon more reliably.

In this experiment 5, 10, 15, and 20 symbols were used by the agents to label the same twenty meanings. Thirty trials were made of each using different random initial network weights but the same no-noise data set. Figure 1 shows the mean values and three standard error bars of these experiments. It can be seen that when only five symbols are available, the negotiation algorithm is a clear winner. However, imitation is as good as negotiation when there are as many symbols as meanings. Once there are more symbols than meanings imitation becomes the clear winner. This is an unexpected result, as the negotiation algorithm was designed to explicitly maximize mutual information in the symbol usages. If negotiation works at all, one would expect it to work in all cases.

A clue to why this occurs can be seen by looking at the standard error bars; for the ten symbol case there is an order of magnitude more standard error in the distribution of negotiation scores than for the five symbol case (0.10 vs. 0.02 bits). In fact, for the ten symbol case four of the 30 negotiation trials produce solutions of more than 3.0 bits of mutual information (maximum possible is $H(C_a; C_b) = 3.3$ bits). On the other hand, there are also three trials of with less than 1.5 bits of mutual information. In comparison, the mutual information values produced by the imitation algorithm all fall between 2.0 and 3.0 bits. So, in fact, negotiation can actually provide a more informative lexicon even in the ten symbol case, but imitation provides a higher *average* score.

This variance can be attributed to the simple fact that the negotiation algorithm cannot backtrack and reassign a new symbol after the agents agree how to label a meaning. Once a particular meaning is assigned to an under-used symbol it may, in fact, cause that symbol to become heavily over-used. If there are many fewer symbols than meanings, there are more chances to rectify any mistakes made in smoothing out the relative occurrence rates of symbols. So, essentially, negotiation degrades less quickly with decreasing lexicon size than imitation only because there are an increased number of chances to fix mistakes made in symbol assignment.

The clear implication is that the negotiation algorithm proposed is not going to be good enough. Imitation will always produce better symbol mutual information on average as long as it's given enough symbols in the lexicon. To get better performance in tasks with difficult perception, like the landmark labeling task presented in section 3.3, it will be necessary to design an algorithm that can either (1) take into account the relative frequency of perceptions as well as that of symbol, or (2) can re-assign symbols even for previously agreed upon meanings.

5. Conclusions and future work

This paper shows that imitation alone is insufficient to develop lexicons useful for a task. Experiments with a simple imitation learning algorithm show that symbol usages will have less average information than is present in the agents' perceptions. The same simulations showed that when perceptual noise is present this effect becomes much worse. An example of lexicon learning on a robotic task presented an extreme case of information loss (only 3 symbols, 1.4 bits mutual information, out of 20 meanings) that motivated the development of a better method.

This paper argues that lexicon learning algorithms should *explicitly* maximize the mutual information be-

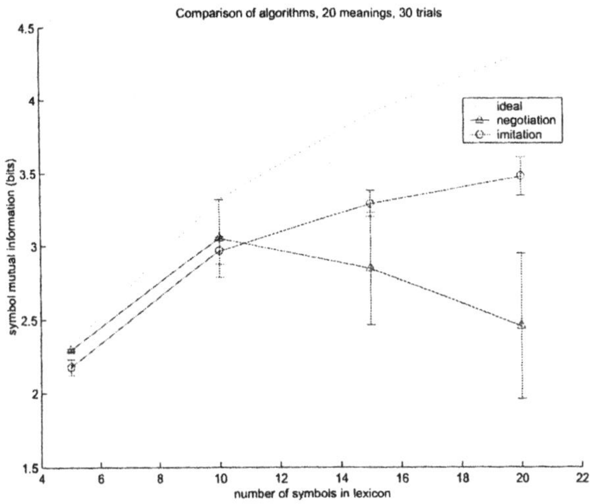

Figure 1: Figure showing how $H(S_a; S_b)$ changes as a function of lexicon size for both imitation and negotiation algorithms. Error bars show three standard errors around the mean value. The ideal line is the mutual information of an equiprobable symbol distribution of the given lexicon size. Plainly negotiation cannot offer better mean mutual information than imitation.

tween symbol usages to prevent this from occurring. Although many lexicon learning algorithms in the literature effectively do this, they do so through hidden or implicate mechanisms, like the example literature analyzed in section 2.3. Because these mechanisms are not explicit they can be difficult to control or have unforeseen consequences in different lexicon learning tasks.

A first step was made toward a useful, information-maximizing algorithm with the proposal of the negotiation algorithm. Unfortunately, the algorithm does not produce better mutual information levels than imitation except in cases where the number of symbols is much less than the number of meanings. This occurs because the algorithm is heuristically approximating the maximization of symbol usage mutual information. Once the algorithm has made a mistake due to that approximation, it cannot re-negotiate a new symbol assignment for that meaning.

The way ahead is therefore clear. A new algorithm should either make a better approximation to mutual information maximization, or it should be able to re-negotiate meaning-symbol assignments that are already agreed upon. The difficulty with the later approach is to be able to guarantee that a lexicon will ever become stable. Therefore, we believe that the former is the best avenue for future investigation. For instance, a new way to calculate an agent's willingness to accept a proposal that uses an estimate of the occurrence rate of the meanings rather than just the occurrence rate of the symbols has the potential to be useful.

Regardless of the method used to achieve it, lexicon learning algorithms would benefit from making information maximization an explicit aim of learning. This approach will hopefully enable the development of minimum performance guarantees and systems which produce higher symbol usage mutual information. In the long term, such research might also shed some light on mechanisms humans and other animals employ to ensure that their symbol usages are informative.

Acknowledgments

The authors thank Christian Freksa at the University of Bremen, Germany for opportunities he provided and Wolfram Burgard at the University of Freiburg, Germany for many helpful discussions about the problems in this paper.

Part of this work was supported by a Canadian & U.S. Citizens' Scholarship and an ATLAS grant, both from the Department of Computer Science at the University of Manchester. Part of this work was supported by a stipend from the International Quality Network on Spatial Cognition, which is supported by the Deutscher Akademischer Austausch Dienst.

References

Balch, T. and Arkin, R. C. (1994). Communication in reactive multiagent robotic systems. *Autonomous Robots*, 1:27–52.

Batali, J. (1998). Computational simulations of the emergence of grammar. In *Approaches to the Evolution of Language*, pages 405–426. Cambridge Univerisity Press.

Cover, T. and Thomas, J. (1991). *Information Theory*. John Whiley, New York.

Fleischer, J. (2004). *Route Communication Between Mobile Robots Using Adaptive Landmark Symbols*. PhD thesis, University of Manchester, England.

Garrod, S. and Doherty, G. (1994). Conversation, coordination, and convention: an empirical investigation of how groups establish linguistic conventions. *Cognition*, 53(3):181–215.

Kohonen, T. (1993). *Self-Organization and Associative Memory, 3rd ed.* Springer, Berlin.

MacLennan, B. (1992). Synthetic ethology: An approach to the study of communication. In Langton, C. G., Taylor, C., and Farmer, J. D., (Eds.), *Artificial Life II*, pages 631–658. Addison-Wesley, Redwood City, CA.

Shannon, C. (1948). A mathematical theory of communication. *The Bell System Technical Journal*, 27:379 - 423, 623 - 656.

Steels, L. (1996). Emergent adaptive lexicons. In *Animals to Animats 4*, pages 562 – 567. MIT Press.

Steels, L. (1998). Synthesizing the origins of language and meaning using coevolution, self-organization, and level formation. In *Approaches to the Evolution of Language*, pages 384–404. Cambridge Univerisity Press.

Vogt, P. (2000). *Lexicon Grounding on Mobile Robots.* PhD thesis, Vrije Universiteit Brussel.

Werner, G. and Dyer, M. (1992). Evolution and communication in artificial organisms. In Langton, C. G., Taylor, C., and Farmer, J. D., (Eds.), *Artificial Life II*, pages 659–687. Addison-Wesley, Redwood City, CA.

The Sensorimotor Bases of Linguistic Structure: Experiments with Grounded Adaptive Agents

Angelo Cangelosi
Adaptive Behaviour & Cognition Research Group
School of Computing, Communication and Electronics
University of Plymouth (UK)
acangelosi@plymouth.ac.uk
www.plymouth.ac.uk/staff/acangelosi

Abstract

This research uses grounded adaptive agents for investigating the evolutionary origins of syntactic categories, such as nouns and verbs. To analyze the sensorimotor bases of linguistic structure, the techniques of categorical perception and of synthetic brain imaging are employed. The simulation uses two different architectures for the adaptive agent's neural controller. Analyses show that the neural processing of verbs is consistently localized in the regions of the networks that perform sensorimotor integration, while nouns are associated with sensory processing areas. The general implications of such model and of the analysis techniques for adaptive behavior and language evolution research are discussed.

1. Modeling language evolution in grounded adaptive agents

Investigations on the emergence of language, both in evolutionary and developmental terms (MacWhinney 1999; Cangelosi & Parisi 2002), have greatly benefited from the use of computational models. This has resulted in an alternative approach to the nature vs. nurture (and biological vs. cultural) dilemma in language emergence. Models demonstrate that language is a complex system that emerges from intricate interactions between various biological and environmental processes. Adaptive behavior and artificial life provide useful modeling methodologies (Kirby 2002; Wagner et al. 2003) for dealing with such a complex system vision of language. In adaptive behavior models, populations of autonomous agents interact via language games to exchange information about the environment. Their coordinated communication system emerges from the direct interaction between agents.

Amongst the various adaptive behavior approaches, some provide a more integrative vision of language and treat it as an integral part of the whole cognitive system. The agent's linguistic abilities are strictly dependent on, and grounded in, other behaviors and skills. Various sensorimotor, cognitive, neural, social and evolutionary factors contribute to the emergence and establishment of communication and language. For example, in these models there exists an intrinsic link between the communication symbols (words) used by the agent and its own cognitive representations (meanings) of the perceptual and sensorimotor interaction with the external world (referents). We call this <u>grounded adaptive agent</u> modeling of the emergence of language. This is consistent with the psychologically-plausible theories of the grounding of language in the organism's perceptual and action systems (Barsalou 1999; Glenberg & Kaschak in press; Joyce *et al.* 2003). Such an approach is in opposition to other adaptive modeling systems that view language as an independent and autonomous capability of the agent, and are subject to the symbol grounding problems (Harnad, 1990).

There are various language models based on grounded adaptive agents. Some use real (or realistic) robots interacting in physical environments, while others use simulated adaptive agents. In robotic models, communication results from the dynamical interaction between the robot's physical body, its cognitive system and the external physical and social environment. Some studies stress the grounding in sensorimotor processes, such as Marocco's *et al.* (2003) model of robotic arms in a physics dynamics simulator and Vogt's (2002) mobile robots. Other robotic models highlight the grounding through social interaction, such as Steels' (2002) talking heads and Kaplan's (2000) interactive toy robot. On the other hand, some studies are based on simulation adaptive agents. They model the agent and its environment with a good degree of detail upon which emergent meanings can be directly constructed. These simulation models have focused on grounding in perceptual experience (Cangelosi *et al.* 2000; Dyer 1994) and in cognitive representations and sensorimotor interactions (Cangelosi & Harnad 2000; Hazlehurst & Hutchins 1998).

Both robotic and simulation models have been used to study the emergence of simple communication systems, and that of more complex syntactic languages. Of the models focusing on syntax, some have specifically addressed the evolutionary transition from non-compositional

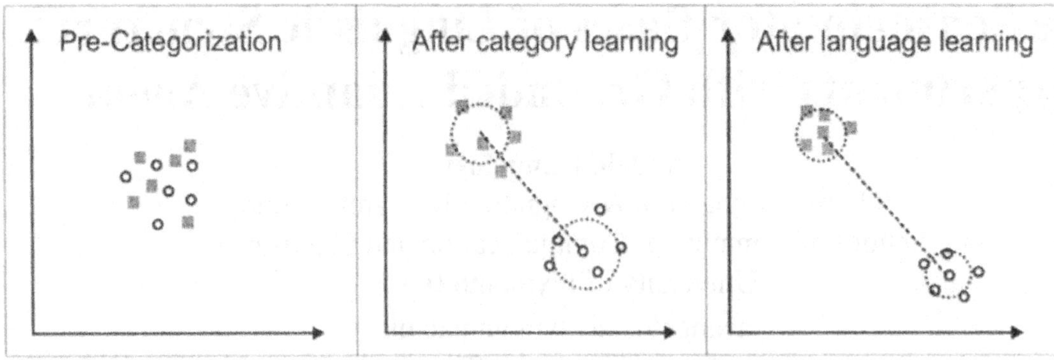

Figure 1: Typical formation of clusters of points (i.e. square and circle categories) during category and language learning. Before category learning (Left), points corresponding to different categories overlap. After categorization (Centre) and language learning (Right) points group in distinct areas. Notice the enhancement of compression and separation (CP effects) between simple categorization and language learning. These diagrams do not correspond to real data, but only provide a sketch of CP effects observed in modeling experiments (e.g. Cangelosi et al. 2000).

communication to syntactic language. This is an important issue in the field of language origins, because it can shed light on the evolutionary emergence and diversification of word categories, such as nouns, verbs and functions words (e.g. prepositions and quantifiers). This topic is also very important for studies on the ontogenetic development of language based on the emergentist/constructivist approach (Tomasello & Brooks, 1999; MacWhinney 1998).

This paper will focus on the use of grounded adaptive agents for investigating the evolutionary origins of syntactic categories, such as nouns and verbs. It will use grounded adaptive agents in which neural networks control all behaviors, including sensorimotor, cognitive and linguistic abilities. In order to better understand how all these behaviors interact, two techniques will be adopted for examining the agents' internal (neural) representations. These analyses can identify and highlight the sensorimotor bases of linguistic structure, such as the role of perceptual and motor knowledge in differentiating the syntactic categories of verbs and nouns. The first methodology considers the phenomena of categorical perception in the hidden representations of neural networks. The second technique applies neuroimaging principles to artificial neural networks for comparing neural representations in different linguistic and behavioral tasks. The results of such analyzes will also be related to our current knowledge of the neural basis of language processing.

2. The neural processing of language and cognition

The neuropsychological and neurocognitive literature on language processing in the brain is quite extensive (Pulvermuller, 2003). This has been possible through the use of a variety of scientific methods, such as neuropsychological experiments of patients with psycholinguistics deficits, brain imaging studies on normal adults and language-impaired patients and artificial neural network models. These methods support detailed investigations of the neural principles and mechanisms behind various linguistic abilities. For example, a recent review paper has analyzed numerous studies on the neural processing of the syntactic categories of nouns and verbs (Cappa & Perani, 2003). This review explicitly includes and compares brain imaging experiments on verb and noun processing (e.g. Martin *et al.*, 1995; Perani *et al.*, 1999). Cappa and Perani (2003) report an overall agreement on the fact that the left temporal neocortex plays a crucial role in lexical-semantic tasks related to the processing of nouns. The processing of verbs, instead, involves additional regions of the left dorsolateral prefrontal cortex.

Such data show that language processing is strictly interconnected with other sensorimotor and cognitive abilities. For example, the prefrontal cortical areas implicated in verb processing are those also involved in motor control. This finding is consistent with other work that illustrates a strong relationship between language and motor development (Greenfield, 1991; Rizzolatti & Arbib, 1998). More generally, in cognitive psychology and neuropsychology it is widely accepted that language is not an autonomous function of the organisms, but it is strictly dependent on/from other cognitive abilities (Gazzaniga, 2000).

In parallel to neuroscientific investigations, neural network models have extensively been used to study the neural control of linguistic behavior (Christiansen & Chater, 2001; Just et al. 1999). However, most of these models tend to study language in isolation. Only a few studies have directly and purposely simulated in the same model linguistic and motor functions (e.g. Reilly, in press). These include grounded adaptive agent models (e.g. Cangelosi 2001; Marocco et al. 2003), where by definition linguistic abilities emerge from sensorimotor and cognitive skills.

2.1 Categorical perception

Category learning has been hypothesized to be one of the core capabilities supporting the evolution and development of language and cognition. In particular, categorical perception (CP) is proposed as the basic mechanism sustaining our ability to build discrete and hierarchically-ordered representations of the environment (i.e. categories). Language and other higher-order cognitive abilities are then grounded upon such categories (Harnad, 1987; Cangelosi & Harnad 2000). CP refers to the effect of "warping" the similarity space of internal categorical representations. This results in the compression of within-category differences between members of the same category and the expansion of between-category distances amongst members of different categories. Such phenomena can be graphically represented through the process of the formation of clusters of points in the similarity space of categories (Figure 1). The first diagram (Fig. 1 Left) shows the undifferentiated similarity space before category acquisition. The other two diagrams represent the formation of quite distinct clusters (categories) after category learning has occurred without (Fig. 1 Centre) and with (Fig. 1 Right) language. These diagrams show the presence of two distinct categories (cluster of squares vs. cluster of circles) in an abstract two-dimensional similarity space. Each dimension may correspond to some classification component (e.g. presence of a feature) or to the hidden unit activation of a neural network. Relative distances in this space can be calculated using Euclidean measures between points. The two dotted circles in each diagram represent the within-category distances, corresponding to the standard deviation of the Euclidean distances between each point and the center of its cluster. The continuous straight line represents the between-category distance, e.g. the Euclidean distance between the centers of the two clusters.

CP has been shown to occur in animals (Zental *et al.* 1986) and human subjects (Goldstone 1994; Andrews, Livingstone & Harnad, 1998). The compression effects have also been analyzed in real neural systems (Kosslyn *et al.* 1989) and in artificial neural networks (Tijsseling & Harnad 1997; Cangelosi, Greco & Harnad 2000; Nakisa & Plunkett 1998). CP effects have been reported in models performing both category and language learning. Specifically, language appears to enhance the CP warping effects. The acquisition of categories via linguistic instruction produces stronger compression effects than those obtained in categories acquired through direct experience with stimuli. This is sketched out in the comparison of Figure 1 Left (after simple category learning) and Right (after category learning via language).

More recently, CP effects have been investigated in different syntactic categories. In a grounded model of the evolutionary acquisition of verbs and nouns (Cangelosi & Parisi 2001), it was reported that the strength of language-enhanced CP effects varies depending on the word type used by agents. Verbs produce stronger warping effects than nouns (Figure 2). These distinct internal representation patterns for the processing of verbs and nouns are also shown to generate differential benefits on the agent. This is because more similar representations of different situations help the agents to respond with the same action, albeit they are used in tasks with more variable environmental contexts (e.g. different objects/category may require the use of the same action, and its corresponding verb). Nouns, on the other end, tend to co-vary with the features of input stimuli (categories) and require less differentiated categorical representations (Cangelosi & Parisi, 2001).

In this paper further investigations of such differential verb/noun effects will be presented. Specifically, CP measurements will be performed in two neural architectures where the integration of sensorimotor information varies depending on the input of motor-related (proprioceptive) signals.

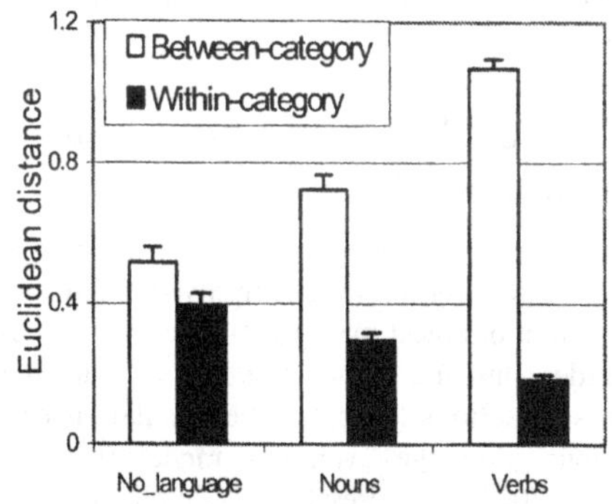

Figure 2: Categorical perception (CP) effects with verbs and nouns. Notice the enhanced CP effects in verbs, compared to the other conditions (adapted from Cangelosi & Parisi 2001)

2.2 Synthetic brain imaging

Experimental brain imaging techniques (also known as neuroimaging) permit the visualization and identification of neural activity in accurate regions of the brain in response to controlled tasks. Neuroimaging is based on comparisons of differences in brain activity (rCBF, regional Cerebral Blood Flow) between tasks. The two most commonly used techniques are Positron Emission Tomography (PET) and functional Magnetic Resonance Imaging (fMRI).

Synthetic brain imaging (SBI) (Arbib et al. 2000; Horwitz et al. 1999; Cangelosi & Parisi, in press) is a modeling technique used to highlight the regions of an artificial neural network that are functionally active in response to specific stimulations and tasks. SBI can be used to compare directly PET and fMRI imaging data in empirical studies and in computational models. This permits the development of more detailed and neurally-plausible models of behavioral functions. In analogy with experimental neuroimaging, SBI is based on differences in artificial neural "activity" between tasks. For example, following Arbib et al.'s (2000) assumptions, in synthetic PET the involvement of a region in a specific task is correlated with local integrated synaptic activity. Synaptic activity is computed by summing together the synaptic inputs arriving to the artificial neurons of a specific area (ignoring the positive/negative sign of the connections). Arbib and colleagues (2002) used synthetic PET to test their FARS model of parietal-premotor interactions in the grasping control of primates. In a study on language processing, Just *et al.* (1999) applied synthetic neuroimaging in a computational model of sentence comprehension. They found good agreement between the number of activated voxels in human fMRI experiments and the predictions of their model for sentence of different complexity.

In this paper SBI techniques will be used to analyze the internal activations in the neural controllers of adaptive agents. Various linguistic conditions will be compared, in particular to show differences in the neural processing of verbs and nouns.

3. The evolution of nouns and verbs: An adaptive behavior model

A grounded simulation model of the evolution of nouns and verbs was employed. This uses a population of object manipulation agents that are able to understand the names of objects (nouns) and the names of actions (verbs)[1]. The simulation extends a previous language evolution model (Cangelosi & Parisi, 2001, in press), by employing different architectures in the agent's neural controller.

3.1 The agent and the tasks

The simulated agent (Figure 3) consists of an organism with a retina and a two-segment arm (e.g. Schlesinger & Barto, 1999). The arm moves in a 2D environment in front of the agent. One of two different objects (object A = vertical bar; object B = horizontal bar) can appear at any time in the environment. The object occupies either three vertical cells or three horizontal cells and is centered in one of the 9 central cells of a 5x5 grid. The pixel values (0 for empty cell, 1 for cell occupied by a portion of the object) will be projected to the 25 units of the agent's retina. The agent's arm is transparent and absent from the retinal image.

During each task, an agent begins by grasping an object with its hand and it has either to pull the object toward itself or to push the object away from itself. The agent may responds to visual stimuli, linguistic instructions (nouns, verbs), or both. There are 11 different tasks which combine these two sensory modalities (Figure 4). The agent always receives proprioceptive input regarding the current position of the two segments of its arm.

In the first task (No_Language), the retina receives a 5x5 image of the object location and shape. No linguistic input is present. When object A is present in the scene, the agent must push it. Instead, object B must be pulled towards the agent's shoulder.

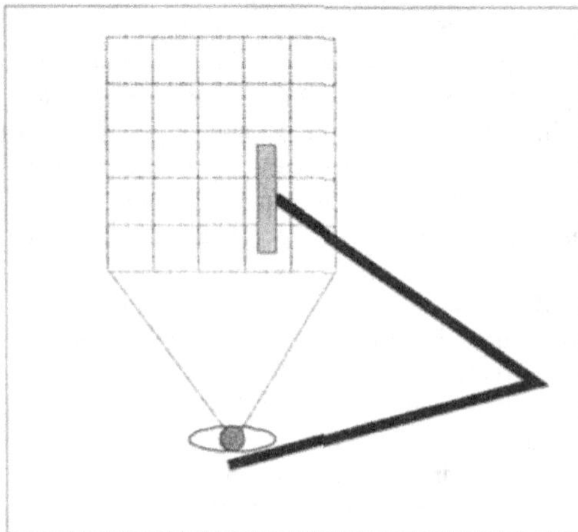

Figure 3: Agent interaction with environment. The two-segment arm grasps the object and has to pull it toward itself or push it away from itself. This depends on the type of object and the linguistic command that the agent may receive.

	OBJECT A		OBJECT B	
	retina	*language*	*retina*	*language*
No_language	▮	+ * *	▬	+ * *
Noun only	▮	+ * A	▬	+ * B
" "	*	+ * A	*	+ * B
Verb only	▮	+ Push *	▬	+ Pull *
" "	*	+ Push *	*	+ Pull *
" "	▮	+ Pull *	▬	+ Push *
" "	*	+ Pull *	*	+ Push *
Noun Verb	▮	+ Push A	▬	+ Pull B
" "	*	+ Push A	*	+ Pull B
" "	▮	+ Pull A	▬	+ Push B
" "	*	+ Pull A	*	+ Push B

Figure 4: List of the retina and language input for the 11 tasks. The * symbol indicates the absence of input (i.e. a series of 0s)

The remaining 10 tasks involve language understanding. In 5 Vision+Language tasks, in addition to the retina and proprioceptive input, the following linguistic instructions are received by the agent: (1) "A" or "B", as the nouns of objects A and B; (2) "Push" or "Pull" for the verbs describing the default action (push for object A; pull for B); (3) "Push" or "Pull" for the verbs describing the opposite action (pull for object A; push for object B); (4) "Push A" or "Pull B" for describing the default actions by combining a noun and a verb; (5) "Push B" or "Pull A" for describing the opposite actions. When

[1] This is a model of language comprehension only. A related model that deals with both linguistic comprehension and production is described in Cangelosi (2001).

the verb is received as input, the agent must do whatever the verb meaning indicates, thus overriding the default action if the verb describes the opposite action. In the 5 Language-only tasks, only the linguistic instruction and the proprioceptive input are provided.

The use of the terms "noun" and "verb" is not intended to correspond to the full blown grammatical categories of verbs and nouns. Nevertheless, the nouns and verbs of the present simulation capture some of the fundamental properties of nouns and verbs of real human languages (Parisi *et al.*, 2002). Nouns co-vary with the input stimuli and can act as attentional cues. Verbs co-vary with the motor action performed by the agents. These simple properties may have been those of primitive proto-nouns and proto-verbs which, through cultural and linguistic evolution, have given way to the complex, full blown grammatical categories. As suggested by Pinker (1994) the use of mental labels for different types of objects and actions has clear adaptive advantages.

3.2 The neural networks

The behavior of the agent is controlled by a feedforward neural network. Two different architectures will be used (Figure 5). They differ in the connection pattern between some input and hidden nodes.

The input layer contains 33 units. The proprioceptive input is encoded in 4 units. These encode the position of the arm through the input of the angles of the two pairs of muscles (extensor and flexor) of each of the two arm's segments (shoulder and elbow). The visual object is perceived through a retina of 25 units corresponding to the 5x5 cells of the environment grid. Four localist-encoding linguistic units are used for language input, two respectively for the nouns "A" and "B" and two for the verbs "Push" and "Pull".

The output layer contains four motor units. These control the extension/contraction of the four arm muscles (a pair of extension/contraction muscles per arm segment). The output activation corresponds to the force that is applied to each muscle.

The network has two layers of hidden units, whose organization varies in the two different architectures. In the first network configuration (Figure 5 top), the first hidden layer has two separate modules: a group of 5 units for the preprocessing of proprioceptive information and a group of 10 units for the preprocessing on linguistic and visual stimuli. In the other network architecture (Figure 5 bottom), all 14 first-layer hidden nodes preprocess the input coming from all input units. In all networks, the second hidden layer contains 5 units with connections to the previous layer of hidden nodes. The motivation behind such manipulations of the neural architecture is to create a first network with a full modular organization, where separate sensory processing modules for motor information (proprioceptive input) and visual information (retina input) exist, in addition to the sensorimotor integration module of the second hidden layer. The second architecture, instead, uses a fully distributed approach where proprioceptive and visual information are integrated since the first hidden layer.

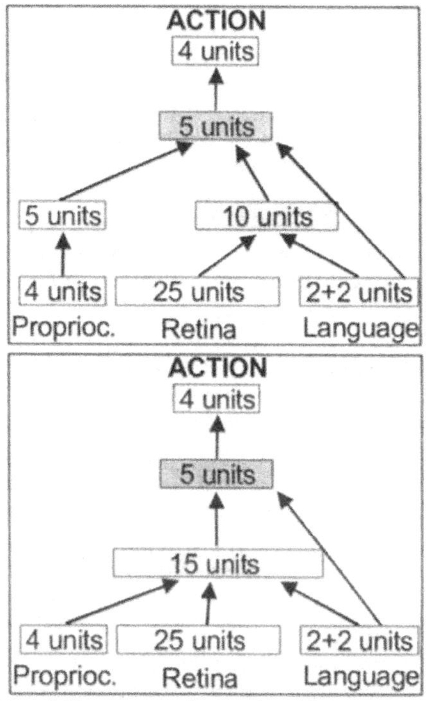

Figure 5: The two neural network architectures used in the simulation: Modular (Top) and distributed (Bottom) architectures. The connection patterns between the proprioceptive input units and the two hidden layers (shaded rectangles) are varied in the two networks.

The two architectures resulting from the different organization of the hidden layers will be used to compare the categorical perception effects in two experimental conditions: (1) use of a modular architecture, with separated sensory-processing and sensorimotor integration modules; (2) use of a fully connected network directly based on sensorimotor integration. These two architectures permit the analysis of the functional organization of verb and noun processing in the networks in relationship to their sensorimotor abilities. In addition, the first architecture will be used to make qualitative comparisons with brain imaging data on modular language processing in humans. In all networks, the input language units are always connected to both hidden layers in the two networks. This is to better control the specific effects of the differential sensorimotor processing levels on linguistic input.

3.3 The evolutionary algorithm

A genetic algorithm is used to evolve the connection weights of the agents' neural networks. During the agent's lifetime, the weights do not change. The motor behavior and language understanding improves (i.e. is learned) during evolution.

During each generation, a population of 80 agents is used. At the beginning of the simulation, 80 genotypes (connection weights) are randomly generated in the range of ±1. Weights are encoded as real numbers. A single task in the lifetime of an individual consists of a total of 18 subtasks (2 objects x 9 positions). Each subtask lasts for 20 input/output cycles, necessary for the arm to move an object from its initial position to the target. The fitness formula computes the total number of subtasks successfully completed by each agent in all conditions. A subtask is considered successful when (a) object A is pulled to a distance of 5 points or less from the agent's shoulder, or (b) object B is pushed to a distance of 45 points or more from the agent's shoulder.

At the end of the generation, agents are ranked on the basis of their fitness formula. The 20 agents with highest fitness are selected for reproduction. Each individual asexually generates 4 offspring with the same genotype of its single parent except for the addition of random changes (mutations) to some of the weights. Five percent of weights are mutated by adding a random quantity between ±1.

In the first 1000 generations, agents only perform the No_language task. These initial generations are needed for evolving a set of connection weights that allows agents to respond appropriately to the two objects (before language is introduced). From generation 1001 to 2000, agents have a longer lifetime and are exposed to all 11 tasks. These are experienced by each individual in a random order. In the fitness formula for the 10 tasks with language, a subtask is successful if the agent pushes or pulls the object according to the linguistic input.

3.4 Fitness results

For each network architecture, 20 replications were performed using different random initial populations. In all replications, agents evolved an ability to respond appropriately to the different input conditions. At the end of generation 2000, over 90% of the objects are responded to with the appropriate pushing or pulling behavior in all tasks.

In this paper the focus is on the neural processing of nouns and verbs, rather than on evolutionary and fitness results. Therefore the following sections will discuss only the categorical perception analyses and the synthetic brain imaging data. For a more detailed discussion of fitness and evolutionary data in this type of model, see Cangelosi & Parisi (2001).

4. Analyses on language processing

4.1 Categorical Perception (CP) results

In previous simulations of this language evolution model (Cangelosi & Parisi, 2001), it was shown that categorical perception (CP) varied in the word categories of verbs and nouns. Verbs produced an enhancement of between-category distances and reduction of within-category sizes when compared with nouns (Figure 2). These results refer to the activation patterns of the second hidden layer where sensorimotor integration happens[2]. The proposed explanation for such phenomenon is that verbs are better than nouns in shaping the internal representations (i.e. enhanced CP effects) because they "prepare the motor output with which the organism must respond to the input" (Cangelosi & Parisi 2001: 174).

This explanatory hypothesis, based on CP data taken from the hidden layer preceding the output units, can have two possible interpretations. The first hypothesis assumes that verbs distinguish themselves from nouns in the hidden layers next to the motor layer. As a consequence, the closer a hidden layer is to the output motor layer, the more action-oriented the representation is (which is the case of verbs that co-vary with the action to be performed). The alternative interpretation is that verbs produce CP effects whenever hidden units play a direct role in shaping the type of motor response through the integration of internal sensory (visual and linguistic) representations and motor (proprioceptive) information. The analyses on the two architectures will resolve such an ambiguity and shed light on the relationship between sensorimotor processing and linguistic categories.

In this study the CP data were computed for both hidden layers[3] of each of the two architectures. Figure 6 reports the between-category data for the modular (top) and the fully distributed (bottom) architecture. Data in Figure 6 were computed and averaged using the hidden unit activations of the best agent in each of the 20 replications. To compute the between-category distances, the sets of N hidden activation values h for the object A ($h1_A, h2_A \ldots hn_A$) and B ($h1_B, h2_B \ldots hn_B$) are necessary. These constitute the coordinates of two points in a N-dimensional Cartesian plane. Then the Pythagoras theorem is applied to obtain the Euclidean distance d_{A-B} between the two points:

$$d_{A-B} = \sqrt{(h1_A - h1_B)^2 + (h2_A - h2_B)^2 + \ldots + (hn_A - hn_B)^2} \quad (1)$$

Results clearly indicate that in the modular network (Figure 6 Top) verbs differ from nouns only in the second hidden layer, i.e. where motor (proprioceptive) information integrates with sensory (visual) input. Instead, in the distributed architecture (Figure 6 Bottom) verbs differentiate from nouns starting from the first sensorimotor hidden layer.

[2] In Cangelosi & Parisi (2001), the network architecture is similar to that of Figure 5 (top). The only difference was in the linguistic input solely connected to the second hidden layer.

[3] The CP data on the proprioception preprocessing layer of the first architecture were not computed because they do not receive linguistic input.

These data supports the hypothesis that verbs produce enhanced CP effects whenever hidden units play a direct role in shaping the motor response through sensorimotor integration. This is independent from the relative proximity to the motor output layer.

The localization of the CP effects also highlights the specific sensorimotor basis of verbs, and the sensorial characterization of nouns. The linguistic structure behind the differentiation of the word categories of verbs and nouns is strictly interconnected with sensorimotor representations.

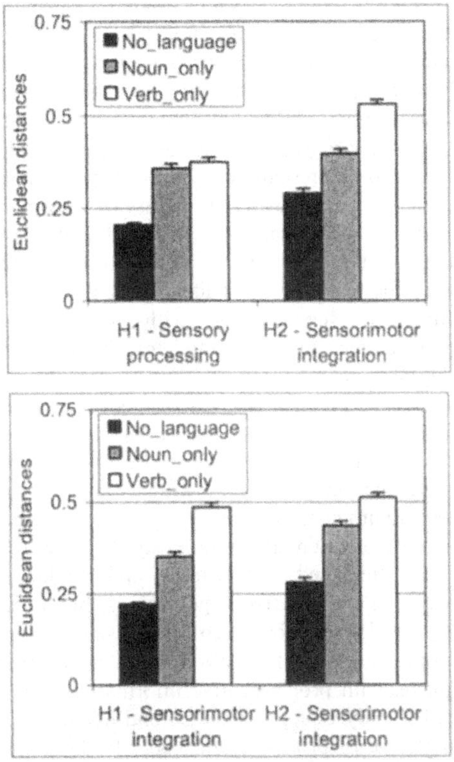

Figure 6: Between-category distances for both layers of the modular network (Top) and the fully-distributed architecture (Bottom). The Euclidean distances in the Y axes are normalized to permit direct comparisons between layers with different number of units.

4.2 Synthetic Brain Imaging (SBI) results

Synthetic brain imaging analyses were performed only on both neural architectures. SBI will be used to make some qualitative comparisons between these artificial models and known data on syntax processing in the human brain.

The SBI technique presented here will compare the neural activity of the two hidden layers in different linguistic tasks. This method is called Integrated Neural Activity (INA) and generally follows Arbib *et al.* (2000) synthetic fMRI procedure. A related synthetic brain imaging method, called Integrated Synaptic Activity (ISA), has also been developed to make comparisons with PET data. ISA will not be described and applied here, but for a comparison of INA and ISA analyses in grounded neural networks see Cangelosi & Parisi (in press).

The overall value of the integrated neural activity $INA_A(1/2)$ reflects the difference of activation between the two tasks 1 and 2 in the region A (e.g. hidden layer). It can be computed using the following formula:

$$INA_A(1/2) = |rINA_A(1) - rINA_A(2)| \tag{2}$$

where the regional Integrated Neural Activity for each task $rINA_A$ is

$$rINA_A = \sum_{i=1}^{N} a_i \tag{3}$$

with N being to the number of units in the hidden layer, and a_i the activation of each individual neuron in the region A.

In small networks like the ones used here, it is possible to consider the $INA_A(1/2)$ as the average of differences between individual neuron activations in the two tasks. In this case, the following formula can be used:

$$INA_A(1/2) = \frac{\sum_{i=1}^{N} |a_i(1) - a_i(2)|}{N} \tag{4}$$

Formula (4) will be used here, due to the small number of units in each hidden layer. Only the first activation value produced at the beginning of each task is used. Observations of the behavior showed that agents produce distinct push/pull movements since the first activation cycle. The INA values for all possible pairs of task subtractions were calculated separately for each hidden layer. Subsequently, all subtractions were averaged for the two linguistic classes of verbs and nouns. For example, the average of the values No_language–Noun_only and Verb_only–Noun_verb contribute to the INA value for Noun processing.

Figure 7 contains the average INA values over the 80 individuals of the 20 populations at generation 2000. Means and standard errors were computed for both the modular architecture (Figure 7 Top) and the fully distributed architecture (Figure 7 Bottom). In general, higher columns correspond to more activity in the specific hidden layer and linguistic task to which they refer.

For each network, an analysis of variance with repeated factors (MANOVA) was computed to compare the average differential activities in the two hidden layers and in the two linguistic tasks of noun and verbs. A sample size of 20 was considered, i.e. one data point for each of the 20 populations. The dependent variable was the INA value averaged over the 80 networks of each population. In computing the INA values for each individual network, only the values when the individual agent correctly manipulates the two objects (over 90% of the cases) were used. The two independent variables used in the MANOVA were HIDDEN_LAYER (2 levels: 1st and 2nd) and NOUN_VERB (2 levels: noun tasks and verb tasks).

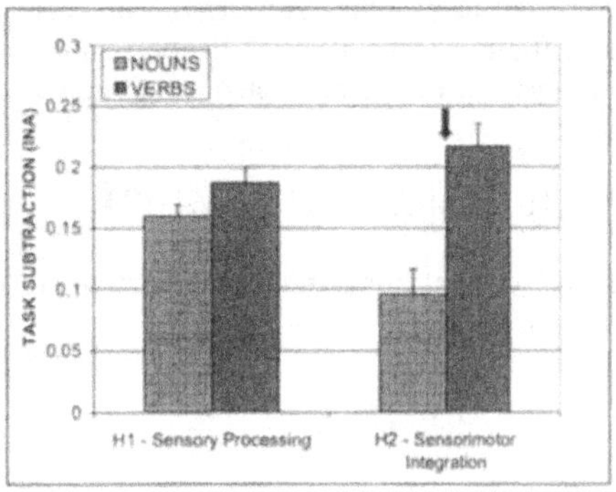

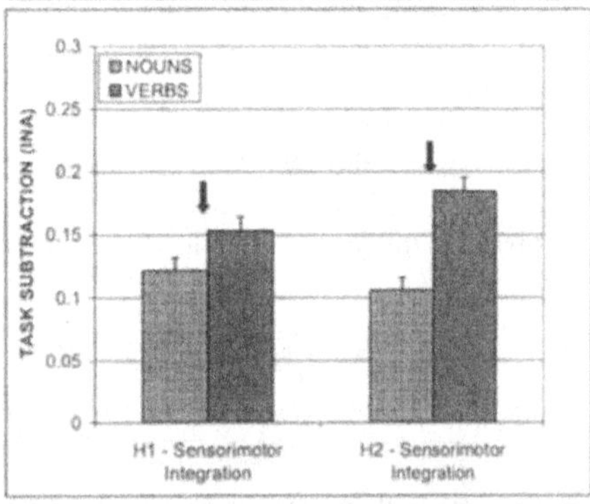

Figure 7: Data on the synthetic brain imaging of verb and noun processing in the two networks. Black arrows indicate significant statistical difference between the two verb/noun columns. The modular network (Top) has no significant differential activity for nouns in the first hidden layer. Verbs are significantly more active in the second hidden layer. In the network with fully distributed architecture (Bottom) verbs are significantly more active in both layers.

Two individual MANOVAs were carried out, one for the modular network and one for the distributed architecture. In both analyses, only the NOUN_VERB factor and the interaction between the two factors were significant. In particular, to identify the specific differences between the 4 means of the interaction, some post hoc test were carried out. In the first network (Figure 7 Top), there is no significant difference of activity between nouns and verbs in the first hidden layer ($p=0.281$). Instead, verbs produce significantly more neural activity than nouns in the second hidden layer ($p=0.009$). In the second network (Figure 7b) verbs are significantly more active than nouns in both layers ($p=0.017$ in the first layer, and $p<0.001$ in the second).

Overall, this pattern of results confirms that verb processing is "localized" in the regions of the network specialized for sensorimotor integration. This is the case of only the second hidden layer in the modular network, and of both layers in the fully distributed architecture. In addition, the results are consistent with the categorical perception analyses in the two neural architectures.

5. Discussion and Conclusions

This research was based on grounded adaptive agents for simulating the evolutionary acquisition of simple lexicons based on verbs and nouns. The use of neural networks that control all behaviors (sensorimotor, cognitive and linguistic) has permitted a better understanding of the interaction between such abilities. The manipulation of the connectivity in the two networks had the scope of designing agents with different neural processing strategies for the control of sensorimotor and linguistic behavior. The CP and SBI analyses have highlighted that the linguistic structure differentiating the two categories of nouns and verbs is strictly dependent on the perceptual and sensorimotor representations evolved by the agents. For example, tasks involving the use of verbs require the activation of the same neurons (and hidden layers) that play the function of integrating sensorial information (e.g. vision for identifying the shape and position of the object) with motor signals (proprioception for planning arm movement) to produce action.

In section 4.1, two alternative mechanisms where hypothesized for explaining differences in the neural processing of verbs and nouns. The first hypothesis suggests that the closer the hidden layer is to the motor output units, the more it will be specialized for verb representation and processing (e.g. verbs will produce enhanced CP effects and higher SBI activity). The alternative hypothesis is that a hidden layer specializes for verb processing when it integrates sensory information, such as vision and language, with proprioceptive information. Both CP and SBI analyses support the second hypothesis. For example, verbs produce enhanced CP effects in the hidden layers that play a direct role in determining the motor response through sensorimotor integration. This is the case of the second hidden layer in the modular network and of both layers in the fully-distributed network. This also contradicts the first interpretative hypothesis on the correlation between specialization for verb processing and relative proximity to the motor layer. In fact, in the distributed architecture, both layers show enhanced CP effects, whit stronger effects in the layer more distant from the output units.

The value of this model in highlighting the sensorimotor grounding of the linguistic categories is strengthened by the similarity between the model performance and data on the neuroscience of language processing. In section 2.2 it was illustrated that experimental brain imaging studies have consistently established that nouns activate more the posterior areas of the brain related to sensory and associative processing, while verbs activate more the anterior motor areas (Cappa & Perani, 2003). Although the aim of this simulation and the SBI analyses is not that of building a neurally-plausible model of language processing in the brain, it is nevertheless remarkable that such a simple neural network model can be constrained to exhibit some functional and architectural similarities with the neural processing of language. This outcome further supports the use of grounded adaptive agent models for testing language evolution hypothesis. For example, the current model could be easily expanded to manipulate known constraints on the animal and human brains that have been hypothesized to affect the evolution of cognitive and linguistics abilities (Deacon, 1997). Simulations would then investigate the validity of such hypotheses.

From a more general perspective on language origins, one of the most important contributions of this simulation is the demonstration that syntax evolution is directly grounded in sensorimotor and cognitive representations. The CP and SBI analyses show that verbs systematically "follow" the sensorimotor integration processing units, regardless of their position in the neural network. Therefore we can say that there is some kind of functional and neural equivalence between understanding the name of an action and performing the action (also regardless of any linguistic input). This coupling of language and sensorimotor processing abilities is consistent with cognitive grammar literature (Langacker 1987). According to such a theoretical approach, the formation of linguistic constructs leads back to cognitive constraints and functions.

This work also has more general methodological implications for research in adaptive agents. The two techniques of CP and SBI have been used here to visualize and analyze the internal representation used by agents in various tasks and behavioral contexts. With these methodologies, the neural and categorical representations can be not only visually represented, but can also be compared using quantitative methods (e.g. Figure 7 with statistically comparisons of means in different hidden layers and linguistic tasks). Information visualization has become an important issue in adaptive behavior and artificial life research (Smith *et al.* 2002; Marocco et al. 2003). Current adaptive behavior models are becoming increasingly more complex in terms of evolutionary, behavioral and structural properties. Simulations and robotic studies can produce huge sets of data that are often difficult to analyze using existing methodologies. The development of information visualization tools, widespread in other areas such as genomic and neuroscience research (Stuart *et al.* 2001), can be fruitfully adopted in adaptive behavior. The two techniques used here can be easily adapted and used to visualize and interpret the internal representations that adaptive agents use to control complex behaviors.

The model is currently being extended to study the emergence of syntactic languages. The lexicon will not be provided by the experimenter, but rather it will self-organize through the communicative interaction between agents (e.g. Cangelosi 1999). In addition, agents will learn language from their own peers and parents, so that the simulation can combine both the evolutionary and developmental acquisition of language. New simulations will focus on the transition from a non-symbolic and non-compositional lexicon to the emergence of combinatorial and syntactic languages. Further manipulations of the neural network architectures will be conducted to determine the modular organizations favoring the transition to syntactic languages.

Acknowledgements

The research was supported by the UK Engineering and Physical Research Council (EPSRC Grant GR/N01118). The author thanks Domenico Parisi for valuable discussions on this model.

References

Andrews J., Livingston K., Harnad S. (1998). Categorical perception effects induced by category learning. *Journal of Experimental Psychology: Learning, Memory, and Cognition*, 24: 732-753

Arbib M.A., Billard A., Iacoboni M., Oztop E. (2000). Synthetic brain imaging: grasping, mirror neurons and imitation. *Neural Networks*, 13: 975-997

Arbib M.A., Fagg A.H., Grafton S.T. (2002). Synthetic PET imaging for grasping: From primate neurophysiology to human behavior. In F. Sommer, A. Wichert (Eds.), *Explorative analysis and data modelling in functional neuroimaging*. Cambridge, MA: The MIT Press.

Barsalou L. (1999) Perceptual symbol systems. *Behavioral and Brain Sciences,* 22: 577-609.

Cangelosi A. (2001). Evolution of communication and language using signals, symbols, and words. *IEEE Transactions on Evolutionary Computation*. 5(2): 93-101

Cangelosi A., Greco A., Harnad S. (2000). From robotic toil to symbolic theft: Grounding transfer from entry-level to higher-level categories. *Connection Science*, 12(2): 143-162

Cangelosi A., Harnad S. (2000). The adaptive advantage of symbolic theft over sensorimotor toil: Grounding language in perceptual categories. *Evolution of Communication*. 4(1): 117-142

Cangelosi A., Parisi D. (in press). The processing of verbs and nouns in neural networks: Insights from Synthetic Brain Imaging. *Brain and Language*.

Cangelosi A., Parisi D. (2002). *Simulating the Evolution of Language*. London: Springer.

Cangelosi A., Parisi D. (2001). How nouns and verbs differentially affect the behavior of artificial organisms. In J.D. Moore, K. Stenning (Eds.), *Proceedings of the 23rd Annual Conference of the Cognitive Science Society*, London: Lawrence Erlbaum Associates, 170-175.

Cappa S.F., Perani, D. (2003). The neural correlates of noun and verb processing. *Journal of Neurolinguistics*, 16(2-3): 183-189

Christiansen M.H., Chater N. (Eds.) (2001). *Connectionist Psycholinguistics*, Westport: Ablex.

Deacon T.W. (1997). *The Symbolic Species.* London: Penguin.

Dyer M.G. (1994). Grounding language in perception. In V. Honavar, L. Uhr (Eds.), *Artificial Intelligence and neural networks: Steps toward principled integration.* Boston: Academic Press.

Gazzaniga M.S. (Ed.) (2000). *The New Cognitive Neurosciences.* Cambridge, MA: MIT Press.

Glenberg A.M., Kaschak M. P. (in press). Grounding language in action. *Psychonomic Bulletin & Review*

Goldstone R. (1994). Influences of categorization of perceptual discrimination. *Journal of Experimental Psychology: General*, 123: 178-200

Greenfield P. (1991). Language, tool and brain: The ontogeny and phylogeny of hierarchically organized sequential behavior. *Behavioral and Brain Sciences*, 14: 531-595.

Harnad S. (Ed.) (1987). *Categorical Perception: The Groundwork of Cognition.* New York: CUP

Harnad S. (1990). The symbol grounding problem. *Physica D*, 42: 335-346

Hazlehurst B., Hutchins E. (1998). The emergence of propositions from the co-ordination of talk and action in a shared world. *Language and Cognitive Processes*, 13: 373-424

Horwitz B., Tagamets M.-A., McIntosh A.R. (1999). Neural modeling, functional brain imaging, and cognition, *Trends in Cognitive Science*, 3: 91-98.

Joyce D., Richards L., Cangelosi A., Coventry K.R. (2003), On the foundations of perceptual symbol systems: Specifying embodied representations via connectionism. *Proceedings of the 5th Intl. Conference on Cognitive Modeling (ICCM 2003).* Bamberg

Just M.A., Carpenter P.A., Varma S. (1999). Computational modeling of high-level cognition and brain function. *Human Brain Mapping*, 8: 128-136.

Kaplan F. (2000). Talking AIBO: First experimentation of verbal interactions with an autonomous four-legged robot. In A. Nijholt, D. Heylen, K. Jokinen (Eds.). *Learning to Behave: Interacting agents. CELE-TWENTE Workshop on Language Technology*, 57-63.

Kirby S. (2002). Natural language and artificial life. *Artificial Life*, 8: 185-215.

Kosslyn S.M., Koenig O., Barrett A., Cave C.B. et al. (1989). Evidence for two types of spatial representations: Hemispheric specialization for categorical and coordinate relations. *Journal of Experimental Psychology: Human Perception & Performance*, 15(4): 723-735.

Langacker R.W. (1987). *Foundations of Cognitive Grammar, Vol.1: Theoretical Prerequisites.* Stanford, Stanford University Press.

MacWhinney B (Ed.) (1999). *Emergence of Language.* Hillsdale, NJ: LEA.

MacWhinney B. (1998). Models of the emergence of language. *Annual Review of Psychology*, 49: 199-227.

Marocco D., Cangelosi A., Nolfi S. (2003), The emergence of communication is evolutionary robots. *Philosophical Transactions of the Royal Society London – A*, 361: 2397-2421

Marocco D., Stuart E.J., Cangelosi A. (2003). Visualisation techniques in artificial life: The method of parallel coordinates. *1st Italian Workshop on Artificial Life*, Universita' della Calabria, Italy.

Martin A., Haxby J.V., Lalonde F.M., Wiggs C.L., Ungerleider L.G. (1995). Discrete cortical regions associated with knowledge of color and knowledge of action. *Science*, 270: 102-105.

Nakisa R.C., Plunkett K. (1998) Evolution of a rapidly learned representation for speech. *Language and Cognitive Processes*, 13: 105-127

Parisi D., Cangelosi A., Falcetta I. (2002). Verbs, nouns and simulated language games. *Italian Journal of Linguistics*, 14(1): 99-114

Perani D., Cappa S.F., Schnur T., Tettamanti M., Collina S., Rosa M.M., Fazio, F. (1999). The neural correlates of verb and noun processing: A PET study. *Brain*, 122: 2337-44.

Pinker S. (1994). *The Language Instinct.* Penguin.

Pulvermuller F. (2003) *The neuroscience of language. On brain circuits of words and serial order.* Cambridge: Cambridge University Press.

Reilly R.G. (in press). The relationship between object manipulation and language development in Broca's area: A connectionist simulation of Greenfield's hypothesis. *Behavioral and Brain Sciences.*

Rizzolatti, G., & Arbib, M. (1998). Language within our grasp. *Trends in Neuroscience*, 21: 188-194.

Smith T., Bullock S., Bird J. (2002). Beyond fitness: Visualising evolution - Workshop overview. *Artificial Life VIII*, Sydney.

Steels L. (2002). Grounding symbols through evolutionary language games. In A. Cangelosi, D. Parisi (Eds.), *Simulating the Evolution of Language.* London: Springer-Verlag.

Stuart L., Walter M., Borisyuk R. (2002). Visualisation of synchronous firing in multi-dimensional spike trains. *BioSystems*, 67: 265-279.

Tijsseling A., Harnad S. (1997). Warping similarity space in category learning by backprop nets. In M. Ramscar et al. (Eds.). *Proceedings of SimCat 1997: Interdisciplinary Workshop on Similarity and Categorization*, Edinburgh University, 263-269.

Tomasello M., Brook P.J. (1999). Early syntactic development: A Construction Grammar approach. In M. Barrett (Ed.), *The Development of Language.* Philadelphia, PA: Psychology Press, 161-190

Vogt P. (2002). The physical symbol grounding problem. *Cognitive Systems Research*, 3: 429-457

Wagner K., Reggia J.A., Uriagereka J., Wilkinson G.S. (2003). Progress in the simulation of emergent communication and language. *Adaptive Behavior*, 11(1): 37-69.

Zentall T.R., Jackson-Smith P., Jagielo J.A., Nallan G.B. (1986). Categorical shape and color coding by pigeons. *Journal of Experimental Psychology: Animal Behavior Processes*, 12(2): 153-159.

APPLIED ADAPTIVE BEHAVIOR

Evolving Opponents for Interesting Interactive Computer Games

Georgios N. Yannakakis*
*Centre for Intelligent Systems
and their Applications
The University of Edinburgh
AT, Crichton Street, EH8 9LE
g.yannakakis@sms.ed.ac.uk

John Hallam**
** Mærsk Mc-Kinney Møller Institute
for Production Technology
University of Southern Denmark
Campusvej 55, DK-5230
john@mip.sdu.dk

Abstract

In this paper we introduce experiments on neuro-evolution mechanisms applied to predator/prey multi-character computer games. Our test-bed is a modified version of the well-known Pac-Man game. By viewing the game from the predators' (i.e. opponents') perspective, we attempt off-line to evolve neural-controlled opponents capable of playing effectively against computer-guided fixed strategy players. However, emergent near-optimal behaviors make the game less interesting to play. We therefore discuss the criteria that make a game interesting and, furthermore, we introduce a generic measure of predator/prey computer games' interest. Given this measure, we present an evolutionary mechanism for opponents that keep learning from a player while playing against it (i.e. on-line) and we demonstrate its efficiency and robustness in increasing and maintaining the game's interest. Computer game opponents following this on-line learning approach show high adaptability to changing player strategies which provides evidence for the approach's effectiveness against human players.

1. Introduction

Machine learning in computer games is nowadays still in its very early stages, where computer games continue to use simple rule-based finite and fuzzy-state machines for nearly all their AI needs (Woodcock, 2001). Therefore, we still meet new released games with the same 20-year old concept in brand new graphics engines. Unfortunately, instead of designing intelligent opponents to play against, game developers mainly concentrate on the graphical presentation of the game.

Predator/prey games is a very popular category of computer games and among its best representatives is the classical Pac-Man released by Namco (Japan) in 1980. Even though Pac-Man's basic concept and graphics are very simple, the game still keeps players interested after so many years, and its basic ideas are still found in many new released games. In Pac-Man, the player's (*PacMan's*) goal is to eat all the pellets appearing in a maze-shaped stage while avoiding being killed by four opponent characters named '*Ghosts*'. On the other hand, *Ghosts* are aiming to kill (by 'touching') *PacMan* as soon as possible. The game is over when either all pellets in the stage are eaten by *PacMan* or *Ghosts* manage to kill *PacMan*. In that case, the game restarts from the same initial positions for all five characters. Since there are four *Ghosts* on the game field, they are designed to be slower than *PacMan* so that the game is fairer to play.

There are some examples, in the Pac-Man domain literature, of researchers attempting to teach a controller to drive *PacMan* in order to acquire as many pellets as possible and to avoid being eaten by *Ghosts*. Koza (1992) considers the problem of controlling an agent in a dynamic non-deterministic environment and, therefore, sees Pac-Man as an interesting multi-agent environment for applying off-line learning techniques based on genetic programming. The same Pac-Man application domain has been used for analyzing size and generality issues in genetic programming (Rosca, 1996).

On the other hand, there are many researchers who use predator/prey domains in order to obtain efficient emergent teamwork behavior of either homogeneous or heterogeneous predators. For example, Luke and Spector (1996), among others, have designed an environment similar to the Pac-Man game (i.e. Serengeti world) in order to examine different breeding strategies and coordination mechanisms for the predators. Finally, there are examples of work (see (Haynes and Sen, 1995) and (Miller and Cliff, 1994)) in which both the predators' and the prey's strategies are co-evolved in continuous or grid-based environments.

Similar to Luke and Spector (1996), we view Pac-Man from the *Ghosts'* perspective. Our first aim is to emerge effective teamwork hunting behaviors by the use of an off-line training approach, based on evolutionary computation techniques, applied to homogeneous neural controlled (Yao, 1999) *Ghosts*. However, playing a computer game like Pac-Man against optimal hunters cannot be interesting because of the fact that you get easily killed. To this end, the primary objective of this work is to introduce an efficient generic measure of interest of any predator/prey game. We believe that the interest of any computer game is directly related to the interest generated by the opponents' behavior rather than to the graphics or even the player's behavior. Thus, when 'interesting game' is mentioned we mainly refer to interesting opponents to play against.

We present a robust on-line neuro-evolution learning mechanism capable of increasing the game's interest (starting from near-optimal off-line trained behaviors) as well as keeping that interest at high levels as long as the game is being played. The arcade version of Pac-Man uses a handful of very simple rules and scripted sequences of actions combined with some random decision-making to make the *Ghosts'* behavior less predictable. The game's interest decreases at the point where *Ghosts* are too fast to beat (Rabin, 2002). In our Pac-Man version we require *Ghosts* to keep learning and constantly adapting to the player's strategy instead of being uninteresting opponents with fixed strategies. In addition, we explore learning procedures that achieve good real-time performance (i.e. low computational effort while playing).

2. The Pac-Man World

The computer game test-bed studied is a modified version of the original Pac-Man computer game released by Namco. The original game includes a number of special pellets called 'power-pills' that allow *PacMan* to 'eat' *Ghosts* for a short period of time. Power-pill is the only feature not included in this modified version of the game. This omission contributes to the simplicity of the game without affecting the game's playing interest.

As previously mentioned, the Pac-Man game is investigated from the viewpoint of *Ghosts* and more specifically how *Ghosts'* emergent adaptive behaviors can contribute to the interest of the game. Pac-Man — as a computer game domain for emerging adaptive behaviors — is a two-dimensional, multi-agent, grid-motion, predator/prey game. The game field (i.e. stage) consists of corridors and walls. Both the game's dimensions and its maze structure are predefined. For the experiments presented in this paper we use the 19×29 grid maze-stage presented in Figure 1.

The characters visualized in the Pac-Man game (as illustrated in Figure 1) are a white circle that represents *PacMan* and 4 ghost-like characters representing the *Ghosts*. Additionally, there are black squares that represent the pellets and dark grey blocks of walls.

Figure 1: Snapshot of the Pac-Man game

PacMan moves at double the *Ghosts'* speed and since there are no dead ends, it is impossible for a single *Ghost* to complete the task of killing it. Since *PacMan* moves faster than a *Ghost*, the only effective way to kill *PacMan* is for a group of *Ghosts* to hunt cooperatively. It is worth mentioning that one of *Ghosts'* properties is permeability. In other words, two or more *Ghosts* can simultaneously occupy the same cell of the game grid.

The simulation procedure of the Pac-Man game is as follows. *PacMan* and *Ghosts* are placed in the game field (initial positions) so that there is a suitably large distance between them. Then, the following occur at each simulation step:

1. Both *PacMan* and *Ghosts* gather information from their environment.

2. *PacMan* and *Ghosts* take a movement decision every simulation step and every second simulation step respectively. (That is how *PacMan* achieves double the *Ghost's* speed.)

3. If the game is over (i.e. all pellets are eaten, *PacMan* is killed, or the simulation step is greater than a predetermined large number), then a new game starts from the same initial positions.

4. Statistical data such as number of pellets eaten, simulation steps to kill *PacMan* as well as the total *Ghosts'* visits to each cell of the game grid are recorded.

2.1 PacMan

Both the difficulty and, to a lesser degree, the interest of the game are directly affected by the intelligence of the *PacMan* player. We chose three fixed *Ghost*-avoidance and pellet-eating strategies for the *PacMan* player, differing in complexity and effectiveness. Each strategy is based on decision making applying a cost or probability approximation to the player's 4 neighbor cells (i.e. up, down, left and right). Even though the initial positions are constant, the non-deterministic motion of *PacMan* provides lots of diversity within games.

- Cost-Based (CB) *PacMan*: The CB *PacMan* moves towards its neighbor cell of minimal cost. Cell costs are assigned as follows: $c_p = 0$, $c_e = 10$, $c_{ng} = 50$, $c_g = 100$, where c_p: cost of a cell with a pellet (pellet cell); c_e: cost of an empty cell; c_g: cost of a cell occupied by a *Ghost* (*Ghost* cell); c_{ng}: cost of a *Ghost's* 4 neighbor cells. Wall cells are not assigned any cost and are ignored by *PacMan*. In case of equal minimal neighbor cell costs (e.g. two neighbor cells with pellets), the CB *PacMan* makes a random decision with equal probabilities among these cells.

  In other words, the CB *PacMan* moves towards a cost minimization path that produces effective *Ghost*-avoidance and (to a lesser degree) pellet-eating behaviors but only in the local neighbor cell area.

- Rule-Based (RB) *PacMan*: The RB *PacMan* is a CB *PacMan* plus an additional rule for more effective and global pellet-eating behavior. This rule can be described as follows. If all *PacMan's* neighbor cells are empty ($c = 10$), then the probability of moving towards each one of the available directions (i.e. not towards wall cells) is inversely proportional to the distance to the closest pellet on that direction.

- Advanced (ADV) *PacMan*: The ADV *PacMan* checks in every visible direction (i.e. no wall cell) for *Ghosts*. If there is at least one *Ghost* in sight, then the probability of moving towards each one of the available directions is directly proportional to the distance to a *Ghost* in that direction. If there is no *Ghost* in sight, then the ADV *PacMan* behaves like a RB *PacMan*.

  The ADV moving strategy is expected to produce a more global *Ghost*-avoidance behavior built upon the RB *PacMan's* good pellet-eating strategy.

2.2 Neural Controlled Ghosts

Neural networks are a suitable host for emergent adaptive behaviors in complex multi-agent environments (Ackley and Littman, 1992). A feedforward neural controller is employed to manage the *Ghosts'* motion and is described in this subsection. Apart from the neural controller, three non-evolving fixed ways of controlling the *Ghosts* are presented in subsection 2.3.

Input

Using their sensors, *Ghosts* inspect the environment from their own point of view and decide their next action. Each *Ghost* receives input information from its environment expressed in the neural network's input array of dimension 4 (see Figure 2). The input array consists of the relative distances from (a) *PacMan* in x ($\Delta_{x,P} = x_g - x_p$) and y ($\Delta_{y,P} = y_g - y_p$) axis and (b) the closest *Ghost* in x ($\Delta_{x,C} = x_g - x_c$) and y ($\Delta_{y,C} = y_g - y_c$) axis; where (x_g, y_g), (x_p, y_p) and (x_c, y_c) are the cartesian coordinates of the current *Ghost's*, *PacMan's* and closest *Ghost's* current position respectively. *Ghost's* input includes information for only one neighbor *Ghost* as this constitutes the minimal information for emerging teamwork cooperative behaviors.

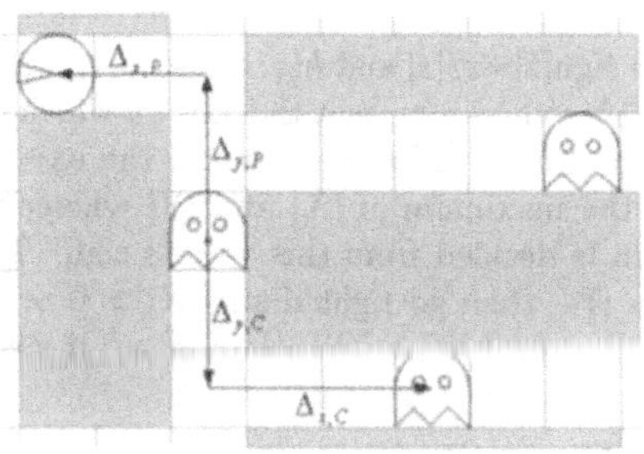

Figure 2: *Ghost's* input data

All input values are linearly normalized into [0, 1] via $0.5[(\Delta_{i,J}/L_i) + 1]$ where $i \in \{x, y\}$, $J \in \{P, C\}$ and L_x, L_y are the width and height of the stage respectively.

Architecture

As previously mentioned, a multi-layered fully connected feedforward neural network has been used for the experiments presented here. The sigmoid function is employed at each neuron.

The connection weights take values from -5 to 5 while the neural network's output is a four-dimensional vector with respective values from 0 to 1 that represents the *Ghost's* four movement options (up, down, left and right respectively). Each *Ghost* moves towards the direction represented by the highest output value.

2.3 Fixed Strategy Ghosts

Apart from the neural controlled *Ghosts*, three additional fixed non-evolving strategies have been tested for controlling the *Ghost's* motion. These strategies are used as baseline behaviors for comparison with any neural controller emerged behavior.

- Random (R): *Ghosts* that randomly (uniform distribution) decide their next available movement. Available movements have equal probabilities to be picked.

- Followers (F): *Ghosts* designed to follow *PacMan* constantly. Their strategy is based on moving so as to reduce the greatest of their relative distances $(\Delta_{x,P}, \Delta_{y,P})$ from *PacMan*.

- Near-Optimal (O): A *Ghost* strategy designed to produce attractive forces between *Ghosts* and *PacMan* as well as repulsive forces among the *Ghosts*. For each *Ghost* X and Y values are calculated as follows.

$$\begin{aligned} X &= \text{sign}[\Delta_{x,P}]h(\Delta_{x,P}, L_x, 0.25) \\ &- \text{sign}[\Delta_{x,C}]h(\Delta_{x,C} - 1, L_x, 10) \quad (1) \\ Y &= \text{sign}[\Delta_{y,P}]h(\Delta_{y,P}, L_y, 0.25) \\ &- \text{sign}[\Delta_{y,C}]h(\Delta_{y,C} - 1, L_y, 10) \quad (2) \end{aligned}$$

where $\text{sign}[z]=z/|z|$ and $h(z, z_m, p) = [1 - (|z|/z_m)]^p$. X and Y values represent the axis on which the near-optimal *Ghost* will move. Hence, the axis is picked from the maximum of $|X|$ and $|Y|$ whereas, the direction is decided from this value's sign. That is, if $|X| > |Y|$, then go right if $\text{sign}[X] > 0$ or go left if $\text{sign}[X] < 0$; if $|Y| > |X|$, then go up if $\text{sign}[Y] > 0$ or go down if $\text{sign}[Y] < 0$.

3. Optimal Behavior

When a predator/prey game is investigated from the predator's viewpoint, optimality can be measured in the predators' ability to kill the prey. Thus, a predator's behavior that always manages to kill the prey in such games is obviously a desired behavior in terms of optimality.

Prey-killing ability is the primary factor that determines how good a behavior is (i.e. its performance) in the Pac-Man game as well. Furthermore, the behavior of preventing *PacMan* from eating pellets constitutes an additional factor of the desired optimal behavior. This behavior also implies a fast-killing behavior, which is also desired from optimal predators. Given these, a measure designed to give an approximation of a group of *Ghosts'* performance over a specific number N of games played, is

$$P = \frac{\alpha\frac{k}{N} + \beta\min\{1 + \frac{(e_{min} - E\{e\})}{e_{max}}, 1\}}{\alpha + \beta} \quad (3)$$

where P is the performance of a *Ghost* group behavior taking values from 0 to 1; k is the number of *PacMan* kills within N games; $E\{e\}$ is the average number of pellets eaten by *PacMan* over the N games; e_{min}, e_{max} are the lower and upper bound of the eaten pellets e respectively (in this paper $e_{min} = 70$ and $e_{max} = 187$); α, β are weight parameters (in this paper $\alpha = \beta = 1$).

4. Interesting Behavior

In order to find an objective (as possible) measure of interest in the Pac-Man computer game we first need to define the criteria that make a game interesting. Then, second, we need to quantify and combine all these criteria in a mathematical formula. The game should be, then, tested by human players and therefore have this formulation of interest cross-validated against the interest the game produces in real conditions. This last part of our investigation constitutes a crucial phase of future work.

In order to simplify this procedure we will ignore the graphics' as well as the player's contribution to the interest of the game and we will concentrate on the *Ghost's* behavior that effects the game's interest. That is because, we believe, the computer-guided opponent character contributes the vast majority of features that make a computer game interesting.

By being as objective and generic as possible, we believe that the criteria that collectively define interest on the Pac-Man game are as follows.

1. *When the game is neither too hard nor too easy.* In other words, the game is interesting when *Ghosts* manage to kill *PacMan* sometimes but not always. In that sense, optimal behaviors are not interesting behaviors and *vice versa*.

2. *When there is diversity in Ghosts' behavior over the games.* That is, when *Ghosts* are able to find different ways of hunting and killing *PacMan* in each game so that their strategy is less predictable.

3. *When Ghosts' behavior is aggressive rather than static.* That is, *Ghosts* that move towards killing *PacMan* but meanwhile, move constantly all over the game field instead of simply following it. This behavior gives player the impression of an intelligent strategic *Ghosts'* plan which increases the game interest.

In order to estimate and quantify each of the aforementioned criteria of the game's interest, we let the examined group of *Ghosts* play the game N times (each game for a sufficiently large evaluation period of t_{max} simulation steps; this period should be enough for *PacMan* to clear the stage of pellets; in the experiments presented here $t_{max} = 300$ simulation steps) and we record the simulation steps t_k taken to kill *PacMan* as well as the total number of *Ghosts'* visits v_i at each cell i of the game field for each game. Given these, the quantifications of the Pac-Man game's three interest criteria can be presented as follows.

1. According to the first criterion, the estimate of how interesting the behavior is, is given by T in (4).

$$T = [1 - (E\{t_k\}/\max\{t_k\})]^{p_1} \quad (4)$$

where

$$t_{min} \leq t_k \leq t_{max} \quad (5)$$

and $E\{t_k\}$ is the average number of simulation steps taken to kill *PacMan* over the N games; $\max\{t_k\}$ is the maximum t_k over the N games; t_{min} is the minimum number of simulation steps required for *Ghosts* to kill *PacMan* (t_{min} is 30 simulation steps in this paper); p_1 is a weighting parameter (for the experiments presented here $p_1 = 0.5$).

The T estimate of interest demonstrates that the greater the difference between the average number of steps taken to kill *PacMan* and the maximum number of steps taken to kill *PacMan*, the higher the interest of the game. Given (4), both poor-killing ('*too easy*') and near-optimal ('*too hard*') behaviors get low interest estimate values (i.e. $E\{t_k\} \simeq \max\{t_k\}$).

2. The interest estimate for the second criterion is given by S in (6).

$$S = \left(s^2/s_{max}^2\right)^{p_2} \tag{6}$$

where

$$s_{max}^2 = \frac{N}{4(N-1)}(t_{max} - t_{min})^2 \tag{7}$$

and s^2 is the sample variance of t_k over the N games; s_{max}^2 is the maximum value of s^2; p_2 is a weighting parameter (for the experiments presented here $p_2 = 1$).

The S estimate of interest demonstrates that the greater the variance of the steps taken to kill *PacMan* over N games, the higher the interest of the behavior. Therefore, by using (6) we promote *Ghosts* that produce high diversity in the time taken to kill *PacMan*.

3. A good measure for quantifying the third interest criterion is through entropy of the *Ghosts'* cell visits in a game, which quantifies the completeness and uniformity with which the *Ghosts* cover the stage. Hence, for each game, the cell visits' entropy H is calculated via (8) and it is normalized into $[0,1]$ via (9).

$$H = -\sum_i \frac{v_i}{V} log\left(\frac{v_i}{V}\right) \tag{8}$$

$$H_n = (H/logV)^{p_3} \tag{9}$$

where V is the total number of visits of all visited cells (i.e. $V = \sum_i v_i$) and p_3 is a weighting parameter (for the experiments presented here $p_3 = 4$).

Given the normalized entropy values for all N games, the interest estimate for the third criterion can be presented as their average value $E\{H_n\}$ over the N games. This implies that the higher the average entropy value, the more interesting the game becomes.

All three criteria are combined linearly (10)

$$I = \frac{\gamma T + \delta S + \epsilon E\{H_n\}}{\gamma + \delta + \epsilon} \tag{10}$$

where I is the interest value of the Pac-Man game; γ, δ and ϵ are criterion weight parameters (for the experiments presented here $\gamma = 1, \delta = 2, \epsilon = 3$).

The measure of the Pac-Man game's interest introduced in (10) can be effectively applied to any predator/prey computer game because it is based on generic features of this category of games. These features include the time required to kill the prey as well as the predators' entropy throughout the game field. We therefore believe that (10) — or a similar measure of the same concepts — constitutes a generic interest approximation of predator/prey computer games.

5. Off-line learning

As previously stressed, our primary aim is to generate emergent *Ghost* behaviors that make the game interesting via an on-line learning mechanism. We therefore use an off-line evolutionary learning approach in order to produce some 'good' (i.e. in terms of performance) initial behaviors for the on-line learning mechanism.

The neural networks that determine the behavior of the *Ghosts* are themselves evolved. In the algorithm presented here, the evolving process is limited to the connection weights of the neural network.

The evolutionary procedure, which is based on previous work by Yannakakis et al. (2003), is as follows. Each *Ghost* has a genome that encodes the connection weights of its neural network. A population of 80 (we keep this number low because of the computational cost) neural networks (*Ghosts*) is initialized randomly with initial uniformly distributed random connection weights that lie within [-5, 5]. Then, at each generation:

Step 1: Every *Ghost* in the population is cloned 4 times. These 4 clones are placed in the Pac-Man game field and play N_t games, each one for an evaluation period of t simulation steps (in the experiments presented in this paper $N_t = 10$ games and $t = 300$ simulation steps). The outcome of these games is to ascertain the time taken to kill *PacMan* t_k for each game.

Step 2: Each *Ghost* is evaluated via (11) for each game and its fitness value is given by $E\{f\}$ over the N_t games.

$$f = [1 - (t_k/t)]^{\frac{1}{4}} \tag{11}$$

By the use of the f fitness function, that takes values from 0 to 1, we promote *PacMan*-killing behaviors capable of achieving high performance values P.

Step 3: A pure elitism selection method is used where only the 10% best fit solutions determine the members of the intermediate population and, therefore, are able to breed.

Step 4: Each parent clones an equal number of offspring in order to replace the non-picked solutions from elitism.

Step 5: Mutation occurs in each gene (connection weight) of each offspring's genome with a small probability p_m (e.g. 0.01). A uniform random distribution is used again to define the mutated value of the connection weight.

The algorithm is terminated when a predetermined number of generations g is achieved (e.g. $g = 1000$) and the best-fit *Ghost's* connections weights are saved.

Ghosts play few games (i.e. $N_t = 10$) when evaluated by the off-line learning mechanism. Even though, this evaluation procedure constitutes an approximation of the examined *Ghost's* overall performance in a greater number of games, it keeps the computational cost in low levels.

6. On-line learning (OLL)

This learning approach is based on the idea of *Ghosts* that learn while they are playing against *PacMan*. In other words, *Ghosts* that are reactive to any player's behavior and learn from its strategy instead of being the predictable and, therefore, uninteresting characters that exist in all versions of this game today. Furthermore, this approach's additional objective is to keep the game's interest at high levels as long as it is being played.

Beginning from any initial group of homogeneous offline trained (OLT) *Ghosts*, the OLL mechanism attempts to transform them into a group of heterogeneous *Ghosts* that are interesting to play against.

The OLL procedure is as follows. An OLT *Ghost* is cloned 4 times and its clones are placed in the Pac-Man game field to play against a selected *PacMan* type of player. Then, at each generation:

Step 1: Each *Ghost* is evaluated every t simulation steps via (12), while the game is played. For the experiments presented in this paper $t = 50$ simulations steps.

$$f' = \sum_{i=1}^{t/2} (d_{P,2i} - d_{P,(2i-1)}) \qquad (12)$$

where $d_{P,i}$ is the distance between the *Ghost* and *PacMan* at the i simulation step. This fitness function promotes *Ghosts* that move towards *PacMan* within an evaluation period of t simulation steps.

Step 2: A pure elitism selection method is used where only the fittest solution is able to breed. The best-fit parent clones an offspring with a probability p_c that is inversely proportional to the normalized cell visits' entropy (i.e. $p_c = 1 - H_n$) given by (9). In other words, the higher the cell visits' entropy of the *Ghosts*, the lower the probability of breeding new solutions. If there is no cloning, then go back to Step 1, else continue to Step 3.

Step 3: Mutation occurs in each gene (connection weight) of each offspring's genome with a small probability p_m (e.g. 0.01). A gaussian random distribution is used to define the mutated value of the connection weight. The mutated value is obtained from (13).

$$w_m = \mathcal{N}(w, 1 - H_n) \qquad (13)$$

where w_m is the mutated connection weight value and w is the connection weight value to be mutated. The gaussian mutation, presented in (13), suggests that the higher the normalized entropy of a group of *Ghosts*, the smaller the variance of the gaussian distribution and therefore, the less disruptive the mutation process as well as the finer the precision of the GA.

Step 4: The cloned offspring is evaluated briefly via (12) in off-line mode, that is, by replacing the worst-fit member of the population and playing an off-line (i.e. no visualization of the actions) short game of t simulation steps. The fitness values of the mutated offspring and the worst-fit *Ghost* are compared and the better one is kept for the next generation. This pre-evaluation procedure for the mutated offspring attempts to minimize the probability of group behavior disruption by low-performance mutants. The fact that each mutant's behavior is not tested in a single-agent environment but within a group of heterogeneous *Ghosts* helps more towards this direction. If the worst-fit *Ghost* is replaced, then the mutated offspring takes its position in the game field as well.

The algorithm is terminated when a predetermined number of games has been played.

We mainly use small simulation periods (i.e. $t = 50$) in order to evaluate *Ghosts* in OLL. The aim of this high frequency of evaluations is to accelerate the on-line evolutionary process. The same short period is used for evaluating mutated offspring. The two primary objectives of this are: 1) to apply a fair comparison between the mutated offspring and the worst-fit *Ghost* (i.e. same evaluation period) and 2) to avoid undesired high computational effort in on-line mode (i.e. while playing). However, the evaluation function (12) constitutes an approximation of the examined *Ghost's* overall performance for large simulation periods. Keeping the right balance between computational effort and performance approximation is one of the key features of this approach. In the experiments presented here, we use minimal evaluation periods capable of achieving good estimation of the *Ghosts'* performance.

7. Results and analysis

Results obtained from experiments applied on the PacMan game are presented in this section. These include, off-line and on-line learning emergent behavior analysis as well as experiments for testing robustness and adaptability of the on-line mechanism proposed.

7.1 Off-line experiments

The experiment presented in this subsection is focused on producing well-behaved *Ghosts* in terms of the performance measure described in section 3. We train *Ghosts* against all three types of *PacMan* players through the off-line learning mechanism presented in section 5. The off-line learning experiment is described as follows.

- Apply the off-line learning mechanism by playing against each type of *PacMan* players separately.
- *Ghosts* trained against a specific type of *PacMan* player are evaluated by playing 100 non-evolution games against the same *PacMan* type.
- In order to minimize the non-deterministic effect of the *PacMan's* strategy on the *Ghost's* performance and interest values as well as to draw a clear picture of these averages' distribution, we apply the following procedure. Using a uniform random distribution we pick 10 different 50-tuples out of the 100 aforementioned games. These 10 samples of data (i.e. e, k, t_k, v_i) from 50 games (i.e. $N = 50$) are used to determine the *Ghosts'* average performance and interest values. The outcome of this experiment is presented in Table 1.

	Trained off-line by playing against					
	CB		RB		ADV	
	P	I	P	I	P	I
R	0.446	0.429	0.4	0.43	0.411	0.424
F	0.776	0.705	0.839	0.733	0.822	0.722
O	1.0	0.607	0.96	0.625	1.0	0.582
B	0.987	0.458	0.79	0.555	0.875	0.535
A	0.724	0.646	0.7	0.614	0.749	0.513
H	0.6	0.297	0.587	0.476	0.464	0.357
$E\{\}$	0.755	0.524	0.713	0.572	0.72	0.522

Table 1: Performance (P) and Interest (I) values (average values of 10 samples of 50 games each) of fixed strategy (R, F, O) and OLT *Ghosts* (B, A, H) playing against all three *PacMan* types (CB, RB, ADV). Average P and I values ($E\{\}$) of all six strategies appear in the bottom row. Experiment Parameters: population size is 80, $g = 1000$, $t = 300$ simulation steps, $N_t = 10$ games, $p_m = 0.01$, 5-hidden neurons controller.

More analytically, in Table 1, both the performance and the interest values of six different *Ghosts'* behaviors against all three different *PacMan* types as well as their average values $E\{\}$ are presented for comparison. In the first three rows of Table 1 the fixed strategy *Ghosts'* (i.e. R:Random, F:Followers, O:Near-Optimal) performance and interest values are presented against each type of *PacMan* player. Furthermore, P and I values of three different emergent behaviors against each *PacMan* type (9 in total) are presented. These behaviors can be described as follows.

- Blocking (B): These are the OLT *Ghosts* that achieve the best performance ($P > 0.75$) against each *PacMan* type. Their behavior is characterized as 'Blocking' because they tend to wait for *PacMan* to enter into a specific area that is easy for them to block and kill.
- Aggressive (A): These are suboptimal OLT *Ghosts* that achieve lower performance ($0.6 < P \leq 0.75$) in comparison to the blockers. Their behavior is characterized as 'Aggressive' because they tend to follow *PacMan* all over the stage in order to kill it.
- Hybrid (H): These are suboptimal OLT *Ghosts* that achieve the lowest performance ($P \leq 0.6$) in comparison to the aforementioned B and A *Ghosts*. Their behavior is characterized as 'Hybrid' because they tend to behave as a Blocking-Aggressive hybrid which proves to be ineffective at killing *PacMan*.

The two main criteria in distinguishing the aforementioned behavior types are their performance values and their behavior seen on screen.

Near-optimal and Blocking behavior *Ghosts* achieve high-performance values against all three *PacMan* types whereas their interest value is not as high as their performance value. This illustrates the compromise between optimality and interest we have to make because, in a predator/prey computer game, optimal killing behaviors cannot be interesting behaviors.

On the other hand, Followers and Aggressive behavior *Ghosts* are likely to produce the two most interesting behaviors (among the behaviors examined in Table 1) for the game. Given the highest achieved interest values (Followers — 0.705 against CB, 0.733 against RB and 0.722 against ADV), we can determine the threshold of $I \geq 0.7$ in accepting a team of *Ghosts'* behavior (i.e. a game) as interesting.

Viewing results presented in Table 1 from the *PacMan* type perspective (i.e. average values in the bottom row of the table), it looks as if the RB is the harder to kill and, meanwhile, the most interesting *PacMan* player for the *Ghosts* to play against. Furthermore, the CB *PacMan* seems to be the easiest to kill whereas, ADV — even though apparently not significantly different from CB — seems to be the least interesting type of *PacMan* for *Ghosts* to play against.

7.2 On-line experiments

As previously mentioned, the off-line learning procedure is a mechanism that produces near-optimal solutions to the problem of killing *PacMan* and minimizing the pellets eaten in a game. These solutions will be the OLL mechanisms' initial points in the search for more interesting games. The OLL experiment is described as follows.

- Pick the nine different emerged *Ghosts'* behaviors produced from the off-line learning experiments presented in subsection 7.1 (i.e. B, A and H behaviors emerged by playing against each *PacMan* type).
- Starting from each OLT behavior, apply the OLL mechanism by playing against each type of *PacMan* player separately. This makes a total of 27 different OLL attempts.
- Calculate the interest of the game every 100 games during each OLL attempt.

In order to calculate the interest of the game we take the *Ghosts'* neural controllers every 100 OLL games and evaluate them by letting them play 100 non-evolution games against the *PacMan* type they were playing against during OLL. We need to minimize the non-deterministic effect of the *PacMan's* strategy on the game's interest and therefore, we use a uniform random distribution to pick 10 different 50-tuples out of these 100 games. These 10 samples of data (i.e. t_k, v_i), of 50 games each (i.e. $N = 50$), are used to determine the games' average as well as boundary values of interest described in section 4. The outcome of this experiment is presented in Table 2 and Figure 3.

Figure 3 illustrates the overall picture of the OLL experiments. The evolution of interest over the OLL games of each one of the nine different OLT behaviors is presented in a sub-figure of Figure 3. For each sub-figure, three lines are illustrated, representing the interest values of the OLL attempt playing against the three different *PacMan* types (a total of 27 different OLL attempts).

As seen from Figure 3, the OLL mechanism manages to find ways of increasing the interest of the game no matter what the initial OLT behavior or the *PacMan* player they play against. In all experiments presented here the learning mechanism is capable of producing games of higher than the initial interest as well as keeping that high interest for a long period. There is obviously a slight probability of disruptive mutations (i.e. the higher the game's interest, through the cell visit's entropy value, the less the probability of mutation — see section 6) that can cause undesired drops in the game's interest. However, as seen from Figure 3, the learning mechanism is robust enough to recover from such disruptive phenomena. When the initial *Ghost* behavior is interesting (see Figures 3(b) and 3(e)) then the mechanism is likely to keep the game at these high, or ever higher, levels of interest. Given an interesting initial behavior (e.g. Aggressive behavior, $I > 0.6$) or even a suboptimal H behavior (see Figures 3(c) and 3(f)), it takes some hundreds of games (i.e. around 500 games in most cases) for the learning mechanism to produce games of high interest. On the other hand, it takes some thousand games to transform an uninteresting near-optimal blocking behavior (see Figures 3(a), 3(d) and 3(g)) into an interesting one. That is because the OLL process requires an initial long period to disrupt the features of an uninteresting blocking behavior in order to be able to increase the interest of the game. This long period of disruption appears when the initial on-line *Ghosts'* behavior (B, A, or H) is emerged by playing against ADV *PacMan* as well (see Figures 3(g), 3(h) and 3(i)). This happens likely because off-line training against ADV *PacMan* seems to produce the least interesting games (see Table 1).

It is obvious that a number in the scale of 10^3 constitutes an unrealistic number of games for a human player to play. In other words, it is very unlikely for a human player to play so many games in order to see the game's interest increasing. The reason for the OLL process being that slow is a matter of keeping the right balance between the process' speed and its 'smoothness' (by 'smoothness' we define the interest's magnitude of change over the games). A solution to this problem is to consider the initial long period of disruption as an off-line learning procedure and start playing as soon as the game's interest is increased. How effective will this mechanism be in a potential change from a fixed strategy to a human *PacMan* player? Subsection 7.3 provides evidence in order to support the answer.

			OLL - Playing against		
			CB	RB	ADV
Initial OLT behaviors	CB	B	0.731	0.711	0.682
		A	0.761	0.714	0.772
		H	0.627	0.418	0.609
	RB	B	0.743	0.727	0.724
		A	0.716	0.688	0.705
		H	0.673	0.772	0.676
	ADV	B	0.701	0.574	0.564
		A	0.706	0.591	0.733
		H	0.763	0.595	0.675

Table 2: Best interest values achieved from on-line learning.

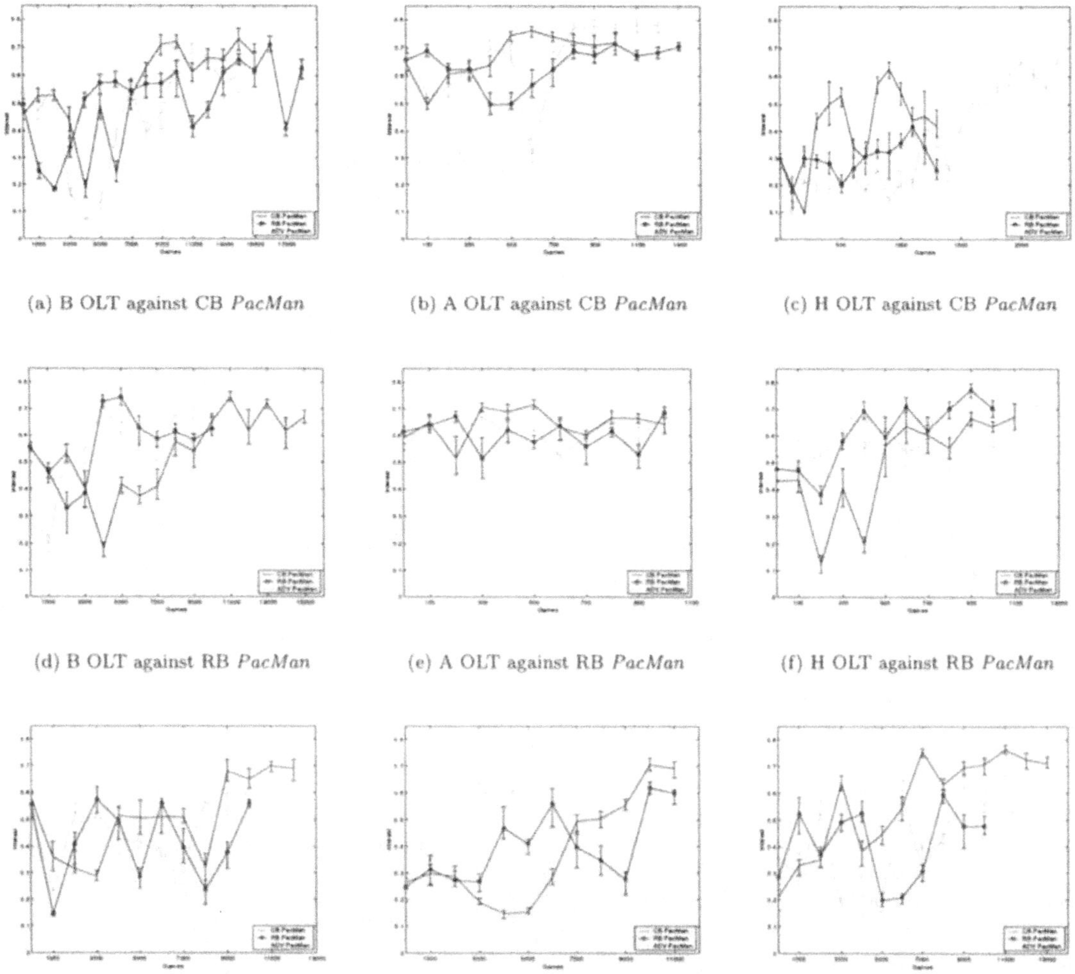

(a) B OLT against CB *PacMan* (b) A OLT against CB *PacMan* (c) H OLT against CB *PacMan*

(d) B OLT against RB *PacMan* (e) A OLT against RB *PacMan* (f) H OLT against RB *PacMan*

(g) B OLT against ADV *PacMan* (h) A OLT against ADV *PacMan* (i) H OLT against ADV *PacMan*

Figure 3: Game interest over the number of OLL games. For reasons of computational effort, the OLL procedure continues for a number of games, large enough to illustrate the mechanism's behavior, after a game of high interest ($I \geq 0.7$) is found. Initial *Ghost* behaviors appear in sub-figure captions. Experiment Parameters: $t = 50$ simulation steps, $p_m = 0.01$, 5-hidden neurons controller.

Table 2 presents the best average interest values obtained from the OLL mechanism. It is clear that the OLL approach constitutes a robust mechanism that, starting from near-optimal or suboptimal *Ghosts*, manages to emerge interesting games (i.e. interesting *Ghosts*) in the majority of cases (i.e. in 15 out of 27 cases $I > 0.7$). It is worth mentioning that in 10 out of 27 different OLL attempts the best interest value is greater than the respective Follower's value (i.e. 0.705 against CB, 0.733 against RB and 0.722 against ADV).

7.3 *Adaptability*

In order to test the OLL mechanism's ability to adapt to a changing environment (i.e. change of *PacMan* strategy), the following experiment is proposed. Beginning from an initial OLT behavior we apply the OLL mechanism against a specific *PacMan* type. During the on-line process we keep changing the type of player as soon as interesting games (i.e. $I > 0.7$) are produced. The process stops when all types of players have played the game.

Since we have three types of players, the total number of different such experiments is 6 (all different player type sequences). These experiments illustrate the over-

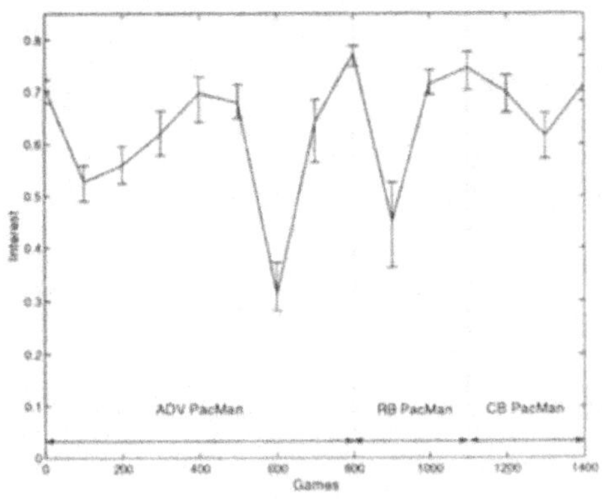

Figure 4: On-line learning *Ghosts* playing against changing types of *PacMan*. Initial behavior: A OLT against CB *PacMan*.

all picture of the mechanism's behavior against any sequence of *PacMan* types. Due to space considerations we present only one (see Figure 4) out of the 6 experiments here. As seen in Figure 4, the OLL mechanism is able to quickly recover a sudden change in the player's strategy (i.e. from ADV to RB and finally to CB *PacMan*) and increase the game's interest at high levels after a number of games has been played (i.e. 100 to 400 games). The mechanism demonstrated a similar adaptive behavior for all 6 different sequences of *PacMan* players which illustrates the mechanism's independence from the sequence of the changing *PacMan's* type.

In addition, by experimenting with all sorts (B, A, H) of initial behaviors, the mechanism's time scale of adaptation to new players is seen to be a function of the initial *Ghosts'* behavior: the more interesting the initial behavior, the faster the adaptation to new players.

Results obtained from this experiment provide evidence for the mechanism's ability to adapt to new types of players as well as its efficiency in producing interesting games against human players.

8. Conclusion

Predator strategies in prey/predator computer games are still nowadays based on simple rules which make the game pretty predictable and, therefore, uninteresting (by the time the player gains more experience and playing skills). A computer game becomes interesting primarily when there is an on-line interaction between the player and its opponents who demonstrate interesting behaviors.

Given some objective criteria for defining interest in predator/prey games presented in this paper we introduced a generic method for measuring interest in such games. We saw that by using the proposed on-line learning mechanism, maximization of the individual simple distance measure (see (12)) coincides with maximization of the game's interest. Apart from being fairly robust, the proposed mechanism demonstrates high adaptability to changing types of player (i.e. playing strategies). Therefore, we believe that such a mechanism will be able to produce interesting interactive opponents (i.e. games) against even the most complex human playing strategy.

On the other hand, for future work, we believe that the methods used need to be tested on more complex PacMan stages in order to provide more evidence for their generality, and the interest measure proposed needs to be cross-validated against human players.

References

Ackley, D. H. and Littman, M. L. (1992). Interactions between learning and evolution. In *Artificial Life II*, pages 478–507, Reading, MA. Sante Fe Institute Studies in the Sciences and Complexity, Addison-Wesley.

Haynes, T. and Sen, S. (1995). Evolving behavioral strategies in predators and prey. In *IJCAI-95 Workshop on Adaptation and Learning in Multiagent Systems*, pages 32–37. Morgan Kaufmann.

Koza, J. (1992). *Genetic Programming: On the programming of computers by means of natural selection.* MIT Press.

Luke, S. and Spector, L. (1996). Evolving teamwork and coordination with genetic programming. In *Genetic Programming 1996: Proceedings of the First Annual Conference*, pages 150–156. MIT Press.

Miller, G. and Cliff, D. (1994). Protean behavior in dynamic games: Arguments for the co-evolution of pursuit-evasion tactics. In *Proceedings of the Third International Conference on Simulation of Adaptive Behavior (SAB-94)*, pages 411–420. MIT Press.

Rabin, S. (2002). *AI Game Programming Wisdom.* Charles River Media, Inc.

Rosca, J. (1996). Generality versus size in genetic programming. In *Genetic Programming 1996: Proceedings of the First Annual Conference*, pages 381–387, Stanford University, CA, USA. MIT Press.

Woodcock, S. (2001). Game AI: The State of the Industry 2000-2001: It's not Just Art, It's Engineering. Game Developer magazine.

Yannakakis, G. N., Levine, J., Hallam, J., and Papageorgiou, M. (2003). Performance, robustness and effort cost comparison of machine learning mechanisms in *FlatLand*. In *Proceedings of the 11th Mediterranean Conference on Control and Automation MED'03*. IEEE.

Yao, X. (1999). Evolving artificial neural networks. In *Proceedings of the IEEE*, volume 87, pages 1423–1447.

Evolvable Recovery Membranes in Self-monitoring Aerospace Vehicles

Peter Wang and Mikhail Prokopenko
Centre for Intelligent Systems Design
CSIRO Information and Communication Technologies Centre
Locked bag 17, North Ryde, NSW 1670, Australia
{peter.wang, mikhail.prokopenko}@csiro.au

Abstract

In this paper we proposed and verified a methodology underlying the design of localised algorithms for complex multi-agent systems, exemplified by self-monitoring aerospace vehicles. In particular, we considered the emergence of self-organising impact boundaries and recovery membranes, separating damaged and/or potentially recovering regions from less affected agents. In order to identify phase transitions in system's dynamics, we investigated graph-theoretic and information-theoretic metrics, and incorporated them within fitness functions for a genetic algorithm. The GA involved a generation gap strategy and targeted a response time as well as spatial connectivity, temporal persistence and size of emergent boundaries and membranes. A variety of stable spatiotemporal patterns were produced under selection pressure, highlighting the potential for the *design at the edge of chaos*.

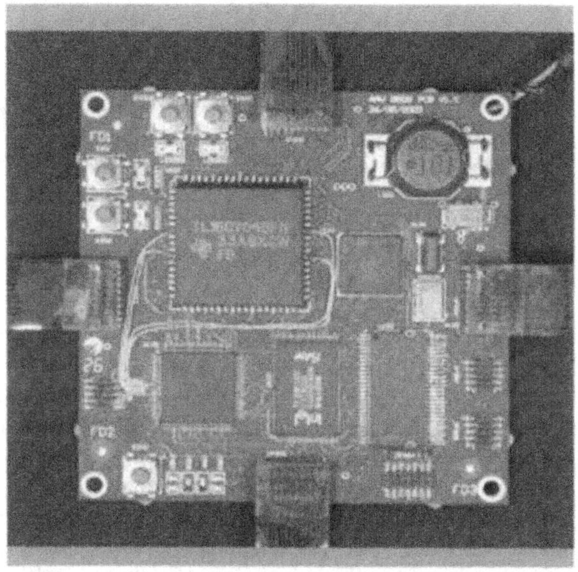

Figure 1: A single square cell with 4 communication ports: Concept Demonstrator view of a printed circuit board.

1. Introduction

Robust or "ageless" aerospace vehicles (AAVs) are expected to be capable of structural self-assessment and repair. The research results presented in this paper were obtained as part of the joint CSIRO-NASA AAV project. At this stage, the project is concerned primarily with the self-assessment, or self-monitoring, functions, but it will progress towards the incorporation of damage prognostic models enabling a predictive damage mitigation decision-making process. The initial goal of the present AAV Concept Demonstrator (Abbott et al., 2002, Abbott et al., 2003) is the detection and characterisation of high velocity impacts caused, for instance, by micro-meteoroids in space.

In general, the self-monitoring functions can be divided between those carried out by distributed sensors and decentralised processing and communication on the skin or within the structure, and those that could be more effectively provided by autonomous robotic non-destructive evaluation (NDE) agents deployed to monitor damage or integrity of the vehicle structure. In this paper, we consider the first type of agents embedded in the skin of the AAV. In particular, we shall consider a multi-cellular sensing and communication network. Each cell in this network has the ability to sense via its 4 piezoelectric sensors, process the inputs, and communicate through its 4 communication ports (Abbott et al., 2003), as shown in Figures 1 and 2.

A two-dimensional array of cells is modelled in a biologically-inspired hybrid Concept Demonstrator, some cells existing in dedicated hardware (a cell per micro-processor) and some residing within interconnected personal computers (a number of cells per PC) (Abbott et al., 2003). We also used a stand-alone Simulator capable of simulating some simple environmental effects such as particle impacts of various energies.

The solution adopted for the problem of handling distributed sensor data and making the self-monitoring

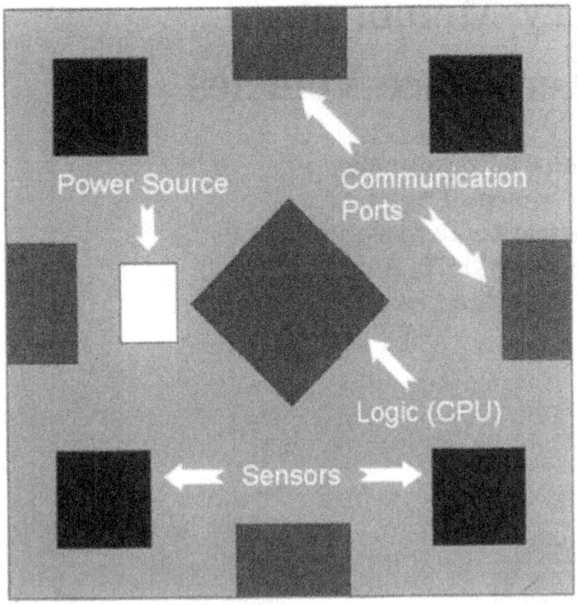

Figure 2: A single square cell with 4 communication ports: AAV Simulator view.

system robust, is to distribute the processing as well, so that most computation takes place near the data source, and there is no single or small number of points of failure. In other words, without centralised controllers, agents (cells) are expected to self-organise and survive on the basis of local, rather than global, information (no single agent has access to information about any others but its neighbours). This approach uses the idea of localised algorithms (Estrin et al., 1999, Macias and Durbeck, 2002, Durbeck and Macias, 2002, Prokopenko et al., 2004), in which simple local behaviours achieve a desired global objective, while communicating only with cells within an immediate neighbourhood. Single cells may need to make fast and automatic responses to sudden damage, while collections of cells may solve more complex tasks, for example, produce an impact boundary with desired characteristics (Lovatt et al., 2003, Foreman et al., 2003) or form a spanning tree connecting cells that detected non-critical impacts (Wang et al., 2003, Prokopenko et al., 2004).

Typically, the desirable emergent behaviour on the system level is not easily predictable from local agents' rules and interactions. Our proposed approach to designing localised algorithms is based on an iterative process including the following steps: a) forward simulation leading to emergent behaviour for a class of localised algorithms; b) quantitative measurement of spatiotemporal stability of the emergent patterns; and c) evolutionary modelling of the algorithms, with the metrics obtained at step b) contributing to the fitness functions. This paper investigates the step c).

The next section will cover some background on measures of emergent behaviour, followed by a description, in section 3., of a localised algorithm producing impact boundaries and emergent recovery membranes. Section 4. will then describe a genetic algorithm and our experiments on evolving the recovery membranes to fit the desired criteria.

2. Background

A very promising direction was investigated by Wright et al. (Wright et al., 2000) who designed *a measure of emergence* of swarming/flocking behaviour in multi-agent systems. The approach borrows from the analysis of dynamical systems and assumes a selection of a set of local state variables (eg., 2-dimensional coordinates of agents/particles and their corresponding velocities). The proposed measure Ω estimates the level of self-organisation in the multi-agent flock via approximation of the dynamical system's *characteristic dimension* — i.e., by determining how well a swarm/flock can be described as a single body. The characteristic dimension is approximated through the Shannon entropy of singular value spectra sampled over time (Wright et al., 2000). The proposed calculation involves a number of approximations and is computationally intensive, but importantly, it associates "a systems' ability to exhibit emergent behaviour with sudden transitions in the Ω measure, relative to smooth changes in system parameters" (Wright et al., 2000).

There are other information-theoretic metrics that target complexity of a multi-agent system. For example, a rule-space of 1-dimensional cellular automata was characterised with the Shannon entropy of rules' frequency distribution (Wuensche, 1999). The input-entropy settles to fairly low values for ordered dynamics, but fluctuates irregularly within a narrow high band for chaotic dynamics. For the complex CA, the input-entropy generally settles onto a short attractor cycle, where order and chaos may predominate at different times causing the entropy to vary. A measure of the variability of the input-entropy curve is its variance or standard deviation, calculated over time. Wuensche has convincingly demonstrated that only complex dynamics exhibits high variance of input-entropy, leading to automatic classification of the rule-space. Importantly, the peak of input-entropy variance points to a phase transition as well, indicating the edge of chaos (complexity).

We would like to point out, however, that both the Ω measure and the input-entropy of the CA rule-space rely purely on local state variables that can be associated with each particle/cell, and do not directly capture *inter-agent* connections and interactions. In other words, the Ω measure estimates degrees of freedom in the dynamical system, rather than complexity of spatial inter-connections in a multi-agent network. Similarly, the input-entropy of the CA rule-space traces diversity of rules used over time and identifies temporally persistent configurations, rather than spatial connectivity among neighboring cells. If one needs to capture emergence and stability of a specific spatial pattern (eg., a continuously connected boundary, a maximally connected sub-graph, a desired geometric or topological shape, etc.), then another type of measure is needed.

It is well-known that graph connectivity can be analysed in terms of the size of the largest connected subgraph (LCS) and its standard deviation obtained across an ensemble of graphs, as suggested by the random graphs theory (Erdös and Renyi, 1961). In particular, critical changes occur in connectivity of a directed graph as the number of edges increases — the size of the LCS rapidly increases as well and fills most of the graph, while the variance in the size of the LCS reaches a maximum at some critical point before decreasing. Thus, a metric based on LCS variance may capture *spatial* aspects of desired emergent patterns (connectivity), while a metric based on entropy of local agent variables (rules, coordinates, velocities, etc.) may capture *temporal* persistence of emergent behaviour.

Importantly, the approach introduced in the work done by (Wright et al., 2000) suggests a feedback using the Ω measure in *evolving* the desired pattern: swarming behaviour as opposed to fully coordinated "crystalline" behaviour or totally uncoordinated dynamics of independent particles. A feedback to evolvable localised algorithms, based on a quantitative measure of global emergent behaviour, is central to our approach as well. Moreover, the identified distinction between spatial and temporal metrics allows us to implement this feedback systematically, focussing on those evolvable aspects that are important for a specific task (spatial connectivity and/or temporal persistence, size of emergent structures, etc.).

3. Impact Boundaries

This section presents a localised algorithm producing *impact boundaries* in the presence of connectivity disruptions and cell failures resulting from impacts. Typically, the damage on the AAV skin caused by a high energy impact is most severe at the point of impact (an epicentre). It will be assumed that not only the cells at the epicentre are destroyed, but the communication capability of the neighbouring cells may be reduced — e.g., the communication damage may propagate out with an exponential decay to a certain radius. In this case, the damage can be characterised by a probability P_i of an error corrupting a message bit i, dependent on proximity of the affected communication port to the epicentre:

$$P_i = \frac{1}{2}\,(1 - \frac{d}{R})^{\alpha},$$

where d is the distance between the involved communication port and the epicentre of the impact with the radius

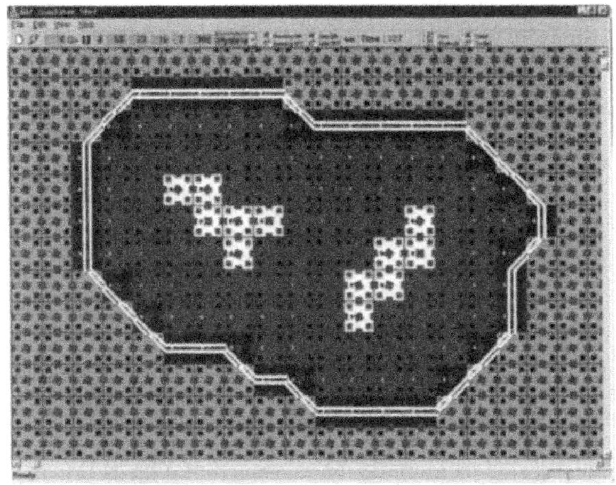

Figure 3: A stable impact boundary: white cells are destroyed, dark-grey cells form "scaffolding", black cells form "frame". Boundary links are shown as white double-lines.

R, and α is the exponential decay of the communication loss. Obviously, multiple impacts result in overlapping damaged regions, and the cumulative bit error probability can be approximated as

$$P_i = \frac{1}{2}\sum_{j=1}^{j=m}(1 - \frac{d_j}{R_j})^{\alpha},$$

where m is the number of impacts. The probability that the whole message containing n bits is corrupted depends on the employed error correction code. For example, the $(12, 8)$ Hamming error correction code (8 data and 4 parity check bits) corrects a single bit error per 12 bits, and leads to the following estimation

$$P_{failure} = \begin{cases} 0 & \text{if } P_i^n \leq r; \\ 1 - P_i^n & \text{if } P_i^n > r, \end{cases}$$

where $r = \frac{1}{12}$ is the error correction rate[1].

The impact boundaries are expected to enclose critically damaged regions and form continuously connected closed circuits (Lovatt et al., 2003). Moreover, impact boundaries must be robust to communication malfunctions caused by proximity to the impact — in other words, in the presence of unreliable communications between cells due to high probabilities $P_{failure}$, and without knowing the distances between each cell and the epicentre. Figure 3 illustrates a typical situation: the white cells are destroyed, normal cells (background colour) do not detect any damage, while cells shown with white double-lines self-organise into the impact boundary. Impact boundaries are intended to ensure a reliable communication pathway around the damaged region, and

[1]A computationally expensive Reed-Solomon error correction code would give a much better error correction rate than $\frac{1}{12}$ in the best case of consecutive "burst" errors.

also create an outline for subsequent self-repair, being inspired by a biological analogy — the clotting of a wound on mammalian skin. Not surprisingly, spatiotemporal stability in impact boundaries is an important concern: even at the periphery of a single impact region the probability $P_{failure}$ may be fairly high. Moreover, multiple impacts often introduce some asymmetry within cell neighbourhoods, so that a cell may be able to transmit but not receive data or vice versa (Lovatt et al., 2003).

3.1 Emergent Membranes and Recovery Mode

In this section, while describing the localised algorithm producing impact boundaries, we shall highlight the role of emergent *recovery membranes* that separate the boundaries from inner impact-surrounding regions, including (possibly recovering) cells that may communicate unreliably. We intend to show that a recovery membrane is an emergent structure, and its emergence is precisely the reason for stabilisation of the impact boundary. In doing so we shall also underscore the parameters used to evolve recovery membranes and stable impact boundaries.

First of all, we consider the following two communication behaviours potentially leading to emergence of an impact boundary, in the absence of sensory readings from cells destroyed by impacts:

(i) At the start of every cycle, every cell sends a *Ping* message to each of its neighbours.

(ii) If a boolean variable *EnableAcks* is true, a cell sends an Acknowledgment reply when it receives a Ping message.

The second behaviour is evolvable, while the first one is pre-determined. Two binary circular arrays are used to store the communication histories for ping messages (*pingArray*) and acknowledgments (*ackArray*) for each communication port. The size of these arrays is called the communication history length.

Every cycle, after receiving communication messages, each cell updates the following parameters:

(1) For each communication port, *PingFailure* will result if the percentage of lost Pings in the *pingArray* is greater than *PingFailThreshold.*

(2) For each communication port, *PingSuccess* will result if the percentage of Pings received in the *pingArray* is greater than or equal to *PingSuccessThreshold.*

(3) For each communication port, *AckSuccess* will result if the percentage of Acknowledgments received in the *AckArray* is greater than or equal to *AckSuccessThreshold.*

The *Failure* and *Success* parameters are hysteretic: they change values only when a sufficient communication history is accumulated. This lagging of an effect behind its cause provides a (temporary) resistance to change and ensures a degree of stability in the treatment of communication connections between any two cells:

(4) A neighbour is considered to be *communicating* when *EnableAcks* is false and *PingSuccess* is true, or both *EnableAcks* is true and *AckSuccess* is true.

(5) *Scaffolding* state S_s will result if there are no *communicating* neighbours.

(6) *Frame* boundary state S_f will result if *PingSuccess* is true for at least one communication port, and *PingFailure* is true for at least one communication port[2].

(7) *Closed* boundary state S_c will result if the cell state is S_f, and there are at least two *communicating* neighbours.

In order for a continuous impact boundary to emerge, the following two communication behaviours were considered:

(iii) if the cell state is S_c, 1) determine a cell α that failed to communicate; 2) determine two communicating neighbour cells β_1 and β_2 nearest to the cell α[3]; 3) map the directions to α, β_1 and β_2 to a direction γ; and 4) send a "Connect(τ, γ)" message to both β_1 and β_2 with a *time to live* parameter τ;

(iv) upon receiving "Connect(τ, γ)" message from a cell α, if the cell state is not S_c, 1) switch to the state S_c, 2) if $\tau > 0$, follow the steps 2) and 3) from (iii) producing new direction γ' and 3) send a "Connect($\tau - 1, \gamma'$)" message to neighbours β_1 and β_2.

The *time to live* τ is an evolvable parameter, and prevents spurious links from persisting. In general, the described policy achieves the desired robustness and continuity of self-organising impact boundaries for a variety of cell shapes (triangular and square) and communication damage probability distributions (Foreman et al., 2003). One particular evolvable behaviour is critical in achieving the desired stability:

(v) If a boolean variable *EnableShutdown* is true, the cell in the *Scaffolding* state S_s will stop transmitting messages.

This is needed in order to break asymmetry between neighboring cells, where a cell is able to transmit data but not receive, or vice versa — it is better in such cases

[2] This condition is similar to the condition of having at least one communicating neighbour and at least one mis-communicating neighbour, but does not assume Acknowledgement messages at all.

[3] When cell δ determines its own neighbour cells nearest to some other neighbour cell α, it chooses the cells β_1 and β_2 on opposite sides (clockwise and counter-clockwise), relative to α.

not to communicate at all. This behaviour is effective in achieving stable impact boundaries. However, the cells that stopped transmitting messages may need to resume communications under certain conditions — for example, when a repair action is initiated, and their neighbours are again ready to receive communications (i.e., when the cause of asymmetry is eliminated). The conditions for resumption of communications have to be precise so that they are not reacted upon prematurely, interfering with boundary formation. A variant of these *recovery* conditions is given below:

(vi) Recovery state S_r will result if all ports have *pingSuccess* for a number of consecutive cycles, where this number is set by *RecoveryThreshold*.

(vii) Recovery state S_r will result if all ports have *pingFailure* for a number of consecutive cycles, where this number is set by *RecoveryThreshold*.

(viii) A cell stays in the Recovery state S_r and may send communication messages during the next *RecoveryInertia* cycles.

The behaviour (vii) enables totally isolated cells to be in a ready recovery state as well — a feature that, understandably, may be needed when a repair is in progress and cells are being transported across a panel. It is precisely the combination of the "shutdown" and "recovery" conditions that makes the problem of designing (evolving) stable impact boundaries particularly difficult. On one hand, "shutdown" condition (v) disables "asymmetric" messages, making the task of boundary formation easier. On the other hand, conditions (vi)-(viii) may enable "asymmetric" messages, potentially destabilising the boundary.

The solution to this problem is provided by emergent spatiotemporal structures — recovery membranes — that separate the boundaries from recovering cells. A recovery membrane always forms on the inside of the closed boundary, and on the outside of the recovering area. Interestingly, unlike scaffolding and frame boundary, the membrane is not a designated state into which a cell can switch. Membrane cells shut down their communications, following behaviour (v) like other scaffolding cells, but do not resume communications because behaviours (vi)—(viii) are not applicable, as there are some (but not all) mis-communicating neighbours. Without a membrane, the cells on the frame boundary would be confused by intermittent messages from scaffolding cells attempting recovery. Figure 4 illustrates a checkered-pattern recovery membrane shown with dark-grey colour, while the recovering cells are shown in darker shade of white. It is quite obvious that the membrane cells separate the recovering scaffolding cells from the frame boundary.

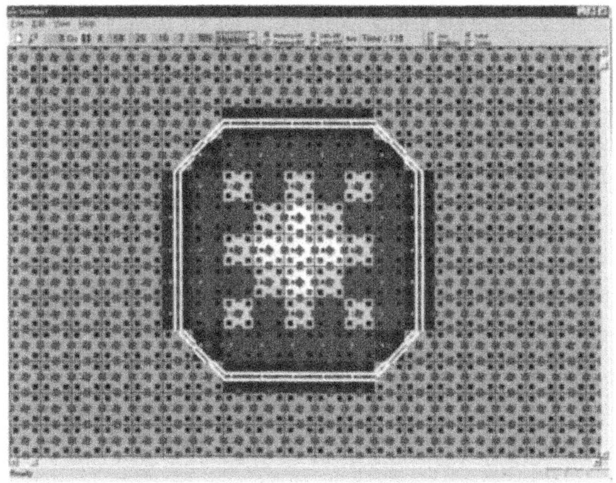

Figure 4: Five white cells at the epicentre are destroyed. A recovery membrane shown in dark-grey "absorbs" scaffolding cells that attempt recovery, shown in darker shade of white, and separates them from the frame, shown in black.

4. Evolving Recovery Membranes

In this section, we describe a Genetic Algorithm (GA) aimed at evolving the parameters used in impact boundary formation. In particular, we examine the interrelationships between communication "shutdown" and "recovery" conditions that give rise to recovery membranes needed for stable impact boundaries. Some of the evolving parameters designate conditionalised branches of the impact boundary algorithm, and some represent various thresholds, so the search space is significantly large. It is well-known that a genetic algorithm, as a problem-solving tool based on biological evolution, works on improving a solution via a search through a process of selection, recombination (crossover) and mutation, and is particularly useful when the search space has many local optima or is too large to use conventional techniques.

4.1 Encoding

We followed a traditional GA encoding: binary strings encoding the chromosome as a structure containing the collection of parameters (genes) and representing a behavioural trait of the individual. Our chromosome contains the following 42 bits:

History, ρ (5 bits) — number of cycles for which a cell remembers the received Pings and Acknowledgements.

PingFailThreshold (7 bits) — a percentage representing the number of Ping messages lost in the last History cycles including the current cycle.

PingSuccessThreshold (7 bits) — a percentage representing the number of Ping messages received in the last *History* cycles including the current cycle.

AckSuccessThreshold (7 bits) — a percentage representing the number of Acknowledgements messages received in the last *History* cycles including the current cycle.

RecoveryThreshold, π (5 bits) — number of cycles the recovery condition must hold for communication recovery to start.

RecoveryInertia (6 bits) — number of cycles in which a cell continues to communicate with its neighbours while recovering.

Time to live, τ (3 bits) — number of times a Connect message is sent before being discarded.

EnableShutdown (1 bit) — a Boolean variable indicating whether a cell shutdown functionality is enabled.

EnableAcks (1 bit) — a Boolean variable indicating whether a cell has the ability to send and receive Acknowledgements messages.

4.2 Fitness/Objective Function

The evolution of recovery membranes is based on spatiotemporal metrics incorporated within a fitness (objective) function. The analysis presented by (Foreman et al., 2003) used two metrics to characterise stability of emergent impact boundaries: spatial and temporal.

The spatial metric is based on the variance in the size of the connected boundary-fragment (CBF). A CBF is simply a set F of cells in the closed state S_c such that every cell in F is connected with at least one other cell in F, and there exists no cell outside F, which is connected to at least one cell in F (an analogue of a maximally connected sub graph or a graph component). We calculate the maximum size $H_{sp}(t)$ of CBF's in self-organising impact boundaries at each cycle. Its variance σ^2_{sp} over time is then used as a spatial metric within the objective function. This metric, as mentioned before, is inspired by random graphs theory and is intended to capture *spatial connectivity* in impact boundaries. A continuous boundary may, however, change its shape over time, without breaking into fragments, while keeping the size of CBF almost constant. Therefore, a temporal metric that measures the diversity of cell transitions (analogously to cellular automata input-entropy or the Ω measure used to characterise flocking behaviour) may be complementary.

In order to analyse *temporal persistence*, we consider state changes in each cell at every time step. Given 6 symmetric boundary links possible in each square cell ("left-right", "top-bottom", "left-top", etc.), there are 2^6 possible boundary states (including "no-boundary"), and $m = 2^{12}$ transitions. The entropy $H_{temp}(t)$ of a particular frequency distribution $S_i(t)$, where t is a time

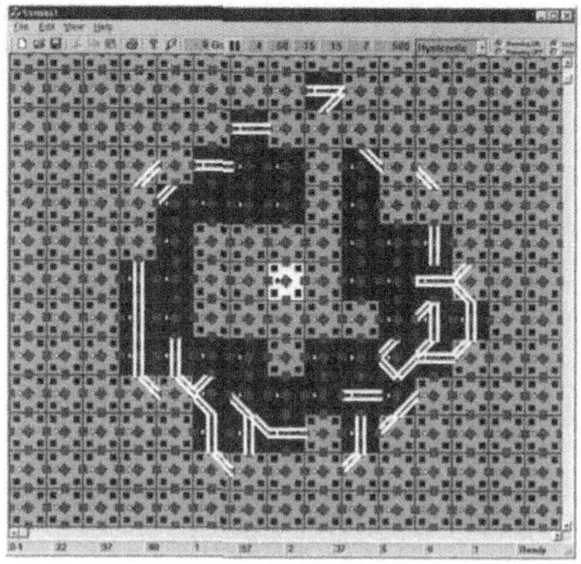

Figure 5: A chaotic boundary with $\overline{H_{sp}} \leq 16$ and zero-length communication ρ. A membrane does not form at all. Both σ^2_{sp} and σ^2_{temp} are low-to-medium.

step, and i is a cell transition index: $1 \leq i \leq m$, can be calculated as follows:

$$H_{temp}(t) = -\sum_{i=1}^{m} \frac{S_i(t)}{n} \, log \frac{S_i(t)}{n} ,$$

where n is the total number of cells, and $S_i(t)$ is the number of times the transition i was used at time t across all cells. Again, the variance σ^2_{temp} of the entropy $H_{temp}(t)$ over time is used as a temporal metric within the objective function.

Our task is complicated by the fact that emergent structures are characterised by a *phase transition* detectable by either σ^2_{sp} or σ^2_{temp}, rather than a particular value range. Therefore, simply rewarding low values for these entropy-based metrics would be insufficient. In particular, it has been observed (Foreman et al., 2003) that both metrics are low-to-medium for algorithms with zero-length communication ρ (tropistic algorithms and chaotic regimes — Figure 5), increase dramatically for ρ in the range $1 \leq \rho \leq \rho_0$, where ρ_0 is a critical value at and below which complex unstable behaviours occur (Figure 6), and undergo a phase transition to very low values when $\rho > \rho_0$ (hysteretic algorithms and ordered regimes).

The critical value ρ_0 is, of course, dependent on all other parameters used by the algorithm. Nevertheless, the chaotic regimes, which are more stable simply due to a small number of connections, can often be identified by a low average $\overline{H_{sp}}$ of the maximum sizes $H_{sp}(t)$ of CBF's in impact boundaries, ruling out at least zero-length histories. In particular, impact boundaries with the average $\overline{H_{sp}} \leq 16$ can be safely ruled out — the

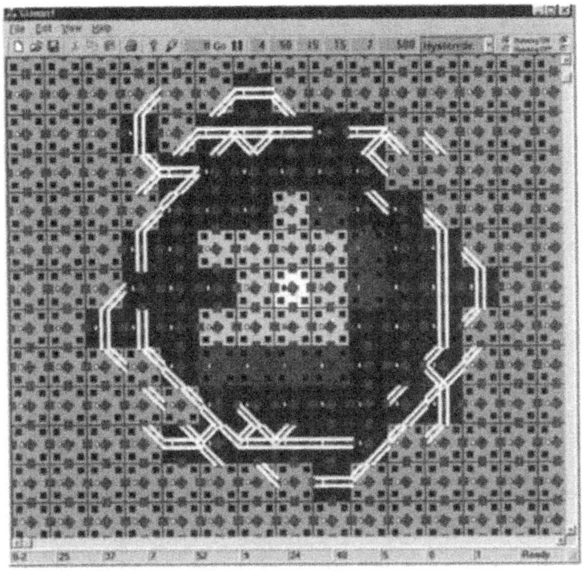

Figure 6: An unstable boundary with ρ close to its critical value. The membrane is fragmentary. Both σ^2_{sp} and σ^2_{temp} are close to their peaks: the phase transition.

resulting chaotic patterns, illustrated in Figure 5, are of no interest.

On the other hand, a preference among ordered regimes towards shorter histories is another useful identifier of a phase transition and the critical value ρ_0. Besides, a shorter communication history ρ enables a quicker response, as do lower values of τ and π. Thus, our first experiment used minimisation of the following objective function:

$$f_{sp}(\beta) = \begin{cases} M & \text{if } \overline{H_{sp}} \leq 16; \\ 4.0\ \sigma^2_{sp} + \rho + \tau + \pi + \beta\ \overline{H_{sp}} & \text{if } \overline{H_{sp}} > 16, \end{cases}$$

where M is the maximal integer value provided by the compiler. The coefficient β reflects the relative importance of the length of impact boundaries in the objective function — sometimes, it may be as important to obtain smallest possible impact perimeter as it is to maintain a shortest possible communication history. We alternated between $\beta_1 = 0.25$ and $\beta_2 = 2.0$.

The second experiment focussed on temporal metric σ^2_{temp} embedded in the objective function $f_{temp}(\beta)$ constructed analogously to the function $f_{sp}(\beta)$:

$$f_{temp}(\beta) = \begin{cases} M & \text{if } \overline{H_{sp}} \leq 16; \\ 10^5\ \sigma^2_{temp} + \rho + \tau + \pi + & \\ \qquad + \beta\ \overline{H_{sp}} & \text{if } \overline{H_{sp}} > 16. \end{cases}$$

Finally, our ultimate objective function is defined as follows:

$$f(\beta) = \begin{cases} M & \text{if } \overline{H_{sp}} \leq 16; \\ \frac{1}{2}\ (4.0\ \sigma^2_{sp} + 10^5\ \sigma^2_{temp}) + & \\ \qquad + \rho + \tau + \pi + \beta\ \overline{H_{sp}} & \text{if } \overline{H_{sp}} > 16 \end{cases}$$

Each experiment involves an impact at a predefined cell, and lasts 500 cycles; the first 30 cycles are excluded from the series $H_{sp}(t)$ and $H_{temp}(t)$ in order not to penalise longer history lengths ρ. We repeat the experiment 3 times for every chromosome and average the objective (fitness) values obtained over these runs.

4.3 Selection

We have chosen a *generation gap* replacement strategy hoping to use better search capabilities offered by the generational replacement and faster convergence typically provided by the steady state selection. Some of this faster convergence, however, may be explained by the stochastic nature of the selection operator: the rate of genetic drift in steady state selection is twice that of generational selection (Rogers and Prügel-Bennett, 1999). In our experiments, we set the generation gap parameter $G = 0.2$. In other words, the entire old population (40 chromosomes) is sorted according to fitness, and we choose the best 20% for direct replication in the next generation, employing an elitist selection mechanism. The selection phase and recombination (crossover) phase can be merged (Thierens and Goldberg, 1994). We follow a similar approach but still explicitly keep the selection phase replicating the elitist offspring. The rest of selection functionality is moved into the crossover.

4.4 Crossover and Mutation

We used a variation of the n-point crossover, where the probability of having n points in the crossover depends on the fitness of the chromosome. We choose this since there are benefits of having a low and high n-point crossover, enabling a better exploration in the search space. Our variation involves three different crossovers, each having an equal probability of contributing to the generation of new chromosomes. In other words, after the elite takes 20% of the new population, for each remaining place we randomly perform a crossover chosen among the following three:

Elitist driven: parent 1 is randomly chosen from the best performing 20% of the old population and parent 2 is randomly chosen from the entire old population, followed by a low 1- to 2-point crossover (the number of points is determined randomly). This low-point crossover is used here since it makes sense to disturb an elitist-driven solution as little as possible.

Mid-range: parent 1 is randomly chosen from the next 20% - 50% of the old population and parent 2 is randomly chosen from the entire old population, followed by a medium 1- to 4-point point crossover (the number of points is determined randomly as well). This medium-point crossover is more applicable when it makes sense to disturb a mediocre solution.

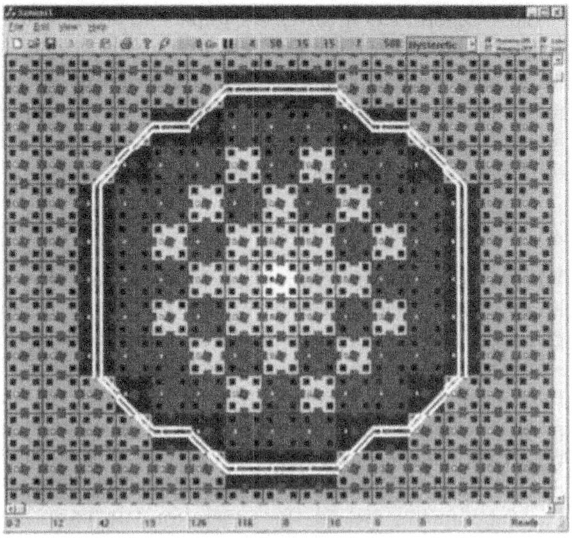

Figure 7: A large checkered-pattern membrane, with short hysteresis, within a morphing but closed and continuous boundary ($\beta = 0.25$). Both σ^2_{sp} and σ^2_{temp} are low.

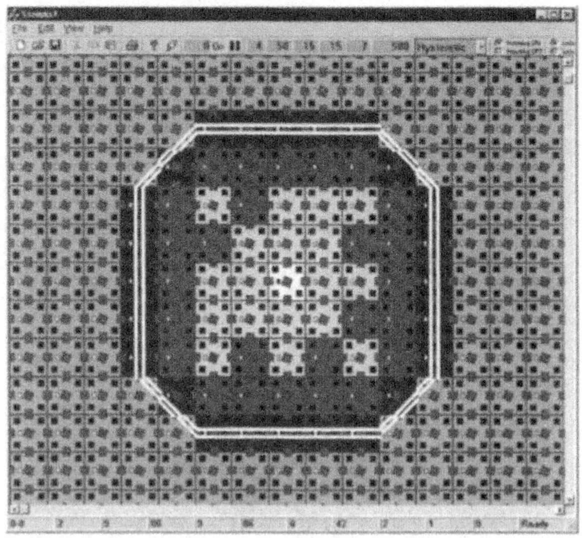

Figure 8: A small membrane, with long hysteresis, within a regular octagonal boundary ($\beta = 2.0$). Both σ^2_{sp} and σ^2_{temp} are very low.

Remainder: parent 1 is randomly chosen from the worst performing 50% of the population and parent 2 is randomly chosen from the entire old population, followed by a high 1- to 7-point crossover, potentially disrupting an under-fit solution a lot in order to enter a new search region.

A feature of this multi-point crossover is that the first gene is always inherited from parent 1. We also ensure that offspring generated from the crossover is unique. We use a slightly higher than typical mutation rate: each bit in the chromosome has a 0.04 probability of being flipped. We also ensure that the mutation results in a unique chromosome by repeating mutation if the produced chromosome already exists in the new population. Mutation is not performed on the chromosome generated via elitist selection.

5. Experimental Results

Our first experiment minimising the objective function $f_{sp}(0.25)$ was mainly concerned with spatially connected and stable impact boundaries, and the form taken by corresponding recovery membranes. The length of the boundary was of lesser importance. Not surprisingly, the evolved solution achieved long robust and continuous impact boundaries with $\overline{H_{sp}} = 40$ (Figure 7), around large impact-surrounding regions, while requiring fairly short hysteresis: $\rho = 2$ and $\pi = 5$. The stabilisation of an impact boundary around a large region occurs at the periphery of the communication damage, where the probability $P_{failure}$ falls to 0 due to the error correction code, and the process has a cascading nature, where the boundary expands to eventually cover all the impact-surrounding region. Interestingly, the evolved continuous boundaries may change their shape and only rarely stabilise as a regular octagon, while keeping their length $\overline{H_{sp}} = 40$ constant. The emergent recovery membrane evolved to separate boundary from the recovering cells. It has a checkered pattern that can be explained by the opposing nature of the conditions (vi) and (vii), and the short hysteresis enabling oscillations between recovery and scaffolding states.

On the other hand, minimisation of $f_{sp}(2.0)$ resulted in more compact impact-surrounding regions ($\overline{H_{sp}} = 32$, Figure 8) and thinner membranes, at the expense of longer hysteresis: $\rho = 6$ and $\pi = 4$. These boundaries morph as well, but generally keep the shape of a regular octagon. Interestingly, the second experiment produced a typical *speciation*, where the longer hysteresis solutions took only one niche, while shorter hysteresis $\rho = 2$ and $\pi = 4$ solutions co-evolved into a separate niche, both niches evolving compact regions with boundaries $\overline{H_{sp}} = 32$. This supports our conjecture that the employed generational gap selection with $G = 0.2$ counteracts the genetic drift to a reasonable degree.

Both solutions favoured $\tau = 1$ as expected for square cells (while triangular cells require at least $\tau = 2$ to achieve continuity). Also, the evolved cells prefer to send Acknowledgements (*EnableAcks* = 1) and stop transmitting in scaffolding state (*EnableShutdown* = 1). Without the latter feature membranes would not emerge, and the recovering cells would disrupt the boundaries.

The second pair of experiments focussed on evolving temporally stable boundaries, minimising $f_{temp}(0.25)$ and $f_{temp}(2.0)$. These experiments produced results very similar to the ones obtained by minimising the spatial

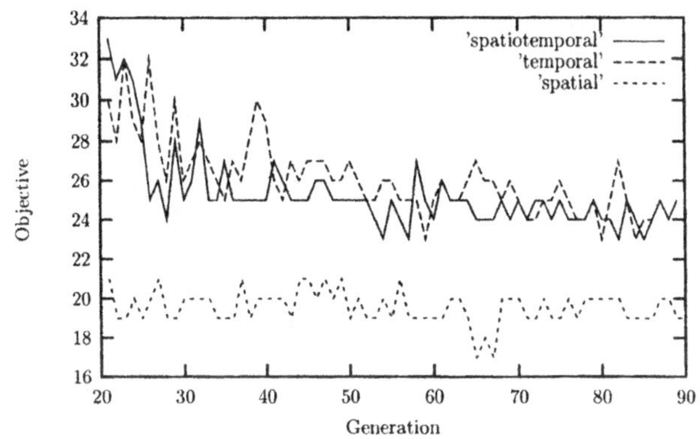

Figure 9: The functions $f_{sp}(0.25)$, $f_{temp}(0.25)$ and $f(0.25)$.

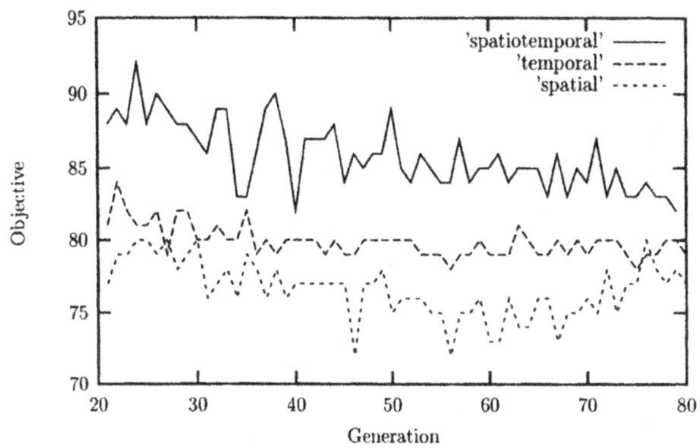

Figure 10: The functions $f_{sp}(2.0)$, $f_{temp}(2.0)$ and $f(2.0)$.

metric. In particular, two types of boundaries evolved, one with a shorter hysteresis ($\rho = 2$ and $\pi = 8$) and longer boundaries $\overline{H_{sp}} = 40$, and the other with a longer hysteresis ($\rho = 6$ and $\pi = 4$) and shorter boundaries $\overline{H_{sp}} = 32$. Both solutions involve membranes, the main difference being that even the shorter hysteresis solutions with longer boundaries tend to reach and retain the regular octagon shape. However, sometimes the boundary fragments. These outcomes are expected for an objective function rewarding temporal stability.

The final pair of experiments combined the spatial and temporal metric, minimising $f(0.25)$ and $f(2.0)$. Again, both types: ($\overline{H_{sp}} = 40$, $\rho = 2$, $\pi = 8$) and ($\overline{H_{sp}} = 32$, $\rho = 6$, $\pi = 8$) were produced. As expected, the evolved membranes and boundaries were more stable, and had a regular octagon shape in both cases, mostly without morphing or fragmenting.

As mentioned above, the case $\beta = 0.25$ results in longer boundaries that are capable of morphing without breaking into fragments. The objective functions $f_{sp}(0.25)$, $f_{temp}(0.25)$ and $f(0.25)$ for the most fit individual in each generation are plotted in Figure 9. All plots exclude the initial period (20 generations) of the rapid decrease typical for GA-based exploration of the search-space. The function $f_{sp}(0.25)$ converges well but does not explore the search-space considerably. This is so because it uses the spatial metric and, therefore, rewards the boundary's continuity, ignoring morphing instability — which is "allowed" in this case. On the contrary, the function $f_{temp}(0.25)$ attempts to minimise morphing instability and has to explore a large part of the space.

The objective functions $f_{sp}(2.0)$, $f_{temp}(2.0)$ and $f(2.0)$ are plotted in Figure 10. The case $\beta = 2.0$ results in shorter boundaries that cannot morph without breaking into fragments, so any instability leads to fragmentation. Consequently, the function $f_{temp}(2.0)$ shows the convergence within a narrower band — it uses the temporal metric and rewards persistence, ignoring occasional fragmentations. Its counterpart, the function $f_{sp}(2.0)$, specifically puts selection pressure on continuity, leading to a wider exploration and poorer convergence.

Importantly, the functions $f(0.25)$ and $f(2.0)$, using spatiotemporal metrics, provided a good compromise in both cases. The experiments confirmed that the choice of an objective function depends on the main task: when the target is not only continuity of impact boundaries but also their shape, then the spatiotemporal metric is more suitable. On the other hand, if morphing is acceptable, then a spatial metric capturing connectivity rather than shape is sufficient. Similarly, a temporal metric may be better suited for a general shape design if sporadic fragmentations are tolerable.

6. Conclusion

In this paper we suggested and verified a methodology underlying design of localised algorithms for complex multi-agent systems, exemplified by self-monitoring aerospace vehicles. We started by briefly describing a multi-agent algorithm leading to emergence of impact boundaries and recovery membranes, followed by quantitative measurement of spatiotemporal stability of the emergent patterns. The graph-theoretic and information-theoretic metrics, capable of identifying phase transitions, contributed to fitness functions for evolutionary modelling of boundaries and membranes. The produced results are promising and demonstrate the possibility for a multi-objective design of localised algorithms. In particular, the desired response time as well as size (and potentially, shape) of impact boundaries and membranes may be specified in advance, leaving the precise logic and parameterisation of the localised algorithms to selection pressures. We believe that the proposed methodology is well suited to the *design at the edge of chaos*, where the design objective (e.g., a specific shape) may be unstable, while other parameters (e.g. the response time) may be optimal. The presented methods should increase the reliability of the design of complex multi-agent systems, accounting for emergent patterns that are not easily predictable by human designers.

Acknowledgements This work was carried under CSIRO contract to NASA. It is a pleasure to record our appreciation to Dr Ed Generazio (NASA Langley Research Center) for his encouragement and support. The authors are grateful to other members of the AAV project for many open and motivating discussions, especially Tony Farmer, Mark Foreman, Mark Hedley, Geoff James, Mark Johnson, Chris Lewis, Geoff Poulton, Don Price, Andrew Scott, Sarath Seneviratne, and Philip Valencia. The authors are also grateful to all members of $GREM^{Lab}$ (a collaborative CSIRO laboratory on Global Response Engineering in Multi-Agent Networks) and the discussion group on "Entropy and self-organisation in multi-agent systems" in CSIRO.

References

Abbott, D., Batten, A., Carpenter, D., Dunlop, J., Edwards, G., Farmer, A., Gaffney, B., Hedley, M., Hoschke, N., Isaacs, P., Johnson, M., Lewis, C., Murdoch, A., Poulton, G., Price, D., Prokopenko, M., Rees, D., Scott, A., Seneviratne, S., Valencia, P., Wang, P., and Whitnall, D. (2003). Development and evaluation of sensor concepts for ageless aerospace vehicles (4): Phase 1 — implementation of the concept demonstrator. In *CSIRO CTIP Report No TIPP 1898*. September 2003.

Abbott, D., Doyle, B., Dunlop, J., Farmer, A., Hedley, M., Herrmann, J., James, G., Johnson, M., Joshi, B., Poulton, G., Price, D., Prokopenko, M., Reda, T., Rees, D., Scott, A., Valencia, P., Ward, D., and Winter, J. (2002). Development and evaluation of sensor concepts for ageless aerospace vehicles. development of concepts for an intelligent sensing system. In *NASA technical report NASA/CR-2002-211773*. Langley Research Center, Hampton, Virginia.

Durbeck, L. and Macias, N. (2002). Defect-tolerant, fine-grained parallel testing of a cell matrix. In Schewel, J., James-Roxby, P., Schmit, H., and McHenry, J., (Eds.), *Proceedings of SPIE ITCom 2002 Series, Vol. 4867*.

Erdös, P. and Renyi, A. (1961). On the strength of connectedness of random graphs. *Acta Mathematica Scientia Hungary*, 12:261–267.

Estrin, D., Govindan, R., Heidemann, J., and Kumar, S. (1999). Next century challenges: Scalable coordination in sensor networks. In *Proceedings of the Fifth Annual ACM/IEEE International Conference on Mobile Computing and Networks*, pages 263–270. ACM Press.

Foreman, M., Prokopenko, M., and Wang, P. (2003). Phase transitions in self-organising sensor networks. In Banzhaf, W., Christaller, T., Dittrich, P., KIm, J., and Ziegler, J., (Eds.), *Advances in Artificial Life - Proceedings of the 7th European Conference on Artificial Life (ECAL)*, volume 2801 of *Lecture Notes in Artificial Intelligence*. Springer Verlag.

Lovatt, H., Poulton, G., Price, D., Prokopenko, M., Valencia, P., and Wang, P. (2003). Self-organising impact boundaries in ageless aerospace vehicles. In Rosenschein, J., Sandholm, T., Wooldridge, M., and Yokoo, M., (Eds.), *Proceedings of the 2nd International Joint Conference on Autonomous Agents and Multi-Agent Systems*, pages 249–256. ACM Press.

Macias, N. and Durbeck, L. (2002). Self-assembling circuits with autonomous fault handling. In Stoica, A., Lohn, J., Katz, R., Keymeulen, D., and Zebulum, R., (Eds.), *Proceedings of NASA/DoD Conference on Evolvable Hardware*. IEEE Computer Society Press.

Prokopenko, M., Wang, P., Foreman, M., Valencia, P., Price, D., and Poulton, G. (2004). On connectivity of reconfigurable impact networks in ageless aerospace vehicles. *Journal of Robotics and Autonomous Systems*, Special Issue:in press.

Rogers, A. and Prügel-Bennett, A. (1999). Genetic drift in genetic algorithm selection schemes. *IEEE Transactions on Evolutionary Computation*, 3(4):298–303.

Thierens, D. and Goldberg, D. (1994). Elitist recombination: an integrated selection recombination ga. In Michalewicz, Z., Schaffer, J., Schwefel, H.-P., Fogel, D., and H., K., (Eds.), *Proceedings of the First IEEE Conference on Evolutionary Computation*, pages 508–512. IEEE Press.

Wang, P., Valencia, P., Prokopenko, M., Price, D., and Poulton, G. (2003). Self-reconfigurable sensor networks in ageless aerospace vehicles. In Nunes, U., de Almeida, A., Bejczy, A., Kosuge, K., and Machado, J., (Eds.), *IEEE Proceedings of the Eleventh International Conference on Advanced Robotics, ICAR 2003*, pages 1098–1103. Coimbra, Portugal.

Wright, W., Smith, R., Danek, M., and Greenway, P. (2000). A measure of emergence in an adapting, multi-agent context. In Meyer, J., Berthoz, A., Floreano, D., Roitblat, H., and Wilson, S., (Eds.), *Proceedings of the Sixth International Conference on the Simulation of Adaptive Behaviour, SAB 2000*, pages 20–27. ISAB Press.

Wuensche, A. (1999). Classifying cellular automata automatically: Finding gliders, filtering, and relating space-time patterns, attractor basins, and the z parameter. *Complexity*, 4(3):47–66.

Learning to Behave: Adaptive Behavior for Planetary Surface Rovers

Terry Huntsberger
Jet Propulsion Laboratory
MS 82-105, 4800 Oak Grove Drive
Pasadena, CA 91109
Terry.Huntsberger@jpl.nasa.gov

Hrand Aghazarian
Jet Propulsion Laboratory
MS 82-105, 4800 Oak Grove Drive
Pasadena, CA 91109
Hrand.Aghazarian@jpl.nasa.gov

Abstract

Robotic missions to planetary surfaces are becoming more ambitious and of longer duration. The nominal mission timeline for the MER (Mars Exploration Rovers) called Spirit and Opportunity currently on the Martian surface is 90 days, with extensions to 180 days and beyond depending on rover health. The upcoming 2009 MSL (Mars Science Laboratory) mission is planned to be 300-500 days, and will possibly involve traverses on the order of multiple kilometers. Due to time delays of up to 40 minutes round-trip for control, the rovers will require a high degree of onboard autonomous behavior that must also adapt to declining health and unknown environmental conditions during a long duration mission. This paper presents an algorithm for onboard adaptive learning of weights within a rover hierarchical behavior control framework called SMART (System for Mobility and Access to Rough Terrain). SMART is based on earlier work in free flow behavior hierarchies for planetary surface rovers (Huntsberger & Rose, 1998; Huntsberger, 2001). We also present the results of some preliminary laboratory and field studies.

1. Introduction

High-value science data acquisition on rough terrain (example shown in Figure 1(a)) is beyond the capabilities of current NASA rover designs. Although the JPL technology prototype rover SRR (Sample Return Rover) shown in Figure 1(b) has the ability to mechanically adapt itself to changing terrain by varying its shoulder angles, such an operation will require a high level of adaptability in the onboard control algorithms in order to maintain the health of the rover. In addition, as the mission progresses, the onboard control must also adapt to degraded performance due to wear-and-tear on components such as the steering and drive mechanisms.

(a) (b)

Figure 1: Planetary surface terrain and technology example for autonomous access to high risk, scientifically interesting regions. (a) Mars cliff-face with signs of water outflows; (b) JPL technology prototype of a terrain-adaptive reconfigurable rover.

We have developed a behavior-based framework called SMART (System for Mobility and Access to Rough Terrain) to address these concerns at the system level by treating rover motion, rover health, and resource management within a free flow behavior hierarchy (Huntsberger & Rose, 1998; Huntsberger, 2001). SMART uses a previously developed control architecture called BISMARC (Biologically Inspired System for Map-based Autonomous Rover Control) for long duration missions (Huntsberger and Rose, 1998; Huntsberger, 2001). It is based on a modified free-flow hierarchy (FFH) similar to the DAMN architecture (Rosenblatt and Payton, 1989; Tyrrell, 1993), and has been used successfully for a number of different simulated mission scenarios including multiple cache retrieval (Huntsberger, 1997), fault tolerance for long duration missions (Huntsberger, 1998), and site preparation (Huntsberger, *et al.*, 1999).

The major limitation in the original implementation of BISMARC in all of our previous studies was the use of fixed weights in the FFH, which effectively made it unable to adapt to situations outside of the original world model. This paper presents an onboard mechanism for learning weights that will adapt not only to the dynamic environment around the rover, but also to the degradation of mechanical components during the mission lifetime. Underlying behaviors and the organization of the FFH are predefined based on mission requirements and rover capabilities. Our goal in this research is not to find the optimal policy, but instead one that is "good enough" to maintain rover health while still achieving its higher levels goals.

The next section briefly describes the organization of the action selection mechanism of BISMARC, followed by a discussion of the learning mechanism of the system. Next, we discuss related work, and close with experimental studies and conclusions.

2. BISMARC Organization

An example of the action selection mechanism used in BISMARC is shown in Figure 2 for a rough terrain navigation mission that is used for the experimental studies reported in this paper. The rectangular boxes represent behaviors and the ovals are sensory inputs (either fixed, direct, or derived). At the top are the high level behaviors including *Don't Tip Over, Go to Goal, Avoid Obstacles, Preserve Motors, Warm Up, Get Power,* and *Sleep at Night.* These goals are related to both task and rover safety. For example, since most planetary surface rovers have only visual sensors for navigation, the sensory input for *Proximity to Night* is derived from knowledge of the sun's position and forces the rover to sleep at night by weighting the input to *Sleep at Night* heavier (4.0) than any other behavior in the hierarchy. The *Avoid Obstacles* behavior uses the output of an onboard local navigation algorithm as recommendations for viable paths. The rovers are equipped with solar panels and the *Rest* behavior allows the batteries to recharge if the sun is up. The *Rest* behavior is also used to cool down the motors for *Preserve Motors* if they are working too hard going up a steep slope, or to stop and turn on the heaters for *Warm Up* if the internal temperature of the rover drops below a safety threshold.

The intermediate level *Change CG* behavior is an example of a sophisticated combination behavior discussed in Section 1 that works to shift the center of gravity of the rover (see Figure 1(b)) much like an animal does in response to traveling up or along a steep slope. This behavior is implemented using a finite state machine based on a well-tested algorithm for pose reconfiguration (Schenker, *et al.*, 2003a-b). The algorithm uses the onboard gyroscopes and accelerometers, which would be equivalent to the inner ear mechanism in mammals for roll and pitch determination. Recommendations for shoulder angle and arm end position changes to help stabilize the rover are generated and passed on to the bottom level behaviors.

The intermediate level behaviors are designed to interact with both the short term memory (STM), which corresponds to perceived sensory stimuli, and the long term memory (LTM), which encodes remembered sensory information. Control loops are prevented through temporal penalties (shown as T-ovals in Figure 2) that constrain the system to only repeat a behavior a predetermined number of times. The bottom level behaviors in the hierarchy fuse the sensory inputs and the activations of the higher level

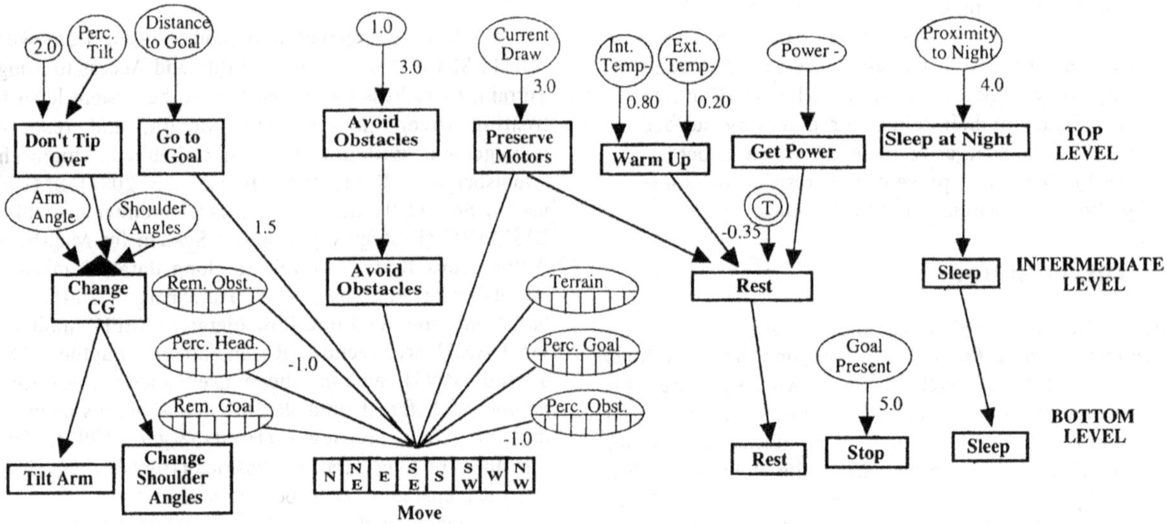

Figure 2: Free-flow hierarchy action selection mechanism for rough terrain navigation mission scenario. Ovals represent inputs derived from sensory stimuli, rectangular boxes are behaviors, and double ovals are temporal penalties. All weights on inputs to behaviors are 1.0 unless otherwise noted. Segmented boxes and ovals represent directional inputs (only cardinal directions shown but in practice continuous coverage). See text for further details.

behaviors in order to select appropriate actions for rover safety and goal achievement. The rover will continue to move until it achieves the goal position as determined by a rover localization algorithm (Hoffman, *et al.*, 1998) shown as the *Goal Present* input to *Stop* in Figure 2, or its health deteriorates due to dead batteries, freezing, burned out motors, or tipping over.

SMART's map-based LTM (Long Term Memory) is similar to hippocampus place cells. Landmarks corresponding to obstacles and goals are extensively mapped and stored for comparison to perceived inputs, with a probabilistic update of memories based on the positional variance of the rover and the match strength of the current perception to memory contents. A LTM landmark is encoded as a four-byte field that includes relative height of the landmark (2 bytes), actions leading to the landmark (1 byte), and accelerometer readings on the robot (1 byte). A similar approach is the coupled goal/representation framework of (Mataric, 1992; Mataric, 1997a). Another alternate approach is an occupancy grid that gives dense coverage of the environment, but doesn't scale well for long duration planetary surface missions (Elfes, 1987).

3. Learning Mechanism

Learning mechanisms for planetary surface rovers have the same requirements as terrestrial robots (Mahadevan & Connell, 1992): (1) noise immunity, (2) fast convergence, (3) incrementality (improving performance while learning), (4) tractability (iterations of algorithm doable in real-time), and (5) groundedness (information limited to onboard sensors). In particular, the fast convergence and tractability requirements are key for planetary surface rovers because they are typically computationally challenged (i.e., MER uses a 27Mhz CPU) due to power constraints. We address (2) and (4) through a behavior decomposition process similar to the use of *heterogeneous reward functions* developed by Mataric (Mataric, 1997b). We give the details of the reward function for updating the weights for the *Move* behavior (see Figure 2) in this section. For point (3) we use the W-learning algorithm of Humphrys (Humphrys, 1997) supplemented with a dynamic reward function directly related to rover health. For (1) we use a sequence memory similar to that of McCallum (McCallum, 1995-96) and Michaud and Mataric (Michaud & Mataric, 1998-99). Finally, we restrict our inputs to onboard sensors only as stipulated in point (5).

The weights on the links between modules are usually heuristically set based on mission goals. These goals are specified at a relatively high level without complete knowledge of the operating environment of the rover. There is however a priority derived from mission risk mitigation requirements explicitly included in the relative size of the weights. The maximum activation of the high level behaviors are weighted to give the highest priorities to rover preservation. In order of highest priority to lowest these are *Sleep at Night*, *Avoid Obstacles*, *Preserve Motors*, *Don't Tip Over*, *Get Power*, and *Warm Up*. In addition, rover health will degrade as the mission progresses, and weights chosen at full health may no longer be appropriate. Rover health is defined in Equation (1) as:

$$\text{rover\_health} = \left[w_p \text{power} + w_{mc}(1 - \text{motor\_current})\right]\left[\frac{(\text{AGE\_MAX} - \text{age})}{\text{AGE\_MAX}}\right] \tag{1}$$
$$w_p + w_{mc} = 1$$

where *power* is the current battery levels, *motor_current* is the current draw on the motors, *AGE_MAX* is the maximum expected lifetime for the rover, *age* is the current age of the rover, and w_p and w_{mc} are weights (currently both set to 0.5 since dead batteries are as lethal as burned-out drive motors). A dynamic reward function is defined in SMART based on changes in rover health and progress towards goal achievement for each step:

$$\text{reward} = w_{rh}\Delta\text{rover\_health} + w_{ga}\Delta\text{goal\_achievement} \tag{2}$$
$$w_{rh} + w_{ga} = 1$$

where Δ is the change, and w_{rh} and w_{ga} are weights (currently set to 0.65 and 0.35 based on the relative importance of health and goal achievement determined experimentally).

Learning is only enabled in the weights on the links feeding into the *Move* behavior at the lowest level in the FFH shown in Figure 2. This is done in order to maintain the rover safety embodied in the relatively high priorities of the *Sleep at Night*, *Get Power*, and *Warm Up* high level behaviors. A modified version of the W-learning algorithm of Humphrys (Humphrys, 1997) is used in SMART to dynamically update the weights. In W-learning, agents suggest their actions with a weight *W* and the maximum weight is chosen as the leader. In our case there are three behaviors vying for control of *Move*, these being *Go to Goal*, *Avoid Obstacles*, and *Preserve Motors*. W-learning uses the difference between the predicted reward $\mathcal{P}$ and the actual reward $\mathcal{A}$ to determine which weights are to be updated (Humphrys, 1997).

Humphrys used a genetic algorithm run off-line to determine his reward functions. We instead use the expression in Equation (2) in order to capture the true change in the rover health through an action (the motor currents and battery levels are read in real-time). Rover behavior is extensively studied prior to launch through both laboratory and field trial studies, so the predicted changes in rover battery levels and motor currents are known for rover movement and in fact are used for resource management planning during the missions.

A small sampling of the predicted rewards are shown in Figure 3 for typical rover behavior. The reward (1) for movement towards the goal on even terrain from a start position is the highest since rover health has a minimal change compared to progress towards the goal. As progressively steeper slopes are attempted, the rewards (2-3) start out being positive since progress towards the goal is still outweighing the impact on rover health, but become more and more negative (4-5) as the steepness increases. Backing-off the slope has a negative reward (6-7) for the relatively benign slopes since the rover health improvement in rover health is outweighed by the lack of progress towards the goal, becoming positive (8-9) for the steeper slopes. The reward (10) for driving sideways is slightly negative since minimal impact on the rover health is outweighed by the movement away from the goal. The reward is less negative than that associated with driving up the steeper slopes which will come into play during the experimental studies. The last reward (11), that of driving away from the goal is a large negative value primarily due to the movement in a direction totally opposite the goal.

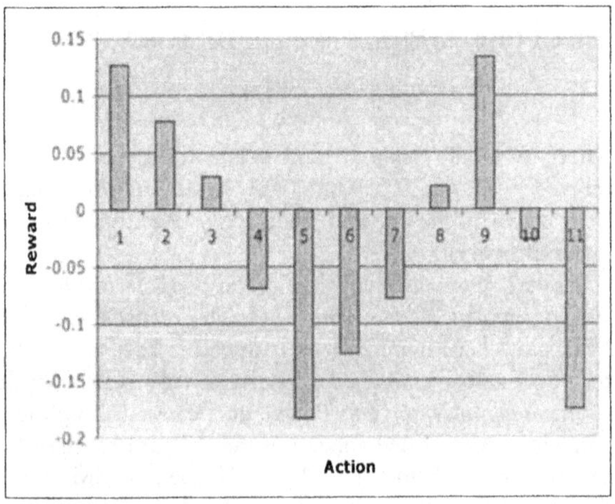

Figure 3: Sample of rewards derived using Equation (2) for a number of different types of actions (in reality the action space is continuous). Reference point is rover health of 1.0 except in cases where the rover is reacting to a current situation (i.e., backing off climbing a slope). All moves are with respect to the goal actions are: (1) normal driving on even terrain, (2) driving up a 5° slope, (3) driving up a 10° slope, (4) driving up a 25° slope, (5) driving up a 45° slope, (6) backing-off a 5° slope drive, (7) backing-off a 10° slope drive, (8) backing-off a 25° slope drive, (9) backing-off a 45° slope drive, (10) driving sideways, and (11) driving backwards. The weights were both set to 0.5 in Equation (1) and to 0.65 and 0.35 respectively for the change in rover health and change in relation to goal in Equation (2).

The *Move/Tilt Arm/Change Shoulder Angles*, *Rest*, *Stop*, and *Sleep* actions at the lowest level in the FFH shown in Figure 2 are mutually exclusive and the action with the maximum activation is chosen using a competitive action selection. The *Tilt Arm, Change Shoulder Angles*, and *Move* actions at the lowest level in the FFH shown in Figure 2 can be done simultaneously, so they are treated as a unit during the action selection process. However, progress towards the goal will be compromised if the rover tips over, so there is a dynamic relationship between the two higher level goals of *Go to Goal* and *Don't Tip Over*. The W-learning algorithm is applied to the links feeding into the *Move* behavior in the hierarchy with a time delay between activations. The *Don't Tip Over* behavior activation occurs in the first time slice, followed by the *Go to Goal* weight updates and activation. This maintains the rover health, while at the same time making progress towards the goal. Another instance where this process is applied is the relative direction that the rover moves. In order to *Preserve the Motors*, the rover will attempt to climb a steep incline, and either back off, go sideways, or rest if the perceived motor currents in the rear wheels are too high. If the weights are not dynamically adjusted, this could lead to dithering where the rover attempts to climb, backs off, and then attempts to climb in the same direction. Adaptive weighting using the W-learning algorithm changes the direction of attack, since progress towards the goal is being compromised by the dithering. For this situation, there is a time delay between application of W-learning to the two incoming links of *Go to Goal* and *Preserve Motors*, with *Preserve Motors* occurring first, followed by *Go to Goal*.

Although our convergence times are typically within 500ms, it is still desirable to limit CPU cycles devoted to learning if it may not be needed. Noise in the sensors can lead to state aliasing where the same sequence of state transitions experienced previously is not recognized. One possible solution to this problem is to provide a memory to the system (McCallum, 1995-96, Michaud & Mataric, 1998-99). We maintain a fixed number of memory traces (currently 100) of limited length (currently 25 steps) of the most recent experiences of the rover. As new experiences come in they are checked for similarity to previous sequences and merged. In the event that the behavior sequence is new, the oldest traces are deleted. These traces are organized using the tree structure developed by Michaud and Mataric (Michaud & Mataric, 1998-99). Rather than use these traces to trigger alternate behaviors as done by Michaud and Mataric, we instead use them to seed the W-learning process with the sequence of expected rewards. In our preliminary studies, we have seen a speedup of a factor of two in our step-wise learning.

4. Related Work

Prior research by Tyrrell (Tyrrell, 1993) and Bryson (Bryson, 2000) demonstrated superior performance of a hierarchical system for action selection over purely reactive systems. In particular, the agents in the Edmund system of Bryson (Bryson, 2000) are built as related sensing and action functions that exhibit selective attention with the payoff of a higher efficiency than the modified Rosenblatt and Payton (RP) mechanisms of Tyrell (Tyrell, 1993). A comprehensive overview of action selection systems can be found in Bryson (Bryson, 2001). Although BISMARC uses the modified RP mechanisms, the nodes in the free flow behavior hierarchy perform operations that are more sophisticated than simple combination. In some sense, they are closer to the *competence* structures of Bryson (Bryson, 2000), in that a collection of plan elements are organized as a prioritized finite state machine whose outputs converge on a specific goal. These nodes have undergone extensive evaluation at the modular level either through field or mission testing.

To date, there has been very little research into learning for hierarchical action selection systems which are typically characterized by multiple, possibly conflicting goals. The dominant learning strategy for single goal achievement such as robotic navigation has been reinforcement learning (RL), an unsupervised method that seeks to maximize a reward signal based on the utility of pairings of input and output states and their subsequent actions (Kaelbling, 1993; Kaebling, *et al.*, 1996; Sutton and Barto, 1998). One of the most popular RL algorithms is Q-learning (Watkins, 1989) and its variations such as Q-PSP (Horiuchi, *et al.*, 1996), and hierarchical Q-learning (Lin, 1993). RL algorithms typically suffer from slow convergence, large state spaces, and difficulties in handling uncertain sensory inputs. Continuous valued versions of the Q-learning algorithm have been developed to address the large state space problem (Gaskett, *et al.*, 1999; Takahashi, *et al.*, 1999; Takeda, *et al.*, 2000). These works used a continuous Q-value derived from neural networks or other function approximation methods. The state space concerns were also addressed for deterministic environments using a forgetting mechanism in a penalty-based hierarchical Q-learning algorithm, which reduces the amount of state information that an agent must maintain by using a low level agent to maintain local state information and a high level agent to maintain global state information (Yen, *et al.*, 2001; Yen and Hickey, 2002). During planetary surface rover operations, the prediction of a state following an action is difficult since it is closer to a non-deterministic process due to interactions with the terrain. Most of the RL studies to date have been confined to simulations and interior navigation in 2-D environments.

An alternate learning system that performs in the presence of a multiple conflicting goals where subtasks are only partially satisfied (Maes, 1991) is W-learning and its variations, which are based on compromise or negotiated decision making between agents (Humphrys, 1997). W-learning is a memory efficient method that is more suited for operation onboard planetary surface rovers than traditional or hierarchical Q-learning systems, and a temporally prioritized modification of it is running under SMART.

5. Experimental Studies

In order to determine the utility of SMART for planetary surface operations in rough terrain, we have run three different types of experimental studies: (1) 2000 simulated rough terrain navigation missions, (2) 50 laboratory sequences with SRR, and (3) 4 sequences with SRR in natural terrain in the Arroyo Seco outside JPL. We have attempted to match the fidelity of the simulation models for terrain and rovers to those used for the laboratory and field studies.

5.1 Simulation studies

The first series of experimental studies used simulated terrain based on MOLA (Mars Orbiter Laser Altimeter) data from the Dao Valis region of Mars, which had slopes of up to 65°. A 200 meter by 200 meter sub-area of the rough terrain dataset is shown in Figure 4 and a view of

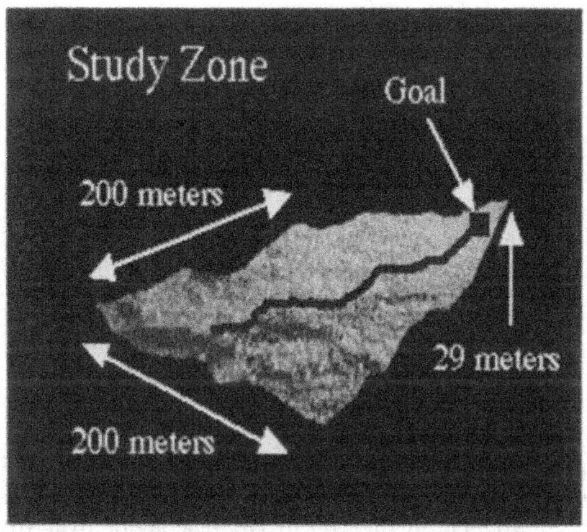

Figure 4: Terrain model for rough terrain navigation mission, with goal position at box and path shown as solid line. Study zone is 200 m by 200 m and terrain variation is from 271 m to 300 m.

the SRR during one of the simulation runs is shown in Figure 5. Mission success was defined as the attainment of the randomly selected goal position without dying due to freezing, dead batteries, burned out motors, or tipping over. The experimental setup included:

- Random starting and goal positions
- Timestep of 0.1s
- 10% loss of traction in rocky terrain
- 1 sq. km study area (5 cm resolution)
- Top speed of 15 cm/sec

The model of SRR matches the physical platform and has two sets of stereo cameras, one body-mounted and one mast mounted, a 3 DOF (degrees of freedom) manipulator and a twelve week battery lifetime supplemented with solar panels.

Figure 5: The SRR climbing a 35° slope in simulated terrain derived from MOLA data in the Dao Valis region of Mars. The model of the rover contained full kinematics and dynamics and used a probabilistic slip assumption. The FFH shown in Figure 2 was used for control and adaptive learning for 2000 simulation runs.

Our studies had a 95.9% mission success with the onboard adaptive learning mechanism, and a 43% success rate without the adaptive learning. The primary failure mode (3.8%) for the system with learning enabled was dead batteries which from a mission standpoint would indicate a need for larger solar panels. An analysis of the 57% of the missions that failed with no learning enabled gives:

- Tipping over - 27%
- Dead batteries - 15%
- Burned out motors - 9%
- Freezing - 6%

Since 27% of the missions failed due to tipping over, the initial weights for inputs to *Move* were set too high, giving an overall bias to the *Get to Goal* behavior over rover safety related behaviors such as *Don't Tip Over*.

5.2 Laboratory studies

The second set of experimental studies was run in the Planetary Robotics Lab (PRL) at JPL and used the JPL technology prototype rover SRR shown in Figure 6. SRR has independently articulated shoulders which allow it to dynamically change its pose and lean much like an animal does on sloped terrain. The full range of shoulder movement is shown in Figure 6. SRR also has independent four wheel drive and independent four wheel steering enabling it to travel sideways.

One of the experimental runs is shown in Figure 7, where we have set up a worse case scenario of opposing hills and valleys for the rover. The SRR (Sample Return Rover) successfully negotiated the course based on a subnet of the full hierarchy shown in Figure 2. This subnet included the *Don't Tip Over, Go to Goal, Avoid Obstacles*, and *Preserve Motors* top level nodes. The *Warm Up, Get Power,* and *Sleep at Night* top level node activation levels were all set to zero since the interior of the lab was warm

Figure 6: Sample Return Rover (SRR) range of hardware adaptation including clockwise from upper left - the lowest range of the shoulder articulation, the highest range of shoulder articulation, and the mid-range of shoulder articulation coupled with extended arm movement.

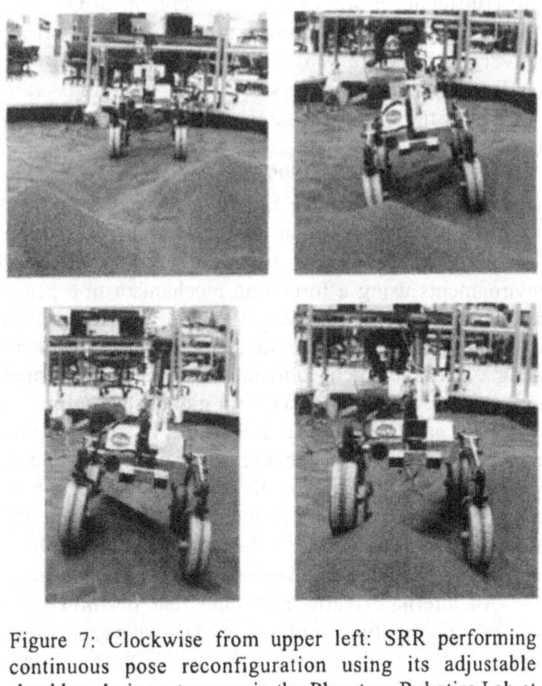

Figure 7: Clockwise from upper left: SRR performing continuous pose reconfiguration using its adjustable shoulders during a traverse in the Planetary Robotics Lab at JPL. The terrain was a set of two opposing hills and valleys, with 45° degree slopes.

and not exposed to the sun.

Another series of laboratory trials used a ramp set at a 65° slope with the rover positioned at the bottom. The goal position was on the other side of the ramp which was beyond SRR's stability capabilities to climb even with shoulder reconfiguration. Initially the rover attempted to climb the slope, but repeatedly backed off and then tried again. This behavior can be traced to the combination of *Go to Goal, Avoid Obstacles,* and *Preserve Motors* using the default weights. The learning algorithm progressively reduced the *Go to Goal* weight from 1.5 to 0.45 while at the same time increasing the weights of *Go to Goal* and *Avoid Obstacles* from 1.0 to a high-water mark of 1.6, which caused the rover to try to skirt the ramp by moving sideways while still maintaining movement towards the goal. Although adaptation of the *Avoid Obstacles* weights lagged behind those of the *Preserve Motors*, the ramp was eventually seen as an obstacle and the obstacle avoidance behavior kicked in. As the rover cleared the side of the ramp it then started movement towards the goal due to the *Go to Goal* behavior output dominating the inputs to *Move* without any obstacles or sloped terrain in front of the rover.

The dynamic weight adaptation seen in the ramp trials is shown in Figure 8, where the weights are shown for the *Go to Goal, Avoid Obstacles,* and *Preserve Motors* behaviors. There are rapid changes in the weights as the rover attempts to climb the ramp, followed by oscillations about a fixed point after numerous backing-off behaviors and then skirting the edge of the ramp. The variability in the weights over the trials is greatest when they are stabilizing to their new values (as seen in the size of the error bars). The eventual outcome of the sequence was that

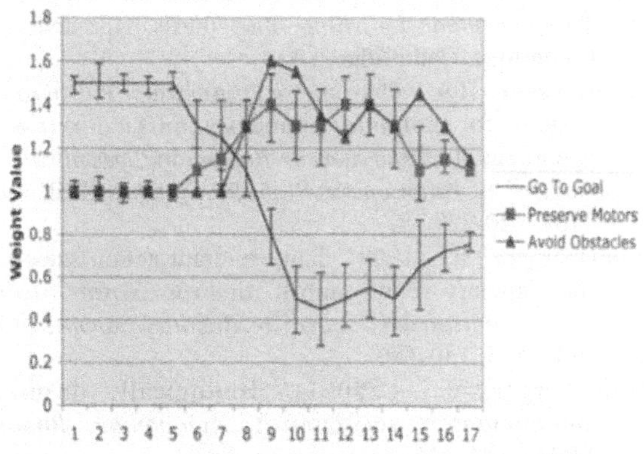

Figure 8: Adaptive learning of weights in the FFH for a rover attempting to go to a goal blocked by a steep ramp. Referring to Figure 2, the inputs to *Move* are *Go To Goal* (plain line), *Preserve Motors* (square line), and *Avoid Obstacles* (triangle line). Points on curves are the average over 20 trials with error bars giving the range of variation (error bars on *Avoid Obstacles* curve omitted for clarity).

the rover learned to treat steeply sloped terrain as an obstacle, while at the same time trying to prevent motor burn-out. In the field, this behavior would be equivalent to the rover trying to find a safe way up a slope to get to the goal, as will be shown in the next sub-section.

5.3 Field studies

The last series of experimental studies was done in the Arroyo Seco, a dry wash that is next to JPL. This site is used for technology prototype rover testing and is characterized by a mixture of benign sand and rocky beds that have been scoured by the periodic water flow bordered by steeply sloped cliffs. An example of the terrain with SRR during a traverse is shown in Figure 9. The learning component of SMART was not fully implemented at the time, so only qualitative results are available at this time.

Figure 9: SRR in the bottom of a rock-strewn gully in the Arroyo Seco outside of JPL. The right shoulder is almost horizontal compared to the left one because the rover just came off of the rock behind the right rear wheel.

We were only able to complete a preliminary series of 4 runs in the Arroyo Seco and will return for more data collection in the spring of 2004 after the winter rains. An example of the skirting behavior along a slope, as previously seen in the laboratory studies discussed in section 4.2, is shown in Figure 10, where the rover approaches the slope in the left frame and is not able to climb, skirts to the side in the middle frame, and finally gets enough traction to climb to the top of the rise and continue on towards the goal.

6. Conclusions

We have developed an autonomous rover control system called SMART for planetary rovers traversing rough and highly sloped terrain. It is based on the previously developed free flow hierarchical action selection of

Figure 10: Skirting behavior of SRR along the length of a slope in the Arroyo Seco wash outside of JPL where the rover initially can not get enough traction to climb so the direction of travel favors lateral motion. Time flows from left to right in the sequence with the left frame being the initial approach almost parallel to the slope, the middle frame showing a change in rover heading more perpendicular to the slope for better traction on the top of the rise, and the right frame showing the rover continuing along its initial heading towards the goal. The mast was fixed in its orientation for this run and would have given the rover more traction if pointed up slope.

BISMARC, coupled with an onboard learning mechanism for changing weights in the hierarchy. The learning mechanism enabled SMART to maintain rover health in both simulated and actual rover studies in rough terrain. Of particular importance for future NASA rover missions was the analysis of the rover failures, indicating that an additional 52.9% of missions would potentially be successful with adaptive learning. We are currently optimizing the memory trace implementation and preparing for further trials in the Arroyo Seco (results should be collected in time for the meeting). We are also starting the integration of the SMART control techniques into the recently developed CAMPOUT (Control Architecture for Multi-robot Planetary Outposts) running on two technology prototype rovers at JPL (Huntsberger, *et al.*, 2003; Schenker, *et al.*, 2003a).

7. Acknowledgments

The authors would like to thank the three anonymous reviewers for their helpful comments on the initial draft of this manuscript. The research described in this paper was carried out at the Jet Propulsion Laboratory, California Institute of Technology, under a contract with the National Aeronautics and Space Administration. The views contained herein are those of the authors and are not meant to in any way imply that they are those of NASA or JPL.

References

Bryson, J.J. (2000). Hierarchy and Sequence vs. Full Parallelism in Action Selection. In *From Animals to Animats 6, Proc. of the Sixth Intern. Conf. on Simulation of Adaptive Behavior (SAB'00),* 147-156.

Bryson, J.J. (2001). *Intelligence by Design.* Ph.D. dissertation, Dept. of Elec. Engineering and Computer Science, MIT, Cambridge, MA, USA.

Elfes, A. (1987). Sonar-based real-world mapping and navigation. *IEEE Journal of Robotics and Automation*, RA-3(3): 249-265.

Gaskett, C., Wettergreen, D., and Zelinsky, A. (1999). Q-learning in continuous state and action spaces. In *Proc. 12th Australian Joint Conf. on AI*, Sydney, Australia.

Hoffman, B.D., Baumgartner, E.T., Huntsberger, T., and Schenker, P.S. (1999). Improved rover state estimation in challenging terrain. *Autonomous Robots,* 6(2): 113-130.

Horiuchi, T., Fujino, A., Katai, O., and Sawaragi, T. (1996). Q-PSP Learning: An exploitation-oriented Q-learning algorithm and its applications. In *Proc. IEEE International Sympos. on Evolutionary Computation*, 76-81.

Humphrys, M. (1997). *Action Selection Methods using Reinforcement Learning*. PhD thesis, University of Cambridge, Cambridge, UK.

Huntsberger, T.L. (1997). Autonomous multi-rover system for complex planetary surface retrieval operations. In *Proc. Sensor Fusion and Decentralized Control in Autonomous Robotic Systems*, SPIE Vol. 3209, 220-229.

Huntsberger, T.L. (1998). Fault-tolerant action selection for planetary rover control. In *Proc. Sensor Fusion and Decentralized Control in Robotic Systems*, SPIE Vol. 3523, 150-156.

Huntsberger, T.L. (2001). Biologically inspired autonomous rover control. *Autonomous Robots,* 11(11): 341-346.

Huntsberger, T.L., Mataric, M.J., and Pirjanian, P. (1999). Action selection within the context of a robotic colony. In *Proc. Sensor Fusion and Decentralized Control in Robotic Systems II*, SPIE Vol. 3839, 84-91.

Huntsberger, T., Aghazarian, H., Cheng, Y., Baumgartner, E.T., Tunstel, E., Leger, C., Trebi-Ollennu, A., and Schenker, P.S. (2002). Rover Autonomy for Long Range Navigation and Science Data Acquisition on

Planetary Surfaces. In *Proc. 2002 IEEE International Conf. on Robotics and Automation (ICRA2002)*, Washington, DC, 3161-3168.

Huntsberger, T., Pirjanian, P., Trebi-Ollennu, A., Nayar, H.D., Aghazarian, H., Ganino, A., Garrett, M., Joshi, S.S., and Schenker, P.S. (2003). CAMPOUT: A Control Architecture for Tightly Coupled Coordination of Multi-Robot Systems for Planetary Surface Exploration. *IEEE Trans. Systems, Man & Cybernetics, Part A: Systems and Humans, Special Issue on Collective Intelligence,* 33(5): 550-559.

Huntsberger, T.L. and Rose, J. (1998). BISMARC. *Neural Networks*, 11(7/8): 1497-1510.

Kaelbling, L.P. (1993). *Learning in Embedded Systems.* MIT Press, Cambridge, MA, USA.

Kaelbling, L.P., Littman, M.L., and Moore, A.W. (1996) Reinforcement Learning: A survey. *J. of Artificial Intelligence Research*, 4: 237-285.

Lin, L.-J. (1993) *Reinforcement Learning for Robots Using Neural Networks*. Ph.D. dissertation, Carnegie Mellon University, Pittsburgh, PA, USA.

Maes, P. (Ed.) (1991). *Designing Autonomous Agents Theory and Practice from Biology to Engineering and Back*, MIT Press.

Mahadevan, S. and Connell, J. (1992) Automatic Programming of Behavior Based Robots Using Reinforcement Learning. Artificial Intelligence, 55: 311-365.

Mataric, M.J. (1992). Integration of representation into goal-driven behavior-based robots. *IEEE Trans. on Robotics and Automation*, 8(3): 304-312.

Mataric, M.J. (1997a). Behavior-based control: Examples from navigation, learning, and group behavior. *Journal of Experimental and Theoretical Artificial Intelligence, Special Issue on Software Architectures for Physical Agents*, 9(2-3): 323-336.

Mataric, M.J. (1997b) Reinforcement Learning in the Multi-Robot Domain. *Autonomous Robots,* 4(1): 73-83.

McCallum, A.K. (1995). *Reinforcement Learning with Selective Perception and Hidden State.* PhD dissertation, Department of Computer Science, Univ. of Rochester.

McCallum, A.K. (1996). Learning to Use Selective Attention and Short-Term Memory in Sequential Tasks. In *From Animals to Animats 4, Proc. Fourth International Conf. on Simulation of Adaptive Behavior, (SAB'96)*, Cape Cod, MA.

Michaud, F. and Mataric, M.J. (1998) Learning from History for Behavior-Based Mobile Robots in Non-stationary Conditions. Joint Special Issue on Learning in Autonomous Robots, *Machine Learning,* 31(1-3): 141-167, and *Autonomous Robots,* 5(3-4): 335-354.

Michaud, F. and Mataric, M.J. (1999) Representation of behavioral history for learning in nonstationary conditions. *Robotics and Autonomous Systems,* 29: 187-200.

Pirjanian, P. (1998). *Multiple Objective Action Selection and Behavior Fusion Using Voting.* PhD dissertation, Laboratory of Image Analysis, Department of Medical Informatics and Image Analysis, Aalborg University, Denmark.

Pirjanian, P. and Mataric, M., (2000). Multiple objective vs. fuzzy behavior coordination. In *Lecture Notes in Computer Science on Fuzzy Logic Techniques for Autonomous Vehicle Navigation*, D. Drainkov and A. Saffiotti, (Eds.), Springer-Verlag, 235-253.

Rosenblatt, J.K. and Payton, D.W. (1989). A fine-grained alternative to the subsumption robot control. In *Proc. IEEE/INNS Joint Conf. on Neural Networks*, 317-324.

Schenker, P.S., Huntsberger, T.L., Pirjanian, P., Baumgartner, E.T., and Tunstel, E. (2003a). Planetary Rover Developments Supporting Mars Exploration, Sample Return and Future Human-Robotic Colonization. *Autonomous Robots*, 14(2/3): 103-126.

Schenker, P.S., Huntsberger, T., Pirjanian, P., Dubowsky, S., Iagnemma, K., and Sujan, V. (2003b). Rovers for Intelligent, Agile Traverse of Challenging Terrain. In *Proc. International Conference on Advanced Robotics (ICAR'03)*, University of Coimba, Portugal, 1683-1692.

Sutton, R.S. and Barto, A.G. (1998). *Reinforcement Learning: An Introduction.* MIT Press, Cambridge, MA.

Takahashi, Y., Takada, M., and Asada, M. (1999). Continuous valued Q-learning for vision-guided behavior acquisition. In *Proc. International Conf. on Multisensor Fusion and Integration for Intelligent Systems*, 716-721.

Takeda, M., Nakamura, T., Imai, M., Ogasawara, T., and Asada, M. (2000). Enhanced continuous valued Q-learning for real autonomous robots. In *From Animals to Animats 6, Proc. of the Sixth Intern. Conf. on Simulation of Adaptive Behavior (SAB '00).*

Tunstel, E. (2001). Ethology as an inspiration for adaptive behavior synthesis in autonomous planetary rovers. *Autonomous Robots,* 11(11): 333-340.

Tyrrell, T. (1993). The use of hierarchies for action selection. *Journal of Adaptive Behavior*, 1(4).

Watkins, C.J. (1989). *Learning from Delayed Rewards.* Ph.D. dissertation, Cambridge Univ., Cambridge, UK.

Yen, G., Yang, F., Hickey, T., and Goldstein, M. (2001). Coordination of exploration and exploitation in a dynamic environment. In *Proc. International Joint Conference on Neural Networks (IJCNN '01), Volume 2*, 1014-1018.

Yen, G. and Hickey, T. (2002). Reinforcement learning algorithms for robotic navigation in dynamic environments. In *Proc. of the 2002 Intern. Joint Conf. on Neural Networks (IJCNN '02), Vol. 2*, 1444-1449.

AUTHOR INDEX

Abt, Roland, 114
Aghazarian, Hrand, 519
Alexander, William H., 13
Arleo, Angelo, 255
Avila-García, Orlando, 243

Baccon, Jean Christophe, 57
Baldassarre, Gianluca, 385
Bargmann, Robert, 193
Bauckhage, Christian, 315
Beer, Randall D., 49, 357
Belpaeme, Tony, 455
Blaesing, Bettina, 173
Boucheny, Christian, 255
Bourjot, Christine, 425
Bovet, Simon, 114, 122
Bryden, John, 183
Buchli, Jonas, 153
Buckley, Christopher, 334
Bullock, Seth, 334
Buffet, Olivier, 223

Cañamero, Lola, 243
Cangelosi, Angelo, 487
Charpillet, François, 223
Chevrier, Vincent, 425
Cohen, Avis H., 201
Cohen, Netta, 183, 334
Crabbe, Frederick L., 233

Daucé, Emmanuel, 305
Davy, Simon, 395
de Boer, Bart, 455
Degris, Thomas, 255
Der, Ralf, 324
Desor, Didier, 425
De Vylder, Bart, 455
Diefenbacher, Marco, 114
Dorigo, Marco, 405
Doya, Kenji, 287
Dutech, Alain, 223

Edelman, Gerald M., 130

Federici, Diego, 373
Fend, Miriam, 114, 122
Filliat, David, 3
Flacher, Fabien, 435
Fleischer, Jason, 477
Franks, Daniel W., 395
Fukuoka, Yasuhiro, 201

Garcowski, Jacob, 49
Gaussier, Philippe, 57
Girard, Benoit, 3
Goldenberg, Eldan, 49
Guillot, Agnès, 3

Hafemeister, Laurence, 57
Hallam, John C. T., 87, 499
Harter, Derek, 264
Hayes, Gillian M., 297
Herrmann, J. Michael, 324
Hoffmann, Heiko, 213
Holicki, Michael, 324
Humphrys, Mark, 39
Huntsberger, Terry, 519
Husbands, Phil, 163

Ijspeert, Auke Jan, 153

Jansen, Bart, 455
John, Roul Sebastian, 23

Kallmann, Marcelo, 193
Khamassi, Mehdi, 3
Kim, DaeEun, 140
Kim, Hyundo, 104
Kimura, Hiroshi, 201
Komatsu, Takanori, 415
Konidaris, George D., 297
Kozma, Robert, 264
Krafft, Martin, 114
Krichmar, Jeffrey L., 130

Lachèze, Loïc, 3, 255
Lau, Boris, 104

Matarić, Maja, 193
McHale, Gary, 163
McKinstry, Jeffrey L., 130
Meyer, Jean-Arcady, 3
Möller, Ralf, 213
Müller, Rolf, 87

Nadel, Jacqueline, 57
Noble, Jason, 395
Nolfi, Stefano, 385

O'Leary, Ciarán, 39

Paine, Rainer W., 274
Parisi, Domenico, 385
Paul, Chandana, 33
Peremans, Herbert, 77
Pfeifer, Rolf, 122
Prepin, Ken, 57
Prokopenko, Mikhail, 509

Quinn, Matt, 282
Quoy, Mathias, 141

Reijniers, Jonas, 77

Sagerer, Gerhard, 315
Salomon, Ralf, 96, 364
Schermerhorn, Paul, 445
Scheutz, Matthias, 445
Seth, Anil K, 130
Seys, Chad W., 357
Shapiro, Jonathan, 477
Sigaud, Olivier, 435
Sporns, Olaf, 13
Steels, Luc, 67
Sugita, Yuuya, 467

Tani, Jun, 274, 467
ten Thij, Tom, 455
Thomas, Vincent, 425
Thurau, Christian, 315
Trianni, Vito, 405
Triesch, Jochen, 104
Tuci, Elio, 405

Uchibe, Eiji, 287

Walshe, Ray, 39
Wang, Peter, 509
Werner, Christian W., 23
Williams, Hywel, 344

Yannakakis, Georgios N., 499

Made in the USA
Monee, IL
13 September 2026

62977352R20299